Optimum and Adaptive Signal Processing

Optimum and Adaptive Signal Processing

S.V. Narasimhan
S. Veena
S. Roopa

National Aerospace Laboratories, Bangalore-560017
Council of Scientific and Industrial Research, New Delhi, INDIA

ISBN: 978-93-89633-03-0

Edition: 2019

Printed at: Shree Maitrey Printech Pvt. Ltd., Noida

। श्री गणेशाय नमः ।

Shri Ganeshaaya Namaha

। श्रद्धाहि परमागतिः ।

Shraddaahi paramaa gatihi

Preface

Optimum and adaptive signal processing has played an important role in real time processing of signals in almost all the fields of science and engineering. The optimum estimators known as parametric signal representation, are generally based on a model (allpole/autoregressive (AR), all-zero/moving average (MA) or pole-zero/ARMA) and when the model matches the signal, provide accurate information about the signal as they have less bias and variance. This is due to either signal/spectral modeling of the signal. On the other hand, the non-parametric estimators based on fixed transforms like, popular Discrete Fourier transform; for a given data length, have a very poor performance as their bias, variance or root mean square error, are very large. Also for a given length of data, they suffer from a tradeoff between bias and variance. But with the optimum approach, even with short data length, the frequency bias (resolution) achieved is much superior to that achieved by non-parametric approach. Also the spectral variance is superior as it fits a model to the spectral data of the signal. The optimum (minimum mean square error) estimators as they represent the signal/spectrum parsimoniously (by few parameters) compared to the non-parametric approach, they facilitate efficient signal compression, storage, transmission and classification.

The desirable features of an optimum estimator are valid for a stationary signal (i.e., whose spectral characteristics do not change with time). However as most of the practical signals are nonstationary, there is an absolute necessity to extend the optimum property of the estimator also to them. This is achieved by moving from one optimum to another and is achieved by adaptation, thus resulting in a *optimum adaptive estimator/ processing*. The adaptation of optimum may involve high computational load which may not be feasible in many applications and this is due to requirement of sophisticated fast DSP platform on which the estimators are implemented. Therefore, there is a necessity to develop simple, computationally efficient optimum adaptive estimators for the changing signal scenario by having fast convergence rate. Further as practical signals are not only nonstationry but also nonlinear, it is also necessary to accommodate the nonlinearity. *Any real time processing should be adaptive and also optimum to get the best desirable results.* Such a requirement is most common in almost all the fields like communication, audio, video, control and biomedical applications. In view of these, the study of the subject, Optimum and adaptive processing is essential for students of most of the branches of engineering which require real time processing.

There are many books available on the subject of adaptive signal processing/filters from information point of view and they excel in one or more topics. For a person with good mathematical or communication engineering background, these books may meet the requirement. But as many engineering fields are becoming multidisciplinary, for a beginner or student from other fields like mechanical, civil, chemical or biological sciences, who is interested in self learning, will find the treatment really hard to understand and apply the concepts in his/her developmental, research and teaching activities. Further, the optimum and adaptive/adaptive signal processing is not only taught at graduate level as a core subject but also as an optional subject at undergraduate level for electrical, electronics, telecommunication and instrumentations engineering students. For such readers, the amount of material in these books may confront the reader what is essential to get a working knowledge, for a specific purpose. In circumventing the above problems, few books have been published. In some books, the approach may be highly mathematical and will suit merited communication or electrical students/professionals/researchers. Further, the rigor involved may appear too much for a beginner making the subject not attractive. Also keeping the required background to a minimum, helps in maintaining the pace of understanding.

In the present book, the subject has been developed in a logical directed manner to get the working knowledge so that the student will not suffer from the limitations mentioned above. To cater for a beginner, the material has been kept to a minimum avoiding any confrontation about the decision "what is essential?" As the subject has been developed systematically from understanding point of view, the background required is minimum (physics and mathematics of *junior degree level)* and the basics of signals and systems are desirable. The ease of understanding *is not at the cost of omission of intricate concepts but by giving all the intermediate steps in a mathematical derivation and their physical meaning in a simple way.* As the student covers the chapters with clear understanding, he gains confidence to get a working knowledge which is the goal of a good book.

The book is *self contained* as the initial 2 - 5 chapters cover general back ground of signal processing, including the required random process theory. The chapter-6 brings out the limitations of nonparametric estimators based on Discrete Fourier transform, their limitations of high bias and high variance and their tradeoff. This chapter also presents recent approach based on group delay processing which provides improved performance. The estimation of bispectrum and recovery of nonminimum system phase are given.

The chapter-7, the parametric spectral estimation treats the optimum estimation using allpole and pole-zero models, in detail covering the common existing algorithms and their implementation structures highlighting their merits and drawbacks. A pole-zero modeling based on pole-zero decomposition in group delay domain which overcomes some of the problems of existing methods is covered. Estimation of moving average model parameters for a given data by spectral factorization and by allpole modeling are given. Bispectrum estimation via AR modeling by solving third order recursion equation is also presented. In chapter-8, the parametric modeling considered for causal signals has been generalized to non-causal signals and has been shown that the physically unrealizable estimator/filter provides the lowest possible error/irreducible error which forms the lower bound for any physically realizable estimator.

The chapter-9, deals with the method of steepest descent which is iterative in nature and achieves the solution over many iterations or samples asymptotically on sample to sample basis instead of handling as a single block of data. The condition for convergence for the mean weight vector has been derived with full mathematical details. The analysis of steepest descent leads to the method of orthogonalization based methods. The Newton method has also been treated and it is interpreted as one where method of steepest descent is applied to orthogonal components of input. The chapter -10 , the Least Mean Square (LMS) algorithm simplifies the method of steepest descent as it does not require any computation or advance information about autocorrelation matrix of input and crosscorrelation vector between the put and the desired signal. This simplicity of LMS algorithm has made it most popular in all fields of engineering. The expression for missadjustment and the condition for convergence in mean square sense (not in sense of mean) have also been derived. Many variations of the LMS algorithm, normalized LMS, variable stepsize LMS, correlation, momentum LMS, leaky LMS, Griffiths LMS, sign-sign LMS and median LMS; are considered. The effect of delay in the LMS loop which affects convergence rate is discussed. Noise (even color noise) robust variable stepsize algorithm which not only uses noise robust stepsize but also noise robust gradient for the weight adaptation without requiring any explicit noise power estimate, is presented and this is based on Griffiths gradient LMS algorithm. For a transversal structure, to reduce the misadjustment by 6dB, the method which uses forward and backward error gradients for weight adaptation is given. The use of conjugate gradient approach to adaptive filtering is considered with some modifications and this provides better convergence performance even with single iteration per sample instead of requiring many iterations for every sample required by the original conjugate gradient method. Post smoothing of weight vector to reduce the misadjustment has been considered when large value of stepsize is required to avoid stalling due to low levels of input. Non minimum phase system adaptive identification based on bispectrum using LMS algorithm is treated.

The Chapters -11 deals with methods which overcome the LMS limitation of slow convergence due to high eigenvalue spread of input autocorrelation matrix by using input orthogonalization approaches like different orthogonal transforms like DFT, DCT and DST and various types of lattice filters with different degrees of success in improving the convergence rate. The transform domain version of forward-backward algorithm which in addition to reduced misadjustment also provides faster convergence, is presented. Further noise robust transform domain variable stepsize algorithm which also uses robust gradient for weight vector adaptation is covered and this is again based on Griffiths LMS algorithm. Transform domain Conjugate gradient (TCG) algorithm for adaptive filtering is considered and this provides improved convergence performance even with single iteration per sample.

The normal lattice (two multiplicative) with the Laguerre structure has been dealt and this reduces the adaptive filter length by about 50% compared to that of lattice without Laguerre structure. The use escalator lattice for adaptive filtering has been illustrated and is appropriate for non stationary signals as it uses either forward or backward error only for adaptation. Further use of Laguerre structure with escalator lattice also reduces the adaptive filter order by 50% preserving its original performance. The simple cascade predictors are used to improve the convergence rate as they reduce the eigenvalue spread. The cascade predictors are extended to the minimization of both forward and backward error energies together to achieve better misadjustment.

In chapter-12, the eigenvalue spread is reduced by decomposing the signal into subbands using a perfect reconstruction filterbank and using an adaptive filter for each subband. As signal is processed (filters are adapted) in subbands of smaller bandwidth compared that of fullband, the decimation can be done and hence more time is available for processing which enables use of slow DSP processors. Also, as decimation expands the subband spectrum, the eigenvalue spread for each subband gets further reduced (in addition to that due to signal decomposition) and hence further improves the convergence rate . The problems like instability due to synthesis filter delay, eigenvalue spread which results due to oversampling (to solve aliasing problem) and errors in synthesis filter bank; are solved by using delayless subband adaptive filter (DLSAF) in the closed loop mode. In the DLSAF realized by polyphase filter bank, the weight transform (constructing overall adaptive filter from sub adaptive filter) is quite involved as it needs tedious spectrum stacking due to complex nature of the DFT of sub adaptive filters. As an extension of subbands, wavelet transform is also considered and as this will have better performance for nonstationary signals due to nonuniform bandwidth of WT. The weight transform becomes simpler if the DFT of subadaptive filters are real. A DCT harmonic analysis filterbank/wavelet transform which has built in decimation, is simple compared to time domain analysis filterbank/ discrete wavelet transform and results in real inputs to subadaptive filters and results in real subadaptive filters. Further mere concatenation of DCT of the subadaptive filters lead to overall adaptive filter and makes the weight transform simple. Thus use of DCT subbands (grouping of DCT coefficients of input) solves basically aliasing and eigenvalue spread tradeoff. Computation of DCT on sample to sample basis efficiently has been achieved by running/recursive DCT. The concept of an adaptive type of signal decomposition into subbands based mutual correlation between input and desired signal as against fixed type of decomposition is treated.

Use of pole-zero model (often known as IIR filters) will not only efficiently extract the general signal characteristics but also reduces the computational load as the orders required both for numerator and denominator are comparatively small than using either allpole or allzero model. The output error pole-zero method suffers from local minima due to nonlinearity and instability due to movements of poles outside the unit circle while adaptation. Equation error method (EQE) solves these problems but suffers from pole polynomial bias due to observation noise and does not minimize the true error. Though the monic normalization removes bias for white observation noise but it fails color observation noise. The Steiglitz-Mc Bride method minimizes true error but for color observation noise, this also suffers from bias problem. To solve the bias problem for both the methods, a bias free approach which takes the estimate of system output as input to the adaptive filter for the denominator, has been covered. These aspects and possible implementation of pole-zero estimators with stability monitoring; are covered in chapter-13.

The chapter-14 deals with the frequency domain block LMS algorithm. This does the adaptation on block to block basis not only reduces the computational load as it implements convolution operations using FFT but also improves the convergence rate as it exploits the readily available orthogonalization of the input. Improved versions of FBLMS, like portioned block LMS and delayless partitioned FBLMS are considered to solve the input block delay problem. The iterated FBLMS which reuses the input data for improved performance is briefly treated.

In chapter -15 deals with the Recursive type of algorithms and. The derivation of standard recursive least square algorithm has been given. For stationary signals, the forgetting factor (λ) can have the value unity and this leads to zero misadjsutment. However for nonstationary signals, for updating the autocorrelation matrix and corsscorrelation vector, λ should be in the range $(0<\lambda<1)$. The effect of value of λ on the speed of convergence and misadjustment are derived. Smaller the forgetting factor faster is the convergence but it results in a larger misadjustment. The RLS and Newton algorithms have similarity for large number of samples. The standard RLS algorithm as it requires N^2 multiplications (N is the order of the filter), is highly computationally intensive. Fast transversal RLS (FTRLS) algorithm which uses the a posteriori forward and backward prediction errors and reduces the computational load from N^2 to $8N$ has been derived. However, the fast transversal RLS (FTRLS) is highly sensitive to finite word length effects and may result in stability problems. For this purpose, fast RLS based on joint lattice structure with error feedback based on a priori forward and backward errors, has been derived. The use of error feedback for coefficient adaptation provides additional immunity to finite word length effects of processor used for implementation. . The derivation of fast lattice RLS algorithms without error feed back for coefficient adaptation based on a posteriori and also on a priori errors, are also derived. The computational complexity of fast RLS lattice is of the order of $18N$ and is higher than that of transversal which is $8N$. But this is tolerable compared to direct RLS implementation of RLS which has computational load of N^2 and also as the fast lattice RLS has high immunity to finite word length effects.

The affine projection (AFP) adaptive filter is also considered and this has a convergence rate in between that of normalized LMS (NLMS) and RLS. This due to the fact that AFP uses 5-10 input vectors instead of a single input vector as used by NLMS algorithm. However its misadjustment will increase as the number of input vectors considered increases. The Kalman estimator which is nothing but recursive way of implementing Wiener filter, is treated in its simplest form and is illustrated for the estimation of the signal form a mixture of signal plus white noise.

The linear adaptive filters cannot account for any nonlinearity present in the input or system as they extract only linear features. For this, adaptive filters should have nonlinear processing element. The neural network or Volterra series based adaptive filters have such provision. Since for many practical signals, their nonlinear feature may be of importance, the neural network based on back propagation algorithm and 2nd order Volterra series, are dealt in chapter-16.. The orthogonalization of the input by lattice or by an orthogonal transform like, DCT; improves the convergence rate compared to that of transversal Volterra algorithm and this is due to reduction in crossterms.

The Chapter -17 covers reasonably a wide variety of applications of adaptive filters/ processing. In communication applications: ADPCM, sigma-delta modulation, decision feedback equalization, adaptive inverse control, adaptive time delay estimation, adaptive coherence estimation, telephone line and data network echo cancellation, optimal and adaptive beamformers for communication are dealt. In speech, audio and acoustics: noise cancellation for speech, cross talk suppression, acoustic echo cancellation and hearing aid feedback cancellation, active low frequency noise control; are treated. In biomedical applications: enhancement of fetus ECG in the presence of mothers ECG, power line interference cancellation from ECG by adaptive notch filter, muscle noise suppression from EEG, suppression of EOG from EEG, adaptive filtering of evoked potentials.

The current research topics in the field of adaptive filtering like state space LMS for single and multichannels, Bayesian adaptive filtering, Adaptive set-membership filtering,

Partial-update and sparse adaptive filtering Kernel adaptive filters, Wavenets and wavelet network and Complex valued (nonlinear) adaptive filters; are briefly introduced.

In realizing this book, the authors have referred to many classical books and the material available in them has been *selectively picked* and presented in a *simple and logical order in a directed way*. I some cases, the authors have developed their own interpretations and derivations which adds to ease of understanding. *The research work done by authors of this book (and their co-researchers) which covers many aspects of adaptive processing , have been included.* For *self-learning*, the Matlab programs used for most of the illustrations, are provided.

The approach followed has been found to be successful as many students at, graduate/master level, undergraduate level and undergraduate faculty who have attended many guest lectures by the authors in many continuing educational programmes; have been benefited. This book is the result of: thesis guidance to 20 Masters students from National Institute of technology, Karnataka (NITK) Surathkal, India; teaching a full semester course (39 hours of lectures) on adaptive signal processing twice during 2004 and 2006 as a guest faculty (first author) at Indian Institute of Science for master and doctoral students, a 4 days intensive course on adaptive signal processing under NAL–University lecture programme in Nov.2008 at National Aerospace laboratories, Bangalore, India.

In spite of many revisions, few typographical/technical corrections may be still required. The authors welcome corrections or suggestions, which improve the quality of the book

Study Guide

(a) For under graduate students (B.E, B.Tech, M.Sc, B.S)
Chapters : 1 – 5, 6.1-6.2, 7.1-7.6, 7.7.1, 7.8-7.10, 7.10.1,7.10.3, 7.10.6,
9, 10.1-10.6, 10.11,10.13, 11,11.4, 11.10,11.10.1,
13.1 -13.4, 15.1,15.2, 17.1,17.2,17.5,17.8,17.8.3, 17.11, 17.13.2

(b) Postgraduate/ Masters (M.S, M.Tech, M.E) with DSP and random process background:
Chapters: 6.1, 6.2, 7.1 to 7.11, 7.14 to 7.16, chapter-8, 9, 10.1to 10.13, 11.1 to11.7, 11.10, 12.1 to 12.15,131 to13.8, 14.1 to 14.7, 15.1 to 15.4, 17 (selected relevant applications)

(c) For research students : Same as for (b) plus other relevant topic sections of interest and section --17.15 Current topics.

S.V. Narasimhan
S. Veena
S. Roop

Acknowledgements

The authors thank the present and previous Directors Mr. Shyam Chetty and Dr. A.R. Upadhya, respectively of CSIR-National Aerospace laboratories (NAL), Bangalore and the Present and previous Heads Mr. K.G. Venkatanaryana and Dr. M.R. Nayak, respectively of the Aerospace Electronics & Systems Division; for all their encouragement and support in completing this book. The first author expresses his sincere gratitude to Human Resources Division, Council of Scientific and Industrial Research (CSIR), New Delhi, India, for awarding the CSIR Emeritus Scientist Fellowship under which the compilation of book is evolved. The authors appreciate the efforts of Dr. J.S. Mathur, Head, Knowledge and Technology management Division, in getting the copyright from CSIR, The knowledge presented in the book is due to practical execution and research of 6 projects carried out on the activity, Active Noise Control (ANC) for which the optimum and adaptive signal processing forms an integral part. Therefore the authors are grateful to the funding agencies: Aeronautical research and development Board(ARDB), National programme on smart materials and systems (NPSM), Aeronautical Development Agency (ADA), National programme on nano materials (NPMASS) (all under Defence research and development organization (DRDO)) and Modernization programme and High Science Technology (both under CSIR) . The author S.V. Narasimhan thanks Prof. M.L.Munjal, department of Mechanical engineering, Indian Institute of Science, Bangalore, for having inspired him to work in the field of ANC. They express the gratitude Dr. T.S. Pralhad and Dr.T.R. Pai to previous directors and Ex Head ALSD, Mr. N.N. Murthy; for their support and encouragement, during the different stages of these projects.

The subject knowledge has been gained by Guiding 15 masters project thesis of Naional Institute of Technology, Karnataka (NITK), Surathkal, India. The first author expresses his gratitude to Prof. Anurag Kumar, Chairman , Department of Electrical Communication Engineering, Indian Institute of Science, Bangalore, India for having given an opportunity to teach the full semester course on Adaptive signal processing twice during 2004 and 2006 for the Masters and Doctoral students, as this enabled to consolidate and organize the material presented in the book. Also the 4 days intensive course on Adaptive signal processing conducted during 4-7 November 2008 at National Aerospace laboratories (NAL), Bangalore under NAL- University Programme, added more strength to the knowledge.

He also thanks Prof. B. Yegnanaraya (III T, Hyderabad, India), Prof. M.N.S. Swamy and Prof. E.I. Plotkin (Concordia University, Montreal, Canada), Prof. N.S. Nagaraja (late), Prof. P.S. Naidu, Prof. D. Narayana Dutt, Prof. V.U. Reddy, Prof. K.R. Ramakrishnan, Prof. V. Krishnan (Indian Institute of Science Bangalore, India), Prof. T.V. Ananthapadmanabha (Director, Voice and Speech Systems Ltd., Bangalore), Dr.V.S.V. Mani (Honeywell Corporation, Bangalore, India) and B.S. Adiga (consultant to Tata consultants, Bangalore, India); whose guidance, encouragement and support helped him in acquiring the knowledge in this field.

The book has become possible mainly due to the exposure of the authors to the rich quality of material available in various books and references. The authors *sincerely* thank and acknowledge the authors of the books and references given in the bibliography.

Further, the interaction with the student community enabled the authors to get to know the necessity of such an effort and the repeated lectures/discussion on this subject evolved the contents of this book. The authors express their gratitude to Dr. Sudesh Kumar Kashyap (Flight Mechanics and Control Division, N.A.L) for reviewing the Neural network part of the chapter on Nonlinear adaptive filters. The authors thank the Masters thesis students: Mr. R. Krishna Dutt, Mr. G.C. Yala,, Mr.K.suresh, Mr. A.V.Kulakarni, K. Radhakrishna, Ms. Malini Nayak, Mr. Harish ,Mr. Vikram, Ms. Savita. K. Shankarling, Mr. K. Nagesh, Ms. Nagamani, Ms. S. Pavitra, Ms. Tulika, Mr. Naveen Kumar. Mr. B. babloo and Yogesh Yadav (all from National Institute of Technology Karnataka, Surathkal, India), Dr. M.K. Sridhar, Ms. S. Roopa (B.M.S. College of Engg., Bangalore, India), Dr. M. Girish Chandra (I.I.T. Madras) of Technology, Madras) and many undergraduate students who carried out their projects with the authors.

The second author thanks her teachers Ms. B.V. Meenakshamma, Ms. Satyabhama, Ms. Ratna, Mr. Gururaj, Mr. Lakshman, Mr. Jagadeesh, Mr. V.P. Shanbagh, Mr. L. Parthasarathy and Mr. K. Anantharam for inculcating the thirst for knowledge during her educational career.

The third Author is grateful to her mentors Dr. S.V. Narasimhan and DSP group members of ALD, NAL for providing valuable guidance. She sincerely thank Prof. Mrs. M.N. Suma (B.M.S. college of Engineering, Bangalore) for her support and encouragement.

The authors sincerely thank and appreciate the efforts the DSP Systems Group members Mr. H. Lokesha (scientist, NAL), Nandini Basu Mallick (presently, Scientist at Central Glass and ceramic Research Institute, Kolcatta, India), S.M. Vaitheeswaran (scientist, NAL), S. Pavithra (Project Assistant, NAL), Ms. V. Roopashree, Ms. S. Suhasini and D.T.Geetha (NAL Graduate trainees).

The authors thank the entire staff of IK International Publishers Pvt. Limited, for their keen interest, enthusiasm, cooperation and their efforts; in bringing out the book in an excellent form.

The first author expresses his gratitude to his parents Sri S. Venkatarangachar and Smt. Leelavatamma and his brother S.V. Srinivasan and sister Mrs. S.V. Sarasamma whose efforts and cooperation has enabled him to acquire the wealth of education. Further, he acknowledges his wife Dr. Mrs. M.V. Chandravadana for all the support and encouragement in addition to taking the family responsibilities and appreciates his son Mr. Mukund Sriram for his understanding during the course of the development of the book. He expreses his gratitude to his best friend Mr. Sai Deep Chand Rathnam and his parents who gave him all the support and encouragement, during his educational period.

The second author expresses her gratitude to her parents Mrs. V. Sharada and Mr. S. Venkateshaiah and her husband Mr. S. Shantharam for their support and encouragement and her daughters Lekha Payasvini and Dyuti Manasvini for their understanding.

The third author is grateful to her parents Mr. B.M. Siddagangaiah and Mrs. T.G. Manjula. Besides she thanks her husband Mr. P. Yogesh, her baby daughter Miss Y. Hima and her in-laws Mr. A.S. Prasad and Mrs. Sumangala for their understanding and support. She appreciates the inspiration provided by her brothers Mr. K.S. Shankar and Mr. S. Mallesh Babu in taking part of this scientific venture.

S.V. Narasimhan
S. Veena
S. Roopa

The Authors

S.V. Narasimhan was a CSIR Emeritus Scientist (2009-2014) at Aerospace Electronics & Systems Division, National Aerospace Laboratories, Bengaluru, At the same institution, he served as a scientist at various levels during 1977-1982 and 1991-2009 and retired as Chief Scientist (2006-2009). He was a Post-Doctoral Fellow (June 89-April 91) and a Visiting Scientist (July-Oct. 1996) at the Dept. of Electrical and Computer. Engineering, Concordia University, Montreal, Canada. He got his B.E. (Electrical Communication Engineering, 1973) and Ph.D. (EEG Signal Processing,1988) from Indian Institute of Science, Bangalore, India and M.Tech. in Electrical Engineering (Radar & Communication, 1977) from Indian Institute of Technology, Madras, India. His research field covers: 1D/2D spectral estimation, coherence estimation, higher order spectra (bispectrum), cyclostationarity processing, time-frequency analysis, group delay processing, adaptive signal processing, active noise control, wavelet transform, ADPCM, channel equalization, EEG signal processing and speech enhancement. He has 45 peer reviewed international journal papers and 30 international/national conference papers. His developmental activities involved mainly active noise control for ducts, helicopter cabins, aircraft cockpits and pilot headsets; which were funded by national agencies. He has guided 2 Ph.D. and 2 M.Sc. (by research) candidates and 42 M.Tech projects. He is the author of *Principles of Signal Processing and Implementation* (2005), *Introduction to Wavelet Transform: A Signal Processing Approach* (2011).

S. Veena is a Principal Scientist at CSIR, National Aerospace Laboratories, Bengaluru. She has about 18 years of research experience at the same organization. She is B.Tech from Bangalore University and M.Sc. (Engg.) from VTU, Belgaum. She is a member of IEEE. Her research interests include digital signal processing, active noise control, speech processing technologies for control applications in aerospace and algorithms for information security. Her team has developed active noise control based quiet helmet for fighter aircraft. She is also consultant for UIET, Panjab for the active noise control activities of the university. She is also involved developing speech activated micro air vehicles. She has about 10 journal publications and 20 conference publications. She is the co-author of *Principles of Signal Processing and Implementation* (2005), *Introduction to Wavelet Transform: A Signal Processing Approach* (2011). She has guided about 12 M.Tech students.

S. Roopa is Assistant Professor, Electronics Department, Siddaganga Institute of Technology, Tumakuru. She holds a gold medal for M.Tech from VTU and is pursuing Ph.D. from the same university. She has contributed to the field of signal processing, as a Research Fellow at CSIR-National Aerospace Laboratories, Bengaluru. She has about 9 journal publications and a conference publication.

The Authors

Contents

Note
CD containing MATLAB programs for most of the illustrations is provided.

Chapter 1

INTRODUCTION

This chapter emphasises the necessity for parametric optimal and adaptive signal processing by providing its evolution history by bringing out the limitations of conventional nonparametric approach based on the Discrete Fourier transform. Further by providing briefly many day-to-day applications in different scenarios where such techniques play the essential role, the motivation for the signal processing community to pursue and improve this subject for further utility, has been attempted.

1.1 INTRODUCTION

Generally, any real signal is considered to be made up of many sinusoids of different frequencies having different magnitudes and added with different phases (starting point). This information about the signal is referred to as frequency content/spectrum of a signal (in the year 1900) which may or may not change with respect to time. Signals whose spectrum does not change with respect to time belong to the class of stationary signals, otherwise they come under the class of nonstationary signals. Hence, for stationary signals, the spectrum of any segment (of same length) of a signal taken at different instants of time will be similar but this is not true for nonstationary signals. The spectrum can be obtained by taking Fourier transform (FT) of the signal as the latter represents the signal by many sinusoids which cover the signal frequency range (ideally from $-\infty$ to ∞) over the signal time duration (ideally from $-\infty$ to ∞). That is, FT assumes that the spectrum will remain same for the entire time interval ($-\infty$ to ∞). Hence, the FT and FT based processing methods are valid for only stationary signals and not for nonstationary signals. For example, consider a signal of 10 seconds duration and if it has a sinusoid of frequency 5Hz up to initial 5 seconds and for the remaining 5 seconds, the sinusoid is of frequency 10 Hz, then the FT of this signal for duration of 10 seconds indicates that *there are two sinusoids of frequencies 5 Hz and 10Hz for the entire duration of 10 seconds* (Fig. 1.1(d)), which is not true. But the FT of the segments of initial 5 seconds and that of later 5 seconds durations, shows a spectral peak only at 5Hz and only at 10Hz, respectively but not two spectral peaks together (Fig. 1.1(a, b)) which is the reality. Also from the FT of the complete signal, by its inverse FT, it is possible to get back the original signal. This indicates that to get proper spectrum, it is necessary to segment the data into stationary segments and apply FT to each of them. Further given a nonstationary signal segment, the FT can *synthesize* the signal by many sinusoids whose *magnitudes remain same throughout the infinite interval.*

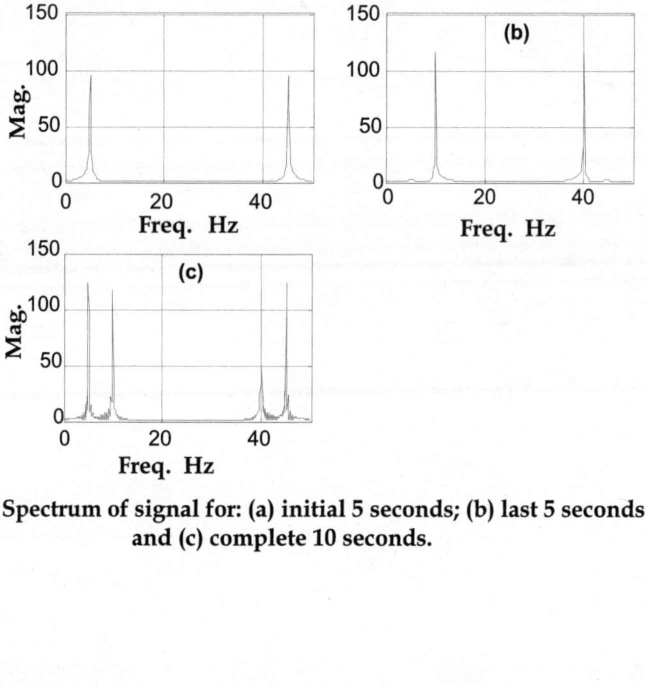

Fig. 1.1 Spectrum of signal for: (a) initial 5 seconds; (b) last 5 seconds and (c) complete 10 seconds.

In practice, most of the natural signals are nonstationary. The most common signal, the speech, is very much nonstationary and it is its nonstationary nature only enables to provide different information. This is so because, the excitation provided by vocal chords gets filtered by the vocal tract whose shape changes with time resulting in different types resonances at different times, hence, the different spectra of the speech signal. For biosignals like ECG, EEG, EMG; their spectra depend on the conditions of the body, which varies with respect to time hence, join the class of nonstationary signals. The line frequency of the electricity produced by a power generating station is not constant as it varies with load and other conditions. Hence, the line frequency signal which appears (due to power supply pickup) as interference in many signals like ECG, has a frequency content that varies with time and is nonstationary.

The segmentation affects the frequency content in a signal as abrupt truncation results in Gibbs effect and leakage effect (will be discussed in detail in a later chapter). The Gibbs effect introduces ripples/variations in a spectrum (more obvious in a flat spectrum) and the leakage will scatter the energy from a high amplitude frequency region to a lower one and vice versa. Such a leakage from a high level peak may even obscure those low level spectral peaks, which are in its frequency neighbourhood. The segment duration also decides the ability of FT to resolve the two sinusoids closely spaced in frequency. Further the Gibbs ripple magnitude varies from segment to segment, depending on the signal amplitude at truncation points, which is random. Thus while getting the frequency content information by applying FT to segmented data/signal, the data segmentation will introduce variability/variance and shift (bias) both in magnitude and frequency. Further, there can be a trades-off between frequency resolution and variance, as the given data can be segmented into more number of smaller segments and their FT magnitudes can be averaged to get a reduced variance.

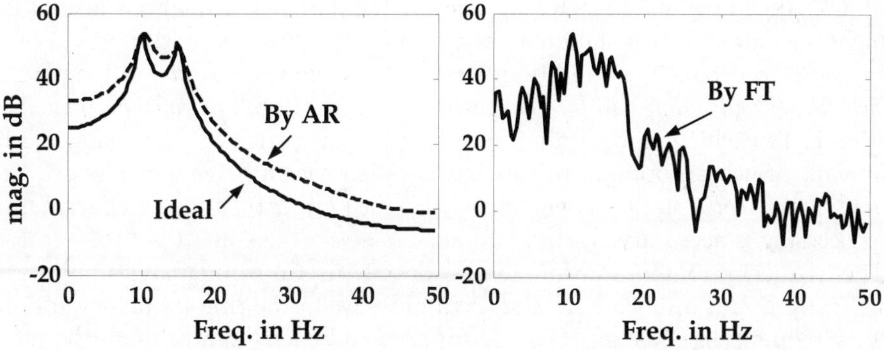

Fig. 1.1 (d) Spectrum of an AR process : ideal, by AR method and by FT.

The spectrum obtained *by using FT* suffers from variance, bias and also by limited frequency resolution problems. Hence, to overcome these problems, it is necessary to process the data/signal in *time domain without going to frequency domain* using FT. As long as the processing in time is confined to available data duration, the spectrum achieved will have less variance, bias and improved frequency resolution. This also will overcome the trade-off between variance and the frequency resolution. This is possible by fitting a time domain model /filter like autoregressive (AR) or autoregressive moving average (ARMA) model to the data/spectrum. The AR and ARMA models are nothing but allpole and pole-zero filters, respectively. The parameters/coefficients of the model/filter are extracted from the signal so that the model will represent the given signal/ its spectral characteristics. That is, the coefficients of the model are chosen to minimize the error energy between the signal and the model, over the signal duration (1935-1945). If the type of the model matches the given data (i.e., the signal is generated by such a model driven by white noise/impulse), the model parameters extracted are *optimal.* This approach is generally known as parametric spectral estimation (modelling) and has significantly less variance, bias and better frequency resolution. Such a spectral estimation based on AR model has been shown to correspond to maximum entropy (Burg, 1967). Figure 1.1(d) shows the spectrum obtained by FT and by AR modelling of an AR process. The noise driving the system mainly contributes to the variance in addition to the other effects already mentioned. The spectrum obtained by parametric method is very smooth and also resolves the closely spaced spectral peaks even with small data length compared to that by FT and hence the former has significantly less variance and better frequency resolution.

In arriving at the model parameters, the minimization of the error energy results in a system of equations which in turn involves statistics of the signal/the input to the model. These statistics are computed based on the assumption that the signal is stationary over the segment interval and hence the validity of optimality is *confined to that interval or block of data.* Outside this interval if the frequency content changes, a different optimum model/filter is required. That is, this approach *is valid only for processing stationary signal segments/blocks and is often known as block parametric method and* it provides better quality spectral estimate. Hence, in the above nonstationary example where the frequency of the sinusoid is changed from 5Hz to 10Hz, to apply the parametric methods also, it is required to segment the data into two stationary segments, one up to 5 seconds and the other from 5 to 10 seconds. To extend the validity of the model beyond the segment length considered, *the model/ filter should* ***adapt*** *to the changes in the frequency content of the signal.* That is, the adaptation has to achieve a new minimum mean square error for the changed situation or environment. Such an optimal processing is called *adaptive signal processing. In real time application (which works on sample to sample basis) even for stationary signals,* such an adaptive signal processing is necessary.

The frequency content information is not only useful from analysis point of view but also has many useful applications. For example, say, in the line frequency interference in ECG, if its frequency is constant, it can be suppressed by a fixed notch filter with its notch at the interference frequency. The frequency content information helps in deciding the notch frequency of the filter. However, if the line interference frequency changes, a notch filter with its notch at a fixed frequency may not effectively suppress the interference, as the notch frequency will not match the actual frequency of line interference. Hence, for proper line frequency interference suppression, it is necessary to know the line frequency variation with time and based on this, the notch position has to be adjusted. The other approach is to estimate the component of interference adaptively based on its some reference and subtract it from the interference contaminated ECG (Widrow, 1965). That is, the suppression operation should ***adapt*** to the varying nature of the interference and hence the suppression or processing becomes adaptive or adaptive signal processing is necessary.

Signals are invariably accompanied by undesired components, which take the role of noise with respect to signal. In the above case considered, the ECG forms the signal and the line interference the noise. Assuming the line frequency interference is stationary, its removal by a fixed notch filter is possible only if its notch occupies a different frequency region other than that of the signal. Otherwise, the signal component at the line frequency also gets suppressed resulting in signal distortion. That is, if the signal and noise have *overlapping frequency regions*; it is not possible to remove the noise without distorting the signal. In such a situation, the noise removal is done again by *optimal filtering also known as Wiener filtering* (Norbert Wiener, 1947). Here, as mentioned above, the filter is chosen such that over the signal interval, the energy of the error between the true signal and its estimate derived from the noise corrupted signal, becomes minimum. Again its validity is confined to that signal interval or block and hence valid *for signals, which are stationary over that interval.* For a nonstationary scenario, *the adaptation from one optimum to another is essential* and the adaptive processing mentioned above serves this purpose.

The optimal parametric methods mainly differ in the manner they utilize the available data and an effective utilization will result in a better solution. Further, efficient solution of system of equations is possible by using the property of the statistics like using the Toeplitz property of the signal autocorrelation matrix resulting in a recursive solution by Levinson (1947)). The solution can also be simplified by using constrained minimization of the error energy (Burg method, 1975). However, the constraint may get reflected in the performance and may affect it also. The characteristics of the structures (normal lattice, Makhoul, 1978 and escalator lattice, Ahmed and Youn, 1980) used for implementation can also be included in the process of minimization and they may aid in further processing. Further, the minimization can be linear or nonlinear depending on the model used and for example, allpole model results in a linear solution and the pole-zero model leads to a nonlinear solution. But for computational simplicity, it is possible to convert a nonlinear solution to a linear one if a suboptimal solution is acceptable.

In the adaptive optimal methods, the parameters of the model are adapted by different approaches like steepest descent gradient method (1950) and its simplified version of least mean square (LMS) algorithm (Widrow, 1959), recursive least square (RLS) method (Plackett, 1950), block LMS approach (Clark and Mitra, 1981) and the Kalman filter (Kalman, 1960). The type of adaptation differs from one approach to the other and hence their performance and computational load. Generally, as expected, there will be a trade-off between computational load and performance. In the LMS gradient algorithm, as the adaptation is based on sample to sample basis without using any statistical quantities, it is simple. From utility point of view, the basic LMS algorithm has been used in other forms like, leaky LMS, Griffiths LMS, momentum LMS, median LMS, variable stepsize LMS, etc. There can be variations of these methods based on the *orthogonalization of the input* by orthogonal transforms like discrete cosine transform (M.J. Narasimha et al., 1981) and implementation structures like lattice filter (Griffiths, 1977), which have a improved convergence performance. For stationary or slow nonstationary signals and which involve long impulse response lengths of the order of 1000 to 2500 samples, adaptation on block to block basis which can exploit efficient frequency domain convolution for its output estimation was proposed. In an attempt to achieve reduction in eigenvalue spread and reduce the computational load per input sampling period, subband adaptive filters which is realized by a perfect reconstruction filter bank came in to existence Most of these approaches use *finite duration impulse response (FIR)/ all zero filters* but they can be extended to pole-zero filters also. However, with pole-zero models, their stability and local minima aspects have to be taken into account. The algorithm and the structure used for its implementation may also differ in their numerical performance to inevitable finite word length arithmetic used.

The conjugate gradient (CG) method introduced by M.R. Hestenes and E. Stiefel in 1952 for solving large system of linear equations in finite difference and finite element methods (for solving partial differential equations, circuit analysis problems and sparse matrix equations) has been modified for the purpose adaptive filtering even with a single iteration per sample (unlike many iterations per sample of original CG). This is similar to steepest descent but the new direction chosen is orthogonal to previous directions and in each direction, the error is minimized so that the convergence occurs in only a number of samples same as the order of the system/ filter (to be identified).

The RLS algorithm is also another attempt for achieving improved convergence performance by recursively computing the statistical quantities, viz., the inverse autocorrelation matrix of the input and the cross-correlation between input and the desired signal, on sample to sample basis and hence it is computationally complex. To reduce the RLS computational complexity from N^2 (N is filter length) to order of N, fast versions of it based both on transversal (FTRLS) and lattice versions (FLRLS) have come into existence. The Kalman filter is also a recursive solution to Wiener filtering and is similar to RLS algorithm but it assumes a model for its input data.

A nonlinear process or system cannot be handled by linear methods. But by neural network which uses back-propagation algorithm to train multilayer perceptrons (Rumelhart, Hinton and Williams in 1986) and Volterra adaptive filters (Gabor, 1954) can handle such a scenario. These have built in nonlinear functions /property and may include algorithms which adapt the higher order statistics (beyond autocorrelation like, triple correlation) by using gradient approach. However, computational load of these can be significantly high, for example, the neural network approach achieves its superior performance with many stages and by adapting a large number of parameters at each stage (similar to adaptation of cascade of many transfer functions).

A sophisticated adaptive algorithm like Kalman filter may provide good performance but their appropriateness and the cost involved in realizing it will decide its utility. Thus, many simple algorithms with their even limited performance may be appropriate in many situations and their performance improvement with marginal increase in cost is of importance. Also for a certain application, only a particular type of optimal/adaptive algorithm may be suitable and may even require some additional tailoring to achieve a good performance. In the forthcoming chapters, a detailed treatment of these optimal and adaptive algorithms, their evolution and their applications with the required background to understand, will be considered.

The adaptive parametric processing is not only important from the point of view of getting the actual and better quality spectral information but as it represents the data by a smaller number of parameters than the actual signal length, it provides *data compression* and hence will be useful for efficient storage, transmission and classification purposes. Such a compression is popular under the name linear predictive coding (LPC) (Atal, 1970) in the field of speech processing. The adaptive filter/signal processing has very vast applications. But here only some of them are considered briefly giving a bird's-eye view of this field to provide motivation. But many applications will be dealt in detail in the chapter 17.

1.2 ADAPTIVE SPECTRAL ESTIMATION/LINE ENHANCER/ADAPTIVE PREDICTION FILTER

In the prediction, the present sample $x(n)$ is approximated as a weighted sum of previous samples,

$$\hat{x}(n) = \sum_{k=1}^{M} w_k(n)\, x(n-k).$$

The weights w_k, $k = 1, 2, \ldots, M$; are chosen such that the energy of the error $e(n)$ between the present sample $x(n)$ and its estimate $\hat{x}(n)$, given by $e(n) = x(n) - \hat{x}(n)$; is minimized. This is known as linear prediction and the coefficients of the adaptive filter are nothing but the denominator polynomial of an IIR filter.

$$X(z) = \frac{\sigma_e}{1 + \sum_{k=1}^{M} a_k z^{-k}}, a_k = -w_k(n), \sigma_e^2 = \frac{1}{L}\sum_{n=0}^{L-1} e^2(n)$$

The frequency response of such an IIR filter provides the spectrum of the signal $x(n)$.

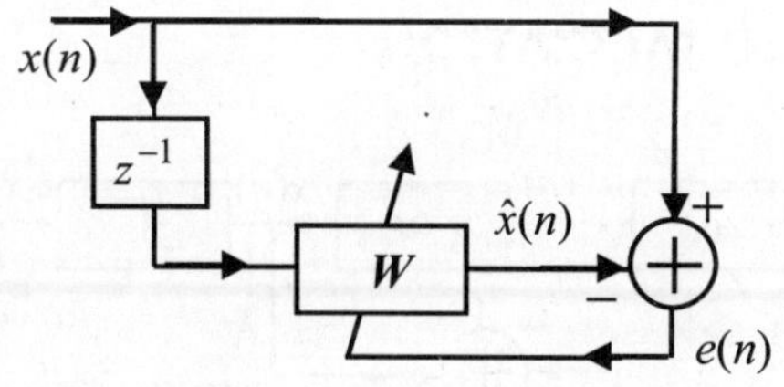

Fig. 1.2 Schematic of spectral estimation by linear prediction.

Instead of delaying the input by a single delay unit, use of a higher value of delay D will provide an estimate of a narrowband signal, when it is associated with a broadband noise and such an application is known *as line enhancer/predictive filtering*. The name prediction filter is due to the fact that the present sample is predicted by P samples, which are D samples away. Since beyond D samples, only the signal can have a correlation but not the broadband noise and hence only signal will contribute to the weighted sum. Thus, the estimate $\hat{x}(n)$ is free from noise and the narrowband signal gets enhanced, hence the name line (spectral line) enhancer. The line enhancer can also be realized using notch filter and this will be discussed in the chapter 17 on applications.

1.3 ADAPTIVE ECHO CANCELLATION

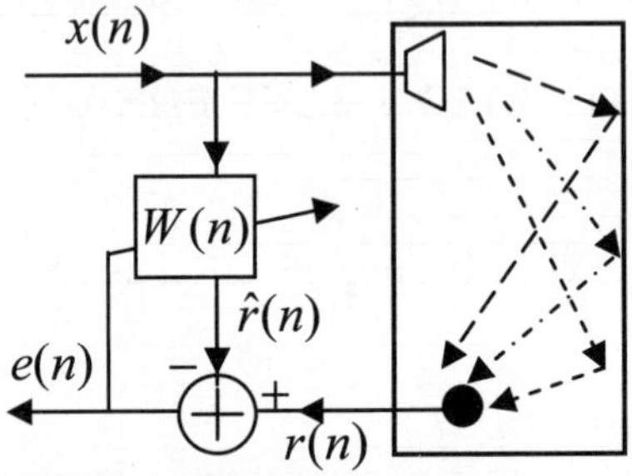

Fig. 1.3 Schematic of adaptive echo cancellation.

In acoustic/line echo cancellation, the acoustic echo path/the line echo path spectral characteristics vary with time. In acoustic/line echo, speech $x(n)$ given out by the loudspeaker (at the far end) itself reaches you with some delay after your speech, as $r(n)$ picked up by the microphone. This is due to the reflection or echo from the room/hall environment at the far end. This will be very annoying to continue the speech conversation and hence need to be cancelled. For this, at the far end, an estimate of the echo $\hat{r}(n)$ is made and is subtracted from the microphone output $r(n)$. In getting an estimate of the echo, the echo path is modeled by the adaptive filter $W(n)$.

1.4 ADAPTIVE CHANNEL EQUALIZATION

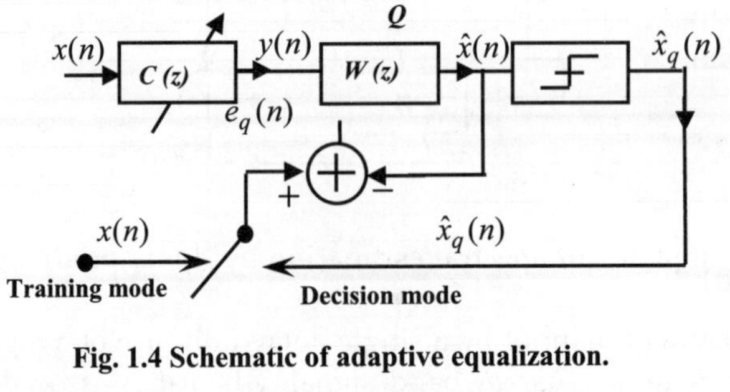

Fig. 1.4 Schematic of adaptive equalization.

The band limited communication channel $C(z)$ distorts the input $x(n)$ by introducing inter-symbol interference. This can be removed by equalizing the channel output by knowing $x(n)$ or considering the quantized adaptive filter output $\hat{x}_q(n)$ (by the quantizer Q) is same as $x(n)$. In such a case, the adaptive filter will model the *inverse of the channel* so that its quantized output provides the input to the channel (1965). The channel $C(z)$ and its inverse model $W(z)$ cancel each other to provide the input $x(n)$ as $C(z)W(z)=1$. If the input is available at the receiver, then it is called the *training mode.* However, the channel spectral characteristics change with time (for example, ionosphere channel during day is different from that of the night) and it may be required to estimate it adaptively for a proper removal of inter-symbol interference. This calls for frequent transmission of the training sequence to estimate the channel, whenever it is changed and its quantized output of the filter is compared with the training sequence. To make it efficient, instead of the input to the channel, the quantized adaptive filter output (assuming it is correct) can be used. This is known as *decision directed mode* but this would take more time for the adaptive filter to work properly. An improved version of equalization is achieved feeding back the previous decisions, known as decision feedback equalizer (Austin, 1967).

1.5 ADAPTIVE DIFFERENTIAL PULSE CODE MODULATION (ADPCM)

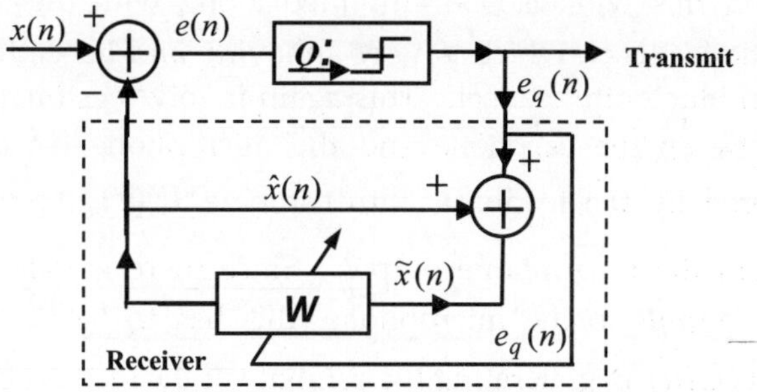

Fig. 1.5 Schematic of adaptive DPCM.

In pulse code modulation (PCM), the number of bits required depends up on the variance of the signal and each sample of the signal is quantized by as many as 8 to 16 bits. However, for efficient transmission, such a representation is not desirable as this will increase the data rate (number of bits per second) which demands a higher channel bandwidth. Hence, it is necessary to represent the signal samples with minimum number of bits with minimum reconstruction error. To meet this in differential pulse code modulation (DPCM), the error $e(n)$, the difference between the signal $x(n)$ and its estimate $\hat{x}(n)$, is quantized and this quantized error $e_q(n)$ is transmitted. Hence, for a good low bit rate transmission, a good estimate of speech is essential. However, as speech is nonstationary, such an estimate of speech can be achieved by an adaptive prediction filter $W(z)$. The sigma-delta modulation which is a simplified extension of ADPCM, also uses adaptive predictor to reduce its quantization noise which is of broadband nature.

1.6 ADAPTIVE FEEDBACK CANCELLATION IN HEARING AIDS

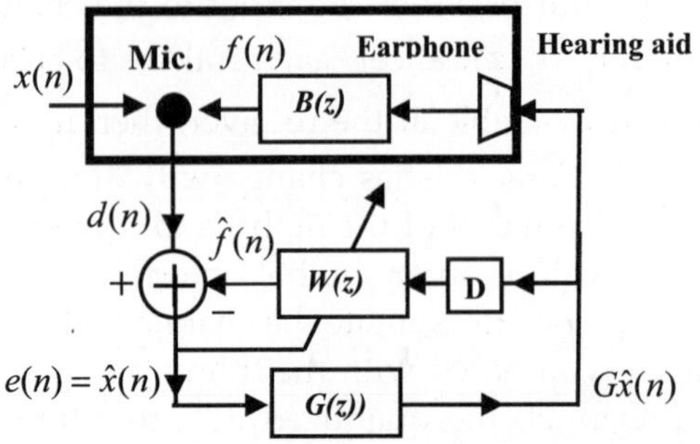

Fig. 1.6 Schematic of hearing aid adaptive feedback cancellation.

In a hearing aid, basically the signal $x(n)$ picked up by the microphone is amplified $G\hat{x}(n)$ and sent to the earphone. However, because of small size of the hearing aid, the close proximity between the microphone and the earphone makes the former to pick up the latter's output $f(n)$ and this will also be amplified along with the actual input $x(n)$, leading to a self acoustic feedback which is very annoying and hence has to be cancelled. This can also happen in bluetooth headsets. This again involves estimation of the acoustic feedback path $B(z)$ between the earphone and the microphone. To overcome this, the earphone input is filtered by the feedback path estimate $W(z)$ and the resulting signal $\hat{f}(n)$ is subtracted from the microphone output. Since there is delay (D) involved in reaching the earphone output to the microphone, this has to be accounted by using a delayed earphone input $G\hat{x}(n)$ in generating the estimate of the feed back signal $\hat{f}(n)$.

1.7 ACTIVE NOISE CANCELLATION (CONTROL)

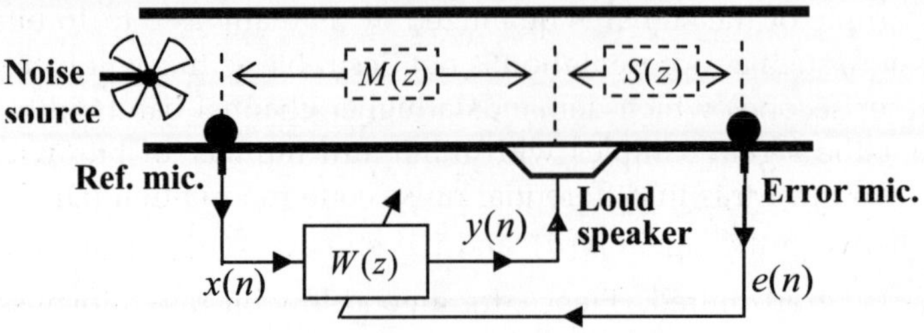

Fig. 1.7 Schematic of active noise cancellation.

In industrial ducts and mobile vehicles it is difficult to reduce the low frequency noise (less than 500Hz) due to rotating machinery like the fan in the duct, engine in a car and propeller in an aircraft/ helicopter. This is due to the requirement of very large size passive absorbers like foam, glass-wool, etc., whose weight does not permit to use. Hence, noise cancellation is done by generating an antinoise adaptively in acoustic domain and superimposing it on the existing noise filed. The residual noise is used as error to correct the quality of antinoise. As the two noise fields interact over a physical space, effectively the error signal is available after it has passed through an acoustic path, called the secondary path, $S(z)$. This path has to be taken into account in generating the antinoise and plays a crucial role in the performance active noise cancellation. Here, basically the adaptive filter $W(z)$ develops the change in magnitude and phase as the sound propagates from the source end to the loudspeaker (main acoustic path $M(z)$)over the reference signal $x(n)$ picked up by the reference microphone. The residual noise $e(n)$ is sensed by the error microphone at the termination end. It is possible to get noise reduction of the order of 15dB for noise field characterized by fundamental and a few of its harmonics. Though this concept was there even in 1937, but till early 1990, the technology was not in a position to cater for the required accuracy and computational load.

1.8 ADAPTIVE BEAMFORMING OF SENSORS

Beamforming is a spatial filtering that enables selective reception of the waves arriving in a particular direction and rejecting from other directions. Such a spatial filtering is realized by an array of sensors by suitably weighing each sensor of the array. The beamforming has applications in radar, sonar, seismology and wireless communication for steering the reception in the desired direction. Further, in speech processing it is used to reduce the ambient noise effect. For a plane wave arriving at an angle with respect to an array of sensors in a line with its elements separated by a uniform distance, there will be a constant delay, in reaching the successive sensors. By compensating for these delays in each sensor, the signals arriving in that specific direction can be time aligned in phase. But as this is not possible for signals coming from other directions, they suffer cancellation/attenuation. Instead of using mere delay for each sensor, a digital filter in addition will provide not only spatial filtering which steers the beam in the desired direction but also temporal filtering. In a changing scenario, an adaptive version of such beamformer will be of immense utility. There are many variations of such a beamformer providing the desired features for different scenarios.

The power/ second order spectrum considered till now, is sufficient to characterize a system or signal completely, only if it is minimum phase or the signal has a Gaussian probability density function. If the process is non-Gaussian or is generated by nonlinear mechanisms, higher-order spectra are required to characterize the process. Bispectrum, the Fourier Transform (FT) of Triple Correlation Sequence (TCS), is one such. The bispectrum enables estimation of the deviation from the Gaussian behavior of a process, identification of a mixed phase system from its output, and characterization of the nonlinear properties of the mechanism that generates the process. It not only provides the spectral magnitude but also the complete spectral phase of the signal or system. Due to its computational complexity and poor quality of the estimate using discrete FT (DFT), its development and applications were limited. But the introduction of parametric approach to bispectrum estimation (C.L. Nikias and M.R. Raghuveer in 1984), gave impetus to its further growth. The bispectrum is used in a wide range of applications like recovery of reflection series in seismology, channel equalization in data communication, nondestructive evaluation using ultrasonic methods, signal or image reconstruction only from phase, and in knowing the history of evolution of the universe using the cosmic microwave background as its deviation from Gaussianity provides the details. Bicoherence, a squared normalized version of the bispectrum, is a convenient measure for quantifying the extent of phase coupling in a signal, monitoring non-Gaussianity and nonlinearity in applications like aircraft structure fatigue monitoring and checking gear teeth in helicopter transmissions. In view of the importance of bispectrum, its estimation by simple parametric and adaptive methods, is briefly considered.

The above applications as stated already are given briefly to bring out the importance of optimal and adaptive processing and to motivate about its diversity and universality. Chapter 17 on the applications of adaptive filters, not only provides detailed treatment of these applications but also of many more.

Chapter 2

SIGNAL APPROXIMATION: The FOURIER TRANSFORM

This chapter deals with the signal approximation by orthogonal basis functions over a finite interval by finding the weights such that the mean square error between the signal and its estimate. Such an approximation is valid over an infinite interval for a periodic signal as the signal in one period repeats infinite times and the representation is discrete in frequency. However, such an approximation, is not directly valid for an arbitrary signal over an infinite interval. Such an attempt of representation of an arbitrary signal over an infinite interval by an orthogonal set of complex exponentials has been considered and is known as Fourier transform which is continuous in frequency. Further, its properties and limitations are considered. The Fourier transforms of important functions which are essential for processing are derived.

2.1 INTRODUCTION

The simplest signal is a sinusoid of a definite frequency f_o with an initial phase θ_0. That is, if $x(t)$ is the signal (Fig. 2.1(a)), then

$$x(t) = \sin(2\pi f_o t + \theta_0)$$

In practice, the signals that one comes across are not as simple as this but in general are complicated. Signals like speech, ECG, EEG communication signal like PCM, music signal are highly complex compared to a sinusoid (Fig. 2.1 (a)). But it is possible to realize these complex signals in terms of many simple sinusoid signals. What is seen as a record of any physical signal may be a combination of many simple sinusoids of different amplitudes. In a record, all these components are *always* present and hence it is *not possible to see them separately* as shown in Fig. 2.1(b), which involves four sinusoids corrupted with noise.

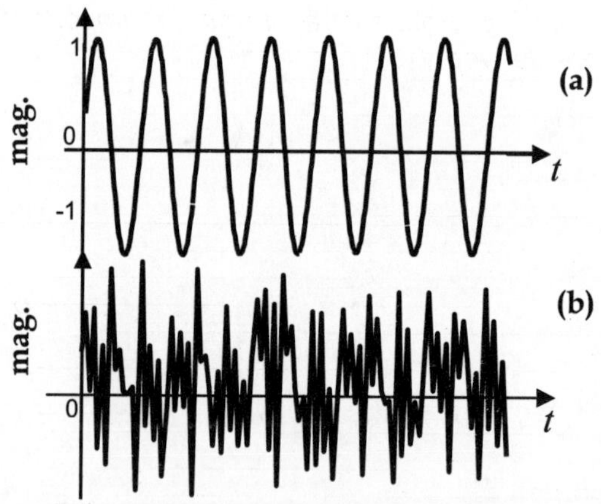

Fig. 2.1 Time domain signal (a) Simple sinusoid and (b) Sum of four sinusoids.

Hence, it is not possible to modify each component, as they are present simultaneously. Any attempt to modify like amplification of any component, may affect all of them. One can think of decomposing the signal into *different types of components not necessarily sinusoids.* To process a given signal selectively, it is required to decompose the given signal into a *desired* set of components. In other words, for a given set of components, how they have been combined to get the desired signal is of importance. Hence, it is of importance to approximate a given signal in terms of a *given set* of component functions and this is called signal approximation, which will be dealt in the subsequent section. The signal approximation not only helps in forming the desired signal but as the components of the signal are known, it also enables selective processing of each component to get desired results in the final signal.

2.2 APPROXIMATION OF VECTORS

Signals are similar to vectors and signal approximation/representation is similar to that of vectors.

A vector V_1 can be approximated by another vector V_2 as (Fig 2.2a)

$$V_1 \approx C_{12} V_2 \text{ and } V_1 = C_{12} V_2 + V_e \tag{2.1}$$

$C_{12}V_2$ is the *component of the vector* V_1 *along the vector* V_2 and is obtained by drawing a perpendicular from the end of V_1 to the vector V_2.

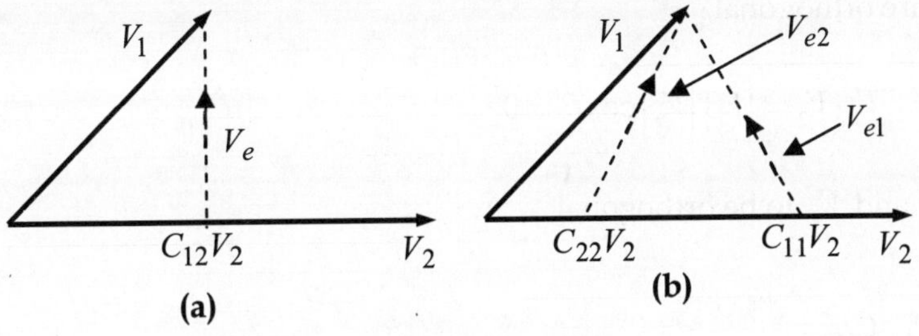

Fig. 2.2 Vector approximation.

The vector V_1 can also be expressed as (Fig. 2.2b)

$$V_1 = C_{11} V_2 + V_{e1} \text{ or } V_1 = C_{22} V_2 + V_{e2}$$

The representation of the vector V_1 by V_2 is made up of two parts, one in terms V_2 and another the error V_e or V_{e1} or V_{e2}. But V_e is unique, as it is the *smallest error* possible. From the geometry, $V_e < V_{e1}$ and also $V_e < V_{e2}$ and for that matter, for any arbitrary $V_{ei}, V_e < V_{ei}$. The component of a vector along another vector is the one for which the *error is minimum.* If the component of a vector is large along another vector that means the two vectors are similar in their direction. C_{12} is a measure of their similarity. If $C_{12} = 0$, then one vector does not have any component on the other vector and the two vectors are perpendicular to each other and are called ***orthogonal vectors***.

If θ is the angle between V_1 and V_2, then their dot product is given by,

$$V_1 \cdot V_2 = |V_1||V_2|\cos\theta \tag{2.2}$$

The magnitude of the component of V_1 along V_2 is,

$$|V_1|\cos\theta = \frac{V_1 \cdot V_2}{|V_2|}$$

The component of vector of V_1 along V_2 is

$$\left(\frac{V_1 \cdot V_2}{|V_2|^2}\right)V_2 = C_{12}V_2$$

$$C_{12} = \frac{V_1 \cdot V_2}{|V_2|^2} \tag{2.3}$$

If V_1 and V_2 are orthogonal,

$$V_1 \cdot V_2 = |V_1||V_2|\cos\theta = 0$$

That is, for V_1 and V_2 to be orthogonal

$$C_{12} = \frac{V_1 \cdot V_2}{|V_2|^2} = 0$$

Similarly, the magnitude of component of V_2 along V_1 is

$$|V_2|\cos\theta = \frac{V_1 \cdot V_2}{|V_1|}$$

2.3 THE SIGNAL APPROXIMATION

The concept of vector approximation can be extended to signals also. A signal $x_1(t)$ can be approximated in terms of another signal $x_2(t)$ over **an interval** $(t_1 \le t \le t_2)$ as

$$x_1(t) \approx C_{12}\, x_2(t) \text{ for } (t_1 \le t \le t_2) \tag{2.4}$$

Similar to vectors, the best approximation is achieved by choosing C_{12} to minimize the error function $x_e(t)$ over the time interval $(t_1 \le t \le t_2)$ where

$$x_e(t) = x_1(t) - C_{12}\, x_2(t)$$

This is achieved by minimizing the average of square of the error ε over the interval $(t_1 \le t \le t_2)$, given by

$$\varepsilon = \frac{1}{(t_1 - t_2)} \int_{t_1}^{t_2} x_e^2 (t)\, dt$$

Hence, C_{12} is chosen such that

$$\frac{d\varepsilon}{dC_{12}} = 0 \text{, i.e., } \frac{d}{dC_{12}} \left\{ \frac{1}{(t_1 - t_2)} \int_{t_1}^{t_2} \left[x_1(t) - C_{12}\, x_2(t) \right]^2 dt \right\} = 0$$

Changing the order of differentiation and integration,

$$\frac{1}{(t_2 - t_1)} \int_{t_1}^{t_2} \frac{d}{dC_{12}} \left[x_1^2(t) - 2C_{12}\, x_1(t)\, x_2(t) + C_{12}^2\, x_2^2(t) \right] dt = 0$$

$$\frac{1}{(t_2 - t_1)} \left[\int_{t_1}^{t_2} \frac{d}{dC_{12}} \left[x_1^2(t) \right] dt - 2\int_{t_1}^{t_2} x_1(t)\, x_2(t) dt + 2C_{12} \int_{t_1}^{t_2} x_2^2(t) dt \right] = 0$$

The first integral is zero. Therefore,

$$C_{12} = \frac{\int_{t_1}^{t_2} x_1(t) x_2(t)\, dt}{\int_{t_1}^{t_2} x_2^2(t) dt} = \frac{1}{K_2} \int_{t_1}^{t_2} x_1(t) x_2(t) dt\,, \quad K_2 = \int_{t_1}^{t_2} x_2^2(t)\, dt$$

This is similar to vectors as

$$V_1 \cdot V_2 \;\Leftrightarrow\; \int_{t_1}^{t_2} x_1(t) x_2(t) dt \text{ and } V_2 \cdot V_2 \;\Leftrightarrow\; \int_{t_1}^{t_2} x_2^2(t)\, dt$$

$$\frac{V_1 \cdot V_2}{|V_2|^2} = \frac{V_1 \cdot V_2}{V_2 \cdot V_2} = C_{12}$$

When $x_1(t)$ and $x_2(t)$ are complex,

$$C_{12} = \frac{1}{K_2} \int_{t_1}^{t_2} x_1(t) x_2^*(t)\, dt \ ,$$

C_{12} represents the magnitude of the component of $x_2(t)$ in the signal $x_1(t)$.

Example: Approximate

$$x(t) = \begin{cases} 1, & 0 < t < \dfrac{\pi}{2} \\ -1, & \dfrac{\pi}{2} < t < \pi \end{cases} \quad \text{by } \cos t$$

$$x(t) \approx C_{12} \cos t \text{ and } C_{12} = \frac{\int_0^{\pi} x(t) \cos t\, dt}{\int_0^{\pi} \cos^2 t\, dt}$$

$$\int_0^{\pi} \cos^2 t\, dt = \int_0^{\pi} \frac{(1+\cos 2t)}{2} dt = \frac{1}{2}\left[t + \frac{\sin 2t}{2}\right]_0^{\pi} = \frac{\pi}{2}$$

$$C_{12} = \frac{\int_0^{\pi} x(t) \cos t\, dt}{\int_0^{\pi} \cos^2 t\, dt} = \frac{2}{\pi}\int_0^{\pi} x(t) \cos t\, dt = \frac{2}{\pi}\left[\int_0^{\pi/2} (1) \cos t\, dt + \int_{\pi/2}^{\pi} (-1) \cos t\, dt\right]$$

$$= \frac{2}{\pi}\left[\sin t\Big|_0^{\pi/2} - \left\{\sin t\Big|_{\pi/2}^{\pi}\right\}\right] = \frac{4}{\pi}$$

Therefore

$$x(t) = \frac{4}{\pi} \cos t .$$

That is, the rectangular function has a component of $\cos t$ and its magnitude is $(4/\pi)$. If two functions *vary at the same rate,* then they have a high similarity and one will contain a large component of the other.

If $C_{12} = 0$, then the signal $x_1(t)$ contains no component of signal $x_2(t)$ and in such a case, these two functions are called *orthogonal* signals. It follows for orthogonal signals,

$$\int_{t_1}^{t_2} x_1(t)\,x_2(t)\,dt = 0 \qquad (2.5)$$

For example, $\sin(n\omega_0 t)$ and $\sin(m\omega_0 t)$ are orthogonal over the interval $\left(t_0, t_0 + 2\pi/\omega_0\right)$ for integer values of m and n. Let

$$I = \int_{t_0}^{t_0+2\pi/\omega_0} \sin(n\omega_0 t)\sin(m\omega_0 t)\,dt$$

$$= \int_{t_0}^{t_0+2\pi/\omega_0} \frac{1}{2}\left[\cos(n-m)\omega_0 t - \cos(n+m)\omega_0 t\right]dt$$

$$= \frac{1}{2}\left[\frac{\sin(n-m)\omega_0 t}{(n-m)\omega_0} - \frac{\sin(n+m)\omega_0 t}{(n+m)\omega_0}\right]_{t_0}^{t_0+2\pi/\omega_0}$$

Since n and m are integers $(n-m)$ and $(n+m)$ are also integers, hence $I = 0$. Therefore, the two functions are orthogonal. Similarly $\{\sin(n\omega_0 t)$ and $\cos(m\omega_0 t)\}$ and $\{\cos(n\omega_0 t)$ and $\cos(m\omega_0 t)\}$ are orthogonal pairs.

For a set of n functions $g_1(t), g_2(t), \cdots g_n(t)$ which are orthogonal to one another over an interval t_1 to t_2

$$\int_{t_1}^{t_2} g_i(t) g_l(t)\,dt = 0, \quad i \neq l$$
$$= K_i, \quad i = l \qquad (2.6)$$

These are analogous to the co-ordinate vectors in two- or three-dimensional space.

Any arbitrary function $x(t)$ can be approximated over an interval (t_1, t_2) by a linear combination of $(n+1)$ mutually orthogonal functions, as

$$x(t) \approx C_0 g_0(t) + C_1 g_1(t) + C_2 g_2(t) + \ldots + C_n g_n(t) \qquad (2.7)$$

For the best approximation, C_r's are chosen to minimize ε, the mean square of $x_e(t)$.

$$x_e(t) = x(t) - \sum_{r=0}^{n} C_r g_r(t)$$

$$\varepsilon = \frac{1}{(t_2 - t_1)} \int_{t_1}^{t_2} \left(x(t) - \sum_{r=0}^{n} C_r g_r(t) \right)^2 dt \tag{2.8}$$

C_r represents component of $x(t)$ over the signal $g_r(t)$ and $x(t)$ can be expressed by them as in Eqn. (2.7). Multiplying $x(t)$ by $g_r(t)$ and integrating,

$$\int_{t_1}^{t_2} x(t) g_r(t)\, dt \approx \int_{t_1}^{t_2} [C_0 g_0 + C_1 g_1(t) + C_2 g_2(t) + \ldots. + C_n g_n(t)] g_r(t)\, dt$$

Since,

$$\int_{t_1}^{t_2} g_i(t)\, g_r(t)\, dt = \begin{cases} 0, & i \neq r \\ \int_{t_1}^{t_2} g_r^2(t)\, dt, & i = r \end{cases}$$

$$C_r \int_{t_1}^{t_2} g_r^2(t) dt = \int_{t_1}^{t_2} x(t)\, g_r(t)\, dt.$$

$$C_r = \frac{\int_{t_1}^{t_2} x(t)\, g_r(t)\, dt.}{\int_{t_1}^{t_2} g_r^2(t) dt} = \frac{1}{K_r} \int_{t_1}^{t_2} x(t)\, g_r(t)\, dt \tag{2.8a}$$

where

$$K_r = \int_{t_1}^{t_2} g_r^2\, (t)\, dt \tag{2.8b}$$

If $g_1, g_2, \ldots\ldots g_n$ are not orthogonal, then

$$-\int_{t_1}^{t_2} x(t) g_i(t)\, dt + C_i \int_{t_1}^{t_2} g_i^2(t)\, dt + \sum_{\substack{r=0 \\ r \neq i}}^{n} C_r \int_{t_1}^{t_2} g_i(t) g_r(t)\, dt = 0\,,\ i = 0, 1, 2 \ldots n$$

In this, since $x(t)$ and the basis members $g_r(t)$, $r = 0, 1, 2, \cdots, n$; are known, each integral can be computed and it is required to solve *(n+1)* simultaneous equations to find the

C_i unknowns. But when $g_r s$ form an orthogonal basis, the above equations get *decoupled* and $C_i s$ can be found independently. The mean square error ε is expanded as

$$\varepsilon = \frac{1}{(t_2 - t_1)} \int_{t_1}^{t_2} \left[x(t) - \sum_{r=0}^{n} C_r\, g_r(t) \right]^2 dt$$

$$= \frac{1}{(t_2 - t_1)} \left[\int_{t_1}^{t_2} x^2(t)dt - 2\sum_{r=0}^{n} C_r \int_{t_1}^{t_2} x(t) g_r(t)dt + \int_{t_1}^{t_2} \sum_{r=0}^{n} C_r^2 g_r^2(t)dt \right]$$

From Eqn. (2.8a),

$$\int_{t_1}^{t_2} x(t)\, g_r(t)\, dt = K_r C_r$$

Therefore,

$$\varepsilon = \frac{1}{(t_2 - t_1)} \left[\int_{t_1}^{t_2} x^2(t)dt - 2\sum_{r=0}^{n} C_r^2 K_r + \sum_{r=0}^{n} C_r^2 K_r \right]$$

$$= \frac{1}{(t_2 - t_1)} \left[\int_{t_1}^{t_2} x^2(t)dt - \sum_{r=0}^{n} C_r^2 K_r \right]$$

By definition, ε is positive, so increasing n to ∞, ε reduces to zero. That is,

$$\int_{t_1}^{t_2} x^2(t)\, dt = \sum_{r=0}^{\infty} C_r^2 K_r \tag{2.9}$$

The set of functions $\{ g_i(t) \}$, which are mutually orthogonal over the interval (t_1, t_2) is said to be a complete or closed set, if there exists no function $\zeta(t)$ for which it is true that

$$\int_{t1}^{t_2} \varsigma(t)\, g_r(t) dt = 0, \quad r = 1, 2, 3, 4, \cdots$$

If the set $\{g_r(t)\}$ forms a complete set, representation of $x(t)$ by mutually orthogonal set of functions is called *generalized Fourier series representation* of $x(t)$. This representation by an infinite series is given by

$$x(t) = C_1 g_1(t) + C_2 g_2(t) + C_3 g_3(t) + \cdots + C_n g_n(t) + \cdots$$

The complete set of mutually orthogonal functions is called *orthogonal basis.* If

$$K_r = \int_{t_1}^{t_2} g_r^2(t)\, dt = 1, \text{ for all } r = 1, \ldots, n.$$

Then the orthogonal basis is called orthonormal basis. In such a case,

$$\int_{t_1}^{t_2} x^2(t)dt = \sum_{r=0}^{n} C_r^2 \tag{2.10}$$

This is known as Parseval's theorem, stated as *the energy in time domain is equal to the energy in the frequency domain.* The error minimization says that the error should be orthogonal to the functions $g_i s$, involved in the approximation. The signal approximation is valid even for the case of non-orthogonal basis.

2.4 THE FOURIER SERIES REPRESENTATION

The trigonometric functions, viz., the sine and cosine functions form an orthogonal basis and the representation of a signal over an interval by this basis leads to the well-known Fourier series:

$$x(t) = \sum_{n=0}^{\infty} \left[a_n \cos(n\omega_o t) + b_n \sin(n\omega_o t) \right], \; t_o < t < t_o + T \tag{2.11}$$

where, $T = 2\pi / \omega_o$,

$$a_n = \frac{\int_{to}^{t_o+T} x(t) \cos(n\omega_o t)\, dt}{\int_{to}^{t_o+T} \cos^2(n\omega_o t)\, dt} = \frac{2}{T} \int_{t_o}^{t_o+T} x(t) \cos(n\omega_o t)\, dt \tag{2.12}$$

$$b_n = \frac{\int_{to}^{t_o+T} x(t) \sin(n\omega_o t)\, dt}{\int_{to}^{t_o+T} \sin^2(n\omega_o t)\, dt} = \frac{2}{T} \int_{t_o}^{t_o+T} x(t) \sin(n\omega_o t)\, dt \tag{2.13}$$

$\cos(n\omega_o t)$ and $\sin(n\omega_o t)$ are orthogonal over the same interval. Therefore, $\cos(n\omega_o t)$ and $\sin(n\omega_o t)$ by themselves do not form a complete set.

The exponential functions $e^{j\omega_o nt}$ form an orthogonal set over an interval $(t_o, t_o + 2\pi/\omega_o)$. An arbitrary function $x(t)$ can be expressed in terms of these functions as

$$x(t) = \sum_{n=-\infty}^{\infty} X_n e^{jn\omega_o t} \tag{2.14}$$

and

$$X_n = \frac{\int_{to}^{t_o+T} x(t)(e^{jn\omega_o t})^* dt}{\int_{t_o}^{t_o+T} \left(e^{jn\omega_o t}\right)\left(e^{jn\omega_o t}\right)^* dt}$$

$$X_n = \frac{1}{T}\int_{t_o}^{t_o+T} x(t)e^{-jn\omega_o t} dt \tag{2.15}$$

This representation is called exponential Fourier series. This representation approximates the signal over a finite interval $(t_o, t_o + T)$. *Outside this interval, the function $x(t)$ and the corresponding Fourier series representation need not be the same. But if the function $x(t)$ is periodic, the series representation applies to the entire interval* $(-\infty < t < \infty)$. This follows from the fact that

$$e^{-jn\omega_o t} = e^{-jn\omega_o (t+T)}, \text{ for } T = 2\pi/\omega_o$$

Relation to trigonometric Fourier series:

$$a_o = x_o, \quad a_n = X_n + X_{-n}, \; b_n = j(X_n - X_{-n})$$

and

$$X_n = (a_n - jb_n)/2, \quad X_{-n} = (a_n + jb_n)/2$$

2.5 THE FOURIER TRANSFORM REPRESENTATION

An arbitrary signal $x(t)$ can be represented over the interval $(-\infty, \infty)$ by the Fourier transform and this can be derived by constructing a periodic signal $x_T(t)$ with period T, from $x(t)$ and representing it by an exponential Fourier series and then allowing the period T to go to ∞.

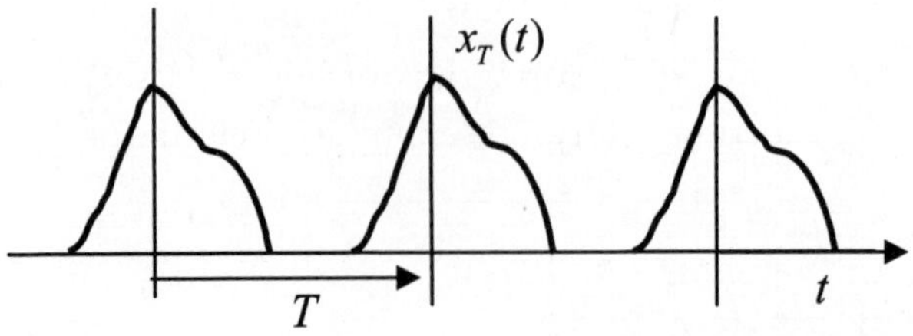

Fig. 2.3 Periodic signal generated from an arbitrary signal.

$$\lim_{T\to\infty} x_T(t) = x(t)$$

$x_T(t)$ can be represented as

$$x_T(t) = \sum_{n=-\infty}^{\infty} X_n e^{j\omega_o nt}, \quad \omega_o = 2\pi / T$$

$$X_n = \frac{1}{T}\int_{-T/2}^{T/2} x(t)\, e^{-jn\omega_0 t} dt$$

X_n is the amplitude of the frequency component $n\omega_o$. As T becomes large, ω_o becomes smaller and the spectrum becomes denser. Also, the amplitudes of the components become smaller, however the shape of the spectrum remains same, as all the frequency components are scaled by the same factor, viz., $(1/T)$. In the limit when $T = \infty$, there will be infinite number of frequency components, that is, the spectrum exists for every ω and the spectrum is continuous unlike for the Fourier series for which the spectrum is discrete.

If $n\omega_o = \omega_n$, then X_n is a function of ω_n i.e.,

$$X_n = X_n(\omega_n).$$

Further if

$$T X_n(\omega_n) = X(\omega_n)$$

Then

$$x_T(t) = \sum_{n=-\infty}^{\infty} X_n e^{j\omega_o nt},$$

becomes

$$x_T(t) = \frac{1}{T}\sum_{n=-\infty}^{\infty} X(\omega_n) e^{j\omega_n t},$$

with

$$X(\omega_n) = T X_n = \int_{-T/2}^{T/2} x_T(t) e^{-j\omega_n t} dt$$

Since $T = \dfrac{2\pi}{\omega_o}$,

$$x_T(t) = \frac{1}{2\pi} \sum_{n=-\infty}^{\infty} X(\omega_n) e^{j\omega_n t} \omega_o$$

This equation states that $x_T(t)$ can be represented by the sum of exponentials, $e^{j\omega_n t}$ of frequencies $\omega_1, \omega_2, \omega_3, \omega_4, \cdots \omega_n, \cdots$, with the magnitude of frequency ω_n equal to $X_n = X(\omega_n)\omega_o / 2\pi$. The amplitude of frequency component ω_n is proportional to $X(\omega_n)$.

The above equation can be interpreted as the area under the curve $X(\omega_n) e^{-j\omega_n t}$; $X(\omega_n) e^{-j\omega_n t} \omega_o$ represents the area of a strip of height $X(\omega_n) e^{-j\omega_n t}$ and width ω_o. Since ω_o becomes small as $T \to \infty$, the evaluation of this area under the curve becomes better. Also, $x_T(t) \to x(t)$ as $T \to \infty$. In such a case , the summation sign can be replaced by an integral, ω_n by ω and ω_o by $d\omega$, which leads to

$$x(t) = \frac{1}{2\pi} \int_{-\infty}^{\infty} X(\omega)\, e^{j\omega t} d\omega \tag{2.16}$$

and

$$X(\omega) = \int_{-\infty}^{\infty} x(t) e^{-j\omega t} dt \tag{2.17}$$

Eqn. (2.17) is known as the *direct Fourier transform* and Eqn. (2.16) as the *inverse Fourier transform.* Eqn. (2.16) is also referred to as Fourier representation /expansion of the signal. Unlike the Fourier series, the spectrum $X(\omega)$ is *continuous.*

For a real signal, since $X(\omega)$ given by Eqn. (2.17), has to be conjugate symmetric,

$$|X(\omega)| = |X(-\omega)| \text{ and } \theta(\omega) = -\theta(-\omega)$$

where $|X(\omega)|$ is the magnitude spectrum and $\theta(\omega)$ is the phase spectrum. The magnitude and phase spectra are symmetric and antisymmetric, respectively with respect to frequency about the origin.

The negative frequency exists because a real function has been expressed in terms of complex exponentials $e^{j\omega t}$. Since $e^{j\omega t} = \cos\omega t + j\sin\omega t$, $\cos\omega t$ *or* $\sin\omega t$ which are real can be realized *only by having the complex conjugate of* $e^{j\omega t}$, $e^{-j\omega t}$ and the negative frequency comes into existence. For the same reason in realizing a real quantity in terms of complex quantity, for a real signal the magnitude spectrum and the phase spectrum are symmetric and antisymmetric, respectively.

2.5.1 Existence of Fourier Transform

If the integral in the Equation,

$$X(\omega) = \int_{-\infty}^{\infty} x(t) e^{-j\omega t} dt$$

is finite, then the FT exists. But since the magnitude of $e^{-j\omega t}$ is unity, the sufficient condition (but not a necessary) for the existence of FT of $x(t)$ is that

$$\int_{-\infty}^{\infty} |x(t)|\, dt < \infty \tag{2.18}$$

This is also known as absolute integrability or Dirichlet's condition.

2.5.2 Fourier Transform of Some Important Functions

1. Gate function

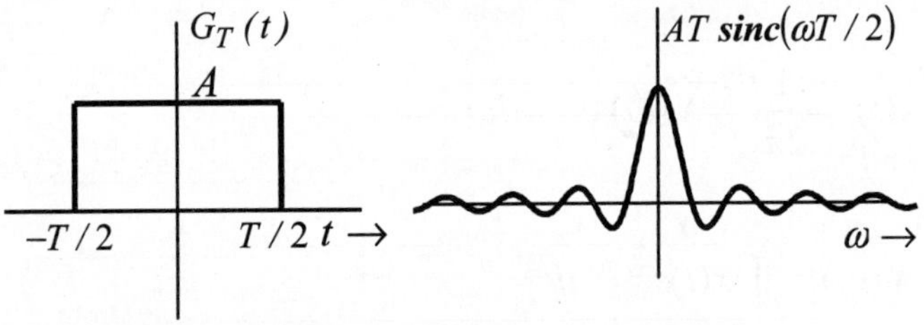

Fig. 2.4 Gate function and its Fourier transform.

A gate function $G_T(t)$ is a rectangular pulse and is defined as

$$G_T(t) = A, \quad |t| < T/2$$

$$= 0, \quad |t| > T/2$$

FT of this function is given by

$$X(\omega) = \int_{-T/2}^{T/2} A\, e^{-j\omega t} dt = \frac{A}{-j\omega}\left[e^{-j\omega T/2} - e^{j\omega T/2}\right] = \frac{2A}{\omega}\sin(\omega T/2)$$

$$= AT\left[\frac{\sin\left(\omega T/2\right)}{\left(\omega T/2\right)}\right] = AT\ \mathrm{sinc}\left(\omega T/2\right)$$

The FT of a gate function is a sinc function.

2. Impulse function

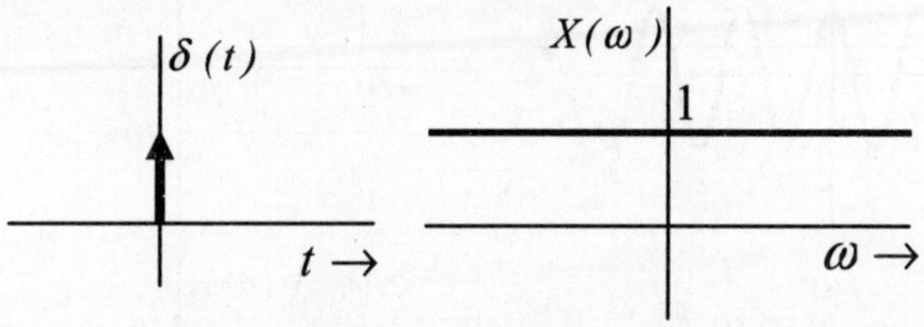

Fig. 2.5 Impulse function and its FT.

Impulse function is defined as

$$\left.\begin{aligned}\delta(t) &= 1,\ t=0\\ &= 0,\ t\neq 0\end{aligned}\right\}\ \text{and} \int_{-\infty}^{\infty}\delta(t)\,dt = 1$$

$$X(\omega) = \int_{-\infty}^{\infty}\delta(t)\,e^{-j\omega t}dt \ = e^{-j\omega(0)}\int_{-\infty}^{\infty}\delta(t)\,dt = 1$$

The FT of an impulse function is unity and constant throughout the frequency range.

3. A constant function

$x(t) = A$, is a constant function. It does not satisfy the absolute integrability. A gate function becomes a constant function, in the limit as T approaches ∞. The FT of a constant function is therefore the FT of a gate function as T tends to ∞. The FT of the gate function is

$$X(\omega) = AT\mathbf{sinc}\left(\omega T/2\right)$$

As $T\to\infty$,

$$X(\omega) = \lim_{T\to\infty} AT\ \mathbf{sinc}\left(\omega T/2\right) = 2\pi A \lim_{T\to\infty}\left((T/2\pi)\mathbf{sinc}\left(\omega T/2\right)\right)$$

$$= 2\pi A\delta(\omega) \quad [\because \lim_{T\to\infty}(T/2\pi)\mathbf{sinc}\left(\omega T/2\right) = \delta(\omega)]$$

The FT of a constant is an impulse at the origin.

4. Trigonometric functions: $\cos\omega_o t$ and $\sin\omega_o t$

A sinusoid/co-sinusoid does not satisfy the condition of absolute integrability; hence its FT is computed as a limiting case.

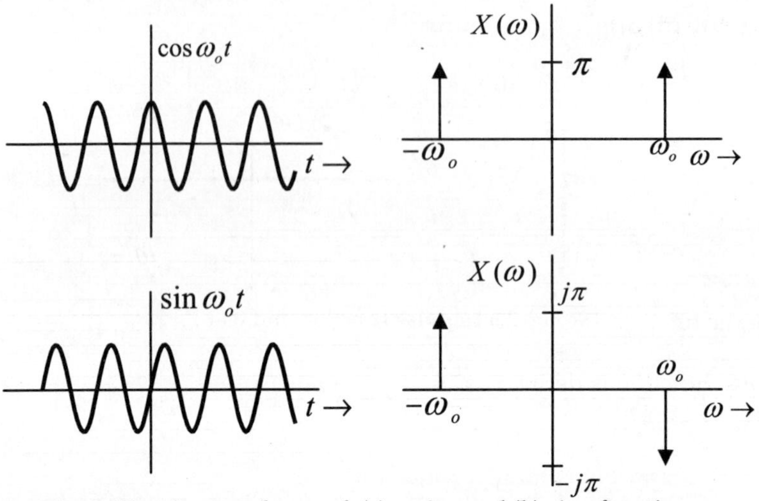

Fig. 2.6 Fourier transforms of (a) cosine and (b) sine functions.

$$F\left[\cos\omega_o t\right] = \lim_{T\to\infty} \int_{-T/2}^{T/2} \cos\omega_o t \, e^{-j\omega t} dt$$

$$= \lim_{T\to\infty}(1/2) \int_{-T/2}^{T/2} \left\{ e^{-j(\omega-\omega_o)t} + e^{-j(\omega+\omega_o)t} \right\} dt$$

$$= \lim_{T\to\infty}(1/2) \left[\frac{e^{-j(\omega-\omega_o)t}}{-j(\omega-\omega_o)} + \frac{e^{-j(\omega+\omega_o)t}}{-j(\omega+\omega_o)} \right]_{-T/2}^{T/2}$$

$$= \pi \lim_{T\to\infty}(T/2\pi) \left[\frac{\sin(\omega-\omega_o)T/2}{(\omega-\omega_o)T/2} + \frac{\sin(\omega+\omega_o)T/2}{(\omega+\omega_o)T/2} \right]$$

$$F\left[\cos\omega_o t\right] = \pi\left[\delta(\omega-\omega_o) + \delta(\omega+\omega_o)\right]$$

Similarly, the FT of $\sin\omega_o t$ is given by

$$F\left[\sin\omega_o t\right] = j\pi\left[\delta(\omega+\omega_o) - \delta(\omega-\omega_o)\right]$$

5. Exponential function

$$e^{j\omega_o t} = \cos\omega_o t + j\sin\omega_o t$$

$$F\left[e^{j\omega_o t}\right] = F\left[\cos\omega_o t\right] + jF\left[\sin\omega_o t\right]$$

$$= \pi\left[\delta(\omega-\omega_o)+\delta(\omega+\omega_o)\right] + j\left[j\pi\left\{\delta(\omega+\omega_o)-\delta(\omega-\omega_o)\right\}\right]$$

$$= \pi\left[\delta(\omega-\omega_o)+\delta(\omega+\omega_o)\right] + \pi\left[\delta(\omega-\omega_o)-\delta(\omega+\omega_o)\right]$$

$$F\left[e^{j\omega_o t}\right] = 2\pi\ \delta(\omega-\omega_o)$$

6. An impulse train

For a periodic function $x(t)$, the Fourier series representation is given by Eqn. (2.14)

$$x(t) = \sum_{n=-\infty}^{\infty} X_n e^{jn\omega_o t}$$

and taking FT on both the sides,

$$F\left[x(t)\right] = F\left[\sum_{n=-\infty}^{\infty} X_n e^{jn\omega_o t}\right],\ X(\omega) = \sum_{n=-\infty}^{\infty} X_n F\left[e^{jn\omega_o t}\right]$$

$$X(\omega) = 2\pi\sum_{n=-\infty}^{\infty} X_n\delta(\omega - n\omega_o) \tag{2.19}$$

The above expression is a mathematical proof for the statement, FT *of a periodic signal is discrete and it is made up of impulses located at the harmonic frequencies of the fundamental frequency* ω_o *of the signal* and its strength is 2π times the value of the corresponding coefficient in the exponential Fourier series.

In case of an impulse train with period *T*, the Fourier series coefficient X_n is given by

$$X_n = \frac{1}{T}\int_{-T/2}^{T/2}\delta(t)e^{-jn\omega_o t}dt = \frac{1}{T}$$

Therefore, from Eqn. (2. 19),

$$X(\omega) = 2\pi\sum_{n=-\infty}^{\infty}\frac{1}{T}\delta(\omega - n\omega_o)$$

$$X(\omega) = \omega_o\sum_{n=-\infty}^{\infty}\delta(\omega - n\omega_o) = \omega_o\delta_{\omega_o}(\omega) \tag{2.20}$$

The FT of an impulse train of period T is also an impulse train with a period $\omega_o = 2\pi / T$ radians.

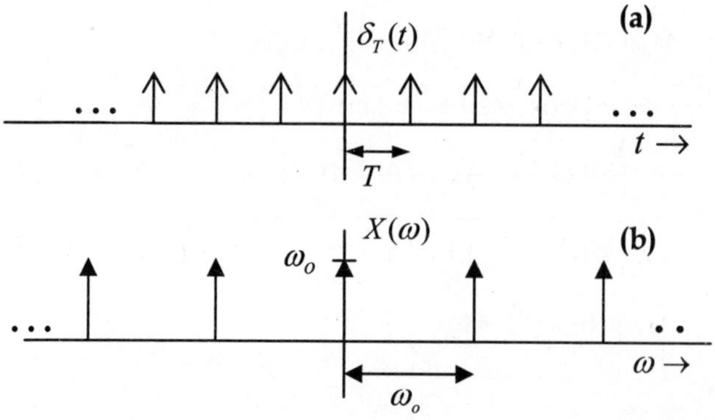

Fig. 2.7 Fourier transform of an impulse train: (a) impulse train and (b) its FT.

For a periodic signal with period $T = 2\pi / \omega_o$, there can be only those frequencies which are integrable multiples of the fundamental frequency ω_o, to ***maintain the zero crossings.*** *Hence, the spectrum of a periodic signal is discrete and occurs only at multiples of the basic frequency ω_o.*

2.6 IMPORTANT PROPERTIES OF FOURIER TRANSFORM

1. Symmetry property

If $x(t) \leftrightarrow X(\omega)$, then $X(t) \leftrightarrow 2\pi\, x(-\omega)$ (2.21)

From Eqn. (2.16),

$$x(t) = \frac{1}{2\pi}\int_{-\infty}^{\infty} X(\omega)\, e^{j\omega t} d\omega \quad \therefore \quad 2\pi\, x(-t) = \int_{-\infty}^{\infty} X(\omega)\, e^{-j\omega t} d\omega$$

$$2\pi\, x(-t) = \int_{-\infty}^{\infty} X(p)\, e^{-jpt}\, dp \quad \text{(changing the variable } \omega \rightarrow p\text{)}$$

$$2\pi\, x(-\omega) = \int_{-\infty}^{\infty} X(p)\, e^{-jp\omega} dp \quad \text{(changing the variable } t \rightarrow \omega\text{)}$$

$$= \int_{-\infty}^{\infty} X(t)\, e^{-j\omega t} dt \quad \text{(changing the variable } p \rightarrow t\text{)}$$

$$2\pi\, x(-\omega) = F\left[X(t)\right]$$

2. Frequency shifting property (modulation)

If $x(t) \leftrightarrow X(\omega)$, then $x(t)e^{j\omega_o t} \leftrightarrow X(\omega - \omega_o)$ (2.22)

$$F\left[x(t)e^{j\omega_o t}\right] = \int_{-\infty}^{\infty} x(t)e^{j\omega_o t}e^{-j\omega t}dt = \int_{-\infty}^{\infty} x(t)e^{-j(\omega-\omega_o)t}dt = X(\omega - \omega_o)$$

Multiplication of function $x(t)$ by $e^{j\omega_o t}$ in the time domain is equivalent to shifting the function in the frequency domain by ω_o.

3. Time shifting property

If $x(t) \leftrightarrow X(\omega)$,

then $x(t - t_o) \leftrightarrow X(\omega)e^{-j\omega t_o}$ (2.23)

$$F[x(t-t_o)] = \int_{-\infty}^{\infty} x(t-t_o)e^{-j\omega t}dt$$

$$= \int_{-\infty}^{\infty} x(p)e^{-j\omega(p+t_o)}dp \qquad (t - t_o = p)$$

$$= e^{-j\omega t_o}\int_{-\infty}^{\infty} x(p)e^{-j\omega p}dp = e^{-j\omega t_o}X(\omega)$$

If a function is shifted in the time domain by t_o seconds, then its magnitude spectrum $|X(\omega)|$ remains unchanged, but the phase spectrum is changed by an amount $(-\omega t_o)$.

4. Fourier transforms of even and odd functions

Any arbitrary function $x(t)$ can be split into even part $x_e(t)$ and odd part $x_o(t)$. For an even function $x_e(t)$, $x_e(t) = x_e(-t)$, i.e., it is symmetrical about the origin. For an odd function $x_o(t)$, $x_o(t) = -x_o(-t)$, i.e., it is antisymmetric about the origin.

$$x(t) = \left[\frac{x(t)}{2} + \frac{x(-t)}{2}\right] + \left[\frac{x(t)}{2} - \frac{x(-t)}{2}\right] = x_e(t) + x_o(t)$$

$$x_e(t) = \left[\frac{x(t)}{2} + \frac{x(-t)}{2}\right]$$, is even as t is changed to $-t$, it remains same.

$$x_o(t) = \left[\frac{x(t)}{2} - \frac{x(-t)}{2}\right]$$, is odd as t is changed to $-t$, it changes its sign.

(a) $x_e(t) \quad \leftrightarrow \quad R_e(\omega) = \int_{-\infty}^{\infty} x_e(t)\cos(\omega t)dt$

$$X(\omega) = \int_{-\infty}^{\infty} x_e(t)e^{-j\omega t}dt = \int_{-\infty}^{\infty} x_e(t)\cos(\omega t)\,dt - j\int_{-\infty}^{\infty} x_e(t)\sin(\omega t)\,dt$$

$$= \int_{-\infty}^{\infty} x_e(t)\cos(\omega t)\,dt \quad = \quad R_e(\omega)$$

The imaginary term is zero as the integrand is an odd function. Also, as $\cos(\omega t)$ is an even function, $x_e(t)\cos(\omega t) = x_e(t)\cos(-\omega t)$. Therefore, $X_e(\omega) = X_e(-\omega)$. Similarly, if $X(\omega)$ is real and even the inversion gives

$$x(t) = \int_{-\infty}^{\infty} X_e(\omega)e^{j\omega t}d\omega = \int_{-\infty}^{\infty} R_e(\omega)e^{j\omega t}d\omega$$

$$= \int_{-\infty}^{\infty} R_e(\omega)\cos(\omega t)d\omega - j\int_{-\infty}^{\infty} R_e(\omega)\sin(\omega t)d\omega$$

$$= \int_{-\infty}^{\infty} R_e(\omega)\cos(\omega t)\,d\omega = x_e(t)$$

(b) $x_o(t) \quad \leftrightarrow \quad jI_o(\omega)$

$$X(\omega) = \int_{-\infty}^{\infty} x_o(t)e^{-j\omega t}dt = \int_{-\infty}^{\infty} x_o(t)\cos(\omega t)\,dt - j\int_{-\infty}^{\infty} x_o(t)\sin(\omega t)\,dt$$

$$= -j\int_{-\infty}^{\infty} x_o(t)\sin(\omega t)\,dt \quad = \quad jI_o(\omega)$$

The real integral is zero as multiplication of an odd function by an even function is an odd function. Also as $\sin(\omega t)$ is an odd function,

$$x_o(t)\sin(\omega t) = -x_o(t)\sin(-\omega t)$$

Therefore, $X_o(\omega) = -X_o(-\omega)$, i.e., the frequency function is odd. Similar to even function case, if $X(\omega)$ is given as odd and imaginary, then the resulting inverse is an odd function in time.

This shows that *the real and imaginary parts of the FT of a (real) signal are the FTs of the even and odd parts of the signal, respectively and vice versa. Further, with respect to frequency, the real and imaginary parts of the FT are symmetric and antisymmetric, respectively.*

5. FT of complex signal

For a complex signal $x(t) = x_r(t) + jx_i(t)$,

$$X(\omega) = \int_{-\infty}^{\infty} x(t)e^{-j\omega t}dt = \int_{-\infty}^{\infty} \left[x_r(t) + jx_i(t)\right]e^{-j\omega t}dt$$

$$X(\omega) = \int_{-\infty}^{\infty} \left[x_r(t)\cos(\omega t) + x_i(t)\sin(\omega t)\right]dt$$

$$- j\int_{-\infty}^{\infty} \left[x_r(t)\sin(\omega t) - x_i(t)\cos(\omega t)\right]dt$$

$$X(\omega) = R(\omega) \quad + jI(\omega)$$

where

$$R(\omega) = \int_{-\infty}^{\infty} \left[x_r(t)\cos(\omega t) + x_i(t)\sin(\omega t)\right]dt,$$

$$I(\omega) = -\int_{-\infty}^{\infty} \left[x_r(t)\sin(\omega t) - x_i(t)\cos(\omega t)\right]dt$$

Inverse FTs are given by

$$x_r(t) = \int_{-\infty}^{\infty} \left[R(\omega)\cos(\omega t) - I(\omega)\sin(\omega t)\right]d\omega$$

$$x_i(t) = -\int_{-\infty}^{\infty} \left[R(\omega)\sin(\omega t) + I(\omega)\cos(\omega t)\right]d\omega$$

If $x(t)$ is a real signal, $x(t) = x_r(t)$, then the real and imaginary parts of FT of $x(t)$ are

$$R_e(\omega) = \int_{-\infty}^{\infty} \left[x_r(t)\cos(\omega t)\right]dt \quad , \quad I_o(\omega) = -\int_{-\infty}^{\infty} \left[x_r(t)\sin(\omega t)\right]dt$$

Further, $R_e(\omega) = R_e(-\omega)$ and $I_o(\omega) = -I_o(-\omega)$ as obtained above for a real signal.

If $x(t)$ is a imaginary signal, $x(t) = x_i(t)$, then the real and imaginary parts of FT of $x(t)$ are

$$R_o(\omega) = \int_{-\infty}^{\infty} [x_i(t)\sin(\omega t)]dt \quad , \quad I_e(\omega) = -\int_{-\infty}^{\infty} [x_i(t)\cos(\omega t)]dt$$

Further, $I_e(\omega) = I_e(-\omega)$ and $R_o(\omega) = -R_o(-\omega)$. For a complex signal under different conditions, the nature of the FT is listed in the Table 2.1.

Table 2.1: Properties of the FT of a complex function

Time domain $x(t)$	Frequency domain $X(\omega)$
Real	Real part even Imaginary part odd
Imaginary	Real part odd Imaginary part even
Real part even Imaginary part odd	Real
Real part odd Imaginary part even	Imaginary
Real and even	Real and even
Real and odd	Imaginary and odd
Imaginary and even	Imaginary and even
Imaginary and odd	Real and odd
Complex and even	Complex and even
Complex and odd	Complex and odd

6. Linearity

$$F[a_1x_1(t) + a_2x_2(t)] = \int_{-\infty}^{\infty} [a_1x_1(t) + a_2x_2(t)]e^{-j\omega t}dt$$

$$= a_1 \int_{-\infty}^{\infty} x_1(t)e^{-j\omega t}dt + a_2 \int_{-\infty}^{\infty} x_2(t)e^{-j\omega t}dt$$

$$F[a_1 x_1(t) + a_2 x_2(t)] = a_1X_1(\omega) + a_2 X_2(\omega)$$

7. Convolution

Convolution is the most commonly encountered operation, in the field of signal processing. The convolution $y(t)$ of two functions/ signals $x_1(t)$ and $x_2(t)$ is given by

$$y(t) = x_1(t) * x_2(t) = \int_{-\infty}^{\infty} x_1(\tau)\, x_2(t-\tau)\, d\tau$$

The operations involved are:

1. The signal $x_2(\tau)$ flipped/ reflected about the origin to get $x_2(-\tau)$

2. The signals $x_1(\tau)$ and $x_2(-\tau)$ are multiplied at every τ

3. To find the convolved signal $y(t_0)$ at a particular $t = t_0$,

$$y(t_p) = \int_{-M+t_0}^{L} x_1(\tau)\, x_2(t_0-\tau)\, d\tau$$

(a) $x_2(-\tau)$ is shifted in time by t_0 to get $x_2(t_0-\tau)$,

(b) The product signal $x_1(\tau)\, x_2(t_0-\tau)$ is computed,

(c) The integral of the product signal $x_1(\tau)x_2(t_0-\tau)$ is computed.

The upper limit of integral is the upper limit of the fixed function $x_1(\tau)$ and the lower limit is decided by the shifted function $x_2(t_0-\tau)$ and is $(-M+t_0)$ for the example shown in the Fig 2.8.

4. To find $y(t)$ at all t, step 3 is repeated for every value of t.

It is important to notice that in computing $y(t)$, 't' forms the *shift parameter*. The duration of $y(t)$ is the sum of the durations of the two signals involved, as the shift of $x_2(-\tau)$ beyond this does not contribute to the integral.

Time convolution theorem

If $x_1(t) \leftrightarrow X_1(\omega)$ and $x_2(t) \leftrightarrow X_2(\omega)$, then,

$$y(t) = x_1(t) * x_2(t) = \int_{-\infty}^{\infty} x_1(\tau)x_2(t-\tau)\, d\tau \leftrightarrow X_1(\omega)\, X_2(\omega)$$

$$F\left[x_1(t) * x_2(t)\right] = \int_{-\infty}^{\infty} e^{-j\omega t} \left(\int_{-\infty}^{\infty} x_1(\tau)\, x_2(t-\tau)\, d\tau \right) dt$$

$$= \int_{-\infty}^{\infty} x_1(\tau) \left(\int_{-\infty}^{\infty} x_2(t-\tau) e^{-j\omega t} dt \right) d\tau = \int_{-\infty}^{\infty} x_1(\tau)\, e^{-j\omega \tau} X_2(\omega)\, d\tau$$

$$F\left[x_1(t) * x_2(t)\right] = X_1(\omega) X_2(\omega) \qquad (2.24a)$$

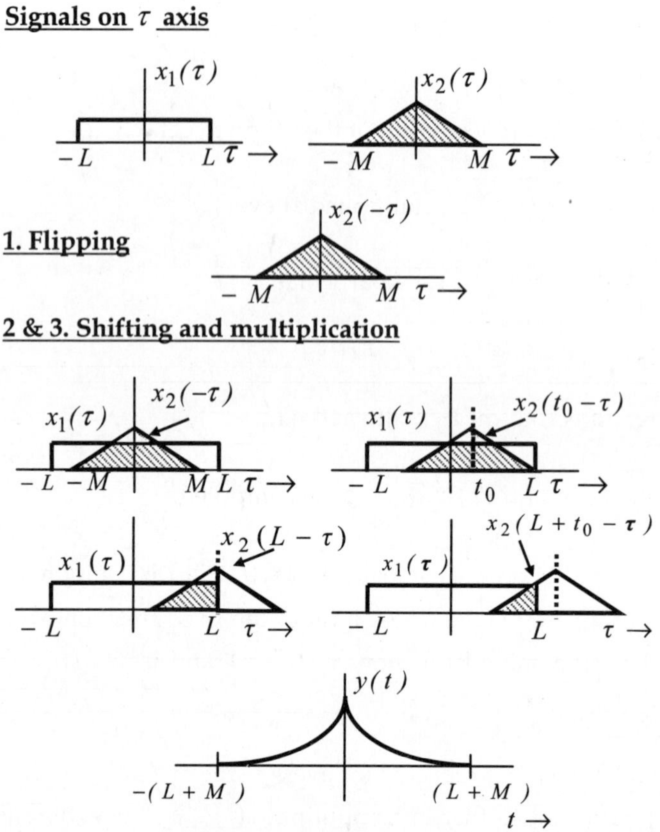

Fig. 2.8 Convolution operation.

That is, *convolving two functions in the time domain is equivalent to the multiplication of their FTs, in the frequency domain.*

The meaning of convolution operation in time is not explicit, but as the convolution theorem states that it is equivalent to the product of the FTs of the signals involved, hence the convolution can be interpreted as *frequency domain weighing. Filtering* selects only the desired frequency components and rejects the other components which is nothing but weighing. Hence, filtering is a convolution operation in time domain between the input and the impulse response of the filter.

Frequency convolution theorem

If $x_1(t) \leftrightarrow X_1(\omega)$ and $x_2(t) \leftrightarrow X_2(\omega)$, then

$$x_1(t)x_2(t) \leftrightarrow \frac{1}{2\pi}\left[X_1(\omega) * X_2(\omega)\right]$$

$$F^{-1}\left\{\frac{1}{2\pi}\left[X_1(\omega) * X_2(\omega)\right]\right\}$$

$$= \frac{1}{2\pi}\left\{\int_{-\infty}^{\infty} \frac{1}{2\pi}\left[\int_{-\infty}^{\infty} X_1(\xi)\, X_2(\omega-\xi) d\xi\right] e^{j\omega t}\, d\omega\right\}$$

$$= \frac{1}{2\pi}\int_{-\infty}^{\infty} X_1(\xi)\left[\frac{1}{2\pi}\int_{-\infty}^{\infty} X_2(\omega-\xi\,)e^{j\omega t} d\omega\right] d\xi$$

$$F^{-1}\left\{\frac{1}{2\pi}\left[X_1(\omega) * X_2(\omega)\right]\right\} = x_2(t)\frac{1}{2\pi}\int_{-\infty}^{\infty} X_1(\xi) e^{j\xi t}\, d\xi$$

$$(\because x(t)\, e^{j\omega_o t} \leftrightarrow X(\omega-\omega_o))$$

$$F^{-1}\left\{\frac{1}{2\pi}\left[X_1(\omega) * X_2(\omega)\right]\right\} = x_1(t)\, x_2(t) \qquad (2.\ 24b)$$

Multiplication of two signals in the time domain is equivalent to convolution of their FTs in the frequency domain.

Convolution of a function with unit impulse function

If $x(t) \leftrightarrow X(\omega)$ then from time convolution theorem,

$$x(t) * \delta(t) \leftrightarrow X(\omega).1 \;\; (\text{as } \delta(t),\ \delta(t) \leftrightarrow 1)$$

Consequently,

$$x(t) * \delta(t) = x(t)$$

The convolution of a function $x(t)$ *with a unit impulse function yields the function* $x(t)$ *itself. This is known as impulse response.* Extending this result,

$$x(t) * \delta(t-T) \leftrightarrow X(\omega) e^{-j\omega T} = x(t-T) \qquad (2.25)$$

Convolution relationships

i. Commutative law

$$x_1(t) * x_2(t) = x_2(t) * x_1(t)$$

$$x_1(t) * x_2(t) = \int_{-\infty}^{\infty} x_1(\tau)\, x_2(t-\tau)\, d\tau$$

$$= \int_{-\infty}^{\infty} x_2(p)\, x_1(t-p)\, dp, \quad p = t-\tau$$

$$= x_2(t) * x_1(t)$$

ii. Distributive law

$$x_1(t) * [x_2(t) + x_3(t)] = x_1(t) * x_2(t) + x_1(t) * x_3(t)$$

iii. Associative law

$$x_1(t) * [x_2(t) * x_3(t)] = [x_1(t) * x_2(t)] * x_3(t)$$

8. Time integration and differentiation

If $x(t) \leftrightarrow X(\omega)$ then $\frac{dx(t)}{dt} \leftrightarrow j\omega\, X(\omega)$ and $\int_{-\infty}^{t} x(\tau)d\tau \leftrightarrow \frac{X(\omega)}{j\omega}$

if $X(\omega)/\omega$ is bounded at $\omega = 0$, i.e., $X(0) = 0$ or $\int_{-\infty}^{\infty} x(\tau)d\tau = 0$

$$x(t) = \frac{1}{2\pi}\int_{-\infty}^{\infty} X(\omega)e^{j\omega t}d\omega$$

$$\frac{dx(t)}{dt} = \frac{1}{2\pi}\frac{d}{dt}\int_{-\infty}^{\infty} X(\omega)e^{j\omega t}d\omega = \frac{1}{2\pi}\int_{-\infty}^{\infty} j\omega\, X(\omega)e^{j\omega t}d\omega$$

$$F\left[\frac{dx(t)}{dt}\right] = j\omega\, X(\omega) \text{ and } F\left[\frac{d^n x(t)}{dt^n}\right] = (j\omega)^n X(\omega)$$

If $$\phi(t) = \int_{-\infty}^{t} x(\tau)\, d\tau, \quad \frac{d\phi(t)}{dt} = x(t)$$

Also if $\phi(t) \leftrightarrow \Phi(\omega)$, $x(t) \leftrightarrow j\omega\,\Phi(\omega)$,

$$\therefore \quad X(\omega) = j\omega\,\Phi(\omega), \qquad \Phi(\omega) = \frac{X(\omega)}{j\omega}$$

That is

$$\int_{-\infty}^{t} x(\tau)d\tau \quad \leftrightarrow \frac{1}{j\omega} \quad X(\omega)$$

In the frequency domain, differentiation and integration in time are equivalent to multiplication and division by $j\omega$, respectively.

9. Frequency differentiation

If $x(t) \leftrightarrow X(\omega)$, then $-jtx(t) \quad \leftrightarrow \quad \dfrac{dX(\omega)}{d\omega}$

$$X(\omega) = \int_{-\infty}^{\infty} x(t)e^{-j\omega t}dt$$

$$\frac{dX(\omega)}{d\omega} = \frac{d}{d\omega}\left[\int_{-\infty}^{\infty} x(t)e^{-j\omega t}dt\right] = \int_{-\infty}^{\infty} -jtx(t)e^{-j\omega t}dt$$

$$F^{-1}\left[\frac{dX(\omega)}{d\omega}\right] = F^{-1}\left[\int_{-\infty}^{\infty} -jtx(t)e^{-j\omega t}dt\right]$$

Therefore, $\dfrac{dX(\omega)}{d\omega} \leftrightarrow -jtx(t)$ or $\dfrac{d^nX(\omega)}{d\omega^n} \leftrightarrow (-jt)^n x(t)$

That is, differentiation in frequency domain is equivalent to multiplication by $(-jt)$, in time domain.

10. Scaling property

If $x(t) \leftrightarrow X(\omega)$, then $x(at) \leftrightarrow \dfrac{1}{|a|}X\left(\dfrac{\omega}{a}\right)$

$$F\left[x(at)\right] = \int_{-\infty}^{\infty} x(at)e^{-j\omega t}dt\,,$$

Let $a > 0$, $\tau = at$, $\quad dt = {d\tau}/{a}$, then

$$F\left[x(at)\right] = \int_{-\infty}^{\infty} x(\tau)e^{-j\omega\left(\frac{\tau}{a}\right)}\frac{d\tau}{a} = \left(\frac{1}{a}\right)\int_{-\infty}^{\infty} x(\tau)e^{-j\left(\frac{\omega}{a}\right)\tau}d\tau$$

$$F\left[x(at)\right] = \left(\frac{1}{a}\right)X\left(\frac{\omega}{a}\right)$$

Let $\quad a < 0,\ a = -b,\ b > 0; \tau = at$

Then $\tau = -bt,\ d\tau = -bdt,\ t \to -\infty, \tau \to \infty$ and $t \to \infty, \tau \to -\infty$

$$F\left[x(at)\right] = F\left[x(-bt)\right]$$

$$= \int_{-\infty}^{\infty} x(-bt)e^{-j\omega t}dt = \int_{\infty}^{-\infty} x(\tau)e^{-j\omega\left(\frac{\tau}{-b}\right)}\frac{d\tau}{-b} = \left(\frac{1}{b}\right)\int_{-\infty}^{\infty} x(\tau)e^{-j\omega\left(\frac{\tau}{-b}\right)}d\tau$$

$$F\left[x(at)\right] = \left(\frac{1}{b}\right)X\left(\frac{\omega}{-b}\right)$$

Combining both cases for $a > 0$ and $a < 0$,

$$F\left[x(at)\right] = \left(\frac{1}{|a|}\right)X\left(\frac{\omega}{a}\right)$$

The function $x(at)$ represents the function $x(t)$ compressed in the time domain by a factor 'a'. Equivalently, in the frequency domain, $X\left(\frac{\omega}{a}\right)$ represents a function $X(\omega)$ expanded by the same factor.

The scaling property states that compression in the time domain is equivalent to the expansion in the frequency domain and vice versa.

Also, if $a = -1$,

$$F\left[x(at)\right] = \left(\frac{1}{|a|}\right)X\left(\frac{\omega}{a}\right),\ F\left[x(-t)\right] = X(-\omega) = X^*(\omega)$$

Therefore,

$$x(-t) \leftrightarrow X^*(\omega)$$

2.7 THE UNCERTAINTY PRINCIPLE

The scaling property *qualitatively* states that an increase in time duration of a signal leads to compression in its frequency spectrum and vice versa. The uncertainty principle provides a *quantitative statement* of the scaling property.

The equivalent bandwidth of a spectral function is given by

$$B_{eq} = \frac{\int_{-\infty}^{\infty} |X(\omega)|\, d\omega}{2|X(\omega)|_{\max}} = \frac{\int_{0}^{\infty} |X(\omega)|\, d\omega}{|X(\omega)|_{\max}}$$

Similarly, the equivalent time interval is given by

$$T_{eq} = \frac{\int_{-\infty}^{\infty} |x(t)|\, dt}{|x(t)|_{\max}}$$

The uncertainty principle states that,

$$B_{eq} T_{eq} \geq \pi \tag{2.26}$$

This is nothing but scaling property.

Proof:

$$B_{eq} T_{eq} = \frac{\int_{-\infty}^{\infty} |X(\omega)|\, dw}{2|X(\omega)|_{\max}} \frac{\int_{-\infty}^{\infty} |x(t)|\, dt}{|x(t)|_{\max}} \geq \pi \tag{2.27}$$

Consider,

$$x(t) = \frac{1}{2\pi} \int_{-\infty}^{\infty} X(\omega) e^{j\omega t} d\omega$$

$$|x(t)| = \frac{1}{2\pi} \left| \int_{-\infty}^{\infty} X(\omega) e^{j\omega t} d\omega \right| \leq \frac{1}{2\pi} \int_{-\infty}^{\infty} |X(\omega)|\, d\omega$$

This is true for all t. Therefore,

$$|x(t)|_{\max} \leq \frac{1}{2\pi} \int_{-\infty}^{\infty} |X(\omega)|\, d\omega \quad \text{or} \quad \frac{\int_{-\infty}^{\infty} |X(\omega)|\, d\omega}{|x(t)|_{\max}} \geq 2\pi$$

$$X(\omega) = \int_{-\infty}^{\infty} x(t) e^{-j\omega t} dt, \quad |X(\omega)| \leq \int_{-\infty}^{\infty} |x(t)|\, dt$$

$$|X(\omega)|_{\max} \leq \int_{-\infty}^{\infty} |x(t)|\, dt \quad \text{or} \quad \frac{\int_{-\infty}^{\infty} |x(t)|\, dt}{|X(\omega)|_{\max}} \geq 1$$

Therefore, Eqn. (2.27) becomes,

$$B_{eq}T_{eq} = \frac{\int_{-\infty}^{\infty}|X(\omega)|\,d\omega}{|x(t)|_{\max}} \; \frac{\int_{-\infty}^{\infty}|x(t)|\,dt}{2|X(\omega)|_{\max}} \geq \{\,2\pi\,\}.\left\{\frac{1}{2}\right\}$$

Or

$$B_{eq}T_{eq} \geq \pi$$

For a given B_{eq}, T_{eq} is $> \pi / B_{eq}$. That is, the frequency resolution B_{eq} can be achieved only by having a signal of time duration T_{eq} greater than π / B_{eq} and vice versa. That is, it is not possible to get time and frequency resolution simultaneously and one has to be sacrificed for the other. The equality is valid only for $x(t) = ae^{-\beta\,t^2}, \beta \geq 0$ which is a Gaussian function whose traditional duration is infinite.

2.8 PARSEVAL'S THEOREM

The Parseval's theorem developed will be proved here. It states that energy in time domain is equal to energy in frequency domain.

$$\int_{-\infty}^{\infty} x^2(t)\,dt = \int_{-\infty}^{\infty} x(t)\left[\frac{1}{2\pi}\int_{-\infty}^{\infty} X(\omega)\,e^{j\omega\,t}d\omega\right]dt$$

$$= \frac{1}{2\pi}\int_{-\infty}^{\infty} X(\omega\,)\left[\int_{-\infty}^{\infty} x(t)\,e^{j\omega\,t}dt\right]d\omega$$

$$\int_{-\infty}^{\infty} x^2(t)\,dt = \frac{1}{2\pi}\int_{-\infty}^{\infty} X(\omega)X(-\omega)\,d\omega$$

$$\int_{-\infty}^{\infty} x^2(t)\,dt = \frac{1}{2\pi}\int_{-\infty}^{\infty} |X(\omega)|^2\,d\omega$$

2.9 GRAM SCHMIDT ORTHOGONALIZATION

In a set of functions $\{\,g_l(t)\}$, $l = 1,2,3,4,\cdots N$, which are mutually orthogonal over the interval (t_1,t_2), any $g_l(t)$ cannot be expressed in terms of other. That is, for a member $g_l(t)$ of the set,

$$g_I(t) = C_1 g_1(t) + C_2 g_2(t) + C_3 g_3(t) + \cdots + C_N g_N(t)$$

is *valid only for* $C_l = 0, l = 1, 2, \cdots N$. Each $g_l(t)$ *cannot* be expressed as linear combination of other functions and is independent and such a set is also known *as linearly independent set.* A minimal set of independent functions that covers a functional space (all possible combinations of set of functions) forms a *basis.*

On the other hand, for a function $g_d(t)$,

$$g_d(t) = C_1 g_1(t) + C_2 g_2(t) + C_3 g_3(t) + \cdots + C_N g_N(t)$$

for which $C_l \neq 0$, at least for some of $l = 1, 2, 3, 4, \cdots N$, $g_d(t)$ is a linear combination of the set of functions considered and is called a *linearly dependent* function and a set of such functions *cannot* form a basis.

An orthogonal set of functions can form a basis. However, a basis need not be orthogonal. As it has been seen it is advantageous to have an orthogonal basis and hence it is necessary to convert a nonorthogonal basis to an orthogonal. The Gram Schmidt procedure facilitates this.

Let $h_l(t), l = 1, 2, 3, 4, \cdots N$, be the desired orthogonal basis and is derived as follows.

Let $h_1(t) = g_1(t)$. Then $h_2(t)$ should be orthogonal to $h_1(t)$.

Therefore,

$$g_2(t) = C_{12}\, g_1(t) + h_2(t)$$

$$C_{12} = \frac{\int_{t_1}^{t_2} h_1(t)\, g_2(t) dt}{\int_{t_1}^{t_2} h_1^2(t)\, dt}, \quad \because g_1(t) = h_1(t)$$

$$h_2(t) = g_2(t) - C_{12} h_1(t)$$

The component of $g_2(t)$ along $h_1(t)$ is $C_{12} h_1(t)$ and the part of $g_2(t)$ which is not in alignment with $h_1(t)$, $g_2(t) - C_{12} h_1(t)$ will be along $h_2(t)$ which is orthogonal to $h_1(t)$.

Similarly, $h_3(t)$ is given by

$$h_3(t) = g_3(t) - \left[C_{13} h_1(t) + C_{23} h_2(t) \right]$$

The component of $g_3(t)$ along $h_1(t)$ is $C_{13} h_1(t)$ and $h_2(t)$ is $C_{23} h_2(t)$ and the part of $g_3(t)$ which is not in alignment with $h_1(t)$ and $h_2(t)$,

$$g_3(t)-\left[C_{13}h_1(t)+C_{23}h_2(t)\right]$$

will be along $h_3(t)$ which is orthogonal to $h_1(t)$ and $h_2(t)$.

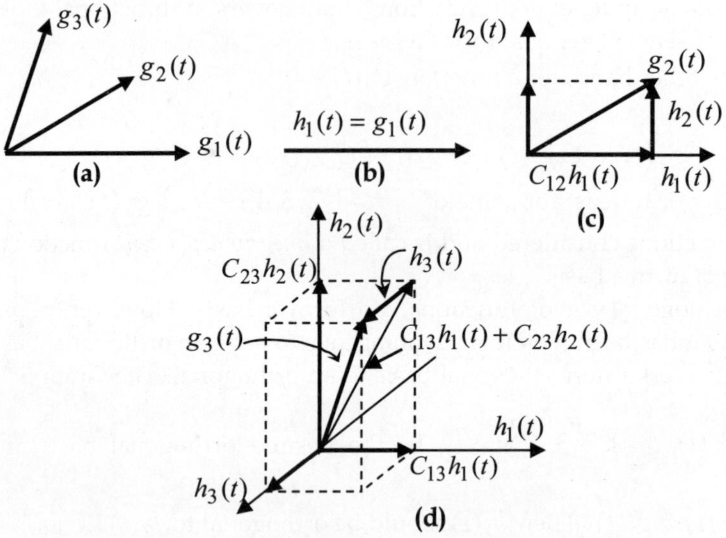

Fig. 2.9 (a) Non-orthogonal vector basis (N=3)
(b), (c) & (d) Gram Schmidt orthogonalization procedure.

Generalizing,

$$h_l\ (t)=g_l\ (t)-\sum_{k=1}^{l-1}C_{kl}\ h_k\ (t)\ , \qquad l=2,3,4,\cdots N$$

where

$$C_{kl}\ =\frac{\int_{t1}^{t_2} h_k\ (t)\ g_l\ (t)dt}{\int_{t1}^{t_2} h_k^2\ (t)\ dt}$$

Thus, given a set of non-orthogonal set of functions, it is possible to convert it to an orthogonal set. In summary, each new vector is expressed in terms of the existing orthogonal vectors and the left-over component is a new vector, which is orthogonal to the existing orthogonal vectors. Thus, we get a complete orthogonal basis.

Chapter 3

SIGNAL TRANSMISSION AND POWER SPECTRAL DENSITY

This chapter deals with the linear time invariant system providing the condition to be obeyed by them for the transmission of signals through them without any distortion. The signal energy representation in frequency domain by power spectral density (PSD) and its relation to time domain correlation are given. The concept of the stationarity of the signal based on PSD, the applicability of Fourier transform (FT) only to stationary signals by the filtering view of FT and the trade-off between the time and frequency resolutions; are described. Further, the concept of white noise is introduced through correlation/PSD.

3.1 LINEAR TIME INVARIANT SYSTEM

A linear system obeys the principle of superposition. That is, if $y_1(t)$ and $y_2(t)$ are the outputs for the inputs $x_1(t)$ and $x_2(t)$, respectively for a linear system, then its output for the input $\{a_1 x_1(t) + a_2 x_2(t)\}$ is given by $y(t)$,

$$y(t) = a_1 y_1(t) + a_2 y_2(t) \tag{3.1}$$

Further, if the system is time invariant, then for the input $x_1(t - t_o)$, the output will be $y_1(t - t_o)$. That is, if the input is delayed by an amount, then the output is delayed by the same amount. This will not be valid for time varying system, in which the parameters of the system will change with time. If the system is linear time invariant, then for the input $\{a_1 x_1(t - t_o) + a_2 x_2(t - t_1)\}$, the output will be

$$y(t) = a_1 y_1(t - t_o) + a_2 y_2(t - t_1) \tag{3.2}$$

3.2 LINEAR SYSTEM CONVOLUTION

The response $Y(\omega)$ of a linear system $H(\omega)$ to a unit impulse input $\delta(t)$ is

$$Y(\omega) = H(\omega).F[\delta(t)] = H(\omega).1 = H(\omega)$$

Therefore,

$$Y(\omega) = H(\omega) \tag{3.3}$$

That is, the output of the system represents the system itself for an impulse input and in time domain $y(t) = h(t)$ and this is called the impulse response of the system. To the output, *only the system contributes and not the input.* This response of the system to impulse input is very characteristic of the system and is known as the *impulse response* of the system. If $h(t)$ is known, then only the complete system is known.

For a time invariant system, for a delayed unit impulse $\delta(t - \tau)$ the output is (from Eqn. 3.2)

$$h(t) * \delta(t - \tau) \leftrightarrow H(\omega)\, e^{-j\omega\tau} = h(t - \tau) \tag{3.4}$$

That is, for a delayed unit impulse, the system output is again the impulse response of the system, but a delayed one corresponding to the input delay.

A linear operator/system input $\{a_1 x_1(t) + a_2 x_2(t)\}$ gives the output $\{a_1 y_1(t) + a_2 y_2(t)\}$, if $y_1(t)$ and $y_2(t)$ are the outputs for inputs $x_1(t)$ and $x_2(t)$, respectively. Therefore, for an input of sum of the K delayed impulses $\delta(t - \tau_k)$ of different amplitudes a_k, the system output will have a sum of delayed impulse responses of amplitudes decided by those of the corresponding impulses in the input. That is,

$$y(t) = \sum_{k=1}^{K} a_k\, h(t - \tau_k) \tag{3.5}$$

For a general input waveform $x(t)$, it can be considered to be made of small elements of width Δ_τ of height $x(n\Delta_\tau)$. These elements can be considered impulses with area $x(n\Delta_\tau)\Delta_\tau$. The output for such an impulse input is $\{x(n\Delta_\tau)\Delta_\tau\, h(t - n\Delta_\tau)\}$.

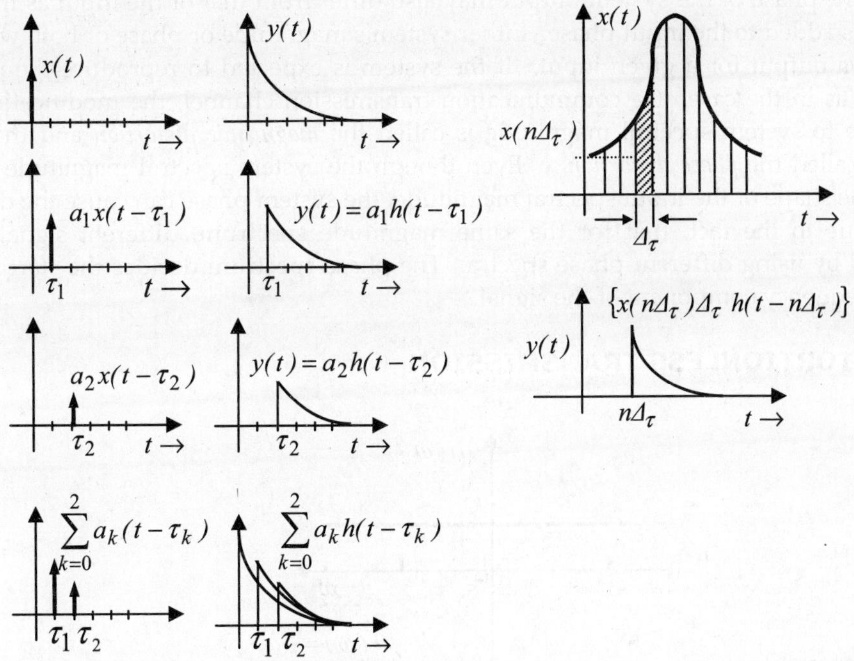

Fig. 3.1 Graphical illustration of linear system convolution.

The output due to all such elements is given by

$$y(t) = \sum_{n=-\infty}^{\infty} \{x(n\Delta_\tau)\Delta_\tau h(t - n\Delta_\tau)\}$$

As $\Delta_\tau \to 0$ and $n \to \infty$, $n\Delta_\tau \to \tau$, then

$$y(t) = \int_{-\infty}^{\infty} x(\tau)\, h(t-\tau)\, d\tau$$

From the time convolution theorem, in the frequency domain,

$$Y(\omega) = H(\omega)X(\omega) = |H(\omega)| e^{j\Phi(\omega)} |X(\omega)| e^{j\theta(\omega)}$$

$$|Y(\omega)| e^{j\Psi(\omega)} = |X(\omega)||H(\omega)| e^{j\{\theta(\omega)+\Phi(\omega)\}}$$

$$|Y(\omega)| = |X(\omega)||H(\omega)| \quad \text{and} \quad \Psi(\omega) = \theta(\omega) + \Phi(\omega)$$

The FT magnitude of the input gets weighed by the FT magnitude of the system impulse response and the system adds its FT phase to phase of the FT of the input. That is, the signal spectral magnitude gets shaped by the system spectral magnitude and this may result in retainment or enhancement or attenuation of the frequency components of the input.

The phase of the system output may also differ from that of the input as the system phase gets added to the input phase. Either system's magnitude or phase or both will decide the system output for a given input. If the system is expected to reproduce the input, for example, as in the case of a communication transmission channel, the modification of the input due to system spectral magnitude is called the *magnitude distortion* and that due to phase is called the *phase distortion*. Even though the system spectral magnitude does not change the shape of the input spectral magnitude, the system phase can cause the distortion. This is due to the fact, that for the same magnitude spectrum, different signals can be generated by using different phase spectra. The phase spectrum decides the starting phase of each frequency component of the signal.

3.3 DISTORTIONLESS TRANSMISSION

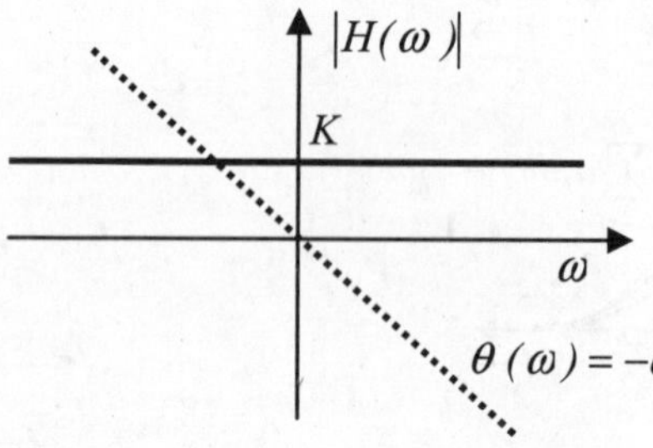

Fig. 3.2 Magnitude and phase spectral characteristics of a distortionless system.

A system will be considered distortionless, if its *output has same shape as the time domain input, even if it is delayed*. That is, the system can have a gain factor and it can introduce a delay. For an input $x(t)$,

$$y(t) = K\,x(t - t_o) \tag{3.6}$$

From the time shifting property of the FT,

$$Y(\omega) = KX(\omega)\,e^{-j\omega t_o}$$

But

$$Y(\omega) = H(\omega)X(\omega)$$

Comparing the above two equations,

$$H(\omega) = Ke^{-j\omega t_o}$$

This is the transfer function of a distortionless system.

$$|H(\omega)|e^{j\theta(\omega)} = Ke^{-j\omega t_o}$$

The magnitude of the transfer function is

$$|H(\omega)| = K$$

That is, the magnitude spectrum of the system is constant for all the frequencies. The phase of a distortionless system is given by

$$\theta(\omega) = -\omega t_o$$

That is, the phase is linearly proportional to frequency of the signal. Different frequency components of a signal have to be shifted in time by the same amount by a system, then their phase changes will be proportional to their frequencies.

Taking the change of sign of the system output into account, the system phase is given by

$$\theta(\omega) = m\pi - \omega t_o, \text{ where } m \text{ is an integer.}$$

In practice, the systems will have flat magnitude spectral characteristic over a limited range of frequencies and this is known as the bandwidth of the system, $B = (\omega_2 - \omega_1)$. The bandwidth is defined as the frequency range over which $|H(\omega)|$ remains within $(1/\sqrt{2})$ times (3dB) its maximum value.

3.4 PHYSICAL REALIZABILITY

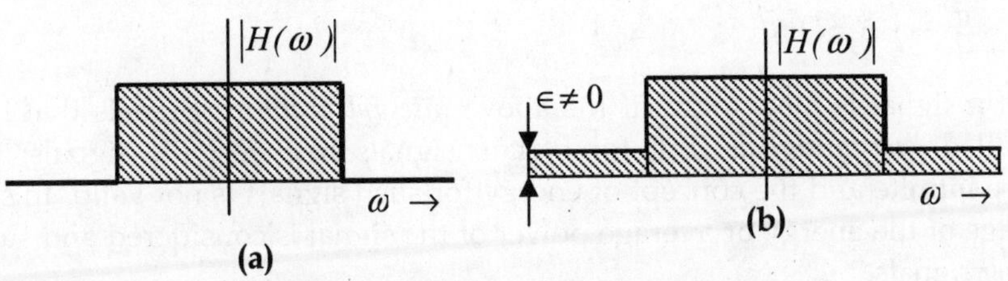

Fig. 3.3 Magnitude spectrum of physically: (a) unrealizable system and (b) realizable system.

A physically realizable system must have its response only after the input is applied and not before and this is known as causality condition. This requires that the impulse response $h(t)$ of a physically realizable system must be causal. That is, $h(t) = 0,\ t < 0$. This is the time domain requirement and its equivalent in frequency domain is

$$\int_{-\infty}^{\infty}\left[\frac{|\ln|H(\omega)||}{(1+\omega^2)}\right]d\omega < \infty \tag{3.7}$$

This is known as **Paley-Wiener (PW) criteria.** This criterion assures that for such a magnitude spectrum, it is possible to find a phase function, which will result in a physically realizable system. The above condition, in turn, requires,

$$\int_{-\infty}^{\infty} |H(\omega)|^2 \, d\omega < \infty \tag{3.8}$$

The PW criterion implies that $|H(\omega)|$ can be zero at some discrete frequencies but not over a finite band of frequencies as the integral becomes infinite (Fig.3.3). Hence, ideal filters cannot be realized physically. According to the PW criteria, the amplitude function cannot fall to zero faster than an exponential order. That is, it is possible to have,

$$|H(\omega)| = Ke^{-\alpha\omega}$$

But it is ***impossible*** to have,

$$|H(\omega)| = Ke^{-\alpha\omega^2}$$

3.5 ENERGY DENSITY SPECTRUM

The energy E of a signal $x(t)$ is defined as the energy dissipated by a voltage $x(t)$ applied across a 1-ohm resistor (or by a current $x(t)$ passing through a 1-ohm resistor). Therefore

$$E = \int_{-\infty}^{\infty} x^2(t)\, dt \tag{3.9}$$

The energy of a signal is meaningful if the above *integral is finite.* Signals that have finite energy are called *energy signals.* But for certain signals like periodic signals, the above integral will be infinite and the concept of energy for such signals is not valid. In such cases, the time average of the energy or average power of the signal is considered and such signals are called *power* signals.

$$x(t) = \frac{1}{2\pi}\int_{-\infty}^{\infty} X(\omega)e^{j\omega t}d\omega$$

$$E = \int_{-\infty}^{\infty} x^2(t)dt = \int_{-\infty}^{\infty} x(t)\left[\frac{1}{2\pi}\int_{-\infty}^{\infty} X(\omega)e^{j\omega t}d\omega\right]dt = \frac{1}{2\pi}\int_{-\infty}^{\infty} X(\omega)\left[\int_{-\infty}^{\infty} x(t)e^{j\omega t}dt\right]d\omega$$

$$= \frac{1}{2\pi} \int_{-\infty}^{\infty} X(\omega)\, X(-\omega)\, d\omega = \frac{1}{2\pi} \int_{-\infty}^{\infty} |X(\omega)|^2\, d\omega$$

since for a real signal,

$$X(\omega)X(-\omega) = |X(\omega)|^2$$

$$E = \frac{1}{2\pi} \int_{-\infty}^{\infty} |X(\omega)|^2\, d\omega = \int_{-\infty}^{\infty} |X(\omega)|^2\, df = \int_{-\infty}^{\infty} \Psi(\omega)\, df$$

The energy is the area under the curve $\Psi(\omega) = |X(\omega)|^2$. Hence, $\Psi(\omega)$ is called the *energy density function, i.e.,* $\Psi(\omega)$ represents the energy per unit bandwidth.

The energy density function of the output $y(t)$ of a system $H(\omega)$ for an input $x(t)$ is given by

$$\Psi_y(\omega) = |Y(\omega)|^2 = |H(\omega)X(\omega)|^2 = |H(\omega)|^2 |X(\omega)|^2 = |H(\omega)|^2 \Psi_X(\omega)$$

$$\Psi_y(\omega) = |H(\omega)|^2 \Psi_X(\omega)$$

That is, *the energy density function of the output is the product of the energy density function of the input and the magnitude square of the system frequency response.*

3.6 POWER DENSITY SPECTRUM

As mentioned already, signals that have infinite energy are called power signals. The average power of a signal $x(t)$ as the power dissipated by a voltage $x(t)$ applied across a 1-ohm resistor (or by a current $x(t)$ passing through a 1-ohm resistor). This is given by

$$P = \lim_{T\to\infty} \frac{1}{T} \int_{-T/2}^{T/2} x^2(t)dt \qquad (3.10)$$

The average power is also the mean square value of $x(t)$. If the truncated function of $x(t)$ during the interval $(-T/2, T/2)$ is represented by $x_T(t)$, then

$$\int_{-\infty}^{\infty} x_T^2(t)\, dt = \frac{1}{2\pi} \int_{-\infty}^{\infty} |X_T(\omega)|^2\, d\omega$$

But

$$\int_{-\infty}^{\infty} x_T^2(t)\, dt = \int_{-T/2}^{T/2} x^2(t)\, dt$$

Therefore,

$$P = \lim_{T\to\infty} \frac{1}{T} \int_{-T/2}^{T/2} x^2(t)\,dt = \frac{1}{2\pi} \int_{-\infty}^{\infty} \lim_{T\to\infty} \frac{1}{T} |X_T(\omega)|^2\,d\omega$$

If $$S_X(\omega) = \lim_{T\to\infty} \frac{1}{T} |X_T(\omega)|^2,$$

then

$$P = \frac{1}{2\pi} \int_{-\infty}^{\infty} S_X(\omega)\,d\omega = \int_{-\infty}^{\infty} S_X(\omega)\,df$$

That is, total power of a signal is the area under the curve of $S_X(\omega)$. Hence, $S_X(\omega)$ represents power per unit bandwidth. Since $S_X(\omega)$ shows how the power is distributed over the frequency range, it is called the *power spectral density* (PSD). Since PSD involves only magnitude square value, it does not have any phase information.

The PSD function of the output $y(t)$ of a system $H(\omega)$ for an input $x(t)$ is given by

$$S_Y(\omega) = \lim_{T\to\infty} \frac{1}{T} |Y_T(\omega)|^2 = \lim_{T\to\infty} \frac{1}{T} |H(\omega) X_T(\omega)|^2 = |H(\omega)|^2 \lim_{T\to\infty} \frac{1}{T} |X_T(\omega)|^2$$

$$S_Y(\omega) = |H(\omega)|^2 S_X(\omega)$$

That is, *the PSD of the system output is* $|H(\omega)|^2$ *times the PSD of the input signal.*

3.7 CORRELATION

The similarity between two signals $x_1(t)$ and $x_2(t)$ is given by

$$C_{12} = \int_{t_1}^{t_2} x_1(t) x_2(t)\,dt$$

That is, if C_{12} has a large magnitude then $x_1(t)$ and $x_2(t)$ are similar to a greater extent over the interval (t_1, t_2). That is, $x_1(t)$ has a large component of $x_2(t)$ and vice versa. This concept can be put on an average basis as

$$C_{12} = \lim_{(t_2 - t_1)\to\infty} \frac{1}{(t_2 - t_1)} \int_{t_1}^{t_2} x_1(t)\, x_2(t)\,dt$$

This definition is called the *cross correlation* between $x_1(t)$ and $x_2(t)$. If the signals involved are the signal and its shifted version, then it is called *autocorrelation* and this indicates how fast the signal changes and is given by

$$C(\tau) = \lim_{(t_2 - t_1) \to \infty} \frac{1}{(t_2 - t_1)} \int_{t_1}^{t_2} x(t)\, x(t+\tau)\, dt$$

Consider

$$F^{-1}\left[X(\omega) X^*(\omega) \right] = F^{-1}\left[X(\omega) \right] \otimes F^{-1}\left[X^*(\omega) \right]$$

Here, $\otimes$ indicates convolution operation.

$$F^{-1}\left[X(\omega) X^*(\omega) \right] = x(\tau) \otimes x(-\tau) = \int_{-\infty}^{\infty} x(\lambda) x(\lambda - (-\tau)) d\lambda = \int_{-\infty}^{\infty} x(\lambda) x(\lambda + \tau) d\lambda$$

$$F^{-1}\left\{ \lim_{T = (t_2 - t_1) \to \infty} \left[X_T(\omega) X_T^*(\omega) \right] \right\} = \lim_{T = (t_2 - t_1) \to \infty} \int_{t_1}^{t_2} x(\lambda) x(\lambda + \tau) d\lambda$$

where $T = (t_2 - t_1)$

Therefore

$$F^{-1}\left\{ \lim_{T \to \infty} \frac{1}{T} \left[X_T(\omega)\, X_T^*(\omega) \right] \right\} = \lim_{(t_2 - t_1) \to \infty} \frac{1}{T} \int_{t_1}^{t_2} x(\lambda)\, x(\lambda + \tau)\, d\lambda$$

$$\lim_{T \to \infty} \frac{1}{T} \int_{t_1}^{t_2} x(t) x(t+\tau) dt \quad \Leftrightarrow \quad \lim_{T \to \infty} \frac{1}{T} \left| X_T(\omega) \right|^2 = S_x(\omega)$$

$$C(\tau) \quad \Leftrightarrow \quad \lim_{T \to \infty} \frac{1}{T} \left| X_T(\omega) \right|^2 = S_x(\omega)$$

That is,

$$S_X(\omega) = \int_{-\infty}^{\infty} C(\tau)\, e^{-j\omega\tau} d\tau \tag{3.11}$$

or

$$C(\tau) = \int_{-\infty}^{\infty} S_X(\omega)\, e^{j\omega\tau} d\omega \tag{3.12}$$

The autocorrelation function and the power spectral density function are Fourier transform pairs and this is called the **Wiener-Khincthine** relation. Further, if $t_2 = -t_1$,

$$C(0) = \lim_{T \to \infty} \frac{1}{T} \int_{-T/2}^{T/2} x(t)\, x(t+0)\, dt$$

$$C(0) = \lim_{T \to \infty} \frac{1}{T} \int_{-T/2}^{T/2} x^2(t)\, dt = \textit{average power}$$

The autocorrelation at zeroth lag gives the average power of the signal and this is the maximum value that the autocorrelation can have. That is,

$$\begin{aligned} C(0) &\geq C(\tau), \quad \tau \neq 0 \\ C(\tau) &\to 0, \quad \tau \to \infty \end{aligned} \tag{3.13}$$

If $x(t)$ is periodic with period T, then $C(\tau)$ is also periodic with the same period T. Further, since

$$F[C(\tau)] = |X(\omega)|^2,$$

the FT of the correlation function is nonnegative.

The autocorrelation function measures the similarity/dependence (dependence leads to similarity) between different parts of the signal as it compares the original signal with its shifted versions. If the signal characteristics of two parts of the signal are almost same/ dependent, then the correlation function will not fall rapidly. But if the characteristics differ, then the correlation will decrease with increase in separation or lag length. A rapid fall in the correlation with lag, calls for a large number of frequency components to be associated and hence for such signals, the FT will have a large bandwidth. An extreme case of this is the correlation function being an *impulse* and the signal will have infinite bandwidth. For a slowly decaying correlation, the bandwidth associated with the signal is small. The correlation basically tells about the *rapidity with which the signal characteristics are changing and the bandwidth associated with signal.*

A positive correlation value between two quantities indicates that the dependence is *in harmony,* i.e., if one increases the other also increases and vice versa. On the other hand, if the correlation value is negative, it indicates that there is dependence but in *anti-harmony,* i.e., if one quantity increases the other decreases and vice versa. But a zero correlation value means there is *no dependence at all.*

Both the impulse and the *white noise* can have their correlation function as an impulse. For an impulse, it is easy to understand that there is no correlation as it occurs only at one point (or over a very short duration) and it is unique at that point and different from other points in the neighborhood. However, with the white noise as this term is newly introduced only in this section, it will be explained. Unlike an impulse, the white noise has values everywhere in time and not just at one point like that of an impulse. But these values are not dependent on one another. Each point is *independent* to take any value and at any point it does not have any dependence on the value at any other point. That is, the correlation at any point is zero with any other point separated by a distance (lag length) greater than zero. It is correlated with itself and not with any other. Such a process, which generates this type of signal, is called *white noise process.* Further, as the neighboring and other samples are independent of their values, they will be random and there is no definite pattern associated with it. Hence, a white noise process is also a *random process.* If a white noise process is passed through a system, the system will introduce its characteristics of inertia/damping into the white noise and gives out as its output. Such an output is often called *colored noise.* This is because for complete independent nature, all the frequencies from

($-\infty$ *to* ∞) are required. But as the physical system cannot change instantaneously, its frequency spectrum is limited and hence it removes those frequencies, which are not there in its spectrum. Thus, the output has only limited number of frequencies hence, it is called colored in analogy with light.

Even though for the white noise the sample values are uncorrelated, for particular type of process in a given interval, on an average the number of sample values gets distributed in particular form. That is, the *histogram* or the sample density function (normalized by the total number of samples) has a specific shape. In a histogram, the complete range of values that the signal can take is divided into finite number of intervals and the number of values falling into different intervals is plotted against the interval values. A histogram normalized with respect to total number of samples considered provides an estimate of the probability density function when the number of samples is sufficiently large. White noises having different types of density functions like Gaussian (Normal), uniform, sine, square, exponential exist.

3.8 STATIONARY PROCESS

The filtering view of FT gives an insight into stationarity and it will be considered here.

3.8.1 Filtering view of Fourier Transform

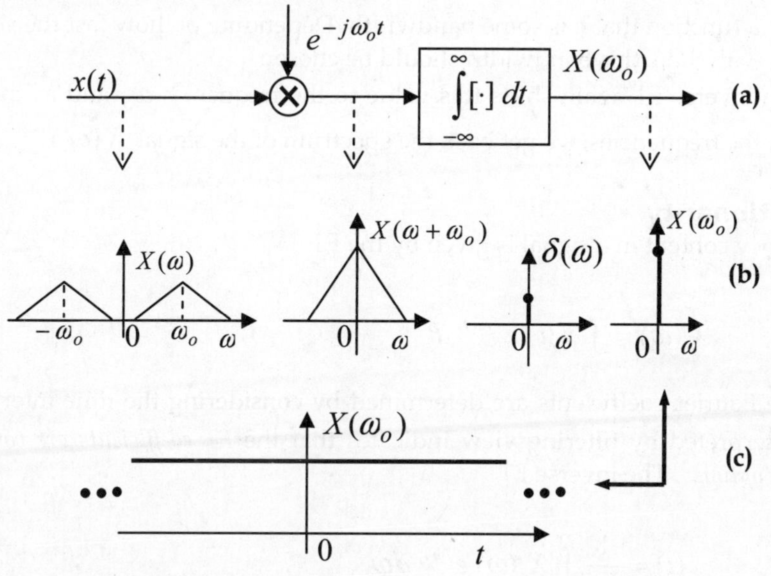

Fig. 3.4. Filtering view of Fourier transform: (a) Schematic of operations; (b) corresponding outputs and (c) Time domain representation of $X(\omega_o)$ at the origin.

The Fourier transform of $x(t)$ is given by

$$X(\omega) = \int_{-\infty}^{\infty} x(t)e^{-j\omega t}dt$$

At frequency ω_o,

$$X(\omega_o) = \int_{-\infty}^{\infty} x(t)e^{-j\omega_o t}dt$$

This can be schematically represented as in Fig. 3.4. The demodulation operation multiplication $x(t)$ by $e^{-j\omega_o t}$, shifts the spectrum $X(\omega)$ to the origin as indicated by $X(\omega+\omega_o)$ in Fig. 3.4(b). Further, the integration/summation operation can be interpreted as the convolution of the function with a constant of unit magnitude, in frequency domain; it is equivalent to multiplication of $X(\omega+\omega_o)$ by an impulse $\delta(\omega)$ at the origin. The effect of this is to pick up the value of $X(\omega)$ at ω_o which is a constant $X(\omega_o)$. That is, the value of the frequency component at ω_o of the signal *will not change with time* and this is shown in Fig. 3.4(c). According to the FT, the value of the signal spectral components will not change with time and this is due to the infinite time limits that result in an impulse function $\delta(\omega)$ at the origin. In order to accommodate the nonstationary nature of the signal, that is, the changing nature of spectrum of the signal, the $\delta(\omega)$ function at the origin should be replaced by a function that has some bandwidth. Depending on how fast the signal spectrum is changing with time, this bandwidth should be chosen.

The inverse FT again shifts this value to the frequency ω_o and as this operation is done for all the frequencies, we get back the spectrum of the signal $X(\omega)$.

3.8.2 Stationarity

The frequency content in a signal is given by the FT,

$$X(\omega) = \int_{-\infty}^{\infty} x(t)\, e^{-j\omega t}\, dt$$

That is, the Fourier coefficients are determined by considering the time interval $(-\infty,\infty)$. This, as interpreted by filtering view indicates that the *FT coefficients are constants and are independent of time.* The inverse FT

$$x(t) = \frac{1}{2\pi}\int_{-\infty}^{\infty} X(\omega)\, e^{j\omega t}\, d\omega$$

also says that the signal can be constructed from the FT coefficients which are constants over the interval. $(-\infty,\infty)$. *A signal whose FT coefficients are time invariant is called a stationary signal.*

Further, since

$$S_x(\omega) = \lim_{T\to\infty} \frac{1}{T} |X_T(\omega)|^2$$

$$C(\tau) = \frac{1}{2\pi} \int_{-\infty}^{\infty} S_x(\omega)\, e^{j\omega\tau} d\omega$$

the invariance of FT coefficients implies that the PSD, $S_x(\omega)$ hence, the autocorrelation function $C(\tau)$ are also time invariant or stationary. That is, the *autocorrelation function pattern is invariant with respect to time.* Autocorrelation tells us the degree to which the signal samples $x(t)$ and $x(t+\tau)$ are separated by τ are correlated. This is computed by collecting all the pairs of samples, which are separated by τ and averaging the product of each pair, over the total number of pairs and this is done for different values of τ.

The component $x_i(t)$ at ω_i of signal $x(t)$ associated with a certain bandwidth σ_i can be represented as

$$x_i(t) = e^{-\sigma_i t} \sin \omega_i t$$

The correlation between two such components $x_i(t)$ and at frequencies ω_i and ω_j is given by

$$C_{ij} = \int_{-\infty}^{\infty} x_i(t)\, x_j(t)\, dt \quad = \int_{-\infty}^{\infty} [e^{-\sigma_i t} \sin \omega_i t][\, e^{-\sigma_j t} \sin \omega_j t]\, dt$$

$$C_{ij} = \int_{-\infty}^{\infty} e^{-(\sigma_i+\sigma_j)t} [\sin \omega_i t][\, \sin \omega_j t]\, dt$$

$$= \frac{1}{2} \int_{-\infty}^{\infty} e^{-(\sigma_i+\sigma_j)t} [\cos(\omega_i - \omega_j)t - \cos(\omega_i + \omega_j)\, t]\, dt$$

Let $\lambda = \left(\sigma_i + \sigma_j\right)$, $\omega_d = (\omega_i - \omega_j)$, $\omega_p = (\omega_i + \omega_j)$

$$C_{ij} = \frac{1}{2} \int_{-\infty}^{\infty} e^{-\lambda t} [\cos \omega_d t - \cos \omega_p t]\, dt$$

$$= \int_{0}^{\infty} e^{-\lambda t} [\cos \omega_d t - \cos \omega_p t]\, dt$$

$$= \int_{0}^{\infty} e^{-\lambda t} \cos \omega_d t\, dt - \int_{0}^{\infty} e^{-\lambda t} \cos \omega_p t\, dt$$

$$= \int_{0}^{\infty} e^{-\lambda t} \left(\frac{e^{j\omega_d t} + e^{-j\omega_d t}}{2} \right) dt - \int_{0}^{\infty} e^{-\lambda t} \left(\frac{e^{j\omega_p t} + e^{-j\omega_p t}}{2} \right) dt$$

$$= \frac{1}{2} \int_{0}^{\infty} \left(e^{-(\lambda - j\omega_d)\, t} + e^{-(\lambda + j\omega_d) t} \right) dt - \frac{1}{2} \int_{0}^{\infty} \left(e^{-(\lambda - j\omega_p)\, t} + e^{-(\lambda + j\omega_p) t} \right) dt$$

$$= \frac{1}{2}\left(\frac{e^{-(\lambda-j\omega_d)\,t}}{-(\lambda-j\omega_d)} + \frac{e^{-(\lambda+j\omega_d)\,t}}{-(\lambda+j\omega_d)}\right)\Bigg|_0^\infty - \frac{1}{2}\left(\frac{e^{-(\lambda-j\omega_p)\,t}}{-(\lambda-j\omega_p)} + \frac{e^{-(\lambda+j\omega_p)\,t}}{-(\lambda+j\omega_p)}\right)\Bigg|_0^\infty$$

$$= \frac{1}{2}\left(\frac{-1}{-(\lambda-j\omega_d)} + \frac{-1}{-(\lambda+j\omega_d)}\right) - \frac{1}{2}\left(\frac{-1}{-(\lambda-j\omega_p)} + \frac{-1}{-(\lambda+j\omega_p)}\right)$$

$$= \frac{1}{2}\left(\frac{2\lambda}{\lambda^2+\omega_d^2}\right) - \frac{1}{2}\left(\frac{2\lambda}{\lambda^2+\omega_p^2}\right)$$

$$C_{ij} = \frac{\lambda\left(\omega_p^2-\omega_d^2\right)}{\left(\lambda^2+\omega_d^2\right)\left(\lambda^2+\omega_p^2\right)} \neq 0 \tag{3.14}$$

If $\lambda = 0$, that is $(\sigma_i + \sigma_j) = 0$ or $\sigma_i = \sigma_j = 0$ (as $\sigma_i \neq -\sigma_j$ since the bandwidth cannot be negative), C_{ij} will be zero as

$$C_{ij} = \int_{-\infty}^{\infty} [\sin \omega_i t][\sin \omega_j t]dt$$

which corresponds to the correlation between the two orthogonal signals $\sin \omega_i t$ and $\sin \omega_j t$.

This implies that the signal Fourier components associated with bandwidth are correlated. Further, as such a bandwidth is associated with signals of nonstationary nature, *for nonstationary signals the Fourier components are correlated.* But *for stationary signals, the Fourier components are orthogonal or uncorrelated.*

The FT *only synthesizes* a nonstationary signal. But it does not tell when spectral changes occurred, as there is no time reckoning in FT. That is, for example, if a sinusoid of frequency ω_1 is added to another existing sinusoid of frequency ω_2 at a later time, according to FT of the record, the two sinusoids of average amplitude existed through out the record, which is not true.

The timing information is in the phase of the FT. However, extracting/computing the phase information accurately is *not a simple task.* This is due to the fact that the phase computed from the FT is the principal value of the phase and not the actual one. The principal phase value is discontinuous and is known as wrapped phase. This lies only in the range $(-\pi, \pi)$ irrespective of its actual value which may exceed this range. It is necessary to add/subtract integral multiple of π to the principal phase value to make it continuous and is known as *phase unwrapping.* Though at the outset it looks simple, it is not so as it depends upon the signal characteristics like location of its roots and hence the frequency resolution required in computing the principal value of the phase is different depending up on the situation. In view of this, the methods/transforms which directly involve *both time and frequency* in signal representation have become popular.

Chapter 4

SAMPLING THEOREM AND DISCRETE FOURIER TRANSFORM (DFT)

This chapter introduces the concept of sampling which is the gateway for the discrete time signal processing and its implication of undesired aliasing due to improper sampling. Based on the sampling theorem, the Fourier transform applicable to discrete time signal, the Discrete Fourier transform (DFT) is considered. Further, in DFT, the effect of the inevitable data truncation, the Gibbs ripple/leakage effect and the method of reducing this ill effect by windowing which introduces the trade-off between frequency resolution and leakage for a fixed length of data are dealt with. In this direction, the use of the discrete cosine transform (DCT) which overcomes the leakage effect significantly and at the same time providing better frequency resolution, is explained.

The in-depth knowledge of such transforms is essential as they play the role of pre-processing (orthogonalization) which adds to the value of for the much beneficial parametric methods (which will be considered in coming chapters). The interrelationships among the Fourier transform, Laplace transform, Z-transform and DFT are described. Further to analyse systems by transfer function approach, the rational transfer function and its variations, minimum phase, maximum phase, mixed phase and minimum phase equivalent systems, are briefly explained.

4.1 THE SAMPLING THEOREM

Sampling is achieved by multiplying the desired signal $x(t)$ by an impulse train $\delta_T(t)$, T being the sample interval or the sampling frequency being $\omega_o = 2\pi / T$.

$$x_S(t) = x(t)\delta_{T_s}(t) \tag{4.1}$$

$$x(t) \leftrightarrow X(\omega) \text{ and } \delta_{T_s}(t) \leftrightarrow \omega_s \delta_{\omega_s}(\omega)$$

This process of sampling results in a convolution of the FT's $X(\omega)$ and $\omega_s \delta_{\omega_s}(\omega)$ of $x(t)$ and $\delta_{T_s}(t)$, respectively.

$$x_S(t) \leftrightarrow \frac{1}{2\pi}\left[X(\omega) * \omega_s \delta_{\omega_s}(\omega)\right]$$

or $$x_S(t) \leftrightarrow \frac{1}{T_s}\left[X(\omega) * \delta_{\omega_s}(\omega)\right] \text{, where } \delta_{\omega_s}(\omega) = \sum_{n=-\infty}^{\infty} \delta(\omega - n\omega_s)$$

$$X_S(\omega) = \frac{1}{T_s}\left[X(\omega) * \delta_{\omega_s}(\omega)\right] = \frac{1}{T_s}\left[X(\omega) * \sum_{n=-\infty}^{\infty} \delta(\omega - n\omega_s)\right]$$

$$= \frac{1}{T_s}\left[\sum_{n=-\infty}^{\infty} X(\omega) * \delta(\omega - n\omega_s)\right]$$

$$X_S(\omega) = \frac{1}{T_s}\left[\sum_{n=-\infty}^{\infty} X(\omega - n\omega_s)\right] \tag{4.2}$$

$X_S(\omega)$ is the spectrum of the sampled signal and in this, the original signal spectrum $X(\omega)$ repeats at every frequency interval of ω_o. This shows that for a sampled signal the spectrum is periodic. [On the other hand, for a periodic signal, the spectrum is sampled or discrete, which has been established already.] In view of this, if $X(\omega)$ is not limited to a maximum frequency $\omega_m \le \omega_o / 2$, the repetition of $X(\omega)$, results in an overlapping and the high frequency components of $X(\omega)$ that repeats get added to the low frequency components of the original $X(\omega)$. This distorts the signal spectrum, which will be recovered by using a low-pass filter $G_{2\omega_m}(\omega)$. Though the low-pass filter removes the images (repetitions) of the original signal spectra, it cannot distinguish and remove those components, which are already added to the components within the signal frequency band.

To recover a proper analog signal from the sampled signal, the sampling frequency must be greater than twice or *at least twice* the maximum frequency content of the original signal.

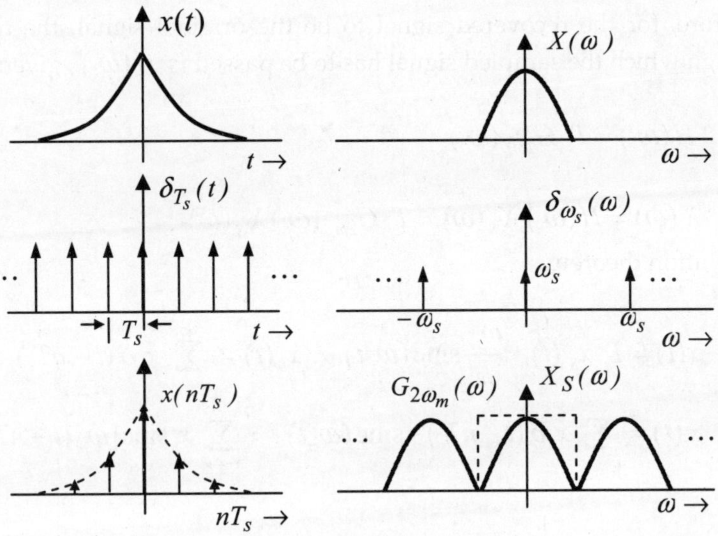

Fig. 4.1 The sampling process.

In such a case, the repetition of the signal spectrum due to sampling will not distort the recovered analog signal, using a low-pass filter. The distortion in the recovered signal, introduced due to insufficient sampling frequency, is called *aliasing*. This terminology is appropriate, since the high frequency components of the signal appear as low-frequency components and get added to the low frequency components (of the original signal) thereby appearing as low frequency components. In practice, to ensures/fix proper sampling frequency, the signal is first band limited by a low-pass filter, which provides sufficient attenuation (at least 40dB) above its cutoff frequency.

Sampling Theorem: *To reconstruct the signal without any distortion, the signal has to be sampled at a frequency ω_o that is greater or at least equal to twice the maximum frequency ω_m contained in the signal. This sampling frequency is known as Nyquist sampling frequency.*

4.2 RECOVERING THE SIGNAL FROM ITS SAMPLES

As stated, the original signal can be recovered by passing the sampled signal through a low-pass filter with a cutoff frequency ω_m. That is

$$\hat{X}(\omega) = X_S(\omega)\, G_{2\omega_m}(\omega) \tag{4.3}$$

Low-pass filtering is achieved by a gate function $G_{2\omega_m}(\omega)$.

$$\hat{X}(\omega) = X_S(\omega) G_{2\omega_m}(\omega) = G_{2\omega_m}(\omega) \frac{1}{T_s}\left[\sum_{n=-\infty}^{\infty} X(\omega - n\omega_s)\right] = \frac{1}{T_s} X(\omega)$$

Therefore, for the recovered signal to be the original signal, the transfer function required through which the sampled signal has to be passed is $H(\omega)$, given by

$$H(\omega) = T_s\, G_{2\omega_m}(\omega)$$

Therefore,

$$X(\omega) = H(\omega) X_S(\omega) = T_s\, G_{2\omega_m}(\omega) X_S(\omega)$$

By time-convolution theorem,

$$x(t) = T_s\, x_S(t) * \frac{\omega_m}{\pi} \operatorname{sinc}(\omega_m t), \quad x_S(t) = \sum_{n=-\infty}^{\infty} x_n \delta(t - nT_s)$$

$$x(t) = \sum_{n=-\infty}^{\infty} x_n \delta(t - nT_s) * \operatorname{sinc}(\omega_m t) = \sum_{n=-\infty}^{\infty} x_n \operatorname{sinc}[\omega_m (t - nT_s)] \qquad (4.4)$$

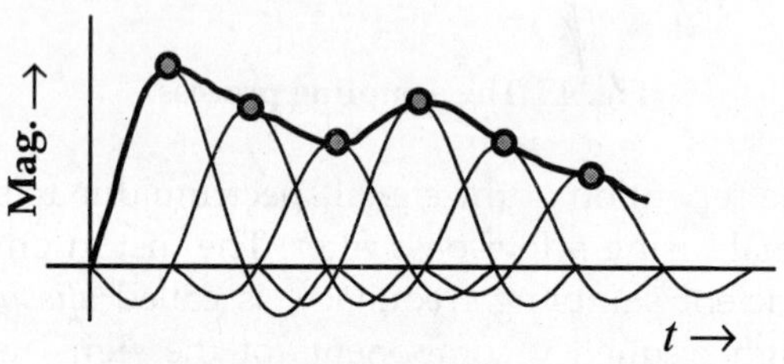

Fig. 4.2 Reconstruction from samples.

That is, the reconstructed signal can be obtained from the samples by multiplying each sample by a sinc function and adding all the resulting waveforms (Fig. 4.2). However, the reconstructed signal obtained from a non-ideal filter, specifically from the finite number of samples of the filter impulse response, will deviate from the original signal. Thus, the filter determines the quality of the signal approximation.

4.3 DISCRETE FOURIER TRANSFORM (DFT)

Here, the development of the DFT is treated briefly giving the main steps involved to get to know the mechanism involved and hence complete derivation is not considered (but can be found in the book by S.V. Narasimhan and S.Veena, 2008).

For a periodic continuous time signal, the FT is discrete in frequency. However, for a general (aperiodic in time) signal $x(t)$, its FT $X(\omega)$ is neither periodic nor discrete (Fig. 4.3a).

The first step in handling the continuous signal is to discretize in time according to the sampling theorem and is given by

$$x(nT_S) = x(t)\,\delta_{T_S}(t) = x(t)\sum_{n=-\infty}^{\infty}\delta\,(t - nT_S)$$

$$x(nT_S) = \sum_{n=-\infty}^{\infty} x(nT_S)\,\delta\,(t - nT_S) \tag{4.5}$$

Since the computer memory is finite and also to carry out the analysis of a specific part of the signal, signal of finite duration has to be considered and this is achieved by applying a rectangular window $w(t)$ (Fig. 4.3d) to the sampled time signal sequence $x(nT_S)$.

$$w(t) = 1, \qquad -T_o/2 < t < T_o/2 \tag{4. 6a}$$
$$= 0, \qquad \text{otherwise.}$$

T_o is the duration of the window. The FT of $w(t)$ is a sinc *function*. The windowed or truncated signal is

$$w(t)\,x\,(nT_S) = w(t)\left[\sum_{n=-\infty}^{\infty} x(nT_S)\,\delta\,(t - nT_S)\right]$$

$$w(t)\,x\,(nT_S) = \sum_{n=0}^{N-1} x\,(nT_S)\,\delta\,(t - nT_S) \tag{4.6b}$$

N is the number of samples in the interval T_o. That is, $N = \dfrac{T_o}{T_S}$. The truncation of the signal in time domain, results in a *rippling effect,* in the FT, $\tilde{X}_{\omega_S}(\omega)$ (Fig. 4.3e). This is the second step.

But for a sampled aperiodic signal $x(nT_S)$, its FT $X_{\omega_o}(\omega)$ is periodic and continuous in frequency ω and not discrete (Fig. 4.3e).

To handle any processing in a computer, it is essential that the quantities involved are discrete. However, unless the time signal is periodic, its FT does not satisfy this criterion hence, its FT should be sampled. Hence, the last step in the DFT development is discretization of the frequency axis. This is achieved by multiplying $\tilde{X}_{\omega_S}(\omega)$ by $\delta_\Delta(\omega)$, an impulse train in the frequency domain and this results in sampled spectrum $\tilde{X}_{\omega_S}(k\Delta)$. The inverse FT of this impulse train $\delta_\Delta(\omega)$, is another impulse train in time $\delta_N(t)$ (Fig. 4.3d). Further, multiplication of $\delta_\Delta(\omega)$ and $\tilde{X}_{\omega_S}(\omega)$ giving $\tilde{X}_{\omega_S}(k\Delta)$, in the frequency domain results in a convolution of the sampled time signal $w(t)x(nT_S)$ and the impulse train $\delta_N(t)$ in time, resulting in $x_N(nT_S)$ which is periodic.

$$\delta_N(t) = T_o\sum_{r=-\infty}^{\infty}\delta\,(t - rT_o) \tag{4.7a}$$

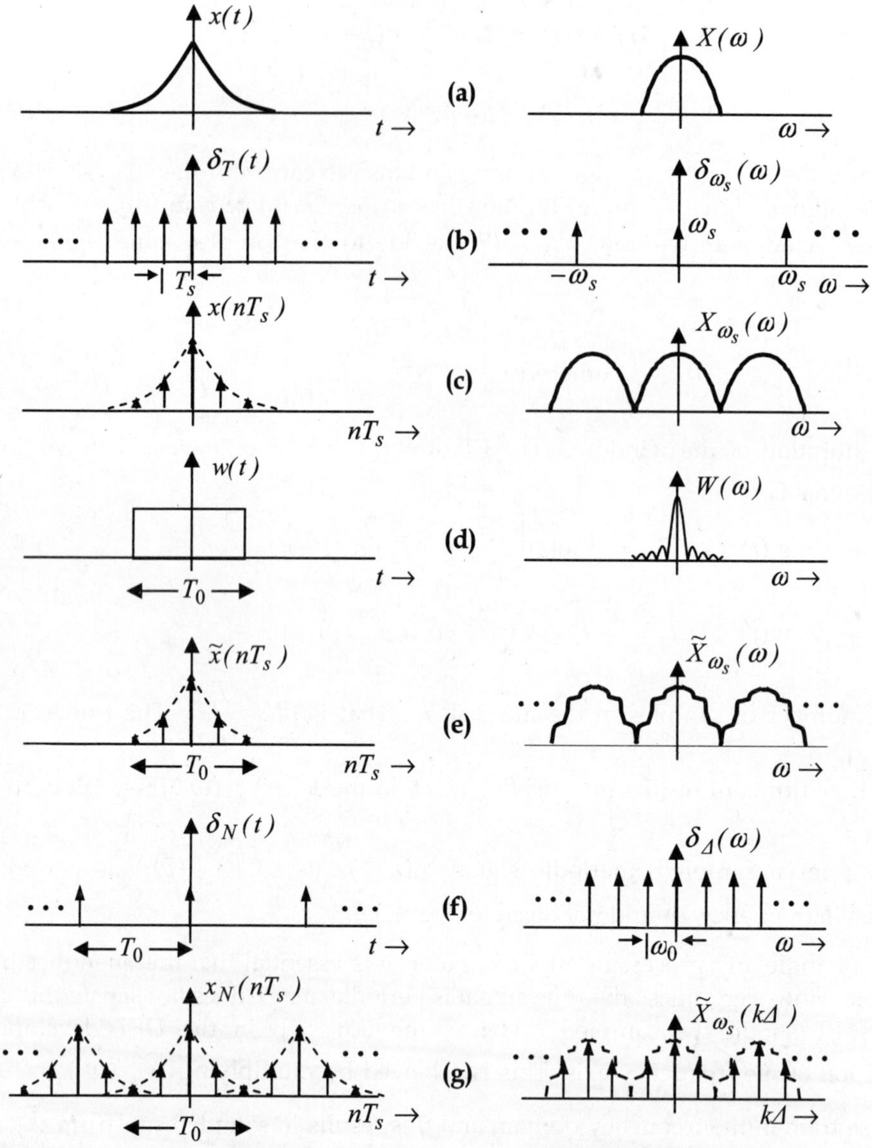

Fig. 4.3 Effect of data truncation on the development of DFT.

Further, multiplication of $\delta_{\Delta}(\omega)$ and $\tilde{X}_{\omega_s}(\omega)$ giving $\tilde{X}_{\omega_s}(k\Delta)$, in the frequency domain results in a convolution of the sampled time signal $w(t)\,x(nT_s)$ and the impulse train $\delta_N(t)$ in time, resulting in $x_N(nT_s)$ which is periodic.

$$\delta_N(t) = T_o \sum_{r=-\infty}^{\infty} \delta\,(t - rT_o) \tag{4.7a}$$

Therefore,

$$x_N(nT_S) = \left[w(t)\,x(nT_S)\right] * \delta_N(t) = \left[\sum_{n=0}^{N-1} x(nT_s)\delta\,(t - nT_S)\right] * \left[T_o \sum_{r=-\infty}^{\infty} \delta\,(t - rT_0)\right] \tag{4.7b}$$

The FT of this sequence gives the required FT.

To recover the time-sampled signal, the frequency sampling should avoid possible time domain aliasing. That is, the signal length decides the frequency-sampling rate. With the permissible minimum frequency-sampling rate, the time signal is periodic with its length as its period.

Such a FT that involves sampling both in time and frequency is called discrete FT (DFT). In contrast to this, the FT of a sampled signal is called Discrete Time FT (DTFT), which specifically says that discretization is *only in time and not in frequency.* In a DFT, the periodicity is *built in both, in time and frequency,* due to sampling (in the other domain) though the original signal may not be periodic.

In similarity to the continuous FT, the DTFT is given by

$$X\,(\omega) = \sum_{n=0}^{N-1} x\,(nT)\,e^{-j\,\omega\,nT}$$

Since ω is discretized corresponding to the signal segment length N, the sampled signal becomes periodic with period N. The Discrete Fourier transform (DFT) is given by

$$X(k) = \sum_{n=0}^{N-1} x\,(nT)\,e^{-j\left(\frac{2\pi}{N}\right)k\,nT}, \quad k = 0, 1, 2, \ldots, (N-1) \tag{4.8a}$$

The FT is sampled at frequency interval $(2\pi / N)$. The Inverse Discrete Fourier transform (IDFT) is

$$x\,(nT) = \sum_{n=0}^{N-1} X\,(k)\,e^{j\left(\frac{2\pi}{N}\right)k\,nT}, \quad n = 0, 1, 2, \ldots, (N-1) \tag{4.8b}$$

4.4 PERFORMANCE OF THE DFT

It is of interest to know the performance of the DFT compared to its continuous counterpart. The truncating window plays a significant role in the estimation of the FT of a signal and this will be considered in this section. In general, it is not possible to compute the true FT of a signal using a computer and the implicit rectangular window will alter the true FT of the signal, on hand The truncation of a signal is nothing but multiplication of the signal by a rectangular window and this results in convolution of the FT of the signal with that of the rectangular window, which is a sinc function. For some signal conditions, this convolution may not affect the true FT of the signal. However, in general, it affects the computation of the FT of the signal, to different degrees. In view of this, the DFT may provide a spectrum significantly different from the true spectrum of the untruncated signal.

If the FT $X(\omega)$ is a gate function representing an ideal low-pass filter then its inverse FT, $x(t)$ is a sinc function of infinite length $(-\infty, \infty)$ and its DFT will significantly differ from gate function. Here also, the DFT introduces periodicity and makes the frequency axis discrete. But in addition to this, the truncation of $x(t)$ (the sinc function) by a window function $w(t)$ in time domain, results in a *ripple effect* (Fig. 4.4b).

The truncation in time results in the convolution between $X_{\omega_s}(\omega)$ and the sinc function $W(\omega)$ giving $\tilde{X}_{\omega_s}(\omega)$ in the frequency domain.

$$\tilde{X}_{\omega_s}(\omega) = \frac{1}{2\pi} \int_{-\omega_s/2}^{\omega_s/2} X_{\omega_s}(\theta) W(\omega - \theta)\, d\theta \tag{4.9}$$

As $W(\omega - \theta)$ slides across the discontinuity of $X_{\omega_s}(\omega)$ with ω variation, the integral *oscillates* as each side lobe of $W(\omega - \theta)$ crosses (Fig. 4.4a) the discontinuity and results in an undesired ripple in $\tilde{X}_{\omega_s}(\omega)$. The magnitude of the oscillation depends on the side lobe magnitude and smaller the side lobe level smaller is the ripple.

However, a longer window length only increases the ripple frequency and not its magnitude (Fig. 4.4b and c). To reduce the ripple magnitude, window with smaller side lobe level has to be used. As the side lobes are nothing but high frequency components of the window, its smoother termination reduces the magnitude of the ripple effect. This ripple effect is known as *Gibbs effect*.

The ripple effect introduces some amount of uncertainty in the magnitude of the components of the spectrum and hence the variance in the estimation of the spectrum. To reduce this variance the only way is to use windows with smooth truncation. The examples of such windows are Hamming, Hanning, Blackman, and Kaiser, etc. However, due to smooth truncation, the *effective length* of the window (area under the window) is reduced, resulting in an increased width of the main lobe of its FT. As finer changes in the spectrum over a frequency width smaller than that of the main lobe cannot be made out, frequency resolution is decreased. That is, reducing the ripple effect or the variance of the spectrum is only at the cost of frequency resolution.

Let $X(\omega)$ corresponds to two sinusoids of different amplitudes (one large and one small). The convolution of the spectrum $X(\omega)$ with $W(\omega)$, due to truncation by the window can result in two spectral peaks may be of the same magnitude. This is due to the fact in the convolution process, when $W(\omega)$ is positioned at the smaller spectral peak, the side lobe at the large spectral peak, gets weighed by the spectral peak magnitude and contributes to the convolution sum value thereby enhancing the magnitude of the smaller peak. That is, the larger spectral peak *leaks through* the side lobe of the FT of the window and contributes to the component at the main lobe position (Fig. 4.5).

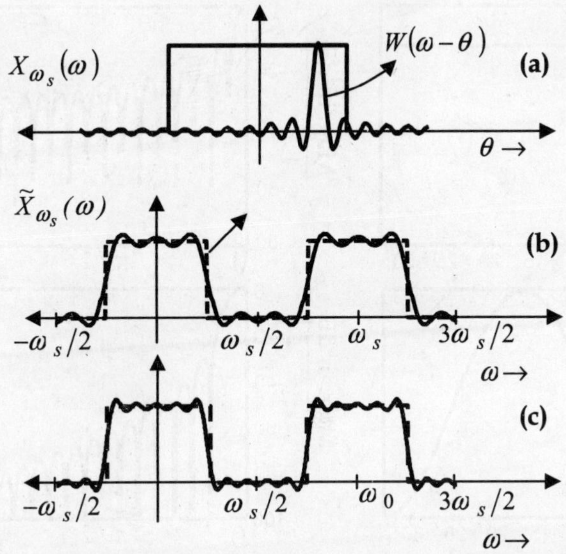

Fig. 4.4 Gibbs phenomenon: (a) Mechanism of Gibbs phenomenon; (b) Gibbs ripple with 7 coefficients and (c) Gibbs ripple with 11 coefficients.

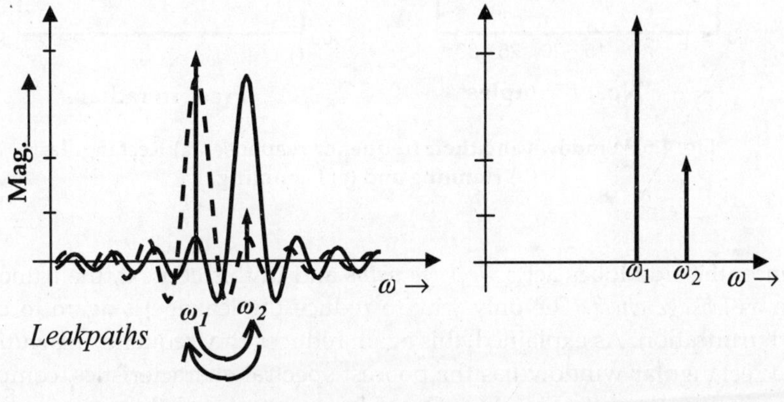

Fig. 4.5 Leakage effect mechanism.

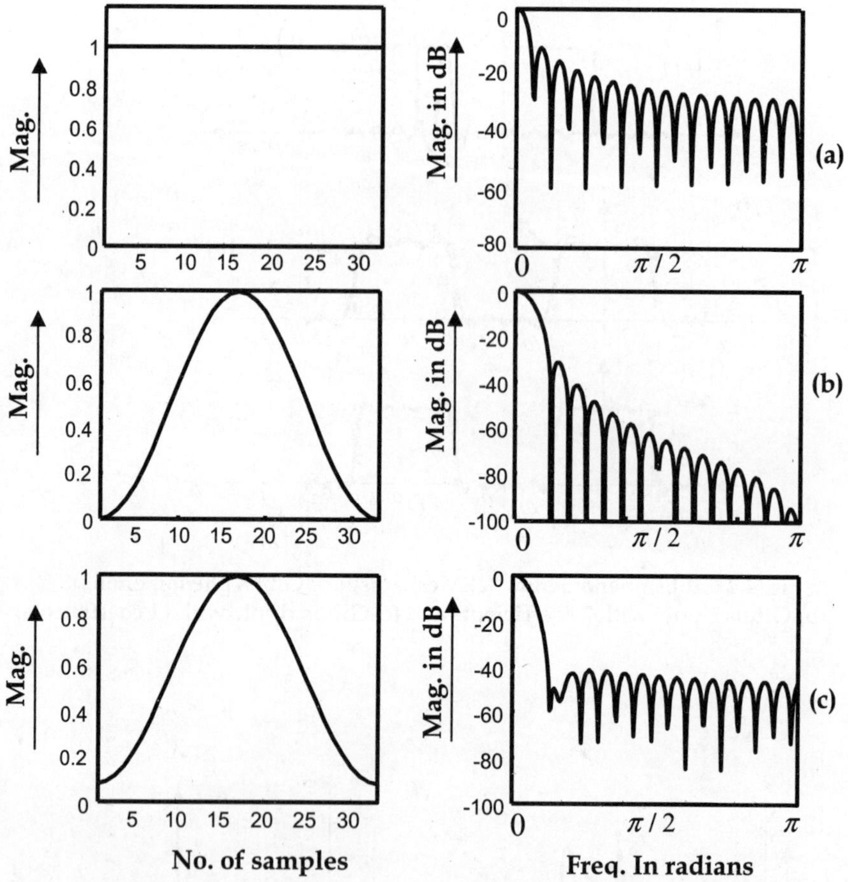

Fig. 4.6 Windows and their frequency response: (a) Rectangular; (b) Hanning and (c) Hamming.

Hence, the side lobes act as *leakage paths* and in this context, the truncation effect is also known as *leakage effect.* The only way to reduce the leakage is again to use a window with smooth truncation. As explained, this again reduces the frequency resolution.

The rectangular window has the poorest spectral characteristics, compared to other windows smoother than it (Fig. 4.6). The FT magnitude of the rectangular window is characterized by: a first side lobe which is 13 dB below the main lobe level and a main lobe which is of smallest width (for a given length of the window N) compared to other windows like Hanning and Hamming.

Hanning window:

$$w(n) = \frac{1}{2}\left[1 - \cos\left(\frac{2\pi n}{N-1}\right)\right], \quad 0 \le n \le N-1$$

Hamming window:

$$w(n) = 0.54 - 0.46\cos\left(\frac{2\pi n}{N-1}\right), \quad 0 \le n \le N-1$$

N is the length of the window.

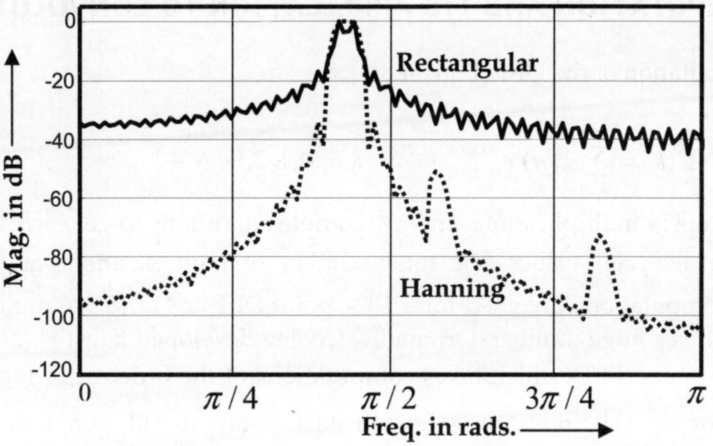

Fig. 4.7 Effect of windowing on leakage.

For these windows, the peak side lobe is significantly below the main lobe level, compared to that of rectangular window (Hanning (-31 dB), Hamming (-41dB)). Though the leakage effect is reduced due to their relatively lower peak side lobe levels, their frequency resolution is inferior as their main lobes are wider than that for the rectangular window. Specifically, the Hamming and Hanning windows have same main lobe width (effective length 0.5N) but they differ in their peak side lobe level and roll off rate. *The Hanning has a better roll off rate (-18dB/octave) than the Hamming window (-6dB/octave)* though it has a poor peak side lobe level (Fig. 4.6(b, c)). Hence, Hanning will have a reduced leakage effect and for this reason, is preferred to Hamming, for spectral analysis

To bring out the effect of leakage in estimating the spectral content of a signal, a composite signal made up of four sinusoids is considered. Their frequencies are $0.4\pi, 0.42\pi, 0.55\pi$ and 0.8π. Their corresponding amplitudes in order are $20.0, 20.0, 0.06$ and 0.005. The first two sinusoids are of equal amplitude and are closely spaced. However, the other two sinusoids are of relatively small magnitude (-50dB and –75dB below) and themselves well separated and also away from the closely spaced sinusoids. For a segment data of 256 samples without any window being applied, it is not possible to get the complete spectral information about the signal (Fig. 4.7). This is because, the leakage effect due to the rectangular window is so high that it masks the low level sinusoids at (0.55π, 0.8π) and is unable to indicate even their presence. But due to its good frequency resolution ability it is able to resolve the two closely spaced spectral lines at (0.4π, 0.42π).

The application of the Hanning window for the same data is able to bring out the two low level spectral lines (-50dB and -75dB) at 0.55π and 0.8π explicitly (Fig. 4.7). However, due to the poor frequency resolution of the Hanning window compared to that of the rectangular window, it is unable to resolve the spectral lines at $(0.4\pi,\ 0.42\pi)$ and indicates them as a single broad peak.

4.5 DFT COMPUTATION AND ITS APPLICATION TO CONVOLUTION

The direct computation of the DFT (Eqn. (4.8a)),

$$X(k) = \sum_{n=0}^{N-1} x(n)\, e^{-j2\pi\, n\, k/N}, \quad k = 0,1,2,\ldots N-1$$

involves N complex multiplications and N complex additions to get each value of $X(k)$. Since there are N such values, the total number of multiplications and additions are N^2 each. The computations involved for a 1024 point DFT are 1,048,576 multiplications and additions, which is a huge number. Originally, Cooley developed a fast algorithm known as *Fast Fourier Transform* (FFT) which takes computations of the order of $N\log_2 N$ instead of N^2. That is, for a 1024 point DFT, the FFT takes only 10,240 computations as against 1,048,576 computations for the direct DFT and the FFT is efficient by more than *two order* magnitude which is significantly remarkable. To compute the DFT for 100 points, it takes 10,000 multiplications and additions each. If the DFT can be calculated twice using half the total number of points, i.e., 50, the computations involved are $2 \times 50^2 = 5000$ which is half as that of direct DFT computation. That is, if the DFT computation of N samples could be broken into two DFT computations of $N/2$ points, the total number of computations will be reduced. This process can be continued from $N/2$ points to $N/4$ and so on, till it end with one point DFT. This is the basic idea of FFT algorithm.

4.5.1 Convolution of two Sequences using DFT

The convolution of $x_1(n)$ with $x_2(n)$ is given by

$$y(n) = \sum_{i=-\infty}^{\infty} x_1(i)\, x_2(n-i) \tag{4.10}$$

The above convolution operation in time domain is computationally demanding as to get the convolution for each time instant, it involves multiplication of the flipped delayed signal with the other and summation of the product.

The convolution of two signals in time results in multiplication of the FTs of the two signals, involved. This concept can be used for efficiently computing the convolution operation by multiplying the FTs of the two signals and taking the inverse FT of the product. This involves two DFT operations and multiplication of the two DFTs only once and one IDFT.

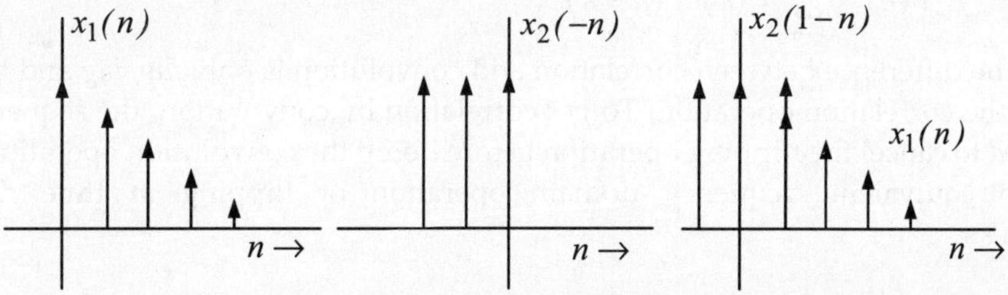

Fig. 4.8 Convolution of two finite length sequences.

If the length of $x_1(n)$ is M and $x_2(n)$ is L, then the convolution operation results in a signal of length $L+M-1$.

$$y(n) = \sum_{i=0}^{M-1} x_1(i)\, x_2(n-i) \tag{4.11}$$

Assuming $M > L$, if the DFT of the two sequences are taken for a length M (by padding zeros to the signal of length L) and the inverse FT of the product $X_1(k)X_2(k), \quad k = 0,1,2,\ldots M-1$, then there will be a *time domain aliasing due to the built in periodicity of the DFT* as the convolved signal will have a length of $(L+M-1)$, which is larger than M. In Fig. 4.8, $M = 5$, $L = 3$ and $(L+M-1) = 7$ which is larger than $M = 5$ and hence there will be time domain aliasing over two points. This convolution is called *circular convolution.*

However, the desired convolution is a linear convolution. To avoid this time domain aliasing of the convolved signal due to circular convolution, the DFT of each signal is computed for a length equal to/greater than $(L+M-1)$ by padding $(L-1)$ or more zeros to $x_1(n)$ and $(M-1)$ or more zeros to $x_2(n)$. With this, the convolved signal though repeats, will not overlap over its adjacent replica and hence there is no time domain aliasing problem. The padding of sufficient number of zeros enables to compute linear convolution from circular convolution.

It has been found that the use of FFT for convolution is computationally advantageous only if the FFT length involved is greater than 32.

4.5.2 Computation of Autocorrelation by Convolution

The discrete correlation is given by

$$y(n) = \sum_{k=0}^{L-1} x(k)\, x(k+n) = \sum_{k=0}^{L-1} x(k)\, x(n+k) \tag{4.12}$$

But the convolution is given by

$$y(n) = \sum_{k=0}^{L-1} x(k)\, x(n-k)$$

Therefore, the difference between correlation and convolution is only *flipping* and there is no flipping in the correlation operation. To get correlation by convolution, the sequence has to be *pre-flipped* to cancel the flipping operation before doing the convolution operation.

The equivalent frequency domain operation of flipping in time domain is conjugation as

$$x(an) \leftrightarrow \frac{1}{|a|} X\left(\frac{\omega}{a}\right) \qquad \text{(Scaling property)}$$

If $a = -1$,

$$x(-n) \leftrightarrow X\left(\frac{\omega}{-1}\right) = X(-\omega) = X^*(\omega)$$

Therefore, to perform convolution between $x(n)$ and $x(-n)$ using frequency domain approach,

$$Y(\omega) = X(\omega)\, X^*(\omega)$$

and the correlation is given by inverse transform of $Y(\omega)$.

Here also, the convolution length has to be considered and since the signal and its flipped versions are convolved, it is required to the FT for a length twice the signal duration by padding zeros equal to the duration of the signal. This is essential to avoid time domain aliasing due to periodic nature of the DFT.

It may be thought as

$$Y(\omega) = X(\omega)\, X^*(\omega) = |X(\omega)|^2$$

by taking the inverse transform of the FT magnitude square of the signal, the correlation can be computed. This is true provided in computing the FT magnitude square the proper zero padding has been done. Otherwise, the correlation function obtained will be in error due to time domain aliasing.

To compute the *cross-correlation* also the same method and considerations are valid and instead of taking the conjugate of the FT of the same function, the conjugate of the FT of the second function involved is taken and the length of the individual FTs should be more than the sum of the duration of the two sequences involved.

The methods of efficient computation of convolution and correlation will find application in computing the wavelet transform, which will be discussed in chapter 12.

4.6 LAPLACE TRANSFORM

Signals that are non-zero in the interval $-\infty \le t \le \infty$ are called two-sided or bilateral signals. Signals which are non-zero in the interval $0 \le t \le \infty$ are called unilateral or one sided signals or causal signals. Laplace transform (LT) is often studied for unilateral signals but this is not a restriction and LT is applicable for bilateral signals also. The Fourier transform (FT) of a unilateral signal is given by

$$X(\omega) = \int_0^{\infty} x(t)\, e^{-j\omega t}\, dt \quad \text{and} \quad x(t) = \frac{1}{2\pi}\int_{-\infty}^{\infty} X(\omega)\, e^{j\omega t}\, d\omega$$

where $x(t) = 0, t < 0$.

The FT may not exist for a function $x(t)$, but if it is multiplied by a function that decays with time, the FT of the resulting function may exist.

The FT of $x(t)e^{-\sigma t}$, where σ is a real number is given by

$$X_\sigma(\omega) = \int_0^{\infty} \left[x(t)e^{-\sigma t}\right] e^{-j\omega t}\, dt$$

If σ is positive and sufficiently large, $x(t)\, e^{-\sigma t}$ will decay rapidly and its FT can exist which is given by

$$X_\sigma(\omega) = \int_0^{\infty} x(t)\, e^{-(\sigma + j\omega)t}\, dt \tag{4.13}$$

The inverse FT is

$$x(t)\, e^{-\sigma t} = \frac{1}{2\pi}\int_{-\infty}^{\infty} X_\sigma(\omega)\, e^{j\omega t}\, d\omega \tag{4.14}$$

$$x(t) = \frac{1}{2\pi}\int_{-\infty}^{\infty} X_\sigma(\omega)\, e^{(\sigma + j\omega)t}\, d\omega$$

Here, the $e^{-\sigma t}$ factor has been moved to the right side as it is a constant factor independent of integration variable ω. Let $s = \sigma + j\omega$ and $X(s) = X_\sigma(\omega)$, then $ds / d\omega = j,\ d\omega = (1/j)ds$. Substituting these in Eqns. (4.13) and (4.14)

$$X(s) = \int_0^{\infty} x(t)\, e^{-(s)t}\, dt \tag{4.15}$$

and

$$x(t) = \frac{1}{2\pi j} \int_{\sigma - j\infty}^{\sigma + j\infty} X(s)\, e^{st}\, ds \tag{4.16}$$

These equations (4.15) and (4.16) form the Laplace Transform pair. At first glance, the interpretation of LT of a function by itself is not easy. But from the above development of the LT, the LT of a function can be interpreted as *the FT of an exponentially weighted function.* Apparently, it looks like that the FT can be obtained from LT by replacing s by $j\omega$. To some extent it is true but there are functions for which LT exists though the FT does not exist.

For the FT to exist,

$$\int_0^{\infty} |x(t)|\, dt < K, \qquad K \text{ is a finite quantity.}$$

For the LT to exist,

$$\int_0^{\infty} |x(t)|\, e^{-\sigma t}\, dt < M, \qquad M \text{ is a finite quantity.}$$

Even though a function may not satisfy the condition for existence of FT, for the LT, a suitable choice of σ will enable this condition to be satisfied.

For example if $x(t) = u(t)\, e^{8t}$, then for a choice of $\sigma > 8$,

$$\int_0^{\infty} |\, e^{8t}\, |\, e^{-\sigma t}\, dt = \int_0^{\infty} |\, e^{(8-\sigma)t}\, |\, dt$$

this integral is finite. *Hence, any function, which has FT, will have LT. However, the converse is not true.*

If the LT exists for some value of say σ_o, then it exists for all values of $\sigma \geq \sigma_o$. In addition to the Laplace transform relation in Eqn. (4.15), the range of values of σ for which the LT exists, also has to be specified. That is, complete specification of LT is

$$X(s) = \int_0^{\infty} x(t)\, e^{-(s)t}\, dt\, ,\ \sigma \geq \sigma_o \tag{4.17}$$

4.6.1 System Analysis by Laplace Transform

Generally, it is required to find the response of a physical system described by the time domain differential equation for a given input. Consider a system described by the differential equation with zero initial conditions

$$\frac{d^2 y(t)}{dt^2} + a\frac{dy(t)}{dt} + by(t) = c\frac{dx(t)}{dt} + dx(t) \tag{4.18}$$

where $x(t)$ and $y(t)$ are input and output of the system.

The LT of the derivative of a function $x(t)$ is given by

$$X'(s) = \int_0^\infty \frac{dx(t)}{dt} e^{-(s)t} dt$$

This integration will be carried out by integration by parts, given by

$$\int u\, dv = uv - \int v\, du$$

If $u = e^{-st}$, $dv = \frac{dx(t)}{dt}$, then

$$\int_0^\infty e^{-st} \frac{dx(t)}{dt} dt = e^{-st} x(t)\Big|_0^\infty - \int_0^\infty x(t)\, e^{-st}(-s)\, dt$$

$$= x(0) + s\int_0^\infty x(t)\, e^{-st} dt,$$

If $x(0) = 0$, then

$$X'(s) = \int_0^\infty \frac{dx(t)}{dt} e^{-st} dt = sX(s) \tag{4.19}$$

Similarly,

$$X''(s) = \int_0^\infty \frac{d^2x(t)}{dt^2} e^{-(s)t} dt = s^2 X(s) \tag{4.20}$$

Taking Laplace transform on both sides of Eqn. (4.18),

$$L\left\{\frac{d^2y(t)}{dt^2}\right\} + aL\left\{\frac{dy(t)}{dt}\right\} + bL\{y(t)\} = cL\left\{\frac{dx(t)}{dt}\right\} + dL\{x(t)\}$$

Applying the above results for the derivatives of a function,

$$s^2Y(s) + as\,Y(s) + bY(s) = cs\,X(s) + dX(s)$$

$$H(s) = \frac{Y(s)}{X(s)} = \frac{cs + d}{s^2 + as + b} \tag{4.21}$$

$X(s)$ and $Y(s)$ are Laplace transforms of $x(t)$ and $y(t)$, respectively. $H(s)$ is known as the *transfer function*. Knowing $X(s)$, $Y(s) = X(s)\,H(s)$ and hence $y(t)$ can be found using inverse Laplace transform.

The Laplace transform is advantageous as *it converts the differential equation into an algebraic equation.*

$h_2(t)$

4.6.2 Poles and Zeros

If

$$H_1(s) = \frac{d/a}{s+b/a} = \frac{G}{s+\sigma_0}, \qquad G = d/a, \sigma_0 = b/a$$

For $s = -\sigma_0$, the denominator becomes zero and $H_1(s)$ becomes infinite. This value of s is called a **pole**. The inverse transform of $H_1(s)$, called the impulse response $h_1(t)$ is given by

$$h_1(t) = G\, e^{-\sigma_0 t}$$

For $s = -\sigma_0$, the pole is located on the LHS of the *s*-plane on the σ-axis (Fig. 4.9a).

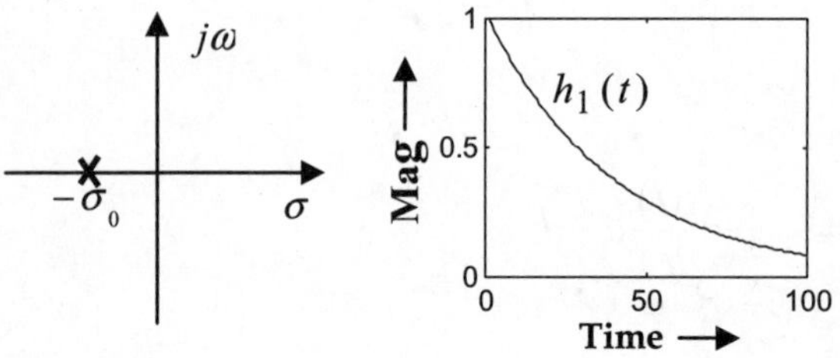

Fig. 4.9(a) *s*-plane representation of $H_1(s)$ and its impulse response.

Consider,

$$H_2(s) = \frac{cs}{s^2 + as + b} = \frac{Gs}{(s+p)\,(s+p^*)}, \quad G = c,\ p = p_R + jp_I$$

At $s = -p$ and $s = -p^*$, $H_2(s)$ is infinity. Hence, $H_2(s)$ has poles at these values of s. The value of s at which the transfer function becomes zero is called its **zero.** Here, the numerator and hence $H_2(s)$ becomes zero at $s = 0$. Therefore, $s = 0$ is the *zero* of $H_2(s)$. The inverse transform of $H_2(s)$ is given by

$$h_2(t) = 2|G|\, e^{-\sigma_0 t} \cos(\omega_0 t + \theta)$$

where $\sigma_0 = p_R$, $\omega_0 = p_I$ and $\theta = \tan^{-1}(p_I / p_R)$.

Here, $h_2(t)$ is a decaying oscillating signal at a frequency ω_0 and decay rate is determined by σ_0 (Fig. 4.9b).

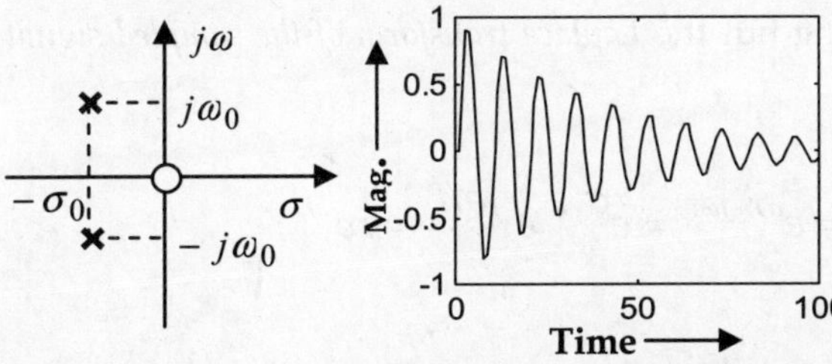

Fig. 4.9(b) S-plane representation of $H_2(s)$ and its impulse response.

Both for $h_1(t)$ and $h_2(t)$ in the s-plane, if the poles are on the left of the $j\omega$-axis, they will decay, otherwise if the poles are on the right of the s-plane, they grow indefinitely. For a bounded input, to get a bounded output, the system poles should be on the *left half of the s-plane.* If the poles are on $j\omega$ - axis, the system will oscillate.

4.7 Z-TRANSFORM

A sampled signal is given by

$$x_s(t) = \sum_{n=0}^{\infty} x\,(t)\,\delta\,(t-nT) = \sum_{n=0}^{\infty} x\,(nT)\,\delta\,(t-nT)$$

Here, the sampling interval is indicated by T and not by T_S to avoid any confusion that may arise as " s " is also used in the definition of Laplace transform.
Laplace transform of $x_s(t)$ is given by

$$L\{x_s(t)\} = X_s(s) = \int_{t=0}^{\infty} x_s(t)\,e^{-st}\,dt$$

$$= \int_{t=0}^{\infty} \sum_{n=0}^{\infty} x(nT)\,\delta\,(t-nT)\,e^{-st}\,dt = \sum_{n=0}^{\infty} x\,(nT)\,e^{-snT}$$

$$X(z) = X_s(s)\Big|_{z=e^{sT}} = \sum_{n=0}^{\infty} x\,(nT)\,z^{-n} \tag{4.22}$$

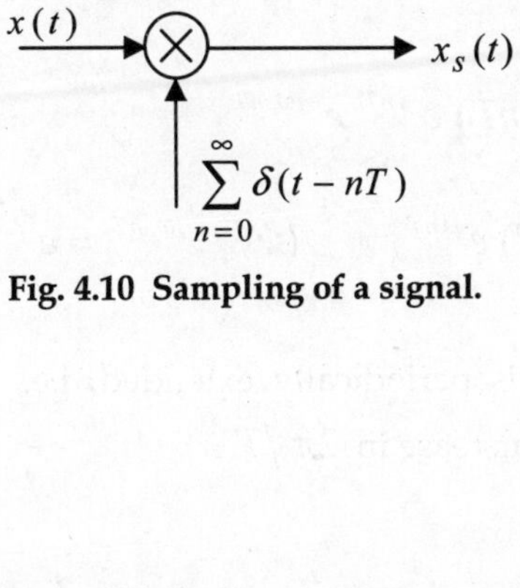

Fig. 4.10 Sampling of a signal.

$X(z)$ is called the Z-transform of the sequence $x(nT)$. Therefore, the Z-transform of a sequence $x(nT)$ is nothing but the *Laplace transform of the sampled signal* $x_s(t)$ *with* $z = e^{sT}$.

Consider

$$z = e^{sT} = e^{\sigma T + j\omega T} = e^{\sigma t} e^{j\omega T} = re^{j\theta}$$

where

$$r = e^{\sigma t} \text{ and } \theta = \omega T \tag{4.23(a, b)}$$

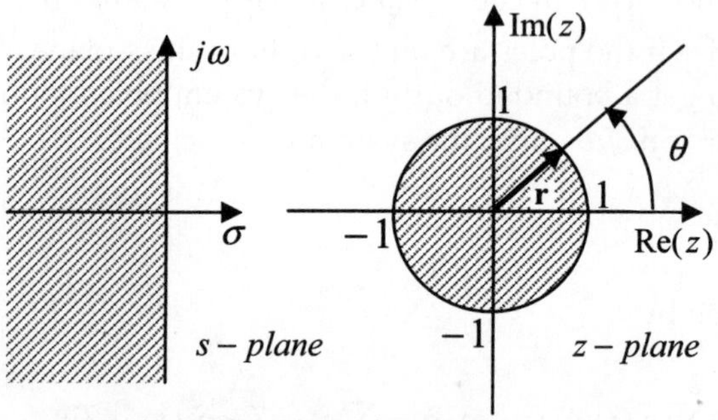

Fig. 4.11 Mapping of *s*-plane to *z*-plane.

$\sigma = 0, r = 1$: The imaginary axis on the *s*-plane is mapped as unit circle in *z*-plane.

$\sigma < 0, r < 1$: The *s*-plane left to the $j\omega$-axis is mapped inside the unit circle.

$\sigma > 0, r > 1$: The *s*-plane right to the $j\omega$-axis is mapped outside the unit circle (Fig. 4.11).

Therefore, for the stability of a system, poles must lie inside the unit circle.

In the *s*-plane as ω increases from 0 to $2\pi / T$, since $\theta = \omega T$, θ makes one rotation in the *z*-plane. Every increase of $2\pi / T$ in ω, in the *s*-plane results in one rotation in the *z*-plane. Thus strips of width $2\pi / T$ in the s-plane correspond to one rotation in the z-plane.

$$X_s(s + j\omega_s) = \sum_{n=0}^{\infty} X(nT)\, e^{-snT}\, e^{-j\omega_s nT}$$

$$= \sum_{n=0}^{\infty} X(nT)\, e^{-snT} \qquad (\because, e^{-j\omega_s nT} = e^{-j\frac{2\pi}{T}nT} = 1)$$

Thus in the *s*-plane, $X_s(s)$ is periodically extended, i.e., the information in the region 0 to $2\pi /T$, repeats for every increase in $2\pi /T$.

The location of poles in the *z*-plane is of importance as with the *s*-plane. The *z*-plane maps the left half of the *s*-plane inside the unit circle, the right half of *s*-plane outside the unit circle and the $j\omega$ axis itself as the unit circle. Further, the strips of width ω_S on the $j\omega$ axis maps to rotations on the unit circle.

For a real pole located inside the unit circle, the time response (TR) is a decaying exponential and is nothing but impulse response of a first order low-pass filter. The rate of decay depends upon the distance of the pole from the unit circle. For a pole away from the unit circle, the TR will decay faster. On the other hand, for a real pole on the unit circle, corresponds to DC, which does not decay at all with time. If the real pole is outside unit circle, this leads to a growing exponential which also corresponds to a pole in the right half of the *s*-plane. Again its rate of growth is more if it is away from the unit circle.

If the real pole is on the negative side of the real axis, it is equivalent to a high-pass filter as the real pole is shifted from the zero frequency to half the sampling frequency or π. That is, the low-pass time response $x(n) = e^{-\sigma_o n}$ is multiplied by

$$e^{-j\pi n} = \cos \pi n - j \sin \pi n = (-1)^n$$

That is,

$$y(n) = (-1)^n e^{-\sigma_o n} = (-1)^n r_o^{-n}$$

Therefore, for a real pole on the negative side of the real axis in the *z*-plane, the TR takes positive and negative values alternately and again its distance from the unit circle determines the decay rate. As in the case of a low-pass filter, if the pole is away from the unit circle, the TR decays faster. For a pole located on the unit circle, i.e., at π, the TR goes positive and negative alternately but without any decay. If the pole is outside the unit circle, the TR grows in a bipolar fashion and the rate again depends up on its distance from the unit circle.

For the complex conjugate pair of poles, the angle θ_o indicates the frequency of oscillation scaled with respect to sampling frequency corresponding to 2π. If the poles are located away from the unit circle, the response decays and farther the location, faster is the decay. For the case where the poles are located on the unit circle, again the frequency of oscillation is decided by θ_o but the oscillation amplitude is sustained without any decay and represents an oscillator. If the pole locations are outside the unit circle, in such a case, the oscillations grow in amplitude and lead to an unstable system.

The decay rate of the TR is determined by the distance of location of poles from the unit circle. This distance decides the bandwidth associated with the center frequency (zero frequency for low-pass or the half the sampling frequency for high-pass). If this distance is more, the bandwidth is more and faster is the decay or growth. If this distance is less or zero, decay (growth) rate is slow or the amplitude is sustained (no decay or growth). For a complex conjugate pair of poles, in the spectrum, the spectral peak will be narrower if the poles are located close to the unit circle and otherwise, the spectral peak will be broader.

For the case when the poles are on the unit circle, the spectral peaks ideally should have zero bandwidth. This is so, for real poles also.

4.7.1 Sampling in the Frequency Domain

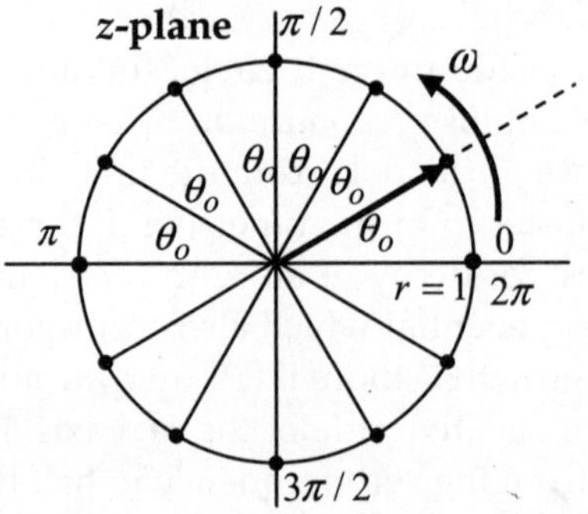

Fig. 4.12 Discrete Fourier transform.

The Z- transform of a sequence $x(nT)$ is given by

$$X(z) = \sum_{n=-\infty}^{\infty} x(nT)\, z^{-n}$$

where

$$z = e^{sT} = e^{(\sigma + j\omega)T} = e^{\sigma T}.\, e^{j\omega T} = r.\, e^{j\omega T}$$

If $\sigma = 0$, then

$$r = e^{\sigma T} = e^{0.T} = 1$$

Therefore,

$$z = re^{j\omega T} = e^{j\omega T}$$

$$X(z) = \sum_{n=-\infty}^{\infty} x(nT)\left(e^{j\omega T}\right)^{-n}$$

$$X(\omega) = \sum_{n=-\infty}^{\infty} x(nT)\, e^{-j\omega nT} \tag{4.23c}$$

This is nothing but Fourier transform of the sampled signal, which is continuous in frequency ω and is the ***Discrete Time Fourier Transform*** **(DTFT)** of the signal $x(t)$. Therefore *the Z- transform of a signal evaluated on the unit circle i.e., $r = 1$, is the* ***Discrete Time Fourier Transform.***

If the DTFT is sampled in frequency, this will result in ***Discrete Fourier Transform (DFT)*** which is *sampled both in time and frequency.* Such a sampling is indicted in the Fig. 4.12.

To summarize, the Z-transform is the Laplace transform of the sampled signal. The Fourier transform is the Laplace transform evaluated on the $j\omega$ axis. In turn, the LT can be considered the FT of an exponentially weighted sequence. The DTFT is the Z-transform evaluated on the unit circle. Finally, the DFT is the Z-transform, sampled in the frequency on the unit circle.

4.7.2 Rational Transfer Function

A general rational transfer function is given by

$$H(z) = G\, z^{q} \frac{\prod_{k=1}^{N_i}(1-a_k z^{-1})\prod_{k=1}^{N_o}(1-c_k z)}{\prod_{k=1}^{M_i}(1-b_k z^{-1})\prod_{k=1}^{M_o}(1-d_k z)} \tag{4.23d}$$

where $|a_k|, |b_k|, |c_k|$ and $|d_k|$ are all less than unity. Then

- a_k and b_k are the zeros and poles inside the unit circle.
- c_k and d_k are the zeros and poles outside the unit circle.
- z^q corresponds to the linear phase shift introduced by the system and G is the gain.

If only a_k and b_k are present, c_k and d_k are zero, then the system has poles and zeros only inside the unit circle. Such a system is called a *minimum phase* system.

If only c_k and d_k are present, a_k and b_k are zero, then the system has poles and zeros only outside the unit circle. Such a system is called a *maximum phase* system.

If some of a_k, b_k, c_k and d_k are present, then the system poles and zeros will lie both inside and outside the unit circle. Such a system is called a *mixed phase* system.

However, for a physically realizable system, the poles cannot be outside the unit circle and only zeros can be either inside or outside the unit circle.

A minimum phase system has the following properties:

- It is causal, that is, its impulse response $h(n) = 0$ for $n < 0$
- It is invertible (i.e., $1/H(z)$ exists)
- A minimum phase system is a minimum delay system, i.e., energy is concentrated towards the origin.

Since the poles and zeros are inside the unit circle and the region of convergence includes $z = \infty$. Hence, a minimum phase system is causal. It is also invertible as the zero on inversion become poles and they are inside the unit circle ensuring a stable system. The poles will become zeros and they will not pose any problem of stability. As the name itself indicates, its phase curve deviates minimum from the frequency axis and hence contributes minimum delay (group delay).

The maximum phase system corresponds to an anti-causal system, i.e. its impulse response $h(n)$ is zero for $n > 0$ and has maximum delay. Such a system is also invertible. But a mixed phase system is not invertible if the zeros lie outside the unit circle.

The magnitude spectrum of a real signal (mixed phase), will be the same irrespective of its roots (poles or zeros) lie inside or outside the unit circle in the z-plane as it does not preserve phase information. Thus, the magnitude spectrum is blind to whether roots are of maximum or minimum phase nature and corresponds to the minimum phase. For a mixed phase system

$$H(z) = \frac{z\,(1-0.35z^{-1})\,(z^{-1}-0.6)}{1-0.3981z^{-1}+0.4149z^{-2}}$$

The roots of the denominator polynomial and the zero at $z = -0.35$ are of minimum phase nature. But the zero $z = 1/0.6$ is a maximum phase zero. Therefore, *the minimum phase equivalent* will have a zero at $z = 0.6$ corresponding to the maximum phase zero $z = 1.666$ and is obtained by reflecting it to the reciprocal conjugate position. Here, the conjugate operation is ineffective as it is a real zero and only reciprocal operation is effective. This zero is in addition to the minimum phase zero at $z = 0.35$ and the complex conjugate pole pair, forming a minimum phase system

$$H_{\min}(z) = \frac{z\,(1-0.35z^{-1})\,(1-0.6z^{-1})}{1-0.3981z^{-1}+0.4149z^{-2}}$$

To get the minimum phase equivalent of a mixed phase signal $h(n)$, its even part of the cepstrum $c_{ev}(n)$ form its spectral magnitude$|H(k)|$ is computed as

$$c_{ev}(n) = IFT\left[\ln|H(k)|\right]$$

$$c_{ev}(n) = \frac{c(n)+c(-n)}{2}$$

Since for a minimum phase signal, $c(-n) = 0$, $c(n)$ is given by

$$\begin{aligned} c(n) &= c_{ev}(n), \quad n = 0 \\ &= 2c_{ev}(n), \; n > 0 \\ &= 0, \qquad\quad n < 0 \end{aligned}$$

Algorithm using DFT/FFT:

1. Compute: $H(k) = DFT\left[h(n)\right],\ \hat{H}(k) = \left\{\ \ln\left|H(k)\right|\ \right\},\ k = 0, 1, 2, \ldots, (N-1).$

2. Compute: $EV\left[\hat{h}_{\min}(n)\right] = IDFT\left[\hat{H}(k)\right]$, where $EV[\ \cdot\ ]$: even part and IDFT : inverse DFT

3. Compute: $\hat{h}_{\min}(n) = l_{\min}(n) EV\left[\hat{h}(n)\right]$

where
$$l_{\min}(n) = \begin{cases} 1, & n = 0, N/2 \\ 2, & n = 1, \ldots, N/2 - 1 \\ 0, & n = N/2 + 1, \ldots, N - 1 \end{cases}$$

4. Compute: $\hat{H}_{\min}(k) = DFT\ [\hat{h}_{\min}(n)]$

$$H_{\min}(k) = e^{\hat{H}_{\min}(k)},\ k = 0, 1, 2, \ldots, N-1$$

$$h_{\min}(n) = IDFT\left[H_{\min}(k)\right]$$

4.8 DISCRETE COSINE TRANSFORM (DCT)

The real part of the FT for a real signal is the FT of the even part of the signal. That is, if the signal is an even function, its FT will be real. It is possible to construct from the given discrete signal an even symmetric function and for this, the DFT will be real and is known as the discrete cosine transform (DCT). The DCT is commonly used for signal processing applications like image coding, adaptive algorithms and for many analyses due to its own properties like energy compactness and computational advantage being real in nature than DFT which is complex. Though there are many versions of the DCT like DCT-1, 2, 3 and 4, DCT-2 is widely used for different applications.

To derive the DCT of a signal, consider a sequence $x(n),\ n = 0, 1, 2, \ldots (N-1)$. Now form a new sequence $y(n)$ of length $2N$ given by

$$y(n) = x(n) + x\,(2N - 1 - n) \tag{4.24}$$

$$y(n) = \begin{cases} x(n), & 0 \le n \le N-1 \\ x(2N-1-n), & N \le n \le 2N-1 \end{cases}$$

This sequence $y(n)$ is symmetric with respect to the point $\left(N - \dfrac{1}{2}\right)$ (Fig. 4.13).

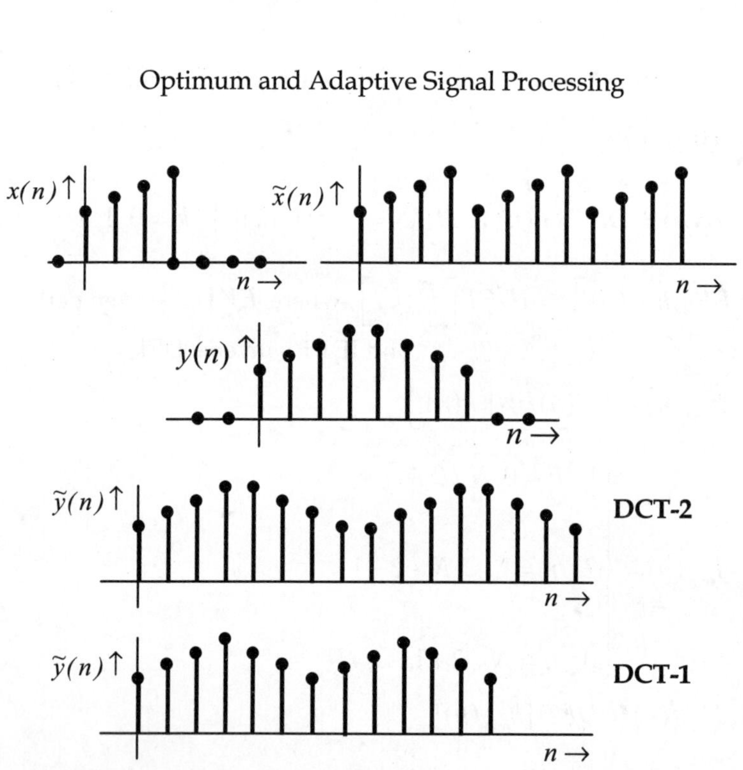

Fig. 4.13 Development of periodic symmetric signal for DCT-2 and DCT-1.

The $2N$ point DFT of $y(n)$ is

$$Y(k) = \sum_{n=0}^{2N-1} y(n)\, e^{-j2\pi k n/2N}, \qquad 0 \le k \le 2N-1$$

$$= \sum_{n=0}^{N-1} x(n)\, e^{-j2\pi k n/2N} + \sum_{n=N}^{2N-1} x(2N-1-n)\, e^{-j2\pi k n/2N}$$

Let $l = 2N-1-n$. Then for $n = N$, $l = N-1$ and for $n = 2N-1$, $l = 0$.

$$Y(k) = \sum_{n=0}^{N-1} x(n)\, e^{-j2\pi k n/2N} + \sum_{l=N-1}^{0} x(l)\, e^{-j2\pi k (2N-1-l)/2N}$$

$$Y(k) = \sum_{n=0}^{N-1} x(n)\, e^{-j2\pi k n/2N} + \sum_{l=0}^{N-1} x(l)\, e^{j2\pi kl/2N} e^{-j2\pi k}\, e^{j2\pi k/2N}$$

$$= \sum_{n=0}^{N-1} x(n)\, e^{-j2\pi k n/2N} + \sum_{l=0}^{N-1} x(l)\, e^{j2\pi kl/2N} e^{j2\pi k/2N} \qquad (\because e^{-j2\pi k} = 1)$$

Multiplying and dividing the RHS by $e^{j2\pi k/4N}$,

$$Y(k) = e^{j2\pi k/4N}\left[\sum_{n=0}^{N-1} x(n)\, e^{-j2\pi kn/2N} e^{-j2\pi k/4N} + \sum_{l=0}^{N-1} x(l)\, e^{j2\pi kl/2N} e^{j2\pi k/2N} e^{-j2\pi k/4N}\right]$$

$$= e^{j2\pi k/4N}\left[\sum_{n=0}^{N-1} x(n)\, e^{-j2\pi k(2n+1)/4N} + \sum_{l=0}^{N-1} x(l)\, e^{j2\pi k(2l+1)/4N}\right]$$

$$= 2e^{j2\pi k/4N}\left[\sum_{n=0}^{N-1} x(n)\left(\frac{e^{-j\pi k(2n+1)/2N} + e^{j\pi k(2n+1)/2N}}{2}\right)\right]$$

$$Y(k) = e^{j2\pi k/4N}\left[\sum_{n=0}^{N-1} 2x(n)\cos\frac{\pi k(2n+1)}{2N}\right],\ 0 \le k \le 2N-1 \tag{4.25}$$

The N point DCT of $x(n)$, $C_x(k)$ is given by

$$C_x(k) = \begin{cases} e^{-j2\pi k/4N} Y(k), & 0 \le k \le N-1 \\ 0, & \text{Otherwise} \end{cases} \tag{4.26}$$

Therefore, the DCT of $x(n)$ is given by

$$C_x(k) = \begin{cases} \sum_{n=0}^{N-1} 2x(n)\cos\frac{\pi k(2n+1)}{2N}, & 0 \le k \le N-1 \\ 0, & \text{otherwise} \end{cases} \tag{4.27a}$$

Consider

$$C_x(2N-k) = \sum_{n=0}^{N-1} 2x(n)\cos\frac{\pi(2N-k)(2n+1)}{2N},$$

As

$$\cos\frac{\pi(2N-k)(2n+1)}{2N} = \cos\left(\pi(2n+1) - \frac{\pi k(2n+1)}{2N}\right) = -\cos\frac{\pi k(2n+1)}{2N}$$

$$C_x(2N-k) = -C_x(k),\ k = 0,1,\ldots,2N-1 \tag{4.27b}$$

This is the *symmetry* property of DCT-2.

To get the inverse relation, that is to get $x(n)$ from $C_x(k)$, it is required to get $y(n)$ from $Y(k)$. From $y(n)$, $x(n)$ can be obtained.

Proceeding in the reverse order, $C_x(k)$ is only one side of the symmetric function $e^{-j2\pi k/4N}Y(k)$. Therefore,

$$Y(k) = e^{j2\pi k/4N}C_x(k), \qquad 0 \le k \le N-1$$

The symmetric $Y(k)$ over $2N$ points (using Eqns. (4.27)), is given by

$$Y(k) = \begin{cases} e^{j2\pi k/4N}C_x(k), & 0 \le k \le N-1 \\ 0, & k = N \\ -e^{j2\pi k/4N}C_x(2N-k), & N+1 \le k \le 2N-1 \end{cases}$$

Therefore,

$$y(n) = \frac{1}{2N}\sum_{k=0}^{2N-1} Y(k)e^{j2\pi nk/2N}, \quad 0 \le n \le 2N-1$$

and

$$x(n) = \begin{cases} y(n), & 0 \le n \le N-1 \\ 0, & otherwise \end{cases}$$

$$y(n) = \frac{1}{2N}\sum_{k=0}^{N-1}\left[e^{j2\pi k/4N}C_x(k)\right]e^{j2\pi nk/2N} + Y(N)e^{j2\pi nN/2N}$$
$$+\frac{1}{2N}\sum_{k=N+1}^{2N-1}\left[-e^{j2\pi k/4N}C_x(2N-k)\right]e^{j2\pi nk/2N}$$

Let $p = 2N-k$, when $k = N+1, p = N-1$ and when $k = 2N-1, p = 1$

$$y(n) = \frac{C_x(0)}{2N} + \frac{1}{2N}\sum_{k=1}^{N-1}\left[e^{j2\pi k/4N}C_x(k)\right]e^{j2\pi nk/2N}$$
$$+\frac{1}{2N}\sum_{p=N-1}^{1}\left[-e^{j2\pi(2N-p)/4N}C_x(p)\right]e^{j2\pi n(2N-p)/2N}$$

In the second summation,

$$-e^{j2\pi(2N-p)/4N}e^{j2\pi n(2N-p)/2N}$$

$$= -e^{j\pi}.e^{-j2\pi\, p/4N} e^{j2\pi\, n} e^{-j2\pi\, n\, p/2N}$$

$$= e^{-j2\pi\, p/4N} e^{-j2\pi\, n\, p/2N}$$

$$= e^{-j2\pi\, p(2n+1)/4N}$$

$$y(n) = \frac{C_x(0)}{2N} + \frac{1}{2N}\sum_{k=1}^{N-1} 2C_x(k)\left[\frac{e^{j2\pi\, k(2n+1)/4N} + e^{-j2\pi\, k(2n+1)/4N}}{2}\right]$$

Therefore,

$$y(n) = \frac{1}{N}\left[\frac{C_x(0)}{2} + \sum_{k=1}^{N-1} C_x(k)\cos\left(\frac{\pi\, k(2n+1)}{2N}\right)\right]$$

and

$$x(n) = \frac{1}{N}\sum_{k=0}^{N-1} U(k)C_x(k)\cos\left(\frac{\pi\, k(2n+1)}{2N}\right), \quad 0 \le n \le N-1 \qquad (4.28)$$

where

$$U(k) = \begin{cases} 1/2, & k = 0 \\ 1, & 1 \le k \le N-1 \end{cases}$$

$$C_x(k) = \begin{cases} \sum_{n=0}^{N-1} 2x(n)\cos\frac{\pi\, k(2\,n+1)}{2N}, & 0 \le k \le N-1 \\ 0, & otherwise \end{cases} \qquad (4.29)$$

These two relations Eqn. (4.28) and Eqn. (4.29) are the *discrete cosine transform pair*. Further this pair is also known as DCT-2. Here, the sequence has been extended so that it has even periodic symmetry about $1/2, (N-1/2), (2N-1/2), \ldots$ and the total number of points $2N$ is even and hence it is called even symmetrical DCT.

4.8.1 The Performance of DCT

In the DFT, a segment of the signal considered is a truncated one and further it becomes periodic in DFT. Thus, the signal on hand for a DFT, in general is a periodic signal with a *periodic discontinuity*. The discontinuity due to truncation, as already explained introduces Gibbs ripple and leakage effect, which contributes to the variance and bias of the estimate. It has been seen that the reduction of the Gibbs and leakage effect is achieved by smooth windows and averaging the DFT of these windowed segments. The segmentation of the given data into smaller segments and further use of a smooth window to each segment, result in a reduced frequency resolution. Thus, there is a trade-off between better frequency resolution and reduction in leakage and variance. Thus, with the DFT based spectral estimation, this trade-off is inevitable.

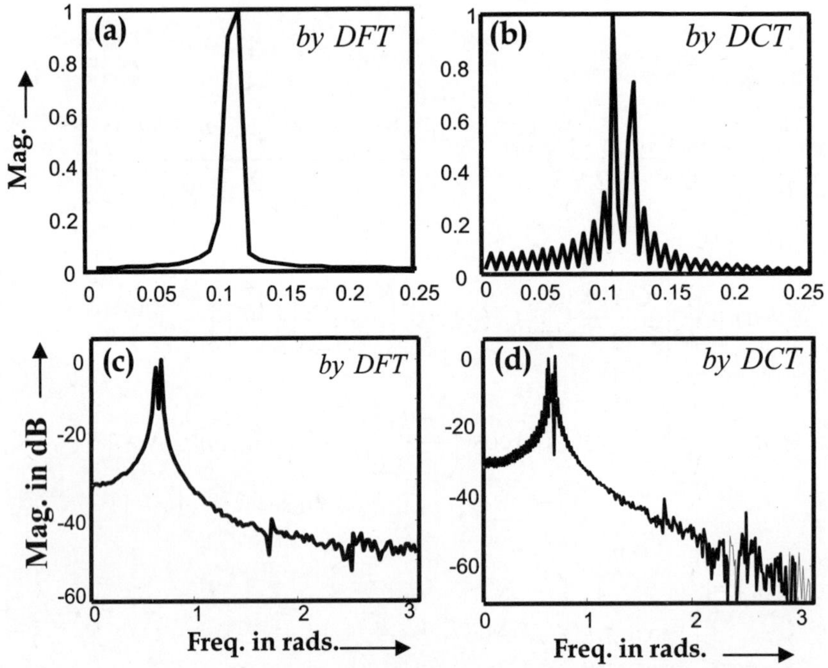

Fig. 4.14 Performance comparison of DFT and DCT Resolution of DFT (a) and DCT (b) Leakage or detectability of DFT (c) and DCT (d).

In DCT, the generation of a periodic even symmetric sequence removes the discontinuity as the symmetric signal moves from one period to the next period smoothly. In the DFT it is the discontinuity that generates the high frequency components, which are not in the original signal. Further, due to the leakage effect, the energy gets leaked into undesired high frequency components from the desired signal components and vice versa. For recovering the signal back, the only way is to *undo the total damage* that has occurred. Thus it is not possible to get rid off the high frequency components by setting them to zero values as they have some part of the low frequency energy. Hence, for a good approximation of the signal, large number of DFT coefficients is necessary and only the last few coefficients can be neglected. In the DCT, on the other hand, due to the absence of the discontinuity, no high frequency components are introduced. Further, due to smooth truncation from one period to another (as if there is no window), there is no leakage into either higher frequency region from signal frequency components. That is the signal energy is confined to only signal frequency components and there is no distribution of the signal energy into high frequency region. This is called the *energy compaction* of the DCT. Hence, the DCT enables representation of the signal only by few coefficients and is unlike the DFT, which requires more coefficients as the signal energy is scattered into high frequencies due to leakage. Effectively, the windowing problem is relatively reduced.

For the DCT, the symmetric signal is of twice its length. This is unlike zero padding where the zeros are padded to the original signal. The zero padding only interpolates the intermediate values improving the readability and there is no improvement in the basic frequency resolution that is decided by the signal length. In the DCT, the symmetric signal length itself is double its original length and hence its frequency resolution is better by a factor of two. Thus, the DCT both from the signal compression point of view and also from the spectral estimation point of view is superior.

The above properties of the DCT are illustrated by examples. First sum of two sinusoids of frequencies 0.1π and 0.11π each of same amplitude are considered. The DFT spectrum of this data of 128 samples cannot resolve the two spectral peaks (Fig. 4.14 (a)). But the DCT of the same data is able to resolve the two spectral peaks clearly (Fig. 4.14 (b)). This shows the superior frequency resolution ability of the DCT over DFT.

In the second case, four sinusoids, two of them are closely spaced in frequency being at 0.4π *or* 1.2566 and at 0.42π *or* 1.3192 with equal amplitudes and the other two sparsely spaced at 0.55π *or* 1.7279 and 0.8π *or* 2.5133. The magnitudes of the last two sinusoids are 40 and 45 dB below that of the first two sinusoids, respectively. It is seen that a DFT of 256 samples is unable to detect the low level spectral peaks clearly (Fig. 4.14(c)) and this is due to leakage effect but the DFT is able to resolve closely spaced spectral peaks. The use of a Hanning window enables the detection of the low level spectral peaks but at the cost of frequency resolution. On the other hand, the DCT of the same data not only resolves the closely spaced spectral peaks but also detects the low level spectral peaks (Fig. 4.14(d)). The detection of the low level spectral peaks is due to its reduced leakage effect due to the absence of discontinuity in the signal and the improved frequency resolution is due to its enhanced length of symmetric signal used for the DCT. The amount of leakage reduction is more than 15dB, as for DFT the spectrum does not go below – 45 dB level whereas for DCT it is below –60dB. This is very significant as it enables the detection of low-level spectral peaks.

Chapter 5

RANDOM SIGNAL THEORY

This chapter introduces the concept of random signal analysis. A signal which cannot be defined by a mathematical relation is a random signal. On the other hand, a signal which can be defined by a mathematical relation is not random but is deterministic. Most of the practical signals like speech and image and system like windmill cannot be represented mathematically as they are influenced by many factors simultaneously. Hence, it is necessary to treat them as random and their processing has to be based on statistical averages. The tools which can handle them are called estimators and it is of interest to find estimators which best represent the random process on hand. The basic estimator like correlation estimator and its performance indices like bias and variance, are considered.

5.1 INTRODUCTION

Most of the signals in nature like speech, weather, seismic, stock market prices, bio-signals like electroencephalograph, etc., are random in nature. Therefore, to handle these signals, a theory based on probabilistic approach will be introduced in this chapter. The required probability basics are briefly reviewed.

5.2 PROBABILITY

In a game or an experiment, if there are many events/outcomes then the probability of an event A, $P(A)$ is given by

$$P(A) = \frac{No.\ of\ favorable\ trials}{Total.\ no.\ of\ trials} = \lim_{N\to\infty} \frac{m}{N} \tag{5.1}$$

All the events/outcomes of an experiment form a *sample space*.

Example 1: If a fair die is thrown, find the probability of getting an odd number.

The sample space is {1, 2, 3, 4, 5, 6}, N=6

The number of favorable events to get an odd number *are* {1, 3, 5}, m=3

Therefore, the probability of getting an odd number is $P = \frac{m}{N} = \frac{3}{6} = \frac{1}{2}$

5.2.1 Joint Probability, Conditional Probability and Independence

Let an experiment having two sets of outcomes be performed N times out of which n_1 sets are favorable for the first set, A and out of these n_1 trials, let n_2 trials, be favorable to the second set, B. Then n_2 is the number of trials favorable for joint event AB. Therefore,

$$P(AB) = \lim_{N\to\infty} \frac{n_2}{N} = \lim_{N\to\infty} \frac{n_1}{N}\frac{n_2}{n_1} \tag{5.2}$$

$$P(A) = \lim_{N\to\infty} \frac{n_1}{N} \quad \text{and} \quad P(B/A) = \lim_{N\to\infty} \frac{n_2}{n_1}$$

In Fig. 5.1, $P(B / A)$ *is the relative probability of B with respect to* ***the reduced sample space*** *A.*

$$P(AB) = P(A)\,P(B/A)$$

Therefore,

$$P(B/A) = P(AB)/P(A) \qquad P(A) \neq 0$$

$$P(A/B) = P(AB)/P(B) \qquad P(B) \neq 0$$

$$P(A/B) = \frac{P(B/A)P(A)}{P(B)} \tag{5.3a}$$

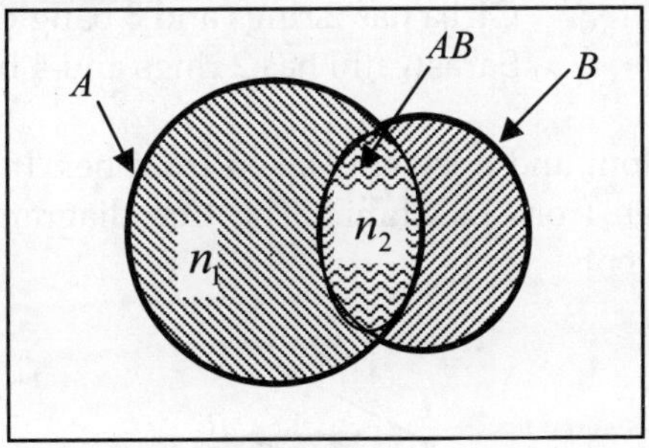

Fig. 5.1 Conditional probability space.

Similarly

$$P(B/A) = \frac{P(A/B)P(B)}{P(A)} \tag{5.3b}$$

These two relations (5.2) and (5.3) are known as *Bayes rule.*

Example 2:

A pair of dice is thrown. What is the probability of getting a 3 and probability of getting a 3 if the sum of numbers on both is 7?

Let B be the event of getting a number 3.

Totally there are 36 combinations and the possibilities of getting a 3 are

{(3,1), (3,2), (3,3), (3,4), (3,5), (3,6), (6,3), (5,3), (4,3), (2,3), (1,3)}.

Therefore, $P(B) = \frac{11}{36}$.

The event A of getting the sum as 7 occurs for the possible combinations

{(1,6), (2,5), (3,4), (4,3), (5,2), (6,1)}

Event B of getting a number 3 is the conditional probability of getting 3 when the sum is 7 is P(B/A). Therefore,

$$P(B/A) = \frac{\text{Number of ways } A \text{ and } B \text{ can occur}}{\text{Number of ways } A \text{ can occur}} = \frac{2}{6} = \frac{1}{3}$$

Example 3:

Three ladies, Lakshmi, Girija and Saraswathi have jewels, viz., rings and bangles as given below:

Lakshmi has 4 rings and 12 bangles,
Girija has 2 rings and 8 bangles and
Saraswathi has 2 rings and 4 bangles

A lady is selected at random and a jewel is taken from her. If the item is a ring what is the probability that it is taken from Lakshmi. The tree diagram shown enables finding the probability of different events.

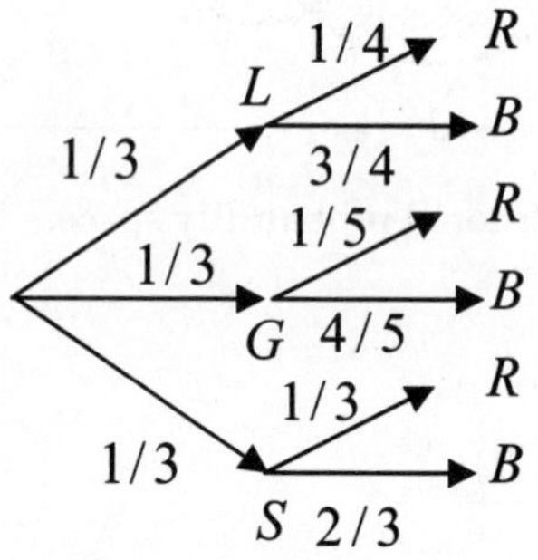

Probability of getting a ring is

$$\begin{aligned} P(R) &= (1/3)(1/4) + (1/3)(1/5) + (1/3)(1/3) \\ &= (1/3)[1/4 + 1/5 + 1/3] \\ &= (1/3)[(15 + 12 + 20)/60 \\ &= (1/3)(47/60) = 47/180 \end{aligned}$$

Probability that Lakshmi is selected and the jewel is a ring is

$$P(LR) = (1/3)(1/4) = 1/12$$

Therefore

$$P(L/R) = P(LR)/P(R) = \frac{(1/12)}{(47/180)} = \frac{1}{12}\frac{180}{47} = \frac{15}{47}$$

5.2.2 Independent Events

If one event is not influenced by the other event, then the two events are called independent events.

If A and B are independent of each other

$$P(B/A) = P(B)$$
$$P(A/B) = P(A)$$

Therefore

$$P(AB) = P(A)\,P(B/A)$$
$$P(AB) = P(A)\,P(B) \tag{5.4}$$

Example 4:

Urn-A contains 4 green and 6 red balls. Urn-B contains 7 blue and 3 green balls. A ball is drawn from each urn. What is the probability of getting a green ball, P (G) from both the urns.

P (GA) = 4/10, P (GB) = 3/10

$$P(G) = P[(GA) \cap (GB)] = (4/10)(3/10) = (2/5)(3/10) = 3/25$$

5.3 STATISTICAL AVERAGES

If each event is associated with a number, then this number is called a random variable. For example, in tossing a fair coin, there are two events/outcomes, head and tail. The random variable X associated with this can have any value associated with these outcomes. For example, X can take with the head, 5 apples and with tail, 0 apples.

5.3.1 Probability Density and Distribution Function

Table 5.1: Probability density function for two dice thrown together

X	2	3	4	5	6	7	8	9	10	11	12
No. of ways X occurs	1	2	3	4	5	6	5	4	3	2	1
$p_X(x)$	$\frac{1}{36}$	$\frac{2}{36}$	$\frac{3}{36}$	$\frac{4}{36}$	$\frac{5}{36}$	$\frac{6}{36}$	$\frac{5}{36}$	$\frac{4}{36}$	$\frac{3}{36}$	$\frac{2}{36}$	$\frac{1}{36}$

Let us consider the example of throwing two dice together and the sum of the two numbers on the two dice as the random variable. The values that the random variable X can take, the number of ways each of these values can appear and hence their probability of occurrence of each event or the value of the random variable, are listed in Table 5.1.

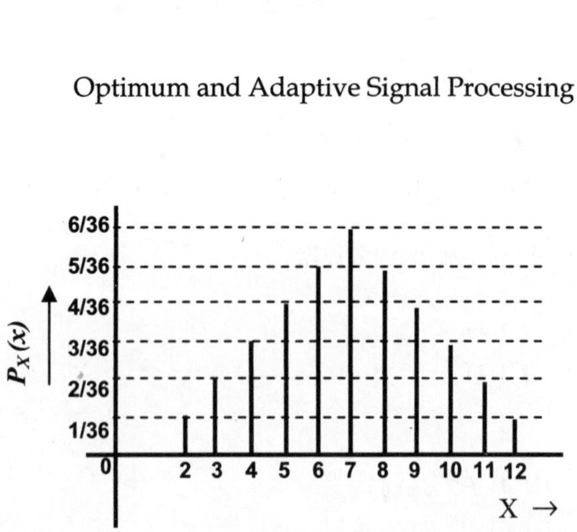

Fig. 5.2 (a) Density function of the r.v. for the sum of the numbers for a pair of dice.

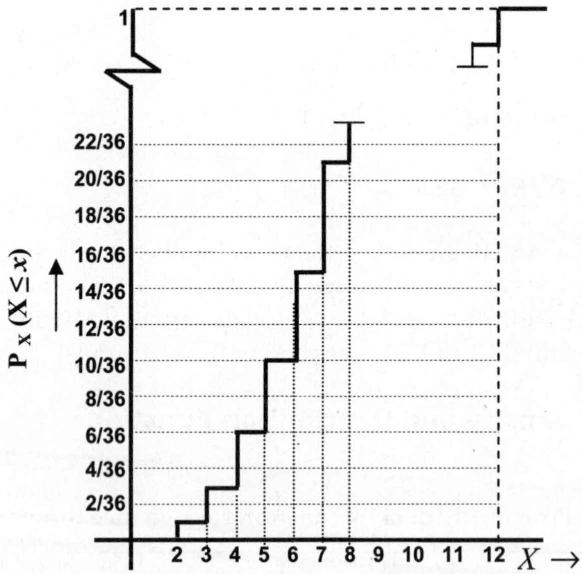

Fig. 5.2 (b) Cumulative distribution function of a r.v. for the sum of the numbers for a pair of dice.

Further, the plot of $p_X(x)$ against X is also shown in the Fig. 5.2(a). The probability of $P_X(X \le x)$ which is the sum of the probabilities of all events $X \le x$ is plotted in Fig. 5.2(b) and is called the *cumulative probability distribution* function and is denoted by $F_X(x)$. For example, $P_X(X \le 4)$ is the cumulative sum of the probabilities $p_X(2), p_X(3)$ and $p_X(4)$ of the occurrence of X equal to 2, 3 and 4. Thus, Fig. 5.2(a) is the derivative of Fig. 5.2(b) and is called the *density function* (also known as probability distribution function) which gives the probability of each event or the value of the random variable and *not up to* that value. Therefore, if the probability of observing a *continuous* random variable X in the interval x and $x+\Delta x$ is $p_X(x)\Delta x$ as $\Delta x \to 0$, then $p_X(x)$ is the density function. $p_X(x)\Delta x$ is the area under the curve $p_X(x)$ over the interval Δx (Fig. 5.3(a)). That is, if $P_X(X \le x)$ is the probability of X taking values $\le x$, then

$$p_X(x) = \frac{\partial}{\partial x} P_X(X \le x) \tag{5.5a}$$

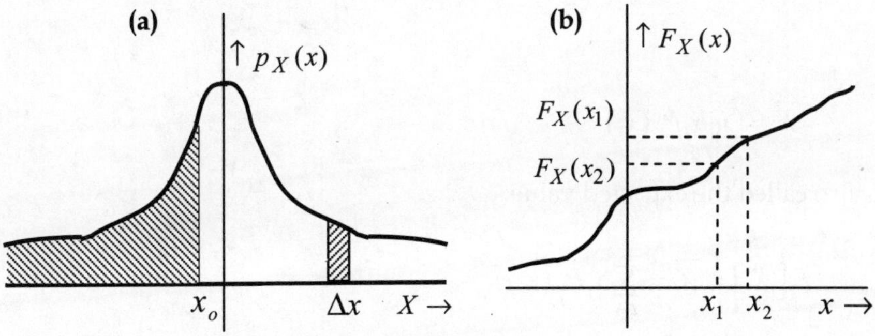

Fig. 5.3 (a) Probability density function and (b) Cumulative probability distribution function of a continuous r.v.

$P(X \le x)$ is also known as the probability distribution denoted by $F_X(x)$ (Fig. 5.3 (b)).

$$F_X(x) = \int_{-\infty}^{x} p_X(x)\, dx \tag{5.5b}$$

Also,

$$P(\alpha_1 \le X \le \alpha_2) = F_X(\alpha_2) - F_X(\alpha_1)$$

If $\alpha_1 \le \alpha_2$, $F_X(\alpha_1) \le F_X(\alpha_2)$. Also, $F_X(-\infty) = 0$ and $F_X(\infty) = 1$

5.3.2 Expectation

Let X be a random variable, which takes n values $x_1, x_2 \ldots x_n$. Let the experiment represented by X be repeated N times, $N \to \infty$ and let $m_1, m_2 \ldots m_n$ be the favorable trials for $x_1, x_2 \ldots x_n$, respectively.

The mean value of X is given by

$$\overline{X} = \frac{1}{N}[m_1 x_1 + m_2 x_2 + \ldots\ldots\ldots + m_n x_n]$$

$$= \frac{m_1}{N} x_1 + \frac{m_2}{N} x_2 + \ldots\ldots\ldots\ldots\ldots + \frac{m_n}{N} x_n$$

As $N \to \infty$,

$$\frac{m_i}{N} \to P_X(x_i)$$

Therefore

$$\overline{X} = \sum_{i=1}^{N} x_i P_X(x_i) \tag{5.6}$$

The mean is also called the expected value.

$$E[X] = \overline{X} = \sum_{i=1}^{N} x_i P_X(x_i) \tag{5.7a}$$

For continuous random variables,

$$E[X] = \int_{-\infty}^{\infty} x\, p_X\,(x)\, dx \tag{5.7b}$$

Mean/expectation of a random variable is its average value.

5.3.3 Moments

The nth moment of a random variable X is given by

$$E\left[x^n\right] = \int_{-\infty}^{\infty} x^n p_X\,(x)\, dx \tag{5.8}$$

The first moment, i.e., for $n = 1$, is same as the mean or the expected value of the random variable.

The nth central moment of a random variable X is its moment about its mean value m and is given by

$$E\left[(x-m)^n\right] = \int_{-\infty}^{\infty} (x-m)^n p_X(x)\, dx \tag{5.9}$$

The second central moment is known as the *variance*, σ_x^2 and is given by

$$\sigma_x^2 = E\left[(x-m)^2\right] = \int_{-\infty}^{\infty} (x-m)^2\, p_X(x)\, dx$$

$$= \int_{-\infty}^{\infty} (x^2 - 2mx + m^2)\, p_X(x)\, dx$$

$$\sigma_x^2 = E\left[x^2\right] - m^2 \qquad \text{(from Eqn. 12.7b)}$$

$$Var(X) = \sigma_x^2 = E\left[x^2\right] - m^2 \tag{5.10}$$

Variance of a random variable gives its *average deviation* σ_x *from its mean value.*

The third central moment of a random variable (r.v.) X is $E\left[(x-m)^3\right]$ and is known as *skewness* and is a *measure of asymmetry of density function about its mean value.* For symmetric density functions, all odd moments are zeros.

The fourth-central moment of a r.v. X is $E\left[(x-m)^4\right]$ and is known as *kurtosis* and is a *measure of peakedness of the density function at its mean.*

The expectation of a function of a random variable, generalizes the above moments.

For a function $g(X)$,

$$E[g(X)] = \int_{-\infty}^{\infty} g(x)\, p_X(x)\, dx$$

For $E\left[(x-m)^n\right]$,

$$g(x) = (x - m_X)^n .$$

5.3.4 Joint Density Function

If $p_{XY}(x,y)\Delta x \Delta y$ as $\Delta x \to 0$ and $\Delta y \to 0$, is the probability of observing X and Y in the interval $(x, x+\Delta x)$ and $(y, y+\Delta y)$, then $p_{XY}(x,y)$ is called the joint probability density function of X and Y. That is

$$p_{XY}(x,y) = \frac{\partial^2}{\partial x \partial y} P_{XY}(X \le x, Y \le y) \tag{5.11}$$

where $P_{XY}(X \le x, Y \le y)$ is the probability that $(X \le x, Y \le y)$ which is also the probability distribution function given by

$$P_{XY}(X \le x, Y \le y) = \int_{-\infty}^{x} \int_{-\infty}^{y} p_{XY}(\xi_1, \xi_2)\, d\xi_1\, d\xi_2 \tag{5.12}$$

$$P_{XY}(X \le \infty, Y \le \infty) = \int_{-\infty}^{\infty} \int_{-\infty}^{\infty} p_{XY}(\xi_1, \xi_2)\, d\xi_1\, d\xi_2 = 1$$

The probability density functions $p_X(x)$ and $p_Y(y)$ can be obtained from the joint density functions and are called the marginal density functions and are given by

$$p_X(x) = \int_{-\infty}^{\infty} p_{XY}(x, y)\, dy \tag{5.13}$$

and

$$p_Y(y) = \int_{-\infty}^{\infty} p_{XY}(x, y)\, dx \tag{5.14}$$

5.3.5 Expectation of a Function of Two Random Variables

$$E[g(X,Y)] = \int_{-\infty}^{\infty} \int_{-\infty}^{\infty} g(x, y) p_{XY}(x, y)\, dx\, dy$$

For $g(X,Y) = X$,

$$E[X] = \int_{-\infty}^{\infty} x \left[\int_{-\infty}^{\infty} p_{XY}(x, y)\, dy \right] dx = \int_{-\infty}^{\infty} x p_X(x)\, dx \tag{5.15a}$$

For $g(X,Y) = X^2$,

$$E[X^2] = \int_{-\infty}^{\infty} x^2 \left[\int_{-\infty}^{\infty} p_{XY}(x, y)\, dy \right] dx = \int_{-\infty}^{\infty} x^2 p_X(x)\, dx \tag{5.15b}$$

For $g(X,Y) = (X - m_X)^2$,

$$E[(X - m_X)^2] = \int_{-\infty}^{\infty} (x - m_X)^2 \left[\int_{-\infty}^{\infty} p_{XY}(x, y)\, dy \right] dx = \int_{-\infty}^{\infty} (x - m_X)^2 p_X(x)\, dx$$

$$= \int_{-\infty}^{\infty} (x^2 - 2x m_X + m_X^2)\, p_X(x)\, dx = E[X^2] - 2m_X^2 + m_X^2$$

$$E[(X-m_X)^2]=E[X^2]-m_X^2 \quad (5.15c)$$

For $g(X,Y)=X^mY^n$,

$$E[X^mY^n]=\int_{-\infty}^{\infty}\int_{-\infty}^{\infty} x^m y^n p_{XY}(x,y)\,dxdy \quad (5.15d)$$

For $g(X,Y)=XY$,

$$E[XY]=\int_{-\infty}^{\infty}\int_{-\infty}^{\infty} xyp_{XY}\ (x,y)\,dx\,dy$$

$E[XY]$ is the mean cross product and is known as correlation between X and Y and will be dealt in detail later.

5.3.6 Conditional probability density functions

Since

$$P_{X/Y}(X\le x/y)=\frac{P_{XY}(X\le x,y)}{P(Y)} \quad (5.16)$$

$$p_X(x/y)=\frac{\partial}{\partial x}P_{X/Y}(X\le x/Y)=\frac{\dfrac{\partial^2}{\partial x\partial y}P(X\le x,Y\le y)}{\dfrac{\partial}{\partial y}P_Y(Y\le y)}$$

Therefore,

$$p_X(x/y)=\frac{p_{XY(x,y)}}{p_Y(y)} \quad (5.17a)$$

Similarly,

$$p_Y(y/x)=\frac{p_{XY(x,y)}}{p_X(x)} \quad (5.17b)$$

Further, if X and Y are independent,

$$p_X(x/y)=\frac{}{x}P_X(X\le x)=p_X(x)$$

Therefore,

$$p_X(x/y)=\frac{p_{XY(x,y)}}{p_Y(y)}=p_X(x)$$

$$p_{XY(x,y)}=p_X(x)\ p_Y\ (y) \quad (5.18)$$

If the random variables X and Y are independent of each other, then their joint probability density function is equal to the product of their individual probability density functions. Further, the expectation of sum of two independent random variables,

$$E\left[X+Y\right]=\int_{-\infty}^{\infty}\int_{-\infty}^{\infty}(X+Y)f_{XY}(x,y)\,dXdY$$

Since X and Y are independent random variables,

$$f_{XY}(x,y)=f_X(x)f_Y(y) \tag{5.19}$$

$$E\left[X+Y\right]=\int_{-\infty}^{\infty}\int_{-\infty}^{\infty}(X+Y)f_X(x)f_Y(y)\,dXdY$$

$$=\int_{-\infty}^{\infty}Xf_X(x)\,dX\int_{-\infty}^{\infty}f_Y(y)\,dY+\int_{-\infty}^{\infty}Yf_Y(y)\,dY\int_{-\infty}^{\infty}f_X(x)\,dX$$

$$E[X+Y]=E\left[X\right]+E\left[Y\right] \tag{5.20}$$

$$V\left[X+Y\right]=\int_{-\infty}^{\infty}\int_{-\infty}^{\infty}(X-m_X+Y-m_Y)^2f_X(x)f_Y(y)\,dXdY$$

$$=\int_{-\infty}^{\infty}(X-m_X)^2f_X(x)\,dX\int_{-\infty}^{\infty}f_Y(y)\,dY+\int_{-\infty}^{\infty}(Y-m_Y)^2f_Y(y)\,dY\int_{-\infty}^{\infty}f_X(x)\,dX$$

$$+2\int_{-\infty}^{\infty}\int_{-\infty}^{\infty}(X-m_X)(Y-m_Y)f_X(x)f_Y(y)\,dXdY$$

$$\int_{-\infty}^{\infty}\int_{-\infty}^{\infty}(X-m_X)(Y-m_Y)f_X(x)f_Y(y)\,dXdY=$$

$$\int_{-\infty}^{\infty}\int_{-\infty}^{\infty}(XY-m_XY-m_YX+m_Xm_Y)f_X(x)f_Y(y)\,dXdY$$

$$=\int_{-\infty}^{\infty}\int_{-\infty}^{\infty}(XYf_X(x)f_Y(y)\,dXdY-m_X\int_{-\infty}^{\infty}Yf_Y(y)\,dY\int_{-\infty}^{\infty}f_X(x)\,dX$$

$$-m_Y\int_{-\infty}^{\infty}Xf_X(x)\,dX\int_{-\infty}^{\infty}f_Y(y)\,dY+m_Xm_Y\int_{-\infty}^{\infty}\int_{-\infty}^{\infty}f_X(x)f_Y(y)\,dXdY$$

$$= \int_{-\infty}^{\infty} Xf_X(x)\, dX \int_{-\infty}^{\infty} Yf_Y(y)\, dY - m_X m_Y - m_X m_Y + m_X m_Y$$

$$= m_X m_Y - m_X m_Y - m_X m_Y + m_X m_Y$$

$$= 0$$

Therefore,

$$V[X+Y] = \int_{-\infty}^{\infty} (X - m_X)^2 f_X(x)\, dX \int_{-\infty}^{\infty} f_Y(y)\, dY$$

$$+ \int_{-\infty}^{\infty} (Y - m_Y)^2 f_Y(y)\, dY \int_{-\infty}^{\infty} f_X(x)\, dX$$

$$V[X+Y] = V[X] + V[Y] \tag{5.21}$$

Therefore, for independent random variables,

$$E[X+Y] = E[X] + E[Y].$$

$$V(X+Y) = V(X) + V(Y)$$

A practical way of obtaining the density function for a given data is by histogram and generally this is called amplitude distribution. There are different density functions like, Gaussian, uniform, exponential, sine, square, etc.

Gaussian type is amenable to theoretical studies/analysis as it can be expressed mathematically and is quite popular. Many natural processes are of Gaussian type. In view of this, if nothing is known about any noise process, it is customary to assume it as of Gaussian distribution. The Gaussian density function $f_X(X)$ is given by

$$f_X(X) = \frac{1}{\sigma\sqrt{2\pi}} e^{-\frac{(X - m_X)^2}{2\sigma^2}} \tag{5.22}$$

The mean is given by

$$E[X] = \int_{-\infty}^{\infty} Xf_X(x)\, dX = \int_{-\infty}^{\infty} X \frac{1}{\sigma\sqrt{2\pi}} e^{-\frac{(X - m_X)^2}{2\sigma^2}}\, dX$$

$$E[X] = \frac{1}{\sigma\sqrt{2\pi}} \int_{-\infty}^{\infty} X e^{-\frac{(X - m_X)^2}{2\sigma^2}}\, dX$$

Let $Y = X - m_X$, then

$$E[X] = \frac{1}{\sigma\sqrt{2\pi}} \int_{-\infty}^{\infty} (Y + m_X) e^{-\frac{Y^2}{2\sigma^2}} dY$$

$$E[X] = \frac{1}{\sigma\sqrt{2\pi}} \left[\int_{-\infty}^{\infty} Y e^{-\frac{Y^2}{2\sigma^2}} dY + m_X \int_{-\infty}^{\infty} e^{-\frac{Y^2}{2\sigma^2}} dY \right]$$

The first integral is zero as the integrand is an odd function and second integral is given by

$$\int_{-\infty}^{\infty} e^{-a^2Y^2} dY = \frac{\sqrt{\pi}}{a} \text{ where } a^2 = \frac{1}{2\sigma^2}$$

$$E[X] = m_X \frac{1}{\sigma\sqrt{2\pi}} \sigma\sqrt{2\pi}$$

$$E[X] = m_X \tag{5.23}$$

$$E[X^2] = \int_{-\infty}^{\infty} X^2 f_X(x) dX = \int_{-\infty}^{\infty} X^2 \frac{1}{\sigma\sqrt{2\pi}} e^{-\frac{(X-m_X)^2}{2\sigma^2}} dX$$

Let $Y = \dfrac{X - m_X}{\sigma}$,

$$E[X^2] = \frac{\sigma}{\sigma\sqrt{2\pi}} \int_{-\infty}^{\infty} (\sigma Y + m_X)^2 e^{-\frac{Y^2}{2}} dY$$

$$E[X^2] = \frac{1}{\sqrt{2\pi}} \int_{-\infty}^{\infty} (\sigma^2 Y^2 + m_X^2 + 2\sigma Y m_X) e^{-\frac{Y^2}{2}} dY$$

$$E[X^2] = \frac{1}{\sqrt{2\pi}} \left[\int_{-\infty}^{\infty} (\sigma^2 Y^2 e^{-\frac{Y^2}{2}} dY + \int_{-\infty}^{\infty} m_X^2 e^{-\frac{Y^2}{2}} dY + 2\sigma \int_{-\infty}^{\infty} Y m_X e^{-\frac{Y^2}{2}} dY) \right]$$

Let

$$I_1 = \int_{-\infty}^{\infty} \sigma^2 Y^2 e^{-\frac{Y^2}{2}} dY$$

Putting

$$q = Y/\sqrt{2},\ Y = q\sqrt{2},\ \frac{dY}{dq} = \sqrt{2}$$

$$I_1 = 2\sqrt{2}\sigma^2 \int_{-\infty}^{\infty} q^2 e^{-q^2}\, dq$$

Since

$$\int_{-\infty}^{\infty} q^2 e^{-q^2}\, dq = \frac{\sqrt{\pi}}{2},\ I_1 = 2\sqrt{2}\sigma^2 \frac{\sqrt{\pi}}{2} = \sqrt{2\pi}\,\sigma^2$$

Let

$$I_2 = \int_{-\infty}^{\infty} m_X^2 e^{-\frac{Y^2}{2}} dY$$

Since

$$\int_{-\infty}^{\infty} e^{-a^2Y^2} dY = \frac{\sqrt{\pi}}{a}, \text{ with } a = 1/\sqrt{2}$$

$$I_2 = m_X^2 \int_{-\infty}^{\infty} e^{-\frac{Y^2}{2}} dY = \sqrt{2\pi}\, m_X^2$$

Let

$$I_3 = 2\sigma \int_{-\infty}^{\infty} Y m_X e^{-\frac{Y^2}{2}} dY = 0, \text{ since the integrand is an odd function.}$$

Therefore,

$$E[X^2] = \frac{1}{\sqrt{2\pi}}\left[\int_{-\infty}^{\infty}\left\{\sigma^2 Y^2 e^{-\frac{Y^2}{2}} dY + \int_{-\infty}^{\infty} m_X^2 e^{-\frac{Y^2}{2}} dY + 2\sigma \int_{-\infty}^{\infty} Y m_X e^{-\frac{Y^2}{2}} dY\right\}\right]$$

$$= \frac{1}{\sqrt{2\pi}}\left(\sqrt{2\pi}\,\sigma^2 + \sqrt{2\pi}\, m_X^2 + 0\right)$$

$$E[X^2] = \sigma^2 + m_X^2 \qquad (5.24)$$

Variance,

$$V[X] = E[(X - m_X)^2] = E[X^2 - 2Xm_X + m_X^2]$$

$$= E[X^2] - 2E[X]m_X + E[m_X^2]$$

$$= E[X^2] - 2m_X^2 + m_X^2$$

$$V[X] = E[X^2] - m_X^2 \tag{5.25}$$

Since

$$E[X^2] = \sigma^2 + m_X^2,$$

$$V[X] = \sigma^2 + m_X^2 - m_X^2 = \sigma^2$$

$$V[X] = \sigma^2 \tag{5.26}$$

$$E[X] = m_X \text{ and } V[X] = \sigma^2.$$

Hence, m_X *is the mean and* σ^2 *is the variance* of the Gaussian random variable X.

The uniform density function of R.V. distributed uniformly between *(a, b)* is given by

$$f_X(X) = \begin{cases} \dfrac{1}{b-a}, a \le X \le b \\ 0, \quad otherwise \end{cases} \tag{5.27}$$

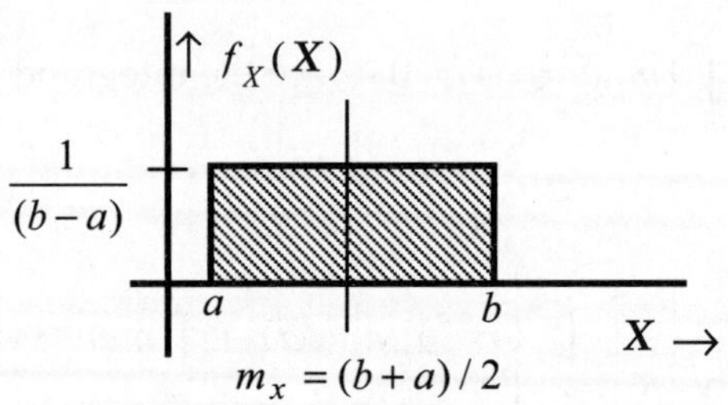

Fig. 5.4 Uniform density function

The mean of the uniformly distributed R.V., X is

$$m_X = E[X] = \int_a^b X \frac{1}{(b-a)} dX = \frac{1}{(b-a)} \left[\frac{X^2}{2} \right]_a^b$$

$$= \frac{1}{(b-a)}\left[\frac{b^2-a^2}{2}\right] = \frac{(b+a)}{2}$$

and its variance is $V_X = E[X^2] - m_X^2$

$$E[X^2] = \int_a^b X^2 \frac{1}{(b-a)} dX = \frac{1}{(b-a)}\left[\frac{X^3}{3}\right]_a^b$$

$$= \frac{1}{(b-a)}\left[\frac{b^3-a^3}{3}\right] = \frac{(b^2+a^2+ab)}{3}$$

$$V_X = E[X^2] - m_X^2 = \frac{(b^2+a^2+ab)}{3} - \left[\frac{(b+a)}{2}\right]^2$$

$$= \frac{4(b^2+a^2+ab) - 3(b^2+a^2+2ab)}{12} = \frac{(b-a)^2}{12}$$

Therefore,

$$m_X = \frac{(b+a)}{2} \tag{5.28a}$$

$$V_X = \frac{(b-a)^2}{12} \tag{5.28b}$$

Uniform noise is also quite common as the quantisation noise is of this type and quantisation noise exists when a signal sample values are quantized.

5.3.7 Characteristic Function of a Random Variable

Characteristic function (CF) of a random variable, $\phi_X(u)$: It is defined as the expected value of $[e^{juX}]$. That is,

$$\phi_X(u) = E[e^{juX}] = \int_{-\infty}^{\infty} e^{juX} f_X(X)\, dX$$

The CF of a random variable is similar to FT of $f_X(X)$ except for the sign in the exponent. Therefore, the CF and the probability density function of random variable form a FT pair. Thus

$$\phi_X(u) = \int_{-\infty}^{\infty} e^{juX} f_X(X)\, dX \tag{5.29a}$$

$$f_X(X) = \frac{1}{2\pi}\int_{-\infty}^{\infty} e^{-juX}\phi_X(u)du \tag{5.29b}$$

Since,

$$\phi_X(u) = \int_{-\infty}^{\infty} e^{juX} f_X(X)\, dX$$

$$\phi_X(u) = \int_{-\infty}^{\infty} f_X(X)\,[1 + juX + \frac{1}{2!}(juX)^2 + \ldots\ldots]\, dX$$

$$= \int_{-\infty}^{\infty} f_X(X)\, dX + ju\int_{-\infty}^{\infty} Xf_X(X)dX + \ldots\ldots + \frac{(ju)^n}{n!}\int_{-\infty}^{\infty} X^n f_X(X)\, dX + \ldots$$

$$= 1 + juE[X] + \ldots\ldots + \frac{(ju)^n}{n!}E[X^n] + \ldots\ldots$$

$$\phi_X(u) = \sum_{n=0}^{\infty}\frac{(ju)^n}{n!}E[X^n] \tag{5.30}$$

where $\phi_X(0) = 1$. The coefficients of $\frac{(ju)^n}{n!}$ of the CF of the random variable X are the central moments of the random variable.

5.3.8 The Probability Density Function of Sum of Two Independent Random Variables

Let Z be the sum of two independent random variables, X and Y. The density function of Z is found using CF of Z.

$$\phi_Z(u) = \int_{-\infty}^{\infty} e^{juZ} f_Z(Z)\, dZ$$

$$= \int_{-\infty}^{\infty}\int_{-\infty}^{\infty} e^{ju(X+Y)} f_X(X) f_Y(Y)\, dXdY$$

$$= \int_{-\infty}^{\infty} e^{juX} f_X(X)\, dX \int_{-\infty}^{\infty} e^{juY} f_Y(Y)\, dY$$

$$\phi_Z(u) = E[e^{juZ}] = E[e^{juX}]E[e^{juY}]$$

$$f_Z(Z) = \frac{1}{2\pi}\int_{-\infty}^{\infty} e^{-juZ}\phi_Z(u)du \quad = \frac{1}{2\pi}\int_{-\infty}^{\infty} e^{-juZ}E[e^{juX}]E[e^{juY}]\; du$$

$$= \frac{1}{2\pi}\int_{-\infty}^{\infty} e^{-juZ}\left[\int_{-\infty}^{\infty} e^{juX} f_X(X)dX \int_{-\infty}^{\infty} e^{juY} f_Y(Y)dY\right] du$$

$$= \int_{-\infty}^{\infty}\int_{-\infty}^{\infty} f_X(X)f_Y(Y)\left[\frac{1}{2\pi}\int_{-\infty}^{\infty} e^{-ju(Z-X-Y)}du\right] dXdY$$

Substituting,

$$\frac{1}{2\pi}\int_{-\infty}^{\infty} e^{-ju(z-X-Y)}du = \delta(Z-X-Y),$$

$$(\because \frac{1}{2\pi}\int_{-\infty}^{\infty} e^{-j\omega t}d\omega = \delta(t))$$

$$f_Z(Z) = \int_{-\infty}^{\infty} f_X(X)\left\{\int_{-\infty}^{\infty} f_Y(Y)\,\delta\,(Z-X-Y)\,dY\right\} dX$$

$$= \int_{-\infty}^{\infty} f_X(X)f_Y(Z-X)\,dX \qquad (\because \int_{-\infty}^{\infty} f(t)\delta(t-t_o)\,dt = f(t_o))$$

$$f_Z(Z) = f_X(z) * f_Y(Z) \tag{5.31}$$

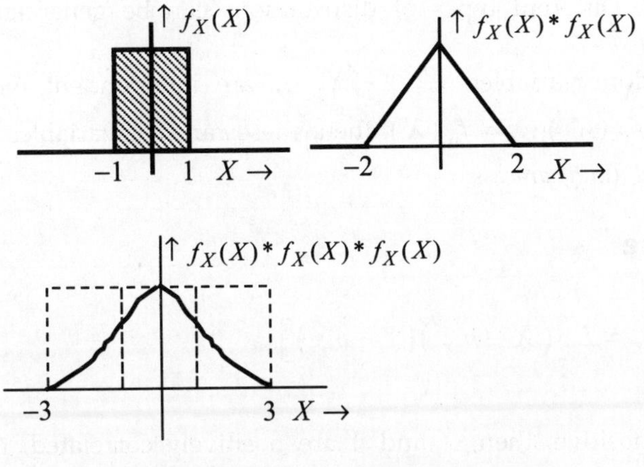

Fig. 5.5 Illustration of central limit theorem.

The convolution of many density functions of different independent random variables leads close to a Gaussian density function. This can be seen even for three variables with uniform distribution, it has become close to the Gaussian. The density function of sum of large number of random variables results in a more exact Gaussian distribution.

If $m_{X_1}, m_{X_2}, m_{X_3} \ldots$ and $\sigma_1^2, \sigma_2^2, \sigma_3^2 \ldots$ are the mean and variances of the variables $X_1, X_2, X_3, \ldots$ with density functions $f_{X_1}(X_1), f_{X_2}(X_2), \ldots$ then, the density function of the sum

$$Z = X_1 + X_2 + X_3, \ldots$$

is given by

$$f_Z(Z) = f_{X_1}(X_1) * f_{X_2}(X_2) * f_{X_3}(X_3) * \ldots f_{X_n}(X_n)$$

$$\lim_{n \to \infty} f_Z(Z) = \frac{1}{\sigma\sqrt{2\pi}} e^{-\frac{(X - m_Z)^2}{2\sigma^2}} \tag{5.32a}$$

where

$$\sigma^2 = \sigma_1^2 + \sigma_2^2 + \sigma_3^2 + \ldots \tag{5.32b}$$

$$m_Z = m_{X_1} + m_{X_2} + m_{X_3} + \ldots \tag{5.32c}$$

This leads to the Central limit theorem, which states that *the sum of or linear combination of a large number of independent random variables of any distribution results in a Gaussian distribution*. This enables to generate Gaussian distribution from say, uniform distribution. A random variable with uniform distribution is easily generated in computers and it has been found that addition of 12 uniform random numbers results in a fairly good Gaussian numbers. Different types of distributions can be generated from the uniform random numbers.

If the random variables $X_1, X_2, X_3, \ldots$ are independent with identical density functions $f_{X_1}(X_1), f_{X_2}(X_2), \ldots = f_X(X)$ then, these random variables form an *independent identically distributed (i.i.d) process.*

5.3.9 Covariance

$$C_{XY} = E\left[(X - m_X)(Y - m_Y)\right] \tag{5.33}$$

If C_{XY} is positive, then X and Y are positively correlated, this implies that the variation of the two variables X and Y are in harmony (in the same direction up or down). Similarly, if C_{XY} is negative, then X and Y are negatively correlated, this implies that the variation of the two variables X and Y are not in harmony. Negative value of C_{XY} does not mean that they are not dependent, they are dependent but in the opposite direction. If $C_{XY} = 0$, then the r.v. X and Y are uncorrelated.

$$C_{XY} = E\left[(X - m_X)(Y - m_Y)\right]$$

$$= E[XY] - E[m_X Y] - E[X m_Y] + E[m_X m_Y]$$

$$= E[XY] - m_X m_Y - m_X m_Y + m_X m_Y$$

$$C_{XY} = E[XY] - m_X m_Y \tag{5.34}$$

If $C_{XY} = 0$,

$$E[XY] = m_X m_Y$$

If X and Y are independent, then

$$E[XY] = m_X m_Y$$

Therefore, $C_{XY} = 0$. Hence independent random variables are uncorrelated. However, uncorrelated random variables are not necessarily independent. For independent random variables the functions of X and Y i.e., $g_1(X)$ and $g_2(Y)$ satisfy

$$E[g_1(X)g_2(Y)] = E[g_1(X)]\,E[g_2(Y)]$$

However, uncorrelatedness satisfies only $E[XY] = m_X m_Y$. But for Gaussian random variables, uncorrelatedness implies independence.

5.3.10 Correlation

The correlation R_{XY} between random variables X and Y is defined as

$$R_{XY} = E[X\,Y] \tag{5.35}$$

Therefore,

$$C_{XY} = R_{XY} - m_X m_Y \tag{5.36}$$

If X and Y are uncorrelated $C_{XY} = 0$ and $R_{XY} = m_X m_Y$. If $R_{XY} = 0$, then X and Y are called as orthogonal. If either one or both of m_X and m_Y are zero then X and Y are uncorrelated. Covariance or correlation is a measure of relatedness of two variables. For a zero mean process, covariance and correlation both are same.

5.4 RANDOM PROCESS

Random process is an extension of the concept of a random variable. In the case of a random variable, a number is assigned to each event or sample point of an experiment. The collection of all sample points or events is generally known as a sample space. In a random process, for each sample point a waveform is assigned and this waveform is called a *sample function*. The probability distribution of the sample functions is decided by the probability distribution of the sample points.

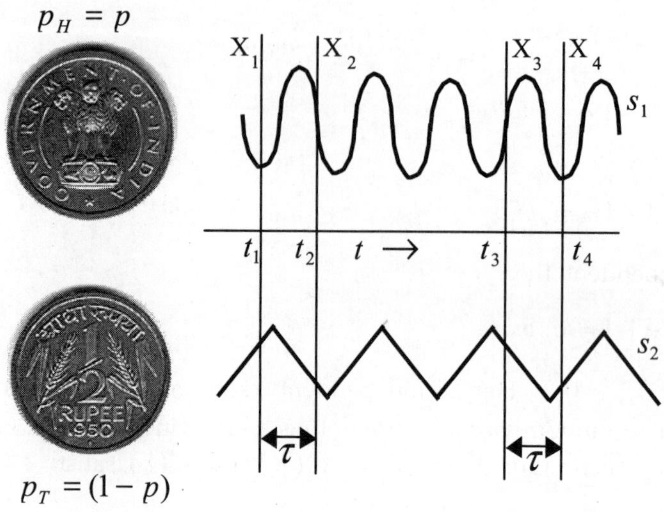

Fig. 5.6 Random process.

The collection of waveforms/sample functions is called an *ensemble. The probability system comprising of the sample space, the ensemble, probability distribution function is called a random process.*

The waveforms themselves are not random but the randomness is associated with the *occurrence* of a particular random function. The occurrence of a sample function is decided by the occurrence of the sample point.

For example, in the coin tossing experiment, the head and tail outcomes are associated with some probabilities say p and (1-p) and the outcomes can be associated with some waveforms like sine (s_1) and triangle (s_2) (Fig. 5.6).

A random process gives a sequence of random variables. Each random variable has its own density function. Further, the occurrence of two or more random variables together is determined by their joint density functions. That is, for the case of two random variables, their occurrence is determined by the joint density of two random variables (second order joint density function).

5.4.1 First Order Statistics

This is concerned with the statistics of a random variable $X(t)$ corresponding to a time instant t. The statistics are decided by the probability density function $P_{X(t)}(x)$ or by $P_X(x)$ i.e., $E\left[X(t)\right]$ is given by

$$E[X(t)] = \int_{-\infty}^{\infty} X(t) P_{X(t)}(x)\, dx \qquad (5.37a)$$

$$E[X^2(t)] = \int_{-\infty}^{\infty} X^2(t) P_{X(t)}(x)\, dx \qquad (5.37b)$$

5.4.2 Second and Higher Order Statistics

The first order statistics gives the amplitude distribution of sample functions for all values of t. However, it cannot give any information about the frequency content of the signal. The correlation between signal values at various time intervals gives this information. The correlation information is obtained by the joint density function where the joint density function (or the second order joint density function) of the two random variables $X(t_1)$ and $X(t_2)$ at instants t_1 and t_2, respectively.

An nth order probability function represents the joint probability density function of n random variables defined at $t = t_1, t_2 ....t_n$. Also, all its lower order density functions can be obtained from it. A random process is completely specified by the joint probability density function of n random variables defined at $t = t_1, t_2 ....t_n$, for all n. However, it is very difficult to find such a joint density function and for the purpose of correlation between two random variables, second order joint density functions are sufficient.

The correlation between random variables $X(t_1)$ and $X(t_2)$ will be denoted by $R_X(t_1, t_2)$. This is called the autocorrelation function of the process $X(t)$, since it is the correlation between two variables defined for the same process. It is a measure of dependence on the amplitudes of the sample functions at $t = t_1$ and on the amplitudes of the same (auto) functions at $t = t_2$. Thus, $R_X(t_1, t_2)$ is given by

$$R_X(t_1, t_2) = E[X(t_1)X(t_2)] = \int_{-\infty}^{\infty}\int_{-\infty}^{\infty} x(t_1)x(t_2)\, P_{X(t_1)X(t_2)}(x(t_1), x(t_2))\, dx(t_1)\, dx(t_2)$$

$$R_X(t_1, t_2) = \int_{-\infty}^{\infty}\int_{-\infty}^{\infty} x_1 x_2 P_{X_1 X_2}(x_1, x_2; t_1, t_2)\, dx_1 dx_2 \qquad (5.38)$$

where $X(t_1)$, $X(t_2)$ are represented by x_1, x_2, respectively and the time dependence is explicitly mentioned in the joint density function by (t_1, t_2).

5.4.3 Stationarity of a Random Process

In general, the process is said to be stationary if all its statistics or all orders of joint probability functions exhibit invariance with time shift. For the first order,

$$P_{X(t)}(x;t) = P_X(x) \tag{5.39a}$$

For the second order if there is a shift in time by T, for all values T, the stationarity requires,

$$P_{X_1X_2}(x_1,x_2;t_1,t_2) = P_{X_1X_2}(x_1,x_2;t_1+T,t_2+T) \tag{5.39b}$$

This is possible if

$$P_{X_1X_2}(x_1,x_2;t_1,t_2) = P_{X_1X_2}(x_1,x_2;\tau) \quad \text{where } \tau = t_2 - t_1.$$

Hence, for a stationary process,

$$R_X(t_1,t_2) = R_X(\tau) \tag{5.39c}$$

That is, autocorrelation between two variables separated by τ is independent of the time origin. Such processes where the second order statistics is independent of time origin are known as *wide sense or weakly stationary.*

Similarly, if the joint density function of more variables say three, four etc., is time invariant, it is possible to have higher order stationarity.

As it is seen for a stationary process, it is required that probability joint density function and the time waveforms associated with the outcomes of the probabilistic game/experiment should not change with time.

For a random process to be *strictly stationary,* the joint density function of all the orders/all higher order moments, should be stationary. But for the Guassian process since the density function is completely determined by the mean μ_X and variance σ^2, the second moment, if it is widely stationary is also strictly stationary.

In Fig. 5.6 for the random variable X_1, the probabilistic average is done over this ensemble for X_1. As long as the distribution p and $(1-p)$, and the corresponding waveforms triangle and sine remain same, the process is stationary. For the random process shown in the Fig. 5.6, to be stationary, the correlation between (X_1,X_2) and that between (X_3,X_4) will be the same as they are separated by the same time distance τ and autocorrelation is given by

$$R_X(\tau) = E\left[X(t)X(t+\tau)\right] \tag{12.40}$$

The FT of the autocorrelation function is the PSD. Therefore, the stationarity of autocorrelation, the second order statistics of a random process implies that the PSD does not change with time. That is, if PSD is computed by taking data segments (of same length) at different time instants, which may be far apart (non-overlapping segments) they are almost similar. This enables to check the stationarity of the data in practice.

The meaning of stationarity in terms of probability distribution function can be got starting from PSD. That is, if the PSD at different time instants are same (similar), then the autocorrelation function will be independent of time (time origin). For a random signal, the autocorrelation function is given by

$$R_X(\tau) = \int_{-\infty}^{\infty}\int_{-\infty}^{\infty} x_1 x_2 \, P_{X_1X_2}(x_1, x_2; t_1, t_2)\, dx_1 dx_2$$

for random variables X_1 at t_1 and X_2 at t_2 with $\tau = (t_2 - t_1)$. For autocorrelation to be independent of time, the joint density function $P_{X_1X_2}(x_1, x_2; t_1, t_2)$ should also be independent of time. This is what exactly required for the stationarity of a random process.

Similarly, if the joint density function of more variables or joint statistics of say three, four etc., is time invariant, it is possible to have higher order stationarity. The joint density of three variables is given by

$$P_{X_1X_2X_3}(x_1, x_2, x_3; t_1, t_2, t_3) = P_{X_1X_2X_3}(x_1, x_2, x_3; \tau_1, \tau_2)$$

where, $\tau_1 = (t_2 - t_1), \tau_2 = (t_3 - t_2)$

The correlation of three variables known as *triple correlation* is given by

$$b_X(\tau_1, \tau_2) = E\left[x(t)\, x(t+\tau_1)\, x(t+\tau_2)\right] \tag{5.41}$$

If this is independent of time, then the process is said to be wide sense stationary up to third order. As the autocorrelation corresponds to PSD, the triple correlation corresponds to bispectrum (M. R. Raghuveer and C.L. Nikias, 1987).

For a stationary process, the power spectral density is given by

$$S_x(\omega) = \lim_{T\to\infty} \frac{E\left[\left|X_T(\omega)\right|^2\right]}{T} \tag{5.42a}$$

$$S_x(\omega) = \lim_{T\to\infty} \frac{E\left[X_T(\omega)X^*(\omega)\right]}{T}$$

$$= \lim_{T\to\infty} \frac{E\left[X_T(\omega)X(-\omega)\right]}{T}$$

The PSD gives *the distribution of power of the signal between two frequencies, which add to zero.* That is, $(\omega_1 + \omega_2) = 0$ or $\omega_1 = -\omega_2$, power distribution between positive and corresponding negative frequencies. Also,

$$S_x(\omega) = \int_{-\infty}^{\infty} R_X(\tau) e^{-j\omega\tau} d\tau \ , \tag{12.42b}$$

where $R_X(\tau) = E\left[X(t)X(t+\tau)\right]$. Further, the bispectrum gives *the distribution of power of the signal among three frequencies that add to zero.* That is, $(\omega_1 + \omega_2 + \omega_3) = 0$ or $\omega_3 = -(\omega_1 + \omega_2)$. The distribution power among three frequencies ω_1, ω_2 and $\omega_3 = -(\omega_1 + \omega_2)$.

Consider $\quad b_X(\tau_1, \tau_2) = E\left[x(t)\, x(t+\tau_1)\, x(t+\tau_2)\right]$

$$= \int_{t=-\infty}^{\infty} x(t)\, x(t+\tau_1)\, x(t+\tau_2)\, dt$$

The Fourier transform of triple correlation is given as,

$$B_x(\omega_1, \omega_2) = \int_{-\infty}^{\infty}\int_{-\infty}^{\infty} b_X(\tau_1, \tau_2) e^{-j(\omega_1\tau_1 + \omega_2\tau_2)} d\tau_1 d\tau_2 \tag{5.42c}$$

$$B_x(\omega_1, \omega_2) = \int_{-\infty}^{\infty}\int_{-\infty}^{\infty}\int_{-\infty}^{\infty} x(t)\, x(t+\tau_1)\, x(t+\tau_2)\, e^{-j\omega_1\tau_1} e^{-j\omega_2\tau_2} d\tau_1 d\tau_2 dt$$

Let $t + \tau_1 = p \Rightarrow \tau_1 = p - t$ and $d\tau_1 = dp$ and $t + \tau_2 = q \Rightarrow \tau_2 = q - t$ and $d\tau_2 = dq$, then

$$B_x(\omega_1, \omega_2) = \int_{-\infty}^{\infty}\int_{-\infty}^{\infty}\int_{-\infty}^{\infty} x(t)\, x(p)\, x(q)\, e^{-j\omega_1 p} e^{-j\omega_2 q} e^{j(\omega_1+\omega_2)t} dp\, dq\, dt$$

$$= \int_{-\infty}^{\infty} x(t)\, e^{j(\omega_1+\omega_2)t}\, dt \int_{-\infty}^{\infty} x(q)\, e^{-j\omega_2 q}\, dq \int_{-\infty}^{\infty} x(p)\, e^{-j\omega_1 p} dp$$

$$B_x(\omega_1, \omega_2) = X^*(\omega_1 + \omega_2) X(\omega_1) X(\omega_2)$$

To compute bispectrum, the spectrum obtained from triple correlation has to be ensemble averaged over segments. Therefore, the magnitude and phase of bispectrum are given by (M. R. Raghuveer and C.L. Nikias, 1987).

$$B_x(\omega_1, \omega_2) = E\left[X(\omega_1) X(\omega_2) X^*(\omega_1 + \omega_2)\right] \tag{5.42d}$$

$$\left|B_x(\omega_1, \omega_2)\right| = E\left[\, |X(\omega_1)|\, |X(\omega_2)|\, |X\ (-(\omega_1 + \omega_2))|\, \right]$$

$$\psi_x(\omega_1, \omega_2) = E\left[\phi_x(\omega_1) + \phi_x(\omega_2) - \phi_x(\omega_1 + \omega_2)\right]$$

The stationarity of triple correlation is one step improvement compared to second order stationarity.

Strictly stationary process

Theorem: The independent random variables with density functions

$$f_X[X(t_1)], f_X[X(t_2)], \ldots = f_X[X(t)\ ]$$

form a strictly stationary random process.

$$f_X[X_1(t), X_2(t), \ldots X_n(t)] = \quad f_X[(X(t_1)]f_X\ [X(t_2)], \ldots f_X\ [X(t_n)]$$

Since $f_X\ [X(t_i)]$ is same for all t_i,

$$f_X\ [X(t_i)] = f_X\ [X(t_i + \tau)] \text{ for all } t_i \tag{5.43}$$

The joint density function $f_X[X_1(t), X_2(t), \ldots X_n(t)]$ will be the same for all t_i and independent of time. Therefore, *an i.i.d process is a strictly stationary process* as it is true for all n. This is a very important result as the white noise process, which is an i.i.d, will be frequently visited in signal processing.

Example 5:

Consider the random process $X(t) = A\cos(\omega_c t + \theta\ )$ where θ is random variable with uniform distribution in the interval $(\ 0, 2\pi\ )$, A and ω_c are constants. Is this process stationary ?

The mean of the process is given by

$$m_X = E[X(t)] = E[A\cos(\omega_c t + \theta)] = A\int_0^{2\pi} \cos(\omega_c t + \theta)\ d\theta$$

$$= A\int_0^{2\pi} [\cos\omega_c t \cos\theta - \ \sin\omega_c t\ \sin\theta\ ]\ d\theta$$

$$m_X = A\cos\omega_c t\int_0^{2\pi} \cos\theta\ d\theta - A\sin\omega_c t\int_0^{2\pi} \sin\theta\ d\theta$$

$$= A\cos\omega_c t\left[\ \sin\theta\ \Big|_0^{2\pi}\ \right] + A\sin\omega_c t\left[\ \cos\theta\ \Big|_0^{2\pi}\ \right]$$

$$= A\cos\omega_c t[0-0] + A\sin\omega_c t[1-1] = 0$$

That is the mean m_X is constant and is not a function of time. The autocorrelation is given by

$$R_X(\tau) = E\left[X(t\ )X(t\ +\tau)\right]$$

$$= E\left[\ A\cos(\omega_c t + \theta\ )A\cos(\ \omega_c\left\{t + \tau\ \right\} + \theta\ )\right]$$

$$= E\left[A^2 \cos(\omega_c t + \theta)\cos(\omega_c \{t + \tau\} + \theta)\right]$$

Since

$$\cos A \cos B = (1/2)[\cos(A - B) + \cos(A + B)],$$

$$R_X(\tau) = (A^2/2)E\Big[\cos\big(\omega_c t + \theta - (\omega_c \{t + \tau\} + \theta)\big)$$

$$+ \cos\big(\omega_c t + \theta + \omega_c \{t + \tau\} + \theta\big)\Big]$$

$$= (A^2/2)E\left[\cos(\omega_c \tau) + \cos(\omega_c (2t + \tau) + 2\theta)\right]$$

$$= (A^2/2)E\left[\cos(\omega_c \tau)\right] + E\left[\cos(\omega_c (2t + \tau) + 2\theta)\right]$$

The first term on R.H.S is independent of θ,

$$R_X(\tau) = (A^2/2)\cos(\omega_c \tau) + E\left[\cos(\omega_c (2t + \tau) + 2\theta)\right]$$

Since $E\left[\cos(\omega_c (2t + \tau) + 2\theta)\right]$ is nothing but the mean of $X(t)$ it is equal to zero. Therefore,

$$R_X(\tau) = (A^2/2)\cos(\omega_c \tau)$$

This is independent of time t. Since the mean m_X and autocorrelation $R_X(\tau)$ of the process $X(t)$ are independent of time, it is a wide sense stationary process.

Example 6: A random process is having sample functions $X_1(t) = 5$ and $X_2(t) = \cos t$. Further, they occur with probability $P[X_1(t)] = 2/3$ and $P[X_2(t)] = 1/3$. The mean of the process is given by

$$E[X(t)] = \frac{2}{3}X_1(t) + \frac{1}{3}X_2(t) = \frac{1}{3}[10 + \cos t]$$

$$E[X(t)X(t+\tau)] = \frac{2}{3}X_1(t)X_1(t+\tau) + \frac{1}{3}X_2(t)X_2(t+\tau)$$

$$= \frac{2}{3}5(5) + \frac{1}{3}[\cos t \cos(t+\tau)]$$

$$= \frac{50}{3} + \frac{1}{3}[\cos t \cos(t+\tau)]$$

Since the mean and the autocorrelation function are functions of time, the process is not stationary.

5.4.4 Gaussian Random Process

For jointly Gaussian random variables $x_1, x_2, \dots x_N$, the joint density function is given by

$$f_{x_1,x_2,x_3\dots x_N}(X_1, X_2, X_3 \dots X_N) =$$

$$\frac{1}{\sqrt{(2\pi)^N |C|}} \exp\left[\frac{-1}{2|C|} \sum_k \sum_l \Delta_{kl}(x_k - m_{x_k})(x_l - m_{x_l})\right] \quad (5.44a)$$

where, C is a covariance matrix

$$C = \begin{bmatrix} c_{11} & c_{12} & \dots & c_{1N} \\ c_{21} & c_{22} & \dots & c_{2N} \\ \vdots & \vdots & \vdots & \vdots \\ c_{N1} & c_{N2} & \dots & c_{NN} \end{bmatrix}, \quad c_{kl} = E\left[(x_k - m_{x_k})(x_l - m_{x_l})\right] \quad (5.44b)$$

$|C| = \det.(C)$, Δ_{kl} is cofactor of the element c_{kl} and $c_{kk} = c_k^2$.

1. The Guassian process is completely determined by the first and second moments, mean and covariances. As for a Gaussian random variable only first two moments μ_X and variance σ^2, are required to specify the density function.
2. If N Gaussian random variables are uncorrelated, they are independent. That is,

$$f_{x_1,x_2,x_3\dots x_N}(X_1, X_2, X_3 \dots X_N) = f_{x_1}(X_1) f_{x_2}(X_2) \dots f_{x_N}(X_N)$$

The joint density function of X_1 and X_2 is

$$f_{x_1,x_2}(X_1, X_2) = \frac{1}{2\pi\sqrt{M}} e^{-[c_{22}x_1^2 + c_{11}x_2^2 - 2c_{12}x_1x_2]/2M}$$

$$c_{11} = E[X_1^2], \quad c_{22} = E[X_2^2], \quad c_{12} = E[X_1 X_2], \quad M = c_{11}c_{22} - c_{12}^2$$

Since X_1 and X_2 are uncorrelated, $c_{12} = E[X_1 X_2] = 0$, hence,

$$f_{x_1,x_2}(X_1, X_2) = \frac{1}{2\pi\sqrt{c_{11}c_{22}}} e^{-[x_1^2 c_{22} + c_{11}x_2^2]/2c_{11}c_{22}}$$

$$= \frac{1}{\sqrt{2\pi c_{11}}} e^{-x_1^2/2c_{11}} \frac{1}{\sqrt{2\pi c_{22}}} e^{-cx_2^2/2c_{22}}$$

$$f_{x_1,x_2}(X_1, X_2) = f_{x_1}(X_1) f_{x_2}(X_2)$$

This indicates but that the joint density function of X_1 and X_2 is the product of the individual density functions of X_1 and X_2.

This is true only for Gaussian case. In other cases, this is not valid. However, independence implies uncorrelatedness but the converse is not true.

3. For the process to be stationary, all the higher order joint density function or the higher order statistics (moments) should be shift invariant. But for a Gaussian process, since only the first and second order statistics determine the joint density function, it is sufficient if the second order statistics are shift invariant, which corresponds to wide sense stationarity. That is, *a Gaussian process is strictly stationary if it is wide sense stationary.*
4. If the input process is Gaussian for a linear system, the output process is also Gaussian, as the output is a linear combination of the input and this is valid by central limit theorem. This is not true with other processes as the linear combination of sum of random variables of any distribution result in Gaussian distribution, hence the identity of the input distribution is not maintained at the output. However, for a nonlinear system, the Gaussian nature of the input can change for the output process and the degree of deviation form the Gaussianity will be used as a measure of non-linearity.
5. The third moment is a *measure of asymmetry* of the density function about its mean value and is zero for a Gaussian process as its density function is symmetric. The triple correlation being the third moment is also zero for a Gaussian process and this is true for any symmetric density function.

5.4.5 Ergodic Process

The probabilistic random process considered which is having its density function associated with the outcomes of the process and its sample functions cannot be identified in a practical scenario. For a random process, only a sequence or stream of output is available. For example, the noise of an electrical power station turbine may be considered as the random process. To get the ensemble (all ensemble members) of the process, it is necessary to switch on the turbine many times. However, it may not be possible to switch on many times and can be done only once. In this scenario, the random process model considered as will not fit in. But in some situations it has been found that there is no necessity to have the complete ensemble of the process and even a single *realization* is sufficient to analyze the process. This is possible only *if a single realization represents the ensemble*, i.e., it has all the possibilities of the ensemble and instead of averaging over the ensemble, it can be done over different segments of a single realization. Such a random process is called an *ergodic process.* For the same reason it is not possible to measure the joint probability density function, hence only the moments like mean (first moment) and autocorrelation (second moment) which can be computed from the data, will be considered for handling a random process analysis and these are also sufficient for this purpose.

For the random process to be stationary, its density function and the sample functions ensemble should not change with time, as these reflect in averages like mean and autocorrelation, these quantities should be stationary even for the ergodic process. Rather, *it is necessary that the process is stationary to be ergodic.* Under the assumption of ergodicity, the probabilistic averages can be replaced by time averages. Hence, the definition for autocorrelation based on time average is also valid for a random process under the assumption that it is ergodic. Therefore, m_x and $R_X(\tau)$ are given by

$$m_x = \overline{X(t)} = \lim_{T\to\infty} \frac{1}{T} \int_{-T/2}^{T/2} X(t)\, dt \tag{5.45}$$

$$R_X(\tau) = \lim_{T\to\infty} \frac{1}{T} \int_{-T/2}^{T/2} X(t)\, X(t+\tau)\, dt \tag{5.46}$$

As already stated in the previous chapter, autocorrelation tells the rapidity with which the signal changes, i.e., it gives information about the frequency content of the signal. For white noise, the autocorrelation has a nonzero value only at the zeroth lag and for a periodic signal it is periodic.

5.4.6 Autoregressive (AR), Moving Average (MA) and Autoregressive Moving average Random Processes

A white noise process is a random process as its correlation function has nonzero value only at the zeroth lag and zero for all other lags. A filtered version of it is also a random process. Here, the randomness is only due to the white noise input and not due to the filter. If the noise process is stationary, that is, its characteristics, viz., its power and amplitude distribution function and the filter parameters are time invariant, then the output process is also a stationary random process. Depending on the type of filter like FIR or IIR or combination of the random processes are respectively known as moving average (MA) process or autoregressive (AR) process or moving average autoregressive (ARMA) process. These terminologies are coined by statisticians. Further, the terminologies are appropriate in the sense, the moving average stands for that the average is local, confined to a small interval around the desired time instant and not global. For autoregressive, auto means self and regressive means dependence, i.e., any sample value depends on its own previous samples. However, in signal processing, the MA process is known as all zero process and the AR as the all pole process and the ARMA as pole-zero process. The ARMA process is the most general process. For example, a white noise process is an extreme case of MA process and can be represented by zeros only. The sinusoidal process is an AR process as sinusoids can be represented by only poles. But sinusoids plus noise is an ARMA process. An ARMA process can be approximated by a higher order AR process. An MA process $x(n)$ is given by

$$x(n) = \sum_{k=-Q}^{Q} b(k) \in (n-k) \tag{5.47}$$

or $$X(z) = H(z)\, E(z)$$

where, $\in(n)$ is a white noise process, $b(n)$ are the coefficients of all zero filter

$$H(z) = \sum_{k=-Q}^{Q} b(k)\, z^{-k} a$$

An AR process $x(n)$ is given by

$$x(n) = \in(n) + \sum_{k=1}^{P} a(k)\, x(n-k) \tag{5.48}$$

where, $\in(n)$ is white noise process and $a(n)$ are the coefficients of an all pole filter

$$H(z) = \frac{X(z)}{\in(z)} = \frac{1}{1-\sum_{k=1}^{P} a(k) z^{-k}}$$

Further,

$$x(n) \approx \sum_{k=1}^{P} a(k)\, x(n-k),$$

if $\in(n)$ is small and can be considered an error. In such a case $x(n)$ can be predicted based on its previous values $x(n-1), x(n-2), \ldots x(n-P)$ by weighing them by the coefficients $a(n)$.

Given the process $x(n)$, $a(n)$ can be found by minimizing the error energy $\varepsilon = \sum_{n=0}^{L} \in^2(n)$ and when the coefficients are best one, then $\in(n)$ will be white, i.e., its correlation with adjacent samples and also its power will be minimum. Since the present sample is a weighted sum of the previous samples, these coefficients $a(n)$s are called prediction filter coefficients and $\in(n)$ the prediction error.

5.4.7 Relationship between Autocorrelation and Power Spectral Density Functions

The power spectral density is given by

$$S_x(\omega) = \lim_{T\to\infty} \frac{E\left[|X_T(\omega)|^2\right]}{T}$$

$$S_x(\omega) = \lim_{T\to\infty} \frac{1}{T} E\left[\int_{-T/2}^{T/2} X_T(t_1) e^{j\omega t_1}\, dt_1 \left[\int_{-T/2}^{T/2} X_T(t_2) e^{-j\omega t_2}\, dt_2\right]\right]$$

$$= \lim_{T\to\infty} \frac{1}{T} E\left[\int_{-T/2}^{T/2} \int_{-T/2}^{T/2} X_T(t_1) X_T(t_2) e^{-j\omega(t_2-t_1)} dt_2 dt_1 \right]$$

$$= \lim_{T\to\infty} \frac{1}{T} \int_{-T/2}^{T/2} \int_{-T/2}^{T/2} E\left[X_T(t_1) X_T(t_2)\right] e^{-j\omega(t_2-t_1)} dt_2 dt_1$$

put $t_1 = t$, then $dt_1 = dt$, $\tau = t_2 - t_1 = t_2 - t$, then $d\tau = dt_2$

$$S_x(\omega) = \lim_{T\to\infty} \frac{1}{T} \int_{-T/2-t}^{T/2-t} \int_{-T/2}^{T/2} R_X(t, t+\tau) e^{-j\omega\tau} dt d\tau$$

$$S_x(\omega) = \int_{-\infty}^{\infty} \left[\lim_{T\to\infty} \frac{1}{T} \int_{-T/2}^{T/2} R_X(t, t+\tau) d\tau \right] e^{-j\omega\tau} dt$$

Here,

$$\lim_{T\to\infty} \frac{1}{T} \int_{-T/2}^{T/2} R_X(t, t+\tau)\, d\tau = \overline{R_X(t, t+\tau)},$$

is the time average of the autocorrelation function.

$$S_x(\omega) = \int_{-\infty}^{\infty} \overline{R_X(t, t+\tau)}\, e^{-j\omega\tau} dt$$

For an ergodic process, $\overline{R_X(t, t+\tau)} = R_X(\tau)$

$$S_x(\omega) = \int_{-\infty}^{\infty} R_X(\tau)\, e^{-j\omega\tau} dt \qquad (5.49a)$$

$$R_X(\tau) = \frac{1}{2\pi} \int_{-\infty}^{\infty} S_x(\omega)\, e^{j\omega\tau} d\omega \qquad (5.49b)$$

5.4.8 Cross Power Density Spectrum

The cross power density spectrum is defined as

$$S_{xy}(\omega) = \lim_{T\to\infty} \frac{E\left[X_T(\omega)\, Y_T^*(\omega) \right]}{T} \qquad (5.50a)$$

$$S_{xy}(\omega) = \int_{-\infty}^{\infty} R_{XY}(\tau)\, e^{-j\omega\tau} dt \tag{5.50b}$$

Properties of autocorrelation function:

1. $|R_X(\tau)| \leq R_X(0)$ (upper bound)
2. $R_X(-\tau) = R_X(\tau)$ (by stationarity)
3. $R_X(0) = E\left[X_T^2\right]$ = Mean square value or power
4. $\lim\limits_{\tau \to \infty} R_X(\tau) = 0$, for a zero mean and non-periodic signal

Properties of power spectral density:

1. $S_x(\omega) \geq 0$
2. $S_x(-\omega) = S_x(\omega)$, for $X(t)$ real.
3. $S_x(\omega)$ is real
4. $\dfrac{1}{2\pi}\int_{-\infty}^{\infty} S_x(\omega)\, d\omega = E\left[X^2(t)\right]$

5.4.9 Linear System Response to Random Inputs

For a wide sense stationary random input $X(t)$, to a system with impulse response $h(t)$ its output $Y(t)$ is given by

$$Y(t) = \int_{-\infty}^{\infty} h(\xi)X(t-\xi)\, d\xi \tag{5.51}$$

The mean value of the response $Y(t)$ is given by

$$E\left[Y(t)\right] = E\left[\int_{-\infty}^{\infty} h(\xi)X(t-\xi)\, d\xi\right]$$

$$= \int_{-\infty}^{\infty} h(\xi)E\left[X(t-\xi)\right] d\xi$$

$$E\left[Y(t)\right] = \overline{X}\int_{-\infty}^{\infty} h(\xi)\, d\xi = \overline{Y} \tag{5.52}$$

The mean value of the output $Y(t)$ is the mean value of the input $X(t)$ times the area under the impulse response.

Mean square value of $Y(t)$:

$$E\left[Y^2(t)\right] = E\left[\int_{-\infty}^{\infty} h(\xi_1)X(t-\xi_1)\,d\xi_1 \int_{-\infty}^{\infty} h(\xi_2)X(t-\xi_2)\,d\xi_2\right]$$

$$E\left[Y^2(t)\right] = \int_{-\infty}^{\infty}\int_{-\infty}^{\infty} E\left[X(t-\xi_1)X(t-\xi_2)\right]h(\xi_1)h(\xi_2)d\xi_1 d\xi_2$$

$$= \int_{-\infty}^{\infty}\int_{-\infty}^{\infty} R_X(\xi_1-\xi_2)\,h(\xi_1)\,h(\xi_2)\,d\xi_1\,d\xi_2$$

$$\overline{Y^2} = E\left[Y^2(t)\right] = \int_{-\infty}^{\infty}\int_{-\infty}^{\infty} R_X(\xi_1-\xi_2)\,h(\xi_1)\,h(\xi_2)\,d\xi_1\,d\xi_2 \tag{5.53}$$

For white noise input, $R_{XX}(\xi_1-\xi_2) = (\mathrm{N}_0/2)\,\delta\,(\xi_1-\xi_2)$ i.e., the autocorrelation function is an impulse at the origin. Therefore

$$\overline{Y^2} = E\left[Y^2(t)\right] = \int_{-\infty}^{\infty}\int_{-\infty}^{\infty} (\mathrm{N}_0/2)\delta(\xi_1-\xi_2)\,h(\xi_1)d\xi_1\,h(\xi_2)d\xi_2$$

$$\overline{Y^2} = (\mathrm{N}_0/2)\int_{-\infty}^{\infty} h^2(\xi_2)\,d\xi_2 \tag{5.54}$$

For this case, the output power becomes proportional to the square of the area under the curve $h(t)$.

5.4.10 Autocorrelation Function of the Response

$$R_Y(t,t+\tau) = E\left[Y(t)Y(t+\tau)\right]$$

$$= E\left[\int_{-\infty}^{\infty} h(\xi_1)X(t-\xi_1)\,d\xi_1 \int_{-\infty}^{\infty} h(\xi_2)\,X\,(t+\tau-\xi_2)\,d\xi_2\right]$$

$$= \int_{-\infty}^{\infty}\int_{-\infty}^{\infty} E\left[X(t-\xi_1)X(t+\tau-\xi_2)\right]h(\xi_1)h(\xi_2)d\xi_1\,d\xi_2$$

$$R_Y(\tau) = \int_{-\infty}^{\infty}\int_{-\infty}^{\infty} R_X(\tau+\xi_1-\xi_2)\,h(\xi_1)\,h(\xi_2)\,d\xi_1\,d\xi_2$$

1. If $X(t)$ is wide sense stationary, then $Y(t)$ is also wide sense stationary since $R_Y(\tau)$ does not depend on t.
2. $R_Y(\tau)$ is a twofold convolution of the input autocorrelation function with the system's impulse response, i.e.,

$$R_Y(\tau) = R_X(\tau) * h(-\tau) * h(\tau)$$

Taking FT on both sides,

$$S_y(\omega) = S_x(\omega)H(-\omega)\, H(\omega)$$

$$S_y(\omega) = S_x(\omega)|H(\omega)|^2 \tag{5.55}$$

The output power spectral density is the input power spectral density multiplied by square of the magnitude of the transfer function of the system.

For the white noise, the autocorrelation function is an impulse at the origin given by $(\mathrm{N}_0/2)\delta(\tau)$. The power spectral density for white noise is the FT of $(\mathrm{N}_0/2)\delta(\tau)$, i.e., $(\mathrm{N}_0/2)$, since the FT of $\delta(\tau)$ is unity.

Similar to this, for the bispectrum of the system output is,

$$B_y(\omega_1,\omega_2) = B_H(\omega_1,\omega_2)\, B_x(\omega_1,\omega_2) \tag{5.56a}$$

$$|B_y(\omega_1,\omega_2)| = |B_x(\omega_1,\omega_2)||B_H(\omega_1,\omega_2)| \tag{5.56b}$$

and $$\Psi_y(\omega_1,\omega_2) = \Psi_x(\omega_1,\omega_2) + \Psi_H(\omega_1,\omega_2) \tag{5.56c}$$

For *non-Gaussian white noise* input, $|B_x(\omega_1,\omega_2)| = \beta$, a constant and $\Psi_x(\omega_1,\omega_2) = 0$. Therefore, the magnitude and phase of the bispectrum of the output are

$$|B_y(\omega_1,\omega_2)| = \beta\,|B_H(\omega_1,\omega_2)| \quad \text{and} \quad \Psi_y(\omega_1,\omega_2) = \Psi_H(\omega_1,\omega_2)$$

Considering the output bispectrum magnitude,

$$|B_y(\omega_1,\omega_2)| = \beta\,|H(\omega_1)||H(\omega_2)||H(-(\omega_1+\omega_2))|$$

At $\omega_2 = 0$,

$$B_y(\omega_1,0) = |H(0)||H(\omega_1)||H(-\omega_1)| = |H(0)||H(\omega_1)|^2$$

$$|H(\omega_1)|^2 = \frac{B_y(\omega_1,0)}{|H(0)|}, \quad \text{for } |H(0)| \neq 0 \tag{5.57}$$

Hence, system spectral magnitude can be obtained from the output bispectrum magnitude. (M. R. Raghuveer and C.L. Nikias, 1987). But the spectral magnitude can be found as well from the PSD itself.

Considering the output bispectrum phase,

$$\psi_y(\omega_1,\omega_2) = \phi(\omega_1) + \phi(\omega_2) - \phi(\omega_1+\omega_2) \tag{5.58}$$

The bispectrum phase of the output contains the system phase information but this has to be extracted from it.

5.4.11 Cross-correlation of Input and Output

The cross-correlation function between $X(t)$ and $Y(t)$ is

$$R_{XY}(t,t+\tau) = E\left[X(t)Y(t+\tau)\right]$$

$$= E\left(X(t)\int_{-\infty}^{\infty} h(\xi)X(t+\tau-\xi)\,d\xi\right)$$

$$= \int_{-\infty}^{\infty} E\left[X(t)X(t+\tau-\xi)\right]h(\xi)d\xi$$

Since $X(t)$ is stationary,

$$R_{XY}(\tau) = \int_{-\infty}^{\infty} R_X(\tau-\xi)\,h(\xi)d\xi$$

which is the convolution of $R_X(\tau)$ with $h(\tau)$. That is,

$$R_{XY}(\tau) = R_X(\tau) * h(\tau) \tag{5.59a}$$

Also

$$R_{YX}(\tau) = R_X(\tau) * h(-\tau) \tag{5.59b}$$

Taking FT on both sides of Eqn.(5.59a),

$$S_{xy}(\omega) = S_x(\omega)\,H(\omega)$$

or

$$H(\omega) = \frac{S_{xy}(\omega)}{S_x(\omega)} \tag{5.60}$$

The system transfer function is given by the ratio of the cross-spectral density $S_{xy}(\omega)$ between the input and the output to the power spectral density of the input.

5.4.12 Coherence Function

Coherence function (CF) is a measure of the bidirectional linearity between two processes. The CF is a normalized cross-spectral density function and the normalization process forces its magnitude square of CF (MSC) to be less than or equal to unity. The coherence between two processes $x(t)$ and $y(t)$, $\gamma_{xy}(\omega)$ is given by

$$\gamma_{xy}(\omega) = \frac{S_{xy}(\omega)}{\sqrt{S_{xx}(\omega)S_{yy}(\omega)}}, \quad 0 \le \left|\gamma_{xy}(\omega)\right|^2 \le 1 \tag{5.61}$$

In the frequency domain, the CF plays the role of correlation function in time domain. The two processes are non-coherent, i.e., uncorrelated at a particular frequency if the MSC at that frequency, is zero. On the other hand if MSC is equal to unity at all frequencies, then the two processes are totally coherent. For a system, at any frequency ω, the MSC is a measure of the portion of power of the output $y(t)$ due to input $x(t)$. For perfect linearity between the output and input, MSC must be unity at all the frequencies and this happens in the absence of any disturbing noise. The deviation of MSC from unity is an indication of the presence of noise, or non-linearity or the two processes do not depend linearly on one another. If the signal power is unity then $\left|\gamma_{xy}(\omega)\right|^2$ gives the portion of power in the signal, hence the noise power is $\left(1-\left|\gamma_{xy}(\omega)\right|^2\right)$. Therefore the signal to noise power ratio (S/N) is

$$(S/N) = \frac{1}{1-\left|\gamma_{xy}(\omega)\right|^2} \tag{5.62}$$

The coherence function and its magnitude play a crucial role in many signal processing applications and its estimation. The methods of their estimation are available in references given in the Bibliography (Y.T. Chen and R.K. Miskowicz, 1984, J.A. Cadzow and O.M. Solomon, Jr., 1987, S.V. Narasimhan, G.R. Reddy, E. I. Plotkin and M. N. S. Swamy,, 1995)

5.5 BASICS OF ESTIMATION

The measurement of a physical quantity/parameter (which is constant) like amplitude or frequency or speed, due to the tolerance of the measuring device or due to any other reason, may show different values at different trials. The histogram of these values gives a rough estimate of the density function. Generally, the values will be spread about a mean and the spread depends up on the measuring device/estimator. The measuring device/estimator can be assessed statistically or probabilistically and this will be considered.

Let ϕ be the parameter to be estimated. If $\hat{\phi}$ is the estimate, then $\overline{\phi}$ is its mean value, that is $E\left[\hat{\phi}\right] = \overline{\phi}$.

If $\phi - \overline{\phi} = 0$, then it is an *unbiased estimate.*

Otherwise if

$$\phi - \overline{\phi} = B \tag{5.63a}$$

the estimate is said to be biased and B is called the bias.
The variance of the estimate is

$$\sigma_{\hat{\phi}}^2 = E\left[\hat{\phi} - \overline{\phi}\right]^2 \tag{5.63b}$$

If the variance and bias B tend to zero as the data size increases, then it is called a *consistent estimate.* The mean square error between the estimate and its true value is the expected value of the squared difference between them.

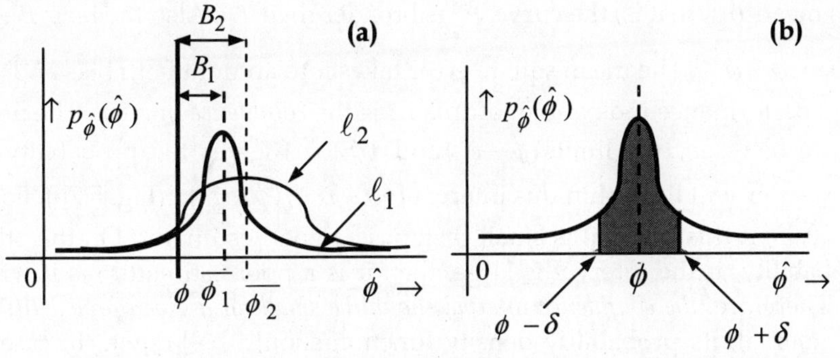

Fig. 5.7 (a) Comparison of estimators ℓ_1 and ℓ_2 of the parameter ϕ and (b) Confidence interval of an estimator.

Mean square error $\overline{\varepsilon}$ is given by

$$\overline{\varepsilon} = E\left[\phi - \hat{\phi}\right]^2 \tag{5.63c}$$

In Eqn. (12.63c), add and subtract $\overline{\phi}$,

$$\overline{\varepsilon} = E\left[\phi - \hat{\phi} + \overline{\phi} - \overline{\phi}\right]^2 = E\left[(\phi - \overline{\phi}) - (\hat{\phi} - \overline{\phi})\right]^2$$

$$= E\left[(\phi - \bar{\phi})\right]^2 + E\left[(\hat{\phi} - \bar{\phi})\right]^2 - 2E\left[\hat{\phi}\,\phi - \hat{\phi}\bar{\phi} - \bar{\phi}\,\phi + \bar{\phi}^2\right]$$

Here, $E\left[(\phi - \bar{\phi})\right] = (\phi - \bar{\phi})$, as it is constant.

$$\bar{\varepsilon} = B^2 + \sigma_\phi^2 - 2\left[\phi\bar{\phi} - \bar{\phi}^2 - \bar{\phi}\,\phi + \bar{\phi}^2\right]$$

$$\because E\left[\hat{\phi}\right] = \bar{\phi},\ E\left[\bar{\phi}\right] = \bar{\phi} \text{ and } E[\phi] = \phi$$

$$\bar{\varepsilon} = B^2 + \sigma_\phi^2 \tag{5.64}$$

For a good estimator the bias and variance or the mean square error must be as small as possible. From this point of view if the density function of the estimator is plotted, it should be concentrating / peaking at its mean value and the mean value should be close to the true value of the parameter being estimated. The narrow curve indicates lesser variance and better performance of the estimator. In Fig. 5.7a, the performance of the estimator ℓ_1 is better than that of ℓ_2 for the parameter ϕ, as the latter has a larger spread in the measured values about its mean $\bar{\phi}_2$, that is, the curve ℓ_2 is broader than ℓ_1. Also the bias $B_2 = \phi - \overline{\phi_2}$ is larger than $B_1 = \phi - \overline{\phi}_1$. The mean square error takes into account both bias and variance.

Another performance index of an estimator is the *confidence interval*. The area under the density curve between two limits $(\phi - \delta)$ and $(\phi + \delta)$ gives the probability that the value of the parameter will lie within this interval $[(\phi - \delta), (\phi + \delta)]$ (Fig. 5.7b). For a given probability, say 0.95, if this interval is small then, it is a good estimator. On the other hand for a given probability, if the interval is large, then it is a poor estimator, as it will have a large variance. Therefore, *the confidence interval should be small for a given probability.* To find the confidence interval, its probability density function should be known. In cases, where this is not known, it is found by computing the Gaussian density function from the mean and variance.

In practice, the mean and variance are computed from the observations/discrete data and are called the sample statistics, sample mean and sample variance.

Based on N observations/data points of a parameter x, its sample mean m_x and sample variance V_x are given by

$$m_x = \frac{1}{N}\sum_{n=1}^{N} x(n) \tag{5.65a}$$

and

$$V_x = \sigma_x^2 = \frac{1}{N}\sum_{n=1}^{N}[x(n) - m_x]^2 \tag{5.65b}$$

and mean square error is

$$\overline{\varepsilon_x} = \frac{1}{N}\sum_{n=1}^{N}[x_o - x(n)]^2 \tag{5.65c}$$

x_o is the true value of x.

5.5.1 Estimates of Autocorrelation and Cross-correlation

From the N observations/data, the estimates of the auto- and cross-correlation $r_x(k)$ and $r_{xy}(k)$, are computed as given below:

Unbiased estimates:

$$\hat{r}_x(k) = \frac{1}{N-|k|}\sum_{n=1}^{N-|k|} x(n)x(n+k)$$

$$\hat{r}_{xy}(k) = \frac{1}{N-|k|}\sum_{n=1}^{N-|k|} x(n)y(n+k) \tag{5.66a}$$

Biased estimates:

$$\hat{r}_x^B(k) = \frac{1}{N}\sum_{n=1}^{N-|k|} x(n)\, x(n+k)$$

$$\hat{r}_{xy}^B(k) = \frac{1}{N}\sum_{n=1}^{N-|k|} x(n)\, y(n+k) \tag{5.66b}$$

Bias of the estimates:

$$E[\hat{r}_x(k)] = \frac{1}{N-|k|}\sum_{n=1}^{N-|k|} E[x(n)\, x(n+k)]$$

$$E[\hat{r}_x(k)] = \frac{1}{N-|k|}\sum_{n=1}^{N-|k|} r_x(k) = r_x(k)$$

$$E[\hat{r}_x(k)] = r_x(k) \tag{5.67a}$$

Similarly, $E[\hat{r}_{xy}(k)] = r_{xy}(k)$. Therefore, $\hat{r}_x(k)$ and $\hat{r}_{xy}(k)$ are unbiased estimates as sum of $(N-|k|)$ quantities is divided by $(N-|k|)$. But $\hat{r}_x^B(k)$ and $\hat{r}_{xy}^B(k)$ are biased estimates as sum of $(N-|k|)$ quantities is divided by N.

Consider

$$\hat{r}_x^B(k) = \frac{1}{N}\sum_{n=1}^{N-|k|} x(n)\, x(n+k)$$

$$E\left[\hat{r}_x^B(k)\right] = \frac{1}{N} E\left[\sum_{n=1}^{N-|k|} x(n)\, x(n+k)\right] = \frac{1}{N}\left[\sum_{n=1}^{N-|k|} E\{x(n)\, x(n+k)\}\right]$$
$$= \frac{1}{N}\left[\sum_{n=1}^{N-|k|} r_x(k)\right]$$

$$E\left[\hat{r}_x^B(k)\right] = \frac{1}{N}\left(N-|k|\right) r_x(k)$$
$$E[\hat{r}_x^B(k)] = \left(1-\frac{|k|}{N}\right) r_x(k)$$

Similarly,

$$E[\hat{r}_{xy}^B(k)] = \left(1-\frac{|k|}{N}\right) r_{xy}(k)$$

Therefore,

$$B\left[\hat{r}_x^B(k)\right] = r_x(k)\left(\frac{k}{N}\right) \qquad \because B\left[\hat{r}_x(k)\right] = 0 \tag{5.67b}$$

Variance of the estimates:

$$V\left[\hat{r}_x(k)\right] = E[\hat{r}_x^2(k)] - \{E[\hat{r}_x(k)]\}^2$$

$$E[\hat{r}_x^2(k)] = \frac{1}{\left(N-|k|\right)^2}\sum_{i=1}^{N-|k|}\sum_{j=1}^{N-|k|} E\left\{x(i+|k|)\, x(i)\, x(j)\, x(j+|k|)\right\}$$

Since the expectation of four random variables is given by

$$E[V_1V_2V_3V_4] = E[V_1V_2]E[V_3V_4] + E[V_1V_3]E[V_2V_4] + E[V_1V_4]E[V_2V_3]$$

$$E\left\{x(i+|k|)\, x(i) x(j)\, x(j+|k|)\right\} =$$
$$E\{x(i+|k|)\, x(i)\}\, E\{x(j)\, x(j+|k|)\} + E\{x(i+|k|) x(j)\}\, E\{x(i)\, x(j+|k|)\}$$
$$+ E\{x(i+|k|)\, x(j+|k|\}\, E\{x(i)\, x(j))\}$$

$$= r_x(k)r_x(k) + E\{x\,(i+|k|)\,x\,(j)\}\,E\{x\,(i)\,x\,(j+|k|)\} + r_x(i-j)r_x(i-j)$$
$$= r_x^{\,2}(k) + E\{x\,(i+|k|)\,x\,(j)\}\,E\{x\,(i)\,x\,(j+|k|)\} + r_x^{\,2}(i-j)$$

Now,

$$V\left[\hat{r}_x(k)\right] = E[\hat{r}_x^{\,2}(k)] - \{E[\hat{r}_x(k)]\}^2$$

$$= r_x^{\,2}(k) + \left\{\frac{1}{\left(N-|k|\right)^2}\sum_{i=1}^{N-|k|}\sum_{j=1}^{N-|k|}[\,E\{x\,(i+|k|)\,x\,(j)\}\,E\{x\,(i)\,x\,(j+|k|)\} + r_x^{\,2}(i-j)] - r_x^{\,2}(k)\right\}$$

$$= \frac{1}{\left(N-|k|\right)^2}\sum_{i=0}^{N-|k|-1}\sum_{j=0}^{N-|k|-1}[\,E\{x\,(i+|k|)\,x\,(j)\}\,E\{x\,(i)\,x\,(j+|k|)\} + r_x^{\,2}(i-j)]$$

Let $q = i - j$, then

$$E\{x\,(i+|k|)\,x\,(j)\}\,E\{x\,(i)\,x\,(j+|k|)\} + r_x^{\,2}(i-j)$$

$$= E\{x\,(i+|k|)\,x\,(i-q)\}\,E\{x\,(i)x(i-q+|k|)\} + r_x^{\,2}(q)$$

$$= r_x(q+|k|)r_x(q-|k|) + r_x^{\,2}(q)$$

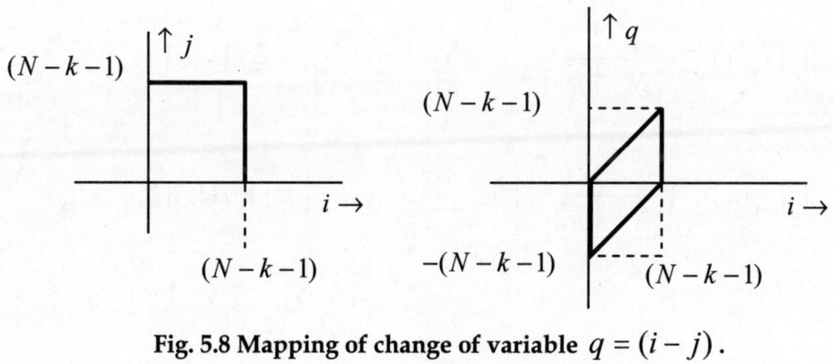

Fig. 5.8 Mapping of change of variable $q = (i - j)$.

$$i = 0, j = 0, q = 0,$$
$$i = 0, j = N - |k| - 1, q = -(N - |k| - 1)$$
$$i = N - |k| - 1, j = 0, q = N - |k| - 1$$
$$i = N - |k| - 1, j = N - |k| - 1, q = 0$$
$$i : 0 \rightarrow N - |k| - 1, q : -(N - |k| - 1) \rightarrow (N - |k| - 1)$$

This mapping is shown in Fig.(5.8). When

$$-(N - |k| - 1) \leq q \leq -1, \quad 0 \leq i \leq N - |k| - 1 + q$$
$$0 \leq q \leq (N - |k| - 1), \qquad q \leq i \leq N - |k| - 1$$

The double summation

$$\sum_{i=0}^{N-|k|-1} \sum_{j=0}^{N-|k|-1} [.] = \left(N - |k|\right)^2$$

$$\left\{ \sum_{q=-(N-|k|-1)}^{-1} \sum_{i=0}^{N-|k|-1+q} [.] + \sum_{q=0}^{(N-|k|-1)} \sum_{i=q}^{N-|k|-1} [.] \right\} = \left(N - |k|\right)^2$$

Therefore,

$$V\left[\hat{r}_x(k)\right] = \frac{1}{[N - |k|]^2} \sum_{q=-(N-|k|-1)}^{N-|k|-1} \sum_{i=0}^{N-|k|-1} \xi\ [k,q]$$

where

$$\xi\ [k,q] = r_x(q + |k|) r_x(q - |k|) + r_x^2(q)$$

$$V\left[\hat{r}_x(k)\right] = \frac{1}{[N - |k|]^2} \left\{ \sum_{q=-(N-|k|-1)}^{-1} \sum_{i=0}^{N-|k|-1+q} \xi\ [k,q] + \sum_{q=0}^{(N-|k|-1)} \sum_{i=q}^{N-|k|-1} \xi\ [k,q] \right\}$$

$$V\left[\hat{r}_x(k)\right] = \frac{1}{[N - |k|]^2} \left\{ \sum_{q=-(N-|k|-1)}^{-1} (N - |k| + q) \xi\ [k,q] \right.$$
$$\left. + \sum_{q=0}^{(N-|k|-1)} (N - |k| - q) \xi\ [k,q] \right\}$$

In the first summation, $q < 0$ and in second $q \geq 0$, hence the two terms can be combined using $| q |$ notation by $(N - | k | - | q |)$, which satisfies both the summations.

$$V[\hat{r}_x(k)] = \frac{1}{[N - | k |]^2} \sum_{q=-(N-|k|-1)}^{(N-|k|-1)} (N - | k | - | q |)\left[r_x(q + | k |)r_x(q - | k |) + r_x^2(q)\right]$$

$$V[\hat{r}_x(k)] = \frac{1}{[N - | k |]} \sum_{q=-(N-|k|-1)}^{(N-|k|-1)} \left(1 - \frac{| q |}{[N - | k |]}\right)\left[r_x(q + | k |)r_x(q - | k |) + r_x^2(q)\right]$$

The variance of $\hat{r}_x(k)$ for large N is given by

$$V\left[\hat{r}_x(k)\right] \approx \frac{N}{[N - | k |]^2} \sum_{i=-\infty}^{\infty} r_x^2(i) + r_x(i + |k|)r_x(i - |k|) \qquad (5.68a)$$

As $k \to N$, $V\left[\hat{r}_x(k)\right]$ becomes very large and this is due to the fact that as $k \to N$, the data contributing to its computation becomes less and less.

For biased case,

$$V\left[\hat{r}_x^B(k)\right] \approx \frac{\left[N - |k|\right]}{N^2} V\left[\hat{r}_x(k)\right] \qquad (5.68b)$$

Therefore, for the biased estimate $\hat{r}_x^B(k)$ as $k \to N$, the variance will not become as large as in the unbiased case. But from Eqn. (5.67b),

$$B\left[\hat{r}_x^B(k)\right] = r_x(k)\left(\frac{k}{N}\right)$$

As $k \to N$, the bias of $\hat{r}_x^B(k)$ approaches $r_x(k)$. That is for the biased estimate, the bias is as big as the quantity to be estimated, $r_x(k)$.

From these results, neither the unbiased nor the biased one appears to be good.

For a given k, as N is increased, for the unbiased case, the variance decreases and for the biased case, both the variance and bias decrease. But the mean square error for the biased over that of the unbiased estimator is less and generally, the unbiased is preferred. Both the estimators can be used by ensuring sufficiently large data lengths as compared to the maximum autocorrelation lag considered.

In practice, as a rule of thumb, a data length *more than ten times* the maximum lag length will be taken.

To bring out the significance of accuracy of the estimation of autocorrelation problem, an example of random gyro (R. G. Brown, Introduction to Random Signal Analysis and Kalman Filtering) is considered.

For a random gyro drift process with an autocorrelation following the relation

$$r(\tau) = Ke^{-|\tau|/\tau_o} \qquad (K \text{ is a constant})$$

for a time constant τ_o equal to 10 *hours*, to get an autocorrelation of 10% accuracy (Fig. 5.9), data of 2000 *hours* is required which is about the mean life time of a gyro.

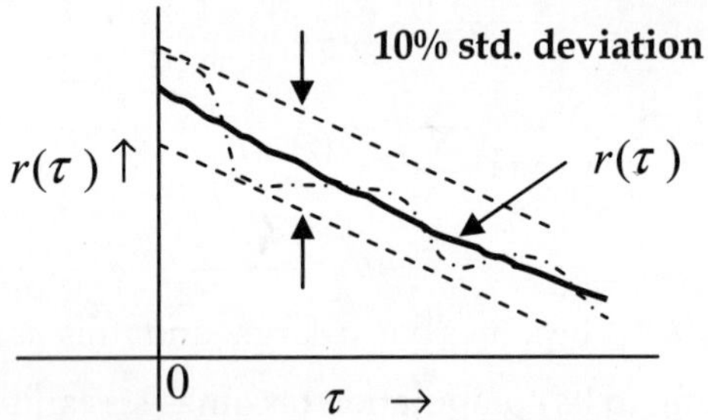

Fig. 5.9 Autocorrelation estimation.

Further, for an accuracy of 1% about **23 *Years!!!!*** of data is required, indicating that an accurate estimation of the autocorrelation function is not so simple.

Chapter 6

NON-PARAMETRIC SPECTRAL ESTIMATION

The spectral estimation of a signal is most important as it reveals the frequency content of the signal. Prior to processing any signal, it is essential to know its contents, hence obtaining this information with minimum error on an average basis, is the goal of spectral estimation. This chapter introduces the concept of random sampling signals. A signal which cannot be defined by a mathematical relation is a random signal. On the other hand a signal which can be defined by a mathematical relation is not random but is deterministic. But most of the practical signals like speech and image and system like windmill cannot be represented mathematically. Hence, it is necessary to treat them as random signals and their processing has to be based on statistical averages. The tools which can handle them are called estimators and it is of interest to find estimators which best represent the random process on hand. The basic estimator like correlation estimator and its performance indices like bias and variance are considered. The spectral estimators like, periodogram, the windowed periodogram, Welch method are treated. These methods mainly suffer from the trade-off between frequency resolution (bias) and variance. In addition to these, the new methods like modified group delay method which reduces the variance without compromising on frequency resolution are introduced. Further, a new DFT having desirable DCT properties like less leakage and double the frequency resolution (of conventional DFT), the analytic DCT is introduced for spectral estimation. The combination of analytic DCT and modified group delay which achieves better frequency resolution, less leakage and reduced variance is considered. Bispectrum and extraction of nonminimum phase from bispectral phase is covered.

6.1 INTRODUCTION

The aim of spectral estimation is to get an accurate estimate of the power spectral density (PSD) or spectral information of a signal on an *average basis*. That is, an accurate estimate of the magnitudes of various spectral components present in a signal, on an average basis. An accurate estimate of the PSD refers to that estimate whose variance and bias or the mean square error is less and has a required frequency resolution. For this purpose, the most important tool is the Fourier transform and from the computational point of view, it is the discrete Fourier transform. The DFT by itself has limitations/problems like leakage and poor frequency resolution. These aspects as they affect the bias and variance of the estimate, have to be taken into account in arriving at a spectral estimator.

In general, a signal is the output of a system for some input. The input can be an impulse/impulse train or a white noise. The examples for impulse-excited type are voiced speech; electrocardiogram and those due to white noise are electroencephalograph, seismic signals (reflections form the earth layers). Further, signals are generally associated with noise. The FT of a segment of noise (white) is not a constant and it will differ from one segment to the other. It is only the average of the FTs of a large number of segments that is constant with respect to frequency. Hence, the spectral information of interest is modified by the noise spectrum, which is about the system or the original signal (i.e., without noise). Since the noise spectrum is different for different segments of noise, the noise mainly introduces the variability or variance. Further, there will be variability of data truncation from segment to segment that leads to leakage of different magnitudes. Further, any extra frequency resolution than required increases the variance as it picks up undesired details. However, to reduce the bias, it is necessary to use a DFT that provides the required frequency resolution.

Here, both Fourier transform based non-parametric and parametric methods of spectral estimation will be discussed.

6.2 TOOLS FOR SPECTRAL ESTIMATION

6.2.1 The Periodogram

It is defined as

$$I(\omega) = \frac{1}{LU}\left|X(\omega)\right|^2 \tag{6.1}$$

where

$$X(\omega) = \sum_{n=0}^{L-1} x(n)\, w(n)\, e^{-j\omega n} \tag{6.2a}$$

and

$$U = \frac{1}{L}\sum_{n=0}^{L-1} w^2(n) \tag{6.2b}$$

The window $w(n)$ selects L samples from the data sequence $x(n)$. The factor U removes the bias due to window in the estimate $I(\omega)$ (for a rectangular window $U = 1$). If $w(n)$ is rectangular, $I(\omega)$ is called periodogram, otherwise it is called a modified periodogram.

In practice, the periodogram (or modified periodogram) is computed using DFT of the sequence $x(n)w(n)$ and is given by

$$I(k) = \frac{1}{LU}|X(k)|^2$$
$$= \frac{1}{LU}\left|\sum_{n=0}^{N-1} g(n)e^{-j\frac{2\pi}{N}kn}\right|^2, \quad k = 0,1,2,\ldots(N-1). \tag{6.3}$$

where

$$g(n) = \begin{cases} [x(n)w(n) - \mu], & n = 0,1,2,\ldots(L-1) \\ 0, \quad n = L,(L+1),\ldots(N-1). \end{cases} \tag{6.4}$$

$$\mu = \frac{1}{L}\sum_{n=0}^{L-1} x(n)\, w(n) \tag{6.5}$$

It is necessary that the mean is removed otherwise it may introduce considerable leakage for the frequencies near the zero frequency.

Again from Eqn. (6.1),

$$I(\omega) = \frac{1}{LU}|X(\omega)|^2$$

$$LU\, I(\omega) = |X(\omega)|^2 = X(\omega)X^*(\omega)$$

$$X(\omega)X^*(\omega) \Leftrightarrow \sum_{n=0}^{L-|k|-1} x(n)\, x(k-(-n)) = \sum_{n=0}^{L-|k|-1} x(n)\, x(n+k)$$

$$I(\omega) \Leftrightarrow \frac{1}{LU}\sum_{n=0}^{L-|k|-1} x(n)\, x(n+k) = \frac{1}{U}\hat{r}_x^B(k)$$

$$E[I(\omega)] \Leftrightarrow \frac{1}{U}E\,[\hat{r}_x^B(k)]$$

Since,

$$E[\hat{r}_x^B(k)] = \left(1 - \frac{|k|}{L}\right) r_x(k),$$

$$U\,E[I(\omega)] \Leftrightarrow \left(1-\frac{|k|}{L}\right) r_x(k)$$

$$E[I(\omega)] = \frac{1}{U}\sum_{k=-(L-1)}^{L-1}\left(1-\frac{|k|}{L}\right) r_x(k)\, e^{-j\omega k}$$

$$E[I(\omega)] = \frac{1}{2\pi}\int_{-\pi}^{\pi} S_X(\theta) W(\omega-\theta)\, d\theta,$$

$$w(k) = \begin{cases} \left[1-|k|/L\right], & |k| \le L-1 \\ 0, & |k| > L-1 \end{cases}$$

$$w(k) \Leftrightarrow W(\omega) = \frac{1}{L}\left(\frac{\sin[\omega L/2]}{\sin[\omega/2]}\right)^2$$

$$E[I(\omega)] = W(\omega) * S_X(\omega) \tag{6.6}$$

where

$$S_X(\omega) \Leftrightarrow r_x(k)$$

Thus, the *mean of the sample spectrum is the convolution of the true PSD with the FT of the* $w(k)$, *the Bartlett* ***window***. The sample PSD is biased for finite L but becomes unbiased as $L \to \infty$.

Variance of the periodogram:

$$LU\, I(\omega) = |X(\omega)|^2 = X(\omega) X^*(\omega)$$

$$= \sum_{m=0}^{L-1} x(m)\, e^{-j\omega m} \sum_{l=0}^{L-1} x(l)\, e^{j\omega l}$$

$$= \sum_{m=0}^{L-1}\sum_{l=0}^{L-1} x(m)\, x(l)\, e^{-j\omega m} e^{j\omega l}$$

The autocorrelation of the periodogram is given by

$$E\left[I(\omega_1) I(\omega_2)\right]$$

$$= \frac{1}{(LU)^2}\sum_{k=0}^{L-1}\sum_{l=0}^{L-1}\sum_{m=0}^{L-1}\sum_{n=0}^{L-1} E[x(k)x(l)x(m)x(n)] e^{j\omega_1 k} e^{-j\omega_2 n} e^{-j\omega_1 l} e^{j\omega_2 m}$$

$$E\left[I(\omega_1) I(\omega_2)\right] = \frac{1}{(LU)^2}\sum_{k=0}^{L-1}\sum_{l=0}^{L-1}\sum_{m=0}^{L-1}\sum_{n=0}^{L-1} E[x(k)\, x(l)\, x(m)\, x(n)]\, e^{j[\omega_1 (k-l)+\omega_2 (m-n)]} \tag{6.7}$$

This expression cannot be simplified even for a white noise process for which $E[x(n)x(n+m)] = \sigma_x^2 \delta(m)$, unless it is Gaussian, since for this case it is possible to write

$$E[x(m)x(l)x(k)x(n)] = E[x(k)x(l)]E[x(m)x(n)] + E[x(k)x(m)]E[x(l)x(n)] + E[x(k)x(n)]E[x(l)x(m)]$$

$$E[x(m)x(l)x(k)x(n)] = \begin{cases} \sigma_x^4, & (k=l, m=n) \text{ or } (k=m, l=n) \text{ or } (k=n, l=m) \\ 0\ , & otherwise \end{cases}$$

Substituting this in Eqn. (6.7)

$$E\left[I(\omega_1)I(\omega_2)\right] = \frac{1}{(LU)^2}\left[\sum_{k=0}^{L-1}\sum_{l=0}^{L-1}\sum_{m=0}^{L-1}\sum_{n=0}^{L-1} E[x(k)x(l)]E[x(m)x(n)]e^{j[\omega_1(k-l)+\omega_2(m-n)]}\right.$$

$$+\sum_{k=0}^{L-1}\sum_{l=0}^{L-1}\sum_{m=0}^{L-1}\sum_{n=0}^{L-1} E[x(k)x(m)]E[x(l)x(n)]e^{j[\omega_1(k-l)+\omega_2(m-n)]}$$

$$\left.+\sum_{k=0}^{L-1}\sum_{l=0}^{L-1}\sum_{m=0}^{L-1}\sum_{n=0}^{L-1} E[x(k)x(n)]E[x(l)x(m)]e^{j[\omega_1(k-l)+\omega_2(m-n)]}\right]$$

The first summation term, T_1: since $(k=l, m=n)$

$$T_1 = \sum_{k=l=0}^{L-1}\sum_{m=n=0}^{L-1}\sigma_x^4 = L^2\sigma_x^4$$

The second summation term, T_2: since $(k=m, l=n)$

$$T_2 = \sum_{k=m=0}^{L-1}\sum_{l=n=0}^{L-1}\sigma_x^4 e^{j[\omega_1(m-n)+\omega_2(m-n)]} = \sum_{m=0}^{L-1}\sum_{n=0}^{L-1}\sigma_x^4 e^{j[(\omega_1+\omega_2)(m-n)]}$$

The third summation term, T_3: since $(k=n, l=m)$

$$T_3 = \sum_{k=n=0}^{L-1}\sum_{l=m=0}^{L-1}\sigma_x^4 e^{j[\omega_1(n-m)+\omega_2(m-n)]} = \sum_{k=n=0}^{L-1}\sum_{l=m=0}^{L-1}\sigma_x^4 e^{j[(\omega_1-\omega_2)(n-m)]}$$

Therefore,

$$E\left[I(\omega_1)I(\omega_2)\right] =$$

$$\frac{\sigma_x^{\ 4}}{(LU)^2}\left[L^2 + \sum_{m=0}^{L-1}\sum_{n=0}^{L-1} e^{j(m-n)(\omega_1+\omega_2)} + \sum_{m=0}^{L-1}\sum_{n=0}^{L-1} e^{j(n-m)(\omega_1-\omega_2)}\right] \tag{6.8}$$

$$\sum_{m=0}^{L-1}\sum_{n=0}^{L-1} e^{j(m-n)(\omega_1+\omega_2)} = \sum_{m=0}^{L-1} e^{jm(\omega_1+\omega_2)} \sum_{n=0}^{L-1} e^{-jn(\omega_1+\omega_2)}$$

$$= \sum_{m=0}^{L-1} e^{jm(\omega_1+\omega_2)} \left[\frac{1-e^{-jL(\omega_1+\omega_2)}}{1-e^{-j(\omega_1+\omega_2)}}\right]$$

$$= \left[\frac{1-e^{jL(\omega_1+\omega_2)}}{1-e^{j(\omega_1+\omega_2)}}\right]\left[\frac{1-e^{-jL(\omega_1+\omega_2)}}{1-e^{-j(\omega_1+\omega_2)}}\right]$$

$$= \frac{2-e^{-jL(\omega_1+\omega_2)} - e^{jL(\omega_1+\omega_2)}}{2-e^{-j(\omega_1+\omega_2)} - e^{j(\omega_1+\omega_2)}}$$

$$\sum_{m=0}^{L-1}\sum_{n=0}^{L-1} e^{j(m-n)(\omega_1+\omega_2)} = \frac{2-2\cos L(\omega_1+\omega_2)}{2-2\cos(\omega_1+\omega_2)}$$

Therefore,

$$\sum_{m=0}^{L-1}\sum_{n=0}^{L-1} e^{j(m-n)(\omega_1+\omega_2)} = \frac{\sin^2\{(\omega_1+\omega_2)L/2\}}{\sin^2\{(\omega_1+\omega_2)/2\}}.$$

Similarly,

$$\sum_{m=0}^{L-1}\sum_{n=0}^{L-1} e^{j(n-m)(\omega_1-\omega_2)} = \frac{\sin^2\{(\omega_1-\omega_2)L/2\}}{\sin^2\{(\omega_1-\omega_2)/2\}}$$

Therefore, Eqn. (6.8) becomes

$$E\left[I(\omega_1)I(\omega_2)\right] = \frac{\sigma_x^{\ 4}}{U^2}\left[1+\left(\frac{\sin\{(\omega_1+\omega_2)L/2\}}{L\sin\{(\omega_1+\omega_2)/2\}}\right)^2 + \left(\frac{\sin\{(\omega_1-\omega_2)L/2\}}{L\sin\{(\omega_1-\omega_2)/2\}}\right)^2\right]$$

since,

$$COV\left[I(\omega_1)I(\omega_2)\right] = E\left[I(\omega_1)I(\omega_2)\right] - E\left[I(\omega_1)\right]E\left[I(\omega_2)\right] \tag{6.9}$$

and for white noise ,

$$E\left[I(\omega_1)\right] = E\left[I(\omega_2)\right] = \frac{\sigma_x^{\ 2}}{U}$$

$$COV\left[I(\omega_1)I(\omega_2)\right] =$$

$$\frac{\sigma_x^{\ 4}}{U^2}\left[\left(\frac{\sin\{(\omega_1+\omega_2)L/2\}}{L\sin\{(\omega_1+\omega_2)/2\}}\right)^2 + \left(\frac{\sin\{(\omega_1-\omega_2)L/2\}}{L\sin\{(\omega_1-\omega_2)/2\}}\right)^2\right]$$

Therefore, variance of the periodogram at a particular frequency is obtained if $\omega = \omega_1 = \omega_2$,

$$V\left[I(\omega)\right] = COV\left[I(\omega)I(\omega)\right]$$

$$V\left[I(\omega)\right] = \frac{\sigma_x^{\ 4}}{U^2}\left[\left(\frac{\sin(\omega L)}{L\sin(\omega)}\right)^2 + 1\right] \tag{6.10}$$

As $L \to \infty$, $V\left[I(\omega)\right]$ *does not approach zero, hence the periodogram is not a consistent estimator.* The variance of $I(\omega)$, $V\left[I(\omega)\right]$ is of the order of $\sigma_x^{\ 4}$ irrespective of the size L.

Again from the expression for the covariance of the periodogram,

$$COV\left[I(\omega_1)I(\omega_2)\right] = \frac{\sigma_x^{\ 4}}{U^2}\left[\left(\frac{\sin\{(\omega_1+\omega_2)L/2\}}{L\sin\{(\omega_1+\omega_2)/2\}}\right)^2 + \left(\frac{\sin\{(\omega_1-\omega_2)L/2\}}{L\sin\{(\omega_1-\omega_2)/2\}}\right)^2\right]$$

If $\omega_1 = 2\pi\, k / L$ and $\omega_2 = 2\pi\, l / L$ for integer values of k and l,

$$COV\left[I(k)I(l)\right] = \frac{\sigma_x^{\ 4}}{U^2}\left[\left(\frac{\sin\{\pi(k+l)\}}{L\sin\{\pi(k+l)/L\}}\right)^2 + \left(\frac{\sin\{\pi(k-l)\}}{L\sin\{\pi(k-l)/L\}}\right)^2\right]$$

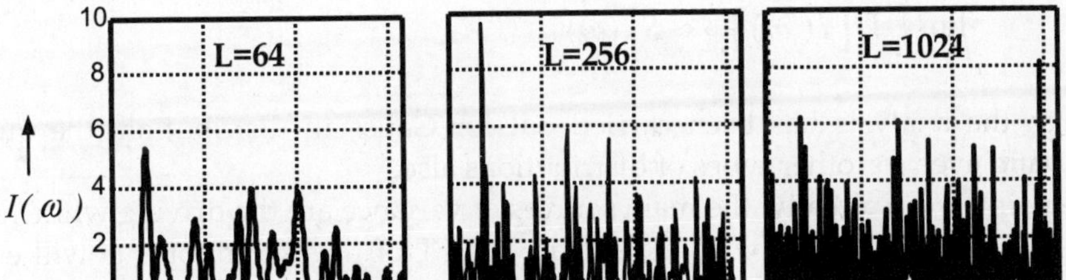

Fig. 6.1 Periodogram for different data lengths, L.

If $k \neq l$, then $\pi(k+l)$ and $\pi(k-l)$ are integer multiples of π, though this will not result in the denominator terms $\sin\{\pi(k+l)/L\}$ and $\sin\{\pi(k-l)/L\}$ equal to zero, but certainly results in the numerator terms $\sin\{\pi(k+l)\}$ and $\sin\{\pi(k-l)\}$ equal to zero. Therefore for $k \neq l$,

$$COV\left[I(k)I(l)\right] = 0$$

That is, the periodogram values at frequencies separated by integer multiples of $(2\pi / L)$ are *uncorrelated.*

As L increases, these frequencies become closer and closer, and however these frequencies are again uncorrelated. That is, the values at these frequencies are random and do not obey any definite pattern. But since the noise considered is white, it is expected that $I(k),\ 0 \le k \le 2\pi$ should be constant or flat. Hence, with the increase in L, the periodogram fluctuates more rapidly, which is *not at all a desired* result. Figure 6.1 shows the periodogram for data lengths 64, 256 and 1024 samples of the noise sequence computed with a FFT of 1024 points. It is seen that the fluctuations increase with data length.

This analysis considering the signal as white Gaussian gives some insight into the variance of the periodogram. For colored noise, since it can be considered the filtered white noise, its variance is given by

$$V\left[I(\omega)\right] = \left|H(\omega)\right|^4 \frac{\sigma_x^4}{U^2}\left[1+\left(\frac{\sin(\omega L)}{L\sin(\omega)}\right)^2\right] \ge E[I^2(\omega)]$$

$$V\left[I(\omega)\right] = \frac{S_X^2(\omega)}{U^2}\left[1+\left(\frac{\sin(\omega L)}{L\sin(\omega)}\right)^2\right] \ge E[I^2(\omega)] \tag{6.11}$$

where

$$S_X(\omega) = \left|H(\omega)\right|^2 \sigma_x^{\,2}$$

and $H(\omega)$ is the filter frequency response.

As L increases, the variance of the periodogram is proportional to square of the spectrum and hence, it is *not a consistent estimate.*

$$\lim_{L\to\infty}\left\{ V\left[I(\omega)\right] \right\} \approx S_X^2(\omega)$$

Even though the analysis has been carried out for Gaussian distribution, the qualitative analysis is valid even for other types of distributions also.

As pointed out already, the main sources of variance are the driving white noise of a system or the associated noise with the signal. The FT of these components will either get multiplied or added to PSD of the signal, hence introduces extra variance to the already existing ones due to Gibbs ripple effect and leakage. Therefore, for a signal associated with noise, the variance is given by

$$V[I(\omega)] = V[I_S(\omega)] + V[I_\eta(\omega)] \tag{6.12}$$

$V[I_S(\omega)]$: Variance of the periodogram of the signal (which includes Gibbs ripple and leakage effects)

$V[I_\eta(\omega)]$: Variance of the periodogram of the associated noise, which is given by the expression (for Gaussian white noise).

For the case of a system driven with white noise, the variance is similar that for a colored noise. Further, the system transfer function corresponds to that of the filter.

A spectrum/PSD, which is smooth, is called a *statistically stable spectrum*. Increasing the length of the data basically increases the frequency resolution and picks up unwanted details, which will vary from segment to segment, hence increases the variance of the PSD estimate. On the other hand, only the required frequency resolution will not do so and will not increase the variance unnecessarily. Hence, it is desirable to have only the required frequency resolution and any extra resolution will only add to the variance, which is undesirable and decreases the statistical stability. The statistical stability is often quantified by the parameter Q called the *quality ratio*, the ratio of the variance of the spectral estimate to the square of the mean of the estimate and is given by

$$Q = \frac{V[I(\omega)]}{\{E[I(\omega)]\}^2} \tag{6.13}$$

It is effectively the reciprocal of signal to noise ratio. Small value of Q (less than unity) indicates that the spectral estimate is smooth with small fluctuations whereas a value greater than unity corresponds to a noisy spectrum with large fluctuations or variance.

6.2.2 Smoothed Periodogram Estimators

The periodogram is not a consistent estimator though its bias decreases, its variance increases with increase in data length. Therefore to get an estimate whose variance decreases with increase in data length, it is necessary to average over many segments of data.

Bartlett method: Averaging the periodograms

In the method proposed by Bartlett, a given record of data is divided into a number of segments each of length sufficient enough to give the required frequency resolution and their periodograms are averaged.

The data record $x(n),\ 0 \le n \le L-1$ is divided into K segments each of length M samples so that $L = KM$. The ith segment is given by

$$x^{(i)}(n) = x(n + M(i-1)),\ 0 \le n \le M-1 \text{ and } 1 \le i \le K$$

The ith periodogram is given by,

$$I^{(i)}(\omega) = \frac{1}{M}\left|\sum_{n=0}^{M-1} x^{(i)}(n)e^{-j\omega n}\right|^2,$$

If $r_x(m)$ is small for $m > M$, then the segments are independent and the averaging will reduce the variance. The spectrum estimate $\bar{I}(\omega)$ is given by

$$\bar{I}(\omega) = \frac{1}{K}\sum_{i=1}^{K} I^{(i)}(\omega)$$

The expected value of $\bar{I}(\omega)$ is

$$E[\bar{I}(\omega)] = \frac{1}{K}\sum_{i=1}^{K} E[I^{(i)}(\omega)] = E[I^{(i)}(\omega)]$$

But

$$E[I^{(i)}(\omega)] = S_X(\omega) * W_B(\omega)\ ,\ W_B(\omega) = \frac{1}{M}\left[\frac{\sin(\omega M/2)}{\sin(\omega/2)}\right]^2$$

$$E[I^{(i)}(\omega)] = \frac{1}{2\pi}\int_{-\pi}^{\pi} S_X(\xi)\, W_B(\omega-\xi)d\xi$$

$$E[\bar{I}(\omega)] = \frac{1}{2\pi M}\int_{-\pi}^{\pi} S_X(\xi)\left(\frac{\sin[(\omega-\xi)M/2]}{\sin[(\omega-\xi)/2]}\right)^2 d\xi$$

That is, the expected value of the Bartlett estimate is the convolution of the true PSD with the FT of the triangular window of length M. Thus, this estimate is a biased estimator. Further, as $M = L/K$, main lobe width of the Bartlett window with M samples is K *times wider* than with that of length L. The narrower the main lobe of $W_B(\omega)$, the closer it is to the impulse and its convolution with the true PSD results in an estimate that is closer to $S_X(\omega)$. Hence, the bias of the Bartlett estimate is *higher than* that of the periodogram (with data length L). But the averaging of K independent periodograms of data length M samples will bring down the variance by a factor of K.

$$V[\bar{I}(\omega)] = \frac{1}{K}V[I(\omega)]$$

$$V[\bar{I}(\omega)] = \frac{1}{K}\left\{\frac{S_X^2(\omega)}{U^2}\left[1+\left(\frac{\sin(\omega M)}{M\sin(\omega)}\right)^2\right]\right\} \tag{6.14}$$

Thus, a decrease in variance is accompanied by an increase in bias. Thus, for a given length of data, there is trade-off between frequency resolution or bias and variance.

The Bartlett window is a built-in window for the periodogram and its ability to smooth the periodogram along the frequency axis is fixed for a given length. But to have additional smoothing to improve the statistical stability, it is necessary to use other windows over and above the built-in Bartlett window. Since to smooth beyond the built-in window, the main lobe of the FT of the applied window should be larger than that of the built-in one. The smoothed periodogram is given by

$$\tilde{I}(\omega) = I(\omega) * W(\omega) = \frac{1}{2\pi}\int_{-\pi}^{\pi} I(\xi)\, W(\omega-\xi)\, d\xi$$

But,

$$E[I(\omega)] = \frac{1}{2\pi}\int_{-\pi}^{\pi} S_X(\xi)\, W_B(\omega-\xi)\, d\xi$$

Further,

$$E\left[\tilde{I}(\omega)\right] = \left[S_X(\omega) * W_B(\omega)\right] * W(\omega)$$

Or

$$E\left[\tilde{I}(\omega)\right] = \sum_{k=-(M-1)}^{(M-1)} r_x(k)\, w_B(k)\, w(k)\, e^{-j\omega k}$$

As mentioned above for smoothing beyond by the built in Bartlett window, the *smoothing window,* $w(n)$ effect should be comparatively higher than that of the built in Bartlet window. In such a case,

$$E\left[\tilde{I}(\omega)\right] \approx \sum_{k=-(M-1)}^{(M-1)} r_x(k)\, w(k)\, e^{-j\omega k} \quad \text{or} \quad E\left[\tilde{I}(\omega)\right] = S_X(\omega) * W(\omega)$$

Since the main lobe width of $W(\omega)$ is larger than that of $W_B(\omega)$, the frequency resolution of the smoothed periodogram is less than that of unsmoothed one. If the main lobe width of $W(\omega)$ is small compared to the desired variations in $S_X(\omega)$, then asymptotically, i.e., as $M \to \infty$, $W(\omega) \to \delta(\omega)$. For which,

$$w(0) = \frac{1}{2\pi}\int_{-\pi}^{\pi} W(\omega)\, d\omega = 1$$

The variance of the periodogram is given by

$$V[I(\omega)] = \frac{S_X^2(\omega)}{U^2}\left[1 + \left(\frac{\sin(\omega L)}{L\sin(\omega)}\right)^2\right] \tag{6.15}$$

Further,

$$V[I(\omega)] = \begin{cases} 2S_X^2(0)/U^2, \omega = 0 \\ 2S_X^2(\pi)/U^2, \omega = \pi \\ S_X^2(\omega)/U^2,\ \omega \neq 0, \omega \neq \pi \end{cases}$$

For the smoothed periodogram the effect of Bartlett window is negligible, hence in this case, the term in square bracket is equal to the unity and the value of $U < 1$ resulting,

$$V[I(\omega)] \approx S_X^2(\omega)/U^2$$

Since $\tilde{I}(\omega) = I(\omega) * W(\omega)$, the variance of the smoothed periodogram with the window $w(n)$ is given by

$$V\left[\tilde{I}(\omega)\right] \approx \frac{1}{L}\left\{V[I(\omega)] * W^2(\omega)\right\}$$

$$V\left[\tilde{I}(\omega)\right] = \frac{1}{2\pi\, LU^2}\int_{-\pi}^{\pi} S_X^2(\xi)\, W^2(\omega - \xi)\, d\xi$$

Further, if the main lobe width $W(\omega)$ is small compared to the variations in $I(\omega)$, that is, the length of smoothing window is sufficient to provide the required frequency resolution then,

$$V\left[\tilde{I}(\omega)\right] = \frac{1}{LU^2} S_X^2(\omega) \frac{1}{2\pi}\int_{-\pi}^{\pi} W^2(\xi)\, d\xi$$

From Parseval's theorem,

$$\frac{1}{2\pi}\int_{-\pi}^{\pi} W^2(\xi)\, d\xi = \sum_{n=-(M-1)}^{M-1} w^2(n)$$

Therefore,

$$V\left[\tilde{I}(\omega)\right] = \frac{S_X^2(\omega)}{U^2}\left[\frac{1}{L}\sum_{n=-(M-1)}^{M-1} w^2(n)\right] \tag{6.16}$$

That is, the variance of the smoothed periodogram depends on the factor in the square bracket and for the *variance to be small, this factor should be less than unity.* To achieve this, the value of M and the shape of the window can be chosen. As the variance of smoothed periodogram and that of the periodogram differ by this factor, the advantage is decided by it.

The estimate of the PSD should be positive and this is valid with the unsmoothed periodogram as $W_B(\omega) > 0$ and hence its convolution with $S_X(\omega)$ does not result in negative values. However, for the smoothed periodogram, if windows like Hanning or Hamming is used, then as $W(\omega)$ can also take negative values,

$$E\left[\tilde{I}(\omega)\right] = S_X(\omega) * W(\omega)$$

cannot be assured to be positive. Hence, *the PSD estimate can take negative values also.*

Welch method: Averaging the windowed periodograms

This method is well suited to compute the spectral estimate directly using the FFT. Here, the data record is divided into $K = L / M$ segments each of M samples. The window is applied directly to the data segment before computing the periodogram. Further, due to windowing there is loss of data (approximately 50%) depending upon the type of window. In order to compensate for this loss, the data segments are overlapped by a factor corresponding to the loss. The data segment windowed by the window $w(n)$ is given by

$$x^{(i)}(n) = w(n)x(n + \lambda i), \quad 0 \le n \le M-1 \text{ and } 0 \le i \le K-1$$

λ is the number of samples ($\lambda \le M$) to be shifted to move from $(i-1)^{th}$ to i^{th} segment in the data record. For an overlap of $(M - \lambda)$ samples, the maximum number of segments K, available for averaging is integer part of $\left((L - M)/\lambda + 1\right)$. The ith periodogram is given by

$$I^{(i)}(\omega) = \frac{1}{MU}\left|\sum_{n=0}^{M-1} x^{(i)}(n)e^{-j\omega n}\right|^2, \; U = \frac{1}{M}\sum_{n=0}^{M-1} w^2(n)$$

The Welch spectrum estimate given by $\bar{I}_W(\omega)$ is

$$\bar{I}_W(\omega) = \frac{1}{K}\sum_{i=0}^{K-1} I^{(i)}(\omega) \tag{6.17}$$

The mean of this estimate is given by

$$E[\bar{I}_W(\omega)] = \frac{1}{2\pi}\int_{-\pi}^{\pi} S_X(\xi)\, W(\omega - \xi)\, d\xi \tag{6.18}$$

where,

$$W(\omega) = \frac{1}{MU}\left|\sum_{n=0}^{M-1} w(n)\, e^{-j\omega n}\right|^2 \tag{6.19}$$

Further, its variance is given by

$$V[\bar{I}_W(\omega)] \approx \frac{S_X^2(\omega)}{K} \tag{6.20}$$

As the overlapping increases the number of segments also increases approximately by a factor of two due to the fact that for many windows the data loss is approximately 50%, the variance improvement by a factor of about two is achieved over the non-overlapping segmented averaging.

An estimate of the spectrum of the given data can be computed either by computing the correlation and taking its Fourier transform, known as *correlogram method* or by *direct method* where it is obtained directly from the FT of the data. Though these two approaches are equivalent, it has to be noted that in the case of correlogram method, its variance depends on how many lags are considered. However, with the direct method, all lags up to half the data length are considered and this makes variance of the spectral estimate greater than that obtained by the correlogram method. This is due to the fact that the correlation at higher lags has larger variance compared to that of lower lags.

Example: (a) An AR process is generated by the difference equation

$$x(n) = 2.7377\,x(n-1) - 3.7476\,x(n-2) + 2.6293\,x(n-3) - 0.9224\,x(n-4) + \in(n)$$

where $\in(n)$ is a zero mean and unit variance white Gaussian noise.

Using data records of length $l = 128$, an ensemble of 50 spectrum estimates are computed. For Welch method a Hanning window of length l is used.

$$NRMSE(k) = 20 * \log_{10}\left[\frac{\sigma(k)}{S(k)}\right],\ 0 \le k \le \frac{N}{2} - 1 \tag{6.21}$$

where, $\sigma(k)$ is the root mean square error (RMSE)

The RMSE is given by

$$\sigma(k) = \left[\frac{1}{L}\sum_{i=1}^{L}\left\{S(k) - \hat{S}_i(k)\right\}^2\right]^{1/2} \quad 0 \le k \le \frac{N}{2} - 1 \tag{6.22}$$

$S(k)$ is the true spectrum and $\hat{S}_i(k)$ is the ith spectrum estimate, L is the number of trials.

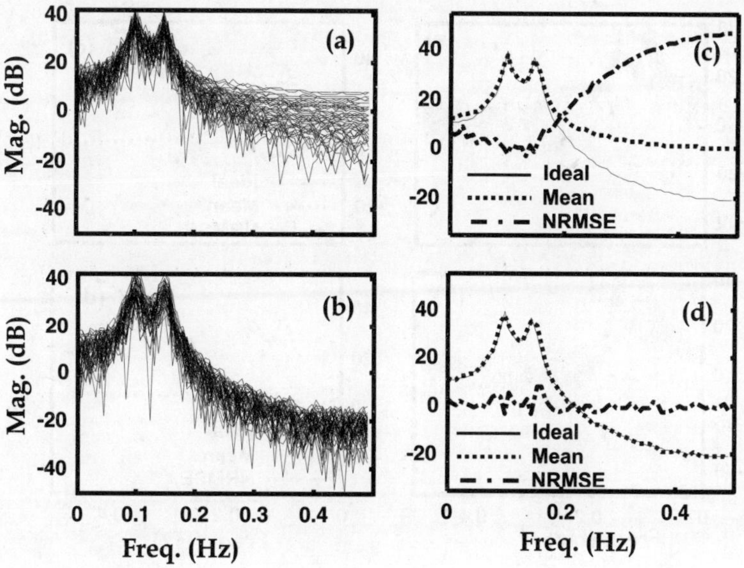

Fig. 6.2 Spectrum estimation of an AR process 50 estimates of : (a) Periodogram; (b) Welch method Ideal, mean, NRMSE ; (c) Periodogram and (d) Welch method.

For the periodogram, the estimates are more scattered or spread is more compared to that for the Welch method (Fig. 6.2 (a) and (b)) indicating that the variance is reduced. This is expected as the Hanning window used reduces the leakage effect. Further, for the Welch method, the mean spectrum follows the true spectrum better than that by the periodogram. This is reflected in the NRMSE curves as for the Welch method; its magnitude is less (Fig. 6.2 (c) and (d)). The parameter summed normalized root mean square error (SNRMSE) is

$$SNRMSE = \sum_{k=0}^{N/2-1} \frac{\sigma(k)}{S(k)}$$

For the Welch method this factor is 70 and for periodogram, 4000.

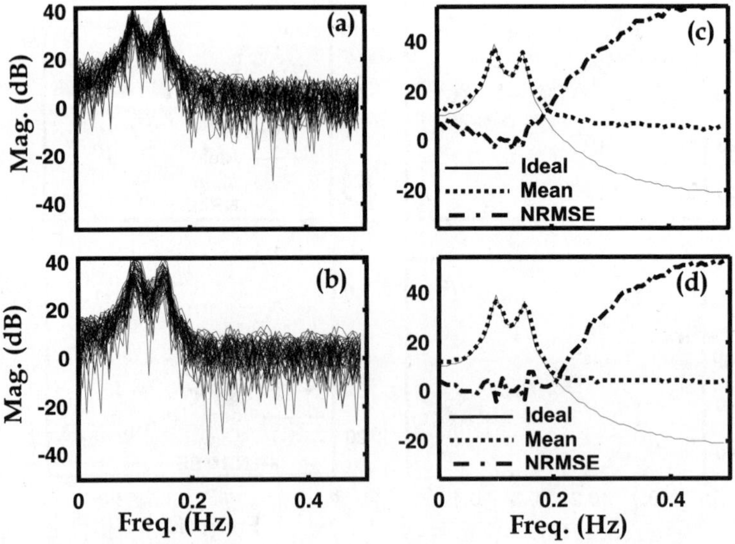

Fig. 6.3 Spectral estimation of AR process with noise, SNR =20dB. 50 estimates of : (a) Periodogram; (b) Welch method Ideal, mean, NRMSE; (c) Periodogram and (d) Welch method.

For the same AR process with an output SNR of 20 dB, curves obtained for periodogram and Welch method are shown in Fig. 6.3. In this case, SNRMSE for Welch and periodogram is 7000 and 10000 respectively.

6.3 SPECTRAL ESTIMATION BY MODIFIED MAGNITUDE GROUP DELAY (MMGD)

The variance of a spectral estimate is due to undesired fine structure. It is of interest to capture the spectral envelope and discard the fine structure without affecting the envelope. The ripples in the magnitude spectrum / the spikes in the group delay (negative derivative of phase spectrum) result in zeros close to the unit circle in the z-plane and contribute significantly to the variance of the spectral estimate. As discussed already, this ripple effect can be removed by using normal smoothing windows but only at the cost of frequency resolution. This is due to the fact that the smoothing window effect results in not only pulling the zeros close to unit circle but also the signal poles, towards the origin in the *z*-plane. The signal truncation effects, input white noise and associated white noise result in spectral ripples and contribute to the variance of the spectral estimate. Thus, the spectral ripples or variance is due to the zeros close to unit circle. Figure 6.4(a) shows the location of zeros for a white Gaussian noise of 50 samples and it is seen that most of them are close to unit circle. On the other hand, the signal/system spectral peaks are due to the denominator/poles. To reduce the variance of a spectral estimate, it is required to reduce the effect of the numerator/ zeros and this can be achieved by dividing the transfer function by an estimate of the numerator. However, in doing so, there can be a problem of singularity (division by zero). The group delay is defined as the negative derivative of the phase of the signal. Presently, the group delay domain, provides a platform to achieve this operation without any singularity as it involves only multiplication of the group delay function by the square of the spectral magnitude estimate of the numerator.

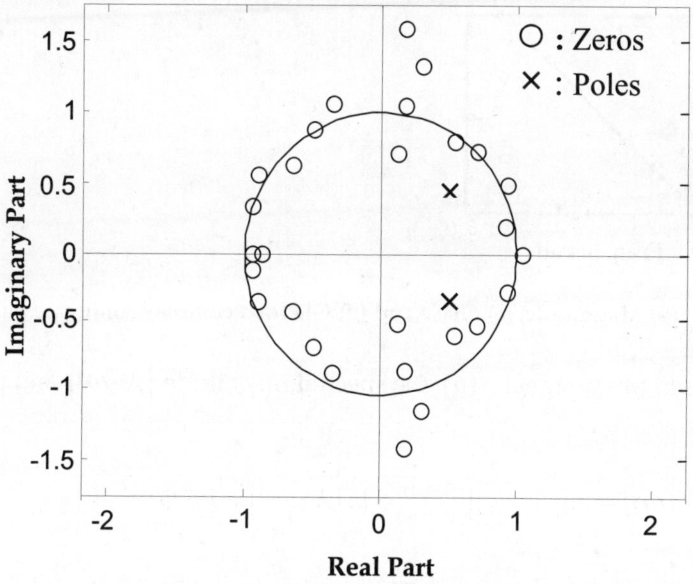

Fig. 6.4 (a) Location of zeros for a white Gaussian noise of 50 samples.

The phase transitions are fast for the roots located close to unit circle. Further, the phase transitions are comparatively faster than magnitude transitions for a given root location from the unit circle (Fig. 6.4(b and c) for a conjugate pair pole location $r = 0.85$ and $\theta = \pm\pi/4$). In view of this, the group delay(GD) is better localized and narrower than the magnitude spectrum (Fig. 6.4(b and d)) and provides better frequency resolution. That is, as the GD is localized better, can resolve (separate) the closely spaced zeros /poles better than the spectral magnitude. The significant values of GD for a zero/pole inside the unit circle are negative/positive. In other words, for a zero /pole inside the unit circle, the GD shows a dip/peak. Since the phases of two cascaded systems get added, their GDs also get added and the GD domain processing provides computational simplicity.

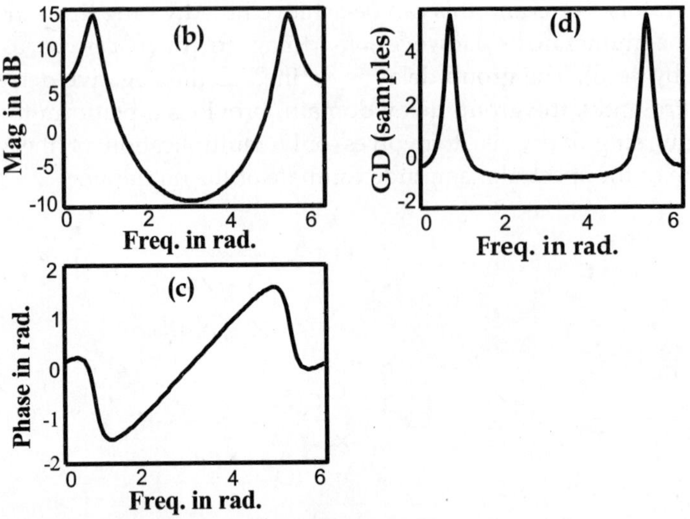

Fig. 6.4 (b) Magnitude; (c) phase and (d) GD for a complex conjugate pole pair.

For a minimum phase signal $x(n)$ the spectral magnitude $|X(\omega)|$ and phase $\theta(\omega)$ are given by.

$$\ln[X(\omega)] = \ln[|X(\omega)|e^{j\theta(\omega)} = \ln|X(\omega)| + j\theta(\omega) \tag{6.23a}$$

$$\ln[X(\omega)] = \sum_{n=0}^{\infty} c(n)e^{-j\omega n} = \sum_{n=0}^{\infty} c(n)\cos\omega n - j\sum_{n=0}^{\infty} c(n)\sin\omega n \tag{6.23b}$$

$c(n)$ are the cepstral coefficients of the signal $x(n)$. Therefore,

$$\ln|X(\omega)| = \sum_{n=0}^{\infty} c(n)\cos\omega\, n \tag{6.24a}$$

$$\theta\,(\omega) = -\sum_{n=0}^{\infty} c(n)\sin\omega\, n \tag{6.24b}$$

where $c(n)$ are the cepstral coefficients (In Eqn. (6.23a), $c(n)$ was derived from power spectral density which is the averaged squared spectral magnitude.) The group delay $\tau\,(\omega)$, the negative derivative of the phase $\theta\,(\omega)$, is given by

$$\tau(\omega) = -\frac{d}{d\omega}\theta(\omega) = \sum_{n=0}^{\infty} nc(n)\cos\omega n \tag{6.24c}$$

If $x(n)$ is a signal generated by an all-pole system driven by a white noise or sinusoids associated with white noise and further, if it can be represented by

$$X(\omega) = N(\omega) / D(\omega)$$

then, $D(\omega)$ corresponds to the system or sinusoids and $N(\omega)$ to the excitation or the associated noise. Its group delay function is given by

$$\tau(\omega) = \tau_N(\omega) - \tau_D(\omega) \tag{6.24d}$$

$\tau_N(\omega)$ and $\tau_D(\omega)$ are group delay functions of $N(\omega)$ and $D(\omega)$, respectively. This can be expressed as

$$\tau(\omega) = \frac{\alpha_N(\omega)}{|N(\omega)|^2} - \frac{\alpha_D(\omega)}{|D(\omega)|^2} \tag{6.24e}$$

$\alpha_N(\omega)$ and $\alpha_D(\omega)$ are the numerator in the expression for $\tau_N(\omega)$ or $\tau_D(\omega)$,

$$\tau_i(\omega) = \frac{P_R(\omega)Q_R(\omega) + P_I(\omega)Q_I(\omega)}{|P(\omega)|^2}, \; i = N \text{ or } D.$$

$P(\omega)$ and $Q(\omega)$ are the FTs of $p(n)$ and $q(n) = np(n)$, respectively. For the zeros close to unit circle due to signal truncation or input driving noise of a system or the associated noise with the signal, the $|N(\omega)|^2$ is small hence from Eqn. (6.24e) the magnitude of $\tau_N(\omega)$ will be very large compared to that of $\tau_D(\omega)$ as the signal poles are away from the unit circle. Hence, the ripples due to truncation or driving noise or the associated noise will mask the signal peaks in the group delay domain.

By multiplying $\tau(\omega)$ by $|N(\omega)|^2$, Eqn. (6) (6.24e) becomes

$$\tau_o(\omega) = \alpha_N(\omega) - \frac{\alpha_D(\omega)}{|D(\omega)|^2}|N(\omega)|^2 \tag{6.24f}$$

This modification reduces the masking effect of ripples due to the zeros close to the unit circle on the spectral peaks due to poles (B. Yegnanarayana and Hema A. Murthy, 1992). To get $\tau_o(\omega)$, which is free from fluctuations, $|N(\omega)|^2$ has to be determined with the given signal. It is important to preserve the values of $|N(\omega)|^2$ around the zeros so that it cancels the small values in the denominator of the first term in Eqn. (6.24e). Therefore, $|N(\omega)|^2$ should retain all the sharp fluctuations of the magnitude spectrum and should have *a flat envelope.* Since $|\hat{N}(\omega)|^2$, the estimate of $|N(\omega)|^2$ has *flat envelope* (constant), the second term of Eqn.(6.24e) will preserve its nature even after multiplying by $|N(\omega)|^2$ (Eqn. (6.26f)). The $|N(\omega)|^2$ estimate, $|\hat{N}(\omega)|^2$ is generally derived using a cepstrally smooth spectrum $|\tilde{X}(\omega)|^2$ obtained by truncating the cepstral coefficient sequence of the signal given by

$$|\hat{N}(\omega)|^2 = \frac{|X(\omega)|^2}{|\tilde{X}(\omega)|^2} \tag{6.24g}$$

The spectral estimate of reduced variance is derived from the modified group delay using the Eqns. (6.24c) and (6.24a). An extension of this to complex signals using group delay for complex signals, has also been investigated (S.V. Narasimhan, E. I. Plotkin and M. N. S. Swamy, 1999).

6.3.1 Spectral Estimation Based on DCT and MMGD

As explained in chapter 4, the DCT spectrum has low spectral leakage / magnitude bias. Further, the low leakage / magnitude bias enables or enhances DCT detectability of a low level spectral peak in the neighborhood of high level one. Also compared to the DFT, the DCT has a better frequency resolution and this enables DCT to resolve the closely spaced spectral peaks which is not possible by the former. Thus, from the spectral estimation point of view, the DCT is superior as it is having lesser bias both in spectral magnitude and frequency hence, a better detectability. However, as the frequency resolution is more, its variance can be more as it captures more spectral details.

To exploit the properties of the DCT for DFT, the DFT coefficients should be derived *directly* from DCT coefficients. The analytic DCT spectrum derived from DCT incorporates the desired properties of the DCT and this forms the desired new DFT. This is similar to (dual to) getting a one sided spectrum from a symmetric spectrum by an analytic signal. For a real signal the spectral magnitude is symmetric. For an analytic signal, the spectrum is one sided which implies that for a one sided signal, the spectrum is analytic. In DCT, the signal is symmetric. To get a one-sided signal from the DCT, its *analytic spectrum must be considered* (S.V. Narasimhan, and M. Harish, 2006). The analytic DCT spectrum, $C_x^a(k)$ is given by

$$C_x^a(k) = C_x(k) + j\,C_{xH}(k) \tag{6.25a}$$

where $C_{xH}(k)$: Hilbert transform of $C_x(k)$. If the desired DFT is represented as $X_d(k)$, then its magnitude $|X_d(k)|$ and phase $\phi_d(k)$ are given by

$$\left|X_d(k)\right| = \frac{1}{2}\left|C_x^a(k)\right| = \frac{1}{2}\sqrt{C_x^2(k) + jC_{xH}^2(k)} \tag{6.25b}$$

and

$$\phi_d(k) \;=\; \tan^{-1}\left[\frac{C_{xH}(k)}{C_x(k)}\right] \tag{6.25c}$$

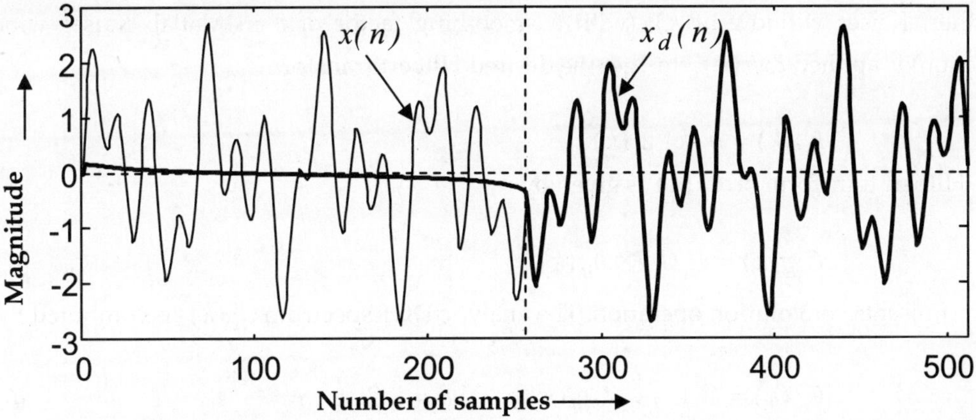

Fig. 6.5 Signal $x_d(n)$ derived from analytic DCT and the original signal $x(n)$.

A signal (made up of three sinusoids) and the signal derived from its analytic DCT spectrum are shown in Fig 6.5. This shows that the one sided signal can be recovered from the magnitude and phase of the analytic DCT spectrum. It is seen that the later half of the derived signal $x_d(n)$ is almost identical to the original signal $x(n)$ and its initial (symmetrical) part is almost zero. The Hilbert transform used in deriving the analytic DCT spectrum determines the accuracy of $x_d(n)$.

The magnitude of the analytic spectrum is the envelope of the DCT spectrum. That is, the DFT coefficients are the envelopes of the DCT coefficients and this is the reason for the frequency resolution of the DFT to be poorer than that of DCT. Further, the DFT coefficients so obtained from the DCT *enjoy the desired properties of the DCT,* viz., the reduced spectral leakage / reduced spectral magnitude bias and better detectability.

These desirable features of the DCT and above proposed new relation to get the equivalent DFT from the DCT have motivated to improve its variance performance using the MGD that has been described.

The procedure is listed in seven steps.

Step 1. Compute the DCT of the given signal, $C_x(k)$.

Step 2. Compute the analytic spectrum $C_x^a(k)$ using Eqn. (6.25a). This involves computation of the Hilbert Transform of the DCT, $C_x(k)$. The Hilbert transformer impulse response is given by

$$h(k) = \frac{2\sin^2[\pi(k - N/2)/2)]}{[(k - N/2)\pi]}, k = 1, 2, \ldots, N \tag{6.25d}$$

Further, Kaiser window $w(k)$ (with a smoothing factor of 8 in Matlab Kaiser window function) is applied to $h(k)$, to get the desired Hilbert transformer,

$$h_H(k) = h\,(k)\,w\,(k)\backslash$$

The Hilbert transform of $C_x(k)$ is given by

$$C_{xH}(k) = C_x(k) \otimes h_H(k) \tag{6.25e}$$

$\otimes$ represents convolution operation. The analytic DCT spectrum $C_x^a(k)$ is computed by

$$C_x^a(k) = C_x(k) + jC_{xH}(k), \qquad k = 1, 2, \ldots, N \tag{6.25f}$$

The accuracy of the recovered one-sided signal is decided by the quality of the Hilbert transformer used. By just making the FT spectral values for the negative frequency equal to zero and scaling the spectral values for positive frequency by a factor 2, the Hilbert transformed signal could have been computed. But once the FT is taken, there will be leakage from negative frequency region to the positive frequency region, which makes the further results erroneous.

Step 3. Find the magnitude of this analytic spectrum using Eqn. (6.25b)

$$|X_d(k)| = \frac{1}{2}|C_x^a(k)| = \frac{1}{2}\sqrt{C_x^2(k) + jC_{xH}^2(k)}\ , k = 1, 2, \ldots, N.$$

Step 4. Build the symmetric DFT magnitude spectrum $|X_{dS}(k)|$ using $|X_d(k)|$.

Step 5. Modified group delay (MGD) method of spectral estimation is applied to $|X_{ds}(k)|$ as follows:

(a) Compute the cepstral coefficients $c(n)$ using Eqn.(6.24a)

(b) Compute the group delay (GD) $\tau(\omega)$ using Eqn (6.24c)

(c) Compute the cepstrally smoothed spectrum,

$$\hat{\tilde{X}}_{ds}(k) = \ln\left|\tilde{X}_{ds}(k)\right| = \sum_{n=0}^{M} c\,(n)\cos\,(k\,n)\,, M \text{ can be in the range } (5\text{ -}10)$$

(d) Compute:

$$\ln \hat{N}(k) = \hat{X}_{ds}(k) - \hat{\tilde{X}}_{ds}(k)\ ,\hat{X}_{ds}(k) = \ln\left|X_{ds}(k)\right|$$

and $$\left|\hat{N}(k)\right|^2 = e^{2\ln \hat{N}(k)}$$

(e) Multiplying $\tau\,(\omega)$ by $\left|\hat{N}(k)\right|^2$ results in modified magnitude group delay function (MGD), $\tau_m(\omega)$.

Step-6. The improved log spectrum is derived from the MGD using Eqns. (6.24c) and (6.24a), i.e., by retracing the GD computation procedure in the reverse order. However, this log spectrum has to be properly scaled for its gain with reference to its original log spectrum to achieve proper match between them.

This spectral estimate obtained from DCT and MGD will have less leakage effect/spectral magnitude bias, lower variance, better frequency resolution (corresponding to main lobe width of a rectangular window) and better detectability than the one derived from either MGD based on DFT or only by DCT.

To illustrate the performance of the proposed DCTMGD, it is applied to sum of two sinusoids in noise and also to AR process associated with noise. For computing the mean and variance, 50 data segments are considered each with a data length of 256 samples. DFT size $N = 256$ and the DCT-II (symmetrical about half point) with $2N$ points are used in all the cases. In case of sinusoids, initial 3 cepstral coefficients are used for DFTMGD and 4 for DCTMGD to get cepstrally-smoothed spectrum. Initial 4 cepstral coefficients are used in case of AR process. DFT magnitude spectrum obtained from the DCT analytic spectrum is of $2N$ points.

For sinusoids with noise, the signal is given by

$$x(n) = \sqrt{10}\sin(0.2\pi\, n) + \sqrt{20}\sin(0.3\pi\, n) + u(n)\,,$$

where $u(n)$ is a zero mean white Gaussian noise.

One of the performance index used for comparison is the average variance per frequency bin (AVPFB) and is given by

$$AVPFB = \frac{1}{N}\sum_{k=1}^{N} V(k), \quad V(k) = \frac{1}{P}\sum_{i=1}^{P}\left[S_i(k) - S_m(k)\right]^2, \; S_m(k) = \frac{1}{P}\sum_{i=1}^{P}\left[S_i(k)\right]$$

$S_i(k)$ are the estimates of individual segments and P is the number of signal segments.

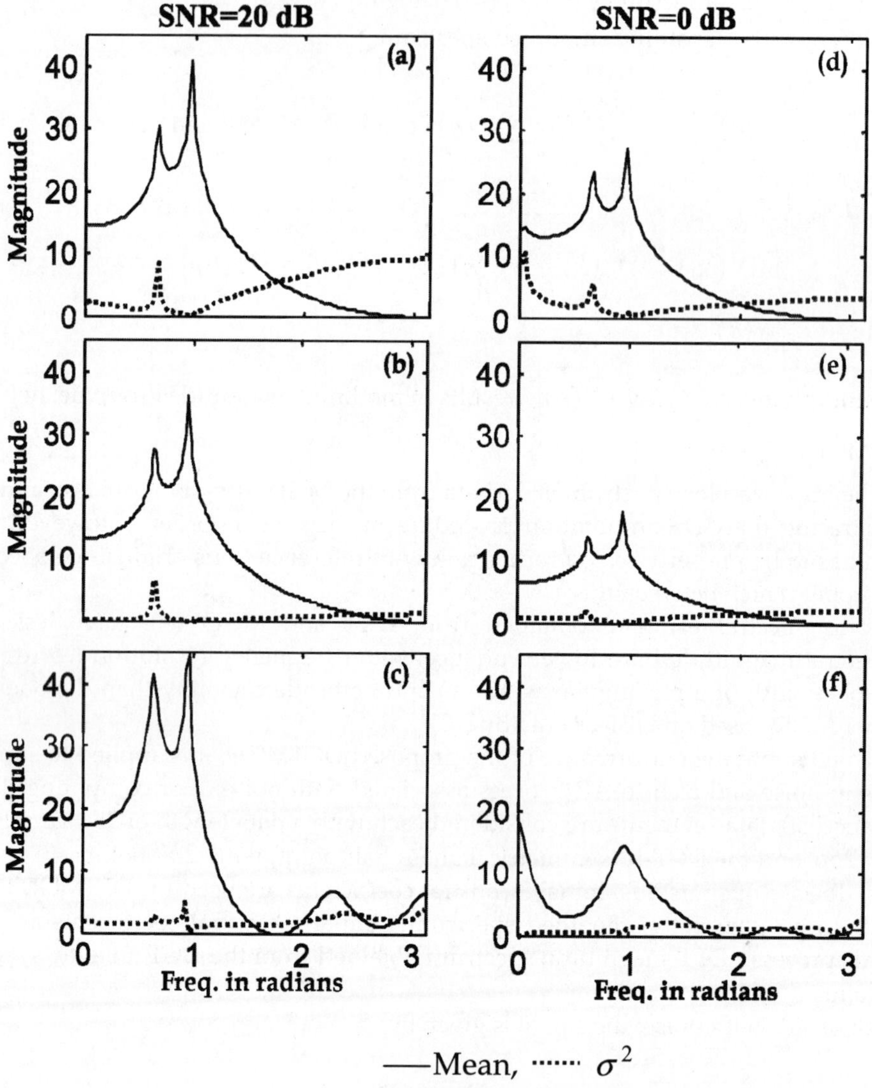

Fig. 6.6 (a) DFTMGD; (b) DCTMGD; (c) Burg estimate; (d) DFTMGD; (e) DCTMGD and (f) Burg estimate

Table 6.1: AVPFB by different methods for sinusoids

SNR in dB	DFTMGD	DCTMGD	Burg
20	0.0988 (0.1409)	0.0171 (80%) (0.0202)	0.0418 (56%)
0	0.0513 (Resolves)	0.0277 (46%) (Resolves)	0.0254 (50%) (Cannnot resolve)

For SNR=20dB, the results indicate that in terms of AVPFB, the proposed method DCTMGD has an improvement of 80% over that of the DFTMGD (Table-6.1, Figs. 6.6(a) and (b)). For SNR=0 dB, this improvement is 46% (Table 6.1, Figs. 6.6(c) & (d)).

It has been observed that there is no significant reduction in AVPFB by analytic DCT alone over DFT especially at a low SNR (less than 20dB).

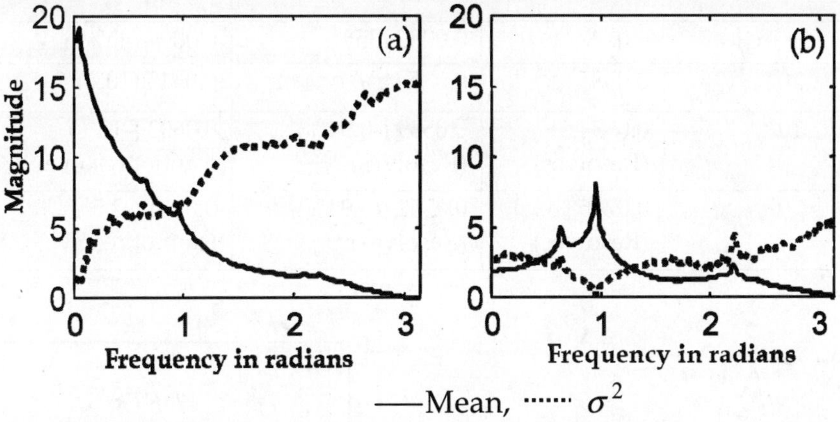

Fig. 6.7 (a) DFTMGD and (b) DCTMGD for sinusoids (SNR= - 13dB).

The performance of DCTMGD is compared with that of Burg method which is a parametric one. The parametric algorithm is based on a model (AR model) and the model parameters are chosen to minimize the error energy between the signal and model output (which does signal modeling or spectral modeling depending on the input to the model, whether it is impulse or white noise, respectively). Since the model parameters are found in time domain without going to frequency domain as in non parametric approach, it will not suffer from leakage and resolution limitations. Hence, the model based approach provides good frequency resolution and good reduction in variance (this will be dealt within detail in chapter 7). However, this model based approach suffers from poor immunity to noise. For a model order 8, at SNR=20 dB, the performance of model based Burg method in terms of AVPFB, is relatively poor compared to that of DCTMGD (Table 6.1 and Fig. 6.6c) but it is able to resolve the two sinusoids. However, at SNR = 0dB, Burg estimate is unable to resolve the spectral peaks though its AVPFB is less compared to that of DCTMGD (Fig. 6.6f, Table 6.1).

The results for SNR = –13dB, are shown in the Fig. 6.6 and the AVPFB is 64% better than the DFTMGD and the DCTMGD could locate and resolve the peaks properly (Fig. 6.7b). However, the DFTMGD is unable to bring out the spectral peaks properly (Fig. 6.6a). This is true even up to SNR = –17dB, whereas DFTMGD was unable to do so even below SNR = – 8 dB.

The AR process y(n) considered is given by

$$y(n) = u(n) + 2.76y(n-1) - 3.809y(n-2) + 2.654y(n-3) - 0.924y(n-4)$$

where $u(n)$ is zero mean white Gaussian noise.

Table 6.2: ARMSE by different methods for AR process

SNR in dB	DFTMGD	DCTMGD	Burg
Clean	0.0127	0.0082 (35%)	0.0094 (25%)
20	0.0385	0.037 (4%)	0.0345 (10%)
10	0.0534 (Resolves)	0.0539 (-0.94%) (Resolves)	0.0602 (-12%) (cannot resolve)
0	0.0745 (Resolves)	0.0752 (-0.94%) (Resolves)	0.0764 (-2.5%) (cannot resolve)

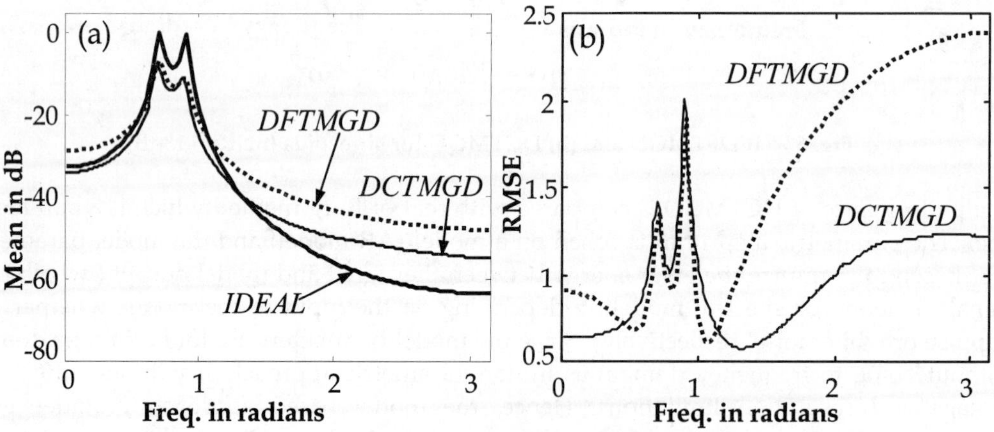

Fig. 6.8 Performance of the proposed algorithm for an AR process (a) Mean and (b) RMSE.

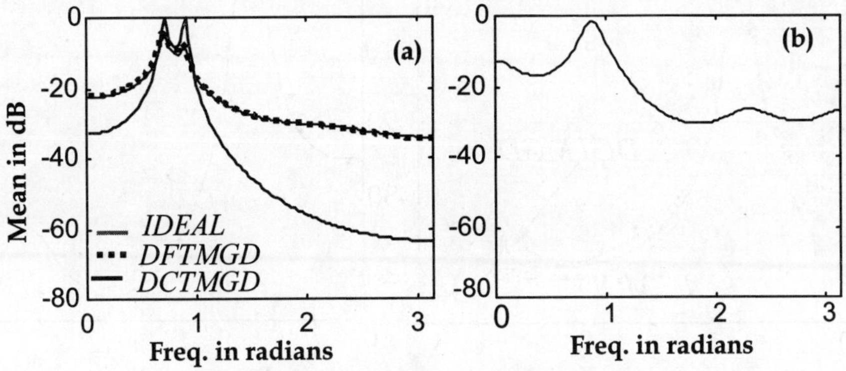

Fig. 6.9 For AR process at SNR=10 dB,
(a) DFTMGD, DCTMGD and (b) Burg estimate.

The average RMSE (ARMSE) is used as a performance index to evaluate the performance of the proposed method for the AR process and it is compared with those of DFTMGD and Burg method for different SNR values. The average RMSE per frequency bin, is defined as

$$\text{ARMSE} = (1/N)\sum_{i=1:}^{N}\sigma_i\,,\;\; \sigma_i \text{ is RMSE of } i^{th} \text{ frequency bin.}$$

For clean AR process, the improvement in ARMSE is 35% (Table 6.2). The DCTMGD spectrum is closer to the ideal compared to that by DFTMGD (Fig. 6.9(a)) indicating its low bias. Further, its ARMSE (Fig. 6.9b) is significantly less compared to that of DFTMGD. At SNR=20dB, the ARMSE (Fig. 6.9b) is significantly less compared to that of DFTMGD. At SNR=20dB, the percentage improvement in ARMSE for the DCTMGD over the DFTMGD is very marginal as it is only 4%.

Further, at SNR =10dB and below, the DFTMGD performs better than DCTMGD as the latter's ARMSE is inferior to that of the former though both are equally good in terms of frequency resolution as they resolve the spectral peaks (Fig. 6.9(a)). But the Burg method works well only when the SNR is 20dB as its ARMSE is poor (Table 6.2) and cannot resolve the peaks at SNR = 10 dB (Fig. 6.9(b)) and 0dB. Hence, at low SNR, the DFTMGD is preferred to the other two methods considered.

For a signal with the above AR process with white Gaussian noise (SNR = 20dB) plus a sinusoid located at 0.8π radians whose spectral peak is about 30dB below the AR peaks, the DCTMGD is able to detect the low level peak better than the DFTMGD (Fig. 6.10(a)). Burg method also detects the low level peak but with a broader peak (Fig. 6.10(b)) as the AR process acts as a noise and the model does not match the signal as far as this low level spectral peak is concerned. Hence, the DCTMGD not only detects the low level spectral peak *significantly better* than DFTMGD but also its *resolution is not affected* by the AR process as it happens for the Burg method.

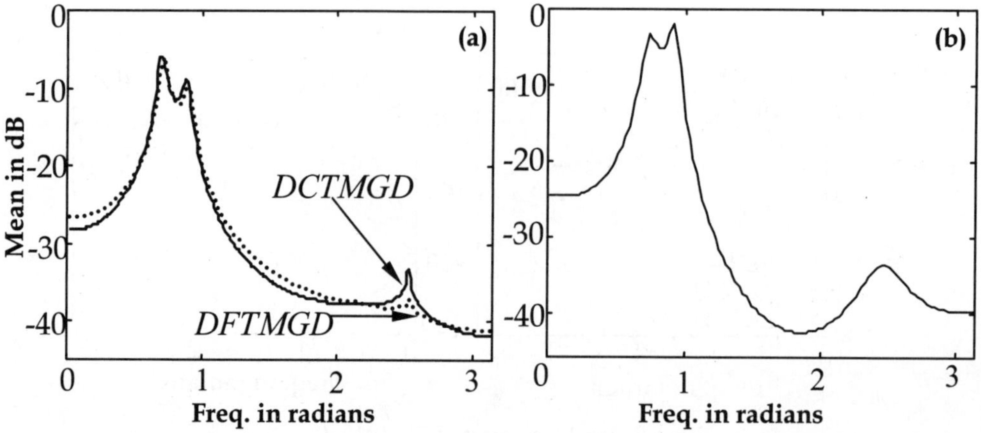

Fig. 6.10 Detectability of a – 30 dB low level sinusoid (at 0.8π radians) embedded in the AR process at SNR=20 dB. (a) DCTMGD and DFTMGD and (b) Burg estimate.

The analytic DCT -MMGD method is able to significantly reduce the bias due to low leakage of DCT and the variance without any loss of frequency resolution due to the ability of MGD to remove the ripples without any windowing. The low bias enables DCTMGD to detect low level peaks in the presence of a higher one.

For sinusoids, the DCTMGD is able to locate and resolve the spectral peaks even at SNR= -17dB while the DFTMGD is unable to do so below -8 dB. It has an improvement in variance measure (Average variance per frequency bin (AVPFB)) of about 80% for a signal to noise ratio (SNR) = 20dB and 46% for SNR=0dB, over those of DFTMGD.

For the clean autoregressive (AR) process, the root mean square error measure (average root mean square error (ARMSE)) reduces by 35% over that of DFTMGD. Further, it has an improved detectability as it brings out a low level (−30dB) sinusoidal peak significantly well in an AR process compared to that of the DFTMGD and Burg methods.

Though for sinusoids the proposed method has a good performance, for AR process at low SNR (less than 10dB), the DFTMGD performs better than the DCTMGD. Also its performance for the examples considered, is found to be better than that obtained by parametric Burg method.

The DCTMGD provides a significant reduction in bias and variance, better signal detection ability with a better frequency resolution; compared with those of based on DFT (DFTMGD). Further, in the presence of noise the performance of DCTMGD is better than parametric AR method like Burg method.

The present method of spectral estimation using modified group delay and analytic DCT, has been extended to 2D signals also using 2D group delay functions and 2D Hilbert transform (P. Sandeep, B.K. Shreyamsha Kumar and S.V. Narasimhan, 2013).

For the variance and bias reduction of the periodogram spectral estimate, use of multi-taper windows also has been explored (D.J. Thomson, 1982, B. Babadi and E.N. Brown,

2014). In this method, different windows are generated from a single window and they are orthogonal to each other. The spectra obtained by using such windows for the same segment of data are averaged to get the improved spectral estimate (Maria Hansson-Sandsten, 2013). Each of the windows derived, can have different properties like reducing the bias or variance, so that the overall spectral estimate has all the desired features at least to some extent.

It is of interest to have a good frequency resolution for the spectral estimate but at the same time to reduce its variance/ fine structure. This is a conflicting requirement. But it can be met to some extent by treating the spectrum which has finer details / higher variance as a signal and by denoising it (P. Moulin, 1994). In this process, the finer details correspond to noise and the gross features to the signal. The undesired finer details are removed as they contribute to the wavelet transform coefficient values, which are of small magnitude in higher scales. But the sharp spectral peaks correspond to the significant values of wavelet transform coefficients of lower scales. Thus, the spectrum reconstructed after denoising (removing the small values below a threshold in each scale) will be an improved one as it will be not only smoother but also retains the desired details (P. Moulin, 1994). Having removed the finer details of the spectra for each data segment, their number required for averaging will be considerably smaller than for the case of without denoising, to arrive at the same performance. Such a denoising is also carried out as a post-processing for a multi-tapered spectrum to reduce its variance.

6.4 SYSTEM PHASE ESTIMATION BY BISPECTRUM

The triple correlation/bispectrum was introduced in chapter 5 in connection with the degree of stationarity and it was stated that if triple correlation/bispectrum of a process is time invariant, its stationarity is of higher degree than that of the second order stationarity. For a *non-Gaussian white noise* input, $|B_x(\omega_1,\omega_2)| = \beta$, a constant and $\Psi_x(\omega_1,\omega_2) = 0$. Therefore, the magnitude and phase of the bispectrum of the output are (C.L. Nikias and Raghuveer M.R., 1987).

$$|B_y(\omega_1,\omega_2)| = \beta\,|B_H(\omega_1,\omega_2)| \quad \text{and} \quad \Psi_y(\omega_1,\omega_2) = \Psi_H(\omega_1,\omega_2)$$

In chapter 5, the system magnitude from the output bispectrum magnitude, was considered and is given by

$$|H(\omega_1)|^2 = \frac{B_y(\omega_1,0)}{|H(0)|}, \quad \text{for } |H(0)| \neq 0 \tag{6.26a}$$

and it was also stated that the system spectral magnitude can be found as well from the PSD of the output. However, it was stated that the system phase has to be extracted from the output bispectrum phase and is achieved by a simple group delay approach. Since the white noise does not contribute to the phase, the output bispectrum phase is equal to system bispectrum phase. Hence,

$$\psi(\omega_1,\omega_2) = \psi_H(\omega_1,\omega_2) = \phi\,(\omega_1) + \phi\,(\omega_2) - \phi\,(\omega_1 + \omega_2) \tag{6.26b}$$

where $\phi\,(\omega)$ is system phase and $H(\omega) = |H(\omega)|e^{j\phi(\omega)}$. The bispectral phase $\psi(\omega_1,\omega_2)$ contains information about the system phase $\phi\,(\omega)$ and extraction of this will be considered (S.V. Narasimhan., Reddy, G.R., Plotkin, E.I., Swamy, M.N.S, 1992) . Taking the derivative with respect to ω_2 and evaluating at $\omega_2 = 0$,

$$\frac{\partial}{\partial\,\omega_2}\left[\psi(\omega_1,\omega_2)\right] = \frac{\partial}{\partial\,\omega_2}\phi(\omega_2) - \frac{\partial}{\partial\,\omega_2}\phi(\omega_1+\omega_2)$$

$$\frac{\partial}{\partial\,\omega_2}\left[\psi(\omega_1,\omega_2)\right]\Bigg|_{\omega_2=0} = \frac{\partial}{\partial\,\omega_2}\phi(\omega_2)\Bigg|_{\omega_2=0} - \frac{\partial}{\partial\,\omega_2}\phi(\omega_1+\omega_2)\Bigg|_{\omega_2=0}$$

But,

$$\frac{\partial}{\partial\,\omega_2}\phi(\omega_1+\omega_2)\Bigg|_{\omega_2=0} = \frac{\partial}{\partial\,\omega_1}\phi(\omega_1+\omega_2)\Bigg|_{\omega_2=0} = \frac{\partial}{\partial\,\omega_1}\phi(\omega_1)$$

Therefore,

$$\psi'_{\omega_2}(\omega_1,0) = \phi'_{\omega_2}(0) - \phi'_{\omega_1}(\omega_1)$$

$$\phi'_{\omega_1}(\omega_1) = -[\psi'_{\omega_2}(\omega_1,0) - \phi'_{\omega_2}(0)]$$

$$\tau_{\omega_1}(\omega_1) = -\left[\,\tau_{\omega_2}(\omega_1,0) - \tau_{\omega_2}(0)\right]$$

$\tau_{\omega_2}(0)$ is a constant with respect to ω_1 and is the mean value of $\tau_{\omega_1}(\omega_1)$ corresponding to linear phase component (which results only time delay which does contribute to system phase). Hence, $\tau_{\omega_1}(\omega_1)$ can be found by removing the mean value $\tau_{\omega_2}(0)$ of $\tau_{\omega_2}(\omega_1,0)$ and further changing sign of resulting function. The required cepstrum sequence $c(n)$ is found by

$$\tau_{\omega_1}(\omega_1) = \sum_{n=0}^{\infty} nc\,(n)\cos\,(\omega_1\,n)$$

Hence, the phase is given by

$$\phi(\omega_1) = -\sum_{n=0}^{\infty} c(n) \sin(\omega_1 n) \tag{6.26c}$$

Therefore, from the phase and magnitude, the system $H(\omega)$ can be obtained by

$$H(\omega) = |H(\omega)| e^{j\phi(\omega)}$$

Thus, the bispectrum enables the phase recovery when the system phase is of non-minimum phase nature. Hence, the complete non-minimum phase system can be identified and hence the process can be deconvolved. Here, only one bispectral slice is used in extracting the system phase. But to get a system phase estimate with reduced variance, all bispectral slices have to be used (Holombe, R.S., Ray, A.K., Basu, T.K, 1996) instead of a single bispectral slice. Further, improving the system phase estimate variance and frequency resolution, have been achieved by using modified group delay approach (S.V. Narasimhan, B. Nandini and Chaitanya, 2008). In some situations, the complete bispectrum estimation is required and this has been achieved by using 2D-modified group delay function (Nandini Basumallick S. V. Narasimhan, 2011).

Chapter 7

PARAMETRIC SPECTRAL ESTIMATION

Some of the non-parametric spectral estimation methods for a given length of data, suffer from the trade-off between frequency resolution and variance. In other cases, the frequency resolution is decided by data length. The parametric methods alleviate the problem of frequency resolution dependence on data length. These methods assume that the data obeys a model and the model parameters can be used to extend the data beyond the available data and hence they provide better frequency resolution. Further, as the model fits a polynomial to the spectral data, it provides a smooth spectrum, thereby reducing the variance. As the data is processed in time domain confining to the available data and without going to frequency domain, the problems of frequency domain like leakage and limited frequency resolution are avoided. However, as the parametric models exploit the signal correlation, their performance is assured only when the signal to noise ratio is high.

Different methods of spectral estimation based on autoregressive (AR) model, like Yule: Walker, covariance, Burg Method and modified covariance method, warped Burg method (which compress the frequency axis there by requiring a lower model order) are considered. Further, these methods are realized by normal lattice which minimizes average of both forward and backward error energies. The lattice orthogonalizes the input which is useful for applications like system identification. But as the escalator lattice minimises only the backward (forward) error energy and hence its orthogonal decomposition is valid even for nonstationary signals.

In a spectrum to represent the valley region also in a better way in addition to peaks, different autoregressive moving average (ARMA) methods like Pade's approximation, Prony's method, Shank's method, Steiglitz-McBride methods which minimize error energy and some of them which make the error zero (for impulse response type of signals) and frequency domain method of pole-zero decomposition in group delay domain are described.

Further, the realization of ARMA models by lattice structure has been conceived by using IIR lattice for the denominator and FIR tapped delay line for the numerator Bispectrum estimation via AR modelling has been treated.

7.1 INTRODUCTION

In the last chapter, nonparametric method of spectral estimation was considered. It has been found that the periodogram and its variations suffer from limited frequency resolution for a given length of data and large variance. Further, there is a trade-off between variance and frequency resolution. To achieve better frequency resolution and small variance, the periodogram needs a stationary signal of long duration. Further, the periodogram by itself is not an efficient form to represent the spectral information contained in it as it forms a large data. *From the point of view of storage, transmission and data classification it is essential to represent the spectral information in an efficient way in parametric form.* That is, a model, described by a few parameters, represents the spectral information. The model due to its nature may provide not only the compact data representation but also a *better frequency resolution,* even for a signal of a short duration. However, the parametric methods as they are based mainly on correlation of the data, may not perform well in the presence of noise and the signal nature is not matched to that of the model. In the present chapter, the parametric spectral modeling is derived from the system identification view as this approach also forms the basis for sequential adaptive techniques.

The process of determining the complete characteristics of an unknown system from its output with input being known is called system identification. The characteristic of the unknown system can be quantified in terms of its impulse response. The method of least squares is one of the techniques for performing system identification. In this method, a model is assumed for the system and the parameters of the model that minimizes the error energy between the outputs of the actual system and that of the model, provides the optimal parameters of the model for identification. Generally, mean square error (MSE) is minimized. The MSE minimization with respect to each parameter yields a set of simultaneous equations known as normal equations. The fundamental property of minimum mean square error is that the error is orthogonal to the input and hence the name normal equations.

There are two specific methods of solving the normal equations for the optimal filter coefficients, each leading to a practical implementation using the acquired data. The first one is a block method, which uses Durbin's algorithm, and the second is a sequential one based on the method of steepest descent leading to its approximation namely, the least mean square adaptive algorithm.

For block processing, Durbin's algorithm provides an efficient method of solving the above system of equations, for the optimal solution. Further, Durbin's method provides a stage-by-stage or order recursive method. The most important benefit of Durbin's algorithm is that it leads to a stable computational linear filter structure known as the *lattice filter*. The lattice has advantages over the transversal filter of the linear predictor, especially when used in source modeling application.

Linear prediction is a special case of system identification. Autoregressive (AR) power spectrum estimation is an application of linear prediction. Linear prediction, as the name indicates enables prediction of future samples based on present samples. That is, the future samples are expressed as weighted sum of present samples. The weights/coefficients are selected such that the prediction error energy is minimized. The prediction coefficients form an IIR or AR model. An AR model whose coefficients are the coefficients of prediction error filter can approximate a wide-sense stationary process of zero mean. By exciting this AR model with a white noise sequence of zero mean and estimated variance equal to the average prediction error power, an approximate realization of the process can be generated. The coefficients of the model and the average prediction error power represent the results of a linear prediction analysis performed on samples of the process. The AR power spectrum estimate for the process represents an estimate of the power spectrum of a wide-sense stationary process, under the assumption that the process may be approximated by an AR model. Such a model offers insight into the structure of the process, such as the number of resonances, and produces a high resolution spectral estimate in a compact form represented by a few parameters. The model parameters can be extracted by various methods such as Yule-Walker, Burg, Covariance and Modified Covariance.

7.2 SYSTEM IDENTIFICATION

In Fig. 7.1, it is desired to learn the structure of an unknown system $P(z)$ from knowledge of its input $x(n)$ and output $d(n)$. The structure of the unknown system can be quantified in terms of its impulse response $p(n)$, which is time invariant.

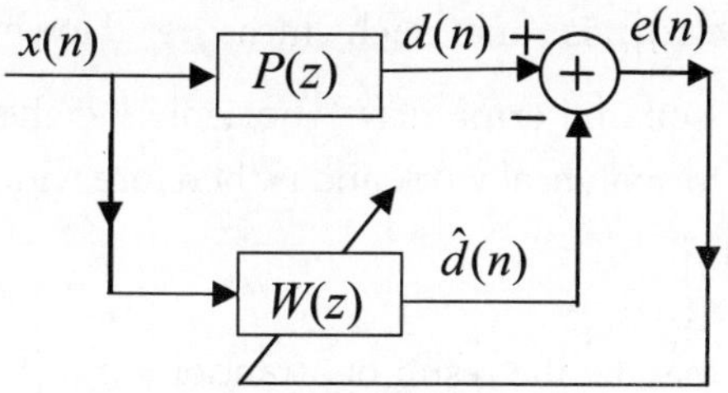

Fig. 7.1 System identification.

The output of the unknown system is given by,

$$d(n) = \sum_{k=0}^{N-1} p_k x(n-k) \tag{7.1}$$

where both the input signal $x(n)$ and the impulse response p have been assumed to be causal. It is also assumed that the p forms a finite impulse response of duration N samples. In system identification, the true impulse response p is unknown and must be obtained.

The estimate of $d(n)$, denoted as $\hat{d}(n)$, can be obtained by using a set of filter coefficients $w_k(n)$

$$\hat{d}(n) = \sum_{k=0}^{N-1} w_k(n)\, x(n-k) = W_N^T(n) X_N(n) \tag{7.2}$$

where

$$W_N^T(n) = [w_0(n), w_1(n), \cdots, w_{N-1}(n)],$$

$$X_N(n) = [x(n), x(n-1), \cdots, x(n-N+1)]^T$$

If each chosen $w_k(n)$ is close to each p_k, then the estimation error,

$$e(n) = d(n) - \hat{d}(n) \tag{7.3}$$

should be small in magnitude. In system identification, if $d(n) \approx \hat{d}(n)$, then $p_k \approx w_k(n)$. Therefore, minimising some measure of $e(n)$ should force $w_k(n)$ to approach p_k, thus identifying the unknown system.

7.3 MEAN SQUARE ERROR (MSE)

Generally, a large value of negative error is as detrimental as that of a positive error. Hence, some performance measure that weighs positive and negative errors equally is desirable. The absolute value of the error $|e(n)|$ is one such measure. However, this measure is mathematically not tractable. The squared error $e^2(n)$ not only weighs positive and negative errors equally but also lends itself to analytical work and is therefore widely used.

The squared error is given by ,

$$e^2(n) = [d(n) - \hat{d}(n)]^2$$

Often, the desired signal may be the result of a random or a stochastic process and in such a case the error signal is also of stochastic nature. Hence, the mean square error (MSE), mean of the squared error sequence is used as a performance measure.

$$\varepsilon(n) = E\{e^2(n)\} \tag{7.4}$$

where $E\{\ .\ \}$ is the mean or expectation operator.

7.4 LINEAR MINIMUM MSE FILTERING

Using the definition of the estimation error $e(n)$, the expression for the MSE becomes

$$\varepsilon(W_N) = E\{[d(n) - \hat{d}(n)]^2\}$$

Expanding

$$\varepsilon(W_N) = E\{d^2(n)\} - 2E\{d(n)\,\hat{d}(n)\} + E\{\hat{d}^2(n)\}$$

Using the definition of $\hat{d}(n)$,

$$\varepsilon\ (W_N) = E\{d^2(n)\} - 2W_N^T(n)E\{d(n)X_N(n)\} + E\left\{\left[W_N^T(n)X_N(n)\right]^2\right\}$$

$$\begin{aligned}\varepsilon\ (W_N) = E\{d^2(n)\} - 2W_N^T(n)E\{d(n)X_N(n)\}\\ + W_N^T(n)E\{X_N(n)X_N^T(n)\}W_N(n)\end{aligned} \quad (7.5)$$

The first expectation is simply the mean square power σ_d^2 of the desired signal and the second one is the cross-correlation of the desired response $d(n)$ and the input $x(n)$. Expressing the second expectation in vector form

$$P_N = E\{d(n)X_N(n)\} \quad (7.6)$$

Expanding

$$P_N = \begin{bmatrix} E\{d(n)x(n)\} \\ E\{d(n)x(n-1)\} \\ \vdots \\ E\{d(n)x(n-N+1)\} \end{bmatrix} = \begin{bmatrix} \phi_{xd}(0) \\ \phi_{xd}(1) \\ \vdots \\ \phi_{xd}(N-1) \end{bmatrix}$$

where $\phi_{xd}(m)$ are the cross-correlation coefficients given by $\phi_{xd}(m) = E\{x(n)\ d\ (n-m)\}$.

Generally, the true value of $\phi_{xd}(m)$ is not known and for an ergodic signal, it is estimated from the data using an estimator.

$$\hat{\phi}_{xd}(m) = \frac{1}{K}\sum_{i=0}^{K-|m|-1} x\ (n-i)\ d\ (n-|m|-i)$$

Similarly, the third expectation in Eqn. (7.5) is the autocorrelation matrix of the input $x(n)$. Denoting

$$R_{NN} = E\{X_N(n)X_N^T(n)\} \quad (7.7)$$

Expanding

$$R_{NN} = E\left\{\begin{bmatrix} x(n) \\ x(n-1) \\ \vdots \\ x(n-N+1) \end{bmatrix}\begin{bmatrix} x(n) & x(n-1) & \cdots & x(n-N+1) \end{bmatrix}\right\}$$

Therefore,

$$R_{NN} = \begin{bmatrix} \phi_x(0) & \phi_x(1) & \ldots\ldots & \ldots\phi_x(N-1) \\ \phi_x(1) & \phi_x(0) & \ldots\ldots & \ldots \quad \phi_x(N-2) \\ \vdots & & & \\ \phi_x(N-1) & \phi_x(N-2) & \ldots\ldots & \ldots\phi_x(0) \end{bmatrix}$$

where $\phi_x(m)$ is the mth autocorrelation coefficient of the stationary signal $x(n)$ given by

$$\phi_x(m) = E\{x(n)\, x(n-m)\}$$

and is estimated from the data using an estimator

$$\hat{\phi}_x(m) = \frac{1}{K} \sum_{i=0}^{K-|m|-1} x(n-i)\, x(n-|m|-i)$$

R_{NN} is called the autocorrelation matrix. Substituting these in Eqn. (7.5)

$$\varepsilon(W_N) = \sigma_d^2 - 2W_N^T P_N + W_N^T R_{NN} W_N \tag{7.8}$$

The MSE is a function of predictor filter coefficients W_N. Any component of W_N is allowed to change then the computed value of MSE also changes. For $N = 2$, the MSE corresponds to a surface and is called the MSE surface.

7.4.1 Properties of the MSE Surface

The most important property of the surface is that it has only one extremum and the extremum is a minimum point. For $N = 2$, the MSE corresponds to an error surface in a three-dimensional space and is called the MSE surface.

For $N = 2$, one characteristic of the surface at the minimum point is that, at this point the slope of the tangents to the surface in directions parallel to w_1 and w_2 axes must be zero. Mathematical formulation of Fig. 7.2 is that the curves are concave up, i.e., second derivative of $\varepsilon(W_N)$ w.r.t w_2 along the line $w_1 = w_1^*$, is positive. Therefore, for a minimum of the MSE surface, its first derivative evaluated at minimum point must be zero and second derivative must be positive.

$$\left[\frac{\partial}{\partial w_2}\varepsilon(W_N)\right]_{W_N = W_N^*} = 0 \tag{7.9}$$

$$\left[\frac{\partial^2}{\partial w_2^2}\varepsilon(W_N)\right]_{W_N = W_N^*} > 0 \tag{7.10}$$

Similarly,

$$\left[\frac{\partial}{\partial w_1}\varepsilon\,(W_N)\right]_{W_N = W_N^*} \quad 0, \qquad \left[\frac{\partial^2}{\partial w_1^2}\varepsilon\,(W_N)\right]_{W_N = W_N^*} > 0 \tag{7.11}$$

where

$$W_N = \left[w_1, w_2\right]^T \text{ and } W_N^* = \left[w_1^*, \; w_2^*\right]^T$$

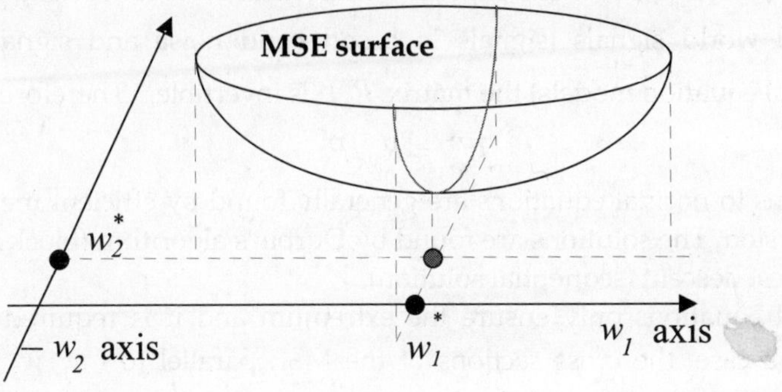

Fig. 7.2 A typical MSE surface with its cross sections with respect to w_1 and w_2

The Eqns. (7.9), (7.10) and (7.11) quantify the geometrical properties of MSE at its minimum. To show that the MSE surface $\varepsilon(W_N)$ satisfies these properties, expand Eqn. (7.8)

$$\varepsilon(W_N) = \sigma_d^2 - 2W_N^T P_N + W_N^T R_{NN} W_N$$

$$\varepsilon\,(W_N) = \varepsilon\,(w_1, w_2) = \sigma_d^2 - 2\,[w_1 \;\; w_2]\begin{bmatrix}\phi_{xd}(0)\\ \phi_{xd}(1)\end{bmatrix} + [w_1 \;\; w_2]\begin{bmatrix}\phi_x(0) & \phi_x(1)\\ \phi_x(1) & \phi_x(0)\end{bmatrix}\begin{bmatrix}w_1\\ w_2\end{bmatrix}$$

$$= \sigma_d^2 - 2w_1\,\phi_{xd}(0) - 2w_2\,\phi_{xd}(1) + w_1^2\,\phi_x(0) + 2w_1 w_2\,\phi_x(1) + w_2^2\,\phi_x(0)$$

Differentiating this w.r.t w_1 and w_2 and at $\left[w_1 \; w_2\right] = \left[w_1^* \; w_2^*\right]$,

$$2w_1^*\,\phi_x(0) + 2w_2^*\,\phi_x(1) - 2\phi_{xd}\,(0) = 0$$
$$2w_1^*\,\phi_x(1) + 2w_2^*\,\phi_x(0) - 2\phi_{xd}\,(1) = 0 \tag{7.12a,b}$$

or

$$\begin{bmatrix} \phi_x(0) & \phi_x(1) \\ \phi_x(1) & \phi_x(0) \end{bmatrix} \begin{bmatrix} w_1^* \\ w_2^* \end{bmatrix} = \begin{bmatrix} \phi_{xd}(0) \\ \phi_{xd}(1) \end{bmatrix}$$

$$R_{NN} W_N^* = P_N \tag{7.13}$$

These equations are often called the "normal equations". If the second-order statistics of the signal are known, the normal equations provide a theoretical basis for solving the optimal filter W_N^*. For most real world signals (signals in broad band noise and signals generated by stochastic differential equation models) the matrix R_{NN} is invertible. Therefore,

$$W_N^* = R_{NN}^{-1} P_N \tag{7.14}$$

The solutions to normal equations are generally found by efficient methods, which do not use matrix inversion. The solutions are found by Durbin's algorithm (block solution) and by the method of steepest descent (sequential solution).

The normal equations only ensure the extremum and it is required that it must be minimum. In such a case, the cross sections of the MSE parallel to (w_1, w_2) plane must be concave upwards. That is, the second derivative must be positive (Eqns. (7.10) and (7.11)).

Differentiating the LHS of the Eqns. (7.12a) and (7.12b) w.r.t w_1 and w_2, respectively

$$\left[\frac{\partial^2}{\partial w_1^2} \varepsilon(W_N)\right]_{W_N = W_N^*} = 2\phi_x(0), \qquad \left[\frac{\partial^2}{\partial w_2^2} \varepsilon(W_N)\right]_{W_N = W_N^*} = 2\phi_x(0)$$

Therefore, the requirement for the error surface to be concave up is that the diagonal elements $\varphi_x(0)$ of R_{NN} must be greater than zero. This is a valid assumption for practical signals, since

$$\phi_x(0) = E[x^2(n)] = \sigma_x^2$$

σ_x^2, the mean square value of the signal is positive. Therefore, error surface at extremum is concave up and is a minimum. That is, the *MSE surface is a bowl shaped curve with its minimum at its bottom.*

7.4.2 Minimum Mean Square Error

The mean square error is given by the Eqn. (7.8),

$$\varepsilon(W_N) = \sigma_d^2 - 2W_N^T P_N + W_N^T R_{NN} W_N$$

At minimum,

$$\varepsilon_{\min} = \varepsilon(W_N^*) \text{ and } W_N^* = R_{NN}^{-1} P_N$$

$$\varepsilon_{\min} = \sigma_d^2 - 2(R_{NN}^{-1}\ P_N)^T P_N + (R_{NN}^{-1}\ P_N)^T R_{NN}(R_{NN}^{-1}\ P_N)$$
$$= \sigma_d^2 - 2\ P_N{}^T (R_{NN}^{-1}\ P_N) + P_N{}^T\ (R_{NN}^{-1}\ P_N)$$
$$= \sigma_d^2 - 2P_N^T\ W_N^* + P_N^T\ W_N^*$$

$$\varepsilon_{\min}(W_N^*) = \sigma_d^2 - P_N^T\ W_N^* \tag{7.15}$$

If $d(n)$ and $x(n)$ are stationary signals with known cross-correlation properties, then using W_N^* as an estimation filter will produce an error sequence which has minimum power $\varepsilon_{\min}$. The optimal filter effectively removes the portions of the desired signal, which are correlated with the data signal $x(n)$. This reduces the error power by a factor $P_N^T W_N^*$ and this is the best a linear mean square error predictor can do, provided the signals $x(n)$ and $d(n)$ are stationary.

For the minimum mean square filtering to have any impact there must be non-zero correlation between the desired signal $d(n)$ and the input $x(n)$. However, the desired signal $d(n)$ and the data $x(n)$ themselves need not be correlated.

7.5 DATA PREDICTION AND ORTHOGONALITY

The normal equations can be derived from the fundamental property of minimum mean square error filters known as orthogonality. To derive these conditions, minimize MSE with respect to W_N:

$$\frac{\partial}{\partial W_N} E\left\{e^2(n)\right\} = E\left\{\frac{\partial}{\partial W_N} e^2(n)\right\} = E\left\{2e(n)\frac{\partial}{\partial W_N} e(n)\right\} = 0 \tag{7.16}$$

where the linearity of the gradient and expectation operators are used to interchange their order. Also,

$$\frac{\partial}{\partial W_N} e(n) = \frac{\partial}{\partial W_N}\left[d(n) - W_N{}^T X_N(n)\right] = -X_N(n)$$

Therefore,

$$E\left\{e(n)X_N(n)\right\} = 0$$

or

$$E\left\{e(n)x(n-i)\right\} = 0 \qquad 0 \le i \le N-1 \tag{7.17}$$

This set of conditions is known as orthogonality conditions and results from the minimization of the mean square error. It states that the expectation of the product of the estimation error with the data used in the estimation is zero. The orthogonality condition must be maintained by the minimum mean square error estimator irrespective of the manner in which the estimator coefficients are computed. The error and the data used in the estimator are uncorrelated.

The normal equations can be derived from the orthogonality condition:

$$E\{X_N(n)\, e\,(n)\}_{W_N=W_N^*} = 0$$

$$E\left\{X_N(n)\left[d(n) - X_N^{\,T}(n)\, W_N\right]\right\}_{W_N=W_N^*} = 0$$

$$E\{X_N(n)\, X_N^{\,T}(n)\}\, W_N^* = E\{X_N(n)\, d\,(n)\}$$

$$R_{NN}\, W_N^* = P_N$$

This is nothing but the solution obtained by minimizing the MSE with respect to the model parameters. Hence, mean square error minimization implies orthogonality between the input signal and the error and vice versa.

There are two specific methods of solving the normal equations for the optimal filter coefficients, each leading to a practical implementation using the acquired data. The first one is a block method, which uses Durbin's algorithm, and the second is a sequential one based on the method of steepest descent leading to its approximation namely, the least mean square (LMS) adaptive algorithm.

7.6 LINEAR PREDICTION

Linear prediction (LP) is a special case of system identification. The condition for orthogonality is

$$E\{e\,(n)\, x\,(n-i)\} = 0 \qquad 0 \le i \le N-1$$

where, N is the predictor order.

$$E\left\{\left[d(n) - \underbrace{\sum_{j=0}^{N-1} w_j x(n-j)}_{\downarrow\, X_N^{\,T} W_N}\right]\left[x(n-i)\right]\right\} = 0, \qquad 0 \le i \le N-1$$

$$X_N^{\,T} = \left[x(n)\; x\,(n-1) \ldots\ldots x(n-N+1)\right], \qquad W_N^{\,T} = \left[w_0\; w_1 \ldots\ldots w_{N-1}\right]$$

For linear prediction, where n^{th} sample $x(n)$ is predicted by N previous samples, $\left[x\,(n-1)\; x\,(n-2) \ldots x(n-N)\right]$, the orthogonality condition is

$$E\left\{\left[x\,(n) - \underline{\sum_{j=1}^{N} w_j x(n-j)}\right]\left[x\,(n-i)\right]\right\} = 0, \qquad 1 \le i \le N \tag{7.18}$$

$$\sum_{j=1}^{N} w_j E\{x(n-j)\,x(n-i)\} = E\{x(n)\,x(n-i)\}$$

$$\sum_{j=1}^{N} w_j \phi_x(j-i) = \phi_x(i), \quad 1 \le i \le N$$

where

$$\phi_x(j-i) = E\{x(n-j)\,x(n-i)\} \quad \text{and} \quad \phi_x(i) = E\{x(n)\,x(n-i)\}$$

For $N = 3$,

$$\begin{aligned} w_1\,\phi_x(0) + w_2\,\phi_x(1) + w_3\,\phi_x(2) &= \phi_x(1) \\ w_1\,\phi_x(-1) + w_2\,\phi_x(0) + w_3\,\phi_x(1) &= \phi_x(2) \\ w_1\,\phi_x(-2) + w_2\,\phi_x(-1) + w_3\,\phi_x(0) &= \phi_x(3) \end{aligned}$$

$$\begin{bmatrix} \phi_x(0) & \phi_x(1) & \phi_x(2) \\ \phi_x(-1) & \phi_x(0) & \phi_x(1) \\ \phi_x(-2) & \phi_x(-1) & \phi_x(0) \end{bmatrix} \begin{bmatrix} w_1 \\ w_2 \\ w_3 \end{bmatrix} = \begin{bmatrix} \phi_x(1) \\ \phi_x(2) \\ \phi_x(3) \end{bmatrix}$$

Generalizing,

$$\underbrace{\begin{bmatrix} \phi_x(0) & \phi_x(1) & \cdots & \phi_x(N-1) \\ \phi_x(-1) & \phi_x(0) & \cdots & \phi_x(N-2) \\ \vdots & \vdots & \vdots & \vdots \\ \phi_x(-(N-1)) & \phi_x(-(N-2)) & \cdots & \phi_x(0) \end{bmatrix}}_{\downarrow R_{NN}} \underbrace{\begin{bmatrix} w_1 \\ w_2 \\ \vdots \\ w_N \end{bmatrix}}_{\downarrow W_N} = \underbrace{\begin{bmatrix} \phi_x(1) \\ \phi_x(2) \\ \vdots \\ \phi_x(N) \end{bmatrix}}_{\downarrow = P_N}$$

Therefore, the solution for the prediction coefficients is given by

$$W_N = R_{NN}^{-1}\,P_N$$

7.6.1 Properties of the Autocorrelation Matrix

1. *The correlation matrix is symmetric, i.e.,* $R_{NN}^T = R_{NN}$.

If $x(n)$ is a widesense stationary process, then $\phi_x(m) = \phi_x(-m)$. Therefore, $R_{NN}^T = R_{NN}$. Also for an ergodic process $x(n)$ which is also stationary, the ensemble average can be replaced by time average and hence the autocorrelation estimate is given by

$$\phi_x(m) = \frac{1}{K}\sum_{n=0}^{K-1} x(n)\, x(n-m), \quad m = 0,1,\ldots, N$$

K is the total number of samples used in the estimate and $\phi_x(m) = \phi_x(-m)$.

2. *Correlation matrix is always non-negative definite, i.e., for any constant vector* a,

$$a^T R_{NN}\, a \geq 0$$

To prove this statement the required background is presented here.

Quadratic form: Quadratic form of a $N \times N$ matrix R_{NN} with respect to a nonzero vector a is defined as

$$Q_R(a) = a^T R_{NN}\, a$$

If the quadratic form of matrix R_{NN} is positive, $Q_R(a) > 0$, then R is said to be *positive definite* and is written as $R_{NN} > 0$.

If the quadratic form is non-negative, $Q_R(a) \geq 0$, then R_{NN} is said to be *positive semi-definite*. Similarly, if $Q_R(a) < 0$ and $Q_R(a) \leq 0$, then R_{NN} is said to be *negative definite and negative semi-definite, respectively. Further,* a matrix that is none of the above is *indefinite*.

For a random process input vector $X(n)$, the inner product $a^T X(n) = X^T(n)a$ is a scalar and random. Since mean square value of a random variable is non-negative,

$$E\left[\left(a^T X\right)\left(X^T a\right)\right] \geq 0 \text{ or } a^T E\left[XX^T\right] a \geq 0$$

$$XX^T = \begin{bmatrix} x(n) \\ x(n-1) \\ \vdots \\ x(n-N+1 \end{bmatrix} \left[x(n) \quad x(n-1) \quad \ldots \quad x(n-N+1)\right]$$

$$= \begin{bmatrix} x(n)x(n) & x(n)x(n-1) & \ldots & x(n)x(n-N+1) \\ x(n-1)x(n) & x(n-1)x(n-1) & \ldots & x(n-1)x(n-N+1) \\ \vdots & \vdots & \ddots & \vdots \\ x(n-N+1)x(n) & x(n-N+1)x(n-1) & \ldots & x(n-N+1)x(n-N+1) \end{bmatrix}$$

Taking expectation of XX^T,

$$R_{NN} = E\left[XX^T\right]$$

$$= \begin{bmatrix} E\left[x(n)x(n)\right] & E\left[x(n)x(n-1)\right] & \dots & E\left[x(n)x(n-N+1)\right] \\ \left[x(n-1)x(n)\right] & E\left[x(n-1)x(n-1)\right] & \dots & E\left[x(n-1)x(n-N+1)\right] \\ \vdots & \vdots & \ddots & \vdots \\ E\left[x(n-N+1)x(n\right]) & E\left[x(n-N+1)x(n-1)\right] & \dots & E\left[x(n-N+1)x(n-N+1)\right] \end{bmatrix}$$

Therefore,

$$a^T R_{NN}\ a \geq 0$$

That is, R_{NN} *is non-negative definite.*

For R_{NN} to be *positive definite,* it should satisfy

$$a^T R_{NN} a > 0$$

The positive definiteness ensures nonsingular nature of the matrix, i.e., the matrix is invertible. This is explained as follows: For a square matrix R_{NN} , its inverse is given by

$$R_{NN}^{-1} = \frac{adj\,[R_{NN}]}{Det\,[R_{NN}]}$$

$adj[R_{NN}]$ is the adjoint matrix of R_{NN}. If Φ_{ij} indicates the cofactor of φ_{ij}, the $(ij)^{th}$ element of R_{NN} in the determinant of the matrix R_{NN}, then $adj[R_{NN}]$ is a square matrix with its $(ij)^{th}$ elements, the cofactors Φ_{ij} and $Det[R_{NN}]$ is the determinant of the matrix R_{NN}. Therefore, for the inverse of a square matrix R_{NN} to exist, $Det[R_{NN}] \neq 0$. Thus, the determinant of the matrix determines whether the matrix is non-singular or singular.

A vector a to be a eigenvector of the square matrix A it should only get scaled by a factor when A operates on it. That is, X should satisfy,

$$Aa = \lambda\, a \quad \text{or} \quad \left[A - \lambda\, I\right] a = 0$$

The associated scale factor λ is the eigenvalue the matrix A.

For a nonzero vector a to be a solution of the above equation, the matrix $\left[A - \lambda I\right]$ must be singular. That is

$$P(\lambda) = Det\left[A - \lambda I\right] = 0$$

The polynomial $P(\lambda)$ is called the characteristic polynomial of matrix A. If A is a $(n \times n)$ square matrix, for the above equation there are n roots λ_k, $k = 1, 2, \dots n$, and also corresponding n vectors a_k, $k = 1, 2, \dots n$ known as eigenvectors of matrix A so that

$$Aa_k = \lambda_k a_k$$

The quadratic form of a real symmetric matrix A, $Q_A(a)$ for a nonzero vector a, i.e. $a \neq 0$ is

$$Q_A(a) = a^T Aa$$

Since for a vector to be eigenvector ,

$$Aa = \lambda a,$$

$$Q_A(a) = a^T Aa = \lambda a^T a = \lambda a^2$$

For positive definiteness,

$$Q_A(a) > 0 \text{ or } a^T Aa > 0$$

Further, since

$$\|a\|^2 > 0 \text{ as } a \neq 0, \ \lambda = \frac{a^T Aa}{a^T a} = \frac{a^T Aa}{\|a\|^2} > 0$$

That is, *if a matrix* A *is positive definite, its eigenvalues are positive.*

Conversely, if the eigenvalues are positive,

$$Q_A(a) = a^T Aa == \lambda a^T a = \lambda \|a\|^2,$$

then

$$Q_A(a) = a^T Aa > 0$$

That is , if the eigenvalues of A are positive, then A is positive definite.

Using the orthonormal transformation for R_{NN} (will be discussed in the next chapter),

$$R_{NN} = M \, \Lambda \, M^T$$

M and Λ are the matrix of eigenvectors or modal matrix and Λ matrix of eigenvalues of R_{NN}

$$\det[R_{NN}] = \det\left[M \, \Lambda \, M^T\right] = \det\left[M^T M \Lambda\right]$$

$$\left(\because \ \det[AB] = \det[BA] = \det[A]\det[B], A = M\Lambda, \quad B = M^T\right)$$

$$(\because M^T M = I)$$

$$\det\left[M^T M \Lambda\right] = \det[I\Lambda],$$

$$\det[R_{NN}] = \det[\Lambda] = \prod_{k=1}^{n} \lambda_k$$

Therefore, the positive nature of eigenvalues of a square matrix R_{NN} , implies,

$$\det[R_{NN}] = \prod_{k=1}^{n} \lambda_k > 0$$

This implies that the positive definiteness of a matrix *ensures its nonsingularity* since as already stated $\det[R_{NN}] = 0$ results in singularity.

A Toeplitz matrix has all its main diagonal elements same and this is true with respect to subdiagonals above and below the main diagonal also. Further, for a symmetric Toeplitz matrix, the corresponding subdiagonals above and below the main diagonal will be same due to symmetry. The autocorrelation matrix satisfies the properties of a symmetric Toeplitz matrix. *This Toeplitz property is due to stationarity of the process under consideration. If the correlation matrix is Toeplitz, it implies that it corresponds to a stationary process.*

Here also, as stated previously, the solution can be obtained by the Durbin's algorithm or by the method of steepest descent. If R_{NN} is symmetric Toeplitz, then Durbin's algorithm provides an efficient method of solving the above system of equations for the optimal solution W_N^*. It requires only $O(N^2)$ instead of $O(N^3)$ arithmetic operations required for solving by the solution of a system of equations method. Further, Durbin's method provides a stage-by-stage or order recursive method, in which an optimal solution for the pth order predictor may be computed from the knowledge of the optimal solution of the $(p-1)^{th}$ order predictor. The most important benefit of Durbin's algorithm is that it leads to a stable computational linear filter structure known as the lattice filter. The lattice has advantages over the transversal filter of the linear predictor, especially when used in source modeling application. The Durbin's algorithm computes a set of reflection coefficients k_p, where $1 \le p \le N$ which are used in the lattice structure.

7.6.2 Durbin's Algorithm

Durbin's algorithm is based on the Toeplitz property of the autocorrelation matrix. In the linear prediction (LP) mode, for a stationary signal, the autocorrelation matrix satisfies this property. In this algorithm, the reflection coefficients provide lattice as an alternative structure for LP filter and also parameterize the information contained in the signal, which is useful for further processing. However, the covariance method does not guarantee Toeplitz property as the elements of the covariance matrix are given by

$$\phi_x(i,j) = \frac{1}{K}\sum_{n=N+1}^{K} x(n-i)x(n-j), \qquad i, j = 1, 2, \ldots, N$$

But the covariance method is accurate as no assumption regarding the data outside the window as in autocorrelation that assumes the data is zero outside the window. However, the covariance method may suffer from matrix ill-conditioning problems, i.e., the covariance matrix may become singular.

Given R_{NN} is Toeplitz, the Durbin's algorithm is derived as below.

Suppose $1 \le p \le N$ and the $(p-1)^{th}$ order prediction is known. Then the normal equations for the $(p-1)^{th}$ order are

$$R_{p-1,p-1} W_{p-1}^{p-1} = P_{p-1} \tag{7.19}$$

where

$$R_{p-1,p-1} = \begin{bmatrix} \phi_x(0) & \phi_x(1) & \cdots & \phi_x(p-2) \\ \phi_x(1) & \phi_x(0) & \cdots & \phi_x(p-3) \\ \vdots & \vdots & & \vdots \\ \phi_x(p-2) & \phi_x(p-3) & \cdots & \phi_x(0) \end{bmatrix}, \quad P_{p-1} = \begin{bmatrix} \phi_x(1) \\ \phi_x(2) \\ \vdots \\ \phi_x(p-1) \end{bmatrix}$$

and the $(p-1)^{th}$ order predictor W_{p-1}^{p-1} is given by

$$W_{p-1}^{p-1} = \left[w_1^{(p-1)}, w_2^{(p-1)} \ldots w_{p-1}^{(p-1)} \right]^T$$

The parenthetical subscript signifies that the coefficient elements are for $(p-1)^{th}$ order vector. The reversed vectors corresponding to W_{p-1}^{p-1} and P_{p-1} are defined as

$$\overline{W}_{p-1}^{p-1} = \left[w_{p-1}^{(p-1)} \; w_{p-2}^{(p-1)} \ldots w_1^{(p-1)} \right]^T$$

$$\overline{P}_{p-1} = \left[\phi_x(p-1), \phi_x(p-2) \ldots \phi_x(2), \phi_x(1) \right]^T$$

Eqn. (7.19) can be written with reversed vectors

$$R_{p-1,p-1} \overline{W}_{p-1}^{p-1} = \overline{P}_{p-1}$$

Also from Eqn. (7.19)

$$W_{p-1}^{p-1} = R_{p-1,\,p-1}^{-1} P_{p-1} \tag{7.20a}$$

$$\overline{W}_{p-1}^{p-1} = R_{p-1,\,p-1}^{-1} \overline{P}_{p-1} \tag{7.20b}$$

The strategy is to derive a method in which optimize p^{th} order predictor by writing it as a function of the $(p-1)^{th}$ order predictor, which is assumed to be known. For this, the pth order normal equations have to be partitioned.

$$\left[\begin{array}{cccc|c} \phi_x(0) & \phi_x(1) & \cdots & \phi_x(p-2) & \phi_x(p-1) \\ \phi_x(1) & \phi_x(0) & \cdots & \phi_x(p-3) & \phi_x(p-2) \\ \vdots & \vdots & \vdots & \vdots & \vdots \\ \phi_x(p-2) & \phi_x(p-3) & \cdots & \phi_x(0) & \phi_x(1) \\ \hline \phi_x(p-1) & \phi_x(p-2) & \cdots & \phi_x(1) & \phi_x(0) \end{array}\right] \left[\begin{array}{c} w_1^{(p)} \\ w_2^{(p)} \\ \vdots \\ w_{p-1}^{(p)} \\ \hline w_p^{(p)} \end{array}\right] = \left[\begin{array}{c} \phi_x(1) \\ \phi_x(2) \\ \vdots \\ \phi_x(p-1) \\ \hline \phi_x(p) \end{array}\right]$$

$$\left[\begin{array}{c|c} R_{p-1,p-1} & \overline{P}_{p-1} \\ \hline \overline{P}_{p-1}^{T} & \phi_x(0) \end{array}\right]\left[\begin{array}{c} W_{p-1}^{p} \\ \hline w_p^{(p)} \end{array}\right] = \left[\begin{array}{c} P_{p-1} \\ \hline \phi_x(p) \end{array}\right] \tag{7.21}$$

This matrix portioning is based on the symmetric Toeplitz structure of the autocorrelation matrix.

Let $w_p^{(p)} = k_p$, k_p is called p^{th} reflection coefficient, W_{p-1}^{p} is a $(p-1)$ length vector consisting of the first $(p-1)$ coefficients of the pth order filter. Determination of k_p is most important part of this approach to linear prediction and its computation provides main advantage of using Durbin's algorithm.

It should be noted that at p^{th} iteration of the algorithm, k_p and W_{p-1}^{p} completely specify the pth order optimal filter.

$$W_p^p = \left[\begin{array}{c} W_{p-1}^{p} \\ \hline k_p \end{array}\right] \quad \text{and} \quad P_p = \left[\begin{array}{c} P_{p-1} \\ \hline \phi_x(p) \end{array}\right]$$

Hence, solution for k_p and W_{p-1}^{p} must be obtained, from Eqn. (7.21)

Vector equation $\quad R_{p-1,p-1}W_{p-1}^{p} + k_p\overline{P}_{p-1} = P_{p-1}$ (7.21a)

Scalar equation $\quad \overline{P}_{p-1}^{T}W_{p-1}^{p} + \phi_x(0)k_p = \phi_x(p)$ (7.21b)

Eqn. (7.21) represents p equations in p unknowns W_{p-1}^{p} and k_p.

To solve these equations, premultiply Eqn. (7.21a) by $R_{p-1,p-1}^{-1}$

$$W_{p-1}^{p} + k_p\, R_{p-1,p-1}^{-1}\, \overline{P}_{p-1} = R_{p-1,p-1}^{-1}\, P_{p-1}$$

$$W_{p-1}^{p} = R_{p-1,p-1}^{-1}\, P_{p-1} - k_p\, R_{p-1,p-1}^{-1}\overline{P}_{p-1}$$

Using $\quad W_{p-1}^{p-1} = R_{p-1,p-1}^{-1}\, P_{p-1}$

$$\overline{W}_{p-1}^{p-1} = R_{p-1,p-1}^{-1}\, \overline{P}_{p-1}$$

$$W_{p-1}^{p} = W_{p-1}^{p-1} - k_p\overline{W}_{p-1}^{p-1} \tag{7.22}$$

This is known *as Levinson recursion relation.* That is, the $(p-1)$ coefficients of the pth order filter are simply a linear combination of the elements of the $(p-1)^{th}$ order optimal filter. The scaling factor controlling the amount of coefficient change in updating from $(p-1)^{th}$ order solution to p^{th} order solution is the reflection coefficient k_p, which is still unknown. Eqn. (7.22) explicitly brings out the importance of reflection coefficient in linear prediction.

To solve for k_p, in Eqn.(7.21b))

$$\bar{P}_{p-1}^{T}\ W_{p-1}^{p} + \phi_x(0)k_p = \phi_x(p),$$

Substitute for W_{p-1}^{p} from the Levinson recursion (Eqn. (7.22))

$$W_{p-1}^{p} = W_{p-1}^{p-1} - k_p \overline{W}_{p-1}^{p-1},$$

$$\bar{P}_{p-1}^{T} W_{p-1}^{p} + \phi_x(0)k_p = \phi_x(p)$$

$$\bar{P}_{p-1}^{T}\left[W_{p-1}^{p-1} - k_p \overline{W}_{p-1}^{p-1}\right] + k_p\phi_x(0) = \phi_x(p)$$

Here all parameters except k_p are assumed to be known. Therefore,

$$k_p\left[\phi_x(0) - \bar{P}_{p-1}^{T}\overline{W}_{p-1}^{p-1}\right] = \phi_x(p) - \bar{P}_{p-1}^{T}W_{p-1}^{p-1}$$

Since

$$\bar{P}_{p-1}^{T}\overline{W}_{p-1}^{p-1} = P_{p-1}^{T}W_{p-1}^{p-1}$$

$$k_p\left[\phi_x(0) - P_{p-1}^{T}\ W_{p-1}^{p-1}\right] = \phi_x(p) - \bar{P}_{p-1}^{T}\ W_{p-1}^{p-1}$$

From Eqn. (7.15)

$$\varepsilon_{\min}(W_N^*) = \sigma_d^2 - P_N^T\ W_N^* \tag{7.23}$$

Therefore, the term in the square bracket is the mean square prediction error ε_{p-1} obtained using the $(p-1)^{th}$ order minimum MSE forward prediction filter, which is given by

$$\varepsilon_{p-1} = \phi_x(0) - P_{p-1}^{T}\ W_{p-1}^{p-1}$$

Therefore, this can be written as

$$k_p = \frac{1}{\varepsilon_{p-1}}\left[\phi_x(p) - \bar{P}_{p-1}^{T}\ W_{p-1}^{p-1}\right] \tag{7.24}$$

Although, ε_p can be computed for each new order p using Eqn. (7.23), a simple recursion can be derived.

Recursion for ε_p *:* From Eqn. (7.23),

$$\varepsilon_p = \phi_x(0) - P_p^T W_p^p$$

From Eqn.(7.21),

$$P_p^T W_p^p = [P_{p-1}^T \ \phi_x(p)] \begin{bmatrix} W_{p-1}^p \\ w_p^{(p)} \end{bmatrix}$$

$$P_p^T W_p^p = P_{p-1}^T W_{p-1}^p + \phi_x(p) w_p^{(p)}$$

Therefore,

$$\varepsilon_p = \phi_x(0) - P_{p-1}^T W_{p-1}^p - \phi_x(p) w_p^{(p)}$$

Since

$$W_{p-1}^p = W_{p-1}^{p-1} - k_p \overline{W}_{p-1}^{p-1}$$

$$\varepsilon_p = \phi_x(0) - P_{p-1}^T \left[W_{p-1}^{p-1} - k_p \overline{W}_{p-1}^{p-1} \right] - K_p \phi_x(p)$$

$$\varepsilon_p = \varepsilon_{p-1} - k_p \left[\phi_x(p) - P_{p-1}^T \overline{W}_{p-1}^{p-1} \right] \qquad (7.25)$$

Substituting for $\phi_x(p)$ from Eqn. (7.21b)

$$\overline{P}_{p-1}^T W_{p-1}^p + \phi_x(0) k_p = \phi_x(p),$$

the term $\phi_x(p) - P_{p-1}^T \overline{W}_{p-1}^{p-1}$ becomes

$$\phi_x(0) k_p + \overline{P}_{p-1}^T W_{p-1}^p - P_{p-1}^T \overline{W}_{p-1}^{p-1}$$

Substituting for W_{p-1}^p from Eqn.(7.22)

$$W_{p-1}^p = W_{p-1}^{p-1} - k_p \overline{W}_{p-1}^{p-1},$$

$$\phi_x(0) k_p + \overline{P}_{p-1}^T \left[W_{p-1}^{p-1} - k_p \overline{W}_{p-1}^{p-1} \right] - P_{p-1}^T \overline{W}_{p-1}^{p-1}$$

$$= k_p \left[\phi_x(0) - \overline{P}_{p-1}^T \overline{W}_{p-1}^{p-1} \right], \qquad \because \overline{P}_{p-1}^T W_{p-1}^{p-1} = P_{p-1}^T \overline{W}_{p-1}^{p-1}$$

$$= k_p \ \varepsilon_{p-1}, \qquad \because \varepsilon_p = \phi_x(0) - P_p^T W_p^p$$

Therefore, Eqn. (7.25)

$$\varepsilon_p = \varepsilon_{p-1} - k_p\left[\phi_x(p) - P_{p-1}^T \overline{W}_{p-1}^{p-1}\right]$$

becomes

$$\varepsilon_p = \varepsilon_{p-1} - k_p^2\, \varepsilon_{p-1}$$

$$\varepsilon_p = \varepsilon_{p-1}\left[1 - k_p^2\right] \tag{7.26}$$

All the quantities required to solve the normal equations have been computed. The steps for the Durbin's algorithm are given below.

1. Initialize

$$\varepsilon_0 = \phi_x(0)$$

2. For $1 \le p \le N$

$$k_p = \frac{1}{\varepsilon_{p-1}}\left[\phi_x(p) - \sum_{j=1}^{p-1} w_j^{(p-1)} \phi_x(p-j)\right], \quad w_p^{(p)} = k_p$$

3. For $1 \le j \le p-1$

$$w_j^{(p)} = w_j^{(p-1)} - k_p w_{(p-j)}^{(p-1)}$$

$$\varepsilon_p = \varepsilon_{p-1}\left[1 - k_p^2\right]$$

At the end of iteration $p = N$, the optimum linear prediction filter is given by

$$w_j^{*(N)} = w_j^{(N)} \qquad 1 \le j \le N$$

7.6.3 Lattice Derivation

For linear prediction, the Lattice filter is an alternative to the transversal filter W_N. The lattice filter has advantages like low round off noise in fixed word length implementation, relatively insensitive to quantisation noise.

Linear prediction filter is used to predict the present sample $x(n)$ by N previous samples $x(n-1), x(n-2), \cdots, x(n-N)$. The prediction is given by,

$$\hat{x}(n) = X_N^T(n)W_N^* = \sum_{i=1}^{N} w_i^{(N)} x(n-i)$$

The prediction error is given by,

$$e_N^f(n) = x(n) - \sum_{i=1}^{N} w_i^{(N)} x(n-i) \tag{7.27}$$

The error $e_N^f(n)$ is called the forward prediction error (FPE), since the predicted value $\hat{x}(n)$ is of a sample $x(n)$ that is forward in time with respect to data used in the prediction. Eqn. (7.27) defines a filter with scalar input $x(n)$ with a scalar output that is the error $e_N^f(n)$. This filter may be considered N^{th} order error filter. For the input $x(n)$, the filter output is only that portion of $x(n)$ that is unpredictable using the Nth order predictor. The set of prediction coefficient in Eqn. (7.27) that produces $e_N^f(n)$ is called FPE filter.

Since in Eqn. (7.27), the discrete input and discrete output are related, taking Z - transform,

$$E_N^f(Z) = \left[1 - \sum_{i=1}^{N} w_i^{(N)} Z^{-i}\right] X(Z)$$

Hence, the Nth order FPE filter transfer function is given by

$$G_N^f(Z) = \frac{E_N^f(Z)}{X(Z)} = 1 - \sum_{i=1}^{N} w_i^{(N)} Z^{-i} \tag{7.28a}$$

Since, Eqn. (7.28a) is for any general filter of order N, there is no loss of generality if N is replaced by any integer p, where $1 \le p \le N$. Hence, the pth order FPE is given by

$$G_p^f(Z) = 1 - \sum_{j=1}^{p} w_j^{(p)} Z^{-j} \tag{7.28b}$$

The first $p-1$ elements of the pth order filter can be written in terms of the previously computed prediction filter W_{p-1} using the Eqn. $W_{p-1}^{p} = W_{p-1}^{p-1} - k_p \bar{W}_{p-1}^{p-1}$

Hence,

$$w_j^{(p)} = w_j^{(p-1)} - k_p w_{p-j}^{(p-1)} \qquad 1 \le j \le p-1$$

To derive the pth order FPE filter in terms of $(p-1)^{th}$ order FPE filter, the above relation can be used.

$$G_{p-1}^{f}(Z) = 1 - \sum_{j=1}^{p-1} w_j^{(p-1)} Z^{-j}$$

$$G_p^{f}(Z) = 1 - \sum_{j=1}^{p-1} w_j^{(p)} Z^{-j} - w_p^{(p)} Z^{-p}$$

$$G_p^{f}(Z) = 1 - w_p^{(p)} Z^{-p} - \sum_{j=1}^{p-1} \left[w_j^{(p-1)} - k_p w_{p-j}^{(p-1)} \right] Z^{-j}$$

$$= \underline{1 - \sum_{j=1}^{p-1} w_j^{(p-1)} Z^{-j}} + \sum_{j=1}^{p-1} \left[k_p w_{p-j}^{(p-1)} \right] Z^{-j} - \underline{w_p^{(p)}} Z^{-p}$$

$$\downarrow k_p$$

Hence,

$$G_p^{f}(Z) = G_{p-1}^{f}(Z) - k_p \left[Z^{-p} - \sum_{j=1}^{p-1} w_{p-j}^{(p-1)} Z^{-j} \right] \tag{7.28c}$$

This recursion for the pth order FPE transfer function in Eqn. (7.28c) may be simplified by considering the bracketed term on the right hand side. This term introduces the concept "backward" prediction error (BPE). It is important to note that the concept of BPE is entirely due to use of Durbin's algorithm and is chosen to describe the operation in the brackets on the right hand side of the Eqn. (7.28c).

For the $(p-1)^{th}$ order FPE transfer function

$$G_{p-1}^{f}(Z) = 1 - \sum_{j=1}^{p-1} w_j^{(p-1)} Z^{-j} = 1 - w_1^{(p-1)} Z^{-1} - w_2^{(p-1)} Z^{-2} - \cdots - w_{p-1}^{(p-1)} Z^{-(p-1)}$$

Hence, FPE filter coefficients for the $(p-1)^{th}$ order are given by,

$$g_{p-1}^{f} = \left[1 \quad -w_1^{(p-1)} \quad -w_2^{(p-1)} \quad \cdots\cdots \quad -w_{p-1}^{(p-1)} \right]^T \tag{7.28d}$$

In forming the prediction $\hat{x}(n)$, the coefficient "1" is multiplied with $x(n)$, $-w_1^{(p-1)}$ with $x(n-1)$ and so on until $-w_{p-1}^{(p-1)}$ with $x(n-p+1)$.

The backward prediction filter g_{p-1}^{b} is a vector of coefficients that is time reversed replica of Eqn. (7.28d).

$$g_{p-1}^{b} = \left[-w_{p-1}^{(p-1)} \quad \cdots\cdots \quad -w_{2}^{(p-1)} \quad -w_{1}^{(p-1)} \quad 1 \right]^{T} \qquad (7.28e)$$

The backward predictor is appropriate since the inner product of g_{p-1}^{b} with p as the length of vector data $\left[x(n)\ x(n-1)\ x(n-2) \cdots x(n-p+1) \right]^{T}$ is given by,

$$x(n-p+1) - w_{1}^{(p-1)} x(n-p+2) - \quad \cdots \quad - w_{p-2}^{(p-1)} x(n-1) - w_{p-1}^{(p-1)} x(n)$$
$$= x(n-p+1) - \hat{x}(n-p+1)$$

where $\hat{x}(n-p+1)$ is the prediction of the sample which is $(p-1)$ samples backward in time (with reference to $x(n)$) . The BPE filter transfer function is therefore given by,

$$G_{p-1}^{b}(Z) = \left[-w_{p-1}^{(p-1)} \cdots \quad - w_{2}^{(p-1)} Z^{-(p-3)} \quad - w_{1}^{(p-1)} Z^{-(p-2)} \quad + Z^{-(p-1)} \right] \qquad (7.28f)$$

From Eqn. (7.28c)

$$G_{p}^{f}(Z) = G_{p-1}^{f}(Z) - k_{p} \left[Z^{-p} - \sum_{j=1}^{p-1} w_{p-j}^{(p-1)} Z^{-j} \right]$$

Expanding,

$$G_{p}^{f}(Z) = G_{p-1}^{f}(Z)$$
$$- k_{p} \left[- w_{p-1}^{(p-1)} Z^{-1} \cdots - w_{2}^{(p-1)} Z^{-(p-2)} \quad - w_{1}^{(p-1)} Z^{-(p-1)} \quad + Z^{-(p)} \right]$$

$$G_{p}^{f}(Z) = G_{p-1}^{f}(Z)$$
$$- Z^{-1} k_{p} \left[- w_{p-1}^{(p-1)} \cdots - w_{2}^{(p-1)} Z^{-(p-3)} \quad - w_{1}^{(p-1)} Z^{-(p-2)} \quad + Z^{-(p-1)} \right]$$

$$G_{p}^{f}(Z) = G_{p-1}^{f}(Z) - Z^{-1} k_{p}\, G_{p-1}^{b}(Z) \qquad (7.29a)$$

This is the Z-domain recursion for the FPE transfer function, which is seen to be function of both $(p-1)^{th}$ order FPE and BPE transfer function. It is important to note that $(p-1)^{th}$ order BPE filter is required to compute the pth order FPE transfer function.

From the definition of BPE transfer function (Eqn. (7.28f)), the pth order BPE transfer function is

$$G_p^b(z) = -w_p^{(p)} - w_{p-1}^{(p)} Z^{-1} \cdots - w_2^{(p)} Z^{-(p-2)} - w_1^{(p)} Z^{-(p-1)} + Z^{-p}$$

$$= Z^{-p} - w_p^{(p)} - \sum_{j=1}^{p-1} w_{p-j}^{(p)} Z^{-j}$$

Using the relation,

$$w_j^{(p)} = w_j^{(p-1)} - k_p w_{p-j}^{(p-1)} \qquad 1 \le j \le p-1$$

$$G_p^b(Z) = Z^{-p} - w_p^{(p)} - \sum_{j=1}^{p-1} \left\{ w_{p-j}^{(p-1)} - k_p\, w_{p-(p-j)}^{(p-1)} \right\} Z^{-j}$$

$$= Z^{-p} - w_p^{(p)} + w_p^{(p)} \sum_{j=1}^{p-1} w_j^{(p-1)} Z^{-j} - \sum_{j=1}^{p-1} w_{p-j}^{(p-1)} Z^{-j}$$

$$= Z^{-p} - w_p^{(p)} \left\{ 1 - \sum_{j=1}^{p-1} w_j^{(p-1)} Z^{-j} \right\} - \sum_{j=1}^{p-1} w_{p-j}^{(p-1)} Z^{-j}$$

$$= Z^{-p} - w_p^{(p)} \left\{ G_{p-1}^f(Z) \right\} - \sum_{j=1}^{p-1} w_{p-j}^{(p-1)} Z^{-j}$$

$$= Z^{-p} - \sum_{j=1}^{p-1} w_{p-j}^{(p-1)} Z^{-j} - w_p^{(p)} G_{p-1}^f(Z)$$

$$= \left\{ -w_{p-1}^{(p-1)} Z^{-1} \quad - w_{p-2}^{(p-1)} Z^{-2} \cdots - w_2^{(p-1)} Z^{-(p-2)} \right.$$

$$\left. -w_1^{(p-1)} Z^{-(p-1)} \quad + Z^{-p} \right\} - w_p^{(p)} G_{p-1}^f(Z)$$

$$G_p^b(Z) = Z^{-1} \left\{ -w_{p-1}^{(p-1)} \quad - w_{p-2}^{(p-1)} Z^{-1} \cdots\cdots - w_2^{(p-1)} Z^{-(p-3)} \right.$$

$$\left. -w_1^{(p-1)} Z^{-(p-2)} \quad + Z^{-(p-1)} \right\} - w_p^{(p)} G_{p-1}^f(Z)$$

$$G_p^b(Z) = Z^{-1} G_{p-1}^b(Z) - k_p G_{p-1}^f(Z) \tag{7.29b}$$

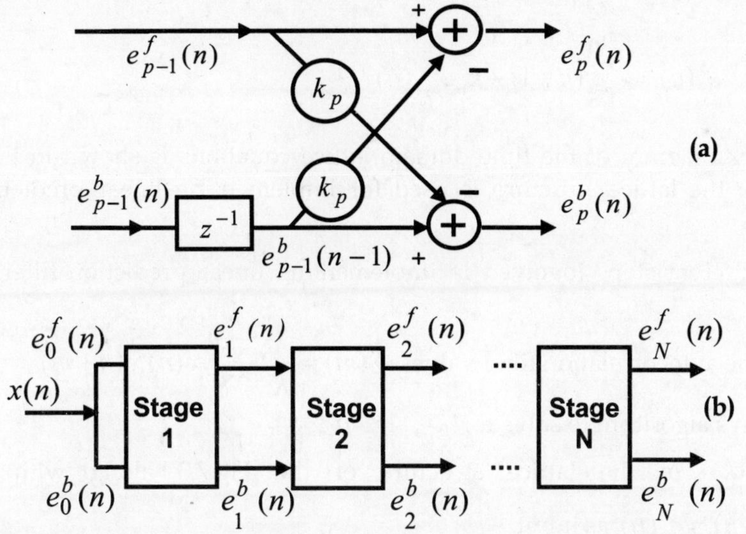

Fig. 7.3 (a) Single stage lattice predictor and (b) Multistage lattice predictor.

Eqns. (7.29a) and (7.29b) completely define the transfer function of the pth order predictor. The lattice filter structure follows directly by examining the time series of the FPE and BPE outputs. If the time signal $x(n)$ is input to the FPE and BPE filter, then the Z-transform of the pth order outputs are

$$\left.\begin{aligned} E^f_p(Z) &= X(Z)G^f_p(Z) \\ E^b_p(Z) &= X(Z)G^b_p(Z) \end{aligned}\right\} \qquad (7.30)$$

Using Eqn. (7.29),

$$E^f_p(Z) = X(Z)G^f_{p-1}(Z) - k_p Z^{-1} X(Z) G^b_{p-1}(Z)$$

$$E^b_p(Z) = Z^{-1} X(Z) G^b_{p-1}(Z) - k_p X(Z) G^f_{p-1}(Z)$$

Using Eqn. (7. 30),

$$E^f_p(Z) = E^f_{p-1}(Z) - k_p Z^{-1} E^b_{p-1}(Z)$$

$$E^b_p(Z) = Z^{-1} E^b_{p-1}(Z) - k_p E^f_{p-1}(Z)$$

Taking inverse Z - transform

$$e_p^f(n) = e_{p-1}^f(n) - k_p e_{p-1}^b(n-1) \tag{7.31a}$$

$$e_p^b(n) = e_{p-1}^b(n-1) - k_p e_{p-1}^f(n) \tag{7.31b}$$

A block diagram of the time domain lattice equations is shown in Fig. 7.3, which illustrates how the lattice structure is used for implementing linear prediction filters on actual data.

The different steps involved in implementing linear prediction filter using lattice structure are:

1. Estimate the autocorrelation filter values $\hat{\phi}_x(m) = \frac{1}{K}\sum_{n=0}^{K-m+1} x(n)\,x(n+m)$
2. Use Durbin's algorithm to solve for $k_p,\ 1 \le p \le N$
3. Use the k_p in the lattice structure of the Fig.7.3 along with data signal $x(n) = e_0^f(n) = e_0^b(n)$ as input.

When used in this manner, the forward error output of the Nth stage $e_N^f(n)$ is equivalent to the forward error of the N coefficient transversal linear predictor. Since $e_N^f(n)$ is now an uncorrelated signal, a linear predictor used in this manner is called *whitening filter*.

The forward error output at the the final stage, the Nth stage, $e_N^f(n)$ is given by

$$e_N^f(n) = e_0^f(n) - \sum_{l=1}^{N} k_l e_l^b(n-1) = x(n) - \sum_{l=1}^{N} k_l e_l^b(n-1)$$

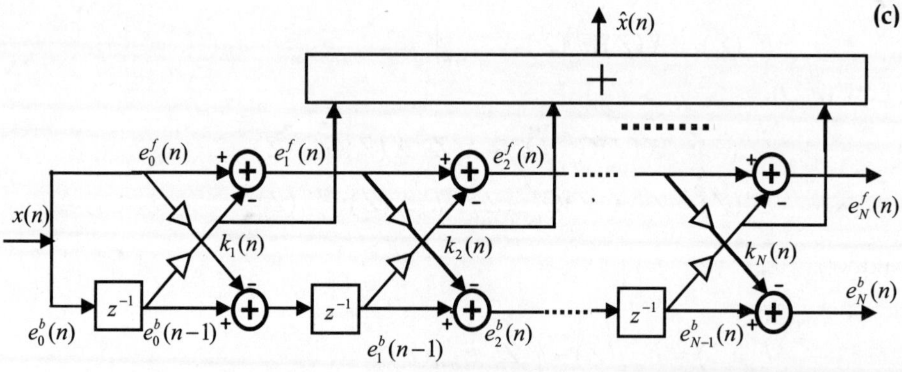

Fig. 7.3 (c) Direct computation of prediction estimate from the lattice structure.

Since

$$e_0^f(n) = x(n),$$

$$e_N^f(n) = x(n) - \sum_{l=1}^{N} k_l e_l^b(n-1)$$

This implies that the second term on R.H.S. involving summation of the weighted backward prediction errrors is nothing but the prediction estimate $\hat{x}(n)$ of $x(n)$.

$$\hat{x}(n) = \sum_{l=1}^{N} k_l\, e_l^b(n-1) \tag{7.31c}$$

It is for this reason in practice, the lattice structure is also referred to as *lattice predictor*. But this is not explictly evident from the lattice structure as its outputs are forward and backward prediction errors. The prediction estimate $\hat{x}(n)$ can also be obtained by computing the forward prediction filter coefficients $w_l^{(N)}, l = 1, 2, \ldots, N$, from the reflection coefficients $k_l, l = 1, 2, \ldots, N$, using the Levison recusion relation as

$$\hat{x}(n) = \sum_{l=1}^{N} w_l^{(N)} x(n-l)$$

However, this approach of finding the prediction estimate is compuationally involved unless the prediction filter coefficients are required for the other purposes like spectrum computaion.

7.6.4 Conversion from Direct Form to Lattice Form

The lattice forward and backward prediction error filter transfer functions (Eqns. (7.29a,b)) are

$$G_p^f(Z) = G_{p-1}^f(Z) - Z^{-1}k_p G_{p-1}^b(Z)$$

$$G_p^b(Z) = Z^{-1}G_{p-1}^b(Z) - k_p G_{p-1}^f(Z)$$

From the above second relation,

$$G_{p-1}^b(Z) = \frac{G_p^b(Z) + k_p G_{p-1}^f(Z)}{Z^{-1}}$$

Substituting this in the first relation,

$$G_p^f(Z) = G_{p-1}^f(Z) - Z^{-1}k_p \left\{ \frac{G_p^b(Z) + k_p G_{p-1}^f(Z)}{Z^{-1}} \right\}$$

$$G_p^f(Z) = G_{p-1}^f(Z)[1 - k_p^2] - k_p G_p^b(Z)$$

$$G_{p-1}^f(Z) = \frac{G_p^f(Z) + k_p G_p^b(Z)}{\left[1 - k_p^2\right]} \tag{7.31d}$$

From Eqn.(7.28b),

$$G_p^f(Z) = 1 - \sum_{j=1}^{p} w_j^{(p)} Z^{-j} \quad \text{or} \quad g_p^f = \left[1 \quad -w_1^{(p)} \quad -w_2^{(p)} \quad \cdots\cdots \quad -w_p^{(p)}\right]^T$$

Since from Eqn.(7.28f),

$$G_p^b(z) = Z^{-p} - w_p^{(p)} - \sum_{j=1}^{p-1} w_{p-j}^{(p)} Z^{-j}$$

or

$$g_p^b = \left[-w_p^{(p)} \quad -w_{p-1}^{(p)} \cdots - w_2^{(p)} \quad -w_1^{(p)} \quad 1\right]^T$$

Eqn. (7.31d)

$$G_{p-1}^f(Z) = \frac{G_p^f(Z) + k_p G_p^b(Z)}{\left[1 - k_p^2\right]}$$

This is valid for all the elements of $g_{p-1}^f(z)$ and hence selectively for the unkown elements (coefficients) can be written as

$$\begin{bmatrix} -w_1^{(p-1)} \\ -w_2^{(p-1)} \\ \vdots \\ -w_{p-1}^{(p-1)} \end{bmatrix} = \frac{1}{\left[1 - k_p^2\right]} \left\{ \begin{bmatrix} -w_1^{(p)} \\ -w_2^{(p)} \\ \vdots \\ -w_{p-1}^{(p)} \end{bmatrix} + k_p \begin{bmatrix} -w_{p-1}^{(p)} \\ -w_{p-2}^{(p)} \\ \vdots \\ -w_1^{(p)} \end{bmatrix} \right\} \tag{7.31e}$$

It sould be noted that the definition of k_p is given by $k_p = w_p^p$ and $k_p \neq -w_p^p$. The derivation of the lattice from the all zero prediction error filter $A_p(z)$ will be illustrated.

$$A_3(z) = 1 - 0.6z^{-1} + 0.5z^{-2} - 0.1z^{-3}$$

$$A_3(z) = 1 - w_1^{(3)} z^{-1} - w_2^{(3)} z^{-2} - w_3^{(3)} z^{-3}$$

Comapring,

$$-w_1^{(3)} = -0.6, \quad -w_2^{(3)} = 0.5 \quad \text{and} \quad -w_3^{(3)} = -0.1.$$

Further, $k_3 = w_3^{(3)} = 0.1$. Therefore Eqn. (7.31e) becomes

$$\begin{bmatrix} -w_1^{(2)} \\ -w_2^{(2)} \end{bmatrix} = \frac{1}{\left[1-k_3^2\right]}\left\{\begin{bmatrix} -w_1^{(3)} \\ -w_2^{(3)} \end{bmatrix} + k_3 \begin{bmatrix} -w_2^{(3)} \\ -w_1^{(3)} \end{bmatrix}\right\}$$

$-w_1^{(3)} = -0.6,\ -w_2^{(3)} = 0.5$ and $k_3 = 0.1$.

$$\begin{bmatrix} -w_1^{(2)} \\ -w_2^{(2)} \end{bmatrix} = \frac{1}{\left[1-0.1^2\right]}\left\{\begin{bmatrix} -0.6 \\ 0.5 \end{bmatrix} + 0.1 \begin{bmatrix} 0.5 \\ -0.6 \end{bmatrix}\right\} = \frac{1}{0.99}\begin{bmatrix} -0.55 \\ 0.44 \end{bmatrix}$$

$$\begin{bmatrix} -w_1^{(2)} \\ -w_2^{(2)} \end{bmatrix} = \begin{bmatrix} -0.5445 \\ 0.4356 \end{bmatrix}$$

$$A_2(z) = 1 - 0.5445z^{-1} + 0.4356z^{-2}$$

$$A_2(z) = 1 - w_1^{(2)}z^{-1} - w_2^{(2)}z^{-2}$$

Therefore,

$-w_1^{(2)} = -0.5445,\ -w_2^{(2)} = 0.4356$ and $k_2 = -0.4356$.

(d)

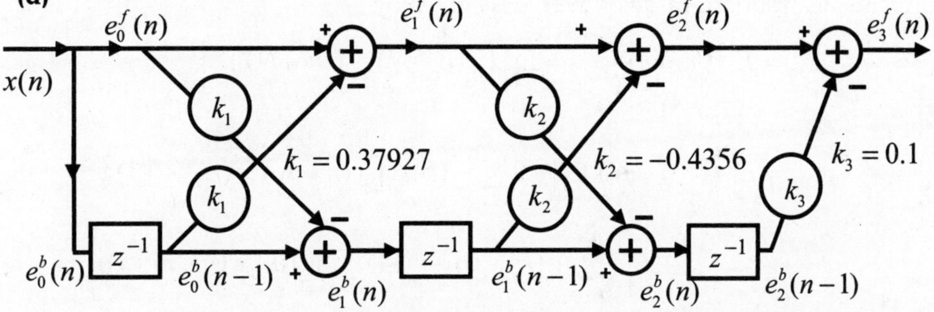

Fig. 7.3d. Lattice structure for the transfer function $A_3(z)$.

$$-w_1^{(1)} = \frac{1}{\left[1-k_2^2\right]}\left\{-w_1^{(2)} + k_2\left[-w_1^{(2)}\right]\right\}$$

$$-w_1^{(1)} = 1.2342\left\{-0.5445 - 0.4356\left[-0.5445\right]\right\}$$
$$= 1.2342\left\{-0.3073\right\} = -0.37927$$

$$A_1(z) = 1 - w_1^{(1)} z^{-1} = 1 - 0.37927 z^{-1}$$
$$k_1 = 0.37927, k_2 = -0.4356, k_3 = 0.1,$$

The lattice structure for the transfer function $A_3(z) = 1 - 0.6z^{-1} + 0.5z^{-2} - 0.1z^{-3}$ is shown in the Fig. (7.3d).

7.7 ORTHOGONALITY OF BACKWARD PREDICTION ERRORS (BPE)

The backward prediction errors resulting at the various stages of the model are orthogonal to each other. That is,

$$E\left[e_i^b(n)\, e_k^b(n) \right] = \begin{cases} \varepsilon_N^b & ; \quad i = k \\ 0 & ; \quad i \le k-1 \end{cases} \tag{7.32a}$$

Proof:
As a part of the proof, the *augmented normal equations for backward linear prediction will be derived.*

The estimate $\hat{x}(n-N)$ of $x(n-N)$ from future samples is given by

$$\hat{x}(n-N) = \sum_{k=1}^{N} w(k)\, x(n-k+1)$$

Therefore, the mean square value of the BPE is given by

$$\varepsilon_M^b = E\left[\left(e_M^b(n) \right)^2 \right] = E\left[\left(x(n-N) - \hat{x}(n-N) \right)^2 \right]$$

$$= E\left[\left(x(n-N) \right)^2 \right] - 2\sum_{k=1}^{N} w(k) E\left[x(n-N)\, x(n-k+1) \right]$$
$$+ E\left[\sum_{k=1}^{N} w(k) x(n-k+1) \sum_{m=1}^{N} w(m) x(n-m+1) \right]$$

$$= E\left[\left(x(n-N) \right)^2 \right] - 2\sum_{k=1}^{N} w(k) E\left[x(n-N)\, x(n-k+1) \right]$$
$$+ E\left[\sum_{k=1}^{N} \sum_{m=1}^{N} w(k) w(m) x(n-k+1) x(n-m+1) \right]$$

$$\varepsilon_N^b = \phi_x(0) - 2\sum_{k=1}^{N} w(k) \phi_x(N-k+1) + \sum_{k=1}^{N} \sum_{m=1}^{N} w(k) w(m) \phi_x(k-m) \tag{7.32b}$$

For ε_M^b to be minimum,

$$\frac{\varepsilon_N^b}{w(k)} = -2\phi_x(N-k+1) + 2\sum_{m=1}^{N} w(m)\phi_x(k-m) = 0, \quad k = 1, 2, \cdots, N$$

Therefore,

$$\sum_{m=1}^{N} w(m)\phi_x(k-m) = \phi_x(N-k+1) \qquad k = 1, 2, \cdots, N \tag{7.32c}$$

These represent *normal equations for backward linear prediction.* From Eqn. (7.32b),

$$\varepsilon_N^b = \phi_x(0) - 2\sum_{k=1}^{N} w(k)\phi_x(N-k+1) + \sum_{k=1}^{N} w(k)\underline{\sum_{m=1}^{N} w(m)\phi_x(k-m)}$$

Substituting for the underlined term from Eqn. (7.32c),

$$\varepsilon_N^b = \phi_x(0) - 2\sum_{k=1}^{N} w(k)\phi_x(N-k+1) + \sum_{k=1}^{N} w(k)\phi_x(N-k+1)$$

$$\varepsilon_N^b = \phi_x(0) - \sum_{k=1}^{N} w(k)\phi_x(N-k+1) \tag{7.32d}$$

Consider the normal Eqn.(7.32c),

$$\sum_{m=1}^{N} w(m)\phi_x(k-m) = \phi_x(N-k+1)$$

$$\phi_x(N-k+1) - \sum_{m=1}^{N} w(m)\phi_x(k-m) = 0 \quad k = 1, 2, \cdots\cdots, N$$

For a stationary input process as $\phi_x(k-m)$ equals $\phi_x(m-k)$,

$$\phi_x(N-k+1) - \sum_{m=1}^{N} w(m)\phi_x(m-k) = 0$$

Defining,

$$a_N(N-k+1) = \begin{cases} 1, & k = N+1 \\ -w(k), & k = 1, 2, \ldots, N \end{cases} \tag{7.32e}$$

Therefore,

$$\sum_{m=1}^{N+1} a_N(N-m+1)\phi_x(m-k) = 0$$

Replacing $(m-1)$ with m,

$$\sum_{m=0}^{N} a_N(N-m)\,\phi_x(m+1-k) = 0$$

Replacing $(k-1)$ with k,

$$\sum_{m=0}^{N} a_N(N-m)\phi_x(m-k) \;=\; 0, \qquad k = 0, 2, \cdots\cdots, N-1 \tag{7.32f}$$

Consider the MSE(Eqn 7.32d),

$$\varepsilon_N^b = \phi_x(0) - \sum_{k=1}^{N} w(k)\phi_x(N-k+1)$$

Since,

$$a_N(N-k+1) = \begin{cases} 1 & ,k = N+1 \\ -w(k), & k = 1, 2, \ldots, N \end{cases}$$

ε_N^b becomes,

$$\varepsilon_N^b = \sum_{k=1}^{N+1} w(k)\,\phi_x(N-k+1)$$

Further,

$$\varepsilon_N^b = \sum_{k=1}^{N+1} a_N(N-k+1)\;\phi_x(N-k+1)$$

Replacing $(k-1)$ with m,

$$\varepsilon_N^b = \sum_{m=0}^{N} a_M(N-m)\;\phi_x(N-m)$$

For a stationary input process $\phi_x(N-m)$ equals $\phi_x(m-N)$, therefore

$$\varepsilon_N^b = \sum_{m=0}^{N} a_N(N-m)\;\phi_x(m-N) \tag{7.32g}$$

Therefore, Eqns. (7.32f) and (7.32g) can be combined as

$$\sum_{m=0}^{N} a_N(N-m)\;\phi_x(m-k) \;=\; \begin{cases} 0 & ; \quad k = 0, 1, \cdots\cdots, N-1 \\ \varepsilon_N^b & ; \quad k = N \end{cases} \tag{7.32h}$$

This set of equations is known as *the augmented normal equations for backward linear prediction.*

In the light of the Eqn.(7.32e), the expression for BPE can be written in a different form. The BPE is given by

$$e_N^b(n) = x(n-N) - \hat{x}\left(n-N\right)$$

where,

$$\hat{x}\left(n-N\right) \;=\; \sum_{k=1}^{N} w(k)\; x(n-k+1)$$

Therefore,

$$e_N^b(n) = x(n-N) \;-\; \sum_{k=1}^{N} w(k)\; x(n-k+1)$$

Since from Eqn. (7.32e),

$$a_N(N-k+1) = \begin{cases} 1 & ,k = N+1 \\ -w(k)\,, & k = 1,2,...,N \end{cases}$$

The BPE can be written as

$$e_N^b(n) = x(n-N) \;+\; \sum_{k=1}^{N} a_N(N-k+1)\; x(n-k+1)$$

Replacing $(k-1)$ by k,

$$e_N^b(n) = x(n-N) + \sum_{k=0}^{N-1} a_N(N-k)\; x(n-k)$$

$$e_N^b(n) \;=\; \sum_{k=0}^{N} a_N(N-k)\; x(n-k) \qquad (7.33a)$$

Expanding the summation on the R.H.S,

$$e_N^b(n) = a_N(N)x(n) + a_N(N-1)x(n-1) + \ldots a_N(0)x(n-N), \quad a_N(0) = 1$$

But

$$e_N^f(n) = a_N(0)x(n) + a_N(1)x(n-1) + \ldots + a_N(N)x(n-N), \quad a_N(0) = 1$$

That is, Eqn.(7.33a) expresses the backwad prediction error in terms of forward prediction filter coefficients. For this, the coefficients order is reversed but the input sequence order is maintained same as for forward prediction error. Further, using Eqn. (7.33a), $e_i^b(n)$ and $e_k^b(n)$ can be written as

$$e_i^b(n) = \sum_{m=0}^{i} a_i(i-m)\; x(n-m) \text{ and } e_k^b(n) = \sum_{p=0}^{k} a_k(k-p)\; x(n-p)$$

Therefore,

$$E\left[\; e_i^b(n)\, e_k^b(n)\right] = E\left[\sum_{m=0}^{i}\;\sum_{p=0}^{k} a_i(i-m)\; a_k(k-p)\; x(n-m)\; x(n-p)\right]$$

$$= \sum_{m=0}^{i}\;\sum_{p=0}^{k} a_i(i-m)\; a_k(k-p)\; E\left[x(n-m)\; x(n-p)\right]$$

$$E\left[\, e_i^b(n)\, e_k^b(n) \right] = \sum_{m=0}^{i} \sum_{p=0}^{k} a_i(i-m)\, a_k(k-p)\, \phi_x(p-m) \tag{7.33b}$$

If $i = k$

$$E\left[\, e_k^b(n)\, e_k^b(n) \right] = E\left[\left(e_k^b(n) \right)^2 \right] = \sum_{m=0}^{k} \sum_{p=0}^{k} a_k(k-m)\, a_k(k-p)\, \phi_x(p-m)$$

$$E\left[\left(e_k^b(n) \right)^2 \right] = \sum_{m=0}^{k} \sum_{p=0}^{k} a_k(k-m)\, a_k(k-p)\, \phi_x(p-m)$$

$$E\left[\left(e_k^b(n) \right)^2 \right] = \sum_{m=0}^{k} a_k(k-m) \sum_{p=0}^{k} a_k(k-p)\, \phi_x(p-m)$$

The MSE of backward prediction error is given by Eqn. (7.32h),

$$\sum_{m=0}^{k} a_k(k-p) \phi_x(p-m) = \begin{cases} 0 & ; \quad m = 0, 1, \cdots, k-1 \\ \varepsilon_N^b & ; \quad m = k \end{cases}$$

Therefore,

$$E\left[\left(e_k^b(n) \right)^2 \right] = a_k(0)\, \varepsilon_k^b$$

Also as $a_k(0) = 1$,

$$E\left[\left(e_k^b(n) \right)^2 \right] = \varepsilon_k^b$$

From Eqn.(7.33b),

$$E\left[\, e_i^b(n)\, e_k^b(n) \right] = \sum_{m=0}^{i} \sum_{p=0}^{k} a_i(i-m)\, a_k(k-p)\, \phi_x(p-m)$$

If $i \le k-1$,

$$E\left[\, e_i^b(n)\, e_k^b(n) \right] = \sum_{m=0}^{i} a_i(i-m) \sum_{p=0}^{k} a_k(k-p)\, \phi_x(p-m)$$

Here, $m \le i$. Since $i \le k-1$, $m \le k-1$.

The MSE of backward prediction error is given by (Eqn. 7.32h),

$$\sum_{m=0}^{k} a_k(k-p) \varphi_x(p-m) = \begin{cases} 0 & ; \quad m = 0, 1, \cdots, k-1 \\ \varepsilon_M^b & ; \quad m = k \end{cases}$$

Therefore,

$$\sum_{p=0}^{k} a_k(k-p)\, \phi_x(p-m) = 0$$

$$E\left[\, e_i^b(n)\, e_k^b(n) \right] = 0$$

$$E\left[e_i^b(n)\, e_k^b(n) \right] = \begin{cases} \varepsilon_k^b \quad ; & i = k \\ 0 \quad ; & i \le k-1 \end{cases}$$

That is, the backward prediction errors $e_i^b(n)$ *and* $e_k^b(n)$ *at stage* i *and at stage* k *(of the lattice) are orthogonal for* $i \neq k$.

7.7.1 Alternate Proof for the Orthogonality of BPE for a Lattice Predictor

Consider Eqn.(7.31a),

$$e_p^f(n) = e_{p-1}^f(n) - K_p e_{p-1}^b(n-1)$$

then,

$$e_{p-1}^f(n) - e_p^f(n) = K_p e_{p-1}^b(n-1) = \rho_p \tag{7.34a}$$

$$e_{l-1}^f(n) - e_l^f(n) = K_l e_{l-1}^b(n-1) = \rho_l \tag{7.34b}$$

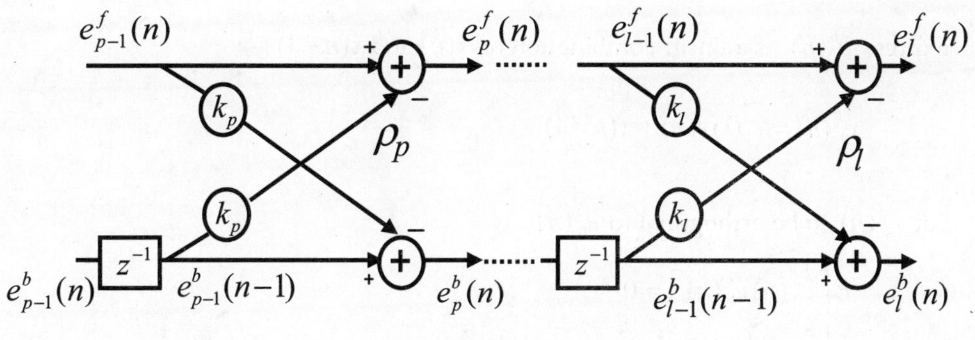

Fig. 7.4 Lattice filter structure.

ρ_p and ρ_l are uncorrelated as these components are removed from the forward prediction errors (FRPE), at p^{th} and l^{th} stage. That is,

$$E\left[\rho_p \rho_l \right] = 0 \tag{7.34c}$$

Or $$E\left[\left(K_p e_{p-1}^b(n-1) \right) \left(K_l e_{l-1}^b(n-1) \right) \right] = 0$$

$$K_p K_l E\left[e_{p-1}^b(n-1)\, e_{l-1}^b(n-1) \right] = 0$$

For $K_p, K_l \neq 0$,

$$E\left[e_{p-1}^b(n-1)e_{l-1}^b(n-1)\right] = 0 \tag{7.34d}$$

That is, backward prediction errors are orthogonal

7.8 CONSTRAINT ON THE REFLECTION COEFFICIENTS

Consider a sequence of correlated random variables denoted by $x(n), x(n-1), ..., x(n-N)$ that are linearly independent in algebraic sense. The sequence is assumed to be of zero mean and known correlation function $\phi_x(k)$ for $0 \le k \le N$. Gram Schmidt orthogonalization is used to transform this sequence into a new sequence of uncorrelated random variables $e_0^b(n)$, $e_1^b(n)$,..., $e_N^b(n)$.

Steps:

1. Let $e_0^b(n) = x(n)$ (7.35a)

2. Express $e_1^b(n)$ as a linear combination of $x(n)$ and $x(n-1)$,

$$e_1^b(n) = c_1(1)x(n) + x(n-1)$$

For $e_1^b(n)$ to be orthogonal to $e_0^b(n)$,

$$E\left[e_0^b(n)e_1^b(n)\right] = 0$$
$$E\left[x(n)\{c_1(1)x(n) + x(n-1)\}\right] = 0$$
$$E\left[c_1(1)x(n)\ x(n) + x(n)x(n-1)\right] = 0$$
$$c_1(1)E\left[x(n)x(n)\right] + \left[x(n)x(n-1)\right] = 0$$
$$c_1(1)\phi_x(0) + \phi_x(1) = 0$$
$$c_1(1) = -\frac{\phi_x(1)}{\phi_x(0)} \tag{7.35b}$$

3. Express $e_2^b(n)$ as

$$e_2^b(n) = c_2(2)x(n) + c_2(1)x(n-1) + x(n-2)$$

For $e_2^b(n)$ to be orthogonal with $e_0^b(n)$,

$$E\left[e_0^b(n)e_2^b(n)\right] = 0 \Rightarrow \quad c_2(2)\phi_x(0) + c_2(1)\phi_x(1) + \phi_x(2) = 0$$

Further, $e_2^b(n)$ is also orthogonal with $e_1^b(n)$,

$$E\left[e_1^b(n)e_2^b(n)\right] = 0$$

$$E\left\{\left[c_1(1)x(n)+x(n-1)\right]\left[c_2(2)x(n)+c_2(1)x(n-1)+x(n-2)\right]\right\}=0$$

$$E\left[c_1(1)c_2(2)x(n)x(n)\right]+E\left[c_2(2)x(n)x(n-1)\right] + E\left[c_1(1)c_2(1)x(n)x(n-1)\right]+E\left[c_2(1)x(n-1)x(n-1)\right] + E\left[c_1(1)x(n)x(n-2)\right]+E\left[x(n-1)x(n-2)\right]=0$$

$$c_1(1)c_2(2)\,\phi_x(0)+c_2(2)\phi_x(1)+c_1(1)c_2(1)\phi_x(1)+c_2(1)\phi_x(0) + c_1(1)\phi_x(2)+\phi_x(1)=0$$

Substituting for $c_1(1)$ from Eqn. (7.35b),

$$c_1(1) = -\frac{\phi_x(1)}{\phi_x(0)}$$

Results

$$-c_2(2)\,\phi_x(0)\frac{\phi_x(1)}{\phi_x(0)}+c_2(2)\,\phi_x(1)-c_2(1)\frac{\phi_x(1)}{\phi_x(0)}\phi_x(1)+c_2(1)\phi_x(0) - \frac{\phi_x(1)}{\phi_x(0)}\phi_x(2)+\phi_x(1)=0$$

The first two terms on the left hand side cancel each other leaving the remaining 4 terms as

$$-c_2(1)\frac{\phi_x^2(1)}{\phi_x(0)}+c_2(1)\phi_x(0)-\frac{\phi_x(1)\phi_x(2)}{\phi_x(0)}+\phi_x(1)=0$$

$$c_2(1)\left[\phi_x(0)-\frac{\phi_x^2(1)}{\phi_x(0)}\right]=\frac{\phi_x(1)\phi_x(2)}{\phi_x(0)}-\phi_x(1)$$

$$c_2(1)\left[\frac{\phi_x^2(0)-\phi_x^2(1)}{\phi_x(0)}\right]=\frac{\phi_x(1)\phi_x(2)-\phi_x(1)\phi_x(0)}{\phi_x(0)}$$

$$c_2(1)=\frac{\phi_x(1)\phi_x(2)-\phi_x(1)\phi_x(0)}{\phi_x^2(0)-\phi_x^2(1)} \tag{7.35c}$$

Again consider

$$c_2(2)\phi_x(0)+c_2(1)\phi_x(1)+\phi_x(2)=0$$

Substituting for $c_2(1)$ from Eqn.(7.35c),

$$c_2(2)\phi_x(0)+\left[\frac{\phi_x(1)\phi_x(2)-\phi_x(1)\phi_x(0)}{\phi_x^2(0)-\phi_x^2(1)}\right]\phi_x(1)+\phi_x(2)=0$$

$$c_2(2)\varphi_x(0)=-\left[\frac{\phi_x(1)\phi_x(2)-\phi_x(1)\phi_x(0)}{\phi_x^2(0)-\phi_x^2(1)}\right]\phi_x(1)-\phi_x(2)$$

$$=\frac{-\phi_x^2(1)\phi_x(2)+\phi_x^2(1)\phi_x(0)-\phi_x^2(0)\phi_x(2)+\phi_x^2(1)\phi_x(2)}{\phi_x^2(0)-\phi_x^2(1)}$$

$$=\frac{\phi_x^2(1)\phi_x(0)-\phi_x^2(0)\phi_x(2)}{\phi_x^2(0)-\phi_x^2(1)}$$

$$c_2(2)=\frac{\phi_x^2(1)-\phi_x(0)\phi_x(2)}{\phi_x^2(0)-\phi_x^2(1)} \tag{7.35d}$$

3. Continue this till the last random variable $e_N^b(n)$ has been computed and determine $\{c_N(N), c_N(N-1), ..., c_N(1)\}$ so that,

$$e_N^b(n)=c_N(N)x(n)+c_N(N-1)x(n-1)+...+x(n-N) \tag{7.35e}$$

This completes the Gram-Schmidt orthogonalization. The above relations for $e_i^b(n)$, $i=0,1,2,....,N$ can be expressed as

$$\begin{bmatrix} e_0^b(n) \\ e_1^b(n) \\ \vdots \\ e_N^b(n) \end{bmatrix}=\begin{bmatrix} 1 & 0 & \cdots & 0 \\ c_1(1) & 1 & \cdots & 0 \\ \vdots & \vdots & & \vdots \\ c_N(N) & c_N(N-1) & \cdots & 1 \end{bmatrix}\begin{bmatrix} x(n) \\ x(n-1) \\ \vdots \\ x(n-N) \end{bmatrix} \tag{7.35f}$$

In the above relation, the coefficient matrix is a *unit lower triangular matrix and* the nonzero coefficients are determined by Gram Schmidt algorithm. Further, the diagonal elements are all equal to 1.

The backward prediction errors $e_k^b(n)$ are given by Eqn.(7.33a) as

$$e_k^b(n) \;=\; \sum_{p=0}^{k} a_k(k-p)\,x(n-p), \quad k=0,1,\dots N$$

Let

$$X(n)=\left[x(n),x(n-1),...,x(n-N)\right]^T \tag{7.36a}$$

$$e^b(n) = \left[e_0^b(n), e_1^b(n), \ldots, e_N^b(n) \right]^T \tag{7.36b}$$

and
$$L = \begin{bmatrix} 1 & 0 & \cdots & 0 \\ a_1(1) & 1 & \cdots & 0 \\ \vdots & \vdots & & \vdots \\ a_N(N) & a_M(N-1) & \cdots & 1 \end{bmatrix} \tag{7.36c}$$

Then

$$\begin{bmatrix} e_0^b(n) \\ e_1^b(n) \\ \vdots \\ e_N^b(n) \end{bmatrix} = \begin{bmatrix} 1 & 0 & \cdots & 0 \\ a_1(1) & 1 & \cdots & 0 \\ \vdots & \vdots & & \vdots \\ a_N(N) & a_N(N-1) & \cdots & 1 \end{bmatrix} \begin{bmatrix} x(n) \\ x(n-1) \\ \vdots \\ x(n-N) \end{bmatrix}$$

This is identical to Eqn.(7.35f) except for the coefficient symbol (c inplace of a) and hence indicates that the lattice does the orthogonalization according to Gram-Schmidt rule and as $e_i^b(n)$, $i = 0,1,2,\ldots, N$ are orthogonal, this also implies that the backward prediction errors are orthogonal to each other. the Further the above equation can be written as

$$e^b(n) = LX(n) \tag{7.36d}$$

That is, the output vector $e^b(n)$ is the input vector transformed by the matrix L.

Let R_x^b be output autocorrelation matrix and this will be of dimension $(N+1)$ by $(N+1)$ as $e^b(n)$ is a vector of dimension $(N+1)$ by 1. Then using Eqn.(7.36d),

$$R_x^b = E\left[\mathrm{e}^b(n) e^{bT}(n) \right] = E\left[LX(n)X^T(n)L^T \right] = LE\left[X(n)X^T(n) \right] L^T$$

$$R_x^b = L\, R_x\, L^T$$

where $R_x = E\left[X(n)X^T(n) \right]$ is of dimension $(N+1)$ by $(N+1)$ as vector $X(n)$ is of dimension $(N+1)$ by 1. Since the output vector $e^b(n)$ is a vector consisting of uncorrelated random variables, the matrix R_x^b has only diagonal elements equal to the mean square values of the respective random variable that constitutes the vector $e^b(n)$.

$$R_x^b = \begin{bmatrix} \varepsilon_0 & 0 & \ldots & 0 \\ 0 & \varepsilon_1 & \ldots & 0 \\ \vdots & \vdots & \ldots & \vdots \\ 0 & 0 & \ldots & \varepsilon_{N+1} \end{bmatrix} \tag{7.36e}$$

where,

$$\varepsilon_i = E\left[\left\{ e_i^b \right\}^2\right] \quad , \quad i = 0,1,\dots,N$$

For a positive definite matrix R_x and for the nonsingular matrix L (as its determinant, $|L| = 1$), both of dimension $(N+1)\times(N+1)$,

$$R_x^b = LR_xL^T > 0$$

That is, R_x^b is positive definite.

Proof: Consider $X^T(L^TR_xL)X$ and this can be written as

$$X^T(L^TR_xL)X = (LX)^T R_x (LX)$$

Let $LX = V$, then

$$X^T(L^TR_xL)X = (V)^T R_x (V)$$

If $R_x > 0$, then $V^TR_xV > 0$, that is,

$$(LX)^T R_x (LX) = X^T\left(L^TR_xL\right)X > 0$$

This implies

$$L^TR_xL > 0$$

That is, $R_x^b = L^TR_xL$ is positive definite. This is possible as the nonsingularity of L (as its determinant, $|L| = 1 \neq 0$),) ensures $V = LX$ is nonzero for any nonzero vector X.

For R_x^b to be positive definite, its determinant should be positive. That is, from Eqn.(7.36e),

$$\det\left[R_x^b\right] = \prod_{k=0}^{M} \varepsilon_k > 0$$

This implies that the diagonal elements of R_x^b (Eqn. (7.36e)) have to be positive. That is,

$$\varepsilon_i > 0 \ , \quad i = 0,1,\dots,M$$

By definition,

$$\varepsilon_0 = E\left[e_0^{b2}(n)\right] = E\left[x^2(n)\right] > 0$$

By Levinson Durbin recursion,

$$\varepsilon_p = \varepsilon_{p-1}\left[1 - K_p^2\right]$$

Therefore,

$$\varepsilon_1 = \varepsilon_0 \left[1 - K_1^2\right]$$

For ε_1 to be positive,

$$\left|K_1\right| < 1$$

Continuing on similar lines,

$$\left|K_i\right| < 1, \qquad i = 1, 2, ..., N \tag{7.36f}$$

Hence, if R_x is positive definite, then all the reflection coefficients $K_i, i = 1, 2, ..., N$ of the lattice predictor have a magnitude less than unity. Conversely, if the all the reflection coefficients have a magnitude value less than unity, then the correlation matrix R_x of the lattice filter is positive definite.

Multistage lattice predictor acts as a Gram-Schmidt Processor for an input vector $[x(n), x(n-1), \ldots x(n-N)]$ and transforms it to component vectors $e_0^b(n), e_1^b(n), \ldots e_N^b(n)$, which are uncorrelated to each other. However, the forward prediction errors, $e_0^f(n), e_1^f(n), \ldots e_N^f(n)$ are *correlated to each other*. This is because at any stage i, the error $e_i^f(n)$ contains the errors $e_m^f(n), m > i$ and the final stage error $e^f{}_N(n)$ is present in all stages.

The forward errors $e_0^f(n), e_1^f(n), \ldots e_N^f(n)$ can also be made uncorrelated. For a prediction filter of sufficient order/length, the final error $e^f{}_N(n)$ is white. To decorrelate the error at any stage with any other stage, the largest decorrelation delay required is N samples and this corresponds to the delay that decorrelates the first error with that of the last stage. Further, for the adjacent stages, the delay decreases by one, in the increasing order direction. That is $[e^f{}_0(n-N), e^f{}_1(n-N+1), \ldots, e^f{}_{N-1}(n-1), e^f{}_N(n)]$ will be uncorrelated and to realize this, additional storage is necessary. But the backward prediction errors do not require any additional storage and hence, preferred.

7.8.1 Minimum Phase Property of Forward Prediction Error (FPE) Filter

The FRPE filter $A_N(z)$ operates on the input $x(n)$ and produces the error , the output $e_N^f(n)$. The FRPE filter $A_N(z)$ is given by

$$A_N(z) = \sum_{k=0}^{N} a_N(k) z^{-k}, \quad a_N(0) = 1 \tag{7.37a}$$

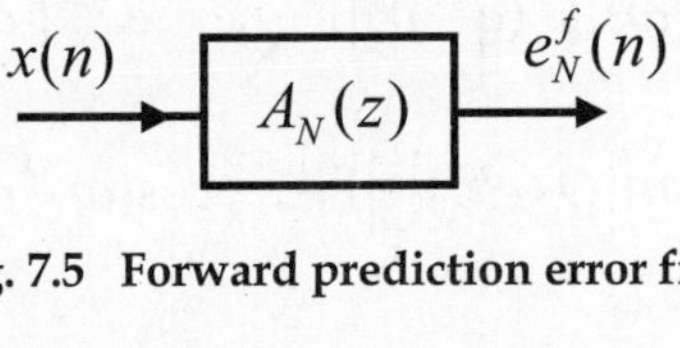

Fig. 7.5 Forward prediction error filter.

Let $S_x(\omega)$ be the PSD of $x(n)$, then $S_f(\omega)$, the PSD of $e_N^f(n)$ is given by

$$S_f(\omega) = \left|A_N(e^{j\omega})\right|^2 S_x(\omega) \tag{7.37b}$$

By Parseval theorem, the mean square value of $e_N^f(n)$ is

$$\varepsilon_N^f = E\left[f_N^2(n)\right] = \frac{1}{2\pi}\int_{-\pi}^{\pi} S_f(\omega)d\omega \;)$$

From Eqn. (7.37b),

$$\varepsilon_N^f = \frac{1}{2\pi}\int_{-\pi}^{\pi} \left|A_N(e^{j\omega})\right|^2 S_x(\omega)d\omega \tag{7.37c}$$

If $z_1, z_2, \cdots z_n$ are the roots of $A_N(z)$

$$A_N(z) = \prod_{m=1}^{N}(1 - z_m z^{-1})$$

If z_i is a root with $z_i = r_i\, e^{j\omega_i}$, then

$$A_N(z) = Q_i(z)(1 - z_i z^{-1}), \quad Q_i(z) = \prod_{\substack{m=1\\ m\neq i}}^{N}(1 - z_m z^{-1})$$

Therefore, Eqn. (7.37c),

$$\varepsilon_N^f = \frac{1}{2\pi}\int_{-\pi}^{\pi} \left|A_N(e^{j\omega})\right|^2 S_x(\omega)\,d\omega$$

becomes

$$\varepsilon_N^f = \frac{1}{2\pi}\int_{-\pi}^{\pi} S_x(\omega)\left|Q_i(e^{j\omega})\right|^2 \left|1 - z_i e^{-j\omega}\right|^2 d\omega$$

$$= \frac{1}{2\pi}\int_{-\pi}^{\pi} S_x(\omega)\left|Q_i(e^{j\omega})\right|^2 \left|1 - r_i e^{j\omega_i} e^{-j\omega}\right|^2 d\omega$$

$$= \frac{1}{2\pi}\int_{-\pi}^{\pi} S_x(\omega)\left|Q_i(e^{j\omega})\right|^2 \left|1 - r_i\left\{\cos(\omega - \omega_i) - j\sin(\omega - \omega_i)\right\}\right|^2 d\omega$$

$$= \frac{1}{2\pi}\int_{-\pi}^{\pi} S_x(\omega)\left|Q_i(e^{j\omega})\right|^2 \left[\{1 - r_i\cos(\omega-\omega_i)\}^2 + r_i^2\sin^2(\omega-\omega_i)\right] d\omega$$

$$= \frac{1}{2\pi}\int_{-\pi}^{\pi} S_x(\omega)\left|Q_i(e^{j\omega})\right|^2$$

$$\left[1 - 2\lambda_i\cos(\omega-\omega_i) + \underline{r_i^2\cos^2(\omega-\omega_i) + r_i^2\sin^2(\omega-\omega_i)}\right] d\omega$$

$$\varepsilon_N^f = \frac{1}{2\pi}\int_{-\pi}^{\pi} S_x(\omega)\left|Q_i(e^{j\omega})\right|^2 \left[1 - 2r_i\cos(\omega-\omega_i) + r_i^2\right] d\omega$$

For the prediction filter to be optimum, $A_M(z)$ should be chosen such that

$$\frac{\partial\, \varepsilon_N^f}{\partial a_i} = 0, \qquad i = 1,2,3\ldots N$$

$$\varepsilon_N^f = \frac{1}{2\pi}\int_{-\pi}^{\pi} S_x(\omega)\left|Q_i(e^{j\omega})\right|^2 \left[1 - 2r_i\cos(\omega-\omega_i) + r_i^2\right] d\omega$$

Therefore,

$$\frac{\partial\, \varepsilon_N^f}{\partial r_i} = 0$$

$$\frac{\partial\, \varepsilon_N^f}{\partial r_i} = \frac{1}{\pi}\int_{-\pi}^{\pi} S_x(\omega)\left|Q_i(\omega)\right|^2 \left[r_i - \cos(\omega-\omega_i)\right] d\omega \qquad (7.37d)$$

Let $r_i > 1$, then z_i lies outside the unit circle. Since $-1 \le \cos(\omega-\omega_i) \le 1$,

$$r_i - \cos(\omega-\omega_i) > 0 \text{ as } r_i > 1.$$

Further, as $S_x(\omega) > 0$ and $\left|Q_i(\omega)\right|^2 > 0$, from Eqn. (7.37d),

$$\frac{\partial\, \varepsilon_N^f}{\partial r_i} > 0, \text{ for } r_i > 1.$$

But *for the filter to be optimum,*

$$\frac{\partial\, \varepsilon_N^f}{\partial r_i} = 0$$

and hence $A_N(z)$ *cannot have zero outside the unit circle.*

For a sinusoidal process,

$$x(n) = \sum_{k=1}^{K} \sqrt{P_k} e^{j(n\omega_k + \theta_k)} \quad (7.37e)$$

θ_k is the phase $(-\pi, \pi)$.

For this process the PSD is discrete and is

$$S_x(\omega) = \sum_{k=1}^{K} P_k \delta(\omega - \omega_k) \quad (7.37f)$$

For a forward prediction filter with $N > K$,

$$\varepsilon_N^f = \frac{1}{2\pi} \int_{-\pi}^{\pi} \left| A_m(e^{j\omega}) \right|^2 S_x(\omega) d\omega$$

$$= \frac{1}{2\pi} \int_{-\pi}^{\pi} \sum_{k=1}^{K} P_k \delta(\omega - \omega_k) \left| A_N(\omega) \right|^2 d\omega = \sum_{k=1}^{K} P_k \left| A_N(\omega_k) \right|^2$$

$$\varepsilon_M^f = P_1 \left| A_N(\omega_1) \right|^2 + P_2 \left| A_N(\omega_2) \right|^2 + \ldots P_K \left| A_N(\omega_K) \right|^2 \quad (7.37g)$$

If K of the N zeros are chosen to lie on the unit circle, then

$$A_N(\omega_1) = A_N(\omega_2) = \ldots = A_N(\omega_k) = 0$$

- For the sinusoidal process $x(n)$ with discrete spectrum with $K < N$, the MSE can be minimized and can be made even zero by choosing K of its N zeros on the unit circle and the remaining $(N - K)$ zero anywhere *inside the unit circle.*
- But for the case $K = N$, all the zero *must be on the unit circle.*

For a process, which does not have exactly discrete spectrum, all the zeros of $A_N(z)$ must be inside the unit circle.

If a zero is on the unit circle, $r_i = 1$, using Eqn.(7.37d),

$$\frac{\partial \varepsilon_N^f}{\partial \rho_i} = \frac{1}{\pi} \int_{-\pi}^{\pi} S_x(\omega) \left| Q_i(\omega) \right|^2 \left[r_i - \cos(\omega - \omega_i) \right] d\omega$$

$$\frac{\partial \varepsilon_N^f}{\partial \rho_i} = \frac{1}{\pi} \int_{-\pi}^{\pi} S_x(\omega) \left| Q_i(\omega) \right|^2 \left[1 - \cos(\omega - \omega_i) \right] d\omega \quad (7.37h)$$

If $S_x(\omega)$ does not have only the δ functions up to M, then as

$1 - \cos(\omega - \omega_i) \geq 0$.

and as $S_x(\omega) \neq 0$,

$$\frac{\partial \varepsilon_M^f}{\partial r_i} > 0$$

Therefore, if the PSD of the input is continuous or mixed with discrete components (up to N), *it is impossible to have a* $A_N(z)$ *to have zero on the unit circle and at the same time optimize* $A_N(z)$ *for minimum mean square error.*

That is, if the power spectrum is not exactly discrete, $A_N(z)$ *has all its zeros inside the unit circle.*

7.9 AUTOREGRESSIVE (AR) MODELING

Modeling a time sequence by a parametric model such as an autoregressive model, offers several advantages over the direct spectral representation by Fourier transform. A model offers insight into the structure of the process, such as the number of resonances, and produces a high resolution spectral estimate in a compact form represented by a few parameters.

A signal $x(n)$ at instant n can be represented by a linear predictive AR model as

$$x(n) = \sum_{i=1}^{p} h_i \, x(n-i) \quad + \quad e(n) \tag{7.38a}$$

where, $x(n)$ is the given signal at instant n

p is the order of the model

h_i are the predictor coefficients

$e(n)$ is the white noise in the signal not modeled by the chosen model.

Therefore, $$e(n) = x(n) \quad - \quad \sum_{i=1}^{p} h_i \, x(n-i)$$

If $$a_0^p = 1, a_i^p = -h_i$$

Then $$e(n) = \sum_{i=0}^{p} a_i^p \, x(n-i) \tag{7.38b}$$

The output power spectral density $S_x(\omega)$ is related to the input noise power spectral density $S_e(\omega) = \varepsilon_p$. This is given by

$$S_x(\omega) = \frac{\varepsilon_p}{|1 + \sum_{i=1}^{p} a_i^p \, e^{-ji\omega}|^2} \tag{7.38c}$$

Eqn. (7.38c) is termed the AR spectral estimate, where a_i^p is the ith AR parameter of the pth order AR process.

As already pointed out *even for short datalength, the AR parametric methods provide higher frequency resolution and lower variance than those by FT based methods.* The parameters have the required *accurate* spectral magnitude information as they have been estimated *in time domain without going to frequency domain.* This is in contrast to the FT magnitude which suffers from leakage and the effect of window applied. Further the model being an IIR, represents the poles/ spectral peaks very well.

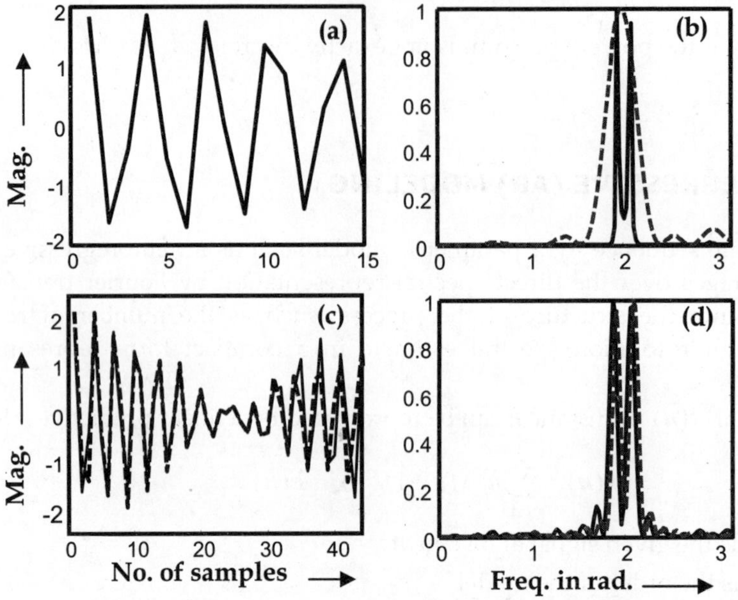

Fig. 7.6 Illustration of high frequency resolution by AR model.
(a) Segment of 15 samples of original data;
(b) Its spectra: FT spectrum - - -, AR model ___;
(c) Original data of length 45 samples ___
Extended data by the AR model - - - - and
(d) Spectra of original data - - - - and extended data ___

For the case of a very short datalength, where the FT cannot resolve the spectral peaks, the AR spectrum does. This can be seen from the ability of the AR model to extend data by prediction. For the prediction of data beyond the accurate spectral magnitude information is in the model parameters and spectral phase is provided by the data (available) by itself. To show this effect, a part of the segment of data available is taken (Fig. 7.6(a)) and its AR parametrs are estimated by a standard method (Burg method to be described) and the data is extended by prediction on either side (Fig. 7.6(c)). The spectrum of the extended data and that of the original data of same length have almost the same magnitude spectrum (Fig. 7.6(d)) but the FT of the short data used for AR parameter estimation is having a very poor frequency resolution (Fig. 7.6(b)). This indicates the power of the AR parameters and shows its ability to extend the data. Hence, the spectrum of AR parameters by itself can give a superior frequency resolution than the FT of the data segment considered(Fig. 7.6(b)). The extension of data by prediction may give a feeling that it is possible to extend the data infinitely and get infinite frequency resolution. But this is not true, as the farther predictions depend upon the predicted data and the accuracy of the prediction reduces resulting in only a better but not an infinite frequency resolution. Further, as the model fits a polynomial to the fluctuating spectral data, it only provides the envelope or gross feature of the spectrum, rejecting the unwanted finer details. This leads to an estimate which has a lower variance than that of the FT based estimate.

Qualitatively, it can be said that by predicting the data by an AR model, the data has been allowed to take its own course and has not been constrained to be zero or periodic. That is, maximum freedom possible or maximum uncertainty has been provided by the AR model by prediction outside the available data. By prediction it implies, using the given data. That is, the data is extended with the utmost uncertainty or freedom satisfying the given data. As the entropy by itself is the measure of total uncertainty/probability, maximum uncertainty corresponds to maximum entropy. Hence, AR model based method of spectral estimation is also called *maximum entropy spectral estimation.*

If some event is known already, it is not an information and only unknown / uncertain event gives information. Hence, any thing that is uncertian provides information and measure of uncertinity is a measure of information. Less probable event gives more information and hence information and probability are inversely related.

The average information or entropy H is given by

$$H = \int_{-\infty}^{\infty} f_x(X)\ln\left[\frac{1}{f_x(X)}\right]dX = -\int_{-\infty}^{\infty} f_x(X)\ln\left[f_x(X)\right]dX = -E\left[\ln f_x(X)\right] \tag{7.39a}$$

The entropy of a random vector of $(P+1)$ random variables of a Gaussian process is given by

$$H = -E\left[\ln f_x(X)\right] = (P+1) + \ln\pi + \ln R_{P,P} \tag{7.39b}$$

The entropy per sample ΔH is related to the power spectrum by

$$\Delta H = -\frac{1}{2\pi}\int_{-\pi}^{\pi} \ln S_x(\omega)d\omega \ + \eta, \quad \eta: \text{constant} \tag{7.39c}$$

Since

$$R_{P,P}(k) = \frac{1}{2\pi}\int_{-\pi}^{\pi} S_x(\omega)e^{j\omega k}d\omega$$

Where $R_{P,P}(k), k = \pm 0, \pm 1, \pm 2, \ldots \pm P$ are the *given* correlation values. The values of $R_{P,P}(k), \quad k > P$ are to be chosen to maximize ΔH. That is,

$$\frac{\partial}{\partial R_{P,P}(k)}\left[\Delta H\right] = \frac{1}{2\pi}\int_{-\pi}^{\pi}\frac{1}{S_x(\omega)}\nabla_{R_{P,P}(k)}\left[S_x(\omega)\right]d\omega = 0, \quad k > P \tag{7.39d}$$

$$\nabla_{R_{P,P}(k)}\left[S_x(\omega)\right] = \nabla_{R_{P,P}(k)}\sum_{k=\infty}^{\infty} R_{P,P}(k)e^{-j\omega k}$$

$$= \nabla_{R_{P,P}(k)}\sum_{k=\infty}^{\infty} R_{P,P}(k)e^{j\omega k} = e^{j\omega k}$$

Let

$$G_x(\omega) = \frac{1}{S_x(\omega)},$$

then,

$$g(k) = \frac{1}{2\pi}\int_{-\pi}^{\pi} G_x(\omega)e^{j\omega k}d\omega = \frac{1}{2\pi}\int_{-\pi}^{\pi}\frac{1}{S_x(\omega)}e^{j\omega k}d\omega$$

According to Eqn. (7.39b), $g(k)$ has finite length and is only non-zero in the interval $[-P, P]$. That is,

$$G_x(\omega) = \frac{1}{S_x(\omega)} = \sum_{k=-P}^{P} g(k)e^{-j\omega k}$$

or

$$S_x(\omega) = \frac{1}{\sum_{k=-P}^{P} g(k)e^{-j\omega k}} = \frac{1}{G(\omega)} \tag{7.39e}$$

In Z- domain,

$$S_x(z) = \frac{1}{\sum_{k=-P}^{P} g(k)z^{-k}} = \frac{1}{G(z)}.$$

Since it is possible to factorize the spectral density function,

$$S_x(\omega) = \frac{K_0}{C(z)C^*(1/z^*)} = \frac{K_0}{\left(\sum_{k=0}^{P} c(k)z^{-k}\right)\left(\sum_{k=0}^{P} c^*(k)z^{k}\right)}$$

$$S_x(\omega) = \frac{1}{|\sum_{k=0}^{P} c(k)z^{-k}|^2}, \quad \text{with } c(0) = 1,\ K_0 \text{ a constant} \tag{7.39f}$$

This implies that the MEM random process is the output of white noise driven linear filter

$H_m(\omega) = \left[\sum_{k=0}^{P} c(k) z^{-k}\right]^{-1}$. Thus, the maximum entropy random process is represented by a *autoregressive/ all-pole model.* The power spectral density of a MEM random process is given by Eqn. (7.41)(7.39f). This is nothing but the spectrum of an AR process with coefficients $c(k)$ and driven by a white noise. Thus the maximum entropy random process is represented by an AR model.

7.10 AR MODELING METHODS

There are various methods for extracting the AR model parameters and they basically minimise the error energy between the present sample and its estimate. For the minimization, the part of the available data and the constraint used determines the method. The method which uses complete data available and uses minimum or no constraints, provides good results. This trend can be seen in the evolution of these methods and are described in the same order.

- Autocorrelation or Yule-Walker method
- Covariance method
- Burg method
- Modified covariance method

7.10.1 Yule-Walker Method

In the autocorrelation method of all-pole modeling, the AR coefficients are found by solving the autocorrelation normal equations. Autocorrelation normal equations can be obtained as follows.

$$e(n) = \sum_{i=0}^{p} a_i^p \; x(n-i).$$

Can be written in vector form as

$$e(n) = X_{p,n}^T A_p \tag{7.40a}$$

where $A_p^T = [1, a_1^p, \cdots, a_p^p]$, $X_{p,n}^T = [x(n), x(n-1), \cdots, x(n-p)]$, and T denotes vector transpose. If each side of Eqn. (7.42) (7.40a) is pre-multiplied by the vector $X_{p,n}$, and the expected value taken, then

$$X_{p,n} e(n) = X_{p,n} X_{p,n}^T A_p$$

$$E\left[X_{p,n}\, e(n)\right] = E\left[X_{p,n}\, X_{p,n}^T\right] A_p$$

$$R_p\, A_p = E_p \tag{7.40b}$$

where

$$R_p = E\left[X_{p,n}\, X_{p,n}^T\right] = \begin{bmatrix} \phi_x(0) & \phi_x(1) & \cdots & \phi_x(p) \\ \phi_x(1) & \phi_x(2) & \cdots & \phi_x(p-1) \\ \vdots & \vdots & \ddots & \vdots \\ \phi_x(p) & \phi_x(p-1) & \cdots & \phi_x(0) \end{bmatrix}$$

is $(p+1)\times(p+1)$ Toeplitz autocorrelation matrix and

$$\varphi_x(i) = \frac{1}{N}\sum_{n=0}^{N-i-1} x(n+i)\, x\,(n) \quad \text{for } i = 0, 1, \cdots\cdots, p \tag{7.40c}$$

is the autocorrelation function at lag i.

$$E_p^T = [\varepsilon_p, 0, \cdots, 0]$$

where $\varepsilon_p = E\left[\,e(n)e(n)\right]$ is white noise power spectral density. The white noise $e(n)$ is uncorrelated with all $x(i)$, for $i < n$.

Thus,

$$E\left[\,e(n)x(n)\right] = E\left[\,e(n)e(n)\right].$$

The set of linear equations represented by Eqns. (7.43) & (7.40b) is called the autocorrelation normal equations or Yule-Walker normal equations.

The spectrum can be estimated by knowing the AR coefficients. Knowing the $p+1$ autocorrelation lags $\varphi_x(0), \cdots\cdots, \varphi_x(p)$, the solution for the AR parameter vector A_p in Eqns. (7.43) & (7.40b) can be computed by the recursive Levinson algorithm, in which the AR parameters for order p may be obtained recursively from the AR parameters previously determined for order $p-1$.

That is,

$$a_k^p = a_k^{(p-1)} + k_p\, a_{(p-k)}^{(p-1)} \quad \text{for} \quad k = 1, \cdots\cdots, p-1 \tag{7.40d}$$

$$a_p^p = k_p \tag{7.40e}$$

where

$$k_p = -\frac{\sum_{k=0}^{p-1} a_k^{(p-1)}\phi(p-k)}{\varepsilon_{(p-1)}} \tag{7.40f}$$

Recursion Eqn. (7.40d) is obtained by taking advantage of the positive definite Toeplitz structure of the autocorrelation matrix. The recursion for the noise spectral density is given by

$$\varepsilon_p = \varepsilon_{(p-1)}\left[1-\left|k_p\right|^2\right] \tag{7.40g}$$

Eqn. (7.40d) and the fact that $\left|k_p\right| \le 1$ are sufficient to guarantee that the roots of the polynomial

$$1 + \sum_{i=1}^{p} a_i^p z^{-i} = 0 \tag{7.40h}$$

are within the unit circle. That is, the modulus of each pole of Eqn. (7.40h) is less than unity. This guarantees that a synthesis of an AR filter using the AR parameters computed by the Levinson recursion will be a stable, minimum phase filter. Also, use of Eqn. (7.40c) for computation of biased form of autocorrelation function, guarantees a positive-definite autocorrelation matrix. *It is important to note that the Yule-Walker method is derived starting from the AR process and hence it assumes the process is an AR process.* However, other methods do not make such an assumption.

Biased form of the autocorrelation estimate ensures positive-definite autocorrelation matrix and hence stable AR model is guaranteed. Although stability is not a critical issue in power spectrum estimation, it is believed that a stable AR model best represents the data. Using unbiased autocorrelation estimates, autocorrelation matrix is not guaranteed to be positive-definite and as a result, the variance of the spectrum estimate tends to become large when the autocorrelation matrix is ill conditioned or singular. Therefore, the biased estimate is generally preferred over the unbiased estimate. Biased autocorrelation estimates reduce the risk of ill conditioning, but at the expense of a degradation of the AR spectral resolution and a shifting of spectral peaks from their true locations. The shift effect is termed frequency estimation bias. Autocorrelation method is rarely used since better resolution may be obtained by the method, which estimates the AR parameters directly from the data.

All-pole modeling with the autocorrelation method forces the signal to zero outside the considered interval by applying a window, even though an all-pole model implies that the signal is infinite in length. Therefore, it results in a decrease in the accuracy of model and generally produces a lower resolution estimate than the approaches that do not window the data, such as the Burg's method, covariance method and modified covariance method. Consequently, for short data records the autocorrelation method is not generally used.

7.10.2 Covariance Method

An all-pole model for the signal $x(n)$ specified for $n = 1, 2, \cdots, N$ can be estimated by minimizing the error energy ε_p, which is a sum of the squares of the error $e(n)$.

That is,

$$\varepsilon_p = \sum_{n=p+1}^{N} e^2(n) \tag{7.41a}$$

where

$$e(n) = \sum_{i=0}^{p} a_i^p \ x(n-i) \tag{7.41b}$$

The limit in Eqn. (7.41a) is because of the fact that in order to evaluate $e(n)$ at time n, it is necessary that the values of $x(n)$ be known at times $n, n-1, \cdots, n-p$ and $x(n)$ is specified only over the interval $[1, N]$, so, the error may only be evaluated for values of n in the interval $[p+1, N]$. Therefore, summation of ε_p includes only those values of $e(n)$ that may be computed from the data. Hence, minimizing ε_p by covariance method avoids any assumption about the data outside the interval $[1, N]$ and does not require that the data be windowed.

Covariance normal equations can be derived as follows: The error energy is

$$\varepsilon_p = \sum_{n=p+1}^{N} e^2(n)$$

From Eqn. (7.41b)

$$\varepsilon_p = \sum_{n=p+1}^{N} \left[\sum_{i=0}^{p} a_i^p \ x(n-i) \right]^2 \qquad i = 1, 2, \cdots, p$$

Therefore, the coefficients that minimize this squared error may be found by setting the partial derivatives of ε_p with respect to $a_i^{*(p)}$ equal to zero as follows

$$\frac{\partial \ \varepsilon_p}{\partial a_i^{(p)}} = \sum_{n=p+1}^{N} \frac{\partial \left[e^2(n) \right]}{\partial a_i^{(p)}} = \sum_{n=p+1}^{N} e(n) \frac{\partial e(n)}{\partial a_i^{(p)}} = 0, \ i = 1, 2, \cdots, p$$

Since the partial derivative of $e(n)$ with respect to $a_i^{(p)}$ is $x(n-i)$, above Eqn. reduces to

$$\sum_{n=p+1}^{N} e(n) \ x(n-i) = 0 \qquad i = 1, 2, \cdots, p \tag{7.41c}$$

This is well known orthogonality principle. Substituting Eqn. (7.41b) for $e(n)$ into Eqn. (7.41c) yields

$$\sum_{n=p+1}^{N} \left[\sum_{l=0}^{p} a_l^p \ x(n-l) \right] x(n-i) = 0, i = 1, 2, \cdots, p$$

or equivalently,

$$\sum_{l=1}^{p} a_l^p \left[\sum_{n=p+1}^{N} x(n-l) \ x(n-i) \right] = - \sum_{n=p+1}^{N} x(n) \ x(n-i) \ , \ i = 1, 2, \cdots, p \tag{7.41d}$$

Define,

$$\phi_x(i,l) = \sum_{n=p+1}^{N} x(n-l)\,x(n-i), \qquad i = 1, 2, \cdots, p$$

which is a symmetric function, $\phi_x(i,l) = \phi_x(l,i)$. In terms of this autocorrelation sequence, Eqn. (7.41d) reduces to

$$\sum_{l=1}^{p} a_l^p\,\phi_x(i,l) = -\phi_x(i,0) \qquad i = 1, 2, \cdots, p \tag{7.41e}$$

which is a set of p linear equations with p unknowns $a_1^p, a_2^p, \cdots, a_p^p$ known as covariance normal equations. In matrix form, the normal equations are

$$\begin{bmatrix} \phi_x(1,1) & \phi_x(1,2) & \cdots & \phi_x(1,p) \\ \phi_x(2,1) & \phi_x(2,2) & \cdots & \phi_x(2,p) \\ \vdots & \vdots & \ddots & \vdots \\ \phi_x(p,1) & \phi_x(p,2) & \cdots & \phi_x(p,p) \end{bmatrix} \begin{bmatrix} a_1^p \\ a_2^p \\ \vdots \\ a_p^p \end{bmatrix} = \begin{bmatrix} \phi_x(1,0) \\ \phi_x(2,0) \\ \vdots \\ \phi_x(p,0) \end{bmatrix} \tag{7.41f}$$

Here for example,

$$\phi_x(3,2) = \sum_{n=p+1}^{N} x[n-2]\,x[n-3] = \sum_{m=p}^{N-1} x(m-1)\,x(m-2)$$

$$\phi_x(3,2) = \phi_x(2,1) - \underline{x(N-1)\,x(N-2) + x(p-1)\,x(p-2)}$$

In general,

$$\phi_x(i+1,l+1) = \sum_{n=p+1}^{N} x[n-(l+1)]\,x[n-(i+1)] = \sum_{m=p}^{N-1} x(m-l)\,x(m-i)$$

$$\phi_x(i+1,l+1) = \phi_x(i,l) - \underline{x(N-l)\,x(N-i) + x(p-l)\,x(p-i)}$$

Therefore, the covariance normal equations are not Toeplitz. The Levinson-Durbin recursion cannot be used to solve Eqn. (7.41f).

Finally, the covariance modeling error is derived as follows

$$\varepsilon_{p(\min)} = \sum_{n=p+1}^{N} [e(n)]^2 = \sum_{n=p+1}^{N} e(n)\,e(n)$$

Substituting

$$e(n) = \sum_{i=0}^{p} a_i^p x(n-i) = x(n) + \sum_{i=1}^{p} a_i^p x(n-i)$$

$$\varepsilon_{p(\min)} = \sum_{n=p+1}^{N} e(n) \left[x(n) + \sum_{i=1}^{p} a_i^p x(n-i) \right]$$

$$= \sum_{n=p+1}^{N} e(n)x(n) + \sum_{n=p+1}^{N} e(n) \sum_{i=1}^{p} a_i^p x(n-i)$$

$$= \sum_{n=p+1}^{N} e(n)x(n) + \sum_{i=1}^{p} a_i^p \left(\sum_{n=p+1}^{N} e(n)x(n-i) \right)$$

$$\varepsilon_{p(\min)} = \sum_{n=p+1}^{N} e(n)x(n) + \sum_{i=1}^{p} a_i^p \left(\sum_{n=p+1}^{N} e(n)x(n-i) \right)$$

$$= \sum_{n=p+1}^{N} \left[x(n) + \sum_{i=1}^{p} a_i^p x(n-i) \right] x(n) + \sum_{i=1}^{p} a_i^p \left(\sum_{n=p+1}^{N} e(n)x(n-i) \right)$$

But, due to orthogonality condition (Eqn.(7.41c),

$$\left(\sum_{n=p+1}^{N} e(n)x(n-i) \right) = 0$$

Therefore,

$$\varepsilon_{p(\min)} = \phi_x(0,0) + \sum_{i=1}^{p} a_i^p \ \phi_x(0,i) \tag{7.41g}$$

Covariance method is also a method for all-pole signal modeling, that does not make any assumptions about the data outside the given observation interval and does not apply a window to the data in the formation of the autocorrelation estimates. As a result, the model obtained with the covariance method is typically more accurate than the autocorrelation method. Therefore, for short data records the covariance method generally produces higher resolution spectrum estimates than the autocorrelation method. However, as the data record length increases and becomes large compared to the model order, the effect of the data window becomes small and the difference between the two approaches becomes negligible. With the covariance method, however, the Toeplitz structure of the normal equations along with the guaranteed stability of the all pole model is lost.

7.10.3 Burg Method

In Burg algorithm, an all-pole model is estimated directly from the data. Also, the reflection coefficients are computed sequentially by minimizing the sum of the forward and backward prediction errors.

Assuming an all-pole stationary stochastic process, the forward linear prediction error

$$e_p^f(n) = x(n) + \sum_{i=1}^{p} a_i^p \ x(n-i) = \sum_{i=0}^{p} a_i^p \ x(n-i), \ p+1 \le n \le N \tag{7.42a}$$

and the backward linear prediction error

$$e_p^b(n) = \sum_{i=0}^{p} a_i^{(p)}\, x(n-p+i),\ p+1 \le n \le N \tag{7.42b}$$

where a_0^p is defined as unity. Since stationarity is assumed, the backward AR coefficients are simply the conjugates of the forward AR coefficients.

In Burg method, AR parameters are estimated by minimizing the sum of the forward and backward prediction error energies

$$\varepsilon_p = \sum_{n=p+1}^{N}\left[e_p^f(n)\right]^2 + \sum_{n=p+1}^{N}\left[e_p^b(n)\right]^2 \tag{7.42c}$$

subject to the constraint that the AR parameters satisfy the Levinson recursion, Eqn. (7.40d, e), for all orders 1 to p. This constraint is motivated to guarantee a stable AR filter (poles within the unit circle). A close evaluation of Eqn. (7.40g) and Eqn. (7.42c) shows that the prediction error energy ε_p at order p is a function of $k_p = a_p^p$, which is unknown, and the AR parameters $a_1^{(p-1)}$ through $a_{(p-1)}^{(p-1)}$, which will always be known from order $(p-1)$ of the recursion. Thus, Eqn. (7.42c) is a function of the single parameter k_p, often termed the reflection coefficient for its physical meaning in seismological applications.

Substituting Eqn. (7.40d, e) into Eqn. (7.42a) and Eqn. (7.42b) yields the following recursive relationships between the forward and backward prediction errors.

$$e_p^f(n) = e_{(p-1)}^f(n) + k_p\, e_{(p-1)}^b(n-1) \tag{7.42d}$$

$$e_p^b(n) = e_{(p-1)}^b(n-1) + k_p\, e_{(p-1)}^f(n), \quad \text{for } p+1 \le n \le N. \tag{7.42e}$$

Substituting Eqn. (7.42d) and Eqn. (7.42e) into Eqn. (7.42c) and minimizing with respect to k_p (set derivative to zero), it reduces to

$$\varepsilon_p = \sum_{n=p+1}^{N}\left[e_p^f(n)\right]^2 + \sum_{n=p+1}^{N}\left[e_p^b(n)\right]^2$$

$$\varepsilon_p = \sum_{n=p+1}^{N}\left[e_{(p-1)}^f(n) + k_p\, e_{(p-1)}^b(n-1)\right]^2$$

$$+ \sum_{n=p+1}^{N}\left[e_{(p-1)}^b(n-1) + k_p\, e_{(p-1)}^f(n)\right]^2$$

$$\frac{\partial\,\varepsilon_p}{\partial k_p} = 0 \quad \Rightarrow \quad \sum_{n=p+1}^{N} 2\left[e^f_{(p-1)}(n) + k_p\, e^b_{(p-1)}(n-1)\right]e^b_{(p-1)}(n-1)$$

$$+ \quad \sum_{n=p+1}^{N} 2\left[e^b_{(p-1)}(n-1) + k_p\, e^f_{(p-1)}(n)\right]e^f_{(p-1)}(n) = 0$$

$$\sum_{n=p+1}^{N} e^b_{(p-1)}(n-1)e^f_{(p-1)}(n) + \sum_{n=p+1}^{N} e^b_{(p-1)}(n-1)e^f_{(p-1)}(n) +$$

$$k_p \sum_{n=p+1}^{N}\left[e^b_{(p-1)}(n-1\right]^2 + k_p \sum_{n=p+1}^{N}\left[e^f_{(p-1)}(n)\right]^2 = 0$$

$$k_p \sum_{n=p+1}^{N}\left[e^b_{(p-1)}(n-1\right]^2 + k_p \sum_{n=p+1}^{N}\left[e^f_{(p-1)}(n\right]^2 = -2\sum_{n=p+1}^{N} e^b_{(p-1)}(n-1)e^f_{(p-1)}(n)$$

$$k_p \quad = \quad -\frac{2\sum_{n=p+1}^{N} e^f_{(p-1)}(n)\ e^b_{(p-1)}(n-1)}{\sum_{n=p+1}^{N}\left[\left[e^f_{(p-1)}(n\right]^2 + \left[e^b_{(p-1)}(n-1)\right]^2\right]} \tag{7.42f}$$

Eqn.(7.42f) ensures that $|k_p| \leq 1$, a necessary condition for a minimum phase linear prediction filter.

Interpretation of Burg estimate k_p : The expression for k_p can be written as

$$k_p \quad = \quad -\frac{\frac{1}{N}\sum_{n=p+1}^{N} e^f_{(p-1)}(n)\ e^b_{(p-1)}(n-1)}{\frac{1}{2}\left\{\frac{1}{N}\sum_{n=p+1}^{N}\left[e^f_{(p-1)}(n\right]^2 + \frac{1}{N}\sum_{n=p+1}^{N}\left[e^b_{(p-1)}(n-1)\right]^2\right\}} \tag{7.43a}$$

1. The numerator represents the averaged cross-correlation between FPE $e^f_{(p-1)}(n)$ and one step delayed BPE $e^b_{(p-1)}(n-1)$.
2. The denominator is the arithmetic mean of time averaged powers of FPE $e^f_{(p-1)}(n)$ and BPE $e^b_{(p-1)}(n-1)$.

The Burg algorithm satisfies certain properties and are considered here.

Property-1: The reflection coefficient of the Burg algorithm is the harmonic mean of the forward and backward reflection coefficients and this is known as the harmonic property of the Burg algorithm.

Proof: The MSE of FPE is

$$\varepsilon_p^f = \sum_{n=p+1}^{N} \left[e_p^f(n) \right]^2 \tag{7.43b}$$

$$e_p^f(n) = e_{p-1}^f(n) - K_p^f e_{p-1}^b(n-1)$$

$$\varepsilon_p^f = \sum_{n=p+1}^{N} \left[e_{p-1}^f(n) - K_p^f e_{p-1}^b(n-1) \right]^2$$

Minimizing the MSE with respect to K_p^f,

$$\frac{\partial \varepsilon_p^f}{\partial K_p^f} = 0, \qquad -2 \sum_{n=p+1}^{N} \left[e_{p-1}^f(n) - K_p^f e_{p-1}^b(n-1) \right] * e_{p-1}^b(n-1) = 0$$

$$\sum_{n=p+1}^{N} e_{p-1}^f(n) e_{p-1}^b(n-1) - K_p^f \sum_{n=p+1}^{N} \left[e_{p-1}^b(n-1) \right]^2 = 0$$

$$K_p^f = \frac{\sum_{n=p+1}^{N} e_{(p-1)}^f(n) \; e_{(p-1)}^b(n-1)}{\sum_{n=p+1}^{N} \left| e_{(p-1)}^b(n-1) \right|^2} \tag{7.43c}$$

Similarly, the MSE of BPE is

$$\varepsilon_p^b = \sum_{n=p+1}^{N} \left[e_p^b(n) \right]^2 \tag{7.43d}$$

$$e_p^b(n) = e_{p-1}^b(n-1) - K_p^b e_{p-1}^f(n)$$

$$\varepsilon_p^b = \sum_{n=p+1}^{N} \left[e_{p-1}^b(n-1) - K_p^b e_{p-1}^f(n) \right]^2$$

Minimizing the MSE with respect to K_p^b,

$$\frac{\partial \varepsilon_p^b}{\partial K_p^b} = 0, \qquad -2 \sum_{n=p+1}^{N} \left[e_{p-1}^b(n-1) - K_p^b e_{p-1}^f(n) \right] * e_{p-1}^f(n) = 0$$

$$\sum_{n=p+1}^{N} e_{p-1}^{f}(n)e_{p-1}^{b}(n-1) - K_p^b \sum_{n=p+1}^{N}\left[e_{p-1}^{f}(n)\right]^2 = 0$$

$$K_p^b = \frac{\sum_{n=p+1}^{N} e_{(p-1)}^{f}(n)\; e_{(p-1)}^{b}(n-1)}{\sum_{n=p+1}^{N}\left[e_{(p-1)}^{f}(n)\right]^2} \tag{7.43e}$$

Now harmonic mean c of variables a and b is defined as

$$\frac{1}{c} = \frac{1}{2}\left[\frac{1}{a}+\frac{1}{b}\right] \tag{7.43f}$$

Therefore,

$$\frac{1}{K_p^f(n)}+\frac{1}{K_p^b(n)} = \left[\frac{\sum_{n=p+1}^{N}\left[e_{(p-1)}^{f}(n\right]^2}{\sum_{n=p+1}^{N} e_{(p-1)}^{f}(n)\; e_{(p-1)}^{b}(n-1)} + \frac{\sum_{n=p+1}^{N}\left[e_{(p-1)}^{b}(n-1)\right]^2}{\sum_{n=p+1}^{N} e_{(p-1)}^{f}(n)\; e_{(p-1)}^{b}(n-1)}\right] = \frac{2}{K_p(n)}$$

$$\frac{1}{K_p(n)} = \frac{1}{2}\left[\frac{1}{K_p^f(n)}+\frac{1}{K_p^b(n)}\right] \tag{7.43g}$$

Thus, the Burg estimate of reflection coefficient $K_p(n)$ is the harmonic mean of the forward and backward reflection coefficients $K_p^f(n)$ and $K_p^b(n)$. *In view of this, Burg algorithm is known as harmonic mean algorithm.*

Property – 2: The Burg estimate of $K_p(n)$ satisfies

$$\left|K_p(n)\right| \le 1, \qquad \textit{for all } p \tag{7.44a}$$

Proof:

The time averaged correlation coefficient ρ_{p-1} of $e_{(p-1)}^{f}(n)$ and $e_{(p-1)}^{b}(n-1)$ is

$$\rho_{p-1} = \frac{\sum_{n=p+1}^{N} e_{(p-1)}^{f}(n)\; e_{(p-1)}^{b}(n-1)}{\sqrt{\sum_{n=p+1}^{N}\left[e_{(p-1)}^{f}(n)\right]^2}\sqrt{\sum_{n=p+1}^{N}\left[e_{(p-1)}^{b}(n-1)\right]^2}}, \quad \left|\rho_{p-1}\right| \le 1 \tag{7.44b}$$

Also let

$$a_{p-1} = \frac{\sqrt{\sum_{n=p+1}^{N}\left[e^{f}_{(p-1)}(n)\right]^2}}{\sqrt{\sum_{n=p+1}^{N}\left[e^{b}_{(p-1)}(n-1)\right]^2}} \tag{7.44c}$$

$$K^{f}_{p} = \frac{\sum_{n=p+1}^{N} e^{f}_{(p-1)}(n)\; e^{b}_{(p-1)}(n-1)}{\sum_{n=p+1}^{N}\left|e^{b}_{(p-1)}(n-1)\right|^2} \Rightarrow k^{f}_{p} = \rho_{p-1}\, a_{p-1} \tag{7.44d}$$

$$K^{b}_{p} = \frac{\sum_{n=p+1}^{N} e^{f}_{(p-1)}(n)\; e^{b}_{(p-1)}(n-1)}{\sum_{n=p+1}^{N}\left[e^{f}_{(p-1)}(n-1)\right]^2} \Rightarrow k^{b}_{p} = \rho_{p-1}\, a_{p-1} \tag{7.44e}$$

From the harmonic property of the Burg reflection coefficient (Eqn. (7.43g)),

$$\frac{1}{K_p} = \frac{1}{2}\left[\frac{1}{K^{f}_{p}} + \frac{1}{K^{b}_{p}}\right] = \frac{1}{2}\left[\frac{1}{\rho_{p-1}\, a_{p-1}} + \frac{a_{p-1}}{\rho_{p-1}}\right]$$

$$\frac{1}{K_p} = \frac{1}{2\rho_{p-1}}\left[\frac{1}{a_{p-1}} + a_{p-1}\right] \tag{7.44f}$$

Substituing for a_{p-1} from Eqn. (7.44c)

$$\frac{1}{a_{p-1}} + a_{p-1} = \frac{\sqrt{\sum_{n=p+1}^{N}\left[e^{b}_{(p-1)}(n-1)\right]^2}}{\sqrt{\sum_{n=p+1}^{N}\left[e^{f}_{(p-1)}(n)\right]^2}} + \frac{\sqrt{\sum_{n=p+1}^{N}\left[e^{f}_{(p-1)}(n)\right]^2}}{\sqrt{\sum_{n=p+1}^{N}\left[e^{b}_{(p-1)}(n-1)\right]^2}}$$

$$\frac{1}{a_{p-1}} + a_{p-1} = \frac{\sum_{n=p+1}^{N}\left[e^{b}_{(p-1)}(n-1)\right]^2 + \sum_{n=p+1}^{N}\left[e^{f}_{(p-1)}(n)\right]^2}{\sqrt{\sum_{n=p+1}^{N}\left[e^{f}_{(p-1)}(n)\right]^2}\sqrt{\sum_{n=p+1}^{N}\left[e^{b}_{(p-1)}(n-1)\right]^2}} \tag{7.44g}$$

Now consider

$$(a+b)^2 = a^2 + 2ab + b^2$$

$$(a+b)^2 - 2ab = a^2 + b^2, \qquad \frac{(a+b)^2}{ab} - 2 = \frac{a^2+b^2}{ab}$$

$$2 - \frac{(a+b)^2}{ab} = -\frac{a^2+b^2}{ab}, \qquad \Rightarrow -\frac{a^2+b^2}{ab} < 2 \qquad \Rightarrow \left|\frac{a^2+b^2}{ab}\right| > 2$$

$$\left|\frac{a^2+b^2}{ab}\right| > 2 \tag{7.44h}$$

Here,

$$a = a_{p-1}, \quad b = 1$$

$$\frac{a_{p-1}^2 + 1}{a_{p-1}} = \frac{1}{a_{p-1}} + a_{p-1} > 2 \tag{7.44i}$$

From Eqn. (7.44f)

$$\frac{1}{K_p} = \frac{1}{2\rho_{p-1}}\left[\frac{1}{a_{p-1}} + a_{p-1}\right],$$

From Eqn. (7.44b) and Eqn. (7.44i)

$$\left(1/\left|\rho_{p-1}\right|\right) \geq 1 \text{ and } \frac{1}{a_{p-1}} + a_{p-1} > 2$$

Therefore,

$$\left|\frac{1}{K_p}\right| = (\geq 1)\left(\frac{1}{2}\right)(>2) \geq 1$$

$$\left|\frac{1}{K_p}\right| \geq 1 \quad or \quad \left|K_p\right| \leq 1$$

This condition is necessary for the linear prediction filter to satisfy the minimum phase property.

Prediction error energy :

$$e_p^f(n) = e_{p-1}^f(n) - K_p^f e_{p-1}^b(n-1)$$

$$e_p^b(n) = e_{p-1}^b(n-1) - K_p^b e_{p-1}^f(n)$$

$$\varepsilon_p^f = \sum_{n=p+1}^{N} \left[e_{p-1}^f(n) - K_p^f e_{p-1}^b(n-1) \right]^2$$

$$\varepsilon_p^f = \sum_{n=p+1}^{N} \left[e_{p-1}^{f^2}(n) + K_p^{f^2} e_{p-1}^{b^2}(n-1) - 2K_p^f e_{p-1}^f(n) e_{p-1}^b(n-1) \right]$$

$$\varepsilon_p^f = \sum_{n=p+1}^{N} \left[\begin{array}{l} e_{p-1}^{f^2}(n) + K_p^2 e_{p-1}^{b^2}(n-1) \\ \quad - 2K_p \dfrac{K_p}{2} \left[e_{(p-1)}^{f^2}(n) + e_{(p-1)}^{b^2}(n-1) \right] \end{array} \right]$$

$$\varepsilon_p^f = \sum_{n=p+1}^{N} e_{p-1}^{f^2}(n) \left[1 - K_p^2 \right]$$

Similarly for,

$$\varepsilon_p^b = \sum_{n=p+1}^{N} \left[e_{p-1}^b(n-1) - K_p^b e_{p-1}^f(n) \right]^2$$

$$\varepsilon_p^b = \sum_{n=p+1}^{N} e_{p-1}^{b^2}(n-1) \left[1 - K_p^2 \right]$$

$$\varepsilon_p = \varepsilon_p^f + \varepsilon_p^b = \sum_{n=p+1}^{N} e_{p-1}^{f^2}(n) \left[1 - K_p^2 \right] + \sum_{n=p+1}^{N} e_{p-1}^{b^2}(n-1) \left[1 - K_p^2 \right]$$

$$\varepsilon_p = \varepsilon_p^f + \varepsilon_p^b = \left[1 - K_p^2 \right] \underline{\sum_{n=p+1}^{N} \left[e_{p-1}^{f^2}(n) + e_{p-1}^{b^2}(n-1) \right]}$$

Therefore,

$$\varepsilon_p = \varepsilon_{(p-1)} \left[1 - |K_p|^2 \right] \tag{7.45}$$

That is, the sum of the forward and backward prediction error energies for p^{th} order filter can be obtained from $(p-1)^{th}$ order filter.

The Burg algorithm finds a set of all-pole model parameters which minimize the sum of the squares of the forward and backward prediction errors. In order to assure that the model is stable, this minimization is performed sequentially with respect to the reflection coefficients. Although Burg algorithm is less accurate than the modified covariance method, it is more accurate than the autocorrelation method since it does not apply any window to the data.

The major advantages of the Burg method for estimating the parameters of the AR model are: it results in high frequency resolution, it yields a stable AR model, and it is computationally efficient. The Burg method is known to have several disadvantages. First, it exhibits spectral line splitting at high signal to noise ratios. By line splitting, it refers to the occurrence of two or more closely spaced spectral peaks in an AR spectral estimate where only one spectral peak should be present. Spectral line splitting occurs most likely, when: the signal to noise ratio is high, the phase offset (initial phase) of sinusoidal components is an odd multiple of $45°$, the time duration of the data sequence is such that sinusoidal components have an odd number of quarter cycles, the model order is a large percentage of the number of data values used for the estimation. For high order models, the method also introduces spurious peaks. The reason for spurious peaks is that, the use of a large model order results in extraneous poles for the transfer function of the all-pole AR model (prediction error filter). Some of these extraneous poles tend to lie close to the unit circle in the Z-plane, thereby giving rise to spurious peaks in the estimated AR power spectrum. Furthermore, for sinusoidal signals in noise, the Burg method exhibits sensitivity to the initial phase of a sinusoid; especially in short data records. This sensitivity manifests as a frequency shift from the true frequency, resulting in a frequency bias that is phase dependent. Burg algorithm utilizes a constrained least squares estimation procedure to obtain the M model parameters from N data samples. The constraint requires the AR parameter estimates to satisfy the Levinson recursion.

For large data block length, Yule-Walker and Burg algorithms yield similar results. For small data block length, the Burg algorithm is preferred over the Yule-Walker algorithm in the sense that it usually yields a better spectral resolution.

Modified Burg method

The line splitting and some of the errors associated with the Burg method can be overcome by using a smoothing window function and is known as modified Burg method. For this, basically, a windowed squared forward and backward prediction errors are used for error energy minimization. The minimization of such energy with respect to reflection coefficient, is carried out in the same manner as in original Burg method. The error energy

$$\varepsilon_p = \sum_{n=p+1}^{N} \left| e_p^f(n) \right|^2 + \sum_{n=p+1}^{N} \left| e_p^b(n) \right|^2$$

is modified as

$$\varepsilon_p = \sum_{n=p+1}^{N} w(n) \left\{ \left[e_p^f(n \right]^2 + \left[e_p^b(n \right]^2 \right\} \tag{7.46a}$$

$w(n)$ is the window function and its length is a function of the filter / model order p due to the lower limit of summation $n = p+1$. Hence

$$\varepsilon_p = \sum_{n=p+1}^{N} w_{p-1}(n) \left[e_{(p-1)}^f(n) + k_p\, e_{(p-1)}^b(n-1) \right]^2$$
$$+ \sum_{n=p+1}^{N} w_{p-1}(n) \left[e_{(p-1)}^b(n-1) + k_p\, e_{(p-1)}^f(n) \right]^2$$

$$\frac{\partial \varepsilon_p}{\partial k_p} = 0 \quad \Rightarrow \quad \sum_{n=p+1}^{N} 2\, w_{p-1}(n)\left[e^f_{(p-1)}(n) + k_p\, e^b_{(p-1)}(n-1) \right] e^b_{(p-1)}(n-1)$$

$$+ \quad \sum_{n=p+1}^{N} 2w_{p-1}(n) \left[e^b_{(p-1)}(n-1) + k_p\, e^f_{(p-1)}(n) \right] e^f_{(p-1)}(n) = 0$$

$$\sum_{n=p+1}^{N} w_{p-1}(n) e^b_{(p-1)}(n-1) e^f_{(p-1)}(n) + \sum_{n=p+1}^{N} w_{p-1}(n) e^b_{(p-1)}(n-1) e^f_{(p-1)}(n) +$$

$$k_p \sum_{n=p+1}^{N} w_{p-1}(n)\left[e^b_{(p-1)}(n-1 \right]^2 + k_p \sum_{n=p+1}^{N} w_{p-1}(n)\left[e^f_{(p-1)}(n) \right]^2 = 0$$

$$k_p \sum_{n=p+1}^{N} w_{p-1}(n)\left\{ \left[e^b_{(p-1)}(n-1 \right]^2 + \left[e^f_{(p-1)}(n \right]^2 \right\} + = -2 \sum_{n=p+1}^{N} w_{p-1}(n) e^b_{(p-1)}(n-1) e^f_{(p-1)}(n)$$

The reflection coefficient is given by

$$k_p \; = \; - \frac{2 \sum_{n=p+1}^{N} w_{p-1}(n)\left[e^f_{(p-1)}(n)\;\; e^b_{(p-1)}(n-1) \right]}{\sum_{n=p+1}^{N} w_{p-1}(n)\left[\left[e^f_{(p-1)}(n \right]^2 + \left[e^b_{(p-1)}(n-1) \right]^2 \right]} \tag{7.46b}$$

A nonnegative window function will not alter the original characteristics of the reflection coefficient i.e., $\left|k_p\right| < 1$ and a Hamming window which has all its values positive can be used. The Hamming window function is

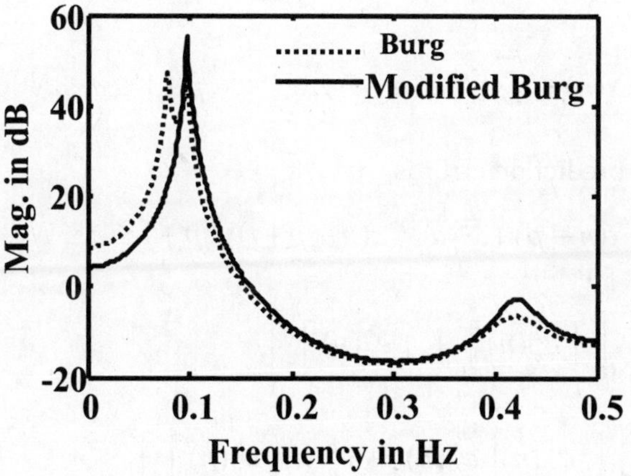

Fig. 7. 7 Comparion of spectral line splitting effect of Burg and Modified Burg algorithms.

$$w_p(n) = 0.54 - 0.46\cos\left(\frac{2\pi(n+1)}{N-p}\right), \quad n = 0, 1, \ldots N-1-p \tag{7.46c}$$

To bring out the performance comparison of Burg and modified Burg methods, an input signal $x(n)$ consisting of a cosinusoid

$$\cos(2\pi f_0 n + \theta) + \sigma_v v(n), \quad \sigma_v^2 = \frac{1}{2}\left(10^{-SNR/10}\right)$$

of frequency $f_0 = 0.1\text{Hz}$ and phase $\theta = -\pi/4$ plus white Gaussian noise to give an $SNR = 19.83\text{dB}$, is considered. The length of the data and the model order, considered are 18 and 6, respectively. Further, the sampling frequency used is 1Hz. It is seen from Fig. 7.7 that the original Burg method suffers from spectral line splitting. But the modified Burg algorithm, which uses a Hamming window, overcomes the undesired effect of line splitting showing a single peak at $f_0 = 0.1\,\text{Hz}$ instead of two peaks as in original Burgs method.

7.10.4 Modified Covariance Method

Modified covariance method or forward-backward algorithm for all-pole signal modeling minimizes the sum of the squares of the forward and backward prediction error energies

$$\varepsilon_p = \varepsilon_p^f + \varepsilon_p^b \tag{7.47a}$$

Here, the minimization is not performed sequentially, i.e., for a pth order model, the modified covariance method finds the AR parameters a_i^p, that minimize ε_p. This is achieved by setting the derivative of ε_p w. r. t $a_i^{(p)}$ equal to zero for $i = 1, 2, \cdots, p$.

Assuming an all-pole stationary stochastic process, the forward linear prediction error is

$$e_p^f(n) = x(n) + \sum_{i=1}^{p} a_i^{(p)}\, x(n-i)\,, \text{for} \quad p+1 \le n \le N \tag{7.47b}$$

and the backward linear prediction error is

$$e_p^b(n) = x(n-p) + \sum_{i=1}^{p} a_i^{(p)}\, x(n-p+i), \quad p+1 \le n \le N \tag{7.47c}$$

$$\varepsilon_p = \sum_{n=p+1}^{N}\left[\left[e_p^f(n)\right]^2 + \left[e_p^b(n)\right]^2\right]$$

$$= \sum_{n=p+1}^{N}\left[\, e_p^f(n)\left[e_p^f(n)\right] + e_p^b(n)\left[e_p^b(n)\right]\right]$$

Then,

$$\frac{\partial \varepsilon_p}{\partial a_i^{(p)}} = \sum_{n=p+1}^{N}\left[e_p^f(n)\frac{\partial\left[e_p^f(n)\right]}{\partial a_i^{(p)}} + \left[e_p^b(n)\right]\frac{\partial e_p^b(n)}{\partial a_i^{(p)}}\right] = 0$$

Using Eqns.(7.47b and 7.47c),

$$\sum_{n=p+1}^{N}\left[e_p^f(n)\, x(n-i) + \left[e_p^b(n)\right]x(n-p+i)\right] = 0\,, i = 1, 2, \cdots, p \tag{7.47d}$$

Substituting Eqn.(7.47b)and Eqn.(7.47c) into Eqn.(7.47d) the normal equations can be obtained as follows

$$\sum_{n=p+1}^{N}\left\{\left[x(n) + \sum_{l=1}^{p} a_l^p\, x(n-l)\right]x(n-i)\right.$$

$$\left. + \left[x(n-p) + \sum_{l=1}^{p} a_l^{(p)}\, x(n-p+l)\right]x(n-p+i)\right\} = 0$$

$$\sum_{n=p+1}^{N} x(n)x(n-i) + \sum_{n=p+1}^{N}\sum_{l=1}^{p} a_l^p x(n-l)x(n-i) +$$

$$\sum_{n=p+1}^{N} x(n-p)x(n-p+i) + \sum_{n=p+1}^{N}\sum_{l=1}^{p} a_l^p x(n-p+l)x(n-p+i) = 0$$

Therefore,

$$\phi_x(i,0) + \sum_{l=1}^{p} a_l^p \phi_x(i,l) + \phi_x(p,p-i) + \sum_{l=1}^{p} a_l^p \phi_x(p-l,p-i) = 0$$

$$\sum_{l=1}^{p}\left[\phi_x(i,l) + \phi_x(p-l,p-i)\right]a_l^p = -\left[\phi_x(i,0) + \phi_x(p,p-i)\right], i = 1, 2, \cdots, p \tag{7.48a}$$

where

$$\phi_x(i,l) = \sum_{n=p+1}^{N} x(n-l)\, x(n-i) \tag{7.48b}$$

Finally, the modified covariance modeling error is derived as follows

$$\varepsilon_{p(\min)} = \sum_{n=p+1}^{N}\left[\left|e_p^f(n)\right|^2 + \left|e_p^b(n)\right|^2\right]$$

$$= \sum_{n=p+1}^{N} \left[e_p^f(n)\left[e_p^f(n)\right] + e_p^b(n)\left[e_p^b(n)\right] \right]$$

$$= \sum_{n=p+1}^{N} \left\{ e_p^f(n)\left[x(n) + \sum_{i=1}^{p} a_i^p \; x(n-i) \right] + \right.$$

$$\left. \left[x(n-p) + \sum_{i=1}^{p} a_i^{(p)} \; x(n-p+i) \right]\left[e_p^b(n)\right] \right\}$$

$$= \sum_{n=p+1}^{N} \left\{ e_p^f(n)x(n) + \sum_{i=1}^{p} a_i^p e_p^f(n) \; x(n-i) + \right.$$

$$\left. x(n-p)\left[e_p^b(n)\right] + \sum_{i=1}^{p} a_i^{(p)} x(n-p+i)\left[e_p^b(n)\right] \right\}$$

$$= \sum_{n=p+1}^{N} \left\{ e_p^f(n)x(n) + x(n-p)\left[e_p^b(n)\right] + \right.$$

$$\left. \sum_{i=1}^{p} a_i^p \left[e_p^f(n) \; x(n-i) + \left[e_p^b(n)\right] x(n-p+i) \right] \right\}$$

But

$$\sum_{n=p+1}^{N} e_p^f(n) \; x(n-i) + \left[e_p^b(n)\right] x(n-p+i) = 0 \qquad \text{(From Eqn. (7.47d))}$$

Therefore,

$$\varepsilon_{p(\min)} = \sum_{n=p+1}^{N} e_p^f(n)x(n) + x(n-p)\left[e_p^b(n)\right]$$

Substituting for $e_p^f(n)$ and $e_p^b(n)$ from Eqn. (7.47b) and Eqn. (7.47c)

$$\varepsilon_{p(\min)} = \sum_{n=p+1}^{N} \left[x(n) + \sum_{i=1}^{p} a_i^p \; x(n-i) \right] x(n)$$

$$+ \, x(n-p)\left[x(n-p) + \sum_{i=1}^{p} a_i^{(p)} \; x(n-p+i) \right]$$

$$= \sum_{n=p+1}^{N} \left\{ x^2(n) + \sum_{i=1}^{p} a_i^p \; x(n-i)x(n) \right.$$

$$\left. + x^2(n-p) + \sum_{i=1}^{p} a_i^{(p)} \; x(n-p)x(n-p+i) \right\}$$

$$= \sum_{n=p+1}^{N} \left\{ x^2(n) + x^2(n-p) \right.$$

$$\left. + \sum_{i=1}^{p} a_i^p \left[x(n-i)x(n) + x(n-p)x(n-p+i) \right] \right\}$$

$$\varepsilon_{p(\min)} = \phi_x(0,0) + \phi_x(p,p) + \sum_{i=1}^{p} a_i^p \left[\phi_x(0,i) + \phi_x(p-i,p) \right] \quad (7.48c)$$

Modified covariance method minimizes the sum of the squares of the forward and backward prediction errors. Here, the minimization is not performed sequentially. Although optimum in the sense of minimizing the Burg error, the model produced with this method is not guaranteed to be stable.

The modified covariance method is similar to the covariance method in that no window is applied to the data. However, instead of finding the autoregressive model that minimizes the sum of the squares of the forward prediction error, the modified covariance method minimizes the sum of the squares of the forward and backward prediction errors. As with the covariance method, the autocorrelation matrix is not Toeplitz. In contrast to other AR spectrum estimation techniques, the modified covariance method appears to give statistically stable spectrum estimates with high resolution. Furthermore, in the spectral analysis of sinusoids in white noise, although the modified covariance method is characterized by a shifting of the peaks from their true locations due to additive noise, this shifting appears to be less pronounced than with other autoregressive estimation techniques. In addition, the peak locations tend to be less sensitive to phase. Finally, unlike the previous methods, modified covariance method is not subjected to spectral line splitting.

7.10.5 Simulation Results

Fig. 7.5 illustrates the AR spectrum estimation techniques to analyze a short data record that is derived from a fourth order narrowband autoregressive process. The AR process is generated by the difference equation

$$x(n) = 2.7377\,x(n-1) - 3.7476\,x(n-2) + 2.6293\,x(n-3) - 0.9224\,x(n-4) + w(n)$$

where $w(n)$ is unit variance white Gaussian noise. Using data records of length $N = 128$, an ensemble of $L = 50$ spectrum estimates are computed by different methods. Figs. 7.8(a) to (d) shows overlay plots of these estimates by Yule-Walker method, Burg method, and covariance method and modified covariance method, respectively. In Figs. 7.8 (e) to (h), the mean and NRMSE computed by using an ensemble of 50 estimates for each method are plotted along with the true AR spectrum. Figs. 7.8 (e) to (h) correspond to Figs. 7.8 (a) to (d), respectively.

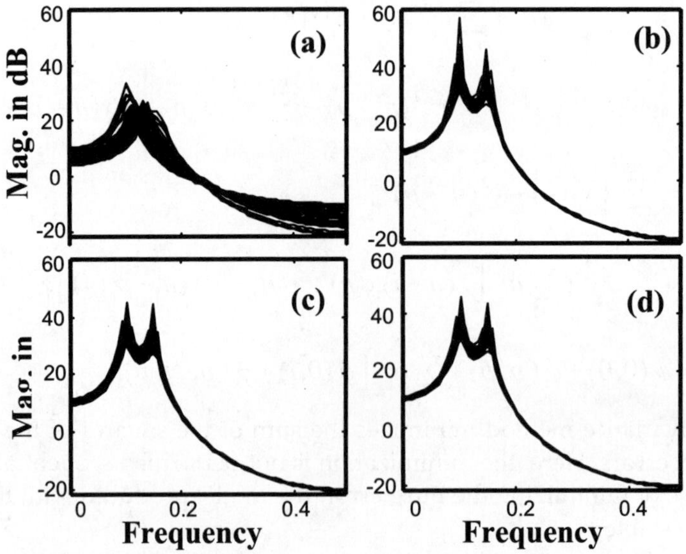

Fig. 7.8 Overlay of 50 spectral estimates of a fourth order autoregressive process by: (a) Yule-Walker; (b) Covariance; (c) Burg and (d) Modified covariance.

For this narrowband process, all the considered methods except the Yule-Walker method, gives less biased estimate with a comparable variance. Also, unlike the other three methods Yule-Walker method is unable to resolve the spectral peaks and has a larger variance. Covariance method (Fig. 7.8(c), (g)) tends to have more spiky estimates near the peaks of the spectrum, than seen in Burg method (Fig. 7.8(b), (f)) and modified covariance method (Fig. 7.8 (d), (h)). Modified covariance method (Fig. 7.8 (d)) exhibits less spread as compared to Burg method (Fig. 7.8 (b)). The normalized root mean square error (NRMSE) is given by

$$NRMSE(k) = 20 * \log_{10} \frac{RMSE(k)}{S(k)} \qquad 0 \le k \le \frac{N}{2} - 1$$

where, RMSE is the root mean square error and is

$$\text{RMSE}(\text{k}) = \sigma(k) = \left[\frac{1}{L} \sum_{i=1}^{L} \left\{ S(k) - \hat{S}_i(k) \right\}^2 \right]^{1/2} \qquad 0 \le k \le \frac{N}{2} - 1$$

$S(k)$ is the true AR spectrum and $\hat{S}_i(k)$ is the ith spectrum estimate, L is the number of trials. The NRMSE plots (broken line) for these methods is also given in Figs. (e) to (h). NRMSE for covariance method (Fig. 7.8 (g)) is more as compared to Burg method (Fig. 7.8(f)) and modified covariance method (Fig.7.8 (h)) especially at the peaks. This justifies that covariance method yields more spiky estimates near the peaks of the spectrum. Table 7.1 gives the percentage normalized sum square mean square error (%NSSMSE) for all the considered methods. %NSSMSE is given by

$$\%\text{NSSMSE} = \frac{\sum_{k=0}^{N/2-1} \sigma^2(k)}{\sum_{k=0}^{N/2-1} S^2(k)} \times 100$$

Table 7.1 % NSSMSE comparison

Method	%NSSMSE
Yule-Walker	69.23
Burg	14.22
Covariance	56.62
Modified covariance	13.42

As seen from the Table 7.1, the %NSSMSE is higher for Yule-Walker and Covariance method, lesser for Burg and modified covariance method. That is, the variance is higher in Yule-Walker and covariance method than the other two methods. Modified covariance method gives least variance.

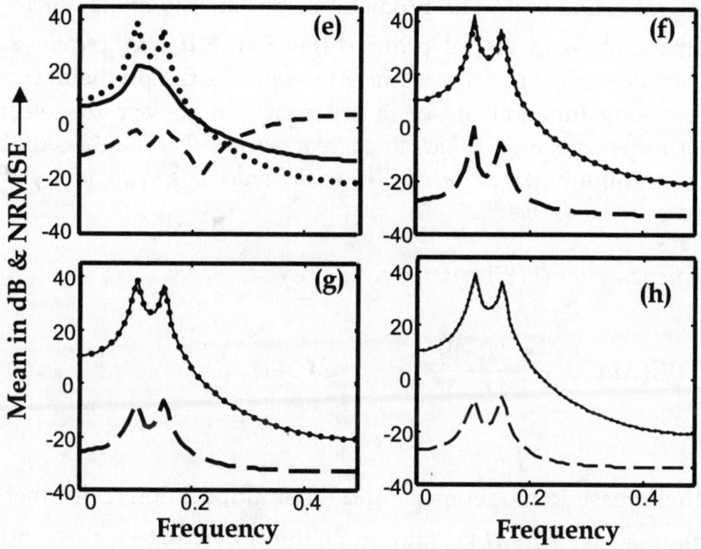

Fig. 7.8(e)-(h): Mean (__) & NRMSE (_ _ _) of estimates of Fig. 14.5 (a)-(d), respectively, compared with true spectrum (......).

The above discussion may give an opinion that the time domain AR based methods are good as they are not affected by data truncation and leakage problems of the periodogram based on DFT. This is valid for clean signals or if the signal power to noise power ratio (SNR) is sufficiently high (say above 20dB). However, if the SNR is low, then AR modeling methods fail severely. This is because, the AR methods are based on correlation of the signal and the presence of noise the disturbs the correlation (dependence between adjacent samples). Higher the noise level, severe is the correlation geting affected and hence the performance of AR model based methods detoriate. But the transform based methods as they are not affected by noise, they form an alternaive to parametric mehods. Different transforms have different properties and can be exploited depending on the situation. For example, DFT and DCT introduced in previous chapters have different properties and can be used when the SNR is poor. It has been shown that the transform based methods work better than AR methods like Burg method, in terms of resolution and detection as given in chapter 6., section 6.3.

7.10.6 Model Order Selection

For a given length of data, an arbitrary model order cannot be used. If the order is low, the spectrum will *have poor frequency resolution and also it may not be sufficient to bring out all the spectral peaks.* If the model is *higher than the required one, the spectral estimate may have spurious spectral peaks.* Particularly, with the Burg method, there can be spectral line splitting. For a given data of length *N*, what is the minimum (optimum) model order *M*, M_{opt} required to result in good spectral estimation? The prediction error energy at M^{th} order ε_M decreases monotonically with increase in model order M (Fig.7.9). *But after certain value of* M, *the decrease is not appreciable and the problem is where to stop?* As the prediction error energy is a monotonically decreasing function of M, a criterion is necessary to detect the optimum point. *By modifying the error energy, which provides a penalty that increases with the order,* it is possible to find the minimum value of *M*. *The model order is specific to a method as the error energy reduction is decided by the method.*

i) Akaike's Final prediction error (FPE) criteria:

$$FPE(M) = \frac{N+M-1}{N-M-1}\varepsilon_M \qquad M = 1, 2, 3, \ldots \tag{7.49a}$$

ε_M decreases with increase in model order and the multiplication factor increases with M. The order that minimizes $FPE(M)$ is chosen as the order. For clean (synthetic) data this criteria provides correct model order. But for ***practical data***/ non-autoregressive data, the order estimated is less than the true one.

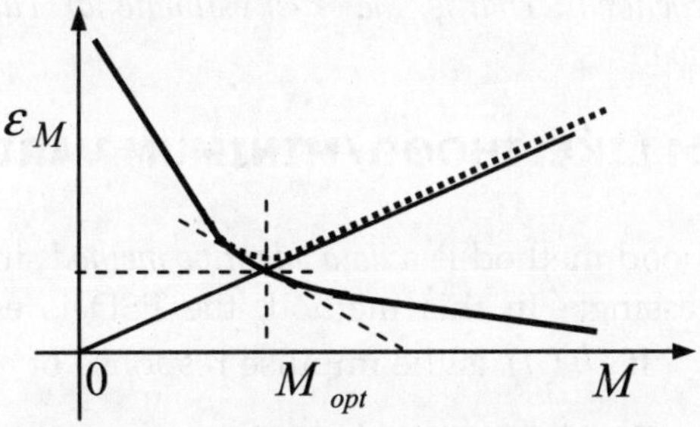

Fig. 7.9 Mechanism of model order selection.

ii) Akaike's information theoretic criterion:

$$AIC(M) = N\ln(\varepsilon_M) + 2M, \qquad M = 1, 2, 3, ... \qquad (7.49b)$$

The term $2M$ is called the penalty for the extra prediction coefficients, which do not reduce the error power significantly. Thus, even if the error energy decreases with increase in order, the penalty term $2M$ increases so that a minimum occurs for some in between order value.

$$\lim_{N\to\infty} \frac{AIC(M)}{N} = \lim_{N\to\infty} FPE(M) \qquad (7.49c)$$

Asymptotically, the two methods are equivalent. AIC also similar to FPE estimates a lower order for practical data. AIC is *not a consistent* estimate as *it over-estimates the order as* $N \to \infty$

iii) Minimum description length (MDL) criterion

In the Akaike's information theoretic criterion, *the penalty is independent of the data length used.* But in the MDL proposed by Rissanen, a larger penalty for large data length is possible by replacing the penalty $2M$ by $M\ln N$.

$$MDL(M) = N\ln(\varepsilon_M) + M\ln N \qquad (7.49d)$$

MDL gives a *consistent estimate* of the order as $M\ln N$ increases faster than $2M$ in AIC. That is , as $N \to \infty$, the order estimate converges to the true order.

iv) Parzen's criterion autoregressive transfer function (CAT)

$$CAT(M) = \frac{1}{N}\left\{\sum_{j=1}^{M} \frac{N-j}{N\varepsilon_j}\right\} - \frac{N-M}{N\varepsilon_M} \qquad (7.49e)$$

Unfortunately, none of the criteria seems to work for short data length. Hence, the order arrived by the above criteria has to be taken as a rough estimate and not the exact model order required. Further, the order estimated is method dependent. That is, the order estimate for Yule-Walker method may be different from that for Burg method.

7.11 CAPON'S MAXIMUM LIKELIHOOD/MINIMUM VARIANCE METHOD

The Capon's maximum likelihood method is a *data adaptive method* similar to Burg, proposed in the context of array processing. In this method, the PSD is estimated by a bank of narrowband bandpass filters. If $h_i(n)$ is the impulse response of an ideal band pass filter with bandwidth $\Delta\,\omega$ and at center frequency ω_i, then

$$H_i(\omega)=\begin{cases}1, & \omega_i-\Delta\,\omega/2\le\omega\le\omega_i-\Delta\,\omega/2\\ 0, & otherwise\end{cases} \tag{7.50a}$$

If $S_x(\omega)$ is the power spectral density of the input, then the PSD of the output of the filter is

$$S_i(\omega)=\left|H_i(\omega)\right|^2 S_x(\omega) \tag{7.50b}$$

and the power output using Parseval's theorem is

$$E\left[\left|y_i(n)\right|^2\right]=\frac{1}{2\pi}\int_{-\pi}^{\pi} S_i(\omega)d\omega$$

$$=\frac{1}{2\pi}\int_{-\pi}^{\pi} S_x(\omega)\left|H_i(\omega)\right|^2 d\omega=\frac{1}{2\pi}\int_{\omega_i-\Delta/2}^{\omega_i+\Delta/2} S_i(\omega)d\omega$$

If $\Delta\,\omega$ is small so that $S_x(\omega)$ is constant over the band $\Delta\,\omega$, then the power output is

$$E\left[\left|y_i(n)\right|^2\right]=S_i(\omega)\frac{\Delta\,\omega}{2\pi} \tag{7.50c}$$

Therefore, the PSD of the process at ω_i is

$$S_x(\omega_i)=E\left(\left|y_i(n)\right|^2\right)\Big/\frac{\Delta\,\omega}{2\pi} \tag{7.50d}$$

The periodogram also computes the PSD similarly and the filter used is

$$H_i(\omega)=\frac{\sin\left[N(\omega-\omega_i)/2\right]}{N\sin\left[(\omega-\omega_i)/2\right]}$$

and the power is divided by the filter bandwidth, $\Delta\,\omega=2\pi/N$. The filter bandwidth is same for all filters and hence the periodogram PSD is independent of data content. If the input signal has significant power in the frequency regions within the side lobes of the bandpass filter, then there will be appreciable leakage through the side lobes and the PSD estimate will be in error. Hence, it is desirable to make the filter *data adaptive* so that *it rejects as much the power from out side the bandpass filter as possible.* The method as discussed at a particular frequency, minimizes the output power of the bandpass filter so that it produces a minimum variance unbiased estimate of the signal in Gaussian noise. The name *maximum likelihood is misnomer* as the method is *nothing to do* with the maximum likelihood and the name *minimum variance method is more appropriate.* Here, a narrowband filter $H_{\omega_o}(\omega)$ centered at frequency ω_o is designed such that its output power is minimized when the input $x(n)$ is applied to it subject to the constraint that the filter has a response

$$H_{\omega_o}(\omega)\Big|_{\omega=\omega_o} = 1 \tag{7.51a}$$

This minimization forces the filter to be *as narrowband as possible.* The spectral estimate of the input at frequency at ω_o is given by the average output power of the filter.

$$P_{ML}(\omega_o) = E\left[\,|y(n)|^2\,\right], \tag{7.51b}$$

$$y(n) = \sum_{k0}^{N-1} h_{\omega_o}(k)x(n-k) = h_o^T X(n) \tag{7.51c}$$

$$h_o^T = \left[\,h_{\omega_o}(0), h_{\omega_o}(1), \ldots h_{\omega_o}(N-1)\,\right] \tag{7.51d}$$

$$W_o^T = \left[1, e^{j\omega_o}, \ldots, e^{j(N-1)\omega_o}\right] \tag{7.51e}$$

and $$X^T(n) = \left[x(n), x(n-1), \ldots, x(n-N+1)\right] \tag{7.51f}$$

Output power of the filter is

$$P_{ML}(\omega_o) = E\left[\,|y(n)|^2\,\right] = E\left[\,h_o^T X(n)X^T(n)h_o\,\right] = h_o^T E\left[\,X(n)X^T(n)\,\right]\,h_o$$

$$P_{ML}(\omega_o) = h_o^T R_x h_o \tag{7.52a}$$

Further the constraint Eqn. (7.51a),

$$H_{\omega_o}(\omega)\Big|_{\omega=\omega_o} = 1$$

is equivalent to

$$y(n) = \sum_{k=0}^{N-1} h_{\omega_o}(k)e^{-j\omega_o k} = W_o^T h_o = 1 \tag{7.52b}$$

The output power minimization (Eqn.(7.52a))with the above constraint (Eqn.(7.52b))is realized by using Lagrange multiplier λ, in the cost function as,

$$Q(h,\lambda) = h_o^T R_x h_o + \lambda\left(1 - h_o^T W_o\right) \tag{7.52c}$$

The minimization is achieved by equating the gradient of $Q(h,\lambda)$ with respect to h, to zero.

$$\nabla_{h_o^T} Q(h,\lambda) = R_x h_o - \lambda W_o = 0$$

$$h_o = \lambda R_x^{-1} W_o \tag{7.52d}$$

The value of the Lagrange multiplier is found setting the derivative of $Q(h,\lambda)$ with respect to λ equal to zero.

$$\frac{\partial Q(h,\lambda)}{\partial \lambda} = 1 - h_o^T W_o = 0, \quad 1 - \left[\lambda R_x^{-1} W_o\right]^T W_o = 0$$

$$1 - \lambda W_o^T R_x^{-1} W_o = 0 \quad \text{Or} \quad \lambda = \frac{1}{W_o^T R_x^{-1} W_o}$$

$$h_o = \lambda R_x^{-1} W_o = \frac{R_x^{-1} W_o}{W_o^T R_x^{-1} W_o}$$

$$h_o == \frac{R_x^{-1} W_o}{W_o^T R_x^{-1} W_o} \tag{7.53a}$$

Hence,

$$P_{ML}(\omega_o) = h_o^T R_x h_o = \left[\frac{R_x^{-1} W_o}{W_o^T R_x^{-1} W_o}\right]^T R_x \left[\frac{R_x^{-1} W_o}{W_o^T R_x^{-1} W_o}\right]$$

$$P_{ML}(\omega_o) = \frac{W_o^T R_x^{-1} R_x R_x^{-1} W_o}{\left(W_o^T R_x^{-1} W_o\right)^2} = \frac{W_o^T R_x^{-1} W_o}{\left(W_o^T R_x^{-1} W_o\right)^2}$$

$$P_{ML}(\omega_o) = \frac{1}{\left(W_o^T R_x^{-1} W_o\right)} \tag{7.53b}$$

Since this is valid for any choice of ω_o, the ML spectral estimate defined by

$$\hat{S}_{ML}(\omega) = \frac{P_{ML}(\omega_o)}{\Delta \omega}$$

becomes,

$$\hat{S}_{ML}(\omega) = \frac{1/\Delta \omega}{\left(W^T R_x^{-1} W\right)}$$

$$\frac{1}{\hat{S}_{ML}(\omega)} = \Delta \omega \left(W^T R_x^{-1} W\right) \tag{7.53c}$$

For a white noise input with variance σ_x^2, $R_x = \sigma_x^2 I$, the filter is

$$h_o = \frac{R_x^{-1} W_o}{W_o^T R_x^{-1} W_o} = \frac{\left[\sigma_x^2 I\right]^{-1} W_o}{W_o^T \left[\sigma_x^2 I\right]^{-1} W_o} = \frac{\sigma_x^{-2} W_o}{\sigma_x^{-2} W_o^T W_o} = \frac{W_o}{N}$$

$$h_o = \frac{W_o}{N}$$

The power estimate at frequency ω_o is,

$$\hat{\sigma}_x^2(\omega) = \frac{1}{\left(W_o^T R_x^{-1} W_o\right)} = \frac{\sigma_x^2}{\left(W_o^T W_o\right)} = \frac{\sigma_x^2}{N}$$

The power output of the filter is independent of ω_o and hence power distribution is constant or uniform. However, as the filter order increases, the filter bandwidth decreases resulting in a decrease in the out of band output power. In addition to the filter power output, the bandwidth of the filter is necessary to compute the PSD, as it is the power per unit bandwidth. To determine this bandwidth, the bandwidth that gives the correct PSD for white noise is used. The power output for white noise is

$$\hat{\sigma}_x^2(\omega) = \frac{\sigma_x^2}{N}$$

By definition PSD is

$$S_x(\omega_i) = \frac{E\left(\left|y_i(n)\right|^2\right)}{\left(\Delta\omega/2\pi\right)} = \frac{\sigma_x^2}{N} \Big/ \frac{\Delta\omega}{2\pi}$$

$$S_x(\omega_i) = \sigma_x^2 \frac{2\pi}{N\Delta\omega}$$

If $\Delta\omega = \frac{2\pi}{N}$,

$$S_x(\omega_i) = \left(\sigma_x^2 \frac{2\pi}{N}\right)\left(\frac{N}{2\pi}\right) = \sigma_x^2$$

Therefore, using this bandwidth, the minimum variance PSD estimate is

$$P_x(\omega_i) = \frac{E\left(\left|y_i(n)\right|^2\right)}{\left(\Delta\omega/2\pi\right)} = \frac{\dfrac{1}{\left(W_o^T R_x^{-1} W_o\right)}}{\left(\Delta\omega/2\pi\right)} = \frac{\dfrac{1}{\left(W_o^T R_x^{-1} W_o\right)}}{\left(\dfrac{2\pi/N}{2\pi}\right)}$$

$$P_x(\omega_i) = \frac{N}{\left(W_o^T R_x^{-1} W_o\right)} \tag{7.53d}$$

This is the *minimum variance spectrum estimate.* The order of the filter determines the performance of the filter as it reduces the variance by rejecting the out of band power. The order of the filter is same as the maximum lag value of the autocorrelation considered and the data length decides the latter. For a good quality autocorrelation estimate, the maximum lag value is much smaller than the data length. Hence, filter length to be used is also much less than the data length.

7.11.1 Relation between Minimum Variance and Maximum Entropy Method

Consider the relation between the autocorrelation matrices of the signal and the backward prediction errors

$$R_b = L\,R_x\,L^T$$

L is lower triangular matrix of filter coefficients and determinant of L, $|L| = 1$. *Since* R_x *is positive definite, both* R_x *and* R_b *are invertible.*

$$\left(L\,R_x L^T\right)^{-1} = R_b^{-1}$$

$$L^{T^{-1}} R_x^{-1} L^{-1} = R_b^{-1}, \quad L^T L^{T^{-1}} R_x^{-1} L^{-1} = L^T R_b^{-1},$$

$$R_x^{-1} L^{-1} = L^T R_b^{-1}, \quad R_x^{-1} L^{-1} L = L^T R_b^{-1} L$$

$$R_x^{-1} = L^T R_b^{-1} L$$

Substituting for R_x^{-1} in,

$$\frac{1}{\hat{S}_{ML}(\omega)} = \Delta\,\omega\left(W^T R_x^{-1} W\right) = \Delta\,\omega\left[W^T\left(L^T R_b^{-1} L\right)W\right]$$

$$\frac{1}{\hat{S}_{ML}(\omega)} = \Delta\,\omega\left[W^T L^T R_b^{-1} L W\right] \tag{7.54}$$

Consider $W^T L^T$ for $N = 2$

$$W^T = \begin{bmatrix} 1 & e^{-j\omega_o} & e^{-j2\omega_o} \end{bmatrix}, \quad L^T = \begin{bmatrix} 1 & a_1(1) & a_2(2) \\ 0 & 1 & a_2(1) \\ 0 & 0 & 1 \end{bmatrix}$$

$$W^T L^T = \begin{bmatrix} 1 & e^{-j\omega_o} & e^{-j2\omega_o} \end{bmatrix} \begin{bmatrix} 1 & a_1(1) & a_2(2) \\ 0 & 1 & a_2(1) \\ 0 & 0 & 1 \end{bmatrix}$$

$$W^T L^T = \begin{bmatrix} 1 & a_1(1) + e^{-j\omega_o} & a_2(2) + a_2(1)e^{-j\omega_o} + e^{-j2\omega_o} \end{bmatrix}$$

Similarly,

$$LW = \begin{bmatrix} 1 & 0 & 0 \\ a_1(1) & 1 & 0 \\ a_2(2) & a_2(1) & 1 \end{bmatrix} \begin{bmatrix} 1 \\ e^{j\omega_o} \\ e^{j2\omega_o} \end{bmatrix}$$

$$LW = \begin{bmatrix} 1 \\ a(1) + e^{j\omega_o} \\ a_2(2) + a_2(1)e^{j\omega_o} + e^{j2\omega_o} \end{bmatrix}$$

Also,

$$R_b = \begin{bmatrix} \varepsilon_0(n) & 0 & 0 \\ 0 & \varepsilon_1(n) & 0 \\ 0 & 0 & \varepsilon_2(n) \end{bmatrix} \Rightarrow R_b^{-1} = \begin{bmatrix} 1/\varepsilon_0(n) & 0 & 0 \\ 0 & 1/\varepsilon_1(n) & 0 \\ 0 & 0 & 1/\varepsilon_2(n) \end{bmatrix}$$

Consider $R_b^{-1}\left(LW\right)$ and substituting for LW

$$R_b^{-1}\left(LW\right) = \begin{bmatrix} 1/\varepsilon_0(n) & 0 & 0 \\ 0 & 1/\varepsilon_1(n) & 0 \\ 0 & 0 & 1/\varepsilon_2(n) \end{bmatrix} \begin{bmatrix} 1 \\ a_1(1) + e^{j\omega_o} \\ a_2(2) + a_2(1)e^{j\omega_o} + e^{j2\omega_o} \end{bmatrix}$$

$$R_b^{-1}\left(LW\right) = \begin{bmatrix} 1/\varepsilon_0(n) \\ \left(1/\varepsilon_1(n)\right)\left(a_1(1) + e^{j\omega_o}\right) \\ \left(1/\varepsilon_2(n)\right)\left(a_2(2) + a_2(1)e^{j\omega_o} + e^{j2\omega_o}\right) \end{bmatrix}$$

$$\left[LW\right]^T = \begin{bmatrix} 1 \\ a_1(1) + e^{j\omega_o} \\ a_2(2) + a_2(1)e^{j\omega_o} + e^{j2\omega_o} \end{bmatrix}^T$$

$$W^T L^T = \begin{bmatrix} 1 & a_1(1) + e^{-j\omega_o} & a_2(2) + a_2(1)e^{-j\omega_o} + e^{-j2\omega_o} \end{bmatrix}$$

$$W^T L^T R_b^{-1}\left(LW\right) = \left[1 \quad a_1(1)+e^{-j\omega_o} \quad a_2(2)+a_2(1)e^{-j\omega_o}+e^{-j2\omega_o}\right] \begin{bmatrix} 1/\varepsilon_0(n) \\ \left(1/\varepsilon_1(n)\right)\left(a_1(1)+e^{j\omega_o}\right) \\ \left(1/\varepsilon_2(n)\right)\left(a_2(2)+a_2(1)e^{j\omega_o}+e^{j2\omega_o}\right) \end{bmatrix}$$

$$\begin{aligned} W^T L^T R_b^{-1}\left(LW\right) = &\left(1/\varepsilon_0(n)\right) \\ &+\left(1/\varepsilon_1(n)\right)\left(a_1(1)+e^{j\omega_o}\right)\left(a_1(1)+e^{-j\omega_o}\right) \\ &+\left(1/\varepsilon_2(n)\right)\left(a_2(2)+a_2(1)e^{j\omega_o}+e^{j2\omega_o}\right) \\ &\quad \left(a_2(2)+a_2(1)e^{-j\omega_o}+e^{-j2\omega_o}\right) \end{aligned}$$

$$\begin{aligned} W^T L^T R_b^{-1}\left(LW\right) = &\left(1/\varepsilon_0(n)\right)+\left(1/\varepsilon_1(n)\right)\left|a_1(1)+e^{+j\omega_o}\right|^2 \\ &+\left(1/\varepsilon_2(n)\right)\left|a_2(2)+a_2(1)e^{+j\omega_o}+e^{+j2\omega_o}\right|^2 \end{aligned}$$

In general,

$$W^T L^T R_b^{-1}\left(LW\right) = \sum_{m=0}^{N-1}\left(1/\varepsilon_m(n)\right)\left|\sum_{k=0}^{m} a_m(k)e^{-j\omega_0 k}\right|^2, \quad a_m(0)=1$$

$$W^T L^T R_b^{-1}\left(LW\right) = \sum_{m=0}^{N-1} \frac{1}{\varepsilon_m(n) \Big/ \left|\sum_{k=0}^{m} a_m(k)e^{-j\omega_0 k}\right|^2}$$

From Eqn. (7.54),

$$\frac{1}{\hat{S}_{ML}(\omega)} = \Delta\,\omega\left[W^T L^T R_b^{-1} LW\right]$$

Therefore,

$$\frac{1}{\hat{S}_{ML}(\omega)} = \Delta\,\omega\left[\sum_{m=0}^{N-1} \frac{1}{\varepsilon_m(n) \Big/ \left|\sum_{k=0}^{m} a_m(k)e^{-j\omega_0 k}\right|^2}\right]$$

But AR / MEM spectrum estimate is given by

$$\hat{S}_{MEM}(\omega) = \frac{\varepsilon_m(n)}{\left|\sum_{k=0}^{m} a_m(k) e^{-j\omega_0 k}\right|^2}$$

Therefore,

$$\frac{1}{P_x(\omega_i)} = \frac{1}{N}\sum_{m=0}^{N-1} \frac{1}{\hat{S}_{MEM}(\omega)} \tag{7.55}$$

This is a very important relation as it relates the minimum variance method and the maximum entropy method. It states that *the reciprocal of minimum variance estimator is equal to the average of the reciprocals of the MEM estimators of orders* 1 *to* N. The averaging operation involved in minimum variance method, reduces its frequency resolution. Here the AR spectrum of highest resolution corresponding to highest order N gets associated with the spectra of different resolutions including the one of poorest of resolution corresponding to the 1st order. In view of this type of averaging, its variance is less than that corresponding to AR spectra of the highest order considered. But its frequency resolution is better than those of other classical spectral estimators, at high signal to noise ratios.

The above relation also facilitates a method of implementation of minimum variance spectral estimator. As stated already the reciprocal of minimum variance estimator is equal to the average of the reciprocals of the MEM estimators of orders 0 *to* $N-1$. According to this, any AR method can be used for this purpose. But the Burg method is most suitable as it provides coefficients of all the lower order filters along with the required order. However, this is not so with other methods and they have to be separately executed for all the different orders starting from order 0 *to* $N-1$ and this may be computationally expensive. The squared magnitude of the Fourier transform of all order filters (divided by their error energies ε_m) of same number of points are computed and their reciprocal spectra are averaged. The reciprocal of this averaged spectrum provides the minimum variance spectrum estimate.

To illustrate this an AR process $x(n)$ generated by the model

$$x(n) = 2.4139x(n-1) - 3.2341x(n-2) + 2.3183x(n-3) - 0.9224x(n-4) + \in(n)$$

where $\in(n)$ is a zero mean white Gaussian noise, is considered .

For the spectral estimation of this process, a 10th order model and 500 samples of data are used. The spectrum is computed from the model by a DFT of 512 points using FFT algorithm. A sampling interval of 1000 Hz is assumed for the AR process. To compute the mean and variance of the spectral estimate, 25 trials are considered.

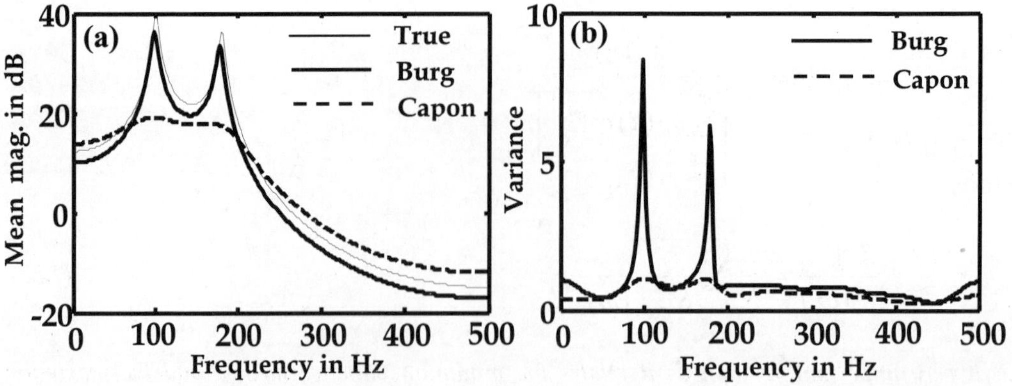

Fig. 7.10 Comparison of spectral estimation by Burg and Capon's methods (a) Mean spectrum and (b) spectrum variance.

The Capon's spectrum is derived from the relation between MEM and Capons method by using Burg spectra of all orders, i.e. , 1 to 10. From Fig. 7.10 (a), it is seen that the mean Capon's spectrum cannot resolved spectral peaks well as it is in some sense the average of all lower order spectra, but the Burg (MEM) is very good in resolving the spectral peaks as there is no averaging in any sense. Also, it follows the true spectrum of the process very well compared to that by Capon's spectrum estimate. However, the variance of the burg spectrum is very high especially at the spectral peaks and somewhat more in the rest of the regions (Fig.7.10 (b)). The variance sum over all frequencies for the Burg estimate is 1.7 times that of the Capon's estimate. This is expected as Capon's estimate has an averaging effect of all the lower order spectra of the Burg method and for the Burg method, it corresponds to only the final order (here, 10th order).

7.12 ESCALATOR LATTICE STRUCTURE

Escalator lattice is an alternative structure to decompose the signal into uncorrelated/ orthogonal components by Gram- Schmidt orthogonalization (N. Ahmed and Youn, 1980, Modern Filters, S. Haykin, Chapter 9). The error at each stage is obtained using the preceding error and hence this structure is known as *escalator*. The basic processing element used for structure is shown in the Fig.7.11(a). It chooses its parameter ρ such that the output, i.e., the error $e(n)$ is uncorrelated to input $x(n)$. That is

$$E\left[e(n)x(n)\right]=0,\quad E\left[\left\{d(n)-\rho\, x(n)\right\}x(n)\right]=0,\quad E\left[d(n)x(n)\right]-\rho E\left[x^2(n)\right]=0$$

$$\rho=\frac{E\left[d(n)x(n)\right]}{E\left[x^2(n)\right]} \tag{7.56a}$$

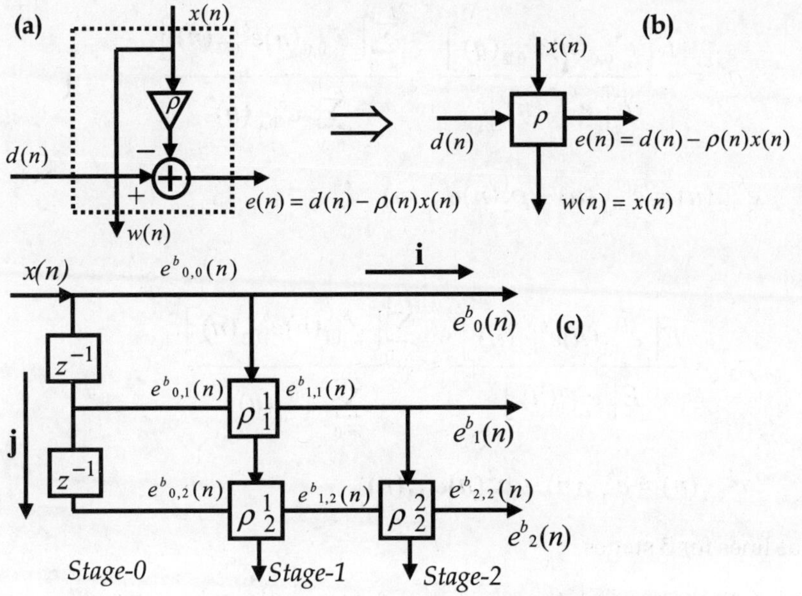

Fig. 7.11(a) Processing element; (b) Equivalent representation and (c) Escalator Structure of order N=2.

where $E[.]$ is the expectation operation. Further, it also transmits the input $x(n)$ as the second output $w(n)$. For simplicity, Fig. 7.9(a) can be represented as in Fig. 7.11(b) without the inner details and this representation will be used in the further treatment.

A two-stage escalator structure using the basic unit (Fig. 7.11(b)) is shown in Fig. 7.11(c). The vertical index "j" corresponds to the order and the horizontal "i" to the stage. The vertical path transmits its input to the next order, at that stage. Since, each element of this structure predicts the previous sample from the present sample, it performs *backward prediction.* As mentioned already, the coefficients are selected such that the backward prediction errors are uncorrelated .

The outputs at each stage (for a 2 stage) is given by

For stage 0:

$$e^b_{0,0}(n) = x(n),\ e^b_{0,1}(n) = x(n-1),\ e^b_{0,2}(n) = x(n-2) \tag{7.56b}$$

$$\rho_1^1 = \frac{E\left[e^b_{0,0}(n)b_{0,1}(n)\right]}{E\left[e^{b2}_{0,0}(n)\right]} \approx \frac{\sum_{n=0}^{L-1}\left[e^b_{0,0}(n)b_{0,1}(n)\right]}{\sum_{n=0}^{L-1} e^{b2}_{0,0}(n)} \tag{7.56c}$$

For stage 1 :

$$e^b_{1,1}(n) = e^b_{0,1}(n) - \rho_1^1(n)e^b_{0,0}(n) \tag{7.56d}$$

$$\rho_2^1 = \frac{E\left[e^b_{0,0}(n)e^b_{0,2}(n)\right]}{E\left[e^{b2}_{0,0}(n)\right]} \approx \frac{\sum_{n=0}^{L-1}\left[e^b_{0,0}(n)e^b_{0,2}(n)\right]}{\sum_{n=0}^{L-1} e^{b2}_{0,0}(n)} \tag{7.56e}$$

$$e^b_{1,2}(n) = e^b_{0,2}(n) - \rho_2^1(n)e^b_{1,1}(n) \tag{7.56f}$$

For stage 2:

$$\rho_2^2 = \frac{E\left[e^b_{1,1}(n)e^b_{1,2}(n)\right]}{E\left[e^{b2}_{1,1}(n)\right]} \approx \frac{\sum_{n=0}^{L-1}\left[e^b_{1,1}(n)e^b_{1,2}(n)\right]}{\sum_{n=0}^{L-1} e^{b2}_{1,1}(n)} \tag{7.56g}$$

$$e^b_{2,2}(n) = e^b_{1,2}(n) - \rho_2^2(n)e^b_{1,1}(n) \tag{7.56h}$$

On the similar lines for 3 stages,

$$e^b_{0,3}(n) = x(n-3)\,;\; e^b_{1,3}(n) = e^b_{0,3}(n) - \rho_3^1(n)e^b_{2,2}(n)$$

$$e^b_{2,3}(n) = e^b_{1,3}(n) - \rho_3^2(n)e^b_{2,2}(n)$$

$$e^b_{3,3}(n) = e^b_{2,3}(n) - \rho_3^3(n)e^b_{2,2}(n) \tag{7.56i}$$

The $e^b_{i,j}(n)$ error at i^{th} stage and j^{th} order is

$$e^b_{i,j}(n) = e^b_{i-1,j}(n) - \rho_j^i(n)e^b_{i-1,i-1}(n),$$

$$i = 1, 2,\ldots, N-1, \quad j = i, i+1,\ldots, N-1 \tag{7.57}$$

where,

$$e^b_{0,i}(n) = x(n-i)\,, \qquad i = 1, 2,\ldots, N-1$$

The prediction errors at the last stage for each order form the outputs of the escalator structure. $e^b_j(n) = e^b_{j,j}(n), \quad j = 0,1,2,\ldots,N-1$. These orthogonal components are used for joint process escalator where they are weighed and summed to get an estimate of the desired signal. Such an estimate is achieved by minimizing the energy of the error, the difference between the estimate and the desired signal. In finding the coefficients $\rho_j^i(n)$, this structure minimizes only the energy of the backward prediction error that is *local* instead of the average energy of the forward and backward errors, which is *global* used by the regular lattice. This implies that, the escalator structure *does not assume the input to be stationary or its autocorrelation matrix is symmetric Toeplitz.* Hence, the escalator is *appropriate* for the orthogonalization of any nonstationary process.

The regular lattice filter has cascaded stages and hence the backward prediction errors are obtained sequentially. But as the escalator has a parallel structure, it produces the backward prediction errors of different orders in parallel. Thus, each stage is independent of its preceding stage. However, this advantage demands higher computational load and is of the order of the order of N^2 which is high compared to those of transversal or lattice which are of the order of N (length of the filter).

7.13 WARPED LINEAR PREDICTION (WLP)

Spectral estimation based on linear prediction provides uniform frequency resolution. In certain situations like audio signals due to the logarithmic frequency response of the human ear, a uniform frequency resolution is not desirable and a bark frequency scale is used. Also in practice, as most of the signals have their energy mainly distributed in the low frequency region, a higher frequency resolution only in that region is required. Further, with a nonuniform / warped frequency resolution, the estimation of the parameters of a signal will be not only superior but also efficient compared to that of uniform frequency resolution. *As the frequency resolution is better in the desired region and coarse in other regions, the estimate discards the unwanted details thereby improving the variance of the estimate. Also as only a certain part of the frequency region has to be resolved better, the order of the predictor required will be small compared to that of having uniform frequency resolution.* Such a nonuniform / warped frequency resolution can be achieved by replacing the unit delay in linear prediction by an allpass function given by

$$D(z) = \frac{z^{-1} - \lambda}{1 - \lambda z^{-1}} \tag{7.58a}$$

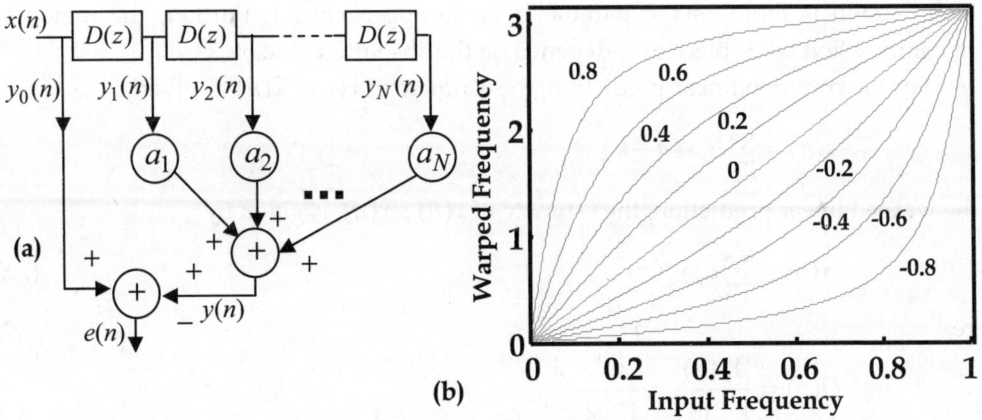

Fig. 7.12. (a) Schematic of warped linear prediction (b) Relation between input and warped frequencies for different values of λ.

where λ, $(-1 < \lambda < 1)$ decides the degree of warping frequency and its region. A chain of allpass unit forms a frequency dependent tapped delay line with magnitude spectrum equal to unity. For $\lambda = 0$, such a chain reduces to a frequency independent ordinary tapped delay line (Fig. 7.12 (a))(A. Harma, 2000).

$$D(z) = \tilde{z}^{-1} = \frac{z^{-1} - \lambda}{1 - \lambda\, z^{-1}}$$

$$e^{-j\tilde{\omega}T} = \frac{e^{-j\omega T} - \lambda}{1 - \lambda\, e^{-j\omega T}} = \frac{\cos\omega T - \lambda - j\sin\omega T}{1 - \lambda os\omega T + j\lambda\sin\omega T}$$

For sampling period $T = 1$,

$$e^{-j\tilde{\omega}} = \frac{(\cos\omega - \lambda - j\sin\omega)(1 - \lambda os\omega - j\lambda\sin\omega)}{(1 - \lambda os\omega)^2 + \lambda^2\sin^2\omega}$$

$$e^{-j\tilde{\omega}} = \frac{(1+\lambda^2)\cos\omega - 2\lambda + j(1-\lambda^2)\sin\omega}{(1 - \lambda os\omega)^2 + \lambda^2\sin^2\omega}$$

$$e^{j\tilde{\omega}} = e^{j\tan^{-1}\left\{\frac{(1-\lambda^2)\sin\omega}{(1+\lambda^2)\cos\omega - 2\lambda}\right\}}$$

Therefore,

$$\tilde{\omega} = \tan^{-1}\left\{\frac{(1-\lambda^2)\sin\omega}{(1+\lambda^2)\cos\omega - 2\lambda}\right\} \tag{7.58b}$$

This provides the relation between the warped frequency and the input frequency. This relation is plotted for different values of the warping parameter λ (Fig.7.12 (b)). It is seen that for values of $\lambda > 0$, the higher frequency region is compressed and the compression is more as its value approaches unity. For $\lambda < 0$, the low frequency region is compressed and the compression is more as the parameter value approaches 1. Further, the region over which compression takes place also depends on the absolute value of λ.

For the common linear prediction, the estimate of $x(n)$, $\hat{x}(n)$ is given by

$$\hat{x}(n) = \sum_{k=1}^{N} a_k x(n-k)$$

For the warped linear prediction, the estimate of $x(n)$, $\hat{x}(n)$ is given by

$$\hat{x}(n) = \sum_{k=1}^{N} a_k y_k(n) \tag{7.58c}$$

where

$$D(z) = \frac{Y_k(z)}{Y_{k-1}(z)} = \frac{z^{-1} - \lambda}{1 - \lambda\, z^{-1}}$$

$$y_k(n) = y_{k-1}(n-1) - \lambda y_{k-1}(n) + \lambda y_k(n-1), \quad k = 1, 2, \ldots, N \tag{7.58d}$$

$$y_0(n) = x(n), \quad k = 0$$

The warped linear prediction coefficients are chosen to minimize the error energy given by

$$\varepsilon = \sum_{n=0}^{L-1} e^2(n) = \sum_{n=0}^{L-1} \{x(n) - \hat{x}(n)\}^2$$

Since the minimization of error energy obeys the orthogonality principle, viz., *the error and the input corresponding to the weight* are orthogonal, the solution for the prediction coefficients cab obtained using this principle. In the present case, for the kth coefficient a_k,

$$\sum_{n=0}^{L-1} y_k(n)e(n) = 0, \quad \sum_{n=0}^{L-1} y_k(n)\{x(n) - \hat{x}(n)\} = 0, \quad \sum_{n=0}^{L-1} y_k(n)\left\{x(n) - \sum_{l=1}^{N} a_l y_l(n)\right\} = 0$$

$$\sum_{l=1}^{N} a_l \sum_{n=0}^{L-1} y_k(n) y_l(n) = \sum_{n=0}^{L-1} y_k(n) x(n)$$

Defining

$$c(k,l) = \frac{1}{L}\sum_{n=0}^{L-1} y_k(n) y_l(n), \quad k,l = 1, 2, \ldots, N-1$$

$$c(k,0) = \frac{1}{L}\sum_{n=0}^{L-1} y_k(n) x(n)$$

$$\sum_{l=1}^{N-1} a_l c(k,l) = c(k,0), \quad k = 1, 2, \ldots, N \tag{7.58e}$$

This will result in a linear system of equations given by

$$\begin{bmatrix} c(0,0) & c(0,1) & \ldots & c(0,N-1) \\ c(1,0) & c(1,1) & \ldots & c(1,N-1) \\ \vdots & \vdots & \vdots & \\ c(N-1,0) & c(N-1,1) & \ldots & c(N-1,N-1) \end{bmatrix} \begin{bmatrix} a_1 \\ a_2 \\ \vdots \\ a_N \end{bmatrix} = \begin{bmatrix} c(1,0) \\ c(2,0) \\ \vdots \\ c(,N,0) \end{bmatrix} \tag{7.58f}$$

or

$$Ca = c$$

where

$$C = \begin{bmatrix} c(0,0) & c(0,1) & \ldots & c(0,N-1) \\ c(1,0) & c(1,1) & \ldots & c(1,N-1) \\ \vdots & \vdots & \vdots & \\ c(N-1,0) & c(N-1,1) & \ldots & c(N-1,N-1) \end{bmatrix}, \quad a = \begin{bmatrix} a_1 \\ a_2 \\ \vdots \\ a_N \end{bmatrix}, \quad c = \begin{bmatrix} c(1,0) \\ c(2,0) \\ \vdots \\ c(,N,0) \end{bmatrix}$$

Hence, the warped linear prediction coefficients are given by

$$a = C^{-1}c \tag{7.58g}$$

Thus, for the solution to exist, the matrix C must be invertible. Here, it has been assumed that the C matrix is a covariance matrix and may not be a correlation matrix which has Toeplitz form. The warped predictor has a cascade of allpass elements and hence, they satisfy,

$$D(z) = D_k(z) / D_{k-1}(z) \quad \text{Or} \quad D_k(z) = D^k(z)$$

Further,

$$D(z) = \frac{z^{-1} - \lambda}{1 - \lambda\, z^{-1}} = \frac{(1/z) - \lambda}{1 - \lambda / z} = \frac{1 - \lambda z}{z - \lambda}$$

But

$$D(z^{-1}) = \frac{z - \lambda}{1 - \lambda\, z}$$

Therefore, for the allpass element,

$$D(z^{-1}) = D^{-1}(z)$$

The covariance element $\tilde{c}(k,l)$ in the expectation sense is given by

$$\tilde{c}(k,l) = E\left\{ y_k(n) y_l(n) \right\} \approx \sum_{n=-\infty}^{\infty 1} y_k(n) y_l(n)$$

Further by Parceval's theorem,

$$\tilde{c}(k,l) = \frac{1}{2\pi i} \oint D^k(z) X(z) D^l(z^{-1}) X(z^{-1}) \frac{dz}{z}$$

Since $\quad D(z^{-1}) = D^{-1}(z)$,

$$\tilde{c}(k,l) = \frac{1}{2\pi i} \oint D^k(z) D^{-l}(z) X(z) X(z^{-1}) \frac{dz}{z} = \frac{1}{2\pi i} \oint D^{k-l}(z) X(z) X(z^{-1}) \frac{dz}{z}$$

$$= \sum_{n=-\infty}^{\infty} y_{k-l} y_0(n)$$

$$\tilde{c}(k,l) = \tilde{c}(k-l,0)$$

This indicates that for the allpass elements, the matrix $\tilde{c}(k,l)$ is Toeplitz. Hence $Ca = c$ is a Yule-Walker equation and can be solved by using Levinson–Durbin algorithm. Thus, even for warped linear prediction using allpass elements instead of delay units, efficient solution is possible. The prediction error $e(n)$ is given by

$$e(n) = x(n) - \hat{x}(n) = x(n) - \sum_{k=1}^{N} a_k y_k(n)$$

In the Z-domain,

$$E(z) = X(z) - \hat{X}(z) = X(z) - \sum_{k=1}^{N} a_k Y_k(z)$$

But $$Y_k(z) = D_k(z) X(z),$$

Therefore,

$$E(z) = X(z)\left[1 - \sum_{k=1}^{N} a_k D_k(z)\right], \qquad X(z) = \frac{E(z)}{1 - \sum_{k=1}^{N} a_k D_k(z)}$$

$$\left|X(e^{j\omega})\right|^2 = \frac{|E(e^{j\omega})|^2}{|1 - \sum_{k=1}^{N} a_k D_k(e^{j\omega})|^2}$$

Representing $S_x(\omega) = |X(e^{j\omega})|^2$ and since $\varepsilon_N = |E(e^{j\omega})|^2$ is a constant,

$$S_x(\omega) = \frac{\varepsilon_N}{|1 - \sum_{k=1}^{N} a_k D_k(e^{j\omega})|^2} \tag{7.59a}$$

In normal linear prediction, it was possible to compute the spectrum of the signal by computing the Fourier transform of the sequence $\{1, -a_1, -a_2, \ldots, -a_N\}$. But for the warped signal, its spectrum cannot be computed so easily using the Fourier transform. Since

$$D(z) = \frac{z^{-1} - \lambda}{1 - \lambda z^{-1}}, \qquad D(e^{j\omega}) = \frac{e^{-j\omega} - \lambda}{1 - \lambda e^{-j\omega}}$$

Therefore,

$$S_x(\omega) = \frac{\varepsilon_N}{\left|1 - \sum_{k=1}^{N} a_k \left(\frac{e^{-j\omega} - \lambda}{1 - \lambda e^{-j\omega}}\right)^k\right|^2} \tag{7.59b}$$

That is, the expression for $S_x(\omega)$ has to be evaluated for ω in steps of $(2\pi / M)$, M is the number of bins in Fourier transform in the frequency range $(0 - 2\pi)$.

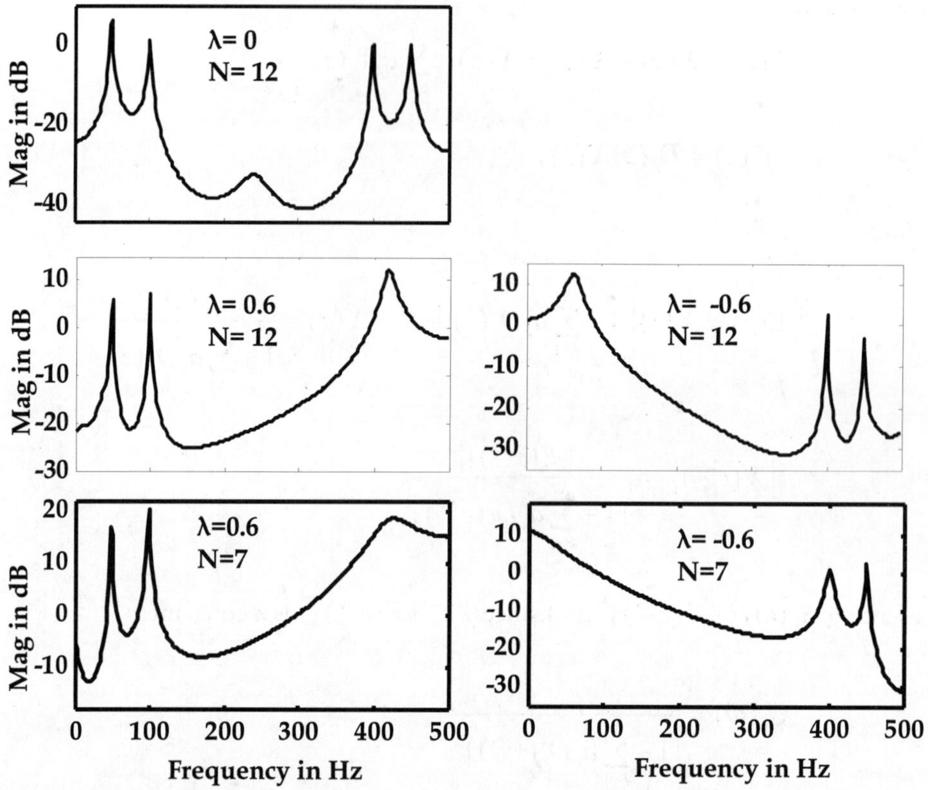

Fig. 7.13 Spectral estimation by Warped linear prediction.

Example :

The performance of warped linear prediction is illustrated for a signal made up of four sinusoids at frequencies $f_1 = 50Hz$, $f_2 = 100Hz$, $f_3 = 400Hz$ and $f_4 = 450Hz$ plus zero mean white Gaussian noise $\eta(n)$ with SNR=26dB. The signal is given by

$$x(t) = 2\sin\left(\frac{2\pi n f_1}{f_s}\right) + \sin\left(\frac{2\pi n f_2}{f_s}\right) + \sin\left(\frac{2\pi n f_3}{f_s}\right) + \sin\left(\frac{2\pi n f_4}{f_s}\right) + \eta(n)$$

where $f_s = 1000$ Hz is the sampling frequency. The data length used is 500 samples and the frequency response is evaluated over 512 bins. With the filter order $N = 12$ and with the warping parameter $\lambda = 0$ which corresponds to normal linear prediction, spectral peaks at all the 4 frequencies are brought out *with uniform frequency resolution through out the frequency range of* $(0-500)$ Hz (Fig. 7.13). But with $\lambda = 0.6$, only the two frequencies $f_1 = 50$ Hz, $f_2 = 100$ Hz, which are in the low frequency range (0-250) Hz are clearly resolved.

But the frequencies $f_3 = 400\,\text{Hz}$, $f_4 = 450\,\text{Hz}$ are not resolved but a single spectral peak indicates their presence. Similarly, for $\lambda = -0.6$, the frequencies f_1 = 50 Hz, f_2 = 100 Hz are not at all resolved but the frequencies f_3 = 400 Hz, f_4 = 450 Hz are well resolved. This is expected as the warping for $\lambda = 0.6$, compress the higher frequency range whereas $\lambda = -0.6$ compress the low frequency range. *It is important to notice that even with reduced the order $N = 7$, almost the same results are achieved for both values of $\lambda = \pm 0.6$*. This indicates that by using warped linear prediction, the required order can be reduced. This is due to the fact that by proper choice of the warping parameter, frequency resolution can be varied in different frequency regions. Generally, to resolve more number of spectral peaks/lines, a higher order is required. But as the resolution is made coarse in some regions depending on the warping parameter, a lower order is sufficient to resolve only those peaks which are in the unwrapped (not compressed) region and there is no such necessity in the warped region where the frequency range is compressed.

7.13.1 Warped Burg Algorithm

The warped Burg algorithm is based on the lattice, which uses an allpass unit $D(z)$ instead of the delay unit z^{-1} (Fig. 7.14). As the allpass units are in the backward prediction branch of the lattice filter, the backward prediction errors get modified. For the lth stage, the error $e_l^b(n)$ gets modified as the warped backward error $\tilde{e}_l^b(n)$ and is given by (Fig. 7.14b)

$$\tilde{e}_l^b(n) = e_l^b(n-1) + \lambda\left\{\tilde{e}_l^b(n-1) - e_l^b(n)\right\}, \quad l = 1, 2, \ldots, N \tag{7.60}$$

where λ is the warping parameter and this is valid for $l = 1, 2, \ldots, N$. Further, the forward and backward prediction errors $e_l^f(n)$ and $e_l^b(n)$ at stage l (Fig. 7.14) are given by

$$e_l^f(n) = e_{l-1}^f(n) + k_l \tilde{e}_{l-1}^b(n) \tag{7.61a}$$

$$e_l^b(n) = \tilde{e}_{l-1}^b(n) + k_l e_{l-1}^f(n), \quad l = 1, 2, \ldots, N \tag{7.61b}$$

The error energy is given by

$$\varepsilon = \frac{1}{2}\sum_{n=l}^{L-1}\left\{\left[e_l^f(n)\right]^2 + \left[e_l^b(n)\right]^2\right\} \tag{7.61c}$$

and the reflection coefficient is derived by minimizing the error energy (Kari Roth, I.Kauppinen, P.A.A.Esquef and V.Valimaki, 2003).
Consider

$$E_d(n) = \frac{\partial}{\partial k_l}\left\{\frac{1}{2}\left[e_l^f(n)\right]^2 + \left[e_l^b(n)\right]^2\right\} = e_l^f(n)\frac{\partial e_l^f(n)}{\partial k_l} + e_l^b(n)\frac{\partial e_l^b(n)}{\partial k_l}$$

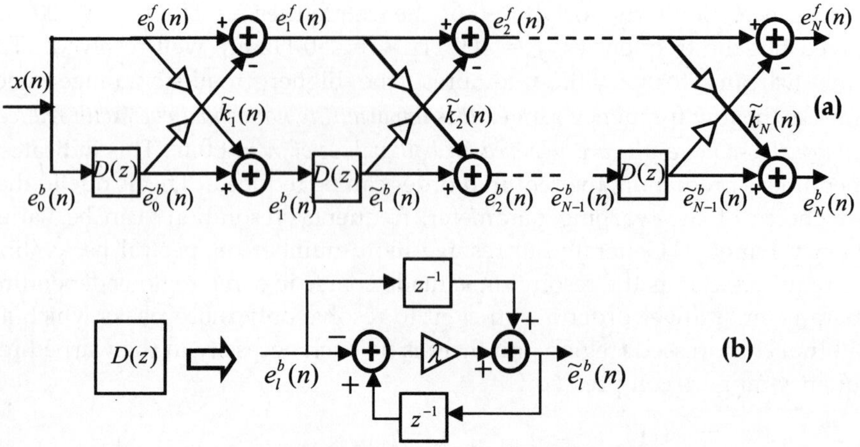

Fig. 7.14 (a) Warped lattice predictor and (b) Warping unit.

Substituting for $e_l^f(n)$ and $e_l^b(n)$ (with in the derivatives),

$$E_d(n) = e_l^f(n)\frac{\partial}{\partial k_l}\left\{e_{l-1}^f(n) + k_l\tilde{e}_{l-1}^b(n)\right\} + e_l^b(n)\frac{\partial}{\partial k_l}\left\{\tilde{e}_{l-1}^b(n) + k_l e_{l-1}^f(n)\right\}$$

$$= e_l^f(n)\tilde{e}_{l-1}^b(n) + e_l^b(n)e_{l-1}^f(n)$$

Again substituting for $e_l^f(n)$ and $e_l^b(n)$,

$$E_d(n) = \left\{e_{l-1}^f(n) + k_l\tilde{e}_{l-1}^b(n)\right\}\tilde{e}_{l-1}^b(n) + \left\{\tilde{e}_{l-1}^b(n) + k_l e_{l-1}^f(n)\right\}e_{l-1}^f(n)$$

$$= 2e_{l-1}^f(n)\tilde{e}_{l-1}^b(n) + k_l\left[\left\{\tilde{e}_{l-1}^b(n)\right\}^2 + \left\{\tilde{e}_{l-1}^f(n)\right\}^2\right]$$

Therefore, for minimization,

$$\frac{\partial\,\varepsilon}{\partial k_l} = 0\,, \quad \frac{\partial\,\varepsilon}{\partial k_l} = \sum_{n=l}^{L-1}\frac{\partial}{\partial k_l}\left\{\frac{1}{2}\left[e_l^f(n)\right]^2 + \left[e_l^b(n)\right]^2\right\} = \sum_{n=l}^{L-1}E_d(n) = 0$$

becomes

$$2\sum_{n=l}^{L-1}e_{l-1}^f(n)\tilde{e}_{l-1}^b(n) + k_l\sum_{n=l}^{L-1}\left[\left\{\tilde{e}_{l-1}^b(n)\right\}^2 + \left\{\tilde{e}_{l-1}^f(n)\right\}^2\right] = 0$$

or

$$k_l = \frac{-2\sum_{n=l}^{L-1} e_{l-1}^{f}(n)\tilde{e}_{l-1}^{b}(n)}{\sum_{n=l}^{L-1}\left[\left\{\tilde{e}_{l-1}^{b}(n)\right\}^2 + \left\{\tilde{e}_{l-1}^{f}(n)\right\}^2\right]}, \quad l = 1, 2, \ldots, N \tag{7.62}$$

The AR coefficients of the lth order can be derived from the $(l-1)^{th}$ order coefficients using the reflection coefficient for the lth order, i.e., k_l by Levinson recursion (Eqn. 7.22) as

$$\begin{aligned} a_m^l &= a_m^{l-1} + k_l a_{l-m}^{l-1}(n), m = 1, 2, \ldots, l \\ a_l^l &= k_l \end{aligned} \qquad \text{with } a_0^0 = 1$$

The warped linear prediction is based on the warped transversal structure. But the Burg algorithm is based on warped lattice structure. As the lattice provides backward prediction errors, which are orthogonal, this can be used for the joint process lattice structure already mentioned. As the IIR structure is superior in modeling efficiency compared to that of FIR and as the warped linear prediction is a combination of IIR (allpass units) and FIR structures, its performance is better than that of a linear prediction that uses only tapped delay line FIR structure. For the same reason the predictor order/ length required will also be less compared to normal linear predictor.

The Burg method is known for its ability to resolve closely spaced spectral peaks even for short data lengths compared to other autoregressive methods. This property is preserved even with warped Burg algorithm.

The performance of warped Burg algorithm is illustrated for a signal made up of four sinusoids at frequencies f_1 = 50 Hz, f_2 = 100 Hz, f_3 = 400 Hz and f_4 = 450 Hz plus zero mean white Gaussian noise $\eta(n)$ with SNR=31dB. The signal is given by

$$x(t) = 2\sin\left(\frac{2\pi n f_1}{f_s}\right) + \sin\left(\frac{2\pi n f_2}{f_s}\right) + \sin\left(\frac{2\pi n f_3}{f_s}\right) + \sin\left(\frac{2\pi n f_4}{f_s}\right) + \eta(n)$$

where f_s = 1000 Hz is the sampling frequency. *The data length used is only 20 samples* and the frequency response is evaluated over 512 bins.

The spectrum achieved by warped linear prediction and warped Burg for the warping parameter $\lambda = 0$ and order $N = 11$, are shown in Fig. 7.15. It is seen that all the four frequencies are brought out by both the methods since there are spectral peaks for each frequency. But the spectral peaks derived by the Burg algorithm are well resolved and enhanced. However, the corresponding peaks derived by the linear prediction are not so well resolved and enhanced.

The performance of warped Burg algorithm and warped linear prediction with warping parameter $\lambda = 0.2$ and order $N = 7$ are shown in Fig. 7.15. Both the algorithms bring out the low frequency peaks (50Hz and 100Hz) and due to warping effect, the two high frequency peaks are represented by a single peak. In this attempt, the Burg brings out the two low frequency peaks very well even though the amplitude of the 100Hz sinusoid is half

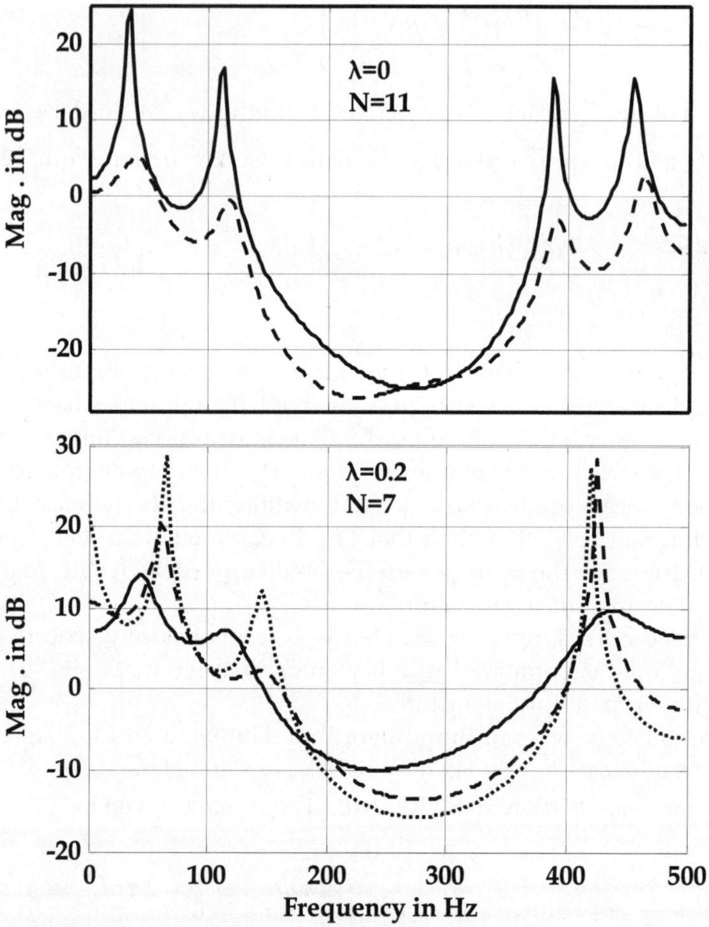

—— Burg, – – · WLP, •••• WLP for 66 data

Fig. 7.15 Comparison of warped linear prediction and warped Burg algorithms for data of 20 samples.

that of 50Hz. However, in the warped linear prediction spectrum, the peak for 100 Hz is about 20 dB below that of 50 Hz. It has been found by experiment that to get a spectrum similar to that of Burg, the warped linear prediction requires a data length of 66 samples, which is almost 3 times that used for the warped Burg method. This indicates that even the warped Burg algorithm performs better than warped linear prediction, for short data lengths.

7.13.2 Warped Escalator Algorithm

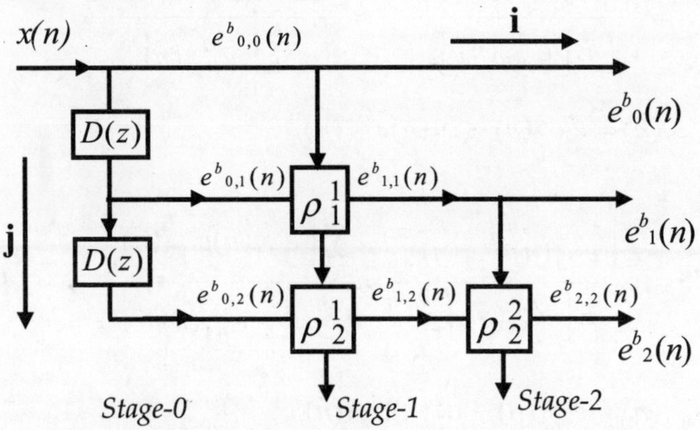

Fig. 7.16 Escalator Structure of order N=2.

Similar to warped lattice, the escalator structure can also be converted to its warped version to utilize the benefits of warping, viz., the performance advantage and the order reduction. Here also, the delay units z^{-1} are replaced by allpass units $D(z)$ as shown in Fig. 7.16 . Over the ordinary escalator structure, in the warped escalator, the introduction of IIR allpass structure provides better modeling performance and order reduction (S.Veena and S.V. Narasimhan, 2007). The orthogonalization provided by such a structure can be advantageously used for the joint process escalator lattice.

The outputs at each stage (for a 2 stage) is given by

For stage 0:

$$e^b_{0,0}(n) = x(n), \tag{7.63a}$$

$$e^b_{0,1}(n) = e^b_{0,0}(n-1) - \lambda e^b_{0,0}(n) + \lambda e^b_{0,1}(n-1) \tag{7.63b}$$

$$e^b_{0,2}(n) = e^b_{0,1}(n-1) - \lambda e^b_{0,1}(n) + \lambda e^b_{0,2}(n-1) \tag{7.63c}$$

$$\rho_1^1 = \frac{E\left[e^b_{0,0}(n)e^b_{0,1}(n)\right]}{E\left[e^{b2}_{0,0}(n)\right]} \approx \frac{\sum_{n=0}^{L-1}\left[e^b_{0,0}(n)e^b_{0,1}(n)\right]}{\sum_{n=0}^{L-1} e^{b2}_{0,0}(n)} \tag{7.63d}$$

For stage 1

$$e^b_{1,1}(n) = e^b_{0,1}(n) - \rho_1^1(n)e^b_{0,0}(n) \tag{7.64a}$$

$$\rho_2^1 = \frac{E\left[e^b{}_{0,0}(n)e^b{}_{0,2}(n)\right]}{E\left[e^{b2}_{0,0}(n)\right]} \approx \frac{\sum_{n=0}^{L-1}\left[e^b{}_{0,0}(n)e^b{}_{0,2}(n)\right]}{\sum_{n=0}^{L-1} e^{b2}_{0,0}(n)} \tag{7.64b}$$

$$e^b{}_{1,2}(n) = e^b{}_{0,2}(n) - \rho_2^1(n)e^b{}_{1,1}(n) \tag{7.64c}$$

For stage 2

$$\rho_2^2 = \frac{E\left[e^b{}_{1,1}(n)e^b{}_{1,2}(n)\right]}{E\left[e^{b2}_{1,1}(n)\right]} \approx \frac{\sum_{n=0}^{L-1}\left[e^b{}_{1,1}(n)e^b{}_{1,2}(n)\right]}{\sum_{n=0}^{L-1} e^{b2}_{1,1}(n)} \tag{7.65a}$$

$$e^b{}_{2,2}(n) = e^b{}_{1,2}(n) - \rho_2^2(n)e^b{}_{1,1}(n) \tag{7.65b}$$

On the similar lines for 3 stages,

$$e^b{}_{0,3}(n) = e^b{}_{0,2}(n-1) - \lambda e^b{}_{0,2}(n) + \lambda e^b{}_{0,3}(n-1) \tag{7.66a}$$

$$e^b{}_{1,3}(n) = e^b{}_{0,3}(n) - \rho_3^1(n)e^b{}_{2,2}(n) \tag{7.66b}$$

$$e^b{}_{2,3}(n) = e^b{}_{1,3}(n) - \rho_3^2(n)e^b{}_{2,2}(n) \tag{7.66c}$$

$$e^b{}_{3,3}(n) = e^b{}_{2,3}(n) - \rho_3^3(n)e^b{}_{2,2}(n) \tag{7.66d}$$

The $e^b{}_{i,j}(n)$ error at i^{th} stage and j^{th} order is

$$e^b{}_{i,j}(n) = e^b{}_{i-1,j}(n) - \rho_j^i(n)e^b{}_{i-1,i-1}(n), \quad i = 1, 2, ..., N-1, \quad j = i, i+1, ..., N-1 \tag{7.67a}$$

where,

$$e^b{}_{0,i}(n) = e^b{}_{0,i-1}(n-1) - \lambda e^b{}_{0,i-1}(n) + \lambda e^b{}_{0,i}(n-1), \quad i = 1, 2, ..., N-1 \tag{7.67b}$$

The warped orthogonal components are

$$e^b{}_0(n) = e^b{}_{0,0}(n), \; e^b{}_1(n) = e^b{}_{1,1}(n), \; e^b{}_2(n) = e^b{}_{2,2}(n)$$

In general,

$$e^b{}_j(n) = e^b{}_{j,j}(n), \; j = 0, 1, 2, ..., N-1 \tag{7.67c}$$

These warped orthogonal components are used for the joint process escalator. Here, the weighted sum of these components (the output) is compared with the desired signal and the weights are chosen to minimize the energy of the error signal, the difference between the output and the desired signal. As already stated, the computational load of escalator is higher than the normal lattice structure. From this point of view, due to warping as the escalator order decreases, warped escalator is advantageous both in terms of computational load and, performance over the ordinary escalator.

7.14 POLE–ZERO/ARMA MODELING

The AR modeling is popular due to its *efficiency and computational simplicity.* For signals with *dominant poles and damped zeros,* a higher order AR model represents the spectral peaks very well and also approximates the signal spectrum to a higher degree. However, the AR model cannot represent predominant zeros. A spectral zero has two effects: a dip in the spectrum and a 12dB/octave rise. A higher order AR produces the 12dB/ *octave rise but fails to create a dip in the spectrum without disturbing the neighborhood of the zero frequency.* Near a zero, this results *in a shift or merging of the spectral peaks.* To approximate the effect of a zero, *many poles are required and this increases the total number of parameters.* For example, for a process **with 4 *poles and 4 zeros,*** a ***25th AR model*** may be required and hence the importance of pole-zero or ARMA model.

Pole-zero estimation: The power spectrum estimation/signal modeling is done by a rational transfer function $H(z)$ given by,

$$H(z) = G\frac{B(z)}{A(z)}, \tag{7.68a}$$

$$A(z) = 1 + \sum_{k=1}^{P} a_k z^{-k}, \quad B(z) = \sum_{k-0}^{Q} b_k z^{-k} \tag{7.68b, c}$$

The problem is to choose a_k and b_k such that the impulse response $h(n)$ approximates the signal $x(n)$ as closely as possible. That is, the error energy ε

$$\varepsilon = \sum_n \left[x(n) - h(n)\right]^2 \tag{7.68d}$$

is minimized by proper choice of coefficients a_k and b_k.

That is,

$$\frac{\partial\, \varepsilon}{\partial(A,B)} = 0 \tag{7.69a}$$

The error energy ε by Parseval's theorem, in frequency domain becomes

$$\varepsilon = \frac{1}{2\pi}\int_{-\pi}^{\pi}\left|X(\omega) - \frac{B(\omega)}{A(\omega)}\right|^2 d\omega$$

$$\frac{\partial\, \varepsilon}{\partial a_k} = \frac{1}{2\pi}\int_{-\pi}^{\pi} 2\,\mathrm{Re}\left\{\frac{-B(\omega)e^{-j\omega k}}{A^2(\omega)}\left[X^*(\omega) - \frac{B^*(\omega)}{A^*(\omega)}\right]\right\}d\omega = 0\,, \quad k = 1,2,\ldots P \tag{7.69b}$$

$$\frac{\partial \varepsilon}{\partial b_k} = \frac{1}{2\pi}\int_{-\pi}^{\pi} 2\,\mathrm{Re}\left\{\frac{e^{-j\omega k}}{A(\omega)}\left[X^*(\omega) - \frac{B^*(\omega)}{A^*(\omega)}\right]\right\} d\omega = 0\ ,\ k = 0,1,2,\ldots Q \quad (7.69c)$$

Eqn. (7.69b) is not linear due to the existence of the squared and cubed terms of a_k, but Eqn. (7.69c) is linear in b_k as it does not appear in higher powers other than unity. In the absence of such terms, problem of pole-zero modeling would have become linear which is simple to solve. Hence, *only linearized versions* of this nonlinear problem are solved.

If $x(n)$ is the signal derived by exciting a system by an impulse then, $h(n)$ can approximate $x(n)$ and in this case it is called *signal modeling. On the other hand* if $x(n)$ is the output of a system driven by a white noise, then only the spectral features of $x(n)$ are represented by $h(n)$ and this is called *spectral modeling,* (as it is not possible to reproduce all the variations of the signal due to random noise.)

7.14.1 Pade's Approximation

The history of pole zero modeling starts from Pade's approximation. Here, the model represents the given signal. That is, the signal is the impulse response of the model for $n = 0, 1, \ldots, P+Q$. That is,

$$X(z) = \frac{B(z)}{A(z)}, \quad \text{with } G = 1 \quad (7.70a)$$

This results in

$$X(z) = \frac{\sum_{k=0}^{Q} b_k z^{-k}}{1 + \sum_{k=1}^{P} a_k z^{-k}}, \quad (7.70b)$$

Assuming that the impulse response of the model $h(n) = 0, n < 0$ with $b_n = 0, n < 0, n > Q$, the above relation becomes,

$$x(n) + \sum_{k=1}^{P} a_k x(n-k) = \begin{cases} b_n, & n = 0, 1, \ldots, Q \\ 0, & n = Q+1, Q+2, \ldots, Q+P \end{cases} \quad (7.70c)$$

Expanding this,

$$\begin{bmatrix} x(0) & 0 & \cdots & 0 \\ x(1) & x(0) & \cdots & 0 \\ \vdots & \vdots & \cdots & \vdots \\ x(Q) & x(Q-1) & \cdots & x(Q-P) \\ x(Q+1) & x(Q) & \cdots & x(Q-P+1) \\ \vdots & \vdots & \cdots & \vdots \\ x(Q+P) & x(Q+P-1) & \cdots & x(Q) \end{bmatrix} \begin{bmatrix} 1 \\ a_1 \\ a_2 \\ \vdots \\ a_P \end{bmatrix} = \begin{bmatrix} b_0 \\ b_1 \\ \vdots \\ b_Q \\ 0 \\ \vdots \\ 0 \end{bmatrix} \quad (7.70d)$$

This can be written as

$$\begin{bmatrix} x(0) & 0 & \cdots & 0 \\ x(1) & x(0) & \cdots & 0 \\ \vdots & \vdots & \cdots & \vdots \\ x(Q) & x(Q-1) & & x(Q-P) \end{bmatrix} \begin{bmatrix} 1 \\ a_1 \\ \vdots \\ a_P \end{bmatrix} = \begin{bmatrix} b_0 \\ b_1 \\ \vdots \\ b_Q \end{bmatrix} \tag{7.70e}$$

and

$$\begin{bmatrix} x(Q+1) & x(Q) & \cdots & x(Q-P+1) \\ x(Q+2) & x(Q+1) & \cdots & x(Q-P+2) \\ \vdots & \vdots & \cdots & \vdots \\ x(Q+P) & x(Q+P-1) & & x(Q) \end{bmatrix} \begin{bmatrix} 1 \\ a_1 \\ \vdots \\ a_P \end{bmatrix} = \begin{bmatrix} 0 \\ 0 \\ \vdots \\ 0 \end{bmatrix}$$

or

$$\begin{bmatrix} x(Q) & x(Q-1) & \cdots & x(Q-P+1) \\ x(Q+1) & x(Q) & \cdots & x(Q-P+2) \\ \vdots & \vdots & \cdots & \vdots \\ x(Q+P-1) & x(Q+P-2) & & x(Q) \end{bmatrix} \begin{bmatrix} 1 \\ a_1 \\ \vdots \\ a_P \end{bmatrix} = - \begin{bmatrix} x(Q+1) \\ x(Q+2) \\ \vdots \\ x(Q+P) \end{bmatrix} \tag{7.70f}$$

Eqns. (7.70e) and (7.70f) form the Pade's approximation method. The problem with this method is, it is valid over the signal interval $(0, P+Q)$ and does not ensure its validity outside this interval, i.e., for values of $n > (P+Q)$. This is because it models the signal exactly over this interval by making the error equal to zero. Instead of making the error zero exactly, by minimizing the squared error with respect to coefficients, it is possible to achieve a better approximation for all values $n > (P+Q)$ and this forms the basis of Prony's method.

7.14.2 Prony's Method

As stated the error between the model and the signal to be modeled is

$$e(n) = x(n) - h(n) \quad \text{or} \quad E(z) = X(z) - H(z) \tag{7.71a}$$

$$E(z) = X(z) - \frac{B(z)}{A(z)} \quad \text{or} \quad E(z)A(z) = A(z)X(z) - B(z)$$

If $E'(z) = E(z)A(z)$, then

$$E'(z) = A(z)X(z) - B(z) \tag{7.71b}$$

In time domain,

$$e'(n) = a_k * x(n) - b_n$$

$e'(n)$ can be called filtered error, the original error filtered by $A(z)$. Since $b_n = 0, n > Q$,

$$e'(n) = \begin{cases} x(n) + \sum_{k=1}^{P} a_k x(n-k) - b_n, & n = 0, 1, \ldots, Q \\ x(n) + \sum_{k=1}^{P} a_k x(n-k), & n > Q \end{cases} \tag{7.71c}$$

Instead of modeling exactly over the interval $(0, P+Q)$, *by making the error zero as in Pade's method, the model to be valid outside the interval,* ***the filtered error energy is minimized***. That is,

$$\varepsilon_{P,Q} = \sum_{n=Q+1}^{\infty} [e'(n)]^2 = \sum_{n=Q+1}^{\infty} \left[x(n) + \sum_{k=1}^{P} a_k x(n-k) - b_n \right]^2 \tag{7.72a}$$

$$\varepsilon_{P,Q} = \sum_{n=Q+1}^{\infty} \left[x(n) + \sum_{k=1}^{P} a_k x(n-k) \right]^2 \text{ as } b_n = 0, n > Q \tag{7.72b}$$

Hence $\varepsilon_{P,Q}$ can be minimized by making

$$\frac{\partial \varepsilon_{P,Q}}{\partial a_k} = 0, \qquad 2 \sum_{n=Q+1}^{\infty} e'(n) \frac{\partial e'(n)}{\partial a_k} = 0, \qquad k = 1, 2, \ldots, P$$

Using Eqn.(7.71c),

$$\sum_{n=Q+1}^{\infty} e'(n) \frac{\partial}{\partial a_k} \left[x(n) + \sum_{k=1}^{P} a_k x(n-k) \right] = 0, \qquad k = 1, 2, \ldots, P$$

$$\sum_{n=Q+1}^{\infty} e'(n) x(n-k) = 0, \qquad k = 1, 2, \ldots, P \tag{7.72c}$$

This is similar to orthogonality principle but here the *input and filtered error are orthogonal to each other.* Substituting for $e'(n)$ from Eqn. (7.71c),

$$\sum_{n=Q+1}^{\infty} \left[x(n) + \sum_{l=1}^{P} a_l x(n-l) \right] x(n-k) = 0, \qquad k = 1, 2, \ldots, P$$

$$\sum_{l=1}^{P} a_l \sum_{n=Q+1}^{\infty} x(n-l) x(n-k) = -\sum_{n=Q+1}^{\infty} x(n) x(n-k), \qquad k = 1, 2, \ldots, P \tag{7.72d}$$

Defining

$$r_x(k,l) = \sum_{n=Q+1}^{\infty} x(n-l) x(n-k) \tag{7.72e}$$

which is symmetric i.e., $r_x(k,l) = r_x(l,k)$, the above equation becomes,

$$\sum_{l=1}^{P} a_l r_x(k,l) = -r_x(k,0), \qquad k = 1, 2, \ldots, P \tag{7.73a}$$

This is a system of P linear equations in P unknowns $a_k, k = 1, 2, \ldots, P$ and is known as *Pronys normal equations.* The filtered error is as already given by (Eqn. 7.71c))

$$e'(n) = \begin{cases} x(n) + \sum_{k=1}^{P} a_k x(n-k) - b_n, & n = 0, 1, \ldots, Q \\ x(n) + \sum_{k=1}^{P} a_k x(n-k), & n > Q \end{cases}$$

The coefficients $b_k, k = 0, 1, 2, \ldots, Q$ are chosen such that $e'(n) = 0$, for $n = 0, 1, 2, \ldots, Q$. That is,

$$x(n) + \sum_{k=1}^{P} a_k x(n-k) - b_n, = 0, \qquad n = 0, 1, \ldots, Q$$

$$b_n = x(n) + \sum_{k=1}^{P} a_k x(n-k), \qquad n = 0, 1, \ldots, Q \tag{7.73b}$$

This is similar to the procedure adopted in Pade's method and the numerator polynomial is given by

$$\begin{bmatrix} x(0) & 0 & \cdots & 0 \\ x(1) & x(0) & \cdots & 0 \\ \vdots & \vdots & \cdots & \vdots \\ x(Q) & x(Q-1) & & x(Q-P) \end{bmatrix} \begin{bmatrix} 1 \\ a_1 \\ \vdots \\ a_P \end{bmatrix} = \begin{bmatrix} b_0 \\ b_1 \\ \vdots \\ b_Q \end{bmatrix} \tag{7.73c}$$

The minimum energy achieved by the Prony's method is

$$\varepsilon_{P,Q} = \sum_{n=Q+1}^{\infty} \left[e'(n)\right]^2 = \sum_{n=Q+1}^{\infty} e'(n)\left[x(n) + \sum_{k=1}^{P} a_k x(n-k) - b_n\right]$$

$$= \sum_{n=Q+1}^{\infty} e'(n)\left[x(n) + \sum_{k=1}^{P} a_k x(n-k)\right], \qquad b_n = 0, \text{ as } n > q$$

$$= \sum_{n=Q+1}^{\infty} e'(n)x(n) + \sum_{k=1}^{P} a_k \sum_{n=Q+1}^{\infty} e'(n)x(n-k)$$

$$= \sum_{n=Q+1}^{\infty} e'(n)x(n) \qquad \left(\text{as } \sum_{n=Q+1}^{\infty} e'(n)\, x(n-k) = 0 \text{ due to orthogonality}\right)$$

$$= \sum_{n=Q+1}^{\infty} x(n)\left[x(n) + \sum_{k=1}^{P} a_k x(n-k) \right] = \sum_{n=Q+1}^{\infty} x^2(n) + \sum_{k=1}^{P} a_k \sum_{n=Q+1}^{\infty} x(n)x(n-k)$$

$$\varepsilon_{P,Q} = r_x(0,0) + \sum_{k=1}^{P} a_k r_x(0,k) \qquad (7.73d)$$

It is important to note that here as the orthogonality between $e'(n)$ and $x(n-k),\ k = 1, 2, \ldots, P$,

$$\sum_{n=Q+1}^{\infty} e'(n)x(n-k) = 0, \qquad k = 1, 2, \ldots, P \qquad (7.73e)$$

is used and it is essential for the coefficients $a_k, k = 1, 2, \ldots, P$ to satisfy this orthogonality.

Allpole modeling by Prony's method:

The Prony's method of pole-zero modeling considers the filtered error $e'(n)$, which is nothing, but the original error $e(n)$ filtered by $A(z)$.

$$E'(z) = A(z)E(z)$$

$$E(z) = X(z) - \frac{B(z)}{A(z)} = X(z) - \frac{\sum_{k=0}^{Q} b_k z^{-k}}{1 + \sum_{k=1}^{P} a_k z^{-k}},$$

$$E'(z) = A(z)X(z) - B(z),$$

$$e'(n) = \begin{cases} x(n) + \sum_{k=1}^{P} a_k x(n-k) - b_n, & n = 0, 1, \ldots, Q \\ x(n) + \sum_{k=1}^{P} a_k x(n-k), & n > Q \end{cases}$$

with $x(n) = 0$ for $n < 0$. For a AR modeling of $x(n)$ as $Q = 0$,

$$X(z) = \frac{b_0}{1 + \sum_{k=1}^{P} a_k z^{-k}}, \qquad (7.74a)$$

That is, *Prony's pole-zero modeling with* $Q = 0$, *results in all-pole modeling.* Therefore, for the Prony's error energy minimization

$$\varepsilon_{P,Q} = \sum_{n=Q+1}^{\infty} \left[e'(n)\right]^2$$

becomes,

$$\varepsilon_{P,0} = \sum_{n=1}^{\infty} \left[e'(n)\right]^2 \qquad (7.74b)$$

with the Prony's error,

$$e'(n) = x(n) + \sum_{k=1}^{P} a_k x(n-k), \qquad n > Q$$

As $x(n) = 0$ for $n < 0$, at $n = 0$, Prony's error is equal to $x(0)$ and does not depend on $a_k, k = 1, 2, \ldots, P$. Hence, the coefficients those minimize $\varepsilon_{P,0}$ are same as those which minimize the error energy

$$\varepsilon_P = \sum_{n=0}^{\infty} \left[e'(n)\right]^2 \tag{7.74c}$$

Therefore, the minimization equations of $\varepsilon_{P,0}$ are same as those for ε_P, and this leads to the the same Prony normal equations (Eqn. 7.73a),

$$\sum_{l=1}^{P} a_l r_x(k,l) = -r_x(k,0), \qquad k = 1, 2, \ldots, P \tag{7.74d}$$

$$r_x(k,l) = \sum_{n=0}^{\infty} x(n-l)x(n-k) \tag{7.74e}$$

It is important that the *lower limit in the definition of* $r_x(k,l)$ *here is zero and not unity* as directly obtained by putting $Q = 0$ in Eqn. (7.72e) of Prony's method of pole-zero modeling. This is due to the use of the error energy ε_P instead of $\varepsilon_{P,0}$. Since for a causal process $x(n) = 0$ *for* $n < 0$, $r_x(k,l)$ depends up on the difference between the indicies "k" and "l", defining

$$r_x(k-l) = r_x(k,l) = \sum_{n=0}^{\infty} x(n-l)x(n-k)$$

and also as $r_x(k,l) = r_x(l,k)$, $r_x(k-l) = r_x(l-k)$, the normal equations become

$$\sum_{l=1}^{P} a_l r_x(k-l) = -r_x(k), \qquad k = 1, 2, \ldots, P \tag{7.74f}$$

That is,

$$\begin{bmatrix} r_x(0) & r_x(1) & \cdots & r_x(P-1) \\ r_x(1) & r_x(0) & \cdots & r_x(P-2) \\ \vdots & \vdots & \cdots & \vdots \\ r_x(P-1) & r_x(P-2) & \cdots & r_x(0) \end{bmatrix} \begin{bmatrix} a_1 \\ a_2 \\ \vdots \\ a_P \end{bmatrix} = -\begin{bmatrix} r_x(1) \\ r_x(2) \\ \vdots \\ r_x(P) \end{bmatrix} \tag{7.74g}$$

Here as the correlation matrix is Toeplitz symmetric, the normal equations can be solved by using Durbin Levinson recursion algorithm. The minimum error energy ε_P for the all-pole modeling is obtained by incorporating the modifications for the case of Prony's pole-zero modeling. The minimum error energy for the pole-zero modeling is given by Eqn.(7.73d),

$$\varepsilon_{P,Q} = r_x(0, 0) + \sum_{l=1}^{P} a_l r_x(0, l)$$

becomes as

$$\varepsilon_P = r_x(0) + \sum_{l=1}^{P} a_l r_x(l) \tag{7.74h}$$

To match the gain of $X(z)$, b_0 is chosen as $b_0 = \sqrt{\varepsilon_P}$ and *this is on the basis that the error* $e(n)$ *will be white.*

Example: The Prony's pole-zero modeling will be applied to speech and impulse responses charcterized both by poles and zeros.

For a speech segment of an utterance "m" which is a nasal having prominent spectral dips sampled at 8.82kHz and of lenth of 256 samples is considered for the illustration of the pole-zero Prony's method. Here, the poles are estimated by Prony's modeling using 10 poles (P=10) and zeros by Pade's method using 8 zeros (Q=8), respectively.

The poles are determined by Prony's pole method using Eqns. (7.72e) and (7.73a)

$$r_x(k,l) = \sum_{n=Q+1}^{\infty} x(n-l)x(n-k)$$

$$\sum_{l=1}^{P} a_l r_x(k,l) = -r_x(k,0), \qquad k = 1, 2, \ldots, P$$

The zeros are determined using Eqn.(7.73c)

$$\begin{bmatrix} x(0) & 0 & \cdots & 0 \\ x(1) & x(0) & \cdots & 0 \\ \vdots & \vdots & \cdots & \vdots \\ x(Q) & x(Q-1) & & x(Q-P) \end{bmatrix} \begin{bmatrix} 1 \\ a_1 \\ \vdots \\ a_P \end{bmatrix} = \begin{bmatrix} b_0 \\ b_1 \\ \vdots \\ b_Q \end{bmatrix}$$

The Eqn. (7.73a) is solved by using the matrix inverse function of Matlab. The pole and zero spectra are shown in Fig. 7.17(a). The pole sectrum emphasies the spectral peaks but not the valleys and the zero spectrum emphasies the dips but not the peaks. The pole-zero spectrum emphasies both spectral peaks and dips as shown in Fig. 7.17(b) and has a good match to the DFT spectrum.

The pole-zero modeling is applied to the impulse response of a transfer function,

$$G(z) = \frac{1 - 1.1289z^{-1} + 1.0058z^{-2} - 0.7617z^{-3}}{1 - 0.59z^{-1} + 0.455z^{-2} - 0.0754z^{-3} + 0.504z^{-4}}.$$

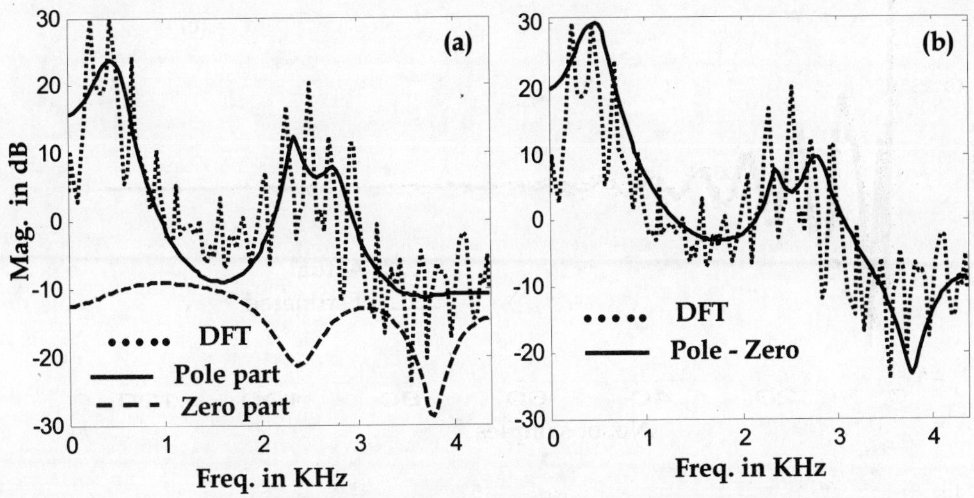

Fig. 7.17 Prony's pole-zero modelling of speech spectrum (a) Pole and zero spectra and (b) Pole-zero spectrum.

The pole polynomial is determined by *Prony's allpole modelling* using Eqns. (7.74e) and (7.74d)

$$r_x(k,l) = \sum_{n=0}^{\infty} x(n-l)x(n-k)$$

$$\sum_{l=1}^{P} a_l r_x(k,l) = -r_x(k,0), \qquad k = 1, 2, \ldots, P$$

but not by Prony's pole modeling. This is because in Prony's pole modeling, the lower limit for Eqn.(7.72e) is not zero but $(Q+1)$ and this removes the initial $(Q+1)$ points of the impulse response. But with Eqn. (7.74e), the lower limit is zero and not $(Q+1)$. Thus, the Prony's pole modelling aims at only impulse response omitting the initial $(Q+1)$ points. But it is required to model the complete impulse response starting from the zeroth point and hence the Prony's allpole modeling should only be used for modeling impulse response (signal modeling). Fig. 7.17(c) shows the estimated impulse response obtained by giving unit impulse the estimated by pole-zero model, poles being determined by Prony's allpole modeling and this has a good match with the actual impulse response computed again by giving unit impulse input to the the actual transfer function $G(z)$.

For spectral modelling, still the Prony's pole modeling can be used in spite of omitting the initial $(Q+1)$ points. This is because the spectral information exists in all parts

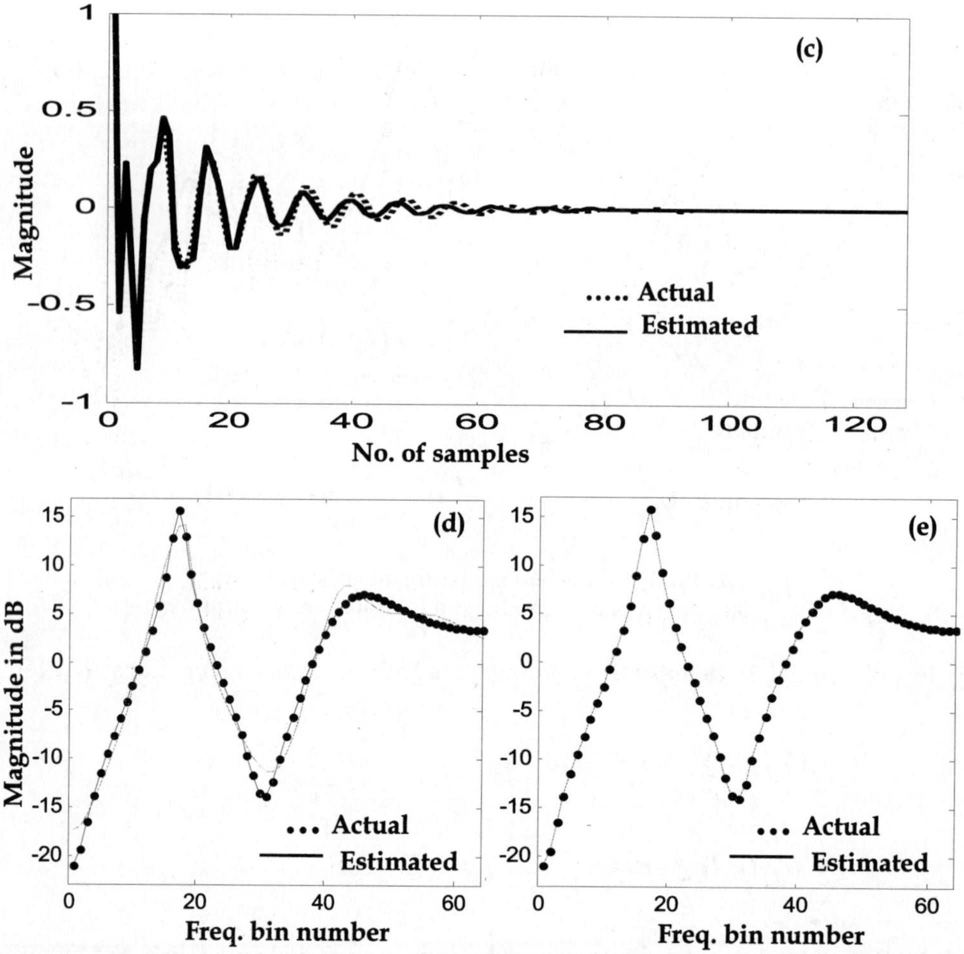

Fig. 7. 17 (c, d, e) : (c) Impulse response estimated by Prony's pole-zero (poles by Prony's allpole) method (d)) Spectrum estimated by Prony's pole-zero method (poles by Prony's allpole estimation) (e) Spectrum estimated by Prony's pole-zero method (poles by Prony's pole estimation).

of the impulse response (before it decays completely to zero). The spectrum obtained by Prony's pole-zero modeling where poles are determined by Prony's pole modeling has a Prony's pole modeling has a good match with the spectrum of transfer function given by $G(z)$ (Fig.(7.17(e)). However, the impulse response computed by giving unit impulse using the estimated model (Prony's pole modeling) blows off (becomes infinite). *Thus, the estimated pole-zero model may provide the spectrum but cannot be used for filtering the input* as it may blow off as it has happened in the present study. It is to be noted that the spectrum estimated by Prony'smethod using allpole modeling poles (Fig. 7.17(d)) (though it provides a impulse response having good match to the actual one) is not so accurate as that obtained using by Prony pole modelling for poles.

7.14.3 Shanks Method

In Prony's method, though the estimation of denominator coefficients $a_n, n = 1, 2, \ldots, P$ are made valid for all n even outside the interval $(0, P+Q)$, the estimated numerator coefficients are still limited (valid) to the interval $(0, Q)$, only. The Shanks method alleviates this problem by considering the original error $e(n)$ and this is given in Z-domain and time domain as

$$E(z) = X(z) - \hat{X}(z)$$

$$e(n) = x(n) - \hat{x}(n) \tag{7.75a}$$

Since

$$\hat{X}(z) = \frac{B(z)}{A(z)},$$

$$E(z) = X(z) - \frac{B(z)}{A(z)}$$

But $\hat{X}(z)$ can be written as

$$\hat{X}(z) = \frac{B(z)}{A(z)} = B(z)\left\{\frac{1}{A(z)}\right\}$$

If $F(z) = 1/A(z)$, then

$$\hat{X}(z) = B(z)F(z)$$

or

$$\hat{x}(n) = \sum_{k=0}^{Q} b_k f(n-k) \tag{7.75b}$$

and

$$f(n) = \delta(n) - \sum_{k=1}^{P} a_k f(n-k) \tag{7.75c}$$

That is, $f(n)$ is the impulse response of the filter $\left[1/A(z)\right]$ and this is further shaped by the filter $\left[B(z)\right]$ to match the signal, in the best possible way. This is achieved by assuming that $\left[1/A(z)\right]$ or its impulse response $f(n)$ is known and the coefficients $b_n, n = 0, 1, 2, \ldots, Q$ are estimated by minimizing the error energy ε_S given by

$$\varepsilon_S = \sum_{n=0}^{\infty} e^2(n)$$

where

$$e(n) = x(n) - \hat{x}(n) = x(n) - \sum_{k=0}^{Q} b_k f(n-k) \tag{7.75d}$$

Therefore, to find b_k, $k = 0, 1, 2, \ldots, Q$,

$$\frac{\partial \varepsilon_S}{\partial b_k} = \frac{\partial}{\partial b_k}\left\{\sum_{n=0}^{\infty}[e(n)]^2\right\}, \qquad k = 0, 1, 2, \ldots, Q$$

$$\frac{\partial \varepsilon_S}{\partial b_k} = 2\sum_{n=0}^{\infty} e(n)\frac{\partial e(n)}{\partial b_k} = 0$$

Since from Eqn. (7.75d)

$$e(n) = x(n) - \sum_{k=0}^{Q} b_k f(n-k),$$

Therefore,

$$\frac{\partial \varepsilon_S}{\partial b_k} = -2\sum_{n=0}^{\infty} e(n) f(n-k) = 0, \quad k = 0, 2, \ldots, Q$$

Also again substituting for $e(n)$,

$$\sum_{n=0}^{\infty}\left[x(n) - \sum_{l=0}^{Q} b_l f(n-l)\right] f(n-k) = 0$$

$$\sum_{l=0}^{Q} b_l\left[\sum_{n=0}^{\infty} f(n-k) f(n-l)\right] = \sum_{n=0}^{\infty} f(n-k)x(n) \tag{7.75e}$$

It is important to note that the minimization has resulted in the above equation which is linear in b_l provided $f(n)$ is known which are constants with respect to b_l. In general defining

$$\phi_{yx}(t,u) = \sum_{n=0}^{L} y(n-t)x(n-u)$$

then,

$$\phi_{ff}(k,l) = \left[\sum_{n=0}^{L} f(n-k) f(n-l)\right], \qquad \phi_{fx}(k,0) = \sum_{n=0}^{L} f(n-k)x(n) \tag{7.75f, g}$$

In defining $\phi_{ff}(k,l)$ and $\phi_{fx}(k,0)$, the upper limit for n is replaced by data length available L instead of ∞. Therefore, Eqn. (7.75e) becomes

$$\sum_{l=0}^{Q} b_l \phi_{ff}(k,l) = \phi_{fx}(k,0), \quad k = 0, 2, \ldots, Q \tag{7.75h}$$

The above equation provides a solution for $b_l, l = 0,1, 2, \ldots, Q$ which is not limited to $n = 0,1, 2, \ldots, Q$ but for all $n = 0,1, 2, \ldots,$.

Till now it was assumed that $[1/A(z)]$ or its impulse response $f(n)$ is known for the estimation of $b_k, k = 0,1, 2, \ldots, Q$. That is, poles and zeros are estimated in two steps. The poles are estimated first to know $f(n)$ and then the zeros are estimated using $f(n)$ which is the information about poles. *Since the poles and zeros are estimated separately, instead of approximating the whole spectrum by an all-pole model by autocorrelation method, the poles are estimated accurately by covariance method* and a_k are given by

$$\sum_{k=1}^{P} a_k \phi_x(k, r) = -\phi_x(0, r) \quad k = 1,2,\ldots P \tag{7.75i}$$

where

$$\phi_x(t, u) = \sum_{n=0}^{L} x(n-t)x(n-u)$$

Though, the poles and zeros can be estimated by the above method, it has a limitation. Since it is based on linear prediction, it requires the signal to be either minimum or maximum phase but not a mixed phase signal. But in practice, generally signals are of mixed phase. The spectral zeros which lie outside the unit circle in the Z-plane makes them of mixed phase nature. These zeros pose problem for modeling the numerator part of the transfer function by LP based methods. It is possible to overcome this problem by converting the mixed phase signal to a minimum phase equivalent. This minimum phase equivalent signal can be used for pole-zero modeling as it satisfies the requirement of linear prediction. The minimum phase equivalent basically reflects the zeros outside the unit circle to its conjugate reciprocal locations inside the unit circle and thus converts it to a minimum phase signal, which is compatible for LP. The minimum phase equivalent signal can be derived from the cepstrum $c(n)$ which is the inverse Fourier transform of the logarithm of the given signal spectrum magnitude $|X(\omega)|$.

$$\ln|X(\omega)| = \sum_{n=0}^{\infty} c(n)\cos \omega n \tag{7.75j}$$

The phase $\theta(\omega)$ of the minimum phase equivalent signal is given by

$$\theta(\omega) = -\sum_{n=0}^{\infty} c(n)\sin \omega n \tag{7.75k}$$

Knowing the spectral magnitude and the phase, the minimum phase equivalent signal can be derived. Such an approach, which combines the homomorphic approach and linear prediction, is called *homomorphic prediction*. The homomorphic approach is applicable to all type of signals including the one characterized by zeros outside the unit circle. The LP on the other hand is applicable to signals represented by poles and zeros inside the unit circle or zeros only outside the unit circle (assuming a physical stable system/ process for which poles are inside). This combination has both generality and the efficiency in extracting the structure of the signal.

7.14.4 Steiglitz-McBride Method: Simultaneous pole–zero modeling/ Iterative Prefiltering Method

In Shank's method, the numerator polynomial is determined by knowing the denominator polynomial by minimizing the energy of the error given by Eqn. (7.75d)

$$e(n) = x(n) - \sum_{k=0}^{Q} b_k f(n-k)$$

where $F(z) = 1/A(z)$ in a single step. In view of this, the model may not be accurate. But the Steiglitz- McBride method aims to achieve this iteratively. Based on the estimation of $A(z)$, new estimates of $A(z)$ and $B(z)$ are obtained till the present estimate of $A(z)$ is almost same as its previous one by linearising the error. The linearization is achieved by prefiltering the given signal $x(n)$ and the impulse $\delta(n)$ by the previous estimate of $\left[1/A(z)\right]$ (T.H. Joo, J.H. McClellan, R.A. Pole, G.S. Mayers and R.S. Lees, 1983).

Here, the rational transfer function $[B(z)/A(z)]$ is chosen so that its impulse response $h(n)$ represents the given signal and is achieved by minimizing the error energy with respect to $A(z)$ and $B(z)$. The error energy is given by

$$E = \sum_{n=0}^{N-1} \left\{x(n) - h(n)\right\}^2 \tag{7.76a}$$

This can be written as

$$E = \sum_{n=0}^{N-1} \left\{\frac{1}{A}\left[A[x(n)] - B[\delta(n)]\right]\right\}^2$$

$A[\bullet]$ and $B[\bullet]$ are convolutional operators for $A(z)$ and $B(z)$, respectively. Further $1/A[\bullet]$ is the corresponding operator for $1/A(z)$. If $A(z)$ is known at the $(i-1)^{th}$ as $A_{i-1}(z)$, then

$$\tilde{E}^{(i)} = \sum_{n=0}^{N-1} \left\{\frac{1}{A_{i-1}}\left[A_i[x(n)] - B_i[\delta(n)]\right]\right\}^2 = \sum_{n=0}^{N-1} \left\{\tilde{e}^{(i)}(n)\right\}^2 \tag{7.76b}$$

$$\tilde{e}^{(i)}(n) = A_i\left[\frac{1}{A_{i-1}}[x(n)]\right] - B_i\left[\frac{1}{A_{i-1}}[\delta(n)]\right]$$

$$\tilde{e}^{(i)}(n) = A_i\left[\tilde{x}(n)\right] - B_i\left[\tilde{y}(n)\right] \tag{7.76c}$$

$\tilde{x}(n)$ and $\tilde{y}(n)$ are the filtered signals corresponding to $x(n)$ and $\delta(n)$ and the filter being $1/A_{i-1}(z)$. The use of these *filtered signals enables linearization of the error with respect to*

$A(z)$ *and* $B(z)$. *During iterations, the minimization occurs close to true error when* $A_i(z) = A_{i-1}(z)$ *as Eqn. (7.76b) becomes close to Eqn. (7.76a).* In practice, the minimization occurs for a small number of iterations, like less than 10. For minimization, (using Eqn.(7.76c).

$$\frac{\partial \tilde{E}^{(i)}}{\partial a_k} = 2\sum_{n=0}^{N-1} \tilde{e}^{(i)}(n)\frac{\partial \tilde{e}^{(i)}(n)}{\partial a_k} = 2\sum_{n=0}^{N-1} \tilde{e}^{(i)}(n)\frac{\partial [A_i[\tilde{x}(n) - B_i[\tilde{y}(n)]]}{\partial a_k} = 0$$

$$\sum_{n=0}^{N-1} \tilde{e}^{(i)}(n)\frac{\partial [A_i[\tilde{x}(n]]}{\partial a_k} = \sum_{n=0}^{N-1} \tilde{e}^{(i)}(n)\frac{\partial}{\partial a_k}[[1+\sum_{l=1}^{P} a_l^{(i)} z^{-l}][\tilde{x}(n)]] = 0$$

$$\sum_{n=0}^{N-1} \tilde{e}^{(i)}(n)\frac{\partial}{\partial a_k}[\tilde{x}(n) + \sum_{l=1}^{P} a_l^{(i)}\tilde{x}(n-l)] = 0$$

$$\sum_{n=0}^{N-1} \tilde{e}^{(i)}(n)\tilde{x}(n-k) = 0, \qquad k == 1, 2, \ldots, P. \tag{7.76d}$$

$$\frac{\partial \tilde{E}^{(i)}}{\partial b_j} = 2\sum_{n=0}^{N-1} \tilde{e}^{(i)}(n)\frac{\partial \tilde{e}^{(i)}(n)}{\partial b_j} = 2\sum_{n=0}^{N-1} \tilde{e}^{(i)}(n)\frac{\partial\, [A[\tilde{x}(n) - B[\tilde{y}(n)])}{\partial b_j} = 0$$

$$\sum_{n=0}^{N-1} \tilde{e}^{(i)}(n)\frac{\partial [-B_i[\tilde{y}(n]]}{\partial b_j} = -\sum_{n=0}^{N-1} \tilde{e}^{(i)}(n)\frac{\partial}{\partial b_j}[[\sum_{m=0}^{Q} b_m^{(i)} z^{-m}][\tilde{y}(n)]] = 0$$

$$\sum_{n=0}^{N-1} \tilde{e}^{(i)}(n)\frac{\partial}{\partial b_j}\left[\sum_{m=0}^{Q} b_m^{(i)}\tilde{y}(n-m)\right] = 0$$

$$\sum_{n=0}^{N-1} \tilde{e}^{(i)}(n)\tilde{y}(n-j) = 0, \qquad j = 0, 1, 2, \ldots, Q. \tag{7.76e}$$

Substituting for

$$\tilde{e}^{(i)}(n) = A_i\left[\tilde{x}(n)\right] - B_i\left[\tilde{y}(n)\right],$$

in Eqn. (7.76d)

$$\sum_{n=0}^{N-1} \tilde{e}^{(i)}(n)\tilde{x}(n-k) = 0, \qquad k == 1, 2, \ldots, P.$$

$$\sum_{n=0}^{N-1} \tilde{x}(n-k)A_i[\tilde{x}(n)] - B_i[\tilde{y}(n)] = 0$$

$$\sum_{n=0}^{N-1} \tilde{x}(n-k)\left\{\tilde{x}(n)+\sum_{l=1}^{P} a_l \tilde{x}(n-l)\right\} - \sum_{n=0}^{N-1} \tilde{x}(n-k)\sum_{m=0}^{Q} b_m \tilde{y}(n-m) = 0,$$

$$k = 1, 2, \ldots, P.$$

$$\sum_{n=0}^{N-1} \tilde{x}(n)\tilde{x}(n-k) + \sum_{l=1}^{P} a_l \sum_{n=0}^{N-1} \tilde{x}(n-l)\tilde{x}(n-k) - \sum_{m=0}^{Q} b_m \sum_{n=0}^{N-1} \tilde{x}(n-k)\tilde{y}(n-m) = 0$$

$$k = 1, 2, \ldots, P.$$

Defining

$$\phi_{fg}(k,l) = \sum_{n=0}^{N-1} f(n-k)g(n-l)$$

$$\sum_{m=0}^{Q} b_m \phi_{\tilde{x}\tilde{y}}(k,m) - \sum_{l=1}^{P} a_l \phi_{\tilde{x}\tilde{x}}(l,k) = \phi_{\tilde{x}\tilde{x}}(0,k), \quad k = 1, 2, \ldots, P. \tag{7.77a}$$

Again substituting for

$\tilde{e}^{(i)}(n) = A_i\left[\tilde{x}(n)\right] - B_i\left[\tilde{y}(n)\right]$ in Eqn.(7.76e),

$$\sum_{n=0}^{N-1} \tilde{y}(n-j)A_i[\tilde{x}(n)] - B_i[\tilde{y}(n)] = 0, \quad j = 0, 1, 2, \ldots, Q.$$

$$\sum_{n=0}^{N-1} \tilde{y}(n-j)\left\{\tilde{x}(n)+\sum_{l=1}^{P} a_l \tilde{x}(n-l)\right\} - \sum_{n=0}^{N-1} \tilde{y}(n-j)\sum_{m=0}^{Q} b_m \tilde{y}(n-m) = 0, \quad j = 0,1, 2, \ldots, Q.$$

$$\sum_{n=0}^{N-1} \tilde{x}(n)\tilde{y}(n-j) + \sum_{l=1}^{P} a_l \sum_{n=0}^{N-1} \tilde{x}(n-l)\tilde{y}(n-j)) - \sum_{m=0}^{Q} b_m \sum_{n=0}^{N-1} \tilde{y}(n-m)\tilde{y}(n-j) = 0$$

$$j = 0,1, 2, \ldots, Q.$$

Since

$$\phi_{fg}(k,l) = \sum_{n=0}^{N-1} f(n-k)g(n-l)$$

$$\phi_{\tilde{x}\tilde{y}}(0,j) + \sum_{l=1}^{P} a_l \phi_{\tilde{x}\tilde{y}}(l,j) - \sum_{m=0}^{Q} b_m \phi_{\tilde{y}\tilde{y}}(m,j) = 0, \quad j = 0,1, 2, \ldots, Q.$$

$$\sum_{m=0}^{Q} b_m \phi_{\tilde{y}\tilde{y}}(m,j) - \sum_{l=1}^{P} a_l \phi_{\tilde{x}\tilde{y}}(l,j) = \phi_{\tilde{x}\tilde{y}}(0,j), \quad j = 0,1, 2, \ldots, Q. \tag{7.77b}$$

From Eqn. (7.77a)

$$\sum_{m=0}^{Q} b_m \phi_{\tilde{x}\tilde{y}}(k,m) - \sum_{l=1}^{P} a_l \phi_{\tilde{x}\tilde{x}}(l,k) = \phi_{\tilde{x}\tilde{x}}(0,k), \quad k = 1, 2, \ldots, P.$$

These equations can be written in matrix form as

$$\begin{bmatrix} \phi_{\tilde{y}\tilde{y}}(m,j) & \phi_{\tilde{x}\tilde{y}}(l,j) \\ \phi_{\tilde{x}\tilde{y}}(k,m) & \phi_{\tilde{x}\tilde{x}}(l,k) \end{bmatrix} \begin{bmatrix} b_m \\ -a_l \end{bmatrix} = \begin{bmatrix} \phi_{\tilde{x}\tilde{y}}(0,j) \\ \phi_{\tilde{x}\tilde{x}}(0,k) \end{bmatrix}, \tag{7.77c}$$

$$m = j = 0, 1, 2, \ldots, Q., \quad l = k = 1, 2, \ldots, P.$$

The solution of this equation gives the coefficients of the numerator and denominator polynomials $A_i(z)$ and $B_i(z)$ at the present i^{th} iteration. The denominator polynomial of this i^{th} iteration $A_i(z)$ is compared with that of the previous iteration $A_{i-1}(z)$ i.e., of $(i-1)^{th}$ iteration. If the sum of the squared difference of the coefficients of the denominator polynomials between i^{th} and $(i-1)^{th}$ becomes least, i.e., $A_i(z)$ and $A_{i-1}(z)$ match best, the iteration is stopped and the coefficients of $A_i(z)$ and $B_i(z)$ are considered valid.

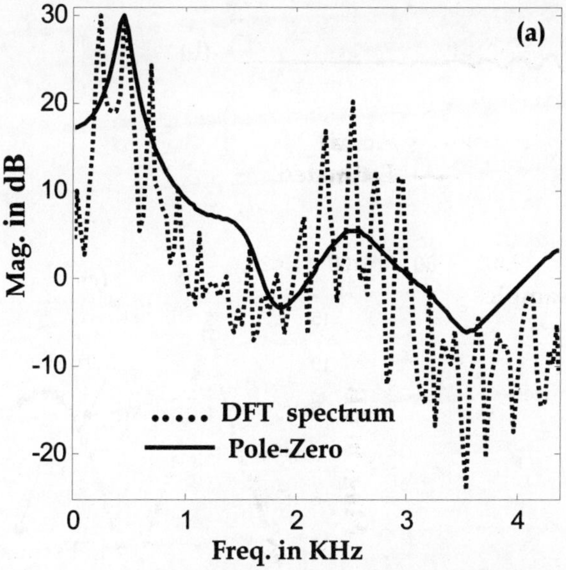

Fig. 7. 18(a) Pole-zero spectrum by Steigliz McBride method for the speech data.

For the same speech data of 256 samples considered previously, the Steigliz McBride method is applied. For this pole and zero polynomial orders P and Q used are 9 and 4 , respectively . Further, the number of iterations used is 6. The initial pole polynomial used is that obtained by Prony's all-pole modeling. Further to invert the matrix of Eqn.(7.77c), the inverse function (of Matlab) resulted in singular solution. Hence, the pseudoinverse function (of Matlab) has been used.

To solve the linear equation $Ax = b$, if A is a $(m \times n)$ matrix with $m \geq n$ and inverse of $A^T A$ exists, then the pseudoinverse of A is $A^+ = \left(A^T A\right)^{-1} A^T$. This is because $A^+ A = \left(A^T A\right)^{-1}\left(A^T A\right) = I$ and hence $A^+ = A^{-1}$.

The spectrum of the speech data achieved by the Steigliz McBride method is shown in the Fig. 7.18(a). The pole and zero polynomial orders and the number of iterations to be used are quite critical in achieving a good spectral match to the DFT spectrum. Also the spectrum obtained by this method is some what poor compared to that of Prony's pole-zero modeling (Fig,7.17b).

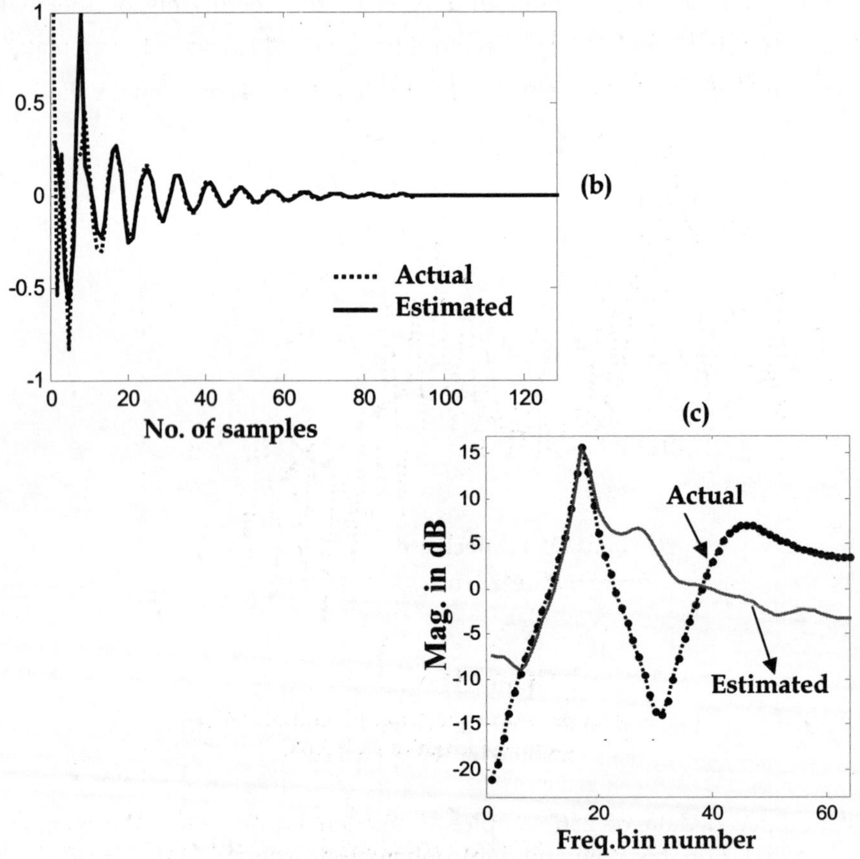

Fig. 7. 18 (b, c) Steigliz McBride method for the impulse response data.
(b) Estimated impulse response and (c) Estimated spectrum

To the same impulse response data considered for the Prony's method of pole-zero modeling, the Steigliz McBride method is applied. Here the pole and zero polynomial orders P and Q used are 15 and 7, respectively. The number of iterations used is 6 and the impulse response data used is 128. Here also, the impulse response estimated (Fig. 7.18(b)) is comparatively poor to that obtained by Prony's pole-zero method (Fig.7.17(c,d)). The spectrum obtained (Fig. 7.18(c)) also does not match well with that of DFT compared to that obtained by Prony's pole-zero method.

Steiglitz-McBride Alternate method (SMA): From Eqn. (7.76a), the error is given by

$$e(n) = x(n) - h(n)$$

This can be written as

$$e(n) = \frac{1}{A}\left[A\left[x(n)\right] - B\left[\delta(n)\right]\right]$$

$A[\bullet]$ and $B[\bullet]$ are convolutional operators for $A(z)$ and $B(z)$, respectively. Further, $(1/A)[\bullet]$ is the corresponding operator for $1/A(z)$. If $A(z)$ is known at the $(i-1)^{th}$ as $A_{i-1}(z)$, then (Eqn. (7,76b))

$$\tilde{e}^{(i)}(n) = A_i\left[\frac{1}{A_{i-1}}\left[x(n)\right]\right] - B_i\left[\frac{1}{A_{i-1}}\left[\delta(n)\right]\right]$$

$$\tilde{e}^{(i)}(n) = A_i\left[\tilde{x}(n)\right] - B_i\left[\tilde{y}(n)\right] \qquad E = \sum_{n=0}^{N-1}\left\{\tilde{e}^{(i)}(n)\right\}^2$$

Instead of estimating $a_l^{(i)}, l = 1,2,\ldots,P$ and $a_m^{(i)}, m = 0,1,2,\ldots,Q$ by minimising the error energy E, the error $\tilde{e}^{(i)}(n)$ can be equated to zero at each ieration. This results in

$$\tilde{x}(n) + \sum_{l=1}^{P} a_l^{(i)}\tilde{x}(n-l) - \sum_{m=0}^{Q} b_m^{(i)}\tilde{y}(n-m) = 0, \quad P \le n \le N, \text{ (assuming in } P \ge Q)$$

If $P = 3, Q = 2$, then

$$a_1^i\tilde{x}(n-1) + a_2^i\tilde{x}(n-2) + a_3^i\tilde{x}(n-3) - b_0^i\tilde{y}(n) - b_1^i\tilde{y}(n-1) - b_2\tilde{y}(n-2) = -\tilde{x}(n)$$

For $n \ge P$,

$$a_1^i\tilde{x}(P-1) + a_2^i\tilde{x}(P-2) + a_3^i\tilde{x}(P-3) - b_0^i\tilde{y}(P) - b_1^i\tilde{y}(P-1) - b_2\tilde{y}(P-2) = -\tilde{x}(P)$$

$$a_1^i\tilde{x}(P) + a_2^i\tilde{x}(P-1) + a_3^i\tilde{x}(P-2) - b_0^i\tilde{y}(P+1) - b_1^i\tilde{y}(P) - b_2\tilde{y}(P-1) = -\tilde{x}(P+1)$$

$$a_1^i\tilde{x}(P+1) + a_2^i\tilde{x}(P) + a_3^i\tilde{x}(P-1) - b_0^i\tilde{y}(P+2) - b_1^i\tilde{y}(P+1) - b_2\tilde{y}(P) = -\tilde{x}(P+2)$$

$$\vdots \qquad\qquad \vdots \qquad\qquad \vdots$$

$$a_1^i\tilde{x}(N-2) + a_2^i\tilde{x}(N-3) + a_3^i\tilde{x}(N-4) - b_0^i\tilde{y}(N-1) - b_1^i\tilde{y}(N-2) - b_2\tilde{y}(N-3) = -\tilde{x}(N-1)$$

$$\begin{bmatrix} \tilde{x}(P-1) & \tilde{x}(P-2) & \tilde{x}(P-3) & \tilde{y}(P) & \tilde{y}(P-1) & \tilde{y}(P-2) \\ \tilde{x}(P) & \tilde{x}(P-1) & \tilde{x}(P-2) & \tilde{y}(P+1) & \tilde{y}(P) & \tilde{y}(P-1) \\ \tilde{x}(P+1) & \tilde{x}(P) & \tilde{x}(P-1) & \tilde{y}(P+2) & \tilde{y}(P+1) & \tilde{y}(P) \\ \vdots & \vdots & \vdots & \vdots & \vdots & \vdots \\ \tilde{x}(N-2) & \tilde{x}(N-3) & \tilde{x}(N-4) & \tilde{y}(N-1) & \tilde{y}(N-2) & \tilde{y}(N-3) \end{bmatrix} \begin{bmatrix} a_1 \\ a_2 \\ a_3 \\ -b_0 \\ -b_1 \\ -b_2 \end{bmatrix} = \begin{bmatrix} -\tilde{x}(P) \\ -\tilde{x}(P+1) \\ -\tilde{x}(P+2) \\ \vdots \\ -\tilde{x}(N-1) \end{bmatrix}$$

For the data $x(n)$ of length $N > P+Q+1$, the above system of equations is an overdetermined one as the number of equations is more than the number of unknowns. Therefore, this system of equations is solved by taking pseudoinverse of the matrix on the left hand side. Since the error is equated to zero instead of minimizing the error energy, the estimated model provides a good fit to the signal which is specifically an impulse response of a system, compared to that of minimizing the error energy.

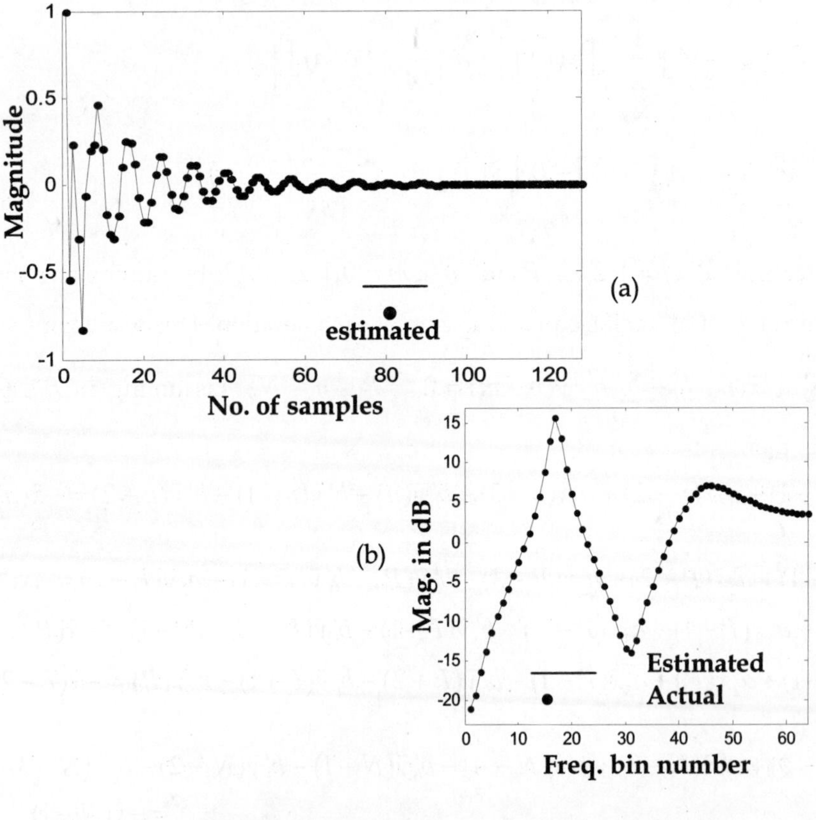

Fig. 7. 19 (a, b) Alternate Steigliz McBride method for the impulse response data. (a) Estimated impulse response and (b) Estimated spectrum.

For the pole-zero system (considered previously)

$$G(z)=\frac{1-1.1289z^{-1}+1.0058z^{-2}-0.7617z^{-3}}{1-0.59z^{-1}+0.455z^{-2}-0.0754z^{-3}+0.504z^{-4}}$$

its impulse response of length 128 samples is modeled by a pole-zero model with $P=10$ and $Q=5$ for 10 iterations by the alternate method. The Prony's all-pole model is used for initial pole polynomial estimation. It is seen from Fig. (7.19a), the estimated impulse response perfectly matches with the actual impulse response. Hence, the spectrum of the model also should match with that of actual impulse response and this is really so as seen from Fig. (7.19b). However, the spectral match of the estimate to the actual one obtained by the error energy minimization method is very poor though for the impulse respone there is some match of the estimate to the actual.

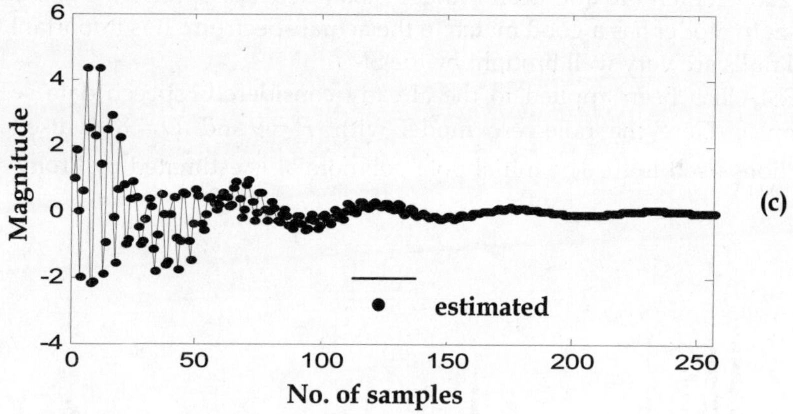

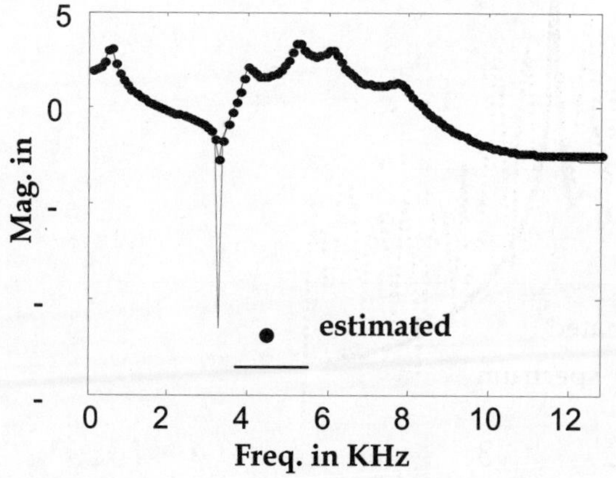

Fig. 7.19 (c, d) Alternate Steigliz McBride method for the impulse responsedata. (c) Estimated impulse response and (d) Estimated spectrum.

Another example for the impulse response which has zeros on the unit circle is considered to illustrate the performance of SMA method. The transfer function of this system is

$$G(z) = \frac{1 - 1.414214z^{-1} + 1z^{-2}}{\begin{bmatrix} 1 - 3.300959z^{-1} + 7.222431z^{-2} - 11.62311z^{-3} + 14.69756z^{-4} - 15.58842z^{-5} \\ + 13.53270z^{-6} - 9.888342z^{-7} + 5.741208\, z^{-8} - 2.461647z^{-9} + 0.7301360z^{-10} \end{bmatrix}}$$

Its impulse response has 256 samples. This is modeled by a pole-zero model with with $P = 10$ and $Q = 5$ for 10 iterations by SMA. The initial pole polynomial is estimated by Prony's all-pole modeling. For this example also, the impulse response estimate by SMA provides very good match to the actual and as expected the sprctrum obtained by the estimated pole-zero model has a good match to the actual spectrum. It is important to notice that the spectral nulls are very well brought by the SMA.

The SMA has been applied to the already considered speech data segment of length 256 samples. Here the pole-zero model with $P = 9$ and $Q = 8$ is used and the number of iterations used is 10. The initial pole polynomial is estimated by Prony's all-pole modeling.

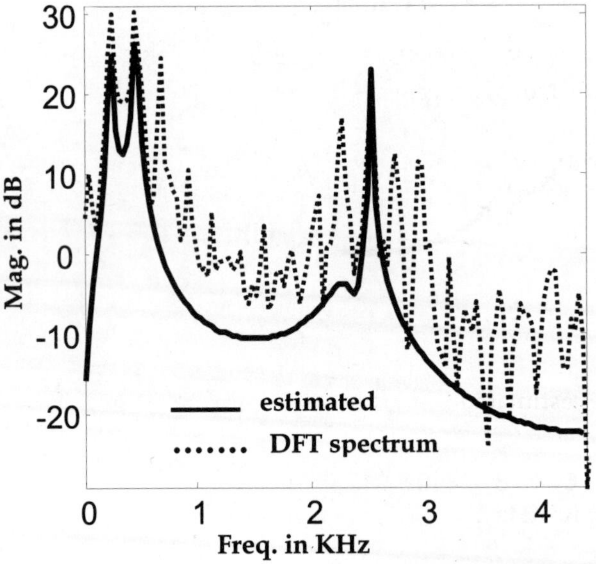

Fig. 7.20 (a) Pole-zero spectrum by SMA method for the speech data.

For the speech data, the SMA model gives a poor spectral match to the DFT sprctrum (Fig. 7.20a) though for an impulse response input data, it gives very good sperctal match to actual spectrum (Fig. 7.19(b,d)). The SMA model spectrum exhibits *very narrow spectral peaks* and also it has picked up peaks corresponding to pitch/ noise peaks instead of showing the gross features bringing out spectral peaks and valleys of the vocal tract. (W.J. Done and C.K. Rushforth, 1979)

The performance of the SMA to speech signal can be improved by making the input close to the impulse response as it works well for an impulse response type of input. For this the minimum phase equivalent of the speech input (256 samples) is found and this is given as input. For this, pole zero model with $P = 3$ and $Q = 10$ is used and the number of iterations used is 2. The initial pole polynomial is estimated by Prony's allpole modeling.

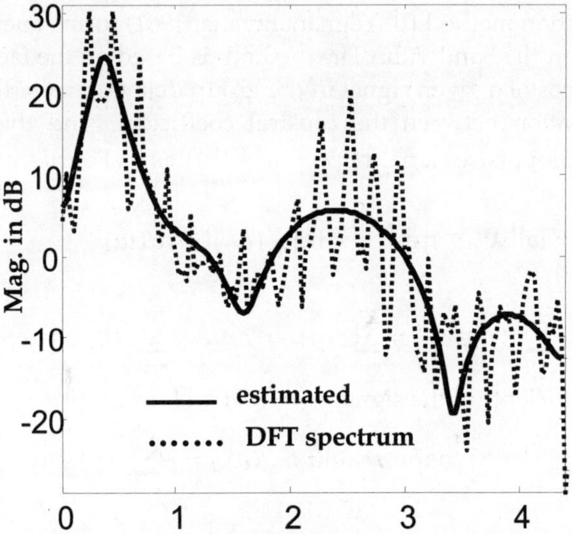

Fig. 7. 20(b) Pole-zero spectrum by SMA method for minimum phase equivalent the speech data.

It is seen that with the application of SMA to minimum phase equivalent, a significantly better fit to DFT speech spectrum has been achieved (Fig.7.20b) to that of applying directly to speech data (Fig.7.20a). Further, this estimated spectrum does not suffer from narrower spectral peaks and does not pick up peaks corresponding to pitch / noise and only represents the gross feature corresponding to vocal tract (W.J. Done and C.K. Rushforth, 1979). In linear prediction based approaches, generally, a higher order pole polynomial is used to get a better spectral fit and the extra poles do not cause any harm. But with iterative filtering, the extra poles may lie outside the unit circle making the iteration to become unstable and hence it is preferable not to over specify the model order (Keneth Steiglitz, 1977).

7.14.5 Pole-zero Modeling by Pole-zero Decomposition

The Steiglitz- McBride method (which equates error to zero, referred to as SMA) has been found to perform well for signals generated by exciting a system by an impulse. But for signals which are outputs of systems excited by impulse train (like voiced speech, ECG and PCG(phonocardiograph))/ white noise process (like / EEG and unvoiced speech) it exhibits sharper and higher spectral peaks with increase in iterations (W.J. Done and C.K. Rushforth, 1979). Further, SM method based on error energy minimization is not satisfactory irrespective of excitation and the estimates also may become unstable. This indicates that the bandwidth information obtained by Steiglitz-McBride method is not accurate for signals generated by impulse train /white noise excitation. This is expected as the method attempts to match the signal with the impulse response of a system in question. But application of SMA to minimum phase equivalent seems to perform satisfactory even for signals generated by impulse train excitation. This necessitates to look for alternatives and the pole-zero decomposition method (B. Yegnanarayana, 1981) forms one such as it provdes a better information about the bandwith. The method is based on the fact that it is possible to separate poles and zeros of a given signal in the group delay domain. Further, it also uses the direct recursive relation between the cepstral coefficients and the linear prediction coefficients. For a minimum phase signal $x(n)$ with $X(\omega)$ as its Fourier transform,

$$\ln[X(\omega)] = \ln[|X(\omega)|e^{j\theta(\omega)} = \ln|X(\omega)| + j\theta(\omega) \tag{7.78a}$$

$$\ln[X(\omega)] = \sum_{n=0}^{\infty} c(n)e^{-j\omega n} = \sum_{n=0}^{\infty} c(n)\cos\omega n - j\sum_{n=0}^{\infty} c(n)\sin\omega n \tag{7.78b}$$

$c(n)$ are the cepstral coefficients of the signal $x(n)$. Therefore,

$$\ln|X(\omega)| = \sum_{n=0}^{\infty} c(n)\cos\omega\, n \quad \text{and} \quad \theta(\omega) = -\sum_{n=0}^{\infty} c(n)\sin\omega\, n \tag{7.78c,d}$$

The signal derived from the magnitude sprctrum and its phase derived from the spectral magnitude from the above relation, result in aminimum phase equivalent signal.

The group delay, the negative derivative of phase is given by

$$\tau(\omega) = -\frac{d\theta(\omega)}{d\omega} = \sum_{n=0}^{\infty} nc(n)\cos\omega\, n \tag{7.78e}$$

For a minimum phase signal, for poles the significant values of group delay are positive and those for the zero are negative. This enables pole-zero decomposition. Representing the positive and negative parts of the group delay $\tau(\omega)$ by $\tau^{+}(\omega)$ and $\tau^{-}(\omega)$ respectively,

$$\tau^{+}(\omega) = C + \sum_{n=0}^{\infty} nc^{+}(n)\cos\omega n \tag{7.78f}$$

and

$$\tau^{-}(\omega) = C + \sum_{n=0}^{\infty} nc^{-}(n)\cos\omega n \tag{7.78g}$$

The sequences $c^{+}(n)$ and $c^{-}(n)$ can be computed using inverse FT. C is the average value that does not contribute to the shape of the spectrum. Further, the relation between LP and cepstral coefficients can be used to get the numerator and denominator polynomials of the transfer function.

The gain G of the transfer function is given by

$$\ln G = c(0)/2 \tag{7.79a}$$

Relation between LP and cepstral coefficients:

If $\ln\left[H(z)\right]$ is related to LP coefficients, then

$$\ln\left[H(z)\right] = \sum_{k=0}^{\infty} c(k) z^{-k} = \ln\left[1/\left\{1 + \sum_{k=1}^{M} a_k z^{-k}\right\}\right] \tag{7.79b}$$

$$\sum_{k=0}^{\infty} c(k) z^{-k} = -\ln\left(1 + \sum_{k=1}^{M} a_k z^{-k}\right)$$

Let $z^{-1} = v$, then

$$\sum_{k=0}^{\infty} c(k) v^{k} = -\ln\left(1 + \sum_{k=1}^{M} a_k v^{k}\right)$$

Differentiating both sides with respect to $z^{-1} = v$,

$$\sum_{k=1}^{\infty} kc(k) v^{k-1} = -\frac{\left(\sum_{k=1}^{M} k a_k v^{k-1}\right)}{\left(1 + \sum_{k=1}^{M} a_k v^{k}\right)}$$

$$\left(\sum_{k=1}^{\infty} kc(k)v^{k-1}\right)\left(1+\sum_{k=1}^{M} a_k v^k\right) = -\sum_{k=1}^{M} ka_k v^{k-1}$$

$$\left(\sum_{k=1}^{\infty} kc(k)\left(z^{-1}\right)^{k-1}\right)\left(1+\sum_{k=1}^{M} a_k \left(z^{-1}\right)^k\right) = -\sum_{k=1}^{M} ka_k \left(z^{-1}\right)^{k-1}$$

$$\left(\sum_{k=1}^{\infty} kc(k)z^{-k+1}\right)\left(1+\sum_{k=1}^{M} a_k z^{-k}\right) = -\sum_{k=1}^{M} ka_k z^{-k+1}$$

Let $M = 3$, for this,

$$\left(c(1)+2c(2)z^{-1}+3c(3)z^{-2}+4c(4)z^{-3}+\ldots\right)\left(1+a_1 z^{-1}+a_2 z^{-2}+a_3 z^{-3}\right)$$
$$= -\left(a_1 + 2a_2 z^{-1} + 3a_3 z^{-2}\right)$$

$$c(1)+\left[c(1)a_1+2c(2)\right]z^{-1}+\left[c(1)a_2+2c(2)a_1+3c(3)\right]z^{-2}+\ldots$$
$$= -\left(a_1 + 2a_2 z^{-1} + 3a_3 z^{-2}\right)$$

Comparing the coefficients of the powers of z^{-1} both sides

$$i = 1, \quad a_1 = -c(1),$$

$$i = 2, \quad 2a_2 = -2c(2) - c(1)a_1,$$

$$i = 3, \quad 3a_3 = -3c(3) - 2c(2)a_1 - c(1)a_2$$

Generalizing,

$$a_1 = -c(1),$$
$$ia_i = -ic(i) - \sum_{k=1}^{i-1} kc(k)a_{i-k}, \quad i = 1,2,3,\ldots M \tag{7.79c}$$

Applying this relation:
For poles:

$$a_1 = -c^+(1), \quad ia_i = -ic^+(i) - \sum_{k=1}^{i-1} kc^+(k)a_{i-k}, \; i = 1,2,3,\ldots M \tag{7.79d}$$

For zeros:

$$b_1 = c^-(1), \quad ib_i = ic^-(i) + \sum_{k=1}^{i-1} kc^-(k)b_{i-k}, \; i = 1,2,3,\ldots N \tag{7.79e}$$

Here, the minimum phase equivalent operation is *built into the method* as the cepstral coefficients are derived from the magnitude spectrum. Further the logarithmic operation compresses/ flattens the spectrum magnitude, the cepstral coefficient can represent a large dynamic range spectrum better than the autocorrelation coefficients and this method is appropriate to such a situation like speech. In the least square methods, the autocorrelation coefficients of the transfer function are matched with those of the signal. But here, the cepstral coefficients of the model are matched with those of the signal.

The positive and negative values of the GD $\tau^+(\omega)$ and $\tau^-(\omega)$ respectively correspond to GD of poles and zeros respectively. The cepstral coefficients for the poles $c^+(n)$ and for zeros $c^-(n)$ are obtained from $\tau^+(\omega)$ and $\tau^-(\omega)$ by taking the inverse FT and dividing by n.

The number of cepstral coefficients for the poles and for the zeros to be used is determined by computing the normalized cepstral error energy sequences.

$$E_p(n) = \frac{\sum_{k=M+1}^{\infty} c^{+2}(k)}{\sum_{k=0}^{\infty} c^{+2}(k)} \tag{7.79f}$$

and $$E_z(n) = \frac{\sum_{k=M+1}^{\infty} c^{-2}(k)}{\sum_{k=0}^{\infty} c^{-2}(k)} \tag{7.79g}$$

Initially, $E_p(n)$ and $E_z(n)$ fall rapidly and as the number of cepstral coefficients increases their decay rate is significantly reduced. The number of coefficients beyond which the decay rate is not significant determines the number of poles and zeros to be used.

Some of the above methods are illustrated for the same a speech segment of 256 samples of an utterance "m" (Fig. 7.21(a)) which is a nasal and is having prominent spectral dips. The signal DFT spectrum with 256 points is also shown in comparison with the pole spectrum and zero spectrum by Shank's method for 10 poles and 5 zeros. The pole spectrum is obtained by covariance method. Further, the zeros by Shank's method knowing the information about poles. The combined pole-zero spectrum and the DFT spectrum are shown in Fig. 7.21c. The pole-zero spectrum provides a better match to the DFT spectrum especially at the zeros / dips. In this example the roots of the denominator polynomial were all inside the unit circle.

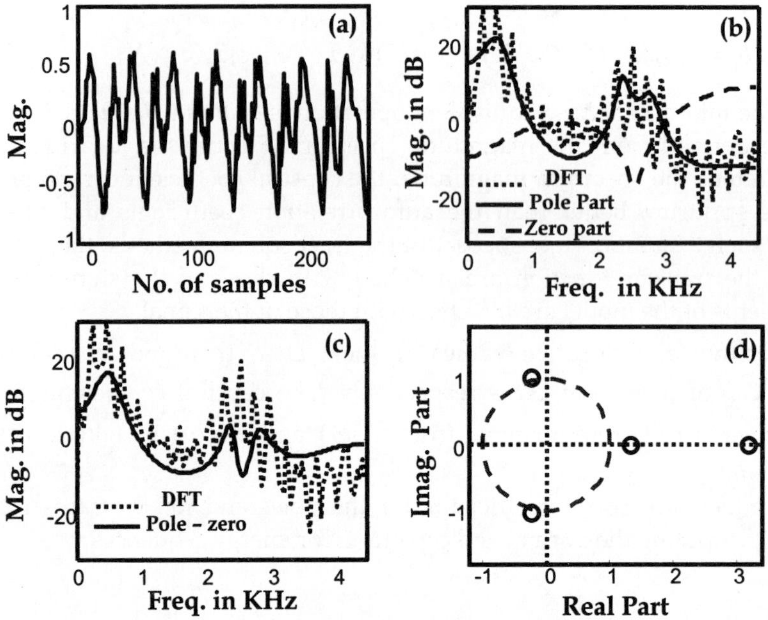

Fig. 7.21 Shanks method of parametric pole-zero modeling of a segment of speech: (a) of utterance "m"; (b) Component spectra of poles and zeros; (c) Pole-zero spectrum and (d) Zero locations on the Z-plane

But generally, in the estimation of poles, the ill conditioning of the covariance matrix may pose stability problem, that is, the inverse may have roots outside the unit circle. In case, if a certain root $r_0 e^{j\theta_0}$ is outside the unit circle, it can be reflected inside by using the reciprocal of its magnitude $(1/r_0)$ (instead of r_0) with its angle θ_0, and computing the denominator polynomial of the new roots. However, the roots of the numerator are as shown in the Fig. 7.21(d) and some of them are well outside the unit circle. In view of this, the use of linear prediction based estimation of the numerator polynomial may not be proper.

To overcome the problem of the signal having mixed phase zeros (some outside the unit circle), the *homomorphic prediction* is applied and for this, the signal is converted to a minimum phase equivalent and the LP based method is applied for the estimation of zero polynomial.

The pole-zero spectrum is shown in the Fig. 7.22(c). The zero locations in the Z-plane Fig. 7.22(d) show that the roots are inside the unit circle unlike for the method where the shank's method is applied directly to the original signal.

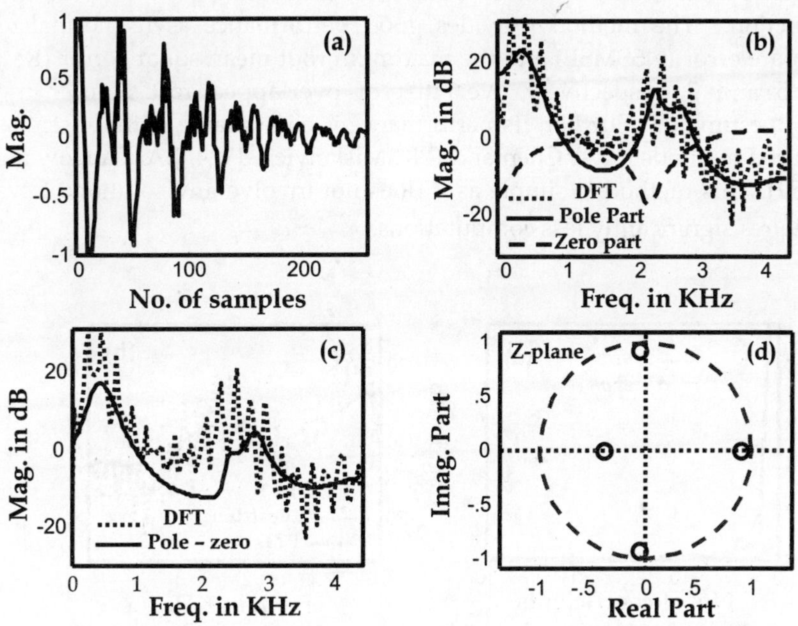

Fig. 7.22 Homomorphic prediction method of parametric pole-zero modeling of a segment of speech of utterance "m": (a) Minimum phase equivalent; (b) Component spectra of poles and zeros; (c) Pole-zero spectrum and (d) Zero locations on the Z-plane.

For the pole-zero decomposition method as mentioned already, the signal is converted to minimum phase and the roots are all ensured to be inside the unit circle. The smoothed group delay from the magnitude information is computed using the first 35 cepstral coefficients and this is decomposed into pole (10) and zero (8)parts by splitting the group delay into two parts by checking its values whether it is positive or negative.

It is seen that the spectral peaks are sharp for linear prediction based methods like Shank and the Burg all-pole modeling as linear prediction represents the poles effectively. Increasing the order though improves the overall spectral fit, it also brings out the peaks due to pitch that is not desirable (Fig. 7.23c). In the pole-zero decomposition the spectral peaks are wider representing the true spectral widths without getting affected by the pitch peaks (showing peaks corresponding to pitch peaks). Further, the notches or valleys are brought out well by the pole-zero decomposition compared to the other methods considered (Fig. 7.23). A detailed study of he application of pole-zero decomposition for speech (B.Yegnanarayana, 1981) and for EEG (S.V. Narasimhan, 1989) has been made. Further, the of pole-zero decomposition in group delay domain, has been used for the estimation of magnitude square coherence (MSC) function (S.V. Narsimhan, G.R. Reddy, E.I. Plotkin and M.N.S. Swamy, 1995). Here, estimation is achieved by modeling the periodogram estimate of MSC and the ARMA model has been achieved by pole-zero decomposition property of the group delay function. The method provides good performance as its normalized sum sample mean square error (NSSMSE) and the maximum root mean square error (RMSE) are less by factors 5 and 2, respectively over that of overlapped fast Fourier transform processing (OFTP) estimator. Further, its performance is comparable with those of existing ARMA model based methods (Y.T. Chen and R.K. Miskowicz, 1984, J.A. Cadzow and O.M. Solomon, Jr., 1987). This method is simple as it does not involve any solution of system of equations and reqires significantly less computations.

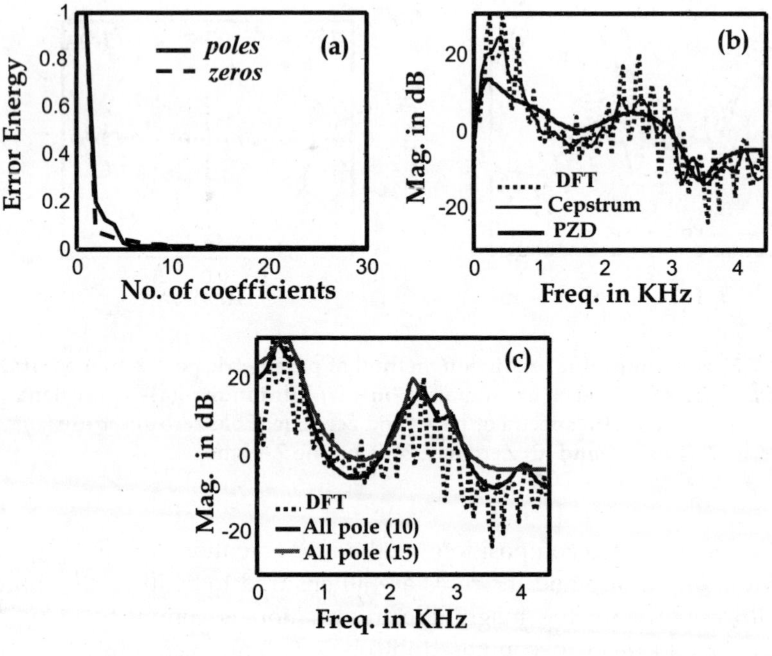

Fig. 7.23 Spectral estimation by pole-zero decomposition: (a) Error energy vs. no of cepstral coefficients; (b) Spectrum by DFT, by truncated cepstrum and PZD and (c) Spectrum by all pole Burg method.

The PZD has been used in the identification of a mixed phase system by AR/ARMA model based on Bispectrum (S.V. Narasimhan, G.R. Reddy, E.L. Plotkin and M.N.S. Swamy, 1992).

Here, the PZD method has been applied to group delays of minmum and maxmum phase components ($\tau_{\min}(\omega)$ and $\tau_{\max}(\omega)$. These group delay components are derived from the group delays derived from system output spectral magnitude as $\tau_m(\omega)$ and system ouput phase as $\tau_p(\omega)$ using the relations, $\tau_{\min}(\omega) = [\tau_p(\omega) + \tau_m(\omega)]/2$ and $\tau_{\max}(\omega) = [\tau_m(\omega) - \tau_p(\omega)]/2$. The cepstral coefficient sequences of $\tau_{\min}(\omega)$ and $\tau_{\max}(\omega)$ can be converted to AR coefficicents using the relationship between cepstral and AR coefficients and this results in AR modeling of non-minimum phase system based on bispectrum. As stated already, for the minimum and maximum phase group delay components, the PZD is applied to get the ARMA model which identifies the non-minimum phase system. A method which solves a third order recursion equations which may be Toeplitz in form but not symmetric has been proposed and the bispectrum derived from these AR coefficients has better variance and frequency resolution (Mysore R.Raghuveer and C.L. Nikias, 1985), will be dealt within detail in Section 7.17.

Further, for channel blind identification based on cyclostationarity, the PZD has also been applied along with the modified group delay (S.V. Narasimhan, M. Hazarathaiah and P.V.S. Giridhar, 2005) and this provides improved results. [Cyclostationary process is a process which is stationary from period to period but within a period, the spectral characteristics change in a definite pattern].

It is important to note that for the purpose of spectral estimation, the roots of denominator or numerator polynomial can be outside the unit circle, as the magnitude square of the FT of the polynomials are considered. That is, for spectral estimation, the stability is *not a necessary* condition. However, if the transfer function determined by such a denominator polynomial is used for filtering an input, it results in an output, which may *blow off resulting in values of magnitude infinity.*

7.14.6 Pole-zero Model Order Selection

As orders for both numerator and denominator have to be chosen, it is difficult for the AR process. For methods which determine both the poles and zeros based on by AR modelling methods, the AR order determination criterion can be extended. For example, for the pole-zero model with P poles and Q zeros, the AIC is given by

$$AIC(P,Q) = N \ln \varepsilon(P,Q) + 2(P+Q) \tag{7.80}$$

7.15 ALL-POLE AND POLE-ZERO LATTICE FILTERS

The realization of all-pole and pole-zero transfer functions using lattice structure will be considered.

7.15.1 All-pole Lattice

The transfer function $G_N^f(z)$, the forward prediction filter of Nth order is an FIR filter and provides the prediction error for a given input $x(n)$. Further, for this lattice structure,

$$e_0^f(n) = e_0^b(n) = x(n)$$

and from Eqn. (7.28a)

$$G_N^f(z) = \frac{E_N^f(z)}{X(z)} = 1 - \sum_{i=1}^{N} w_i^{(N)} z^{-i}$$

Therefore,

$$G_N^f(z) = \frac{E_N^f(z)}{E_0^f(z)} = 1 + \sum_{i=1}^{N} a_i^{(N)} z^{-i}$$

where

$$-w_i^{(N)} = a_i^{(N)}$$

$$G_N^f(z) = 1 + a_1^{(N)} z^{-1} + a_2^{(N)} z^{-2} + \ldots + a_N^{(N)} z^{-N}$$

The transfer function $G_N^b(z)$, the backward prediction filter of Nth order is an FIR filter is

$$G_N^b(z) = Z^{-N} - w_N^{(N)} - \sum_{j=1}^{N-1} w_{N-j}^{(N)} Z^{-j}$$

$$= Z^{-N} + a_N^{(N)} + \sum_{j=1}^{N-1} a_{N-j}^{(N)} Z^{-j}$$

$$= Z^{-N}[1 + a_N^{(N)} Z^N + \sum_{j=1}^{N-1} a_{N-j}^{(N)} Z^{N-j}]$$

$$G_N^b(z) = Z^{-N}[1 + a_N^{(N)} Z^N + a_{N-1}^{(N)} z^{N-1} + a_{N-2}^{(N)} z^{N-2} + \ldots a_1^{(N)} z\ ]$$

$$= Z^{-N}[1 + a_1^{(N)} z\ + a_2^{(N)} z^2 + \ldots + a_{N-2}^{(N)} z^{N-2} + a_{N-1}^{(N)} z^{N-1} + a_N^{(N)} Z^N]$$

$$G_N^b(z) = Z^{-N} G_N^f(z^{-1})$$

Defining

$$G_N^b(z) = 1 + \sum_{i=1}^{N} b_i^{(N)}(i) z^{-i}$$

$$G_N^b(z) = 1 + b_1^{(N)} z^{-1} + b_2^{(N)} z^{-2} + \ldots + b_{N-2}^{(N)} z^{-(N-2)} + b_{N-1}^{(N)} z^{-(N-1)} + b_N^{(N)} Z^{-N} \quad \text{(7.81a) (7.1)}$$

But

$$G_N^b(z) = [Z^{-N} + a_1^{(N)} z^{-(N-1)} + a_2^{(N)} z^{-(N-)2} + \ldots + a_{N-2}^{(N)} z^{-2} + a_{N-1}^{(N)} z^{-1} + a_N^{(N)}]$$

$$G_N^b(z) = [a_N^{(N)} + a_{N-1}^{(N)} z^{-1} + a_{N-2}^{(N)} z^{-2} \ldots a_2^{(N)} z^{-(N-)2} + a_1^{(N)} z^{-(N-1)} + Z^{-N}] \quad (7.81b)$$

Comparing the coefficients of same powers of z^{-1} in Eqn. (7.81a) with Eqn. (7.81b),

$$b_i^{(N)} = a_{N-i}^{(N)} \text{ and } b_N^{(N)} = 1 \text{ or } b_m^{(m)} = 1 \quad (7.81c)$$

An all-pole filter corresponding to $G_N^f(z)$ is

$$\frac{1}{G_N^f(z)} = \frac{E_0^f(z)}{E_N^f(z)} = \frac{1}{1 + \sum_{i=1}^{N} a_i^{(N)} z^{-i}} \quad (7.81d)$$

It states that $e_0^f(n)$ can be obtained from $e_N^f(n)$ or the lower order prediction errors can be derived from higher order errors. Also for the all-pole filter, the roles of given $e_0^f(n)$ (or $x(n)$) and $e_N^f(n)$ in the all-zero filter are interchanged. For the all-pole filter, $e_N^f(n) = x(n)$ instead of $e_0^f(n) = x(n)$, as in the all zero FPEF. For the all-pole FPEF, the input becomes the output and vice versa. For a single stage all-pole lattice, the Fig. 7.24 (a) becomes as in Fig. 7.24(b). It is important to note that except for the interchange of input and output (structure is flipped from right to left), the lattice structure remains the same and hence the Eqn. (7.31a),

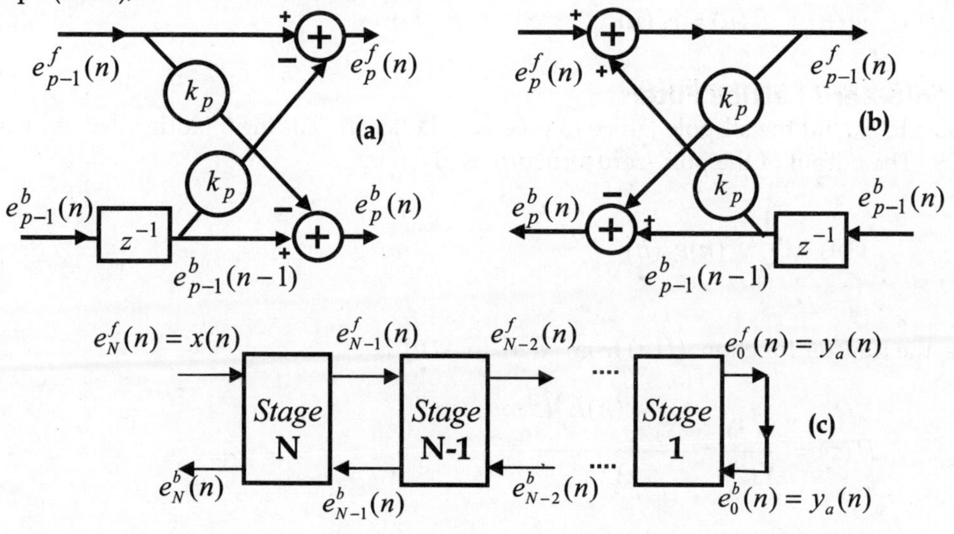

Fig. 7.24 (a) Single stage all zero lattice (b) Single satge all-pole lattice and (c) multistage all-pole lattice filter.

$$e_p^f(n) = e_{p-1}^f(n) - k_p e_{p-1}^b(n-1)$$

needs to be written as

$$e_{p-1}^f(n) = e_p^f(n) + k_p e_{p-1}^b(n-1) \tag{7.81e}$$

This states that the FPE at the $(p-1)^{th}$ order can be derived from that of pth order. However, the BPE relation need not be changed and the higher order BPEs are derived from lower order BPE as in all zero lattice filter and is

$$e_p^b(n) = e_{p-1}^b(n-1) - k_p e_{p-1}^f(n) \tag{7.81f}$$

The Eqns. 7.81e and 7.81f define the single stage all-pole lattice filter. A cascade of N single all-pole lattice stages result in the realization of a N stage lattice all-pole filter (Fig. 7.24 (c)). Here the input $x(n)$ $x(n)$ becomes $e_N^f(n)$. Therefore, the equations for all-pole lattice are

$$e_N^f(n) = x(n) \tag{7.82a}$$

$$e_{p-1}^f(n) = e_p^f(n) + k_p e_{p-1}^b(n-1) \tag{7.82b}$$

$$e_p^b(n) = e_{p-1}^b(n-1) - k_p e_{p-1}^f(n) \tag{7.82c}$$

The output of this all-pole filter is $y_a(n)$ and is given by

$$y_a(n) = e_0^f(n) = e_0^b(n) \tag{7.82d}$$

7.15.2 Pole-Zero Lattice Filter

It is possible to extend the all-pole lattice to pole-zero lattice, by adding a ladder structure as in Fig. 7.25. The output of the pole–zero structure is given by

$$y(n) = \sum_{p=0}^{N} v_p(n) e_p^b(n) \tag{7.83a}$$

Therefore, the transfer function $H(z)$ from $x(n)$ to $y(n)$ is

$$H(z) = \frac{Y(z)}{X(z)} = \frac{\sum_{p=0}^{N} v_p(n) E_p^b(z)}{X(z)}$$

$$H(z) = \frac{\sum_{p=0}^{N} v_p(n) E_p^b(z)}{E_N^f(z)}, \quad \because E_N^f(z) = x(z) \tag{7.83b}$$

Since $E_0^f(z) = E_0^b(z)$, $H(z)$ can be written as,

$$H(z) = \left(\frac{E_0^f(z)}{E_0^b(z)}\right)\frac{\sum_{p=0}^{N} v_p(n)E_p^b(z)}{E_N^f(z)} = \sum_{p=0}^{N} v_p(n)\left(\frac{E_p^b(z)}{E_0^b(z)}\right)\left(\frac{E_0^f(z)}{E_N^f(z)}\right)$$

Since

$$G_N^f(z) = \frac{E_N^f(z)}{E_0^f(z)} \text{ and } G_p^b(Z) = \frac{E_p^b(Z)}{E_0^b(Z)},$$

$$H(z) = \frac{\sum_{p=0}^{N} v_p(n)G_p^b(z)}{G_N^f(z)} \tag{7.83c}$$

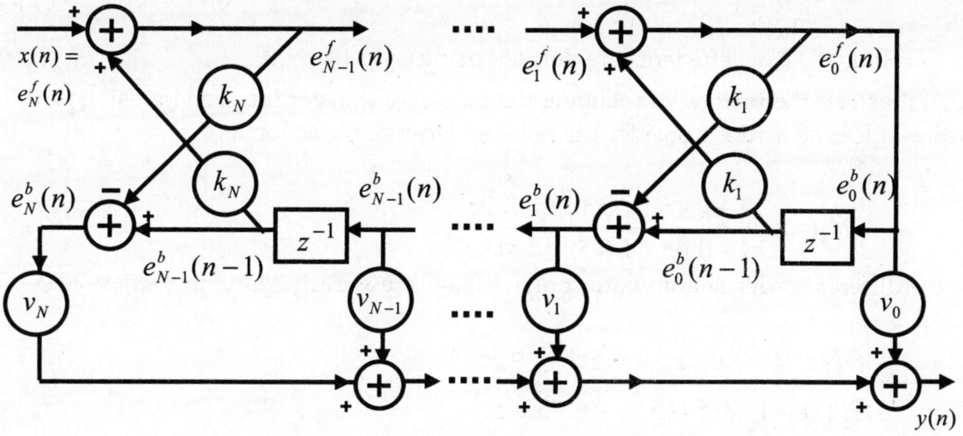

Fig. 7.25 Lattice ladder structure for pole-zero model.

If

$$C_N(z) = \sum_{p=0}^{N} v_p(n)G_p^b(z)$$

$$H(z) = \frac{C_N(z)}{G_N^f(z)} \tag{7.83d}$$

Defining

$$C^{(m)}(z) = \sum_{i=0}^{m} v_i(n)G_i^b(z), \tag{7.83e}$$

$$C^{(m)}(z) = C^{(m-1)}(z) + v_m(n)G_m^b(z)$$

$$C^{(m-1)}(z) = C^{(m)}(z) - v_m(n)G_m^b(z) \tag{7.83f}$$

Since from Eqn. (7.81c),

$$b_m^{(m)} = 1$$

Therefore from Eqn. (7.83f),

$$C^{(m-1)}(z) = C^{(m)}(z) - v_m(n)G_m^b(z)$$

$$v_m = C_m^{(m)} \tag{7.83g}$$

Therefore, the starting will be with

$$v_N = C_N^{(N)} \tag{7.83h}$$

and working backwards using ,

$$C^{(m-1)}(z) = C^{(m)}(z) - v_m(n)G_m^b(z) \tag{7.83i}$$

at each stage with

$$C_{m-1}^{(m-1)} = v_{m-1}, \tag{7.83j}$$

all the v_i, $i = 0, 1, \ldots, N$ coefficients are derived using $G_m^b(z)$.

To illustrate the method of realizing the pole-zero transfer function by using the all-spole lattice-ladder structure, consider the pole-zero transfer function

$$H_3(z) = \frac{1 - 0.1z^{-1} + 0.1z^{-2} + 0.2z^{-3}}{1 - 0.6z^{-1} + 0.5z^{-2} - 0.1z^{-3}}$$

For this, the different order denominator polynomials are already found in Section-7.7.3 and are:

$$G_3^f(z) = 1 - 0.6z^{-1} + 0.5z^{-2} - 0.1z^{-3}$$

$$G_2^f(z) = 1 - 0.5445z^{-1} + 0.4356z^{-2}$$

$$G_1^f(z) = 1 - 0.37927z^{-1}$$

Therefore,

$$G_3^b(z) = -0.1 + 0.5z^{-1} - 0.6z^{-2} + z^{-3}$$

$$G_2^b(z) = 0.4356 - 0.5445z^{-1} + z^{-2}$$

$$G_1^b(z) = -0.37927 + z^{-1}$$

$$C^{(3)}(z) = 1 - 0.1z^{-1} + 0.1z^{-2} + 0.2z^{-3}$$

$$v_3 = C_3^{(3)} = 0.2$$

$$C^{(m-1)}(z) = C^{(m)}(z) - v_m(n)G_m^b(z)$$

$$C^{(2)}(z) = C^{(3)}(z) - v_3 G_3^b(z)$$

$$C^{(2)}(z) = \left[1 - 0.1z^{-1} + 0.1z^{-2} + 0.2z^{-3}\right] - 0.2\left[-0.1 + 0.5z^{-1} - 0.6z^{-2} + z^{-3}\right]$$

$$C^{(2)}(z) = 1.02 - 0.2z^{-1} - 0.22z^{-2}$$

$$v_2 = C_2^{(2)} = -0.22$$

$$C^{(1)}(z) = C^{(2)}(z) - v_2 G_2^b(z)$$

$$C^{(1)}(z) = \left[1.02 - 0.2z^{-1} - 0.22z^{-2}\right] - (-0.22)\left[0.4356 - 0.5445z^{-1} + z^{-2}\right]$$

$$C^{(1)}(z) = 1.115832 - 0.31979z^{-1}$$

$$v_1 = C_1^{(1)} = -0.31979$$

$$C^{(0)}(z) = C^{(1)}(z) - v_1 G_1^b(z)$$

$$C^{(0)}(z) = \left[1.115832 - 0.31979z^{-1}\right] - (-0.31979)\left[-0.37927 + z^{-1}\right]$$

$$C^{(0)}(z) = 0.994545$$

$$v_0 = C_0^{(0)} = 0.994545$$

7.16 MOVING AVERAGE (MA)/ALL ZERO MODELING

Generally, a signal is mainly represented by the spectral peaks of its spectrum and these peaks correspond to poles. The varies due to zeros only enhance the peaks. In view of this, the all-pole (AR) or pole-zero (ARMA) model s have become very popular. Further, the all-pole model with a higher order is reasonably efficient in representing the spectrum of a general signal. Unless the given signal is due to pure moving average process, its requirement is not called for and it is not as popular as AR/all-pole or ARMA/pole-zero mode. But as it is known for its stability, this has enabled its utility in adaptive filters compared to the other two models mentioned. The MA process is given by

$$x(n) = \sum_{k=0}^{Q} b(k)v(n-k) \tag{7.84a}$$

where $v(n)$ is a white noise with zero mean unit variance. The autocorrelation $r_x(k)$ at lag k, of the model is given by

$$r_x(k) = b(k) * b^*(-k) = \sum_{l=0}^{Q-|k|} b(l+|k|)b^*(l) \tag{7.84b}$$

Since the PSD of the signal is the Fourier transform of its autocorrelation, the PSD is given by

$$S_x(z) = \sum_{k=-Q}^{Q} r_x(k)z^{-k} \tag{7.84c}$$

Since for a MA process of order Q, $r_x(k)=0, |k|>Q$. Thus, to find the *spectrum of the MA process*, there is no necessity to find the parameters of the process. But for the purpose of identification, the parameters of the process are necessary. This can be achieved by spectral factorization.

Moving average modeling by spectral factorization: The spectral factorization is given by

$$S_x(k)=\sigma^2 B'(z)B'^*(1/z^*) \tag{7.84d}$$

where σ^2 is the input white noise variance. The spectral factorization can be carried out easily in cepstral domain. The factors $B'(z)$ and $B'^*(1/z^*)$ correspond to minimum phase and maximum phase components. If

$$\ln[S_x(k)]=\sum_{n=-Q}^{Q} c(n)z^{-n} \tag{7.84e}$$

where $c(n)$ are the cepstral coefficients of $S_x(k)$. They can be thought of as the Fourier coefficients of $\ln[S_x(k)]$. The minimum phase component corresponding to $\sigma\, B'(z)$ is given by $c_{\min}(n)$. (As the Fourier transform of $r_x(k)$ form PSD, to get good quality PSD, sufficient data length $\geq 10Q$, Q the model order which corresponds to maximum lag of $r_x(l)$) should be considered). Futher, as $r_x(l), |l|>Q$ will have large variance and hence their contribution should be avoided. Thus

$$S_x(k)=\{DFT[r_x(l)]\},\ r_x(l)=0,|l|>Q \tag{7.84f}$$

$$c_{\min}(n)=\begin{cases} c(0)/2, & n=0 \\ c(n), & n>0 \end{cases} \tag{7.84g}$$

$$\hat{S}_{\min}(z)=\ln[B(z)]=\sum_{n=0}^{Q} c_{\min}(n)z^{-n}, \qquad B(z)=e^{\hat{S}_{\min}(z)}, \tag{7.84h}$$

$$B(z)=b(0)+b(1)z^{-1}+b(2)z^{-2}+\cdots\cdots b(Q)z^{-Q} \tag{7.84i}$$

If $b(0)=1$, for this

$$c_{\min}(n)=\begin{cases} 0, & n=0 \\ c(n), & n>0 \end{cases} \tag{7.84j}$$

The inverse FFT of $B(e^{j\omega})$ gives the moving average model parameter values.

Moving average modeling using all-pole modeling

The MA process modeling can be achieved by using all-pole modeling. Here, a higher order AR model (P^{th} *order) will model* the given MA data of Qth order and $P >> Q$. Then by considering the AR coefficient sequence obtained as data, it is modeled by a Qth order AR model. The coefficients of this Q^{th} order AR polynomial form the MA coefficients of the MA process. This is because the higher order AR models the MA signal spectrum. Since the resulting AR polymomial coefficients are modeled by another AR model of order of Q, this model represents the reciprocal spectrum the original AR polynomial and hence correponds to the numerator polynomial, that is the MA model coefficients. An MA process is given by

$$x(n) = \sum_{k=0}^{Q} b(k)v(n-k) \tag{7.85a}$$

This is modeled by AR process of order $P > Q$. Therefore,

$$B(z) = \frac{1}{A(z)} = \frac{1}{a_0 + \sum_{k=1}^{P} a(k)z^{-k}} \tag{7.85b}$$

But as $A(z)$ is modeled by a Qth order AR model,

$$A(z) = \frac{1}{B(z)} = \frac{1}{b_0 + \sum_{k=1}^{Q} b(k)z^{-k}} \tag{7.85c}$$

Therefore, $B(z)$ represents the coefficients of the moving average process and as this was developed by Durbin, it is known as Durbin method.

Example: Consider the MA model , b=[1.0000 -1.6251 2.0908 -1.3424 0.7310]; driven by a zero mean unit variance white Gaussian noise.

The power spectral density of the process is computed by considering 512 moving average process data points and taking Fourier transform of size 1024 points to overcome the aliaing problem, for the autorelation function. The squared FT magnitude gives an estimate of PSD. The cepstrum (complex cepstrum) for the PSD is found by taking the natural logarithm of PSD and its inverse FT enables the minimum phse spectral factorization which is achieved by considering the complex cepstrum for $n \geq 0$ and the gain unityfor the MA model estimate by choosing $c_{\min}(0) = 0$.

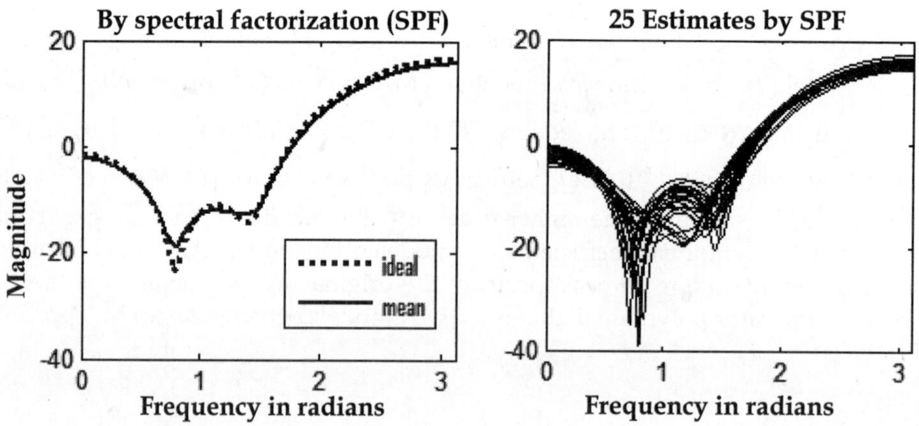

Fig. 7.26 (a) MA model estimation by SPF.

The FT of the $c_{\min}(n)$, its exponentiation and further its inverse transform provides the MA model parameters estimate (first 5 coefficents). By estimating the MA parameters of 25 segments, the mean spectrum of the process is computed. Further, it is compared with the spectrum computed from the true MA parameters. Further, the individual 25 sprectral estimates are also overlaid.

It is seen that there is a good match between the true spectrum and mean of the estimates except for the fact the the null depths are somewhat less compared to those of true spectra (Fig.(7.26a)). But the spectral estmates are spread wide indicating its variance is more. This is expected as the MA parameters estimates are based on DFT whose varance is large.

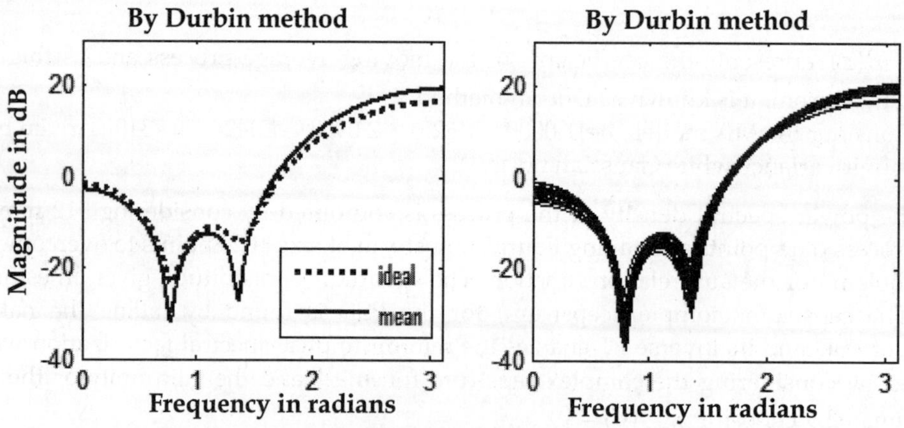

Fig. 7.26 (b) MA model estimation by Durbin method.

For the Durbin method, the same data length of 512 points is considered. The spectrum is AR modeled by Burg method with order 20. The AR sequence is AR modeled by Burg method with order 5. Here also for 25 segments of data each of 512 points using 1024 point FFT, the mean spectrum is computed and is compared with the spectrum obtained by true MA parameters. Further, the individal 25 sprectral estimates are also overlaid. The mean spectrum follows the true spectrum but its nulls overshoot over the true spectrum ((Fig. 7.26(b))). This is expected as the Burg (AR) method is used and its spectrum is characterized by sharper spectral peaks. The spectral estimates for Durbin method are better concentrated indicating its variance is less than that by spectral factorization method.

7.17 BISPECTRUM ESTIMATION VIA AR MODELING

Consider a Pth order AR process

$$x(n) = -\sum_{i=1}^{P} a(i)x(n-i) + w(n) \tag{7.86a}$$

Here, $w(n)$ is a i.i.d zero mean non-Gaussian process with $E[w^3(n)] = \beta$. The bispectrum of the process is given by

$$B_x(\omega_1,\omega_2) = \beta H(\omega_1)H(\omega_2)H^*(\omega_1+\omega_2) \tag{7.86b}$$

where

$$H(\omega), = \frac{1}{1+\sum_{i=1}^{P} a(i)e^{-j\omega i}} \tag{7.86c}$$

Under the assumption $x(n)$ is stationary, its third moment sequence $g(k,l)$ is given by

$$g(k,l) = E[x(n)x(n+k)x(n+l) \tag{7.86d}$$

and it obeys the third order recursion (TOR) relation

$$g(-k,-l) + \sum_{i}^{P} a(i)g(i-k,i-l) = \delta(k,l), \qquad k,l \geq 0 \tag{7.86e}$$

and its proof is considered. Consider Eqn. (7.86a)

$$x(n) = -\sum_{i=1}^{P} a(i)x(n-i) + w(n)$$

Then

$$E[x(n)x(n-k)x(n-l)$$
$$= -\sum_{i=1}^{P} a(i)E[x(n-i)x(n-k)x(n-l)] + E[w(n)x(n-k)x(n-l)]$$

Case 1: For $k, l > 0$
Since

$$E[w(n)x(n-k)x(n-l)] = 0, \qquad k, l > 0$$

Therefore

$$E[x(n)x(n-k)x(n-l) + \sum_{i=1}^{P} a(i)E[x(n-i)x(n-k)x(n-l)] = 0$$

$$g(-k), -l) + \sum_{i=1}^{P} a(i)g(-i, -l) = 0, \quad k, l > 0 \quad \text{B(4)} \tag{7.86f}$$

Case 2: For $k = 0, l > 0$

$$E[x(n)x(n-k)x(n-l)$$
$$= -\sum_{i=1}^{P} a(i)E[x(n-i)x(n-k)x(n-l)] + E[w(n)x(n-k)x(n-l)]$$

Becomes

$$E[x^2(n)x(n-l)$$
$$= -\sum_{i=1}^{P} a(i)E[x(n-i)x(n)x(n-l)] + E[w(n)x(n)x(n-l)] \tag{7.86g}$$

Now multiplying

$$x(n) = -\sum_{i=1}^{P} a(i)x(n-i) + w(n)$$

by $x(n-l)w(n)$ and taking expectation,

$$E[x(n)x(n-l)w(n)] +$$
$$\sum_{i=1}^{P} a(i)\underline{E[x(n-i)x(n-l)w(n)]} = \underline{E[w^2(n)x(n-l)]}$$
$$\downarrow = 0 \qquad\qquad \downarrow = 0, \quad \textit{for } k = 0, l > 0$$

$$E[x(n)x(n-l)w(n)] = 0 \tag{7.86h}$$

Therefore, Eqn. (7.86g), becomes,

$$E[x^2(n)x(n-l)+\sum_{i=1}^{P}a(i)E[x(n-i)x(n)x(n-l)]=0,$$

$$g(-k),-l)+\sum_{i=1}^{P}a(i)g(-i,-l)=0, \quad for\ k=0, l>0 \tag{7.86i}$$

Since bispectrum follows the symmetry relation

$$g(k,l)=g(l,k)=g(-k,l-k)=g(k-l,-l),$$

Eqn. (7.86i) is valid for $for\ k>0, l=0$.
Case 3:, $for\ k=0, l=0$
Rom Eqn. (7.86g),

$$E[x^2(n)x(n-l)$$
$$=-\sum_{i=1}^{P}a(i)E[x(n-i)x(n)x(n-l)]+E[w(n)x(n)x(n-l)]$$

Setting $l=0$,

$$E[x^3(n)]$$
$$=-\sum_{i=1}^{P}a(i)E[x(n-i)x^2(n)]+E[w(n)x^2(n)]$$

Multiplying Eqn. (7.86a)

$$x(n)=-\sum_{i=1}^{P}a(i)x(n-i)+w(n)$$

by $x(n)w(n)$ and taking expectations,

$$E[x^2(n)w(n)$$
$$=-\sum_{i=1}^{P}a(i)\underline{E[x(n-i)x(n)w(n)]}+E[x(n)w^2(n)]$$
$$\downarrow=0$$

$$E[x^2(n)w(n)=E[x(n)w^2(n)] \tag{7.86j}$$

But multiplying Eqn. (7.86a)

$$x(n), = -\sum_{i=1}^{P} a(i)x(n-i) + w(n)$$

By $w^2(n)$ and taking expectations,

$$E[x(n)w^2(n)]$$

$$= -\sum_{i=1}^{P} a(i)\underline{E[x(n-i)w^2(n)]} + E[w3(n)]$$

$$\downarrow = 0$$

$$E[x(n)w^2(n)] = E[w^3(n)] \tag{7.86k}$$

From Eqns. (7.86k), (7.86j) and (7.86i)

$$g(-k), -l) + \sum_{i=1}^{P} a(i)g(-i, -l) = \beta, \quad for\ k = 0, l = 0 \tag{7.86l}$$

The above three cases considered prove that for an AR process, the TOR Eqn. (7.86e)

$$g(-k, -l) + \sum_{i}^{P} a(i)g(i-k, i-l) = \delta(k, l), \qquad k, l \geq 0$$

is satisfied.

To approximate the third moment sequence of an arbitrary process by an AR process driven by a non-Gaussian white noise, the third moment sequence of the process

$$r(k, k), \quad k, = \pm 0, \pm 1, \pm 2, \ldots\ldots, \pm P \tag{7.87a}$$

are matched to those of AR process,

$$g(k, k), \quad k, = \pm 0, \pm 1, \pm 2, \ldots\ldots, \pm P \tag{7.87b}$$

That is, for the AR process,

$$g(k, k) = r(k, k), \quad k, = \pm 0, \pm 1, \pm 2, \ldots\ldots, \pm P \tag{7.87c}$$

For this, the TOR becomes

$$r(-k, -k) + \sum_{i}^{P} \hat{a}(i)r(i-k, i-k) = \beta\ \delta(k, k), \qquad k = 0, 1, 2, \ldots\ldots, P$$

This in matrix form,

$$R\bar{a} = \bar{b} \tag{7.87d}$$

where

$$R = \begin{bmatrix} r(0,0) & r(1,1) & \cdots & r(P,P) \\ r(-1,-1) & r(0,0) & \cdots & r(P-1,P-1) \\ \vdots & \vdots & \cdots & \vdots \\ r(-P,-P) & r(1-P,1-P) & \cdots & r(0,0) \end{bmatrix}$$

$$\bar{a} = [1, \hat{a}(1), \hat{a}(2), \ldots\ldots \hat{a}(P)], \qquad \bar{b} = [\beta, 0, 0, \ldots\ldots, 0]$$

Here, R will be non-symmetric Toeplitz matrix and Eqn. (7.87d)) can be solved for $\bar{a}$ and β. If the third order moment sequence $r(k, k), \quad k, = \pm 0, \pm 1, \pm 2, \ldots\ldots, \pm P$ are those of a Pth order AR process driven by a non-Gaussian white noise, then the estimated parameters match with those of true process. The AR model for the process is given by

$$\hat{H}(\omega) = \frac{1}{1 + \sum_{i=1}^{P} \hat{a}(i) e^{-j\omega i}} \tag{7.87e}$$

and the bispectrum of the process can be obtained by

$$\hat{B}_x(\omega_1, \omega_2) = \beta \hat{H}(\omega_1) \hat{H}(\omega_2) \hat{H}^*(\omega_1 + \omega_2) \tag{7.87f}$$

The degree of match between the estimated bispectrum with the true one depends on the match between the given process to the Pth order AR non-Gaussian process.

In above method, it is required to find

$$r(k, k), \quad k, = \pm 0, \pm 1, \pm 2, \ldots\ldots, \pm P$$

from the given data $x(n), \quad n = 1, 2, \ldots\ldots, N = KM$. For this, divide the data into K records each of M samples. Then for each record, compute the biased third moment estimate.

$$r^{(i)}(k,k) = \frac{1}{M} \sum_{l=1}^{M-k} x^{(i)}(l) x^{(i)}(l+k), \tag{7.87g}$$

$$k = 1, 2, \ldots\ldots, P, = KM, i = 1, 2, \ldots\ldots, K$$

Here, $x^{(i)}(l)$ l^{th} sample in the ith record. The final third estimate to be used is given by

$$r(k,k) = \frac{1}{K}\sum_{l=1}^{K} r^{(i)}(k,k), \quad k = 1, 2, \ldots\ldots, P, \tag{7.87h}$$

Group delay approach for blind system identification by AR/ARMA models based on bispectrum:

Non-minimum phase system identification/bispectrum estimation by AR/ARMA model can be achieved using group delay approach (S.V. Narasimhan, G.R. Reddy, E.L. Plotkin and M.N.S. Swamy, 1992) and this already has been briefly dealt with while stating the applications of PZD method. However, its frequency resolution and variance depend on the data length used for DFT and number of segments used in computing the output bispecutm which is used in computing the magnitude and phase group delay functions involved. Also its variance depends on the way in which the system phase group delay is derived from the bispectrum phase either by using single bispectrum slice or all the bispectrum slices.

CHAPTER 8

PARAMETRIC APPROACH IN ITS GENERAL FORM: THE WIENER FILTER

In this chapter the parametric approach in its general form which is well known as Wiener filter is considered. Here the data till the infinite past (i.e., from the starting point) will be considered. This consideration permits only positive lags for the crosscorrelation (between input and desired signal) term involved in Weiner-Hopf equation and hence does not permit a simple solution in Z-transform domain. This has been overcome by pre-whitening approach which whitens the input and makes the solution possible as a cascade of the whitening filter and the Wiener filter for whitened input. The pre-whitening filter has been derived by the canonical spectral factorization of the input (original) power spectral density. By considering that the crosscorrelation (between input and desired signal) for both positive and negative lags is possible, an unrealizable Wiener filter which cannot be realized in real time has also been considered. The unrealizable Wiener filter can be considered a physically realizable with an infinite delay in the desired response. For unrealizable filter, in estimating the desired response at the present time, the entire past and future input values are used. The performance of an unrealizable filter can be achieved by a realizable one by waiting for future input values in estimating the desired response at a later time. Further the Wiener filter for finite past data which is the only realizable one in real time in practice has been treated. The minimum mean square error for the three cases indicate that this index for unrealizable Wiener filter has the least value compared to the other two cases.

8.1 INTRODUCTION

The parametric approach discussed in the previous chapter, in its most general form is known as *Wiener filter and* will be now considered. This deals with the estimation of a signal $s(n)$ in the presence of additive noise $v(n)$, which commonly occurs in scenarios of communication and many other scientific and engineering fields. Here also, by minimizing the *mean square error* ε between the estimator output $y(n)$ and the desired response $d(n)$, with respect to coefficients of the estimator/filter $W(n)$, the optimum estimation is achieved. Though previously the optimum estimators were considered to be of finite length, presently *the estimator* $W(n)$ will be considered to be *of same length as that of the input* $x(n)$. This implies that if the input is of finite duration, the estimator is also of finite duration. On the other hand, if the input extends to infinite past, the optimum filter is also of infinite duration.

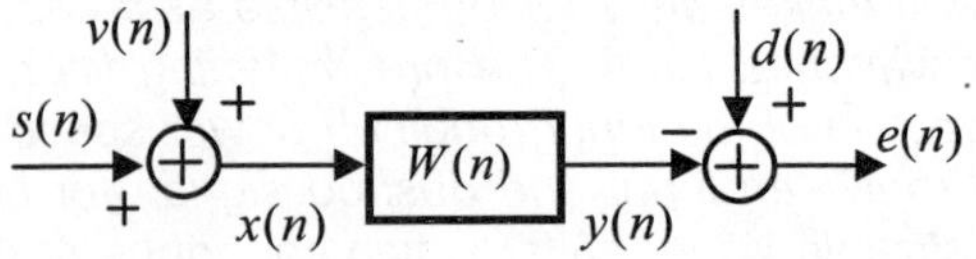

Fig. 8.1 The problem of Wiener filter.

Based on the desired signal $d(n)$, the estimator performs different functions.

1. If $d(n) = s(n)$, the estimator acts as a filter.
2. If $d(n) = s(n+\Delta), \Delta > 0$, the estimator acts as predictor.
3. If $d(n) = x(n-\Delta), \Delta > 0$, the estimator acts as smoothing function.

Assuming a general desired signal $d(n)$, the Wiener filter will be derived for the input extending into infinite past.

8.2 WIENER FILTER FOR INPUT EXTENDING TO INFINITE PAST

It is required to make the error $e(n) = d(n) - y(n)$ as small as possible in statistical sense. That is, $W(n) = \{w(0), w(1), \ldots w(\infty)\}$ should be chosen to make $\varepsilon = E[e^2(n)]$ minimum. The output of the estimators $y(n)$ is given by

$$y(n) = \sum_{k=0}^{\infty} w(k)x(n-k)$$

$$\varepsilon = E\left[e^2(n)\right] = E\left[\{d(n)-y(n)\}^2\right] = E\left[d^2(n)\right] - 2E\left[d(n)y(n)\right] + E\left[y^2(n)\right]$$

$$\varepsilon = E\left[d^2(n)\right] - 2\sum_{k=0}^{\infty} w(k)E\left[d(n)x(n-k)\right] + \sum_{k=0}^{\infty}\sum_{m=0}^{\infty} w(k)w(m)E\left[x(n-m)x(n-k)\right]$$

Here again

$E\left[d^2(n)\right] = \phi_d(0)$: desired signal power,

$E\left[d(n)x(n-k)\right] = \phi_{dx}(k), \quad k = 0,1,2,\ldots\infty$: cross-correlation between input and the desired signal

$E\left[x(n-m)x(n-k)\right] = \phi_x(k-m)$: autocorrelation of the input for lag $(k-m)$,

Therefore,

$$\varepsilon = \phi_d(0) - 2\sum_{k=0}^{\infty} w(k)\,\phi_{dx}(k) + \sum_{k=0}^{\infty}\sum_{m=0}^{\infty} w(k)w(m)\,\phi_x(k-m) \tag{8.1}$$

Minimizing ε with respect to $w(k)$,

$$\frac{\partial\,\varepsilon}{\partial w(k)} = -2\,\phi_{dx}(k) + 2\sum_{m=0}^{\infty} w(m)\,\phi_x(k-m) = 0, \quad k = 0,1,2,\ldots\infty$$

$$\sum_{m=0}^{\infty} w^*(m)\phi_x(k-m) = \phi_{dx}(k), \quad k = 0,1,2,\ldots\infty \tag{8.2}$$

Eqn. (8.2) is the *Wiener-Hopf Equation in its general form and involves* $\phi_x(k-m)$*, the autocorrelation of the input and* $\phi_{dx}(k)$*, the cross-correlation between input and the desired signal.* This is similar to previously derived *Wiener-Hopf Equation* where its L. H. S is a convolution of $W_o(k)$ and $\phi_x(k)$ which are *infinitely long but the R. H. S,* $\phi_{dx}(k)$ exists *both for positive and negative values of* k *with* $\phi_{dx}(k) \neq \phi_{dx}(-k)$. However, the Wiener-Hopf equation is *valid only for* $k = 0, 1, 2, \ldots\infty$ *and not for negative values of* k. Hence, its direct solution by taking the Z-transform of Eqn. (8.2), is not possible and a pre-whitening approach which simplifies the solution, will be considered. This involves conversion of the input $x(n)$ to a white noise sequence $v(n)$ by passing it through a filter.

8.3 WIENER-HOPF EQUATION SOLUTION

Assuming that the conversion of the input $x(n)$ to a white noise sequence $v(n)$ is done by a filter $G_I(z)$, then for the white noise $v(n)$,

$$\phi_v(k) = E\left[v(n)v(n-k)\right]$$

$$\phi_v(k-m) = \begin{cases} \sigma_v^2, & m = k \\ 0, & otherwise \end{cases}, \tag{8.3}$$

$\phi_v(0) = \sigma_v^2$ is variance of $v(n)$. Substituting this in the Wiener-Hopf equation (Eqn. (8.2)),

$$\sum_{m=0}^{\infty} w^*(m)\phi_x(k-m) = \phi_{dx}(k), \; k = 0, 1, 2, \ldots \infty$$

For $m = k$,

$$w_o^*(k)\sigma_v^2 = \begin{cases} \phi_{dv}(k), & k = 0,1,2\ldots\infty \\ 0, & k < 0 \end{cases}$$

$$w_o^*(k) = \begin{cases} \phi_{dv}(k)/\sigma_v^2, & k = 0,1,2\ldots\infty \\ = 0, & k < 0 \end{cases}, \tag{8.4}$$

where

$$\phi_{dv}(k) = E\left[d(n)v(n-k)\right]$$

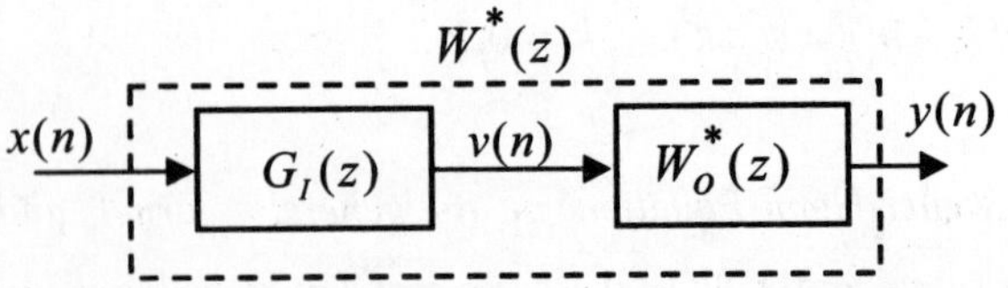

Fig. 8.2 Whitening approach for Wiener-Hopf equation.

Thus, in this approach *there are two filters (Fig.8.2), the whitening filter* $G_I(z)$ and $W_o^*(z)$ *that operates on the white noise sequence* $v(n)$. The filter $G_I(z)$ which converts is input $x(n)$ to white noise $v(n)$, will be discussed later. The Z-transform of $w_o^*(k)$ is

$$W_o^*(z) = \sum_{k=0}^{\infty} w_o^*(k) z^{-k} \tag{8.5}$$

It is important to note that *the lower limit starts from* $k = 0$ and this is due to the definition of $w_o^*(k)$ (Eqn. (8.4)). Therefore,

$$W_o^*(z) = \frac{1}{\sigma_v^2} \sum_{k=0}^{\infty} \phi_{dv}(k) z^{-k} \tag{8.6}$$

But the cross-correlation $\phi_{dv}(k)$ is defined *both for positive and negative values of* k *with* $\phi_{dv}(k) \neq \phi_{dv}(-k)$ *and*

$$S_{dv}(z) = \sum_{k=-\infty}^{\infty} \phi_{dv}(k) z^{-k} \tag{8.7}$$

For $z = e^{j\omega}$, $S_{dv}(z)$ is the cross power spectrum of $d(n)$ and $v(n)$ and $S_{dv}(z)$ is a *two sided z-transform as* k *takes values* $-\infty \leq k \leq \infty$. But as stated already (Eqn.8.4), $W_o^*(z)$ uses *only that part of* $S_{dv}(z)$ *which is valid for values of* $k = 0,1,2,\ldots\infty$ and is defined by,

$$\left[S_{dv}(z)\right]_+ = \sum_{k=0}^{\infty} \phi_{dv}(k) z^{-k} \tag{8.8}$$

Therefore,

$$W_o^*(z) = \left(1/\sigma_v^2\right)\left[S_{dv}(z)\right]_+ \tag{8.9}$$

This requires $S_{dv}(z)$ *and* σ_v^2 *which can be obtained from* $v(n)$ *the whitened version of the input* $x(n)$.

As already stated that the whitening of the input $x(n)$ is done by a filter $G_I(z)$ and $v(n)$ is given by

$$v(n) = \sum_{l=0}^{\infty} g_I(l) x(n-l), \tag{8.10}$$

where, $g_I(l), l = 0, 1, 2, \ldots \infty$; is the impulse response of the *whitening filter* given by,

$$G_I(z) = 1/G(z) = \sum_{l=0}^{\infty} g_I(l) z^{-l} \tag{8.11}$$

Now using Eqn. (8.10) for $v(n)$,

$$\phi_{dv}(k) = E\left[d(n)v(n-k)\right] = E\left[d(n)\sum_{l=0}^{\infty} g_I(l)x(n-k-l)\right]$$

$$= \sum_{l=0}^{\infty} g_I(l) E\left[d(n)x(n-k-l)\right]$$

$$\phi_{dv}(k) = \sum_{l=0}^{\infty} g_I(l)\phi_{dx}(k+l) \tag{8.12}$$

$\phi_{dx}(k+l)$ is the cross-correlation between $d(n)$ and $x(n)$ for *lag* $(k+l)$ and this *is assumed to be known*. Therefore, from Eqn.(8.12),

$$S_{dv}(z) = \sum_{k=-\infty}^{\infty} \phi_{dv}(k)\, z^{-k} = \sum_{k=-\infty}^{\infty}\sum_{l=0}^{\infty} g_I(l)\phi_{dx}(k+l)z^{-k}$$

$$= \sum_{l=0}^{\infty} g_I(l) \sum_{p=-\infty}^{\infty} \phi_{dx}(p) z^{-(p-l)}, \quad k+l=p,$$

$$S_{dv}(z) = \sum_{l=0}^{\infty} g_I(l) z^{l} \sum_{p=-\infty}^{\infty} \phi_{dx}(p) z^{-p} \tag{8.13}$$

Since from Eqn. (8.11)

$$\sum_{l=0}^{\infty} g_I(l) z^{-l} = 1/G(z),$$

$$\sum_{l=0}^{\infty} g_I(l) z^{l} = G_I(z^{-1}) = 1/G(z^{-1})$$

Also

$$\sum_{p=-\infty}^{\infty} \phi_{dx}(p) z^{-p} = S_{dx}(z)$$

Therefore, Eqn. (8.13) becomes,

$$S_{dv}(z) = \frac{S_{dx}(z)}{G(z^{-1})}$$

Substituting this in the expression for the filter $W_o^*(z)$ (Eqn.(8.9)),

$$W_o^*(z) = \frac{1}{\sigma_v^2}\left[S_{dv}(z)\right]_+$$

$$W_o^*(z) = \frac{1}{\sigma_v^2}\left[\frac{S_{dx}(z)}{G(z^{-1})}\right]_+ \tag{8.14}$$

The term $\left[\frac{S_{dx}(z)}{G(z^{-1})}\right]_+$ corresponds to *poles and zeros inside the unit circle.*

The Wiener filter $W_o(z)$ *is a cascade of* $G_I(z) = 1/G(z)$ *and* $W_o^*(z)$. That is,

$$W_o(z) = W_o^*(z) \bullet \frac{1}{G(z)}$$

Therefore, from Eqn. (8.14),

$$W_o(z) = \frac{1}{\sigma_v^2 G(z)}\left[\frac{S_{dx}(z)}{G(z^{-1})}\right]_+ \tag{8.15}$$

This is the Z-transform version of the solution to Wiener-Hopf equation.

8.4 DERIVING THE WHITENING FILTER $G_I(z)$ FROM THE INPUT

In the above treatment, it has been assumed that the filter $G_I(z) = 1/G(z)$ will whiten the input $x(n)$ to get $v(n)$ which is a white noise process. The theory behind achieving such a filter will be considered.

A stochastic process $x(n)$ is generated by filtering a white noise process $v(n)$ by a system / filter $G(e^{j\omega})$. For the stochastic process $x(n)$, such an associated process $v(n)$ is called the *innovation process.* The name "*innovation process*" is because being a white noise process, *its samples at different instants are uncorrelated, and its each sample brings new information.* As $v(n)$ *samples are uncorrelated, it is a much simpler process than* $x(n)$ whose samples are correlated. This is the motivation for converting the original process to an innovation process. The innovation process $v(n)$ can be generated by filtering the stochastic process $x(n)$ by $G_I(e^{j\omega})$ that is *causal and invertible.*

$$G_I(e^{j\omega}) \leftrightarrow \begin{cases} g_I(n) & n \geq 0 \\ 0 & n < 0 \end{cases} \tag{8.16}$$

The causality of $G_I(e^{j\omega})$ enables *real time implementation* and its *causally invertible property facilitates* the property,

$$G(e^{j\omega}) = \frac{1}{G_I(e^{j\omega})} \tag{8.17}$$

Further, $G(e^{j\omega})$ *is causal,* i.e., $g(n) = 0, \quad n < 0$. For the filter $G_I(e^{j\omega})$ *to whiten the input* $x(n)$ with $S_x(e^{j\omega})$ as its power spectral density,

$$S_x(e^{j\omega})\left|G_I(e^{j\omega})\right|^2 = \sigma_v^2 \tag{8.18}$$

σ_v^2 is the variance of the innovation process. Therefore,

$$\left|G_I(e^{j\omega})\right|^2 = \frac{\sigma_v^2}{S_x(e^{j\omega})}$$

$G_I(e^{j\omega})$ is known as the *whitening filter.* Further, due to Eqn. (8.17), it is required that

$$S_x(\omega) = \sigma_v^2 \left| G(e^{j\omega}) \right|^2 \tag{8.19}$$

This is the *canonical innovation representation of the wide sense stationary stochastic process (WSSP)* $x(n)$. *The stochastic process* $x(n)$ *and the innovation process have the same statistical information as it is possible to go from one to the other. Also the innovation process is simpler than its stochastic process.*

8.4.1 Canonical Factorization of Power Spectral Density $S_x(\omega)$ of a Stochastic Process

In the above Eqn. (8.19), for a stochastic process with its PSD $S_x(e^{j\omega})$, the problem is how to get $G(e^{j\omega})$. The $S_x(z)$ can be expressed in terms of its autocorrelation function $\phi_x(k)$ as

$$S_x(z) = \sum_{k=-\infty}^{\infty} \phi_x(k) z^{-k} \tag{8.20}$$

and $S_x(z)$ is a two sided z-transform. If $ln\left[S_x(z)\right]$ *is analytic* (i.e., $ln\left[S_x(z)\right]$ and all its derivatives are continuous) in $\rho < |z| < 1/\rho$, $\rho < 1$, then

$$ln\left[S_x(z)\right] = \sum_{k=-\infty}^{\infty} C_k z^{-k} = \sum_{k=-\infty}^{\infty} C_k e^{-j\omega k} \qquad (\text{as } z = e^{j\omega}) \tag{8.21a}$$

C_k*'s are the Fourier coefficients of In $S_x(z)$ which is periodic in* ω *with period* 2π and are given by,

$$C_k = \frac{1}{2\pi} \int_{-\pi}^{\pi} \ln S_x(\omega) e^{j\omega k} d\omega \text{ and } C_0 = \frac{1}{2\pi} \int_{-\pi}^{\pi} ln\, S_x(\omega) d\omega \tag{8.21b}$$

Since $S_x(\omega)$ *is real and even function of* ω, C_k *is also even, i.e.,* $C_k = C_{-k}$. Further Eqn. (8.21a)

$$ln\left[S_x(z)\right] = \sum_{k=-\infty}^{\infty} C_k z^{-k}$$

can be written as

$$ln\left[S_x(e^{j\omega})\right] = C_0 + \left[\sum_{k=1}^{\infty} C_k z^{-k}\right] + \left[\sum_{k=-\infty}^{-1} C_k z^{-k}\right]$$

or

$$S_x(e^{j\omega}) = e^{C_0} \exp\left[\sum_{k=1}^{\infty} C_k z^{-k}\right] \exp\left[\sum_{k=-\infty}^{-1} C_k z^{-k}\right] \tag{8.22}$$

Defining

$$\sigma_v^2 = e^{C_0}, \tag{8.23a}$$

$$G(z) = \exp\left[\sum_{k=1}^{\infty} C_k z^{-k}\right], \quad |z| > \rho \tag{8.23b}$$

and

$$G(1/z) = \exp\left[\sum_{k=-\infty}^{-1} C_k z^{-k}\right],$$

Replacing k by $-k$ and using $C_k = C_{-k}$,

$$G(1/z) = \exp\left[\sum_{k=1}^{\infty} C_k z^{k}\right], \quad |z| < 1/\rho \tag{8.23c}$$

Therefore, from Eqns. (8.23a),(8.23b) and (8.23c),Eqn. (8.22) becomes

$$S_x(z) = \sigma_v^2 G(z) G(1/z), \tag{8.23d}$$

For $z = e^{j\omega}$,

$$S_x(e^{j\omega}) = \sigma_v^2 G(e^{j\omega}) G\left(e^{-j\omega}\right)$$

$$S_x(e^{j\omega}) = \sigma_v^2 \left|G(e^{j\omega})\right|^2 \tag{8.24}$$

$G(z)$ is normalized to have $G(\infty) = 1$ or g(0)=1. The Eqn. (8.24) is same as Eqn. (8.19) representing *canonical representation of a WSSP*. The representation by Eqn. (8.23d) is called *canonical factorization of the PSD. The* $G(z)$ *has poles and zeros inside the unit circle* hence, $G(z)$ represents a *minimum phase* filter that is stable and invertible. But $G(1/z)$ has poles and zeros outside the unit circle and is a *maximum phase filter*. Further, $G(1/z)$ and $1/G(1/z)$ are anti-causal.

From the above treatment, it is clear that $G(z)$ can be obtained from the given $S_x(e^{j\omega})$, since

$$G(z) = \exp\left(\sum_{k=1}^{\infty} C_k z^{-k}\right), \quad C_k = \frac{1}{2\pi}\int_{-\pi}^{\pi} e^{j\omega k} \ln\left[S_x(\omega)\right] d\omega \tag{8.25}$$

and $g(k)$ is a causal sequence with $g(0) = 1$. For a process $x(n)$ that can be considered the output of a filter $G(z)$ to a white noise input $v(n)$, Eqn. (8.24) is satisfied. That is, for a WSSP, the *canonical spectral factorization is equivalent to modeling of the process as the output of a discrete time linear filter* $g(n)$ *for a white noise input process* $v(n)$ and is given by

$$x(n) = \sum_{k=0}^{\infty} g(k)v(n-k) \tag{8.26}$$

Such a WSSP modeling is called *canonical innovation representation of a WSSP*. Such a WSSP which is the output of a causal MA system possibly of infinite order to a white noise input is also known as *Wold decomposition / representation* (G. Zelniker and F.J. Taylor, chapter 5, 1994 and J.G. Praokis et al. Chapter 3, 2002) Further, for the reverse approach,

$$v(n) = \sum_{k=0}^{\infty} g_I(k)x(n-k), \tag{8.27}$$

where $G_I(z) = 1/G(z)$ with $g_I(0) = 1$ or $G_I(\infty) = 1/G(\infty) = 1$. This indicates that the associated *white noise process for a given* WSSP *can be represented by a linear combination of correlated samples of the* WSSP. This looks strange but it is possible to remove the *correlations in the stochastic process by this inverse operation which* again involves linear combination of samples.

For a power spectral density $S_x(e^{j\omega})$ which is real, even and is nonnegative and satisfies the conditions,

$$\int_{-\pi}^{\pi} S_x(e^{j\omega})\,d\omega < \infty \qquad \text{and} \qquad \exp\left[\frac{1}{2\pi}\int_{-\pi}^{\pi} \ln\left(S_x(e^{j\omega})\right)d\omega\right] > 0$$

there exists a causal, finite energy sequence $g(n)$ such that

$$S_x(e^{j\omega}) = \left|G(e^{j\omega})\right|^2$$

This is known as *Kolmogorov-Szego theorem.* The Wold decomposition implies that AR and ARMA processes can be represented by MA model possibly of infinite order. The *Kolmogorov-Szego theorem* implies that ARMA and MA can be represented by all-pole model possibly of infinite order. Thus, these two representations provide the flexibility of representing one type of process by the other. Practically, this is important as the process type will be unknown, it is possible to get reasonably good approximation even with the choice of a wrong model (G. Zelniker and F.J. Taylor, chapter 5, 1994).

Steps involved in obtaining the Wiener-Hopf equation solution:

1. Factorize $S_x(z)$ as

   $$S_x(z) = \sigma_v^2 G(z)G(z^{-1})$$

   For this, assign all the poles and zeros of $S_x(z)$ inside the unit circle to $G(z)$ and assign all the poles and zeros of $S_x(z)$ outside the unit circle to $G(z^{-1})$.

2. Obtain $S_{dx}(z)$ as *a two-sided z- transform of* $\phi_{dx}(k)$. Further, divide $S_{dx}(z)$ by $G(z^{-1})$. The causal part of $w_o^*(k)$, $w_o^*(k)$, $k \geq 0$ is given by

$$W_o^*(z) = \left[\frac{S_{dx}(z)}{G(z^{-1})} \right]_+$$

3. Obtain the Wiener filter transfer function by multiplying $W_o^*(z)$ by the whitening filter $1/G(z)$ (except for the scale factor σ_v^2)

$$W^*(z) = \frac{1}{G(z)} \left[\frac{S_{dx}(z)}{G(z^{-1})} \right]_+$$

This treatment of the Wiener filter as two parts is only conceptual, but in practice, it is a single entity.

8.5 MINIMUM MEAN SQUARE ERROR

The mean square error is given by (Eqn. 8.1)

$$\varepsilon = \phi_d(0) - 2\sum_{k=0}^{\infty} w(k)\phi_{dx}(k) + \sum_{k=0}^{\infty}\sum_{m=0}^{\infty} w(k)\,w(m)\,\phi_x(k-m)$$

Wiener-Hopf solution (Eqn. 8.2) is

$$\sum_{m=0}^{\infty} w^*(m)\phi_x(k-m) = \phi_{dx}(k)\ , \quad k = 0,1,2,\ldots\infty$$

Therefore,

$$\varepsilon_{\min} = \phi_d(0) - 2\sum_{k=0}^{\infty} w^*(k)\phi_{dx}(k) + \sum_{k=0}^{\infty} w^*(k)\phi_{dx}(k)$$

$$\varepsilon_{\min} = \phi_d(0) - \sum_{k=0}^{\infty} w^*(k)\phi_{dx}(k)$$

Since $w^*(k)$ *is zero for* $k < 0$,

$$\varepsilon_{\min} = \phi_d(0) - \sum_{k=-\infty}^{\infty} w^*(k)\phi_{dx}(k) \tag{8.28}$$

Minimum MSE in Frequency domain : By Parseval's theorem,

$$\sum_{n=-\infty}^{\infty} x(m)y(n) = \frac{1}{2\pi j}\oint_c X(z)Y(z^{-1})\frac{dz}{z} \tag{8.29}$$

Therefore, the second term on the R.H.S of Eqn. (8.28) becomes

$$\sum w^{*}(k)\phi_{dx}(k) = \frac{1}{2\pi j}\oint_{c} W^{*}(z)S_{dx}(z^{-1})\frac{dz}{z} \tag{8.30}$$

(Note that *the superscript "*" in* $W^{*}(z)$, *refers to optimum filter and does not refer to usual conjugate operation*). Further,

$$S_{d}(z) = \sum_{k=-\infty}^{\infty} \phi_{d}(k)z^{-k} \quad \text{and} \quad \phi_{d}(k) = \frac{1}{2\pi j}\oint_{c} S_{d}(z)z^{k}\frac{dz}{z}, \quad k = 0, \pm 1, \pm 2, \ldots \infty$$

and

$$\phi_{d}(0) = \frac{1}{2\pi j}\oint_{c} S_{d}(z)\frac{dz}{z} \tag{8.31}$$

Using Eqn. (8.30) and Eqn. (8.31), minimum MSE (Eqn. (8.28))

$$\varepsilon_{\min} = \phi_{d}(0) - \sum_{k=-\infty}^{\infty} w^{*}(k)\phi_{dx}(k)$$

becomes

$$\varepsilon_{\min} = \frac{1}{2\pi j}\oint_{c}\left[S_{d}(z) - W^{*}(z)S_{dx}(z^{-1})\right]\frac{dz}{z} \tag{8.32}$$

If

$$F(z) = \frac{1}{z}\left[S_{d}(z) - W^{*}(z)S_{dx}(z^{-1})\right]$$

and $z_{1,}z_{2},\ldots z_{m}$ are the simple poles of $F(z)$ *inside the unit circle*, then by Cauchy's residue theorem,

$$\oint_{c} F(z)dz = 2\pi j\sum_{k=1}^{M}\text{Res}\left[F(z), z_{k}\right] \tag{8.33}$$

$$\text{Res}\left[F(z), z_{k}\right] = \underset{z\to z_{k}}{Lt}(z - z_{k})F(z) : \text{residue of } F(z) \text{ at } z = z_{k} \tag{8.34}$$

Therefore,

$$\varepsilon_{\min} = \sum_{k=1}^{M}\text{Res}\left[F(z), z_{k}\right], \tag{8.35}$$

Thus, by computing the residues of $F(z)$, the $\varepsilon_{\min}$ can be computed.

8.6. UNREALIZABLE WIENER FILTER

Till now, it has been assumed that the optimum filter is *physically realizable in real time.* That is, $w(k) = 0, \quad k < 0$. If $w(k) \neq 0$ for $k < 0$, then the filter is not causal and is *physically unrealizable* for real time operation. But this will be of use for *non-real time operation.* For this case,

$$\sum_{m=-\infty}^{\infty} w_u^*(m)\phi_x(k-m) = \phi_{dx}(k) \;, \quad k = 0, \pm 1, \pm 2 \ldots \tag{8.36}$$

Mean square error is

$$\varepsilon_u = \phi_d(0) - \sum_{k=-\infty}^{\infty} w_u^*(k)\phi_{dx}(k) \tag{8.37}$$

$w_u^*(k)$ is a *unrealizable Wiener filter.*

It is important to note that the time variable k *is allowed to take negative values. Since Eqn. (8.36) is valid for all* k *, it can be solved by taking z-transform both sides of the equation.* That is,

$$W_u^*(z)S_x(z) = S_{dx}(z) \;\text{ Or }\; W_u^*(z) = \frac{S_{dx}(z)}{S_x(z)}$$

$$W_u^*(z) = \sum_{k=-\infty}^{\infty} w_u^*(k)z^{-k} \,,$$

$$S_x(z) = \sum_{k=-\infty}^{\infty} \varphi_x(k)z^{-k} \,,$$

and

$$S_{dx}(z) = \sum_{k=-\infty}^{\infty} \varphi_{dx}(k)z^{-k}$$

8.6.1 MSE of Realizable and Unrealizable Wiener Filter

Minimum MSE for unrealizable case is

$$\varepsilon_u = \phi_d(0) - \sum_{k=-\infty}^{\infty} w_u^*(k)\phi_{dx}(k)$$

For realizable case, the lower limit $k = -\infty$ *becomes zero.* In the frequency domain the MSE is

$$\varepsilon_u = \frac{1}{2\pi j}\oint_c \left[S_d(z) - W_u^*(z) S_{dx}(z^{-1}) \right] \frac{dz}{z} \tag{8.38}$$

ε_u is similar in form for realizable Wiener filter (Eqn.8.32), *but it uses unrealizable Wiener filter.* As $w_u^*(k)$ is the best filter possible, its MSE ε_u corresponds *to lower bound attainable* by a physically realizable filter hence, it is called *irreducible error*. The unrealizable Wiener filter can be considered a physically realizable one with *an infinite delay in the desired response*. For unrealizable filter, in estimating the desired response at the present time, the entire past and future input values are used. The performance of an unrealizable filter can be achieved by a realizable one *by waiting for future input values in estimating the desired response at a later time.* Thus, the error ε can be made close to ε_u *by increasing the delay in the desired response.*

8.7. WIENER FILTER FOR DATA EXTENDING TO FINITE PAST

The finite data case has been considered already in chapter 7. Hence, this will be treated briefly for the sake of completion. The Eqn. (8.2)

$$\sum_{m=0}^{\infty} w_o(m)\phi_x(k-m) = \phi_{dx}(k), \quad k = 0,1,2,\ldots\infty$$

becomes

$$\sum_{m=0}^{N-1} w_o(m)\phi_x(k-m) = \phi_{dx}(k), \quad k = 0,1,2,\ldots N-1$$

In matrix form

$$W_{fN}^* = R_{NN}^{-1} \ P_N \tag{8.39a}$$

$$W_{fN}^* = \begin{bmatrix} w_f^*(0) \\ w_f^*(1) \\ \vdots \\ w_f^*(N-1) \end{bmatrix} \quad R_{NN} = \begin{bmatrix} \phi_x(0) & \phi_x(1) & \ldots\ldots & \ldots\phi_x(N-1) \\ \phi_x(1) & \phi_x(0) & \ldots\ldots & \ldots\phi_x(N-2) \\ \vdots & & & \\ \phi_x(N-1) & \phi_x(N-2) & \ldots\ldots & \ldots\phi_x(0) \end{bmatrix},$$

$$\phi_x(k-m) = E\left[x(n-m)x(n-k)\right]$$

$$P_N = \begin{bmatrix} \phi_{xd}(0) \\ \phi_{xd}(1) \\ \vdots \\ \phi_{xd}(N-1) \end{bmatrix}, \quad \phi_{dx}(k) = E\left[d(n)x(n-k)\right]$$

Mean square error is

$$\varepsilon_{\min}^{f} = \phi_d(0) - \sum_{k=0}^{N-1} w_f^*(k)\phi_{dx}(k)$$

$$\varepsilon_{\min}^{f} = \phi_d(0) - P_N^T W_{fN}^*$$

$$\varepsilon_{\min}^{f} = \phi_d(0) - P_N^T R_{NN}^{-1} P_N \qquad (8.39b)$$

If the length of the filter is sufficiently large, then the minimum mean square error achieved by the finite length case will be close to that of Wiener filter of infinite length and this will become clear from the illustration considered in the next section.

8.8. EXAMPLES FOR ILLUSTRATION

Consider a signal $x(n)$ consisting of a first order auto regressive process $d(n)$ and additive white noise process $v(n)$ of unit variance.

$$x(n) = d(n) + v(n) \quad n = 0, 1, 2, \ldots$$

The power spectrum of the AR processes is

$$S_d(\omega) = \frac{0.4}{1.25 - \cos(\omega)}$$

Find the Wiener filter required to estimate $d(n)$.

Solution: Substituting for

$$\cos(\omega) = (e^{j\omega} + e^{-j\omega}) / 2$$

$$S_d(\omega) = \frac{0.4}{1.25 - \left[\dfrac{e^{j\omega} + e^{-j\omega}}{2}\right]} = \frac{0.4}{1.25 - 0.5\left[\, e^{j\omega} + e^{-j\omega}\right]}$$

Now

$$x(n) = d(n) + v(n) \text{ and } S_x(\omega) = S_d(\omega) + S_v(\omega)$$

$$S_x(\omega) = \frac{0.4}{1.25 - 0.5e^{j\omega} - 0.5e^{-j\omega}} + 1 \quad \text{as} \quad S_v(\omega) = 1$$

$$S_x(\omega) = \frac{1.65 - 0.5e^{j\omega} - 0.50.5e^{-j\omega}}{1.25 - 0.5e^{j\omega} - 0.5e^{-j\omega}}$$

As $z = e^{j\omega}$,

$$S_x(z) = \frac{1.65 - 0.5z - 0.5z^{-1}}{1.25 - 0.5z - 0.5z^{-1}} = \frac{-z\left\{0.5z^{-2} - 1.65z^{-1} + 0.5)\right\}}{-z\left\{0.5z^{-2} - 1.25z^{-1} + 0.5)\right\}}$$

Factorizing,

$$S_x(z) = \frac{\left(z^{-1} - 2.96244\right)\left(z^{-1} - 0.33756\right)}{\left(z^{-1} - 0.5\right)\left(z^{-1} - 2\right)} = \frac{(-2.96244)(1 - 0.33756z)z^{-1}(1 - 0.33756z)}{z^{-1}(1 - 0.5z)(-2)(1 - 0.5z^{-1})}$$

$$S_x(z) = 1.48122\frac{(1 - 0.33756z^{-1})(1 - 0.337561z)}{(1 - 0.5z^{-1})(1 - 0.5z)}$$

***Wiener filter for data extending to infinite past*:** Comparing the above $S_x(z)$ with

$$S_x(z) = \sigma^2 G(z)G(z^{-1})$$

$$\sigma_v^2 = 1.48122,\ G(z) = \frac{(1 - 0.55376z^{-1})}{(1 - 0.5z^{-1})} \text{ and } G(z^{-1}) = \frac{(1 - 0.33756z)}{(1 - 0.5z)}$$

The cross-correlation between $d(n)$ and $x(n)$ is

$$\phi_{dx}(k) = E\left[d(n)x(n+k)\right] = E\left[\,d(n)\left\{d(n+k) + v(n+k)\right\}\right]$$

$$\phi_{dx}(k) = \phi_d(k) \quad \text{for } k > 0$$

as $d(n)$ and $v(n)$ are uncorrelated with each other. Further, this implies,

$$S_d(\omega) = \frac{0.4}{1.25 - 0.5z - 0.5z^{-1}}$$

$$S_{dx}(z) = S_d(z) = \frac{0.4}{1.25 - 0.5z - 05z^{-1})} = \frac{0.4}{(1-0.5z^{-1})(1-0.5z)}$$

$$\left[\frac{S_{dx}(z)}{G(z^{-1})}\right]_+ = \left[\frac{0.4}{(1-0.5z^{-1})(1-0.5z)} \frac{(1-0.5z)}{(1-0.33756z)}\right]_+$$

$$\left[\frac{S_{dx}(z)}{G(z^{-1})}\right]_+ = \left[\frac{0.4}{(1-0.5z^{-1})(1-0.33756z)}\right]_+$$

The causal and anticausal parts are separated by partial fraction,

$$\left[\frac{S_{dx}(z)}{G(z^{-1})}\right]_+ = \left[\frac{A}{(1-0.5z^{-1})} + \frac{Bz}{(1-0.33756z)}\right]_+ = \left[\frac{0.4}{(1-0.5z^{-1})(1-0.33756z)}\right]_+$$

$$A(1-0.33756z) + Bz(1-0.5z^{-1}) = 0.4$$
$$A + 0.5Bz = 0.4, \quad -0.33756A + B = 0$$
$$B = -0.162441, \quad A = 0.4812205$$

Therefore,

$$\left[\frac{S_{dx}(z)}{G(z^{-1})}\right]_+ = \left[\frac{0.4812205}{(1-0.5z^{-1})} - \frac{0.162441z}{(1-0.33756z)}\right]_+$$

$$\left[\frac{S_{dx}(z)}{G(z^{-1})}\right]_+ \text{ is the } \textit{causal part} \text{ and is } = \frac{0.4812205}{(1-0.5z^{-1})}$$

The transfer function of the Weiner filter is

$$W^*(z) = \left[\frac{S_{dx}(z)}{G(z^{-1})}\right]_+ \cdot \frac{1}{\sigma^2 G(z)}$$

$$\sigma_v^2 = 1.48122,\ G(z) = \frac{(1-0.33576z^{-1})}{(1-0.5z^{-1})} \text{ and } \left[\frac{S_{dx}(z)}{G(z^{-1})}\right]_+ = \frac{0.4812205}{(1-0.5z^{-1})}$$

Therefore,

$$W^*(z) = \left[\frac{0.4812205}{1-0.5z^{-1}}\right] \frac{1}{1.4812205} \frac{(1-0.5z^{-1})}{(1-0.33756z^{-1})}$$

$$W^*(z) = \frac{0.3248812}{1-0.33756z^{-1}} \quad \text{Or} \quad \frac{Y(z)}{X(z)} = 0.3248812 \frac{1}{1-0.33756z^{-1}}$$

or $$y(n) = 0.3248812\, x(n) + 0.33756\, y(n-1)$$

Therefore,

$$w(n) = 0.3248812(0.33756)^n,\ n = 0,1,2\ldots$$

$$w^*(0) = 0.3248812,\ w^*(1) = 0.109667,\ w^*(2) = 0.037$$

Minimum mean square error:

$$S_{dx}(z) = S_d(z) = \frac{0.4}{(1-0.5z^{-1})(1-0.5z)},\quad W^*(z) = \frac{0.3248812}{1-0.33756z^{-1}}$$

$$\varepsilon_{\min} = \frac{1}{2\pi j}\oint \left[S_d(z) - W^*(z) S_{dx}(z^{-1}) \right] \frac{dz}{z}$$

$$F(z) = \frac{1}{z}\left[S_d(z) - W^*(z) S_{dx}(z^{-1}) \right]$$

$$\varepsilon_{\min} = \sum_{k=1}^{M} \operatorname{Res}\left[F(z), Z_k \right]$$

$$\operatorname{Res}\left[F(z), Z_k \right] = \lim_{z \to z_k} (z - z_k) F(z)$$

$$F(z) = \frac{1}{z}\left[\frac{0.40}{(1-0.5z^{-1})(1-0.5z)} - \left(\frac{0.3248812}{1-0.133756261z^{-1}} \right)\left(\frac{0.40}{(1-0.5z)(1-0.5z^{-1})} \right) \right]$$

$$F(z) = \frac{1}{z}\left[\frac{0.40}{(1-0.5z^{-1})(1-0.5z)} \left(1 - \frac{0.3248812}{1-0.133756z^{-1}} \right) \right]$$

$$= \frac{1}{z}\left[\frac{0.40}{(1-0.5z^{-1})(1-0.5z)} \left(\frac{1-0.3248812-0.33756z^{-1}}{1-0.33756z^{-1}} \right) \right]$$

$$= \frac{1}{z}\left[\frac{0.4(0.6751188)(1-0.5z^{-1})}{(1-0.5z^{-1})(1-0.5z)(1-0.33756z^{-1})} \right] = \frac{1}{z}\left[\frac{0.27004752}{(1-0.5z)(1-0.33756z^{-1})} \right]$$

$$= \frac{1}{z}\left[\frac{2(0.27004752)Z}{(2-z)(z-0.33756)} \right]$$

$$F(z) = \left[\frac{0.540085}{(2-z)(z-0.33756)} \right]$$

The root of interest that is inside the unit circle and this is the pole at $z = 0.33756$.

Therefore,

$$\mathrm{Re}\,s\left[F(z), 0.33756\right] = \lim_{z\to 0.33756}(z-0.33756)F(z)$$

$$= \lim_{z\to 0.33756}\frac{0.540085}{(2-z)} = \frac{0.540085}{(2-0.33576)} = \frac{0.540085}{1.66244} = 0.32482$$

$$\varepsilon_{\min} = \sum_{k=1}^{M}\mathrm{Re}\,s\left[F(z), Z_k\right],$$

Therefore,

$$\varepsilon_{\min} = 0.32482$$

Unrealizable Wiener filter:

$$S_d(z) = S_{dx}(z) = \frac{0.4}{(1-0.5z^{-1})(1-0.5z)}$$

$$S_x(z) = 1.48122\frac{(1-0.33756z^{-1})(1-0.337561z)}{(1-0.5z^{-1})(1-0.5z)}$$

$$W_u^*(z) = \frac{S_{dx}(z)}{S_x(z)}$$

$$S_d(z) = S_{dx}(z) = \frac{0.4}{(1-0.5z^{-1})(1-0.5z)}\,\frac{(1-0.5z^{-1})(1-0.5z)}{1.48122(1-0.33756z^{-1})(1-0.337561z)}$$

$$W_u^*(z) = \frac{0.27005}{(1-0.33756z^{-1})(1-0.337561z)}\ ,\ W^*(z) = \frac{0.27005}{(1-0.33756z^{-1})}$$

Due to pole at $z = 1/0.33756$ *outside the unit circle, the filter is unstable and unrealizable.*
To calculate the error:

$$S_d(z) = S_{dx}(z) = \frac{0.4}{(1-0.5z^{-1})(1-0.5z)}$$

$$W_u^*(z) = \frac{0.27005}{(1-0.33756z^{-1})(1-0.337561z)}$$

$$F_u(z) = \frac{1}{z}\left[S_d(z) - W_u^*(z)S_{dx}(z^{-1})\right]$$

Therefore,

$$F_u(z)=\frac{1}{z}\left[\frac{0.4}{(1-0.5z^{-1})(1-0.5z)}\left\{1-\frac{0.27005}{(1-0.33756z^{-1})\,(1-0.337561z)}\right\}\right]$$

$$=\frac{1}{z}\left[\frac{0.4}{(1-0.5z^{-1})(1-0.5z)}\left\{\frac{0.8438967-0.33756z^{-1}-0.33756z}{(1-0.33756z^{-1})(1-0.337561z)}\right\}\right]$$

$$=\frac{1}{z}\left[\frac{-0.4z}{(1-0.5z^{-1})(1-0.5z)}\left\{\frac{0.33756z^{-2}-0.8438967z^{-1}+0.33756}{(1-0.33756z^{-1})(1-0.337561z)}\right\}\right]$$

$$=\frac{1}{z}\left[\frac{-0.4z}{(1-0.5z^{-1})(1-0.5z)}\left\{\frac{0.33756\left(z^{-1}-2\right)\left(z^{-1}-0.5\right)}{(1-0.33756z^{-1})(1-0.337561z)}\right\}\right]$$

$$F_u(z)=\frac{1}{z}\left[\frac{-0.135024z}{(1-0.5z^{-1})(1-0.5z)}\left\{\frac{-2z^{-1}\left(1-0.5z^{-1}\right)\left(1-0.5z\right)}{(1-0.33756z^{-1})(1-0.337561z)}\right\}\right]$$

$$F_u(z)=\frac{1}{z}\left[\left\{\frac{0.270048}{(1-0.33756z^{-1})(1-0.33756z)}\right\}\right]$$

$$F_u(z)=\frac{1}{z}\left[\left\{\frac{-0.270048z\left(2.9624363\right)}{(z-0.33756)(2.9624363-z)}\right\}\right]$$

$$F_u(z)=\frac{-0.8}{(z-0.33756)(z-2.9624363)}$$

Thus, $F_u(z)$ has a simple pole inside the unit circle at $z=0.33756$ and outside the unit circle at $z=2.9624363$. Using the unit circle as contour of integration, the mean square error is given by finding the residue of $F_u(z)$ at $z=0.33756$. The residue at $z=0.33756$ is,

$$\mathrm{Res}\left[F_u(z),Z_k=0.33756\right]=\lim_{z\to 0.33756}(z-0.33756)F_u(z)$$

$$\lim_{z\to 0.33756}\frac{-0.8}{(z-2.9624363)}=\frac{-0.8}{(0.33756-2.9624363)}=\frac{-0.8}{-2.6248763}=0.3048$$

$$\varepsilon_u=0.3048$$

Therefore,

$$\varepsilon_u=0.3048 \text{ and } \quad \varepsilon_u<\varepsilon_{\min}=0.32482$$

That is, the mean square error for the unrealizable filter is less than that of the realizable one. Hence, the mean square error of the unrealizable filter forms the lower bound.

Wiener filter for data extending to finite past

If the desired signal correlation decreases rapidly for higher lags, then performance of a finite length filter can be closer to that provided by the recursive solution. For the example considered, the signal $d(n)$ associated with white noise $v(n)$ of unit variance,

$$x(n) = d(n) + v(n)$$

$$S_d(z) = S_{dx}(z) = \frac{0.4}{(1-0.5z^{-1})(1-0.5z)} = \frac{0.4}{1.25-0.5(z+z^{-1})}$$

As $z = e^{j\omega}$,

$$S_d(\omega) = \frac{0.4}{1.25-\left[\dfrac{e^{j\omega}+e^{-j\omega}}{2}\right]} = \frac{0.4}{1.25-\cos\omega}$$

For $\phi_d(k) = \sigma^2(\rho)^{|k|}, k = 0, \pm1, \pm2, \ldots$

its Fourier transform is given by

$$S(\omega) = \sum_{k=-\infty}^{\infty} \sigma^2 \rho^k e^{-j\omega k} = \sigma^2 \sum_{k=-\infty}^{\infty} \left[\rho e^{-j\omega}\right]^k$$

$$S(\omega) = \sigma^2 \sum_{k=0}^{\infty} \left[\rho e^{-j\omega}\right]^k + \sigma^2 \sum_{k=-\infty}^{-1} \left[\rho e^{j\omega}\right]^{-k}$$

$$\sigma^2 \sum_{k=0}^{\infty} \left[\rho e^{-j\omega}\right]^k = \frac{\sigma^2}{(1-\rho e^{-j\omega})}$$

$$\sigma^2 \left\{\sum_{k=0}^{\infty} \left[\rho e^{j\omega}\right]^k - 1\right\} = \sigma^2 \left\{\frac{1}{(1-\rho e^{j\omega})} - 1\right\} = \sigma^2 \left\{\frac{\rho e^{j\omega}}{(1-\rho e^{j\omega})}\right\}$$

$$S(\omega) = \sum_{k=-\infty}^{\infty} \sigma^2 \rho^k e^{-j\omega k} = \frac{\sigma^2}{(1-\rho e^{-j\omega})} + \frac{\sigma^2 \rho e^{j\omega}}{(1-\rho e^{j\omega})}$$

$$= \frac{\sigma^2(1-\rho e^{j\omega}) + \sigma^2 \rho e^{j\omega}(1-\rho e^{-j\omega})}{(1-\rho e^{-j\omega})(1-\rho e^{j\omega})}$$

$$= \frac{\sigma^2 - \sigma^2\rho^2}{(1-\rho e^{-j\omega})(1-\rho e^{j\omega})}$$

$$S(\omega) = \frac{\sigma^2(1-\rho^2)}{(1-\rho(e^{j\omega}+e^{-j\omega})+\rho^2}$$

$$S(\omega) = \frac{\sigma^2(1-\rho^2)}{1+\rho^2-2\rho\cos\omega}$$

Therefore, comparing this with the given power spectral density

$$S_d(\omega) = \frac{0.4}{1.25-\cos\omega},$$

$$\sigma^2(1-\rho^2) = 0.4\ ,\ 2\rho = 1$$

Therefore,

$$\rho = 0.5,\ \sigma^2(1-1/4) = 0.4, \sigma^2 = 0.53333$$

$$\phi_d(k) = 0.53333(0.5)^{|k|},\ k = 0, \pm 1, \pm 2, \ldots$$

$$\phi_x(k) = \phi_d(k) + \phi_v(k),\quad k = 0, \pm 1, \pm 2, \ldots$$

As $d(n)$ and $v(n)$ are uncorrelated, and further as $v(n)$ is white, it contribution to $r_x(k)$ is only at 0th lag and other lags of $\phi_x(k)$ is zero. Hence, for $M = 2$,

$$\phi_x(0) = 0.53333 + 1 = 1.53333$$

and for $k \neq 0$,

$$\phi_x(k) = 0.53333(0.5)^{|k|}$$

Hence,

$$\phi_x(1) = 0.53333(0.5)^{|1|} = 0.26667$$

Therefore, R_{NN} with lags $k = 0, 1$; is given by

$$R_{NN} = \begin{bmatrix} 1.53333 & 0.26667 \\ 0.26667 & 1.53333 \end{bmatrix}$$

Further, $\phi_{dx}(k) = \phi_d(k)$ as $d(n)$ and $v(n)$ are uncorrelated. Therefore, the cross-correlation is

$$\phi_{dx}(k) = 0.53333\,(0.5)^{|k|}$$

and cross-correlation vector P is given by

$$P_N = \begin{bmatrix} \phi_{dx}(0) \\ \phi_{dx}(1) \end{bmatrix} = \begin{bmatrix} 0.53333 \\ 0.26667 \end{bmatrix}$$

$$W_f^{*T}(n) = [w_f^*(0), w_f^*(1)]^T, \quad W_{fN}^* = R_{NN}^{-1} P_N$$

$$W_{fN}^*(n) = \begin{bmatrix} w_f^*(0) \\ w_f^*(1) \end{bmatrix} = \begin{bmatrix} 1.53333 & 0.26667 \\ 0.26667 & 1.53333 \end{bmatrix}^{-1} \begin{bmatrix} 0.53333 \\ 0.26667 \end{bmatrix}$$

$$Det(R_{NN}) = \begin{vmatrix} 1.53333 & 0.26667 \\ 0.26667 & 1.53333 \end{vmatrix} = 2.27999$$

Therefore,

$$R_{NN}^{-1} = \frac{1}{2.27999} \begin{bmatrix} 1.53333 & -0.26667 \\ -0.26667 & 1.53333 \end{bmatrix} = \begin{bmatrix} 0.67252 & -0.11696 \\ -0.11696 & 0.67252 \end{bmatrix}$$

$$W_{fN}^*(n) = \begin{bmatrix} w_f^*(0) \\ w_o(1) \end{bmatrix} = \begin{bmatrix} 0.67252 & -0.11696 \\ -0.11696 & 0.67252 \end{bmatrix} \begin{bmatrix} 0.53333 \\ 0.26667 \end{bmatrix} = \begin{bmatrix} 0.327485 \\ 0.116962 \end{bmatrix}$$

$$W_{fN}^*(n) = \begin{bmatrix} 0.327485 \\ 0.116962 \end{bmatrix}$$

The minimum mean square error $\varepsilon_{\min}^f$ achieved by the optimum filter using finite past data is

$$\varepsilon_{\min}^f = \phi_d(0) - P_N^T W_{fN}^*$$

$$\phi_d(0) = 0.53333 \qquad P_N = \begin{bmatrix} 0.53333 \\ 0.26667 \end{bmatrix}$$

Therefore,

$$\varepsilon_{\min}^f = 0.53333 - \begin{bmatrix} 0.53333 \\ 0.26667 \end{bmatrix}^T \begin{bmatrix} 0.327485 \\ 0.116962 \end{bmatrix}$$

$$= 0.53333 - \begin{bmatrix} 0.53333 & 0.26667 \end{bmatrix} \begin{bmatrix} 0.327485 \\ 0.116962 \end{bmatrix}$$

$$\varepsilon_{\min}^f = 0.327481$$

$$w_f^*(0) = 0.327485, \quad w_f^*(1) = 0.116962$$

$$\varepsilon_u = 0.3048 < \varepsilon_{\min} = 0.32482$$

$$\varepsilon_{\min}^f = 0.327481 > \varepsilon_{\min} = 0.32482$$

That is, the performance of the optimum filter using the finite data length is approximately equal to that using data extending to infinite past. This may be due to fast decay of the autocorrelation of the desired response $d(n)$.

$$\phi_d(k) = 0.53333(0.5)^{|k|}, \quad k = 0, \pm 1, \pm 2, \ldots$$

Therefore,

$$\phi_d(0) = 0.53333, \ \phi_d(1) = 0.26667, \phi_d(2) = 0.1333325, \phi_d(3) = 0.06667, \phi_d(4) = 0.033335$$

That is, $\phi_d(4)$ drops to 6.25 % of $\phi_d(0)$. Hence, in $d(n)$ *samples separated by 5 or more samples are uncorrelated.* Hence, a 5-coefficient finite length filter can be sufficient and its performance will be close to that of the recursive Wiener filter. Hence, it is important to note that the *order (length) of the filter is determined by·the correlation function of the desired signal.* For highly correlated signal, a large filter is required and vice versa.

The Wiener filter can be applied provided the statistical quantities, autocorrelation coefficient of the input and the cross-correlation of the input and the desired signal are known. Out of the cases considered, the finite data length case appears to be feasible in a practical environment. However, this also required both the statistical quantities mentioned. The next two chapters 9 and 10 attempt to remove these limitations of Wiener filtering and make its realization simple and practically feasible.

Chapter 9

ITERATIVE SOLUTION TO NORMAL EQUATIONS: METHOD OF STEEPEST DESCENT

The parametric methods operate on a block of data. But in real time processing where action is called for based on the processing in the previous instant or sample, the block parametric methods are not appropriate as it involves delay and also assumes the data to be stationary over the block interval. Hence, the data has to be processed on a sample to sample basis as and when the samples arrive. This is achieved in a simple way by the steepest descent method (MSD) and this requires the autocorrelation matrix of the input and cross-correlation vector between input and the desired signal. The most important parameter of MSD is the upper bound on the stepsize for the mean weight vector to converge to the optimal solution and its complete mathematical derivation has been considered giving the necessary linear algebra background. This bound is the reciprocal of the eigenvalue of the autocorrelation matrix of the input. Further as the convergence rate is decided by the eigenvalue spread (ratio of maximum to minimum eigenvalue) of the autocorrelation matrix of the input and for a large eigenvalue spread, the SDM suffers from slow convergence rate.

9.1 INTRODUCTION

In chapters 7 and 8 solution of the normal equation by treating the data as a block was considered. However, in a practical scenario, the data will be arriving sequentially and it will be required to get the solution at every sample point. Meeting this requirement by block method involves solution of the normal equations at the arrival of each sample, which is computationally very complex. In view of this, it is preferred to obtain the optimum solution sequentially on a sample-to-sample basis. This will also reflect the changes in the signal characteristics of non-stationary signals, which are invariably common in practice. Further, where action is called for depending on the situation, real time processing is required and this is achieved by sequential solution. In this chapter, the solution of normal equations by the method of steepest descent related algorithms will be discussed.

9.2 THE METHOD OF STEEPEST DESCENT

This method provides an iterative procedure for obtaining the solution of the normal equations and is the basis for many currently used adaptive signal processing techniques. In practical adaptive signal processing systems, the method of the steepest descent itself is rarely used but its approximations are common.

If a marble were placed at a point on the mean square error (MSE) surface (Fig. 9.1) corresponding to the initial filter setting $W_N(0)$ and then released, it would roll down towards the bottom/minimum of the surface. It would initially roll in a direction, opposite to the direction of maximum rate of change of the surface. This direction is opposite to that of the gradient or, equivalently, this direction is that of "steepest descent". In the process of rolling towards the bottom of the MSE surface, the marble might overshoot or roll back and forth for a while (provided there is no change in the shape of the bowl), but will eventually settle at the *bottom of the bowl*, which corresponds to minimum MSE. This concept of rolling the marble towards the bottom of the MSE surface is *mathematically mimicked* by the method of steepest descent for solving the normal equations.

Consider Fig. 9.1, which shows MSE surface for $N = 2$. At the point $\left[W_N(0), \varepsilon(0)\right]$ on the MSE surface, it is described by directional derivatives of the surface at that point. The directional derivatives describe the rate of change of MSE surface with respect to $W_N \cdot \frac{\partial \varepsilon}{\partial W_1}, \frac{\partial \varepsilon}{\partial W_2}$ at $\varepsilon(0)$.

For $N = 2$,

$$\left[\frac{\partial \varepsilon}{\partial W_N}\right]_{\varepsilon=\varepsilon(0)} = \begin{bmatrix} \dfrac{\partial \varepsilon}{\partial W_1} \\ \dfrac{\partial \varepsilon}{\partial W_2} \end{bmatrix}_{\varepsilon=\varepsilon(0)}$$

The direction of steepest descent is opposite to the direction of maximum rate of change of the surface at the point of interest.

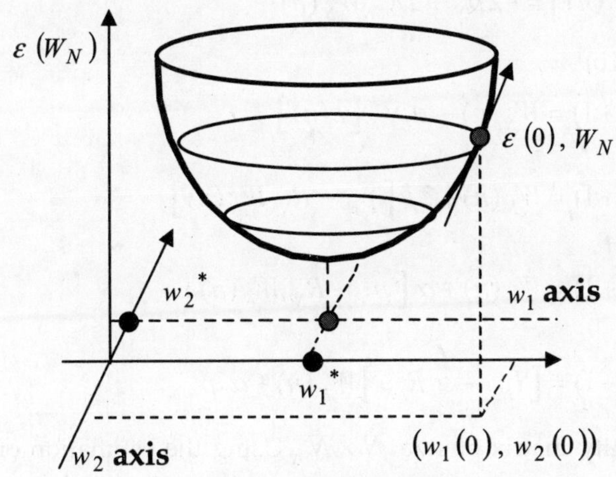

Fig. 9.1 MSE surface.

Given the marble position on the MSE surface at time n, then its position at time $(n+1)$ is achieved by moving along in a direction opposite to the gradient, along the surface. This may be mathematically treated as follows. If the initial position on the MSE surface is $\left[W_N(0), \varepsilon\,(0)\right]$, then the next position $\left[W_N(1), \varepsilon(1)\right]$ would be related to the initial position by

$$\begin{bmatrix} W_N(1) \\ \varepsilon(1) \end{bmatrix} = \begin{bmatrix} W_N(0) - \mu\, \nabla_w \left[\varepsilon(0)\right] \\ \varepsilon(0) - \delta \end{bmatrix} \tag{9.1a}$$

where μ is a constant and δ is the change in MSE which should be positive to reach the optimum value/ bottom of the error surface. To solve for the normal equations, consider

$$W_N(1) = W_N(0) - \mu\, \nabla_w \left[\varepsilon(0)\right]$$
$$W_N(2) = W_N(1) - \mu\, \nabla_w \left[\varepsilon(1)\right]$$
$$\vdots$$
$$W_N(n+1) = W_N(n) - \mu\, \nabla_w \left[\varepsilon(n)\right] \tag{9.1b}$$

The new weight vector is a function of old weight vector plus the correction term dependent upon the properties of the error surface at the old weight vector position. The MSE is given by

$$\varepsilon\,(W_N) = \sigma_d^2 - 2\, W_N^T\, p_N + W_N^T\, R_{NN}\, W_N \tag{9.1c}$$

Since

$$\varepsilon\,(n) = \varepsilon\left[W_N(n)\right]$$
$$\varepsilon\,(n) = \sigma_d^2 - 2\, W_N^T(n)\, p_N + W_N^T(n)\, R_{NN}\, W_N(n)$$

$$\nabla_W \left[\varepsilon(n) \right] = -2p_N + 2R_{NN} W_N(n) \tag{9.1d}$$

Substituting in Eqn.(9.1b)

$$W_N(n+1) = W_N(n) - \mu \nabla_w \left[\varepsilon(n) \right]$$

$$W_N(n+1) = W_N(n) + 2\mu \left[p_N - R_{NN} W_N(n) \right]$$

If $\alpha = 2\mu$,

$$W_N(n+1) = W_N(n) + \alpha \left[p_N - R_{NN} W_N(n) \right]$$

$$W_N(n+1) = \left[I_{NN} - \alpha R_{NN} \right] W_N(n) + \alpha p_N \tag{9.1e}$$

where I_{NN} is an identity matrix of size $N \times N$. Using the estimation of P_N and R_{NN}, the successive iterations could be performed to arrive at the solution.

9.2.1 Solution of Difference Equation

Eqn. (9.1e) can be evaluated directly.

At $n = 0$,

$$W_N(1) = \left[I_{NN} - \alpha R_{NN} \right] W_N(0) + \alpha P_N$$

At $n = 1$,

$$W_N(2) = \left[I_{NN} - \alpha R_{NN} \right] W_N(1) + \alpha P_N$$

$$W_N(2) = \left[I_{NN} - \alpha R_{NN} \right] \left\{ \left[I_{NN} - \alpha R_{NN} \right] W_N(0) + \alpha P_N \right\} + \alpha P_N$$

$$W_N(2) = \left[I_{NN} - \alpha R_{NN} \right]^2 W_N(0) + \alpha P_N + \alpha p_N \left[I_{NN} - \alpha R_{NN} \right]$$

At $n = 2$,

$$W_N(3) = \left[I_{NN} - \alpha R_{NN} \right] W_N(2) + \alpha P_N$$

$$= \left[I_{NN} - \alpha R_{NN} \right] \left\{ \left[I_{NN} - \alpha R_{NN} \right]^2 W_N(0) + \alpha P_N + \alpha P_N \left[I_{NN} - \alpha R_{NN} \right] \right\} + \alpha P_N$$

$$= \left[I_{NN} - \alpha R_{NN} \right]^3 W_N(0) + \alpha P_N + \alpha P_N \left[I_{NN} - \alpha R_{NN} \right] + \alpha P_N \left[I_{NN} - \alpha R_{NN} \right]^2$$

$$= \left[I_{NN} - \alpha R_{NN} \right]^3 W_N(0) + \alpha P_N \sum_{j=0}^{2} \alpha P_N \left[I_{NN} - \alpha R_{NN} \right]^j$$

In general,

$$W_N(n) = \left[I_{NN} - \alpha R_{NN} \right]^n W_N(0) + \alpha P_N \sum_{j=0}^{n-1} \alpha P_N \left[I_{NN} - \alpha R_{NN} \right]^j \tag{9.1f}$$

Given P_N and R_{NN}, the weight vector W_N can be computed at any iteration n. It involves matrix exponentials and running sum of matrix exponentials, hence it is very laborious computationally. Further, no insight is achieved into the convergence of W_N to W_N^*.

9.2.2 Simplifications using Linear Transformations

Consider Eqn.(9.1e)

$$W_N(n+1) = \left[I_{NN} - \alpha\, R_{NN} \right] W_N(n) + \alpha\, p_N$$

$$W_N(n+1) = W_N(n) + \alpha \left[P_N - R_{NN} W_N(n) \right] \tag{9.2}$$

Subtracting the optimum weight vector, W_N^* from both sides

$$W_N(n+1) - W_N^* = W_N(n) - W_N^* + \alpha \left[P_N - R_{NN} W_N(n) \right]$$

If

$$V_N(n) = W_N(n) - W_N^*$$

then,

$$V_N(n+1) = V_N(n) + \alpha \left[P_N - R_{NN} W_N(n) \right]$$

Adding and subtracting $\alpha\, R_{NN}\, W_N^*$

$$\begin{aligned} V_N(n+1) &= V_N(n) + \alpha\, P_N - \alpha\, R_{NN} W_N(n) + \alpha\, R_{NN} W_N^* - \alpha\, R_{NN} W_N^* \\ &= V_N(n) + \alpha\, P_N - \alpha\, R_{NN} \left[\, W_N(n) - W_N^* \,\right] - \alpha\, R_{NN} W_N^* \\ &= V_N(n) + \alpha \left[\, P_N - R_{NN} W_N^* \,\right] - \alpha\, R_{NN} V_N(n) \\ &= V_N(n) + \alpha[0] - \alpha\, R_{NN} V_N(n) \qquad \because R_{NN} W_N^* = P_N \end{aligned}$$

Therefore,

$$V_N(n+1) = \left[I_{NN} - \alpha\, R_{NN} \right] V_N(n) \tag{9.3}$$

and this is easier to evaluate.

For $n = 0$,

$$V_N(1) = \left[I_{NN} - \alpha\, R_{NN} \right] V_N(0)$$

For $n = 1$,

$$V_N(2) = [I_{NN} - \alpha R_{NN}] V_N(1) = [I_{NN} - \alpha R_{NN}]^2 V_N(0)$$

Therefore, the general solution will be

$$V_N(n) = [I_{NN} - \alpha R_{NN}]^n V_N(0) \tag{9.4a}$$

$$W_N(n) = W_N^* + [I_{NN} - \alpha R_{NN}]^n V_N(0) \tag{9.4b}$$

This is an improved version for determining the dynamic behavior of $W_N(n)$. However, it is not clear whether the second term in the RHS will vanish as $n \to \infty$. For $N = 2$ and at $n = 1$,

$$\begin{bmatrix} W_1(1) \\ W_2(1) \end{bmatrix} = \begin{bmatrix} W_1^* \\ W_2^* \end{bmatrix} + \left\{ \begin{bmatrix} 1 & 0 \\ 0 & 1 \end{bmatrix} - \alpha \begin{bmatrix} \phi_x(0) & \phi_x(1) \\ \phi_x(1) & \phi_x(0) \end{bmatrix} \right\} \begin{bmatrix} V_1(0) \\ V_2(0) \end{bmatrix}$$

$$\begin{bmatrix} W_1(1) \\ W_2(1) \end{bmatrix} = \begin{bmatrix} W_1^* \\ W_2^* \end{bmatrix} + \begin{bmatrix} 1 - \alpha\, \phi_x(0) & -\alpha\, \phi_x(1) \\ -\alpha\, \phi_x(1) & 1 - \alpha\, \phi_x(0) \end{bmatrix} \begin{bmatrix} V_1(0) \\ V_2(0) \end{bmatrix}$$

$$w_1(1) = w_1^* + [1 - \alpha\, \phi_x(0)] V_1(0) - \alpha\, \phi_x(1) V_2(0) \tag{9.5a}$$

$$w_2(1) = w_2^* + [1 - \alpha\, \phi_x(0)] V_2(0) - \alpha\, \phi_x(1)\, V_1(0) \tag{9.5b}$$

In Eqn. (9.5a), $w_1(1)$ is not only a function of $V_1(0)$ but also of $V_2(0)$. Similarly, in Eqn. (9.5b), $w_2(1)$ is not only a function of $V_2(0)$ but also of $V_1(0)$. Each weight value *affects the solution of the other weight due to the cross-coupling through* $V_1(0)$ and $V_2(0)$. This situation gets compounded for the general N length vector W_N, as $w_1(n)$ is a function of $w_2(n), w_3(n), \ldots, w_{N-1}(n)$, the weight $w_2(n)$ is a function of $w_1(n), w_3(n), \ldots, w_{N-1}(n)$, and so forth. This *cross-coupling severely prohibits a deeper insight into the dynamics* of $W_N(n)$. Further, it is not clear how $W_N(n)$ converges to $W_N^*(n)$ or if it converges at all. This difficulty is due *to* $[I_{NN} - \alpha R_{NN}]$ *being a nondiagonal matrix* and the question is, is it possible to diagonalize this term? To find out the answer to this question, it is necessary to know the requirements of the diagonalization.

9.3 ORTHONORMAL DIAGONALIZATION

The properties of the autocorrelation matrix were considered in chapter 7. But for the purpose of orthonormal diagonalization, additional properties are essential and they will be dealt presently.

9.3.1 Property: Eigenvectors of an autocorrelation/Hermitian Matrix are Orthogonal

Let $q_1, q_2, ..., q_M$ respectively, be the distinct eigenvectors corresponding to the eigenvalues $\lambda_1, \lambda_2, ..., \lambda_M$ of the $M \times M$ correlation matrix R_x. A set of vectors $\{q_1, q_2, ..., q_p\}$ is said to be orthogonal set if

$$\begin{aligned} q_i \cdot q_j &= 0, \quad i \neq j \\ &= K_{ij}, \quad i = j \end{aligned}$$

If $K_{ij} = 1$, the set is orthonormal set.

Let q_i and q_j denote any two eigenvectors of the correlation matrix R_x for its eigenvalues λ_i and λ_j. Then,

$$R_x q_i = \lambda_i \, q_i \quad \text{and} \quad R_x q_j = \lambda_j \, q_j \tag{9.6a,b}$$

Pre-multiplying Eqn.(9.6a) by q_j^H

$$q_j^H R_x q_i = \lambda_i \, q_j^H q_i \tag{9.7a}$$

Taking Hermitian transpose for, Eqn. (9.6b)

$$\left(R_x q_j\right)^H = \left(\lambda_j \, q_j\right)^H \quad \Rightarrow \quad q_j^H R_x^H = \lambda_j \, q_j^H$$

Since autocorrelation matrix is symmetric/Hermitian, i.e.,

$$R_x^H = R_x,$$

Therefore,

$$q_j^H R_x = \lambda_j \, q_j^H$$

Post multiplying the above equation by q_i,

$$q_j^H R_x q_i = \lambda_j \; q_j^H q_i \tag{9.7b}$$

Subtract Eqn. (9.7b) from Eqn. (9.7a),

$$q_j^H R_x q_i - q_j^H R_x q_i = (\lambda_i - \lambda_j)\, q_j^H q_i$$

$$(\lambda_i - \lambda_j)\, q_j^H q_i = 0$$

Since eigenvalues of the correlation matrix R_x are assumed to be distinct, $\lambda_i \neq \lambda_j$. Therefore,

$$q_j^H q_i = 0 \tag{9.7c}$$

That is, q_i and q_j, *the eigenvectors of the autocorrelation matrix are orthogonal to each other.* In the above treatment, for the orthogonality of eigenvectors of the autocorrelation matrix, the eigenvalues are assumed to be distinct. But such an assumption is not necessary and the orthogonality is valid even for the case of equal eigenvalues. For example, for the identity matrix,

$$I = \begin{bmatrix} 1 & 0 \\ 0 & 1 \end{bmatrix}$$

Its eigenvalues are same and are equal to unity. Any vector is an eigenvector of I and

$$q_1 = [1 \quad 0]^T, \quad q_2 = [0 \quad 1]^T$$

are one such set and they are orthogonal as

$$q_1^T q_2 = 0.$$

9.3.2 Property: If $Q = \{q_1, q_2, ..., q_M\}$ is Orthogonal, then it is Linearly Independent

If $Q = \{q_1, q_2, ..., q_M\}$ is not linearly independent, then

$$q_r = \{c_1 q_1 + c_2 q_2, ... + c_M q_M\} \tag{9.8a}$$

and c_i are found by taking the inner product of q_i with q_r,

$$q_i^T q_r = \{c_1 q_i^T q_1 + c_2 q_i^T q_2, ... + c_M q_i^T q_M\} \tag{9.8b}$$

Since $Q = \{q_1, q_2, ..., q_M\}$ is orthogonal,

$$q_i^T q_r = 0 \tag{9.8c}$$

$$q_i^T q_r = \{c_1 q_i^T q_1 + c_2 q_i^T q_2, ... + c_M q_i^T q_M\} = 0$$

and

$$q_i^T q_j = 0, \textit{ for any } i \neq j \tag{9.8d}$$

Therefore,

$$c_i q_i^T q_i = 0$$

But as

$$q_i^T q_i \neq 0,$$

therefore,

$$c_i = 0. \tag{9.8e}$$

This, is true for any i, therefore

$$q_r = \{c_1 q_1 + c_2 q_2, ... + c_M q_M\}$$

is valid only for $c_i = 0$, for any i.

Therefore, $Q = \{q_1, q_2, ..., q_M\}$ is linearly independent. That is, a set of orthogonal vectors is linearly independent.

9.3.3 Similarity Transformation

If A and B are $n \times n$ matrices, A is said to be similar to B, if there is an invertible matrix P such that

$$\mathrm{B} = \mathrm{P}^{-1}\mathrm{AP} \quad \text{or} \quad \mathrm{A} = \mathrm{PBP}^{-1} \tag{9.9a}$$

If $\mathrm{Q} = \mathrm{P}^{-1}$, then $\mathrm{A} = \mathrm{Q}^{-1}\mathrm{BQ}$ hence, B similar to A, i.e., A and B are said to be similar matrices. *Changing* A *to* B *by* $\mathrm{P}^{-1}\mathrm{AP}$, *is known as similarity transformation.*

Property: Similar matrices A *and* B *of dimension*($n \times n$), *have the same eigenvalues.*

If $\mathrm{B} = \mathrm{P}^{-1}\mathrm{AP}$ then

$$\mathrm{B} - \lambda\mathrm{I} = \mathrm{P}^{-1}\mathrm{AP} - \lambda\mathrm{P}^{-1}\mathrm{P} = \mathrm{P}^{-1}[\mathrm{A} - \lambda\mathrm{I}]\mathrm{P}$$

$$\det(\mathrm{B} - \lambda\mathrm{I}) = \det(\mathrm{P}^{-1}[\mathrm{A} - \lambda\mathrm{I}]\mathrm{P})$$

$$= \det(\mathrm{P}^{-1})\det([\mathrm{A} - \lambda\mathrm{I}])\det(\mathrm{P})$$

$$= \det(\mathrm{P}^{-1}\mathrm{P})\det([\mathrm{A} - \lambda\mathrm{I}])$$

$$= \det([\mathrm{A} - \lambda\mathrm{I}])$$

$$\det(\mathrm{B} - \lambda\mathrm{I}) = \det([\mathrm{A} - \lambda\mathrm{I}]) \tag{9.9b}$$

That is, the characteristic polynomials of A *and* B *are identical, hence their eigenvalues.*

9.3.4 Diagonalization Theorem

- *A square matrix* A *is said to be diagonalizable if* A *is similar to a diagonal matrix* D. That is, $A = PDP^{-1}$, for some invertible matrix P.

- A $n \times n$ matrix A is diagonalizable, if and only if A has n *linearly independent eigenvectors.*

- $A = PDP^{-1}$ with D a diagonal matrix is valid only if the columns of P are n linearly independent eigenvectors of A. In this case, the diagonal entries of D are eigenvalues of A corresponding to the eigenvector in P.

- As it has been shown that for a Hermitian / autocorrelation matrix, orthogonality of its eigenvectors implies their independence and also its eigenvectors are orthogonal, it is diagonizable.

Orthonormal matrices: A $m \times n$ *matrix* Q *has orthonormal columns if and only if*

$$Q^T Q = I \tag{9.10a}$$

Let $Q = \{q_1, q_2, q_3\}$, then

$$Q^T Q = \begin{bmatrix} q_1^T \\ q_2^T \\ q_3^T \end{bmatrix} \begin{bmatrix} q_1 & q_2 & q_3 \end{bmatrix} = \begin{bmatrix} q_1^T q_1 & q_1^T q_2 & q_1^T q_3 \\ q_2^T q_1 & q_2^T q_2 & q_2^T q_3 \\ q_3^T q_1 & q_3^T q_2 & q_3^T u_3 \end{bmatrix} = I \tag{9.10b}$$

That is,

$$q_1^T q_2 = q_2^T q_1 = 0,\ q_1^T q_3 = q_3^T q_1 = 0,\ q_2^T q_3 = q_3^T q_2 = 0 \tag{9.10c}$$

which is also the condition for orthogonality of Q. Further, columns U of all have unit length if $q_1^T q_1 = q_2^T q_2 = q_3^T q_3 = 1$, then

$$Q^T Q = I \text{ or } Q^{-1} = Q^{T} \tag{9.10d}$$

- A matrix A is *orthonormally diagonizable* if there is an *orthonormal matrix* P (i.e., $P^{-1} = P^T$) and a *diagonal matrix* D such that

$$A = PDP^{-1} = PDP^T \text{ or } P^T AP = D.$$

- To orthonormally diagonalize a $n \times n$ matrix it is required to find its n linearly independent orthonormal eigenvectors.

- If A is *orthogonally diagonizable,* then

$$A^T = \left(PDP^T\right)^T = \left(P^T\right)^T D^T P^T = PD^T P^T = PDP^T = A \qquad (9.10e)$$

That is, ***A is symmetric.***

A $m \times n$ matrix A *is orthogonally diagonizable only if* A *is symmetric.* Since R_{NN} is symmetric, *it is diagonizable.* It is possible to have an orthonormal P, as the eigenvectors of R_{NN} is orthonormal.

9.4 ORTHONORMAL DIAGONALIZATION OF $[I_{NN} - \alpha R_{NN}]$

From the above discussion since R_{NN} is symmetric, it is orthogonally diagonalizable. The transformation required to diagonalize A is given by $P^T AP = D$. Here, P is the matrix of eigenvectors of A and D is the diagonal matrix of eigenvalues of A. For the autocorrelation matrix R_{NN}, the matrix P and D are correspondingly represented by M_{NN} and Λ_{NN}, respectively for notational consistency. Hence, the diagonalization transformation is given by,

$$M_{NN}^T R_{NN} M_{NN} = \Lambda_{NN}$$

and

$$R_{NN} = M_{NN} \Lambda_{NN} M_{NN}^T \qquad (9.11a)$$

where

M_{NN} - Modal matrix whose columns are eigenvectors of R_{NN}

and

$$M_{NN} M_{NN}^T = M_{NN}^T M_{NN} = I \qquad (9.11b)$$

Λ_{NN} - $N \times N$ diagonal matrix of eigenvalues λ_i

Therefore Eqn. (9.1e),

$$\mathrm{W}_N(n+1) = \left[\mathrm{I}_{NN} - \alpha\,\mathrm{R}_{NN}\right]\ \mathrm{W}_N(n) + \alpha\,\mathrm{P}_N$$

$$= \left[\mathrm{I}_{NN} - \alpha\,\mathrm{M}_{NN}\Lambda_{NN}\mathrm{M}_{NN}^T\right]\mathrm{W}_N(n) + \alpha\,\mathrm{P}_N$$

$$= \left[\mathrm{M}_{NN}\mathrm{M}_{NN}^T - \alpha\,\mathrm{M}_{NN}\Lambda_{NN}\mathrm{M}_{NN}^T\right]\mathrm{W}_N(n) + \alpha\,\mathrm{P}_N$$

$$(\because\ \mathrm{M}_{NN}\mathrm{M}_{NN}^T = \mathrm{I}_{NN})$$

$$\mathrm{W}_N(n+1) = \mathrm{M}_{NN}\left[\mathrm{I}_{NN} - \alpha\,\Lambda_{NN}\right]\mathrm{M}_{NN}^T\mathrm{W}_N(n) + \alpha\,\mathrm{P}_N \tag{9.11c}$$

Let the uncoupled weight vector $\mathrm{W}_N'(n)$ and uncoupled crosscorrelation be defined as

$$\mathrm{W}_N'(n) = \mathrm{M}_{NN}^T\mathrm{W}_N(n),\quad \mathrm{P}_N' = \mathrm{M}_{NN}^T\mathrm{P}_N$$

On pre-multiplying Eqn. (9.11c) by M_{NN}^T

$$\mathrm{W}_N'(n+1) = \mathrm{M}_{NN}^T\mathrm{M}_{NN}\left[\mathrm{I}_{NN} - \alpha\Lambda_{NN}\right]\mathrm{M}_{NN}^T\mathrm{W}_N(n) + \alpha\mathrm{M}_{NN}^T\mathrm{P}_N$$

Since

$$\mathrm{M}_{NN}^T\mathrm{M}_{NN} = \mathrm{I}_{NN},$$

$$\mathrm{W}_N'(n+1) = \left[\mathrm{I}_{NN} - \alpha\,\Lambda_{NN}\right]\mathrm{W}_N'(n) + \alpha\,\mathrm{P}_N', \tag{9.12a}$$

Since now, $\left[\mathrm{I}_{NN} - \alpha\Lambda_{NN}\right]$ is diagonal, it can be expanded as

$$\begin{bmatrix} w_1'(n+1) \\ w_2'(n+1) \\ \vdots \\ w_N'(n+1) \end{bmatrix} = \begin{bmatrix} 1-\alpha\,\lambda_1 & 0 & \cdots & 0 \\ 0 & 1-\alpha\,\lambda_2 & \cdots & 0 \\ \vdots & \vdots & \ddots & \vdots \\ 0 & 0 & \cdots & 1-\alpha\,\lambda_N \end{bmatrix}\begin{bmatrix} w_1'(n) \\ w_2'(n) \\ \vdots \\ w_N'(n) \end{bmatrix} + \begin{bmatrix} \alpha\,P_1' \\ \alpha\,P_2' \\ \vdots \\ \alpha\,P_N' \end{bmatrix} \tag{9.12b}$$

$$w_i'(n+1) = \left(1-\alpha\,\lambda_i\right)w_i'(n) + \alpha\,P_i' \quad \text{for } 0 \le i \le N \tag{9.12c}$$

Each $w_i'(n+1)$ is only a function of $w_i'(n)$ and not a function of any other $w_j'(n),\ j \neq i$. This is the property needed to write the above equation as a set of N *scalar uncoupled equations*. Since this same scalar difference equation holds for all $w_i'(n)$, it can be solved for a general case and the entire set of solutions are obtained for $0 \le i \le N$. This equation may be evaluated for successive values of n and a general pattern may be recognized.

Consider Eqn. (9.12c),

$$w_i'(n+1) = \left(1-\alpha\,\lambda_i\right)w_i'(n) + \alpha\,P_i'$$

$\alpha\,P_i'$ is a constant.

For $n = 0$,

$$w_i'(1) = \gamma_i\,w_i'(0) + \alpha\,P_i', \qquad \gamma_i = \left(1-\alpha\,\lambda_i\right)$$

For $n = 1$, $\quad w_i'(2) = \gamma_i w_i'(1) + \alpha\,P_i'$

$$= \gamma_i\left[\gamma_i w_i'(0) + \alpha\,P_i'\right] + \alpha\,P_i'$$

$$= \gamma_i^{\,2} w_i'(0) + \alpha\,P_i'\sum_{j=0}^{1}\gamma_i^{\,j}$$

In general,

$$w_i'(n) = \gamma_i^{\,n} w_i'(0) + \alpha\,P_i'\left[\sum_{j=0}^{n-1}\gamma_i^{\,j}\right] \tag{9.13a}$$

From algebra, for $|\gamma_i| < 1$,

$$\sum_{j=0}^{n-1}\gamma_i^j = \frac{1-\gamma_i^n}{1-\gamma_i}, \; 0 \le i \le N$$

Therefore,

$$w_i'(n) = \gamma_i^n w_i'(0) + \alpha\,P_i'\left[\frac{1-\gamma_i^n}{1-\gamma}\right] \tag{9.13b}$$

Solution by z-transform: The same solution for the Eqn. (9.12c),

$$w_i'(n+1) = \gamma_i w_i'(n) + \alpha\,P_i'$$

can be achieved by using z-transform. Applying z–transform on both sides,

$$Z\left[w_i'(n+1)\right] = \gamma_i\,Z\left[w_i'(n)\right] + Z\left[\alpha\,P_i'\right] \tag{9.13c}$$

The L.H.S of Eqn. (9.13c) is given by

$$zW_i'(z) = zW_i'(z) - zw_i'(0),$$

where $w_i'(0)$ is the initial weight $w_i'(n)$ and $Z\left[w_i'(n)\right] = W_i'(z)$. Further the term $Z\left[\alpha\,P_i'\right]$ on the R.H.S of Eqn. (9.13c) is

$$Z\left[\alpha\, P_i'\right] = \alpha\, P_i' \sum_{i=0}^{\infty} z^{-i} = \frac{\alpha\, P_i'}{1-z^{-1}}$$

Therefore, Eqn. (9.13c) becomes

$$zW_i'(z) - zw_i'(0) = \gamma_i W_i'(z) + \alpha\, P_i' / \left(1-z^{-1}\right)$$

$$zW_i'(z)\left[1-\gamma_i z^{-1}\right] = zw_i'(0) + \alpha\, P_i' / \left(1-z^{-1}\right)$$

$$W_i'(z) = zw_i'(0) / z\left[1-\gamma_i z^{-1}\right] + \alpha\, P_i' / z\left[1-\gamma_i z^{-1}\right]\left(1-z^{-1}\right)$$

$$W_i'(z) = \frac{w_i'(0)}{\left[1-\gamma_i z^{-1}\right]} + \frac{\alpha\, P_i' z^{-1}}{\left[1-\gamma_i z^{-1}\right]\left(1-z^{-1}\right)}$$

On taking inverse z-transform,

$$w_i'(n) = \gamma_i^n w_i'(0) + \alpha\, P_i'\left[\frac{1-\gamma_i^n}{1-\gamma_i}\right]$$

As expected, this solution is same as that obtained by direct evaluation Eqn. (9.13b).

9.5 CONVERGENCE CRITERION FOR THE METHOD OF DESCENT STEEPEST

Consider Eqn. (9.13b),

$$w_i'(n) = \gamma_i^n w_i'(0) + \alpha\, P_i'\left[\frac{1-\gamma_i^n}{1-\gamma_i}\right], \quad |\gamma_i| < 1$$

Since γ_i is raised to an increasing positive power as time increases, i.e., $n \to \infty$, the requirement for a bounded solution for $w_i'(n)$ is

$$|\gamma_i| < 1 \qquad \text{for} \qquad 0 \le i \le N \tag{9.13d}$$

$$|1-\alpha\,\lambda_i| < 1$$

i.e., $-1 < (1-\alpha\,\lambda_i) < 1$

i.e., $-1 < (1-\alpha\,\lambda_i)$ and $(1-\alpha\,\lambda_i) < 1$

i.e., $\alpha < \dfrac{2}{\lambda_i}$ and $0 < \alpha$

$$0 < \alpha < \frac{2}{\lambda_i} \quad \text{for} \quad 0 \le i \le N \tag{9.13e}$$

The lowest upper bound for α is determined by the maximum value of λ_i, i.e., $\lambda_{\max}$, the maximum eigenvalue of the autocorrelation matrix R_{NN}. Hence, the maximum range of α to ensure convergence is

$$0 < \alpha < \frac{2}{\lambda_{\max}} \tag{9.14}$$

For α with this range, $|\gamma_i| < 1$ and weight vector $w_i'(n)$ is given by (Eqn. (9.13b))

$$w_i'(n) = \gamma_i^n w_i'(0) + \alpha\, P_i' \left[\frac{1-\gamma_i^n}{1-\gamma_i}\right]$$

For convergence, the weights $w_i'(n)$ should reach $w_i'^*$ as the number of samples increase to ∞.

$$\lim_{n\to\infty} w_i'(n) = \lim_{n\to\infty} \gamma_i^n w_i'(0) + \alpha\, P_i' \lim_{n\to\infty} \left[\frac{1-\gamma_i^n}{1-\gamma_i}\right]$$

$$\lim_{n\to\infty} w_i'(n) = 0 + \alpha\, P_i' \left[\frac{1}{1-\gamma_i}\right] = \alpha\, P_i' \left[\frac{1}{1-(1-\alpha\,\lambda_i)}\right] = \frac{P_i'}{\alpha\,\lambda_i}$$

$$\lim_{n\to\infty} w_i'(n) = \frac{P_i'}{\lambda_i} \tag{9.15a}$$

Consider the uncoupled optimum weight vector,

$$W_N'^* = M_{NN}^T W_N^* = M_{NN}^T R_{NN}^{-1} P_N = M_{NN}^T \left(M_{NN}\, \Lambda_{NN}^{-1}\, M_{NN}^T \right) P_N$$

$$\text{as} \quad R_{NN}^{-1} = \left(M_{NN}^T\, \Lambda_{NN}\, M_{NN} \right)^{-1} = M_{NN}\, \Lambda_{NN}^{-1}\, M_{NN}^T$$

$$W_N'^*(n) = \Lambda_{NN}^{-1}\, M_{NN}^T\, P_N = \Lambda_{NN}^{-1} P_N',$$

$$\text{as} \quad P_N' = M_{NN}^T\, P_N \quad I_N = M_{NN}^T\, M_{NN}$$

$$W_N'^*(n) = \Lambda_{NN}^{-1} P_N' \tag{9.15b}$$

This is similar to $W_N^*(n) = R_{NN}^{-1} P_N$ except it is in terms of uncoupled weight vector. Further, as Λ_{NN} is a diagonal matrix, its inverse is also diagonal, with elements $\left(\Lambda_{NN}^{-1}\right)_{ii} = 1/\lambda_i$. Therefore, each component of weight vector $W_N'^*(n)$ is

$$W_N'^* = \begin{bmatrix} 1/\lambda_1 & \cdots & 0 \\ \vdots & \ddots & \vdots \\ 0 & \cdots & 1/\lambda_N \end{bmatrix} \begin{bmatrix} P_1' \\ \vdots \\ P_N' \end{bmatrix}$$

For individual weights,

$$w_i'^* = \frac{P_i'}{\lambda_i} \tag{9.15c}$$

Therefore, from (9.15a) and (9.15c),

$$\lim_{n\to\infty} w_i'(n) = w_i'^*, \tag{9.15d}$$

i.e., weights by steepest descent converge to same set of weights, obtained by solution of normal equations, if $0 < \alpha < (2/\lambda_{\max})$.

9.6 DYNAMICS OF WEIGHT CONVERGENCE PROCESS

The solution for the weight vector (Eqn. (9.13b) is

$$w_i'(n) = \gamma_i^n w_i'(0) + \alpha\, P_i' \left[\frac{1-\gamma_i^n}{1-\gamma}\right]$$

Subtracting $w_i'^*$ from both sides,

$$w_i'(n) - w_i'^* = \gamma_i^n w_i'(0) + \alpha\, P_i' \left[\frac{1-\gamma_i^n}{1-\gamma_i}\right] - w_i'^*$$

Using the Eqn. (9.15c) $w_i'^* = P_i'/\lambda_i$,

$$v_i'(n) = \gamma_i^n w_i'(0) + \frac{\alpha\, P_i'}{1-\gamma_i} - \frac{\alpha\, P_i' \gamma_i^n}{1-\gamma_i} - \frac{P_i'}{\lambda_i}$$

$$= \gamma_i^n \left[w_i'(0) - \frac{\alpha\, P_i'}{1-\gamma_i} \right] + \frac{\alpha\, P_i'}{1-\gamma_i} - \frac{P_i'}{\lambda_i}$$

Since $\gamma_i = 1 - \alpha\lambda_i$,

$$v_i'(n) = \gamma_i^n \left[w_i'(0) - \frac{\alpha P_i'}{1-(1-\alpha\lambda_i)} \right] + \frac{\alpha P_i'}{1-(1-\alpha\lambda_i)} - \frac{P_i'}{\lambda_i}$$

$$v_i'(n) = \gamma_i^n \left[w_i'(0) - \frac{\alpha P_i'}{\alpha\lambda_i)} \right] + \frac{\alpha P_i'}{\alpha\lambda_i} - \frac{P_i'}{\lambda_i}$$

$$v_i'(n) = (1-\alpha\lambda_i)^n \left[w_i'(0) - \frac{P_i'}{\lambda_i} \right]$$

Further, if

$$v_i'(0) = w_i'(0) - \frac{P_i'}{\lambda_i}$$

then $v_i'(0)$ is a constant, once the initial weight $w_i'(0)$ is chosen. Therefore,

$$v_i'(n) = v_i'(0)(1-\alpha\lambda_i)^n \tag{9.15e}$$

Since the eigenvalues of R_{NN} are positive and real, $v_i'(n)$ decays exponentially from any $v_i'(0)$ as long as α is in the range, $0 < \alpha < 2/\lambda_{\max}$ and the rate of decay is determined by step size α. This implies that $W_N(n)$ also converges to $W_N^*(n)$ in the same manner as $V_N(n) = W_N(n) - W_N^*(n)$. As $v_i'(n)$ decays to zero irrespective of its initial value $v_i'(0)$, irrespective of the choice of starting value $W_N(0)$, $W_N(n)$ will converge to $W_N^*(n)$.

9.7 ILLUSTRATION OF CONVERGENCE

To illustrate the convergence behavior of the method of steepest descent, the a weight vector with two weights is considered. For this, for the input $x(n)$ and output $d(n)$, the autocorrelation and cross-correlation of the estimated values used are

$$\hat{\phi}_x(0) = 1,\ \hat{\phi}_x(1) = 0.9,\ \hat{\phi}_{xd}(0) = 0.9,\ \hat{\phi}_{xd}(1) = 0.6.$$

Therefore, the autocorrelation matrix $\hat{R}_{NN}$ and the cross-correlation vector $\hat{P}_N$ are given by

$$\hat{R}_{NN} = \begin{bmatrix} 1 & 0.9 \\ 0.9 & 1 \end{bmatrix} \quad \text{and} \quad \hat{P}_N = \begin{bmatrix} 0.9 \\ 0.6 \end{bmatrix}$$

The eigenvalues of $\hat{R}_{NN}$ are found by solving $|R_{NN} - \lambda I| = 0$ and are

$$[\lambda_1 \quad \lambda_2] = [0.1 \quad 1.9]$$

and the eigenvectors are found by $[R_{NN} - \lambda I]m = 0$ and the normalized eigenvectors are

$$\mathrm{m}_1 = \frac{1}{\sqrt{2}}[1 \quad 1]^T \text{ and } \mathrm{m}_2 = \frac{1}{\sqrt{2}}[1 \quad -1]^T.$$

For the stability of steepest descent, the step size α should be less than $(2/\lambda_{\max})$. As $\lambda_{\max} = 1.9$, α should be less than $(2/1.9 = 1.0526)$ Hence, the α value is chosen as unity and the initial value of the weights as zero. That is, $\alpha = 1$ and $w_0(0) = 0, w_1(0) = 0$. The weight vector solution is

$$w_i'(n) = \gamma_i^n w_i'(0) + \alpha\, P_i'\left[\frac{1-\gamma_i^n}{1-\gamma}\right]$$

As $W_N(0) = 0$, the term $\gamma_i^n w_i'(0)$ becomes zero. Therefore,

$$w_i'(n) = \frac{P_i'}{\lambda_i}\left[1-(1-\alpha\,\lambda_i)^n\right]$$

The modal matrix is given by

$$M_{NN} = \frac{1}{\sqrt{2}}\begin{bmatrix} 1 & 1 \\ 1 & -1 \end{bmatrix} \qquad M_{NN}^T = \frac{1}{\sqrt{2}}\begin{bmatrix} 1 & 1 \\ 1 & -1 \end{bmatrix}$$

Further, as $P_N' = M_{NN}^T P_N$,

$$P_N' = \frac{1}{\sqrt{2}}\begin{bmatrix} 1 & 1 \\ 1 & -1 \end{bmatrix}\begin{bmatrix} 0.9 \\ 0.6 \end{bmatrix} = \frac{1}{\sqrt{2}}\begin{bmatrix} 1.5 \\ -0.3 \end{bmatrix}$$

Therefore,

$$w_1'(n) = \frac{1.5}{\sqrt{2}\times 1.9}\left[1-(-0.9)^n\right] \quad \text{and} \quad w_2'(n) = \frac{-0.3}{\sqrt{2}\times 0.1}\left[1-(-0.9)^n\right]$$

Since

$$W_N'(n) = M_{NN}^T W_N(n)$$

$$W_N(n) = \left(M_{NN}^T\right)^{-1} W_N'(n) = M_{NN} W_N'(n)$$

$$\begin{bmatrix} w_1(n) \\ w_2(n) \end{bmatrix} = W_N(n) = \frac{1}{\sqrt{2}}\begin{bmatrix} 1 & 1 \\ 1 & -1 \end{bmatrix}\begin{bmatrix} w_1'(n) \\ w_2'(n) \end{bmatrix} = \frac{1}{\sqrt{2}}\begin{bmatrix} w_1'(n) + w_2'(n) \\ w_1'(n) - w_2'(n) \end{bmatrix}$$

$$w_1(n) = \frac{1}{\sqrt{2}}\left\{\frac{1.5}{\sqrt{2}\times 1.9}\left[1-(-0.9)^n\right]+\frac{-0.3}{\sqrt{2}\times 0.1}\left[1-(0.9)^n\right]\right\}$$

$$w_1(n) = \frac{1.5}{2\times 1.9}\left[1-(-0.9)^n\right]+\frac{-0.3}{2\times 0.1}\left[1-(0.9)^n\right]$$

$$w_1(n) = -1.1053 - 0.3947\ (-0.9)^n + 1.5\ (0.9)^n$$

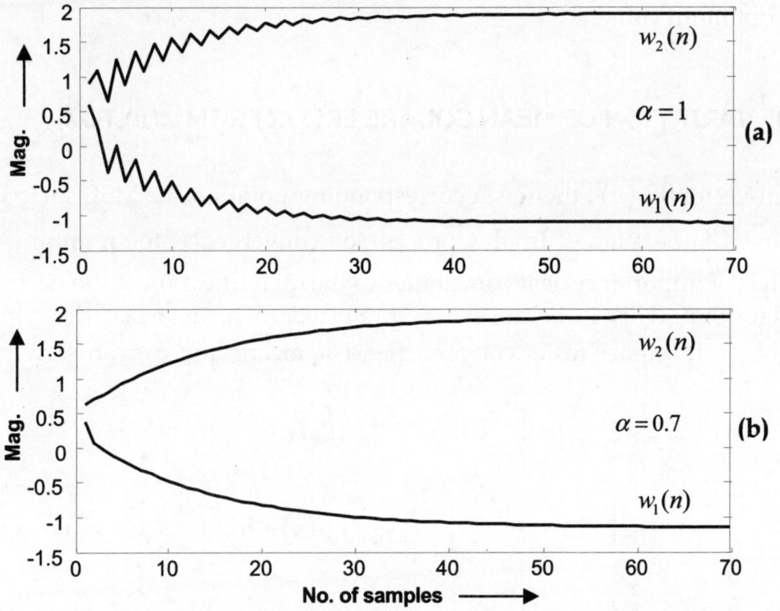

Fig. 9.2 (a, b) Convergence of the adaptive filter weights for different stepsize values.

Similarly,

$$w_2(n) = 1.8947 - 0.3947\ (-0.9)^n - 1.5\ (0.9)^n$$

Therefore, the weight vector solution for $\alpha = 1$, is

$$w_1(n) = -1.1053 - 0.3947\ (-0.9)^n + 1.5\ (0.9)^n$$

$$w_2(n) = 1.8947 - 0.3947\ (-0.9)^n - 1.5\ (0.9)^n$$

For $\alpha = 0.7$, the weight vector solution is

$$w_1(n) = -1.1053 - 0.3947\ (0.05)^n + 1.5\ (0.95)^n$$

$$w_2(n) = 1.8947 - 0.3947\ (0.05)^n - 1.5\ (0.95)^n$$

The dynamics of convergence is shown in Fig. 9.2. The convergence is initially little oscillatory but the oscillations will die out with increase in iterations and finally converge to $[w_1(n) = -1.1053,\ w_2(n) = 1.8947]$. This is nothing but the optimal solution obtained by matrix inversion. That is, weights will not converge monotonically to the optimal solution. With $\alpha = 0.7$, the weight trajectories are smoother than with $\alpha = 1.0$. But it takes longer to converge to the optimum value.

9.8 NATURE OF VARIATION OF MEAN SQUARE ERROR FROM $\varepsilon(n)$ TO $\varepsilon_{\min}$

For every weight vector $W_N(n)$, there is a corresponding point on the MSE surface, i.e., at a height $\varepsilon(n)$ above W_N - plane. In the process of convergence to optimum value, its variation pattern is of importance as it contributes to the quantification of the performance of the adaptive algorithm. This is because it is preferred to have a fast decaying pattern from $\varepsilon(n)$ to $\varepsilon_{\min}$ as it implies faster convergence. The mean square error is given by (Eqn. (9.1c))

$$\varepsilon(n) = \sigma_d^2 - 2W_N^T(n)P_N + W_N^T(n)R_{NN}W_N(n)$$

$$= \sigma_d^2 - 2\left[V_N(n) + W_N^*\right]^T P_N + \left[V_N(n) + W_N^*\right]^T R_{NN}\left[V_N(n) + W_N^*\right]$$

$$\left(\because\quad V_N(n) = W_N(n) - W_N^*\right)$$

$$= \sigma_d^2 - 2V_N^T(n)P_N - 2\underline{W_N^{*T} P_N} + V_N^T(n)R_{NN}V_N(n)$$
$$+ W_N^{*T} R_{NN}V_N(n) + V_N^T(n)\underbrace{R_{NN}W_N^*}_{=(P_N)} + W_N^{*T}\underbrace{R_{NN}W_N^*}_{=(P_N)}$$

$$= \sigma_d^2 - 2V_N^T(n)P_N - W_N^{*T} P_N + V_N^T(n)R_{NN}V_N(n) + (\underbrace{R_{NN}W_N^*}_{=(P_N)})^T V_N(n) + V_N^T(n)P_N$$

$$= \sigma_d^2 - 2V_N^T(n)P_N - W_N^{*T} P_N + V_N^T(n)R_{NN}V_N(n) + \underline{P_N^T V_N(n) + V_N^T(n)P_N}$$

$$= \underbrace{\sigma_d^2 - W_N^{*T} P_N}_{=\varepsilon_{\min}} + V_N^T(n)R_{NN}V_N(n)$$

$$\varepsilon(n) = \varepsilon_{\min} + V_N^T(n)R_{NN}V_N(n) \tag{9.16a}$$

Substituting for R_{NN} as

$$R_{NN} = M_{NN}\Lambda_{NN}M_{NN}^T$$

$$\varepsilon(n) = \varepsilon_{\min} + V_N^T(n)\left(M_{NN}\Lambda_{NN}M_{NN}^T\right)V_N(n)$$

$$= \varepsilon_{\min} + \left(M_{NN}^T V_N(n)\right)^T \Lambda_{NN}\, M_{NN}^T\, V_N(n)$$

$$\varepsilon(n) = \varepsilon_{\min} + V'^T(n)\Lambda_{NN}\, V'(n) \tag{9.16b}$$

At convergence since the weight vector $W_N(n)$ approaches W_N^*, the difference weight vector $V_N(n)$ approaches zero. Therefore, the uncoupled difference weight vector $V'(n)$ also approaches zero at convergence. Therefore, in the expression for $\varepsilon(n)$ (Eqn. 9.16b), the second term containing $V'(n)$ vanishes as $n \to \infty$. Hence, error for the method of descent, the steady state mean square $\varepsilon(n)$, converges to $\varepsilon_{\min}$ which is achieved by the normal equations. This only ensures that the mean square error will converge to the optimal solution. However, the nature in which it has been achieved is of importance and is yet to be considered.

From the expression for mean square error $\varepsilon(n)$ (Eqn. (9.16b)),

$$\varepsilon(n) = \varepsilon_{\min} + V'^T(n)\Lambda_{NN}\, V'(n)$$

it is clear that it is a function of $V'(n)$ and hence the variation of $\varepsilon(n)$ with respect to time is determined by the variation of $V'(n)$. The variation of $v_i'(n)$ is given by (Eqn. (9.15e))

$$v_i'(n) = v_i'(0)\left[1 - \alpha\,\lambda_i\right]^n$$

$$v_i'(n) = v_i'(0)\gamma_i^n \qquad \gamma_i = 1 - \alpha\,\lambda_i$$

Considering $V'^T(n)\Lambda_{NN}V'(n)$ for $N = 2$ and substituting for $v_i'(n)$, $i = 1, 2$

$$V_2'^T(n)\Lambda_{22}V_2'(n) = \begin{bmatrix} v_1'(0)\gamma_1^n & v_2'(0)\gamma_2^n \end{bmatrix}\begin{bmatrix} \lambda_1 & 0 \\ 0 & \lambda_2 \end{bmatrix}\begin{bmatrix} v_1'(0)\gamma_1^n \\ v_2'(0)\gamma_2^n \end{bmatrix} = \sum_{i=1}^{2} \lambda_i v_i'^2(0)\gamma_i^{2n}$$

Therefore,

$$\varepsilon(n) = \varepsilon_{\min} + \sum_{i=1}^{N} \lambda_i v_i'^2(0)\gamma_i^{2n} \tag{9.16c}$$

Since $\lambda_i v_i'^2(0)$ is a constant, the convergence pattern of $\varepsilon(n)$ is governed by γ_i^{2n} and

$$\lim_{n\to\infty} \gamma_i^{2n} = \lim_{n\to\infty} (1 - \alpha\,\lambda_i)^{2n} = 0 \tag{9.16d}$$

That is, γ_i^{2n} decays to zero exponentially as long as $|\gamma_i| < 1$ or $0 < \alpha < 2/\lambda_{max}$. Hence, for the method of steepest descent, the MSE $\varepsilon(n)$ *converges exponentially to* ε_{min} *as* n *increases.*

9.8.1 Time Constant

The variation of uncoupled difference weight is given by (Eqn. (9.15e))

$$v_i'(n) = v_i'(0)(1-\alpha\lambda_i)^n$$

Here, once the initial weight $w_i'(0)$ is chosen,

$$v_i'(0) = w_i'(0) - \frac{P_i'}{\lambda_i}$$

is a constant. Since the *eigenvalues of* R_{NN} *are positive and real,* $v_i'(n)$ *decays exponentially* as long as $(1-\alpha\lambda_i) < 1$ or $0 < \alpha < 2/\lambda_{max}$. Time constant τ_i is the time taken for a particular mode (λ_i) to decay to $1/e$ of its initial value and is given by,

$$\frac{v_i'(n)}{v_i'(0)} = \frac{1}{e} = (1-\alpha\lambda_i)^{\tau_i}, \quad \ln(1/e) = \tau_i \ln(1-\alpha\lambda_i),$$

$$\tau_i = -\frac{1}{\ln(1-\alpha\lambda_i)} \tag{9.17a}$$

For $\alpha << 1$, $l\alpha\,\lambda_i << 1$. Further as

$$\ln(1+x)) \approx x, \quad for\ x << 1,$$

$$\tau_i = -\frac{1}{-\alpha\lambda_i}$$

$$\tau_i = \frac{1}{\alpha\lambda_i} \tag{9.17b}$$

That is, smaller the step size parameter α, slower will be the decay rate hence, slower will be the convergence.

The maximum convergence rate possible is decided by the largest time constant, and from Eqn. (9.17b), it corresponds to minimum eigenvalue λ_{min} with α *near its upper limit.*

That is, for, $\lambda_i = \lambda_{\min}$,

$$\tau_{\lambda_{\min}} = \frac{1}{\alpha\,\lambda_{\min}}$$

and for α at its upper limit, $\alpha = 2/\lambda_{\max}$. Therefore,

$$\tau_{\lambda_{\min}} = \frac{1}{\lambda_{\min}\left(\dfrac{2}{\lambda_{\max}}\right)}$$

$$\tau_{\lambda_{\min}} = \frac{1}{2}\left(\frac{\lambda_{\max}}{\lambda_{\min}}\right) \tag{9.17c}$$

The convergence time is decided by the *ratio of maximum eigenvalue to the minimum eigenvalue or the eigenvalue spread and is inversely proportional to it*. The parameter α is chosen to ensure convergence.

9.8.2 Relationship between Eigenvalues and PSD

In practice, computation of $\lambda_{\max}$ and $\lambda_{\min}$ of R_{NN} or the ratio $\lambda_{\max}/\lambda_{\min}$ is not easy. However, $\lambda_{\max}$ and $\lambda_{\min}$ are related to the power spectral density, $S_x(\omega)$ of the signal. The relation between the eigenvalues and the autocorrelation matrix through the quadratic form has already been derived (Chapter 7, Section 7.7) as

$$\lambda_i = \frac{q_i^H R_x q_i}{q_i^H q_i}, \qquad i = 1,2,\ldots,N \tag{9.18a}$$

Expanding the numerator, $q_i^H R_x q_i$

$$q_i^H R_x q_i = \sum_{k=1}^{N}\sum_{l=1}^{N} q_{ik}^{*}\, r(l-k)\, q_{il} \tag{9.18b}$$

Here q_{ik}^{*} is the kth element of the row vector q_i^H, $r(l-k)$ is the $(l,k)^{th}$ element of the matrix R_{NN} and q_{il} is the lth element of the row vector q_i.

Using the relation between power spectral density $S_x(\omega)$ and autocorrelation $r(l-k)$ of a process $x(n)$,

$$r(l-k) = \frac{1}{2\pi}\int_{-\pi}^{\pi} S(\omega)\, e^{j\omega(l-k)}\, d\omega$$

$$q_i^H R_x q_i = \sum_{k=1}^{M}\sum_{l=1}^{M} q_{ik}^* q_{il} \frac{1}{2\pi}\int_{-\pi}^{\pi} S(\omega) e^{j\omega(l-k)} d\omega$$

$$q_i^H R_x q_i = \frac{1}{2\pi}\int_{-\pi}^{\pi} S(\omega) \sum_{k=1}^{M}\sum_{l=1}^{M} q_{ik}^* q_{il} e^{j\omega(l-k)} d\omega \tag{9.18c}$$

Let the DFT of sequence $q_{1k}^*, q_{2k}^*, \ldots, q_{Mk}^*$ be

$$Q_i'(e^{j\omega}) = \sum_{k=1}^{M} q_{ik}^*\, e^{-j\omega k}$$

Substituting in Eqn. (9.18c),

$$q_i^H R_x q_i = = \frac{1}{2\pi}\int_{-\pi}^{\pi} S(\omega) \left| Q_i'(e^{j\omega}) \right|^2 d\omega \tag{9.18d}$$

On similar lines,

$$q_i^H q_i = \frac{1}{2\pi}\int_{-\pi}^{\pi} \sum_{k=1}^{M}\sum_{l=1}^{M} q_{ik}^*\, q_{il}\, e^{j\omega(l-k)}\, d\omega$$

$$q_i^H q_i = \frac{1}{2\pi}\int_{-\pi}^{\pi} \left| Q_i'(e^{j\omega}) \right|^2 d\omega \tag{9.18e}$$

Therefore,

$$\lambda_i = \frac{q_i^H R_x q_i}{q_i^H q_i} = \frac{\dfrac{1}{2\pi}\displaystyle\int_{-\pi}^{\pi} S(\omega) \left| Q_i'(e^{j\omega}) \right|^2 d\omega}{\dfrac{1}{2\pi}\displaystyle\int_{-\pi}^{\pi} \left| Q_i'(e^{j\omega}) \right|^2 d\omega} \tag{9.18f}$$

If $S_{\max}$ and $S_{\min}$ are absolute maximum and minimum values of power spectral density $S(\omega)$, then

$$\int_{-\pi}^{\pi} S(\omega) \left| Q_i'(e^{j\omega}) \right|^2 d\omega \geq S_{\min} \int_{-\pi}^{\pi} \left| Q_i'(e^{j\omega}) \right|^2 d\omega \tag{9.18g}$$

$$\int_{-\pi}^{\pi} S(\omega)\left| Q_i'(e^{j\omega})\right|^2 d\omega \le S_{\max}\int_{-\pi}^{\pi}\left| Q_i'(e^{j\omega})\right|^2 d\omega \tag{9.18h}$$

Therefore,

$$\frac{\frac{1}{2\pi}\int_{-\pi}^{\pi} S(\omega)\left| Q_i'(e^{j\omega})\right|^2 d\omega}{\frac{1}{2\pi}\int_{-\pi}^{\pi}\left| Q_i'(e^{j\omega})\right|^2 d\omega} \ge S_{\min} \text{ and } \frac{\frac{1}{2\pi}\int_{-\pi}^{\pi} S(\omega)\left| Q_i'(e^{j\omega})\right|^2 d\omega}{\frac{1}{2\pi}\int_{-\pi}^{\pi}\left| Q_i'(e^{j\omega})\right|^2 d\omega} \le S_{\max}$$

From Eqn. (9.18f),

$$\lambda_i \ge S_{\min} \quad \text{and} \quad \lambda_i \le S_{\max}$$

These are true for any λ_i therefore,

$$\lambda_{\min} \ge S_{\min} \quad \text{and} \quad \lambda_{\max} \le S_{\max}$$

As any λ_i lies between $\lambda_{\min}$ and $\lambda_{\max}$,

$$S_{\min} \le \lambda_i \le S_{\max}$$

and the eigenvalue spread $\chi(R)$ is

$$\chi(R) = \left(\frac{\lambda_{\max}}{\lambda_{\min}}\right) \le \left(\frac{S_{\max}}{S_{\min}}\right)$$

That is a spectrum with wide dynamic range has a large eigenvalue spread and as speech signal has large range of spectral values, forms an example for a large eigenvalue spread. The $\lambda_{\max}$ and $\lambda_{\min}$ of R_{NN} cannot be computed easily. But as the *ratio of* $S_{\max}$ *to* $S_{\min}$ *forms the upper bound for the eigenvalue spread,* power spectral density or spectrum can be used as a measure for eigenvalue spread of the signal autocorrelation matrix.
Since from the relation (Eqn. (9.17c)),

$$\tau_{\lambda_{\min}} = \frac{1}{2}\left(\frac{\lambda_{\max}}{\lambda_{\min}}\right)$$

and from Eqn. (9.18g),

$$\tau_{\lambda_{\min}} = K\left(\frac{S_{\max}}{S_{\min}}\right), \quad K \text{ is a constant}$$

That is, the maximum time constant is proportional to eigenvalue spread and the method of steepest descent will converge slowly for an input signal, with a wide dynamic range spectrum. The convergence rate is maximum for a signal with unity eigenvalue spread, that is for a white noise signal whose spectrum is flat. Generally, *narrow band signals* characterized by a *few spectral peaks will have high eigenvalue spread,* that is, for narrowband or colored noise, *the steepest descent has a slow convergence rate.*

9.9 A GEOMETRICAL VIEW OF ERROR SURFACE AND OF AUTOCORRELATION MATRIX

The *orientation* of the error surface is determined by the *statistics of the signal,* the autocorrelation R_{NN} hence, its *eigenvalues and eigenvectors.*
The error surface is given by(Eqn.(9.1c))

$$\varepsilon = \sigma_d^2 - 2\mathrm{W}_N^T\mathrm{P}_N + \mathrm{W}_N^T\mathrm{R}_{NN}\mathrm{W}_N$$

or

$$\mathrm{W}_N^T\mathrm{R}_{NN}\mathrm{W}_N - 2\mathrm{W}_N^T\mathrm{P}_N - (\varepsilon - \sigma_d^2) = 0$$

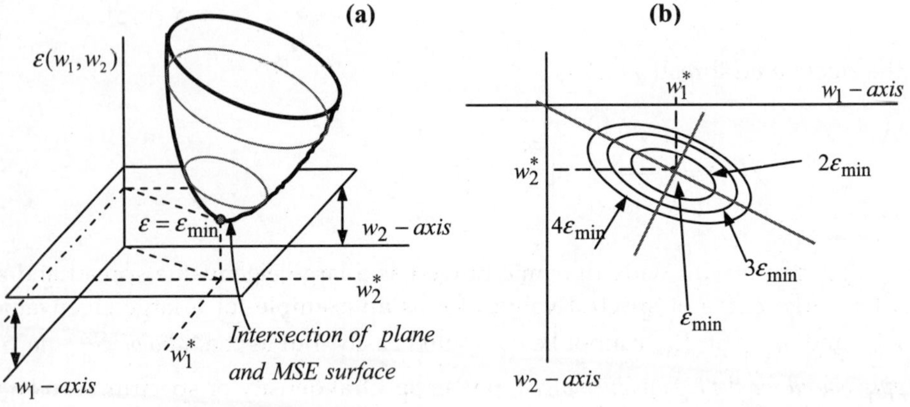

Fig. 9.3 (a) Error surface intersected by a plane parallel to $w_1 - w_2$ plane at $\varepsilon = \varepsilon_{\min}$ (b) Contours of the error surface at different heights.

For simplicity, let $N = 2$. If the MSE surface is intersected by a plane parallel to W_N - plane at a height $\varepsilon = \varepsilon_{\min}$, its projection on the $w_1 - w_2$ plane will be a point corresponding to the optimal filter vector $\left(w_1^*, w_2^*\right)$ (Fig. 9.3(a, b)). If a plane at $\varepsilon = 2\varepsilon_{\min}$ intersects the error surface, then its projections on $w_1 - w_2$ plane will result in a contour with its centre corresponding to $\varepsilon_{\min}$, i.e., $\left(w_1^*, w_2^*\right)$. Similarly, for $\varepsilon = 3\varepsilon_{\min}, \varepsilon = 4\varepsilon_{\min}$ and so on, result in concentric contours

(Fig. (9.3b)). If the successive contours are *very close/crowded together*, then the MSE surface is *very steep* indicating that that the MSE error is *very sensitive to the choice of filter coefficients.* If the successive contours are *widely separated, the MSE is not very sensitive* to choice of the filter coefficients.

The translated filter coefficients are given by

$$\mathbf{V}_N = \mathbf{W}_N - \mathbf{W}_N^* = \mathbf{W}_N - \mathbf{R}_{NN}^{-1}\mathbf{P}_N$$

$$\mathbf{W}_N = \mathbf{V}_N + \mathbf{R}_{NN}^{-1}\mathbf{P}_N$$

Substituting this in the error surface equation,

$$\mathbf{W}_N^T\mathbf{R}_{NN}\mathbf{W}_N - 2\mathbf{W}_N^T\mathbf{P}_N - (\varepsilon - \sigma_d^2) = 0$$

$$\left(\mathbf{V}_N + \mathbf{R}_{NN}^{-1}\mathbf{P}_N\right)^T\mathbf{R}_{NN}\left(\mathbf{V}_N + \mathbf{R}_{NN}^{-1}\mathbf{P}_N\right) - 2\mathbf{P}_N^T\left(\mathbf{V}_N + \mathbf{R}_{NN}^{-1}\mathbf{P}_N\right) - \left(2\varepsilon_{\min} - \sigma_d^2\right) = 0$$

$$\left(\mathbf{V}_N + \mathbf{R}_{NN}^{-1}\mathbf{P}_N\right)^T\left(\mathbf{R}_{NN}\mathbf{V}_N + \mathbf{P}_N\right) - 2\mathbf{P}_N^T\mathbf{V}_N - 2\mathbf{P}_N^T\mathbf{R}_{NN}^{-1}\mathbf{P}_N - \left(2\varepsilon_{\min} - \sigma_d^2\right) = 0$$

$$\mathbf{V}_N^T\mathbf{R}_{NN}\mathbf{V}_N + \mathbf{P}_N^T\mathbf{R}_{NN}^{-1}\mathbf{R}_{NN}\mathbf{V}_N + \mathbf{V}_N^T\mathbf{P}_N + \mathbf{P}_N^T\mathbf{R}_{NN}^{-1}\mathbf{P}_N - 2\mathbf{P}_N^T\mathbf{V}_N$$
$$- 2\mathbf{P}_N^T\mathbf{R}_{NN}^{-1}\mathbf{P}_N - \left(2\varepsilon_{\min} - \sigma_d^2\right) = 0$$

$$\mathbf{V}_N^T\mathbf{R}_{NN}\mathbf{V}_N - 2\varepsilon_{\min} + \underbrace{\sigma_d^2 - \mathbf{P}_N^T\mathbf{W}_N^*}_{\downarrow = \varepsilon_{\min}} = 0, \qquad \because \mathbf{R}_{NN}^{-1}\mathbf{P}_N = \mathbf{W}_N^*$$

$$\mathbf{V}_N^T\mathbf{R}_{NN}\mathbf{V}_N - \varepsilon_{\min} = 0$$

$$\mathbf{V}_N^T\mathbf{R}_{NN}\mathbf{V}_N = \varepsilon_{\min} \tag{9.19a}$$

<u>Example</u>

Let

$$\mathbf{R} = \begin{bmatrix} 5 & -2 \\ -2 & 5 \end{bmatrix}, \qquad \mathbf{V} = \begin{bmatrix} v_1 \\ v_2 \end{bmatrix} \qquad \text{and} \qquad \varepsilon_{\min} = 48$$

$$V^T RV = \begin{bmatrix} v_1 & v_2 \end{bmatrix} \begin{bmatrix} 5 & -2 \\ -2 & 5 \end{bmatrix} \begin{bmatrix} v_1 \\ v_2 \end{bmatrix} = 48$$

$$\begin{bmatrix} 5v_1 - 2v_2 & -2v_1 + 5v_2 \end{bmatrix} \begin{bmatrix} v_1 \\ v_2 \end{bmatrix} = 48$$

$$5v_1^2 - 2v_1v_2 - 2v_1v_2 + 5v_2^2 = 48$$
$$5v_1^2 - 4v_1v_2 + 5v_2^2 = 48$$

The eigen values of R are 3 and 7, eigen vectors are

$$m_1 = \begin{bmatrix} 1/\sqrt{2} \\ 1/\sqrt{2} \end{bmatrix}, \; m_2 = \begin{bmatrix} -1/\sqrt{2} \\ 1/\sqrt{2} \end{bmatrix}$$

and the modal matrix

$$M = \begin{bmatrix} m_1 & m_2 \end{bmatrix} = \begin{bmatrix} 1/\sqrt{2} & -1/\sqrt{2} \\ 1/\sqrt{2} & 1/\sqrt{2} \end{bmatrix}$$

Diagonalizing the matrix R orthonormally by M as it was done previously and substituting $V = MV'$ in $V^T RV$,

$$V^T RV = (MV')^T R\, MV' = V'^T \underbrace{M^T R\, M}_{\downarrow = \Lambda} V'$$

Since M is orthonormal, it diagonalizes R. Therefore,

$$V'^T \Lambda V' = 48, \quad \begin{bmatrix} v_1' & v_2' \end{bmatrix} \begin{bmatrix} 3 & 0 \\ 0 & 7 \end{bmatrix} \begin{bmatrix} v_1' \\ v_2' \end{bmatrix} = 48$$

$$3(v_1')^2 + 7(v_2')^2 = 48, \quad \frac{3}{48}(v_1')^2 + \frac{7}{48}(v_2')^2 = 1$$

$$\frac{(v_1')^2}{(48/3)} + \frac{(v_2')^2}{(48/7)} = 1 \qquad (9.19b)$$

This is of the form

$$\frac{(v_1')^2}{a^2} + \frac{(v_2')^2}{b^2} = 1 \qquad (9.19c)$$

where

$$a = \sqrt{\frac{48}{3}} = \sqrt{\frac{\varepsilon_{min}}{\lambda_{min}}} \quad \text{and} \quad b = \sqrt{\frac{48}{7}} == \sqrt{\frac{\varepsilon_{min}}{\lambda_{max}}} \qquad (9.19d, e)$$

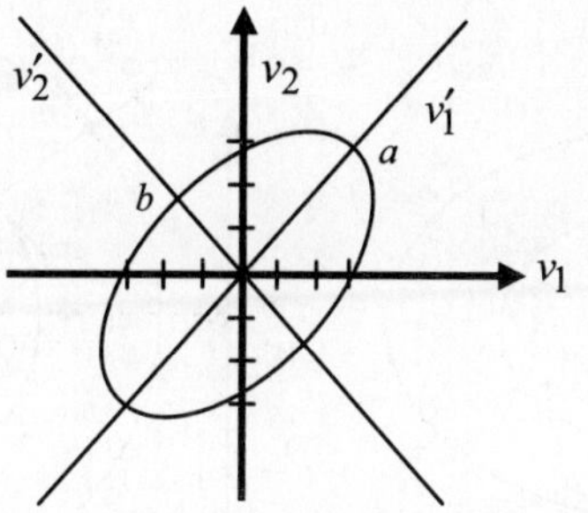

Fig. 9.3(c) Effect of orthonormal diagonalization on the MSE surface contours.

This represents an ellipse with a and b as semi-major and semi-minor axes, respectively. The MSE contour on V'-plane forms an ellipse centred at the origin with its major and minor axes aligned with the V' co-ordinate space (Fig. (9.3c)). The contours intersect the V' - axes at values *dependent on the eigenvalues of* R_{NN} and *specific mean square error value chosen. The general shape of the error contours in the original* W_N *- plane and new* V' *- plane is the same(Fig. (9.3c)). The orthonormal diagonalization rotates* $v_1 - v_2$ to $v_1' - v_2'$, *which is aligned with the principal axes and this provides a simpler mathematical form. The principal axes of the ellipse are nothing but the eigenvectors. The values of a and b are determined by the eigenvalues* λ_{min} and λ_{max}, respectively Eqn. (9.19d, e)). As already stated if the consecutive contours corresponding to equal increments in ε are well separated, then the error surface will be shallow. *If the surface is a hemisphere, then these contours will be equally spaced and the steepness of the surface will be minimum, i.e., which corresponds to* $a = b$ *and the eigenvalue spread equal to one. For large eigenvalue spread or large* (a / b), *the error surface will be very steep and sensitive to the filter coefficients.*

The orthogonalization of the autocorrelation matrix has been used only for the analysis purpose to show that the method of steepest descent algorithm converges to the optimum solution exponentially, provided the step-size is within the permissible limit. However, orthogonalization of the autocorrelation matrix has not been used in the algorithm. In the next section, steepest descent algorithm that utilizes such an orthogonalization will be considered.

The weight vector $W(n)$ takes different values as it starts from an arbitrary point $W(0)$ in the $W(n)$ coordinate space. The curve joining these weight vector values shows the *weight vector path. For a given point* $W(0)$ *on the error surface contour, the gradient is perpendicular to the tangent to the contour at that point. Hence, the movement form one contour to the other in approaching the minimum of the cost function, the weight vector moves in a direction opposite to the gradient at each point (on the contour) and these positions in the* $W(n)$ *space (indicated by black dots in the Fig. 9.3d) form the weight vector path.*

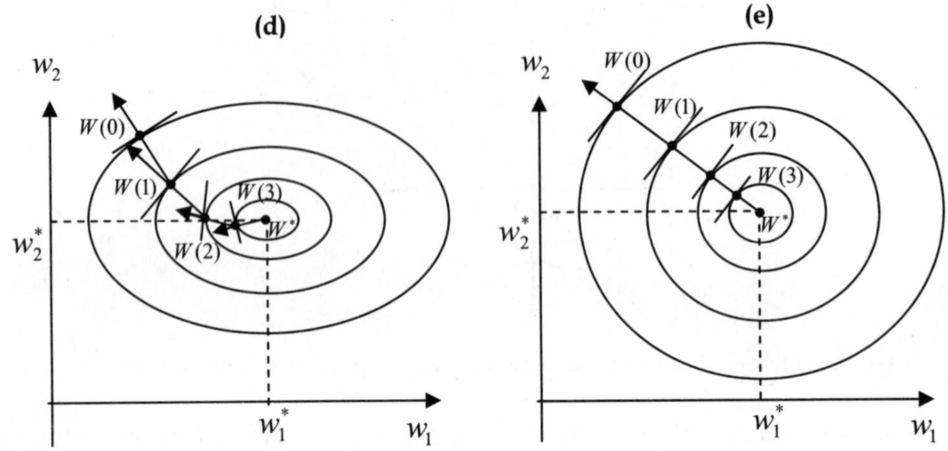

Fig. 9.3 (d, e) Weight vector paths for high (d) and unity eigenvalue spread (e).

The gradient direction is sensitive to the eccentricity of the MSE performance surface (eigenvalue spread). For the case of eigenvalues being significantly different (high eigenvalue spread compared to unity), ***the weight vector path is a curve*** as it depends on the direction (opposite) of the gradients at each point on the contour. Also the directions opposite to the gradients at different positions considered, *are not directed towards the minimum of the MSE error surface.* But for the unity eigenvalue spread, the semi-minor and semi-major axes of the elliptical contours are equal and make the contours *circular.* For these circular contours, the directions opposite to gradients at different positions on the contours *are directed towards the minimum of the MSE error surface* and hence *the weight vector path is* ***a straight line*** *directed towards the minimum and not a curve (Fig. 9.3e).*

For the high eigenvalue spread, the slow convergence rate is due to *the weight vector path being a curve which is a longer path and not a straight line directed towards the minimum point, as in the case of unity eigenvalue spread. This is an additional reason to the one that the stepsize value (for a path chosen) should be smaller for the* high eigenvalue spread compared to that of unity eigenvalue spread.

It should be mentioned that if the starting point of the weight vector is on the semi-minor axis of any contour which correspond to the largest eigenvalue, then the distance to be covered in reaching the minimum point is the shortest compared to any other point on that contour. Further, if weight vector is on the semi-major axis for the contour considered, then this weight vector path is the longest and is the worst in terms of convergence time. Hence, it is not only the step-size that is used but also the weight vector path selected decides the convergence rate. The weight vector path towards minimum can be made better by choosing at each step a proper direction which is orthogonal to its previous directions and is known as *conjugate gradient method* and this will be considered in detail in Section 9.11.

9.10 NEWTON'S METHOD

Steepest descent suffers from slow convergence rate due to large eigenvalue spread of R_{NN}. If this eigenvalue spread is avoided, the convergence rate will improve. The Newton's method does this by replacing the scalar stepsize μ by $\mu\, R_{NN}^{-1}$.

The gradient adaptation is according the rule (Eqn. (9.1b)),

$$W_N(n+1) = W_N(n) + \nabla\ \varepsilon[W_N(n)]$$

$\nabla\,\varepsilon[W_N(n)]$ is the gradient of the error surface with respect to weight vector $W_N(n)$.

For steepest descent,

$$\nabla\,\varepsilon[W_N(n)] = -\mu\,\frac{\partial E\left\{e^2(n)\right\}}{\partial W_N(n)}$$

But for Newton's method,

$$\nabla\,\varepsilon[W_N(n) = -\mu\, R_{NN}^{-1}\,\frac{\partial E\left\{e^2(n)\right\}}{\partial W_N(n)} \tag{9.20a}$$

Therefore,

$$W_N(n+1) = W_N(n) - \mu\, R_{NN}^{-1}\,\frac{\partial E\left\{e^2(n)\right\}}{\partial W_N(n)} \tag{9.20b}$$

$$e(n) = d(n) - y(n)\,,\quad y(n) = W_N^{\,T}(n)X_N(n)$$

$$e^2(n) = \left[\, d(n) - W_N^{\,T}X_N(n)\right]^2$$

$$\frac{\partial E\left\{e^2(n)\right\}}{\partial W_N(n)} = 2E\left\{\left[\, d(n) - W_N^{\,T}(n)X_N(n)\right]\left(-X_N(n)\right)\right\}$$

$$= E\left\{\left[\, -X_N(n)d(n) + W_N^{\,T}(n)X_N(n)X_N^{\,T}(n)\right]\right\}$$

$$\frac{\partial E\left\{e^2(n)\right\}}{\partial W_N(n)} = -2P_N + 2R_{NN}W(n)$$

Substituting this in the update equation (Eqn.(9.20b)),

$$W_N(n+1) = W_N(n) - \mu\, R_{NN}^{-1}\,\frac{\partial E\left\{e^2(n)\right\}}{\partial W_N(n)}$$

$$W_N(n+1) = W_N(n) - 2\mu\, R_{NN}^{-1}\left[-P_N + R_{NN}W_N(n)\right]$$

$$W_N(n+1) = W_N(n) + 2\mu\, R_{NN}^{-1}\, P_N - 2\mu\, RW_N(n)$$

Since

$$W_N^* = R_{NN}^{-1}P_N$$

$$W_N(n+1) = \left[1-2\mu\right]W_N(n) + 2\mu\, W_N^* \tag{9.20c}$$

Evaluating this for different values of n,

For $n = 0$, $\quad W_N(1) = \left[1-2\mu\right]W_N(0) + 2\mu\, W_N^*$

For $n = 1$, $\quad W_N(2) = \left[1-2\mu\right]W_N(1) + 2\mu\, W_N^*$

$$= \left[1-2\mu\right]\left(\left[1-2\mu\right]W_N(0) + 2\mu\, W_N^*\right) + 2\mu\, W_N^*$$

$$= \left[I-2\mu\right]^2 W_N(0) + \left[I-2\mu\right]2\mu\, W_N^* + 2\mu\, W_N^*$$

$$= \left[1-2\mu\right]^2 W_N(0) + 2\mu\, W_N^* - 4\mu^2\, W_N^* + 2\mu W_N^*$$

$$= \left[1-2\mu\right]^2 W_N(0) + 4\mu\, W_N^* - 4\mu^2\, W_N^*$$

$$= \left[1-2\mu\right]^2 W_N(0) - \left[4\mu^2 - 4\mu + 1 - 1\right]W_N^*$$

$$= \left[1-2\mu\right]^2 W_N(0) - \left[(1-2\mu)^2 - 1\right]W_N^*$$

$$= \left[1-2\mu\right]^2 W_N(0) - (1-2\mu)^2\, W_N^* + W_N^*$$

$$W_N(2) = W_N^* + \left[1-2\mu\right]^2\left[W_N(0) - W_N^*\right]$$

In general,

$$W_N(n) = W_N^* + \left[1-2\mu\right]^n\left[W_N(0) - W_N^*\right] \tag{9.20d}$$

For convergence,

$$\left(1-2\mu\right) < 1 \quad \text{or} \quad 0 < \mu < 1 \tag{9.20e}$$

For $\mu = 0.5$,

$$W_N(1) = \left[1 - 2\times\frac{1}{2}\right]\left[W_N(0) - W_N^*\right] + W_N^*$$

$$W_N(1) = W_N^* \tag{9.20f}$$

That is, the weight vector jumps from initial $W_N(0)$ *to the optimum* W_N^* *in a single step* (Fig. 9.4).

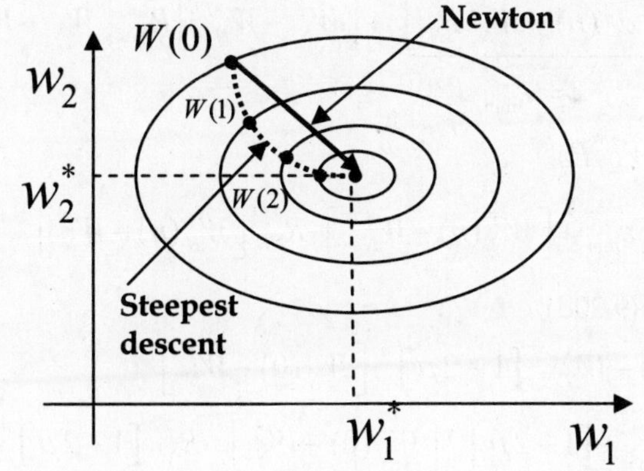

Fig. 9.4 Weight vector path for method of steepest descent and Newton's method.

But in a practical setup, exact values of R_{NN}^{-1} and $\dfrac{\partial E\left\{e^2(n)\right\}}{\partial W_N(n)}$ are not available and have to be estimated. This necessitates a step-size smaller than 0.5.

Consider the mean square error (Eqn. (9.1b)),

$$\varepsilon = E\left[d^2(n)\right] - 2P_N^T W_N + W_N^T R_{NN} W_N$$

$$= E\left[d^2(n)\right] - W_N^T P_N - P_N^T W_N + W_N^T R_{NN} W_N$$

$$= E\left[d^2(n)\right] + W_N^T R_{NN} W_N - W_N^T R_{NN} W_N^* - W_N^{*T} R_{NN} W_N,$$

$\because P_N = R_{NN} W_N$

Adding and subtracting the term $W_N^{*T} R_{NN} W_N^*$,

$$\varepsilon = E\left[d^2(n)\right] + W_N^T R_{NN} W_N - W_N^T R_{NN} W_N^* - W_N^{*T} R_{NN} W_N + W_N^{*T} R_{NN} W_N^* - W_N^{*T} R_{NN} W_N^*$$

$$= E\left[d^2(n)\right] + W_N^T R_{NN}\left[W_N - W_N^*\right] - W_N^{*T} R_{NN}\left[W_N - W_N^*\right] - W_N^{*T} R_{NN} W_N^*$$

$$= E\left[d^2(n)\right] + \left[W_N^T - W_N^{*T}\right] R_{NN}\left[W_N - W_N^*\right] - W_N^{*T}\underbrace{\left(R_{NN} W_N^*\right)}_{\downarrow = p_N}$$

$$= \underline{E\left[d^2(n)\right]} + \left[W_N^T - W_N^{*T}\right] R_{NN}\left[W_N - W_N^*\right] \quad \underline{-W_N^{*T} P_N}$$

$$= \underbrace{E\left[d^2(n) - W_N^{*T} P_N\right]}_{\downarrow = \varepsilon_{\min}} + \left[W_N^T - W_N^{*T}\right] R_{NN} \left[W_N - W_N^*\right],$$

$$\because \varepsilon_{\min} = E\left[d^2(n)\right] - W_N^{*T} P_N$$

$$\varepsilon(n) = \varepsilon_{\min} + \left[W_N^T(n) - W_N^{*T}\right] R_{NN} \left[W_N(n) - W_N^*\right],$$

Substituting from Eqn. (9.20d),

$$W_N(n) - W_N^* = [1 - 2\mu]^n \left[W_N(0) - W_N^*\right]$$

$$\varepsilon = \varepsilon_{\min} + [1 - 2\mu]^n \left[W_N(0) - W_N^*\right]^T R_{NN} [1 - 2\mu]^n \left[W_N(0) - W_N^*\right]$$

$$\varepsilon = \varepsilon_{\min} + [1 - 2\mu]^{2n} \left[W_N(0) - W_N^*\right]^T R_{NN} \left[W_N(0) - W_N^*\right] \quad (9.20g)$$

Now for the method of steepest descent,

$$\varepsilon(n) = \varepsilon_{\min} + \left[W_N^T(n) - W_N^{*T}\right] R_{NN} \left[W_N(n) - W_N^*\right],$$

Therefore,

$$\varepsilon(0) = \varepsilon_{\min} + \left[W_N^T(0) - W_N^{*T}\right] R_{NN} \left[W_N(0) - W_N^*\right],$$

$$\varepsilon(0) - \varepsilon_{\min} = \left[W_N^T(0) - W_N^{*T}\right] R_{NN} \left[W_N(0) - W_N^*\right],$$

Using this relation in Eqn. (9.20g),

$$\varepsilon(n) = \varepsilon_{\min} + [1 - 2\mu]^{2n} \left[\varepsilon(0) - \varepsilon_{\min}\right]$$

Further,

$$\left(\varepsilon(n) - \varepsilon_{\min}\right) = [1 - 2\mu]^{2n} \left[\varepsilon(0) - \varepsilon_{\min}\right]$$

$$\varepsilon_v(n) = [1 - 2\mu]^{2n} \varepsilon_v(0) \quad (9.20h)$$

where,

$$\varepsilon_v(n) = \left(\varepsilon(n) - \varepsilon_{\min}\right), \ \varepsilon_v(0) = \left(\varepsilon(0) - \varepsilon_{\min}\right)$$

The time constant for the adaptation is time required for the difference error to decrease from its initial value $\varepsilon_v(0)$ to $\left(\frac{1}{e}\right)\varepsilon_v(0)$. Therefore,

$$\frac{\varepsilon_v(n)}{\varepsilon_v(0)} = \frac{1}{e} = [1 - 2\mu]^{2\tau}$$

or

$$-1 = \ln[1 - 2\mu]^{2\tau}, \quad 2\tau = \frac{-1}{\ln[1 - 2\mu]}$$

If $2\mu << 1$,

$$\ln\left[1-2\mu\right] \approx -2\mu$$

$$\tau = \frac{1}{4\mu} \qquad (9.20i)$$

Here, there is only one mode of convergence which is solely determined by the step size μ and this is another deviation from that of the method of steepest descent which is characterized by many modes.

9.10.1 Interpretation of Newton's Method

Let,

$$X'_N(n) = M_{NN}^T X_N(n)$$

and the modal matrix of R_{NN} be

$$M_{NN}^T = \left[m_0 \;\; m_1, \ldots, m_{N-1}\right]$$

then,

$$X'_N(n) = \left[x'_0(n), x'_1(n)\; x'_2(n), \ldots x'_{N-1}(n)\right]$$

whose elements $x'_i(n) = 1), 2, \ldots N-1$ are uncorrelated random variables. The eigenvalues of R_{NN} are given by

$$E\left[x_i'^2(n)\right] = \lambda_i, \quad i = 0, 1, 2, \ldots (N-1)$$

Let,

$$X''_N(n) = \left[x''_0(n), x''_1(n)\; x''_2(n), \ldots x''_{N-1}(n)\right]$$

where,

$$x''_i(n) = \lambda_i^{-1/2} x'_i(n), \quad i = 0, 1, 2, \ldots, n-1 \qquad (9.21a)$$

$X''_N(n)$ is a vector of scaled uncorrelated variables. Therefore,

$$R''_{NN} = E\left[X''_N(n) X_N''^T\right] = I,$$

$$W''_N = \Lambda^{1/2} M_{NN}^T W_N$$

$$W''_N X''_N = W_N^T M_{NN} \Lambda^{(1/2)} \Lambda^{-(1/2)} M_{NN}^T X_N$$

$$W''_N X''_N = W_N^T X_N = \hat{d}(n)$$

For the method of steepest descent,

$$W_N(n+1) = \left[I - 2\mu\, R_{NN}\right] W_N(n) + 2\mu\, P_N$$

$$W_N''(n+1) = \left[I - 2\mu\, R_{NN}'' \right] W_N''(n) + 2\mu\, P_N'' \qquad (9.21b)$$

where,

$$P''_{NN} = E\left[X''^{T} d(n) \right]$$

and since,

$$R''_{NN} = I, \qquad (9.21c)$$

Since

$$W_N^{*''} = R_N''^{-1} P_N'' = I''^{-1} P_N'' = P_N'',$$

$$W_N^{*''} = P_N'' \qquad (9.21d)$$

The equation,

$$W_N''(n+1) = \left[I - 2\mu \;\right] W_N''(n) + 2\mu\, P_N''$$

becomes

$$W_N''(n+1) = \left[I - 2\mu \;\right] W_N''(n) + 2\mu\, W_N^{*''} \qquad (9.21e)$$

Newton's algorithm is,

$$W_N(n+1) = \left[1 - 2\mu \right] W_N(n) + 2\mu\, W_N^{*}$$

That is, the Newton's algorithm is identical to the one derived by transforming the input by an orthonormal matrix of R_{NN} and normalized by the eigenvalues of R_{NN}. That is, the input is decorrelated and normalized. Newton's method can be viewed as the steepest descent method applied to a transformed signal whose components are uncorrelated to each other. Further, the eigenvalue-spread is overcome by the power normalizations of its components. It is characterized by a single time constant and is determined by μ, the step size ($\tau = 1/4\mu$).

In the further study of adaptive processing, the concepts of the method of steepest descent play a crucial role as its simpler version, the least mean square algorithm (dealt in next chapter) is extensively used in developing different types of adaptive algorithms which are advantageous in different applications.

Chapter 10

LEAST MEAN SQUARE (LMS) ALGORITHM

The steepest descent method (SDM) though provides optimal solution, it needs the statistics, viz., the autocorrelation of the input and the cross-correlation between the input and the desired signal. In a nonstationary signal scenario, these two statistics change from sample to sample and their computation is involved. To get over this computational problem especially for a real-time situation, the least mean square algorithm (LMS) has been introduced. Here, the statistical quantities are replaced by their instantaneous values and this results in a highly computationally simple algorithm. However, this simplicity is at the cost of excess error over the minimum error as the weight vector jitters about the optimum weight vector. The ratio of excess error to optimum error is known as misadjustment and an expression for this has been derived. The excess error increases with step size and filter length. Further, the SDM noise immunity is lost due to the absence of averaging in the LMS algorithm. Similar to SDM, the mean weight vector of LMS also suffers from slow convergence when the eigenvalue spread of the input autocorrelation matrix is large. The effect of delay between the input and the error on the adaptation stepsize has been derived. Many variations of the LMS algorithm are considered to improve its convergence rate (variable step size LMS, momentum LMS), convergence error magnitude (forward backward error LMS, variable step size LMS), its immunity to noise (Griffiths' LMS, median LMS), stability (normalized LMS and Leaky LMS, filtered input LMS to overcome the stability ill effect due to delay between the error and input involved in computing the gradient) and computational simplicity (like Laguerre transversal LMS, sign error and sign data LMS).

To avoid stalling due to low input magnitude, use of larger step size is suggested and the resulting higher misadjustment is controlled by post-smoothing of the weight vector

To improve the convergence performance of LMS, use of conjugate gradient method (CGM) is considered. In this in approaching the minimum, the new descent direction chosen is orthogonal to its previous directions and the error is totally minimized in the new direction by choosing a proper step size. Thus, the CGM converges within a number of iterations determined by the dimension of the input signal autocorrelation space.

For adaptive filtering as the autocorrelation and cross-correlation change from sample to sample, the CGM which involves many iterations has to be repeated for every sample and this is computationally highly expensive. To overcome this, some modifications for CGM are considered which aim at only one iteration per sample but still has a performance significantly better than that of LMS.

LMS adaptive identification of a non-minimum phase system based on bispectrum is also covered.

10.1 THE LEAST MEAN SQUARE (LMS) ALGORITHM

The method of steepest descent and the Newton's method introduced in the previous chapter requires the signals $x(n)$ and $d(n)$ to be stationary as *second-order statistics* R_{NN} *and* P_N *have to be known*. Further, Newton's method requires R_{NN}^{-1}. For the method of steepest descent (Eqn.(9.2)),

$$W_N(n+1) = W_N(n) + 2\mu\left[p_N - R_{NN}W_N(n)\right]$$

Finding the estimates P_N and R_{NN} is computationally expensive. Also as these estimates have to be updated frequently in a *non-stationary environment depending on its short time stationary nature,* it is not suitable in such environments. The estimation of these correlation values can cause a delay which my result in errors. The computation of R_{NN}^{-1} may not be easy and some time may result in singular solutions depending upon the data and its manipulation.

An alternative to the above methods is the *least mean square algorithm (LMS).* It is a stochastic adaptive algorithm as it uses only *a noisy approximation of the gradient of the MSE surface.* However, the method of steepest descent and Newton's methods use the *exact gradient of the MSE surface.*

10.2 LMS ALGORITHM

The update for the steepest descent algorithm (Eqn. (9.2)) is

$$W_N(n+1) = W_N(n) + 2\mu\left[p_N - R_{NN}W_N(n)\right]$$

Expanding R_{NN} and P_N,

$$W_N(n+1) = W_N(n) + 2\mu\left[E\{X_N(n)d(n)\} - E\{X_N(n)X_N^T(n)\}W_N(n)\right]$$

As the LMS uses only the noisy gradient, it requires the expectation operation involved in R_{NN} and P_N, to be removed to avoid the smoothing/averaging effect. Hence, the above equation becomes,

$$W_N(n+1) = W_N(n) + 2\mu\left[X_N(n)d(n) - X_N(n)X_N^T(n)W_N(n)\right]$$

$$W_N(n+1) = W_N(n) + 2\mu\left[X_N(n)\{d(n) - X_N^T(n)W_N(n)\}\right]$$

But

$$\hat{d}(n) = W_N^T(n)X_N(n) = X_N^T(n)W_N(n)$$

Therefore,

$$W_N(n+1) = W_N(n) + 2\mu\left[X_N(n)\{d(n) - \hat{d}(n)\}\right]$$

Since as

$$e(n) = d(n) - \hat{d}(n)$$

$$W_N(n+1) = W_N(n) + 2\mu\left[X_N(n)e(n)\right] \tag{10.1a}$$

If $\alpha = 2\mu$ then, the weight vector adaptation is given by,

$$W_N(n+1) = W_N(n) + \alpha\, e(n)X_N(n) \tag{10.1b}$$

Expanding the above vector relation

$$\begin{bmatrix} w_0(n+1) \\ w_2(n+1) \\ \vdots \\ w_{N-1}(n+1) \end{bmatrix} = \begin{bmatrix} w_0(n) \\ w_2(n) \\ \vdots \\ w_{N-1}(n) \end{bmatrix} + \alpha e(n) \begin{bmatrix} x(n) \\ x(n-1) \\ \vdots \\ x(n-N+1) \end{bmatrix}$$

Therefore, the individual weight adaptation is

$$w_i(n+1) = w_i(n) + \alpha \underline{e(n)x(n-i)}\ ,\ 0 \le i \le N-1 \tag{10.2}$$

There is a stringent ***time alignment*** *between the error* $e(n)$ *and input* $x(n)$. *This is of importance, since any deviation from this alignment may results in divergence.* For the steepest descent to converge, *the statistical orthogonality* between the error and the input is required. That is,

$$E\left\{e(n)x(n-i)\right\} = 0$$

In LMS algorithm for $w_i(n)$ to converge exactly to w_i^* ,

$$e(n)x(n-i) = 0, \quad 0 \le i \le N-1, \tag{10.3}$$

is obtained by forcing $w_i(n+1) = w_i(n)$ *for all* n. This is a much ***more severe constraint*** than that for the method of steepest descent.

Example

The LMS algorithm is illustrated for a predictor for the autoregressive process

$$x(n) = 0.4x(n-1) - 0.8x(n-2) + r(n)$$

$r(n)$ is a white Gaussian noise with zero mean and unit variance.

$$\hat{x}(n) = \sum_{i=1}^{2} w_i(n)x(n-i+1)$$

$$R_{NN} = \begin{bmatrix} \phi_x(0) & \phi_x(1) \\ \phi_x(1) & \phi_x(0) \end{bmatrix} \text{ and } P_N = \begin{bmatrix} \phi_x(1) \\ \phi_x(2) \end{bmatrix}$$

For a data length of 800 samples,

$$\phi_x(0) = 1.0000, \phi_x(1) = 0.2302, \phi_x(2) = -0.6964$$

$$R_{NN} = \begin{bmatrix} 1.0 & 0.2302 \\ 0.2302 & 1.0 \end{bmatrix} \text{ and } P_N = \begin{bmatrix} 0.2302 \\ -0.6964 \end{bmatrix}$$

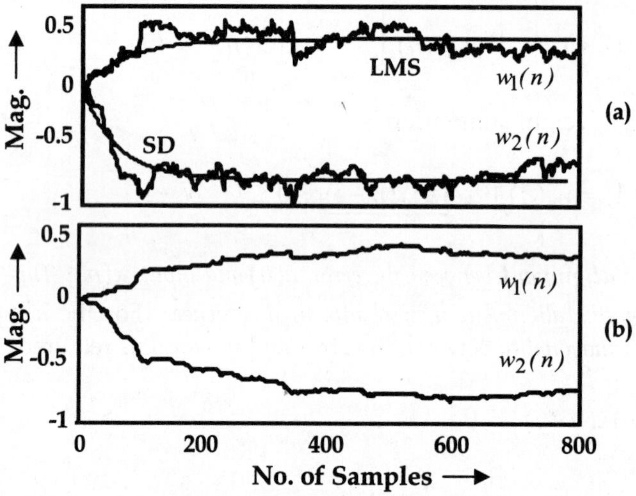

Fig. 10.1 Convergence of LMS algorithm for (a) α = 0.02 and (b) α = 0.004

The eigenvalues of R_{NN} are 0.7698 and 1.2302. The eigenvectors of R_{NN} are

$$\mathrm{m}_1^T = \begin{bmatrix} -0.7071 & 0.7071 \end{bmatrix} \text{ and } \mathrm{m}_2^T = \begin{bmatrix} 0.7071 & 0.7071 \end{bmatrix}.$$

Therefore,

$$P' = M^T P = \begin{bmatrix} -0.6552 \\ -0.3297 \end{bmatrix}$$

For $\alpha = 0.02$, Since, $\gamma_i = 1 - \alpha\,\lambda_i$

$$\gamma_1 = 0.9846 \text{ and } \gamma_2 = 0.9754$$

Compute

$$w_i'(n) = \gamma_i^n w_i'(0) + \alpha\, p_i' \left[\frac{1-\gamma_i^n}{1-\gamma_i}\right] \qquad i = 1,2 \qquad \gamma_i = 1-\alpha\,\lambda_i$$

For $\alpha = 0.02$,

$$\gamma_1 = (1-\alpha\,\lambda_1) = 0.9938 \text{ and } \gamma_2 = (1-\alpha\,\lambda_2) = 0.9262,$$

$$W'(n) = M^T W(n), \quad W(n) = MW'(n)$$

It is seen from Fig. 10.1 (a) that by the method of SD algorithm, the weights converge to their true values after 200 samples and remain constant. But by the LMS algorithm, the estimated weights oscillate about their true values. The oscillations are of significantly large magnitude for $\alpha = 0.02$ compared to those for $\alpha = 0.004$ (Fig. 10.1(b)). But the small value of $\alpha = 0.004$ makes the convergence to occur only after 400 samples (Fig. 10.1(b)) hence, has a slower convergence rate compared to that for $\alpha = 0.02$. Thus there is a trade-off between parameter variance and convergence rate. A larger step size results in a faster convergence but results in a larger variance in weight values and vice versa.

10.3 CONVERGENCE OF THE LMS ALGORITHM

The LMS adaptation is (Eqn.(10.1b))

$$W_N(n+1) = W_N(n) + \alpha\, e(n) X_N(n)$$

$$W_N(n+1) = W_N(n) + \alpha\, X_N(n)\left[d(n) - X_N^T(n) W_N(n)\right]$$

$$W_N(n+1) = \left[I_{NN} - \alpha\, X_N(n) X_N^T\right] W_N(n) + \alpha\, d(n) X_N(n)$$

For the method of steepest descent (Eqn.(9.1e)),

$$W_N(n+1) = \left[I_{NN} - \alpha\, R_{NN}\right] W_N(n) + \alpha\, P_N(n) \tag{10.4}$$

These two equations are identical except for $X_N(n)X_N^T(n)$ in place of R_{NN} and $d(n)X_N(n)$ in place of P_N are present, in the LMS algorithm. As the LMS algorithm *approximates expectations by their instantaneous quantities,* only the *mean weight vector* of the LMS algorithm converges to optimum solution for (Eqn. (9.14)),

$$0 < \alpha < \frac{2}{\lambda_{\max}}$$

which is the criterion for the method of steepest descent to converge to the optimum solution.

If M is an orthonormal matrix with eigenvectors of R_{NN} and λ_i, $i = 1, 2, \ldots N$ are the eigenvalues of R_{NN}, then

$$\lambda_{\max} \leq \sum_{i=1}^{N} \lambda_i \text{ and } R_{NN} = M \, \Lambda \, M^T,$$

$$tr\left[R_{NN}\right] = tr\left[M \, \Lambda \, M^T\right] = tr\left[M^T M \Lambda\right] (\because \; tr\left[AB\right] = tr\left[BA\right], A = M\Lambda, \quad B = M$$

$$tr\left[M^T M \Lambda\right] = tr\left[I \Lambda\right], \qquad (\because \; M^T M = I)$$

$$tr\left[R_{NN}\right] = tr\left[\Lambda\right] = \sum_{i=1}^{N} \lambda_i$$

Therefore,

$$\lambda_{\max} \leq tr\left[R_{NN}\right] = N\phi_x(0) = N\sigma_x^2$$

Therefore, the condition for LMS mean weight vector to converge is (Eqn.(9.14))

$$0 < \alpha < \frac{2}{\lambda_{\max}}$$

becomes

$$0 < \alpha < \frac{2}{N\sigma_x^2} \tag{10.5}$$

This preserves the simplicity of the LMS algorithm as only by computing σ_x^2, the power of the input signal, it is possible to determine the step size that ensures convergence in the mean sense. Similar to the method of steepest descent, the *LMS algorithm also exhibits slow convergence when the eigenvalue spread is large.*

10.4 MISADJUSTMENT

For the method of steepest descent (Eqn.(10.4)),

$$W_N(n+1) = W_N(n) + \alpha\left[P_N - R_{NN} \, W_N\right]$$

At convergence,

$$W_N(n) = W_N^* \text{ and } W_N^* = R_{NN}^{-1} P_N \text{ or } \left[P_N - R_{NN} \, W_N^*\right] = 0$$

Therefore,

$$W_N(n+1) = W_N^*(n) + 0$$

That is, no more weight vector adaptation takes place hence, it stays at the optimum solution W_N^*.

If the expectation operation had not been removed, the Eqn. (10.1a)

$$W_N(n+1) = W_N(n) + 2\mu\left[X_N(n)e(n)\right]$$

would have become

$$W_N(n+1) = W_N(n) + 2\mu E\left[X_N(n)e(n)\right]$$

It is known that the minimization of mean square results in the orthogonality of *error and the input* which is nothing but

$$E\left[X_N(n)e(n)\right] = 0$$

However, in the absence of expectation as in LMS algorithm,

$$X_N(n)e(n) \neq 0$$

Hence, at convergence for LMS algorithm,

$$W_N(n+1) = W_N^*(n) + 2\mu X_N(n)e(n)$$

or

$$w_i(n+1) = w_i^*(n) + 2\mu\, x(n-i)e(n)$$

$$w_i(n+1) = w_i^*(n) + \delta w_i(n), \quad \delta w_i(n) = 2\mu x(n-i)e(n) \tag{10.6}$$

That is *about the optimum value* w_i^*, *the LMS weights vary/ jitter by the quantity* $\delta w_i(n)$ (Fig. 10.2).

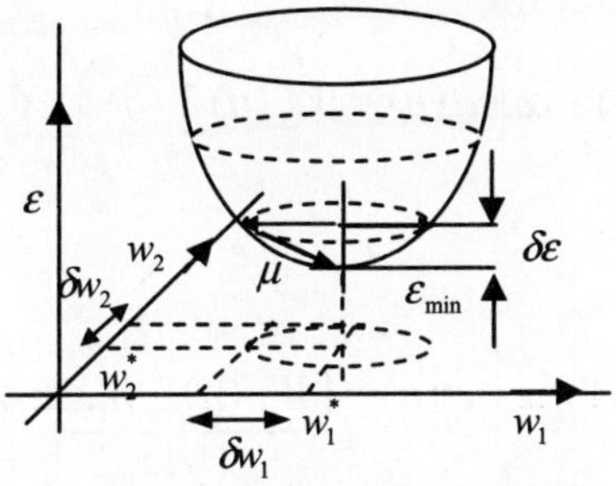

Fig. 10.2 LMS misadjustment.

This causes is an *instantaneous error in the LMS weight value* by δw_i and results in an *excess error* $\delta\ \varepsilon$ compared to the optimum error $\varepsilon_{\min}$ that is obtained by the method of steepest descent. The ratio of the excess error to the optimal error is called *misadjustment,* M .

10.4.1 Expression for LMS Misadjustment

Let

$$V_N(n) = W_N(n) - W_N^*, \qquad W_N(n) = V_N(n) + W_N^* \tag{10.7}$$

Therefore,

$$e(n) = d(n) - W_N^T X(n)$$

becomes

$$e(n) = d(n) - W_N^{*T} X(n) - V_N^T(n)X(n)$$

If

$$e_0(n) = d(n) - W_N^{*T} X(n) \tag{10.8a}$$

Then

$$e(n) = e_0(n) - V_N^T(n)X(n) \tag{10.8b}$$

The LMS adaptation rule(Eqn.10.1) is

$$W_N(n+1) = W_N(n) + \alpha\, X_N(n)e(n)$$

Adding and subtracting W_N^* both sides,

$$W_N(n+1) - W_N^* = W_N(n) - W_N^* + \alpha\, X_N(n)e(n)$$

Using Eqn. (10.7)

$$W_N(n) = V_N(n) + W_N^*$$

$$V_N(n+1) = V_N(n) + \alpha\left[d(n) - W^T(n)X(n)\right] X_N(n)$$

Add and subtract W_N^* on the RHS,

$$V_N(n+1) = V_N(n) + \alpha\left[d(n) - \left(\underbrace{W_N(n) - W_N^*}_{\downarrow = V_N} + W_N^*\right)^T X_N(n)\right] X_N(n)$$

$$= V_N(n) + \alpha\left[\underbrace{d(n) - W_N^{*T} X(n)}_{\downarrow = e_0(n)} - V_N^T X_N(n)\right] X_N(n) \quad \text{(using Eqn.(10.8a))}$$

$$= V_N(n) - \alpha\left[V_N^T X_N(n)\right] X_N(n) + \alpha\, e_0(n) X_N(n)$$

$$= V_N(n) - \alpha\, X^T(n) V_N X_N(n) + \alpha\, e_0(n) X_N(n)$$

$$V_N(n+1) = \left[I_{NN} - \alpha\, X^T(n) X_N(n)\right] V_N + \alpha\, e_0(n) X_N(n) \tag{10.9}$$

Defining,

$$K(n) = E\left\{ V_N(n)\, V_N^T(n) \right\} \tag{10.10}$$

$$K(n+1) = E\left\{ V_N(n+1)\, V_N^T(n+1) \right\}$$

Using Eqn. (10.9),

$$K(n+1) = E\Big\{ \Big(\big[I_{NN} - \alpha\, X_N(n) X_N^T(n) \big]\; V_N(n) + \alpha\, e_0(n) X_N(n) \Big)$$

$$\Big(\big[I_{NN} - \alpha\, X_N(n) X_N^T \big]\; V_N(n) + \alpha\, e_0(n) X_N(n) \Big)^T \Big\}$$

$$K(n+1) = E\Big\{ \big[I_{NN} - \alpha\, X_N(n) X_N^T \big]\; V_N(n) V_N^T(n) \big[I_{NN} - \alpha\, X_N^T(n) X_N(n) \big]$$

$$+ \alpha\, e_0(n) X_N(n)\, V_N^T(n) \big[I_{NN} - \alpha\, X_N^T(n) X_N(n) \big]$$

$$+ \alpha\, e_0(n) X_N^T(n) \big[I_{NN} - \alpha\, X_N(n) X_N^T \big]\; V_N(n) + \alpha^2\, e_0^2(n) X_N^T(n) X_N(n) \Big\}$$

$$K(n+1) = E\Big\{ \big[I_{NN} - \alpha\, X_N(n) X_N^T \big]\; V_N(n) V_N^T(n) \big[I_{NN} - \alpha\, X_N^T(n) X_N(n) \big] \Big\}$$

$$+ E\Big\{ \alpha\, e_0(n) X_N(n)\, V_N^T(n) \big[I_{NN} - \alpha\, X_N^T(n) X_N(n) \big] \Big\}$$

$$+ E\Big\{ \alpha\, e_0(n) X_N^T(n) \big[I_{NN} - \alpha\, X_N(n) X_N^T \big]\; V_N(n) \Big\} \qquad (10.11)$$

$$+ E\big\{ \alpha^2\, e_0^2(n)(n) X_N^T(n) X_N(n) \big\}$$

Now for the optimal solution, the error is (Eqn. (10.8a))

$$e_0(n) = d(n) - W_N^{*T} X(n)$$

and this will be orthogonal to the input. Therefore,

$$E\{\; e_0(n)\; X(n)\; \} = 0$$

Therefore, Eqn. (10.11) becomes,

$$K(n+1) = E\Big\{ \big[I_{NN} - \alpha\, X_N(n) X_N^T \big]\; V_N(n) V_N^T(n) \big[I_{NN} - \alpha\, X_N^T(n) X_N(n) \big] \Big\}$$

$$+ E\big\{ \alpha^2\, e_0^2(n) X_N^T(n) X_N(n) \big\}$$

$$K(n+1) = \big[I_{NN} - \alpha\, R_{NN} \big]\; K(n) \big[I_{NN} - \alpha\, R_{NN} \big] + \alpha^2\, \varepsilon_{\min} R_{NN}$$

$$\because \quad E\big\{ e_0^2(n) \big\} = \varepsilon_{\min}$$

$$K(n+1) = \big[I_{NN} - \alpha\, R_{NN} \big]\; K(n) \big[I_{NN} - \alpha\, R_{NN} \big] + \alpha^2\, \varepsilon_{\min} R_{NN}$$

Applying orthonormal transformation for R_{NN} and $K(n)$ as,

$$R_{NN} = M_{NN}\; \Lambda\; M_{NN}^T, \;\; K(n) = M_{NN}\; S(n)\; M_{NN}^T \qquad (10.12)$$

$$K(n+1) = \left[I_{NN} - \alpha\, M_{NN}\, \Lambda\, M_{NN}^T \right] M_{NN}\, S(n)\, M_{NN}^T$$
$$\bullet \left[I_{NN} - \alpha M_{NN} \Lambda M_{NN}^T \right] + \alpha^2 \varepsilon_{\min} M_{NN} \Lambda M_{NN}^T$$

$$= \left[M_{NN} M_{NN}^T - \alpha\, M_{NN}\, \Lambda\, M_{NN}^T \right] M_{NN}\, S(n)\, M_{NN}^T$$
$$\bullet \left[M_{NN} M_{NN}^T - \alpha M_{NN} \Lambda M_{NN}^T \right] + \alpha^2 \varepsilon_{\min} M_{NN} \Lambda M_{NN}^T$$

$$= \left[M_{NN} - \alpha\, M_{NN}\, \Lambda \right] M_{NN}^T\, M_{NN}\, S(n)\, M_{NN}^T\, M_{NN}$$
$$\bullet \left[M_{NN}^T - \alpha\, \Lambda M_{NN}^T \right] + \alpha^2 \varepsilon_{\min} M_{NN} \Lambda M_{NN}^T$$

$$K(n+1) = \left[M_{NN} - \alpha\, M_{NN}\, \Lambda \right] S(n) \left[M_{NN}^T - \alpha\, \Lambda\, M_{NN}^T \right] + \alpha^2 \varepsilon_{\min} M_{NN} \Lambda M_{NN}^T$$

$$K(n+1) = M_{NN} \left[I_{NN} - \alpha\, \Lambda \right] S(n) \left[I_{NN} - \alpha\, \Lambda \right] M_{NN}^T + \alpha^2 \varepsilon_{\min} M_{NN} \Lambda M_{NN}^T$$

Pre-multiply by M_{NN}^T,

$$M_{NN}^T K(n+1) = \left[I_{NN} - \alpha\, \Lambda \right] S(n) \left[I_{NN} - \alpha\, \Lambda \right] M_{NN}^T + \alpha^2 \varepsilon_{\min} \Lambda M_{NN}^T$$

Post-multiply by M_{NN},

$$M_{NN}^T K(n+1) M_{NN} = \left[I_{NN} - \alpha\, \Lambda \right] S(n) \left[I_{NN} - \alpha\, \Lambda \right] + \alpha^2 \varepsilon_{\min} \Lambda$$

Since

$$K(n) = M_{NN}\, S(n)\, M_{NN}^T, \quad S(n) = M_{NN}^T\, K(n)\, M_{NN}$$

Therefore

$$S(n+1) = \left[I_{NN} - \alpha\, \Lambda \right] S(n) \left[I_{NN} - \alpha\, \Lambda \right] + \alpha^2\, \varepsilon_{\min} \Lambda \tag{10.13a}$$

For each i,

$$S_i(n+1) = \left(1 - \alpha\, \lambda_i \right)^2 S_i(n) + \alpha^2\, \varepsilon_{\min}\, \lambda_i \tag{10.13b}$$

At convergence,

$$S_i(n+1) = S_i(n)$$

Therefore,

$$S_i(n) = \left(1 - \alpha\, \lambda_i \right)^2 S_i(n) + \alpha^2\, \varepsilon_{\min}\, \lambda_i$$

$$S_i(n) \left[1 - \left(1 - \alpha\, \lambda_i \right)^2 \right] = \alpha^2\, \varepsilon_{\min}\, \lambda_i$$

$$S_i(n) = \frac{\alpha^2 \lambda_i \varepsilon_{\min}}{\left[1-\left(1-\alpha \lambda_i\right)^2\right]} = \frac{\alpha^2 \lambda_i \varepsilon_{\min}}{\left[1-1+2\alpha \lambda_i - \alpha^2 \lambda_i^2\right]}$$

$$= \frac{\alpha^2 \lambda_i \varepsilon_{\min}}{2\alpha \lambda_i - \alpha^2 \lambda_i^2}$$

$$S_i(n) = \frac{\alpha \varepsilon_{\min}}{2-\alpha \lambda_i} \tag{10.14}$$

The expression for error is (Eqn. (10.8b)),

$$e(n) = e_0(n) - V_N^T(n)X(n)$$

The mean square error,

$$\varepsilon(n) = E\left\{e^2(n)\right\} = E\left\{ \left(e_0(n) - V_N^T(n)X(n)\right)\left(e_0(n) - X^T(n)V_N(n)\right) \right\}$$

$$\varepsilon(n) = E\left\{e_0^2(n)\right\} - e_0(n)E\left\{\underline{V_N^T(n)X(n) - X^T(n)V_N(n)}\right\} + E\left\{V_N^T(n)X(n)X^T(n)V_N\right\}$$

$$\downarrow = 0$$

$$\varepsilon(n) = E\left\{e_0^2(n)\right\} + E\left\{V_N^T(n)X(n)X^T(n)V_N\right\}$$

$$\varepsilon(n) = \varepsilon_{\min} + \varepsilon_{ex}(n) \tag{10.15a}$$

where

$$\varepsilon_{ex}(n) = E\left\{V_N^T(n)X(n)X^T(n)V_N\right\} \tag{10.15b}$$

Since $\varepsilon_{ex}(n)$ is a scalar,

$$\varepsilon_{ex}(n) = E\left\{ V_N^T(n)X(n)X^T(n)V_N(n) \right\}$$

$$= E\left\{ tr\left[V_N^T(n)X(n)X^T(n)V_N(n)\right]\right\}$$

$$= E\left\{ tr\left[X(n)X^T(n)V_N(n)V_N^T(n)\right]\right\}$$

$$= tr\left\{ E\left[X(n)X^T(n)V_N(n)V_N^T(n)\right]\right\}$$

Using independence property between $X(n)$ and $V_N(n)$ and Eqn. (10.10)

$$K(n) = E\left\{ V_N(n)\, V_N^T(n) \right\},$$

$$\varepsilon_{ex}(n) = tr\left\{ E\left[X(n)X^T(n)\right] E\left[V_N(n)V_N^T(n)\right] \right\} = tr\left[R_{NN}K(n)\right]$$

Therefore,

$$\varepsilon(n) = \varepsilon_{\min} + \varepsilon_{ex}(n) = \varepsilon_{\min} + tr\left[R_{NN}K(n)\right]$$

Applying similarity transform,

$$\varepsilon_{ex}(n) = tr\left[R_{NN}K(n)\right] = tr\left[M_{NN}\ \Lambda\ M_{NN}^T\ M_{NN}\ S(n)\ M_{NN}^T\right]$$

$$\varepsilon_{ex}(n) = tr\left[M_{NN}\ \Lambda\ S(n)\ M_{NN}^T\right] = tr\left[M_{NN}M_{NN}^T\ \Lambda\ S(n)\right]$$

$$\varepsilon_{ex}(n) = tr\left[\Lambda\ S(n)\right] = \sum_{i=1}^{N}\lambda_i s_i(n)$$

Substituting for $S_i(n)$ from Eqn. (10.14),

$$S_i(n) = \frac{\alpha\ \varepsilon_{\min}}{2-\alpha\ \lambda_i}$$

then,

$$\varepsilon_{ex}(n) = \varepsilon_{\min}\sum_{i=1}^{N}\frac{\alpha\ \lambda_i}{2-\alpha\ \lambda_i}$$

Also,

$$\lim_{n\to\infty}\ \varepsilon_{ex}(n) = \varepsilon_{\min}\sum_{i=1}^{N}\frac{\alpha\ \lambda_i}{2-\alpha\ \lambda_i}$$

The LMS misadjustment is,

$$M = \frac{\lim_{n\to\infty}\varepsilon_{ex}(n)}{\varepsilon_{\min}} = \sum_{i=1}^{N}\frac{\alpha\ \lambda_i}{2-\alpha\ \lambda_i}$$

$$M = \sum_{i=1}^{N}\frac{\alpha\ \lambda_i}{2-\alpha\ \lambda_i} \tag{10.16}$$

For small values of α and $M(=0.01,\ 10\%)$, $\alpha\ \lambda_i << 2$. Then

$$M = \sum_{i=1}^{N}\frac{\alpha\ \lambda_i}{2} = \frac{\alpha}{2}\sum_{i=1}^{N}\lambda_i = \frac{\alpha}{2}tr\left[R_{NN}\right] = \frac{\alpha}{2}N\varphi(0) = \frac{\alpha}{2}N\sigma_x^2$$

$$M = \frac{\alpha}{2}N\sigma_x^2 \tag{10.17}$$

The misadjustment is proportional to step size and filter length. The larger step size will result in larger variance or jitter in the weight vector. As the number of parameters (N) increases, the accuracy of estimation becomes poor and hence the misadjustment.

10.5 STABILITY OF LMS ALGORITHM

The stability of the LMS algorithm will be poorer than that of the method of steepest descent as the expectation operation is removed in the former and the weights are freer to take any values. That is, the variance of the weight vector for LMS will be more than that for method of steepest descent. Hence, it is expected that the LMS stability constraint to be much more severe than that for the method of steepest descent and this stability condition for LMS will be derived. From Eqn. (10.16),

$$M = \sum_{i=1}^{N} \frac{\alpha\,\lambda_i}{2-\alpha\,\lambda_i}$$

The desired value of M is in the range, $0 < M < 1$. Smaller the value of M better is the performance. Hence considering the maximum value of α that can be tolerated corresponds to $M = 1$, the value of α is given by,

$$M = \sum_{i=1}^{N} \frac{\alpha\,\lambda_i}{2-\alpha\,\lambda_i} = 1$$

$$\sum_{i=1}^{N} \frac{2\mu\,\lambda_i}{2-2\mu\,\lambda_i} = 1, \qquad \because \alpha = 2\mu$$

$$\sum_{i=1}^{N} \frac{\mu\,\lambda_i}{1-\mu\,\lambda_i} = 1$$

But

$$0 < \mu < \frac{1}{\sum_{i=1}^{N} \lambda_i}$$

Further, if $A + B < 1$, then

$$\frac{A}{1-A} + \frac{B}{1-B} < \frac{A+B}{1-(A+B)}$$

Using this property,

$$\sum_{i=1}^{N} \frac{\mu\,\lambda_i}{1-\mu\,\lambda_i} \le \frac{\mu\sum_{i=1}^{N}\lambda_i}{1-\mu\sum_{i=1}^{N}\lambda_i}$$

The value of μ , $\mu_{\max}$ that satisfies the equation

$$\frac{\mu_{\max}\sum_{i=1}^{N}\lambda_i}{\left(1-\mu_{\max}\sum_{i=1}^{N}\lambda_i\right)} = 1 \qquad (10.18)$$

also satisfies

$$\frac{\mu_{\max}\sum_{i=1}^{N}\lambda_i}{\left(1-\mu_{\max}\sum_{i=1}^{N}\lambda_i\right)} \le 1$$

Therefore, from Eqn. (10.18),

$$\mu_{\max}\sum_{i=1}^{N}\lambda_i = 1-\mu_{\max}\sum_{i=1}^{N}\lambda_i$$

$$\mu_{\max} = \frac{1}{2\sum_{i=1}^{N}\lambda_i} = \frac{1}{2tr\left[R_{NN}\right]}$$

As

$$tr\left[R_{NN}\right] = N\sigma_x^2$$

$$\mu_{\max} = \frac{1}{2N\sigma_x^2}$$

Since $\alpha = 2\mu$,

$$\alpha_{\max} = \frac{1}{N\sigma_x^2}$$

Therefore

$$0 < \alpha < \frac{1}{N\sigma_x^2} \qquad (10.19)$$

For the convergence of the *mean weight vector* of LMS (Eqn.105)),

$$0 < \alpha < \frac{2}{N\sigma_x^2}$$

Thus, *the maximum limit on the step size for the LMS is half that of the method of steepest descent(LMS mean weight vector convergence).* This is due to the fact that the jitter in the LMS gradient makes its convergence / stability more sensitive to the step size parameter α. Hence, for LMS, the upper bound for α should be much lower than that for method of steepest descent. In practice, to ensure convergence, the step size α is chosen such that

$$\frac{1}{100\,N\sigma_x^2} < \alpha < \frac{1}{10\,N\sigma_x^2} \qquad (10.20)$$

The steps involved in the LMS algorithm are:

1. $$\hat{d}(n) = \sum_{i=0}^{N-1} w_i(n)x(n-i)$$

2. $$e(n) = d(n) - \hat{d}(n)$$

3. $$w_i(n+1) = w_i(n) + \alpha\, e(n)x(n-i), \quad i = 0,1,\ldots\ldots, N-1$$

4. The step size α is chosen such that

$$0 < \alpha < \frac{1}{N\sigma_x^2}$$

The LMS algorithm will be modified for its robust operation, to achieve both faster convergence and better misadjustment and ease of implementation. The modifications include normalized LMS, noise robust Griffiths' LMS, variable stepsize LMS, delayed LMS, stable leaky LMS, filtered input LMS (FXLMS), better quality forward-backward LMS, efficient Laguerre LMS, linearly constrained LMS, and simple to implement version like sign-LMS.

10.6 NORMALIZED LMS ALGORITHM

For LMS, the weight adaptation rule (Eqn. (10.2) is

$$w_i(n+1) = w_i(n) + \alpha\, e(n)x(n-i), \quad i = 0, 1, \ldots, N-1$$

and the step size is given by (Eqn. 10.5)

$$0 < \alpha < \frac{2}{N\sigma_x^2}$$

The step size α can be fixed according to the criteria given above knowing σ_x^2, the input signal power and the order of the filter N. However, if the input signal characteristics / statistics change, then the value of α chosen may be no more appropriate and may lead to either too slow a convergence or even divergence. If the input signal statistics change in such a way that σ_x^2 increases, then the initially chosen value of α may become too high and the LMS algorithm can diverge. On the other hand, if σ_x^2 decreases, then the initially chosen step size α may become too small resulting in a slow convergence. Hence, it is essential to make the step size as a function of instantaneous input signal power $\sigma_x^2(n)$ and inversely proportional to it. That is, the step size should be defined as,

$$\alpha(n) = \frac{\alpha_0}{N\sigma_x^2(n)}, \qquad 0 < \alpha_0 < 1$$

Hence, the LMS adaptation rule is,

$$w_i(n+1) = w_i(n) + \frac{\alpha_0}{N\sigma_x^2(n)} e(n)x(n-i) \quad i = 0, 1, \dots, N-1 \tag{10.21}$$

This is known as *normalized LMS (NLMS) algorithm* as the step size is normalized by the instantaneous input signal power $\sigma_x^2(n)$. Thus, the NLMS algorithm is *robust to variations of the statistics of the input.* The LMS algorithm requires *(2N+1)* multiplications per iterations. In this, *N* multiplications are required for the output estimate and *(N+1)* multiplications for weight adaptation. Further, for NLMS, the computations for normalization depend on the type of estimation of input power used for normalization. The input signal power $\sigma_x^2(n)$ can be estimated either on block basis or on recursive approach or by exponential power estimation method (EPE).

(a) Block estimation of the input power

In this, the input power is given by

$$\sigma_x^2(n) = \frac{1}{N}\sum_{k=0}^{N-1} x^2(n-k) = \sigma_x^2(n-1) + \frac{x^2(n) - x^2(n-N)}{N} \tag{10.22a}$$

However, this requires $(N+1)$ storage locations.

(b) Recursive estimation of the input power

In this, the input power is given by

$$\sigma_x^2(n) = \beta\, \sigma_x^2(n-1) + (1-\beta)x^2(k), \quad 0 < \beta < 1 \tag{10.22b}$$

If

$$q(n) = x^2(n), p_x(n) = \sigma_x^2(n),$$
$$p_x(n) = \beta p_x(n-1) + (1-\beta)q(k), \quad 0 < \beta < 1$$

Taking z-transform both sides,

$$P_x(z) = \beta\, P_x(z)\, z^{-1} + (1-\beta)\, Q(z)$$
$$\frac{P_x(z)}{Q(z)} = \frac{(1-\beta)}{1-\beta\ z^{-1}}$$

This transfer function represents a first order lowpass filter with unity gain at DC, which averages $q(n)$. Closer the value of β to unity, better is the averaging as it moves the cut-off to a lower frequency. The equivalence between block and recursive estimations is given by $N \approx 1/(1-\beta)$.

In the weight vector adaptation (Eqn. 10.21), the variable stepsize is decided by α_0 and the input power $\sigma_x^2(n)$ as

$$\alpha(n) = \frac{\alpha_0}{N\sigma_x^2(n)}, \qquad 0 < \alpha_0 < 1$$

Defining

$$\alpha_x(n) = \frac{\alpha_0}{\sigma_x^2(n)}$$

The time constant for the ith mode is given by Eqn. (9.17b)

$$\tau_i = \frac{1}{\alpha_x \lambda_i}$$

λ_i is the ith eigenvalue of the autocorrelation matrix of the input $x(n)$. If λ_{av} is the average of the eigenvalues, then the average time constant τ_{av} is given by

$$\tau_{av} = \frac{1}{\alpha_x \lambda_{av}}$$

Further, as $\lambda_{av} = \sigma_x^2(n)$ and substituting for $\alpha_x = \alpha_0 / \sigma_x^2(n)$,

$$\tau_{av} = \frac{1}{\alpha_x \sigma_x^2(n)} = \frac{1}{\left(\dfrac{\alpha_0}{\sigma_x^2(n)}\right)\sigma_x^2(n)}$$

$$\tau_{av} = \frac{1}{\alpha_0}$$

Computation of the input power $\sigma_x^2(n)$ recursively involves a time constant that is given by

$$\tau_\sigma = \frac{1}{(1-\beta)}$$

Since $\sigma_x^2(n)$ estimation is a part of the LMS algorithm, it is better to have its time constant equal to that of the LMS algorithm, i.e., $\tau_\sigma = \tau_{av}$. This results in,

$$\frac{1}{(1-\beta)} = \frac{1}{\alpha_0}, \qquad \beta = (1-\alpha_0)$$

Thus, the time constant of power estimation is equal to that of the LMS algorithm and this has been found to have better dynamic adaptation for nonstationary signals.

(c) Exponential Power Estimation (EPE)

A good power estimator has the following features: (a) the estimate should follow the major changes of the input signal, (b) the estimation should not add additional computations and (c) the use of the estimator should not result in new complexity in existing setup like adaptive filter. From these points of view, both the block and recursive power estimators make the adaptive filter more complex and may have their own negative influence on the adaptive filter. But the EPE introduced by M. Bellanger and C.C. Evci in 1985 and its modified version are preferred as they do not involve any multiplication or division and their use results only in the complexity reduction as the estimate is represented as power of 2 (i.e., $2^{i(n)}$, $i(n)$ is an integer). Further, its exponential nature, leads to sufficiently fast response compared to those of block and recursive estimators (due to their additive nature) and also the EPE response can be controlled easily.

The EPE aims at to track the changing magnitudes of the input sequence by updating a variable and this is achieved by the following steps.

1. Compute

$$N(n) = |x(n)| 2^{[\lfloor i(n-1)/2 \rfloor}$$

2.

$$\begin{aligned} T(n) &= T(n-1)(1+2^{IT}), \quad N(n) > 1 \\ &= T(n-1)(1-2^{DT}), \quad N(n) < 1 \\ &= T(n-1), \qquad\qquad\quad N(n) = 1 \end{aligned}$$

In step 1, the magnitude of the present input sample $x(n)$ is divided by the magnitude estimate of the previous sample. In step 2, if the resulting ratio in step 1, is greater or less than unity, correspondingly the internal variable $i(n)$ is incremented or decremented.

3. Bound $T(n)$ as,

$$T_{\min} \le T(n) \le T_{\max}$$

4.

$$i(n) = \lfloor 2T(n) \rfloor$$

5.

$$\sigma_x^2(n) = 2^{i(n)} \tag{10.22c}$$

6.

$$\sigma_x(n) = 2^{\lfloor i(n)/2 \rfloor} \tag{10.22d}$$

The internal variable $T(n)$ is updated to track the envelope of the magnitude of the input sequence, $x(n)$ and $i(n)$ is an integer variable. The constants IT and DT are negative integers and determine the amounts of increment, $T(n-1)2^{IT}$ and decrement, $T(n-1)2^{DT}$ of $T(n)$ and hence control the speed of updating. The bounds $T_{\min}$ and $T_{\max}$ limit $T(n)$. The power and magnitude estimates of $x(n)$ are respectively given by $P_x(n)$ and $M_x(n)$. The upper limit $T_{\max}$ is decided by the maximum input signal level using the relation, $|x_{\max}(n)| = 2^{T_{\max}}$ and T is initialized to $T_{\max}$. Further, the lower limit $T_{\min}$ is decided by signal dynamic range (DR) and is given by $DR_{dB} = 6[T_{\max} - T_{\min}]$. The values of IT and DT are selected depending on the required response rate.

The ratio of increment to decrement (I/D) is decided by IT and DT. Generally, (I/D) is chosen to be high and can be 4 or 8.

The LMS adaptation is given by (Eqn. (10.21))

$$wi(n+1) = w_i(n) + \frac{\alpha_0}{N\sigma_x^2(n)} e(n)x(n-i), \quad i = 0,1,\dots N-1,$$

Let

$$S(n) = \frac{\alpha_0}{N\sigma_x^2(n)} \text{ and } \alpha_0 < 2.$$

Since $\sigma_x^2(n)$ is a power of 2, choosing (α_0 / N) also a power of 2, makes $S(n)$ to be expressed as $S(n) = 2^{-q(n)}$, $q(n)$ a positive integer. Therefore, the LMS adaptation can be written as

$$w_k(n+1) = w_k(n) + \left[2^{-q(n)} 2^{\lfloor i(n-k)/2 \rfloor} \right] e(n)\,\text{sgn}\left[x(n-k) \right], \quad k = 0,1,\dots N-1 \tag{10.23}$$

Thus, the simplicity of EPE provides a saving of *2(N + 1)* multiplications plus one division and *(N + 5)* multiplications one division, per sample; over the block and recursive methods, respectively.

The exponential nature of the power estimator together with the computational simplicity involved in implementing the adaptive filter makes it advantageous in terms of tracking the fast changing input signal envelope (like in speech) over that of constant step size LMS and the normalized LMS algorithms which use block and recursive power estimators.

10.6.1 Normalized LMS Algorithm by Step Size Optimisation

By optimizing the step size for which the error energy is minimum, the NLMS algorithm can be derived. Also by minimizing the Euclidian norm of the incremental weight vector, the NLMS can be derived. These two formulations for the NLMS will be considered.

For the weight vector adaptation for gradient algorithm is,

$$W_N(n+1) = W_N(n) - \mu \nabla_w \left[\varepsilon(n) \right]$$

where

$$\nabla_w \left[\varepsilon(n) \right] = -2[P_N - R_{NN} W_N(n)]$$

Therefore,

$$W_N(n+1) = W_N(n) + 2\mu(n) \left[P_N - R_{NN} W_N(n) \right]$$

The correction term is $\mu \nabla_w \left[\varepsilon(n) \right]$. If μ is large, the update suffers from gradient noise amplification. The normalised LMS algorithm seeks to avoid gradient noise amplification. To account for this, the step size is made varying $\mu(n)$ and optimised to minimise the error.

Now,

$$\varepsilon(n+1) = E\left\{ e^2(n+1) \right\}$$

Since

$$e(n+1) = d(n+1) - W_N^T(n+1)X_N(n+1),$$

$$\varepsilon(n+1) = E\Big(\big\{d(n+1) - W_N^T(n+1)X_N(n+1)\big)$$

$$\big(d(n+1) - X_N^T(n+1)W_N(n+1)\big)\Big\}$$

$$\varepsilon(n+1) = E\left\{ d^2(n+1) \right\} - E\left\{ W_N^T(n+1)X_N(n+1)d(n+1) \right\}$$

$$-E\left\{ X_N^T(n+1)W_N(n+1)d(n+1) \right\} + E\left\{ W_N^T(n+1)X_N(n+1)X_N^T(n+1)W_N(n+1) \right\}$$

$$\varepsilon(n+1) = E\left\{ d^2(n+1) \right\} - E\left\{ W_N^T(n+1)\underline{X_N(n+1)d(n+1)} \right\}$$

$$- E\left\{ \underline{d(n+1)X_N^T(n+1)}W_N(n+1) \right\} \tag{10.24}$$

$$+ E\left\{ W_N^T(n+1)X_N(n+1)X_N^T(n+1)W_N(n+1) \right\}$$

Since

$$E\left\{d(n+1)X_N^T(n+1)\right\} = E\left\{d(n+1)X_N(n+1)\right\} = P_N,$$

$$\sigma_d^2(n+1) = E\left\{ d^2(n+1) \right\}$$

Eqn.(10.24) becomes

$$\varepsilon(n+1) = \sigma_d^2(n+1) - W_N^T(n+1)P_N - P_N^T W_N(n+1) + W_N^T(n+1)R_{NN}W_N(n+1)$$

Substituting for,

$$W_N(n+1) = W_N(n) + 2\mu(n)\left[P_N - R_{NN}W_N(n)\right]$$

$$\varepsilon(n+1) = \sigma_d^2(n+1) - \left[W_N^T(n) + 2\mu(n)\left(P_N^T - R_{NN}W_N^T(n)\right) \right] P_N$$

$$- P_N^T\left[W_N(n) + 2\mu(n)\left(P_N - R_{NN}W_N(n)\right) \right]$$

$$+\left[W_N^T(n) + 2\mu(n)\left(P_N^T - R_{NN}W_N^T(n)\right) \right] \bullet R_{NN}\left[W_N(n) + 2\mu(n)\left(P_N - R_{NN}W_N(n)\right)\right]$$

As

$$\nabla\ \varepsilon(n) = -2[P_N - R_{NN}W_N(n)]$$

Therefore,

$$\varepsilon(n+1) = \sigma_d^2(n+1) - \left[W_N^T(n) - \mu(n)\nabla\ \varepsilon^T(n) \right] P_N - P_N^T\left[W_N(n) - \mu(n)\nabla\ \varepsilon(n)\right]$$

$$+\left[W_N^T(n) - \mu(n)\nabla\ \varepsilon^T(n) \right] R_{NN}\left[W_N(n) - \mu(n)\nabla\ \varepsilon(n)\right]$$

$$\begin{aligned}\varepsilon(n+1) = \sigma_d^2(n+1) &- \left[W_N^T(n) - \mu(n)\,\nabla\varepsilon^T(n) \right] P_N \\ &- P_N^T \left[W_N(n) - \mu(n)\,\nabla\varepsilon(n) \right] + W_N^T(n) R_{NN} W_N(n) \\ &- \mu(n)\,\nabla\varepsilon^T(n)\, R_{NN} W_N(n) - \mu(n)\, W_N^T(n) R_{NN} \nabla\varepsilon(n) \\ &+ \mu^2(n)\,\nabla\varepsilon^T(n) R_{NN}\ \nabla\varepsilon(n)\end{aligned} \tag{10.25}$$

Differentiating $\varepsilon(n+1)$ with respect to $\mu(n)$ and equating to zero,

$$\nabla\varepsilon^T(n)\left[P_N - R_{NN}W_N(n) \right] + \nabla\varepsilon(n)\left[P_N^T - W_N^T(n)R_{NN} \right] = -2\,\mu(n)\ \nabla\varepsilon^T(n)R_{NN}\nabla\varepsilon(n)$$

Again using

$$\nabla\varepsilon(n) = -2(P_N - R_{NN}W_N(n)),$$

$$-\nabla\varepsilon^T(n)\nabla\varepsilon(n)/2 - \nabla\varepsilon(n)\nabla\ \varepsilon^T(n)/2 = -2\,\mu(n)\ \nabla\varepsilon^T(n)R_{NN}\nabla\varepsilon(n)$$

$$\mu(n) = \frac{\nabla\varepsilon^T(n)\nabla\varepsilon(n) + \nabla\varepsilon(n)\nabla\varepsilon^T(n)}{4\nabla\varepsilon^T(n)R_{NN}\nabla\varepsilon(n)} \tag{10.26}$$

Now instantaneous autocorrelation is,

$$\hat{R}_{NN} = X_N(n)X_N^T(n)$$

and instantaneous gradient error,

$$\begin{aligned}\nabla\hat{\varepsilon}(n) &= -2\left[X_N(n)d(n) - X_N(n)X_N^T(n)W_N(n) \right] \\ &= -2\left[X_N(n)d(n) - X_N(n)\hat{d}(n) \right] \\ &= -2X_N(n)\left[d(n) - \hat{d}(n) \right] = -2X_N(n)e(n)\end{aligned}$$

$$\nabla\hat{\varepsilon}(n) = -2X_N(n)e(n)$$

Substituting these in Eqn.(10.26),

$$\mu_0(n) = \frac{4e^2(n)X^T(n)X(n) + 4e^2(n)X(n)X^T(n)}{4\left\{-2e(n)X^T(n)\right\}\left\{X_N(n)X_N^T(n)\right\}\left\{-2e(n)X(n)\right\}}$$

$$\mu_0(n) = \frac{e^2(n)X^T(n)X(n)}{2e^2(n)\left[X^T(n)X_N(n) \right]^2} = \frac{1}{2\left[X^T(n)X_N(n) \right]}$$

$$\mu_0(n) = \frac{1}{2\|\ X_N(n)\ \|^2} \tag{10.27}$$

The estimation of $P_x(n) = \|\ X_N(n)\ \|^2$ can be done by the two methods already considered.

The weight update relation is

$$W_N(n+1) = W_N(n) + 2\mu(n)\, X_N(n)e(n)$$

Further using the optimised value of $\mu(n) = \mu_0(n)$,

$$\mu_0(n) = \frac{1}{2\| X_N(n) \|^2}$$

The normalized LMS adaptation rule is

$$W_N(n+1) = W_N(n) + \frac{1}{\| X_N(n) \|^2} X_N(n)e(n), \tag{10.28}$$

Thus, the normalized LMS algorithm corresponds to *an optimum stepsize* algorithm. The step size is optimised in the sense that it minimizes the instantaneous error energy.

To avoid problems when $\| X_N(n) \|^2 \approx 0$, an offset δ (a small positive quantity) can be added and the normalized LMS algorithm becomes

$$W_N(n+1) = W_N(n) + \frac{\alpha}{\delta + \| X_N(n) \|^2} X_N(n)e(n) \tag{10.29}$$

The parameter α is introduced to have control over the step size. When the recursively estimated $\sigma_x^2(n)$ is used in the denominator instead of the squared norm of the vector $X_N(n)$, there is no necessity of use the offset δ as such an estimate does not become zero due to its memory.

10.6.2 NLMS by Euclidian Norm of Incremental Weight Vector Minimization

The normalized LMS algorithm also corresponds to minimization of the Euclidian norm of incremental change in the weight vector, i.e., $\delta W_N(n+1)$ in its successive updates with the constraint

$$d(n) = \sum_{k=0}^{N-1} w_k(n+1)x(n-k). \tag{10.30}$$

The weight incremental norm is given by

$$\|\delta W_N(n+1)\| = \sum_{k=0}^{N-1} \left[w_k(n+1) - w_k(n) \right]^2 \tag{10.31}$$

Such a constrained minimization is achieved by the Lagrange multiplier method. The cost function is formulated such that both the requirements are satisfied by its minimization and is given by

$$J(W_N(n+1),\lambda) = \|\delta W_N(n+1)\| + \lambda\left\{d(n) - \sum_{k=0}^{N-1} w_k(n+1)x(n-k)\right\} \quad (10.32)$$

$$J(W_N(n+1),\lambda) = \sum_{k=0}^{N-1}\left[w_k(n+1) - w_k(n)\right]^2 + \lambda\left\{d(n) - \sum_{k=0}^{N-1} w_k(n+1)x(n-k)\right\}$$

where λ is Lagrange multiplier. Here $W_N(n+1)$ is chosen such that $J(W_N(n+1),\lambda)$ is minimum. That is,

$$\frac{\partial J(W(n+1),\lambda)}{\partial W(n+1)} = 0$$

This results in

$$2\left[w_k(n+1) - w_k(n)\right] + \lambda x(n-k) = 0, \quad k = 0, 1, , \ldots, N-1$$

$$w_k(n+1) = w_k(n) + \frac{\lambda}{2}x(n-k), \quad k = 0, 1, , \ldots, N-1 \quad (10.33)$$

Multiplying by $x(n-k)$ summing over all k both sides,

$$\sum_{k=0}^{N-1} w_k(n+1)x(n-k) = \sum_{k=0}^{N-1} w_k(n)x(n-k) + \frac{\lambda}{2}\sum_{k=0}^{N-1} x^2(n-k)$$

Using the relation,

$$d(n) = \sum_{k=0}^{N-1} w_k(n+1)x(n-k)$$

it becomes

$$d(n) = \sum_{k=0}^{N-1} w_k(n)x(n-k) + \frac{\lambda}{2}\sum_{k=0}^{N-1} x^2(n-k)$$

$$\lambda = \frac{2\left[d(n) - \sum_{k=0}^{N-1} w_k(n)x(n-k)\right]}{\sum_{k=0}^{N-1} x^2(n-k)} = \frac{2e(n)}{\sum_{k=0}^{N-1} x^2(n-k)}$$

Substituting this value of λ in Eqn. (10.33),

$$w_k(n+1) = w_k(n) + \frac{1}{2}\frac{2e(n)}{\sum_{k=0}^{N-1} x^2(n-k)} x(n-k), \quad k = 0, 1, , \ldots, N-1$$

Inserting the parameter α to have control over the stepsize,

$$w_k(n+1) = w_k(n) + \frac{\alpha\, e(n)x(n-k)}{\sum_{k=0}^{N-1} x^2(n-k)}, \quad k = 0, 1, , \ldots, N-1 \tag{10.34}$$

This is nothing but the normalized LMS algorithm.

10.6.3 Convergence of NLMS Algorithm

The NLMS adaptation is given by

$$W_N(n+1) = W_N(n) + \frac{\alpha}{\| X_N(n) \|^2} X_N(n)e(n)$$

Substituting

$$e(n) = d(n) - W_N^T(n)X_N(n),$$

$$W_N(n+1) = W_N(n) + \frac{\alpha}{\| X_N(n) \|^2} X_N(n)\left[d(n) - W_N^T(n)X_N(n)\right]$$

$$W_N(n+1) = W_N(n) + \frac{\alpha}{\| X_N(n) \|^2} X_N(n)\left[d(n) - X_N^T(n)W_N(n)\right]$$

$$W_N(n+1) = \left[I_{NN} - \frac{\alpha\, X_N(n)X_N^T(n)}{\| X_N(n) \|^2} \right] W_N(n) + \frac{\alpha\, X_N(n)d(n)}{\| X_N(n) \|^2} \tag{10.35a}$$

In the above relation for NLMS, the terms are normalised by the input power. However, for LMS, it is not so and

$$W_N(n+1) = \left[I_{NN} - \alpha\, X_N(n)X_N^T(n) \right]\, W_N(n) + \alpha X_N(n)d(n) \tag{10.35b}$$

For LMS for stability(from Eqn.10.35a),

$$0 < \alpha < \frac{1}{tr\left[E\left\{ X_N(n)X_N^T(n) \right\} \right]}$$

From the analogy between LMS and NLMS (from Eqn.(10.35b)),

$$0 < \alpha < \frac{1}{tr\left[E\left\{\dfrac{X_N(n)X_N^T(n)}{\| X_N(n) \|^2}\right\}\right]}$$

But

$$tr\left[E\left\{\frac{X_N(n)X_N^T(n)}{X_N(n)X_N^T(n)}\right\}\right] = 1$$

Therefore, for NLMS

$$0 < \alpha < 1 \tag{10.36}$$

That is, *the NLMS convergence criterion is simpler than that for LMS* and *NLMS generally converges faster than the LMS algorithm since depending on the input power, it adjusts its step size.*

10.7 GRIFFITHS' LMS ALGORITHM (GLMS)

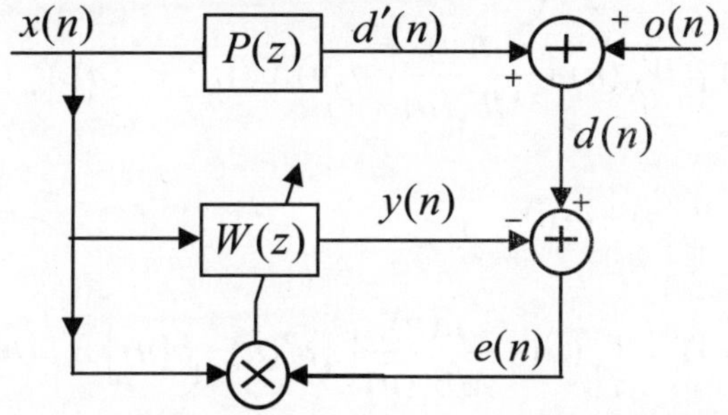

Fig. 10.3 System identification in the presence of observation noise.

In the adaptive filter algorithms considered, it has been assumed that the desired signal $d(n)$ the output of the system $P(z)$ for the input $x(n)$, is available (Fig. 10.3, assuming $o(n) = 0$) and using this the required error $e(n)$ can be formed. However, in certain applications (like active noise control), the desired signal is not available. Hence, it may not be possible to form $e(n)$ and apply the adaptation algorithm to such cases. But if the statistics, viz., the correlation between the input and the desired signal is available, then for such a case to make the LMS algorithm applicable, Griffiths' suggested a modification (Griffiths', 1967). The LMS adaptation is

$$W_N(n+1) = W_N(n) + \frac{\mu(n)}{ME_x(n)}\left[\, e(n)X_N^T(n) \,\right]$$

Since

$$e(n) = d(n) - y(n)$$

$$W_N(n+1) = W_N(n) + \frac{\mu(n)}{ME_x(n)}\left[d(n)X_N^T(n) - y(n)X_N^T(n) \right] \quad (10.37)$$

Replacing $d(n)X_N^T(n)$ by its corresponding correlation $P = E\left\{d(n)X_N^T(n)\right\}$, the Griffiths' algorithm is achieved and is

$$W_N(n+1) = W_N(n) + \frac{\mu(n)}{ME_x(n)}\left[P - y(n)X_N^T(n) \right] \quad (10.38)$$

That is, even in the absence of the desired signal $d(n)$, but by knowing the cross-correlation P between $d(n)$ and $x(n)$, the LMS algorithm can be used.

Presently, for the system identification, the Griffiths' algorithm will be used to suppress the effect of observation noise appearing in the gradient of the LMS adaptation. This algorithm will be referred to as GLMS algorithm. In Fig. 10.3, $d'(n)$ is the output of the system $P(z)$ to be identified with the input $x(n)$. The observation noise $o(n)$ gets added to the output $d'(n)$ to produce the desired signal $d(n)$. The input $x(n)$ is also fed to the model $W(z)$ that generates the output $y(n)$. The LMS adaptation is given by (Eqn.10.37)

$$W_N(n+1) = W_N(n) + \frac{\mu(n)}{ME_x(n)}\left[d(n)X_N^T(n) - y(n)X_N^T(n) \right]$$

Since

$$d(n) = d'(n) + o(n)$$

$$W_N(n+1) = W_N(n) + \frac{\mu(n)}{ME_x(n)}\left[\{d'(n) + o(n)\} X_N^T(n) - y(n)X_N^T(n) \right]$$

$$W_N(n+1) = W_N(n) + \frac{\mu(n)}{ME_x(n)}\left[e'(n)X_N^T(n) \right] + \frac{\mu(n)}{ME_x(n)}\left[o(n)X_N^T(n) \right] \quad (10.39)$$

In the above equation, on the right hand side only the first two terms should have been present. The second term is due to the true error $e'(n) = d'(n) - y(n)$. The third term is undesirable as its is due to observation noise $o(n)$. Since in LMS algorithm only instantaneous values are considered, this third term will not vanish. Further, this term will disturb the convergence process and may even lead to divergence. *Thus the effect of observation noise is to cause damage to the convergence process.* But as $x(n)$ and $o(n)$ are uncorrelated, their expectation will become zero. This is the motivation to use the idea of Griffiths' algorithm of using the correlation rather than instantaneous values of the product between input $x(n)$ and the desired signal $d(n)$. Such a modification reduces the effect of observation noise in the LMS gradient.

On taking expectation,

$$E[W_N(n+1)] = E[W_N(n)] + \frac{\mu(n)}{ME_x(n)}\left\{ E\left[d(n)X_N^T(n) \right] - E\left[y(n)X_N^T(n) \right]\right\} \quad (10.40)$$

$$E_x(n) = \beta E_x(n-1) + (1-\beta)x^2(n), \qquad 0 < \beta < 1$$

The first quantity within the flower brackets (Eqn. (10.40)) is the cross-correlation between the desired signal and the input. Denoting this term by $G(n)$ and substituting for $d(n)$ as,

$$d(n) = d'(n) + o(n)$$

where $d'(n)$ is the system output for the input $x(n)$ and $o(n)$ is the observation noise at the system output. Therefore

$$G(n) = E\left[d(n)X_N^T(n) \right] = E\left[d'(n)X_N^T(n) \right] + E\left[o(n)\, X_N^T(n) \right]$$

Since the input $x(n)$ and observation noise $o(n)$ are independent, $E\left[o(n)\, X_N^T(n) \right] = 0$. Therefore,

$$G(n) = E\left[d'(n)\, X_N^T(n) \right] \quad (10.41a)$$

. By considering

$$E\left[e(n)\, X_N^T(n) \right] = G(n) - E\left[y(n)\, X_N^T(n) \right],$$

the observation noise effect could have been removed, however this will make second the term $y(n)\, X_N^T(n)$ on RHS of the above equation, involving the output of the adaptive filter $y(n)$, sluggish due to its computation by using recursive filtering similar to Eqn. (10.41c)which is not desirable. Further, as instantaneous LMS algorithm, uses instantaneous quantities and also as in the second term $y(n)\, X_N^T(n)$, the observation noise is not directly present as in the first term $G(n) = E\left[d(n)X_N^T(n) \right]$, the expectation operation for the second term can be removed and the LMS algorithm becomes,

$$W_N(n+1) = W_N(n) + \frac{\mu(n)}{ME_x(n)}\left\{ G(n) - y(n)X_N^T(n)\right\} \quad (10.41b)$$

or

$$w_i(n+1) = w_i(n) + \frac{\mu(n)}{ME_x(n)}\left\{ G_i(n) - y(n)x(n-i)\right\}, \quad i = 0, 1, \ldots, N-1$$

$$G_i(n) = \beta_g\, G_i(n-1) + \left(1-\beta_g\right) d(n)x(n-i), \quad 0 < \beta_g < 1 \tag{10.41c}$$

It is important to note that $G(n) = E[d(n)X_N^T(n)]$ or $G_i(n) = E\left[d(n)x(n-i)\right]$ is not readily available and is estimated on a sample to sample basis recursively using Eqn.(10.41c). The value of β_g in the range $(0 < \beta_g < 1)$ decides the averaging effect and the degree of noise rejection. The closer the value of β_g to unity better is the noise rejection. Also if the process involved $d(n)$ and $x(n)$ are nonstationary, then β_g value must be less than unity and if they are stationary, its value can be close to unity giving better noise rejection.

Further defining,

$$\rho_G(n) = \mathbf{G}(n) - y(n)X^T(n) \tag{10.41d}$$

where $G(n) = E[d(n)X_N^T(n)]$, $\rho_G(n)$ is referred to as *Griffiths' cross-correlation which is free from the effect of observation noise component,* $o(n)$.

$$W_N(n+1) = W_N(n) + \frac{\mu(n)\rho_G(n)}{ME_x(n)}$$

Thus the use of Griffiths' correlation reduces the effect of the observation noise on the gradient for the LMS algorithm (S.V. Narasimhan, S. Veena and Lokesha. H, 2010)

10.8 VARIABLE STEP SIZE LMS ALGORITHM

Generally, prior to the convergence a relatively larger step size is required to reach quickly the optimal point on the error surface. However, after convergence as a large step size results in a large convergence error/ misadjustment, there is a necessity to reduce the step size. Hence, for a given step size, there is a trade-off between convergence rate and error. To have a faster convergence, prior to convergence the step size should be large and have to a tolerable convergence error, the step size has to be reduced as the convergence is reached. That is, a variable step size is desirable to a fixed step size. Prior to convergence, the sign of the gradient will remain same and it will not change. But after convergence, its sign continuously alters as the weight vector jitters about its optimum value. This concept was used as a basis in arriving at a variable step size algorithm. For this, the weight adaptation rule is given by

$$W_N(n+1) = W_N(n) + U(n)\; X_N(n)\; e(n) \tag{10.42}$$

where

$$U(n) = \begin{bmatrix} \mu_0(n) & 0 & \cdots & 0 \\ 0 & \mu_1(n) & \cdots & 0 \\ \vdots & \vdots & \ddots & \vdots \\ 0 & 0 & \cdots & \mu_{N-1}(n) \end{bmatrix}$$

$\mu_i(n)$ is increased or decreased by a factor γ depending on the gradient $\nabla_i = x(n-i)e(n)$. If the sign of ∇_i alternates for m_0 successive samples, then $\mu_i(n)$ is decreased by a factor γ. If there is no sign change of ∇_i over m_1 successive samples, then $\mu_i(n)$ is increased by a factor γ. Otherwise $\mu_i(n)$ is left unaltered.

$$\mu_i(n) = \begin{cases} \gamma^{-1}\ \mu_i(n-1) & \textit{sign alternates} \\ \gamma\ \mu_i(n-1) & \textit{sign remains same} \\ \mu_i(n-1) & \textit{otherwise} \end{cases} \tag{10.43a}$$

Initially,

$$\left.\begin{aligned} &\mu_i(0) = \mu_{\max}, \\ &0 < \mu_{\max} < \frac{2}{\lambda_{\max}} \\ &\textit{and}\ \ \mu_{\min} < \mu_i(n) < \mu_{\max} \end{aligned}\right\} \tag{10.43b}$$

γ is generally chosen as powers of 2, to avoid multiplications. Further, $m_0 = 3$ and $m_1 = 3$ are chosen and can be changed depending on the application. The value of the step size is bound by the maximum and minimum values. The limit on the maximum value avoids possible divergence. The limit on the minimum step size avoids possible adaptation stalling (sleep mode).

The above variable step size algorithm is not only computationally complex but also requires more memory. It is also sensitive to observation noise (in a system identification case) and other interfering signals. This is because the observation noise and interferences can change the sign of the gradient, which may not be really due to the actual error. In view of this to take care of the observation noise and the interferences, many variable step size algorithms have been proposed. In this direction, the error energy is taken as an index to change the step size and it is given by

$$\mu(n) = \beta\ \mu(n-1) + \gamma\ e^2(n) \qquad 0 < \beta, \gamma < 1 \tag{10.44a}$$

However, this also suffers from the problem of observation noise and cannot reflect the actual error magnitude. The correlation of the error at unity lag was used as in index and the stepsize is given by

$$\mu(n) = \alpha\ \mu(n-1) + \gamma\ p^2(n), \qquad 0 < \gamma < 1 \tag{10.44b}$$

$$p(n) = \beta\, p(n-1) + (1-\beta)e(n)e(n-1), \qquad 0 < \beta < 1$$

This will provide immunity only for the white observation noise and fails for coloured noise, which has correlation at higher lags also.

To cater to both for the white and coloured observation noise, the *correlation between the input and error,*

$$\rho = E\left[x(n)\, e(n)\right] \tag{10.44c}$$

was considered. This correlation is immune to observation noise and interference which get associated with the error as they do not have any correlation with the input. Further, ρ decreases as the convergence improves and on convergence as the input and the error are orthogonal, it becomes zero. Hence, the step size was made proportional to ρ. That is, when $W_N(n)$ is away from the optimal solution W_N^*, the correlation ρ is high and the adaptive filter is said to be in *active state* with a relatively large step size. But after convergence, the input and the error are nearly uncorrelated and the step size is set close to zero, putting the adaptation algorithm into a *sleep state* also known as stall mode. The LMS algorithm realized using an estimate of correlation between the input and the error signal is given by

$$W_N(n+1) = W_N(n) + \mu(n)\ e(n)X_N(n)$$

$$\mu(n+1) = \alpha\, \rho(n) \tag{10.44d}$$

The correlation $\rho(n)$ is computed recursively as

$$\rho(n) = \beta\, \rho(n-1) + \left(1-\beta\right) e(n)x(n), \quad 0 < \beta < 1 \tag{10.44e}$$

This algorithm is generally known as *correlation LMS* and provides the variable step size behaviour which *is robust to observation noise and interferences* (T.J. Shan and T.Kailath,1988). But its main disadvantage is that the *step size may become negative and* may go to *stall mode* where the adaptation stops due to a very small step size less than that can be detected by the finite word length arithmetic.

Okello suggested a modification for the correlation LMS algorithm. Here *the correlation corresponding to individual weights instead of overall correlation* is considered. The correlation for i^{th} weight is given by

$$\rho_i(n) = \beta\, \rho_i(n-1) + \left(1-\beta\right) e(n)x(n-i), \quad 0 < \beta < 1 \tag{10.45a}$$

Further, the Okello's step size adaptation is given by (Okello, 1998)

$$\mu(n+1) = \alpha\, \mu(n-1) + \sigma \sum_{i=0}^{N-1} \rho_i^2, \quad 0 < \alpha, \sigma < 1 \tag{10.45b}$$

The *sum of the squared values of* $\rho_i(n)$ *avoids the negative step size* and the term $\alpha\ \mu(n-1)$ *avoids stalling*. The mechanism behind this method will be considered

$$\rho(n) = E[e(n)X^T(n)] = E[\{d(n) - y(n)\}X_N^T(n)]$$

$$\rho(n) = [\rho_0(n), \rho_1(n), \cdots, \rho_{M-1}(n)],$$

$$\rho_i(n) = E\left[e(n)x(n-i)\right] \tag{10.45c}$$

Here, the desired signal is $d(n) = d'(n) + o(n)$, $d'(n)$ is system output and $o(n)$ is the observation noise. As $x(n)$ and $o(n)$ are independent,

$$E[o(n)\ X^T(n)] = 0,$$

$$\rho(n) = E[\{d'(n) - y(n)\}X_N^T(n)].$$

Therefore $\rho(n)$ *is free from observation noise.*

If W^* is the optimal solution and assuming that it matches the actual system $P(z)$, then

$$d(n) = W_N^{*T} X_N(n) + o(n)$$

and

$$\rho(n) = E[\{d(n) - y(n)\}X_N^T(n)]$$

becomes

$$\rho(n) = E\left[\{W_N^{*T} X_N(n) + o(n) - y(n)\}X_N^T(n)\right]$$

$$\rho(n) = E\left[\{W_N^{*T} X_N(n) + o(n) - W_N^T(n)X_N(n)\}X_N^T(n)\right]$$

Using

$$E[o(n)\ X_N^T(n)] = 0,$$

$$\rho(n) = E\left[\{W_N^{*T} X_N(n) - W_N^T(n)X_N(n)\}X_N^T(n)\right]$$

$$= E\{W_N^{*T} - W_N^T(n)\}E\{X_N(n)X_N^T(n)\}$$

If the expected value of difference weight vector is $\boldsymbol{\theta}(n)$ given by

$$\boldsymbol{\theta}(n) = E[W_N^* - W_N(n)]$$

then,

$$\rho(n) = \theta^T R_{NN}$$

where $R_{NN}(n)$ is the autocorrelation matrix of $X_N(n)$,

$$R_{NN}(n) == E[X_N(n)X_N^T(n)]$$

Now

$$\mu(n+1) = \alpha\ \mu(n-1) + \sigma\sum_{i=0}^{N-1}\rho_i^2, \quad 0 < \alpha, \sigma < 1$$

can be written as

$$\mu(n) = \alpha\ \mu(n-1) + \sigma\ \rho(n)\rho^T(n)$$

Substituting for

$$\rho(n) = \theta^T R_{NN},$$

$$\mu(n) = \alpha\ \mu(n-1) + \sigma\ \theta^T(n)R_{NN}(n)R_{NN}(n)\theta(n) \tag{10.45d}$$

As convergence is reached, the difference weight vector $\boldsymbol{\theta}(n)$ decreases and the magnitude of the elements $\rho(n)$ will approach zero value. Thus, the cross-correlation $\rho(n)$ enables to achieve a smaller step size as convergence is approached and this ensures reduced misadjustment.

This being a NLMS algorithm, for stability the initial value for $\mu(n)$ is set close to unity and the lower limit for $\mu(n)$ is fixed to prevent adaptation stalling and to provide minimum misadjustment tolerable.

10.8.1 Griffiths' Variable Step Size LMS Algorithm

The GLMS algorithm discussed in section 10.6 takes care of the effect of the observation noise on the *weight vector gradient*. On the other hand, the Okello's variable stepsize LMS (VLMS) algorithm of section 10.7 estimates the *scale factor for the step size* which is less affected by observation noise. The present method to be described combines these two algorithms to achieve an algorithm whose gradient and scale factor for step size, both are less affected by the observation noise.

The Griffiths' cross-correlation $\rho_G(n)$ (Eqn.(10.41d)) is given by,

$$\rho_G(n) = G(n) - y(n)X_N^T(n) \tag{10.46a}$$

where

$$G(n) = E\left[d(n)X_N^T(n)\right]$$

The crosscorrelation $\rho(n)$ between in put $x(n)$ and the desired signal $e(n)$ is given by

$$\rho(n) = E\left[e(n)X_N^T(n)\right] = E\left[\{d(n)-y(n)\}X_N^T(n)\right]$$

$$\rho(n) = E\left[d(n)X_N^T(n)\right] - E\left[y(n)X_N^T(n)\right]$$

$$\rho(n) = G(n) - E\left[y(n)X_N^T(n)\right] \qquad (10.46b)$$

Comparing Eqns.(10.46a) and (10.46b), they differ only in their second term. In Eqn.(10.46b), for the second term, the expectation can be removed due to use of instantaneous quantities in the LMS algorithm and also as noise is not directly present in it. This permits to use the Griffiths' correlation both for the adaptation of the gradient and the step size. It is important to note that as $y(n)$ is a time varying quantity and averaging quantity which contains this in $\rho(n)$ makes the adaptation process sluggish justifying the use of $\rho_G(n)$ instead of $\rho(n)$. Therefore, the LMS adaptation rule becomes

$$\hat{w}_i(n+1) = \hat{w}_i(n) + \frac{\mu(n)}{ME_x(n)}\{G_i(n) - y(n)x(n-i)\} \qquad (10.46c)$$

and

$$\mu(n) = \alpha\,\mu(n-1) + \sigma\sum_{i=0}^{N-1}\left[G_i(n) - y(n)x(n-i)\right]^2, 0 < \alpha, \sigma < 1 \qquad (10.46d)$$

$$G_i(n) = \beta_g\, G_i(n-1) + \left(1-\beta_g\right) d(n)x(n-i), \quad 0 < \beta_g < 1 \qquad (10.46e)$$

Here β_g decides the cut-off frequency and the degree of observation noise rejection.
The Eqn.(10.46c) can be expressed as

$$\hat{w}_i(n+1) = \hat{w}_i(n) + \mu(n)\nabla_i(n) \qquad (10.46f)$$

where

$$\nabla_i(n) = \frac{\rho_{Gi}(n)}{ME_v(n)}, \quad \rho_{Gi} = \left[\hat{G}_i(n) - y(n)x(n-i)\right]$$

$$\mu(n) = \alpha\,\mu(n-1) + \sigma\sum_{i=0}^{N-1}\left[\rho_{Gi}(n)\right]^2 \quad 0 < \alpha, \sigma < 1$$

The second term of Eqn. (10.46f) is the product of step size $\mu(n)$ and the weight vector adaptation gradient $\nabla(n)$. *This algorithm not only uses noise-free step size $\mu(n)$ but also noise-free* *gradient* (S.V. Narasimhan, S. Veena and Lokesha. H, 2010). *However, other variable step size algorithms have used only noise free step-size and not the noise free gradient* $\nabla(n)$, *though the latter is available.* In this algorithm, a high value of σ close to unity and a small value of α enable the step size to track the changes in the error at a faster rate for nonstationary signals, whereas a small σ value and large α are acceptable for stationary signals. The noise-free $\rho_G(n)$ used both for the gradient and step size adaptation in this algorithm significantly improves the convergence performance. This algorithm does not require any explicit estimation of noise power, which is necessary in some other algorithms.

10.8.2 System Identification Example

For illustration, a system $S(z)$ of impulse response length 128 samples for an acoustic path (Fig.10.4a) is used. Here, the SNR is with reference to clean desired signal (system output) and the observation noise. The performance of the algorithms is evaluated based on the following measures:

(a) The *estimation error* $\Delta S(n)$, deviation of the estimate $\hat{S}(z)$ from $S(z)$ is given by

$$\Delta S(n) = \left\| \left(\hat{\mathrm{S}}(n) - \mathrm{S} \right)^2 \right\| / \left\| \mathrm{S}^2 \right\| = \sum_{k=0}^{M-1} \left(\hat{s}_k(n) - s_k \right)^2 / \sum_{k=0}^{M-1} s_k^2$$

Ensemble average of $\Delta S(n)$ expressed in dB,

$$\Delta S_{AV}(n)\Big|_{dB} = 10\log_{10}\left\{ \frac{1}{K} \sum_{i=1}^{K} \Delta S_i(n) \right\}$$

(b) The *smoothed ensemble average square error* (SEASE) expressed in dB, $\xi_{Sd}(n)$

$$\xi_{Sd}(n) = 10\log_{10}(\xi_S(n)),$$

$$\xi_S(n) = \lambda\, \xi_S(n-1) + (1-\lambda)\xi(n),$$

$$\xi(n) = \frac{1}{K} \sum_{i=1}^{K} e_i^2(n), \quad 0 < \lambda < 1$$

$e_i(n)$ - Error for i^{th} realization, Number of realizations $K = 50$ and $\lambda = 0.99$ are considered.

Performance comparison of NLMS, Okello's and VGLMS algorithms:

Table 10.1a: Parameters for system identification algorithms at SNR= - 10dB

Parameter	μ	$\mu_{\max}$, $\mu_{\min}$	β_v	β_{vs}	β_g	α, σ
NLMS	1 and 0.005		0.99			
Okello		1, 0.005	0.99	0.9999		0.99, 0.99
VGLMS		1, 0.005	0.99		0.9	0.99, 0.0001

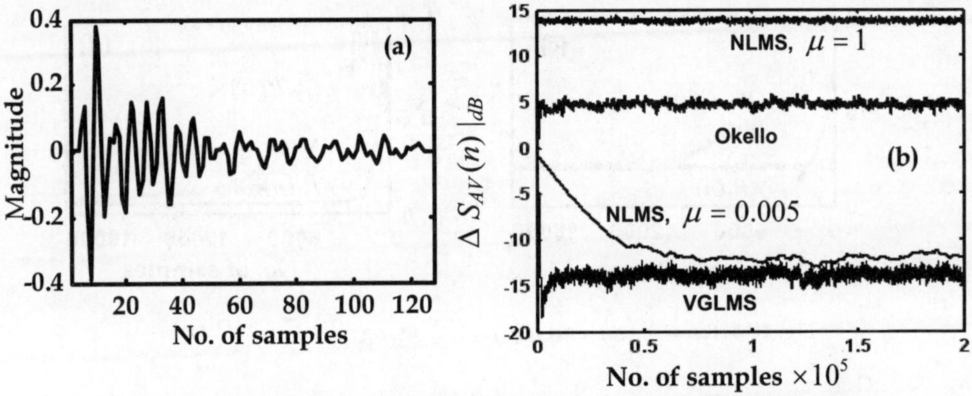

Fig. 10.4 (a) The impulse response of acoustic path and (b) Comparison of SEASE for NLMS, Okello's and VGLMS algorithms.

For this, the input $v(n)$ is a zero mean unit variance Gaussian white noise and the colored observation noise $o(n)$ is a sum of three sinusoids at 100Hz, 200Hz and 300Hz with additive Gaussian noise. The performance of various algorithms is evaluated for an *SNR* of *-10dB* and using the parameters given in Table 10.1a.

The NLMS algorithm with $\mu = 1$ is considered for faster convergence rate and $\mu = 0.005$ for minimum misadjustment and these are compared with VGLMS algorithm. The NLMS with $\mu = 0.005$ is able to achieve better, $\Delta S_{AV}(n)\big|_{dB}$ but at the cost of convergence rate. The VGLMS has a faster convergence rate and lesser estimation error compared to NLMS algorithm (Fig. 10.4b). In Okello's algorithm due to the use of noisy gradient, the step size is unable to reach $\mu_{\min} = 0.005$ and this makes the performance of Okello's algorithm poorer than VGLMS.

Performance comparison with $\mathbf{P}(n)$ *and* $\mathbf{P}_G(n)$ *in VGLMS algorithm*

For this, the input $v(n)$ is a zero mean unit variance Gaussian white noise and the observation noise is a colored noise (50 Hz fundamental and harmonics at 100Hz, 150Hz, 200Hz, 250Hz and additive random noise) and the SNR is 10dB. The parameters used are given in Table 10.1b.

Table 10.1b: Parameters for using $\mathbf{P}(n)$ and $\mathbf{P}_G(n)$ at SNR = 10dB

Parameter	μ_{max}, μ_{min}	β_v	β_{vs}	β_g	α, σ
for $\mathbf{P}(n)$	1, 0.01	0.99	0.9999		0.99, 0.99
for $\mathbf{P}_G(n)$	1, 0.01	0.99		0.9	0.99, 0.0001

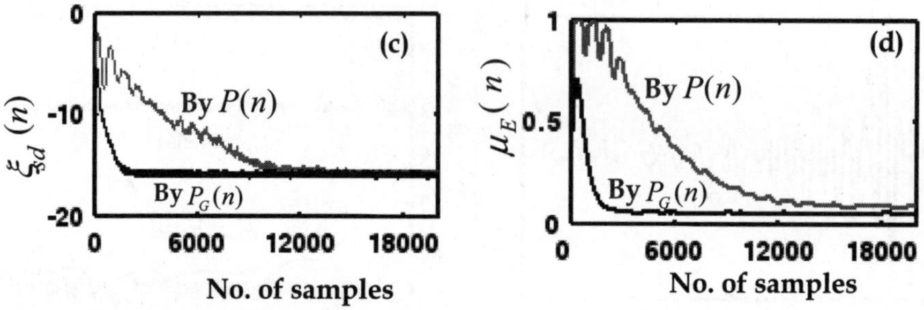

Fig. 10.4 (c) $\xi_{sd}(n)$ and (d) $\mu_E(n)$ using $P(n)$ and $P_G(n)$

The indices $\xi_{sd}(n)$ and $\mu_E(n)$ (the averaged step-size over 50 trials), indicate that the use of $P(n)$ makes the convergence process sluggish (Fig. 10.4 (c) and Fig. 10.4 (d)) compared to using $P_G(n)$ justifying the choice of $P_G(n)$ for VGLMS algorithm.

10.9 EFFECT OF DELAY ON LMS ALGORITHM

The system identification situation where the output of the adaptive filter $W(z)$ is available only after it has been delayed by k samples is considered (Fig. 10.5). Let the delayed output of the adaptive filter be represented by $y(n)$. The error signal $e(n)$ that controls the adaptive filter is given by

$$e(n) = d(n) - y(n)$$

Further,

$$e(n) = d(n) - W_N^T(n-k)X_N(n-k)$$

The goal is to minimize the mean square error defined as

$$\varepsilon(n) = E\left[e(n)^2\right] \tag{10.47a}$$

$$\varepsilon(n) = E\left[d^2(n) + \left(W_N^T(n-k)X_N(n-k) \right)^2 - 2d(n)W_N^T(n-k)X_N(n-k) \right]$$

$$= E\left[d^2(n) \right] + E\left[\left(W_N^T(n-k)X_N(n-k) \right) \right]^2 - 2E\left[d(n)W_N^T(n-k)X_N(n-k) \right]$$

$$= E\left[d^2(n) \right] + W_N^T(n-k)E\left[X_N(n-k)X_N^T(n-k) \right] W_N(n-k) - 2W_N^T(n-k)E\left[d(n)X_N(n-k) \right]$$

$$\varepsilon(n) = E\left[d^2(n) \right] + W_N^T(n-k)R_{NN}W_N(n-k) - 2W_N^T(n-k)P_N \quad (10.47b)$$

Where,

$$R_{NN} = \left[X_N(n-k)X_N^T(n-k) \right], \quad P_N = \left[d(n)X_N(n-k) \right]$$

The optimum weight vector that produces the minimum mean square error can be found by equating the gradient of the mean square error to zero.

$$\nabla\ \varepsilon(n) = \partial\ \varepsilon(n) / \partial W_N(n-k)$$

$$\nabla\ \varepsilon(n) = 2R_{NN}W_N(n-k) - 2P_N \quad (10.47c)$$

Thus, the optimum weight vector, W^* is

$$W^* = R_{NN}^{-1}P_N$$

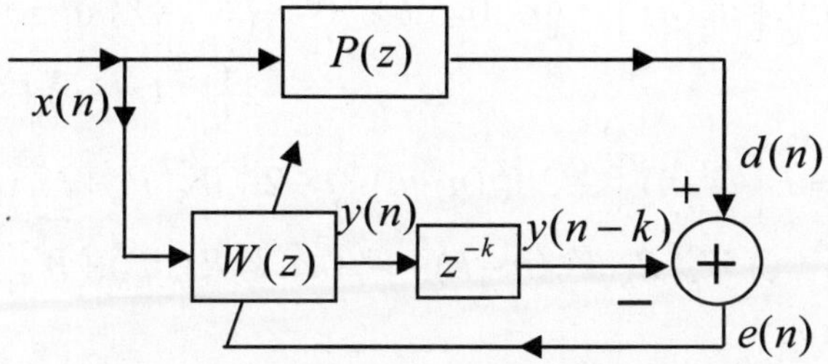

Fig. 10.5 LMS adaptive filter with a delay.

The expression for the mean square error

$$\varepsilon(n) = E\left[d^2(n) \right] + W_N^T(n-k)R_{NN}W_N(n-k) - 2W_N^T(n-k)P_N$$

Thus, the optimum weight vector, W^* is

$$W^* = R_{NN}^{-1} P_N$$

The expression for the mean square error

$$\varepsilon(n) = E\left[d^2(n)\right] + W_N^T(n-k)R_{NN}W_N(n-k) - 2W_N^T(n-k)P_N$$

For the optimum weight vector

$$W_N(n-k) = W_N^* \text{ and } \varepsilon(n) = \varepsilon_{\min}(n)$$

Therefore,

$$\varepsilon_{\min} = E\left[d^2(n)\right] + W_N^{*T} R_{NN} W_N^* - 2P_N^T W_N^*$$

$$= E\left[d^2(n)\right] + W_N^* P_N^T - 2P_N^T W_N^*$$

$$= E\left[d^2(n)\right] - P_N^T W_N^* \tag{10.47d}$$

Let

$$V_N(n-k) = \left[\, W_N(n-k) - W_N^* \,\right]$$

$$\varepsilon(n) = E\left[d^2(n)\right] + W_N^T(n-k)R_{NN}W_N(n-k) - 2W_N^T(n-k)P_N$$

becomes

$$\varepsilon(n) = E\left[d^2(n)\right] + \left\{V_N^T(n-k) + W_N^{*T}\right\} R_{NN} \left\{V_N(n-k) + W_N^*\right\} - 2\left\{V_N^T(n-k) + W_N^{*T}\right\} P_N$$

$$\varepsilon(n) = E\left[d^2(n)\right] - 2V_N^T(n-k)P_N - 2 + W_N^{*T} P_N + V_N^T(n-k)R_{NN}V_N(n-k) + W_N^{*T} R_{NN} V_N(n-k) + V_N^T(n-k)\underline{R_{NN}W_N^*} + W_N^{*T}\, \underline{R_{NN}W_N^*}$$

$$= E\left[d^2(n)\right] - 2V_N^T(n-k)P_N - 2 + W_N^{*T} P_N + V_N^T(n-k)R_{NN}V_N(n-k) + W_N^{*T} R_{NN} V_N(n-k) + V_N^T(n-k)P_N + W_N^{*T} P_N \quad \because P_N = R_{NN}W_N^*$$

$$\varepsilon(n) = E\left[d^2(n)\right] - 2V_N^T(n-k)P_N - W_N^{*T} P_N + V_N^T(n-k)R_{NN}V_N(n-k) + W_N^{*T} R_{NN} V_N(n-k) + V_N^T(n-k)P_N$$

$$= E\left[d^2(n)\right] - 2V_N^T(n-k)P_N - W_N^{*T}P_N + V_N^T(n-k)R_{NN}V_N(n-k)$$
$$+\left(\underline{R_{NN}W_N^*}\right)^T V_N(n-k) + V_N^T(n-k)P_N$$

$$= E\left[d^2(n)\right] - \underline{2V_N^T(n-k)P_N} - W_N^{*T}P_N + V_N^T(n-k)R_{NN}V_N(n-k)$$
$$+\underline{P_N^T V_N(n-k)} + \underline{V_N^T(n-k)P_N}$$

As

$$P_N^T V_N(n-k) = V_N^T(n-k)P_N$$

$$\varepsilon(n) = E\left[d^2(n)\right] - W_N^{*T}P_N + V_N^T(n-k)R_{NN}V_N(n-k)$$

Since,

$$\varepsilon_{\min} = E\left[d^2(n)\right] - W_N^{*T}P_N$$

$$\varepsilon(n) = \varepsilon_{\min} + V_N^T(n-k)R_{NN}V_N(n-k)$$

The adaptive filter coefficient updating equation for the gradient descent algorithm is

$$W_N(n+1) = W_N(n) - \mu\,\nabla\,\varepsilon(n) \tag{10.47e}$$

From Eqn. (10.47c),

$$\nabla\varepsilon(n) = 2R_{NN}W_N(n-k) - 2P_N$$

Eqn.(10.47c) becomes

$$W_N(n+1) = W_N(n) + 2\mu\left[P_N - R_{NN}W_N(n-k)\right]$$

As the input autocorrelation matrix, R_{NN} is symmetric, it can be diagonalized as

$$R_{NN} = M_N\Lambda_{NN}M_N^T,$$

where M is the orthonormal modal matrix of R_{NN} (the matrix of eigen vectors) and Λ_{NN} is the diagonal matrix of eigen values of R_{NN}.

Defining

$$V_N'(n) = M_N^T V_N(n),$$

the mean square error

$$\varepsilon(n) = \varepsilon_{\min} + V_N^T(n-k)R_{NN}V_N(n-k)$$

becomes

$$\varepsilon(n) = \varepsilon_{\min} + V_N^T(n-k)\, M_N\, \Lambda_{NN}\, M_N^T\, V_N(n-k)$$

Since

$$\left[V_N'(n)\right]^T = \left[M_N^T V_N(n)\right]^T = V_N^T(n)M_N,$$

$$\varepsilon_n = \varepsilon^* + V_N'^T(n-k)\Lambda_{NN}V_N'(n-k)$$

With the definitions

$$W_N' = M_N^T W_N \text{ and } W_N^{*'} = M_N^T W_N^*$$

By noting,

$$W_N^* = R_{NN}^{-1}P_N \qquad \text{and} \qquad M_N^T = M_N^{-1}$$

$W_N^{*'}$ can also be written as

$$R_{NN}^{-1} = \left(M_N\Lambda_{NN}M_N^T\right)^{-1} = M_N\Lambda_{NN}^{-1}M_N^T$$

$$\begin{aligned} W_N^{*'} &= M_N^T R_{NN}^{-1}P_N \\ &= M_N^T\left(M_N\Lambda_{NN}^{-1}M_N^T\right)P_N \\ &= \Lambda_{NN}^{-1}M_N^T P_N \qquad \because M_N^T M_N = I \end{aligned}$$

$$W_N^{*'} = \Lambda_{NN}^{-1}P_N'$$

So the weight vector adaptation is

$$W_N(n+1) = W_N(n) + 2\mu\left[P_N - R_{NN}W_N(n-k)\right]$$

$$W_N(n+1) = W_N(n) + 2\mu\left[P_N - M_N\Lambda_{NN}M_N^T W_N(n-k)\right] \tag{10.48a}$$

Multiplying both sides by M_N^T,

$$M_N^T W_N(n+1) = M_N^T W_N(n) + 2\mu\left[M_N^T P_N - M_N^T M_N\, \Lambda_{NN}\, M_N^T W_N(n-k)\right]$$

$$W_N'(n+1) = W_N'(n) + 2\mu\left[P_N' - \Lambda_{NN}W_N'(n-k)\right], \qquad \because M_N^T M_N = I_N$$

Subtracting $W_N^{*\prime}$ from both sides

$$W'_N(n+1) - W_N^{*\prime} = W'_N(n) - W_N^{*\prime} + 2\mu\left[P'_N - \Lambda_{NN} W'_N(n-k) \right]$$

Now substituting for P'_N from the relation

$$W_N^{*\prime} = \Lambda_{NN}^{-1} P'_N ,$$

as

$$P'_N = \Lambda_{NN} W_N^{*\prime}$$

$$W'_N(n+1) - W_N^{*\prime} = W'_N(n) - W_N^{*\prime} + 2\mu\left[\Lambda_{NN} W_N^{*\prime} - \Lambda_{NN} W'_N(n-k) \right]$$

$$W'_N(n+1) - W_N^{*\prime} = W'_N(n) - W_N^{*\prime} + 2\mu\Lambda_{NN}\left[W_N^{*\prime} - W'_N(n-k) \right]$$

In terms of difference weight vector $V'_N(n+1) = W'_N(n+1) - W_N^{*\prime}$, the above expression becomes

$$V'_N(n+1) = V'_N(n) - 2\mu\,\Lambda_{NN} V'_N(n-k) \tag{10.48b}$$

This equation for the individual weights can be written as

$$v'_i(n+1) = v'_i(n) - 2\mu\ \lambda_i v'_i(n-k), \qquad i = 0, 1, \ldots, N-1,$$

where λ_i are the elements of the diagonal matrix Λ. Taking the z-transform with the assumption $v'_i(n) = 0$ for all $n < 0$ and as

$$Z\left[x(n+1)\right] = \sum_{n=0}^{\infty} x(n+1) z^{-n} = z\left[X(z) - x(0)\right],$$

$$zV'_i(z) - zv(0) = V'_i(z) - 2\mu\ \lambda_i\left[z^{-k} V'_i(z)\right]$$

$$V'_i(z)\left[z - 1 + 2\mu\,\lambda_i z^{-k} \right] = z\,v'_i(0)$$

$$V'_i(z) = \frac{z\ v'_i(0)}{z - 1 + 2\mu\,\lambda_i z^{-k}}$$

$$V_i'(z) = \frac{z^{k+1} v_i'(0)}{(z^{k+1} - z^k + 2\mu\lambda_i)}$$

For the algorithm to be stable, the poles of the above equation must be within the unit circle.

The values of $2\mu\lambda_i$ for which the characteristic equation has roots on the unit circle can be found by substituting $e^{j\varphi}$ for Z $\left(z = re^{j\varphi}\right)$ with $r = 1$ and setting the equation equal to zero. This results in

$$2\mu\lambda_i = e^{jk\varphi} - e^{j(k+1)\varphi} \tag{10.48c}$$

$$2\mu\lambda_i = \cos k\phi + j\sin k\phi - [\cos(k+1)\phi + j\sin(k+1)\phi]$$
$$= \cos k\phi - \cos(k+1)\phi + j[\sin k\phi - \sin(k+1)\phi]$$

The solution to this equation can be obtained by equating real and imaginary parts,

$$2\mu\lambda_i = \cos(k\phi) - \cos\left[(k+1)\phi\right] \tag{10.48d}$$

$$0 = \sin(k\phi) - \sin\left[(k+1)\phi\right] \tag{10.48e}$$

Eqn. (10.48e) can be simplified by noting

$$\sin(a+b) - \sin(a-b) = 2\cos a \sin b$$

$$(a-b) = (k+1)\phi$$

$$(a+b) = k\phi$$

On solving above equations,

$$a = \frac{(2k+1)\phi}{2}, \quad b = -\frac{\phi}{2}$$

Now equation (10.48e) can be rewritten as

$$2\cos\left[{(2k+1)\phi}/{2}\right]\sin\left({\phi}/{2}\right) = 0$$

the above condition is met for either

$$\phi = 0 \;\; or \;\; (2k+1)\phi/2 = \pi/2$$

The solution $\phi = 0$ doesn't satisfy Eqn. (3.48d), hence the only possible solution is

$$(2k+1)\phi/2 = \pi/2 \quad \text{or} \quad \phi = \pi/(2k+1).$$

Eqn. (10.48d) gives

$$\cos\left[(k+1)\phi\right] - \cos(k\phi) = -2\mu\,\lambda_i$$

$$2\sin\left[(2k+1)\phi/2\right]\sin(\phi/2) = 2\mu\,\lambda_i$$

$$\sin(\pi/2)\sin\left(\pi/_{2(2k+1)}\right) = \mu\,\lambda_i$$

$$\mu = (1/\lambda_i)\sin\left[\pi/(2k+1)2\right] \tag{10.48f}$$

As the bounds of stability are dependent on the maximum eigenvalue of R_{NN}, the limiting values of the convergence coefficient becomes

$$0 < \mu < (1/\lambda_i)\sin\left[\pi/(2k+1)2\right]$$

Since R_{NN} is an autocorrelation matrix and its eigenvalues are positive, an upper bound on $\lambda_{\max}$ is given by

$$\lambda_{\max} \le \sum_{i=1}^{N}\lambda_i = trace..R_{NN} = N\phi_x(0)$$

$\phi_x(0)$ is the input signal power. The bounds on the step size becomes,

$$0 < \mu < \left[1/(N * input\ power)\right]\sin\left[\pi/(2k+1)2\right] \tag{10.48g}$$

Comparing this result with LMS algorithm without time delay,

$$0 < \mu < \left[1/(N * input\ power)\right]$$

It can be seen that the time delay reduces the maximum allowed value of the stepsize by a factor $\sin\left[\pi/(2k+1)2\right]$, which in turn reduces the rate of convergence of the algorithm (Snyder. S.D and Hansen. C. H, 1990).

10.10 LEAKY LMS ALGORITHM

The eigenvalues of an autocorrelation matrix can become zero when there is no sufficient/persistent excitation. For these modes with zero eigenvalues, the adaptive filter will not converge and due to these marginally stable modes the adaptive filter can diverge/become unstable especially during its practical implementation with finite word length arithmetic or input data is quantized. Such a situation has been found in ADPCM, adaptive equalization and side lobe canceller systems. This unstability/ divergence can be solved by using leaky LMS algorithm. Consider the cost function

$$J(n) = e^2(n) + \gamma W_N^T(n)W_N(n), \quad \gamma \geq 0 \tag{10.49a}$$

The recursive algorithm for the minimization of this cost function can be achieved by the gradient approach. That is,

$$W_N(n+1) = W_N(n) - \mu \frac{\partial J(n)}{\partial W_N(n)} \tag{10.49b}$$

The gradient of the cost function with respect to the weight vector is given by

$$\frac{\partial J(n)}{\partial W_N(n)} = -2e(n)X_N(n) + 2\gamma W_N(n) \qquad \text{as } e(n) = d(n) - X_N^T(n)W_N(n)$$

Therefore, Eqn. (10.49b) becomes

$$W_N(n+1) = W_N(n) - \mu\left[-2e(n)X_N(n) + 2\gamma W_N(n)\right]$$

$$W_N(n+1) = \left[1 - 2\,\mu\,\gamma\right]W_N(n) + 2\,\mu X_N(n)e(n) \tag{10.49c}$$

Eqn. (10.49c) is known as leaky LMS algorithm and γ is the leak parameter. *When the input is removed, the LMS algorithm stalls. But for the leaky LMS algorithm, the weight vector leaks out to zero. This enables stability for finite precision implementation, removes effect of non-persistent excitation and reduces the effect of stalling, bursts, etc.*

The LMS algorithm has fast convergence rate for the highest eigenvalue but has slowest convergence rate for the minimum eigenvalue. This disparity increases with the eigenvalue spread which corresponds to the eccentricity of the elliptical contours of the mean square error surface. Therefore, the LMS algorithm convergence can be improved by decreasing the eigenvalue spread. Substituting for

$$e(n) = d(n) - X_N^T(n)W_N(n)$$

in the above weight vector adaptation Eqn. (10.49c),

$$W_N(n+1) = \left[1 - 2\,\mu\,\gamma\right]W_N(n) + 2\,\mu X_N(n)e(n)$$

$$W_N(n+1) = (1 - 2\,\mu\,\gamma)W_N(n) + 2\,\mu X_N(n)\left[d(n) - X_N^T(n)W_N(n)\right]$$

$$= (1 - 2\,\mu\,\gamma)W_N(n) + 2\,\mu\,X_N(n)d(n) - 2\,\mu X_N(n)X_N^T(n)W_N(n)$$

$$= \left[I - 2\,\mu\left\{X_N(n)X_N^T(n) + \gamma I\right\}\right] W_N(n) + 2\,\mu\,X_N(n)d(n)$$

Taking expectation both sides,

$$E\{W_N(n+1)\} = \left[I - 2\,\mu E\{X_N(n)X_N^T(n) + \gamma I\}\right]\ E\{W_N(n)\} + 2\,\mu\, E\{X_N(n)d(n)\}$$

$$E\{W_N(n+1)\} = [I - 2\,\mu[R_{NN}(n) + \gamma I]]\ E\{W_N(n)\} + 2\,\mu\, P_N(n) \qquad (10.49d)$$

Comparing Eqn.(10.49d) with the method of steepest descent adaptation algorithm,

$$E\{W_N(n+1)\} = [I - 2\,\mu R_{NN}(n)]E\{W_N(n)\} + 2\,\mu\, P_N(n),$$

it is inferred that due to the presence of the leak parameter γ, the autocorrelation matrix $R_{NN}(n)$ of the input is changed to $R_{NN}(n) + \gamma I$.

For the method of steepest descent, the kth uncoupled difference weight given by

$$v_k^{'}(n) = (1 - 2\,\mu\,\lambda_k)^n v_k^{'}(0)$$

As long as $(1 - 2\,\mu\,\lambda_k) < 1$ or $0 < \mu < 1/\lambda_k$, that particular weight will converge. If λ_k is small, then $(1 - 2\,\mu\,\lambda_k)$ will be sufficiently close to unity hence, it takes more time for that weight to converge. But if λ_k is large then, as $(1 - 2\,\mu\,\lambda_k)$ will be much smaller compared to unity and the concerned difference weight will decay faster. Presently, since with the leaky LMS the autocorrelation matrix is $R_{NN}(n) + \gamma I$, and its kth eigenvalue $\lambda_k^{'}$ is given by

$$\lambda_k^{'} = \lambda_k + \gamma \qquad (10.49e)$$

Hence, $(1 - 2\,\mu\,\lambda_k^{'})$ will be much smaller than $(1 - 2\,\mu\,\lambda_k)$ for $\gamma > 0$. Hence $v_k^{'}(n)$ will decay to zero much faster as long as $0 < \mu < 1/(\lambda_k + \gamma)$.

The same analysis can be viewed in an alternate way in terms of eigenvalue spread. With the leaky LMS as the autocorrelation matrix is $R_{NN}(n) + \gamma I$, the eigenvalue spread is

$$\frac{\lambda_{\max}^{'}}{\lambda_{\min}^{'}} = \frac{\lambda_{\max} + \gamma}{\lambda_{\min} + \gamma} \le \frac{\lambda_{\max}}{\lambda_{\min}}, \quad for\ \gamma \ge 0 \qquad (10.49f)$$

This is so since the numerator will not get affected as $\lambda_{\max} >> \gamma$ but the denominator gets affected very much as $\lambda_{\min}$ is small, *resulting in significant reduction in eigenvalue spread.* Therefore, the largest time constant that corresponds to minimum eigenvalue is given by

$$\tau_{\lambda_{\min}^{'}} = \frac{1}{(\alpha\,\lambda_{\min}^{'})}.$$

For α at its maximum limit,

$$\alpha' = \frac{2}{\lambda'_{\max}},$$

The largest time constant which determines the convergence rate is given by

$$\tau_{\lambda'_{\min}} = \frac{1}{\lambda'_{\min}\left(2/\lambda'_{\max}\right)}, \quad \tau_{\lambda'_{\min}} = \frac{1}{2}\left(\frac{\lambda'_{\max}}{\lambda'_{\min}}\right)$$

Since from Eqn.(10.49f)

$$\frac{\lambda'_{\max}}{\lambda'_{\min}} \le \frac{\lambda_{\max}}{\lambda_{\min}}, \quad \textit{for } \gamma \ge 0$$

$$\tau_{\lambda'_{\min}} \le \tau_{\lambda\min} \tag{10.49g}$$

That is the largest time constant or the convergence time with leakage is relatively smaller than for without leakage. *Thus, leakage reduces the eigenvalue spread thereby making the convergence faster.*

As already stated $n \to \infty$, the difference weight vector decays to zero and the adaptation stops making $W_N(n+1) \approx W_N(n)$. Therefore, (Eqn. (10.49d)),

$$E\{W_N(n+1)\} = [I - 2\,\mu[R_{NN}(n) + \gamma I]]\; E\{W_N(n)\} + 2\,\mu\, P_N(n)$$

becomes,

$$E\{W_N(n)\} = [I - 2\,\mu[R_{NN}(n) + \gamma I]]\; E\{W_N(n)\} + 2\,\mu\, P_N(n)$$

$$-2\,\mu[R_{NN}(n) + \gamma I]E\{W_N(n)\} + 2\,\mu\, P_N(n) = 0$$

$$E\{W_N(n)\} = [R_{NN}(n) + \gamma I]^{-1}\; P_N(n) \tag{10.49h}$$

But for no leakage, the solution of steepest descent is

$$E\{W_N(n)\} = R_{NN}^{-1}(n)\; P_N(n)$$

This indicates that for the LMS with leakage, the solution is biased and it is equivalent to adding white noise of variance γ to the input. This increases the uncertainty of the input signal and this hinders growth of the coefficient magnitude. Thus, the leakage introduced for the weight vector though enhances the convergence rate, it converges to a biased optimal weight vector, which is not desirable. Thus, in choosing the leakage factor there is a trade-off between convergence rate and bias in convergence solution. It is essential that till convergence, the convergence should be faster but after convergence, an unbiased solution is required. This motivates to have a variable leakage factor γ which can be high prior to convergence (to make $\left[1-2\mu\ \gamma\right] << 1$) and as convergence is approached γ should be close to zero (to make $\left[1-2\mu\ \gamma\right] \approx 1$) so that the bias in converged solution is tolerable. The bias in the weight vector gets reflected in the excess error and to keep the excess error low, bias in weight vector should be less, that is $\gamma << 1$ (M. Kamenetsky and B.Widrow, 2004). For stability, the step size should satisfy

$$0 < \mu < \frac{1}{(\lambda_{\max} + \gamma)} < \frac{1}{N(\sigma_x^2 + \gamma)}$$

Here, σ_x^2 is the power of the input and from the limits on μ, the convergence improvement is not explicit. But the reduction in eigenvalue spread results in spectral flattening, leading to a biased weight vector which is equivalent to adding white noise of variance γ to input. This white noise reduces the correlation length of input to N' as it significantly destroys correlation at higher lags. For the improvement in convergence speed,

$$N'(\sigma_x^2 + \gamma) < N\sigma_x^2\ ,$$

$$N' < \frac{N\sigma_x^2}{(\sigma_x^2 + \gamma)}$$

Since $\gamma > 0$, the above relation is valid and $N' < N$ and the stepsize with leaky LMS is increased which improves the convergence rate. But as N' is not known using N, the stepsize can be increased by using a higher value of $\alpha_0,\ 0 < \alpha_0 < 1$ in the expression for stepsize

$$\alpha(n) = \frac{\alpha_0}{N\sigma_x^2(n)}, \qquad 0 < \alpha_0 < 1$$

10.11 SIMPLIFIED SIGN LMS ALGORITHMS

Sign-error LMS algorithm: The LMS algorithm of order N requires $2N$ additions and $2N$ multiplications per sample for the computation of adaptive filter output and for updating the weight vector. To reduce this computational load, simplified algorithms which avoid multiplications by using only the sign of the quantities involved have been developed. For the LMS adaptation that uses sign of the error is,

$$W_N(n+1) = W_N(n) + \mu(n)\, X_N(n)\, \text{sgn}\left[e(n)\right] \tag{10.50a}$$

$$w_i(n+1) = w_i(n) + \mu(n)\, x(n-i)\, \text{sgn}\left[e(n)\right], \qquad i = 0, 1, \ldots, N-1 \tag{10.50b}$$

$$e(n) = d(n) - W_N^T(n) X_N(n)$$

$$\text{sgn}\left[u\right] = \begin{cases} 1, & u > 0 \\ 0, & u = 0 \\ -1, & u < 0 \end{cases}$$

This still requires N multiplications and $N-1$ additions to compute the adaptive filter output. But it simplifies the weight vector adaptation. If the step size μ is chosen to be a power of 2, the product of μ and $x(n-i)$ for $x(n-i)$ the ith weight can be done by left shifting $x(n-i)$. Further, the adaptation i^{th} weight is achieved by adding or subtracting depending on the sign of the error $e(n)$ to the product $\mu\, x(n-i)$. Thus, the adaptation is realized by mere N shifts and N additions rather than by N multiplications and N additions. This may improve the computational speed but this is at the expense of performance as only sign of the error is taken into account neglecting its magnitude. This simplification gets reflected as noise in the gradient which, in turn, appears as variance in the weight vector. Thus, the misadjustment of sign-error algorithm may be significantly more than that of its corresponding LMS algorithm without any approximation for the error. Further, to improve its misadjustment, it may be required to reduce the step size to accommodate the noise (inaccuracies) in the weight vector due to due approximation of the error to its sign from its actual value. Such an attempt will make the adaptation towards convergence slower.

Signed data (Sign regressor) LMS algorithm: Here, instead of using the input values, only their signs are considered and this algorithm is

$$W_N(n+1) = W_N(n) + \mu(n)\, \text{sgn}\left[X_N(n)\right] e(n) \tag{10.50c}$$

Similar to sign-error algorithm for weight vector adaptation this also requires N shifts and N additions. Further, this algorithm also suffers from poor convergence performance like high value for misadjustment and slower convergence rate to due to use of smaller step size to tolerate the noise in the weight vector due to the approximation of the input by its sign. In certain situations, the sign-data algorithm has been found to perform better than the sign-error algorithm. However, there is an argument in favour of sign error algorithm. The argument is in sign-error, the approximation only scales the gradient but does not change the direction of the gradient. However, as the approximation in sign-data changes the direction of the gradient, it may lead to divergence, which is more serious.

Sign-Sign LMS algorithm: Here instead of the error and the input values only their signs are considered.

$$W_N(n+1) = W_N(n) + \mu(n)\, \text{sgn}\left[X_N(n)\right] \text{sgn}\left[e(n)\right] \tag{10.50d}$$

Compared to sign-error and sign-data algorithms, this is much simpler as it does the weight vector adaptation without any multiplications. As both the data and the error are approximated instead of any one of them, the resulting gradient of this algorithm will be very noisy and as expected may lead to very inferior performance. In spite of this, its over simplification has tempted to use it in different scenarios. This algorithm may also suffer from divergence problem due to change in the direction of gradient due to the noise introduced in data approximation.

The sign algorithms are suitable for VLSI architectures, but with DSP processors, the determination of sign can break up the instruction pipeline and degrade the performance. All these simplified versions may converge to optimal solution. But in general, this may not be necessarily true. This is because for example, the sign-sign algorithm attempts to satisfy

$$E\left\{\text{sgn}\left[X_N(n)(e(n)\right]\right\} = 0$$

whereas the optimal solution satisfies $E\left\{X_N(n)(e(n)\right\} = 0$, which is much different.

10.12 FILTERED-X LMS (FXLMS) ALGORITHM

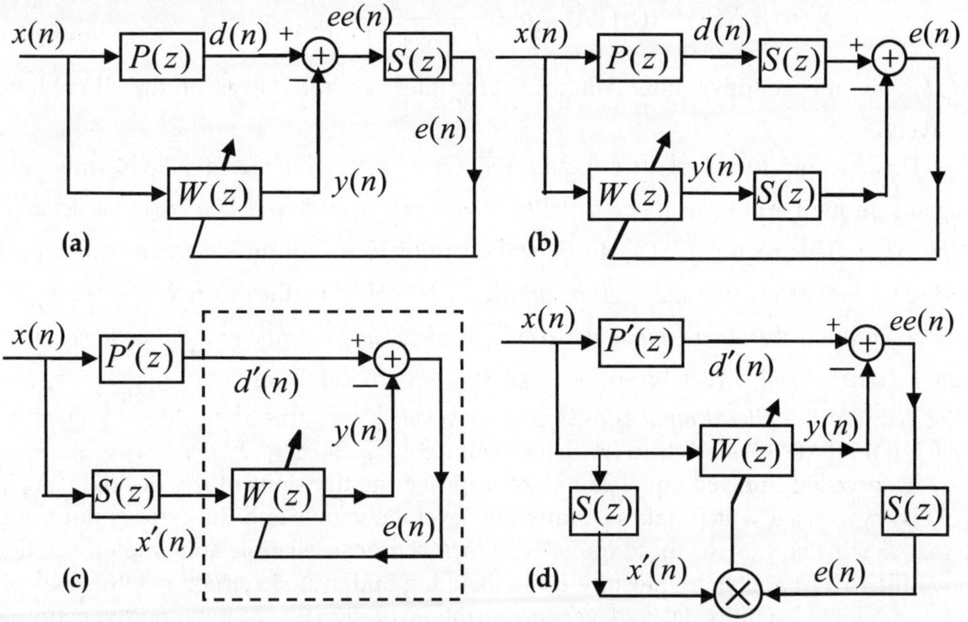

Fig. 10.6 (a) System identification when the error is not directly available; (b–d) Development of FXLMS algorithm.

For Fig. (10.6a), the LMS weight update equation is,

$$w_i(n+1) = w_i(n) + \alpha \, \underline{ee(n)\; x(n-i)}, \quad 0 < i < N-1 \tag{10.51a}$$

This has a *stringent time alignment between the input* $x(n)$ *and the error* $ee(n)$. However, if error signal $ee(n)$ is available only after it passes through the transfer function $S(z)$, as $e(n)$, *the alignment between* $ee(n)$ *and* $x(n)$ *for the LMS adaptation algorithm gets disturbed hence, can lead to divergence. This is supported by the fact the delay due to presence of* $S(z)$ *will reduce limit the range of value of step size and crossing this limit will result in divergence as discussed in the Section 10.8, effect of delay on LMS algorithm.* Hence, the *presence* $S(z)$ *requires modification of the LMS algorithm.*

The transfer function $S(z)$ can be shifted prior to the summing junction (Fig.10.6b). Further, $S(z)$ can be combined with the plant $P(z)$ to form a new plant $P'(z)$ and $S(z)$ in the $W(z)$ branch can be shifted prior to $W(z)$ as shown in Fig.10.6c (assuming $W(z)$ to be time invariant). For the part shown *inside the dotted rectangle* (Fig.10.6c), the error signal $e(n)$ is *directly* available hence, the LMS algorithm can be applied as (L.J. Eriksson, 1991),

$$w_i(n+1) = w_i(n) + \mu\, e(n)\, x'(n-i), \qquad 0 < i < N-1 \tag{10.51b}$$

where

$$x'(n) = S(n) * x(n), \; 0 < \mu < \frac{1}{P_{x'}(N+M)}$$

$N, M, P_{x'}$ are the adaptive filter length, $S(z)$ filter length, power of the filtered input, respectively.

This means, to adapt $W(z)$ with the error $e(n)$ which is available only after it propagates through $S(z)$, *the input signal to be used is a filtered version* $x'(n)$ *of the actual input* $x(n)$ *by* $S(z)$. This requires $S(z)$ to be shifted prior to the summing point to form a new plant $P'(z)$. However, this may *not be practically possible.* But the effect of $S(z)$ can still be overcome, *by using the filtered input* $x'(n)$ *only in the adaptation rule and the input for* $W(z)$ *is retained as* $x(n)$ *itself,* which does not call for any modification in the practical system (Fig.10.6d). Since the *filtered input is used for the adaptation rule,* this algorithm is known as the ***filtered-X LMS*** (FXLMS) algorithm (Widrow et al, 1981; J.C. Burgess, 1981).

In decision directed equalization, to improve the decisions of the desired signal, the Viterbi coding is used which takes duration of several symbols and the error is not available until that much delay. Also in adaptive reference echo cancellation and implementation of adaptive filters by parallel architecture like pipelining and systolic structures, the coefficients have to be adapted with a delayed version of the error signal. The modified version of the algorithm which handles the delayed error, is known as the " Delayed LMS algorithm" (DLMS) and is given by

$$\begin{aligned} &w_i(n+1) = w_i(n) + \mu\, ee(n-M)\, x(n-M-i), \qquad 0 < i < N-1 \\ &ee(n-M) = d(n-M) - y(n-M) \\ &0 < \mu < \frac{1}{P_x(N+M)} \end{aligned} \tag{10.51c}$$

where P_x, N, M are the input power, adaptive filter length and the delay involved for the error, respectively.

The filtered-X LMS algorithm becomes a delayed LMS (DLMS) algorithm when the system $S(z)$ becomes a pure delay *i.e.,* $S(z) = z^{-D}$

10.13 COMPLEX LMS ALGORITHM

In situations like array processing/ beam forming where the magnitude and direction both are sensed or in communication applications which deal with analytic signals the nature of the signals will be complex rather than real. To use the LMS algorithm for such scenarios, it is necessary to extend its ability to complex signals.

For a complex input vector $X_N(n)$,

$$X_N(n) = X_{NR}(n) + i\,X_{NI}(n) = \begin{bmatrix} x_{1R}(n) \\ x_{2R}(n) \\ \cdot \\ x_{NR}(n) \end{bmatrix} + i \begin{bmatrix} x_{1I}(n) \\ x_{2I}(n) \\ \cdot \\ x_{NI}(n) \end{bmatrix} \tag{10.52a}$$

The complex weight vector $W_N(n)$ is

$$W_N(n) = W_{NR}(n) + i\,W_{NI}(n) = \begin{bmatrix} w_{1R}(n) \\ w_{2R}(n) \\ \cdot \\ w_{NR}(n) \end{bmatrix} + i \begin{bmatrix} w_{1I}(n) \\ w_{2I}(n) \\ \cdot \\ w_{NI}(n) \end{bmatrix} \tag{10.52b}$$

where R and I refer to real and imaginary parts. The desired complex signal $d(n)$ and output $y(n)$ of the adaptive filter are

$$d(n) = d_R(n) + id_I(n)\,,\; y(n) = y_R(n) + iy_I(n) \tag{10.52c}$$

$$y(n) = X_N^T(n)W(n) = W_N^T(n)X_N(n)$$

The complex error is

$$e(n) = e_R(n) + ie_I(n) \tag{10.52d}$$

$$e(n) = d(n) - y(n) = d(n) - W_N^T(n)X_N(n) = d(n) - X_N^T(n)W_N(n)$$

The complex LMS algorithm must adapt the real and imaginary parts of $W_N(n)$ *simultaneously minimizing both* $e_R(n)$ *and* $e_I(n)$*. This is done by minimizing the average total error power*

$$E\left[e(n)\,e^*(n)\right] = E\left[e_R^2(n) + e_I^2(n)\right] = E\left[e_R^2(n)\right] + E\left[e_I^2(n)\right] \tag{10.52e}$$

$$e^*(n) = d^*(n) - \left(W_N^T(n)\right)^* X_N^*(n) = d^*(n) - \left(X_N^T(n)\right)^* W_N^*(n)$$

Though the two components of the error are in quadrature to each other, they cannot be minimized independently. The instantaneous gradient of $e(n)e^*(n)$ with respect to the real component of the weight vector is

$$\nabla_R\left(e(n)\,e^*(n)\right) = \begin{bmatrix} \dfrac{\partial\left(e(n)\,e^*(n)\right)}{\partial w_{1R}} \\ \cdot \\ \cdot \\ \dfrac{\partial\left(e(n)\,e^*(n)\right)}{\partial w_{NR}} \end{bmatrix} \tag{10.52f}$$

$$\begin{aligned}
\nabla_R\left(e(n)\,e^*(n)\right) &= e(n)\,\nabla_R\left(e^*(n)\right) + e^*(n)\nabla_R\left(e(n)\right) \\
&= e(n)\,\nabla_R\left(d(n) - W_N^T(n)X_N(n)\right)^* + e^*(n)\nabla_R\left(d(n) - W_N^T(n)X_N(n)\right) \\
&= e(n)\nabla_R\left(d^*(n) - (\underline{W_{NR}(n)} - i\,W_{NI}(n))^T(n)X_N^*(n)\right) \\
&\qquad + e^*(n)\nabla_R\left(d(n) - (\underline{W_{NR}(n)} + i\,W_{NI}(n))^T X_N(n)\right)
\end{aligned}$$

$$\nabla_R\left(e(n)\,e^*(n)\right) = e(n)\left(-X_N^*(n)\right) + e^*(n)\left(-X_N(n)\right) \tag{10.52g}$$

The instantaneous gradient with respect to the imaginary component of the weight vector is

$$\nabla_I\left(e(n)\,e^*(n)\right) = e(n)\,\nabla_I\left(e^*(n)\right) + e^*(n)\nabla_I\left(e(n)\right)$$

$$= e(n)\,\nabla_I\left(d(n) - W_N^T(n)X_N(n)\right)^* + e^*(n)\nabla_I\left(d(n) - W_N^T(n)X_N(n)\right)$$

$$= e(n)\nabla_I\left(d^*(n) - (W_{NR}(n) - i\,\underline{W_{NI}(n))^T}(n)X_N^*(n)\right)$$

$$+ e^*(n)\nabla_I\left(d(n) - (W_{NR}(n) + i\,\underline{W_{NI}(n))^T}\,X_N(n)\right)$$

$$\nabla_I\left(e(n)\,e^*(n)\right) = e(n)\left(iX_N^*(n)\right) + e^*(n)\left(-iX_N(n)\right) \tag{10.52h}$$

Applying the LMS algorithm to the real and imaginary parts of the weight vector by changing them along the respective negative gradient estimates,

$$W_{NR}(n+1) = W_{NR}(n) - \mu\,\nabla_R\left(e(n)\,e^*(n)\right),\quad W_{NI}(n+1) = W_{NI}(n) - \mu\,\nabla_I\left(e(n)\,e^*(n)\right)$$

Since the complex weight vector is

$$W_N(n) = W_{NR}(n) + i\,W_{NI}(n),$$

Therefore, the complex weight iteration rule is

$$W_N(n+1) = W_N(n) - \alpha\left[\nabla_R\left(e(n)\,e^*(n)\right) + i\nabla_I\left(e(n)\,e^*(n)\right)\right]$$

Using Eqns.(10.52g) and (10.52h),

$$W_N(n+1) = W_N(n)$$

$$-\alpha\left[e(n)\left(-X_N^*(n)\right) + e^*(n)\left(-X_N(n)\right) + i\,e(n)\left(i\,X_N^*(n)\right) + e^*(n)\left(-i\,X_N(n)\right)\right]$$

$$= W_N(n) - \alpha\left[-e(n)X_N^*(n) - e^*(n)\,X_N(n) - e(n)\,X_N^*(n) + e^*(n)X_N(n)\right]$$

$$W(n+1) = W(n) + 2\alpha\,e(n)\,X^*(n) \tag{10.52i}$$

Thus, for the complex input, its conjugate is used as input for the LMS adaptation.

10.14 FORWARD BACKWARD LMS PREDICTOR

For the predictor that predicts its future samples from the previous samples (or previous samples from its future samples), the LMS algorithm uses only forward prediction errors. But for this purpose instead of using the forward prediction error alone, both the forward and backward prediction errors can be used. The motivation for this is that it gives a reduction in misadjustment over the conventional LMS transversal predictor that uses only forward prediction error (Y.C. Lim and C.C. Ko., 1990). However, its computational complexity is doubled than that of using only the forward error. The forward and the backward errors are given by (Fig.10.7).

$$e^f(n) = x(n) - X_N^{f^T} W_N(n), \tag{10.53a}$$

$$e^b(n) = X_N(n-N) - X^{bT}_N(n)\, W_N(n) \tag{10.53b}$$

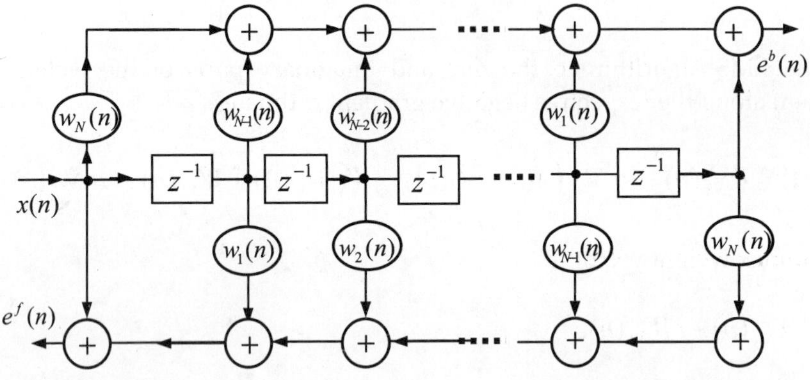

Fig.10. 7 Forward backward LMS predictor.

where

$$X_N^{f^T}(n) = [\ x(n-1), x(n-2)\ldots X(n-N)] \tag{10.53c}$$

$$X_N^{b^T}(n) = [X(n-N+1), X(n-N+2),\ldots, x(n)] \tag{10.53d}$$

The gradient adaptation rule is

$$W_N(n+1) = W_N(n) - \mu\, \nabla[\ \varepsilon(n)]$$

The error energy to be minimized is

$$\varepsilon(n) = \frac{\left[e^f(n)\right]^2 + \left[e^b(n)\right]^2}{2} \tag{10.53e}$$

Therefore, the gradient to be considered is

$$\nabla[\ \varepsilon(n)] = \frac{1}{2}\left\{2e^f(n)\frac{\partial e^f(n)}{\partial W_N(n)} + 2e^b(n)\frac{\partial e^f(n)}{\partial W_N(n)}\right\}$$

$$\nabla\left[\ e^2(n)\right] = -\left\{e^f(n)X_N^f(n) + e^b(n)X_N^b(n)\right\}$$

Therefore, the LMS weight vector adaptation is

$$W_N(n+1) = W_N(n) + \mu\left\{e^f(n)X_N^f(n) + e^b(n)X_N^b(n)\right\} \tag{10.53f}$$

10.14.1 Convergence Criterion for FBLMS Algorithm

The necessary convergence criterion for the FBLMS algorithm will be considered. For this in the FBLMS adaptation rule (Eqn. (10.35f))

$$W_N(n+1) = W_N(n) + \mu\left\{e^f(n)X_N^f(n) + e^b(n)X_N^b(n)\right\}$$

Substituting for (Eqns. 10.53a,b)

$$e^f(n) = x(n) - X_N^{f^T}\ W_N(n),\quad e^b(n) = x(n-N) - X^{bT}{}_N(n)\, W_N(n)$$

$$W_N(n+1) = W_N(n) + \mu X_N^f(n)\left\{\ x(n) - X_N^{f^T}\ W_N(n)\right\}$$
$$+\ \mu X_N^b(n)\left\{x(n-N) - X^{bT}{}_N(n)\, W_N(n)\right\}$$

$$E\left\{W_N(n+1)\right\} = E\left\{W_N(n)\right\}\ + \mu E\left\{X_N^f(n)x(n) - X_N^f(n)X_N^{f^T}\ W_N(n)\right\}$$
$$+\ \mu E\left\{X_N^b(n)x(n-N) - X_N^b(n)X^{bT}{}_N(n)\, W_N(n)\right\}$$

Since for a stationary signal,

$$P_N = E\left\{X_N^f(n)x(n)\right\} = E\left\{X_N^b(n)X_N(n-N)\right\}$$

$$R_{NN} = E\left\{X_N^f(n)X_N^{f^T}(n)\right\} = E\left\{X_N^b(n)X^{bT}{}_N(n)\right\}$$

$$E\left\{W_N(n+1)\right\} = E\left\{W_N(n)\right\}\ + \mu\left[P_N - R_{NN}E\left\{\ W_N(n)\right\}\right] + \mu\left[P_N - R_{NN}E\left\{\ W_N(n)\right\}\right]$$

$$E\left\{W_N(n+1)\right\} = E\left\{W_N(n)\right\}\ + 2\mu\left[P_N - R_{NN}E\left\{\ W_N(n)\right\}\right]$$

Putting $\alpha = 2\mu$,

$$E\left\{W_N(n+1)\right\} = \ \left[I_{NN} - \alpha R_{NN}\right]E\left\{\ W_N(n)\right\} + \alpha P_N$$

For simplicity let $E\left\{W_N(n)\right\}$ be replaced by $W_N(n)$ then

$$W_N(n+1) = \left[I_{NN} - \alpha R_{NN}\right]W_N(n) + \alpha P_N \tag{10.54a}$$

This equation is similar to that seen in chapter 9 and can be treated in similar way by using orthonormal diagonalization using

$$M_{NN}^T R_{NN} M_{NN} = \Lambda_{NN}, \quad R_{NN} = M_{NN}\Lambda_{NN}M_{NN}^T$$

M_{NN}, Λ_{NN} are the modal matrices whose columns are eigenvectors of R_{NN} and diagonal matrix of eigenvalues of R_{NN}, respectively. Therefore,

$$W_N(n+1) = \left[I_{NN} - \alpha M_{NN}\Lambda_{NN}M_{NN}^T\right] W_N(n) + \alpha P_N$$

$$= \left[M_{NN}M_{NN}^T - \alpha M_{NN}\Lambda_{NN}M_{NN}^T\right] W_N(n) + \alpha P_N$$

$$= M_{NN} \left[I_{NN} - \alpha\,\Lambda_{NN}\right] M_{NN}^T W_N(n) + \alpha P_N$$

$$M_{NN}^T W_N(n+1) = M_{NN}^T M_{NN} \left[I_{NN} - \alpha\,\Lambda_{NN}\right] M_{NN}^T W_N(n) + \alpha M_{NN}^T P_N \tag{10.54b}$$

Defining

$$W_N'(n) = M_{NN}^T W_N(n),\; P_N'(n) = M_{NN}^T P_N(n)$$

and

$$W_N'^* = M_{NN}^T W_N^* = M_{NN}^T R_{NN}^{-1}(n) P_N$$

$$= M_{NN}^T \left\{M_{NN}\Lambda_{NN}M_{NN}^T\right\}^{-1} P_N = M_{NN}^T M_{NN}\Lambda_{NN}^{-1}M_{NN}^T P_N$$

$$W_N'^* = \Lambda_{NN}^{-1}P_N',\; w_i'^* = p_i' / \lambda_i \tag{10.54c}$$

Eqn. (10.54b) becomes

$$W_N'(n+1) = \left[I_{NN} - \alpha\,\Lambda_{NN}\right] W_N'(n) + \alpha P_N'$$

Or

$$w_i'(n+1) = (1 - \alpha\,\lambda_i) w_i'(n) + \alpha p_i' \tag{10.54d}$$

The solution of this equation is

$$w_i'(n) = (1 - \alpha\,\lambda_i)^n w_i'(0) + \alpha p_i' \sum_{j=0}^{n-1} (1 - \alpha\,\lambda_i)^j$$

Putting $\gamma_i = (1-\alpha\,\lambda_i)$,

$$w_i'(n) = \gamma_i^n w_i'(0) + \alpha p_i' \sum_{j=0}^{n-1} \gamma_i^j$$

For $|\gamma_i| < 1$,

$$w_i'(n) = \gamma_i^n w_i'(0) + \alpha p_i' \frac{1-\gamma_i^n}{1-\gamma_i} \tag{10.54e}$$

This solution is possible only for

$$|\gamma_i| < 1 \text{ or } |1-2\alpha\,\lambda_i| < 1 \text{ or } 0 < \alpha < 2/\lambda_i \text{ or } 0 < \alpha < 2/\lambda_{\max}$$

Since

$$\lambda_{\max} < \sum_{i=0}^{N-1} \lambda_i = trace\{R_{NN}\} = N\sigma_x^2,$$

$$0 < \alpha < \frac{2}{N\sigma_x^2} \tag{10.54f}$$

This convergence criterion is same as that used for the algorithm that minimizes only forward prediction error energy. Further as

$$\lim_{n\to\infty} w_i'(n) = \alpha p_i' \frac{1}{1-\gamma_i} = \alpha p_i' \frac{1}{1-(1-\alpha\,\lambda_i)}$$

$$\lim_{n\to\infty} w_i'(n) = \frac{p_i'}{\lambda_i} = w_i^{*\prime}(n)$$

That is, the weight elements approach their corresponding optimum values as iterations become large.

10.14.2 Convergence speed of FBLMS

Consider Eqn.(10.54e)

$$w_i'(n) = \gamma_i^n w_i'(0) + \alpha p_i' \frac{1-\gamma_i^n}{1-\gamma_i}$$

$$w_i'(n) - w_i^{*\prime}(n) = \gamma_i^n w_i'(0) + \alpha p_i' \frac{1-\gamma_i^n}{1-\gamma_i} - w_i^{*\prime}(n)$$

But

$$w_i^{*\prime}(n) = \frac{p_i'}{\lambda_i} \text{ and } \gamma_i = \quad (1-\alpha\,\lambda_i)$$

and putting $v_i'(n) = w_i'(n) - w_i^{*\prime}(n)$,

$$v_i'(n) = \quad \gamma_i^{\,n}\left[w_i'(0) - \frac{\alpha p_i'}{1-(1-\alpha\,\lambda_i)}\right] + \frac{\alpha p_i'}{1-(1-\alpha\,\lambda_i)} - \frac{p_i'}{\lambda_i}$$

$$v_i'(n) = (1-\alpha\,\lambda_i)^n\left[w_i'(0) - \frac{p_i'}{\lambda_i}\right]$$

$$v_i'(n) = (1-\alpha\,\lambda_i)^n v_i'(0), \quad v_i'(0) = \left[w_i'(0) - \frac{p_i'}{\lambda_i}\right]$$

Therefore, the time constant is given by

$$\frac{v_i'(n)}{v_i'(0)} = \frac{1}{e} = (1-\alpha\,\lambda_i)^{\tau_i}, \qquad \tau_i = \frac{-1}{\ln(1-\alpha\,\lambda_i)}$$

For $\alpha << 1$,

$$\tau_i = \frac{1}{\alpha\,\lambda_i}$$

The largest time constant corresponds to $,\lambda_{\min}$ $,\alpha = \alpha_{\max} = 2/\lambda_{\min}$. Therefore, the maximum convergence speed is decided by

$$\tau_{\max} = \frac{1}{2}\left(\frac{\lambda_{\max}}{\lambda_{\min}}\right) \tag{10.54g}$$

The forward backward algorithm speed of convergence is also decided by the eigenvalue spread $(\lambda_{\max}/\lambda_{\min})$ of the autocorrelation matrix of the input and this similar to that for forward prediction error energy minimization.

10.14.3 Misadjustment of FBLMS

It has been seen that for the FBLMS, the convergence criterion and convergence speed, are similar to those of conventional LMS algorithm. But as stated the major advantage of FBLMS is that its misadjustment will be less than (which is half that of) that of conventional LMS algorithm and the expression for the misadjustment will be derived.

From Eqn. (10.53a),

$$e^f(n) = x(n) - X_N^{f^T}(n)\, W_N(n),$$

Therefore, the mean square error is given by

$$\varepsilon^f = E\left\{e^{f2}(n)\right\} = E\left\{\left[\, x(n) - X_N^{f^T}(n)\, W_N(n)\right]^2\right\},$$

$$= E\left\{x^2(n) - 2x(n)X_N^{f^T}(n)W_N(n) + W_N^T(n)X_N^f(n)X_N^{f^T}(n)\, W_N(n)\right\}$$

$$= E\left\{x^2(n)\right\} - 2E\left\{x(n)X_N^{f^T}(n)\right\}E\left\{W_N(n)\right\} + E\left\{W_N^T(n)\right\}E\left\{X_N^f(n)X_N^{f^T}(n)\right\}E\left\{W_N(n)\right\}$$

$$\varepsilon^f = E\left\{x^2(n)\right\} - 2P_N E\left\{W_N(n)\right\} + E\left\{W_N^T(n)\right\}R_{NN}\, E\left\{W_N(n)\right\} \tag{10.55a}$$

After convergence, $W_N(n)$ can be expressed as

$$W_N(n) = W_N^*(n) + \delta_N(n) \tag{10.55b}$$

$\delta(n)$ is the deviation in $W_N(n)$ from its optimum value $W_N^*(n)$ and it has been shown that as $n \to \infty$, $E\left\{\delta_N(n)\right\} \to 0$. Therefore,

$$\varepsilon^f = E\left\{x^2(n)\right\} - 2P_N E\left\{W_N^*(n) + \delta_N(n)\right\} + E\left\{W_N^{*T}(n) + \delta_N^T(n)\right\}R_{NN}\, E\left\{W_N^*(n) + \delta_N(n)\right\}$$

$$= E\left\{x^2(n)\right\} - 2P_N W_N^* + W_N^{*T} R_{NN}\, W_N^* + E\left\{\delta_N^T(n)R_{NN}\, \delta_N(n)\right\}$$

Since, $W_N^* = R_{NN}^{-1}P_N$

$$\varepsilon^f = E\left\{x^2(n)\right\} - 2P_N W_N^* + W_N^{*T} R_{NN}\, R_{NN}^{-1}P_N + E\left\{\delta_N^T(n)R_{NN}\, \delta_N(n)\right\}$$

$$= E\left\{x^2(n)\right\} - 2P_N W_N^* + W_N^{*T} P_N + E\left\{\delta_N^T(n)R_{NN}\, \delta_N(n)\right\}$$

$$= E\left\{x^2(n)\right\} - P_N W_N^* + E\left\{\delta_N^T(n)R_{NN}\, \delta_N(n)\right\}$$

But $\varepsilon_{\min} = E\left\{x^2(n)\right\} - P_N W_N^*$,

$$\varepsilon^f = \varepsilon_{\min} + E\left\{\delta_N^T(n)R_{NN}\ \delta_N(n)\right\} \tag{10.55c}$$

The second term on the RHS is the excess error power above that of the optimim/minimum and is due to jittering of the weight vector about its optimum value. The misadjustment is

$$M = \frac{excess\ error\ power}{optimum\ error\ power}, \quad M = \frac{E\left\{\delta_N^T(n)R_{NN}\ \delta_N(n)\right\}}{\varepsilon_{\min}} \tag{10.55d}$$

Therefore,

$$\varepsilon^f = \varepsilon_{\min}\left[1 + \frac{E\left\{\delta_N^T(n)R_{NN}\ \delta_N(n)\right\}}{\varepsilon_{\min}}\right]$$

$$\varepsilon^f = \varepsilon_{\min}\left[1 + M\right] \tag{10.55e}$$

Consider Eqn. (10.53f)

$$W_N(n+1) = W_N(n) + \mu\left\{e^f(n)X_N^f(n) + e^b(n)X_N^b(n)\right\}$$

Subtracting W_N^* both sides,

$$W_N(n+1) - W_N^* = W_N(n) - W_N^* + \mu\left\{e^f(n)X_N^f(n) + e^b(n)X_N^b(n)\right\}$$

$$\delta_N(n+1) = \delta_N(n) + \mu\left\{e^f(n)X_N^f(n) + e^b(n)X_N^b(n)\right\}$$

Consider the product

$$\delta_N^T(n+1)\delta_N(n+1) = \left[\delta_N(n) + \mu\left\{e^f(n)X_N^f(n) + e^b(n)X_N^b(n)\right\}\right]^T \left[\delta_N(n) + \mu\left\{e^f(n)X_N^f(n) + e^b(n)X_N^b(n)\right\}\right]$$

$$\delta_N^T(n+1)\delta_N(n+1) = \left[\delta_N^T(n) + \mu\left\{e^f(n)X_N^f(n) + e^b(n)X_N^b(n)\right\}^T\right] \left[\delta_N(n) + \mu\left\{e^f(n)X_N^f(n) + e^b(n)X_N^b(n)\right\}\right]$$

$$\begin{aligned} &= \delta_N^T(n)\delta_N(n) + \delta_N^T(n)\mu\left\{e^f(n)X_N^f(n) + e^b(n)X_N^b(n)\right\} \\ &\quad + \mu\left\{e^f(n)X_N^f(n) + e^b(n)X_N^b(n)\right\}^T \delta_N(n) \\ &\quad + \mu^2\left\{e^f(n)X_N^f(n) + e^b(n)X_N^b(n)\right\}^T \left\{e^f(n)X_N^f(n) + e^b(n)X_N^b(n)\right\} \end{aligned}$$

$$\delta_N^T(n+1)\delta_N(n+1) - \delta_N^T(n)\delta_N(n) =$$
$$\mu\Big\{\delta_N^T(n)\left\{e^f(n)X_N^f(n)+e^b(n)X_N^b(n)\right\}$$
$$+x\left\{e^f(n)X_N^f(n)+e^b(n)X_N^b(n)\right\}^T\delta_N(n)$$
$$+\mu\left\{e^f(n)X_N^f(n)+e^b(n)X_N^b(n)\right\}^T\left\{e^f(n)X_N^f(n)+e^b(n)X_N^b(n)\right\}\Big\}$$

$$\delta_N^T(n+1)\delta_N(n+1) - \delta_N^T(n)\delta_N(n) =$$
$$2\mu\Big\{\delta_N^T(n)\left\{e^f(n)X_N^f(n)+e^b(n)X_N^b(n)\right\}$$
$$+\frac{\mu}{2}\left\{e^f(n)X_N^f(n)+e^b(n)X_N^b(n)\right\}^T\left\{e^f(n)X_N^f(n)+e^b(n)X_N^b(n)\right\}\Big\}$$

Taking expectations on both sides and as after convergence

$$\delta_N^T(n+1)\delta_N(n+1) = \delta_N^T(n)\delta_N(n),$$

$$2\mu\Big\{E\left\{\delta_N^T(n)\left\{e^f(n)X_N^f(n)+e^b(n)X_N^b(n)\right\}\right\}$$
$$+\frac{\mu}{2}E\left[\left\{e^f(n)X_N^f(n)+e^b(n)X_N^b(n)\right\}^T\left\{e^f(n)X_N^f(n)+e^b(n)X_N^b(n)\right\}\right]\Big\} = 0 \tag{10.55f}$$

Let

$$T_1 = E\left\{\delta_N^T(n)\left\{e^f(n)X_N^f(n)+e^b(n)X_N^b(n)\right\}\right\} \tag{10.55g}$$

$$T_2 = E\left[\left\{e^f(n)X_N^f(n)+e^b(n)X_N^b(n)\right\}^T\left\{e^f(n)X_N^f(n)+e^b(n)X_N^b(n)\right\}\right] \tag{10.55h}$$

then,

$$2\mu\left[T_1+\frac{\mu}{2}T_2\right]=0 \tag{10.55i}$$

Consider the term T_1 (Eqn.10.55g) and substituting for

$$e^f(n) = x(n) - X_N^{f^T}(n)W_N(n), \qquad e^b(n) = x(n-N) - X_N^{bT}(n)W_N(n),$$

$$E\left[\delta_N^T(n)\left\{\left(x(n)-X_N^{f^T}(n)W_N(n)\right)X_N^f(n)+\left(x(n-N)-X_N^{b^T}(n)W_N(n)\right)X_N^b(n)\right\}\right]$$

$$= E\left[x(n)\delta_N^T(n)X_N^f(n)\right] - E\left[\delta_N^T(n)X_N^{f^T}(n)W_N(n)X_N^f(n)\right]$$
$$+E\left[\delta_N^T(n)x(n-N)X_N^b(n)\right] - E\left[\delta_N^T(n)X_N^{bT}(n)X_N^b(n)W_N(n)\right]$$

Since $\delta_N(n)$ is uncorrelated with $X_N^f(n)$ and $X_N^b(n)$,

$$E\left[x(n)\delta_N^T(n)X_N^f(n)\right] = 0, \quad E\left[\delta_N^T(n)x(n-N)X_N^b(n)\right] = 0$$

Therefore,

$$T_1 = -E\left[\delta_N^T(n)X_N^{f^T}(n)W_N(n)X_N^f(n)\right] - E\left[\delta_N^T(n)X_N^{bT}(n)X_N^b(n)W_N(n)\right]$$

Substituting for (Eqn.(10.55b))

$$W_N(n) = W_N^*(n) + \delta_N(n)$$

$$T_1 = -E\left[\delta_N^T(n)\left\{\delta_N(n) + W_N^*(n)\right\}X_N^{f^T}(n)X_N^f(n)\right]$$
$$- E\left[\delta_N^T(n)X_N^{bT}(n)X_N^b(n)\left\{\delta_N(n) + W_N^*(n)\right\}\right]$$
$$= -E\left[\delta_N^T(n)X_N^{f^T}(n)X_N^f(n)\delta_N(n) + \delta_N^T W_N^*(n)X_N^{f^T}(n)X_N^f(n)\right]$$
$$- E\left[\delta_N^T(n)X_N^{bT}(n)X_N^b(n)\delta_N(n) + \delta_N^T(n)W_N^*(n)X_N^{bT}(n)X_N^b(n)\right]$$

$$T_1 = -E\left[\delta_N^T(n)R_{NN}\delta_N(n) + \delta_N^T(n)R_{NN}W_N^*(n)\right]$$
$$- E\left[\delta_N^T(n)R_{NN}\delta_N(n) + \delta_N^T(n)R_{NN}W_N^*(n)\right]$$

Since $E\left[\delta_N(n)\right] = 0$,

$$T_1 = -2E\left[\delta_N^T(n)R_{NN}(n)\delta_N(n)\right] \tag{10.56a}$$

But from the relation (Eqn. (10.55d))

$$M = \frac{E\left\{\delta_N^T(n)R_{NN}\ \delta_N(n)\right\}}{\varepsilon_{\min}}, \tag{10.56b}$$

Therefore, Eqn.(10.55g)

$$T_1 = E\left\{\delta_N^T(n)\left\{e^f(n)X_N^f(n) + e^b(n)X_N^b(n)\right\}\right\}$$

By Eqns. (10.56a) and (10.56b), becomes

$$T_1 = E\left\{\delta_N^T(n)\left\{e^f(n)X_N^f(n) + e^b(n)X_N^b(n)\right\}\right\} = -2M\varepsilon_{\min} \tag{10.56c}$$

Consider the term T_2 (Eqn. (10.55h)),

$$T_2 = E\left[\left\{e^f(n)X_N^f(n) + e^b(n)X_N^b(n)\right\}^T \left\{e^f(n)X_N^f(n) + e^b(n)X_N^b(n)\right\}\right]$$

$$= E\left[\left\{e^f(n)X_N^{f^T}(n) + e^b(n)X_N^{b^T}(n)\right\}\left\{e^f(n)X_N^f(n) + e^b(n)X_N^b(n)\right\}\right]$$

$$= E\left\{e^f(n)X_N^f(n)X_N^{f^T}(n)e^f(n) + e^b(n)X_N^{b^T}(n)e^f(n)X_N^f(n)\right\}$$

$$+ E\left\{e^f(n)X_N^{f^T}(n)e^b(n)X_N^b(n) + e^b(n)X_N^b(n)X_N^{b^T}(n)e^b(n)\right\}$$

Since the left hand side is a scalar,

$$T_2 = E\left[e^{f^2}(n)\right] Trace(R_{NN}) + E\left[e^b(n)e^f(n)\right] E\left[X_N^{b^T}(n)X_N^f(n)\right]$$

$$+ E\left[e^f(n)e^b(n)\right]E\left[X_N^{f^T}(n)X_N^b(n)\right] + E\left[e^{b^2}(n)\right]trace(R_{NN})$$

$$T_2 = 2\left\{E\left[e^{f^2}(n)\right] Trace(R_{NN}) + E\left[e^b(n)e^f(n)\right] E\left[X_N^{b^T}(n)X_N^f(n)\right]\right\}$$

Since from Eqn. (10.55e),

$$\varepsilon^f = E\left[e^{f^2}(n)\right] = \varepsilon_{\min}\left[1 + M\right]$$

$$T_2 = 2\left\{\varepsilon_{\min}(1+M)Trace(R_{NN}) + E\left[e^b(n)e^f(n)\right]E\left[X_N^{b^T}(n)X_N^f(n)\right]\right\} \quad (10.56d)$$

Now from Eqns. (10.53c) and (10.53d)

$$X_N^{f^T}(n) = \left[\,x(n-1),\, x(n-2)\ldots X(n-N)\right]$$

$$X_N^{b^T}(n) = \left[X(n-N+1),\, X(n-N+2),\ldots,\, x(n)\right]$$

Consider

$$E\left[X_N^{b^T}(n)X_N^f(n)\right] = E\left[x(n-(N-1)x(n-1)\right] + E\left[x(n-(N-2)x(n-2)\right]$$

$$+\ldots E\left[x(n-(N)x(n)\right]$$

$$E\left[X_N^{b^T}(n)X_N^f(n)\right] = \sum_{m=1}^{N} E\left[x(n-(N-m)x(n-m)\right]$$

$$E\left[X_N^{b^T}(n)X_N^f(n)\right] = \sum_{m=1}^{N} E\left[x(n-m+m+(N-m)x(n-m)\right]$$

$$E\left[X_N^{b^T}(n)X_N^f(n)\right] = \sum_{m=1}^{N} E\left[x(n-m-(N-2m))x(n-m)\right]$$

Replacing $(n-m)$ by n,

$$E\left[X_N^{b^T}(n)X_N^f(n)\right]=\sum_{m=1}^{N}E\left[x(n-(N-2m))x(n)\right]=\sum_{m=1}^{N}R(N-2m)$$

Therefore, Eqn. (10.56d)

$$T_2=2\left\{\varepsilon_{\min}(1+M)Trace(R_{NN})+E\left[e^b(n)e^f(n)\right]E\left[X_N^{b^T}(n)X_N^f(n)\right]\right\}$$

becomes

$$T_2=2\left\{\varepsilon_{\min}(1+M)Trace(R_{NN})+E\left[e^b(n)e^f(n)\right]\sum_{m=1}^{N}r(N-2m)\right\}$$

Now consider the ratio,

$$Q=\frac{E\left[e^b(n)e^f(n)\right]\sum_{m=1}^{N}r(N-2m)}{\varepsilon_{\min}(1+M)Trace(R_{NN})}$$

Since from Eqn.(10.55e),

$$\varepsilon^f=E\left[\left\{e^f(n)\right\}^2\right]=\varepsilon_{\min}\left[1+M\right]$$

$$Q=\frac{E\left[e^b(n)e^f(n)\right]\sum_{m=1}^{N}r(N-2m)}{E\left[\left\{e^f(n)\right\}^2\right]Trace(R_{NN})}$$

Further as $Nr(0)=Trace(R_{NN})$,

$$Q=\left|\frac{E\left[e^b(n)e^f(n)\right]}{E\left[\left\{e^f(n)\right\}^2\right]}\right|\left|\frac{\sum_{m=1}^{N}r(N-2m)}{Nr(0)}\right|$$

Since $Nr(0)>>r(k), k>>0$,

$$\left|\frac{\sum_{m=1}^{N}r(N-2m)}{Nr(0)}\right|<<1$$

Therefore,

$$Q \le \left| \frac{E\left[e^b(n) e^f(n) \right]}{\varepsilon^f} \right|$$

Here, $e^f(n)$ and $e^b(n)$ are the errors in estimating $x(n)$ and $x(n-N)$ respectively. For example, for an AR process with model order equal to less than N, after convergence $e^f(n)$ and $e^b(n)$ will be white and hence $E\left[e^b(n)e^f(n) \right]$ which is correlation of two white noise processes will be zero and the ratio $E\left[e^b(n)e^f(n) \right] / \varepsilon^f$ will also be zero. In general, $E\left[e^b(n)e^f(n) \right] / \varepsilon^f$ will be much less than unity. Therefore, in the term T_2 ,

$$T_2 = 2\left\{ \varepsilon_{\min}(1+M)Trace\,(R_{NN}) + E\left[e^b(n)e^f(n) \right] \sum_{m=1}^{N} r(N-2m) \right\},$$

In the above expression for T_2 , the term $E\left[e^b(n)e^f(n) \right] \sum_{m=1}^{N} r(N-2m)$ can be neglected compared to the term $\varepsilon_{\min}(1+M)Trace(R_{NN})$ and hence,

$$T_2 = 2\varepsilon_{\min}(1+M)Trace(R_{NN})$$

Substituting for T_2 and T_1 (Eqn. (10.56c)),

$$T_1 = E\left\{ \delta_N^T(n)\left\{ e^f(n)X_N^f(n) + e^b(n)X_N^b(n) \right\} \right\} = -2M\varepsilon_{\min}$$

in (Eqn. (10.55i)),

$$2\mu\left[T_1 + \frac{\mu}{2}T_2 \right] = 0$$

$$2\mu\left[-2M\varepsilon_{\min} + 2\mu\;\varepsilon_{\min}(1+M)Trace(R_{NN}) \right] = 0$$

$$\left[-2M + \mu(1+M)Trace(R_{NN}) \right] = 0$$

$$\frac{M}{1+M} = \frac{1}{2}\mu Trace(R_{NN})$$

For small $M << 1$,

$$M = \frac{1}{2}\mu\; Trace(R_{NN})$$

Since $\alpha = 2\mu$,

$$M = \frac{\alpha}{4} Trace\,(R_{NN}) \tag{10.56e}$$

For the LMS algorithm which minimizes either forward or backward error energy, the misadjustment is given by

$$M = \frac{\alpha}{2} Trace(R_{NN})$$

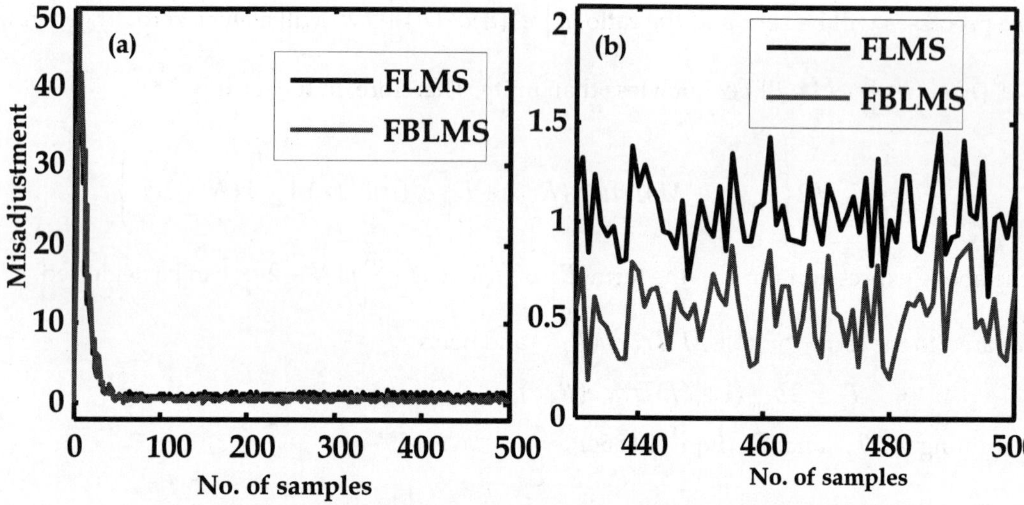

Fig. 10.8 (a) Averaged excess energy for FLMS and FBLMS algorithms and (b) Expanded view for the last part of the sequence (430 to 500 samples).

Therefore, for *the FBLMS, the misadjustment is half that of that of l LMS algorithm which minimizes either only forward or backward error energy. That is, the FBLMS predictor is advantageous in terms of misadjustment but its convergence speed is same that of the conventional LMS that minimizes forward/ backward error energy.*

The FBLMS misadjustment is illustrated for a sinusoid signal with Gaussian noise given by

$$x(n) = \sin[2\pi n(f_0 / f_s)] + \in(n), \quad (f_0 / f_s) = 1/6,$$

The sinusoid has a SNR=10.63 dB. The prediction filter has 6 taps. The step size and forgetting factor used are respectively, 0.2 and 0.99. The optimal error energy is computed using Yule-Walkers method for the FLMS by computing the prediction error for 500 samples. For the FBLMS, the optimal error energy is computed using Burg method by computing the error for the same 500 samples of data. The excess error energy is averaged over 100 trials. It is seen that the excess error energy for FBLMS is about half that of FLMS.

10.15 LAGUERRE TRANSVERSAL LMS ADAPTIVE ALGORITHM

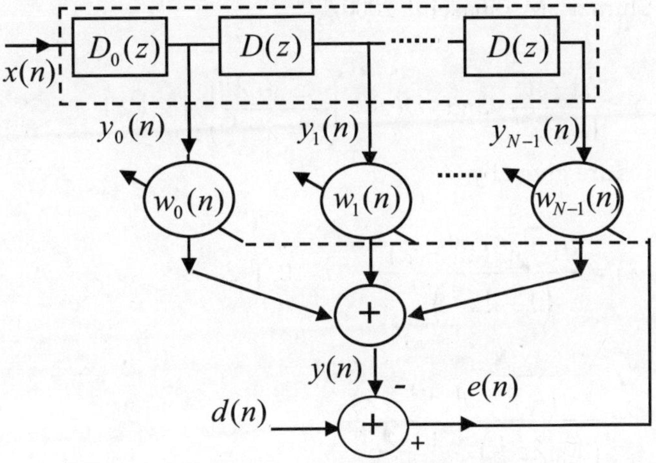

Fig. 10.9 Adaptive Laguerre transversal filter structure.

In applications like acoustic echo cancellation, equalization for underwater acoustic channels and digital control applications, the impulse response of very large length has to be handled due to high sampling rate involved. In such cases, the adaptation of the filter coefficients will make the computational load significantly high. To overcome this, IIR filters, which are having better modeling efficiency, can be used. However, the IIR filters suffer from the mean square error surface local minima and instability/divergence during coefficient adaptation. Approximating the impulse response in terms of truncated orthonormal sequences, the required model filter order can be reduced. The Laguerre sequences form such orthonormal sequences and are generated by the Laguerre structure characterised by IIR response but with all its poles inside the unit circle avoiding stability problem. This is due to the fact that the Laguerre structure retains the FIR structure but with its each delay unit replaced by a first order allpass unit and the allpass chain is fed by a first order lowpass filter (Fig. 10.9). Further, the responses of the allpass units and the lowpass filter are controlled by a single parameter, the Laguerre parameter. For a proper choice of this parameter value, the new structure can achieve the same modeling with a significant reduction in the order of the filter indicating that it is more efficient than the normal FIR structure that corresponds to the zero value of the Laguerre parameter. Such a structure is also known as constrained IIR as all its pole locations are constrained to be same and hence it is less efficient than a normal unconstrained IIR filter. The Laguerre adaptive filter does not suffer from either local minima or instability as only the FIR coefficients are updated which ensures global minimum and as the allpass and input lowpass unit's poles are fixed inside the unit circle by the choice of the Laguerre parameter.

A Laguerre transversal adaptive filter structure is shown in Fig. 10.9. A Laguerre structure is obtained by inserting a single pole low-pass filter $D_0(z)$ at the input and by using a first order allpass filters $D(z)$ instead of delay unit in normal transversal adaptive filter (M.A. Masnadi-Shirazi*, M. Ghasemi, 2000).

$$D_0(z) = \frac{\sqrt{1-\lambda^2}}{1-\lambda z^{-1}} \text{ and } D(z) = \left(\frac{z^{-1}-\lambda}{1-\lambda z^{-1}}\right), |\lambda| < 1 \tag{10.57a,b}$$

The Laguerre sequences are given by

$$D_i(z,\lambda) = \frac{\sqrt{1-\lambda^2}(z^{-1}-\lambda)^i}{\left(1-\lambda z^{-1}\right)^{i+1}}, \quad i = 0, 1, 2, \ldots$$

$$= \frac{\sqrt{1-\lambda^2}}{\left(1-\lambda z^{-1}\right)}\left(\frac{z^{-1}-\lambda)}{1-\lambda z^{-1}}\right)^i$$

$$D_i(z,\lambda) = D_0(z)D^i(z) \tag{10.57c}$$

The Laguerre sequences are the inverse transform of $D_i(z,\lambda)$, $d_i(n,\lambda)$ form a complete orthonormal set for $|\lambda| < 1$ satisfying

$$\sum_{n=-\infty}^{\infty} d_i(n,\lambda)d_j(n,\lambda) = \begin{cases} 1, i = j \\ 0, i \neq j \end{cases} \tag{10.57d}$$

Hence, any impulse response can be $h(n)$ can be expressed in untruncated form in terms of $d_i(n,\lambda)$ (as $d_i(n,\lambda)$ is an IIR filter)

$$h(n) = \sum_{i=-\infty}^{\infty} w_i(\lambda)d_i(n,\lambda) \tag{10.57e}$$

where

$$w_i(\lambda) = \sum_{n=-\infty}^{\infty} h(n)d_i(n,\lambda) \tag{10.57f}$$

(S. Hayakin, 2004) Here, 'λ' is known as the Laguerre parameter, denotes a predetermined pole location for the low-pass filter $D_0(z)$ and for the allpass filter $D(z)$. When $\lambda = 0$, the block $D(z)$ reduces to a unit delay z^{-1}, making the whole structure a normal transversal one. For the input $x(n)$, let the low-pass filter $D_0(z)$ output and the output of $D(z)$ at the ith tap, respectively be $y_0(n)$ and $y_i(n)$. The low-pass output is given by

$$Y_0(z) = D_0(z)X(Z)$$

$$y_0(n) = \lambda y_0(n-1) + \sqrt{1-\lambda^2}\,x(n) \tag{10.58a}$$

The allpass outputs are given by

$$Y_i(z) = D(z)Y_{(i-1)}(Z)$$

$$y_i(n) = y_{i-1}(n-1) - \lambda[y_{i-1}(n) - y_i(n-1)], \quad i = 0, 1, 2, \ldots, N-1 \tag{10.58b}$$

The output of the adaptive filter is

$$y(n) = \sum_{i=0}^{N-1} w_i(n) y_i(n) \tag{10.59a}$$

The error is

$$e(n) = d(n) - y(n)\,. \tag{10.59b}$$

The weight update equation is given by

$$w_i(n+1) = w_i(n) + \mu(n)\, e(n)\, y_i(n)\,, \; i = 0,1,\ldots, N-1 \tag{10.59c}$$

$\mu(n)$ is the adaptation step size is given by

$$\mu(n) = \frac{\mu_0}{P_0(n)}, \qquad 0 < \mu_0 < 1 \tag{10.59d}$$

$P_0(n)$ is the power of the signal $y_0(n)$ and is given by

$$P_0(n) = \beta_0 P_0(n-1) + (1-\beta_0) y_0^2(n)\,, \quad 0 < \beta_0 < 1 \tag{10.59e}$$

The choice of Laguerre parameter decides the order and the overall performance of the adaptive filter. The Laguerre parameter can be in the range of $|\lambda| < 1$. It has been reported that the error surface with respect to λ is quite shallow and a value of $|\lambda|$ close *to* 0.5, can be the choice for many applications. The Laguerre parameter can be estimated on online basis by using the gradient rule with an initial value of $|\lambda| = 0.5$.

For a transversal filter $W(n)$, the error energy between the impulse response of the system $h(n)$ and its estimate is,

$$\psi(\lambda) = \sum_{n=0}^{\infty}\left[h(n) - \sum_{i=0}^{N} w_i(\lambda) l_i(n,\lambda) \right]^2$$

where $w_i(\lambda)$ are weights of a transversal filter of length $N+1$

Using the orthonormal property of the Laguerre sequences and considering real $w_i(\lambda)$, then by minimizing the error surface

$$\psi(\lambda) = \sum_{n=0}^{\infty} h^2(n) - \sum_{i=0}^{N} w_i^2(\lambda)$$

$$\frac{\partial \psi(\lambda)}{\partial \lambda} = -2\sum_{l=0}^{N} w_i(\lambda)\frac{\partial w_i(\lambda)}{\partial \lambda}$$

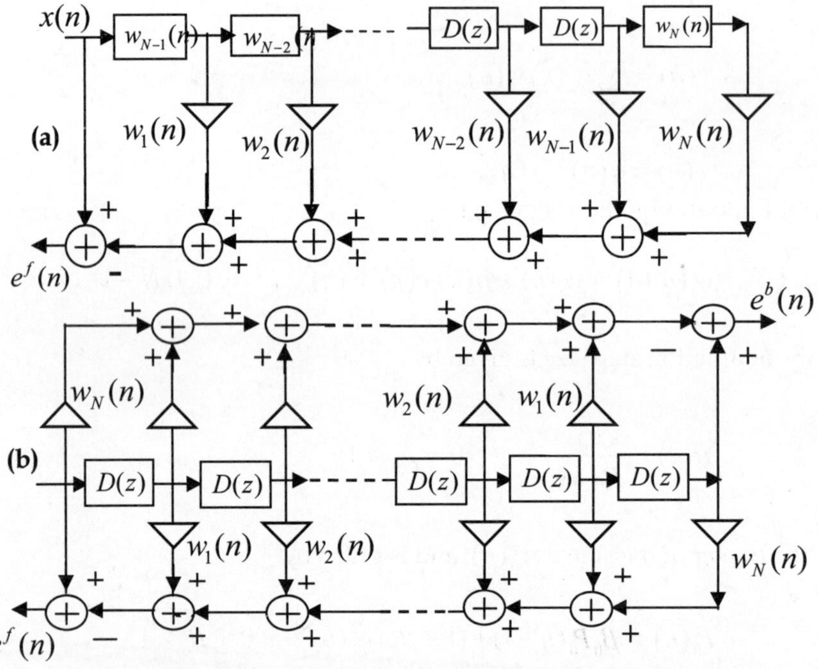

Fig. 10.10 Schematic of WFLMS (a) and WFBLMS (b) predictors.

It has been shown that in the literature (M.A. Masnadi-Shirazi, M. Ghasemi, 2000)

$$\sum_{i=0}^{N} w_i(\lambda)\frac{\partial w_i(\lambda)}{\partial \lambda} = \frac{N+1}{1-\lambda^2} w_{N-1}(\lambda) w_N(\lambda)$$

The adaptation for the Laguerre parameter using the gradient rule is given by

$$\lambda(n+1) = \lambda(n) + 2\mu\sum_{i=0}^{N} w_i(\lambda)\frac{\partial w_i(\lambda)}{\partial \lambda}$$

Therefore ,

$$\lambda(n+1) = \lambda(n) + 2\alpha \frac{N+1}{1-\lambda^2(n)} w_N(\lambda) w_{N+1}(\lambda), \quad 0 < \alpha < 1 \tag{10.60}$$

There is a similarity between warped transversal predictor and the Laguerre transversal adaptive filter (in a predictor configuration) as both have allpass units instead of delay units. This is true, however the Laguerre structure in addition to the allpass units has a first order low-pass filter whose pole location is also decided by the Laguerre parameter. The lowpass filter is essential in forming Laguerre orthonormal set of sequences, which are used in approximating the impulse response to be modeled. In both cases, the allpass units warp the frequency axis depending on the parameter λ value. However, the low-pass filter does not contribute to frequency warping. For its positive value, the high frequency region gets compressed and for its negative value, the low frequency region gets compressed. It is this frequency compression that facilitates the reduction in the order of the filter, as only a part of the frequency region has to be represented well by the model than the full frequency range uniformly.

For the common linear prediction, the estimate of $x(n)$, $\hat{x}(n)$ is given by

$$\hat{x}(n) = \sum_{i=1}^{N} w_i(n) x(n-i)$$

For the warped linear prediction, the estimate of $x(n)$, $\hat{x}(n)$ is given by

$$\hat{x}(n) = \sum_{i=1}^{N} w_i(n) y_i(n)$$

$$D(z) = \frac{Y_i(z)}{Y_{i-1}(z)} = \frac{z^{-1} - \lambda}{1 - \lambda z^{-1}}$$

Or

$$y_i(n) = y_{i-1}(n-1) - \lambda y_{i-1}(n) + \lambda y_i(n-1), \quad 0 < \mu < 1, \quad i = 1, 2, , ..., N$$

$$y_0(n) = x(n), \quad i = 0$$

The weight update for WFLMS (Fig.10.10a) is

$$w_i(n+1) = w_i(n) + \frac{\mu}{P(n)} e^f(n) y_i(n), \quad i = 1, 2, ..., N \tag{10.61a}$$

$$P(n) = \beta P(n-1) + (1-\beta) x^2(n), \quad 0 < \beta < 1$$

The weight update for WFBLMS (Fig.10.10b) is (S. Pavithra, 2011)

$$\left.\begin{aligned} w_i(n+1) &= w_i(n) + \frac{\mu}{P(n)}\left[e^f(n)y_i(n) + e^b(n)y_{N-i}(n)\right], \\ &\qquad\qquad\qquad\qquad i = 1,2,\ldots,N-1 \\ w_N(n+1) &= w_N(n) + \frac{\mu}{P(n)}\left[e^f(n)y_N(n) + e^b(n)x(n)\right], \quad i = N \end{aligned}\right\} \tag{10.61b}$$

$$S_x(\omega) = \frac{1}{\left|1 - \sum_{k=1}^{N} w_k \left(\frac{e^{-j\omega} - \lambda}{1 - \lambda\, e^{-j\omega}}\right)^k\right|^2} \tag{10.61c}$$

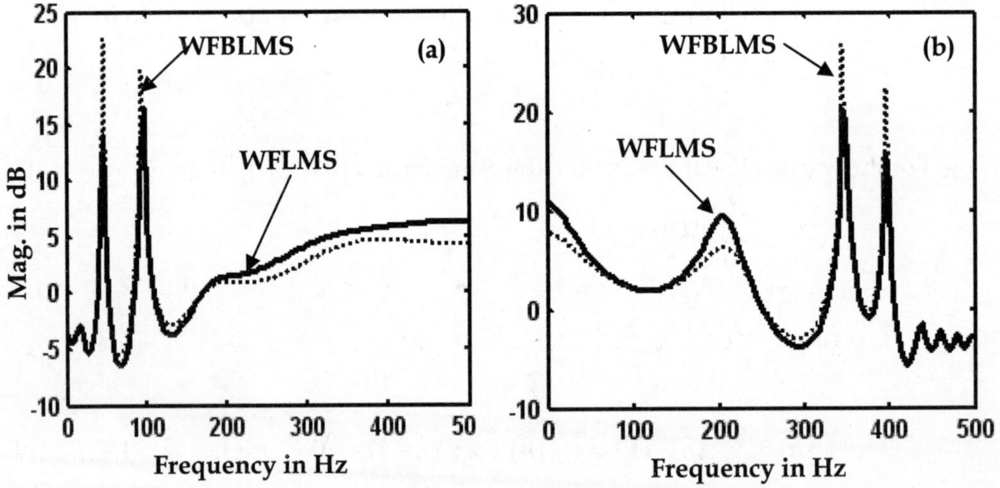

Fig. 10. 11 Spectrum estimated by WFLMS and WFBLMS predictors for signals in low (a) and high frequency (b) regions.

To bring out the performance of the warped FBLMS over that of warped FMLS, two signals are considered. For the first signal, sinusoids are in the low frequency region and are at 50Hz and 100Hz. For the second signal, the sinusoids are in the high frequency region and are at 350Hz and 400Hz. Further, for the two signals, zero mean Gaussian white noise is added to get an SNR of 20dB. The two signals are sampled at 1KHz. For the first signal the warping parameter used is 0.6 so that it compresses the high frequency region. But for the second, it is -0.6.so that it compresses the low frequency region. The compression in unwanted region enables to perform well in the desired region even with a smaller order predictor. Here, for both the signals, the parameters used are : order of the predictors is 15, step size is 0.02 and the forgetting factor for input power estimation is 0.999. The number of data samples used is 5000 and the parameters estimated after convergence is used for driving the spectrum.

It is seen that the WFBLMS enhance the desired signal peaks both for the low and high frequency regions compared to that achieved by WFLMS algorithm. This is due to lower misadjustment of WFBLMS compared to that of WFLMS. The warping helps in reducing the order of the filter required as it compresses the undesired frequency region and the minimization of the error energy is mainly over the desired frequency region.

10.16 LINEARLY CONSTRAINED LMS ALGORITHM

In chapter 7 for minimum variance spectral estimation (Capon's Method) constrained minimization was considered. Also in section 10.6.2 of this chapter, the NLMS was derived by constrained minimization of the Euclidian norm of incremental change in the weight vector in its successive updates. An adaptive version of minimum variance algorithm leads to linearly constrained LMS algorithm that uses a cost function which has a general linear constraint. The cost function given by Eqn. (7.81) is

$$Q(h,\lambda) = h_o^T R_x h_o + \lambda\left(1 - h_o^T W_o\right)$$

Its generalized version can be written as

$$Q(W,\lambda) = W^T R_x W + \lambda\left(W^T C - f\right) \tag{10.62a}$$

where, $W_o^T C = 1$ is changed to $W^T C = C^T W = f$, f is scalar and W_o is replaced by W .

To maintain the consistency of the notations of adaptive algorithms, symbols h_o and W_o , are respectively, replaced by W and C . The constrained LMS algorithm for the weight vector $W(n)$ can be achieved by using the gradient of the instantaneous error energy. The constrained cost function is given by

$$Q(W(n),\lambda(n)) = e^2(n) + \lambda(n)\left(W^T(n)C - f\right) \tag{10.62b}$$

The gradient of the cost function with respect to weight vector is

$$\nabla_W Q\left(W,\lambda(n)\right) = -2e(n)X(n) + \lambda(n)C \tag{10.62c}$$

The weight adaptation is given by

$$W(n+1) = W(n) - \mu \nabla_w Q\left(W,\lambda(n)\right)$$

Therefore,

$$W(n+1) = W(n) - \mu\left[-2e(n)X_N(n) + \lambda(n)C\right]$$

$$W(n+1) = W(n) + 2\mu X_N(n)e(n) - \mu\ \lambda(n)C \tag{10.62d}$$

Applying the constraint $W^T C\ = C^T W = f$ to $W(n+1)$,

$$C^T W(n+1) = C^T W(n) + 2\mu C^T X_N(n)e(n) - \mu\ \lambda(n)C^T C = f$$

Therefore,

$$\mu\,\lambda(n) = \frac{C^T W(n) + 2\mu C^T X_N(n)e(n) - f}{C^T C}$$

Therefore substituting for $\mu\,\lambda(n)$ in Eqn. (10.62d)

$$W(n+1) = W(n) + 2\mu X_N(n)e(n) - \mu\ \lambda(n)C$$

$$W(n+1) = W(n) + 2\mu X_N(n)e(n) + \frac{f - C^T\{W(n) + 2\mu X_N(n)e(n)\}}{C^T C}C$$

Defining

$$W^+(n) = W(n) + 2\mu X_N(n)e(n) \tag{10.62e}$$

$$W(n+1) = W^+(n) + \left[\frac{f - C^T W^+(n)}{C^T C}\right]C \tag{10.62f}$$

Thus, the linear constrained LMS adaptation can be implemented in two stages. At each iteration, first the intermediate weight vector $W^+(n)$ and then the final adaptation weight vector $W(n+1)$, are computed.

10.17 MOMENTUM LMS (MLMS) ALGORITHM

The motivation for momentum LMS algorithm is to improve the convergence speed of the conventional LMS algorithm and it was used for high speed adaptive equalization for digital communication channels. The MLMS algorithm is realized by adding to the gradient, a scaled difference between two weight vectors corresponding to successive previous instants (Rajesh Sharma; W.A. Sethares and J.A. Bucklew, 1998). That is,

$$W(n+1) = W(n) + \alpha e(n)X(n) + \gamma\left[W(n) - W(n-1)\right], \quad 0 < \alpha < 1, |\gamma| < 1 \tag{10.63a}$$

or

$$w_k(n+1) = w_k(n) + \alpha e(n)x(n-k) + \gamma\left[w_k(n) - w_k(n-1)\right], \quad 0 < \alpha < 1, \ |\gamma| < 1 \tag{10.63b}$$

The term in the square bracket is the momentum term. If the previous weight change, i.e., $\left[W(n) - W(n-1)\right]$ is large then adding a fraction of this change (determined by γ) to the present update will accelerate the descent process towards the minimum. This will enable a smoother and faster convergence of the weight vector. But for $\gamma < 0$, the MLMS has a slower convergence and lower misadjustment. Further as $|\gamma| \to 1$, the MLMS algorithm becomes unstable. The MLMS requires additional storage for $W(n-1)$ and a multiplication of a scalar with a vector. For the MLMS, the largest time constant is given by

$$\tau_{\min} = \frac{(1-\gamma)}{\alpha\, \lambda_{\min}} \tag{10.63c}$$

Thus for $\gamma > 0$, the largest time constant is smaller by a factor $(1-\gamma)$ hence, this provides a faster convergence compared to that of LMS. This is due to increased gradient size. However, after convergence as the gradient size is more than that for LMS, the convergence error/ misadjustment will be more. The misadjustment is given by

$$M = \frac{\alpha}{2}\frac{N\sigma_x^2}{(1-\gamma)}, \qquad \gamma > 0 \tag{10.63d}$$

Thus, the misadjustment is more by a factor $\left[1/(1-\gamma)\right]$ compared to that of conventional LMS algorithm. This can be overcome by removing the additional term $\left[W(n) - W(n-1)\right]$ by making $\gamma = 0$ when the difference weight vector term becomes small and this converts MLMS to LMS after convergence. However, using a constant γ value both prior to and after convergence is *not of any use* as the reduced convergence error due to improved convergence rate is eaten away by the increased convergence error, as convergence progresses. *But the MLMS provides a smoothing effect to the mean squared error for the cases where for LMS, its MSE has bursts.* Also for nonlinear criterion error surfaces, the MLMS due to its momentum term helps to escape the local minima (like in neural network learning). Hence, to take advantage of MLMS, the value of γ should be varied from high value (< 1) to a low value as the convergence progresses (M.A. Tugay and Y. Tanik, 1989).

10.18 MEDIAN LMS (MEDLMS) ALGORITHM

The LMS algorithm though simple suffers from slow convergence rate. Further, its performance may get seriously affected by the presence of an impulsive interference either in the input or in the desired signal and may distort the gradient in the weight vector update equation resulting in long term disturbance in the output which my even lead to instability. (P.M. Clarkson and T.I. Haweel, 1989). This problem to some extent can be controlled by reducing the adaptation step size but this will result in reduced convergence rate. Also averaging the gradient over *(N+1)* samples,

$$W(n+1) = W(n) + \alpha \sum_{j=n-N/2}^{n+N/2} e(j)X(n-j) \quad 0 < \alpha < 1 \tag{10.64a}$$

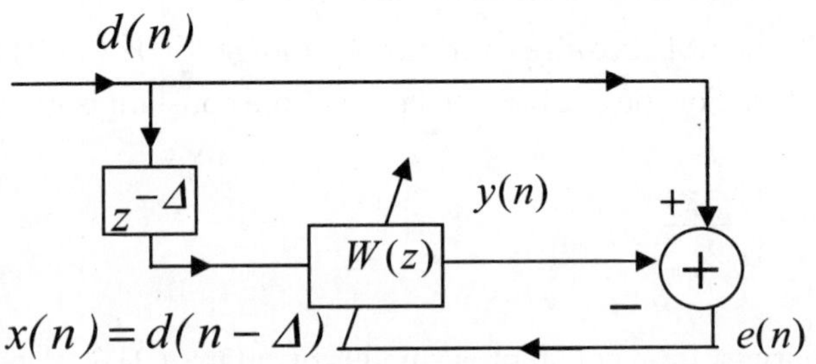

Fig. 10.12 (a) Median LMS adaptive filter with a delay.

may also reduce the severity of the problem but again at the cost convergence speed. But an appropriate solution is to use the median of the gradient, instead other solutions like averaging the gradient or reduction of stepsize. That is,

$$W(n+1) = W(n) + \alpha \; med\left[e(n)X(n)\right]_N \quad 0 < \alpha < 1 \tag{10.64b}$$

where $med\left[e(n)X(n)\right]_N$ is sample median of gradient estimates

$$e(n)X(n),\, e(n-1)X(n-1),\, \ldots\ldots,\, e(n-N+1)X(n-N+1) \tag{10.64c}$$

(The median operation basically arranges the numbers in the increasing /decreasing order and picks up the middle value and thus, it enables rejection of extreme values both on higher and lower sides.) Such a median LMS adaptive filter removes the effect of impulsive interference and its performance is similar to that of ordinary LMS in the absence of impulsive interference. The median operation arranges the sequence of numbers considered in monotonically increasing/ decreasing order and picks up the middle number. However, the LMS and averaged gradient LMS suffer severely after the occurrence of the impulsive interference.

To illustrate the performance of MEDLMS in comparison with that of conventional LMS, the filtering of a sinusoid associated with Gaussian noise with a large amplitude impulse located at the middle point to bring out the effect of impulsive interference and is given by

$$d(n) = \cos \omega_0 n + w(n) + 50\,\delta(n - 10000)$$

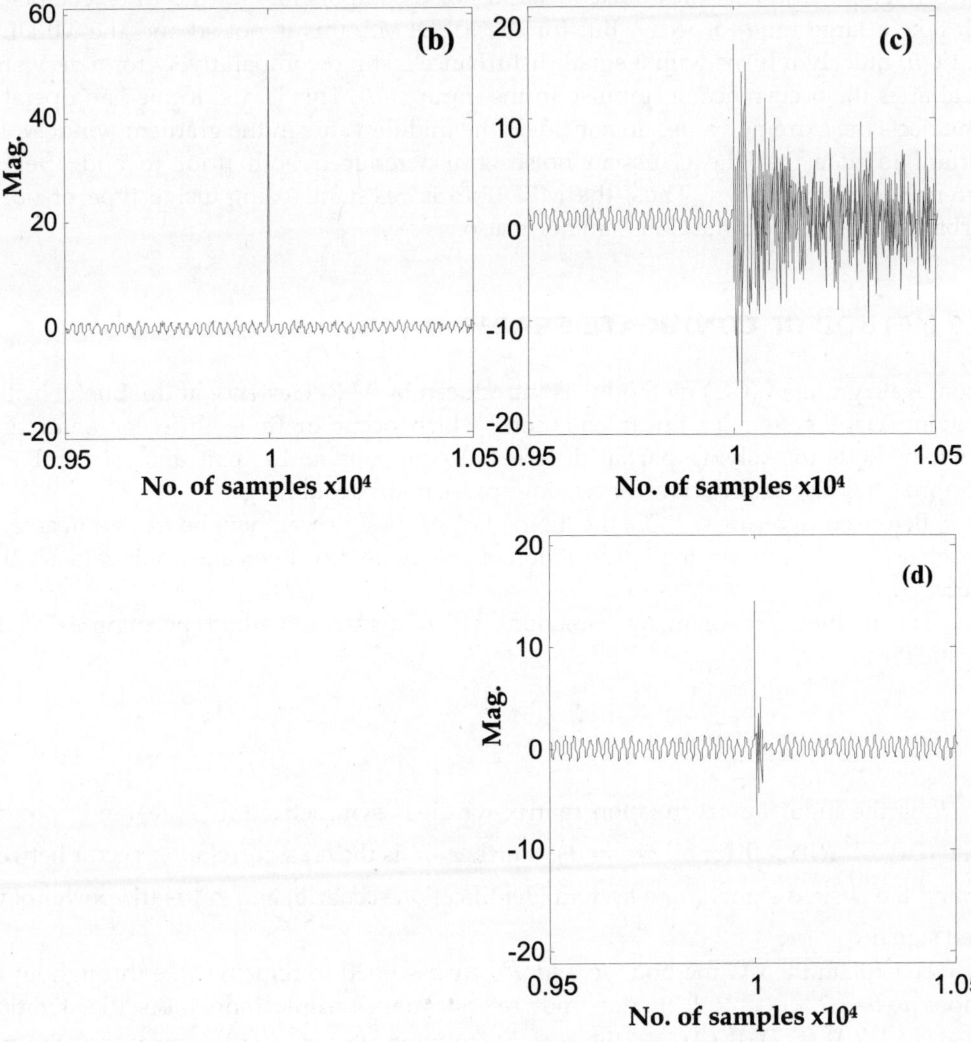

Fig. 10.12 (b, c, d) Adaptive median filtering performance; (b) Input signal having sinusoid with Gaussian noise plus impulse; (c) Filtering by conventional LMS and (d) Filtering by MEDLMS.

where $w(n)$ is a zero-mean unit variance Gaussian white noise. The frequency of the cosinusoid is 500 Hz and the sampling frequency used is 10 KHz, The SNR for sinusoid to Gaussian noise is 11dB The number of samples considered for processing is 20000 (Fig. 10.12b). The filtering schematic is as shown in Fig. 10.12(a).

The parameters used are : delay $\Delta = 1$ and for the MLMS the step size used is 0.01, the filter length $L = 20$ and the median widow length used is $N = 5$.

Prior to occurrence of impulse, the LMS algorithm reduces the Gaussian noise (Fig. 10.12c). But the LMS algorithm experiences severe distortion following the onset of the impulse (Fig. 10.12c). This introduces a large noise component into the filter coefficients which take a large time to decay. But for the MEDLMS, this is not so and the MEDLMS output *y(n)* quickly returns with a small disturbance lasting comparatively for a very short interval after the occurrence of impulse in the input. *d(n)*. This is due to median operation which rejects the extreme values ad considers the middle value in the gradient window. It is important to note that the Gaussian noise is also reduced both prior to and after the occurrence of the impulse. Thus, the MEDLMS is resistant to impulsive type of noise/ disturbance and suppress the Gaussian noise also.

10.19 METHOD OF CONJUGATE GRADIENT (CG)

The conjugate gradient (CG) method was introduced by M.R. Hestenes and E Stiefel in 1952 for solving large system of linear equations which occur in finite difference and finite element methods for solving partial differential equations and circuit analysis problems. This approach is also appropriate for solving sparse matrix equations.

Before considering the CG, the method of steepest descent will be treated in a vector approach and this simplifies the introduction of conjugate directions and conjugate gradient methods.

The method of conjugate directions is an iterative method of minimizing the quadratic form

$$f(W) = \frac{1}{2}W^T RW - P^T W + \sigma_d^2 \tag{10.65a}$$

Here, R is the input autocorrelation matrix which is symmetric ($R^T = R$) and positive definite (i.e., $W^T RW > 0$ for all $W \neq 0$). Further, P is the crosscorrelation vector between input and the desired signal (for a system identification scenario) and σ_d^2 is the power of the desired signal.

To explain the CG method, R and P, are assumed to remain same throughout the iterations *n=0. 1,2, ………, L* (*n* does not refer to data sample index). As the iterations proceed as $W(0), W(1), W(2), \ldots$, the weight/solution vector $W(n)$ approaches the true solution W_*. Here, the error $\in(n)$ for the solution vector at n^{th} iteration $W(n)$, with reference to optimum solution W_* is defined as

$$\in(n) = W(n) - W_* \tag{10.65b}$$

The norm of the error $\|\in(n)\| = \sqrt{\in_1^2(n) + \in_2^2(n) + \cdots + \in_N^2(n)}$ indicates how far away the estimated solution $W(n)$ is from the true/ optimum solution W_*. But unfortunately, as the true/optimum solution W_* is not known, $\in(n)$ cannot be calculated. The minimum of $f(W)$ (Eqn. (10.65a)) corresponds to that $W(n)$, where its derivative $f'(W)$ with respect to W, vanishes. That is,

$$\frac{\partial f(W)}{\partial W} = \frac{\partial}{\partial W}\left[\frac{1}{2}W^T RW - PW^T + \sigma_d^2\right] = 0$$

$$f'(W) = RW_* - P = 0$$

This implies that $f(W)$ minimization is equivalent to the solution of the equation $RW_* = P$, where W_* is the optimal solution. Thus, as already stated this approach can also be used to solve the matrix equation $RW_* - P = 0$. Also for n^{th} iteration,

$$f'(W(n)) = RW(n) - P$$

Since

$$RW_* = P$$

$$f'(W(n)) = RW(n) - RW_*$$

Therefore, using Eqn. (10.65b),

$$f'(W(n)) = R \in (n), \tag{10.65c}$$

The residual at n^{th} iteration $\xi(n)$ is defined as

$$\xi(n) = P - RW(n) \tag{10.65d}$$

Since $RW_* = P$ and using Eqn. (10.65c)

$$\xi(n) = RW_* - RW(n) = -R(\{W(n) - W_*\}$$

$$\xi(n) = -R \in (n)$$

$$\xi(n) = -f'(W(n)) \tag{10.65e}$$

Since P, R and $W(n)$ are known/ estimated, unlike the error $\in(n)$, the residue $\xi(n)$ can be calculated.

In the method of steepest descent, the starting is from an arbitrary point and is moved further in many steps till a satisfactory solution is reached. But every time taking a new step, it is taken in a direction in which $f(W(n))$ deceases quickly, that is, $-f'(W(n)) = P - RW(n) = \xi(n)$. ***Thus, the residue $\xi(n)$ is the direction of steepest descent.***

In the method of steepest descent, the new weight vector is equal to the old vector plus correction. The correction direction is opposite to gradient of the cost function $f(W(n))$ and its magnitude is the gradient scaled by the stepsize $\alpha(n)$ which is given by

$$W(n+1) = W(n) - \alpha(n) f'(W(n))$$

Using Eqn. (10.65e)

$$\xi(n) = -f'(W(n))$$

$$W(n+1) = W(n) + \alpha(n)\xi(n) \tag{10.65f}$$

The magnitude of the stepsize $\alpha(n)$ is chosen by a *line search procedure* which minimizes $f(W(n+1))$ with respect to stepsize $\alpha(n)$ along the line of gradient, $-f'(W(n))$. That is,

$$\nabla_{\alpha(n)} f\{W(n+1)\} = 0, \quad i.e., \quad \frac{\partial f\{W(n+1)\}}{\partial\, \alpha(n)} = 0$$

From chain rule of differentiation,

$$\frac{\partial f^T\{W(n+1)\}}{\partial W(n+1)} \frac{\partial \{W(n+1)\}}{\partial\, \alpha} = 0$$

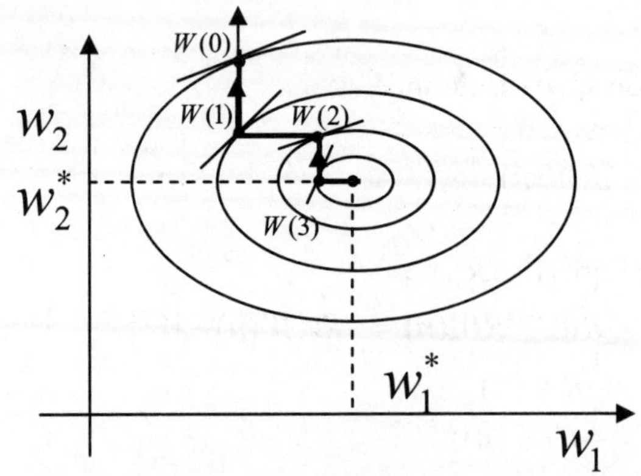

Fig. 10.13 Weight vector path for the line search method.

Further substituting for $W(n+1)$,

$$\nabla^T_{(W(n+1))} f\{W(n+1)\} \frac{\partial[W(n) - \alpha(n)\nabla_{(W(n)} f(W(n))]}{\partial\, \alpha(n)} = 0$$

$$-\nabla^T_{W(n+1)} \left[f(W(n+1)) \right] \nabla_{(W(n)} \left[f(W(n)) \right] = 0$$

$$\nabla^T_{W(n+1)} \left[f(W(n+1)) \right] \nabla_{(W(n)} \left[f(W(n)) \right] = 0 \qquad (10.65g)$$

$$f'^T(W(n+1)) f'(W(n)) = 0$$

From Eqn. (10.65e)

$$\xi(n) = -f'(W(n))$$

Therefore,

$$\xi^T(n+1)\xi(n)) = 0$$

$$\left[P - RW(n+1) \right]^T \xi(n) = 0, \quad \because \quad P - RW(n) = \xi(n), \quad \text{Eqn}.\ (10.65d)$$

From Eqn. (10.65f)

$$W(n+1)) = W(n) + \alpha\ \xi(n)$$

Therefore,

$$\left[P - R\{W(n) + \alpha(n)\xi(n)\} \right]^T \xi(n) = 0$$

$$\left[\{P - RW(n)\}^T - \alpha(n)\xi^T(n)R \right] \xi(n) = 0$$

$$\xi^T(n)\xi(n) - \alpha(n)\xi^T(n)R\xi(n) = 0$$

$$\alpha_o(n) = \frac{\xi^T(n)\xi(n)}{\xi^T(n)R\xi(n)} \qquad (10.65h)$$

Using this optimum step size $\alpha_o(n)$, the new weight vector that best minimizes $f(W(n+1))$ .is given by

$$W(n+1) = W(n) - \alpha_o(n)\nabla f(W(n))$$

From Eqn. (10.65g)

$$\nabla^T_{W(n+1))}\left[f(W(n+1))\right]\nabla_{W(n)} f\left[W(n)\right] = 0$$

This indicates that for optimized step size for the sample instant $(n+1)$, *the gradient is orthogonal to that at sample instant* n *and this makes the path of optimal step size steepest descent, zigzag* (Fig. 10.13). Though this method is simple and easy to apply, it is slow in converging to minimum point especially when the eigenvalue spread is large, i.e., the quadratic surface is tall and narrow. Further, the convergence not only becomes slower as it approaches the minimum point *but also depends on the starting point.* The algorithm is given as below:

Step 1: The quantities R and P are assumed to be known or estimated sequentially and $W(0) = 0$.

$$\xi(n) = P - RW(n)$$

Step 2:

$$\alpha(n) = \frac{\xi^T(n)\xi(n)}{\xi^T(n)R\xi(n)}$$

For these two steps, two matrix vector multiplications are required.

Step 3

$$W(n+1) = W(n) + \alpha(n)\xi(n)$$

This implies

$$\in(n+1) = \in(n) + \alpha(n)\xi(n)$$

To avoid one matrix vector multiplication,

$$\xi(n+1) = P - RW(n+1) = P - R[W(n) + \alpha(n)\xi(n)]$$
$$\xi(n+1) = \xi(n) - \alpha(n)R\xi(n)]$$

can be used instead of $\xi(n+1) = P - RW(n+1)$. *But this does not indicate the status of* $W(n)$ *whether it has reached the true solution* $W_*(n)$ *and the round-off errors may result in an estimated solution which may be only close to the true one but not the true solution.* Though the error $\in(n)$ provides this information, as already stated it cannot be calculated.

With this $\xi(n+1)$ for the second iteration (subsequent iterations) , step 2 and step 3 can be computed.

In the SD method, the initial weight vector used for further adaptation is important and if nothing is known about it, it can be taken as a zero vector. Also as already stated the eigenvalue spread of the input autocorrelation matrix R decides the convergence speed and larger the eigenvalue spread, slower is the convergence.

10.19.1 Method of Conjugate Directions

In the method of steepest descent, there may be retracing of the previous path as the directions need not satisfy any condition like orthogonality (in general). Due to this, the error from previous directions may enter into the new directions and the error may even grow in new directions. Also for the optimum step size, the successive directions will be perpendicular to each other which makes the weight vector path a zigzag path. The weight vector path towards minimum can be made better by choosing a path that pierces through the zigzag path. These undesired possibilities can be overcome if the search directions $V(k), \quad k = 0,1,2,\ldots,N-1$, are orthogonal and in each direction the component of the error is completely minimized by using appropriate stepsize (Arghya Deb, 2012, Dragika Vasileska 2010, Martin Zachariasen, DIKU 2001e). In general, this leads to exact solution/minimization of error in N steps and in some cases, the error becomes small after a few iterations. For this, choose a set of orthogonal search directions, $V(k), \quad k = 0,1,2,\ldots,N-1$ and in each search direction, a step is taken so that the weight adaptation is given by

$$W(n+1) = W(n) + \alpha(n)V(n) \tag{10.66a}$$

with the value of $\alpha(n)$ such that, *the error in the present direction should not have the error of previous direction* (as stated already) *and the error component should be minimized in each direction.* That is, according to the first constraint,

$$\left(W(n+1) - W_*\right) \bullet V(n) = \in(n+1) \bullet V(n) = 0 \tag{10.66b}$$

The symbol $\bullet$ indicates vector dot product operation. That is, the error at $(n+1)^{th}$ sample instant does not have any component along $V(n)$, the direction of previous sample instant n.

The error vector $\in(n+1)$ at $(n+1)^{th}$ sample instant can be written in terms of previous directions $V(i), i = 0,1,2,\ldots$

$$\begin{aligned}\in(n+1) &= W(n+1) - W_* \\ &= W(n) + \alpha(n)V(n) - W_* \qquad \text{from Eqn.(10.66a)} \\ &= W(n) - W_* + \alpha(n)V(n) = \in(n) + \alpha(n)V(n)\end{aligned}$$

$$\in(n+1) = \in(n) + \alpha(n)V(n) \tag{10.66c}$$

$$\begin{aligned}\in(n+1) &= W(n) - W_* + \alpha(n)V(n)\\ &= W(n-1) + \alpha(n-1)V(n-1) - W_*\\ &= \in(n-1) + \alpha(n-1)V(n-1) + \alpha(n)V(n)\end{aligned}$$

$$\in(n+1) = \in(n-1) + \alpha(n-1)V(n-1) + \alpha(n)V(n)$$

$$\vdots \qquad \vdots$$

$$\in(n+1) = \in(0) + \sum_{k=0}^{n} \alpha(k)V(k) \qquad (10.66d)$$

To explain the meaning of the above equation, let $\eta(k) = -\alpha(k)$. Then,

$$\in(n+1) = \in(0) - \sum_{k=0}^{n} \eta(k)V(k)$$

This states that the error at any sample instant is equal to the original error $\in(0)$ plus sum of the components of the error in the directions which are already covered. As sample instants increase, more and more components ($\eta(k)V(k)$) are covered. Thus, with increase in the number of error components, they ***eat away*** (this is indicted by the negative sign associated with the term $-\eta(k)V(k)$) more and more from the original error $\in(0)$ making the error $\in(n)$ to decrease. Finally, at the sample instant N, the error reaches minimum, as all the N components of error are covered.

The Eqn. (10.66b) is

$$V^T(n) \in (n+1) = 0 \quad or \quad \in^T(n+1)V(n) = 0$$

Using Eqn. (2c), (9.23c)

$$\in(n+1) = \in(n) + \alpha(n)V(n)$$

$$V^T(n)\left[\in(n) + \alpha(n)V(n)\right] = 0\,, \qquad \alpha(n) = -\frac{V^T(n)\in(n)}{V^T(n)V(n)}$$

But as $\in(n)$ ***is not known*** (since $\in(n) = W(n) - W_*$ and as true solution W_* is not known), $\alpha(n)$ ***cannot be calculated***. *Hence, to overcome this problem, instead of choosing the search directions orthogonal, the orthogonality with respect to matrix* R *(*R *is symmetric and positive definite) is used and this is known as* R***-orthogonality or R-conjugate*** and is given by

$$V^T(n)R \in (n+1) = 0 \quad or \quad \in^T(n+1)RV(n) = 0 \qquad (10.66e)$$

But using Eqn. (10.66c)

$$V^T(n)R\{\in(n)+\alpha(n)V(n)\}=0, \quad \alpha(n)=-\frac{V^T(n)R\in(n)}{V^T(n)RV(n)}$$

But as

$$R\in(n)=-\xi(n),$$

$$\alpha(n)=\frac{V^T(n)\xi(n)}{V^T(n)RV(n)} \qquad (10.66f)$$

Since all the quantities on the R.H.S of above equation are known, $\alpha(n)$ ***can be computed***

If the search directions/ vectors are the residuals then, Eqn. (10.66f) for step size will be identical to that method of steepest descent. Eqn. (10.66g)

$$\alpha(n)=\frac{\xi^T(n)\xi(n)}{\xi^T(n)R\xi(n)}$$

10.19.2 Minimization of the Error by *R*-Orthogonality of Search Directions

It can be shown that the condition $\in^T(n+1)RV(n)=0$ results in minimization of the error in the direction $V(n)$. This will be shown by proving that the minimization of the error leads to A-orthogonality. *Minimizing the error in search direction is equivalent to minimizing the function.* The function minimization at $(n+1)^{th}$ iteration requires

$$\frac{df(W(n+1))}{d\alpha(n)}=0, \quad \frac{df(W(n+1))}{dW(n+1)}\frac{dW(n+1)}{d\alpha(n)}=0$$

$$\frac{df(W(n+1))}{dW(n+1)}\frac{d(W(n)+\alpha(n)V(n))}{d\alpha(n)}=0$$

$$\nabla^T_{W(n+1)}[f(W(n+1))]\,V(n)=0$$

For the quadratic form,

$$f(W)=\frac{1}{2}W^TRW-P^TW+\sigma_d^2$$

$$\nabla f(W)=RW-P=-\xi$$

Further,

$$W(n+1) - W_* = \in (n+1),$$
$$RW(n+1) - RW_* = R \in (n+1)$$
$$RW(n+1) - P = R \in (n+1) = -\xi(n+1)$$
$$RW(n+1) - P = \nabla^T_{(W(n+1))} f(W(n+1)),\ R^T = R$$

For minimization as stated above,

$$\nabla^T_{(W(n+!)} f(W(n+1))\ V(n) = 0$$

Since

$$RW(n+1) - P = \nabla^T_{(W(n+1)} f(W(n+1))$$

Therefore,

$$-\xi^T (n+1) V(n) = 0$$

But

$$RW(n+1) - P = R \in (n+1) = -\xi(n+1)$$

Therefore,

$$\in^T (n+1) RV(n) = 0 \quad or \quad V^T(n) R \in (n+1) = 0$$

But this is the condition required for the R-orthogonalization of directions $V(n)$ in successive iterations. Hence R-orthogonalization of directions $V(n)$, leads to minimization of error.

10.19.3 Minimization in N Steps

The error $\in (0)$ can be expressed in terms of the orthogonal search directions $V(n), n = 0,1,\ldots,N-1$ as the form its basis.

$$\in (0) = \sum_{j=0}^{N-1} \delta(j) V(j) \tag{10.67a}$$

Pre-multiplying by $V^T(k)R$,

$$V^T(k) R \in (0) = \sum_{j=0}^{N-1} \delta(j) V^T(k) RV(j) = \delta(k) V^T(k) RV(k),$$

since

$$V^T(k)\sum_{\substack{j=0\\ j\neq k}}^{N-1} \delta(j)V(j) = 0,$$

Therefore,

$$\delta(k) = \frac{V^T(k)R \in (0)}{V^T(k)RV(k)} \tag{10.67b}$$

But as

$$V^T(k)\sum_{j=0}^{k-1} \alpha(j)V(j) = 0,$$

$$\delta(k) = \frac{V^T(k)R\left\{\in (0) + \sum_{j=0}^{k-1} \alpha(J)V(j)\right\}}{V^T(k)RV(k)}, \tag{10.67c}$$

From Eqn. (10.66a),

$$W(n+1) = W(n) + \alpha(n)V(n)$$

$$W(n) = W(n-1) + \alpha(n-1)V(n-1)$$

$$W(n) - W_* = W(n-1) - W_* + \alpha(n-1)V(n-1)$$

$$\in (n) = \in (n-1) + \alpha(n-1)V(n-1)$$

$$\in (n-1) = \in (n-2) + \alpha(n-2)V(n-2)$$

$$\in (n) = \in (n-2) + \alpha(n-2)V(n-2 + \alpha(n-1)V(n-1)$$

Thus,

$$\in (n) = \in (0) + \sum_{j=0}^{n-1} \alpha(j)V(j) \tag{10.67d}$$

Therefore, Eqn. (10.67c) becomes as

$$\delta(k) = \frac{V^T(k)R \in (k)}{V^T(k)RV(k)}$$

But from Eqn. (10.66f),

$$\alpha(n) = \frac{V^T(n)\xi(n)}{V^T(n)RV(n)} \text{ with } R \in (n) = -\xi(n)$$

$$\alpha_k = -\frac{V^T(k)R \in (k)}{V^T(k)RV(k)}$$

Thus,

$$\alpha(k) = -\delta(k) \tag{10.67e}$$

and the development of the solution as $W(n+1) = W(n) + \alpha(n)V(n)$ in steps, results in reduction of the error component by component. But from Eqn. (10.67a)

$$\in (0) = \sum_{j=0}^{N-1} \delta(j)V(j)$$

Therefore,

$$\in (n) = \sum_{j=0}^{N-1} \delta(j)V(j) + \sum_{j=0}^{n-1} \alpha(j)V(j)$$

Since from Eqn. (10.67e)

$$\alpha(j) = -\delta(j),$$

$$\in (n) = \sum_{j=0}^{N-1} \delta(j)V(j) - \sum_{j=0}^{n-1} \delta(j)V(j)$$

$$\in (n) = \sum_{j=n}^{N-1} \delta(j)V(j) \tag{10.67f}$$

At each iteration, as one component of the error becomes zero, the total error will become zero in N iterations.

10.19.4 Method of Choosing Search Directions

This is based on well known Gram-Schmidt orthogonalization which converts a linearly independent basis to an orthogonal basis. For a set of linearly independent vectors $U(i), = 0, 1, 2, \ldots, N-1$, the orthogonal basis $V(i), i = 0, 1, 2, \ldots, N-1$ is given by

$$V(0) = U(0)$$

$$V(n) = U(n) - \sum_{k=0}^{n-1} \beta(n,k)V(k), \tag{10.68a}$$

$\beta(n,k)$ are the projections of $U(n)$ on previous $V(k)$ directions. Pre-multiplying Eqn. (10.68a) by $V^T(n)R$

$$\begin{aligned} V^T(n)RV(j) &= U^T(n)RV(j) - \sum_{k=0}^{n-1} \beta(n,k)V^T(k)RV(j) \\ &= U^T(n)RV(j) - \beta(n,j)V^T(j)RV(j), \quad \textit{for } n > j \end{aligned}$$

Hence, from the R-orthogonality property of V vectors, $V^T(n)RV(j) = 0, \; n > j$

$$U^T(n)RV(j) - \beta(n,j)V^T(j)RV(j) = 0$$

$$U^T(n)RV(j) - \beta(n,j)V^T(j)RV(j) = 0$$

$$\beta(n,j) = \frac{U^T(n)RV(j)}{V^T(j)RV(j)} \tag{10.68b}$$

This demands knowledge of previous search directions and this is due to the fact that $V(n)$ is given by Eqn. (10.68a)

$$V(n) = U(n) - \sum_{k=0}^{n-1} \beta(n,k)V(k)$$

Hence, there is a necessity to store all the previous directions which is not practically convenient and requires order of N^3 operations. In view of these, this method is not favourable for use.

10.19.5 The Method of Conjugate Gradient

The conjugate gradient method overcomes the problem of storing the previous directions required by the method of conjugate directions, considered.

The important property of conjugate directions is that *the residue at i^{th} iteration $\xi(n)$ is orthogonal to the space spanned by all the previous (n-1) search directions $V(k), k = 0.1, 2, \ldots n-1$*. The error $\in(n)$ is given by (Eqn. (10.67f))

$$\in(n) = \sum_{j=n}^{N-1} \delta(j) V(j)$$

Pre-multiplying by $V^T(k)R$

$$-V^T(k)R \in(n) = -\sum_{j=n}^{N-1} \delta(j) V^T(k) R V(j) = 0, \quad for \quad k < n$$

But as $R \in(n) = -\xi(n)$,

$$V^T(k)\xi(n) = 0 \quad for \quad k < n \tag{10.69a}$$

That is, the residual $\xi(n)$ at n^{th} iteration is orthogonal to previous search directions. Further from Gram-Schmidt orthogonalization (Eqn. (10.68a)

$$V(n) = U(n) - \sum_{k=0}^{n-1} \beta(n,k) V(k)$$

Substituting this in,

$$V^T(n)\xi(j) = 0, \quad for \quad n < j$$

$$V^T(n)\xi(j) = \left[U(n) - \sum_{k=0}^{n-1} \beta(n,k) V(k) \right]^T \xi(j) = 0, \quad for \quad n < j$$

$$V^T(n)\xi(j) = U^T(n)\xi(j) - \sum_{k=0}^{n-1} \beta(n,k) V(k) \xi(j) = 0, \quad for \quad n < j$$

$$V^T(n)\xi(j) = U^T(n)\xi(j) = 0 \tag{10.69b}$$

This states that the previous search directions $V(n)$ are orthogonal to the residues $\xi(j), n < j$.

Any gradient method to be efficient should avoid retracing its path and this is true with the conjugate directions method considered, as its search directions are *R-orthogonal* which not only ensures mutual independence of search directions but also minimizes the cost function in each search direction enabling convergence in N iterations. It is necessary that the new search directions are orthogonal to previous search directions. As the residuals are orthogonal to previous search directions, the new search directions formed by residuals will also be orthogonal to previous search directions. *Hence, the choice of $U(n)$ to generate $V(n)$ by Gram-Schmidt procedure* ***cannot be arbitrary*** *and the residuals form this unique choice as they result in new search directions orthogonal to previous search directions (as discussed already) hence,* $U(n) = \xi(n)$. (The condition of the residual becoming zero, corresponds to reaching the minimum). Therefore, Eqn.(10.68b)

$$\beta(n,j) = \frac{U^T(n)RV(j)}{V^T(n)RV(j)}$$

becomes

$$\beta(n,j) = \frac{\xi^T(n)RV(j)_j}{V^T(n)RV(j)} \tag{10.69c}$$

Thus, the conjugate gradient method is the method of conjugate direction where the search directions are constructed by conjugation of the residuals, i.e., $U(n) = \xi(n)$. Hence, Eqn. (10.68a),

$$V(n) = U(n) - \sum_{k=0}^{n-1} \beta(n,k)V(k)$$

becomes

$$V(n) = \xi(n) - \sum_{k=0}^{n-1} \beta(n,k)V(k) \tag{10.69d}$$

Further, Eqn. (10.69b)

$$V^T(n)\xi(j) = U^T(n)\xi(j) = 0$$

becomes

$$V^T(n)\xi(j) = \xi^T(n)\xi(j) = 0 \tag{10.69e}$$

For $n = j$,

$$V^T(n)\xi(n) = \xi^T(n)\xi(n) \tag{10.69f}$$

Further, as the search directions are constructed using residues, the subspace spanned by $\{\xi(1)\xi(2),\ldots,\xi(n-1)\}$ is same as the subspace spanned by search directions $\{V(1),V(2),\ldots,V(n-1)\}$. It was shown that $\xi(j)$ is orthogonal to all previous search directions $V(n)$, $n < j$.

$$V^T(n)\xi(j) = 0, \quad n < j$$

Further,

$$\xi^T(n)\xi(j) = 0, \quad n < j$$

This is true as $\xi(j)$ *is orthogonal to all previous search directions* $V(n)$ *and the space spanned by* $V(n)$ *is same as spanned by* $\xi(n)$. In fact

$$\xi^T(n)\xi(j) = 0, \quad n \neq j$$

This is due to the fact that the *subsequent residuals should be orthogonal to space spanned by previous search vectors which is nothing but the space spanned by previous residuals.* Now,

$$\begin{aligned}\xi(n) = -R \in (n) = -R(W(n) - W_*) &= -R\{W(n-1) + \alpha(n-1)V(n-1) - W_*\} \\ &= -R\{W(n-1) - W_*\} - \alpha(n-1)RV(n-1) \\ &= -R \in (n-1) - \alpha(n-1)RV(n-1)\end{aligned}$$

$$\xi(n) = \xi(n-1) - \alpha(n-1)RV(n-1)$$

Thus, the residual is made up of previous residue and previous scaled modified search vector $RV(n-1)$. Similarly,

$$\xi(j+1) = \xi(j) - \alpha(j)RV(j)$$

$$\xi^T(n)\xi(j+1) = \xi^T(n)\xi(j) - \alpha(j)\xi^T(n)RV(j)$$

$$\xi^T(n)RV(j) = \frac{-1}{\alpha(j)}\left[\xi^T(n)\xi(j+1) - \xi^T(n)\xi(j)\right]$$

$$\left.\begin{aligned}\xi^T(n)RV(j) &= \frac{\xi^T(n)\xi(n)}{\alpha(n)} \quad n = j \\ &= \frac{-\xi^T(n)\xi(n)}{\alpha(n)}, \quad n = j+1 \\ &= 0, \quad \textit{otherwise } (i.e., n > j+1)\end{aligned}\right\} \tag{10.69g}$$

For choosing the new search direction, the coefficient is given by Eqn. (10.69c)

$$\beta(n,j)=\frac{\xi^T(n)RV(j)}{V^T(j)RV(j)}$$

Therefore, substituting for $\xi^T(n)RV(j)$ from Eqn. (10.69g),

$$\beta(n,j)=\begin{cases}\dfrac{-1}{\alpha(n-1)}\dfrac{\xi^T(n)\xi(n)}{V^T(n-1)RV(n-1)}, & n=j+1\\ 0, & n>j+1\end{cases} \tag{10.69h}$$

Therefore, the new search direction given by Eqn. (10.69d),

$$V(n)=\xi(n)-\sum_{j}^{n-1}\beta(n,j)V(j)$$

becomes

$$V(n)=\xi(n)-\beta(n,n-1)V(n-1) \tag{10.69i}$$

Hence, the new search direction is a linear combination of previous search direction and the new residual. From Eqn. (10.66d)

$$\alpha(n)=\frac{V^T(n)\xi(n)}{V^T(n)RV(n)} \quad \text{or} \quad \alpha(n-1)=\frac{V^T(n-1)\xi(n-1)}{V^T(n-1)RV(n-1)}$$

But from Eqn. (10.69f)

$$V^T(n)\xi(n)=\xi^T(n)\xi(n) \text{ or } V^T(n-1)\xi(n-1)=\xi^T(n-1)\xi(n-1)$$

Therefore,

$$\alpha(n)=\frac{\xi^T(n)\xi(n)}{V^T(n)RV(n)} \quad \text{or} \quad \alpha(n-1)=\frac{\xi^T(n-1)\xi(n-1)}{V^T(n-1)RV(n-1)}$$

Substituting for $\alpha(n-1)$ in Eqn. (10.69h)

$$\beta(n,n-1)=\frac{-1}{\alpha(n-1)}\frac{\xi^T(n)\xi(n)}{V^T(n-1)RV(n-1)}$$

$$= \frac{-V^T(n-1)RV(n-1)}{\xi^T(n-1)\xi(n-1)} \frac{\xi^T(n)\xi(n)}{V^T(n-1)RV(n-1)}$$

$$\left.\begin{aligned} &\beta(n) = \beta(n,n-1) = \frac{-\xi^T(n)\xi(n)}{\xi^T(n-1)\xi(n-1)} \\ &Or \\ &\beta(n+1) = \beta(n+1,n) = \frac{-\xi^T(n+1)\xi(n+1)}{\xi^T(n)\xi(n)} \end{aligned}\right\} \quad (10.69j)$$

Thus, the new search direction given by Eqn. (10.69i) can also be written as

$$V(n+1) = \xi(n+1) - \beta(n+1,n)V(n) \quad (10.69k)$$

To ensure conjugation of search directions, only $\beta(n,n-1)$ *is required and there is no necessity to store the all old search directions.* This results in reduction of significant storage and computations involved in finding new search directions. The computational complexity is reduced from $O(N^2)$ to $O(m)$, m being number of non-zero entries in R.

The conjugate gradient algorithm:

Step 1 It is assumed R and P are known/ computed from the input and desired signals data.

$$W(0) = 0, \quad \xi(0) = P - RW(0),\ V(0) = \xi(0)$$

Step 2

$$\alpha(n) = \frac{\xi^T(n)\xi(n)}{V^T(n)RV(n)}$$

Step 3

$$W(n+1) = W(n) + \alpha(n)V(n), \qquad \xi(n+1) = \xi(n) - \alpha(n)RV(n)$$

Step 4

$$\beta(n+1,n) = \frac{\xi^T(n+1)\xi(n+1)}{\xi^T(n)\xi(n)}$$

Step 5 $\qquad V(n+1) = \xi(n+1) - \beta(n+1,n)V(n)$

If $\chi = \lambda_{max} / \lambda_{min}$ then the number of iterations n required for the error norm $\|\in(n)\|$ to be less than $\delta\|\in(0)\|$, δ is a small quantity desired, then for SD, $n \le 0.5\chi \ln(1/\delta)$ and for CG it is $n \le 0.5\sqrt{\chi}\ln(2/\delta)$. Thus, CG has a faster convergence than SD. The time complexity for SD is $O(m\chi)$ and for CG is $O(m\sqrt{\chi})$.

One of the advantages CG is that in RLS and Newton's method, the estimate of R^{-1} is required and if this estimate is not positive definite, the algorithm will diverge. Further, to reduce the computational complexity of RLS, fast RLS algorithm has been proposed and however the latter has a tendency for numerical instability. In CG, as the R^{-1} estimate is not used, its ill effect will not lead to divergence. The CG is an intermediate method between steepest descent and Newton's method. The CG accelerates the slow convergence of the method of steepest descent but its computational complexity is relatively less than that of the Newton's method. It has been pointed out the convergence takes place in N iterations for an N-dimensional problem. But for this, the orthogonality of the residuals and hence the *R-orthogonality* of the search directions have to be preserved. This is possible if exact arithmetic is used for implementation. However, if finite precision arithmetic is used, the rounding off of the errors results in gradual degradation of orthogonality of residuals and the convergence in a small fixed finite number of iterations may not be possible. The method has a limitation in cases where the solution/convergence is required in less than such a fixed number of iterations.

Momentum LMS as CG : The momentum LMS can be considered a simplified version of CG (Sumit Roy, and John. J. Shynk, 1990). For CG,

$$W(n+1) = W(n) + \alpha(n)V(n) \tag{10.70a}$$

From Eqn. (10.66f)

$$\alpha(n) = \frac{V^T(n)\xi(n)}{V^T(n)RV(n)} = -\frac{V^T(n)g(n)}{V^T(n)RV(n)}, \quad g(n) = f'(W(n)) = -\xi(n)$$

$$\xi(n) = P - RW \tag{10.70b}$$

From Eqn. (10.69i),

$$V(n) = \xi(n) - \beta(n, n-1)V(n-1)$$

From Eqn. (10.69c)

$$\beta(n, j) = \frac{\xi^T(n)RV(j)}{V^T(j)RV(j)},$$

$$\gamma(n, n-1) = -\frac{g^T(n)RV(n-1)}{V^T(n-1)RV(n-1)},$$

$$\gamma(n, n-1) = -\beta(n, n-1)$$

With this definition of $\gamma(n, n-1)$,

$$V(n) = \xi(n) + \gamma(n, n-1)V(n-1) \tag{10.70c}$$

Substituting Eqn. (10.70c) in Eqn. (10.70a)

$$W(n+1) = W(n) + \alpha(n)V(n)$$

$$W(n+1) = W(n) + \alpha(n)\left[\xi(n) + \gamma(n, n-1)V(n-1)\right] \tag{10.70d}$$

From Eqn. (10.70a),

$$W(n+1) = W(n) + \alpha(n)V(n), \quad W(n) = W(n-1) + \alpha(n-1)V(n-1)$$

$$V(n-1) = \frac{\left[W(n) - W(n-1)\right]}{\alpha(n-1)}$$

Substituting this in Eqn. (10.70d),

$$W(n+1) = W(n) + \alpha(n)\left[\xi(n) + \gamma(n, n-1)\frac{\left[W(n) - W(n-1)\right]}{\alpha(n-1)}\right]$$

$$W(n+1) = W(n) - \alpha(n)g(n) + \frac{\alpha(n)}{\alpha(n-1)}\gamma(n, n-1)\left[W(n) - W(n-1)\right]$$

Instead of using the optimum values for $\alpha(n)$ and $\gamma(n, n-1)$ as in CG, if they are kept fixed as $\alpha(n) = \alpha$, $\gamma(n, n-1) = \gamma$ and using instantaneous gradient $g(n) = -2e(n)X(n)$ instead of exact gradient $\xi(n) = P - RW$;

$$W(n+1) = W(n) + 2\alpha e(n)X(n) + \gamma\left[W(n) - W(n-1)\right] \tag{10.70e}$$

With these simplifications, the above adaptation is nothing but momentum LMS algorithm. But due to simplifications, it no longer retains the properties of the CG algorithm.

10.20 ADAPTIVE FILTERING USING CONJUGATE GRADIENT

The CG provides an alternative method of solving $RW = P$ without inverting the matrix R. If a finite sliding window is used to estimate R and P, then CG can be considered a stochastic gradient method and the termination of iterations is done after *min*{*N, M*} In many applications, it is required to adapt the weight vector on a sample to sample basis. One direct way to use the above developed CG to a sample to sample adaptation case is to perform many iterations for every sample. In this, the CG assumes R and P as constants throughout the iterations done for each sample and this is computationally very expensive.

For sample to sample adaptation, it is desirable to make only one iteration per sample. But for this, $\xi^T(n)V(k) = 0,\ k < n$ is not valid, that is, the present residue is not orthogonal to previous search directions and does not satisfy the requirement of CG method. Another simplification is use of constant step size and this does not choose $\alpha(n)$ to minimize the cost function

$$f(W(n+1)) = f\left(W(n) + \alpha(n)V(n)\right)$$

and the cost function is not completely minimized. Yet another simplification is to use non-constant R and P at each iteration which also deteriorates the performance of the CG method. An algorithm which gives reasonably good match to the performance of CG, called the *modified CG,* will be considered.

10.20.1 Modified CG Algorithm

To apply CG to the case of sample to sample adaptation case some modifications are required to get a solution close to that of CG (P.S. Chang and A.N. Wilson, 2000),. Here, R and P are estimated on sample to sample basis using

$$R(n) = \lambda R(n-1) + X(n)X^T(n)\,, P(n) = \lambda P(n-1) + d(n)X(n)\,, \qquad (10.71\text{a,b})$$

The forgetting factor λ has to satisfy $0 < \lambda < 1$. For sample to sample basis, the residual vector is given by

$$\begin{aligned} \xi(n) &= P(n) - R(n)W(n) \\ \xi(n) &= \lambda\ \xi(n-1) - \alpha(n)R(n)V(n) + X(n)[d(n) - X^T(n)W(n-1)] \end{aligned} \qquad (10.71\text{c})$$

For recursive computation of $R(n)$ *and* $P(n)$,

$$\alpha(n) = \frac{\eta V^T(n)\xi(n-1)}{V^T(n)R(n)V(n)}, \quad \lambda - 0.5 \le \eta \le \lambda \qquad (10.71\text{d})$$

But for CG method,

$$\alpha(n) = \frac{\xi^T(n-1)\xi(n-1)}{V^T(n)R(n)V(n)}, \quad \lambda - 0.5 \le \eta \le \lambda$$

Here, it is to be noticed that $\xi^T(n-1)\xi(n-1)$ for CG has been replaced by $V^T(n)\xi(n-1)$ for the modified CG applicable for sample to sample adaptation. Here, $\alpha(n)$ minimizes the cost function along the search direction and it is used to update the weight vector $W(n)$, $\xi(n)$ is the residual and points to the direction of steepest descent. The new weight vector $W(n)$ is computed as a linear combination of old weight vector and the search direction.

The direction vector $V(n)$ has to be periodically reset to the true steepest descent gradient, for the algorithm to achieve global convergence on a sample to sample basis. For a non-reset type of sample to sample adaptation

$$\beta(n) = \frac{[\xi(n) - \xi(n-1)]^T \xi(n)}{\xi^T(n-1)\xi(n-1)} \tag{10.71e}$$

This is because for sample to sample basis, $\xi^T(n-1)\ \xi(n) \neq 0$ and use of $\beta(n)$ given above performs better than that used for CG given by

$$\beta(n) = \frac{\xi(n)^T \xi(n)}{\xi^T(n-1)\xi(n-1)} \tag{10.71f}$$

and the new $\beta(n)$ (for sample to sample) ensures that the R-orthogonality is preserved between the new search directions.

CG algorithm for sample to sample adaptation (5 steps):

Step 1 $\quad R(n) = \lambda R(n-1) + X(n)X^T(n)$, $P(n) = \lambda P(n-1) + d(n)X(n)$, $0 < \lambda < 1$.

$$W(0) = 0, \quad \xi(0) = P(0),, \ V(1) = \xi(0), n{=}1$$

Step 2

$$\alpha(n) = \frac{V^T(n)\xi(n-1)}{V^T(n)RV(n)}$$

Step 3

$$W(n) = W(n-1) + \alpha(n)V(n),$$

$$\xi(n) = \lambda\ \xi(n-1) - \alpha(n)R(n)V(n) + X(n)[d(n) - X^T(n)W(n-1)]$$

Step 4

$$\beta(n) = \frac{[\xi(n) - \xi(n-1)]^T \xi(n)}{\xi^T (n-1)\xi(n-1)}$$

Step 5 $$V(n+1) = \xi(n) - \beta(n)V(n)$$

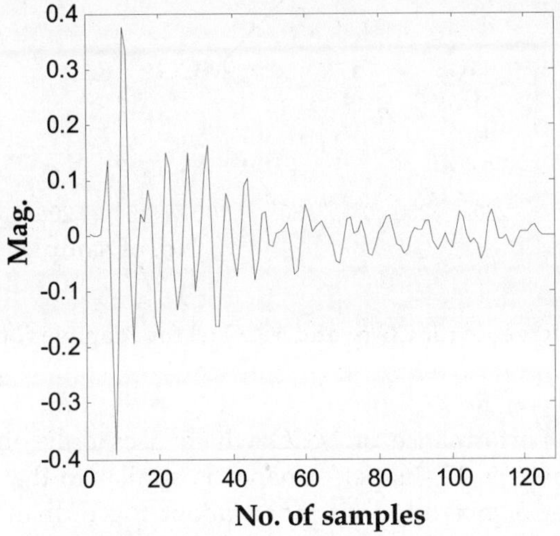

Fig. 10.14 (a) System impulse response used for CG and MCG.

To illustrate the performance of CG and modified CG for adaptive filtering , an acoustic path of length2′ and cross-section 1′x1′ is considered. It impulse response which has 128 samples is shown in Fig. 10.14(a). For the direct application of CG to adaptive filtering on sample to sample basis, 12 iterations and a filter length of 12 , are considered. The parameter for modified CG, $\eta = 0.5$ is used and only one iteration is used per input sample. Further, the filter length for MCG is also 12. An input consisting of sum of three sinusoids at 50 Hz, 100 Hz, and 150 Hz of amplitudes 3, 2,and 1 respectively; is used.

From Fig. 10.14(b), it is seen that the CG converges within about 30 samples and its initial magnitude of error is significantly small with respect to that of time domain LMS which uses 128 coefficients and converges in about 125 samples. For the LMS , the fixed step size of 0.001 is used to get the maximum convergence rate and beyond that the algorithm diverges. Further for CG, the convergence error is almost zero, indicating that CG provides both good convergence rate and error. It is required that initial 12 samples are required for the computations of autocorrelation matrix and cross-correlation vector. Afterwards, these are updated on sample to sample basis recursively using the forgetting factor, $\lambda = 0.99$.

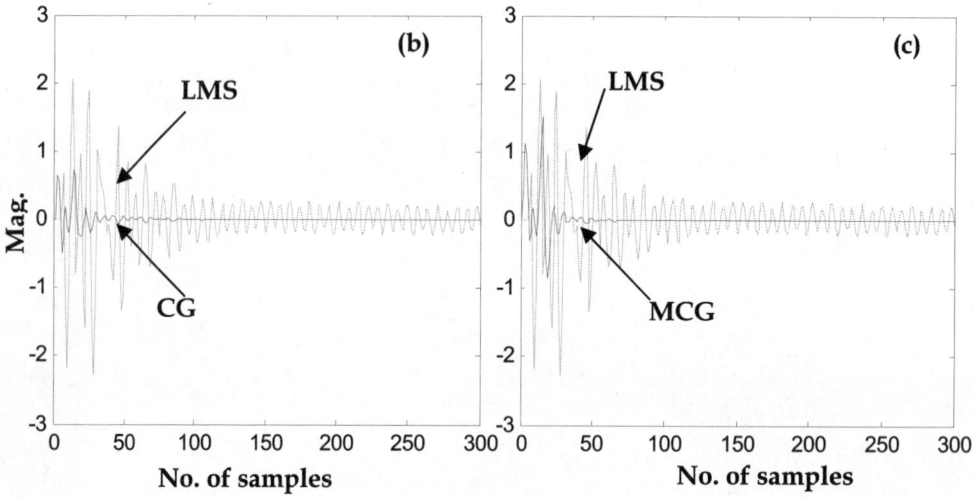

Fig. 10.14(b, c) Error convergence for CG (b)and MCG (c) for adaptive filtering.

Fig. (10.14c) shows the performance of MCG which uses only one iteration per sample. Here also filter length used is 12. Its performance is similar to that of CG with 12 iterations except that its initial error magnitude is more (about twice) than that of CG but after convergence, the error is almost zero as with that of CG. But the MCG convergence rate is same as that for CG and converges in about 30 samples. It is important to note that the MCG has a significantly superior performance in terms of convergence rate and error magnitude than the LMS algorithm though MCG uses only single iteration per sample.

10.21 POST-SMOOTHING LMS

The problem of adaptation stalling occurs when the input magnitude falls below the lowest significant bit specifically. While implementing the algorithm on a *fixed point processor*. This problem can be alleviated by using a larger adaptation step size. However, the increase in step size results in undesirable higher misadjustment. The use of higher step size makes the weight vector at different time instants to be more uncorrelated.

The LMS adaptation for the weight vector $W(n)$ with its elements $w_l(n)$ and input $x(n)$ is (dotted box in Fig. 10.15)

$$w_l(n+1) = w_l(n) + \mu x(n-l)\,e(n), \qquad l = 0,1,\ldots\ldots N-1$$

$$0 < \mu < \alpha / NP_x(n),\; 0 < \alpha < 1,$$

$$P_x(n),+\gamma\, P_x(n-1) + (1-\gamma)x^2(n),\; 0 < \gamma < 1$$

$e(n)$ is the error, the difference between the desired signal $d(n)$ and adaptive filter output $y(n)$. μ : adaptation step size and $P_x(n)$: power of the input. After convergence, $W(n)$ *oscillates* about the optimum the optimum solution and the convergence error are decided by μ. However, a

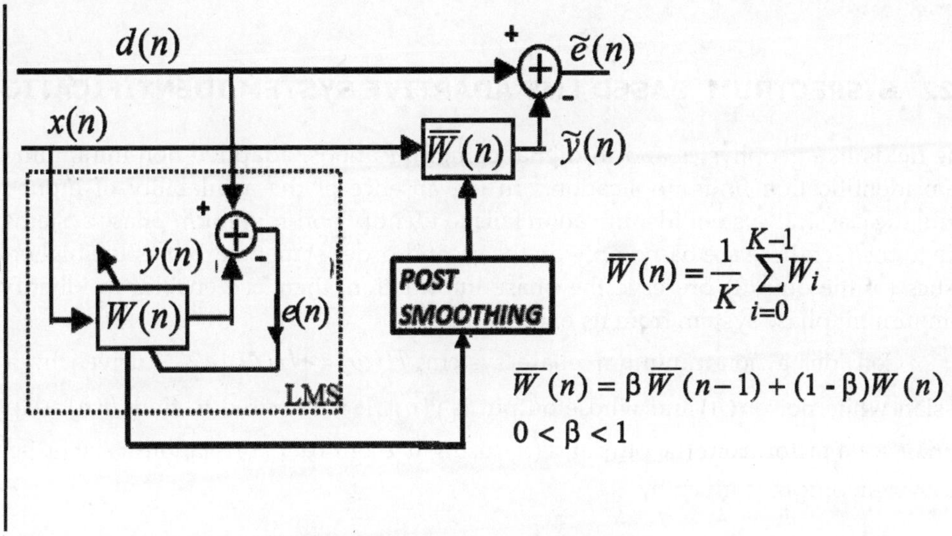

Fig. 10.15 Post Smoothing LMS.

smaller step size results in a lower convergence speed. Since in LMS algorithm, it is the *mean weight vector* that converges to the optimum at any instant, the averaging of the weight vectors corresponding to different iterations in the neighborhood, results in a weight vector close to the optimum. The online averaging of weight vectors is called *online post-smoothing (OPS)* (Y.H. Chung, C.K. Tzou and N.J. Bershad, 1991) and is useful for further processing like filtering and spectral estimation. But it is required to leave the weight vector used in the LMS algorithm *undisturbed* and the smoothing has to be done *outside* the LMS loop [3] (Fig.10.15), outside the dotted box). $\tilde{y}(n)$ and $\tilde{e}(n)$ correspond to $y(n)$ and $e(n)$. For a large step size (within the convergence limit), the OPS significantly reduces the convergence error but for the same step size, the post-smoothed weight vector $\overline{W}(n)$ converges slowly compared to $W(n)$ and the use of a larger step size overcomes the adaptation stalling.

Linear averaging of K weight vectors, involves their storage and K decides the amount of smoothing. Exponential averaging, $\overline{W}(n) = \beta\,\overline{W}(n-1) + (1-\beta)\,W(n)$, requires storage for only one weight vector and for better smoothing, β should be close to unity. Proper choice of β and step size can result in a compromise between convergence rate and misadjustment of $\overline{W}(n)$. In a noise canceller, the averaged weight vector enables better estimate of noise $\tilde{y}(n)$, and the error $\tilde{e}(n)$ forms the output which has reduced noise component than that is obtained by using the direct weight vector without averaging.

Thus, this configuration solves the problem of stalling due to small input value due to use of higher magnitude step size and the resulting higher misadjustment is overcome by averaging the weight vectors at different instants of time. Further, this configuration can also be used in situations where better filtering of noise that is associated with the signal, is required.

10.22 BISPECTRUM BASED LMS ADAPTIVE SYSTEM IDENTIFICATION

In the fields like geophysics and in digital communications, adaptive non-minimum phase system identification finds application. In the absence of the availability of input to the system, the classical system identification fails to identify non-minimum phase systems. This is due to the fact these are based only on the second-order statistics. But as the higher order statistics (of the output) preserve the phase information, their use enables in identifying a non-minimum phase system from its output.

Consider a non-minimum phase system $H(\omega) = |H(\omega)|e^{j\varphi(\omega)}$ driven by a non-Gaussian white noise $x(n)$ and whose output is $y(n)$. It is required to find $h(n)$ adaptively within a scale factor from the output $y(n)$ using higher order correlations. The bispectrum of the system output is given by

$$B_y(\omega_1,\omega_2) = B_H(\omega_1,\omega_2)B_x(\omega_1,\omega_2) \tag{10.72a}$$

For *non-Gaussian white noise* input,

$$|B_x(\omega_1,\omega_2)| = \beta, \quad \Psi_x(\omega_1,\omega_2) = 0$$

Therefore,

$$|B_y(\omega_1,\omega_2)| = \beta\,|B_H(\omega_1,\omega_2)| \quad \text{and} \quad \Psi_y(\omega_1,\omega_2) = \Psi_H(\omega_1,\omega_2)$$

But as

$$B_H(\omega_1,\omega_2) = H(\omega_1)H(\omega_2)H^*(\omega_1+\omega_2),$$

$$B_y(\omega_1,\omega_2) = \beta\, H(\omega_1)H(\omega_2)H^*(\omega_1+\omega_2)$$

Let $\hat{h}(n)$ be the complex cepstrum of $h(n)$ and assuming there are no zeros of $H(z)$ on the unit circle,

$$\hat{h}(n) = F^{-1}[\ln(H(\omega)]$$

Let $g(n)$ be defined as

$$g(n) = F^{-1}[(G(\omega)] = F^{-1}\left[B_y(\omega_1, \omega_2)\,|_{\omega_1=\omega_2=\omega} \right] \tag{10.72b}$$

Since $G(\omega)$ is the diagonal slice of $B_y(\omega_1, \omega_2)$ (Saleh A. Alshebeili, A. Enis Cetin and A. N. Venetsanopoulos, 1991)

$$g(n) = \sum_k r_y(k, n-k) \tag{10.72c}$$

Assuming $G(z)$ no zeros on the unit circle, $\hat{g}(n)$ be the complex cepstrum of $g(n)$ is

$$\hat{g}(n) = F^{-1}[\ln(G(\omega)]$$

Since

$$\ln[B_y(\omega_1, \omega_2)] = \ln\beta + \ln[H(\omega_1)] + \ln[H(\omega_2)] + \ln[H^*(\omega_1 + \omega_2)]$$

Assuming $\beta = 1$,

$$\ln[G(\omega)] = \ln[B_y(\omega_1, \omega_2)\Big|_{\omega_1=\omega_2=\omega}] = \ln[H(\omega)] + \ln[H(\omega)] + \ln[H^*(2\omega)]$$

$$\hat{G}(z) = \ln[G(\omega)] = 2\ln[H(\omega)] + \ln[H^*(2\omega)]$$

The inverse FT of this gives the complex cepstrum $\hat{g}(n)$ of $g(n)$,

$$\hat{g}(n) = \begin{cases} 2\hat{h}(n) + \hat{h}(-n/2), \; for\ n\ even \\ 2\hat{h}(n),, \; for\ n\ odd \end{cases} \tag{10.72d}$$

From this, the complex cepstrum $\hat{h}(n)$ of the impulse response $h(n)$ can be computed recursively from $\hat{g}(n)$.

$$\hat{h}(0) = \hat{g}(0)/3, \quad \hat{h}(1) = \hat{g}(1)/2, \quad \hat{h}(-1) = \hat{g}(-1)/2,$$
$$\hat{h}(2) = \left\{\hat{g}(2) - \hat{h}(-1)\right\}/2, \quad \hat{h}(-2) = \left\{\hat{g}(-2) - \hat{h}(1)\right\}/2,$$
$$\hat{h}(3) = \hat{g}(3)/2, \quad \hat{h}(-3) = \hat{g}(-3)/2,$$

The cepstral sequence $\hat{h}(n)$ of a finite sequence $h(n)$, is of infinite extent. But as $\hat{h}(n)$ decays with a scaling factor of $(1/n)$, for finite length sequence without zeros on the unit circle, $\hat{h}(n)$ can be considered to be limited to a sufficiently long range. That is, $\hat{h}(n)$ has values in the range $-P \le n \le Q, \quad (p,Q) >> 0$. Now,

$$\hat{G}(z) = \ln[G(\omega)]$$

$$\frac{d}{dz}[\hat{G}(z)] = \frac{d}{dz}\{\ln[G(\omega)]\} = \frac{1}{G(z)}\frac{d}{dz}\{G(\omega)\}$$

$$\frac{d}{dz}[\hat{G}(z)] = \frac{1}{G(z)}\frac{d}{dz}\{G(\omega)\}$$

$$zG(z)\frac{d}{dz}[\hat{G}(z)] = z\frac{d}{dz}\{G(\omega)\}$$

Taking the inverse -Z transform both sides,

$$\sum_{k=-\infty}^{\infty} k\hat{g}(k)g(n-k) = ng(n)$$

As $\hat{g}(n)$ also decays with a scaling factor $(1/n)$,

$$g(n) = \sum_{k=-P}^{Q}\left(\frac{k}{n}\right)g(n-k)\hat{g}(k) \tag{10.72e}$$

This forms the crucial relation for the adaptive algorithm and this in matrix form is

$$\overline{g} = \Phi\,\overline{\hat{g}}$$

where

$$\overline{g} = g(-K), g(-K+1), \ldots\ldots, g(K)$$

$$\overline{\hat{g}} = \hat{g}(-P), \hat{g}(-P+1), \ldots\ldots, g(Q),\ K > (P+Q)$$

Φ is a matrix of size $\{2K \times (P+Q)\}$ and its elements are determined by Eqn.(10.72e).

Bispectrum based adaptive system identification algorithm:

Step 1: At each time instant m, compute the triple correlation sequence of the output $r_m(l_1,l_2), \quad -L \le l_1, l_2 \le L$ from a finite length of data available.

Step 2: Compute

$$g_m(n), = \sum_k r_m(n, n-k)$$

Step 3: Adaptive estimation of cepstrum $\hat{h}(n)$ of $h(n)$: The coefficients of $\hat{h}_m(n)$ are updated using the gradient LMS algorithm and for this, the error is given by

$$\bar{e}(m) = \bar{g}_{(m)} - \Phi(m)\bar{\hat{g}}(m)$$

The cepstral coefficients are updated as

$$\bar{\hat{g}}(m+1) = \bar{\hat{g}}(m) + \mu\, \Phi^T(m)\bar{e}(m),$$

$$0 < \mu < \frac{2}{trace[\Phi^T \Phi]}$$

After convergence, at $m = M$, $\hat{h}_M(n)$ are obtained from

$$\bar{\hat{g}}(M) = [\hat{g}_M(-P), \hat{g}_M(-P+1), \ldots\ldots \hat{g}_M(Q)]^T$$

using the Eqn.(10.72d).

Since

$$H_M(\omega) = e^{FT[\hat{h}_M(n)]}$$

Therefore,

$$h_M(n) = FT^{-1}[H_M(\omega)]$$

A non-minimum phase system is given by

$$H(z) = \gamma\, H_+(z) H_-(z) z^{-d}$$

Here $H_+(z)$ and $H_-(z)$ correspond to polynomials whose roots are inside and outside the unit circle, respectively. γ is the system gain and d is the delay introduced by the system.

The minimum and maximum phase components of $h(n)$ can be computed from their respective cepstra,

$$\hat{h}_+(n) = [\ldots 0.0,\, 0.0, \hat{h}(0)/2,\, \hat{h}(1),\, \hat{h}(2), \ldots\ldots]$$

and $$\hat{h}_-(n) = [\ldots\ldots \hat{h}(-2),\, \hat{h}(-1),\, \hat{h}(0)/2, ,), 0.0,\, 0.0,\, 0.0 \ldots\ldots].$$

$$H_+(\omega) = e^{FT[\hat{h}_+(n)]} \text{ and } H_-(\omega) = e^{FT[\hat{h}_-(n)]}$$

Since the complex cepstra of $h(n)$ and $\gamma\, h(n-K)$, are same (γ is a real number), it is only possible to identify the system within a scale factor and delay . In this method, it has been assumed that $H(\omega) \neq 0, -\pi \leq \omega \leq \pi$.

Chapter 11

ORTHOGONALIZATION BASED LEAST MEAN SQUARE ALGORITHM

The main problem with the method of steepest descent or LMS algorithm is its slow convergence when the eigenvalue spread of the autocorrelation matrix of the input signal is large. This situation arises when the input spectrum is coloured or characterised by peaks and valleys or fundamental and its harmonics. The way to reduce the eigenvalue spread is to whiten orthogonalize the input. This is achieved by using an orthogonal transform whose outputs are orthogonal to each other. For each orthogonal component, a single weight LMS algorithm is used and the weighted sum of the orthogonal inputs result in the final output. As for each single coefficient normalized LMS algorithm, the bin input power is very small compared to the power of the input signal and also the filter length is unity instead of total filter length, the step size used gets enhanced, resulting in a faster convergence. The other approach is by using lattice filter for which the backward prediction errors are readily uncorrelated. For each backward error component a single coefficient normalized LMS adaptation is used (known as joint lattice) and similar to the transform domain case, here also the reduced input power and filter length, make the step size larger there by improving the convergence rate. Thus the input orthogonalization which aligns the weight vector axes with the eigenvectors of the autocorrelation matrix is not sufficient but the step size scaling by the power of the orthogonal component necessary is to achieve improved convergence rate. However, this faster convergence rate is at the cost of higher convergence error and variable step size approach is required to achieve both fast convergence rate and small convergence error. The transform domain forward-backward predictor not only provides faster convergence but also reduces the error by a factor of half.

The escalator lattice also produces mutually uncorrelated prediction errors by realizing the Gram-Schmidt orthogonalization algorithm. Further, unlike the regular lattice, it produces the prediction errors in a parallel way and it does not assume stationarity of the input signal as it minimizes only backward (or forward) prediction error energy and is useful for nonstationary input.

Laguerre versions of the regular and escalator lattice predictors are useful as they reduce the adaptive filter length by almost 50% and hence reducing the computational load. The cascade of small order predictors not only provide faster convergence but also computationally simple.

The conjugate gradient method (CGM) improves the convergence performance, but requires many iterations per sample. But the use of an orthogonal transform to orthogonalize the input enables the CGM to require only one iteration per input sample.

Noise robust variable step size both in time and transform domain are important and are considered.

11.1 INTRODUCTION

In chapter 9, it was shown that contours of mean square error surface on W_N - plane were ellipses (for a two elements weight vector) and this nature of contours remained same even after going to uncoupled difference weight vector V'_N - plane. The transformation from W_N - plane to V'_N -plane was realized by translation ($W_N \rightarrow V_N$) and rotation ($V_N \rightarrow V'_N$). This transformation from $W_N \rightarrow V'_N$ gave better insight for analysis as it provided a simpler mathematical form for the solution of the weight vector. For the two elements weight vector case, any specific contour was an ellipse with a and b as its semi-major and semi-minor axes, respectively. Further, a and b are determined by minimum and maximum eigenvalues, respectively. As long as a and b are different, the eigenvalue spread is not unity and higher the ratio a / b, the eigenvalue spread is higher and the contours are more elliptic. This will result in crowding of the contours (less separation) indicating the mean square error surface is a steeper hyper paraboloid. For this, the error surface is very sensitive to filter coefficients limiting the step size to a small value, which results in a slow convergence. On the other hand if a and b are approximately same, the eigenvalue spread is reduced making the error surface close to a hemisphere. For this, the contours will be approximately equally spaced circles indicating the steepness of the surface is reduced and will become minimum when the error surface is a perfect hemisphere.

To overcome the slow convergence due to eigenvalue spread, no action has been taken either in the method of steepest descent or with the *basic LMS* algorithm. The latter case is due to the fact that LMS is an approximation to the method of steepest descent as it uses the gradient of the instantaneous error energy. But the Newton's LMS algorithm already considered has a very fast convergence rate as it converges in a single iteration. In Newton's method, this has become possible due to replacement of μ by μR_{NN}^{-1}. Also, the interpretation of the Newton's algorithm indicated *that it could be viewed as the steepest descent method applied to an orthogonal transformed signal whose components are uncorrelated to each other. Further, the eigenvalue-spread is reduced by power normalization of its components.* Thus, in Newton's method, both the orthogonalization of the input and normalization of the transformed components take place *indirectly* due to use of μR_{NN}^{-1} in place of μ. This seems to be remarkable however its requirement of the estimate of R_{NN}^{-1}, it is problematic and suffers from the similar implementation limitations of method of steepest descent. To achieve faster convergence, unlike in the method of Newton, *the orthogonalization of the input into its components and their power normalization operation can be realized explicitly,* which forms the principle of the transform and other orthogonalization methods like lattice and will be considered in this chapter.

11.2 TRANSFORM DOMAIN ADAPTIVE FILTER (TDAF)

Let

$$X_N(n) = \left[x(n) \quad x(n-1) \quad \ldots \quad \ldots \quad x(n-N+1)\right]^T, \tag{11.1a}$$

be the input vector. When the unitary/ orthogonal transform M_{NN} operates on $X_N(n)$, let $X'(n)$ be the resulting vector given by

$$X'_N(n) = M_{NN}\left[x(n)\right] = \left[x'_1(n)\; x'_2(n) \quad \ldots \quad \ldots \quad x'_N(n)\right]^T \tag{11.1b}$$

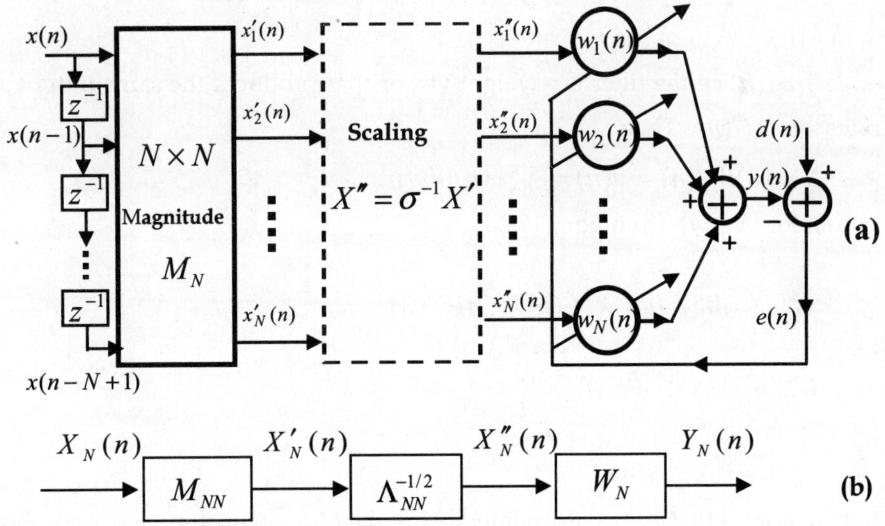

Fig. 11.1 (a) Transform domain adaptive filter and (b) Simplified schematic of transform domain adaptive filter.

If

$$W_N(n) = \left[w_1(n) \quad w_2(n) \quad \ldots \quad \ldots \quad w_N(n)\right]^T \tag{11.1c}$$

then the filter output $y(n)$ (in the absence of the dotted box)

$$y(n) = W_N^{\,T}(n)X'_N(n) = W_N^{\,T}(n)M_{NN}\left[X_N(n)\right] \tag{11.1d}$$

Therefore, the error is

$$e(n) = d(n) - y(n) \tag{11.1e}$$

The adaptation using complex normalized LMS algorithm (S. Shankara Narayan, A.M. Peterson and M.J. Narasimha, 1983) is

$$W_N(n+1) = W_N(n) + \alpha\, e(n)\, \Lambda_{NN}^{-1}\, X_N'^{*}(n) \tag{11.1f}$$

where,

$$\Lambda_{NN} = diag(\sigma_1^2\ \sigma_2^2 \cdots \sigma_N^2),\ \sigma_i^2 \approx E[|\,x_i(n)\,|^2],\ * : \text{conjugate operation}$$

This adaptation equation will be derived.

The simplified schematic of the transform domain adaptive filter is shown in the Fig. (11.1b) and let

$$X_N''(n) = \Lambda_{NN}^{-1/2} M_{NN}\left[x(n)\right] = \Lambda_{NN}^{-1/2} X_N'(n)$$

$$X_N''(n) = \Lambda_{NN}^{-1/2} X_N'(n) \tag{11.2a}$$

Further, let $W_N''(n)$ be the filter tap weight vector that produces the same output $y(n)$ when it operates on $X_N''(n)$. Then

$$X_N'^{T}(n) W_N(n) = y(n) = X_N''^{T}(n) W_N''(n) \tag{11.2b}$$

Substituting for $X_N''(n)$ from Eqn.(11.2a),

$$X_N'^{T}(n) W_N(n) = X_N'^{T}(n) \Lambda_{NN}^{-1/2} W''(n)$$

Therefore,

$$W_N(n) = \Lambda_{NN}^{-1/2} W_N''(n)$$

or

$$W_N''(n) = \Lambda_{NN}^{1/2}\, W(n) \tag{11.2c}$$

If the tap weight vector $W_N''(n)$ is updated using the LMS algorithm,

$$W''(n+1) = W''(n) + \alpha\, e(n)\, X''^{*}(n) \tag{11.2d}$$

Substituting for $W_N''(n+1)$ from Eqn. (11.2c) as

$$W_N''(n+1) = \Lambda_{NN}^{1/2}\, W(n+1)$$

The adaptation becomes

$$W(n+1) = \Lambda_{NN}^{-1/2} W''(n+1)$$

$$= \Lambda_{NN}^{-1/2}\left[W''(n) + \alpha\, e(n)\, X''^{*}(n)\right]$$

$$= \Lambda_{NN}^{-1/2} W''(n) + \alpha\, e(n) \Lambda_{NN}^{-1/2}\, X_N''^{*}(n)$$

Again using Eqn. (11.2c),

$$W_N(n+1) = W_N(n) + \alpha\, e(n) \Lambda_{NN}^{-1/2} X_N''^{*}(n)$$

Substituting for $X''_N(n)$ from Eqn. (11.2a) as

$$X''_N(n) = \Lambda_{NN}^{-1/2} X'_N(n)$$

gives

$$W_N(n+1) = W(n) + \alpha\, e(n) \Lambda^{-1} X''^*(n) \tag{11.2e}$$

This equation is same as the Eqn. (11.1f).

The expression for the mean square error is (Eqn. (9.16a),

$$\varepsilon = E[|\, e(n)\,|^2] = \varepsilon_{\min} + V_N^T R_{NN} V_N, \tag{11.3a}$$

where $V_N = W_N(n) - W_N^*$ is the difference weight vector and W^* is the optimal weight vector. Applying orthonormal transformation,

$$V'_N = M_{NN}^T V_N(n)$$

$$\varepsilon = \varepsilon_{\min} + V_N{}^T \left[M_{NN} \Lambda_{NN} M_{NN}^T \right] V_N$$

$$= \varepsilon_{\min} + \left(M_{NN}^T V_N \right)^T \Lambda_{NN} \left(M_{NN}^T V_N \right)$$

$$\varepsilon = \varepsilon_{\min} + V_N'^T \Lambda V'_N \tag{11.3b}$$

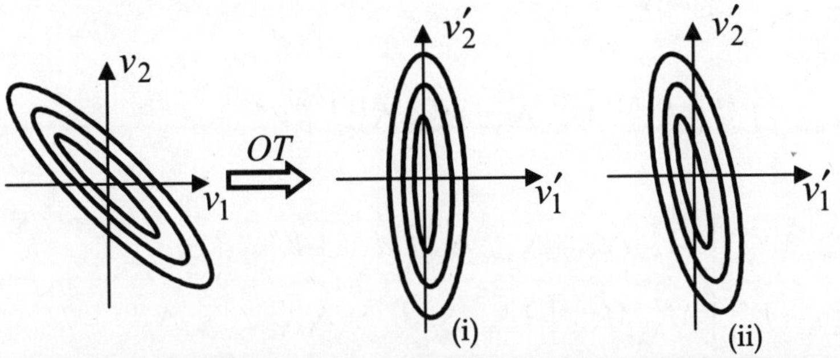

Fig. 11.1(c) Contour plots of the error surface before and after orthogonal transformation (i) satisfactory transform; (ii) unsatisfactory transform.

The orthogonal transformation *aligns the weight co-ordinate axes with the principal axes of the contours (Fig. 11.1c). However, the eccentricity of the surface is unaffected and hence convergence rate of the system is unchanged.* The signal power levels at the adaptive taps have changed after transformation. The quality of the transform determines the extent of alignment of the weight co-ordinate axes with the principal axes (eigenvectors) of the contours. If the alignment is perfect as in Fig. 11.1(c (i)), the transform is satisfactory/good (with respect to a specific type of input signal). However, if the alignment is poor as in

Fig. 11.1(c (ii)), the transform is unsatisfactory (D.F Marshall, W.K. Jenkins and J.J. Murphy,1989; Beaufays F. and Widrow B,1995).

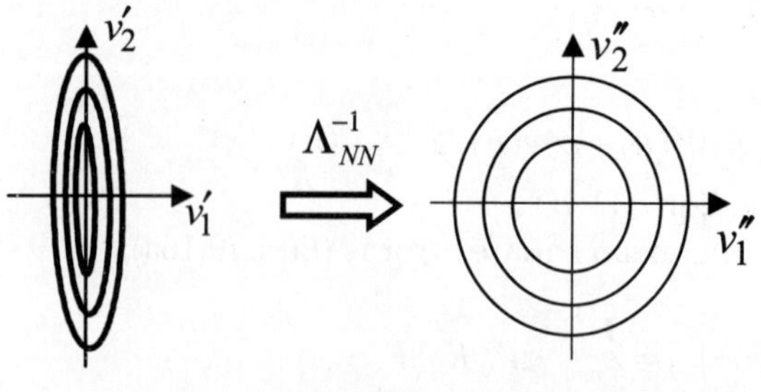

Fig. 11.1(d) Effect of scaling after orthogonal transformation on error surface contours.

Since from Eqn. (11.2c)

$$W_N''(n) = \Lambda_{NN}^{1/2} W_N(n)$$

For the difference weight vector $V_N = W_N(n) - W_N^*$,

$$V_N'' = \Lambda_{NN}^{1/2} V_N \text{ Or } V_N = \Lambda_{NN}^{-1/2} V_N'' \quad (11.3c)$$

Therefore, the mean square error (Eqn. (11.3a)),

$$\varepsilon = \varepsilon_{\min} + V_N^T R V_N$$

Becomes

$$\varepsilon = \varepsilon_{\min} + V''^T \left[\Lambda_{NN}^{-1/2} M_{NN} \Lambda_{NN} M_{NN}^T \Lambda_{NN}^{-1/2} \right] V''$$

$$= \varepsilon_{\min} + V''^T \left[M_{NN}^T \Lambda_{NN}^{-1/2} \Lambda_{NN} \Lambda_{NN}^{-1/2} M_{NN} \right] V''$$

$$= \varepsilon_{\min} + V''^T \left[M_{NN}^T M_{NN} \right] V'', \qquad \because \Lambda_{NN}^{-1/2} \Lambda_{NN} \Lambda_{NN}^{-1/2} = I$$

$$\varepsilon = \varepsilon_{\min} + V''^T V'' \quad (11.3d)$$

For $N = 2$,

$$\varepsilon = \varepsilon_{\min} + \begin{bmatrix} v_1'' & v_2'' \end{bmatrix} \begin{bmatrix} v_1'' \\ v_2'' \end{bmatrix}$$

$$v_1''^2 + v_2''^2 = \varepsilon - \varepsilon_{\min}$$

For a particular value of ε, $\varepsilon - \varepsilon_{\min}$ is a constant and say $\rho^2 = \varepsilon - \varepsilon_{\min}$. Then

$$v_1''^2 + v_2''^2 = \rho^2 \quad (11.3e)$$

Eqn. (11.3d) is nothing but a circle with radius ρ. Thus, the transformation Eqn. (11.2a) and Eqn.(11.2c) convert the elliptical contours to circles (Fig. 11.1d) This, in turn, makes the error surface from a paraboloid to a hemisphere which is having minimum steepness enabling a maximum step size that can by used to achieve best possible convergence rate. It is important to note that *the orthogonal transformation only enables orthogonal components of the signal resulting in rotation of the contours of the MSE surface but in addition to this the scaling of these components by their respective magnitudes (eigenvalues) that makes the paraboloid error surface to a hemisphere.*

For the example considered in Chapter 9, which has

$$\boldsymbol{R} = \begin{bmatrix} 5 & -2 \\ -2 & 5 \end{bmatrix}, \boldsymbol{V} = \begin{bmatrix} v_1 \\ v_2 \end{bmatrix} \quad \text{and} \quad \varepsilon_{\min} = 48,$$

it was shown that

$$\frac{(v_1')^2}{(48/3)} + \frac{(v_2')^2}{(48/7)} = 1$$

$$V'^T \Lambda V' = 48, \quad [v_1' \;\; v_2'] \begin{bmatrix} 3 & 0 \\ 0 & 7 \end{bmatrix} \begin{bmatrix} v_1' \\ v_2' \end{bmatrix} = 48,$$

$$\Lambda = \begin{bmatrix} 3 & 0 \\ 0 & 7 \end{bmatrix}, \; \varepsilon_{\min=48},$$

From Eqn. (11.3b),

$$\varepsilon = \varepsilon_{\min} + V'^T \Lambda V'$$

From Eqn. (11.3c),

$$V'' = \Lambda^{1/2} V' \text{ Or } V' = \Lambda^{-1/2} V'', \quad V' = \begin{bmatrix} 1/\sqrt{3} & 0 \\ 0 & 1/\sqrt{7} \end{bmatrix} V''$$

Therefore,

$$\varepsilon = 48 + V''^T \begin{bmatrix} 1/\sqrt{3} & 0 \\ 0 & 1/\sqrt{7} \end{bmatrix} \begin{bmatrix} 3 & 0 \\ 0 & 7 \end{bmatrix} \begin{bmatrix} 1/\sqrt{3} & 0 \\ 0 & 1/\sqrt{7} \end{bmatrix} V''$$

$$\varepsilon = 48 + V''^T \begin{bmatrix} 1/\sqrt{3} & 0 \\ 0 & 1/\sqrt{7} \end{bmatrix} \begin{bmatrix} \sqrt{3} & 0 \\ 0 & \sqrt{7} \end{bmatrix} V'' = 48 + V''^T \begin{bmatrix} 1 & 0 \\ 0 & 1 \end{bmatrix} V''$$

$$\varepsilon = 48 + [v_1'' \;\; v_2''] \begin{bmatrix} 1 & 0 \\ 0 & 1 \end{bmatrix} \begin{bmatrix} v_1'' \\ v_2'' \end{bmatrix} = 48 + [v_1'' \;\; v_2''] \begin{bmatrix} v_1'' \\ v_2'' \end{bmatrix}$$

$$v_1''^2 + v_2''^2 = \varepsilon - 48 = \text{constant}, \quad \Rightarrow \textit{Circle}$$

Thus, the action of the TDAF system is to rotate the filter coefficient space axes using a unitary rotation matrix M_{NN} and then scale these axes by σ_i^2, so that the error surface contours appear *hyper-spherical*. The actual eccentricity of the error surface contour is reduced by this scaling, and faster convergence is obtained.

11.3 THE MINIMUM MEAN SQUARE ERROR OF TRANSFORM DOMAIN ADAPTIVE FILTER

From Fig. 11.1(a, b),

$$y(n) = X_N''^T(n)W_N(n) = \left\{ \Lambda_{NN}^{-1/2} M_{NN}\left[X(n)\right] \right\}^T W_N(n)$$

$$e(n) = d(n) - y(n)$$

The mean square error for the transform domain adaptive filter, ε_O is

$$\varepsilon_O(n) = E\left\{e^2(n)\right\} = E\left\{ \left(d(n) - y(n)\right)^2 \right\} = E\left\{ d^2(n) \right\} - 2E\left\{ d(n)y(n)\right\} + E\left\{y^2(n)\right\}$$

Since

$$y(n) = X_N''^T(n)W_N(n)$$

$$\varepsilon_O(n) = E\left\{ d^2(n) \right\} - 2E\left\{ d(n)X_N''^T(n)W_N(n)\right\} + E\left\{W_N^T(n)X_N''(n)X_N''^T(n)W_N(n)\right\},$$

$$\varepsilon_O(n)(n) = E\left\{ d^2(n) \right\} - 2W_N^T(n)(n)P'' + W_N^T(n)R_{NN}''W_N(n) \tag{11.4a}$$

where

$$P_N''(n) = E\left\{d(n)X_N''(n)\right\}, \quad R_N''(n) = E\left\{X_N''(n)X_N''^T(n)\right\}$$

Setting the derivative of $\varepsilon_O(n)$ w. r. to $W_N(n)$ equal to zero,

$$W_N^* = R_{NN}''^{-1}{}_{NN}P_N''$$

Substituting this in the expression for $\varepsilon(n)$,

$$\varepsilon_{O\min} = E\left\{ d^2(n) \right\} - 2W^{*T}(n)P'' + W^{*T}(n)R_{NN}''W^*(n)$$

$$= E\left\{ d^2(n) \right\} - 2P_N'^{T}R_{NN}''^{-1}(n)P_N''(n) + P_N'^{T}R_{NN}''^{-1}(n)R_{NN}''W^*(n)$$

$$= E\left\{ d^2(n) \right\} - 2P_N'^{T}R_{NN}''^{-1}(n)P_N''(n) + P_N'^{T}R_{NN}''^{-1}(n)P_N''$$

$$\varepsilon_{O\min} = E\left\{ d^2(n) \right\} - P_N'^{T}R_{NN}''^{-1}(n)P_N'' \tag{11.4b}$$

Now,

$$R''_{NN} = E\left\{V''_N V''^T_N\right\} = E\left\{\left(\Lambda_{NN}^{-1/2} M_{NN}^T X_N(n)\right)\left(\Lambda_{NN}^{-1/2} M_{NN}^T X_N(n)\right)^T\right\}$$

$$= E\left\{\left(\Lambda_{NN}^{-1/} M_{NN}^T X_N(n) X_N^T(n) M_{NN}\right)\right\} = \Lambda_{NN}^{-1/} M_{NN}^T E\left\{X_N(n) X_N^T(n)\right\} M_{NN}$$

Since for time domain transversal filter $R_{NN}(n)$ is given by

$$R_{NN}(n) = E\left\{X_N(n) X_N^T(n)\right\}$$

Therefore,

$$R''_{NN} = \Lambda_{NN}^{-1/} M_{NN}^T R_{NN} M_{NN} \tag{11.4c}$$

and

$$P''_N = E\left\{d(n) V''_N(n)\right\} = E\left\{d(n)\, \Lambda_{NN}^{-1/2} M_{NN}^T X_N(n)\right\} = \Lambda_{NN}^{-1/2} M_{NN}^T E\left\{d(n)\, X_N(n)\right\}$$

Also since for time domain transversal filter, $P_N(n)$ is given by

$$P_N(n) = E\left\{d(n) X_N(n)\right\}$$

Therefore,

$$P''_N = \Lambda_{NN}^{-1/2} M_{NN}^T P_N \tag{11.4d}$$

Substituting for P''_N and R''_{NN} from Eqns. (11.4c) and (11.4d) respectively in the expression for minimum mean square error (Eqn. (11.4b)),

$$\varepsilon_{O\min} = E\left\{ d^2(n) \right\} - P''^T_N R''^{-1}_{NN}(n) P''_N$$

$$= E\left\{ d^2(n) \right\} - [\Lambda_{NN}^{-1/2} M_{NN}^T P_N]^T [\Lambda_{NN}^{-1} M_{NN}^T R_{NN} M_{NN}]^{-1} \Lambda_{NN}^{-1/2} M_{NN}^T P_N$$

$$= E\left\{ d^2(n) \right\} - [\Lambda_{NN}^{-1/2} M_{NN}^T P_N]^T [\Lambda_{NN}^{-1} M_{NN}^T R_{NN} M_{NN}]^{-1} \Lambda_{NN}^{-1/2} M_{NN}^T P_N$$

$$= E\left\{ d^2(n) \right\} - P_N^T M_{NN} \Lambda_{NN}^{-1/2} M_{NN}^T R_{NN}^{-1} M_{NN} \Lambda_{NN}^{-1/2} M_{NN}^T P_N$$

$$= E\left\{ d^2(n) \right\} - P_N^T R_{NN}^{-1} P_N$$

If W_{tN} the weight vector for transversal filter, then its corresponding optimum vector W_{tN}^* is

$$W_{tN}^* = R_{NN}^{-1} P_N$$

Therefore,

$$\varepsilon_{O\min} = E\left\{ d^2(n) \right\} - P_N^T W_{tN}^* \tag{11.4e}$$

The time domain transversal filter's minimum MSE $\varepsilon_{t\min}$ given by

$$\varepsilon_{t\min} = E\left\{ d^2(n) \right\} - P_N^T W_{tN}^*$$

Hence, it is evident that the optimum /minimum mean square error of the transform domain adaptive filter is same as that of time domain transversal filter.

11.3.1 Relation between Time Domain Transversal and Transform Domain Weight Vectors

For the time domain transversal filter, its output $y_t(n)$ is given by

$$y_t(n) = W_{tN}^{\ T}(n)X(n).$$

The output of the transform domain filter is

$$y(n) = W_N^T(n)X''(n) = W_N^T(n)\left\{\Lambda^{-1/2}M_{NN}\left[X(n)\right]\right\}$$

For $y_t(n) = y(n)$,

$$W_{tN}^T(n)X(n) = W_N^T(n)(n)\Lambda^{-1/2}M_{NN}^T X(n)$$

$$W_{tN}^T(n) = W_N^T(n)O_n, \quad O_n = \Lambda^{-1/2}M_{NN}^T$$

$$W_{tN}(n)(n) = O_n^T W_N(n) \tag{11.4f}$$

Where, $O_n = \Lambda^{-1/2}M_{NN}^T$. The time domain weight vector and the transform domain vector are related by the Eqn. (11.4f).

11.3.2 Misadjustment of TADF

For the time domain LMS algorithm, the misadjustment, M_t is given by

$$M_t = \frac{\alpha}{2}\left\{ Tr\left[R_{NN} \right]\right\}$$

Therefore, for the transform domain, the misadjustment, M_o is given by

$$M_o = \frac{\alpha}{2}\left\{ Tr\left[R''_{NN} \right]\right\} \tag{11.4g}$$

But from Eqn.(11.4c),

$$R''_{NN} = \Lambda^{-1}M_{NN}^T R_{NN} M_{NN}$$

$$Tr[R''_{NN}] = Tr\left[\Lambda^{-1}M_{NN}^T R_{NN} M_{NN}\right] = Tr\left[\Lambda^{-1}\Lambda\right] = Tr\left[I\right] = N$$

Substituting this in Eqn. (11.4g),

$$M_0 = \frac{\alpha}{2}\left\{ Tr\left[R''_{NN} \right]\right\} = \frac{\alpha}{2}N$$

$$M_0 = \frac{N\alpha}{2} \tag{11.4h}$$

For the transversal LMS algorithm, the misadjustment is given by

$$M_t = \frac{\alpha}{2}\left\{ Tr\left[R_{NN} \right] \right\} = \frac{\alpha}{2} N\sigma_x^2 , \qquad \because Tr\left[R_{NN} \right] = Nr(0) = N\sigma_x^2$$

For a transversal normalized LMS algorithm as $\sigma_x^2 = 1$,

$$M_t = \frac{\alpha}{2} N \tag{11.4i}$$

The transform domain filter has the same misadjustment as that of normalized LMS transversal filter but its convergence rate is faster. This latter property is due to the fact that for each mode/component, the step size is scaled by its own component's power (much smaller compared to total power of the input signal) and not by total power of the signal. Also for a higher error, the transform domain permits a larger step size for adaptation, than the LMS.

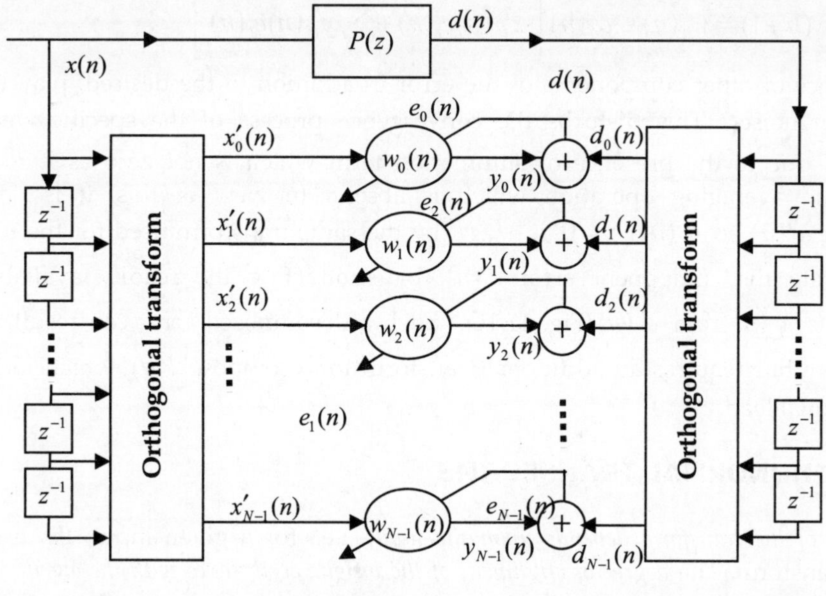

Fig. 11.1(e) Transform domain adaptive filter with the desired signal decomposition.

In the algorithm considered, the adaptation of individual weights is done by Eqn. (11.2e),

$$w_i(n+1) = w_i(n) + \frac{\alpha}{\sigma_{x_i}^2} e(n) x_i'^*(n)$$

$$\sigma_{x_i}^2\left(n+1\right) = \beta\, \sigma_{x_i}^2\left(n\right) + (1-\beta)\left|x_i'^*(n)\right|^2, \qquad i = 0, 1, \cdots, N-1$$

Here the error used is given by

$$e(n) = d(n) - y(n) = d(n) - \sum_{i=0}^{N-1} w_i(n) \quad x_i^{*}(n)$$

However, for the adaptation of the weight $w_i(n)$, the error required is $e_i(n)$ and is given by

$$e_i(n) = d_i(n) - y_i(n) = d_i(n) - w_i(n) \quad x_i^{*}(n)\,,\; i = 0, 1, \cdots, N-1 \tag{11.4j}$$

where, $d_i(n)$ are the components obtained after transforming the desired signal $d(n)$.
The weight adaptation using the required error component is

$$w_i(n+1) = w_i(n) + \alpha_i(n)\, x_i'\,(n) e_i(n)$$

If total error is used then

$$w_i(n+1) = w_i(n) + \alpha_i(n) \left[x_i'\,(n) e_i(n) + \sum_{l \neq i} x_i'\,(n) e_l(n) \right]$$

The presence of other components of the error in addition to the desired, play the role of observation noise. This disturbs the convergence process of the specific weight $w_i(n)$ adaptation due to the presence of summation term which is not zero as it requires the expectation/ averaging operation which is absent. In view of this, it is desirable to decompose $d(n)$ to $d_i(n)$, $i = 0, 1, \cdots, N$ by the same transform used for the input $x(n)$ and generate the component errors $e_i(n)$ by subtracting the readily available output components $y_i(n)$ from $d_i(n)$ (Fig. 11.1e). Such a decomposed error can result in a faster convergence but requires an additional transform to decompose $d(n)$ which increases the computational load.

11.4 ORTHONORMAL TRANSFORMS

The *choice of the transform depends upon the input,* i.e., for a given input, there can be an optimal transform. The *degree of alignment of the weight vector axes with the eigenvectors of the autocorrelation matrix depends on the transform used.* On comparing with the time domain algorithm, the frequency domain algorithm facilitates *rotation and scaling which enhances the convergence rate, which is not possible in time domain approach for the same error energy.* The commonly used different orthogonal transforms are :

1. Karhunen - Loe've Transform (KLT)
2. Discrete Fourier Transform (DFT)
3. Discrete Cosine Transform (DCT)
4. Discrete Sine Transform (DST)
5. Discrete Hartley Transform (DHT)
6. Discrete Walsh Hadmard Transform (WHT)

11.4.1 Karhunen–Love've Transform (KLT)

This transform deals with the representation of a *random vector* in terms of eigenvectors and eigenvalues. Consider a set of N orthonormal vectors $m_1, m_2, \ldots, m_N$ satisfying,

$$m_k^T m_i = \begin{cases} 1 & i = k \\ 0 & i \neq k \end{cases}$$

Let $X(n) = \left[x(n) \quad x(n-1) \quad \cdots \quad \cdots \quad x(n-N+1) \right]^T$ denote a random vector having zero mean and autocorrelation matrix R_{NN}. If $X(n)$ is expressed as a linear combination of $m_1, m_2, \cdots m_N$, then

$$X(n) = \sum_{l=1}^{N} x_l(n) m_l \tag{11.5a}$$

where

$$x_l(n) = m_l^T X(n) \text{ or } X^T(n) m_l \quad l = 1, 2, \cdots N \tag{11.5b}$$

Eqn. (11.5a) and (11.5b) are called discrete time version of Karhunen-Lov`ve expansion. Since $X(n)$ is a vector of random variables, $x_i(n)$ is also a random variable and $E\left[x_i(n)\right] = 0$ as $X(n)$ is of zero mean. For coefficients $x_i(n)$ of Eqn. (11.5a) to be uncorrelated,

$$Cov\left[x_k(n), x_i(n)\right] = E\left[x_k(n) x_i(n)\right] = \begin{cases} \lambda_i & , k = i \\ 0 & , k \neq i \end{cases} \tag{11.5c}$$

since

$$Cov\left[x_k(n), x_i(n)\right] = E\left[x_k(n) - E\left[x_k(n)\right]\right] E\left[x_i(n) - E\left[x_i(n)\right]\right] = E\left[x_k(n) x_i(n)\right]$$

and as

$$E\left[x_l(n)\right] = 0, \; l = 1, 2, \ldots N$$

λ_i is a constant to be specified.

From Eqn. (11.5b),

$$x_l(n) = m_l^T X(n) = X^T(n) m_l, \qquad l = 1, 2, \ldots N$$
$$x_i(n) = m_i^T X(n) = X^T(n) m_i,$$
$$x_k(n) = m_k^T X(n) = X^T(n) m_k$$

Hence,

$$Cov\left[x_k(n) x_i(k)\right] = E\left[m_k^T X(n) X^T(n) m_i\right] = m_k^T E\left[X(n) X^T(n)\right] m_i$$

$$Cov\left[x_k(n)x_i(k)\right] = m_k^T R_{NN} m_i \tag{11.5d}$$

Comparing Eqns. (11.5c) and (11.5d),

$$m_k^T R_{NN} m_i = \begin{cases} \lambda_i & i = k \\ 0 & i \neq k \end{cases} \tag{11.5e}$$

Consider the property of vectors of an orthogonal space

$$m_k^T m_i = \begin{cases} 1 & k = i \\ 0 & k \neq i \end{cases}$$

Multiply both sides by λ_i,

$$\lambda_i m_k^T m_i = \begin{cases} \lambda_i & k = i \\ 0 & k \neq i \end{cases} \tag{11.5f}$$

Comparing this with (11.5e), if Eqn. (11.5e) holds good for all k and a particular i,

$$R_{NN} m_i = \lambda_i m_i, \quad i = 1, 2, \cdots, N \tag{11.5g}$$

This is in the from $Aq = \lambda q$ for which λ and q are eigenvalue and its corresponding eigenvector, respectively. Therefore, in Eqn. (11.5g)), λ_i is eigenvalue corresponding to the eigenvector m_i of R_{NN}. Thus, using Karhuen-Loe've expansion, the random vector $X(n)$ can be expressed in terms of its eigenvectors of autocorrelation matrix.

$$X(n) = \sum_{l=1}^{N} x_l(n) m_l,$$

$$x_l(n) = m_l^T X(n) = X^T(n) m_l, \quad l = 1, 2, \ldots N$$

The eigenvectors $m_1, m_2, \cdots m_N$ constitute an orthonormal set and hence forms an N dimensional space. The random vector $X(n)$ has projection $x_1, x_2, \cdots x_N$ on these axes. Consider

$$X^T(n)X(n) = \left[x(n) \quad x(n-1) \quad \cdots \quad x(n-N+1)\right] \begin{bmatrix} x(n) \\ x(n-1) \\ \vdots \\ x(n-N+1) \end{bmatrix}$$

$$X^T(n)X(n) = \sum_{k=0}^{N-1} x^2(n-k), \tag{11.5h}$$

The quantity $X^T(n)X(n)$ represents energy of $X(n)$. From Eqn. (11.5a) and orthonormal property,

$$X(n) = \sum_{l=1}^{N} x_l(n)m_l\,, \qquad m_k^T m_i = \begin{cases} 1 & k = i \\ 0 & k \neq i \end{cases},$$

Therefore,

$$X^T(n)X(n) = \left(\sum_{k=1}^{N} x_k^T(n)m_k^T\right)\left(\sum_{k=1}^{N} x_k(n)m_k\right) = \sum_{k=1}^{N} x_k^2(n) \tag{11.5i}$$

From Eqns. (11.5h) and (11.5i),

$$\sum_{k=1}^{N} x_k^2(n) = \sum_{k=0}^{N-1} x^2(n-k)$$

Hence, $x_i^2(n)$ *is the energy in the random vector* $X(n)$ *along the* i^{th} *orthonormal vector.*

From Eqn. (11.5c)

$$E\left[x_k(n)x_i(n)\right] = \begin{cases} \lambda_i, & k = i \\ 0\,, & k \neq i \end{cases} \qquad i = 1, 2, \cdots, N$$

Therefore,

$$E\left\{x_i^2(n)\right\} = \lambda_i\,,$$

is the ith eigenvalue , the average energy along i^{th} *eigenvector*. Hence,

$$E\left\{\sum_{k=0}^{N-1} x^2(n-k)\right\} = E\left\{\sum_{k=1}^{N} x_k^2(n)\right\} = \sum_{k=0}^{N-1} E\left\{x_k^2(n)\right\} = \sum_{k=0}^{N-1} \lambda_k$$

$$E\left\{\sum_{k=0}^{N-1} x^2(n-k)\right\} = \sum_{k=0}^{N-1} \lambda_k \tag{11.5j}$$

Total energy of the signal is sum of the eigenvalues of the autocorrelation matrix. KLT provides most desirable representation of the given data in terms of eigenvalues and eigenvectors. However, as it is data dependent and involves more computations, it is not preferred to a fixed type of transform.

11.4.2. Discrete Fourier Transform (DFT)

For a signal $x(n)$ of length N, the direct discrete Fourier transform (DFT) $X(k)$ and its inverse, inverse DFT (IDFT) $\tilde{x}(n)$ are given by

$$X(k) = \sum_{n=0}^{N-1} x(n)e^{-\frac{j2\pi kn}{N}}\,, \qquad k = 0, 1, \ldots., N-1 \tag{11.6a}$$

$$\tilde{x}(n) = \frac{1}{N}\sum_{k=0}^{N-1} X(k)e^{\frac{j2\pi kn}{N}}\,, \qquad n = 0, 1, \ldots, N-1 \tag{11.6b}$$

The number of multiplications required for DFT is N^2 but this gets reduced to $N\log_2 N$ by using the Fast Fourier transform (FFT) algorithm. For adaptive applications, for every new sample, computing DFT even by FFT is not economical.

11.4.3. Discrete Cosine Transform

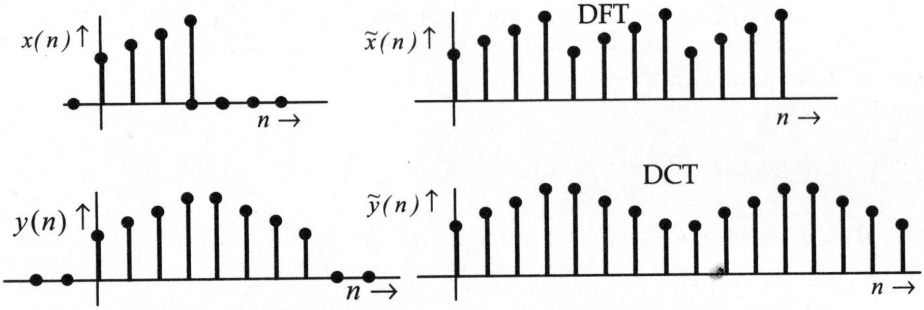

Fig. 11.2 Waveforms associated with DFT and DCT.

To derive the DCT of a signal $x(n)$, $n = 0,1,2,\ldots(N-1)$, form a new sequence $y(n)$ of length $2N$ (Fig. 11.2).

$$y(n) = x(n) + x(2N-1-n)$$
$$= \begin{cases} x(n), & 0 \le n \le N-1 \\ x(2N-1-n), & N \le n \le 2N-1 \end{cases} \tag{11.7a}$$

This sequence $y(n)$ *is symmetric with respect to the point* $(N-1/2)$. The DCT with this type of symmetry is known as DCT-II. The N point DCT of $x(n)$, $C_x(k)$ is given by

$$C_x(k) = \begin{cases} e^{-j2\pi k/4N}Y(k), & 0 \le k \le N-1 \\ 0, & \textit{Otherwise} \end{cases} \tag{11.7b}$$

Therefore, the DCT of $x(n)$ is given by

$$C_x(k) = \begin{cases} \sum_{n=0}^{N-1} 2x(n)\cos\dfrac{\pi k(2n+1)}{2N}, & 0 \le k \le N-1 \\ 0, & \textit{otherwise} \end{cases} \tag{11.7c}$$

and the inverse DCT (IDCT) is given by

$$\tilde{x}(n) = \frac{1}{N}\sum_{k=0}^{N-1} U(k)C_x(k)\cos\left(\frac{\pi k(2n+1)}{2N}\right), \quad 0 \le n \le N-1, \tag{11.7d}$$

$$U(k) = \begin{cases} 1/2, k = 0 \\ 1, \quad 1 \le k \le N-1 \end{cases}$$

In DFT, its associated waveform $\tilde{x}(n)$ *suffers from discontinuity and hence suffers from severe leakage problem as explained already in Chapter 4. But in DCT, its associated waveform has a significantly reduced discontinuity as there is smooth transition from one period to the other. This is due to symmetric extension involved in DCT which makes the DCT leakage significantly less compared to that of DFT.* This makes the quality of orthogonalization by DCT superior to that by DFT. For a first order AR process, the DCT asymptotically approaches KLT as the data length increases. Also unlike KLT, the DCT has fixed basic function that makes it efficient to KLT. Discrete sine transform of a signal $x(n)$ can be obtained by computing the DCT of a signal $y(n) = x(n)(-1)^n$ except for a scale factor. The **DST** is given by

$$S(k) = \frac{D(k)}{N} \sum_{n=0}^{N-1} 2y(n)\cos\left[(2n+1)\frac{k\pi}{2N}\right], \quad k = 0, 1, \ldots, N-1$$

where $y(n) = (-1)^n x(n)$ and $D(k) = \begin{cases} \frac{1}{\sqrt{2}}, & k = 0 \\ 1, & k \neq 0 \end{cases}$,

11.4.4. Discrete Hartley Transform (DHT)

This transform is similar to DFT but it is real transform and is found to require less computational load than the DFT . Its direct and inverse versions are given by

$$\text{DHT:} \quad X(k) = \frac{1}{\sqrt{N}} \sum_{n=0}^{N-1} x(n)\left\{\cos\left(\frac{j2\pi k n}{N}\right) + \sin\left(\frac{j2\pi k n}{N}\right)\right\}, \; k = 0, 1, \ldots., N-1 \quad (11.8a)$$

$$\text{IDHT:} \quad \tilde{x}(n) = \frac{1}{\sqrt{N}} \sum_{k=0}^{N-1} X(k)\left\{\cos\left(\frac{j2\pi k n}{N}\right) + \sin\left(\frac{j2\pi k n}{N}\right)\right\}, \; n = 0, 1, \ldots., N-1 \quad (11.8b)$$

11.4.5. Walsh Hadmard Transform (WHT)

The WHT is given by

$$X(k) = \frac{1}{\sqrt{N}} \sum_{n=0}^{N-1} x(n)\, w_{kn}, \qquad n = 0, 1, \ldots, N-1 \quad (11.9)$$

$$w_{kn} = \prod_{p=0}^{m-1} (-1)^{b_p(k) b_{m-1-p}(n)}, \; m = \log_2 N$$

$b_p(k)$ is the pth bit of binary representation of k. ($p = 0$, corresponds to least significant bit). WHT is simple, as it involves no multiplication, as all of its coefficients are ± 1.

11.5 PERFORMANCE OF DIFFERENT TRANSFORMS

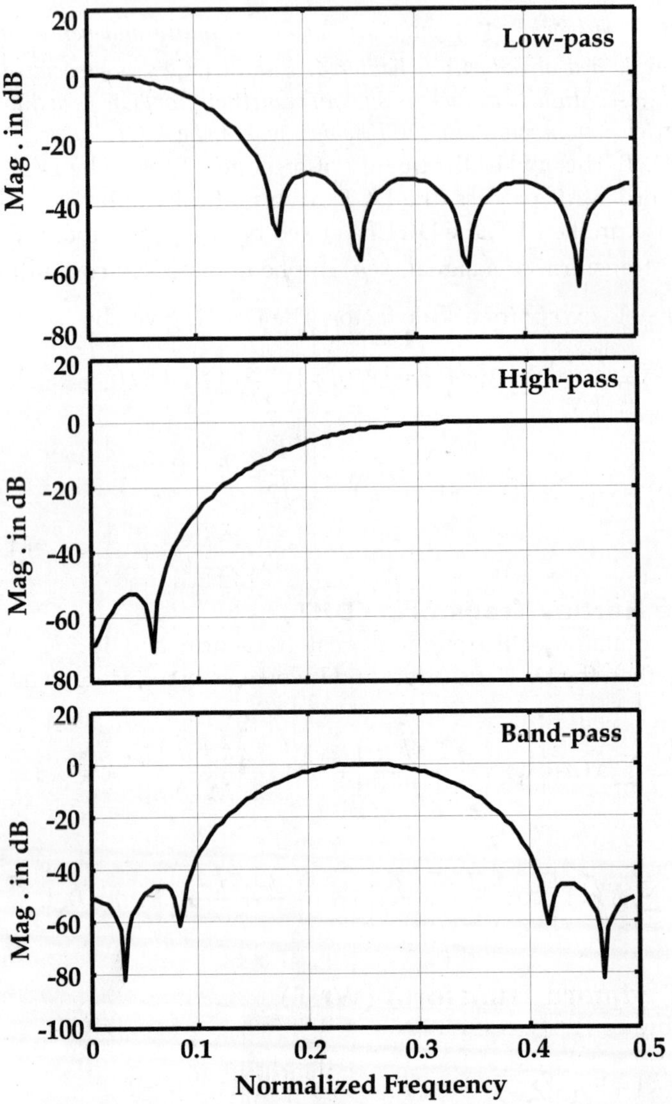

Fig. 11.3(a) Frequency responses of filters used to generate different types of inputs.

The above transforms are useful and effective in reducing the eigenvalue spread depending on the nature of the input signal. But presently in this direction, the effectiveness of commonly used transforms, DCT, DST and DFT will be considered. For this, low-pass, high-pass and band-pass signals are generated by passing the Gaussian white noise, through the respective filters. The low-pass filter has an impulse response length of 11 samples of a *sinc* function and is windowed by a Kaiser Window with smoothing factor of 3.9. The low-pass filter has a cut-off frequency of 50 Hz (0.1π) and sampling frequency of 1k Hz (2π).

For the high-pass filter, the cut-off frequency used is 235 Hz (0.47π). The impulse response length, sampling frequency, the window and its smoothing factor used are as those for the low-pass filter. For the band-pass filter, the impulse response is derived by using a Kaiser window of length 15 and a smoothing factor of 3.9. The band-pass filter has a center frequency of 250 Hz (0.5π) and its bandwidth is 100 Hz (0.1π).

For the low-pass, high-pass and band-pass signals generated, their autocorrelation matrices with 8 lags are computed and their eigenvalue spread are 418.70, 687.88 and 653.57; respectively. Further, for the transformed and scaled signals the autocorrelation matrices R''_{NN} given by $R''_{NN} = E\left[X''(n)X''^T(n)\right]$ (Fig. 11.1b) are computed. The expectation is realized by averaging $\left\{X''(n)X''^T(n)\right\}$ over an ensemble of 100 obtained for the corresponding 100 input samples. The eigenvalue spread of R''_{NN} for the DCT, DST and DFT transforms for different type of signals are tabulated in Table 11.1.

Table 11.1 Eigenvalue spread reduction by different transforms for different inputs

Type of input	Eigenvalue Spread			
	Original	DCT	DST	DFT
Low-pass	418.70	**9.11**	78.14	90.66
High-pass	687.88	50.55	**3.41**	29.28
Band-pass	653.57	118.99	**28.74**	99.20

For the low-pass signal, its eigenvalue spread of 418.7 is reduced to 9.11, 78.14 and 90.66 by DCT, DST and DFT, respectively. For the high-pass signal, its eigenvalue of 687.88, is reduced to 50.55, 3.41 and 29.28; respectively by DCT,DST and DFT. For the band-pass signal, its eigenvalue spread of 653.57 is reduced to 118.99, 28.74 and 28.74; respectively by DCT, DST and DFT. Hence for the low-pass signal, DCT and for high-pass and band-pass, the DST; are suitable. The DFT performs better than DCT for high-pass signal but it is not suitable for low-pass and band-pass signals.

The above performance of different transforms can be explained as follows. The output power of a filter is given by the Parseval's theorem

$$E[y^2(n) = \frac{1}{2\pi}\int_0^{2\pi} S_x(\omega)\,|\,F(\omega)\,|^2\, d\omega$$

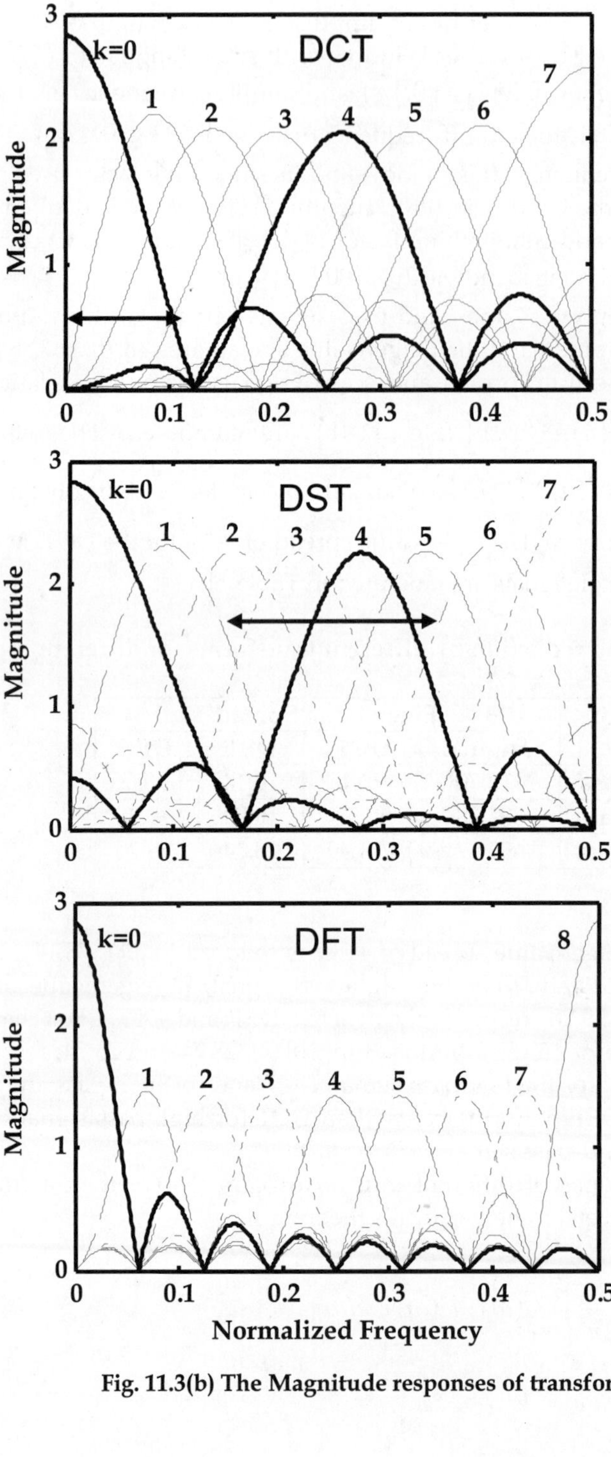

Fig. 11.3(b) The Magnitude responses of transform filters.

Here, $F(\omega)$ is the FT of the filter coefficients, which are the row elements of the transformation matrix M_{NN} and $y(n)$ is the output of the filter for the input $x(n)$ with power spectral density $S_x(\omega)$. For an ideal transform

$$M_{NN}^T M_{NN} = I,$$

which is equivalent to

$$\frac{1}{2\pi}\int_0^{2\pi} |F(\omega)|^2 \, d\omega = 1$$

The filters of transformation are supposed to extract the energy of the signal in the band of the corresponding filter and without altering the signal energy in that frequency band. This is possible only if the filters of the transform are ideal, i.e., has again of unity in the desired band and zero in other regions. However, the filters of different transforms other than KLT will not satisfy this condition. For a given transform, its individual filter has not only mainlobe but also sidelobes. The sidelobe structure differs from transform to transform. For a low-pass signal, if the mainlobe of the filter lies in the region of larger values of $S_x(\omega)$ and the sidelobes due to filters in other regions are of small magnitude, its output power is not affected by the sidelobes, like in DCT (Fig. 11.3a). For a band-pass signal, the DST filters have the desired characteristics as they have mainlobes at the mid-frequency region and the sidelobes due to filters from other regions have sidelobes which are of smaller magnitude (Fig. 11.3b). Hence, DST has a better performance than DCT for a band-pass signal.

The DFT filters on the other hand have sidelobes of same magnitude throughout the frequency range. In view of this, its performance is poor for all the types of signals compared to DCT and DST. In Fig. 11.3b, a transform length of 8 is used for the DCT and DST. But transform length of 16 is used for DFT so that the number of filters is same for all the transforms in the frequency range of $(0, \pi)$.

To be effective the orthogonal transform should align the weight vector axes with the axes (eigenvectors) of the contours of the error surface. The magnitudes of the sidelobes of the filters of the transformation make the orthogonalization approximate. Hence, *transforms to be useful for orthogonalization should have their filters characterised by smaller sidelobe levels.*

11.6 PERFORMANCE INDEX FOR THE COMPARISON OF TRANSFORMS USED FOR ORTHOGONALIZATION

In the above discussion on the utility of transforms for adaptive filters, it is the eigenvalue spread reduction that forms the index of performance. But computing the eigenvalues and their spread is not computationally simple. Hence, it is desirable to have a measure, which not only reflects accurate spread, but also that can be computed easily. Such an index, which can be computed from the autocorrelation matrix directly and easily, is used in practice and this will be presently considered. This is based on the fact that *the ratio of arithmetic mean to geometric mean of the eigenvalues reflects the eigenvalue spread.* Hence, to compare the performance of different transforms/methods, an eigenvalue spread index, ξ has been defined as a function of the ratio of arithmetic mean to geometric mean of eigenvalues and is given by

$$\xi = \left(\frac{\lambda_a}{\lambda g}\right)^N \tag{11.10a}$$

Arithmetic mean of eigenvalues of R_{NN} :

$$\lambda_a = \frac{1}{N}\sum_{i-1}^{N}\lambda_i \tag{11.10b}$$

Geometric mean of eigenvalues of R_{NN} :

$$\lambda_g = \left[\prod_{i=1}^{N}\lambda_i\right]^{1/N} \tag{11.10c}$$

For a given set of numbers $\lambda_i, i = 1, 2, \ldots, N$, since $\lambda_a \geq \lambda_g$, the value of $\xi \geq 1$. The distribution of eigenvalues of R_{NN} decides the value of ξ and ξ is always greater than or equal to unity. For equal eigenvalues, $\xi = 1$ and for any spread of eigenvalues, $\xi > 1$ and further ξ has a higher value for the larger spread. Substituting for λ_a and λ_g,

$$\xi = \left(\frac{\lambda_a}{\lambda g}\right)^N, = \left\{\frac{\left(\frac{1}{N}\sum_{i-1}^{N}\lambda_i\right)}{\left[\prod_{i=1}^{N}\lambda_i\right]^{1/N}}\right\}^N$$

It is known that

$$\sum_{i-1}^{N}\lambda_i = Tr[R_{NN}] \quad and \quad \prod_{i=1}^{N}\lambda_i = Det[R_{NN}]$$

$$\xi = \frac{\left(\frac{1}{N}Tr[R_{NN}]\right)^{1/N}}{Det[R_{NN}]} \tag{11.10d}$$

It is seen that this index involves for the autocorrelation matrix, only its trace and determinant instead of computing its minimum and maximum eigenvalues. Thus, computation of the index ξ is significantly simple compared to that of direct eigenvalue spread. Further, to compare the performance of the two transforms, based on the index ξ an improvement factor I_ξ is used in practice and is given by

$$I_\xi = \ln \xi_1 - \ln \xi_2 \qquad (11.10e)$$

where ξ_1 and ξ_2 are the values of ξ for the transforms -1 and 2, respectively. If I_ξ is positive, then algorithm -2 is better than 1 and vice versa. Let the eigenvalue spread index ξ for a signal with autocorrelation matrix R_{NN}, be ξ_C and that for the orthogonal transformed signal/autocorrelation matrix be ξ_0. Then the index ξ_C is given by (Eqn. 11.10d)

$$\xi_C = \frac{\left(\frac{1}{N} Tr[R_{NN}]\right)^{1/N}}{Det\left(R_{NN}\right)} \qquad (11.10f)$$

Since R_{NN} is transformed as $M_{NN} R_{NN} M_{NN}^T$, then

$$\xi_o = \frac{\left(\frac{1}{N} Tr[M_{NN} R_{NN} M_{NN}^T]\right)^N}{Det\left(M_{NN} R_{NN} M_{NN}^T\right)},$$

Since for matrices A and B,

$$Tr(AB) = TR(BA), \qquad Det(AB) = Det(BA),$$

$$\xi_o = \frac{\left(\frac{1}{N} Tr[M_{NN}^T M_{NN} R_{NN}]\right)^N}{Det\left(M_{NN}^T M_{NN} R_{NN}\right)} = \frac{\left(\frac{1}{N} Tr[R_{NN}]\right)^N}{Det\left(R_{NN}\right)} \quad (as \;\; M_{NN}^T M_{NN} = I)$$

$$\xi_O = \frac{\left(\frac{1}{N} Tr[R_{NN}]\right)^{1/N}}{Det\left(R_{NN}\right)} \qquad (11.10g)$$

Comparing Eqns. (11.10f) and (11.10g),

$$\xi_o = \xi_C \qquad (11.10h)$$

Hence,

$$I_\xi = \ln \xi_C - \ln \xi_o = \ln \xi_C - \ln \xi_C = 0$$

This implies that there is no improvement in eigenvalue spread by mere transformation (without normalization). Hence, it is of interest to consider improvement factor for the case of transformation with scaling.

Let ξ_{os} correspond to ξ of orthogonalized scaled input.

$$Tr[R''_{NN}] = Tr\left[\Lambda^{-1}M^T_{NN}R_{NN}M_{NN}\right] = Tr\left[\Lambda^{-1}\Lambda\right] = Tr\left[I\right] = N$$

$$Det[R''_{NN}] = Det\left[\Lambda^{-1}R'_{NN}\right] = Det\left[\Lambda^{-1}\right]Det\left[R'_{NN}\right]$$

$$\xi_{os} = \frac{\left(\frac{1}{N}Tr[R''_{NN}]\right)^{1/N}}{Det\left(R''_{NN}\right)} = \frac{1}{Det(\Lambda^{-1})Det\left(R'_{NN}\right)} = \frac{Det(\Lambda\ )}{Det\left(R'_{NN}\right)}$$

Since

$$\Lambda\ = dia[R'_{NN}],$$

$$\xi_{os} = \frac{Det(\Lambda\ )}{Det\left(R'_{NN}\right)} = \frac{Det(dia[R'_{NN}])}{Det\left(R'_{NN}\right)},$$

Multiplying and dividing by $Tr[R'_{NN}]$,

$$\xi_{os} = \frac{Det(dia[R'_{NN}])}{Tr[R'_{NN}]}\frac{Tr[R'_{NN}]}{Det\left(R'_{NN}\right)},$$

Since

$$Tr[R'_{NN}] = Tr[dia(R'_{NN})],$$

$$\xi_{os} = \frac{Det(dia[R'_{NN}])}{\{Tr[dia(R'_{NN})]/N\}^N}\frac{\{Tr[R'_{NN}]/N\}^N}{Det\left(R'_{NN}\right)}$$

Let

$$\xi_{odia} = \frac{\{Tr[dia(R'_{NN})]/N\}^N}{Det(dia[R'_{NN}])}$$

Then

$$\xi_{os} = \frac{\xi_o}{\xi_{odia}}, \qquad \ln\xi_{os} = \ln\xi_o - \ln\xi_{odia}$$

Therefore,

$$I^{os}_{\xi} = \ln\xi_C - \ln\xi_{os} = \ln\xi_C - \ln\xi_o + \ln\xi_{odia}$$

Since from Eqn. (11.10h), $\ln \xi_o = \ln \xi_C$,

$$I_{\xi}^{os} = \ln \xi_{odia}$$

The improvement of factor $\ln \xi_{odia}$ is always non-negative and the performance of the transform domain algorithm will not be inferior to ordinary LMS algorithm

For the Newton's algorithm, let ξ_{NT} represent the value of ξ. But for Newton's algorithm, $R_{NN} = I$. Therefore,

$$\xi_{NT} = \frac{\left(\frac{1}{N}Tr[R_{NN}]\right)^{1/N}}{Det\left(R_{NN}\right)} = \frac{\left(\frac{1}{N}Tr[I]\right)^{1/N}}{Det\left(I\right)} = 1$$

Therefore, the improvement factor is

$$I_{\xi} = \ln \xi_C - \ln \xi_{NT} = \ln \xi_C - \ln 1 = \ln \xi_C$$

$$I_{\xi} = \ln \xi_C \qquad (11.10i)$$

This is the maximum possible improvement over ordinary LMS algorithm. The quantity $\ln \xi_i$, quantifies the ability of orthogonalization of the transform and smaller its value better is the orthogonalization. Hence, a larger value for $\ln \xi_i$ indicates a poorer performance compared to the Newton's method, the ideal.

For the same types signals considered previously, their improvement factors obtained by different transforms viz., DCT, DST and DFT are listed in the Table 11.2. For the low-pass signal, the improvement factor I_{ξ} obtained by DCT has the highest value of 14.91. For high-pass and band-pass signals, the index obtained by DST is having highest values of 7.34 and 8.58, respectively. These results are in agreement with the eigenvalue spread reduction computed directly from the eigenvalues (Table 11.1).

This indicates that the improvement factor reflects the performance of the transform for a given type of input signal and can be used without much involved computations. Though the values of I_{ξ} for different transforms differ by small amount, but it should be noted that this is due to logarithmic operation and actual improvement corresponds to exponential power of these values which will be significantly high.

Table 11.2 Improvement factors for different transforms for different inputs

Type of input	Improvement Factor		
	DCT	DST	DFT
Low-pass	**14.91**	9.79	8.31
High-pass	4.10	**7.34**	5.61
Band-pass	6.11	**8.58**	7.20

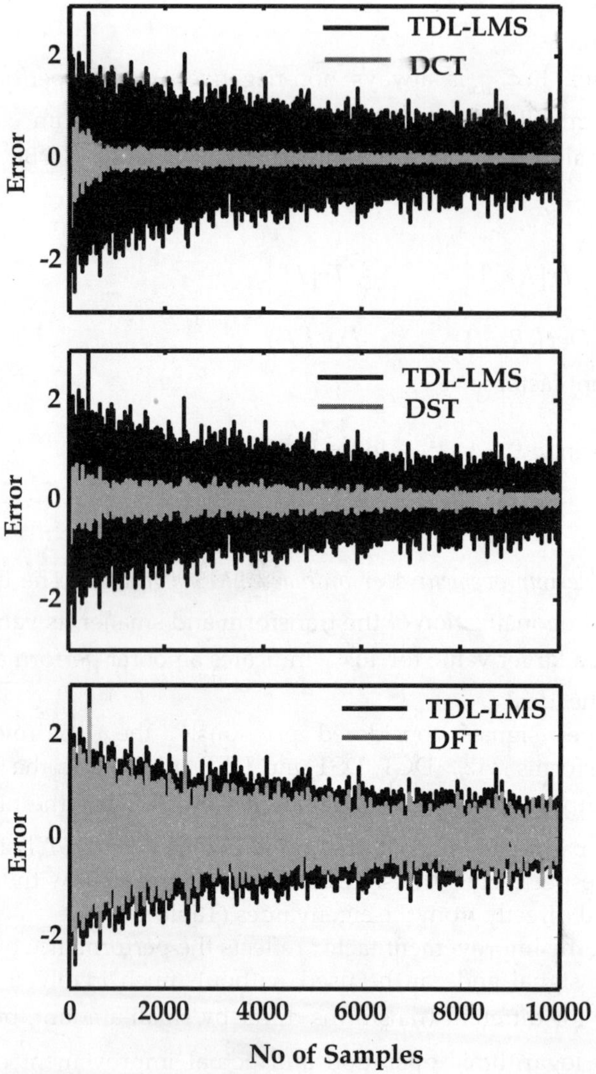

Fig. 11.4 (a) Convergence performance of DCTLMS, DSTLMS, DFTLMS and time domain LMS (TDL-LMS); for a low-pass input.

Example: The performance of DCTLMS, DSTLMS and DFTLMS trnasforms for different types of inputs low-pass, high-pass and band-pass; for system identification ofthe sytem $P(z)$ is cosidered.

$$P(z) = \frac{1 - z^{-1} + 0.8z^{-2} - 0.7z^{-3}}{1 + 0.6z^{-1} - 0.5z^{-2} + 0.1z^{-3} + 0.5z^{-4}}$$

The input low-pass signal is made up of sinusoids with frequencies at 10Hz, 20Hz and 30Hz plus zero mean white Gaussian noise (SNR=26dB). The sampling frequency considered is 1000Hz. A transform length of 32, step size value of 0.004 and the forgetting factor of 0.99 for power computation; are used. The error for 10,000 samples are shown in Fig. 11.4(a) .

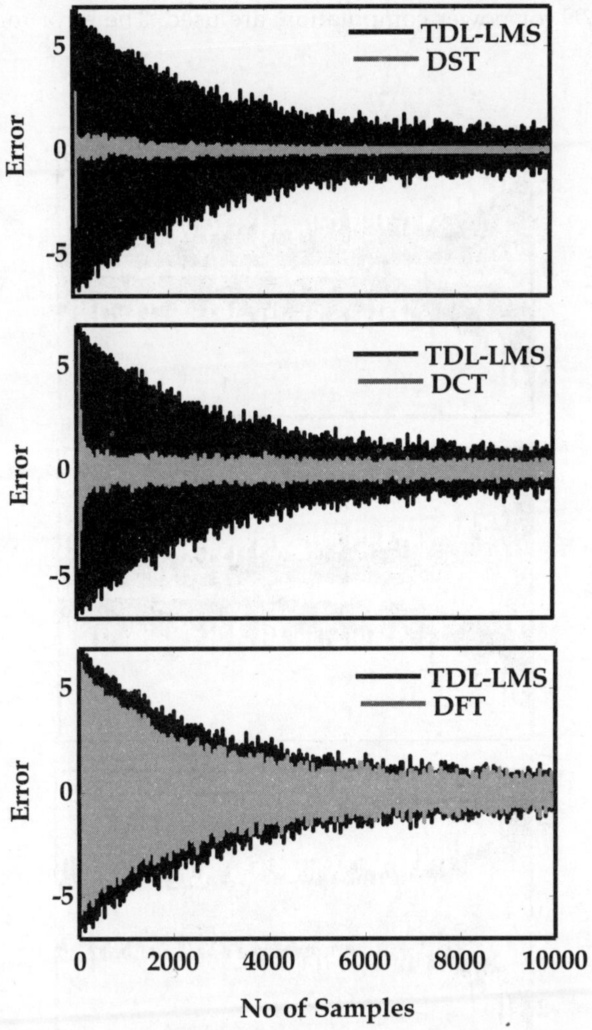

Fig. 11. 4(b) Convergence performance of DSTLMS, DCTLMS, DFTLMS and time domain LMS (TDL); for high-pass input signal.

It is seen that the DCT transform domain LMS exhibits good covergence rate and convergence error compared to those of time domain LMS algorithm. The performance of DSTLMS is inferiror to that of DCTLMS but it is still better than that of time domain LMS algorithm. However the performance of DFTLMS is significantly poorer than those of DCTLMS and DSTLMS and it is amlost similar to that time domain LMS algorithm.

For the high-pass input, the signal which has three sinusoids at frequencies 350Hz, 400Hz and 450Hz plus zero mean white Gaussian noise (SNR=26dB) is used . The sampling frequency considered is 1000Hz. A transform length of 64, step size value of 0.001 and the forgetting factor of 0.99 for power computation; are used. The error for 10,000 samples are shown in Fig. 11.4(b).

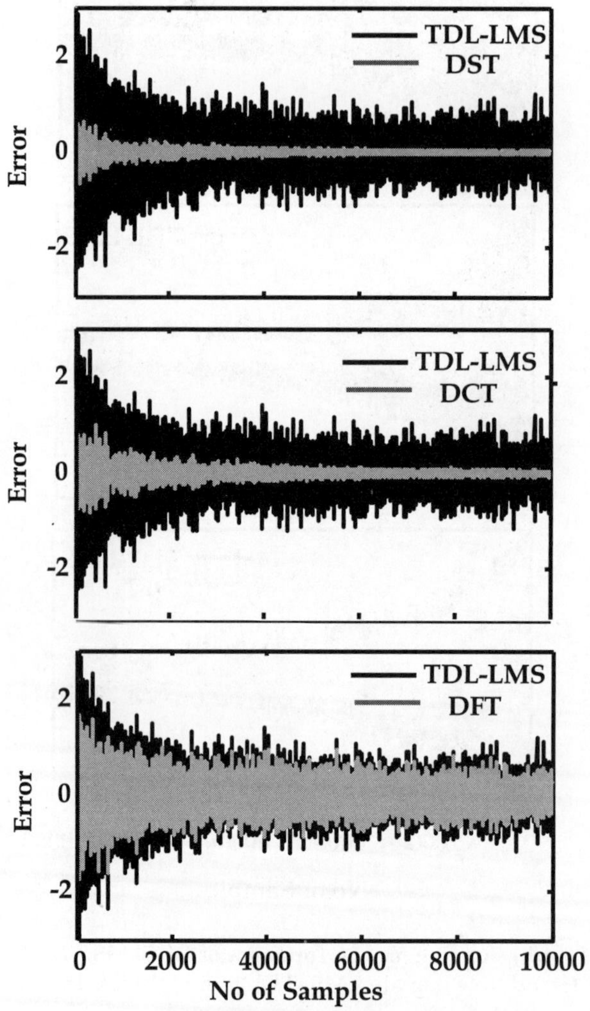

Fig. 11.4(c) Convergence performance of DSTLMS, DCTLMS, DFTLMS and time domain LMS (TDL-LMS) for a bandpass input.

Here, the performance of DSTLMS is significantly better than that of DCTLMS both in terms of convergence rate and error. But the performance of DCTLMS is a superior to those of DFTLMS and time domain LMS algorithms. The DFTLMS performance is almost similar to that of time domain LMS and their convergence rate and error are poor.

For the band-pass input, three sinusoids at frequencies 210Hz, 250Hz and 280Hz plus zero mean Gaussian white noise with SNR=26db, is considered. A sampling frequency of 1000Hz , transform size of 64 points, step size of 0.005 and for the power estimation a forgetting factor of 0.99; are used.

For band-pass input also, the DSTLMS performs maginally better than DCTLMS and its performance is significanly superior to that of DFTLMS and time domain LMS algoritms (Fig. 11.4(c)).

Thus, it is seen that for all types of signals (low-pass, high-pass and band-pass considered), the DCTLMS and DSTLMS perform better compared to those of DFTLMS and time domain LMS algorithms. Further, for low-pass, DCTLMS and for high-pass and band-pass, the DSTLMS are prefered.

Example:

This shows for system identification at specific frequencies, the effect of decomposition of the desired signal over that of using complete desired signal which implies using the total error instead of individual errors. For this, the system considered is an acoustic path having impulse response as shown in Fig. 11.4(d). The input to this system is sum of three sinusoids with frequencies, 300Hz, 350Hz and 400 Hz. The sampling rate used is 1000 Hz. For the observation noise both Gaussian white noise and colored noises are considered (SNR= -10dB) . The colored noise is generated by filtering Gaussian white noise by a system given by

$$H(z) = \frac{1}{1 + z^{-1} - 0.8z^{-2}}$$

The transform used is DCT of for data vector of size 128 points. The step size used is 0.01 and the forgetting factor for power computation used is 0.99. The data length used is 50,000 points. The performance index used is smoothed ensemble average square error (SEASE). $\xi_{Sd}(n)$ is used for comparison and is given by

$$e_i(n) = d_i'(n) - y_i(n) \;, \quad \xi(n) = \frac{1}{k}\sum_{i=1}^{k} e_i^2(n), \; \xi_{Sd}(n) = 10\log_{10}(\xi_S(n))$$

$e_i(n)$:error for i^{th} trial, k : number of trials and $\xi_S(n)$ is the smoothed version of $\xi(n)$ by a moving average filter. The number of trials used $k = 30$. A 12 point moving average filter is used for smoothing $\xi(n)$.

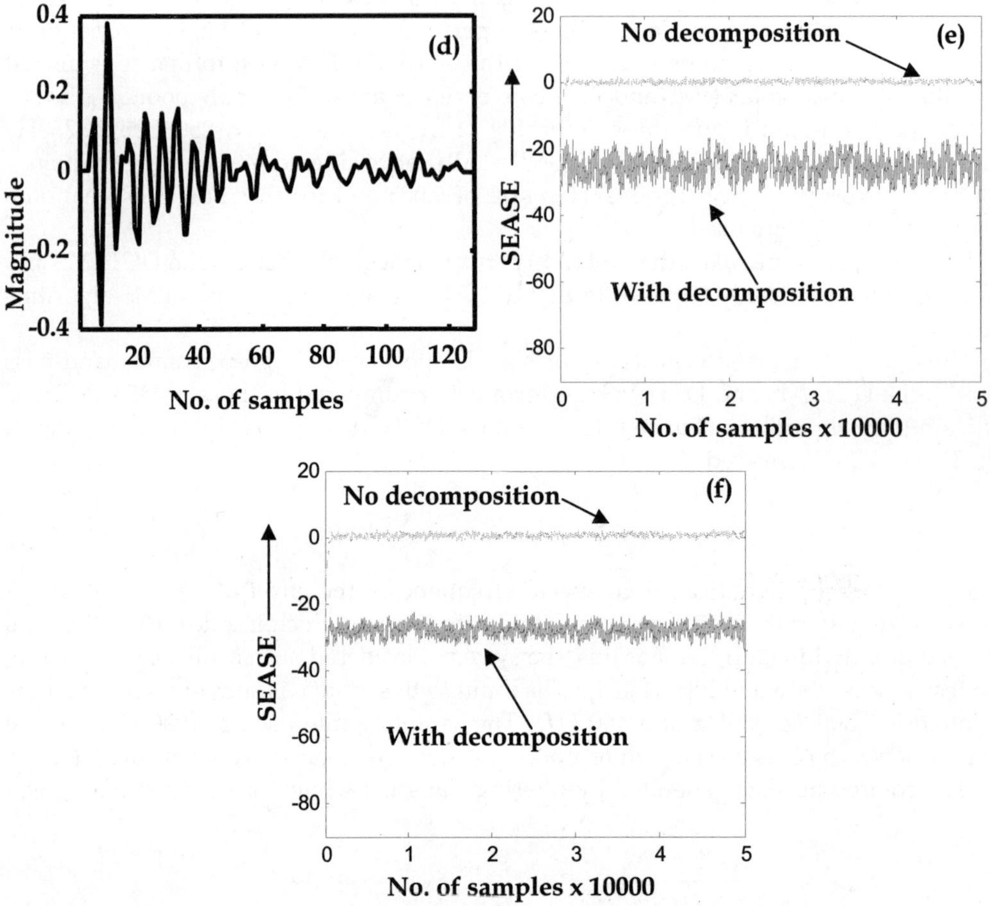

Fig. 11.4 (d, e, f) Effect of desired signal decomposition on the performance index SEASE for system identification. (d) Impulse response of the system, (e) SEASE for white observation noise and (f) SEASE for colour observation noise.

It is seen from Fig. 11.4(e, f) that both for white and coloured observation noise, the desired signal decomposition significantly improves the performance as the index SEASE is about 20 dB and 30dB respectively, below than that for the case of without desired signal decomposition.

11.7 IMPLEMENTATION OF TRANSFORM DOMAIN ALGORITHMS

Many transforms have been considered for orthogonalization. Since the transform is required for a block of data on sample to sample basis, the block transform algorithm has to be repeated for every sample even though only one new sample is added and the last sample in the block (previous) is omitted. For this small change in the block from sample to sample, computing altogether a fresh transform is not justified. Hence, algorithms which only account for the inclusion of the new sample and removal the last sample in the block/window, which are computationally simpler than block transforms, have been developed and are known as *sliding /running/ recursive* transforms. Few important sliding transforms will be considered.

11.7.1 Sliding DFT

The general equation for DFT,

$$z_n(k) = \sum_{p=0}^{N-1} x(n-p) \quad e^{-j\left(\frac{2\pi}{N}\right)pk} \qquad k = 0, 1, 2, \ldots, N-1 \tag{11.11a}$$

where $z_n(k)$ is the N-point discrete Fourier transform of the input signal $x(n-p)$.

Consider

$$Z_n(k) = \sum_{p=0}^{N-1} x(n-p) \quad e^{-j\left(\frac{2\pi}{N}\right)pk}$$

$$Z_n(k) = x(n) + x(n-1)\, e^{-j\frac{2\pi}{N}k} + x(n-2)\, e^{-j\frac{4\pi}{N}k} +$$

$$\ldots + x(n-N+2)\, \underline{e^{-j\frac{2\pi}{N}(N-2)k}} + x(n-N+1)\, \underline{\underline{e^{-j\frac{2\pi}{N}(N-1)k}}}$$

$$= x(n) + x(n-1)e^{-j\frac{2\pi}{N}k} + x(n-2)e^{-j\frac{4\pi}{N}k} +$$

$$\cdots + x(n-N+2)\, \underline{e^{-j\frac{2\pi}{N}Nk}\; e^{j\frac{4\pi}{N}k}} + x(n-N+1)\, \underline{\underline{e^{-j\frac{2\pi}{N}Nk}\; e^{j\frac{2\pi}{N}k}}}$$

Splitting the exponents of underlined terms and using $e^{-j2\pi k} = 1$,

$$Z_n(k) = x(n) + x(n-1)e^{-j\frac{2\pi}{N}k} + x(n-2)e^{-j\frac{4\pi}{N}k} + \cdots + x(n-N+2)\, e^{j\frac{4\pi}{N}k} + x(n-N+1)\, e^{j\frac{2\pi}{N}k} \tag{11.11b}$$

Consider

$$Z_{(n-1)}(k) = \sum_{p=0}^{N-1} x(n-1-p) \quad e^{-j\left(\frac{2\pi}{N}\right)pk}$$

$$Z_{(n-1)}(k) = x(n-1) + x(n-2)\, e^{-j\frac{2\pi}{N}k} + x(n-3)\, e^{-j\frac{4\pi}{N}k} + \cdots$$

$$+ x(n-1-N+1)\, \underline{\underline{e^{-j\frac{2\pi}{N}(N-1)k}}}$$

Splitting exponent of underlined term,

$$Z_{(n-1)}(k) = x(n-1) + x(n-2)\, e^{-j\frac{2\pi}{N}k} + x(n-3)\, e^{-j\frac{4\pi}{N}k} + \cdots$$
$$+ x(n-N)\, e^{-j\frac{2\pi}{N}Nk}\, e^{j\frac{2\pi}{N}k}$$

Using $e^{-j2\pi k} = 1$,

$$Z_{(n-1)}(k) = x(n-1) + x(n-2)\, e^{-j\frac{2\pi}{N}k} + x(n-3)\, e^{-j\frac{4\pi}{N}k} + \cdots + x(n-N+1)\, e^{j\frac{4\pi}{N}k} + x(n-N)\, e^{j\frac{2\pi}{N}k} \quad (11.11c)$$

Consider now Eqn. (11.11b),

$$Z_n(k) = x(n) + x(n-1)e^{-j\frac{2\pi}{N}k} + x(n-2)e^{-j\frac{4\pi}{N}k} + \cdots$$
$$+ x(n-N+2)\, e^{j\frac{4\pi}{N}k} + x(n-N+1)\, e^{j\frac{2\pi}{N}k}$$

$$Z_n(k) = x(n) + \underline{\underline{e^{-j\frac{2\pi}{N}k}}}$$
$$\left[x(n-1) + x(n-2)e^{-j\frac{2\pi}{N}k} + \cdots + x(n-N+2)\, e^{j\frac{6\pi}{N}k} + x(n-N+1)\, e^{j\frac{4\pi}{N}k} \right]$$

Add and subtract $x(n-N)\, e^{j\frac{2\pi}{N}k}$ to the bracketed term,

$$Z_n(k) = x(n) + e^{-j\frac{2\pi}{N}k} \left[x(n-1) + x(n-2)e^{-j\frac{2\pi}{N}k} + \cdots \right.$$
$$\left. + x(n-N+2)\, e^{j\frac{6\pi}{N}k} + x(n-N+1)\, e^{j\frac{4\pi}{N}k} + x(n-N)\, e^{j\frac{2\pi}{N}k} \right] - x(n-N)$$

$$Z_n(k) = + e^{-j\frac{2\pi}{N}k} \left[x(n-1) + x(n-2)e^{-j\frac{2\pi}{N}k} + \cdots + x(n-N+2)\, e^{j\frac{6\pi}{N}k} \right.$$
$$\left. + x(n-N+1)\, e^{j\frac{4\pi}{N}k} + x(n-N)\, e^{j\frac{2\pi}{N}k} \right] + x(n) - x(n-N)$$

Eqn. (11.11c) indicates that the term in the bracket is nothing but $Z_{(n-1)}(k)$, therefore,

$$z_n(k) = z_{(n-1)}(k)\, e^{-j\left(\frac{2\pi}{N}\right)k} + x(n) - x(n-N) \quad (11.11d)$$

Sliding DFT algorithm with zero padding

The time domain convolution of a L point sequence with another M point sequence produces an output whose length is $L+M-1$ points. The calculation of this convolution using the DFT can only correspond to the linear convolution if each of the transform is zero padded to a minimum length of $N = L + M - 1$. Therefore, sliding DFT with zero padding is of interest.

The sliding DFT is given by (Eqn.(11.11d)),

$$z_n(k) = z_{(n-1)}(k)e^{-j\left(\frac{2\pi}{N}\right)k} + x(n) - x(n-N)$$

Substituting for $z_{(n-1)}(k)$ using Eqn.(11.11a),

$$z_n(k) = \sum_{p=0}^{N-1} x(n-p) \; e^{-j\left(\frac{2\pi}{N}\right)pk}$$

$$z_n(k) = \sum_{p=0}^{N-1} x(n-p-1) \; e^{-j(p+1)\left(\frac{2\pi}{N}\right)k} + x(n) - x(n-N)$$

As signal is present only for $n = 0$ to $N/2 - 1$

$$z_n(k) = \sum_{p=0}^{N/2-1} x(n-p-1) \; e^{-j(p+1)\left(\frac{2\pi}{N}\right)k} + x(n) - x(n-N/2)$$

$$z_n(k) = x(n) + x(n-1)\,e^{-j\left(\frac{2\pi}{N}\right)k} + \cdots + x\left(n-\frac{N}{2}\right)e^{-j\pi k} - x\left(n-\frac{N}{2}\right)$$

$$z_n(k) = x(n) + x(n-1)\,e^{-j\left(\frac{2\pi}{N}\right)k} + \cdots + x\left(n-\frac{N}{2}\right)e^{-j\pi k} - x\left(n-\frac{N}{2}\right)$$

As $e^{-j\pi k} = (-1)^k$, the term $x\left(n-\frac{N}{2}\right)e^{-j\pi k}$ becomes $-x(n-N/2)$ for odd values of k and $x(n-N/2)$ for even values of k. The modified equation for computing sliding DFT for zero padded sequence is,

$$z_n(k) = z_{(n-1)}(k)e^{-j\left(\frac{2\pi}{N}\right)k} + x(n) - x\left(n-\frac{N}{2}\right), \qquad k = 0, 2, \ldots..N/2$$

$$z_n(k) = z_{(n-1)}(k)e^{-j\left(\frac{2\pi}{N}\right)k} + x(n) + x\left(n-\frac{N}{2}\right), \quad k = 1, 3, \ldots..(N/2)-1 \tag{11.11e}$$

11.7.2 Sliding DCT

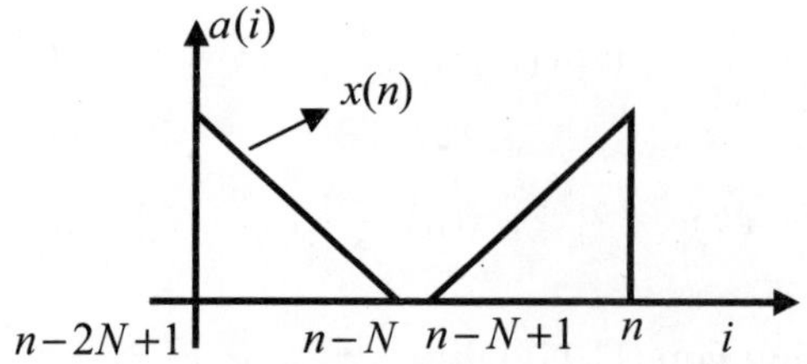

Fig. 11.5(a) Symmetrically extended function.

Consider a sequence of N samples denoted by $x(n), x(n-1),\ldots, x(n-N+1)$. Then an extended sequence $a(i)$ that is symmetric about the point $n-N+\frac{1}{2}$ (Fig. 11.5a) is given by

$$a(i)=\begin{cases} x(i), & i=n,n-1,\cdots,n-N+1 \\ x(-i+2n-2N+1), & i=n-N,n-N-1,\cdots,n-2N+1 \end{cases} \tag{11.12a}$$

Let

$$W_{2N}=\exp\left(\frac{-j2\pi}{2N}\right) \tag{11.12b}$$

The mth element of the $2N$-point DFT of $a(i)$ (Eqn.11.12a) is

$$A_m(n)=\sum_{i=n-2N+1}^{n} a(i)W_{2N}^{m(n-i)}$$

Splitting the summation into two ranges,

$$A_m(n)=\sum_{i=n-N+1}^{n} a(i)W_{2N}^{m(n-i)}+\sum_{i=n-2N+1}^{n-N} a(i)W_{2N}^{m(n-i)} \tag{11.12c}$$

Substituting for $a(i)$ from Eqn. (11,12a),

$$A_m(n)=\sum_{i=n-N+1}^{n} a(i)W_{2N}^{m(n-i)}+\sum_{j=n-2N+1}^{n-N} a(j)W_{2N}^{m(n-j)}$$

$$A_m(n)=\sum_{i=n-N+1}^{n} x(i)W_{2N}^{m(n-i)}+\sum_{j=n-2N+1}^{n-N} x(-j+2n-2N+1)W_{2N}^{m(n-j)} \tag{11.12d}$$

Consider the second sum and let $i = (-j + 2n - 2N + 1)$ then the limits will become

$$j = n - 2N + 1\,,\; i = -n + 2N - 1 + 2n - 2N + 1 = n$$

and

$$j = n - N\,,\; i = -n + N + 2n - 2N + 1 = n - N + 1$$

Therefore,

$$A_m(n) = \sum_{i=n-N+1}^{n} x(i) W_{2N}^{m(n-i)} + \sum_{i=n-N+1}^{n} x(i) W_{2N}^{m(i-n+2N-1)} \qquad (11.12e)$$

On multiplying and dividing the term in the first summation by $W_{2N}^{m\left(N-\frac{1}{2}\right)}$ and splitting the exponent in the second summation,

$$A_m(n) = \sum_{i=n-N+1}^{n} x(i)\, W_{2N}^{m(n-i)}\, W_{2N}^{m\left(N-\frac{1}{2}\right)}\, W_{2N}^{-m\left(N-\frac{1}{2}\right)}$$

$$+ \sum_{i=n-N+1}^{n} x(i) W_{2N}^{m(i-n)} \underline{W_{2N}^{m(2N-1)}}$$

Further, splitting the term $W_{2N}^{m(2N-1)}$ in the second summation,

$$A_m(n) = \sum_{i=n-N+1}^{n} x(i)\, W_{2N}^{m(n-i)}\, W_{2N}^{m\left(N-\frac{1}{2}\right)}\, W_{2N}^{-m\left(N-\frac{1}{2}\right)} + \sum_{i=n-N+1}^{n} x(i) W_{2N}^{m(i-n)} W_{2N}^{m\left(N-\frac{1}{2}\right)}\, W_{2N}^{m\left(N-\frac{1}{2}\right)}$$

Taking common factor $W_{2N}^{m\left(N-\frac{1}{2}\right)}$,

$$A_m(n) = W_{2N}^{m\left(N-\frac{1}{2}\right)} \left[\sum_{i=n-N+1}^{n} x(i)\, W_{2N}^{m(n-i)}\, W_{2N}^{-m\left(N-\frac{1}{2}\right)} + \sum_{i=n-N+1}^{n} x(i)\, W_{2N}^{m(i-n)}\; W_{2N}^{m\left(N-\frac{1}{2}\right)} \right]$$

$$A_m(n) = W_{2N}^{m\left(N-\frac{1}{2}\right)} \left[\sum_{i=n-N+1}^{n} x(i) \left\{ W_{2N}^{-m\left(i-n+N-\frac{1}{2}\right)} + W_{2N}^{m\left(i-n+N-\frac{1}{2}\right)} \right\} \right]$$

Since

$$W_{2N} = e^{-\frac{j2\pi}{2N}} \quad \text{and} \quad W_{2N}^{mN} = e^{-\frac{j\pi\, mN}{N}} = (-1)^m,$$

$$A_m(n) = 2(-1)^m\, W_{2N}^{m\left(N-\frac{1}{2}\right)} \left[\sum_{i=n-N+1}^{n} x(i) \left\{ \frac{\cos\big(m(i - n + N - 1/2)\big)\pi}{N} \right\} \right]$$

$$A_m(n) = 2(-1)^m W_{2N}^{-m/2}\left[\sum_{i=n-N+1}^{n} x(i)\left\{\frac{\cos\left(m(i-n+N-1/2)\right)\pi}{N}\right\}\right] \tag{11.12f}$$

The DCT of the sequence $x(n)$ at time n is given by

$$C_m(n) = k_m \sum_{i=n-N+1}^{n} x(i)\cos\left(\frac{m(i-n+N-1/2)\pi}{N}\right),$$

$$k_m = \begin{cases} 1/\sqrt{2}, & m=0 \\ 1, & otherwise \end{cases} \tag{11.12g}$$

Comparing this with the expression for $A_m(n)$ (Eqn. (11.12f)),

$$C_m(n) = \frac{1}{2}k_m(-1)^m W_{2N}^{m/2} A_m(n), \qquad m = 0,1,\ldots\ldots,N-1 \tag{11.12h}$$

Consider Eqn.(11.12e),

$$A_m(n) = \sum_{i=n-N+1}^{n} x(i)W_{2N}^{m(n-i)} + \sum_{i=n-N+1}^{n} x(i)W_{2N}^{m(i-n+2N-1)}, \quad W_{2N} = e^{-\frac{j2\pi}{2N}}$$

Let

$$A_m^{(1)}(n) = \sum_{i=n-N+1}^{n} x(i)W_{2N}^{m(n-i)}, \tag{11.12i}$$

$$A_m^{(2)}(n) = \sum_{i=n-N+1}^{n} x(i)W_{2N}^{m(i-n+2N-1)} \tag{11.12j}$$

Then

$$A_m(n) = A_m^{(1)}(n) + A_m^{(2)}(n)$$

In Eqn (11.12i), taking out the last term from the summation,

$$A_m^{(1)}(n) = x(n) + \sum_{i=n-N+1}^{n-1} x(i)W_{2N}^{m(n-i)} \tag{11.12k}$$

Also from Eqn. (11.12i), $A_m^{(1)}(n-1)$ is

$$A_m^{(1)}(n-1) = \sum_{i=n-N}^{n-1} x(i)W_{2N}^{m(n-i-1)}$$

Taking out the first term,

$$A_m^{(1)}(n-1) = W_{2N}^{m(n-n+N-1)} x(n-N) + \sum_{i=n-N+1}^{n-1} x(i) W_{2N}^{m(n-i-1)}$$

Since $W_{2N}^{mN} = e^{-\frac{j\pi}{N}mN} = (-1)^m$,

Therefore,

$$A_m^{(1)}(n-1) = (-1)^m W_{2N}^{m} x(n-N) + W_{2N}^{-m} \sum_{i=n-N+1}^{n-1} x(i) W_{2N}^{m(n-i)}$$

Multiplying by W_{2N}^{m}

$$A_m^{(1)}(n-1) W_{2N}^{m} = (-1)^m x(n-N) + \sum_{i=n-N+1}^{n-1} x(i) W_{2N}^{m(n-i)} \tag{11.12l}$$

Subtracting Eqn. (11.12*l*) from Eqn. (11.12k)

$$A_m^{(1)}(n) = A_m^{(1)}(n-1) W_{2N}^{m} + x(n) - (-1)^m x(n-N) \tag{11.12m}$$

Consider Eqn. (11.12j)

$$A_m^{(2)}(n) = \sum_{i=n-N+1}^{n} x(i) W_{2N}^{m(i-n+2N-1)}$$

Taking out the last term from the summation,

$$A_m^{(2)}(n) = x(n) W_{2N}^{m(2N-1)} + W_{2N}^{2mN} W_{2N}^{-m} \sum_{i=n-N+1}^{n-1} x(i) W_{2N}^{m(i-n)}$$

Since $W_{2N}^{2mN} = 1$,

$$A_m^{(2)}(n) = x(n) W_{2N}^{-m} + W_{2N}^{-m} \sum_{i=n-N+1}^{n-1} x(i) W_{2N}^{m(i-n)} \tag{11.13a}$$

From Eqn. (11.12j), $A_m^{(2)}(n-1)$ is

$$A_m^{(2)}(n-1) = \sum_{i=n-N}^{n-1} x(i) W_{2N}^{m(i-n+2N)} = \sum_{i=n-N}^{n-1} x(i) W_{2N}^{m(i-n)} \quad \because \; W_{2N}^{2mN} = 1$$

Taking the first term out of summation,

$$A_m^{(2)}(n-1) = x(n-N) W_{2N}^{mN} + \sum_{i=n-N+1}^{n-1} x(i) W_{2N}^{m(i-n)}$$

$$A_m^{(2)}(n-1) = (-1)^m x(n-N) + \sum_{i=n-N+1}^{n-1} x(i) W_{2N}^{m(i-n)} \tag{11.13b}$$

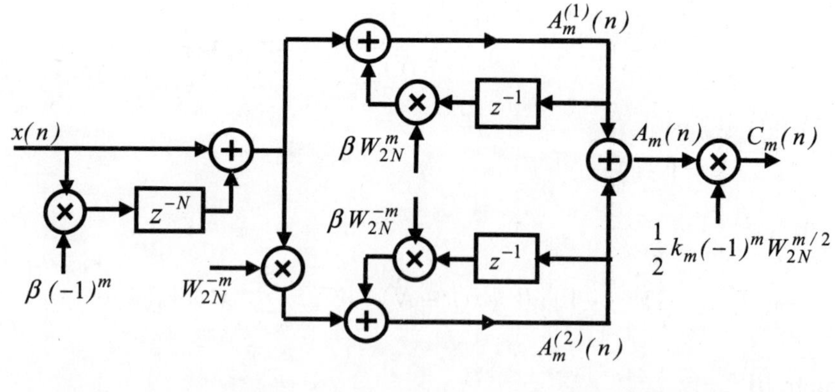

Fig. 11.5(b) Schematic of the implementation of the sliding DCT.

Multiplying Eqn. (11.13b) by W_{2N}^{-m},

$$A_m^{(2)}(n-1)W_{2N}^{-m} = (-1)^m x(n-N)W_{2N}^{-m} + W_{2N}^{-m}\sum_{i=n-N+1}^{n-1} x(i)W_{2N}^{m(i-n)} \tag{11.13c}$$

Subtracting Eqn.(11.13c) from Eqn. (11.13a),

$$A_m^{(2)}(n) - A_m^{(2)}(n-1)W_{2N}^{-m} = x(n)W_{2N}^{-m} - (-1)^m x(n-N)W_{2N}^{-m}$$

$$A_m^{(2)}(n) = A_m^{(2)}(n-1)W_{2N}^{-m} + W_{2N}^{-m}\left[x(n) - (-1)^m x(n-N)\right]$$

Consider,

$$A_m^{(1)}(n) = \beta A_m^{(1)}(n-1)W_{2N}^{-m} + x(n) - (-1)^m x(n-N)$$

and for $\beta = 1$, this will be same as Eqn.(11.12m)

$$A_m^{(1)}(n) = A_m^{(1)}(n-1)W_{2N}^{-m} + x(n) - (-1)^m x(n-N)$$

Taking z-transform

$$A_m^{(1)}(z) = A_m^{(1)}(z)z^{-1}W_{2N}^{-m} + X(z) - (-1)^m X(z)z^{-N}$$

$$A_m^{(1)}(z)\left(1 - z^{-1}W_{2N}^{-m}\right) = X(z) - (-1)^m X(z)z^{-N}$$

$$A_m^{(1)}(z) = \frac{X(z) - (-1)^m X(z) z^{-N}}{\left(1 - z^{-1} W_{2N}^{-m}\right)}$$

The roots of the denominator polynomial are given by

$$\left(1 - z^{-1} W_{2N}^{-m}\right) = 0 \qquad \left(1 - z^{-1} e^{-jm2\pi/2N}\right) = 0 \quad z = e^{-jm2\pi/2N} \qquad |\, z \,| = 1$$

Due to round-off errors, the poles may go out of unit circle making the filter unstable. Hence, a forgetting factor $0 < \beta < 1$ is used to ensure stability and hence $A_m^{(1)}(n)$ is given by

$$A_m^{(1)}(n) = \beta A_m^{(1)}(n-1) W_{2N}^{-m} + x(n) - (-1)^m x(n-N) \tag{11.13d}$$

The same is true for $A_m^{(2)}(n)$ hence, is given by

$$A_m^{(2)}(n) = \beta A_m^{(2)}(n-1) W_{2N}^{-m} + W_{2N}^{-m} \left[x(n) - (-1)^m x(n-N) \right] \tag{11.13e}$$

Implementation of these equations together and obtaining $C_m(n)$ are shown in Fig.(11.5b) as

$$C_m(n) = \frac{1}{2} k_m (-1)^m W_{2N}^{m/2} A_m(n), \qquad m = 0, 1, \ldots\ldots, N-1 \tag{11.13f}$$

where,
$$k_m = \begin{cases} 1/\sqrt{2}, & m = 0 \\ 1, & otherwise \end{cases}$$

$$A_m(n) = A_m^{(1)}(n) + A_m^{(2)}(n) \tag{11.13g}$$

11.8 TRANSFORM DOMAIN FORWARD-BACKWARD LMS ADAPTIVE PREDICTOR

The performance of an adaptive filter is governed by its misadjustment and convergence rate. The LMS forward - backward predictor was introduced in the previous chapter and the analysis indicated improved misadjustment by a factor of 2. However, its convergence speed is same as that of normal LMS algorithm as it is based on tapped delay line structure. Hence it is of interest by using the transform approach to improve the FBLMS predictor's convergence speed, which is already having improved misadjustment (A.O. Ogunfunmi and C. Pham, 1991).

Let the input vectors be (Fig. 11.6a)

$$X_{fN}^{T}(n) = \left[\ x(n-1),\ x(n-2) \ldots X(n-N)\right] \tag{11.14a}$$

$$X_{bN}^{T}(n) = \left[X(n-N+1),\ X(n-N+2), \ldots,\ x(n)\right] \tag{11.14b}$$

When the unitary/orthogonal transform M_{NN} operates on $X_{fN}(n)$ and $X_{bN}(n)$, let $X'_{fN}(n)$ and $X'_{bN}(n)$ be the respective resulting vectors given by

$$X'_{fN}(n) = M_{NN}\left[X^T_{fN}(n)\right] = \left[x'_{f1}(n)\ x'_{f2}(n)\ \ \ldots\ \ldots\ \ x'_{fN}(n)\right]^T \tag{11.14c}$$

$$X'_{bN}(n) = M_{NN}\left[X^T_{bN}(n)\right] = \left[x'_{b1}(n)\ x'_{b2}(n)\ \ \ldots\ \ldots\ \ x'_{bN}(n)\right]^T \tag{11.14d}$$

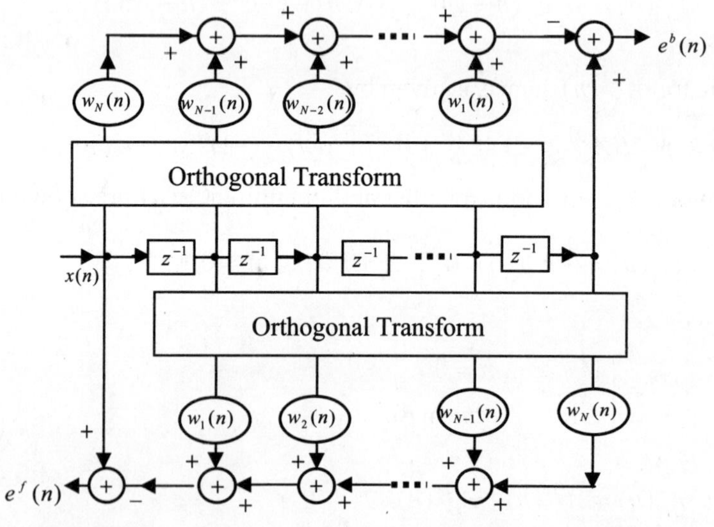

Fig. 11.6(a) Transform domain forward-backward predictor.

If the adaptive filter weight vector is

$$W_N(n) = \left[w_1(n)\quad w_2(n)\quad \ldots\ \ldots\quad w_N(n)\right]^T \tag{11.14e}$$

then the filter outputs $y_f(n)$ and $y_b(n)$ are given by

$$y_f(n) = W_N^{\,T}(n)X'_{fN}(n) = W_N^{\,T}(n)M_{NN}\left[X_{fN}(n)\right] \tag{11.14f}$$

$$y_b(n) = W_N^{\,T}(n)X'_{bN}(n) = W_N^{\,T}(n)M_{NN}\left[X_{bN}(n)\right] \tag{11.14g}$$

The gradient adaptation rule is given by

$$W_N(n+1) = W_N(n) - \mu\,\nabla\left[\ \varepsilon(n)\right]$$

The error energy to be minimized is

$$\varepsilon(n) = \frac{\left[e^f(n)\right]^2 + \left[e^b(n)\right]^2}{2}$$

where,

$$e^f(n) = x(n) - X_{fN}'^T W_N(n), \tag{11.14h}$$

$$e^b(n) = X_N(n-N) - X_{bN}'^T(n)\, W_N(n) \tag{11.14i}$$

Therefore the gradient to be considered is

$$\nabla\left[\ \varepsilon(n)\right] = \frac{1}{2}\left\{2e^f(n)\frac{\partial e^f(n)}{\partial W_N(n)} + 2e^b(n)e^f(n)\frac{\partial e^f(n)}{\partial W_N(n)}\right\}$$

$$\nabla\left[\ e^2(n)\right] = -\left\{e^f(n)X_{fN}'(n) + e^b(n)X_{bN}'(n)\right\}$$

Therefore, the LMS weight vector adaptation is

$$W_N(n+1) = W_N(n) + \mu\left\{\frac{e^f(n)X_{fN}'(n))}{P_f(n)} + \frac{e^b(n)X_{bN}'(n)}{P_b(n)}\right\} \tag{11.14j}$$

where, μ is the gradient stepsize parameter and

$$P_f(n) = [p_{f1}(n), p_{f2}(n), \ldots, p_{fN}(n)],\ P_b(n) = [p_{b1}(n), p_{b2}(n), \ldots, p_{bN}(n)]$$

The individual weight adaptation is given by

$$w_i(n+1) = w_i(n) + \mu\left\{\frac{e^f(n)x_{fi}'(n))}{p_{fi}(n)} + \frac{e^b(n)x_{bi}'(n)}{p_{bi}(n)}\right\} \tag{11.14k}$$

$$p_{fi}(n) = \beta\, p_{fi}(n-1) + (1-\beta)x_{fi}'^{\,2}(n),$$

$$p_{bi}(n) = \beta\, p_{bi}(n-1) + (1-\beta)x_{bi}'^{\,2}(n),$$

$$0 < \beta < 1$$

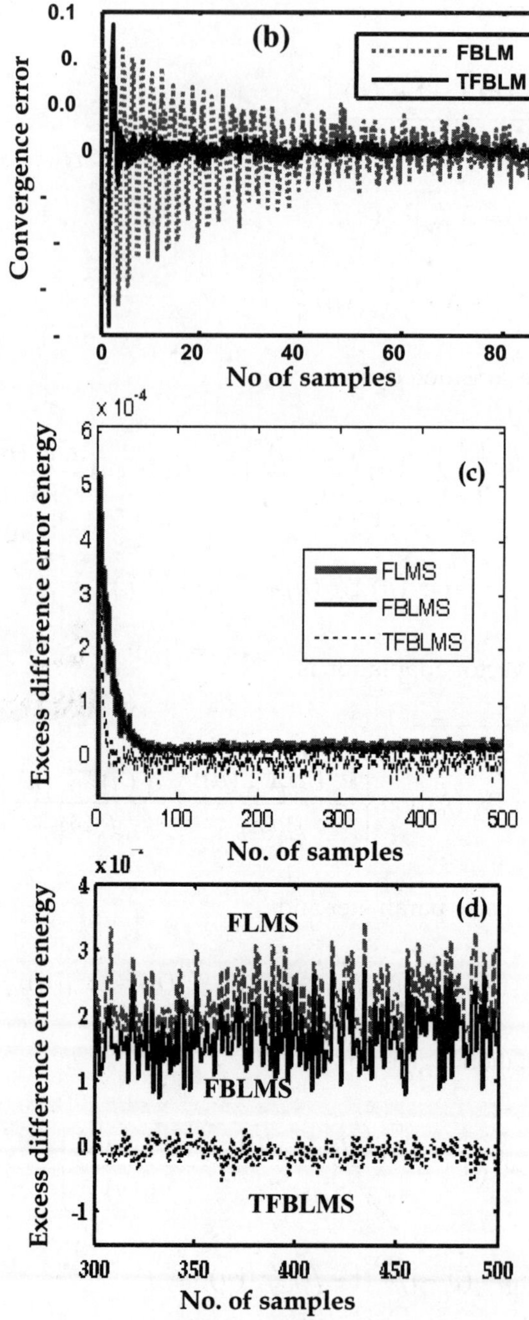

Fig. 11.6(b, c, d) (b) Comparison of error convergence for TBLMS and FBLMS; (c) Excess difference error energy over the optimal for different algorithms and (d) expanded view of last 200 points of (c).

The individual weights of the transform domain adaptive filter will converge faster than that of time domain case as the step size μ is normalized by the respective individual powers of the inputs to the single coefficient LMS adaptive filters than by the total power of the input signal prior to transformation which is significantly higher than the powers of transformed components. Also as it was shown that misadjustment of the transform domain and time domain adaptive filters are same, the misadjustment of forward-backward predictor both in transform domain and time domain will be same.

Thus, for the forward-backward predictor, the transform domain provides improved convergence rate and this is an added desirable feature to its already reduced misadjustment.

The performance of the TFBLMS predictor is illustrated for a signal consisting of three sinusoids 50Hz, 100Hz and 150Hz plus white Gaussian noise (SNR=21.8 dB). The parameters used are : filter length = 16, step size = 0.05 and forgetting factor for power estimation = 0.99 and number of samples =1000. It is seen that the TFBLMS converges significantly faster (8 times) than the time domain FBLMS algorithm (Fig.11.6b). Hence, the convergence error magnitude of FBLMS approaches that of TFBLMS, after it converges taking more time compared to that of TFBLMS. The convergence error magnitude of TFBLMS is slightly lower than that of the time domain FBLMS. The excess difference error energy over the optimal for TFBLMS algorithm is smaller than that of FBLMS (Fig. 11.6 (c, d)).

11.9 TRANSFORM DOMAIN ROBUST VARIABLE STEP SIZE GRIFFITHS' LMS ALGORITHM (TVGLMS)

To overcome the problem of trade-off between convergence rate and misadjustment the variable step size algorithm was introduced. Also the effect of the observation noise which is associated with the error on variable step size algorithm was also overcome to a significant extent by using the Griffiths LMS algorithm. Here, not only the scale factor for the step size but also the gradient, which are free from observation noise effect, are used for the LMS weight adaptation. This as stated was achieved by taking the cross-correlation between the input and the desired signal. Such a variable step size Griffiths' LMS algorithm (VGLMS) for a given maximum step size, has a slower convergence rate due to replacement of instantaneous correlation between input and error (desired signal) by the Griffiths averaged gradient. This motivates the application of VGLMS to algorithms, which have faster convergence rate for inputs with high eigenvalue spread. As the transform domain LMS algorithm is having a faster convergence rate even for inputs with large eigenvalue spread, the application of Griffiths' algorithm to this will not only provide additional convergence speed over that of time domain LMS algorithm but also a smaller misadjustment.

The basic transform domain LMS algorithm using discrete cosine transform is shown in Fig. 11.7a. For the input $x(n)$, the $2N$ point DCT of produces orthogonal components $X_k(n), k = 0, \dots, N-1$ (the remaining N points are symmetrical) and these form inputs to single coefficient adaptive filters $W_k(n), k = 0, \dots, N-1$. The output of these adaptive filters $y_k(n), k = 0, \dots, N-1$ are summed (with a gain factor of 2) to take care of symmetrical components) on sample to sample basis to get the estimate $y(n)$ of the desired signal $d(n)$.

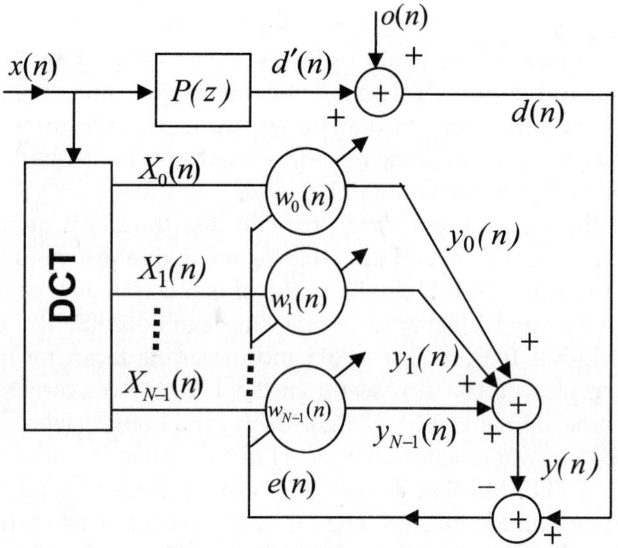

Fig. 11.7(a) DCT based transform domain LMS adaptive filter for system identification.

The error signal $e(n) = d(n) - y(n)$ is used for the adaptation of the weights $W_k(n),\ k = 0, \ldots, N-1$ using the normalized LMS algorithm as

$$w_k(n+1) = w_k(n) + \frac{2\mu(n)}{\sigma_k^2(n)} e(n) X_k(n), \quad k = 0,1,2, \ldots M-1 \tag{11.15a}$$

$$\sigma_k^2(n) = \beta\, \sigma_k^2(n-1) + (1-\beta) X_k^2(n) \qquad 0 < \beta < 1 \tag{11.15b}$$

$\mu(n)$ is the step size used for the adaptation. $\sigma_k^2(n)$ is the input power to adaptive filter $w_k(n),\ k = 0, \ldots, M-1$ and is estimated recursively. The transform domain provides faster convergence compared to transversal filter since

$$\sigma_k^2(n) << \sum_{k=0}^{M-1} \sigma_k^2(n)$$

and the gradient step size will be relatively large for the former. However, due to large step size the convergence error of the transform domain LMS can be larger than that for transversal filter. Since $e(n) = d(n) - y(n)$, Eqn. (11.15a) can be written as

$$w_k(n+1) = w_k(n) + \frac{2\mu(n)}{\sigma_k^2(n)} \left[d(n) - y(n) \right] X_k(n)$$

Further, as

$$d(n) = d'(n) + o(n),$$

$$w_k(n+1) = w_k(n) + \frac{2\mu(n)}{\sigma_k^2(n)}\left[d'(n) - y(n)\right]X_k(n) + \frac{2\mu}{\sigma_k^2(n)}o(n)X_k(n) \tag{11.15c}$$

The adaptation for the weight $w_k(n)$ gets affected by the last term on the right-hand side of Eqn.(11.15c). Applying the Griffiths' algorithm to this case,

$$w_k(n+1) = w_k(n) + \frac{2\mu(n)}{\sigma_k^2(n)}\left[G_k(n) - y(n)X_k(n)\right] \tag{11.15d}$$

where,

$$G_k(n) = E[d(n)X_k(n)]$$

$$= E[d'(n)X_k(n)] + E[o(n)X_k(n)], \qquad k = 0, \ldots, M-1$$

$E[o(n)X_k(n)] = 0$ as the observation noise $o(n)$ and input component $X_k(n)$ are uncorrelated. Hence use of Griffiths gradient as in Eqn.(11.15d) will remove the effect of observation noise on the weight adaptation process. Further, VGLMS can be used for reducing the step size as converges approaches. Defining

$$Q_k(n) = \left[G_k(n) - y(n)X_k(n)\right], \qquad k = 0, \ldots, M-1 \tag{11.15e}$$

$Q_k(n)$ is Griffiths cross-correlation in the context of transform domain LMS algorithm.

Further, $G_k(n)$ is estimated recursively as

$$\hat{G}_k(n+1) = \beta_g\, \hat{G}_k(n) + \left(1 - \beta_g\right) d(n)X_k(n), \quad 0 < \beta_g < 1 \tag{11.15f}$$

Similar to VGLMS algorithm, the step size $\mu(n)$ can be adapted as

$$\mu(n) = \alpha\, \mu(n-1) + \gamma \sum_{k=0}^{M-1} Q_k^2(n) \qquad 0 < (\alpha, \gamma) < 1 \tag{11.15g}$$

(S. Roopa and S.V. Narasimhan, on December 2014)

11.9.1 TVGLMS with Desired Signal Decomposition (DTVGLMS)

For adaptation the weights $w_k(n)$, if the desired signal is orthogonally decomposed by transform, then individual errors $e_k(n)$ will be directly available . Use of these errors $e_k(n)$ is advantageous to that of overall error $e(n)$. This is because, the overall error $e(n)$ will contain not only the required error component but also other error components, which act as observation noise for the weight adaptation under consideration. This not only facilitates faster convergence but also smaller convergence error/ misadjustment. In view of this, the application of VGLMS algorithm to the case where the desired signal is decomposed is of interest.

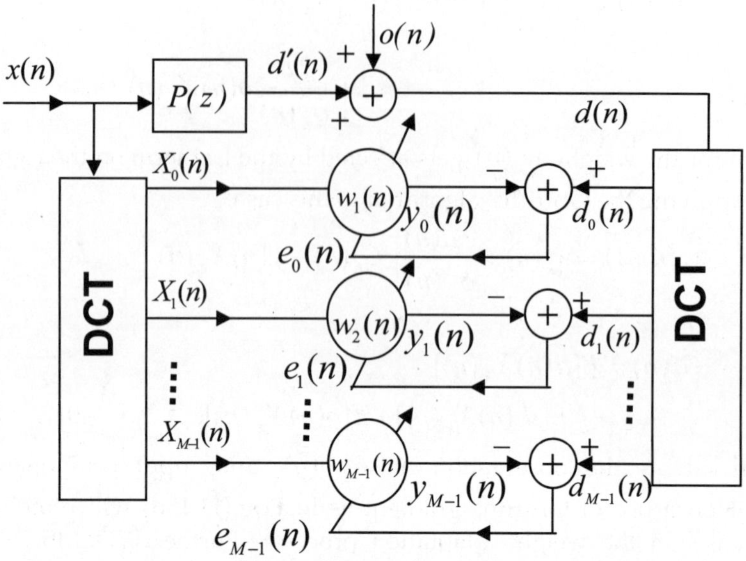

Fig. 11.7(b) DCT based transform domain LMS adaptive filter for system identification with desired signal decomposition.

Fig. 11.7(b) shows the schematic for TVGLMS with desired signal decomposition. Here $d(n)$ is decomposed by another DCT of same number of points $(2N)$ to get the components $d_k(n), k = 0, \ldots, N-1$. Further, the outputs of the adaptive filters $y_k(n)$ are not summed but are remained to get the individual errors $e_k(n)$ as

$$e_k(n) = d_k(n) - y_k(n), \; k = 0, \ldots, M-1$$

The TVGLMS algorithm (Eqns. (11.15d), (11.15e) and)(11.15f) for this case is

$$w_k(n+1) = w_k(n) + \frac{2\mu(n)}{\sigma_k^2(n)}\left[G_k(n) - y_k(n)X_k(n) \right] \tag{11.15h}$$

$$Q_k(n) = \left[G_k(n) - y_k(n)X_k(n) \right], k = 0, \ldots, M-1 \tag{11.15i}$$

$$\hat{G}_k(n+1) = \beta_g \, \hat{G}_k(n) + \left(1 - \beta_g\right) d_k(n)X_k(n), \quad 0 < \beta_g < 1 \tag{11.15j}$$

The step size $\mu(n)$ is adapted as

$$\mu(n) = \alpha \, \mu(n-1) + \gamma \sum_{k=0}^{M-1} Q_k^2(n) \;\; 0 < \left(\alpha, \gamma\right) < 1 \tag{11.15k}$$

This TVGLMS algorithm using components of the desire signal in addition to the components of the input signal, will be referred to as DTVGLMS algorithm. Both TVGLMS (Eqns.(11.15d-g)) and DTVGLMS (Eqns. (11.15h-k)) algorithms, not only use a gradient but also use a variable step size which are robust to observation noise.

For both algorithms, a high value of γ enables better tracking for non-stationary signals. But for stationary signals, a small γ is acceptable. Further, the noise-free gradient and step size, significantly improve the convergence performance. The DTVGLMS algorithm may perform better than that of TVGLMS, due to use of appropriate errors for the individual weight adaptations. (S.Roopa and S.V.Narasimhan, December 2013, on online).

11.9.2 Transform Domain Variable Step Size Algorithm (TDVSS)

Here, an algorithm which uses the local error energy to achieve variable step size adaptation is considered (Radu Ciprian Bilcu, Pauli Kuosmanen and Karen Egiazarian, 2002). In this algorithm, output error energy is used to update the stepsize of each filter tap resulting in a significant improvement of the convergence speed. The step size adaptation is given by

$$A(n) = \alpha\, \mu(n) + \frac{\gamma}{L} \sum_{i=n-L+1}^{n} e(i)^2$$

$$\mu(n+1) = \begin{cases} A(n),\ if\ n = pL\ and\ A(n)\ \varepsilon\ (\mu_{\min}, \mu_{\max}) \\ \mu_{\max},\ if\ n = pL\ and\ A(n) \geq \mu_{\max} \\ \mu_{\min},\ if\ n = pL\ and\ A(n) \leq \mu_{\min} \\ \mu(n),\ if\ n \neq pL \end{cases}, \quad p = 1, 2, 3, \ldots \qquad (11.15l)$$

For a large magnitude output error $e(n)$, the step size takes a higher value so that the convergence speed is increased to reach the convergence quickly. But for a smaller magnitude output error, the step size takes a smaller step size value and this results in a smaller misadjustment. The value of parameter L in Eqn. (11.15l) must be sufficiently large so that the behaviour of the error is obtained by the averaging process. But at the same time the value of L must be small enough to have a sufficient magnitude step size for adaptation. Typical value of L can be 10 and the choice of value of γ should be small to meet the misadjustment. Further, the update of filter coefficient $w_k(n)$ is given by

$$w_k(n+1) = w_k(n) + \mu_k(n) e(n) X_k(n)$$

$$\mu_k(n) = \frac{\mu(n)}{M\sigma_k^2(n)}, \qquad k = 0, 1, 2, \ldots, N-1$$

The parameter $\sigma_k^2(n)$ is the power of the transform coefficient $X_k(n)$ of the input $x(n)$ (Fig. 11.7(a)) and is given by

$$\sigma_k^2(n) = \beta\, \sigma_k^2(n-1) + (1-\beta) X_k^2(n),\ 0 < \beta < 1$$

In this TDVSS algorithm as no provision is made to suppress the effect of observation noise, its performance will be poor and even may diverge when the noise levels are high compared to the system output.

Example: The performance of the proposed algorithms is illustrated for the system identification. The performance index used is the smoothed ensemble average square error (SEASE) expressed in dB, $\xi_{Sd}(n)$

$$e_i(n) = d_i'(n) - y_i(n), \quad \xi(n) = \frac{1}{K}\sum_{i=1}^{K} e_i^2(n), \quad \xi_{Sd}(n) = 10\log_{10}(\xi_S(n))$$

$\xi_S(n)$ is smoothed by a 5-point moving average filter, $e_i(n)$ is the error sequence for the ith trial and K is the number of trials used in computing $\xi(n)$.

In computing the SEASE performance index, the presence of observation noise in the error obscures the actual reduction in error. Hence, to overcome this, the observation noise is removed from the error as it is known in the simulation study.

The plant chosen is an acoustic path of impulse response length $N = 128$. The DCT used is of $2N$ points and $K = 50$. The observation noises considered are Gaussian white (SNR=-8dB) and colored noise (SNR= -4dB). The colored noise is generated by filtering white noise by a transfer function

$$H(z) = \frac{1}{1 + z^{-1} - 0.8z^{-2}}$$

The input used is sum of three sinusoids of frequencies 50, 100 and 150 Hz and additive random noise (SNR =48dB). The parameters used for TVGLMS and DTVGLMS are:

$\mu = 0.001$, $\mu_{\max} = 0.001$ (This is the maximum permissible step size value without divergence), $\mu_{\min} = 0.00001$, forgetting factor $\beta = 0.99$, $\alpha = 0.9999$, $\sigma = 10^{-11}$, $\beta_g = 0.99$.

The parameters used for the TDVSS are: $\mu = 0.001, \mu_{\max} = 0.001, \mu_{\min} = 10^{-5}$, $\beta = 0.99, \alpha = 0.9999$, $L = 10$, $\gamma = 10^{-8}$.

For white observation noise, the performance index SEASE of the TVGLMS and DTVGLMS algorithms is reduced approximately by 12dB and 20dB, respectively over that of the TDVSS algorithm (Fig. 11.8 (a)).

For colored observation noise, the performance index SEASE of the proposed TVGLMS algorithm without and with desired signal decomposition is reduced approximately by 7dB and 14 dB, respectively over that of the TDVSS algorithm (Fig. 11.8(b)). However for the white observation noise, the convergence rate of the proposed algorithms is somewhat reduced due to the averaging process involved in gradient computation compared to that of existing TDVSS. But for the colored observation noise, the proposed algorithms have better convergence rate than that of TDVSS and this may be due to the severe adaptation disturbance of the colored noise compared to that of white noise. It is important to note that the convergence rate of DTVGLMS is faster compared to that of TVGLMS.

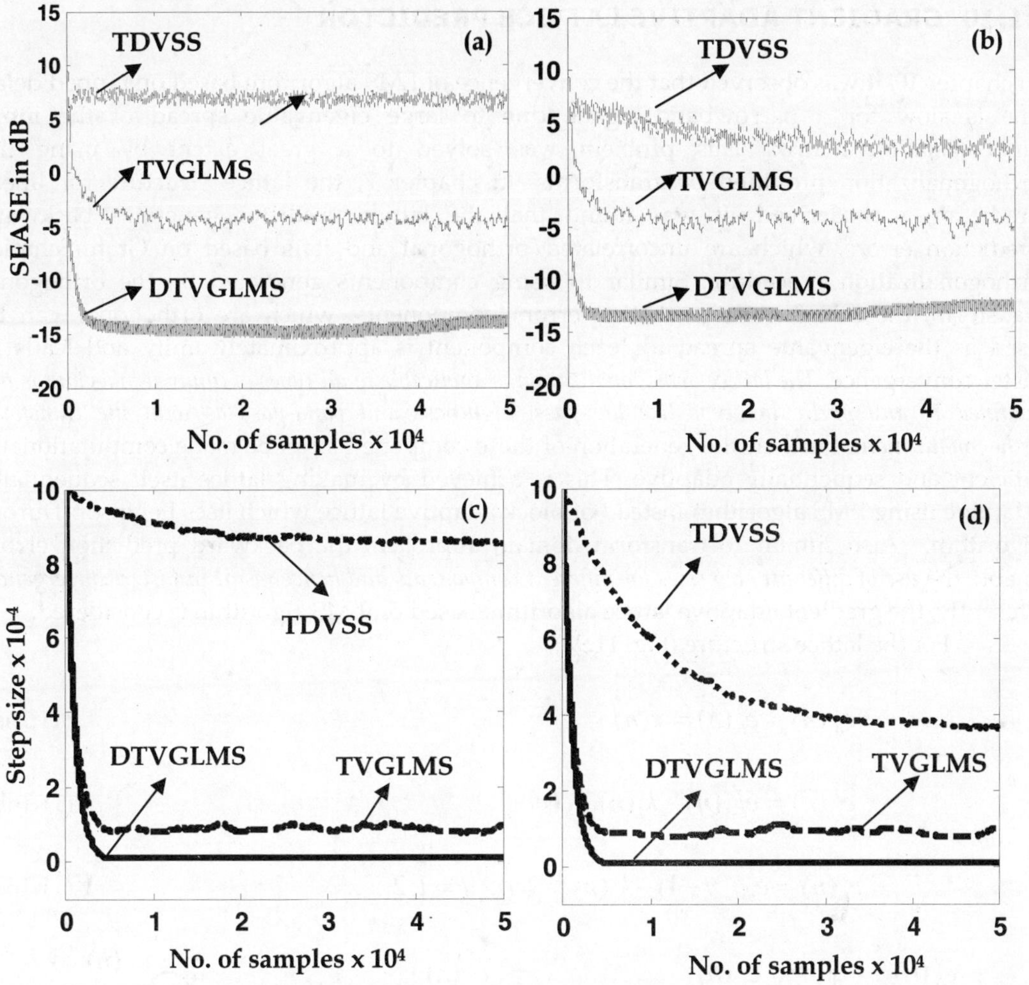

Fig. 11.8. SEASE for observation (a) white noise, (b) colored noise. Variation of step size for (c) white noise, (d) colored noise.

The step size variation for DTVGLMS, TVGLMS and TDVSS for white and colored observation noise are shown in Figs. 11.8(c) and 11.8(d), respectively. To have a faster convergence it is reasonable to start the step size from a large value and for a smaller misadjustment after convergence, the step size has to be reduced. This concept has been followed in the simulation study in getting the above results. It is seen that the step size variation from large value to a small value is faster for the proposed algorithms than that for TDVSS. Further the final step size value for the DTVGLMS is smaller than that for TVGLMS. Also these step size after convergence of the proposed algorithms are significantly lower than that for TDVSS and this supports their lower values of SEASE (misadjustment).

11.10 GRADIENT ADAPTIVE LATTICE PREDICTOR

In chapter 10, it was observed that the convergence of LMS algorithm based on tapped delay line is slow for a narrowband signal, due to large eigenvalue spread of the input autocorrelation matrix. This problem was solved to a great extent by using the orthogonalization provided by transforms. In chapter 7, the lattice structure for linear prediction was derived. It was found that the lattice predictor generates backward prediction errors, which are uncorrelated/orthogonal and it is based on Gram-Schmidt orthogonalization procedure. Similar to using components generated by the orthogonal transform, these backward prediction errors/components, which are orthogonal, can be used as the eigenvalue spread for each component is approximately unity and leads to faster convergence. *The lattice orthogonalization is applicable to all types of input signals and is not confined to any particular type like low-pass, high-pass and band-pass as with the transform orthogonalization.* Further, the generation of these components can be made computationally efficient and sequentially adaptive. This is achieved by making lattice itself sequentially adaptive using LMS algorithm instead of block adaptive lattice which uses Levinson-Durbin algorithm. Also similar to transform domain approach, the backward prediction errors enable the *use of different step sizes for different components and hence result in faster convergence.* Presently, the gradient adaptive lattice algorithm based on LMS algorithm is considered.

For the lattice structure (Fig. 11.9a),

$$e_0^f(n) = e_0^b(n) = x(n) \tag{11.16a}$$

$$e_l^f(n) = e_{l-1}^f(n) - k_l(n)e_{l-1}^b(n-1) \tag{11.16b}$$

$$e_l^b(n) = e_{l-1}^b(n-1) - k_l(n)e_{l-1}^f(n), \quad l = 1,2,\ldots,N \tag{11.16c}$$

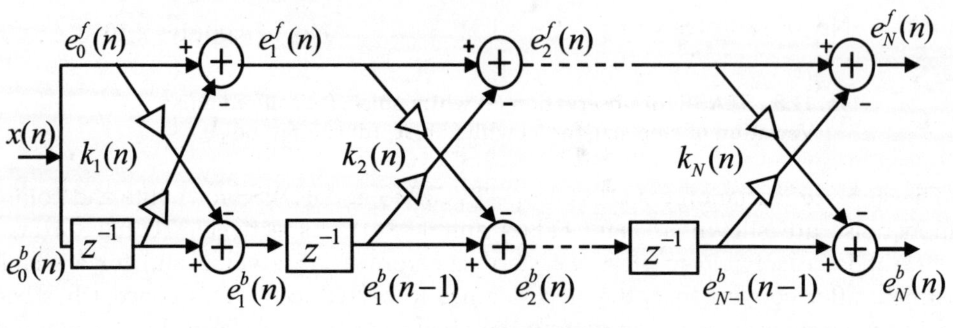

Fig. 11.9(a) Lattice predictor.

At n time and l^{th} stage, $e_l^f(n)$ and $e_l^b(n)$ are the forward and backward prediction errors, respectively and $k_l(n)$ is the corresponding reflection coefficient. The prediction filter coefficients $a_l(n), l = 1, 2, \ldots, N$ are related to reflection coefficients $k_l(n)$ by Levinson recursion relation as

$$a_0^{(0)} = 1$$

$$a_l^{(p)} = a_l^{(p-1)} + K_l \, a_{(p-l)}^{(p-1)} \qquad l = 1, \cdots\cdots, p-1 \tag{11.16d}$$

$$a_p^{(p)} = K_p$$

This is repeated for $p = 1, \cdots\cdots, N$. Hence, the prediction filter coefficients $a_l^{(N)}(n), l = 1, 2, \ldots, N$ are determined by the reflection coefficients $k_l, l = 1, 2, \ldots. N$. Therefore, the prediction filter coefficients are obtained by minimising the average of the square of the forward and backward prediction errors at l^{th} stage with respect to the reflection coefficient, $k_l(n)$ using the LMS algorithm. If

$$\varepsilon_l(n) = \frac{1}{2}\left[\left\{e_l^f(n)\right\}^2 + \left\{e_l^b(n)\right\}^2\right]$$

Using Eqns. (11.16b) and (11.16c)

$$\varepsilon_l(n) = \frac{1}{2}\left\{\left[e_{l-1}^f(n) - k_l(n)e_{l-1}^b(n-1)\right]^2 + \left[e_{l-1}^b(n-1) - k_l(n)e_{l-1}^f(n)\right]^2\right\}$$

The instentaneous gradiant of $\varepsilon_l(n)$ with respect to reflection coefficients $k_l(n)$ is

$$\nabla \varepsilon_l(n) = \frac{1}{2}\left\{2\left[e_{l-1}^f(n) - k_l(n)e_{l-1}^b(n-1)\right]\left[-e_{l-1}^b(n-1)\right] + 2\left[e_{l-1}^b(n-1) - k_l(n)e_{l-1}^f(n)\right]\left[-e_{l-1}^f(n)\right]\right\}$$

$$\nabla \varepsilon_l(n) = -\left[e_l^f(n)e_{l-1}^b(n-1) + e_l^b(n)e_{l-1}^f(n)\right] \tag{11.16e}$$

The reflection coefficient is adapted as

$$k_l(n+1) = k_l(n) - \alpha_l(n)\nabla \xi_l(n)$$

$$k_l(n+1) = k_l(n) + \alpha_l(n)\left[e_l^f(n)e_{l-1}^b(n-1) + e_l^b(n)e_{l-1}^f(n)\right] \tag{11.16f}$$

where,

$$\alpha_l(n) = \frac{\alpha}{E_{l-1}(n)}, \quad 0 < \alpha < 1 \tag{11.16g}$$

$$E_l(n) = \gamma \; E_l(n-1) + (1-\gamma)\left[f_l^2(n) + b_l^2(n-1)\right], 0 < \gamma << 1 \tag{11.16h}$$

$\alpha_l(n)$ and $E_l(n)$ are the step size and the power estimate of the input, for the lth stage and at n^{th} time instant, respectively. The input power for successive stages will reduce monotonically *as each stage absorbs one component of the signal and reducing the power of that component for the succeeding stage.* As a consequence, *the step sizes used for adapting the reflection coefficients at different stages are different.* The *later stages converge faster than the earlier stages.* This results in a faster overall convergence.

The prediction estimate can be achieved by summing the scaled backward prediction errors by their respective reflection coefficients as explained in chapter 7 (section-7.7.2) and is given by (Eqn. (7.31c)),

$$\hat{x}(n) = \sum_{i=1}^{p} k_i(n) e_{i-1}^b(n-1) \tag{11.16i}$$

The above lattice structure is based on one step prediction. This is valid when the signal is noise free. However, if the signal is associated with white/coloured noise, the prediction should filter the noise. This is achieved by exploiting the difference in correlation properties of noise and signal. Such an arrangement for the tapped delay line and for lattice structures (Fig. 11.9(b, c)) and their details will be considered.

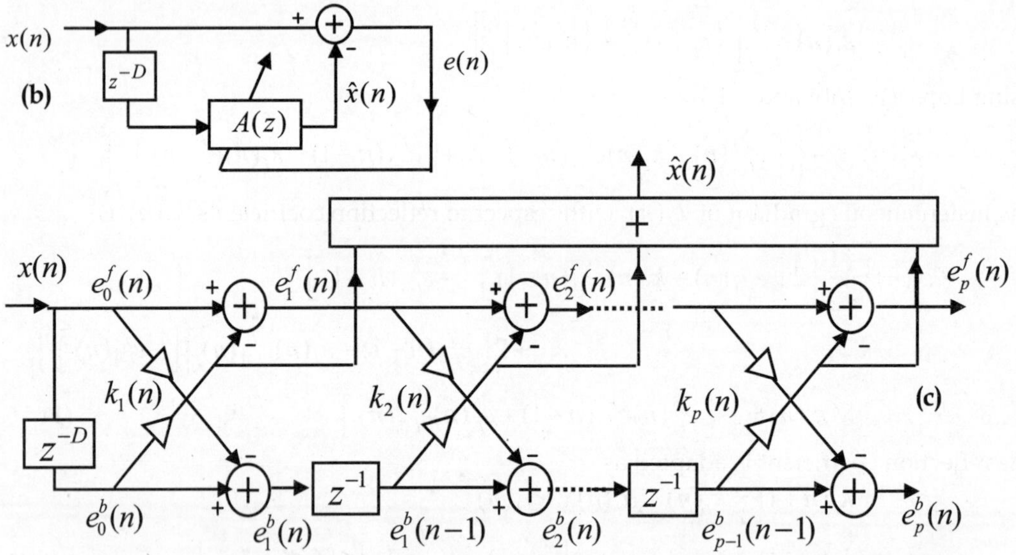

Fig. 11.9 (b, c) (b)Transversal predictor with delay and (c) Lattice predictor with delay.

As $x(n) = s(n) + v(n)$, $s(n)$ is signal and $v(n)$ is noise, the tapped delay line adaptive predictor output is given by

$$\hat{x}(n) = \sum_{i=1}^{p} a_i x(n-i-D+1) = \sum_{i=1}^{p} a_i s(n-i-D+1) + \sum_{i=1}^{p} a_i v(n-i-D+1) \tag{11.16j}$$

For the adaptive filter W_N as $W_N = R_{NN}^{-1} P_N$, it is necessary that there should be correlation (P_N) between the input and the desired signal. If the correlation length of the noise $v(n)$ is less than D (less self correlated), then the samples of $v(n)$ which are D samples away from n, do not contribute to $\hat{x}(n)$, the estimate at instant n and hence the second summation in Eqn. (11.16j) is approximately zero. *Here, the samples, which are D samples away* *from the reference input for the estimation of* $s(n)$. *Since there is no correlation between* $x(n)$ *and noise samples* $\{v(n-D), v(n-D-1),\ldots v(n-D-P)\}$ *and as there is correlation between* $x(n)$ *and the samples* $\{s(n-D), s(n-D-1),\ldots s(n-D-P)\}$, *only the signal samples contribute to signal estimate.* Hence, the estimate $\hat{x}(n)$ is mainly due to the signal $s(n)$ and is given by

$$\hat{x}(n) \approx \sum_{i=1}^{p} a_i s(n-i-D+1) \tag{11.17a}$$

Hence the error $e(n)$ is

$$e(n) = x(n) - \hat{x}(n) \tag{11.17b}$$

The adaptation of the weights $a_k(n)$ is

$$a_k(n+1) = a_k(n) + \frac{\alpha}{E(n)} e(n)x(n-k-D+1) \tag{11.17c}$$

$$E(n+1) = \beta E(n) + (1-\beta)x^2(n-k-D+1) \quad 0 < \gamma << 1 \tag{11.17d}$$

For the lattice predictor, the delay is introduced as in Fig.11.9c and the relations for the lattice remain same except for

$$e_0^f(n) = x(n), \quad e_0^b(n) = x(n-D) \tag{11.17e,f}$$

The actual filter coefficients $a_l(n), l = 1, 2, \ldots, p$ can be obtained by

$$\begin{aligned} a_l^p &= a_l^{(p-1)} + K_p\, a_{(p-l)}^{*(p-1)} \qquad l = 1, \cdots\cdots, p \\ a_l^l &= K_l \qquad l = 1, 2, \ldots. p \\ a_l^l &= K_l \end{aligned} \tag{11.17g}$$

and the spectrum of the predicted signal can be computed by

$$S_x(\omega) = \frac{1}{|A(z)|^2} = \frac{1}{\left|1 + \sum_{l=1}^{p} a_l^p\, e^{-j\omega l}\right|^2} \tag{11.17h}$$

Example: The prediction of a signal (one step) consisting of three sinusoids at 100Hz, 200Hz and 300Hz plus zero mean Gaussian noise SNR =26 dB with a sampling rate of 1000 Hz, is considered. For this, the order of the predictor used is 12, the step size used is 0.001 and the forgetting factor for computation of input powers used is 0.99.

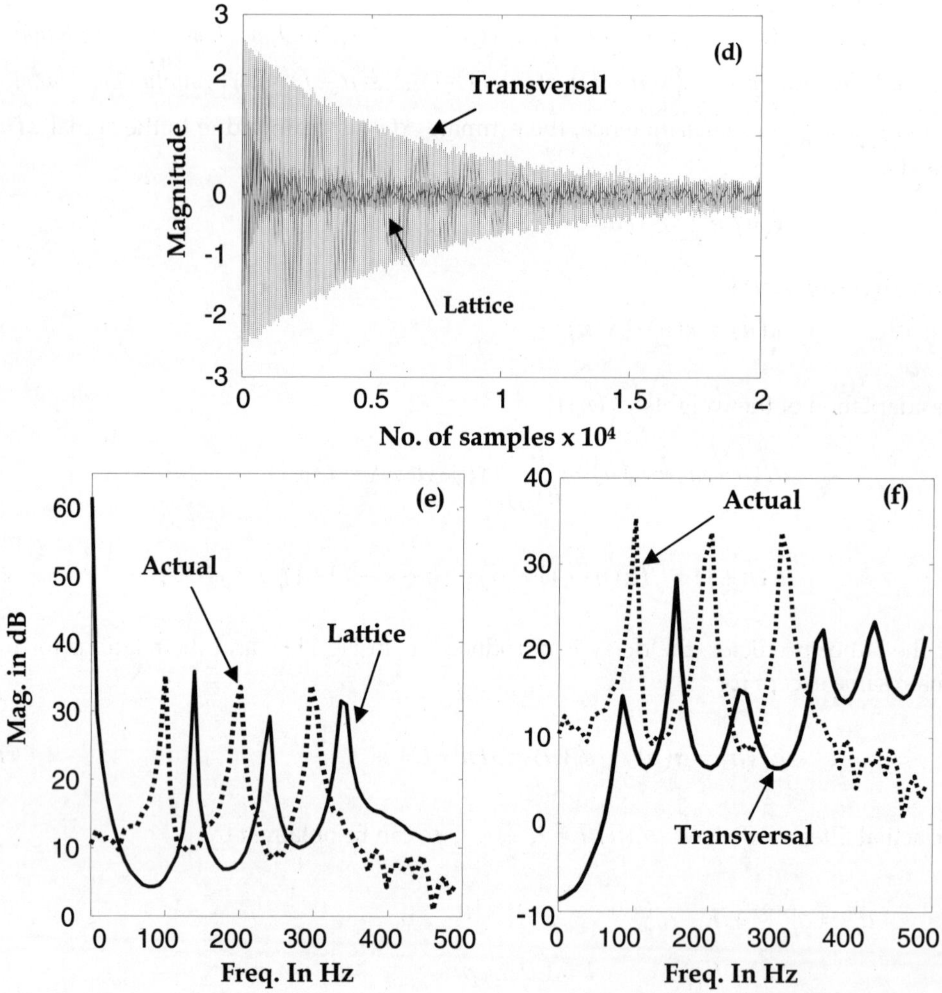

Fig. 11.9(d, e, f): (d) Error convergence of transversal and lattice predictors, Spectrum computed after convergence from the prediction coefficients for (e) lattice and (f) for transversal predictors.

The prediction estimate is directly calculated using Eqn. (11.16i). It is seen that the prediction error obtained by the lattice converges approximately 3 times faster than that by transversal filter (Fig. 11.9d). This is due to the fact that reflection coefficients of the lattice later stages converge faster whereas for the transversal filter all the coefficients converge at the minimum rate decided by step size and its complete input power. There can be situations where the final error magnitude of the lattice can be more than that of transversal due to use of increased step size for later stages of lattice.

The spectrum computed from the prediction coefficients derived from reflection coefficients of the lattice and that from transversal prediction coefficients, are shown in Figs. 11.9(e, f). The prediction coefficients correspond to the last sample (20000th) of the data considered and by this time convergence has taken place. It is seen that the spectrum derived from lattice shows the three spectra peaks clearly even though their frequencies are not accurate. However, the transversal filter not only shows spurious peaks but also the peaks corresponding to input are not brought out well above the noise level.

The prediction filtering of a noisy signal which involves decorrelation delay (more than one step) will be dealt in chapter 17 as an application.

11.10.1 Joint Process Lattice Adaptive Filter (for System Identification)

It has been shown in Chapter 7 that the backward prediction errors are derived by Gram-Schmidt orthogonalization of the input and are given by

$$\begin{bmatrix} e_0^b(n) \\ e_1^b(n) \\ \vdots \\ e_{N-1}^b(n) \end{bmatrix} = \begin{bmatrix} 1 & 0 & \cdots & 0 \\ a_1(1) & 1 & \cdots & 0 \\ \vdots & \vdots & & \vdots \\ a_{N-1}(N-1) & a_{N-1}(N-2) & \cdots & 1 \end{bmatrix} \begin{bmatrix} x(n) \\ x(n-1) \\ \vdots \\ x(n-N+1) \end{bmatrix} \qquad (11.18a)$$

$$e^b(n) = LX(n)$$

where,

$$X(n) = \left[x(n), x(n-1), \ldots, x(n-N+1)\right]^T, \qquad (11.18b)$$

$$e^b(n) = \left[e_0^b(n), e_1^b(n), \ldots, e_{N-1}^b(n)\right]^T \qquad (11.18c)$$

And $$L = \begin{bmatrix} 1 & 0 & \cdots & 0 \\ a_1(1) & 1 & \cdots & 0 \\ \vdots & \vdots & & \vdots \\ a_{N-1}(N-1) & a_{N-1}(N-2) & \cdots & 1 \end{bmatrix} \qquad (11.18d)$$

Since the matrix L is invertible as it is a lower triangular matrix and the inverse is also a lower triangular matrix,

$$X(n) = L^{-1}e^b(n) \qquad (11.18e)$$

The estimate of the desired signal, $y(n)$ which is the output of a filter, can be written as

$$y(n) = C^T X(n) \qquad (11.18f)$$

where, C is the filter weight vector given by

$$C = \begin{bmatrix} c_0 & c_1 & \cdots & c_{N-1} \end{bmatrix}^T \tag{11.18g}$$

Substituting for $X(n)$ from Eqn. (11.18e)

$$y(n) = C^T L^{-1} e^b(n) \tag{11.18h}$$

Defining

$$W = L^{-T} C, \tag{11.18i}$$

$$y(n) = W^T e^b(n) \tag{11.18j}$$

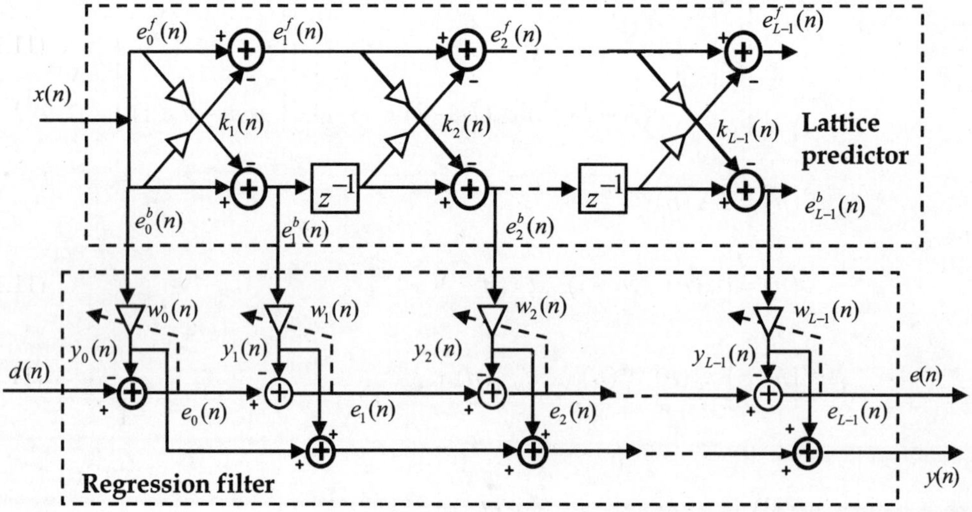

Fig. 11.9(g) Joint process lattice estimator.

This indicates that the output $y(n)$ of the transversal filter C for the input $X(n)$ can also be obtained as the linear combination of the backward prediction errors $e^b(n)$ of the input $x(n)$ instead of the input $x(n)$ itself using the weight vector $W = L^{-T}C$. Such a structure is referred to as the *joint process estimator* as two processes are involved. The first *being* the input process and the second is the orthogonal backward prediction error process generated (converted) by the *lattice predictor* (using the former) which forms input to the *regression filter/ linear combiner* W . The regression filter produces the estimate of the desired output by linearly combining the backward prediction errors.

For the regression filter (Fig. 11.9g), the output estimate is

$$y(n) = \sum_{l=0}^{L-1} w_l(n) e_l^b(n) \tag{11.18k}$$

Therefore, the error $e(n)$ is

$$e(n) = d(n) - y(n) \tag{11.18l}$$

and the LMS adaptation of the weights of the regression filter is given by

$$w_l(n+1) = w_l(n) + \frac{\mu_u}{\hat{G}_l(n)} b_l(n) e_l(n)\,,\ l = 0,1,\ldots\ldots\ldots, L-1 \tag{11.18m}$$

$$e_0(n) = d(n) - y_0(n) = d(n) - w_0(n) e_0^b(n) \ = d(n) - w_0(n) x(n)$$

$$e_l(n) = e_{l-1}(n) - y_l(n) = e_{l-1}(n) - w_l(n) e_l^b(n), \quad l = 1, 2, \ldots, L-1$$

$$\hat{G}_l(n) = \left(1 - \beta_u\right) \hat{G}_l(n-1) + \beta_u \left[e_l^b(n) \right]^2, \qquad 0 < \beta_u \le 1 \tag{11.18n}$$

μ_u is the stepsize and $0 < \mu_u < 1$.

It has been indicated in the literature that the transform domain orthogonalization does not preserve the entropy of the input signal hence, there will be a loss of information. But the lattice orthogonalization of the input is not so and it preserves the entropy of the input hence, it is preferred where there cannot be any loss of information.

Example:
To illustrate the performance of joint lattice for identification of the impulse response frequencies 100, 200 and 300 Hz , of a system given below is considered.

$$G(z) = \frac{1 - 1.1289 z^{-1} + 1.0058 z^{-2} - 0.7617 z^{-3}}{1 - 0.59 z^{-1} + 0.455 z^{-2} - 0.0754 z^{-3} + 0.504 z^{-4}}$$

For this, the input has sinusoids at these frequencies plus white Gaussian noise with SNR=26 dB. The impulse response of the system is shown in Fig. 11.9h. It is seen that the impulse response has significant values up to 64 samples hence, a 64th order regression filter is considered and this implies that the order of the lattice orthogonalizer predictor is also of the same order. For the purpose of comparison, the same adaptation step size is chosen for transversal and regression filters and is 0.005. For the lattice orthogonalizer the step size used is 0.005. For power computation, the forgetting factor used is 0.99.

From Fig 11.9(i), it is evident that the lattice predictor converges very fast as it takes only about 400 samples indicating that orthogonalization is complete within this time.

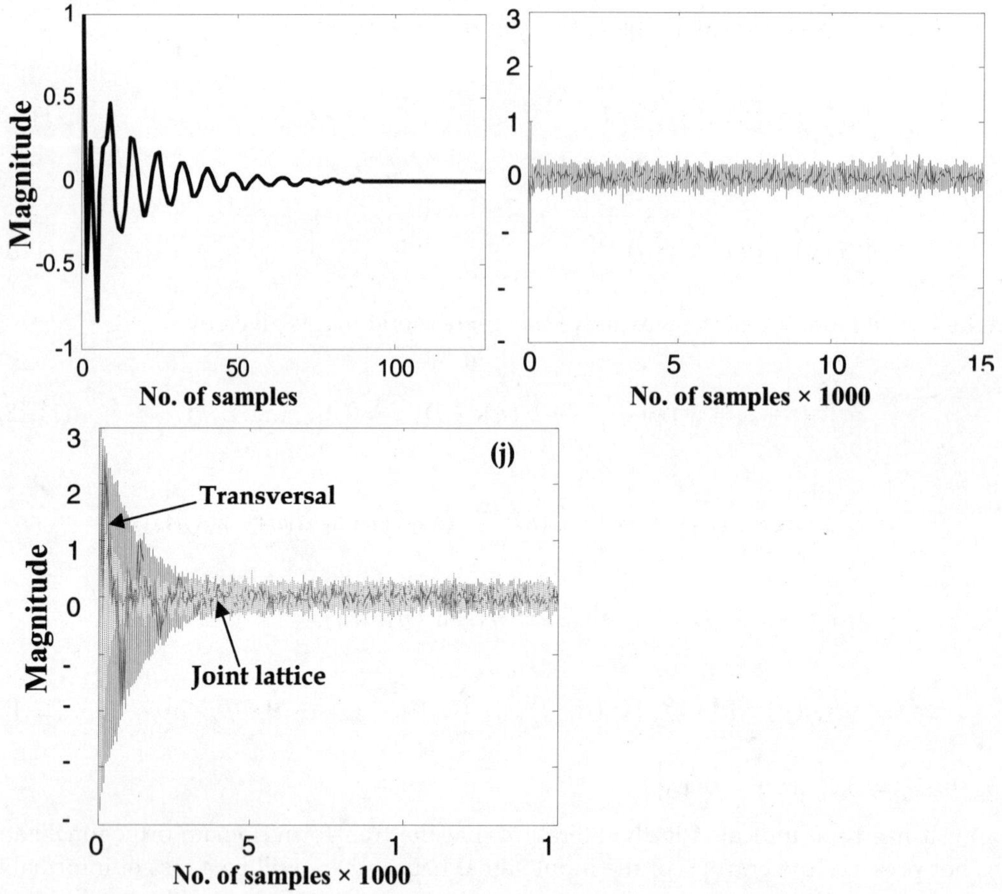

Fig. 11.9(h, i, j) Performance of joint lattice algorithm; (h) System impulse response; (i) Error convergence of lattice orthogonalizer; (j) Error convergence of transversal and joint lattice adaptive filters.

It is seen that the error of the joint lattice converges faster by about 3.5 times than that for transversal filter and this improvement in convergence rate is due to orthogonalization by lattice predictor.

11.10.2 Lattice Adaptation based on Burg Approach

In this, the adaptation is based on learning the present characteristics of the signal and forgetting its past. This is achieved by recursive approach by incorporating a forgetting factor. Hence, this method is different from gradient method which was considered till now.

In the Burg algorithm, the reflection coefficients $k_l(n), l = 1, 2, \ldots, N$ are found by minimising the cost function $\varepsilon_l(n)$ at l^{th} stage with respect to $k_l(n)$. The cost function is

$$\varepsilon_l(n) = \frac{E\left[\left\{e_l^f(n)\right\}^2\right] + E\left[\left\{e_l^b(n)\right\}^2\right]}{2} \tag{11.18o}$$

Here $E[.]$ represents expectation operation and

$$e_l^f(n) = e_{l-1}^f(n) - k_l(n)e_{l-1}^b(n-1)$$

$$e_l^b(n) = e_{l-1}^b(n-1) - k_l(n)e_{l-1}^f(n), \quad l = 1,2,\ldots,N$$

$$\varepsilon_l(n) = (1/2)\left\{E\left[\left\{e_{l-1}^f(n) - k_l(n)e_{l-1}^b(n-1)\right\}^2\right] + E\left[\left\{e_{l-1}^b(n-1) - k_l(n)e_{l-1}^f(n)\right\}^2\right]\right\}$$

$$= (1/2)\left\{ E\left[\left\{e_{l-1}^f(n)\right\}^2 - 2k_l(n)e_{l-1}^f(n)e_{l-1}^b(n-1) + \left\{k_l(n)e_{l-1}^b(n-1)\right\}^2\right]\right.$$

$$\left. +E\left[\left\{e_{l-1}^b(n-1)\right\}^2 - 2k_l(n)e_{l-1}^b(n-1)e_{l-1}^f(n) + \left\{k_l(n)e_{l-1}^f(n)\right\}^2\right] \right\}$$

$$= (1/2)E\left[\begin{array}{l}\left\{e_{l-1}^f(n)\right\}^2 + \left\{e_{l-1}^b(n-1)\right\}^2 + \left\{k_l(n)e_{l-1}^f(n)\right\}^2 + \\ \qquad \left\{k_l(n)e_{l-1}^b(n-1)\right\}^2 - 4k_l(n)e_{l-1}^f(n)e_{l-1}^b(n-1)\end{array}\right]$$

$$\varepsilon_l(n) = (1/2)E\left[\left(\left\{e_{l-1}^f(n)\right\}^2 + \left\{e_{l-1}^b(n-1)\right\}^2\right)\left(k_l^2(n)+1\right) - 4k_l(n)e_{l-1}^f(n)e_{l-1}^b(n-1)\right]$$

Minimising $\varepsilon_l(n)$ with respect to $k_l(n)$, i.e., $\frac{\partial\, \varepsilon_l(n)}{\partial k_l(n)} = 0$. Therefore,

$$(1/2)E\left[\left(\left\{e_{l-1}^f(n)\right\}^2 + \left\{e_{l-1}^b(n-1)\right\}^2\right)\left(2k_l(n)\right) - 4e_{l-1}^f(n)e_{l-1}^b(n-1)\right] = 0$$

$$E\left[\left(\left\{e_{l-1}^f(n)\right\}^2 + \left\{e_{l-1}^b(n-1)\right\}^2\right)\left(k_l(n)\right) - 2e_{l-1}^f(n)e_{l-1}^b(n-1)\right] = 0$$

$$k_l(n) = \frac{2E\left[e_{l-1}^f(n)e_{l-1}^b(n-1)\right]}{E\left[\left\{e_{l-1}^f(n)\right\}^2 + \left\{e_{l-1}^b(n-1)\right\}^2\right]}$$

Replacing the ensemble averages by time averages,

$$k_l(n) = \frac{2\left\{\frac{1}{L}\sum_{n=l+1}^{L-1} e_{l-1}^f(n)e_{l-1}^b(n-1)\right\}}{\frac{1}{L}\sum_{n=l+1}^{L-1}\left\{e_{l-1}^f(n)\right\}^2 + \left\{e_{l-1}^b(n-1)\right\}^2} \tag{11.18p}$$

where, L is the data length considered which has relevance for block/offline processing.

If

$$C_{l-1}(n) = 2\sum_{n=l+1}^{n} e_{l-1}^{f}(n)e_{l-1}^{b}(n-1)$$

and

$$D_{l-1}(n) = \sum_{n=l+1}^{n} \left\{ e_{l-1}^{f}(n)\right\}^2 + \left\{ e_{l-1}^{b}(n-1)\right\}^2$$

then

$$C_{l-1}(n) = C_{l-1}(n-1) + 2e_{l-1}^{f}(n)e_{l-1}^{b}(n-1) \tag{11.18q}$$

and

$$D_{l-1}(n) = D_{l-1}(n-1) + \left\{ e_{l-1}^{f}(n)\right\}^2 + \left\{ e_{l-1}^{b}(n-1)\right\}^2 \tag{11.18r}$$

with

$$e_0^f(n) = e_0^b(n) = x(n)$$

then

$$k_l(n) = \frac{C_{l-1}(n)}{D_{l-1}(n)} \tag{11.18s}$$

and $k_1(n)$ can be computed using $e_0^f(n)$ and $e_0^b(n)$. Further $e_1^f(n)$ and $e_1^b(n)$ can be computed using

$$e_l^f(n) = e_{l-1}^f(n) - k_l(n)e_{l-1}^b(n-1)$$
$$e_l^b(n) = e_{l-1}^b(n-1) - k_l(n)e_{l-1}^f(n),$$

These enable computation of $k_2(n)$ and this process can be continued till $l = N$.

On comparison of this expression for $k_l(n)$ with that obtained by gradient lattice, at the outset, they look different. But it is not so and this will be shown. Let

$$k_l(n+1) = \frac{C_{l-1}(n)}{D_{l-1}(n)}$$

Here, it has been assumed that $k_l(n+1)$ is given by the same expression as that for $k_l(n)$. This enables the correction required to get $k_l(n+1)$ form $k_l(n)$, to depend *only on the forward and backward prediction errors available till time* "n". This approximation is valid for slow adaptation, as $k_l(n+1)$ will not differ much form $k_l(n)$. Let

$$k_l(n+1) = k_l(n) + \Delta_l(n)$$

where, $\Delta_l(n)$ is the correction to be applied for $k_l(n)$ to get $k_l(n+1)$.

Now,

$$\Delta_l(n) = k_l(n+1) - k_l(n)$$

$$\Delta_l(n) = \frac{C_{l-1}(n)}{D_{l-1}(n)} - k_l(n)$$

$$\Delta_l(n) = \frac{1}{D_{l-1}(n)}[C_{l-1}(n) - k_l(n)D_{l-1}(n)]$$

Substituting for $C_{l-1}(n)$ and $D_{l-1}(n)$,

$$\Delta_l(n) = \frac{1}{D_{l-1}(n)}\left[C_{l-1}(n-1) + 2e^f_{l-1}(n)e^b_{l-1}(n-1) - k_l(n)\left\{ D_{l-1}(n-1) + \left\{ e^f_{l-1}(n)\right\}^2 + \left\{ e^b_{l-1}(n-1)\right\}^2 \right\} \right]$$

Since,

$$k_l(n+1) = \frac{C_{l-1}(n)}{D_{l-1}(n)}, \qquad C_{l-1}(n-1) = k_l(n)D_{l-1}(n-1)$$

Therefore,

$$\Delta_l(n) = \frac{1}{D_{l-1}(n)}\Big[k_l(n)C_{l-1}(n-1) + 2e^f_{l-1}(n)e^b_{l-1}(n-1)$$

$$- k_l(n)\left\{ D_{l-1}(n-1) + \left\{ e^f_{l-1}(n)\right\}^2 + \left\{ e^b_{l-1}(n-1)\right\}^2 \right\}$$

$$= \frac{1}{D_{l-1}(n)}\left[2e^f_{l-1}(n)e^b_{l-1}(n-1) - k_l(n)\left\{\left\{ e^f_{l-1}(n)\right\}^2 + \left\{ e^b_{l-1}(n-1)\right\}^2 \right\} \right]$$

$$\Delta_l(n) = \frac{1}{D_{l-1}(n)}\left[e^f_{l-1}(n)e^b_{l-1}(n-1) - k_l(n)\left\{ e^f_{l-1}(n)\right\}^2 + e^f_{l-1}(n)e^b_{l-1}(n-1) - k_l(n)\left\{ e^b_{l-1}(n-1)\right\}^2 \right]$$

$$= \frac{1}{D_{l-1}(n)}\left[e^f_{l-1}(n)\left\{ e^b_{l-1}(n-1) - k_l(n)e^f_{l-1}(n)\right\} + e^b_{l-1}(n-1)\left\{ e^f_{l-1}(n) - k_l(n)e^b_{l-1}(n-1)\right\} \right]$$

As

$$e^f_l(n) = e^f_{l-1}(n) - k_l(n)e^b_{l-1}(n-1) \text{ and } e^b_l(n) = e^b_{l-1}(n-1) - k_l(n)e^f_{l-1}(n),$$

$$\Delta_l(n) = \frac{1}{D_{l-1}(n)}\left[e^f_{l-1}(n)\, e^b_l(n) + e^b_{l-1}(n-1)e^f_l(n) \right]$$

Since,

$$k_l(n+1) = k_l(n) + \Delta_l(n)$$

Therefore,

$$k_l(n+1) = k_l(n) + \frac{1}{D_{l-1}(n)}\left[e_{l-1}^f(n)\, e_l^b(n) + e_{l-1}^b(n-1) e_l^f(n) \right]$$

This is similar to that of gradient lattice predictor given by Eqns. (11.16f, g, h). This is important as the Eqn. (11.18p) appears to have only one term for the numerator in the expression for $k_l(n)$.

For the adaptation of $k_l(n)$ on sample to sample basis, $C_l(n)$ and $D_l(n)$ are estimated recursively using a forgetting factor as

$$C_{l-1}(n) = \beta C_{l-1}(n-1) + 2e_{l-1}^f(n)e_{l-1}^b(n-1)$$

$$D_{l-1}(n) = \beta D_{l-1}(n-1) + \left\{e_{l-1}^f(n)\right\}^2 + \left\{e_{l-1}^b(n-1)\right\}^2, \quad 0 < \beta < 1$$

Here, β is a forgetting factor. For $\beta << 1$, it forgets the previous spectral characteristics and adapts fast to the new spectral changes in the signal which is suitable for non-sationary signals. But if β is close to unity, it takes more time to forget the past spectral characteristics which is appropriate for stationary signals. Thus, the leaning rate is decided by the value of forgetting factor β. This type of adaptation of the lattice parameter will be suitable in certain scenarios.

11.10.3 Sensitivity of the Lattice Structure

The importance of lattice filter structure is its tolerance to finite word length effects in its implementation. Also as long as the reflection coefficient magnitude at all stages is less than unity, it ensures stability. For the lattice predictor,

$$E_i^f(z) = E_{i-1}^f(z) + z^{-1}k_i E_{i-1}^b(z)\,, \quad E_0^f(z) = 1 \tag{11.19a}$$

$$E_i^b(z) = z^{-1}E_{i-1}^b(z) + k_n E_{i-1}^f(z)\,, \quad E_0^b(z) = 1 \tag{11.19b}$$

$$E_i^f(z) = 1 + \sum_{j=1}^{n} e_j^f(n) z^{-j}$$

The relationship between forward and backward prediction errors is,

$$E_i^b(z) = E_i^f(z^{-1})\, z^{-i} \tag{11.19c}$$

Because $E_N^f(z^{-1})$ has real valued coefficients, these zeros must be real or in complex conjugate pairs. *In lattice filter, the first reflection coefficient is the representative of the first few coefficients and the last is that of the last few. Hence, the sensitivity is determined by the first and the last reflection coefficients.* The partial derivatives of Z_1 with respect to the *last coefficient* is

$$\frac{\partial Z_1}{\partial K_p} = \frac{\left(1-Z_1^2\right)\left(1-|Z_1|^2\right)}{\left(1-K_p^2\right)\left(Z_1-Z_1^*\right)} \prod_{p=3}^{i} \frac{\left[Z_1^{-1}-Z_p\right]}{\left[1-Z_1^{-1}Z_p\right]} \tag{11.19d}$$

That is, as the zero approaches the unit circle, the sensitivity of the distance of movement of the reflection coefficient approaches zero. That is *the zeros do not migrate at all.*

The partial derivatives of Z_1 with respect to the *first coefficient* is

$$\frac{\partial Z_1}{\partial K_j} = \left(\frac{e^{j\theta}}{j}\right) \frac{\prod_{p=2}^{N}\left(1-K_p^2\right)}{2\sin(\theta)\left(1-|Z_1|^2\right)} \times \frac{1}{\prod_{p=3}^{i}\left|1-Z_1Z_p\right|^2} \tag{11.19e}$$

where, θ is the angle of Z_1. Since all the terms in this expression are real valued except for the terms in the brackets, that is $\left(\frac{e^{j\theta}}{j}\right)$. But as

$$\left(\frac{e^{j\theta}}{j}\right) = e^{j\left(\theta+\frac{\pi}{2}\right)}, \tag{11.19f}$$

the angle of partial derivative of Z_1 is perpendicular to θ.

This indicates that the direction of movement of Z_1 for small perturbations of K_1 is tangent to the unit circle and not along the radial direction. That is the frequency of the zeros may change, but the bandwidth of the zeros will not decrease, hence stability is guaranteed (M.L. Honig and D.G. Messerschmitt, 1984).

11.11 ADAPTIVE ESCALATOR LATTICE

The escalator lattice was introduced in the chapter 7, and it was shown that it produces mutually uncorrelated prediction errors by realizing the Gram-Schmidt orthogonalization algorithm. Further, unlike the regular lattice, it produces the prediction errors in a parallel way and it does not assume stationarity of the input signal as it minimizes only backward prediction error energy. The error at i^{th} stage and j^{th} order is (Fig. 11.10(a))

$$e_{i,j}^{b}(n) = e_{i-1,j}^{b}(n) - \rho_j^i(n)e_{i-1,j-1}^{b}(n)\,, i = 1, 2,..., N-1,, j = i, i+1,..., N-1 \tag{11.20a}$$

where,

$$e_{0,i}^{b}(n) = x(n-i), \qquad i = 1, 2, \ldots, N-1 \tag{11.20b}$$

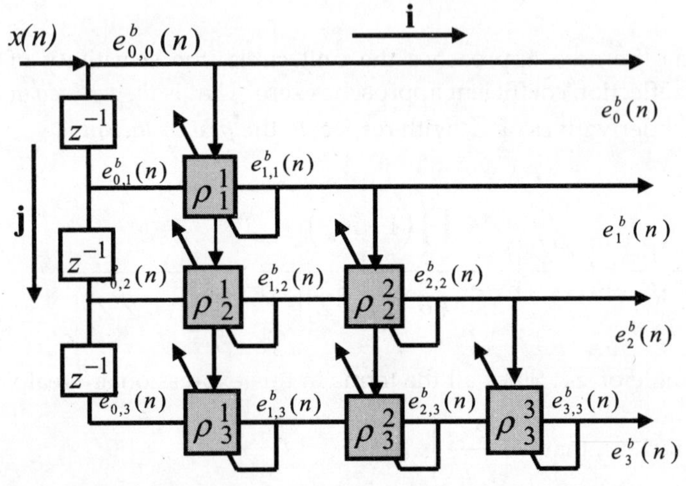

Fig. 11.10(a) Adaptive escalator structure of order N = 3.

The escalator adaptation is based on the gradient LMS algorithm (N. Ahmed, D.H. Youn, 1980, V.N. Parihk, A.Z. Baraniecki, 1998) and here it minimizes the instantaneous square errors and exploits the fact that the prediction errors $b_{i,j}(n)$ at each stage are *local*. Thus,

$$\rho_j^i(n+1) = \rho_j^i(n) - \mu \frac{\partial\left[e_{i,j}^{b}(n)\right]^2}{\partial \rho_j^i(n)}$$

$$\rho_j^i(n+1) = \rho_j^i(n) + \mu_{es}(n)e_{i-1,i-1}^{b}(n)e_{i,j}^{b}(n) \tag{11.20c}$$

where,

$$\mu_{es}(n) = \frac{2\mu}{\sigma_{b_{i-1,i-1}}^2(n)} \tag{11.20d}$$

$$\sigma^2_{b_{i-1,i-1}}(n) = \beta\, \sigma^2_{b_{i-1,i-1}}(n-1) + (1-\beta)\left[e^b_{i-1,i-1}(n) \right]^2, 0 < \beta < 1 \quad (11.20e)$$

and μ is the step size parameter. This algorithm yields N mutually uncorrelated output signals and are given by

$$e^b_i(n) = e^b_{i,i}(n),\ i = 0,1,2,\dots, N-1 \quad (11.20f)$$

11.11.1 Joint Escalator Adaptive Filter

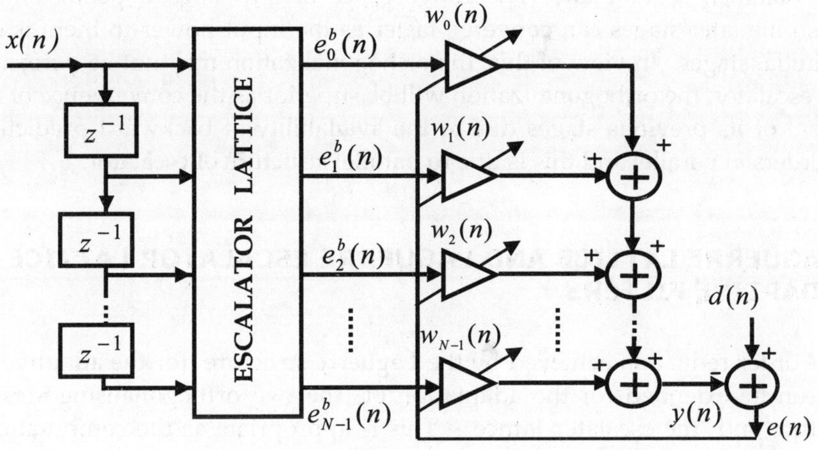

Fig. 11.10(b) Joint Escalator structure.

The escalator structure produces the orthogonal components $\left\{e^b_0(n), e^b_1(n), \dots, e^b_{N-1}(n)\right\}$ and the regression filter, $W(n) = \left\{w_0(n), w_1(n), \dots\ w_{N-1}(n)\right\}$ combines these components to obtain the estimate $y(n)$ of the desired signal $d(n)$ (Fig. 11.10 (b)).

$$y(n) = \sum_{l=0}^{N-1} w_l(n) e^b_l(n) \quad (11.21a)$$

$$e(n) = d(n) - y(n)) \quad (11.21b)$$

The regression coefficients are updated by

$$w_l(n+1) = w_l(n) + \mu_l(n) e^b_l(n) e(n) \quad (11.21c)$$

where,

$$\mu_l(n) = \frac{\mu}{\hat{Q}_l(n)}, \qquad 0 < \mu < 1) \tag{11.21d}$$

$$\hat{Q}_l(n) = \beta \hat{Q}_l(n-1) + (1-\beta)[e_l^b(n)]^2, \quad 0 < \beta < 1 \tag{11.21e}$$

$\hat{Q}_l(n)$ is the estimate of the backward prediction error powers at stage l. This combination of escalator and regression (transversal) filter will be referred as joint escalator structure (N. Ahmed, D.H. Youn, 1980, V.N. Parihk, A.Z. Baraniecki, 1998).

The joint process estimator based on escalator orthogonalization is better than that obtained by regular lattice. In regular lattice as stages are cascaded, the backward prediction errors are obtained sequentially. The convergence of any stage depends on its previous stages. Also the later stages can converge faster as the input power to them is smaller than those for initial stages. In view of this, the orthogonalization may not be perfect at all times. But in the escalator, the orthogonalization will be superior as the convergence of each stage is independent of its previous stages due to the availability of backward prediction errors of different orders in parallel and this is due to parallel structure of escalator.

11.12 LAGUERRE LATTICE AND LAGUERRE ESCALATOR LATTICE ADAPTIVE FILTERS

The model order reduction achieved by the Laguerre structure for the adaptive transversal structure can be extended for the adaptation of the two orthogonalising structures, the regular lattice and the escalator lattice. This is appropriate as the computational load of these two structures are relatively very high compared to that of transversal structure. Further, this is very much essential for the escalator as it requires more computation than the regular lattice. In addition to the order reduction due to the IIR nature of the all-pass characteristic and the frequency compression involved in the Laguerre part, the performance of these structures also will get improved.

11.12.1 Adaptive Laguerre Lattice Structure

The basic Laguerre lattice structure is shown in Fig. 11.10(c). A normal lattice is converted to a Laguerre lattice by inserting is a single pole low-pass filter, $D_0(z)$ at the input and an all-pass filter $D(z)$ after each stage of the lattice, in the backward error path. $D_0(z)$ and $D(z)$ are given by (Zoran Fejzo and Hanoch Lev-Ari, 1997, S. Veena and S.V. Narasimhan, 2004)

$$D_0(z) = \frac{\sqrt{1-\lambda^2}}{1-\lambda z^{-1}} \tag{11.22a}$$

and

$$D(z) = \left(\frac{z^{-1} - \lambda}{1 - \lambda z^{-1}}\right), \quad |\lambda| < 1 \tag{11.22b}$$

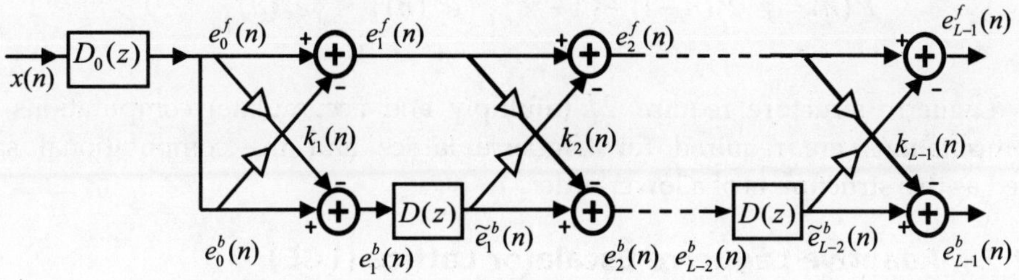

Fig. 11.10(c) Adaptive Laguerre lattice predictor.

Here, the parameter 'λ' is the Laguerre parameter, which denotes a predetermined pole location. When $\lambda = 0$, the block $D(z)$ reduces to an unit delay z^{-1}, making the whole structure, the normal lattice. The choice of Laguerre parameter decides the order and the overall performance of the adaptive filter.

$$E_0^f(z) = E_0^b(z) = D_0(z)X(z)$$

or

$$e_0^b(n) = \lambda e_0^b(n-1) + \sqrt{(1-\lambda^2} x(n) \tag{11.22c}$$

The Laguerre output $\tilde{e}_k^b(n)$, is given by,

$$\tilde{E}_k^b(z) = D(z)E_k^b(z)$$

or

$$\tilde{e}_j^b(n) = e_j^b(n-1) - \lambda\left[e_j^b(n) - \tilde{e}_j^b(n-1)\right] \tag{11.22d}$$

For Laguerre lattice,

$$e_l^f(n) = e_{l-1}^f(n) - k_l(n)\tilde{e}_{l-1}^b(n) \tag{11.22e}$$

$$e_l^b(n) = \tilde{e}_{l-1}^b(n) - k_l(n)e_{l-1}^f(n), \quad l = 1, 2, \ldots, N-1 \tag{11.22f}$$

$$k_l(n+1) = k_l(n) + \alpha_l(n)\left[e_l^f(n)\tilde{e}_{l-1}^b(n) + \tilde{e}_l^b(n)e_{l-1}^f(n)\right] \tag{11.22g}$$

$$\alpha_l(n) = \frac{\alpha}{\hat{P}_l(n)}, \quad 0 < \alpha < 1 \tag{11.22h}$$

$$\hat{P}_l(n) = \gamma\ \hat{P}_l(n-1) + (1-\gamma)\left[\left\{e_l^f(n)\right\}^2 + \left\{\tilde{e}_l^b(n)\right\}^2\right], 0 < \gamma << 1 \tag{11.22i}$$

Laguerre structure require $2L$ (multiply and accumulate) computations . Hence $14L$ computations are required for Laguerre lattice and the computational saving is achieved as this structure is of a lower order.

11.12.2 Adaptive Laguerre Escalator Lattice (LGE)

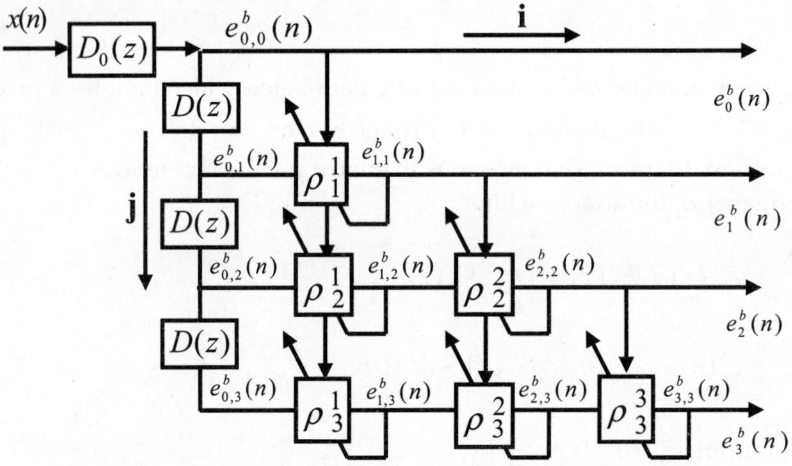

Fig. 11.11 (a) Adaptive Laguerre escalator structure of order N = 3.

The escalator lattice and the Laguerre structure can be combined to form the *Laguerre escalator lattice* (LGE) (S. Veena and S.V. Narasimhan,2007) The error at i^{th} stage and j^{th} order is

$$e^b_{i,j}(n) = e^b_{i-1,j}(n) - \rho^i_j(n) e^b_{i-1,j-1}(n) \tag{11.23a}$$

$$i = 1, 2, \ldots, N-1, j = i, i+1, \ldots, N-1$$

$$e^b_{0,0}(n) = \lambda\, e^b_{0,0}(n-1) + \sqrt{1-\lambda^2}\ x(n) \tag{11.23b}$$

$$e^b_{0,i}(n) = e^b_{0,i-1}(n) - \lambda\left[\, e^b_{0,i-1}(n) - e^b_{0,i}(n-1) \right], \qquad i = 1,2,\ldots,N-1 \tag{11.23c}$$

The escalator adaptation is based on the gradient LMS algorithm and here it minimizes the instantaneous square errors and exploits the fact that the prediction errors $e^b_{i,j}(n)$ at each stage are *local*. Thus,

$$\rho^i_j(n+1) = \rho^i_j(n) - \mu \frac{\partial \left[e^b_{i,j}(n) \right]^2}{\partial \rho^i_j(n)}$$

$$\rho^i_j(n+1) = \rho^i_j(n) + \mu_{es}(n) e^b_{i-1,i-1}(n) e^b_{i,j}(n) \tag{11.23d}$$

where, $\mu_{es}(n) = 2\dfrac{\mu}{\sigma^2_{b_{i-1,i-1}}(n)}$

$$\sigma^2_{b_{i-1,i-1}}(n) = \beta\, \sigma^2_{b_{i-1,i-1}}(n-1) + (1-\beta)\left[e^b_{i-1,i-1}(n) \right]^2, 0 < \beta < 1 \tag{11.23e}$$

and μ is the step size parameter. This algorithm yields N mutually uncorrelated output signals and are given by

$$e^b_i(n) = e^b_{i,i}(n), i = 0,1,2,\ldots,N-1. \tag{11.23f}$$

11.12.3 Joint Laguerre Escalator Structure (JLGE)

The orthogonal components $\left\{ e^b_0(n), e^b_1(n), \ldots, e^b_{N-1}(n) \right\}$ obtained by the Laguerre lattice structure are combined by a regression filter (Fig. 11.11(b)), $W(n) = \left\{ w_0(n), w_1(n), \ldots\, w_{L-1}(n) \right\}$ to obtain the estimate $y(n)$ of the desired signal $d(n)$ (S. Veena and S.V. Narasimhan, 2007)

$$y(n) = \sum_{l=0}^{N-1} w_l(n) e^b_l(n) \tag{11.24a}$$

The regression coefficients are updated by

$$w_l(n+1) = w_l(n) + \mu_l(n) e^b_l(n) e(n) \tag{11.24b}$$

$$\mu_l(n) = \frac{\mu}{\hat{Q}_l(n)}, \quad 0 < \mu \ll 1 \tag{11.24c}$$

$$\hat{Q}_l(n) = \beta \hat{Q}_l(n-1) + (1-\beta)\left[e_l^b(n)\right]^2, \qquad 0 < \beta \ll 1 \tag{11.24d}$$

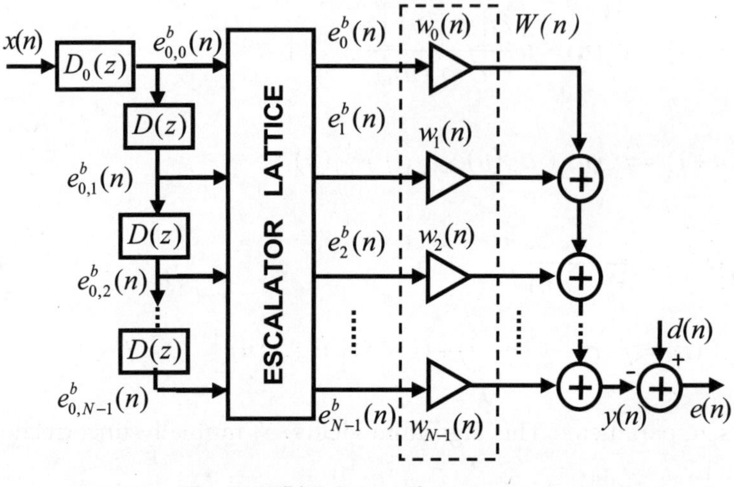

Fig. 11.11(b) Joint escalator Laguerre structure.

$\hat{Q}_l(n)$ is the estimate of the backward prediction error powers at stage l. This combination of escalator Laguerre and regression (transversal) filter will be referred as joint Laguerre escalator (JLGE) structure.

Example: The impulse response of a system

$$P(z) = \frac{1 - z^{-1} + 0.8z^{-2} - 0.7z^{-3}}{1 + 0.6z^{-1} - 0.5z^{-2} + 0.1z^{-3} + 0.5z^{-4}}$$

is considered for system identification and 32 samples of the impulse response are taken for illustrating the performance of different lattice algorithms considered. The input is a stationary input which is a sum of 4 sinusoids with peaks at 50, 100, 400 and 450 Hz and with an additive white Gaussian noise with an SNR of -25.7 dB. For the joint lattice, for orthogonalization of the input, the step size and the forgetting factor in smoothing power used for step size normalization are 0.0001 and 0.999, respectively. The same parameters used for linear combiner are 0.001 and 0.999. The adaptive filter order/length used is 10. For evaluating the performance of different algorithms, the index used is

$$\xi_{sd}(n) = 10\log 10\xi_s(n), \quad \xi_s(n) = \beta\,\xi_s(n) + 1 - \beta)\xi(n),$$

$$\xi(n) = \frac{1}{K}\sum_{i=1}^{K} e_i^2(n)$$

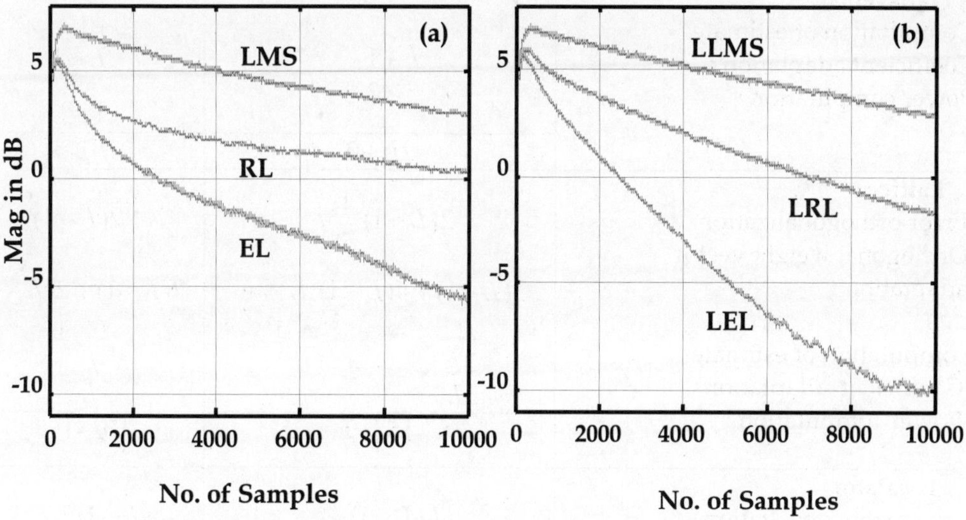

Fig. 11.12 Error convergence for system identification: (a) for LMS, regular lattice(RL), escalator lattice(EL) with L=10 and (b) for Laguerre versions of LMS, RL, and EL with L=6.

Here, K is the number of trials used and is equal to 50 and $e_i(n)$ is the error sequence used for the ith trial. The values for Laguerre parameter λ and the parameter β for smoothing $\xi(n)$, used are 0.5 and 0.999, respectively.

From Fig. 11.12 (a), it is evident that orthogonalization improves convergence rate and error. Further, since orthogonalization is better with escalator lattice (EL) than with regular lattice (RL) as already mentioned, the former shows significant improvement both in terms of convergence rate and error.

The LLMS, LRL and LEL are the Laguerre versions of LMS, RL and EL. For these, the filter length used is 6 and is less compared to those of LMS, RL and EL which is 10. As the filter length is reduced to 50% for the Laguerre versions, to get similar misadjustment, their step size is also reduced to 50%. The LMS and LLMS with reduced filter length, have similar performance (Fig. 11.12(b)). The LRL and LEL with reduced filter length have better performance than those of RL and EL. Further, the improvement with LEL over EL is significant.

11.12.4 Computational Complexity of Transversal, Lattice and Escalator

Table 11.3 Computational complexity of adaptive filter structures

Operation	No. of Multiplications	No. of Additions
Transversal		
Computation of estimate:	L	$L-1$
Coefficient adaptation :	$L+1$	$L-1$
Power computation :	2	1
	$2L+3$	$2L$-1
Lattice		
Error orthogonalization:	$2(L-1)$	$2(L-1)$
Orthogonal weight vector adaptation :	$3(L-1)+4(L-1)\ L$	$2(L-1)+2(L-1)$
Computation of estimate:	L	$L-1$
Coefficient adaptation :	$2L$	L
Power computation :	$3L$	L
	$15L$-9	$10L$-1
Escalator		
Error orthogonalization	$L(L-1)/2$	$L(L-1)/2$
Orthogonal weight vector adaptation	$L(L-1)/2+(L-1)$ +3$(L-1)$	$L(L-1)/2+(L-1)$
Computation of estimate	L	$L-1$
Coefficient adaptation	$2L$	L
Power computation	$3L$	L
	L^2+9L-4	L^2+3L-2
Laguerre outputs	$L+2$	$2L+1$

The computational complexity of transversal, lattice and escalator are given in Table 11.3. The complexity of escalator structure is of the order of L^2, which is very high compared to that of regular lattice structure (order of L). Even though the Laguerre structure increases the number of computations by $L+2$ multiplications and $2L+1$ additions, since the Laguerre structure requires only *about half the filter length* (Zoran Fejzo and Hanoch Lev-Ari, 1997), there will be computational savings in achieving the same performance.

11.13 CASCADE ADAPTIVE LINEAR PREDICTOR

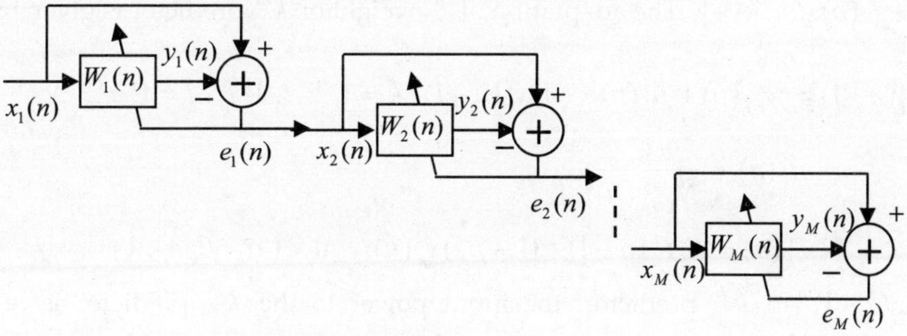

Fig. 11.13(a) Adaptive cascade predictor.

The gradient LMS transversal adaptive predictor is the simplest among all others considered. However, it has been shown that its main problem is slow convergence when the eigenvalue spread of its input autocorrelation matrix is large (chapter 10). Also to handle ARMA type of input, higher order predictors are required and also the eigenvalue spread increases with the order. Further, for higher order predictors, the gradient step size will decrease making the convergence slow. In view of this, many modifications like use of orthogonal transforms, normal lattice and escalator lattice; were considered which decrease the eigenvalue spread to overcome the slow convergence. However, these are quite involved both computationally and conceptually, compared to the basic LMS predictor. *In order to preserve the simplicity of the transversal LMS algorithm while overcoming the problem of slow convergence due to eigenvalue spread, the adaptive cascade structure has come into existence* (P. Padroni and M.Vetterli, 1998). This structure consists of many smaller order independent adaptive predictors in cascade, each minimizing its respective instantaneous error energy. Due to individual predictor's small filter order, they can use larger step sizes compared to that of a single large order adaptive predictor as the step size is normalized by a factor, the order times the input power and this enables faster convergence. The cascade predictor basically decouples individual/groups of components of the input signal as the stages progress. For example, in going from first to second stage, the component $y_1(n)$ and from second to third stage, the component $y_2(n)$ and so on, are removed reducing the eigenvalue spread(Fig. 11.13(a)).

The adaptive one-step cascade predictor consists of M independent adaptive one-step predictors with N_k and $e_k(n)$ being the order and the error of the kth predictor, respectively. The error for the kth predictor is given by

$$e_k(n) = x_k(n) - y_k(n) = x_k(n) - \sum_{l=1}^{N_k} w_{k,l}(n) x_k(n-l) \tag{11.25a}$$

where $w_{k,l}(n)$ are the weights for the kth predictor $W_k(n)$. Further, $x_1(n) = x(n)$ and $x_k(n) = e_{k-1}(n)$ *for* $k > 1$. The adaptation of lth weight of kth predictor is given by

$$w_{k,l}(n+1) = w_{k,l}(n) + \mu_k(n)e_k(n)x_k(n-l),\ k = 1,2,\ldots M, \quad l = 1,2,\ldots N_k \tag{11.25b}$$

$$\mu_k(n) = \frac{\alpha_k}{N_k P_k(n)}, \tag{11.25c}$$

$$P_k(n) = \beta_k P_k(n-1) + (1-\beta_k)x_k^2(n), \quad 0 < (\alpha_k, \beta_k) < 1) \tag{11.25d}$$

Since for $(k-1)^{th}$ to k^{th} predictor, the input power to the kth predictor is reduced, i.e., $P_k(n) < P_{k-1}(n)$, the step size $\mu_k(n)$ can increase monotonically as k increases. That is, higher adaptation step sizes $\mu_k(n)$ can be used for succeeding predictors. Further, this is true as the order of individual predictor's order, N_k is much less than the order N of the single overall predictor, i.e., $N_k << N$. The overall predictor $A(z,n)$ is given by

$$A(z,n) = \prod_{k=1}^{M} A_k(z,n)\ ,\ A_k(z,n) = 1 - \sum_{l=1}^{N_k} w_{k,l}(n)z^{-l}, \quad k = 1, 2, \ldots, M \tag{11.26a}$$

The individual predictors converge to their optimal solution given by

$$a_k(n) = R_{N_k N_k}^{-1} r_k, \quad k = 1, 2, \ldots, M; \qquad a_k(n) = \{a_{k,1}, a_{k,2}, \ldots a_{k,N_k}\},\ k = 1, 2, \ldots, M \tag{11.26b}$$

$$R_{N_k N_k} = \begin{bmatrix} \phi_k(0) & \phi_k(1) & \ldots & \phi_k(N_k-1) \\ \phi_k(1) & \phi_k(0) & \ldots & \phi_k(N_k-2) \\ \vdots & \vdots & \vdots & \vdots \\ \phi_k(N_k-1) & \phi_k(N_k-2) & \ldots & \phi_k(0) \end{bmatrix}, \quad r_k = \begin{bmatrix} \phi_k(1) \\ \phi_k(2) \\ \vdots \\ \phi_k(N_k) \end{bmatrix} \tag{11.26c}$$

$$\phi_k(l) = \frac{1}{L_K} \sum_{n=0}^{L_K-l-1} x_k(n)x_k(n+l), \quad l = 0,1,\ldots,N_K, \quad k = 1, 2, \ldots, M \tag{11.26d}$$

The individual predictors can have even a single coefficient or more coefficients. As pointed out, the cascade predictor provides better convergence rate than a single LMS transversal/lattice predictor of large order. This is due to the fact the error energy minimization will be effective when the order/number of parameters is small. Hence, the convergence error magnitude (misadjustment) will be smaller than that of a single LMS transversal/lattice predictor which is a highly desirable feature. Even with transversal predictors, the cascade structure will be less affected by the coefficient round of errors than that with a single large order predictor as the errors will get localized than being transmitted for the complete predictor, which is the basic feature of cascaded systems.

11.13.1 Cascade Forward Backward LMS Predictor (CFBLMS)

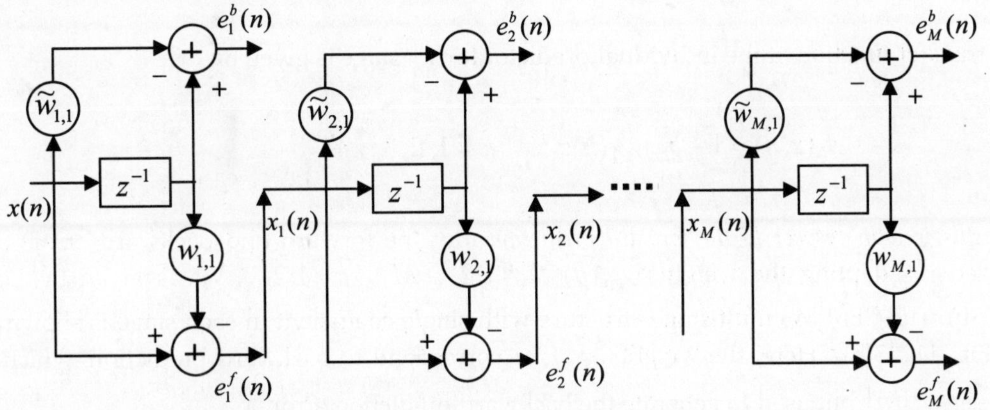

Fig. 11.13(b) Single coefficient multistage CFBLMS predictor.

In the cascade LMS forward backward predictor (Fig. 11.13(b)), the predictor coefficients of each predictor is chosen such that the instantaneous average of forward and backward error energies, is minimized (H.G. Yeh and Martin Vaklev, 2006, Pavthra, 2011). That is for the kth predictor $W_k(n)$,

$$\varepsilon_k(n) = \frac{\left\{e_k^f(n)\right\}^2 + \left\{e_k^b(n)\right\}^2}{2}$$

is minimized.

$$e_k^f(n) = x_k(n) - \sum_{l=1}^{N_k} w_{k,l}(n)x_k(n-l)$$

$$e_k^b(n) = x_k(n-N_k) - \sum_{l=1}^{N_k} w_{k,l}(n)x_k(n-N_k+l) \tag{11.27a}$$

The adaptation of lth weight of kth predictor is given by

$$w_{k,l}(n+1) = w_{k,l}(n) + \mu_k(n)\left[e_k^f(n)x_k(n-l) + e_k^b(n)x_k(n-N_k+l)\right], \tag{11.27b}$$

$$k = 1,2,\ldots M, \quad l = 1,2,\ldots N_k$$

$$\mu_k(n) = \frac{\alpha_k}{N_k P_k(n)},$$

$$P_k(n) = \beta_k P_k(n-1) + (1-\beta_k)x_k^2(n), \quad 0 < (\alpha_k, \beta_k) < 1$$

The overall forward predictor transfer function $A(z,n)$ is given by

$$A(z,n) = \prod_{k=1}^{M} A_k(z,n),$$

The transfer function of the individual predictor for k^{th} stage is given by

$$A_k(z,n) = 1 - \sum_{l=1}^{N_k} w_{k,l}(n) z^{-l}, \quad k = 1, 2, \ldots, M$$

and *this is with respect to the forward prediction.* But the forward and backward errors are utilized in adapting the weights $w_{k,l}(n)$ $k = 1,2,\ldots M,$ $l = 1,2,\ldots N_k$ as given by Eqn. (11.27b). The CFBLMS multistage structure with single coefficient in each stage is shown in the Fig. 11. 13(b). Here, the weights $\tilde{w}_{k,l}(n)$ correspond to backward prediction, which is same as $w_{k,l}(n)$ but used to generate the backward prediction error.

The lattice filter by itself uses both forward and backward prediction errors to update its refection coefficients. Hence, the advantage of using both the errors has already been exploited. But still a cascade structure is possible by considering, for example, a second order two-stage lattice as the basic building block. From the reflection coefficients, the prediction filter coefficients can be derived using the inverse Levinson relation. If k_1 and k_2 are the reflection coefficients of the two-stage lattice, then forward $H^f(z)$ and backward $H^b(z)$ prediction error filters are given by

$$H^f(z) = 1 + \left(k_1 + k_1 k_2\right) z^{-1} + + k_2 z^{-2}$$

$$H^b(z) = k_2 + \left(k_1 + k_1 k_2\right) z^{-1} + z^{-2}$$

Due to lattice lower sensitivity of its coefficients to round off errors as already discussed, a cascade forward-backward adaptive filter based on it, provides a lower mean square error when the reflection coefficients are represented by a small word length (say six bits).

11.14 TRANSFORM DOMAIN CONJUGATE GRADIENT (TCG) ALGORITHM FOR ADAPTIVE FILTERING

For the equation $R(n)W(n) = P(n)$, it is known that with CG, the solution for $W(n)$ can be achieved by $k_{\max}$ (say) iterations. Hence, it is not possible to get exact solution just by a single iteration for every sample, for sample to sample adaptation. *But if the signal is white, then it is possible to get with one iteration per sample, the exact solution for* $W(n)$. This is the motivation for using normalized orthogonal transform to whiten the colored input and improve the performance of CG with single iteration for every sample. Thus, the normalized orthogonal transform can be used to improve the performance of the sample to sample adaptation CG with single iteration for colored input. (Zou Yuexian, S.C. Chan and T.S. Ng, 2000)

For transform domain, the transformed input vector is given by

$$\bar{X}(n) = U[X(n)] = [\bar{x}_1(n), \ldots \bar{x}_N(n)]^T \tag{11.28a}$$

$\bar{X}(n)$, $X(n)$ are the transformed input vector and time domain input vectors. U is a NxN *unitary transform matrix satisfying* $UU^T = I$, I is a unit identity matrix. The autocorrelation of the transformed input vector $\bar{X}(n)$ is

$$\bar{R} = E[\bar{X}(n)\bar{X}^T(n)] \tag{11.28b}$$

The normalized transform domain algorithm is given by

$$\bar{e}(n) = d(n) - \bar{W}^T(n)\bar{X}(n), \tag{11.28c}$$

$$\bar{W}(n+1) = \bar{W}(n) + 2\mu\,\Lambda^{-2}(n)\bar{e}(n)\bar{X}(n) \tag{11.28d}$$

$$\Lambda^2 = dia[\sigma_1^2(n), \sigma_2^2(n), \ldots\ldots \sigma_N^2(n)], \tag{11.28e}$$

$$\sigma_i^2(n) = \lambda\ \sigma_i^2(n-1) + (1-\lambda)\bar{x}_i^2(n),\ 0 < \lambda < 1 \tag{11.28f}$$

Though the fixed transform application results in an approximate alignment of weight vectors along with principal axes of the error surface contours and also the powers of different components of the input are also approximate but these enable the CG to find effective descent direction and step size for the filter to adapt to optimal solution, on sample to sample basis with single iteration.

Normalized transform domain sample to sample CG algorithm

Initialization:

$$\bar{W}(N) = 0,$$

$$\xi(N) = -d(n)\bar{X}(n),\ \bar{R}(N-1) = 0,$$

$$V(N) = -\xi(N),\ n = N$$

For $n < L - N$, (L is the length of the data)

$$\vec{R}(n) = \lambda\vec{R}(n-1) + \bar{X}(n)\bar{X}^T(n) \tag{11.28g}$$

$$\alpha(n) = \frac{V^T(n)\xi(n)}{V^T(n)\vec{R}(n)V(n)} \tag{11.28h}$$

$$\bar{W}(n+1) = \bar{W}(n) + \alpha(n)V(n) \tag{11.28i}$$

$$\xi(n+1) = \xi(n) - \alpha(n)\vec{R}(n)V(n) \tag{11.28j}$$

$$\beta(n) = \frac{\xi^T(n+1)\Lambda^{-2}\xi\ (n+1)}{\xi^T(n)\Lambda^{-2}\xi\ (n)} \tag{11.28k}$$

$$V(n+1) = \Lambda^{-2}\xi\ (n+1) + \beta(n)V(n) \tag{11.28l}$$

End

For a stationary signal, as $V(n)$ always in a direction to decrease the cost function, the convergence is assured. Therefore when $\vec{R}(n)$ converges to $\vec{R}$, $\bar{W}(n)$ converges to $\bar{W}_{opt}$ .The transform domain CG for sample to sample basis with single iteration per sample provides faster convergence and lower error than those obtained by quasi-Newton LMS algorithm.

To illustrate the performance of TCG, the same system with the impulse response of 128 samples, shown in Fig. (10.14a) is used. An input consisting of sum of three sinusoids at 50 Hz, 100 Hz, and 150 Hz of amplitudes 3, 2, and 1 respectively, is used. For the LMS adaptive filter of length 128, the stepsize chosen is 0.001 (maximum stepsize permitted to avoid divergence). The DCT transform size is 128 (to get better orthogonalization and smaller size may not orthogonalize well) with bin power estimation forgetting factor $\beta = 0.99$. For the TCG, the autocorrelation and cross-correlation are adapted using a forgetting factor of $\lambda = 0.99$ For each bin, only one adaptive weight is used and the dimension space of autocorrelation is unity due to orthogonalization. Hence, *only one iteration per input sample* is used for updating the bin weight.

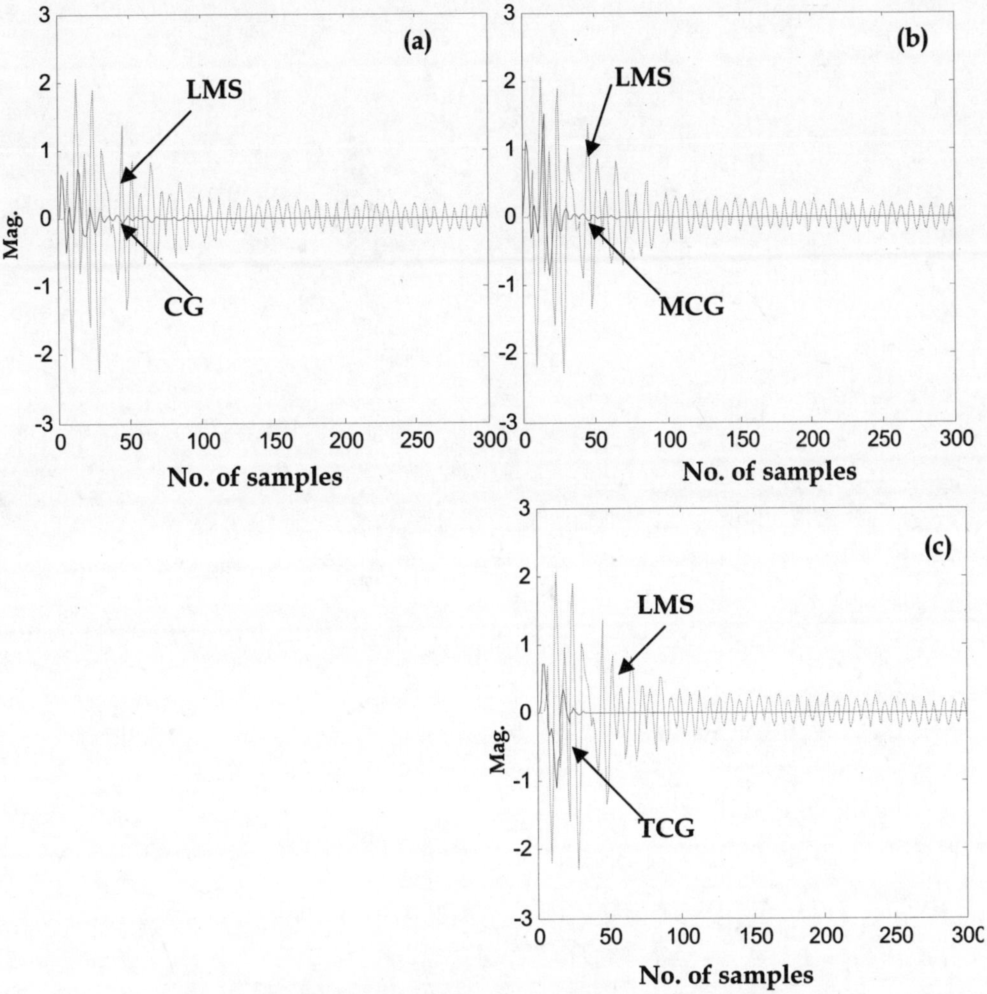

Fig.11.14 Comparison of adaptive filtering error convergence of LMS with those of (a) CG; (b) MCG and (c) TCG.

It is seen that the convergence rate and error of TCG are superior to that of LMS (Fig. 11.14(c)). Further, the initial error magnitude is almost same as that of CG with 12 iterations per input sample (Fig. 11.14 (a, c)). The TCG converges within 30 samples compared to that of LMS which takes 130 samples. The TCG gives an initial error of smaller magnitude and faster convergence compared to that of MCG (Fig. 11.14 (b, c)). Thus, the TCG even with one CG iteration per sample gives an equivalent performance of CG with 12 iterations per sample and further its performance is superior to that of MCG.

Chapter 12

SUBBAND AND WAVELET ADAPTIVE FILTERS

Orthogonalization based on orthogonal transforms and lattice filters involve additional computational load as processing is done at the input sampling rate. Decomposing the signal into different subbands by an analysis filter bank and having a time domain adaptive filter for each subband not only provides reduction in eigenvalue spread for that adaptive filter but also its adaptation can be carried out at a lower sampling/decimated rate as the individual bandwidth of subbands are relatively smaller than that of full band. This enables use of even slower DSP processors in parallel. Further, as decimation expands the spectrum, the eigenvalue spread of the decimated signal is further reduced and hence this enables use of enhanced step size in addition to that due to reduced input power for each adaptive filter and their reduced filter length (which decide the step size). The outputs of the adaptive filters are added by synthesis filter bank which brings the subband sampling rate to the input sampling rate by interpolation. As the filter bank should not introduce additional errors in the desired signal estimate (other than due to adaptive filters), it has to be a perfect reconstruction filter bank (PRFB) and this is realized by a polyphase DFT filter bank where both polyphase structure and DFT matrix are exploited to achieve computational efficiency. In view of this, the theory and design of such a filter bank are considered.

If the synthesis filter bank is used to generate the error (closed loop), the inaccuracies (if any in the PRFB) in the synthesis bank are accounted but there will be delay between the desired signal and its estimate which limits the convergence rate and may even lead to instability. On the other hand if individual errors are generated by decomposing the desired signal (open loop), the errors in the synthesis bank are not accounted. Further, at the band edges, there is a trade-off between aliasing and eigenvalue spread. The method used in overcoming this trade-off still suffers from delay problem. The delay problem of closed loop approach is solved by keeping the complete filter bank out side the LMS loop and the estimate of overall impulse response of the system required to obtain the error (overall) is achieved in frequency domain by stacking Fourier transform of individual sub-adaptive filters (known as weight transform) and is known as delayless subband adaptive filter (DLSAF). The weight transform is quite complicated and further this and the polyphase DFT analysis filter bank, require complex arithmetic. In making the weight transform and the PRFB real and simple, DCT based subband harmonic decomposition filter bank is considered.

Further as the wavelet transform (WT) which is an extension of filterbank with nonuniform bandwidth and which provides additional advantages like handling nonstationary signals and perception, the required theory of WT is covered. Further the DLSAF based on the DCT harmonic WT has been presented.

12.1 INTRODUCTION

In the transform domain adaptive filters, the *improvement in convergence rate is due to decomposing the signal into orthogonal components followed by normalization*. A single transform is not suitable for all types of signals. The DCT is useful for low-pass and the DST for band-pass signals. The reason is that the FT of orthogonal vectors of a transform has poor side lobe levels in some frequency region. The side lobes are relatively small in the low frequency region for DCT and that for the DST in the mid band region. *The high sidelobe levels result in improper orthogonalization, hence* their magnitude reduction is essential and this is achieved by using *filters with good stop band attenuation by the subband approach*. The signal decomposition into narrow band signals *enables decimation or reduction in sampling rate/ increase in sampling interval*. The decimation/compression in time expands the frequency spectrum thereby reducing the eigenvalue spread resulting in a better convergence rate. Also the decimation as it increases the sampling interval, it facilitates their implementation even by low speed DSP processors in parallel. To restore the original signal from these components, it is *necessary to restore their original sampling rate* and add them together. This reverse process of down sampling, is called the *interpolation or up sampling*.

The subband based adaptive filters enable improved convergence rate and implementation on low speed DSP processors. But they will not exploit human perception or nature of signal characteristics. This is because in nature, the signals have broader bandwidth for higher frequencies and smaller bandwidth for lower frequencies. The same rule is followed by human perception also. That is, by simply splitting the signal frequency spectrum into subbands of same width does not exploit the bandwidth change with frequency. Hence, if the frequency range of subbands is made nonuniform to obey the nature's property, the benefit due to it can be utilized. Such a signal decomposition is existing in the literature and is known as wavelet transform and can be utilized for adaptive filters.

12.2 DECIMATION

Reducing the sampling rate is called down sampling. That is, for a signal sampled at Nyquist rate, to reduce the sampling rate by a factor M requires the removal of $(M-1)$ samples and retaining every M^{th} sample.

$$\begin{aligned} x_M(n) &= x(n), \qquad n = 0, \pm M, \pm 2M, \ldots \\ &= 0, \qquad \textit{otherwise} \end{aligned} \tag{12.1a}$$

Here, the sampling interval is increased by a factor M. A fullband signal is passed through a filter (lowpass/bandpass/highpass) and its output is sub-sampled to get the decimated version of the input. As already discussed, typically a lowpass filter is used to restrict the bandwidth to avoid aliasing. However, decimation operation is different from this and involves an additional operation of *renumbering* the retained samples.

$$y(n) = x_M(Mn)$$

$$= x(Mn) = x(n'), \quad n' = MN \tag{12.1b}$$

From this point of view, the signal is compressed and is equivalent to reducing the signal duration. That is, what was happening at n^{th} sample now happens at $(n/M)^{th}$ sample. Thus, the decimation is different from down sampling and as this compresses the signal it is also called *compression*, which is more appropriate.

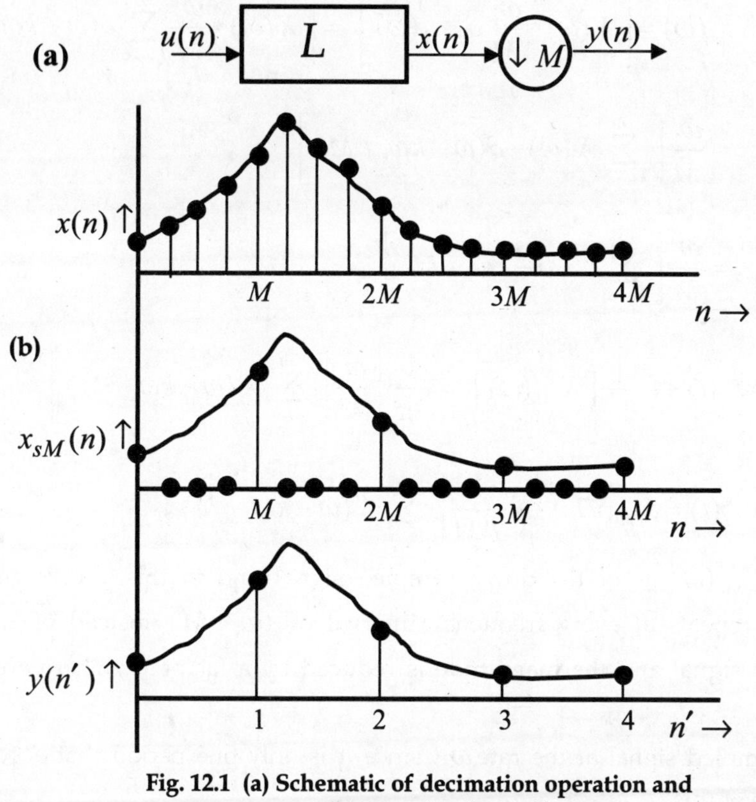

Fig. 12.1 (a) Schematic of decimation operation and (b) Associated waveforms.

As explained already, the sampling is achieved by multiplying $x(t)$ by an impulse train $\delta_{T_s}(t)$, T_s being the sampling interval and $\omega_s = 2\pi / T_s$.

$$x_S(t) = x(t)\delta_{T_s}(t)$$

If

$$x(t) \leftrightarrow X(\omega) \;,\; \delta_{T_s}(t) \leftrightarrow \omega_s \delta_{\omega_s}(\omega)$$

then

$$x_S(t) \leftrightarrow \frac{1}{2\pi}\left[X(\omega) * \omega_s \delta_{\omega_s}(\omega)\right]$$

where

$$\delta_{\omega_s}(\omega) = \sum_{k=-\infty}^{\infty} \delta(\omega - k\omega_s)$$

If this sampled signal is down sampled by a factor M,

$$X_{SM}(\omega) = \left[X(\omega) * \frac{\omega_s}{M}\delta_{\omega_s/M}(\omega)\right] = \left[X(\omega) * \frac{\omega_s}{M}\sum_{k=-\infty}^{\infty}\delta(\omega - k\omega_s / M)\right]$$

$$= \frac{\omega_s}{M}\left[\sum_{k=-\infty}^{\infty} X(\omega) * \delta(\omega - k\omega_s / M)\right]$$

$$= \frac{\omega_s}{M}\left[\sum_{k=-\infty}^{\infty} X(\omega - k\omega_s / M)\right]$$

$$x_{SM}(t) \leftrightarrow \frac{1}{2\pi}\left[X_{sM}(\omega)\right] = \frac{1}{2\pi}\left\{\frac{\omega_s}{M}\left[\sum_{k=-\infty}^{\infty} X(\omega - k\omega_s / M)\right]\right\}$$

$$x_{SM}(t) = x_{SM}(nT_s) \leftrightarrow \frac{1}{T_s M}\left[\sum_{k=-\infty}^{\infty} X(\omega - k\omega_s / M)\right] \tag{12.1c}$$

The spectrum $X_{SM}(\omega)$ is of the down sampled signal and in this, the original signal spectrum $X(\omega)$ repeats at every frequency interval of (ω_s / M) instead of ω_s for the original sampled signal and the magnitude is reduced by a factor M. There are $(M-1)$ replicas in ω_s.

For a sampled signal at the rate ω_s, since it is only one period that is required as others are replica of this,

$$x_{SM}(n) \leftrightarrow \frac{1}{T_s M}\left[\sum_{k=0}^{M-1} X(\omega - k\omega_s / M)\right]$$

That is, in one period $(0-\omega_s)$, there are M replicas of the original spectrum $X(\omega)$.

Further, the down sampled signal is *renumbered* and this results in compression of the signal by a factor M in time domain. Since according to scaling property of the Fourier transform, *compression in one domain leads to expansion in the other domain,* the renumbering of time axis results in the expansion of the frequency axis, that is $\omega \to (\omega / M)$.

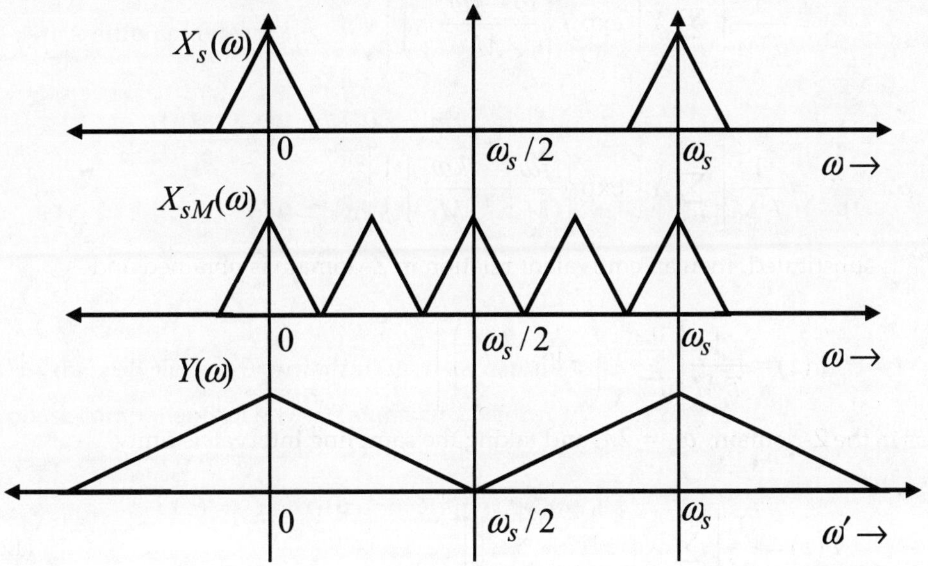

Fig. 12.2 Frequency spectra of signals in decimation for $M = 4$.

$$y(n) \leftrightarrow \frac{1}{T_s M}\left[\sum_{k=0}^{M-1} X(\omega / M - k\omega_s / M)\right] = \frac{1}{T_s M}\left[\sum_{k=0}^{M-1} X\left(\frac{\omega - k\omega_s}{M}\right)\right]$$

$$y(n) \leftrightarrow \frac{1}{T_s M}\left[\sum_{n=-\infty}^{\infty} X(\omega' - k\omega_s)\right] \tag{12.1d}$$

Hence, in the expanded frequency scale $\omega' = \omega / M$, the repeating interval (ω_s / M) becomes ω_s. That is, (ω_s / M) in the original signal spectrum has expanded to ω_s in the new frequency scale ω'.

As the removal of M samples in the decimation process will lead to repetition of the signal spectrum, unless the signal spectrum is band limited to $(\omega_s / 2M)$ or (π / M), the aliasing will take place. In order to avoid this, as shown in the Fig. (12.1a), the input signal $x(n)$ to the decimator should be bandlimited by a *lowpass filter* with cut-off frequency $(\omega_s / 2M)$.

The expression for the FT of $y(n)$ as above is given by

$$Y(\omega) = \frac{1}{T_s M}\left[\sum_{k=0}^{M-1} X\left(\frac{\omega - k\omega_s}{M}\right)\right]$$

$$= \frac{1}{T_s M}\left[\sum_{k=0}^{M-1} X\left(\exp j\left\{\frac{\omega - k\omega_s}{M}\right\}\right)\right]$$

$$= \frac{1}{T_s M}\left[\sum_{k=0}^{M-1} X\left(\exp\left\{\frac{j\omega}{M} - \frac{jk\omega_s}{M}\right\}\right)\right]$$

If $z = e^{j\omega}$ is substituted, then an equivalent relation in Z-domain is obtained and is

$$Y(z) = \frac{1}{T_s M}\left[\sum_{k=0}^{M-1} X\left(z^{\frac{1}{M}} e^{-\frac{jk\omega_s}{M}}\right)\right]$$

Further as in the Z-domain, $\omega_s = 2\pi$ and taking the sampling interval as unity,

$$Y(z) = \frac{1}{M}\left[\sum_{k=0}^{M-1} X\left(z^{\frac{1}{M}} e^{-\frac{j2\pi k}{M}}\right)\right] \qquad (12.1e)$$

For $M = 2$,

$$Y(z) = \frac{1}{2}\left[\sum_{k=0}^{1} X\left(z^{\frac{1}{2}} e^{-j\pi k}\right)\right]$$

$$Y(z) = \frac{1}{2}\left[X(z^{\frac{1}{2}}) + X(-z^{\frac{1}{2}})\right]$$

The compression is a *time variant operation* and this is due to the renumbering operation involved.

Let $x(n)$ be the input sequence to the decimator (M=2) given below.

n	0	1	2	3	4	5	6	7	8	9	10	11	12
$x(n)$	4	3	2	1	4	3	2	1	4	3	2	1	4
$x_{sM}(n)$	4		2		4		2		4		2		4
$y(n)$	4	2	4	2	4	2	4						

$x_{sM}(n)$ is the down sampled sequence where the sampling interval is twice that of the original sequence.

$y(n)$ is the decimated sequence where alternate samples are retained but the sampling interval is same as original one.
For the delayed input by a sample:

$x(n-1)$	0	4	3	2	1	4	3	2	1	4	3	2	1	4
$y(n-1)$	0	3	1	3	1	3	. . . .							

But the delayed output of the decimator for input $x(n)$ is

$y_d(n)$ 2 4 2 4 2 4. . . . .

For a time invariant system/operation, for the delayed input $x(n-1)$ the output $y(n-1)$ should be a delayed version of the output for the input without delay, that is $y_d(n)$. But $y_d(n)$ is not same as $y(n-1)$ hence, the decimation operation is a time variant operation.
In the time domain, the decimator can be (Fig. 12.1b) expressed as

$$y(n) = \sum_{k=-\infty}^{\infty} h(k)u(Mn-k) \tag{12.1f}$$

That is, instead of computing the filter output at every sample instant and then omitting every $(M-1)$ sample, it is computationally advantageous to compute the filter output once in every M samples as given by above equation.

12.3 INTERPOLATION

Interpolation is the reverse process of decimation. Interpolation of a given sequence by a factor M involves introducing $(M-1)$ additional samples in the existing sequence. This can be achieved by first introducing $(M-1)$ zeros between points, renumbering the time axis and then passing this modified signal through a low-pass pass filter with cutoff frequency $(\omega_s / 2M)$ as shown in the Fig. 12.3. The insertion of the $(M-1)$ zeros can be represented by

$$x_{SM}(n) = x(n/M), \qquad n = 0, \pm M, \pm 2M, \ldots \tag{12.2a}$$

$$= 0, \qquad otherwise$$

What used to happen at n^{th} sample now happens at $(Mn)^{th}$ sample in $x_{SM}(n)$. That is, this stretches the time axis of $x(n)$ by a factor M and the frequency axis gets compressed by the same factor M. Let $k = n/M$, $n = kM$,

$$X_{SM}(\omega) = \sum_{k=-\infty}^{\infty} x(n/M)e^{-j\omega kM}$$

$$X_{SM}(\omega) = \sum_{k=-\infty}^{\infty} x(k)\left(e^{jM\omega}\right)^{-k} \tag{12.2b}$$

$$X_{SM}(\omega) = X(M\omega)$$

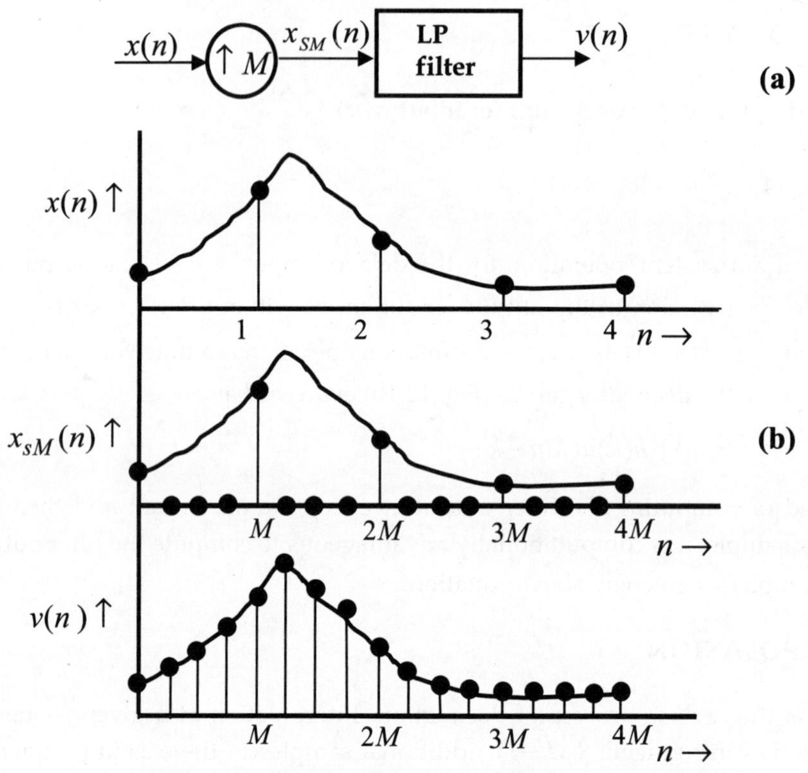

Fig. 12.3 (a) Schematic of interpolation operation and (b) Associated waveforms.

In the frequency domain, whatever happens at ω_k in $X(\omega)$ happens at (ω_k / M) in $X_{SM}(\omega)$. Hence, in the frequency range of $(0-\omega_s)$ there will be $(M-1)$ compressed replica of $X(\omega)$ each of width (ω_s / M) and these are referred to as images (Fig. 12.4). That is, the insertion of $(M-1)$ zeros between the sample points in $x(n)$, results in these images. In other words, the images (are high frequency components) result in forcing the signal to pass through the zero value between the samples of $x(n)$. It is required for the interpolated signal, the $(M-1)$ additional samples to take in between values of $x(n)$ and not zeros. This is achieved by passing the signal $x_M(n)$ through a low-pass filter with cut-off frequency $(\omega_s / 2M)$ (Fig. 12.4).

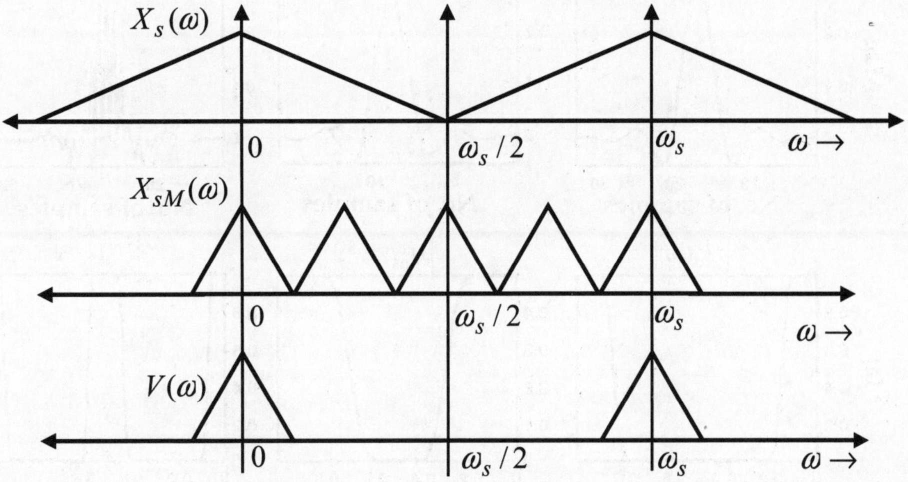

Fig. 12.4 Frequency spectra of signals in interpolation for $M = 4$.

In the relation Eqn. (12.2b), by putting $z = e^{j\omega}$, an equivalent relation in Z-domain is obtained and is

$$X_{SM}(z) = X(z^M)$$

For $M = 2$,

$$X_{SM}(z)\big|_{M=2} = X(z^2)$$

Like compression, the interpolation is also a *time variant operation* and this is due to the renumbering operation involved. In the time domain, computation of the interpolation is given by

$$v(n) = \sum_{l=-\infty}^{\infty} x_{SM}(l)g(n-l) = \sum_{l=-\infty}^{\infty} x(l/M)g(n-l)$$

Let $(l/M) = m$, then

$$v(n) = \sum_{m=-\infty}^{\infty} x(m)g(n-Mm) \tag{12.2c}$$

This is the expression for the time domain implementation of interpolation.

***Example*:** For illustration of the effect of decimation and interpolation, the impulse responses of the low-pass, high-pass and band-pass filters themselves are regarded as data.

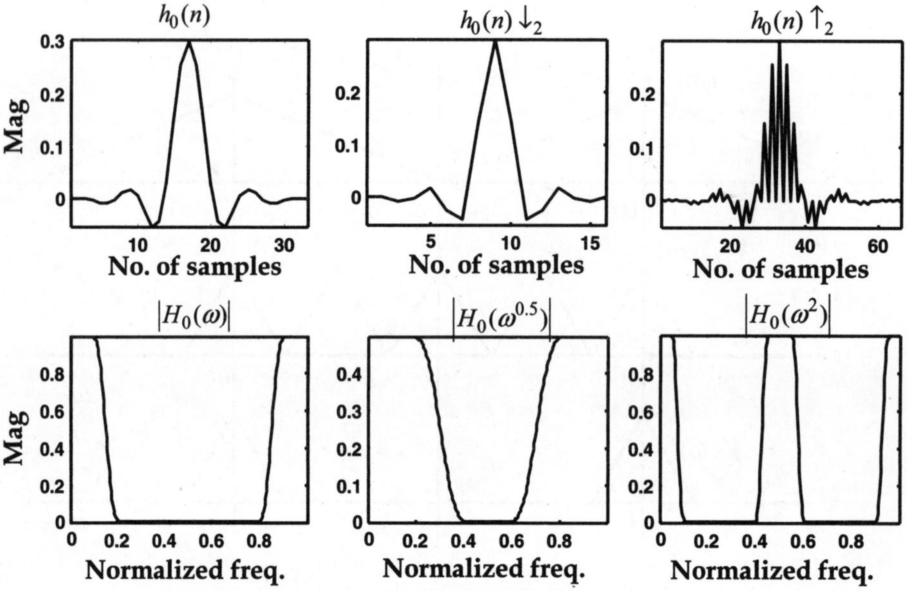

Fig. 12.5(a) Effect of decimation and interpolation on low-pass filter characteristics.

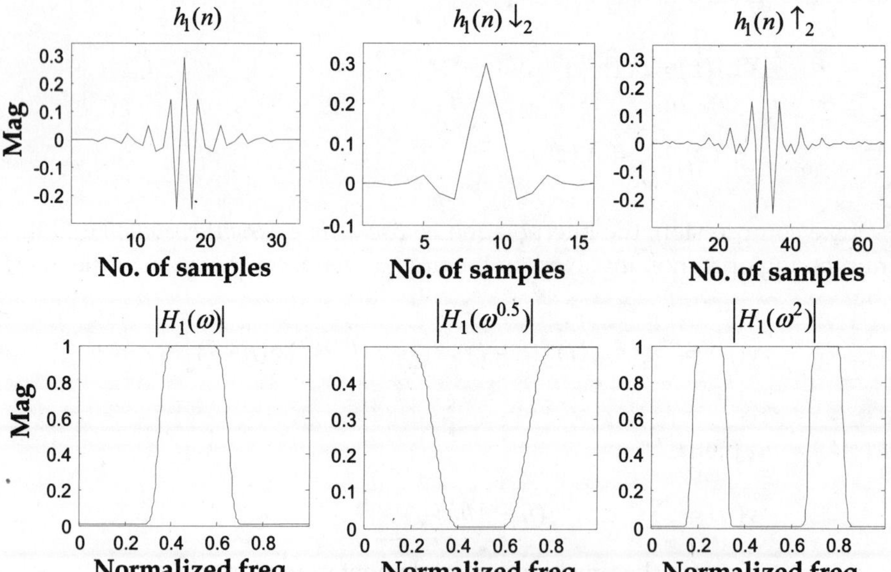

Fig. 12.5(b) Effect of decimation and interpolation on high-pass filter characteristics.

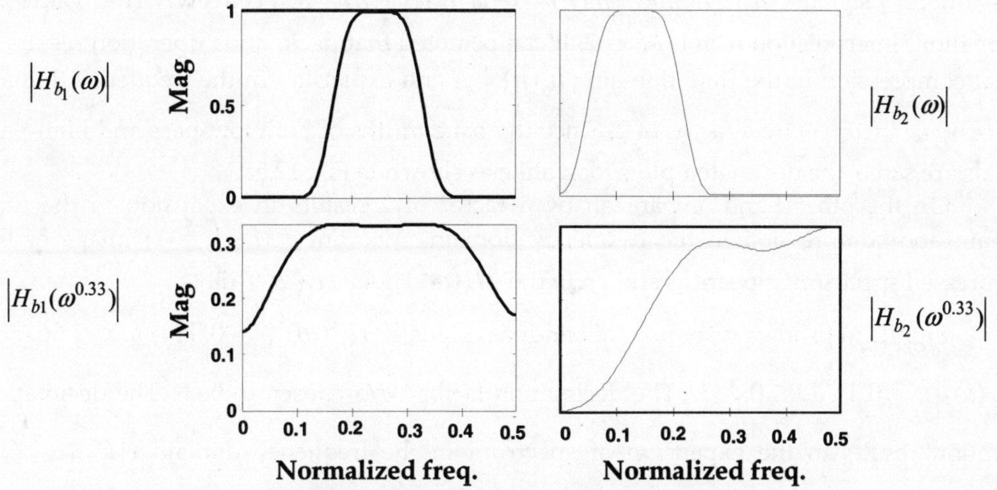

Fig. 12.5(c) Decimation operation on band-pass signal.

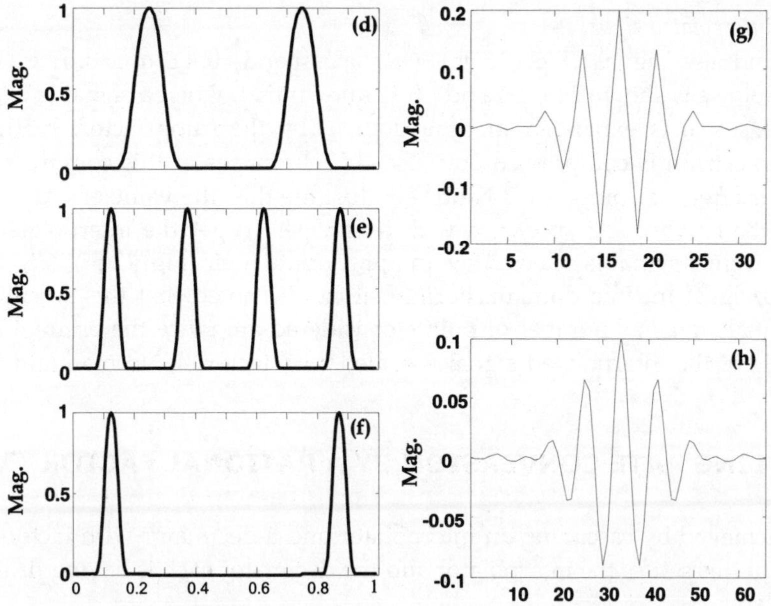

Fig. 12.5 (d - h) Recovery of the interpolated signal: (d) Original signal spectrum; (e) Interpolated spectrum showing images; (f) Frequency response of the image removal Filter; (g) time domain signal and (h) Interpolated signal.

Fig. (12.5a) shows the decimation and interpolation operation on the low-pass and high-pass filters/ signals $h_0(n)$ and $h_1(n) = (-1)^n h_0(N-n)$, respectively. The value of decimation/interpolation factor M is 2. It can be noted that decimation operation results in signal compression in the time domain ($h_0(n)\downarrow_2$) and expansion in the frequency domain ($H_0(\omega^{0.5}), H_1(\omega^{0.5})$) by a factor of 2. Since the bandwidths of both low-pass and high-pass signals are same, the decimated plots look alike as shown in Fig. 12.5a.

On the other hand, expansion by a factor of 2 results in expansion in the time domain and compression in the frequency domain. Also one ($M-1$ = 1) image of the compressed spectrum appears as shown in the $H_0(\omega^2)$ and $H_1(\omega^2)$ plots.

Fig. 12.5(c) shows responses of bandpass signals $\left|H_{b_1}(\omega)\right|$, $0.33 \le \omega \le 0.66$ and $\left|H_{b_2}(\omega)\right|$, $0.1 \le \omega \le 0.433$. The decimation factor M is chosen to be 3. The decimation operation results in the expansion of spectrum in the frequency domain ($\left|H_{b_1}(\omega^{0.33})\right|$, $\left|H_{b_2}(\omega^{0.33})\right|$). Depending on the bandwidth, the decimated frequency response may exhibit aliasing as in the case of $\left|H_{b_2}(\omega^{0.33})\right|$.

Recovering of interpolated signal

A bandpass signal (Fig. 12.5g) with passband $0.4 \le \omega \le 0.6$ is selected. Its frequency response is shown in Fig.12.5d. It is known that, when a signal is expanded by a factor M, the signal is expanded in time domain by the same factor. In the frequency domain, the spectrum is compressed and also $M-1$ images of this compressed spectrum version appear after the compressed band. To illustrate this, the value of M used is 2. Fig. 12.5e shows the compressed spectrum with its image. To get the interpolated signal, the compressed original signal is selected by an appropriate filter (Fig. 12.5f). Fig. 12.5h shows the expanded signal in time domain. Further, it can be noted that this signal is smoother than the original signal as number of points considered are twice the original number. But the magnitude of the interpolated signal is scaled by a factor M to maintain the energy a constant.

12.4 SAMPLING RATE CONVERSION BY A RATIONAL FACTOR (N/M)

This can be achieved by cascading an interpolator and a decimator with factors N and M, respectively. In this setup, the interpolator and the decimator each is having its own lowpass filters separately.

A lowpass filter is required for the interpolator after the zeros are being introduced to remove the spectral images. But for the decimator, a lowpass filter is necessary to remove the high frequency components before the signal being down sampled to avoid aliasing. These two filters can be replaced by a single filter to get a sampling rate conversion by a factor (N/M). For this the frequency response of the lowpass filter should satisfy

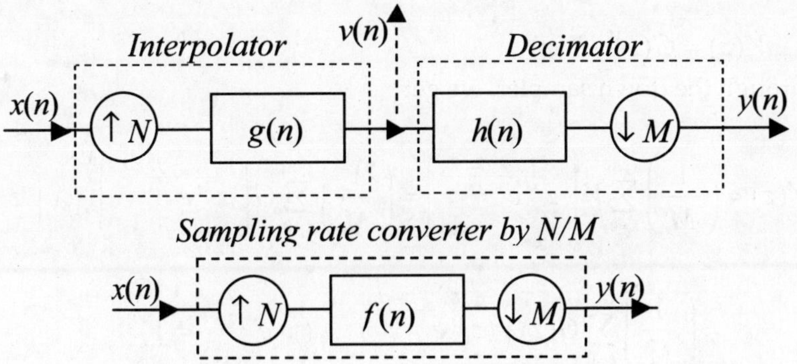

Fig. 12.6 Sampling rate converter by a rational factor.

$$f(\omega_v) = \begin{cases} N, & 0 \le \omega_v \le \min\left(\dfrac{\pi}{M}, \dfrac{\pi}{N}\right), \ \omega_v = \dfrac{\omega_x}{N} \\ 0 & \end{cases} \tag{12.3}$$

ω_x and ω_v correspond to frequency axis of the signals $x(n)$ and $v(n)$.

12.5 NOBLE IDENTITIES

The relations used in manipulating the transfer functions across the up and down samplers are called noble identities.

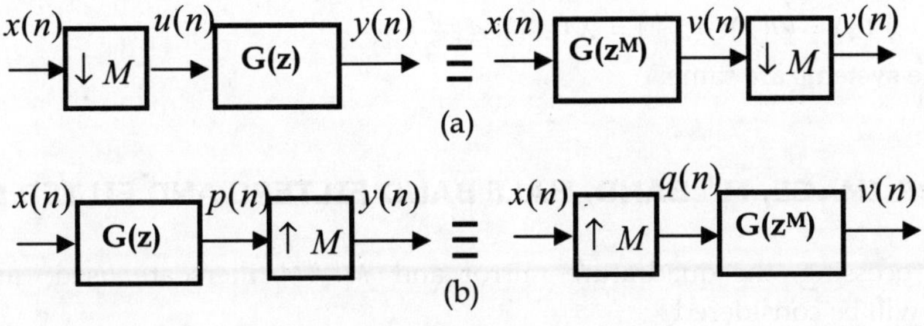

Fig. 12.7 Equivalent structures.

Decimation operation: The output of a down sampler (LHS of Fig. 12.7a) is

$$Y(z) = G(z)U(z) = G(z)\left(\frac{1}{M}\right)\sum_{k=0}^{M-1} X\left(z^{\frac{1}{M}} e^{\frac{-j2\pi k}{M}}\right) \tag{12.4a}$$

For the structure to the right,

$$V(z) = G(z^M)X(z)$$

Passing this through the down sampler, we get

$$Y(z) = \left(\frac{1}{M}\right)\sum_{k=0}^{M-1} V\left(z^{\frac{1}{M}} e^{\frac{-j2\pi k}{M}}\right) = \left(\frac{1}{M}\right)\sum_{k=0}^{M-1} G\left(z^{\frac{1}{M}} e^{\frac{-j2\pi k}{M}}\right)^M X\left(z^{\frac{1}{M}} e^{\frac{-j2\pi k}{M}}\right)$$

$$= \left(\frac{1}{M}\right)\sum_{k=0}^{M-1} G(z)X(z^{\frac{1}{M}} e^{\frac{-j2\pi k}{M}}) \quad (\because e^{-j2\pi k} = 1)$$

$$= \left(\frac{1}{M}\right) G(z)\sum_{k=0}^{M-1} X\left(z^{\frac{1}{M}} e^{\frac{-j2\pi k}{M}}\right) \tag{12.4b}$$

This is the same as $Y(z)$ of Eqn. (12.4a).

Interpolation operation: The output of the interpolator of the LHS (Fig. 12.7b) is

$$P(z) = G(z)X(z)$$

$$Y(z) = P(z^M) = G(z^M)X(z^M) \tag{12.4c}$$

For the structure on the RHS,

$$Q(z) = X(z^M)$$

$$Y(z) = Q(z)G(z^M) = X(z^M)G(z^M) \tag{12.4d}$$

Hence, both the systems are same.

12.6 MIRROR IMAGE, Mth BAND, HALF BAND FILTERS AND FILTER BANKS

In subband processing, the quadrature mirror and $M^{th\ band}$ filters are used and hence presently they will be considered .

12.6.1. Mirror Image Filters

The mirror image filter $h_m(n)$ of a FIR filter $h(n)$ is defined as

$$h_m(n) = (-1)^n h(n) \tag{12.5a}$$

If $H(e^{j\omega})$ is the frequency domain representation of $h(n)$ then,

$$H(e^{j\omega}) = \sum_{n=-\infty}^{\infty} h(n)e^{-j\omega n} \tag{12.5b}$$

Similarly,

$$H_m(e^{j\omega}) = \sum_{n=-\infty}^{\infty} h_m(n)e^{-j\omega n}$$

$$= \sum_{n=-\infty}^{\infty} (-1)^n h\,(n)e^{-j\omega n}$$

$$= \sum_{n=-\infty}^{\infty} e^{j\pi\, n} h\,(n)e^{-j\omega n}, \qquad \because e^{j\pi} = -1$$

$$= \sum_{n=-\infty}^{\infty} h(n)e^{-j(\omega-\pi)n}$$

$$H_m(e^{j\omega}) = H\left[e^{j(\omega-\pi)}\right]$$

$$H_m(e^{j\omega}) = H(-z) \tag{12.5c}$$

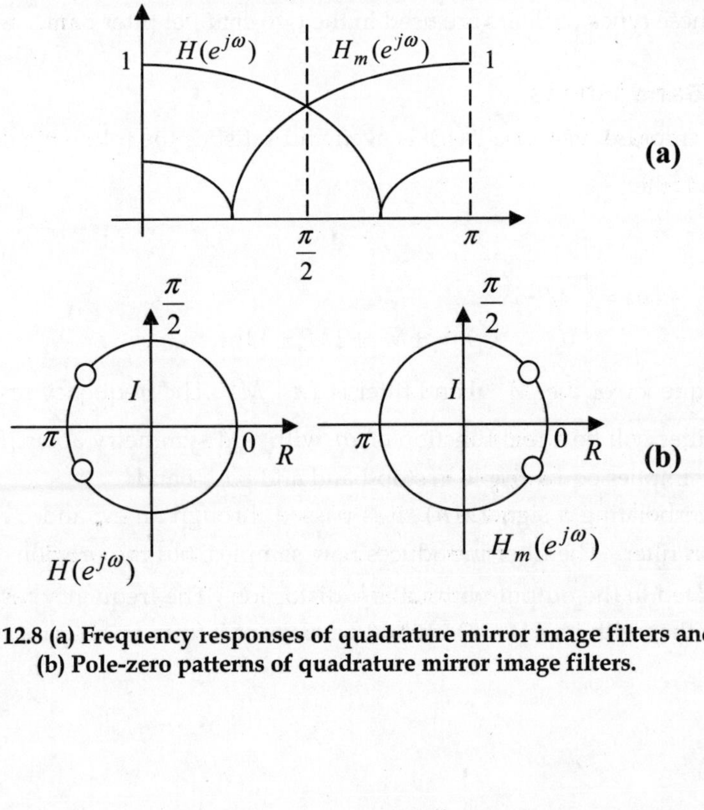

Fig. 12.8 (a) Frequency responses of quadrature mirror image filters and (b) Pole-zero patterns of quadrature mirror image filters.

Substituting for $\omega = \left(\frac{\pi}{2} - \omega\right)$,

$$(\omega - \pi) = \frac{\pi}{2} - \omega - \pi = -\left(\omega + \frac{\pi}{2}\right)$$

Since

$$H_m(e^{j\omega}) = H\left[e^{j(\omega-\pi)}\right],$$

$$\left|H_m\left(e^{j(\pi/2-\omega)}\right)\right| = \left|H\left(e^{-j(\pi/2+\omega)}\right)\right| = \left|H\left(e^{j(\pi/2+\omega)}\right)\right|$$

since *the magnitude is an even function of* ω. Therefore,

$$\left|H_m\left(e^{j(\pi/2-\omega)}\right)\right| = \left|H\left(e^{j(\pi/2+\omega)}\right)\right| \tag{12.5d}$$

the filter responses $H(e^{j\omega})$ and $H_m(e^{j\omega})$ are mirror images about the frequency $\left(\frac{\pi}{2}\right)$. Therefore, they are called quadrature mirror filters (QMF) (Fig. 12.8a).

The quadrature mirror image property implies that the pole zero patterns of the two filters are also mirror images about the imaginary axis i.e., the frequency $\frac{\pi}{2}$ of the z plane (Fig. 12.8b). These types of filters are used in the two channel filter bank sub-band coders.

12.6.2. M^{th} Band Filters

A filter whose impulse response $h(n)$ is even and satisfies the following condition is called as an Mth band filter.

$$h(n) = \begin{cases} \dfrac{1}{M}, & n = 0 \\ 0, & n = \pm M, \pm 2M, \pm 3M \ldots \end{cases} \tag{12.6a}$$

The cut-off frequency of the Mth band filter is (π / M). The frequency response $H(e^{j\omega})$ of an Mth band filter will be a real function of ω with odd symmetry about $[(\pi / M, (1/M)]$. This symmetry implies equal ripple in stop-band and pass-band.

For interpolating a signal $x(n)$, it is passed through an expander and then through an interpolation filter. The filter introduces new samples, but the previous existing samples are communicated to the output without any distortion. The frequency response of an Mth band filter together with its $(M-1)$ shifted versions of it form an all-pass system.

12.6.3 Half-band Filters

An Mth band filter with $M = 2$ is called a half band filter. This filter will have an impulse response, which is even and satisfies

$$h(n) = \begin{cases} 1/2, & n = 0 \\ 0, & n = \pm 2, \pm 4, \pm 6, \dots \end{cases} \tag{12.6b}$$

Its frequency response $H(e^{j\omega})$ has a bandwidth of $(\pi/2)$ and it exhibits anti-symmetry about $[(\pi/2),(1/2)]$ hence, the name half band filter. The peak pass-band and stop-band errors δ_1 and δ_2 are equal, and the band edges ω_p and ω_s are equally spaced from $(\pi/2)$.

Also,

$$H(e^{j\omega}) + H(e^{j(\pi-\omega)}) = 1 \tag{12.6c}$$

Proof:

Let

$$s(n) = h(n) + (-1)^n h(n)$$

Since from the definition of half-band filter, $h(n) = 1/2$ for $n = 0$, therefore $s(n) = 1$ for $n = 0$

Also as $s(n)$ is zero for n odd and n even $(n \neq 0)$ as $h(n)$ itself zero for $n = \pm 2, \pm 4, \pm 6 \dots\dots$,

$$s(n) = \delta(n)$$

i.e.,

$$h(n) + (-1)^n h(n) = \delta(n) \quad \text{or} \quad H(z) + H(-z) = 1$$

or

$$H(e^{j\omega}) + H(e^{j(\pi-\omega)}) = 1$$

A generalization of this to M^{th} band filter, leads to the property stated above that the Mth band filter and its $(M-1)$ shifted versions form an all-pass filter (already mentioned).

12.6.4 Length and Order Consideration

For an FIR filter, the length of the filter and the order N are related by $L = N + 1$. Let k be the index of the highest non-zero coefficient. Since $h(2n) = 0$ for a half band filter, k must be odd. The half band filter is symmetric, and therefore length of the filter is $2k + 1$.

The length L of the impulse response of the half band filter is

$$L = 2(k) + 1 = 2(2n + 1) + 1$$

(k being odd, it can be expressed in terms of n, where n can take odd or even values)

$$L = 4n + 2 + 1 \tag{12.6d}$$

The order of the half band filter is given by $N = 4n + 2$. That is, the order of half band filter is always twice that of an odd number.

12.6.5 Digital Filter Banks

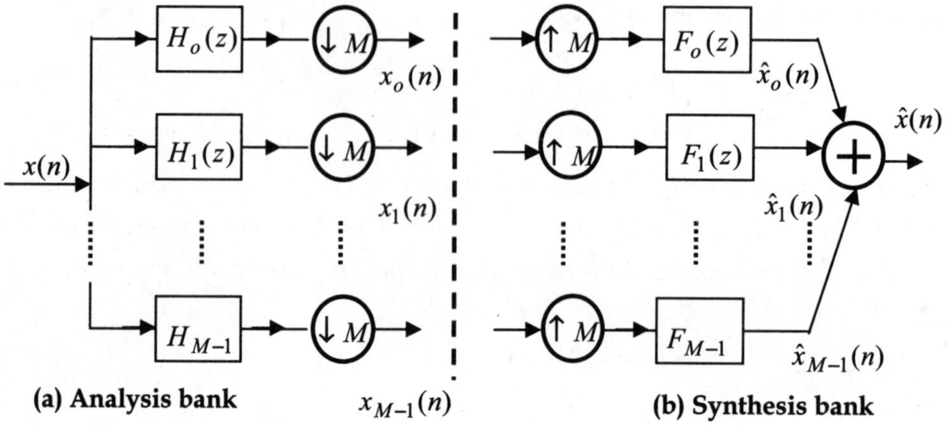

Fig. 12.9(a,b) Digital filter bank.

A digital filter bank is a collection of digital filters with common input or common output. Fig. (12.9a) depicts an analysis filter bank. The filters $H_o(z), H_1(z), ...H_{M-1}$ are called analysis filters. The input signal $x(n)$ is split into M signals $x_o(n), x_1(n), ...x_{M-1}(n)$ which are down sampled by a factor M and these signals are called subband signals. Typical synthesis bank is shown in Fig. (12.9b). The filters $F_o(z), F_1(z), ...F_{M-1}$ are called synthesis filters. The M subband signals are interpolated by a factor M and then are fed to the synthesis filters whose outputs are added to get the reconstructed signal $\hat{x}(n)$ as the output. The cascade of these two filter banks shown in Fig. (12.9a) and (12.9b) results in a M channel filter bank.

As stated in chapter 3, a signal delayed and scaled, is considered a distortionless signal. Hence, for the present context, if the reconstructed/ synthesized signal is delayed and scaled version of the input to the filter bank, then such a system can be treated as a perfect reconstruction system. The transfer function of a cascaded filter bank satisfying the perfect reconstruction property is $T(z) = Cz^{-k}$, where, k is a constant.

12.6.6 Filter Bank Structures

An M-channel filter bank can be realized by using a single prototype M^{th} band filter as a basic unit and translating it to other frequencies. Such a filter bank is called DFT filter bank as the prototype filter translation can be realized using IDFT matrix.

A filter bank with two channels is called a two-channel filter bank. The first filter of this filter bank is a low-pass filter and the second filter is a high-pass filter. The subband tree structure can be realized by using a two-channel filter bank as a building block. A two-channel filter bank is realized by using the QMF.

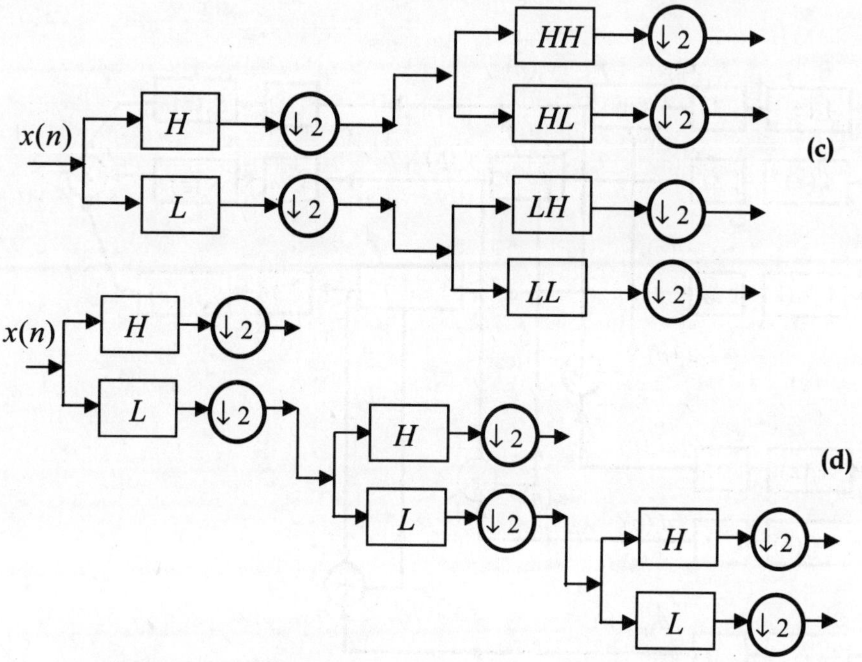

Fig. 12.9(c, d) (c) Regular binary filter bank and (d) Dyadic tree filter bank.

(a) Regular binary

A two-channel filter bank divides the input spectrum into two equal subbands, yielding the low (L) and high (H) bands. These bands can further be divided into quarter bands (LL), (LH), (HL), (HH) by using the same filter bank (Fig. 12.9a). When this process is repeated L times we get 2^L equal width subbands. Such decomposition provides the desired requirements of subband adaptive filters.

(b) Dyadic tree

Here, at any level of the tree, only the lower half of the spectrum after decimation by a factor of 2, is divided into two equal bands (Fig. 12.9(b). Also, at any level of the tree, the higher half band component of the signal is only decimated by a factor of 2 but not decomposed further. Such a filter bank is called dyadic tree filter bank and this structure provides wavelet transform coefficients which can form as inputs to the adaptive filters.

12.7 SUBBAND ADAPTIVE FILTER STRUCTURES

In this section, two basic types of subband adaptive filters, the synthesis independent/open loop and synthesis dependent/closed loop, will be considered (K.A. Lee, W.S. GAn and S.M. Kuo, John Wiley and Sons, 2009). Further, their relative merits in terms of convergence rate and accuracy (misadjustment) will be discussed.

12.7.1 Synthesis Independent/Open Loop Subband Adaptive Filter

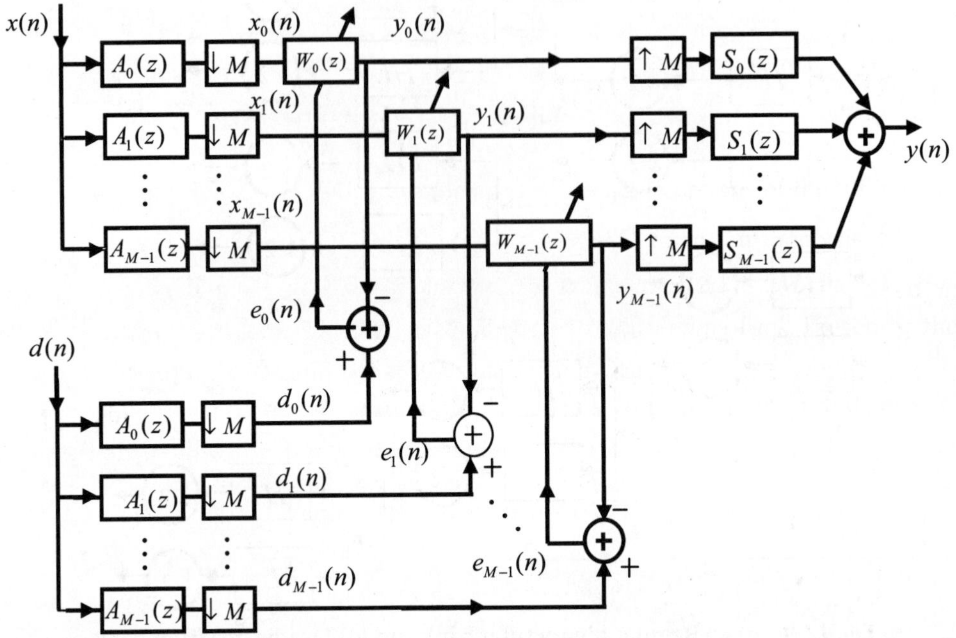

Fig. 12.10 (a) Synthesis independent /open loop subband adaptive structure.

Fig. 12.10(a) shows a subband adaptive filter structure (synthesis in independent). The input $x(n)$ and the desired signal $d(n)$ are split into M subband components by two identical filter banks $A_0(z), A_1(z), \ldots A_{M-1}(z)$. Both the input and desired signal components are down sampled by a factor M to achieve $x_i(n)$ and $d_i(n)$, $i = 0, 1, 2, \ldots M-1$. Each down sampled signal $x_i(n)$ is fed to an adaptive filter $W_i(z)$. The required errors $e_i(n)$ for adaptation are generated from the desired signal components $d_i(n)$ and the outputs $y_i(n)$ of the adaptive filters $W_i(z)$. Using the inputs $x_i(n)$ and the errors $e_i(n)$, the adaptive filters $W_i(z)$ are adapted by NLMS algorithm. A synthesis filter bank interpolates the adaptive filter outputs $y_i(n)$ by a factor M and these are added to get the estimate $y(n)$ of $d(n)$.

The adaptation rule for the ith subband adaptive filter $W_i(n)$ is given by

$$W_i(n+1) = W_i(n+1) + \mu_k(n)\, e_i(n) X_i(n), \qquad i = 0, 1, \ldots M-1 \qquad (12.7a)$$

$$W_i(n) = \left[\, w_{i0}(n), w_{i1}(n), \ldots, w_{iN_i-1}(n) \,\right],$$

$$X_i(n) = \left[x_i(n), x_i(n-1), \ldots, x_i(n-N_i+1)\right]$$

and N_i is the length of the ith subband adaptive filter and its output $y_i(n)$ and error $e_i(n)$ are

$$y_i(n) = \sum_{k=0}^{Ni-1} w_{ik} x_i(n-k)$$

$$e_i(n) = d_i(n) - y_i(n)$$

In terms of individual weight elements, for the kth weight of the ith adaptive filter,

$$w_{ik}(n+1) = w_{ik}(n) + \alpha_i(n)e_i(n)x_i(n-k), \quad k = 0, 1, \ldots N_i - 1 \tag{12.7b}$$

where $\alpha_i(n)$ is the step size for the ith adaptive filter and is given by

$$\alpha_i(n) = \frac{\mu_i}{N_i P_i(n)}, \quad 0 < \mu_i < 1 \tag{12.7c}$$

and $P_i(n)$ is input power for the ith adaptive filter and is recursively computed as

$$P_i(n) = \beta_i P_i(n-1) + (1-\beta_i)x_i^2(n), \quad 0 < \beta_i < 1 \tag{12.7d}$$

The above adaptation is valid for $i = 0, 1, \ldots M-1$.

In Fig. 12.10(a), in the subband adaptive filter loop, the synthesis filter bank is not included and the adaptation is independent of the synthesis filtering. Hence, this structure is known as ***synthesis independent SAF.*** In this, the errors introduced by the synthesis filter bank are left unattended by the adaptation.

12.7.2 Synthesis Dependent/Closed Loop Subband Adaptive Filter

In the synthesis dependent SAF, the errors due to synthesis filters can be reduced by including them in the *adaptive loop.* Here, input is decomposed $x(n)$ is decomposed into components by an analysis bank as in synthesis independent SAF. Instead of computing the individual errors from the desired signal subband components, the fullband error $e(n)$ is computed using the synthesis filter bank output $y(n)$ and the desired signal $d(n)$ (Fig. 12.10b).

In the synthesis-dependent SAF, the errors due to synthesis filters can be reduced by including them in the *adaptive loop.* Here, input is decomposed $x(n)$ is decomposed into components by an analysis bank as in synthesis independent SAF. Instead of computing the individual errors from the desired signal subband components, the fullband error $e(n)$ is computed using the synthesis filter bank output $y(n)$ and the desired signal $d(n)$ (Fig. 12.10b). Further, this error $e(n)$ is decomposed by an identical analysis bank as used for the input $x(n)$ to get the component errors $e_i(n)$ and their decimated versions are used for adaptive filters for each subband. As the synthesis dependent adaptive filter error takes into account the errors due to synthesis filters by feeding back and adjusting the adaptive filter weights to minimize this distortion, it provides a very low distortion in its final output $y(n)$.

For the adaptive filters with respects to the inputs $x_i(n)$, the required errors $e_i(n)$ get delayed by the delay due to synthesis filters $S_i(z)$. This is due to the fact that the analysis filters delay is common to both to inputs $x_i(n)$ and errors $e_i(n)$ and the synthesis filter delay is actual delay Δ for the adaptation assuming the all the synthesis filters are of same length.

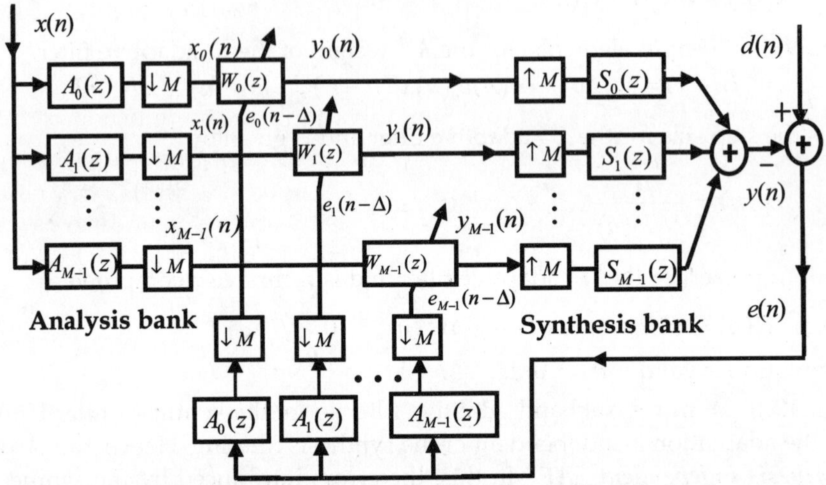

Fig. 12.10(b) Synthesis-dependent /closed loop subband adaptive structure.

The adaptation rule for the ith subband adaptive filter $W_i(n)$ is given by

$$W_i(n+1) = W_i(n+1) + \mu_k(n)\, e_i(n-\Delta) X_i(n), \qquad i = 0, 1, \ldots M-1 \tag{12.8a}$$

The fullband error is

$$e(n) = d(n) - y(n) \tag{12.8b}$$

The individual errors $e_i(n)$ for the subbands are obtained by passing the error $e(n)$ through an analysis filter bank identical to that for the fullband input $x(n)$. In terms of individual weight elements, for the kth weight of the ith adaptive filter,

$$w_{ik}(n+1) = w_{ik}(n) + \alpha_i(n) e_i(n-\Delta) x_i(n-k), \quad k = 0, 1, \ldots N_i - 1 \tag{12.8c}$$

As discussed in the chapter 10, (section 10.8), the step-size is given by

$$\alpha_i(n) = \frac{\mu_i}{N_i P_i(n)} \sin\left(\frac{\pi}{2(2\Delta+1)}\right) \qquad 0 < \mu_i < 1 \tag{12.8d}$$

Since for all values of delay $\Delta \neq 0$, the factor involving *sine function is less than unity* and only for delay $\Delta = 0$, the value of $\alpha_i(n)$ becomes same as that of normal step size,

$$\alpha_i(n) = \frac{\mu_i}{N_i P_i(n)},$$

the delay reduces the range of values that the step size can take *and hence permits only a smaller step size reducing the speed of convergence.* In synthesis independent structure, the delay corresponding to synthesis bank is not there and hence it *converges faster than the synthesis dependent one.* But it suffers from distortion due to aliasing that may be present in the synthesis filters. The *synthesis independent is preferred to the dependent* as they will not suffer from poor convergence and as the aliasing distortion of the synthesis filters can be reduced by proper filter design.

12.8 EFFECT OF DECIMATION ON CONVERGENCE RATE OF SUBBAND ADAPTIVE FILTERS

The subband adaptive filter's *attractive property* is *its faster convergence over that of full band.* As the analysis filters $A_0(z), A_1(z), \ldots A_{M-1}(z)$ split the signal fullband into M non-overlapping subbands (ideally), *the dynamic of range of the input spectrum within each subband is significantly reduced, for large* M, the number of subbands.

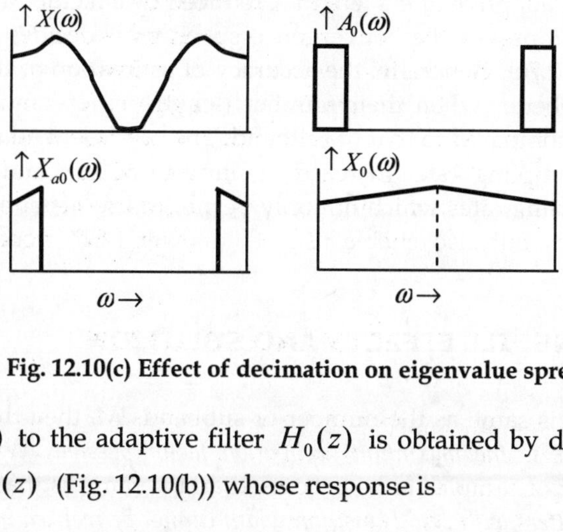

Fig. 12.10(c) Effect of decimation on eigenvalue spread.

The input $x_0(n)$ to the adaptive filter $H_0(z)$ is obtained by decimating the output of the analysis filter $A_0(z)$ (Fig. 12.10(b)) whose response is

$$A_0(e^{j\omega}) = \begin{cases} 1, & |\omega| < \pi / M \\ 0, & otherwise \end{cases}, \qquad X_0(e^{j\omega}) = \frac{1}{M} X_{a0}(e^{j\omega/M})$$

$X_{a0}(\omega)$: Spectrum of output of analysis filter $A_0(z)$ for input $X(\omega)$
$X_0(\omega)$: Spectrum of the decimated output of $A_0(z)$

The spectrum $X_0(\omega)$ *is flatter than* $X(\omega)$ (Fig. 12.10(c)) which implies that $x_0(n)$ *is whiter than* $x(n)$. For large M, this flatness will improve resulting in a much smaller eigenvalue spread of the autocorrelation matrix of $x_0(n)$ compared to that of full-band signal. From the analysis of method of steepest descent adaptation algorithm (chapter 9), the largest time constant which decides the possible fastest convergence speed is proportional to eigenvalue spread. Since eigenvalue spread for each subband is reduced, its convergence speed is increased. The analysis subband decomposes original input signal into M components $x_i(n)$, $i = 0, \ldots, M-1$, which feed the M adaptive filters. The power of these individual components is less than the power of the original input $x(n)$.

Use of normalized LMS algorithm for each of the subband adaptive filter, divides the step size μ_i by the respective component power $P_i(n)$ (Eqn. 12.7c).

$$\alpha_i(n) = \frac{\mu_i}{N_i P_i(n)}, \quad 0 < \mu_i < 1$$

Hence, the step size used for each component is larger than that used for a fullband signal in a single transversal adaptive filter and this results in higher speed of convergence for each subband and consequently that for the overall subband adaptive filter. This will be explicit if $\mu_i(n) = \mu$ and step size of the fullband adaptive filter is μ. Further, due to decimation, the subband adaptive filters length is reduced by a factor M compared to that of fullband and this also improves the convergence speed as it decides the step size of the individual filters (Eqn. 12.7c). Generally, the accuracy of estimation of the coefficients of the adaptive filters will be better when their number (length) is less and this is true for the subband adaptive filter compared to that of fullband. The coefficient adaptation and filtering are done at a lower sampling rate. Increase in number of subbands M will result in significantly lower sampling rates which not only permit to use better algorithms which are computationally complex but also enable use of slower DSP processors whose cost is comparative low.

12.9 OVERSAMPLING: ILL EFFECTS AND SOLUTION

If the decimation factor is same as the number of subbands M, then the sampling is called *critical sampling and this is the maximum decimation factor possible.* With this, there is no increase in total number of samples to be processed and is same as that of original fullband signal. However, FIR *filters with zero transition width cannot be realized in practice (due to their infinite length impulse response) use of critical sampling is not feasible.* Non-overlapping filters remove aliasing but result in spectral gaps, which may affect the performance as they attenuate signal components, which exist in those gap regions. An adaptive line enhancer realized by such a subband filter gets affected if the spectral line to be enhanced falls into one of the spectral gaps. Though spectral gaps can be covered by use of additional filters, it is not easy to implement. Hence, overlapping subband filters is an inevitable choice and this suffers from aliasing due to overlap in the transition region (Fig. 12.10(d)). But as these overlapping filters can be designed with sufficient stopband attenuation, *oversampling* which uses a smaller decimation factor L than the number of subbands M can be used to reduce aliasing. A sufficiently small $L < M$, will significantly reduce aliasing and the residual aliasing can be limited by proper filter design. Even with oversampling, the computational reduction achieved, will be comparable to that of critical sampling.

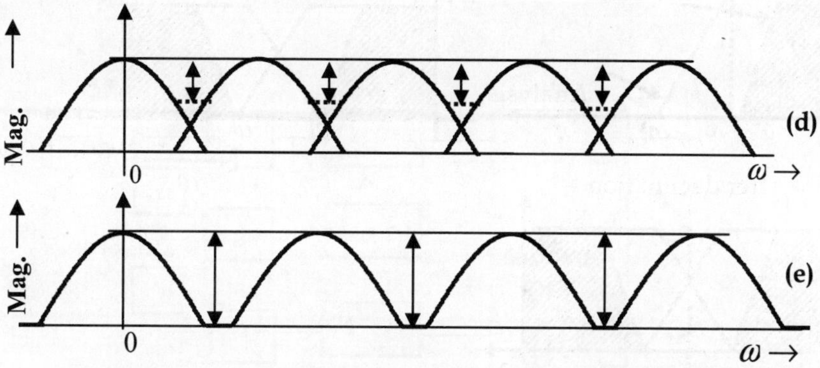

Fig. 12.10 (d, e) Trade-off between aliasing and eigenvalue spread: (d) Low eigenvalue spread due to more aliasing and (e) High eigenvalue spread due to oversampling to avoid aliasing.

Use of oversampling to avoid aliasing makes the adaptive filter input to have a spectrum that varies from a high value within the desired band to a low value at its band ends (Fig. 12.10(d, e)). (This is somewhat similar to non-overlapping filters but does not affect severely due to absence of wide spectral gaps.) *This effectively results in a high eigenvalue spread, which leads to poor convergence of the adaptive filter.* Though the individual filter characteristics of analysis and synthesis filter banks can be of same nature, use of such filter banks cannot solve the problem of eigenvalue spread due to oversampling. The aliasing though undesirable, it reduces the spectral dynamic range and hence reduces the eigenvalue spread. Thus, there is a trade-off between aliasing reduction and eigenvalue spread.

The flexibility of having different filter characteristics for analysis and synthesis banks can be exploited to solve the eigenvalue spread problem. Here, the analysis filter has a flat magnitude and linear phase response from zero to ω_{pa} (passband edge of analysis filter) and $\omega_{pa} > \omega_{ss}$, ω_{ss} is starting of the stopband edge of the synthesis filter. The cascade of such analysis and synthesis filter banks overcome the eigenvalue spread problem as multiplication of analysis and synthesis filters result in a filter that is same as the synthesis filter. That is, the overall frequency response is same as that of synthesis filter since the frequency response of the analysis filter is flat even beyond ω_{ss} as $\omega_{pa} > \omega_{ss}$ and frequencies beyond ω_{ss} are removed by the synthesis filters. The contribution of analysis filters is only a fixed delay to the overall response. Hence, the analysis filters preserve the spectral flatness of the input in each band. Thus, for *each adaptive filter, the input is having a flat spectrum even beyond the required range* ω_{ss} *hence, the problem of large eigenvalue spread at the edges of the subbands is removed.* (For the cascade of analysis and synthesis filter banks to provide perfect reconstruction, it should satisfy power complementary/power symmetric property which will be dealt later).

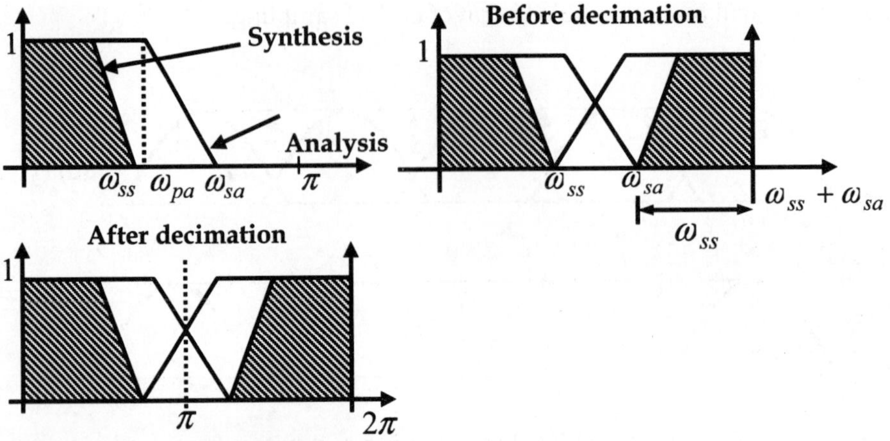

Fig. 12.11 Prototype filters for analysis and synthesis filter bank and the effect of decimation on them.

The maximum frequency range that $(\omega_{ss}+\omega_{sa})$ can be expanded without aliasing of the signal spectra over the bands of interest selected by synthesis filters is given by

$$(\omega_{ss}+\omega_{sa})L_{\max} = 2\pi$$

$$L_{\max} = \left\lfloor \frac{2\pi}{\omega_{ss}+\omega_{sa}} \right\rfloor \tag{12.9}$$

$\lfloor x \rfloor$ denotes largest integer smaller than or equal to x. This choice of L results in an aliasing in the output of the analysis but these aliased frequencies are filtered by the synthesis filters, hence, will not affect the fullband output of the filter.

It is preferred to have a large number of subbands as it reduces the eigenvalue spread with in the subbands, which improves the convergence rate. However, the delay due to analysis and synthesis filters increases with the number of subbands M and also with the oversampling factor L (as its choice is decided by M). This is due to the fact that the samples after decimation correspond to more number of original input sampling intervals and this will increase with M or L. Further, L decides the nature of the analysis and synthesis filters as their transition band's width should be narrow and the stopband attenuation should be large for higher values of M or L. The computational complexity decreases with increase in L as processing is reduced. But as there is limit on tolerable delay in certain applications (echo cancellation (16 milliseconds one way) and active noise control), satisfying the delay limit M and L are chosen as large as possible.

12.9.1 Computational Complexity Reduction of Subband Adaptive Filters

Consider an example with a system having fullband impulse response with N samples/taps, number of subbands M and decimation factor L. Then the number of taps required for each subband, $N_s = N / L$. The subband signals appear as complex conjugate pairs except at 0th and $(M/2)^{nd}$ bands. However, these two bands can be considered a single complex band. Then since it is equivalent having $(M/2)$ subbands, the computational complexity in terms of complex valued subband taps is

$$\frac{N}{L} \times \frac{M}{2} = \frac{NM}{2L}$$

The processing of subbands is at a rate $(1/L)$ of fullband rate. Therefore, the computational complexity in terms of taps becomes,

$$\frac{NM}{2L} \times \frac{1}{L} = \frac{NM}{2L^2}$$

Further, the adaptation is given by

$$W(n+1) = W(n) + 2\alpha e(n) X^*(n)$$

$$W_R(n+1) = W_R(n) + 2\alpha \{e_R(n) + je_I(n))\} \left[X_R(n) + jX_I(n) \right]^*$$

Hence, 1 complex valued tap is equal to 4 real valued taps and the computational. complexity of subband adaptive filter is

$$\frac{4NM}{2L^2} = \frac{2NM}{L^2}$$

The computational complexity of a fullband filter is N. Therefore, the computational advantage achieved by the subband adaptive filter over that of fullband is

$$\frac{\textit{Complexity of fullband filter}}{\textit{Complexity of subband filter}} = \frac{N}{\left(\frac{2NM}{L^2}\right)} = \frac{L^2}{2M}$$

For critical sampling, $L = M$ and the computational advantage becomes $(M/2)$. That is, for a 8 subband system, the computational advantage achieved is 4 over that of fullband. But for a oversampling case of $L = (M/2)$, which is generally used, the advantage is given by $M/8 = 8/8 = 1$ and hence there is no computational advantage.

For the real analysis filter bank, the length of subadaptive filters is also (N/L). Therefore, total number of taps for the M subband adaptive filter is (MN/L). Further, the processing of subbands is at a rate $(1/L)$ of full-band rate. Therefore, the computational complexity becomes (MN/L^2). Therefore, the computational advantage achieved by the subband adaptive filter over that of fullband is

$$\frac{\text{Complexity of fullband filter}}{\text{Complexity of subband filter}} = \frac{N}{\left(\dfrac{NM}{L^2}\right)} = \frac{L^2}{M}$$

For critical sampling, $L = M$ and the computational advantage becomes M. That is, for a 8 channel system, the computational advantage achieved is $L^2/M = M^2/M = M = 8$ over the fullband. But for a oversampling case of $L = (M/2)$, which is generally used, the advantage is given by $(L^2/M) = (M^2/4M) = (M/4) = (8/4) = 2$ and hence there is an advantage in computational load in real filter bank which is not there for a complex analysis filter bank. Thus, the real filter bank is advantageous even for oversampling case of $L = (M/2)$ *hence, real filter banks are preferred to complex one.*

In the above computational complexity analysis, complexity of analysis synthesis banks are not taken into account. The reason for this is that SAF is used for large N, that is for $N > 1000$, and for this, the complexity of analysis and synthesis banks will not be significant and it may be of the order of only 20% of the total computational load required for SAF implementation.

12.10 POLYPHASE REPRESENTATION

The filter bank can be realized efficiently by using the polyphase structure approach. Now polyphase approach will be considered in detail. For a given signal $x(n)$, its Z transform is given by

$$X(z) = \sum_{n=-\infty}^{\infty} x(n)z^{-n}$$

Separating the even and odd indexed terms,

$$X(z) = \sum_{n=-\infty}^{\infty} x(n)z^{-n} = \sum_{n=-\infty}^{\infty} x(2n)z^{-2n} + \sum_{n=-\infty}^{\infty} x(2n+1)z^{-(2n+1)}$$

$$X(z) = \sum_{n=-\infty}^{\infty} x(2n)z^{-2n} + z^{-1}\sum_{n=-\infty}^{\infty} x(2n+1)z^{-2n} \tag{12.10a}$$

Defining

$$x_0(n) = x(2n) \text{ and } x_1(n) = x(2n+1),$$

Then

$$\sum_{n=-\infty}^{\infty} x(2n)z^{-2n} = \sum_{n=-\infty}^{\infty} x_0(n)z^{-2n} = X_0(z^2)$$

$$\sum_{n=-\infty}^{\infty} x(2n+1)z^{-2n} = \sum_{n=-\infty}^{\infty} x_1(n)z^{-2n} = X_1(z^2).$$

Therefore,

$$X(z) = X_0(z^2) + z^{-1}X_1(z^2) \tag{12.10b}$$

In this case $x(2n)$ and $x(2n+1)$ were considered. As an extension if $x(3n)$, $x(3n+1)$ and $x(3n+2)$ are considered, then

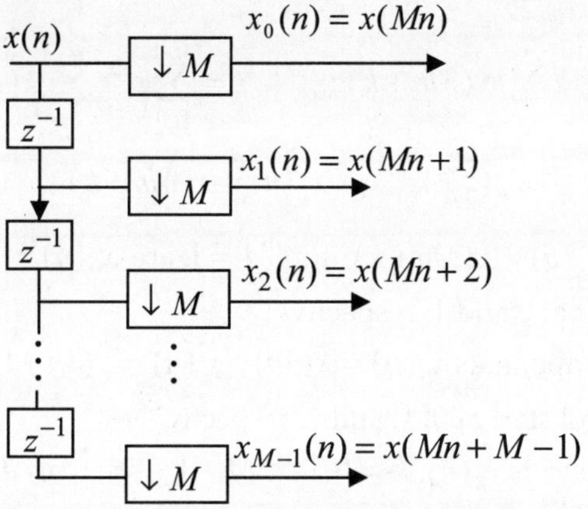

Fig. 12.12 Generation of polyphase components of a signal.

$$x_0(n) = x(3n),\; x_1(n) = x(3n+1) \text{ and } x_2(n) = x(3n+2)$$

which are given by

$$x_0(n) = x(3n) = x(0), x(3), x(6), x(9), x(12)\ldots, \sum_{n=-\infty}^{\infty} x_0(n)z^{-n} = \sum_{n=-\infty}^{\infty} x(3n)z^{-3n} = X_0(z^3)$$

$$x_1(n) = x(3n+1) = x(1), x(4), x(7), x(10), x(13)\ldots, \sum_{n=-\infty}^{\infty} x_1(n)z^{-n} = \sum_{n=-\infty}^{\infty} x(3n+1)z^{-3n} = X_1(z^3)$$

$$x_2(n) = x(3n+2) = x(2), x(5), x(8), x(11), x(14)\ldots, \sum_{n=-\infty}^{\infty} x_2(n)z^{-n} = \sum_{n=-\infty}^{\infty} x(3n+2)z^{-3n} = X_2(z^3)$$

$$X(z) = \sum_{n=-\infty}^{\infty} x(n)z^{-n} = \sum_{n=-\infty}^{\infty} x(3n)z^{-3n} + \sum_{n=-\infty}^{\infty} x(3n+1)z^{-(3n+1)} + \sum_{n=-\infty}^{\infty} x(3n+2)z^{-(3n+2)}$$

$$= \sum_{n=-\infty}^{\infty} x(3n)z^{-3n} + z^{-1}\sum_{n=-\infty}^{\infty} x(3n+1)z^{-3n)} + z^{-2}\sum_{n=-\infty}^{\infty} x(3n+2)z^{-3n}$$

Hence,

$$X(z) = X_0(z^3) + z^{-1}X_1(z^3) + z^{-2}X_2(z^3) \tag{12.10c}$$

Generalizing,

$$\sum_{n=-\infty}^{\infty} x_k(n)z^{-n} = \sum_{n=-\infty}^{\infty} x(Mn+k)z^{-n} = X_k(z^M), \quad k = 0,1,\cdots M-1 \tag{12.10d}$$

$$X(z) = \sum_{k=0}^{M-1}\sum_{n=-\infty}^{\infty} x(Mn+k)z^{-n}$$

$$= \sum_{k=0}^{M-1}\sum_{n=-\infty}^{\infty} x(Mn+k)z^{-(M\,n+k)} = \sum_{k=0}^{M-1} z^{-k}\sum_{n=-\infty}^{\infty} x(Mn+k)z^{-M\,n}$$

$$X(z) = \sum_{k=0}^{M-1} z^{-k}X_k(z^M), \qquad x_k(n) = x(Mn+k) \tag{12.10e}$$

The components $x_k(n) = x(Mn+k)$, for $M = 1$, are $x_0(n) = x(2n)$ $x_1(n) = x(2n+1)$ correspond to $k = 0, 1$ start at 0 and 1; respectively.

For $M = 2$, the components $x_0(n) = x(3n)$, $x_1(n) = x(3n+1)$ and $x_2(n) = x(3n+2)$ correspond to $k = 0, 1, 2$ and start at 0, 1, and 2; respectively.

That is, the components $x_k(n)$, $k = 0,1,\ldots M-1$; start at different instants of time $0,1,\ldots M-1$ hence, are called ***polyphase components.***

The polyphase components of a signal can be realized by a schematic as shown in Fig.12.12. To get M polyphase components, the signal is delayed by $(M-1)$ unit delay elements and each delayed output including the input (which has zero delay) are decimated by a factor M. These delayed decimated outputs are the M polyphase components.

The basis for such a generation is

$$x_k(n) = x(Mn+k), \quad k = 0,1,\cdots M-1$$

$$X(z) = X_0(z^M) + z^{-1}X_1(z^M) + z^{-2}X_2(z^M) + \ldots\ldots z^{-(M-1)}X_{M-1}(z^M)$$

Or

$$X(z) = \sum_{k=0}^{M-1} z^{-k}X_k(z^M) = \begin{bmatrix} 1 & z^{-1} & \cdots & z^{-(M-1)} \end{bmatrix} \begin{bmatrix} X_0(z^M) \\ X_1(z^M) \\ \vdots \\ X_{M-1}(z^M) \end{bmatrix} \tag{12.10f}$$

12.10.1 Polyphase Structure for an FIR Filter

For implementing the lowpass filter $H(z)$ efficiently using polyphase concepts, the first step is to decompose the filter impulse response $h(n)$ into M polyphase components $d_k(n) = h(Mn+k), k = 0,1,\ldots M-1$ and let their transfer functions be represented by $D_k(z^M), k = 0,1,\ldots M-1$. Then, the overall transfer function of the low-pass filter $H(z)$ is

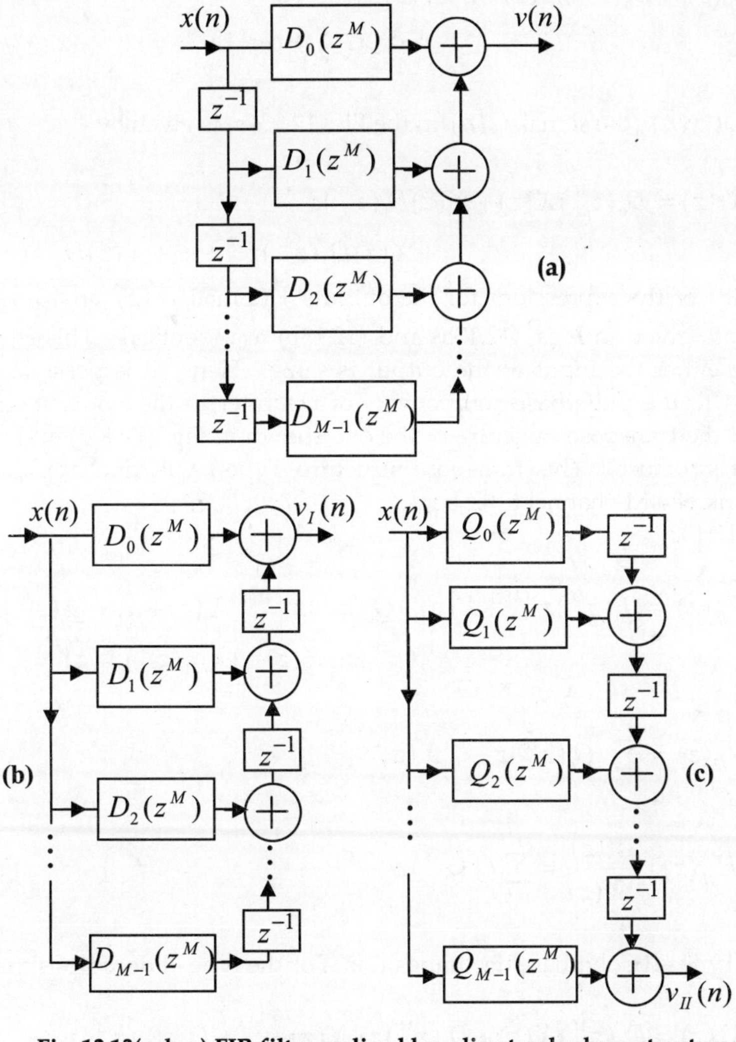

Fig. 12.13(a, b, c) FIR filter realized by : direct polyphase structure (a), transpose polyphase structures Type-I (b)and Type-II (c).

$$H(z) = D_0(z^M) + z^{-1}D_1(z^M) + z^{-2}D_2(z^M) + \ldots + z^{-(M-1)}D_{M-1}(z^M)$$

where

$$D_k(z^M) = \sum_{n=-\infty}^{\infty} d_k(n) z^{-Mn} = \sum_{n=-\infty}^{\infty} h(Mn+k) z^{-Mn}$$

This direct form of realization is shown in Fig. 12.13(a). For an input $x(n)$, the filter output $v(n)$ is given by

$$V(z) = D_0(z^M)X(z) + z^{-1}X(z)D_1(z^M) + z^{-2}X(z)D_2(z^M) + \ldots + z^{-(M-1)}X(z)D_{M-1}(z^M)$$

Again for the input $x(n)$, the signal $v_I(n)$ in the Fig. 12.13(b) is given by

$$V_I(z) = D_0(z^M)X(z) + X(z)D_1(z^M)z^{-1} + X(z)D_2(z^M)z^{-2} + \ldots + X(z)D_{M-1}(z^M)z^{-(M-1)}$$

The right hand side of the expressions for the outputs $V(z)$ and $V_I(z)$ are same and hence the configurations given in Figs. (12.13a) and (12.13b) are identical. This is because the effect of delaying either the input or the output is same. Hence it is possible to shift the delays from input to the polyphase components of the filter to their output side. Such a structure is called the transpose structure of the one shown in Fig. (12.13a) and this is also known as Type-I structure. This transpose structure Type-I will find application in the synthesis filter bank of a M-channel filter bank.

For Fig. (12.13c),

$$V_{II}(z) = Q_0(z^M)z^{-(M-1)}X(z) + Q_1(z^M)z^{-(M-2)}X(z) + \ldots + Q_l(z^M)z^{-l}X(z) + \ldots\ldots + Q_{M-2}(z^M)z^{-1}X(z) + Q_{M-1}(z^M)X(z)$$

$$V_{II}(z) = \sum_{l=0}^{M-1} Q_l(z^M) z^{M-1-l} X(z)$$

Or
$$H_{II}(z) = \frac{V(z)}{X(z)} = \sum_{l=0}^{M-1} Q_l(z^M) z^{-M-1-l}$$

This is known as Type -II polyphase decomposition. For the type -I decomposition

$$V_I(z) = D_0(z^M)X(z) + D_1(z^M)z^{-1}X(z) + \ldots + D_{M-1-l}(z^M)z^{-l} + \ldots \ldots\ldots D_{M-2}(z^M)z^{-(M-2)}X(z) + D_{M-1}(z^M)z^{-(M-1)}X(z)$$

$$V_I(z) = \sum_{l=0}^{M-1} D_l(z^M)z^{-l}X(z)$$

$$H_I(z) = \frac{V(z)}{X(z)} = \sum_{l=0}^{M-1} D_l(z^M)z^{-l}$$

For $H_{II}(z)$ and $H_I(z)$ to be equal and comparing the coefficients of same powers of z in the expressions for $V_I(z)$ and $V_{II}(z)$,

$$Q_l(z^M) = D_{M-1-l}(z^M), \quad 0 \le l \le M-1$$

The $D_l(z)$ of a filter with impulse response $h(n)$ is given by

$$D_l(z) = \sum_{n=-\infty}^{\infty} h(Mn+l)z^{-n}, \quad 0 \le l \le M-1$$

or

$$d_l(n) = h(Mn+l), \quad 0 \le l \le M-1$$

($d_l(n)$ is generally represented as $h_l(n)$). Therefore,

$$q_l(n) = d_{M-1-l}(n) = h_{M-1-l}(n) = h(Mn+M-1-l)$$

$$q_l(n) = h(Mn+M-1-l)$$

That is, for the Type-II representation, $q_l(n)$ is the time domain version of the lth polyphase component of $h(n)$. It is important to note that the transpose type-I and Type-II versions differ in the way the polyphase component outputs are summed to get the final output. In type-I, the delay increases for the first component $D_0(z^M)$ output to last component $D_{M-1}(z^M)$ output. However, for Type-II, it is reverse and the first component $Q_0(z^M)$ output has maximum delay and the last $Q_{M-1}(z^M)$ has the least, i.e., zero delay. This is due to the relation between $Q_l(z^M)$ and $D_l(z^M)$ given above. The transpose structures are used mainly for interpolation in synthesis filter banks and depending upon the convenience, Type-I or Type-II structures are used.

12.10.2 Polyphase Representation of a Decimator

The basic decimator schematic is shown in the Fig. 12.14(a). Here, the lowpass filter $H(z)$ band limits the signal $x(n)$ and then for a decimation factor of M, once in every sample $M-1$ samples is retained. In the direct implementation of the lowpass filter of impulse response length N, N - multiplications and $N-1$ additions/output sample are required.

For the decimator, the lowpass filter $H(z)$ can be realized using polyphase concepts. For this, the lowpass filter realization given in Fig.12.13(a) can be used and the decimator with this is shown in Fig. 12.14(b). For this case also, N multiplications and $N-1$ additions must be done in one sampling period of the input. For succeeding $M-1$ samples the arithmetic units are idle.

For an efficient polyphase implementation of a decimator, the decimation is moved prior to $H_k(z^M)$ using noblé identity and as a consequence Fig. 12.14(b) becomes Fig. 12.14(c).

If N_l multiplications and $(N_l - 1)$ additions per output sample are required by the lth polyphase component filter $D_l(z)$, then the overall structure per decimator output needs,

$\sum_{l=0}^{M-1} N_l = N$ multiplications and $\sum_{l=0}^{M-1}(N_l - 1) + (M-1) = (N-1)$ additions

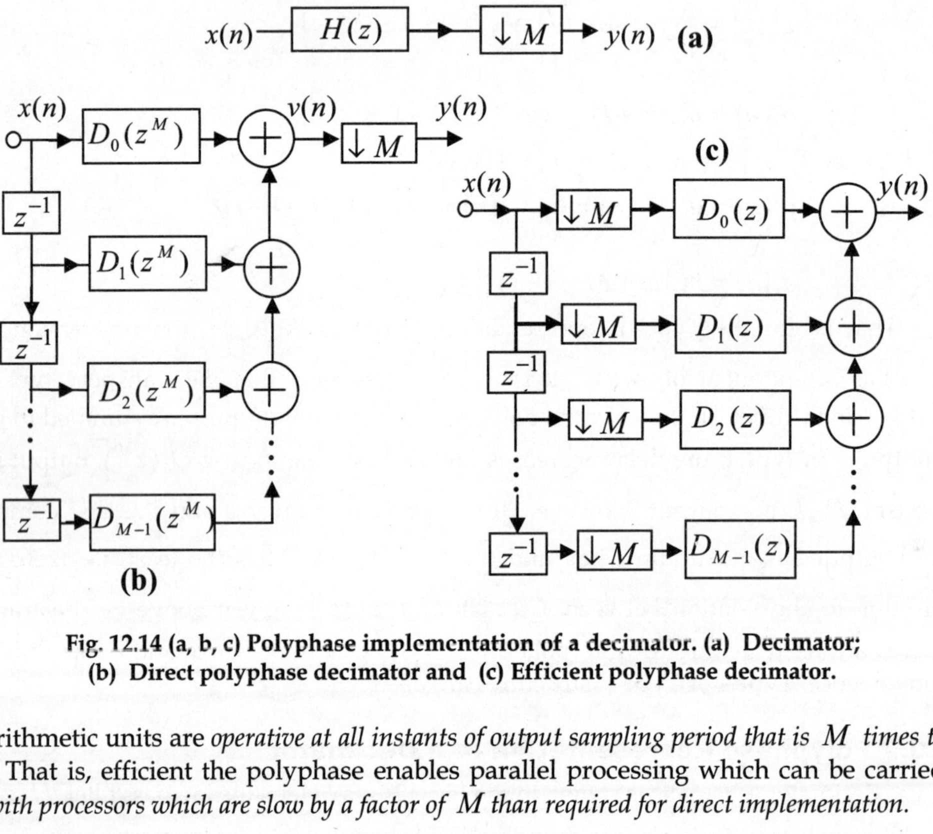

Fig. 12.14 (a, b, c) Polyphase implementation of a decimator. (a) Decimator; (b) Direct polyphase decimator and (c) Efficient polyphase decimator.

The arithmetic units are *operative at all instants of output sampling period that is M times that of input.* That is, efficient the polyphase enables parallel processing which can be carried out *even with processors which are slow by a factor of M than required for direct implementation.*

12.10.3 Polyphase Representation of an Interpolator

The normal interpolator is shown in Fig. 12.15(a). Here, also similar to decimator, the low-pass filter block is implemented by a polyphase structure as in Fig. 12.15(b). In Fig. 12.15(c) the delay elements are shifted after the transfer functions $R_k(z^M), k = 0,1,2\ldots(M-1)$ and this does not modify the functionality. To get efficient implementation, using the noble

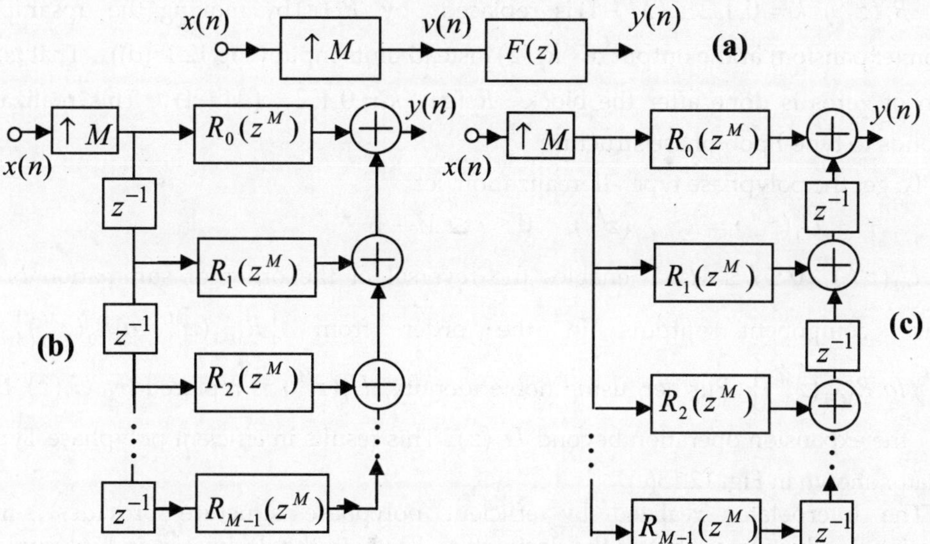

Fig. 12.15(a, b, c) Polyphase interpolator implementation: (a) Normal interpolator; (b) By direct polyphase structure and (c) Transpose polyphase structure.

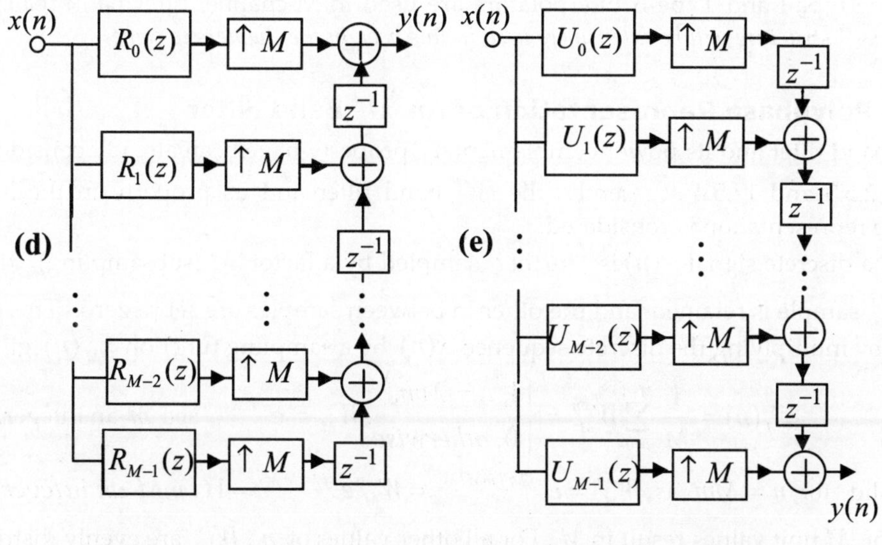

Fig. 12.15(d, e) Polyphase interpolator implementation: (d) Efficient type-I polyphase structure and (e) Efficient type-II polyphase structure.

identity $R_k(z^M), k = 0,1,2\ldots(M-1)$ is replaced by $R_k(z)$ by moving the upsampling operation (expansion) at the output of $R_k(z)$ instead of its input (Fig.12.15(d)). That is, the insertion of zeros is done after the blocks $R_k(z),\ k = 0,1,2\ldots(M-1)$. This realization corresponds to type-I polyphase structure.

To get the polyphase type –II realization, let

$$U_l(z^M) = R_{M-1-l}(z^M), \quad 0 \le l \le M-1$$

Use of $U_l(z^M),\ 0 \le l \le M-1$ enables the reversal of the order of summation of the polyphase component outputs in the order from $\{R_{M-1}(z^M)\ to\ R_0(z^M)\}$ to $\{R_0(z^M)\ to\ R_{M-1}(z^M)\}$. Further, using noble identity, $U_l(z^M)$ is replaced by $U_l(z)$ by is shifting the expansion operation beyond $U_l(z)$. This results in efficient polyphase Type-II interpolator shown in Fig. 12.15(e).

The interpolator realized by efficient polyphase structure provides similar computational reduction as that of the decimator. For both the decimator and interpolator, further reduction in computations is possible in specific cases like linear phase filters by taking advantage of symmetry property. Thus, the implementation of decimator/interpolator by polyphase structure enables parallel processing using even processors which are slow by a factor M.

The type-I and Type-II interpolators are used in M-channel filter bank realization. But *the type-II structure will be extensively used in this book in the chapters to come.*

12.10.4 Polyphase Representation of An Mth Band Filter

The half-band filter and its power complementary property have been already considered in sections 12.5.3 and 12.5.4 Presently, the Mth band filter and its property in the light of polyphase representation is considered.

If a discrete signal $x(n)$ is further sampled by a factor M (subsampling), then its every M^{th} sample is retained and the other in between samples are set to zero. This can be achieved by multiplying the discrete sequence $x(n)$ by a sampling function $\xi_M(n)$ given by

$$\xi_M(n) = \frac{1}{M}\sum_{l=0}^{M-1} W_M^{ln} = \begin{cases} 1, n = Mm, \\ 0, otherwise \end{cases}, W_M = e^{-j2\pi/M} \text{ and } m \text{ an integer.}$$

This is valid for $n = Mm$ as $W_M^{ln} = e^{-j2\pi lmM/M} = W_M^{ln} e^{-j2\pi lmM} = 1(l\,mM\ an\ integer)$ and addition of M unit values result in M. For all other values of n, W_M^{ln} are evenly distributed on the unit circle and their sum is zero. This can be extended for values of n, $n = Mm + k$, k is an offset. For this,

$$\xi_M(n-k) = \frac{1}{M}\sum_{l=0}^{M-1} W_M^{l(n-k)} = \begin{cases} 1, n = Mm + k, \\ 0, otherwise \end{cases}, W_M = e^{-j2\pi/M}$$

and for a given M, the signal $x(n)$ can be reconstructed by M different sampled signals each with a different offset k.

$$x(n) = x_0(n) + x_1(n) + \dots + x_{M-1}(n)$$

$$= x(n)\xi_M(n) + x(n)\xi_M(n-1) + \dots + x(n)\xi_M(n-M+1)$$

$$x(n) = \sum_{k=0}^{M-1} x_k(n) = \sum_{k=0}^{M-1} x(n)\xi_M(n-k)$$

This is the *polyphase representation in time domain* and $x_k(n), k = 0,1,\dots M-1$; are the polyphase components of $x(n)$. Taking *z*-transform,

$$X(z) = \sum_{n=-\infty}^{\infty} x(n)z^{-n} = \sum_{k=0}^{M-1}\sum_{m=-\infty}^{\infty} x(Mm+k)z^{-(Mm+k)}$$

$$= \sum_{k=0}^{M-1}\left[\sum_{m=-\infty}^{\infty} x(Mm+k)z^{-Mm}\right]z^{-k}$$

If

$$X_k(z^M) = \sum_{m=-\infty}^{\infty} x(Mm+k)z^{-Mm},$$

then,

$$X(z) = \sum_{k=0}^{M-1} X_k(z^M)z^{-k},$$

This implies that

$$z^{-k}X_k(z^M) = \sum_{n=-\infty}^{\infty} x_k(n)z^{-n}, \quad k = 0,2,\dots M-1 \tag{12.11a}$$

That is, $X_k(z^M)z^{-k}$ is the Z-transform of $x_k(n)$ represented as

$$X_k(z^M)z^{-k} \Leftrightarrow x_k(n), \quad k = 0,2,\dots M-1.$$

But

$$x_k(n) = x(n)\xi_M(n-k)$$

$$x_k(n) = x(n)\frac{1}{M}\sum_{r=0}^{M-1} W_M^{r(k-n)}, \qquad \because \xi_M(n-k) = \xi_M(k-n) \text{ as } \xi_M(.) = 1 \text{ or } 0$$

Substituting for $x_k(n)$ in Eqn. (12.11a),

$$X_k(z^M)z^{-k} = \frac{1}{M}\sum_{r=0}^{M-1}\sum_{n=-\infty}^{\infty} x(n)z^{-n}W_M^{r(k-n)}$$

$$= \frac{1}{M}\sum_{r=0}^{M-1}\sum_{n=-\infty}^{\infty 1} x(n)W_M^{r(k-n)}z^{-n}$$

$$= \frac{1}{M}\sum_{r=0}^{M-1}\sum_{n=-\infty}^{\infty 1} x(n)[zW_M^r]^{-n}W_M^{kr}$$

$$X_k(z^M)z^{-k} = \frac{1}{M}\sum_{r=0}^{M-1} X_r^{(m)}(z)W_M^{kr} \tag{12.11b}$$

where,

$$X_r^{(m)}(z) = \sum_{n=-\infty}^{\infty} x(n)[zW_M^r]^{-n} \quad \text{or} \quad X_r^{(m)}(z) = X[zW_M^r]$$

That is, the polyphase filter components are obtained from the modulation components using DFT.
The polyphase representation of a lowpass filter $H(z)$ is given by

$$H(z) = \sum_{k=0}^{M-1} z^{-k}D_k(z^M)$$

where $$D_k(z) = \sum_{n=\infty}^{\infty} h(Mn+k)z^{-n}$$

$D_k(z)$ is k^{th} polyphase component of $H(z)$. From Eqn. (12.11b),

$$D_k(z^M)z^{-k} = \frac{1}{M}\sum_{r=0}^{M-1} H_r^{(m)}(z)W_M^{kr}$$

For the 0th polyphase component, $k = 0$,

$$D_0(z^M) = \frac{1}{M}\sum_{r=0}^{M-1} H_r^{(m)}(z)$$

and the 0th polyphase component corresponds to $k = 0$,

$$D_0(z) = \sum_{n=\infty}^{\infty} h(Mn)z^{-n} \tag{12.11c}$$

For an Mth band filter,

$$h(Mn) = \begin{cases} 1/M, & n = 0 \\ 0, & n \neq 0 \end{cases}$$

That is, the 0th value of $h(n)$ is a nonzero constant and those whose indices are multiples of M, are zero. Therefore, the z-transform of the Mth band filter $H_{mb}(z)$ given by

$$H_{mb}(z) = \sum_{\lambda n=\infty}^{\infty} h(Mn)z^{-n} \tag{12.11d}$$

and this becomes

$$H_{mb}(z) = 1/M \tag{12.11e}$$

As the RHS of Eqns. (12.11c) and (12.11d) are same and further from Eqn. (12.11e),

$$D_0(z) = H_{mb}(z) = \frac{1}{M}$$

$$D_0(z^M) = \frac{1}{M} \tag{12.11f}$$

From Eqn. (12.11b),

$$D_k(z^M)z^{-k} = \frac{1}{M}\sum_{r=0}^{M-1} H_r^{(m)}(z)W_M^{kr}$$

For the 0th polyphase component,

$$D_0(z^M) = \frac{1}{M}\sum_{r=0}^{M-1} H_r^{(m)}(z)$$

From Eqn. (12.11f),

$$\frac{1}{M}\sum_{r=0}^{M-1} H_r^{(m)}(z) = \frac{1}{M}$$

From Eqn. (12.11b),

$$H_r^{(m)}(z) = H(zW^r)$$

Therefore,

$$\sum_{r=0}^{M-1} H(zW^r) = 1$$

Putting $z = e^{j\omega}$,

$$\sum_{r=0}^{M-1} H(e^{j(\omega-2\pi r/M)}) = 1 \tag{12.11g}$$

That is, the shifted versions of $H(e^{j\omega})$ by $(2\pi / M)$ will add to unity. This is the property of power complementary filter $H^c(e^{j\omega})$, where, $H(e^{j\omega}) = \left|H^c(e^{j\omega})\right|^2$. Hence, the spectral factor of an Mth band filter is a power complementary filter. From Eqn. (12.11f), the transfer function cannot have powers which are multiples of M. Hence for the Mth band filter,

$$h_{mb}(n) = 0,\ n = \pm iM, \quad i = 1,2,3,\ldots$$

This M^{th} band filter plays a major role in the coming chapters dealing with filter banks. For an interpolator $H(z)$ with a factor M,

$$Y(z) = H(z)X(z^M)$$

$H(z)$ can be realized by its M polyphase components as

$$H(z) = R_0(z^M) + z^{-1}R_1(z^M) + z^{-2}R_2(z^M) + \ldots\ldots$$
$$+ z^{-k}R_k(z^M) + \ldots\ldots + z^{-(M-1)}R_{(M-1)}(z^M)$$

If the kth polyphase components of $H(z)$ is a constant, i.e., $R_k(z) = \alpha$ then, the output of the interpolator is given by

$$Y(z) = \alpha z^{-k}X(z^M) + \sum_{\substack{l=0 \\ l\neq k}}^{M-1} z^{-l}R_l(z^M)X(z^M)$$

For $n = k, k \pm M, k \pm 2M, \ldots,$

$$y(Mn) = \alpha x(n-k)$$

That is, at these instants n, the output samples are same as those of input but in between $M-1$ samples are interpolated. This type filters where some of the polyphase components are constants, are called *Nyquist filters*. Many of their impulse response sample values are zero and this makes them computationally efficient. If the 0th *polyphase component of the filter is a constant,* this corresponds to M^{th} band filter and hence it is special case of a Nyquist filter. For an Mth band filter, every M^{th} sample value of the impulse response is zero.

12.11 GENERAL M-CHANNEL FILTER BANK

The input signal $x(n)$ is decomposed into M components $x_k(n)$ which occupy contiguous non-overlapping frequency bands each of width $2\pi / M$ (for a real input and real impulse response of a filter, the bandwidth is π / M) by passing it through an analysis filter bank

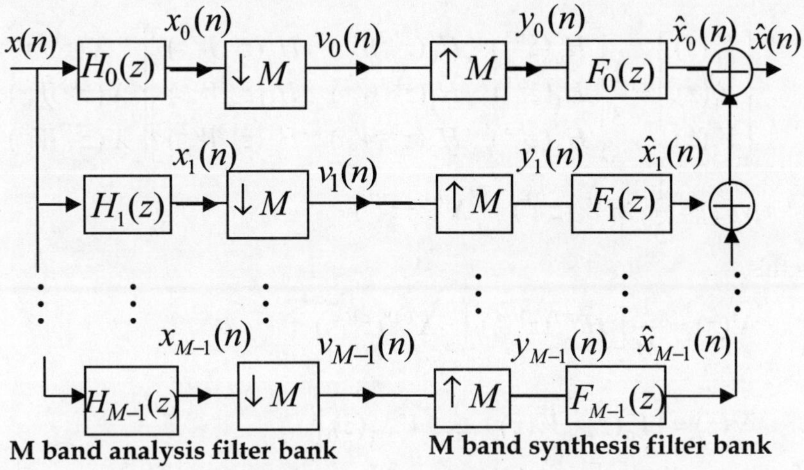

Fig. 12.16 Analysis and synthesis filter.

having M band-pass filters $H_l(z)$ (Fig. 12.16). Further, as their bandwidths are reduced by a factor M, they are decimated by the same factor to get the analysis bank output signals $v_l(n)$. The analysis filter bank outputs $v_k(n)$ are interpolated by a factor M to get signals $y_l(n)$. For the synthesis of the original signal, to remove the images of the interpolated signals $y_k(n)$, they are passed through a set of band-pass filters $F_l(z)$. The outputs $\hat{x}_l(n)$ are added to get the synthesized signal $\hat{x}(n)$. Under certain conditions, $\hat{x}(n)$ can be scaled and shifted version of $x(n)$ in which case the filter bank satisfies the perfect reconstruction property.

Starting from the input side and considering a particular channel, say l, the analysis filter output $X_l(z) = H_l(z)X(z)$ is decimated by a factor M as bandwidth of $H_l(z)$ is (π / M).

Therefore,

$$V_l(z) = \frac{1}{M}\sum_{k=0}^{M-1} H_l(z^{1/M}W_M^k)X(z^{1/M}W_M^k),\ l = 0, 1, \ldots, M-1, \tag{12.12a}$$

where $W_M = e^{-j2\pi/M}$ To obtain Eqn. ((12.12a) using vector notation, it is expanded for $M = 3$ as,

$$V_0(z) = \frac{1}{3}\left[H_0(z^{1/3})X(z^{1/3}) + H_0(z^{1/3}W_3)X(z^{1/3}W_3) + H_0(z^{1/3}W_3^2)X(z^{1/3}W_3^2)\right]$$

$$V_1(z) = \frac{1}{3}\left[H_1(z^{1/3})X(z^{1/3}) + H_1(z^{1/3}W_3)X(z^{1/3}W_3) + H_1(z^{1/3}W_3^2)X(z^{1/3}W_3^2)\right]$$

$$V_2(z) = \frac{1}{3}\left[H_2(z^{1/3})X(z^{1/3}) + H_2(z^{1/3}W_3)X(z^{1/3}W_3) + H_2(z^{1/3}W_3^2)X(z^{1/3}W_3^2)\right]$$

$$\begin{bmatrix} V_0(z) \\ V_1(z) \\ V_2(z) \end{bmatrix} = \frac{1}{3} \begin{bmatrix} H_0(z^{1/3}) & H_0(z^{1/3}W_3) & H_0(z^{1/3}W_3^2) \\ H_1(z^{1/3}) & H_1(z^{1/3}W_3) & H_1(z^{1/3}W_3^2) \\ H_2(z^{1/3}) & H_2(z^{1/3}W_3) & H_2(z^{1/3}W_3^2) \end{bmatrix} \begin{bmatrix} X(z^{1/3}) \\ X(z^{1/3}W_3) \\ X(z^{1/3}W_3^2) \end{bmatrix}$$

$$\mathrm{V}(z) = \frac{1}{3}\left[\boldsymbol{H}^{(m)}(z^{1/3})\right]^T \boldsymbol{X}^{(m)}(z^{1/3})$$

Generalizing this,

$$\mathrm{V}(z) = \frac{1}{M}\left[\boldsymbol{H}^{(m)}(z^{1/M})\right]^T \boldsymbol{X}^{(m)}(z^{1/M}) \tag{12.12b}$$

where

$$\mathrm{V}(z) = \left[V_0(z) \quad V_1(z) \quad \dots \quad V_{M-1}(z)\right]^T \tag{12.12c}$$

$$\mathrm{X}^{(m)}(z^{1/M}) = \left[X_0(z^{1/M}) \quad X_1(z^{1/M}) \quad \dots \quad X_{M-1}(z^{1/M})\right]^T \tag{12.12d}$$

$$\mathrm{X}^{(m)}(z^{1/M}) = \left[X(z^{1/M}) \quad X(z^{1/M}W_M) \quad \dots \quad X(z^{1/M}W_M^{M-1})\right]^T \tag{12.12e}$$

is the *modulation vector of* $x(n)$ and

$$\boldsymbol{H}_{AC}(z) = \boldsymbol{H}^{(m)}(z) = \begin{bmatrix} H_0(z) & H_1(z) & \dots & H_{M-1}(z) \\ H_0(zW_M) & H_1(zW_M & \dots & H_{M-1}(zW \\ \vdots & \vdots & \dots & \vdots \\ H_0(zW_M^{M-1}) & H_1(zW_M^{M-1}) & \dots & H_{M-1}(zW_M^{M-1}) \end{bmatrix} \tag{12.12f}$$

is the *modulation matrix of the analysis filter bank* . This matrix is also known as *alias component (AC)* matrix.

Next, the outputs of the analysis filters $v_l(n)$ are upsampled by a factor M prior to giving as input to synthesis filters. The output of the lth interpolator is given by

$$Y_l(z) = V_l(z^M) = \frac{1}{M}\sum_{k=0}^{M-1} H_l(z\, e^{-j2\pi\, k/M})X(ze^{-j2\pi\, k/M}) \tag{12.13a}$$

$$l = 0, 1, \dots, M-1$$

or

$$\mathrm{V}(z^M) = \frac{1}{M}\left[\boldsymbol{H}^{(m)}(z)\right]^T \boldsymbol{X}^{(m)}(z) \tag{12.13b}$$

$\hat{X}_l(z)$ is the obtained by filtering $y_l(n)$ through synthesis filter $F_l(z)$ as,

$$\hat{X}_l(z) = F_l(z)\, Y_l(z) = F_l(z)\, V_l(z^M), \quad l = 0, 1, \dots, M-1 \tag{12.13c}$$

The output of the synthesis filter bank is the sum of all $\hat{x}_l(n),\ l = 0,1,\cdots,M-1$.

$$\hat{X}(z) = \sum_{l=0}^{M-1} F_l(z)\, Y_l(z) = \sum_{l=0}^{M-1} F_l(z)\, V_l(z^M) \tag{12.13d}$$

$$\hat{X}(z) = \mathrm{F}^T(z)\mathrm{V}(z^M) = \mathrm{F}^T(z)\mathrm{Y}(z) \tag{12.13e}$$

where,

$$\mathrm{F}(z) = \begin{bmatrix} F_0(z) & F_1(z) & \dots & F_{M-1}(z) \end{bmatrix}^T \text{ and } \mathrm{Y}(z) = \begin{bmatrix} Y_0(z) & Y_1(z) & \dots & Y_{M-1}(z) \end{bmatrix}^T$$

Further, the modulated version of the synthesis filter bank output is

$$\hat{X}(zW_M^k) = \mathrm{F}^T(zW_M^k)\mathrm{V}(z^M), \qquad k = 0, 1, \dots, M-1$$

Let the modulated outputs be represented by a vector $\hat{\mathrm{X}}^{(m)}(z)$. Then

$$\hat{\mathrm{X}}^{(m)}(z) = \begin{bmatrix} \hat{X}(z) & \hat{X}(zW_M) & \dots & \hat{X}(zW_M^{M-1}) \end{bmatrix}^T \tag{12.13f}$$

For $M = 3$,

$$\begin{bmatrix} \hat{X}(z) \\ \hat{X}(zW_3) \\ \hat{X}(zW_3^2) \end{bmatrix} = \begin{bmatrix} F_0(z) & F_1(z) & F_2(z) \\ F_0(zW_3) & F_1(zW_3) & F_2(zW_3) \\ F_0(zW_3^2) & F_1(zW_3^2) & F_2(zW_3^2) \end{bmatrix} \begin{bmatrix} V_0(z^3) \\ V_1(z^3) \\ V_2(z^3) \end{bmatrix}$$

Generalizing this,

$$\hat{\mathrm{X}}^{(m)}(z) = \mathrm{F}^{(m)}(z)\, \mathrm{V}(z^M) \tag{12.13g}$$

where,

$$\mathrm{F}^{(m)}(z) = \begin{bmatrix} F_0(z) & F_1(z) & \dots & F_{M-1}(z) \\ F_0(zW_M) & F_0(zW_M) & \dots & F_{M-1}(zW_M) \\ \vdots & \vdots & \vdots & \vdots \\ F_0(zW_M^{M-1}) & F_1(zW_M^{M-1}) & \dots & F_{M-1}(zW_M^{M-1}) \end{bmatrix}$$

is the *modulation matrix of the synthesis filter bank.* Substituting for $\mathrm{V}(z)$ from Eqn. (12.12b) in Eqn. (12.13g),

$$\hat{\mathrm{X}}^{(m)}(z) = \frac{1}{M}\left\{ \mathrm{F}^{(m)}(z) \left[\mathrm{H}^{(m)}(z) \right]^T \right\} \mathrm{X}^{(m)}(z) \tag{12.14a}$$

$$\hat{\mathrm{X}}^{(m)}(z) = \mathrm{T}(z)\, \mathrm{X}^{(m)}(z) \tag{12.14b}$$

where,

$$\mathrm{T}(z) = \frac{1}{M}\left\{ \mathrm{F}^{(m)}(z) \left[\mathrm{H}^{(m)}(z) \right]^T \right\} \tag{12.14c}$$

The matrix $T(z)$ is the transfer matrix and is the product of modulation matrix of the synthesis filter bank and the transpose of the modulation matrix of the analysis filter bank. The input $X(z)$ and its frequency shifted versions are related through $T(z)$ to the output $\hat{X}(z)$ and its frequency shifted versions. For the perfect reconstruction, the output signal $\hat{x}(n)$ can only be delayed and scaled version of the input. That is, $\hat{X}(z) = K\, z^{-n_0} X(z)$.

12.12 POLYPHASE FILTER BANK REPRESENTATION

In this section, both the analysis and the synthesis filter banks are realized by polyphase structures. Further, these are combined to achieve the complete filter bank. The polyphase structures are modified to get computational efficiency and for the resulting structure, the condition required for perfect reconstruction is discussed.

12.12.1 Analysis Filter Bank

In the analysis filter bank of Fig. 12.17(a), the lth filter can be represented as

$$H_l(z) = \sum_{k=0}^{M-1} z^{-k} D_{lk}(z^M) \qquad (12.15a)$$

where $D_{lk}(z)$ is the polyphase component formed by picking the every M^{th} coefficient of $H_l(z)$ starting from the 0th element i.e., $d_{lk}(n) = h_l(nM + k)$, $\quad k = 0, 1, 2, \ldots\ M-1$.

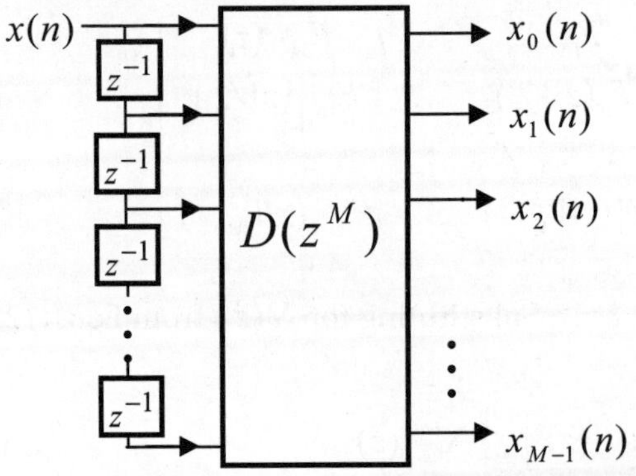

Fig. 12.17 (a) Polyphase representation of analysis filter bank.

Expanding for M=4,

$$H_l(z) = D_{l0}(z^4) + z^{-1}D_{l1}(z^4) + z^{-2}D_{l2}(z^4) + z^{-3}D_{l3}(z^4)$$

or $$H_l(z) = \begin{bmatrix} D_{l0}(z^4) & D_{l1}(z^4) & D_{l2}(z^4) & D_{l3}(z^4) \end{bmatrix} \begin{bmatrix} 1 \\ z^{-1} \\ z^{-2} \\ z^{-3} \end{bmatrix}$$

Now,

$$\begin{bmatrix} H_0(z) \\ H_1(z) \\ H_2(z) \\ H_3(z) \end{bmatrix} = \begin{bmatrix} D_{00}(z^4) & D_{01}(z^4) & D_{02}(z^4) & D_{03}(z^4) \\ D_{10}(z^4) & D_{11}(z^4) & D_{12}(z^4) & D_{13}(z^4) \\ D_{20}(z^4) & D_{21}(z^4) & D_{22}(z^4) & D_{23}(z^4) \\ D_{30}(z^4) & D_{31}(z^4) & D_{32}(z^4) & D_{33}(z^4) \end{bmatrix} \begin{bmatrix} 1 \\ z^{-1} \\ z^{-2} \\ z^{-3} \end{bmatrix}, \; l = 0, 1, 2, 3$$

$$\mathrm{H}(z) = \mathrm{D}(z^4)\mathrm{Z}_d(z)$$

where,

$$\mathrm{H}(z) = \begin{bmatrix} H_0(z) & H_1(z) & H_2(z) & H_3(z) \end{bmatrix}^T \text{ : Analysis filters}$$

$$\mathrm{Z}_d(z) = \begin{bmatrix} 1 & z^{-1} & z^{-2} & z^{-3} \end{bmatrix}^T \text{: Delay chain and}$$

$$\mathrm{D}(z^4) = \begin{bmatrix} D_{00}(z^4) & D_{01}(z^4) & D_{02}(z^4) & D_{03}(z^4) \\ D_{10}(z^4) & D_{11}(z^4) & D_{12}(z^4) & D_{13}(z^4) \\ D_{20}(z^4) & D_{21}(z^4) & D_{22}(z^4) & D_{23}(z^4) \\ D_{30}(z^4) & D_{31}(z^4) & D_{32}(z^4) & D_{33}(z^4) \end{bmatrix}$$

This is polyphase matrix of analysis bank. In general,

$$\mathrm{H}(z) = \mathrm{D}(z^M)\mathrm{Z}_d(z) \tag{12.15b}$$

The output of the analysis filter bank is given by

$$\mathrm{X}^{(m)}(z) = \mathrm{H}(z)\mathrm{X}(z) = \mathrm{D}(z^M)\mathrm{Z}_d(z)\mathrm{X}(z) \tag{12.15c}$$

For $M = 4$,

$$\mathrm{X}^{(m)}(z) = \mathrm{H}(z)\mathrm{X}(z) = \mathrm{D}(z^4)\mathrm{Z}_d(z)\mathrm{X}(z)$$

$$\mathrm{X}^{(m)}(z) = \mathrm{D}(z^4)\left[\mathrm{Z}_d(z)\mathrm{X}(z)\right]$$

That is, $\mathrm{Z}_d(z)$ operates on $\mathrm{X}(z)$ to get the delayed versions of $x(n)$, i.e., $x(n), x(n-1), x(n-2), x(n-3)$ and on these, $\mathrm{D}(z^4)$ operates to get $\mathrm{x}^{(m)}(n) = \begin{bmatrix} x_0(n) & x_1(n) & x_2(n) & x_3(n) \end{bmatrix}^T$. These operations are shown in the Fig. 12.17(a).

12.12.2 Synthesis Filter Bank

Using polyphase components, the lth synthesis filter can be represented as

$$F_l(z) = \sum_{k=0}^{M-1} z^{-(M-1-k)} R_{kl}(z^M) \tag{12.16a}$$

where,

$$r_{kl}(n) = f_l(nM + M - 1 - k), \quad k = 0, 1, 2, \ldots M-1$$

Expanding, for $M = 4$,

$$F_l(z) = z^{-3}R_{0l}(z^4) + z^{-2}R_{1l}(z^4) + z^{-1}R_{2l}(z^4) + R_{3l}(z^4)$$

For the interpolator or for the synthesis side, the delays should appear after the transfer function. To achieve this,

$$F_l(z) = \begin{bmatrix} z^{-3} & z^{-2} & z^{-1} & 1 \end{bmatrix} \begin{bmatrix} R_{0l}(z^4) \\ R_{1l}(z^4) \\ R_{2l}(z^4) \\ R_{3l}(z^4) \end{bmatrix}, \; l = 0, 1, 2, 3$$

$$\begin{bmatrix} F_0 & F_1 & F_2 & F_3 \end{bmatrix} = \begin{bmatrix} z^{-3} & z^{-2} & z^{-1} & 1 \end{bmatrix} \begin{bmatrix} R_{00}(z^4) & R_{01}(z^4) & R_{02}(z^4) & R_{03}(z^4) \\ R_{10}(z^4) & R_{11}(z^4) & R_{12}(z^4) & R_{13}(z^4) \\ R_{20}(z^4) & R_{21}(z^4) & R_{22}(z^4) & R_{23}(z^4) \\ R_{30}(z^4) & R_{31}(z^4) & R_{32}(z^4) & D_{33}(z^4) \end{bmatrix}$$

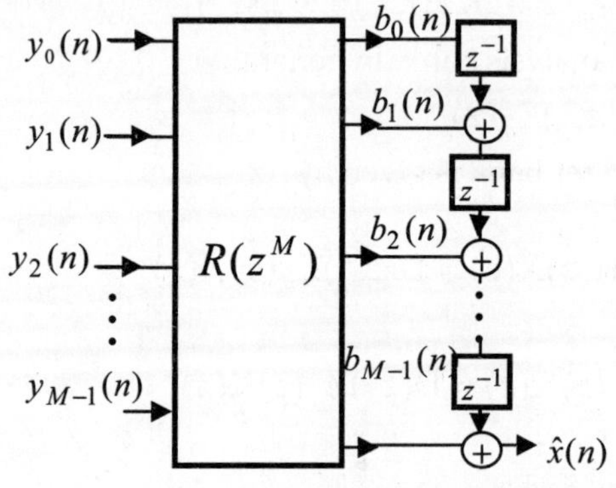

Fig. 12.17(b) Polyphase representation of synthesis filter bank.

$$\mathbf{F}^T(z) = \mathbf{Z}_I^T(z)\mathbf{R}(z^4)$$

where,

$$\text{F}(z) = \left[F_0(z), F_1(z), F_2(z), F_3(z)\right]^T : \text{Synthesis filter bank}$$

$$Z_I = \begin{bmatrix} z^{-3} & z^{-2} & z^{-1} & 1 \end{bmatrix}^T : \text{Delay chain}$$

$$R(z^4) = \begin{bmatrix} R_{00}(z^4) & R_{01}(z^4) & R_{02}(z^4) & R_{03}(z^4) \\ R_{10}(z^4) & R_{11}(z^4) & R_{12}(z^4) & R_{13}(z^4) \\ R_{20}(z^4) & R_{21}(z^4) & R_{22}(z^4) & R_{23}(z^4) \\ R_{30}(z^4) & R_{31}(z^4) & R_{32}(z^4) & D_{33}(z^4) \end{bmatrix} : \text{Synthesis matrix}$$

In general,

$$\mathbf{F}^T(z) = \mathbf{Z}_I^T(z)R(z^M) \tag{12.16b}$$

From Eqn. (12.13e),

$$\hat{X}(z) = F^T(z)Y(z),$$

i.e., $R(z^4)$ operates on $Y(z) = \begin{bmatrix} Y_0(z) & Y_1(z) & Y_2(z) & Y_3(z) \end{bmatrix}^T$ to get the signals (Fig. 12.17(b))

$$B(z) = \begin{bmatrix} R(z^4)Y(z) \end{bmatrix} = [B_0(z) \quad B_1(z) \quad B_2(z) \quad B_3(z)]^T$$

and the resulting signals are delayed by the delay chain Z_I^T to get

$$\hat{X}(z) = Z_I^T B(z) = \begin{bmatrix} z^{-3} & z^{-2} & z^{-1} & 1 \end{bmatrix} \begin{bmatrix} B_0(z) \\ B_1(z) \\ B_2(z) \\ B_3(z) \end{bmatrix}$$

In general,

$$\hat{X}(z) = Z_I^T(z)R(z^M)Y(z) \tag{12.16c}$$

12.12.3 M-channel Polyphase Filter Bank

The analysis and synthesis filter banks realized by polyphase structures can be combined to get the complete filter bank (Fig. 12.17c). Using the noble identities, the down sampling and up sampling units can be moved prior to analysis polyphase matrix $\boldsymbol{D}(z)$ and after the synthesis polyphase matrix $\boldsymbol{R}(z)$, respectively as in Fig. 12.17(d). The polyphase representation of the filter bank *enables the analysis and synthesis filters to operate over longer sampling intervals.* Further, $\boldsymbol{D}(z)$ and $R(z)$ can be combined as $\boldsymbol{C}(z)$ (Fig. 12.17e) with

$$\boldsymbol{C}(z) = \boldsymbol{D}(z)\boldsymbol{R}(z) \tag{12.16d}$$

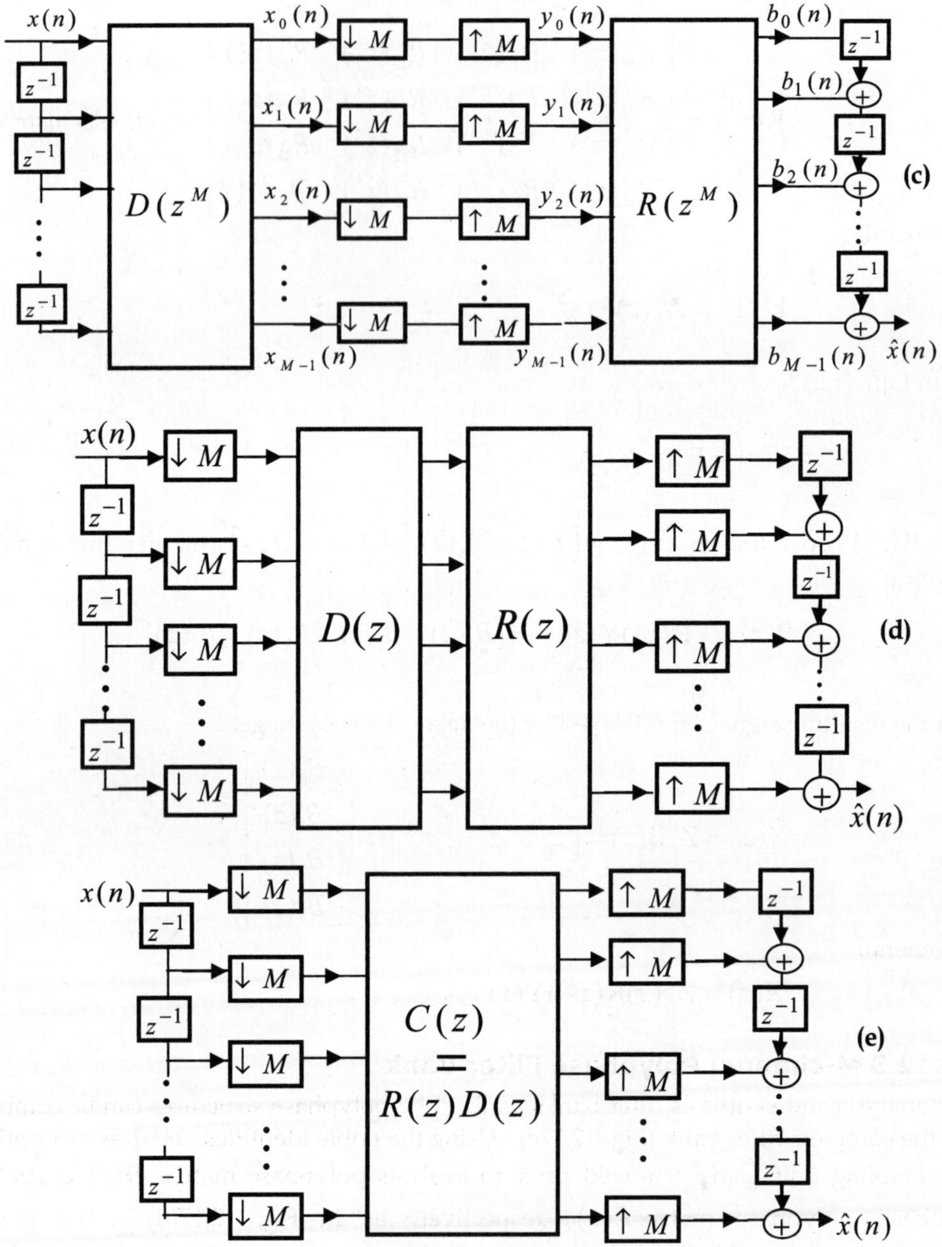

Fig. 12.17(c, d, e) : (c) Complete filter bank representation using polyphase analysis and synthesis banks; (d) Complete efficient polyphase filter bank and (e) Simplified complete efficient polyphase filter bank.

12.12.4 Perfect Reconstruction Requirements

If $C(z) = \mathrm{I}$, i.e., an identity matrix, then the output signal $\hat{x}(n)$ will be equal to the input $x(n)$ and the filter bank realized using polyphase structure provides a perfect reconstruction. This demands that given the analysis polyphase matrix $\boldsymbol{D}(z)$, the synthesis polyphase matrix $\boldsymbol{R}(z)$ should be chosen such that it is the inverse of $\boldsymbol{D}(z)$. However, this requires $\boldsymbol{D}(z)$ to be an invertible one. This condition $C(z) = \mathrm{I}$, is sufficient for perfect reconstruction for both FIR and IIR transfer functions. The condition on $C(z)$ can be further relaxed as

$$C(z) = Kz^{-n_d}\mathrm{I} \tag{12.16e}$$

which implies

$$\boldsymbol{R}(z) = Kz^{-n_d}\boldsymbol{D}^{-1}(z) \tag{12.16f}$$

Here, K is a constant not equal to zero and n_d is a positive integer. Thus, the polyphase implementation enables realization of the perfect reconstruction filter bank. Further, the synthesis polyphase matrix $\boldsymbol{R}(z)$ can be scaled delayed inverse of the analysis polyphase matrix $\mathrm{D}(z)$. Since

$$\boldsymbol{C}(z) = \boldsymbol{D}(z)\boldsymbol{R}(z),$$

$$Det[\mathrm{C}(z)] = Det\left[\mathrm{D}(z)\mathrm{R}(z)\right] = Det[\mathrm{D}(z)]\, Det[\mathrm{R}(z)]$$

Further, for perfect reconstruction,

$$\boldsymbol{C}(z) = \boldsymbol{D}(z)\boldsymbol{R}(z) = Kz^{-n_d}\boldsymbol{I}$$

$$\boldsymbol{R}(z) = \boldsymbol{D}^{-1}(z)Kz^{-n_d}\boldsymbol{I} = \frac{Adj[\mathrm{D}(z)]}{Det[\mathrm{D}(z)]}Kz^{-n_d}\mathrm{I} \tag{12.16g}$$

The elements of the adjoint matrix are cofactors of $\boldsymbol{D}(z)$ which are product and sums of FIR polynomials and hence $Adj[\mathrm{D}(z)]$ is FIR. Also the $Det\left[\boldsymbol{D}(z)\right]$ that involves products of FIR and their sums or differences though is an FIR, *since it occurs in the denominator, in general, makes the filters of $\boldsymbol{R}(z)$ as IIR.* Since perfect reconstruction filter banks are desired to have $H_k(z)$ and $F_k(z)$ as FIR filters, this requires $\boldsymbol{D}(z)$ and $\boldsymbol{R}(z)$ also to be FIR. The only possibility for $\boldsymbol{R}(z)$ to be FIR is that

$$\det\left[\boldsymbol{D}(z)\right] = \alpha z^{-k_d}, \alpha \neq 0,\ k_d \text{ :integer} \tag{12.16h}$$

so that

$$\boldsymbol{R}(z) = \frac{Adj[\boldsymbol{D}(z)]}{Det[\boldsymbol{D}(z)]}Kz^{-n_d}\boldsymbol{I} = Adj[\boldsymbol{D}(z)]\alpha Kz^{-n_d-k_d}\boldsymbol{I}$$

$$\boldsymbol{R}(z) = K'Adj[\boldsymbol{D}(z)]z^{-n_d-k_d}\boldsymbol{I},\quad \mathrm{K}' = \alpha K \tag{12.16i}$$

The FIR nature of $\boldsymbol{D}(z)$ and $\boldsymbol{R}(z)$ implies that their determinants are also of FIR nature. If their product being a constant multiplied by a delay, $\det\left[\boldsymbol{R}(z)\right]$ will be of similar form.

12.13 UNIFORM DISCRETE FOURIER TRANSFORM (UDFT) FILTER BANK

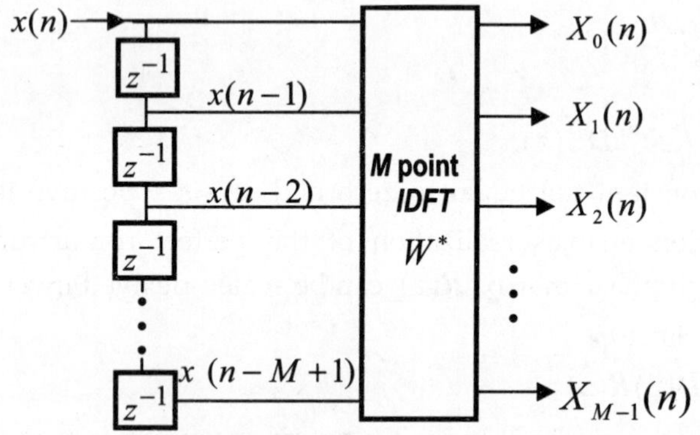

Fig. 12.18 (a) Uniform DFT filter bank.

The DFT of sequence $x(n)$, $X_k(n)$ is given by

$$X_k(n) = \sum_{i=0}^{M-1} x(n-i)e^{\frac{j2\pi ik}{M}}, \quad k = 0, 1, \ldots, M-1,$$

$$X_k(n) = \sum_{i=0}^{M-1} x(n-i)W_M^{-ik}, \quad e^{-\frac{j2\pi}{M}} = W_M = W \tag{12.17a}$$

This can be expressed in matrix form as,

$$\mathrm{X}(n) = \mathrm{W_M x}(n) \tag{12.17b}$$

where,

$$\mathrm{X}(n) = \left[X_0(n) \quad X_1(n) \quad \cdots \quad X_{M-1}(n)\right]^T, \; \mathrm{x}(n) = \left[x(n) \quad x(n-1) \quad \cdots \quad x(n-M+1)\right]^T$$

$$\mathrm{W_M} = \begin{bmatrix} 1 & 1 & \cdots & 1 \\ 1 & W_M & \cdots & W_M^{M-1} \\ 1 & \vdots & \ddots & \vdots \\ 1 & W_M^{M-1} & \cdots & W_M^{(M-1)(M-1)} \end{bmatrix} : \text{DFT matrix.}$$

The Eqn. (12.17a) states that for every time index n, $X_k(n), k = 0, 1, 2, \ldots, M-1$ are produced from $x(n-i), i = 0, 1, 2, \ldots, M-1$ (Fig. 12.18(a)).

Taking Z-transform both sides with respect to n for the Eqn. (12.17a),

$$X_k(n) = \sum_{i=0}^{M-1} x(n-i)W_M^{-ik}, \qquad X_k(z) = \sum_{i=0}^{M-1} z^{-i}W_M^{-ik}X(z), \qquad e^{-\frac{j2\pi}{M}} = W_M$$

$$X_k(z) = \sum_{i=0}^{M-1}\left(zW_M^{k}\right)^{-i} X(z)$$

Representing $X_k(z)$ as

$$X_k(z) = H_k(z)X(z),$$

$H_k(z)$ becomes,

$$H_k(z) = \sum_{i=0}^{M-1}\left(zW_M^{k}\right)^{-i}$$

For $k = 0$,

$$H_0(z) = \sum_{i=0}^{M-1} z^{-i} \tag{12.17c}$$

Therefore,

$$H_k(z) = H_0\left(zW_M^{k}\right) \tag{12.17d}$$

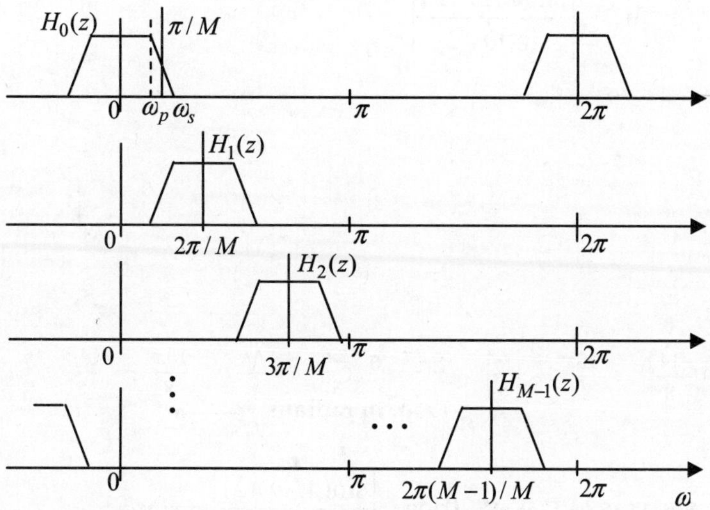

Fig. 12.18 (b) Frequency responses of filters in the analysis bank.

Substituting for $z = e^{j\omega}$ and $W_M = e^{\frac{-j2\pi}{M}}$ in Eqns. (12.17c) and (12.17d),

$$H_0(z) = \sum_{i=0}^{M-1} e^{-j\omega i} \tag{12.17e}$$

$$H_k(e^{j\omega}) = H_0\left(e^{j\omega}\, e^{-\frac{j2\pi k}{M}}\right) = H_0\left(e^{j\left(\omega - \frac{2\pi k}{M}\right)}\right) \tag{12.17f}$$

This indicates that the filters $H_k(e^{j\omega})$ are the *uniformly shifted* versions of the filter $H_0(e^{j\omega})$, *the prototype filter* and $H_k(e^{j\omega})$ for $k = 0, 1, \ldots, M-1$ form an analysis filter bank having M filters. Further, *as successive filters have the same shift of* $(2\pi / M)$ *in their frequency location* (Fig.12.18(b)), $H_k(z)$ is known as *uniform DFT (UDFT) filter bank.* Presently, the aim of UDFT is its use in the realization of a filter bank using DFT.

In this discussion, the prototype filter $H_0(z)$ is a very simple filter and corresponds to a rectangular window function with M samples. That is,

$$H_0(z) = \sum_{i=0}^{M-1} z^{-i} = \frac{1 - z^{-M}}{1 - z^{-1}} = \frac{1 - e^{-j\omega M}}{1 - e^{-j\omega}} = \frac{e^{-j\omega M/2}\left[e^{j\omega M/2} - e^{-j\omega M/2}\right]}{e^{-j\omega/2}\left[e^{j\omega/2} - e^{-j\omega/2}\right]}$$

$$H_0(z) = e^{-j\omega(M-1)/2}\,\frac{\sin(M\omega/2)}{\sin(\omega/2)}$$

$$\left|H_0(e^{j\omega})\right| = \left|\frac{\sin(M\omega/2)}{\sin(\omega/2)}\right| \tag{12.17g}$$

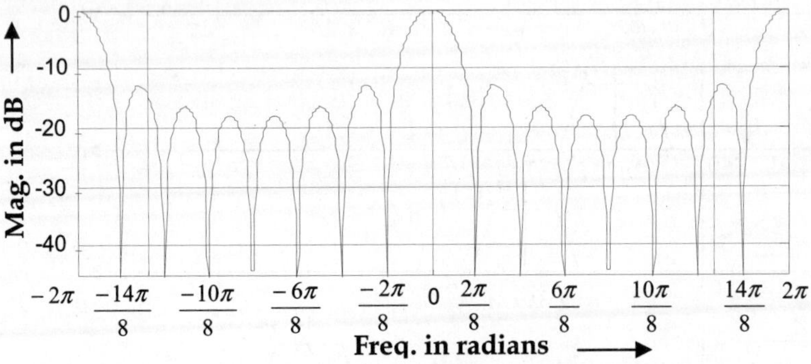

Fig. 12.18 (c) Plot of $20\log_{10}\left|\frac{\sin(M\omega/2)}{\sin(\omega/2)}\right|$ **for** $M = 8$.

The plot of $|\sin(M\omega/2)/\sin(\omega/2)|$ shown in the Fig. 12.18(c) indicates that it has a maximum stopband attenuation of about –13dB (Fig. 12.18(c)). Hence, the filters $H_k(z)$ derived by frequency shifting in increments of $2\pi/M$ (for $M=8$, $2\pi/8$) will have a high level of overlap. This is not at all desirable as the interest is to pick up frequency components within the passband of the filter $H_k(z)$ and totally reject which are outside. But by using a better prototype filter having good stopband and sharp cut-off characteristics, the filter bank can be improved. It is to be noted that the filters $H_k(z)$ (Fig.12.16) obey Eqn. (12.17d) and this property is used in derivation of important results.

12.13.1 Criterion for a Perfect Reconstruction DFT Filter Bank

The DFT filter bank is realized based on the DFT (Fig. 12.19(a)). The signal spectrum $X(\omega)$ is shifted so that its centre of frequency of kth band is brought to the 0th frequency by multiplying the input by $e^{-j2\pi kn/M}$ and further filtered by a lowpass filter $H(z)$ to limit its bandwidth to $(2\pi/M)$. This band limited component is decimated by a factor M to get the output $x_k(n)$. Thus, the components $x_k(n)$, $k=0,1,\ldots(M-1)$ form the analysis filter bank outputs. For the synthesis bank, the analysis bank outputs $x_k(n), k=0,1,\ldots(M-1)$ are upsampled by a factor M (zeros insertion) and the images are filtered by the filters $F(z)$ of bandwidth $(2\pi/M)$ to get the outputs $y_k(n), k=0,1,\ldots(M-1)$. These baseband signals are shifted to their respective positions by multiplying the kth band signal by $e^{j2\pi kn/M}$. These demodulated components are added to get back the original signal.

The kth filter output $x_k(n)$ for DFT filter bank for analysis filter / window $w(i)$ at time n is given by

$$x_k(n) = \sum_{i=-\infty}^{\infty} x(i)w(i-n)e^{\frac{-j2\pi ik}{M}}, k = 0,1,\ldots M-1 \tag{12.18a}$$

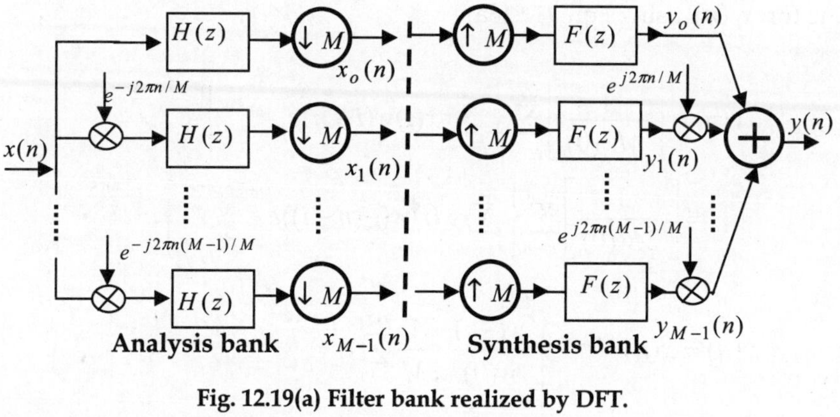

Fig. 12.19(a) Filter bank realized by DFT.

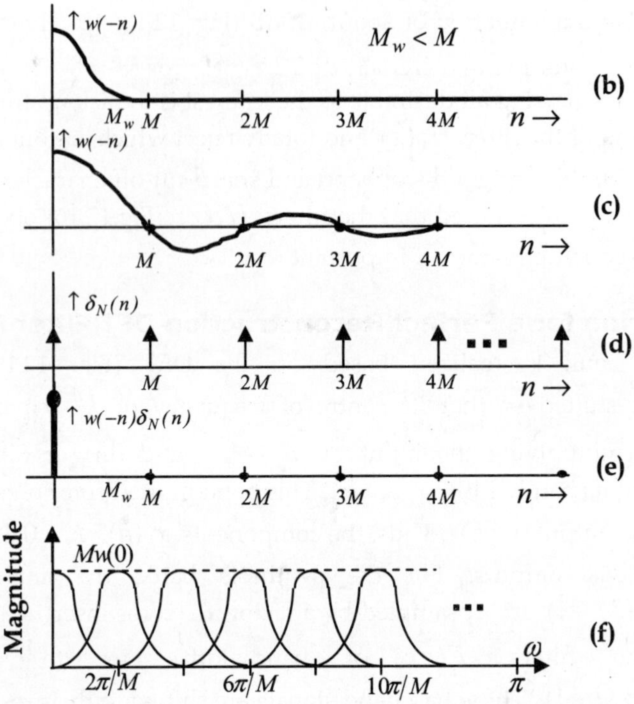

Fig. 12.19 (b-f) Perfect reconstruction criterion for M channel DFT filter bank.

The inverse/reconstructed is the scaled sum of all components $x_k(n), k = 0,1,\ldots M-1$,

$$y(n) = \left[\frac{1}{Mw(0)}\right]\sum_{k=0}^{M-1} x_k(n)e^{\frac{j2\pi nk}{M}} \tag{12.18b}$$

Substituting for $x_k(n)$ using Eqn. (12.18a)

$$y(n) = \left[\frac{1}{Mw(0)}\right]\sum_{k=0}^{M-1}\left\{\sum_{i=-\infty}^{\infty} x(i)w(i-n)e^{\frac{-j2\pi ik}{M}}\right\}e^{\frac{j2\pi nk}{M}}$$

$$= \left[\frac{1}{Mw(0)}\right]\sum_{k=0}^{M-1}\left\{\sum_{i=-\infty}^{\infty} x(i)w(-(n-i))e^{\frac{j2\pi k(n-i)}{M}}\right\}e^{\frac{j2\pi nk}{N}}$$

$$y(n) = x(n) \quad * \quad \left[\frac{w(-n)}{w(0)}\right]\frac{1}{M}\sum_{k=0}^{M-1} e^{\frac{j2\pi nk}{M}} \tag{12.18c}$$

But

$$\left(\frac{1}{M}\right)\sum_{k=0}^{M-1} e^{\frac{j2\pi nk}{M}} = \sum_{r=-\infty}^{\infty} \delta\,(n-rM) = p(n)$$

$p(n)$ is an impulse train with period M. Therefore Eqn. (12.18c) becomes

$$y(n) = x(n) \quad * \quad \left\{\left[\frac{w(-n)}{w(0)}\right]\sum_{r=-\infty}^{\infty}\delta(n-rM)\right\}$$

For $y(n) = x(n)$, the quantity in the flower brackets should be an impulse delta function. That is,

$$\frac{w(-n)}{w(0)}\sum_{r=-\infty}^{\infty}\delta\,(n-rM) = \delta\,(n)$$

$$w(-n)\sum_{r=-\infty}^{\infty}\delta\,(n-rM) = w(0)\delta\,(n) \qquad (12.18d)$$

This is satisfied by any causal filter of length $M_w < M$, M being the number of analysis filters. This is possible as the filter function becomes zero even within M samples and the product $w(-n)\sum_{r=-\infty}^{\infty}\delta(n-rM)$ will be equal to scaled $\delta\,(n)$ (Fig.12.19(b, d)). If the number of filters is less than filter length, $M < M_w$ the above condition is satisfied provided the filter is chosen to have its *every* M^{th} ***sample equal to zero.*** That is,

$$w(rM) = 0, \qquad r = \pm 1, \pm 2, \cdots \qquad (12.18e)$$

the requirement of an Mth *band filter.* For this, from Fig. 12.19 (c, d, e) Eqn. (12.18d) is satisfied and the signal is recovered. The filter bank that *satisfies* M^{th} *band filter condition is called a power complementary filter bank* The L.H.S of Eqn. (12.18d) is the multiplication of the window function $w(-n)$ with an impulse train (in time domain) is in frequency domain equivalent to convolution of the FT of the window function $W(\omega)$ with that of the impulse train which is again an impulse train, $\sum_{r=-\infty}^{\infty}\delta\left(\omega - \frac{2\pi}{N}r\right)$. That is,

$$W(\omega) * \sum_{r=-\infty}^{\infty}\delta\left(\omega - \frac{2\pi}{M}r\right) = Mw(0)$$

$$\sum_{r=-\infty}^{\infty} W\left(\omega - \frac{2\pi\, r}{M}\right) = Mw(0) \qquad (12.18f)$$

This states that for a perfect reconstruction filter bank, the frequency responses of the filters should sum to a constant across the bandwidth which is also the property of a Mth *band filter.*

12.13.2 Polyphase Representation of UDFT Analysis Bank

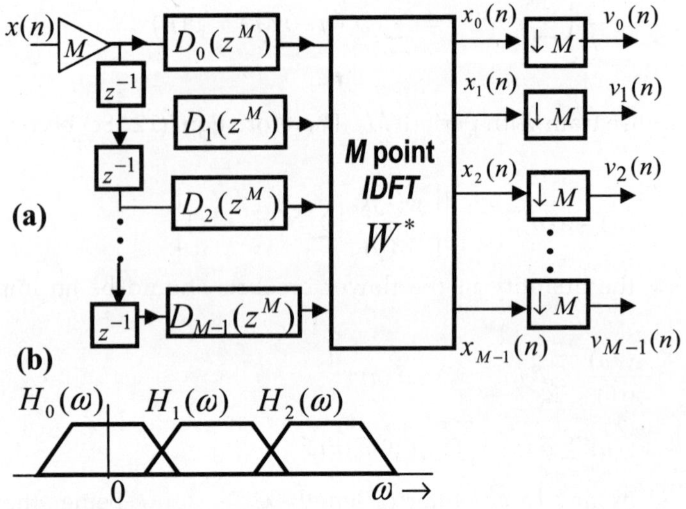

Fig. 12.20(a) Polyphase uniform DFT analysis filter bank and (b) Its frequency response.

The prototype filter $H_0(z)$ can be realized by a polyphase structure with its M components (from Eqn. (12.15a)) as

$$H_0(z) = \sum_{l=0}^{M-1} z^{-l}\, D_l(z^M)$$

where, $D_l(z^M)$ is the lth polyphase filter and

$$D_l(z) = h_0(l) + h_0(l+M)z^{-1} + h_0(l+2M)z^{-2} + \ldots\ h_0(l+(M-1)M)z^{-(M-1)}$$

$$D_l(z) = \sum_{m=-\infty}^{\infty} h_0(Mm+l)z^{-m}$$

Note that since basic filter is $H_0(z)$ and others are derived from it, the polyphase components are represented by $D_l(z)$, instead of $D_{lk}(z)$ as discussed in Section 12.11.1.

In UDFT filter bank, for a prototype filter $H_0(z)$, the filters $H_k(z)$ are (Eqn. (12.17d)),

$$H_k(z) = H_0\left(zW_M{}^k\right) \qquad k = 0, 1, \ldots, M-1$$

Substituting for $H_0(z)$ in terms of its polyphase components,

$$H_k(z) = \sum_{l=0}^{M-1} \left(zW^k\right)^{-l} D_l\left((zW^k)^M\right) = \sum_{l=0}^{M-1} \left(z^{-1}W^{-k}\right)^l D_l\left(z^M W^{k\,M}\right)$$

$$= \sum_{l=0}^{M-1} \left(z^{-1}W^{-k}\right)^l D_l\left(z^M\right) \quad \text{since} \quad W^{kM} = e^{-\frac{j2\pi k}{M}M} = e^{-j2\pi k} = 1$$

$$H_k(z) = \sum_{l=0}^{M-1} z^{-l}\, W^{-k\,l}\, D_l\left(z^M\right) \tag{12.19a}$$

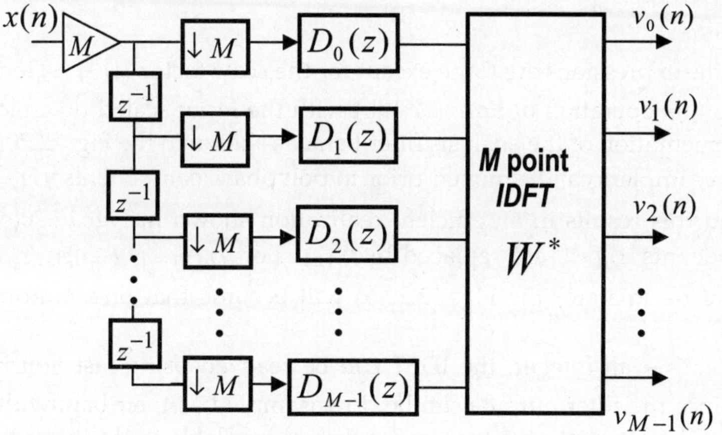

Fig. 12.20(c) Efficient polyphase uniform DFT analysis filter bank.

Expanding this for each value of k,

$$H_0(z) = D_0(z^M) + z^{-1}D_1(z^M) + \ldots + z^{-(M-1)}D_{M-1}(z^M)$$
$$H_1(z) = D_0(z^M) + z^{-1}D_1(z^M)W^{-1} + \ldots + z^{-(M-1)}D_{M-1}(z^M)W^{-(M-1)}$$
$$\vdots \qquad \vdots$$
$$H_{M-1}(z) = D_0(z^M) + z^{-1}D_1(z^M)W^{-(M-1)} + \ldots + z^{-(M-1)}D_{M-1}(z^M)W^{-(M-1)(M-1)}$$

Representing,

$$\boldsymbol{H}(z) = \left[H_0(z) \quad H_1(z) \quad \cdots \quad H_{M-1}(z)\right]^T,$$

$$\boldsymbol{D}_z(z^M) = \left[D_0(z^M) \quad z^{-1}D_1(z^M) \quad \cdots \quad z^{-(M-1)}D_{M-1}(z^M)\right]^T$$

The filters $\boldsymbol{H}(z)$ can be expressed as

$$\boldsymbol{H}(z) = \boldsymbol{W}^* \boldsymbol{D}_z(z^M) \tag{12.19b}$$

Comparing Eqn. (12.17a) ,

$$X_k(n) = \sum_{i=0}^{M-1} x(n-i) W_M^{-ik}$$

with the expression for inverse DFT given by,

$$x(n) = \frac{1}{M}\sum_{k=0}^{M-1} X_k e^{\frac{j2\pi nk}{M}} = \frac{1}{M}\sum_{k=0}^{M-1} X_k W^{-nk}, \qquad n = 0, 1, \ldots, M-1$$

reveals that both the expressions are same except for the scale factor $(1/M)$. Hence, IDFT can be used to realize this operation of Eqn. (12.19b) with the input scaled by a factor M. This leads to the implementation of the analysis filter bank as shown in the Fig. 12.20(a).

The down samplers can be moved prior to polyphase components $D_l(z^M)$ using the noble identity and this results in an efficient realization shown in Fig. 12.20(c). Here, the poly-phase components $D_l(z^M)$ are replaced by $D_l(z)$. For $D_l(z) = 1,\ l = 0, 1, \ldots, M-1$ (which implies absence of decimators) the Fig. 12.20(a) will become that of a uniform DFT filter bank (Fig. 12.18(a)).

With $M = 2^m, m$ an integer, the IDFT can be realized using fast Fourier transform (FFT). Further, since the filter outputs can be downsampled as their bandwidths are small by a factor M, for the efficient structure, more time is available to do the same number of arithmetic operations due to decimation. Hence, it requires M times lesser multiplications and additions per input unit sample interval. That is, indirectly this is equivalent to reduction in computations by the decimation factor.

For a filter $H_0(z)$ of order N' (i.e., number of coefficients $\mathrm{N} = N'+1$), each polyphase filter has $(N+1)/M$ coefficients hence, number of real multiplications for this is $M(N+1)/M = N+1$. Further for the IDFT using radix-2 FFT and the scale factor require $\mathrm{M}\log_2(M)$ and 1 real multiplications, respectively. With the efficient structure, as the computations are effectively reduced by the factor M (due to availability of M times input sampling interval), the total number of real multiplications are

$$\frac{\mathrm{N} + \mathrm{M}\log_2(M)}{M} + 1 = \frac{\mathrm{N}}{M} + \log_2(M) + 1 \quad \mathrm{N} = N' + 1$$

in contrast to directly implementing filter bank, $2M(N+1)$.

For $\mathrm{N} = 128$, direct implementation requires 2048 real multiplications. But polyphase DFT method requires only 20 real multiplications giving a benefit by a factor of about 102. For $\mathrm{N} = 256, M = 16$; the gain is 390. For $\mathrm{N} = 256, M = 32$; the gain is 1170. Hence, this gain improves with increase in number of filters and filter length.

12.13.3 Polyphase Representation of UDFT Synthesis Bank

The synthesis filters $F_k(z)$ for $k = 0, \ldots, M-1$, are given by

$$F_k(z) = \sum_{l=0}^{M-1} z^{-(M-1-l)} W^{lk}$$

The filter $F_0(z)$ corresponds to $k = 0$ and is given by

$$F_0(z) = \sum_{l=0}^{M-1} z^{-(M-1-l)}$$

Hence, replacing z by zW^k in $F_0(z)$,

$$F_0(zW^k) = \sum_{l=0}^{M-1} \left(zW^k\right)^{-(M-1-l)} = \sum_{l=0}^{M-1} z^{-(M-1-l)} W^{-kM} W^k W^{lk}$$

$$= W^k \left[\sum_{l=0}^{M-1} z^{-(M-1-l)} W^{lk}\right], \qquad \because W^{-kM} = 1$$

The quantity in the square brackets on the R.H.S, represents $F_k(z)$, hence

$$F_0(zW^k) = W^k F_k(z)$$

Therefore,

$$F_k(z) = W^{-k} F_0(zW^k)$$

Polyphase representation:

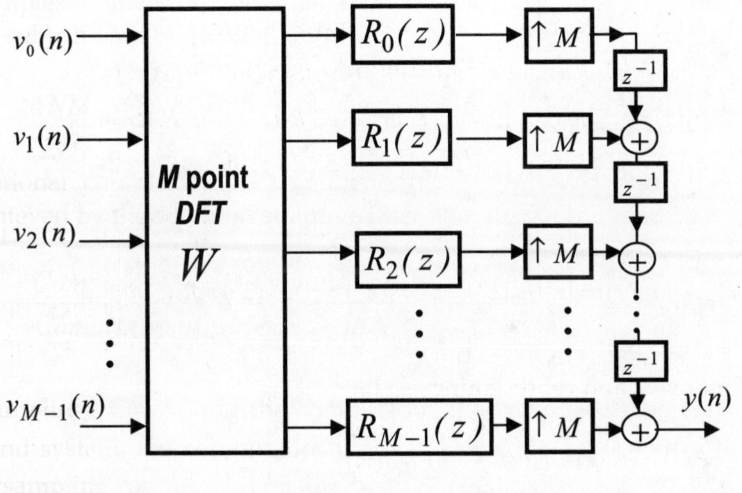

Fig. 12.20(d) Efficient polyphase uniform DFT synthesis filter bank.

The prototype filter $F_0(z)$ can be expressed in terms of its polyphase components as

$$F_0(z) = \sum_{l=0}^{M-1} z^{-(M-1-l)} R_l(z^M)$$

where $R_l(z^M)$ is the lth polyphase filter and is given by

$$R_l(z) = \sum_{m=0}^{M-1} f_0(mM + M - 1 - l))z^{-m}$$

Given a prototype filter $F_0(z)$, the filters $F_k(z)$ are obtained by

$$F_k(z) = W^{-k} F_0(zW^k)$$

Substituting for $F_0(z)$ in terms of its polyphase components,

$$F_k(z) = W^{-k} \sum_{l=0}^{M-1} \left(zW^k\right)^{-(M-1-l)} R_l\left((zW^k)^M\right)$$

$$= W^{-k} \sum_{l=0}^{M-1} z^{-(M-1-l)} \left(W^k\right)^{-(M-1-l)} R_l(z^M)$$

$$F_k(z) = \sum_{l=0}^{M-1} z^{-(M-1-l)} W^{kl} R_l(z^M) \tag{12.20a}$$

Expanding this for each value of k and $M = 4$,

$$F_0(z) = z^{-3}R_0(z^4) + z^{-2}R_1(z^4) + z^{-1}R_2(z^4) + R_3(z^4)$$
$$F_1(z) = \left\{z^{-3}R_0(z^4) + z^{-2}R_1(z^4)W^1 + z^{-1}R_2(z^4)W^2 + R_3(z^4)W^3\right\}$$
$$F_2(z) = \left\{z^{-3}R_0(z^4) + z^{-2}R_1(z^4)W^2 + z^{-1}R_2(z^4)W^4 + R_3(z^4)W^6\right\}$$
$$F_3(z) = \left\{z^{-3}R_0(z^4) + z^{-2}R_1(z^4)W^3 + z^{-1}R_2(z^4)W^6 + R_3(z^4)W^9\right\}$$

$$\left[F_0(z) \;\; F_1(z) \;\; F_2(z) \;\; F_3(z)\right] = \left[z^{-3}R_0(z^4) \;\; z^{-2}R_1(z^4) \;\; z^{-1}R_2(z^4) \;\; R_3(z^4)\right] \begin{bmatrix} 1 & 1 & 1 & 1 \\ 1 & W^1 & W^2 & W3 \\ 1 & W^2 & W^4 & W^6 \\ 1 & W^3 & W^6 & W^9 \end{bmatrix}$$

Generalizing for any M and representing,

$$\boldsymbol{F}(z) = \left[F_0(z) \;\; F_1(z) \;\; \cdots \;\; F_{M-1}(z)\right]^T$$

$$R(z^M) = \left[z^{-(M-1)}R_0(z^M) \quad z^{-(M-2)}R_1(z^M) \quad \cdots \quad R_{M-1}(z^M)\right]^T$$

$$\mathbf{W}_M = \begin{bmatrix} 1 & 1 & \dots & 1 \\ 1 & W^1 & \dots & W^{M-1} \\ \vdots & \vdots & \vdots & \vdots \\ 1 & W^{M-1} & \dots & W^{(M-1)^2} \end{bmatrix}$$

The filters $F(z)$ can be expressed as

$$F^T(z) = R^T(z^M)W_M \tag{12.20b}$$

This realization of the of the synthesis UDFT filter bank using polyphase structure is shown in the Fig. 12.20(d). This also incorporates the modification for the computational efficiency by moving the expanders (upsamplers) to the right of the polyphase components of the filter $R_l(z), l = 0,1,\dots, M-1$ using the noble identity that replaces $R_l(z^M)$ by $R_l(z)$. The computational savings are similar to that of analysis polyphase UDFT filter bank.

The complete UDFT polyphase filter bank is shown in the Fig.12.20(e) where the computationally efficient polyphase UDFT analysis and synthesis banks are cascaded.

Further, the presence of $D_l(z) \neq 1,\ l = 0,1,\dots, M-1$ *enables the prototype filter* $H_0(z)$ *to have an order* N *higher compared to number of filters* M *unlike in the case of UDFT filter bank for which* N *should be equal to* M. *Hence, the polyphase realization* $D_l(z) \neq 1,\ l = 0,1,\dots, M-1$ *permits sharper cut-off filter and also depending upon the nature of* $H_0(z)$, *better stopband attenuation* (Fig. 12.20(b)). These enable to have *less overlap between adjacent filters which is highly desirable.*

In the UDFT filter bank as *filter order is same as the number of filters, the sharpness of cut-off is decided by the number of filters. In the polyphase DFT filter bank, the presence of polyphase components removes this coupling between filter order/length and the number of filters.* Further, as $H_0(z) = F_0(z)$, the same discussion holds good for synthesis bank also.

Since , $W^*W = MI$ (as the DFT matrix W is orthonormal) , Fig. 12.20(e) becomes same as Fig. 12.17.(d) and further from Eqn. (12.16d),

$$C(z) = D(z)R(z)$$

it is equivalent Fig. 12.17(e). It is clear that for perfect reconstruction, a choice of polyphase analysis matrix $D(z) = W^*$, requires a polyphase synthesis matrix $R(z) = W$. That is,

$$C(z) = D(z)R(z) = W^*W = MI$$

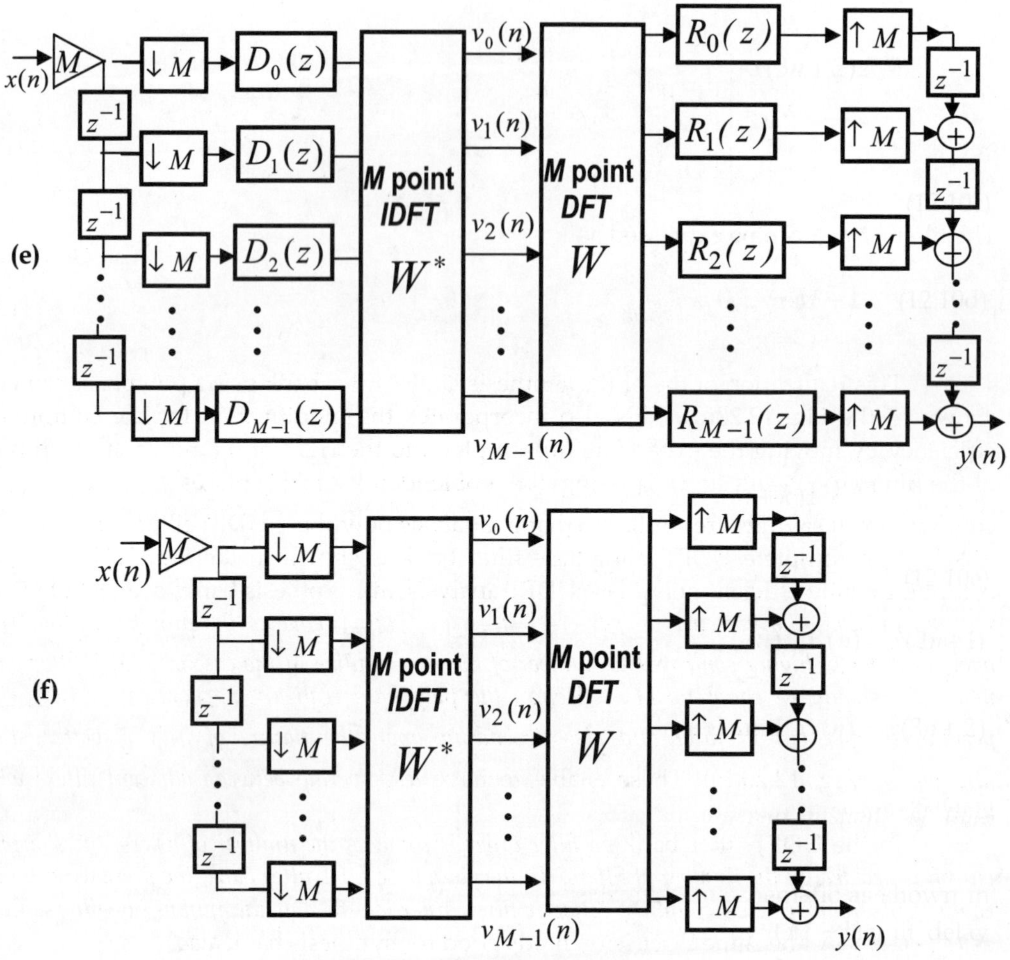

Fig. 12.20 (e) Complete polyphase UDFT filter bank and (f) Perfect reconstruction with UDFT filter bank.

and this is shown in Fig. 12.20(f) and for this the output is given by

$$y(n) = M^3 x(n - M + 1) \tag{12.20c}$$

Hence, for this case $\boldsymbol{D(z) = W^*}$, the aliasing problem due to overlap of frequency response of filters in the UDFT filter bank is cancelled completely.

12.13.4 M-channel Polyphase DFT Filter Bank Design

In the M-channel DFT filter bank, M decides the number of channels and the prototype filter decides the performance of the filter bank. The prototype filter should satisfy the requirement that for $\omega > 2\pi/M$, the attenuation should be large enough so that the aliasing effect is negligible and also this depends on application.

Here, the method given by Fliege for the square root raised cosine filter has been generalized. To start with $G(e^{j\omega}) = \left|H(e^{j\omega})\right|^2$, a zero phase M^{th} band lowpass filter is designed with cutoff frequency $\omega_c = \pi/M$. This satisfies

$$G(e^{j\omega}) + G(e^{j(\omega-2\pi/M)}) = 1\ ,\ 0 \le \omega \le 2\pi/M$$

(Fig. 12.21a) and ensures the power complementary nature of $H(e^{j\omega}), 0 \le \omega \le 2\pi/M$ which is the root spectral factor of $G(e^{j\omega})$. *This property is applicable for its shifted versions in frequency.* That is, for frequency ranges $2\pi k/M \le \omega \le 2\pi(k+1)/M$, $k = 0,1,\ldots(M-1)$ (which includes $0 \le \omega \le 2\pi/M$ also) the aliased components add to unity providing spectral flatness which is essential for perfect reconstruction.

The impulse response of such a $H(e^{j\omega})$ is non-causal and is of infinite duration. But to have a realizable $h(n)$, it is required to consider a shifted version of it by $(N-1)/2$ points and having $(N-1)/2$ points symmetrically on either side. Such a truncation results in a finite stopband attenuation introducing aliasing. In essence, $G(e^{j\omega})$ becomes approximately M^{th} band having some linear distortion. This can be overcome to some extent by choosing a proper larger filter length N.

The root filter $H(e^{j\omega})$ can be achieved from $G(e^{j\omega})$ by cepstral approach using the relation

$$\hat{H}(e^{j\omega}) = 0.5\ln[G(e^{j\omega}) + \delta],$$

δ ensures the positivity of $G(e^{j\omega})$ and is essential as $g(n)$, is impulses response of the Mth band filter obtained by windowing and can make $G(e^{j\omega})$ negative.

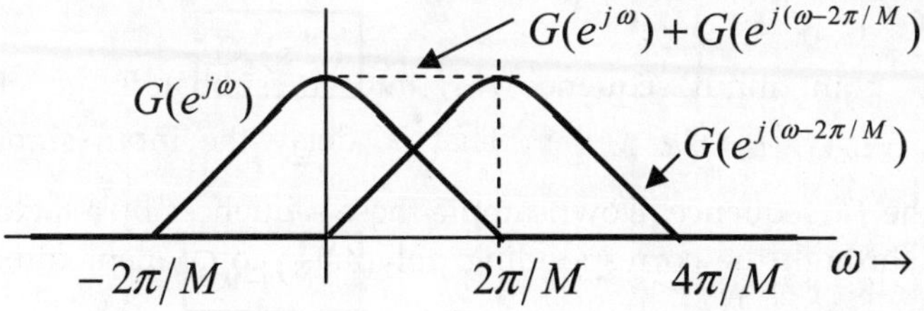

Fig. 12.21(a) Power complementary nature of frequency shifted filters.

$$h(n) = IFT[\exp\hat{H}(e^{j\omega})]\ ,\qquad IFT : \text{inverse Fourier transform.}$$

The use of a Mth band zero phase square root raised cosine filter will avoid the error introduced in getting the spectral factor as the mathematical expression is readily available and is given by

$$h_{src}(n) = \frac{4rn\cos[\pi n(1+r)/M] + M\sin[\pi n(1-r)/M]}{\left[1-(4\pi n/M)^2\right] \; \pi nM},$$

If $H_{src}(z)$ is the z- transform of $h_{src}(n)$, then

$$H_{rc}(z) = H_{src}(z)H_{src}(z)$$

$H_{rc}(z)$ is the z-transform of $h_{rc}(n)$, the impulse response of the Mth band zero phase raised cosine filter given by

$$h_{rc}(n) = \frac{1}{M}\frac{\sin[\pi n/M]}{[\pi n/M]}\frac{\cos[r\pi n/M]}{\left[1-(2rn/M)^2\right]}$$

For $r = 0$, the impulse response of raised cosine filter becomes the ideal lowpass filter. The factor, r controls the rate of decay of the impulse response with respect to index n. Multiplying with a symmetrical rectangular window and shifting it by half the window length results in a realizable causal M^{th} band filter which only approximates the raised cosine filter with a finite stopband attenuation in the range $(1+r)\pi / M \le \omega \le \pi$.

Implementation algorithm for a polyphase complex modulated filterbank:

1. Design an M-band root filter $h(n)$ using the method described above.

2. Compute the polyphase components $d_l(n)$ of analysis filter bank using the relation $d_l(n) = h_0(l + nM)$, $l = 0$ *to* $M - 1$, with $h_0(n) = h(n)$.

3. Since $f_0(n) = h_0(n)$ and $r_l(n) = h_0(nM + M - 1 - l))$, the polyphase components $r_l(n)$ of synthesis bank can be found using $r_l(n) = d_{M-1-l}(n)$.

4. For decomposing an N point input sequence $x(n)$, into M subbands, generate M sequences using $x_{pl}(n) = x(n-l)$, $l = 0$ *to* $M - 1$. That is, delay the input signal by l samples for generating the lth sequence. Downsample these sequences by a factor of M and filter each of them through the corresponding polyphase component filter $d_l(n)$. Form a $M \times (N/M)$ matrix, with the lth row corresponding to the output of the lth polyphase component filter; take an M - point columnwise IDFT of this matrix and multiply by a factor of M^2 (Fig. 12.20(c)). The lth row in the resulting $M \times (N/M)$

matrix corresponds to the lth subband signal. (The multiplication by M^2 is to account for the factor of $1/M$ introduced both by the decimation and IDFT each individually.)

5. For reconstructing the signal, apply an M-point columnwise DFT to the matrix of the subband signals. Filter each row of the resulting matrix by the corresponding filter $r_l(n)$ to obtain M sequences and upsample by a factor of M (introduce $(M-1)$ zeros in between two samples). Delay the lth sequence by $M-l-1$ samples. This has to be carried out for $l = 0$ *to* $M-1$. Add these sequences and remove the delay (filter length) due to filtering, to obtain the reconstructed signal.

Example for open loop polyphase subband adaptive filter

For this, identification of an impulse response of a system

$$G(z)=\frac{1-1.1289z^{-1}+1.0058z^{-2}-0.7617z^{-3}}{1-0.59z^{-1}+0.455z^{-2}-0.0754z^{-3}+0.504z^{-4}}$$

is considered and the length of impulse response is 128 samples.

(b)

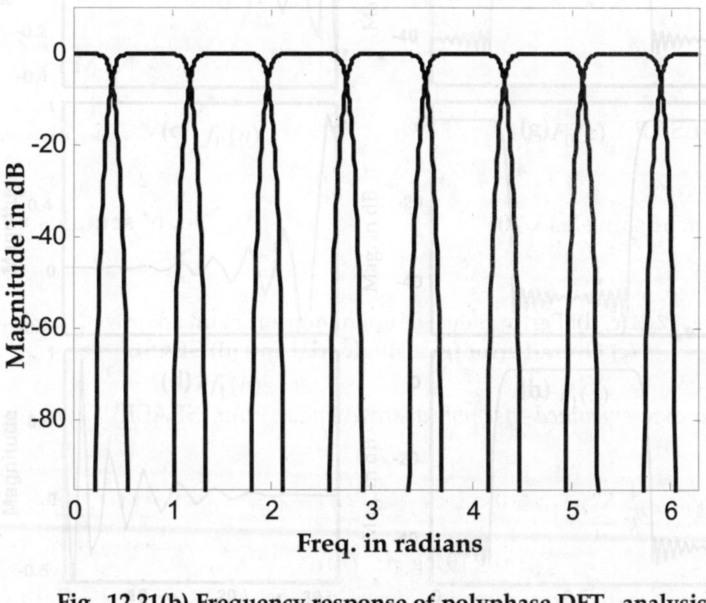

Fig. 12.21(b) Frequency response of polyphase DFT analysis bank for channels.

For the filter bank of 8 channels, a prototype impulse response of zero phase lowpass filter of length 136 is derived using a Kaiser window with smoothing factor of 10 on Matlab platform. The analysis filter bank realized by polyphase DFT filter bank is shown in Fig. 12.21(b). As the system impulse response is 128, the subadaptive filters are of length 32 samples taking into account a oversampling factor of 2 (i.e., 32=2(128/8)). The adaptation step size and the forgetting factor (used computing the input power) are 0.001 and 0.99, respectively.

The input considered is a narrowband signal with frequencies of 200, 400, 1100, 1300 Hz plus Gaussian white noise with SNR=29 dB. The complete polyphase filter bank introduces a delay of $(M-1)=(8-1)=7$ samples (Eqn. (12.20c) hence, the desired signal is delayed by 7 samples for computing the overall error (which is not used for adaptation). But for computing the individual errors for each channel as the input and the desired signals are passed through identical analysis filter bank no delay adjustment is involved. The overall error for the fist trial is shown in Fig. 12.21(c).

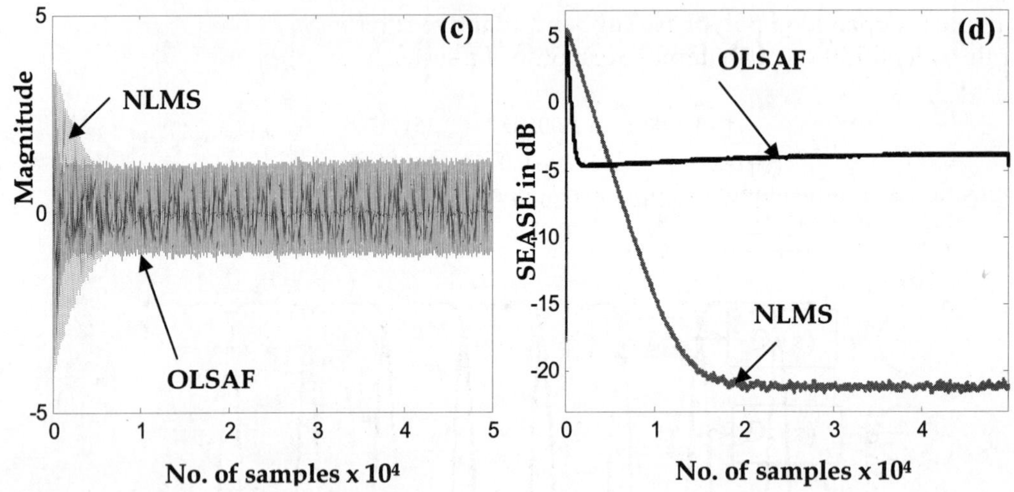

Fig. 12.21(c, d): Performance of open loop subband adaptive filter (c) Overall error for a single trial and (d) SEASE.

The performance index, *smoothed ensemble average square error* (SEASE) $\xi_{Sd}(n)$ is given by

$$\xi(n)=\frac{1}{K}\sum_{i=1}^{K}e_i^2(n),\ 0<\lambda<1,\quad \xi_S(n)=\lambda\,\xi_S(n-1)+(1-\lambda)\xi(n),$$

$$\xi_{Sd}(n)=10\log_{10}(\xi_S(n))$$

$e_i(n)$ - Error for i^{th} realization, number of realizations $K=30$ and $\lambda=0.99$.

The overall error plot indicates that for the OLSAF, the convergence is significantly fast compared to that of the normalized LMS algorithm. The SEASE plot also indicates that the OLASF has a faster convergence but its magnitude of the error is significantly large compared to that of NLMS. The large convergence error is due to large step size that the individual subadaptive filters take due to relatively smaller input power and smaller filter length, compared to those used in NLMS algorithm. The errors due to synthesis filter bank also contribute to the overall error as it is not taken into account.

Example for close loop polyphase subband adaptive filter

For this, the system to be identified, the analysis and synthesis filter banks and the input signal considered for open loop case, are used. The type and length of prototype filter for filter bank used are also same as those of open loop case. In OLSAF, the overall error was computed by delaying the desired signal as the filter bank introduced a delay of $(M-1)=(8-1)=7$ samples and this error was used for only evaluating performance. But this overall error is used for CLSAF, as this error is decomposed into its components (instead of desired signal) and they are used for the adaptation of subfilters. This takes into account the inaccuracies involved in the synthesis filter bank. The step size used is 0.0001.

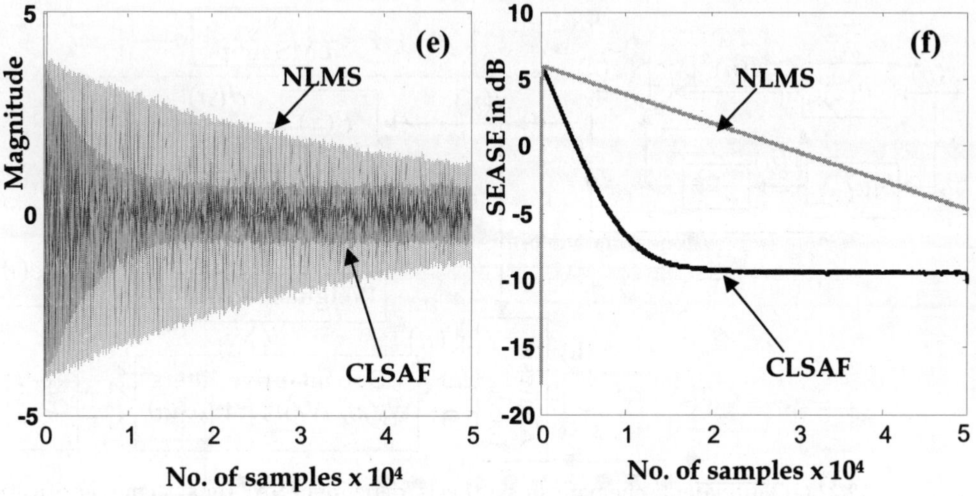

Fig. 12.21(e, f) : Performance of closed loop subband adaptive filter (e)Overall error for a single trial, (f) SEASE.

The overall error plot indicates that for the CLSAF, the convergence is significantly fast compared to that of the normalized LMS algorithm. The SEASE plot also indicates that the CLASF has a faster convergence and that the NLMS takes more time to reduce its SEASE value (error magnitude) equal to that of CLSAF. The error magnitude of NLMS after convergence can also become smaller than that of CLSAF. This is due to large step size that the individual subadaptive filters take due to relatively smaller input and smaller filter length compared those used in NLMS algorithm. The errors due to synthesis filter bank also contributes to the overall error but in CLSAF, this part of error is taken into account. This can be seen from Fig.12.21(e) as the CLSAF error magnitude is smaller than that of NLMS but this is not so with OLSAF (Fig. 2.21(c)). Not only the synthesis filter errors, but also the larger step size (0.001) used for OLASF than that used by CLSAF (0.0001) can also be a reason for poor convergence error of OLSAF.

12.14 DELAYLESS SUBBAND ADAPTIVE FILTER (DLSAF)

Till now, the two types of subband adaptive filters and their parameters, the decimation factor M, oversampling rate L, the delay introduced by these filters and its effect on the SAF performance; were considered. In both cases, the aliasing problem is reduced by using oversampling, $L < M$. The consequence of oversampling, the eigenvalue spread was removed by using different types of analysis and synthesis filters. Unless accuracy (misadjustment) is important it is preferable to use synthesis independent SAF as it provides faster convergence due to absence of delay due to synthesis filter bank in the adaptation loop.

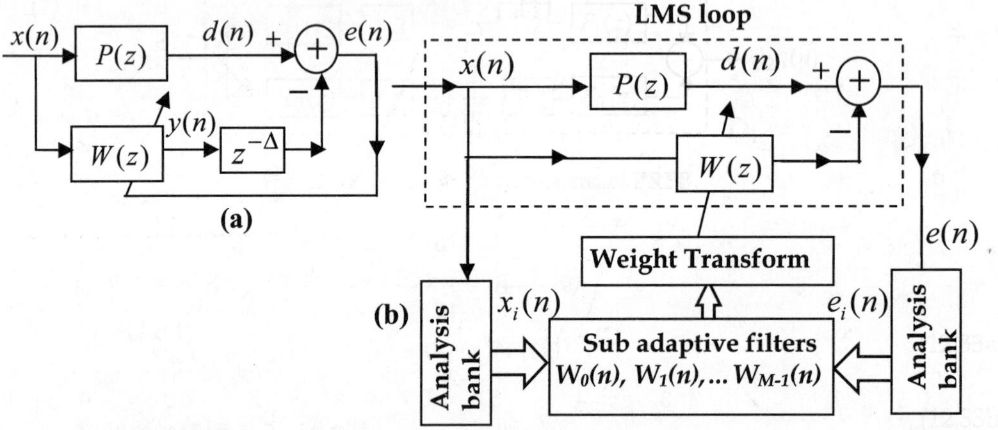

Fig. 12.22 (a) Equivalent schematic of synthesis dependent SAF for system identification and (b) Development of DLSAF for system identification.

But if accuracy is important, it is preferable to go for synthesis dependent SAF which suffers from delay due to both analysis and synthesis filter banks. As stated already, the main problem with the delay is that it not only makes its convergence slow but also not suitable for applications where only limited delay is tolerable. Hence, it is of interest to explore new methods which can provide the accuracy of synthesis dependent SAF and also solve the delay problem. Such a structure is called delayless subband adaptive filter (DLSAF) (D.R.Morgan and J.C.Thi, 1995) and is nothing but modified version of synthesis dependent structure.

In the synthesis dependent structure when used for the system identification, its equivalent can be represented as in Fig. 12.22a. Here $W(z)$ or $W(n)$ is the effective / overall adaptive filter realized by *the synthesis dependent* SAF and $z^{-\Delta}$ is the delay due to analysis and synthesis filter banks. This delay is not desirable in the LMS loop and hence has to be removed. Here, the input and the *overall error* $e(n)$ are decomposed outside the LMS adaptation loop as shown in dotted line box (Fig. 12.22b). Further, the synthesis filter bank is totally removed but its effect is realized by a different approach. The overall error $e(n)$ required is achieved by obtaining the desired signal estimate $y(n)$ by filtering the input $x(n)$ by the complete impulse response estimate $W(n)$ of the plant $P(n)$.

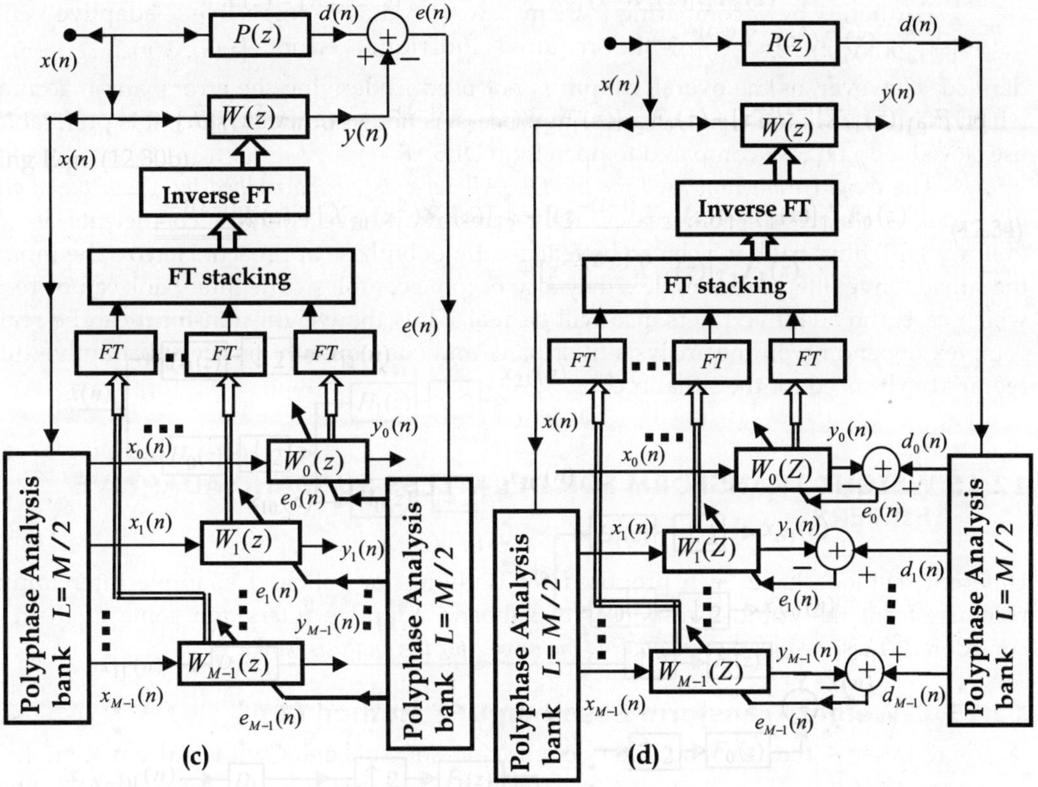

Fig. 12.22 (c, d) System identification by: DLSAF using weight transform for (c) Closed loop and (d) Open loop SAF.

The $W(n)$ is obtained by combining the individual subadaptive filter impulse responses $\{W_0(n), W_1(n), \ldots, W_{M-1}(n)\}$. Since the different subadaptive filters occupy different frequency regions of the overall adaptive filter $W(n)$, to get its transform $W(z)$, the Fourier transform (FT) of the individual impulse responses $\{W_0(z), W_1(z), \ldots, W_{M-1}(z)\}$ are properly stacked taking into account the DFT symmetry and its inverse transform provides $W(n)$. This method of getting the overall weight vector form individual sub weight vectors is known as *weight transform*. The complete DLSAF is shown in the Fig. 12.22(c). It is important to note that this approach does not disturb the *time alignment between the error* $e(n)$ *and the input* $x(n)$ of the LMS adaptation loop.

Even for the open loop SAF, from the subadaptive filter impulse responses, it is possible to obtain the overall impulse response of the system by the weight transform (Fig. 12.22(d)). Further, this enables to get the overall output without any synthesis filter bank. For the open loop SAF, the desired signal is decomposed into its components $\{d_0(n), d_1(n), ..., d_{M-1}(n)\}$by an analysis bank identical to that used for the input signal $x(n)$.

Further, by comparing them with their respective adaptive filter outputs$\{y_0(n), y_1(n), ..., y_{M-1}(n)\}$, the required individual errors $\{e_0(n), e_1(n), ..., e_{M-1}(n)\}$are derived. However, as the overall output is not used in deriving the errors, any inaccuracy, which may occur in the weight transform process, is not accounted. Thus, it is preferable to use closed loop DLSAF compared to open loop DLSAF.

The *weight transform used for the DLSAF depends on the method by which the analysis filter bank is realized.* In some cases, filter bank may have complex coefficients as with uniform DFT filter bank which can be realized by polyphase approach. Here as the input to the subadaptive filters are complex, they also become complex. The filter bank can be real in which case, the adaptive filters also will be real. Thus the weight transform can be real or complex depending on the analysis filter bank and consequently on the subadaptive filters required to be used for the situation.

12.15 WEIGHT TRANSFORM FOR DELAYLESS SUBBAND ADAPTIVE FILTERS

Different methods have been proposed for deriving the fullband adaptive filter impulse response from the impulse responses of subband adaptive filters and some of them are considered. *These assume a oversampling factor of 2 for the analysis bank.*

12.15.1 Weight Transform Based on DFT Method (1)

If N_F represents the fullband filter length, for M subbands with a decimation factor $L = M/2$. *That is, a oversampling factor of 2 is assumed.* Then, the length of each subband adaptive filter $N_i = N_S = 2N_F/M$. Further, let $W_i(k), k = 0, 1, ..., N_S - 1$ represent the N_S DFT coefficients for each subband adaptive filter $W_i(n)$ $n = 0, 1, ..., N_S - 1$. By stacking these M subband DFT coefficient vectors properly to an fullband filter $W(k), k = 0, 1, ... N_F$ and taking its inverse DFT, the desired overall impulse response $W(n), n = 0, 1, ... N_F$ is obtained (D.R. Morgan and J.C. Thi, 1995). The stacking procedure required for this, has been presented in the form of an algorithm (L.-O. Larson, J.M. de Haan and I. Claesson, 2002).

Step 1: For the 0th adaptive filter, $i = 0$

$$W(k) = W_0(k), 0 \le k \le N_S/4 - 1$$

Step 2: For $1 \le i \le M-1$,

If i is odd, then

$$W(k+(i-1)N_S/2) = W_i(k), \quad N_S/4 \le k \le 3N_S/4-1$$

If i even, then

$$W(k+(i-2)N_S/2) = W_i(k), \quad 3N_S/4 \le k \le N_S-1$$

and

$$W(k+iN_S/2) = W_i(k), \qquad 0 \le k \le N_S/4-1$$

Step 3: The last $N_S/4$ points, $W(k)$ are obtained by

$$W(k+(M-2)N_S/2) = W_0(k), 3N_S/4 \le k \le N_S-1$$

Step 4: The inverse DFT of $W(k), k = 0, 1, \ldots N_F$; gives the fullband impulse response $W(n)$.

12.15.2 Weight Transform Based on DFT Method (2)

This method also assumes a oversampling factor of 2. This algorithm is similar to DFT-1 method but overcomes the problem of finite blocking effects/ nulls in the frequency response of derived fullband filter. The weight transform is equivalent to passing /filtering the subband adaptive filter coefficient sequences through the filters of synthesis bank. That is

$$W(z) = \sum_{i=0}^{M-1} S_i(z) W_i(z^L)$$

Thus, the errors in DFT-1 algorithm in realising the fullband impulse response $W(n)$, are due to the circular convolution instead of the required linear convolution. To avoid this in the DFT-2 algorithm, for each subadaptive filters, their impulse responses are zero padded to a length of $2N_S$ instead of N_S (used in the DFT-1 method). The $2N_S$ point DFT coefficients obtained by transforming the zero padded impulse responses are properly stacked to get the fullband DFT sequence $W'(k)$ of length $2N_F$. In the inverse DFT of $W'(k)$, $W'(n)$, the first N_F points form the estimate of the desired fullband impulse response (L.-O. Larson, J.M. de Haan, and I. Claesson, 2002).

For the 0th adaptive filter, $i = 0$ and

$$W'(k) = W_0'(k), 0 \le k \le N/2-1$$

For $1 \le i = M-1$:

For i is odd,

$$W'(k+(i-1)N) = W_i'(k), \quad N/2 \le k \le 3N/2-1$$

For i even,

$$W'(k+(i-2)N/2) = W_i'(k), \quad 3N/2 \le k \le 2N-1$$

and

$$W'(k+iN) = W_i'(k), \qquad 0 \le k \le N/2-1$$

The last $N/2$ points of $W'(k)$ are obtained by

$$W'(k+(M-2)N) = W_0'(k), 3N/2 \le k \le 2N-1$$

Then the inverse DFT of gives the overall impulse response $W'(n)$.

In the above algorithms, it has been assumed that for fullband adaptive filter the contribution is only due to the subadaptive filters considering the analysis and synthesis filter banks as ideal (the DFT-2 method reduces some part of the synthesis bank errors that is present in the DFT-1 method). However, in practice, the analysis and synthesis banks are not ideal providing perfect reconstruction due to errors like aliasing. In view of this, a method of deriving the fullband impulse response taking into account the analysis and synthesis filters has also been attempted. But the above two methods may be generally sufficient for many applications as the errors in delayless synthesis filter bank is considered in synthesis dependent subband adaptive filter implementation.

The computational requirement of DLSAF with M channel polyphase FFT analysis filter bank and DFT-1weight transform for real input and real output, as given by Morgan and Thi , 1995, is considered. This has mainly three components: 1) subband filtering, 2) adaptive weight update and weight transform. The computational complexity considered is in terms of number of multiplications per input sample. Further, a oversampling factor of 2 is assumed.

For the M channel polyphase analysis bank with a oversampling factor of 2, each polyphase component has (L/M) coefficients, L being the length of prototype filter used. Therefore, the number of multiplications for M channels is (L/M)M = L. For the M components the real FFT is requires $\mathrm{M}\log_2 M$. Further, as these computations are done at a reduced rate of M/2, the number of computations required for the analysis bank is $[(\mathrm{L}+\mathrm{M}\log_2 M]/(M/2) = 2(L/M) + 2\log_2 M$.

For the adaptation, due to conjugate symmetry of real input signal, only half of M subbands need to be considered. Further, each subadaptive filter has (2N/M) weights, N is the overall filter order required for the system to be identified. As each complex weight multiplication requires 4 real multiplications, the number of multiplications for every adaptation becomes $(M/2)(2N/M)4 = 4N$. Further, as this is done at a decimated rate of $M/2$, the effective number of multiplications for every adaptation is $4N/(M/2) = 8(N/M)$.

For the weight transform, a (2N/M) point complex FFT is required for each of the M/2 lower subbands and real inverse FFT of N points are required. For a complex FFT of (2N/M) points and for the real FFT of N points, $(4N/M)\log_2(2N/M)$ and $N\log_2 N$ multiplications are respectively, required. Thus for the M/2 subbands,

$$(M/2)[(4N/M)\log_2(2N/M)] = 2N\log_2(2N/M)]$$

multiplications are required. Thus, the total numer multiplications for the weight transform per input sample is $(M/2)[(4N/M)\log_2(2N/M)] = N\{\log_2(2N/M) + \log_2(N)\}$. Since the overall filter output cannot change faster than its impulse response length, it only required to do the weight transform once in every (N/J) input samples. Hence number real multiplications for the weight transform per input sample is $\{\log_2(2N/M) + \log_2(N)\}J$ and thus, the number of real multiplications per input sample for the three parts of DLSAF is $\{(2(L/M) + 2\log_2 M\} + \{8N/M\} + \{\log_2(2N/M) + \log_2(N)\}J$, J can be in the range of 1 to 8.

Thus, the number of real multiplications per input sample for the three parts, viz., M channel subband analysis bank filtering, subadaptive filter weight adaptation and weight transformation of a DLSAF is

$$\{(2(L/M) + 2\log_2 M\} + \{8N/M\} + \{\log_2(2N/M) + \log_2(N)\}J,$$

J can be in the range of 1 to 8.

Example: To illustrate the effect of different weight transformations on the performance of system identification using DLSAF, the same system used previously is considered.

$$G(z) = \frac{1 - 1.1289z^{-1} + 1.0058z^{-2} - 0.7617z^{-3}}{1 - 0.59z^{-1} + 0.455z^{-2} - 0.0754z^{-3} + 0.504z^{-4}}$$

The filter bank with 8 channels is derived using a prototype raised cosine type filter of length 136 samples. A Kaiser window with smoothing factor 10 is used to design the lowpass prototype filter. The frequency response of filter bank derived is shown in the Fig. 12.22 (e). The length of the impulse response of this system considered is also 136. Therefore the subadaptive filters are of length 34 samples for a oversampling factor of 2. The length of DFT used in realising the weight transforms DFT-1 and DFT-2 are respectively, 128 and 256 . To find the effect of weight transforms DFT-1 and DFT-2, the system identification is done using zero mean unit variance Gaussian white noise input.

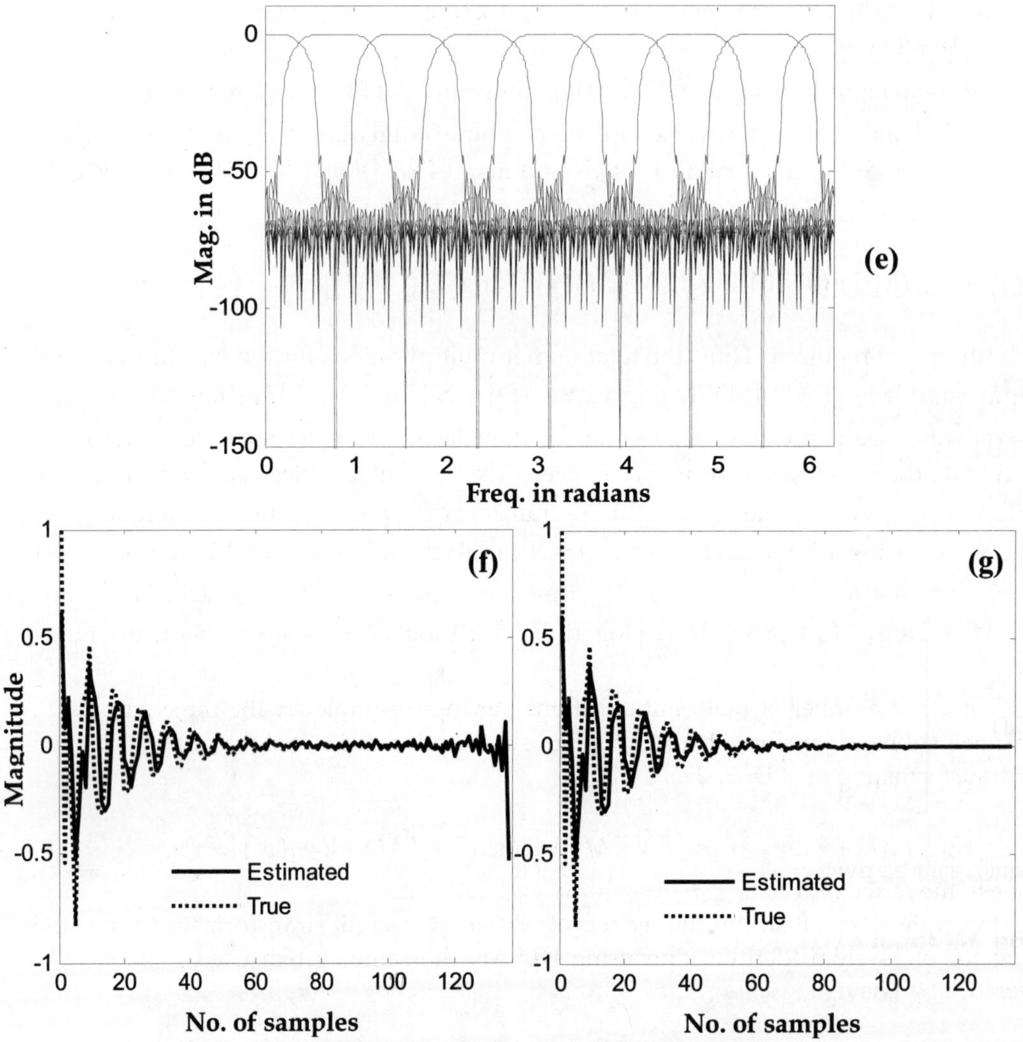

Fig. 12.22(e) Frequency response of 8 channel filter bank derived using raised cosine prototype filter impulse response, The system impulse response by DLSAF using weight transform DFT-1 (f) and DFT-2 (g).

The step size used for subadaptive filters is 0.00001 and the forgetting factor used for computing the power for step size normalization is 0.99. A good weight transform results in an estimate of the overall impulse response which is close to that actual one. The overall impulse response derived by DFT-1 and DFT-2 are shown in Fig.12.22 (f, g). It is seen that the DFT-2 weight transform which overcomes the problem of circular convolution, gives better result as it is close to the actual overall system impulse response, specifically at the far end, it has zero values as against the nonzero values present for DFT-1. Here, no delay adjustment is required for the desired signal, as the delay in the LMS loop is removed.

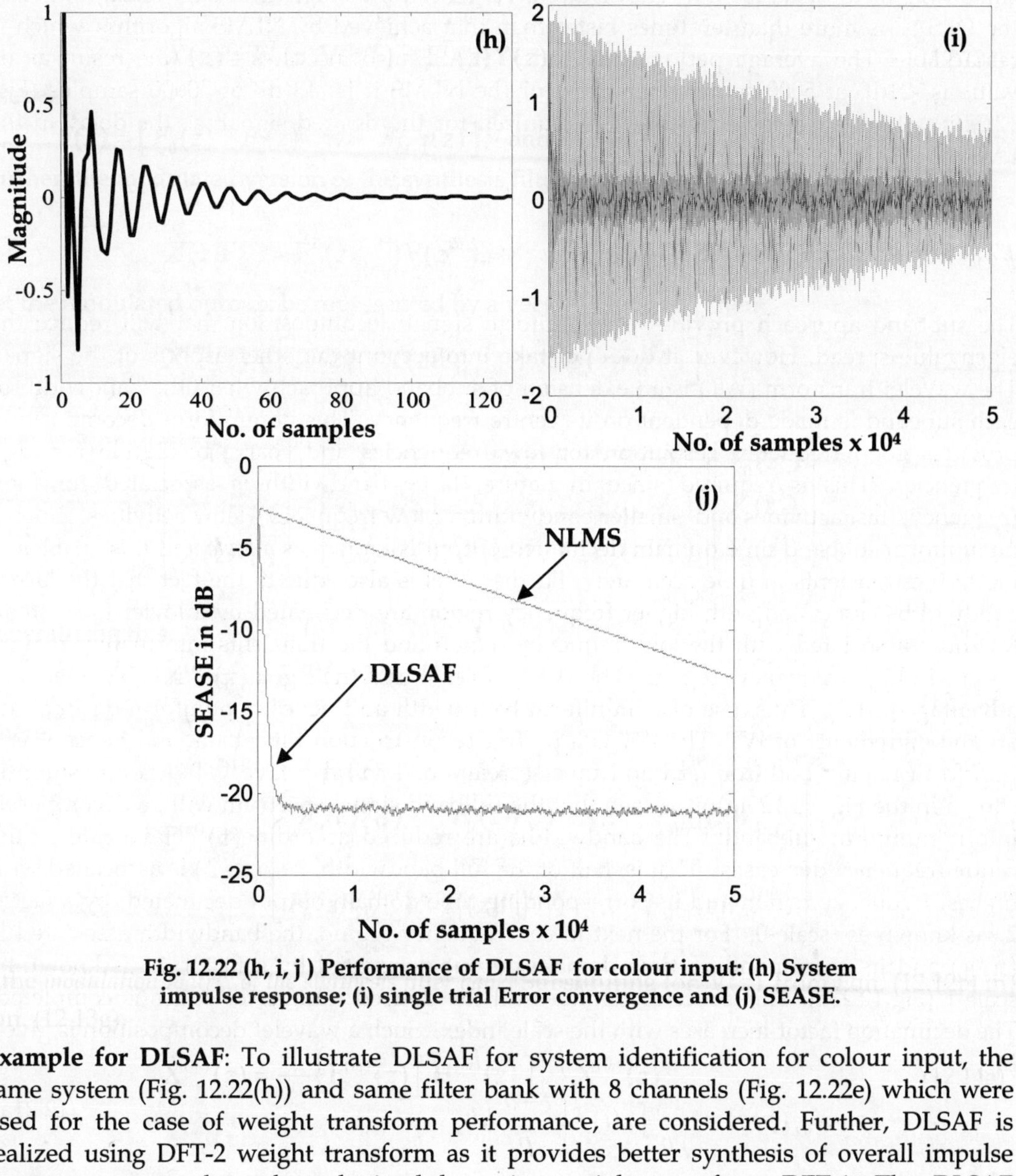

Fig. 12.22 (h, i, j) Performance of DLSAF for colour input: (h) System impulse response; (i) single trial Error convergence and (j) SEASE.

Example for DLSAF: To illustrate DLSAF for system identification for colour input, the same system (Fig. 12.22(h)) and same filter bank with 8 channels (Fig. 12.22e) which were used for the case of weight transform performance, are considered. Further, DLSAF is realized using DFT-2 weight transform as it provides better synthesis of overall impulse response compared to that obtained by using weight transform DFT-1. The DLSAF considered is a close loop system which takes into account the synthesis errors as it decomposes the overall error instead of the desired signal. The subadaptive filters are also same except for step size, which is 0.0001.

The input is made up of sinusoids with frequencies of 100 Hz, 200 Hz, 300 Hz and 900 Hz plus Gaussian white noise to get an SNR=29dB. DFT-2 stacking is used for weight transformation. To compute the performance index SEASE 30 trials are used and the smoothing factor used is 0.99. The error convergence for a single trail (Fig. 12.22i) indicates the DLSAF is more than ten times faster than that achieved by NLMS algorithm which is remarkable. The average performance index SEASE also indicates the same result as its value is -21dB at 5000 samples whereas for the NLMS it is -13 dB at 50000 samples (Fig. 12.22(j)). Here, no delay adjustment is required for the desired signal, as the delay in the LMS loop is removed.

12.16 WAVELET TRANSFORM

The subband approach provides only uniform signal decomposition that will reduce the eigenvalue spread. However, it does not take into account the other aspects of the signal. The wavelet transform (WT) is an extension of subband approach where the bandwidth of each subband is made dependent on its centre frequency. This nonuniform decomposition provides good frequency resolution for low frequencies and coarse resolution for high frequencies. This is required since in nature, large bandwidth is associated for high frequency/ fast activities and smaller bandwidth for low frequency/ slow activities. Such a nonuniform subband time domain decimated output is known as a *scale* and this enables to locate the transients in time accurately. Further, this is also due to the fact that the larger bandwidths associated with higher frequency region are decimated by a lower factor than for those associated with the low frequency region and the transients information mainly will be in high frequency region. The WT enables to use the signal spectral characteristics advantageously. Thus, use of nonuniform bandwidth and hence nonuniform decimation, are the ingredients of WT. The WT is a perfect reconstruction filter bank and hence it is a transform having both forward and inverse relation. Typical octave WT decomposition is shown in the Fig. 12.23(a). It is seen that the fullband signal spectrum will be decomposed into nonuniform subbands. The bandwidths are reduced according to octave rule as the centre frequency decreases. That is, half of the full bandwidth, i.e., $(\pi/2)$ is associated with highest frequency region and its corresponding time domain output decimated by a factor 2, is known as scale-0. For the next lower frequency region, the bandwidth associated is $(\pi/4)$ and its corresponding time domain output decimated by a factor 4 , is known as scale-1 (Fig.12.23a). This process continues till the desired number of scales is completed. The decimation factor increases with the scale index. Such a wavelet decomposition is given by

$$X_{WT}(p,q) = \frac{1}{\sqrt{p}} \int_{-\infty}^{\infty} x(t) f\left(\frac{t-q}{p}\right) dt \tag{12.21a}$$

where

$$p = a^k, \quad q = nTa^k, \quad f(t) = h(-t).$$

Then

$$X_{WT}(k,n) = a^{-\left(\frac{k}{2}\right)} \int_{-\infty}^{\infty} x(t)\, h\left(\frac{nTa^{k} - t}{a^{k}}\right) dt$$

$$X_{WT}(k,n) = a^{-\left(\frac{k}{2}\right)} \int_{-\infty}^{\infty} x(t)\, h\left(nT - a^{-k} t\right) dt$$

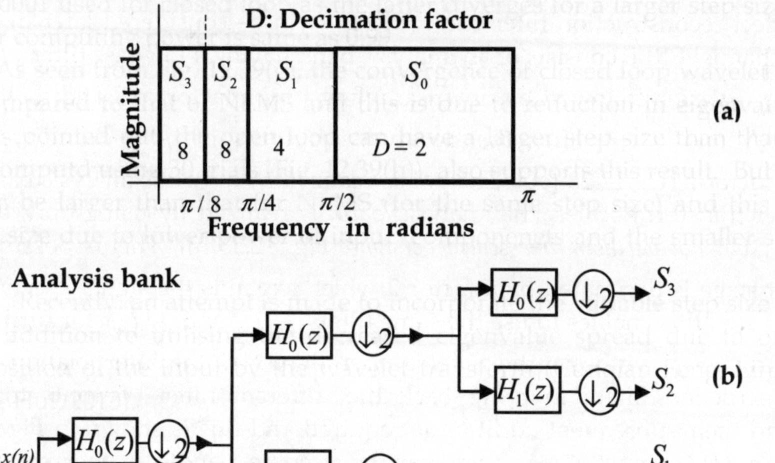

Fig. 12.23 (a) Wavelet transform obeying octave rule; (b) Analysis bank and (c) Synthesis bank.

If $a = 2$, then

$$X_{WT}(k,n) = 2^{-\left(\frac{k}{2}\right)} \int_{-\infty}^{\infty} x(t)\, h\left(nT - 2^{-k} t\right) dt \tag{12.21b}$$

This $X_{WT}(k,n)$ is nothing but the decimated output by a factor 2^k of kth filter whose impulse response is $h(-t)$ time scaled by a factor 2^k. That is, for the scale-0 or $k=1$, the impulse response is expanded by a factor 2 and its output is decimated by a factor 2 which was considered above. The factor $2^{-\left(\frac{k}{2}\right)}$ is the normalization factor which makes the energy independent of k. This illustrates that for $a=2$, the octave filter bank decimated outputs considered represent the wavelet transform. For the realization of the wavelet transform a two channel perfect reconstruction filter bank (PRFB) (will be considered later in detail) is used. This two channel PRFB consists of an analysis bank having a lowpass filter $H_0(z)$ and highpass filter $H_1(z)$ each followed by downsampling by a factor of 2 and the synthesis bank having $F_0(z)$ and $F_1(z)$ after the decimated outputs of $H_0(z)$ and $H_1(z)$ are upsampled by a factor 2 (Fig.12 23 (c)). The WT is realized by a dyadic analysis bank. Here to get all wavelet transform scales, the analysis bank is realized by using only the filters $H_0(z)$ and $H_1(z)$ *but not by their time scaled versions* (Fig. 12.23(b)). This is possible as the input to the next stage is decimated by a factor 2 which expands the signal spectrum from $(0,\pi/2)$ to $(0,\pi)$ and this enables to use the same filters $H_0(z)$ and $H_1(z)$ for further scales also. For reconstruction/ synthesis, the reverse operation, viz., the interpolation with the associated reconstruction filters to bring back the different filter outputs to original sampling rate and their subsequent addition is required and this is shown in Fig. 12.23(c). Here also the same $F_0(z)$ and $F_1(z)$ filters can be used to handle further scales as the upsampling compresses the range from $(0,\pi)$ to $(0,\pi/2)$. The dyadic structure provides the required octave signal spectrum decomposition (Fig. 12.23 (a)) as lowpass filter $H_0(z)$ output at each stage is split into lowpass and highpass and the highpass output is left without splitting (Fig.12.23(b)). The dyadic synthesis bank shown in Fig. 12.23(c) reconstructs the original signal. Presently, for the purpose of wavelet based adaptive filters, the signal decomposition will be confined to only octave rule though other rules are possible. As mentioned, the two channel PRFB filter bank forms the basic unit for the WT realization and it will be considered in detail.

12.17 TWO-CHANNEL PERFECT RECONSTRUCTION FILTER BANK

A band-limited analog signal is sampled at the Nyquist rate to generate a sequence $x(n)$. The spectral content of such a signal extends up to half the sampling frequency, $(0 \ to \ \pi)$. This signal is called fullband signal. A two-channel filter bank, shown in Fig. 12.24, operates on a fullband signal. It divides the input spectrum $X(e^{j\omega}), 0 \le \omega \le \pi$ into two equal subbands. The lowpass and highpass filters $H_0(z)$ and $H_1(z)$ respectively behave as anti-aliasing filters. These output signals $x_L(n)$ and $x_H(n)$ are then downsampled by a factor of 2 to provide decimated versions $V_L(n)$ and $V_H(n)$, respectively. Further, they are processed according to the requirement. On the synthesis side, the subband signals $V_L(n)$ and $V_H(n)$ are passed through the expanders to restore their original data rate.

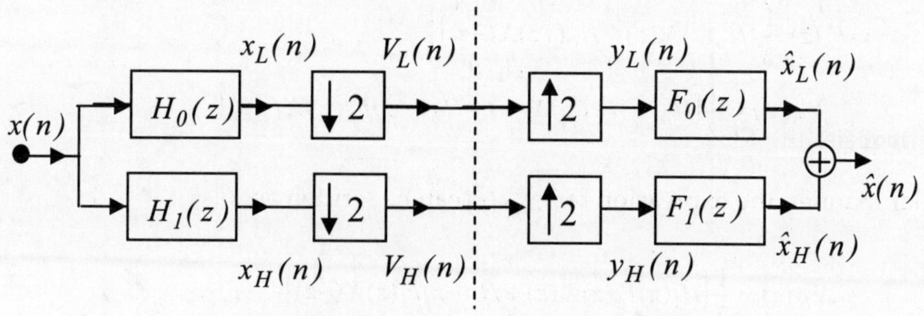

Fig. 12.24 Two-channel filter bank.

Further, the synthesis filters $F_o(z)$ and $F_1(z)$ are used to reject the images that result due to the expanding operation. Finally, the synthesis filter outputs are added to get the reconstructed signal $\hat{x}(n)$. In this chapter, the selection and design of the filters $H_o(z), H_1(z), F_o(z)$ and $F_1(z)$ so as to recover the original signal $x(n)$ will be presented.

12.17.1 Expression for the Transfer Function of a Two Channel Filter Bank

Tracing the low pass subband signal path,

$$X_L(z) = H_o(z)X(z)$$

For $V_L(z)$ obtained by decimating $X_L(z)$, from the theory of decimation,

$$V_L(z) = \frac{1}{2}[X_L(z^{0.5}) + X_L(-z^{0.5})]$$

$$V_L(z) = \frac{1}{2}[H_o(z^{0.5})X(z^{0.5}) + H_o(-z^{0.5})X(-z^{0.5})]$$

For $Y_L(z)$ which is the interpolated version of $V_L(z)$, from the theory of interpolation,

$$Y_L(z) = V_L(z^2) = \frac{1}{2}[H_o(z)X(z) + H_o(-z)X(-z)]$$

By filtering $Y_L(z)$ through $F_o(z)$,

$$\hat{X}_L(z) = Y_L(z)F_o(z)$$

$$= F_o(z)\left[\frac{1}{2}[H_o(z)X(z) + H_o(-z)X(-z)]\right]$$

$$\hat{X}_L(z) = \frac{1}{2}[H_o(z)F_o(z)X(z) + H_o(-z)F_o(z)X(-z)] \tag{12.22a}$$

In a similar manner, the expression for $\hat{X}_H(z)$ can be written as

$$\hat{X}_H(z) = \frac{1}{2}[H_1(z)F_1(z)X(z) + H_1(-z)F_1(z)X(-z)] \tag{12.22b}$$

The reconstructed signal in the z-domain, $\hat{X}(z)$ is obtained by adding $\hat{X}_L(z)$ and $\hat{X}_H(z)$.

$$\hat{X}(z) = \hat{X}_L(z) + \hat{X}_H(z) \tag{12.22c}$$

$$= \frac{1}{2}[H_0(z)F_0(z)X(z) + H_0(-z)F_0(z)X(-z)$$

$$+ H_1(z)F_1(z)X(z) + H_1(-z)F_1(z)X(-z)]$$

Rearranging the terms,

$$\hat{X}(z) = \frac{1}{2}[(H_o(z)F_o(z) + H_1(z)F_1(z))X(z) + (H_o(-z)F_o(z) + H_1(-z)F_1(z))X(-z)]$$

$$\hat{X}(z) = T(z)X(z) + S(z)X(-z) \tag{12.22d}$$

where,

$$T(z) = \frac{1}{2}[H_o(z)F_o(z) + H_1(z)F_1(z)] \tag{12.22e}$$

and

$$S(z) = \frac{1}{2}[H_o(-z)F_o(z) + H_1(-z)F_1(z)] \tag{12.22f}$$

$S(z)$, the coefficient of $X(-z)$ is due to aliasing caused by decimation and imaging caused by expansion. The term $S(z)$ is referred to as an alias term or alias component. In the matrix form, the equation for $\hat{X}(z)$ is

$$\hat{X}(z) = [X(z) \quad X(-z)]\begin{bmatrix} T(z) \\ S(z) \end{bmatrix} = \frac{1}{2}[X(z) \quad X(-z)]\begin{bmatrix} H_o(z)F_o(z) + H_1(z)F_1(z) \\ H_o(-z)F_o(z) + H_1(-z)F_1(z) \end{bmatrix}$$

$$\hat{X}(z) = \frac{1}{2}[X(z) \quad X(-z)]\begin{bmatrix} H_o(z) & H_1(z) \\ H_o(-z) & H_1(-z) \end{bmatrix}\begin{bmatrix} F_o(z) \\ F_1(z) \end{bmatrix} \tag{12.22g}$$

12.18 ERRORS CREATED IN THE FILTER BANK

The two channel filter bank suffers from: aliasing, amplitude distortion and phase distortion errors.

12.18.1 Aliasing

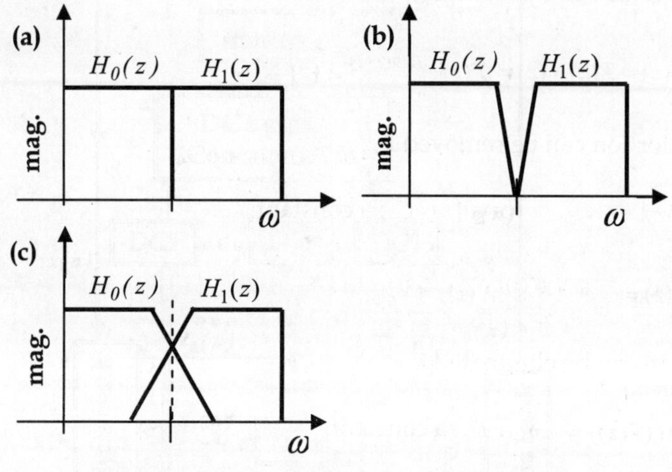

Fig. 12.25(a, b, c) Frequency responses of the filters (a) Ideal; (b) Non-overlapping and (c) Overlapping.

The ideal analysis filters require brick wall frequency response (Fig. 12.25a). However, practical implementation of such a filter is difficult as the filter impulse response to be considered will be very long. For the same reason, even a filter with a very narrow transition width is not a practical one. Hence, filters with a narrow transition width are not used. As an alternative, filters can be non-overlapping, however this creates severe amplitude distortion around ($\pi/2$) (Fig. 12.25b). Though this distortion can be compensated (on the synthesis side) by appropriately boosting the gain in this frequency range, it is not desirable as this will also amplify the noise (due to transmission) severely. Hence, usually, marginally overlapping filters (Fig. 12.25c) are used. For a two-channel filter bank, the bandwidth of each filter is ($\pi/2$). But as substantial energy of the signal outside the intended bandwidth will be present (due to overlap) in the filtered output, aliasing errors are inevitable on maximal decimation.

12.18.2 Amplitude and Phase Distortions

For the reconstructed signal $\hat{X}(z)$ (Eqn. 12.22d),

$$\hat{X}(z) = T(z)X(z) + S(z)X(-z),$$

amplitude distortion occurs *if* $T(z)$ *is not a constant for all frequencies.* If the transfer function $T(z)$ *has non-linear phase, the reconstructed signal suffers from phase distortion.*

12.18.3 Removal of Errors in 2-channel Filter Bank

Consider the Eqn. (12.22d),

$$\hat{X}(z) = T(z)X(z) + S(z)X(-z),$$

the aliasing can be removed by forcing $S(z) = 0$. Then

$$\hat{X}(z) = T(z)X(z) = |T(z)|e^{j\arg(T(z))}X(z)$$

The amplitude distortion can be removed if,

$$|T(z)| = C, \quad \text{for all } \omega,\ C : \text{a constant.}$$

Then

$$\hat{X}(z) = C\, e^{j\arg(T(z))}X(z)$$

The phase distortion can be eliminated if

$$\arg(T(z)) = -m\omega,\ m : \text{a constant.}$$

That is for no distortion,

$$\hat{X}(z) = Ce^{-j\omega m}X(z)$$

or $$\hat{X}(z) = Cz^{-m}X(z) \qquad (12.23a)$$

As stated above, for no aliasing distortion, $S(z) = 0$ and this implies from Eqn. (12.22f)

$$S(z) = \frac{1}{2}[H_o(-z)F_o(z) + H_1(-z)F_1(z)] = 0$$

$$H_o(-z)F_o(z) + H_1(-z)F_1(z) = 0 \qquad (12.23b)$$

The expression for $T(z)$ of Eqn. (12.22e),

$$T(z) = \frac{1}{2}[H_o(z)F_o(z) + H_1(z)F_1(z)]$$

From the perfect reconstruction requirement given by Eqn. (12.23a),

$$T(z) = \frac{1}{2}[H_o(z)F_o(z) + H_1(z)F_1(z)] = z^{-m} \text{ with } C = 1$$

or, $$[H_o(z)F_o(z) + H_1(z)F_1(z)] = 2z^{-m} \qquad (12.23c)$$

Solving Eqn. (12.23c) with Eqn.(12.23b), by Cramer's rule,

$$\begin{bmatrix} F_0(z) \\ F_1(z) \end{bmatrix} = \frac{2z^{-m}}{[H_0(z)H_1(-z) - H_0(-z)H_1(z)]} \begin{bmatrix} H_1(-z) \\ -H_0(-z) \end{bmatrix} \tag{12.23d}$$

The bracketed term of the denominator is nothing but the determinant of the matrix of Eqn. (12.22g). The selection of the filters $H_0(z), H_1(z), F_0(z)$ and $F_1(z)$ to meet the perfect reconstruction requirement of the above equation is the objective of filter bank design.

12.19 FILTER BANK DESIGN

12.19.1 Power Symmetric Filters

If a filter $H(z)$ satisfies,

$$\tilde{H}(z)H(z) + \tilde{H}(-z)H(-z) = 1 \tag{12.24a}$$

such a filter is called a power symmetric filter. Here $\tilde{H}(z)$ represents paraconjugation that is obtained by conjugating the coefficients of $H(z)$ and further replacing z^{-1} by z. This is equivalent to a transfer function whose zeros (poles) are located at conjugate reciprocal locations. Let

$$H(z) = 1 - a\,z^{-1}.$$

Its roots are

$$z^{-1}(z - a) = 0, \quad z = a, \ \infty$$

Then, $\tilde{H}(z)$ is obtained by conjugating the coefficients and replacing z^{-1} by z.

$$\tilde{H}(z) = 1 - a^*\,z$$

Its roots are

$$\left(1 - a^*\,z\right) = 0, \quad z = 1/a^*$$

As stated the root corresponding to 'a' in $H(z)$ becomes $(1/a^*)$. If a is real then,

$$\tilde{H}(z) = 1 - a\ z,\ H(z) = 1 - a\,z^{-1},$$

$$H(e^{j\omega}) = 1 - a\,e^{-j\omega},\ \tilde{H}(e^{j\omega}) = 1 - a\,e^{j\omega} = H(e^{-j\omega})$$

$$\tilde{H}(e^{j\omega}) = H(e^{-j\omega})$$

Therefore, for $H(z)$ with real coefficients,

$$\tilde{H}(z) = H(e^{-j\omega}) \quad \text{and} \quad \tilde{H}(-z) = H(-e^{-j\omega})$$

Therefore,

$$\tilde{H}(z)H(z) + \tilde{H}(-z)H(-z) = 1$$

$$H(e^{-j\omega})H(e^{j\omega}) + H(-e^{-j\omega})H(-e^{j\omega}) = 1 \quad (12.24b)$$

But,

$$H(-e^{j\omega}) = H(e^{j(\omega-\pi)}), \text{ since } -1 = e^{-j\pi}$$

$$H(e^{-j\omega})H(e^{j\omega}) + H(e^{-j(\omega-\pi)})H(e^{j(\omega-\pi)}) = 1$$

Or

$$\left|H(e^{j\omega})\right|^2 + \left|H(e^{j(\pi-\omega)})\right|^2 = 1 \quad \text{as } H(e^{j\omega})H(e^{-j\omega}) = \left|H(e^{j\omega})\right|^2$$

Put $\omega = \left(\frac{\pi}{2} + \theta\right)$, then

$$(\pi - \omega) = \left(\frac{\pi}{2} - \theta\right)$$

and therefore,

$$\left|H\left(e^{j(\pi/2+\theta)}\right)\right|^2 + \left|H\left(e^{j(\pi/2-\theta)}\right)\right|^2 = 1, \qquad \text{for any real } \theta$$

That is, *the magnitude squared function exhibits anti symmetry with respect to* $(\pi/2)$. This is nothing but the quadrature mirror property described in Section-12.5.1.
In the Eqn (12.24a),

$$\tilde{H}(z)H(z) + \tilde{H}(-z)H(-z) = 1,$$

If $\left[\tilde{H}(z)H(z)\right]$ is represented by $G(z)$ then

$$G(z) + G(-z) = 1 \quad (12.24c)$$

Further, $g(n)$ is *even* as it is the inverse FT of $\left|H(e^{j\omega})\right|^2$. Hence, $G(z)$ *satisfies the requirement of a half band filter*. So, *the power symmetric property of* $H(z)$ *is equivalent to saying that* $\left[\tilde{H}(z)H(z)\right]$ *is a half band filter.*

The order of a half band filter $2N$ is always twice that of an odd number N and N is the order of the lowpass filter $H(z)$ (discussed in Section-12.5.4).

12.19.2 Causal Filter Bank

The causal lowpass filter $H(z)$ is derived from the halfband filter is $G(z) = \tilde{H}(z)H(z)$ by spectral factorization (discussed in Section – 12.19). Then, the filters of the analysis side are selected as,

$$H_0(z) = H(z) \quad (12.25a)$$

$$H_1(z) = -z^{-N}\tilde{H}(-z) = -z^{-N}\tilde{H}_0(-z) \tag{12.25b}$$

i.e., the two filters satisfy pseudoquadrature mirror property. With this the denominator of Eqn. (12.23d) becomes,

$$H_0(z)H_1(-z) - H_0(-z)H_1(z)$$

$$= H(z)\{-(-z)^{-N}\tilde{H}(-z)\} + \{-H(-z)\}\{-z^{-N}\tilde{H}(-z)\}$$

(Since N is odd, $(-1)^{-N} = -1$ and $-(-z)^{-N} = z^{-N}$)

$$= z^{-N}\tilde{H}(z)H(z) + z^{-N}H(-z)\tilde{H}(z)$$

$$H_0(z)H_1(-z) - H_0(-z)H_1(z) = z^{-N}\left[H(z)\tilde{H}(z) + H(-z)\tilde{H}(z)\right]$$

If $H(z)$ is power symmetric, then the term within the bracket reduces to a constant. Therefore,

$$H_0(z)H_1(-z) - H_0(-z)H_1(z) = z^{-N} \tag{12.25c}$$

Further, the *delay in the Eqn. (12.25b) is essential to restore causality of* $H_1(z)$. Otherwise, the sequence $\tilde{H}(-z)$ becomes non-causal, if $H(z)$ is causal. Consider Eqn. (12.23d),

$$\begin{bmatrix} F_0(z) \\ F_1(z) \end{bmatrix} = \frac{2z^{-m}}{[H_0(z)H_1(-z) - H_0(-z)H_1(z)]}\begin{bmatrix} H_1(-z) \\ -H_0(-z) \end{bmatrix}$$

This becomes

$$\begin{bmatrix} F_0(z) \\ F_1(z) \end{bmatrix} = \frac{2z^{-m}}{z^{-N}}\begin{bmatrix} H_1(-z) \\ -H_0(-z) \end{bmatrix}$$

Further, by choosing $m = N$,

$$F_0(z) = 2H_1(-z) \tag{12.25d}$$

$$F_1(z) = -2H_0(-z) \tag{12.25e}$$

Time domain expressions: Since

$$H_1(z) = -z^{-N}\tilde{H}_0(-z) = -z^{-N}H_0(-z^{-1}) \Rightarrow -e^{-j\omega N}H_0(e^{-j(\omega-\pi)})$$

$$h_1(n) = (-1)^n h_0(N-n)$$

Similarly,

$$H_1(z) = -z^{-N}\tilde{H}(-z),\ H_1(-z) = z^{-N}\tilde{H}(z)$$

$$F_0(z) = 2H_1(-z) = 2z^{-N}\tilde{H}_0(z)$$

$$F_o(z) = 2z^{-N}\tilde{H}_0(z) \Rightarrow 2e^{-j\omega N}H_0(e^{-j\omega})$$
$$f_0(n) = 2h_0(N-n)$$

and $$F_1(z) = -2H_0(-z), \quad f_1(n) = -2(-1)^n h_0(n)$$

Summarizing,

$$h_0(n) = h(n) \tag{12.26a}$$
$$h_1(n) = (-1)^n h_0(N-n) \tag{12.26b}$$
$$f_0(n) = 2h_0(N-n) \tag{12.26c}$$
$$f_1(n) = -2(-1)^n h_0(n) \tag{12.26d}$$

Summarizing, if $h(n)$ is a power symmetric filter, the choice of other filters as in Eqn. (12.26) results in a perfect reconstruction system satisfying $\hat{x}(n) = x(n-N+1)$.

Directly writing the relationship between the analysis filters $(h_0(n), h_1(n))$ and synthesis filters $(f_0(n), f_1(n))$,

$$F_o(z) = 2z^{-N}\tilde{H}_0(z) \Leftrightarrow f_0(n) = 2h_0^*(N-n)$$
$$H_1(z) = -z^{-N}\tilde{H}_0(-z), \quad H_0(-z) = -z^{-N}\tilde{H}_1(z),$$

Therefore,

$$F_1(z) = -2H_0(-z) = 2z^{-N}\tilde{H}_1(z) \Leftrightarrow f_1(n) = 2h_1(N-n)$$

The synthesis filter $f_0(n)$ is nothing but a time reversed and delayed version of $h_0(n)$, the delay being N samples and $f_1(n)$ is flipped and delayed version of $h_1(n)$.

The dyadic or regular binary tree structure derived from the basic two-channel filter bank also satisfies PR property. The delay offered by such a filter bank depends on the number of stages in the structure.

12.19.3 Orthogonality Conditions to be Satisfied by the Analysis and Reconstruction Filters

$$G(z) + G(-z) = K = 2 \text{ (say)}$$
$$\tilde{H}_0(z)H_0(z) + \tilde{H}_0(-z)H_0(-z) = 2$$
$$H_0(z^{-1})H_0(z) + H_0(-z^{-1})H_0(-z) = 2 \quad \text{(assuming } h_0(n) \text{ to be real)}$$

$$\sum_n h_0(n)z^{-n}\sum_m h_0(m)\, z^m + \sum_n (-1)^{-n}h_0(n)z^{-n}\sum_m (-1)^{-m} f_0(m)\, z^m = 2$$

$$\sum_n\sum_m h_0(n)h_0(m)z^{-n}z^m + \sum_n\sum_m (-1)^{-(n+m)}h_0(n)h_0(m)\, z^{-n}z^m = 2$$

The two terms get cancelled for odd values of $(n+m)$ and since both the terms are equal, they get added for even values of $(n+m)$. Therefore,

$$\sum_n \sum_m h_0(n)h_0(m)z^{-n}z^m = 1 \tag{12.27}$$

Consider the autocorrelation given by

$$r(l) = \sum_k h_0(k)h_0(k+l)$$

Taking z-transform

$$R(z) = \sum_l \left(\sum_k h_0(k)h_0(k+l) \right) z^{-l} = \sum_k h_0(k) \sum_n (h_0(k+l))z^{-l}$$

Let $p = k + l$,

$$R(z) = \sum_k h_0(k) \sum_m h_0(p) z^{-(p-k)}$$

$$R(z) = \sum_k \sum_m h_0(p)h_0(k)z^{-p}z^k$$

Comparing with this, the L.H.S. of (12.27) is the Z-transform of the autocorrelation function. Therefore,

$$G(z) = \sum_n \sum_k h_0(k)h_0(k+n)z^{-n}$$

Since for a half band filter $G(z)$, the even numbered samples vanish,

$$r(2n) = \sum_k h_0(k)h_0(k+2n) = 0$$

That is,

$$r(2n) = \sum_k h_0(k)h_0(k+2n) = \delta(n)$$

$$\sum_k h_0(k)h_0(k+2n) = \delta_n$$

Similarly, using $h_1(n) = (-1)^n h_0(N-n)$,

$$\sum_k h_1(k)h_1(k+2n) = \delta_n,$$

$$\sum_k h_0(k)h_1(k+2n) = 0$$

12.20 DESIGN OF A TWO-CHANNEL QUADRATURE MIRROR FILTER (QMF) BANK

A zero phase half-band filter $G(z)$ satisfying the stopband attenuation and transition width is designed. A spectral factor of such a halfband filter will provide a filter that satisfies the power symmetric property. Then, all other filters of the filter bank can be derived from this filter. $G(e^{j\omega})$ *has to be non-negative* (as it can be represented by its spectral factors). The design is summarized in the following steps.

1. Design a zero phase FIR half band filter $G_w(z)$ of order 2N.
2. Then define $G(z) = G_w(z) + \delta$, where δ is the peak stop band ripple (half of the peak to peak ripple amplitude of $G_w(e^{j\omega})$. This ensures that $G(e^{j\omega}) \geq 0$. This is essential as $G(z)$ has to be factorized. Here, $G(e^{j\omega})$ is equivalent to a power spectral density function and $g(n)$, the impulse response equivalent to an autocorrelation function.
3. Compute the spectral factors $H(z)$ and $\tilde{H}(z)$ of the filter $G(z)$.

12.20.1 Design of a Zero Phase FIR Half-band Filter

By window method:

The window method provides a simple way of designing a good low pass FIR M^{th} band filter, with cutoff frequency (π / M). For an ideal low pass filter $G_I(z)$ with cutoff at (π / M), the impulse response is

$$g_I(n) = \left(\frac{2\pi}{M}\right)\frac{\sin(n\pi / M)}{(n\pi / M)}, -\infty \leq n \leq \infty \qquad (12.28a)$$

Since $sin(n\pi / M)$ is zero for n equal to non-zero multiple of M, $g_I(n)$ satisfies the condition of a Mth band filter.

The ideal filter impulse response is non-causal, does not satisfy the BIBO (bounded input bounded output) requirement and the response exists for infinite length. The impulse response length can be made finite by truncating it. However, as explained in chapter 4, an abrupt truncation, i.e., use of a rectangular results in Gibbs ripple effect, which not only distorts the constant passband gain but also affects the stopband attenuation. The latter effect is due to the fact that a larger ripple magnitude leads to lower stopband attenuation. The *type of truncation* decides the amplitude of the Gibbs ripple and not the truncation length. The smoother the truncation, lower is the ripple magnitude as it is decided by the relative heights of main lobe to side lobe levels. The windows with different smoothness result in different magnitudes of the ripple. Hence, by applying a suitable smooth window to the ideal impulse response, the desired passband ripple/stopband attenuation and

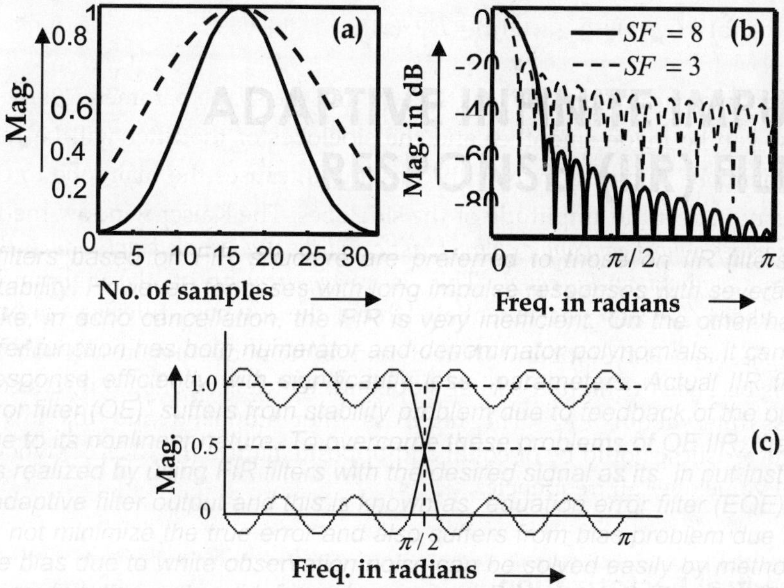

Fig. 12.26(a) Kaiser window; (b) its frequency response for different smoothing factors (SF) and (c) Illustration of equiripple requirement of half band filter.

transition width; can be obtained. The length and shape of the window are the two important parameters: as the former controls the filter transition width and the latter the stopband attenuation/passband ripple. Longer window length results in narrower transition width and smoother window shape gives higher stop band attenuation (due to smaller peak sidelobe level of the window spectrum). However, smoother the window shape, longer window is required as the effective length of the window is reduced resulting in a larger transition width (due to larger mainlobe width of the window spectrum). This trade-off between the mainlobe width and sidelobe level of the window has been nearly optimized in the Kaiser window (Fig. 12.26 a, b). The Kaiser window is given by

$$w(n) = \begin{cases} I_0\left[\beta\left\{1-(n/0.5N)^2\right\}^{0.5}\right]/I_0(\beta), & \frac{N}{2} \le n \le \frac{N}{2} \\ 0, & \textit{otherwise} \end{cases} \tag{12.28b}$$

where, $I_0(x)$ is the zeroth order Bessel function and can be computed from the power series

$$I_0(x) = 1 + \sum_{k=1}^{\infty} \left\{(0.5x)^k / k!\right\}^2 \tag{12.28c}$$

The first 15 terms are sufficient for calculating $I_o[x]$ for most applications.

$$g_w(n) = g_I(n)\, w(n)$$

$$G(z) = G_w(z) + \delta \quad ; \quad g(n) = F\{G(z)\}^{-1}$$

This window has two parameters, the length $L = N$ and the shape parameter β. By increasing β, the window can be made smoother and the sidelobes of the filter will be reduced but with a wider mainlobe. Increase in L with constant β, causes the mainlobe to decrease in width but does not affect the amplitude of the sidelobes. The Kaiser window method of FIR filter design *provides almost equiripple characteristic and the peak ripple amplitude in the passband and stopband are approximately equal.* This is a *requirement* of the halfband filter as antisymmetry about $(\pi/2, 0.5)$ is essential. Unless the lowpass filter is equiripple with respect to the passband and stopband characteristics, the perfect reconstruction will not be satisfied. This is because the highpass filter stopband, which is nothing but reflected version of the lowpass filter about $\pi/2$. The highpass filter stopband and the lowpass filter passband characteristics should be of equal and opposite nature (Fig. 12.26c) when added to give a constant value over all frequencies.

12.20.2 Spectral Factorization

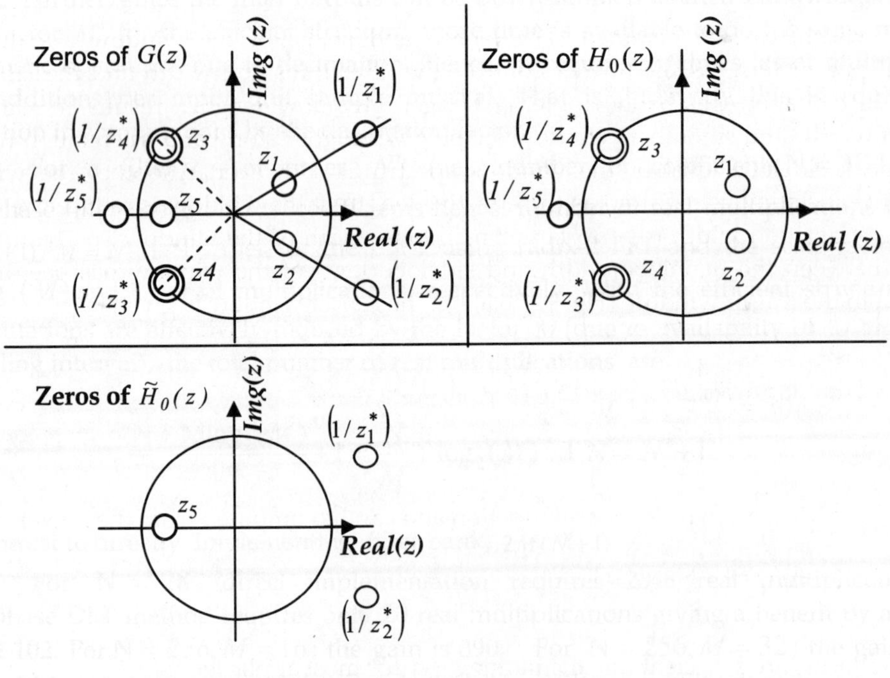

Fig. 12.27 (a) Spectral factorization of G (z).

Any zero phase FIR transfer function $G(z)$ with order $2M$ can be factorized in the form of $G(e^{j\omega}) = |H(e^{j\omega})|^2$, *if* $|G(e^{j\omega})| \geq 0$ *for all* ω. The zero phase property of $G(z)$ implies that if z_k is a zero then $(1/z_k^*)$ is also a zero. If $G(e^{j\omega}) \geq 0$, zeros on the unit circle will be of even multiplicity. In Fig. 12.27a, $G(z)$ which has zeros at z_k, $k = 1, 2, 3, 4, 5$ inside or on the unit circle, because of its zero phase or symmetry property, has zeros at corresponding conjugate reciprocal locations at $(1/z_k^*)$, $k = 1, 2, 3, 4, 5$. In the z-domain, $G(z) = \tilde{H}(z)H(z)$, where $H(z)$ is a FIR with order M. The filter $H(z)$ is called as spectral factor of $G(z)$.

Among the zeros of $G(z)$ any arbitrary set of zeros which have their conjugate counterparts (to form a real signal) form a spectral factor $H(z)$ of $G(z)$. One such is shown in the Fig. 12.27a. In this spectral factor, $H(z)$, z_k, $k = 1, 2, 3,4$ and $(1/z_k^*)$, $k = 3,4,5$ are chosen. Further, z_k, $k = 3, 4$ are on the unit circle and so also $(1/z_k^*)$, $k = 3,4$. But $(1/z_k^*)$, $k = 5$ is a real zero. The other corresponding factor $\tilde{H}_0(z)$ is having zeros z_k, $k = 5$ and $(1/z_k^*)$, $k = 1, 2$, which again form a real signal. The spectral factor $H(z)$ of $G(z)$ is formed by considering either z_k or $(1/z_k^*)$ for all k. If z_k is assigned to $H(z)$, then $(1/z_k^*)$ is assigned to $\tilde{H}(z)$.

Mathematically,

$$H(z) = c\prod_{k=1}^{M}(1 - z^{-1}z_k) \quad \text{and} \quad \tilde{H}(z) = c^*\prod_{k=1}^{M}(1 - z\, z_k^*)$$

(Either z_k or $(1/z_k^*)$ can lie inside or outside the unit circle.)

The most obvious technique of computing a spectral factor is to find the $2M$ zeros of $G(z)$ and to pick an appropriate subset of M zeros to define $H(z)$. In general, the spectral factor $H(z)$ is not unique because $(1 - z^{-1}z_k)$ in $H(z)$ can be replaced with $(1 - z^{-1}/z_k^*)$ and c can be readjusted so that the above equation continues to hold.

If all the zeros of a filter lie within the unit circle then such a filter is referred to as *strictly minimum phase filter*. On the other hand, if all the zeros lie outside the unit circle, such a filter is called *strictly maximum phase filter*. If the zeros lie on and or inside/outside the unit circle, then the filter is referred to as "minimum / maximum phase" filter.

If $G(z)$ has all zeros on the unit circle, then its spectral factor is unique. The minimum phase factor and maximum phase factor are also unique.

12.20.3 Spectral Factorization for Causal Filter Bank Design

Usually, a minimum phase $H(z)$ as shown in Fig. (12.27b) is selected. For filters *of higher order*, finding the zeros of the transfer function is computationally cumbersome process and hence an efficient method based on cepstrum suggested by Main and Nainer is used. This method provides a minimum phase spectral factor that is unique. However *to handle zeros which may lie on or very close to the unit circle*, this approach may pose problems and this can be overcome by making the critical zeros to move away from the unit circle by

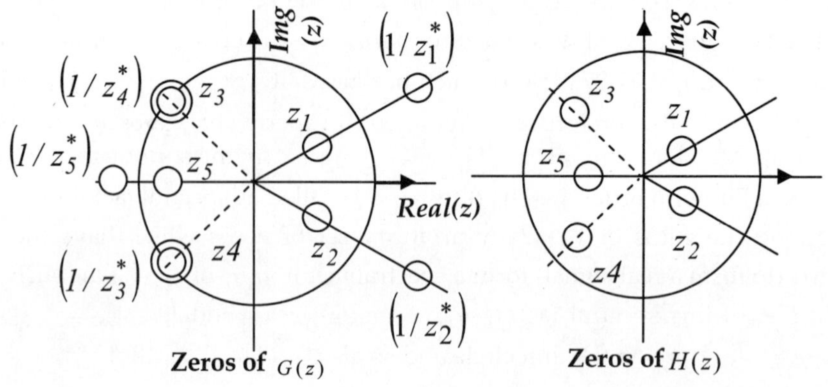

Fig. 12.27(b) Minimum phase spectral factorization of $G(z)$.

adding a positive quantity to the zeroth sample of the impulse response $g(n)$ or equivalently rising the spectral floor level of $G(z)$.

Spectral factorization by cepstrum

The minimum phase spectral factor of a sequence $g(n)$ can be derived from the cepstrum technique. The theory behind the technique is reviewed here. Let $g(n)$ be a sequence with z-transform $G(z)$, its cepstrum (complex cepstrum) $\hat{g}(n)$ is given by

$$\hat{g}(n) = \left(\frac{1}{2\pi}\right)\int_{-\pi}^{\pi}\hat{G}(e^{j\omega})e^{j\omega n}d\omega \tag{12.29a}$$

where

$$\hat{G}(\omega) = \ln[|G(\omega)|] + j\theta(\omega), \tag{12.29b}$$

$G(\omega)$ is the Fourier transform of $g(n)$ and $\theta(\omega)$ is the phase of $G(\omega)$. The phase $\theta(\omega)$ is the unwrapped phase of $G(\omega)$ and generally the angle computed by using *arctan* function is the wrapped phase $\theta_P(\omega)$ which is discontinuous as it lies between $-\pi$ and π and integral multiples of π have to be added to make $\theta_P(\omega)$ continuous and this process is called phase unwrapping.

Consider the function

$$\hat{G}_e(z) = \ln[|G(z)|] \quad \text{Or} \quad \hat{G}_e(e^{j\omega}) = \ln\left|G(e^{j\omega})\right| \tag{12.29c}$$

The even part $\hat{g}_e(n)$ of the cepstrum $\hat{g}(n)$ is given by the inverse Fourier transform of $\hat{G}_e(e^{j\omega})$, i.e.,

$$\hat{g}_e(n) = \left(\frac{1}{2\pi}\right)\int_{-\pi}^{\pi} \hat{G}_e(e^{j\omega})e^{j\omega n}d\omega \tag{12.29d}$$

If $G(z)$ is the z-transform of a minimum phase signal, then the cepstrum $\hat{g}(n)$ is zero for negative time. That is,

$$\hat{g}(n) = 0 \qquad for \quad n < 0 \tag{12.29e}$$

For such a signal, $\hat{g}(n)$ is completely determined by the real part of the Fourier transform, $\ln\left|G(e^{j\omega})\right|$. The cepstrum of the minimum phase signal $\hat{g}(n)$ can be obtained by the logarithm of the magnitude of the Fourier transform alone. The cepstrum of the minimum phase signal is obtained from the even part of the cepstrum and is presently considered.

$$\hat{g}_e(n) = \frac{[\hat{g}(n) + \hat{g}(-n)]}{2} \tag{12.29f}$$

Since $\hat{g}(n)$ is causal, i.e., $\hat{g}(-n) = 0,\ n < 0$, it is given by

$$\hat{h}(n) = U(n)[\hat{g}_e(n)]\ , \quad U(n) = \begin{cases} 1, & n = 0 \\ 2, & n > 0 \\ 0, & n < 0 \end{cases} \tag{12.29g}$$

Presently, it is required to get the minimum phase spectral factor $H(z)$ of the halfband filter $g(n)$. In the DFT domain of size N,

$$\begin{aligned} U(n) &= 1, \quad n = 0, N/2 \\ &= 2, \quad n = \ \ 1 \le n \le N/2 - 1 \\ &= 0, \quad n = N/2 + 1 \le n \le N - 1 \end{aligned}$$

For this,

$$\hat{G}_e(z) = (1/2)\ln[|G(z)|] \tag{12.29h}$$

and cepstrum of the spectral factor $\hat{h}(n)$ is obtained as explained above (Eqn. 12.29g). Then

$$\hat{H}(\omega) = FT[\hat{h}\ (n)] \tag{12.29i}$$

$$H(\omega) = e^{\hat{H}(\omega)} \tag{12.29j}$$

The zeros corresponding to minimum phase spectral factorization are shown in the Fig. 12.27(b). To handle the zeros on the unit circle as stated the floor level is raised by adding a small positive quantity to the zeroth coefficient of $g(n)$ and this results in the *movement of all the zeros radially* towards the origin as the bandwidth associated with the zeros increases. Otherwise, the representation of the zero locations may require a very large size DFT to meet the required frequency resolution as it increases with the proximity of the zero to the unit circle. Hence, this enables the application of the cepstrum computed from a reasonable size DFT, for spectral factorization. However, as this movement of zeros may result in decreased stopband attenuation or filter transition width or both. But these can be met by choosing proper window parameters.

12.20.4 Illustration of Spectral Factorization Technique and PRFB Design

Example 1: Minimum phase spectral factorization by root method

$$G_i(z) = \frac{1}{2} + \sum_{n=1}^{i} g_i(2n-1)\left(z^{-2n+1} + z^{2n-1}\right)$$

The coefficients $h_i(n)$ are determined using the Legrangian interpolation formula,

$$g_i(2n-1) = \frac{(-1)^{n+i-1}\prod_{k=1}^{2i}(i-k+1/2)}{(i-n)!\,(i-1+n)!\,(2n-1)}$$

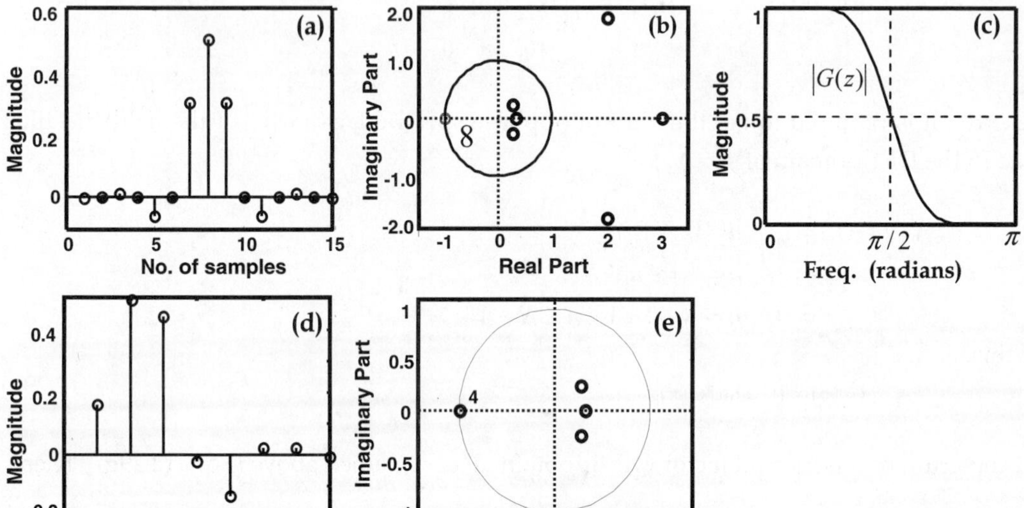

Fig. 12.28 (a) Impulse response; (b) plot of roots of $G(z) = H(z)\widetilde{H}(z)$; (c) Frequency response of $G(z)$. For $H(z)$; (d) Impulse response and (e) its plot of roots.

This filter has $2i$ fold zero at $z = -1$, $i-1$ zeros inside the unit circle and $i-1$ zeros outside the unit circle. The zeros inside and outside the unit circle are mirror images of each other. It is therefore suited for spectral factorization. The spectral factorization procedure is demonstrated for $i=4$

The filter $G(z)$ is of order 14 and length 15. It has eight zeros at $z_i=-1, i=1,2,...,8$ and other 6 roots at

$$z_9 = 0.2841 + i\,0.2432,\ 1/z_9, z_9^*, 1/z_9^* \text{ and } z_{10} = 0.3289, 1/z_{10}.$$

Fig. (12.28c) shows the frequency response of the half band filter $G(z)$.

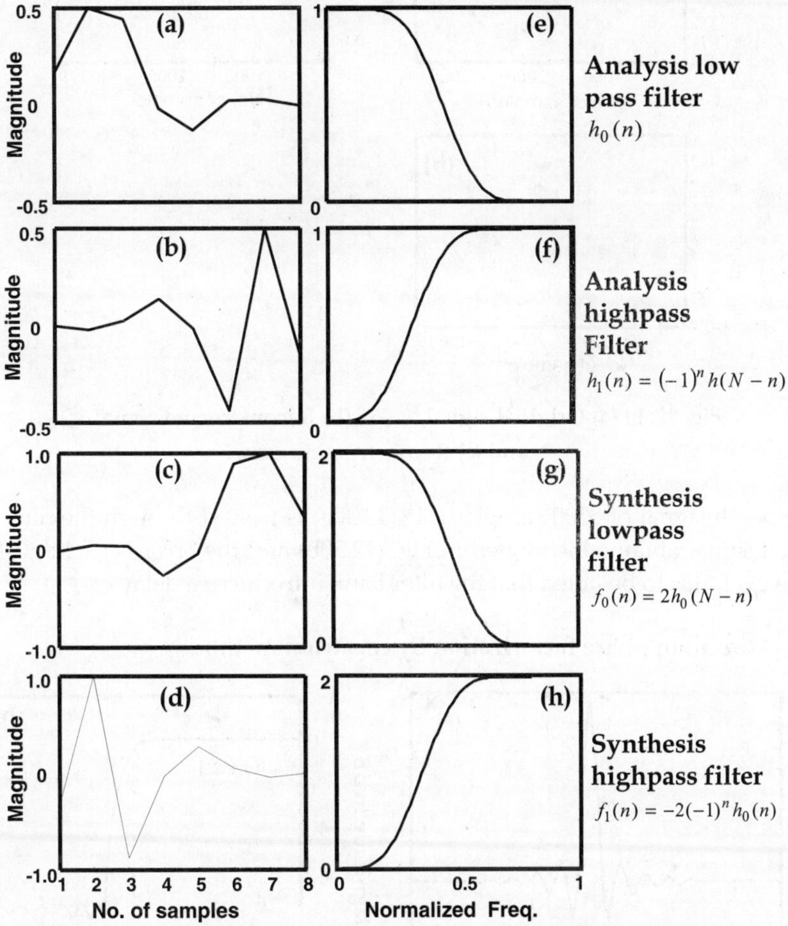

Fig. 12.29 (a, b, c, d) Impulse responses and (e, f, g, h) magnitude responses of the filters in (a,b,c,d), respectively.

To achieve the minimum phase spectral factor $H(z)$, the zeros within the unit circle and half the zeros on the unit circle are selected as shown in Fig. 12.28 (e). The causal filter $h(n)$ obtained by these roots are shown in Fig. 12.28(d). The order of the filter is $N=7$ and its length is $L = 8$. Starting with this $h(n)$, the analysis and synthesis filters of the causal filter bank are derived using the relations of Eqn. (12.26). Fig. 12.29 shows the impulse and frequency responses of the analysis and synthesis filters.

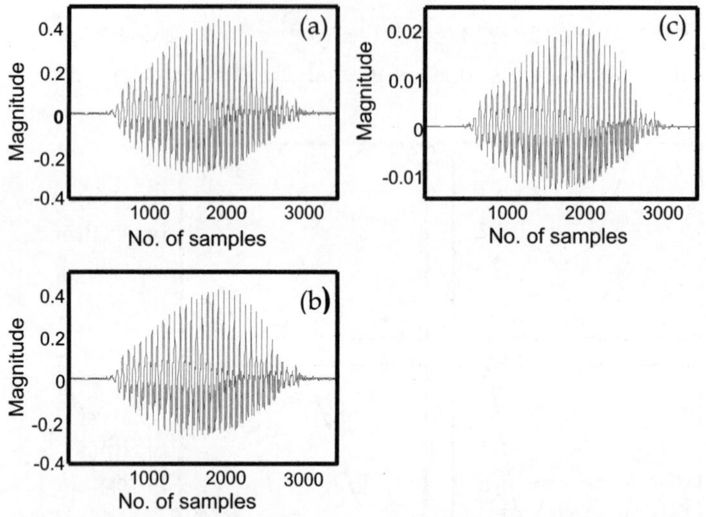

Fig. 12.30 (a) Original signal, $x(n)$; (b) Reconstructed signal, $\hat{x}(n)$ and (c) Reconstruction error.

A speech signal of 3500 samples (Fig. 12.30a) is passed through the filter bank and the reconstructed signal obtained is plotted in Fig. (12.30b) and the error is 0.02 in 0.4, i.e , 5% (Fig. 12.30c). However, it is to be noted that the filter bank introduces a delay of $L-1=7$ samples.

Example 2: Minimum phase factorization by cepstrum method

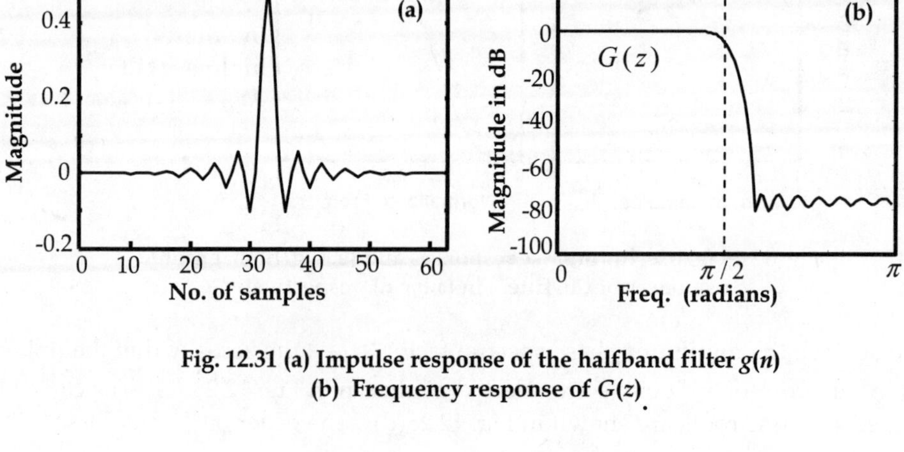

Fig. 12.31 (a) Impulse response of the halfband filter $g(n)$ (b) Frequency response of $G(z)$.

A halfband filter of odd length (63 taps) and order 62 is designed by the window method as discussed in section 6.5. The value of β used for the Kaiser window is 8. The impulse response of the halfband filter $g(n)$ is shown in Fig. 12.31(a). The magnitude response exhibits cut-off at $\pi/2$ and a stopband attenuation of about 80dB (Fig. 12.31(b)).

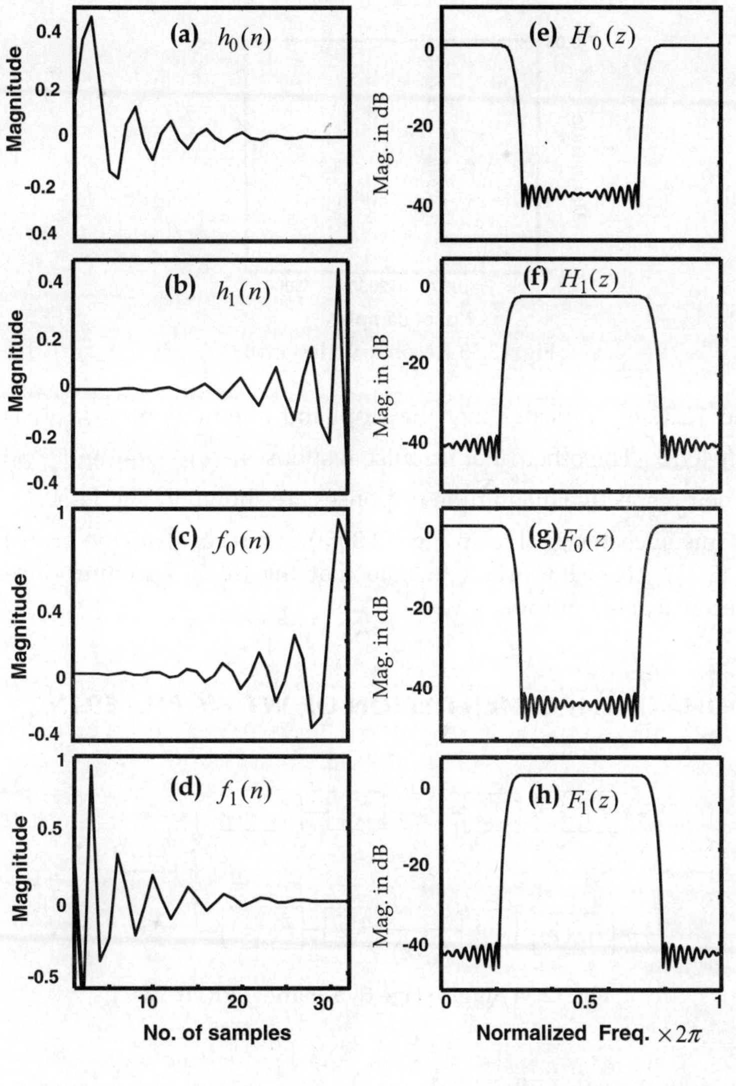

Fig. 12.32 (a, b, c, d) Impulse responses of the filters $H_0(z), H_1(z), F_0(z)$ and $F_1(z)$ and (e, f, g, h) are their magnitude responses, respectively.

The cepstrum of $g(n)$ is computed to perform spectral factorization. For this a DFT of length 256 is used. The zeroth coefficient of $g(n)$ is multiplied with a factor 1.0003 to move the zeros away from the unit circle. The impulse response $h_0(n)$, corresponding to the spectral factor and its magnitude response are shown in Fig. 12.32 (a) and Fig. 12.32(e), respectively. The length of the filter $h_0(n)$ is $L = 32$ and its order is $N = 31$. Comparison of

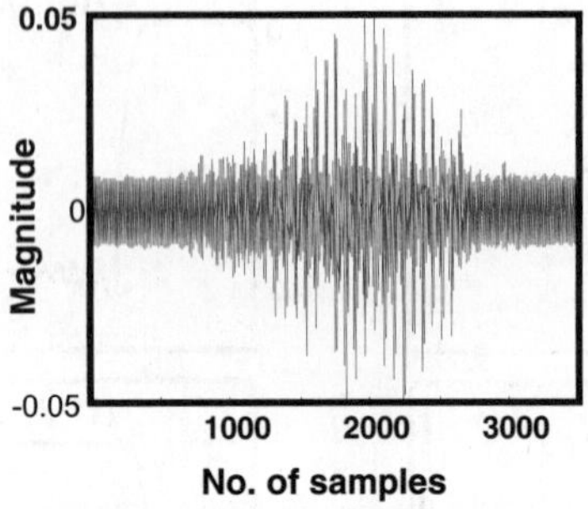

Fig. 12.33 Reconstruction error.

Figs. 12.31(b) and 12.32(e), indicates that the stopband attenuation of $H_0(z)$ is half that of $G(z)$ that is about 40dB. The other filter impulse responses $h_1(n)$, $f_0(n)$ and $f_1(n)$ are derived from $h_0(n)$ and their respective magnitude responses are shown in Fig. 12.32.

For the same speech signal as in Fig. 12.30(a), the reconstruction error is 0.05 in 0.4, i.e., 12.5% (Fig. 12.33). It is important to note that the filter bank introduces a delay of $L-1=31$ in the reconstructed output.

12.21 TIME DOMAIN IMPLEMENTATION OF WT BY FILTERING

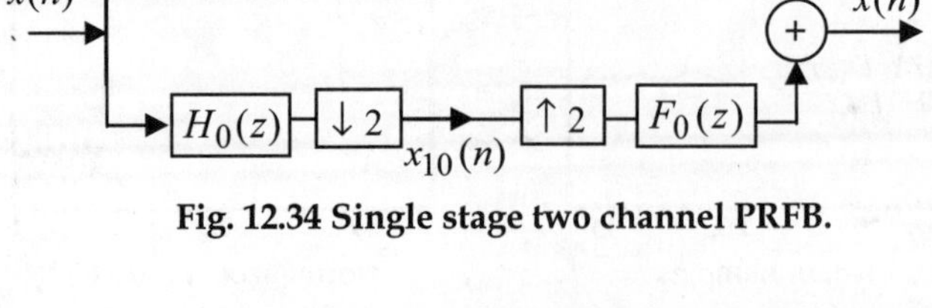

Fig. 12.34 Single stage two channel PRFB.

For a PRFB, the output of the filter bank should be a scaled and delayed version of the input. Since the synthesis filters are made causal by introducing a delay, this delay unless compensated properly, results in an error in the reconstructed signal. This compensation is done by introducing a delay at the output of each scale. The amount of delay that has to be added at the output of each scale depends on the number of stages in the PRFB. After the signal is reconstructed, the accumulated delay is removed to get perfect reconstruction

(Delay analysis in a dyadic two-channel filter bank S.V.Narasimhan, Nandini BasuMallik and S. Veena, , 2011, chapter 12*)*: Consider a two-channel PRFB with input $x(n)$ and equal length filters of even length, L (Fig. 12.34). In the first stage, $x(n)$ is decomposed into two subband signals and after decimation, they represent the low and high scales $x_{10}(n)$ and $x_{11}(n)$, respectively.

$$\hat{X}(z) = X_{10}(z^2)F_0(z) + X_{11}(z^2)F_1(z) \tag{12.30a}$$

$$\hat{X}(z) = z^{-(L-1)}X(z) \tag{12.30b}$$

$$X(z) = z^{(L-1)}\hat{X}(z) \tag{12.30c}$$

For a two-stage, two-channel PRFB (Fig.12.35), the $x_{10}(n)$ channel is split such that the low and high scales, after decimation are $x_{20}(n)$ and $x_{21}(n)$, respectively.

$$\hat{X}_{10}(z) = X_{20}(z^2)F_0(z) + X_{21}(z^2)F_1(z) \tag{12.31a}$$

$$\hat{X}_{10}(z) = z^{-(L-1)}X_{10}(z) \tag{12.31b}$$

$$X_{10}(z) = z^{(L-1)}\hat{X}_{10}(z) \tag{12.31c}$$

Substituting Eqn. (12.31c) in Eqn. (12.30a)

$$\hat{X}(z) = [z^{(L-1)}\hat{X}_{10}(z)]_{\uparrow 2}\, F_0(z) + X_{11}(z^2)F_1(z)$$

$$or,\quad \hat{X}(z) = z^{2(L-1)}[\,\hat{X}_{10}(z^2)F_0(z) + z^{-2(L-1)}X_{11}(z^2)F_1(z)\,]$$

$$or,\quad z^{-2(L-1)}\hat{X}(z) = \hat{X}_{10}(z^2)F_0(z) + [z^{-(L-1)}X_{11}(z)]_{\uparrow 2}F_1(z) \tag{12.32a}$$

Substituting for $\hat{X}(z)$ (from Eqn.(12.30b)) in Eqn.(12.32a),

$$\underline{\underline{z^{-3(L-1)}}}X(z) = \hat{X}_{10}(z^2)F_0(z) + [\underline{z^{-(L-1)}}X_{11}(z)]_{\uparrow 2}F_1(z) \tag{12.32b}$$

Thus $x_{11}(n)$ should be delayed by $D_1 = (L-1)$ before filtering by the synthesis filter bank; and the overall delay is $3\ (L-1)$.

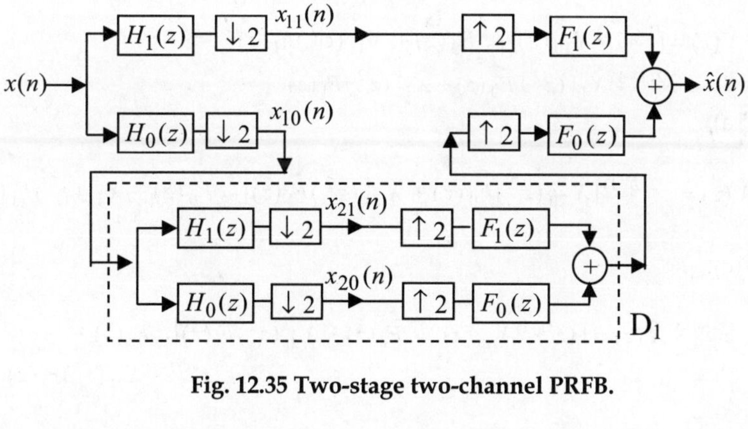

Fig. 12.35 Two-stage two-channel PRFB.

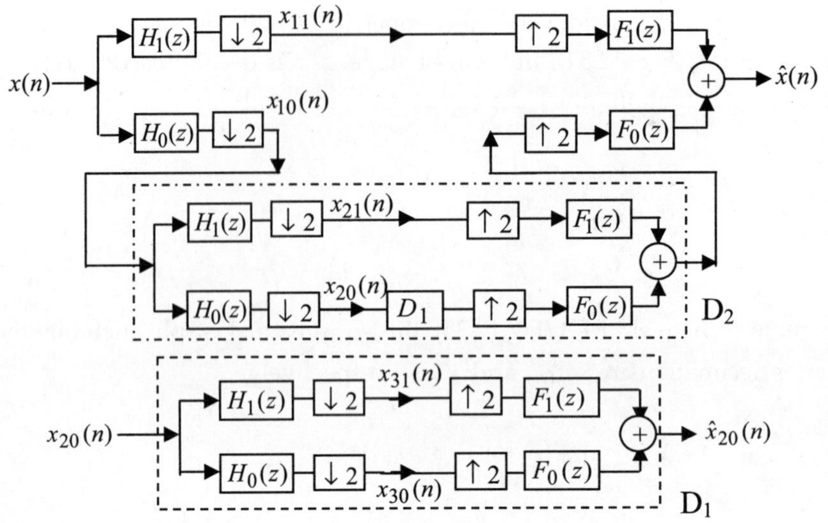

Fig. 12.36 Three-stage two-channel PRFB.

For a three-stage, two-channel PRFB (Fig.12.36),

$$\hat{X}_{20}(z) = X_{30}(z^2)F_0(z) + X_{31}(z^2)F_1(z) \tag{12.33a}$$

$$\hat{X}_{20}(z) = z^{-(L-1)}X_{20}(z) \tag{12.33b}$$

$$X_{20}(z) = z^{(L-1)}\hat{X}_{20}(z) \tag{12.33c}$$

Substituting Eqn. (12.31c) in Eqn. (12.30a),

$$\hat{X}(z) = [z^{(L-1)}\hat{X}_{10}(z)]_{\uparrow 2} F_0(z) + X_{11}(z^2)F_1(z)$$
$$= z^{2(L-1)}\hat{X}_{10}(z^2)F_0(z) + X_{11}(z^2)F_1(z)$$

Using Eqn. (12.31a),

$$\hat{X}(z) = z^{2(L-1)}[X_{20}(z^2)F_0(z) + X_{21}(z^2)F_1(z)]_{\uparrow 2} F_0(z) + X_{11}(z^2)F_1(z)$$

Using Eqn. (12.31c),

$$z^{-2(L-1)}\hat{X}(z) = [(z^{(L-1)}\hat{X}_{20}(z))_{\uparrow 2}F_0(z) + X_{21}(z^2)F_1(z)]_{\uparrow 2}F_0(z) +$$
$$z^{-2(L-1)}X_{11}(z^2)F_1(z)$$

$$= [z^{2(L-1)}\hat{X}_{20}(z^2)F_0(z) + X_{21}(z^2)F_1(z)]_{\uparrow 2} F_0(z) + z^{-2(L-1)} X_{11}(z^2)F_1(z)$$

$$= z^{4(L-1)}[\hat{X}_{20}(z^2)F_0(z) + z^{-2(L-1)} X_{21}(z^2)F_1(z)]_{\uparrow 2} F_0(z) + z^{-2(L-1)} X_{11}(z^2)F_1(z)$$

$$z^{-6(L-1)}\hat{X}(z) = [\hat{X}_{20}(z^2)F_0(z) + z^{-2(L-1)} X_{21}(z^2)F_1(z)]_{\uparrow 2} F_0(z) + z^{-6(L-1)} X_{11}(z^2)F_1(z)$$

$$= [\hat{X}_{20}(z^2)F_0(z) + (z^{-(L-1)} X_{21}(z))_{\uparrow 2} F_1(z)]_{\uparrow 2} F_0(z) + [z^{-3(L-1)} X_{11}(z)]_{\uparrow 2} F_1(z)$$

Using Eqn. (12.30b),

$$\underline{\underline{z^{-7(L-1)}}} X(z) = [\hat{X}_{20}(z^2)F_0(z)]_{\uparrow 2} + [(\underline{\underline{z^{-(L-1)}}} X_{21}(z))_{\uparrow 2} F_1(z)]_{\uparrow 2} F_0(z) + [\underline{\underline{z^{-3(L-1)}}} X_{11}(z)]_{\uparrow 2} F_1(z) \quad (12.34)$$

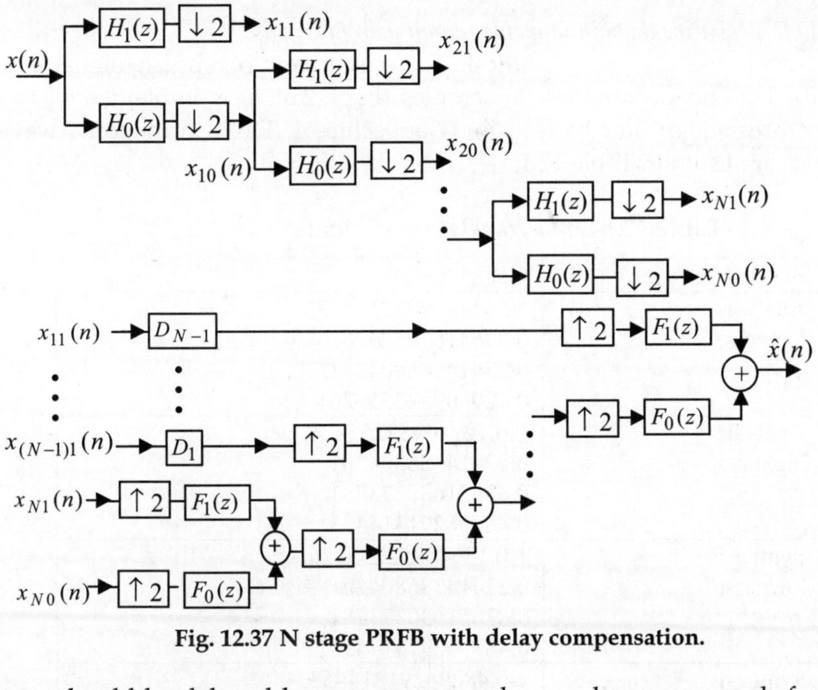

Fig. 12.37 N stage PRFB with delay compensation.

Thus, $x_{11}(n)$ should be delayed by $D_2 = 3(L-1)$ and $x_{21}(n)$ by $D_1 = (L-1)$, before filtering by the synthesis filter bank; and the overall delay is $7(L-1)$.

Generalizing for an 'N' stage PRFB,

$$D_1 = (L-1), \quad D_2 = 2D_1 + (L-1) = 3(L-1), \ldots\ldots$$
$$\ldots\ldots, D_{N-1} = 2D_{N-2} + (L-1) = (2^{N-1} - 1)(L-1) \quad (12.35)$$

Here, all delays are with respect to the sampling rate of the original signal. The same approach can be followed for a PRFB with unequal length filters, with the only difference being the amount of delay added. If the filter lengths are unequal and odd, the delays at different stages are given by

$$D_1 = (L), \quad D_2 = 2D_1 + (L) = 3\,(L), \ldots\ldots, D_{N-1} = 2D_{N-2} + (L) = (2^{N-1} - 1)\,(L),$$

where L is the length of the smaller length filter.

If the filter lengths are unequal and even, the delays at different stages are given by

$$D_1 = (L+1), \quad D_2 = 2D_1 + (L+1) = 3(L+1), \ldots\ldots, D_{N-1}$$
$$= 2D_{N-2} + (L+1) = (2^{N-1} - 1)\,(L+1),$$

where L is the length of the smaller length filter.

The DTWT analysis and synthesis of a signal using an N-stage PRFB with delay compensation is shown in Fig. 12.37.

Example for 1-D DTWT time domain implementation using filtering:

The DTWT analysis and synthesis for a segment of speech signal (vowel 'A') sampled at 10 KHz and of length 4096 samples (Fig. 12.38a), is implemented by filtering method, for a three-stage filter bank using Daubechies-4 (DB-4) orthogonal wavelet. The filter coefficients are listed in Table 12.1.

Table 12.1 DB-4 wavelet coefficients

Filter	Coefficients
Analysis lowpass, h_0	{0.48296291314453, 0.83651630373781, 0.22414386804201, - 0.12940952255126}
Analysis highpass, h_1	{-0.12940952255126, - 0.22414386804201, 0.83651630373781, - 0.48296291314453}
Synthesis lowpass, f_0	{-0.12940952255126, 0.22414386804201, 0.83651630373781, 0.48296291314453}
Synthesis highpass, f_1	{-0.48296291314453, 0.83651630373781, - 0.22414386804201, - 0.12940952255126}

[Note: DB-4 means Daubechies wavelet filter with 4 coefficients. The number of coefficients is related to number of zeros at the digital frequency π. More the number of zeros, the flatter will be the lowpass filter characteristic around the frequency π. This implies more attenuation for the high frequencies and smoother is the wavelet as it is generated by the lowpass filter. Thus, DB-6 is a better wavelet than DB-4.]

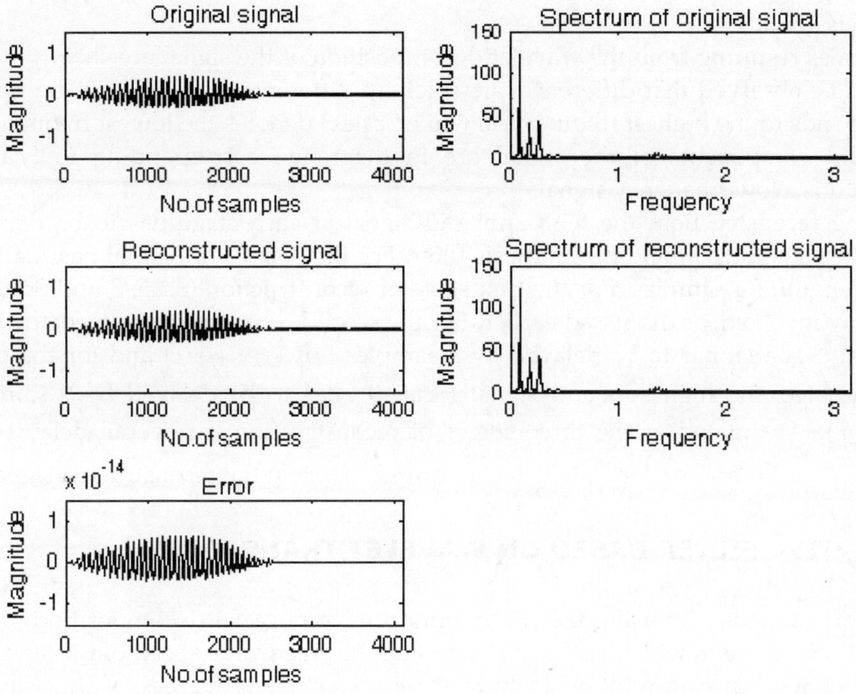

Fig. 12.38(a) Signal reconstruction by filtering, for a speech signal using DB4 wavelet coefficients.

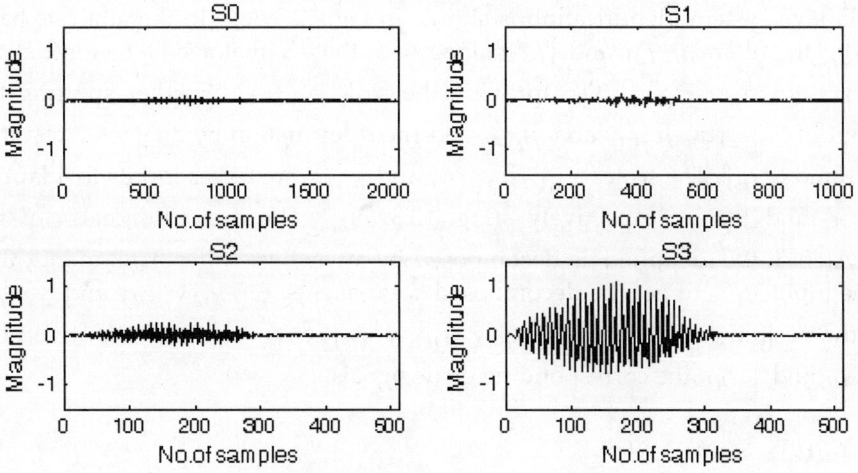

Fig. 12.38(b) Scales for the signal shown in Fig. 12.38(a).

The reconstructed signal is very close to the original input speech and its spectrum is also similar to that of the original (Fig. 12.38a). The magnitude of the error or the difference between the input and reconstructed signal, is very small (10^{-14}), indicating that the reconstruction is good.

The scales resulting from the wavelet decomposition of the signal are shown in Fig. 12.38(b). It can be observed that different scales pick up different frequencies in the signal. Scale-0 corresponds to the highest frequencies, while Scale-3 denotes the lowest frequencies. Most significant components of the signal are in the range corresponding to Scale-3, indicating that it is a low frequency signal.

During reconstruction, the high subband signal at each stage has to be delayed before filtering by the corresponding synthesis filter. For the DB-4 orthogonal basis, all the filters have a length of 4 samples. For the first stage of reconstruction (Scale-3 and Scale-2), there is no delay involved, as discussed earlier. For the second reconstruction stage, the high subband signal (Scale-1), has to be delayed by 3 samples ($D_1 = (L-1) = 3$) and for the third reconstruction stage, the high subband signal (Scale-0), has to be delayed by 9 samples ($D_2 = 3(L-1) = 9$). At the end of the three stages of reconstruction, an overall delay of 21 samples ($7(L-1) = 21$) is accumulated, which is removed to obtain the reconstructed signal.

12.22 ADAPTIVE FILTER BASED ON WAVELET TRANSFORM

The initial efforts aimed at utilising the the nonuniform orthogonal decomposition of the signs in the wavelet domain which reults in decreasing the eigenvalue speed of the signal to achieve faster convergence in addtion to having the percetual properties of the wavelet transform (Erdol N, Basbug F. 1996; Hosur S, Tewfik H. 1997; Liu changyun, Chen changxing, Jia gui, 2002). A three-scale wavelet transform (S_{x0}, S_{x1}, S_{x2}) based synthesis independent adaptive system identification is shown in Fig. 12.39(c). In the analysis bank for the input $x(n)$, the filters $H_0(z)$ and $H_1(z)$ along with the decimator by a factor 2 at the output of $H_1(z)$ (shown in Fig. 12.39(a))provide the scale $S_{x0}(n)$. Further splitting the decimated output of $H_0(z)$ by $H_0(z)$ and $H_1(z)$ and their decimation by a factor 2 result in the scales $S_{x1}(n)$ and $S_{x2}(n)$. The scales $S_{x0}(n), S_{x1}(n)$ and $S_{x2}(n)$ are fed as inputs to adaptive filters $W_0(z)$, $W_1(z)$ and $W_2(z)$, respectively (Fig. 12.39(c)) The desired signal $d(n)$, the output of the system to be identified, is decomposed by an analysis filter bank identical to that used for the input $x(n)$, to get the decomposed components $S_{d0}(n), S_{d1}(n)$ and $S_{d2}(n)$. Further, using the desired signal components $S_{d0}(n), S_{d1}(n)$, $S_{d2}(n)$ and the adaptive filter outputs $y_0(n)$, $y_1(n)$ and $y_2(n)$, the corresponding error signals

$$S_{e0}(n) = S_{d0}(n) - y_0(n),$$
$$S_{e1}(n) = S_{d1}(n) - y_1(n),$$
$$S_{e2}(n) = S_{d2}(n) - y_2(n)$$

are generated. These error signals $S_{e0}(n), S_{e1}(n)$ and $S_{e2}(n)$ are used for updating adaptive filters $W_0(z)$, $W_1(z)$ and $W_2(z)$, respectively. The outputs $y_1(n)$ and $y_2(n)$ of adaptive filters $W_1(z)$ and $W_2(z)$ are interpolated by a factor 2 using image rejection filters $F_1(z)$ and $F_0(z)$, respectively (Fig 12.39(b)). These signals are added and again interpolated by a factor of 2. To this, the interpolated (by a factor 2) version of output $y_0(n)$ of $W_0(z)$ is added to get the estimate $y(n)$ of the desired signal $d(n)$. Prior to $y_0(n)$ interpolation, it has to be delayed by the delay corresponding to the delay of the two channel PRFB formed by the filters $H_0(z)$, $H_1(z)$, $F_0(z)$ and $F_1(z)$. These operations form the output synthesis bank (Fig. 12 39(b)).

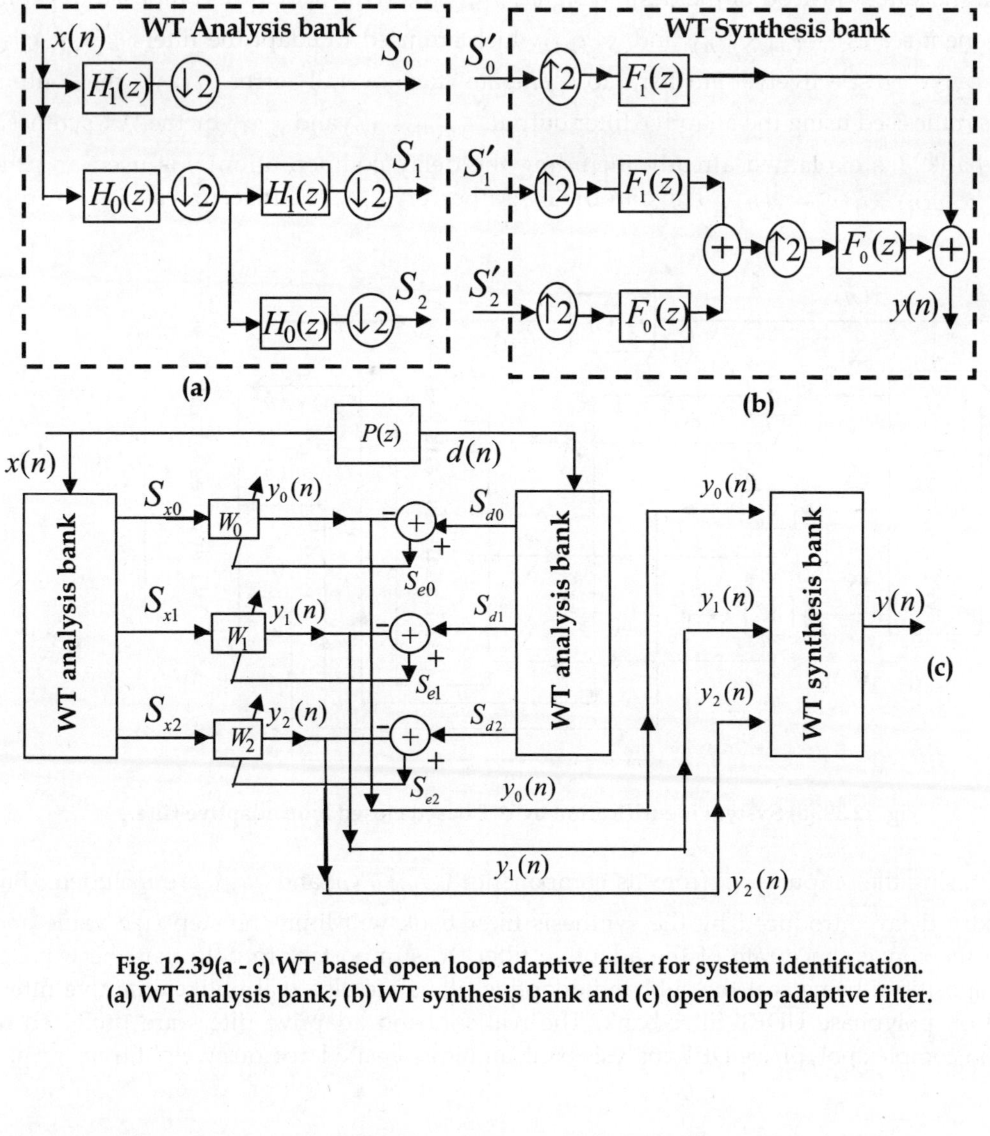

Fig. 12.39(a - c) WT based open loop adaptive filter for system identification. (a) WT analysis bank; (b) WT synthesis bank and (c) open loop adaptive filter.

The delay compensation for synthesis bank considered is simple due to consideration of only 3 scales. But this will be quite tedious and the method already dealt (in section -12.20) should be followed. The errors used for the adaptive filters do not include the errors due to synthesis bank as it is not included in the loop. Only the component outputs $y_0(n)$, $y_1(n)$ and $y_2(n)$ contribute to/decide the error. Hence, the error in the synthesis bank if any is not taken into account by the adaptive filter structure (Fig. 12.39c). It is assumed that the error in synthesis filter bank is tolerable and this structure falls in the category of synthesis independent structure.

In Fig. 12.39(d), synthesis dependent WT based adaptive filter for system identification is given. The only difference between synthesis independent and dependent is that in the latter, instead of the desired signal $d(n)$, the error $e(n)$ is decomposed to get its component scales $S_{e0}(n), S_{e1}(n)$ and $S_{e2}(n)$ which are used to adapt the filters $W_0(z)$, $W_1(z)$ and $W_2(z)$, respectively. Further, to generate the overall error $e(n)$, the output $y(n)$ synthesised using the adaptive filter outputs $y_0(n)$, $y_1(n)$ and $y_2(n)$ by the WT synthesis filter-bank (as explained already including the delay compensation) is used and the error is $e(n) = d(n) - y(n)$. Thus, in the overall error $e(n)$, the inaccuracies involved in

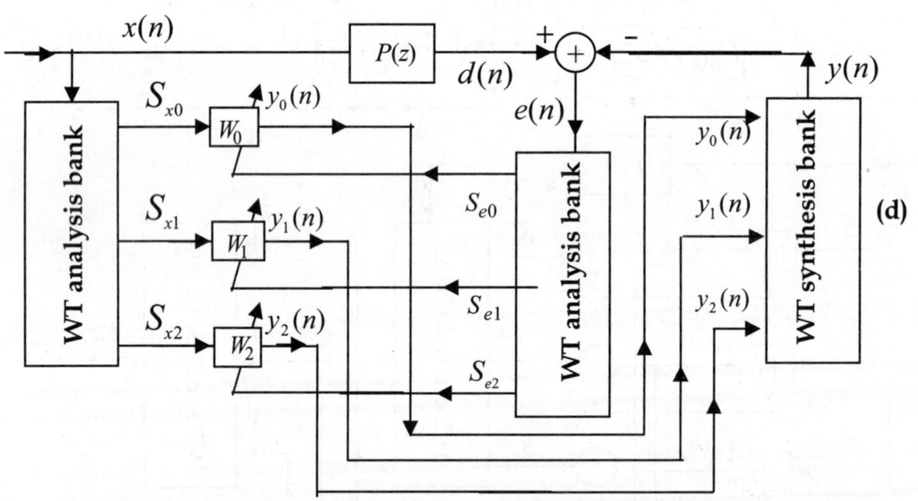

Fig. 12.39 (d) System identification by WT based closed loop adaptive filter.

synthesising the output $y(n)$ from its components $y_0(n)$, $y_1(n)$ and $y_2(n)$ are included. But the extra delay introduced by the synthesis filter bank will limit the step size value and hence the convergence rate of the adaptive filter. It is important to note that the wavelet decompositions being real the subband adaptive filters are also real unlike adaptive filters based on polyphase UDFT filter bank. The real subband adaptive filters are preferred to base on complex polyphase DFT analysis bank and is associated tedious weight transform.

To illustrate the wavelet transform for adaptive filtering a 2 stage 2-channel filter bank which results in 3 scales (0, 1 and 2) using DB-4 wavelet coefficients are used for analysis and synthesis filter bank for the identification of the system

$$G(z) = \frac{1 - 1.1289z^{-1} + 1.0058z^{-2} - 0.7617z^{-3}}{1 - 0.59z^{-1} + 0.455z^{-2} - 0.0754z^{-3} + 0.504z^{-4}}$$

with the input sinusoids of 300Hz, 900Hz and 1800Hz plus Gaussian white noise of SNR=27.7 dB. The sampling frequency used is 4 KHz. The system impulse response length considered which is also the overall adaptive filter length is 128. Hence, the subadaptive filter lengths for scales 0, 1 and 2 are respectively, 64, 32 and 32 samples. For the computation of performance index SEASE,the number of trials and smoothing factor used are 30 and 0.99.

Closed loop wavelet adaptive filter

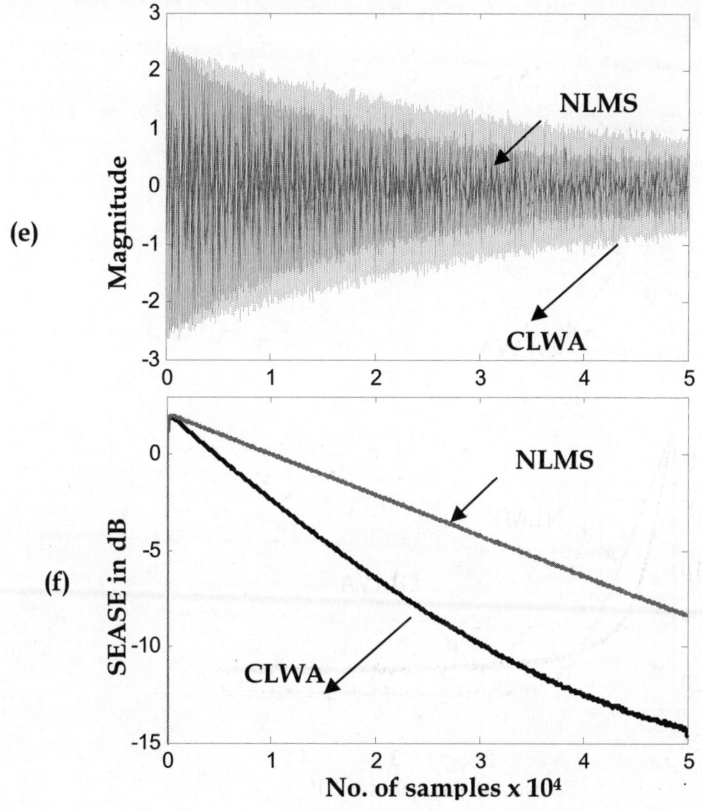

Fig. 12.39 (e, f) Closed loop system identification with 3 scales with DB4 wavelet basis (e) error (f) SEASE.

Here, the error derived by the synthesis bank output and the desired signal, is decomposed by another analysis filter bank which is identical to that used for input decomposition. The delay in synthesising output between scales 2 and 1 is L-1=4-1=3, L is the legth of low/highpass filter used in the filter bank. Hence, scale 0 has to be dalyed by 3 samples and then interpolated and filtered and this is added to synthesised output due to scales a and 2. .The overall delay between the input and outpout for the filter bank is 3(L-1)=9. Thus, the desired signal has to be delayed by 9 samples and then compared with the synthesis bank output to get the overall error. The step size used is 0.0001 and the fogetting factor for computingthe power is 0.99.

As expected the error for the wavelet based adaptive filter converges faster compared to that of NLMS (Fig. 12.39(e)). For more iterations, the convergence error for the NLMS can become less than that of wavelet based adaptive filter and this is due to larger step size used due to use of smaller input power and filter lengths used for different sacles involved. This result is also supported by the SEASE plots shown in Fig. (12.39(f)).

Open loop wavelet adaptive filter

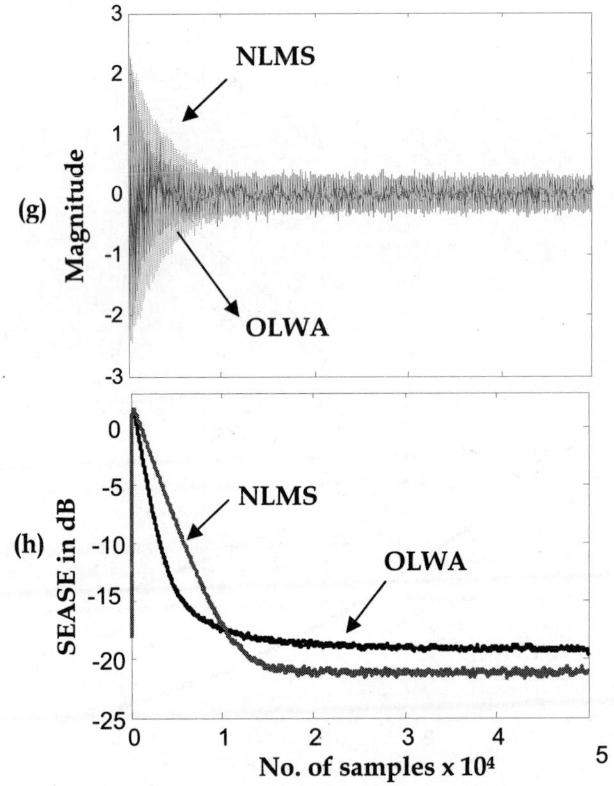

Fig. 12.39(g, h) Open loop system identification with 3 scales with DB4 wavelet basis (e) error (f) SEASE.

For this, the same example for closed loop is considered. Here, instead of the overall error, the desired is decomposed by a wavelet filter bank identical to that used for the input. The WT has scales 0, 1 and 2 and is based on DB4 wavelet. The errors for adapting the subadaptive filters are generated by comparing their outputs with the corresponding desired signal components. Thus, the outputs and desired signals are properly synchronised. For computing the **overall** error, the desired signal is delayed by 9 samples as in closed loop as the synthesis filter bank comes into picture. The step size used fis 0.00.1 as aginst 0.0001 used for closed loop as the latter diverges for a larger step size. The forgetting factor for computing power is same as 0.99.

As seen from Fig. 12.39(g), the convergence of closed loop wavelet adaptive filter is faster compared to that of NLMS and this is due to reduction in eigenvalue spread of the input. As pointed out, the open loop can have a larger step size than that for closed. The SEASE computd using 30 trials (Fig. 12.39(h)), also supports this result. But the convergence error can be larger than that for NLMS (for the same step size) and this is due to larger effective size due to lower power of input componenets and the smaller subadaptivr filter lengths.

Recently, an attempt is made to incorporate the variable step size concept for each scale in addition to utilising the decreased eigenvalue spread due to orthogonal signal decomposition of the input by the wavelet transform (Cunqian Feng, Linrang Zhang and Xiaoping Hui, 2013).

12.23 DELAYLESS SUBBAND ADAPTIVE FILTER BASED ON DISCRETE COSINE HARMONIC WAVELET TRANSFORM (DCHWT)

The delayless subband approach has solved the delay problem and oversampling has solved the aliasing. However, oversampling increases the computational load. The use of delayless polyphase uniform DFT filter bank makes the filter coefficients complex. Hence, the realization of a filter bank with real coefficients is of importance to reduce the computational load. Further, the eigenvalue spread problem at the edges of the passband can be solved by using critical sampling but at the cost of aliasing. By having a larger bandwidth analysis filter compared to that of the synthesis filter, the eigenvalue spread problem has been solved and this again needs oversampling. In view of these comments, to overcome the conflict between oversampling and eigenvalue spread using filter banks with real coefficients incorporating the wavelet signal decomposition nature is a challenging problem. The explicit decimation, interpolation and associated filtering of the filter bank forms 20% of overall computational load of a subband adaptive filter. Further, compared to uniform subband adaptive filter, the wavelet based adaptive filter requires tedious delay compensation also for reconstruction.

12.23.1 Discrete Cosine Harmonic Wavelet Transform (DCHWT)

The harmonic wavelet transform (HWT) is simple due to its built-in generation of different scales (decimated subband signals) by grouping of DFT coefficients and taking their inverse transform of each group. This is in contrast to explicit filtering and decimation.

Its simplicity is also due to signal reconstruction/synthesis which is achieved by taking the inverse DFT of the mere concatenation DFT coefficients groups of different scales (Newland, D.E. , 1993). However, the lack of real symmetry while grouping the DFT coefficients makes the WT coefficients of complex nature. Also the severe DFT leakage results in poor quality WT scales. The realization of HWT using DCT (DCHWT) instead of DFT(DFHWT) makes the DCHWT not only real (instead of complex nature) but also of better quality compared to that by DFT. The real nature and better quality of DCHWT are due to the real nature of DCT and its reduced leakage due to symmetrical data extension, respectively (S.V. Narasimhan, M. Harish and A.R. Haripriya and Nandini Basumallick, 2009).

In Fig. 20.40(a) DCTHWT for a DCT size of 16 is illustrated and the only one side of the symmetrical coefficient sequence is shown, i.e. (0-7). The last half part of the coefficients $C_x(4)$ *to* $C_x(7)$ correspond to scale-0, *i.e.,* C_0. The lower half of the coefficients $C_x(0)$ *to* $C_x(3)$ are split into two groups and the upper group $[C_x(2), C_x(3)]$ corresponds to scale-1, i.e., C_1. Further, the group $[C_x(0), C_x(1)]$ is split into two groups C_2 and C_3 each having single coefficients $C_x(1)$ and $C_x(0)$, respectively. The inverse DCT of each of the groups C_0, C_1, C_2, C_3; results in WT coefficients of the scales $0, 1, 2, 3$, respectively. For $Q = 32$, the scales C_0, C_1, C_2, C_3, C_4 correspond to grouping of coefficients as $\{C_x(8)\ to\ C_x(15)\}, \{C_x(4)\ to\ C_x(7)\}, \{C_x(2)\ to\ C_x(3)\}, \{C_x(1)\}, \{C_x(0)\}$, respectively and this process continues.

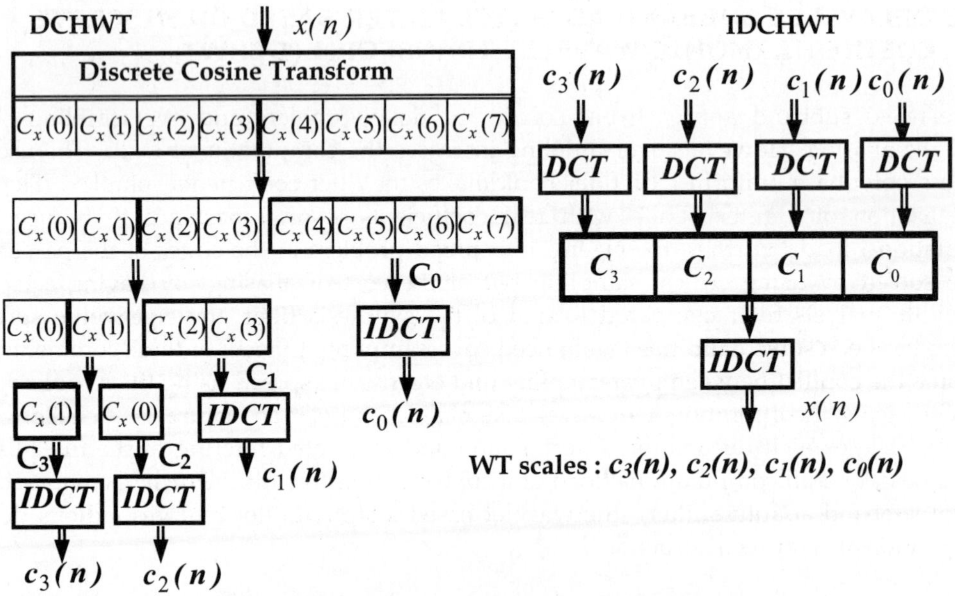

Fig. 12.40(a) DCHWT with 4 scales and its IDCHWT.

For the reconstruction, the DCTs of the subband signals are concatenated to get the DCT of the fullband signal. For the first stage of inverse DCHWT illustrated in Fig. 12.40(a), the DCTs of the subband signals corresponding to groups C_3 and C_4 are concatenated. The resulting group of coefficients is concatenated with the DCT of subband signal corresponding to group C_2, in the next stage. Again, the resulting group of coefficients is concatenated with the DCT of subband signal corresponding to group C_1, to form the DCT of the fullband signal. The WT based on DCT (DCHWT) is very attractive due to its simplicity, real nature and its good quality scales compared to that of DFT. Hence, the possibility of realizing delayless subband adaptive filter based on DCHWT is of interest. (S.V. Narasimhan and S. Veena, 2010) The DSAF as already seen involves signal decomposition into subband components, adaptation of subband filters and weight transform which derives overall adaptive filter impulse response from the individual adaptive filters.

12.23.2 Signal Decomposition into Subband Components

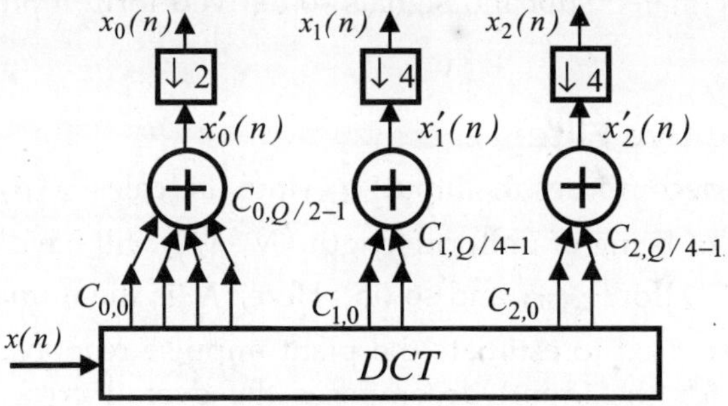

Fig. 12.40(b) DCHWT analysis bank.

The new subband adaptive filter structure based on DCHWT in general is similar to that considered already. But its analysis filter bank is different as it based on DCTHWT and not on the UDFT filter bank. As it is derived by DCTHWT, it also provides dyadic structure and there is no necessity that it should be uniform. Here, the M subband components $\{x_0(n), x_1(n), \ldots, x_{M-1}(n)\}$ are derived by grouping the DCT coefficients of the input $x(n)$ and adding the coefficients of each group on an input sample to sample basis and further decimating the sequence resulting from addition and this is indicated as the *DCT analysis bank* (Fig. 12.40 (b)).

$$x'_0(n) = \sum_{l=0}^{Q/2-1} C_{0l}(n) = \sum_{p=0}^{Q/2-1} C_p(n), \tag{12.36a}$$

$$x'_1(n) = \sum_{l=0}^{Q/4-1} C_{1l}(n) = \sum_{p=Q/2}^{Q/2+Q/4-1} C_p(n), \tag{12.36b}$$

$$x'_2(n) = \sum_{l=0}^{Q/4-1} C_{1l}(n) = \sum_{p=3Q/4}^{Q-1} C_p(n), \tag{12.36c}$$

The decimated subband signals are given by

$$x_0(n)=x_0'(2n),\ x_1(n)=x_1'(4n) \text{ and } x_2(n)=x_2'(4n) \tag{12.36d}$$

$C_p(n),\ p=0,1,\ldots,Q-1$; are the DCT coefficients of the input signal and the length of DCT (II) size is of $2Q$ points. The addition operation is equivalent to IDCT operation on a sample to sample basis rather than on a block basis. It is important to note that as the IDCT is replaced by simple addition and it is computationally advantageous. But here, the built in decimation of HWCT is not used and hence it requires *only explicit subsampling and no filtering.*

For a DCT with half the number coefficients equal to 8, such a derivation of subband components is shown in Fig.12.40(b). This even simplifies the DCHWT as no IDCT is required and involves only addition operation. Using dyadic type of wavelet rule, the lowest scale $x_0(n)$ is decimated by a factor 2 and the subsequent scales $x_1(n)$ and $x_2(n)$ by factors 4 and 8, respectively. The decimated subband signals so derived form inputs to the adaptive filters.

12.23.3 Subband Adaptive Filter

A DCTHWT analysis bank decomposes the input $x(n)$ into its scales $x_i(n), i=0,1,\ldots M-1$. The adaptive filters instead of being of uniform length (N/M), will have lengths $(N/8)$ for $x_0(n)$, $(N/4)$ for $x_1(n)$, $(N/2)$ for $x_2(n)$ and so on. Here, N is the number of tap weights of the overall adaptive filter used to estimate the plant impulse response. A similar DCT analysis filter bank (used for the input) decomposes the overall error $e(n)$ to generate individual errors $\{e_0(n), e_1(n),\ldots, e_{M-1}(n)\}$ for the adaptive filters $W_i(n), i=0,1,\ldots M-1$ (Fig. 12.40(c)).

$$e(n)=d(n)-y(n) \tag{12.36e}$$

where, $y(n)$ is the overall estimate of the plant output,

$$y(n)=\sum_{l=0}^{N-1} w_l x(n-l) \tag{12.36f}$$

Using the subband component inputs $\{x_0(n), x_1(n),\ldots, x_{M-1}(n)\}$ and the error components $\{e_0(n), e_1(n),\ldots, e_{M-1}(n)\}$, the adaptive filter coefficients are updated using normalized LMS algorithm at the respective decimated rates and not at the same rate. The adaptation rule for the ith subband and is given by

$$W_i(n+1)=W_i(n+1)+\mu_k(n)\,e_i(n)X_i(n), \quad i=0,1,\ldots M-1 \tag{12.36g}$$

$$W_i(n)=\left[w_{i0}(n), w_{i1}(n),\ldots, w_{iN_i-1}(n)\right],$$

$$X_i(n)=\left[x_i(n), x_i(n-1),\ldots, x_i(n-N_i+1)\right]$$

and N_i is the length of the ith subband adaptive filter.

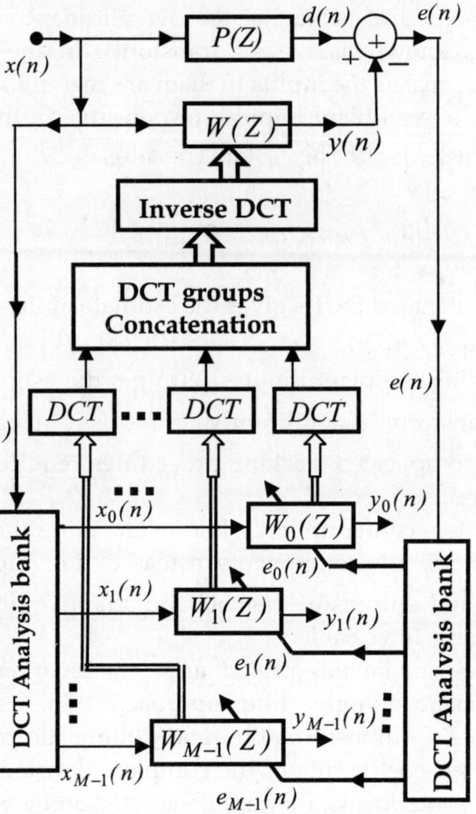

Fig. 12.40(c) Delayless subband adaptive filter based on DCHWT.

In terms of individual weight elements, for the kth weight of the ith adaptive filter,

$$w_{ik}(n+1) = w_{ik}(n) + \mu_i(n) e_i(n) x_i(n-k), \quad k = 0, 1, \ldots N_i - 1 \tag{12.36h}$$

where, $\mu_i(n)$ is the step size for the ith adaptive filter and is given by

$$\mu_i(n) = \frac{\alpha_i}{N_i P_i(n)}, \quad 0 < \alpha_i < 1 \tag{12.36i}$$

and $P_i(n)$ is input power for the ith adaptive filter and is recursively computed as

$$P_i(n) = \beta_i P_i(n-1) + (1-\beta_i) x_i^2(n), \quad 0 < \beta_i < 1 \tag{12.36j}$$

The above adaptation is valid for $i = 0, 1, \ldots M - 1$

12.23.4 Weight Transform

As already indicated the method of achieving the overall adaptive filter from individual subband adaptive filters is known as weight transform. In the present approach, the adaptive filter coefficients are real as the inputs to them are real unlike in a UDFT filter bank. In achieving the estimate of overall plant impulse response $W(n)$, the DCTs of the adaptive filters coefficient sequences $W_i(f) = DCT[W_i(n)]$ are concatenated

$$W(f) = [W_0(f), W_1(f), \ldots, W_{M-1}(f)] \tag{12.36k}$$

The inverse DCT of the concatenated DCT s gives the estimate of the plant impulse response

$$W(n) = [w_0(n), w_1(n), \ldots, w_{N-1}(n)] = IDCT[W(f)] \tag{12.36l}$$

This filter $W(n)$ is used to filter the plant input $x(n)$ to get the estimate $y(n)$ of the desired signal $d(n)$. This weight transform is significantly simple as it involves mere concatenation of DCTs compared to the complicated stacking procedures required for UDFT filter bank and the other delayless methods.

In the DHWT delayless configuration while overcoming the delay problem, as the overall error $e(n)$ is decomposed into components instead of deriving them directly from the desired signal components and outputs of the subband adaptive filters, it compensates for the inaccuracies in the synthesis filter bank.

The DCHWT DSAF can be interpreted as a combination of transform domain method and the usual subband adaptive filter approach. This is because, the subband components are derived by DCT transformation and adding the coefficients with in each group instead of the weighted coefficients of the complete transform coefficients. Further after obtaining the subband components, they are decimated and given to subband adaptive filters which is similar to subband approach. In the proposed method unlike subband approach, the subband components are *critically sampled and no oversampling is required* to overcome aliasing effects due to overlapping of the subbands. This is due to band splitting in the frequency domain and not in the time domain. This *also avoids necessity of cross adaptive filters* which may be used in subband adaptive filters to recover the frequency components which are in transition zone from one band to another (aliased components) due to critical sampling. The cross filters will add to the computation load and the performance of such subband adaptive filters will be same as that of a fullband one and hence reduces the convergence speed. This method is expected to work better than transform method for nonstationary signals as the component signals are associated with bandwidth based on their frequency ranges unlike zero bandwidth used in transform domain which limits its utility to only stationary signals (up to transform length). Such a nonuniform frequency dependent bandwidth is same as that of wavelet transform and hence also desirable from the point of view of perception compared to treatment of fullband in transform domain or uniform subbands in subband adaptive filters. This also provides computational advantage as processing is done at different decimated rates which are less than or equal to half the input sampling rate and also as the decimation factor increases as the order of scales, compared to transform domain which works only at the input sampling rate and those of uniform subband processing.

The performance of the delayless subband adaptive filter based on DCHWT (DSACHWT) is evaluated in system identification configuration for an input comprising five sinusoids at frequencies 0.05, 0.1, 0.15, 0.2 and 0.25 having varying magnitudes (1, 0.7, 0.5, 0.3 and 0.03) and phase, with an additive white Gaussian noise of SNR 30dB. The impulse response of system considered is shown in Fig. (12.41(a).

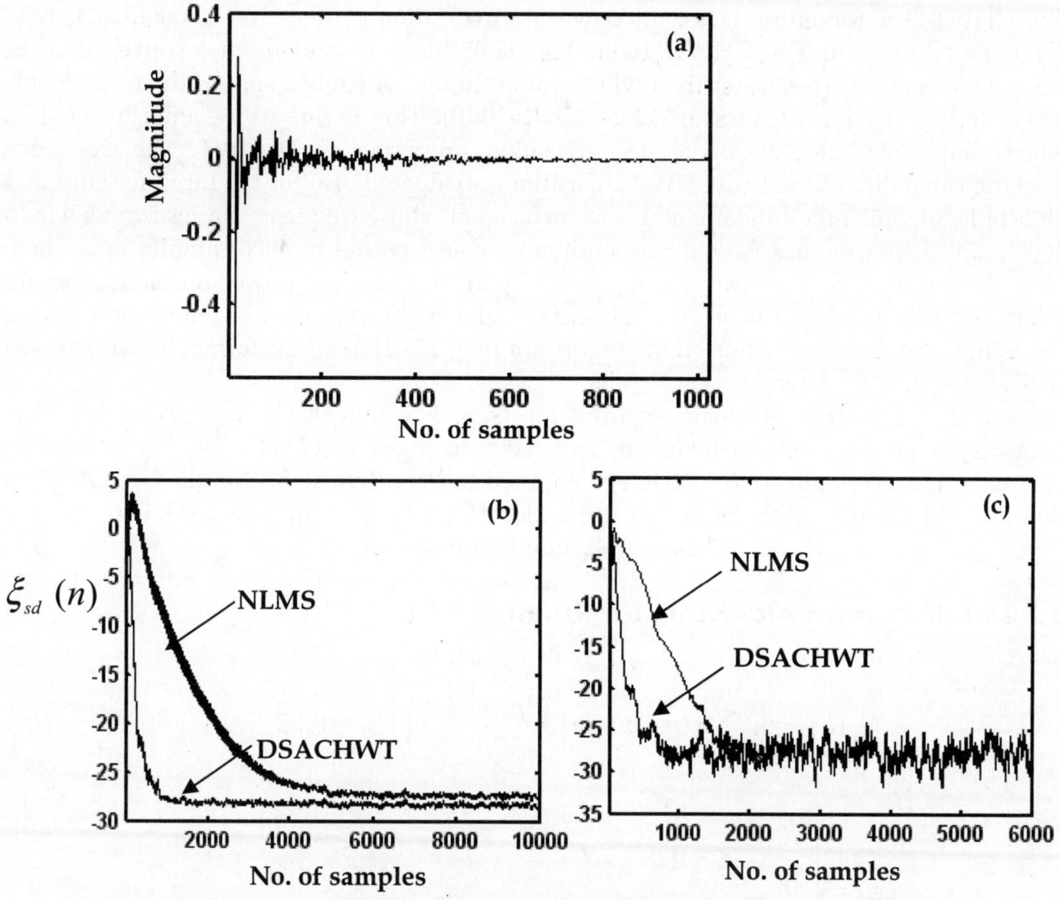

Fig. 12.41 (a) System impulse response, Convergence for ; (b) Sinusoids plus noise and (c) For chirp signal.

The proposed DSACHWT algorithm is compared with fullband normalized LMS (NLMS) algorithm. For both the algorithms, a modeling length of 256 is used. The index $\xi_{sd}(n)$, the *smoothed ensemble average square error* (SEASE) expressed in dB is used as the performance measure and is given by

$$\xi_{Sd}(n) = 10\log_{10}(\xi_S(n)) \quad \xi_S(n) = \sigma\,\xi_S\,(n-1) + (1-\sigma)\xi(n),\ 0 < \sigma < 1$$

and
$$\xi(n) = \frac{1}{K}\sum_{i=1}^{K} e_i^{\,2}(n)$$

where K - number of realizations and $e_i(n)$ - Error for the ith realization. 50 ensembles are considered for the measure. For each trial the phase of the sinusoids are varied randomly and the additive noise sequence is changed.

For the simulation studies, a step-size of 0.01 is used for both NLMS and DSACHWT. The forgetting factor for power normalization of step sizes is taken as 0.9999, for both NLMS and DSACHWT. From Fig. 12.41(b), it is evident that convergence of DSACHWT is five times faster than NLMS for an input, which is a sum of three sinusoids associated with white Gaussian noise (SNR=30dB). This is due to the improvement in eigenvalue spread due to signal decomposition properties of DCHWT. The frequency tracking capability of the DSACHWT algorithm is evaluated also for the input for sum of 3 sinusoids, in which the fundamental is a chirp signal whose frequency varies from 50Hz to 100Hz and its harmonics vary correspondingly, over a period of 4000 samples and there after stationary. For comparison step size of 0.01 and beta for power normalization are taken as 0.9999, respectively. The DSACHW algorithm has better frequency tracking capability compared to that of NLMS algorithm (Fig. 12.41(c)) as the former converges two times faster than the later.

The computational load required for DLSAF using DCHWT (discussed in S.V. Narasimhan and S. Veena, 2010) can be reduced by using *sliding DCT (SDCT)* (discussed in chapter 11) or *the running* DCT (RDCT) computed by recursive algorithm (Z. Lin, J.Z. Chen and R.W. McCallum, 1993) instead of using block DCT used in analysis bank both for the input and the error. The RDCT algoritm is briefly considered.

12.23.5 Recursive Algorithm for Running DCT

The DCT of an input $[x(n), x(n-1), \ldots\ldots x(n-N+1)]$ is

$$C_n(k) = \epsilon_k \sum_{m=0}^{N-1} x(n-m)\cos\left(\frac{(2m+1)k\pi}{2N}\right), \quad k = 0,1,2,\ldots N-1 \tag{12.37a}$$

where,

$$\epsilon_k = \begin{cases} \dfrac{\sqrt{2}}{N}, & k = 0 \\ \dfrac{2}{N} & k = 1,2,\ldots N-1 \end{cases}$$

N: input data vector length. At a given sample time instant n, $C_n(k)$ represents the kth DCT coefficient for the data $[x(n), x(n-1), \ldots\ldots x(n-N+1)]$. For the each new sample time instant, the data vector has to be updated by one sample and such a DCT is called running DCT (Z. Lin, J.Z. Chen and R.W. McCallum, 1993). Now from Eqn. (12.37a)

$$C_n(k) = \in_k \sum_{m=-1}^{N-2} x(n-m)\cos\left(\frac{(2m+3)k\pi}{2N}\right)$$

$$= C_{n-1}(k)\cos(k\pi/N) - \in_k \sum_{m=0}^{N-1} x(n-m-1)\sin\left(\frac{(2m+1)k\pi}{2N}\right)\sin\left(\frac{k\pi}{N}\right)$$

$$+\in_k \cos(k\pi/2N)[x(n)-(-1)^k x(n-N) \quad (12.37b)$$

Similarly, from Eqn.(12.37a),

$$C_{n-2}(k) = \in_k \sum_{m=1}^{N} x(n-m-1)\cos\left(\frac{(2m+1)k\pi}{2N}\right)$$

$$= C_{n-1}(k)\cos(k\pi/N) + \in_k \sum_{m=0}^{N-1} x(n-m-1)\sin\left(\frac{(2m+1)k\pi}{2N}\right)\sin\left(\frac{k\pi}{N}\right)$$

$$-\in_k \cos(k\pi/2N)[x(n-1)-(-1)^k x(n-N-1) \quad (12.37c)$$

Adding Eqns.(12.37b) and (12.37c),

$$C_n(k) = 2\cos(k\pi/N)C_{n-1}(k) - C_{n-2}(k)$$

$$+\in_k \cos\left(\frac{k\pi}{2N}\right)\sum_{m=0}^{N-1} x(n-m-1)\sin\left(\frac{(2m+1)k\pi}{2N}\right) - \in_k \cos(k\pi/2N$$

$$[x(n)-x(n-1)-(-1)^k x(n-N)-(-1)^k x(n-N-1)] \quad (12.37d)$$

This forms the recursive algorithm for running DCT. In this, two real multiplications are required to compute each $C_n(k), k = 0,1,2,\ldots N-1$ and hence $2N$ real multiplications are required to compute N point DCT for each segment of data. But as for direct computation of DCT, N^2 multiplications are required, the reduction factor achieved is $2N/N^2 = 2/N$.

12.23.6 Effect of Subband Processing from Spectral Estimation View

Any processing on the fullband spectrum of a signal is not very effective. This is due to the fact that the processing will give preference to the component of large magnitude and then to others, in order. Hence, the processing is less effective for weak components. Further, it also depends on the proximity of the weaker components to stronger ones and closer the proximity lesser the effectiveness of processing and vice versa. The separation of components/ signal decomposition is a solution to make the processing effective and is achieved by subband processing. Further, signal decomposition into subbands enables decimation which, in turn, results in a spectral expansion or improved frequency resolution (it also reduces the eigenvalue spread as already explained). Further, as it either puts the components into different bands or increases their distance of separation, the component's mutual effect is either totally removed or reduced. In view of this, the subband processing will improve the *noise immunity in addition to improving frequency resolution* (S. Rao and Pearlman, 1996, S.V. Narasimhan and M. Harish, 2004, S.V. Narasimhan, M. Harish and A.R. Haripriya and Nandini Basumallick 2009). The wavelet transform as it separates the signal into different scales, these two advantages will follow automatically for any processing in wavelet transform domain. For example, autoregressive (AR) spectral estimation carried out for each scale will provide better results than applying it to the original fullband signal. This is because, the AR spectral estimation is highly sensitive to noise and its performance deteriorates as signal to noise ratio decreases. Further, as the signal is decomposed in a desired fashion rather than in an arbitrary way like subbands of uniform width, it will be advantageous even from the point of view of perception.

12.24 SIGNAL DEPENDENT WAVELET PACKET BASED ADAPTIVE FILTER

In addition to the limited attenuation of the signal decomposing analysis subband filter which may result in aliasing, the fixed type of decomposition of the input irrespective of its (input) spectral content, poses a major problem. This may occur in an adaptive notch filter or in an adaptive line enhancer. If the sinusoid falls in between two subbands, the sinusoid may not get cancelled or it may not be enhanced. To overcome this, an input signal dependent subband partition is essential.

The wavelet transform can be extended to wavelet packets. The choice of wavelet packet structure, for a particular application may provide further advantage. In a wavelet packet, its tree structure is chosen such that it compresses the complete energy of the signal in a minimum number of coefficients. This is achieved based on entropy considerations at different levels and is valid for applications like to compression and denoising. But for an adaptive filter which is required to handle the input and the desired signal, a band partition based on a different type of criterion which takes into account both the signals, the *mutual wavelet packet criterion* is used (C.B. Lambrecht and M. Karrakchou, 1993).

The Wiener filter, W that best matches the given two signals $x(n)$ (input) and $d(n)$ (desired signal) is given by $W_N = R_{NN}^{-1} P_N$. Here, R_{NN} is autocorrelation matrix of the input $x(n)$ and P_N is the cross-correlation vector between $x(n)$ and $d(n)$. This relation states that by passing the input $x(n)$ through the filter W_N, the resulting signal $y(n)$ will have best matching to the desired signal $d(n)$, in the minimum mean square error sense, as W_N is derived based on this criterion. Hence, in the scenario of adaptive filtering, *it is of interest to decompose both the signal spectra in such a way that their cross-correlation becomes maximum.* Hence, the measure $M(x,d)$ based on cross-correlation between $x(n)$ and $d(n)$ is given by

$$M(x,d) = \sum_{k=0}^{N-1} |\phi_{xd}(k)|,$$

N is the length of the filter W_N and the elements of the vector P_N are given by

$$\phi_{xd}(k) = \frac{1}{L}\sum_{n=0}^{L-1} x(n)d(n+k)\cdot$$

The basic philosophy of the Wiener filter is that the desired signal $d(n)$ and the input $x(n)$ should be correlated though they by themselves need not be correlated. The absence of such a cross-correlation results in a filter W_N with elemental values of zero. An ordinary spectral decomposition of the two signals $x(n)$ and $d(n)$ may not be appropriate as no consideration is given regarding its appropriateness for the purpose for which it is done. But the maximization of the cross-correlation measure takes into account the purpose of decomposition and provides improved results. This is true especially when the signals $x(n)$ and $d(n)$ are colored like narrowband noise or interferences. Particularly the adaptive filters used for the purpose of notch filters and line enhancers to get better performance should satisfy the mutual wavelet packet criterion for their spectral band splitting. This is essential as the signals involved may fall into the gap between the bands and this will deteriorate the performance. However, if the signals involved are of broadband nature, then there may not be much improvement for ordinary band splitting and that by mutual wavelet packet criterion. Here also the decomposition from parent to children splitting is done in a similar way as that of best basis selection but with the criterion of maximizing the cross-correlation between the input and the desired signal.

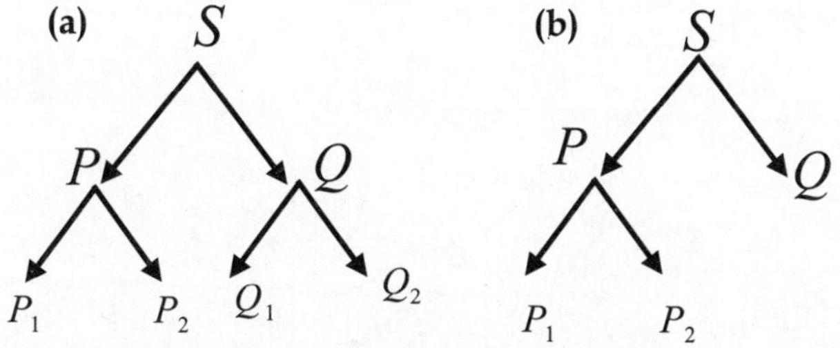

Fig. 12.42(a, b) Mutual wavelet packet decomposition
(a) Ordinary wavelet packet and (b) Mutual wavelet packet.

With reference to the Fig. 12.42(a), the mutual wavelet packet will be illustrated. Here S represents the parent, (P,Q) its children. Further, (P_1,P_2) are children of P and (Q_1,Q_2) are those of Q'.

If it is assumed that $M_S(x,d) \le M_P(x,d) + M_Q(x,d)$, the splitting from $S \to (P,Q)$ is justified as the cross-correlation has to be maximized.

Similarly, if it is assumed that $M_P(x,d) \le M_{P1}(x,d) + M_{P2}(x,d)$, the splitting from $P \to (P1, P2)$ is justified as the cross-correlation has to be maximized.

However, if it is assumed $M_Q(x,d) > M_{Q1}(x,d) + M_{Q2}(x,d)$, the splitting from $Q \to (Q1, Q2)$ is not justified, as this will not maximize the cross-correlation. Hence, the resulting mutual wavelet packet configuration is as shown in Fig. 12.42(b). The application of band splitting , holds to both input and the desired signal.

In view of the above discussion, for the adaptive filtering based on wavelet transform (Fig.12.42), the filter bank used for decomposing the input and the desired signal are identical and can be designed based on the mutual wavelet packet criterion.

Chapter 13

ADAPTIVE INFINITE IMPULSE RESPONSE (IIR) FILTERS

Adaptive filters based on FIR structure are preferred to those on IIR filters, due to their ensured stability. However, for cases with long impulse responses with several thousands of samples like, in echo cancellation, the FIR is very inefficient. On the other hand as the IIR filter transfer function has both numerator and denominator polynomials, it can represent the impulse response efficiently with significantly less parameters. Actual IIR filter known as "output error filter (OE)" suffers from stability problem due to feedback of the output and local minima due to its nonlinear nature. To overcome these problems of OE IIR, the denominator estimate is realized by using FIR filters with the desired signal as its in put instead of feeding back the adaptive filter output and this is known as equation error filter (EQE). However the EQE does not minimize the true error and also suffers from bias problem due to observation noise. The bias due to white observation noise can be solved easily by methods like monic normalization but this not valid for colour noise. Based on Griffiths' LMS and Okello's algorithms, a new bias-free (both for white and colour noise) EQE based on Griffiths' LMS algorithm with variable step size to achieve fast and small convergence has been considered.

It is of interest to have the IIR filter to have EQE nature prior to convergence and that of OE after convergence. In this direction, the Steiglitz-McBride method (SMM) is a composite method having features of both EQE and OE. But it is free from bias only for white observation noise. Based on Griffiths' LMS and Okello's algorithms, a new bias-free (both for white and colour noise) SMM with variable step size which achieves fast and small convergence has also been considered and this minimises true error.

Implementation of IIR filters poses problem of stability and its monitoring is of importance. The lattice structure offers an easy and computationally efficient solution as its reflection coefficients can be computed from the denominator polynomial of the IIR adaptive filter. Further, if the reflection coefficients magnitude is less than unity, the transversal adaptation is valid and otherwise the process continues with previous weight vector. With such a realization, as the filters used are effectively transversal filter with known input and the desired signal, it is possible to overcome the bias problem due to observation white/ colour noise using the Griffiths' algorithm. In case the transversal filter suffers from stability problem due to finite word length effects, it can be realized by an equivalent lattice structure.

Though joint lattice, with IIR lattice for the denominator can be used in realising the IIR (pole-zero) adaptive filter but it suffers from bias in the presence of observation noise to observation noise and in this structure, it is not possible to use the Griffiths' algorithm as the actual inputs to the lattice structure are different from those used for adaptation. Thus, it is preferable to have transversal adaptive filter to both denominator and numerator with stability monitoring.

13.1 INTRODUCTION

In practice, in realising the adaptive filters, the Finite Impulse Response (FIR) filter has become the invariable workhorse. This is mainly due to its ensured stability. However, it suffers from modeling inefficiency for a long impulse response like in acoustic echo cancellation where the effective acoustic path has a length of 2000 samples. In such a case, the FIR adaptive filter length requires an equal length making the computational load very large. This is because the FIR filters can accurately model only the numerator of a rational transfer function easily but can only approximate the effect of the denominator.

In an attempt to improve the efficiency of the FIR filters, the Laguerre adaptive structure approach was considered (in chapters 7, 10 and 11), which is a combination of FIR and IIR. Here in the normal FIR adaptive filter, the delay units are replaced by all-pass units which are IIR and this combination resulted in a reduction of the adaptive filter length by about 50%. The compression is due to the allpass units which enable compression of the frequency range in the undesired region. In the Laguerre structure, as only the FIR part is made adaptive and the IIR part, the allpass filters are nonadaptive, its stability is ensured and hence the efficiency is not at the cost of stability. Thus, the Laguerre structure is a step in improving the efficiency of the FIR adaptive filters.

Generally, IIR adaptive filter is referred to that which can model *a rational transfer function characterised by both poles (denominator polynomial) and zeros (numerator polynomial).* (This is different from the terminology IIR used where it refers to a transfer function characterised by only the denominator and the numerator being a constant). Hence, compared to a FIR filter, IIR filter is *efficient in modeling a rational transfer function as FIR does not approximate the denominator and the denominator is not used to approximate the numerator also.* For example, for a line enhancer (which in a spectrum, enhances the spectral peak compared to its neighbourhood), *if IIR requires 4 coefficients, FIR requires 30 coefficients.* The computational efficiency of the IIR structure will be much more than that of Laguerre structure as it is totally IIR. But IIR filters can easily become *unstable* since their poles may get shifted out of the unit circle during the adaptation process and hence requires stability monitoring. The performance/ objective function / the error surface of an IIR filter usually has *local minima points due to its nonlinear nature and hence may lead to suboptimal solution. Also, the adaptation step size can take only small values to avoid landing into local minima and this effectively results in slow convergence.*

There are basically two methods of realising an IIR adaptive filter, viz., the output error method (OE) and the equation error method (EQE). The OE minimises the true error but suffers from local minima and stability problem. The equation error minimises a modified error which is different from true error but is free from local minima and stability problems. The OE, EQE and their modified versions developed in overcoming their limitations, will be discussed in detail.

13.2 OUTPUT ERROR (OE) METHOD

Let

$$b(z,n)=\sum_{i=0}^{N} b_i(n)z^{-i} \quad \text{and} \quad A'(z,n)=\sum_{i=1}^{M} a_i(n)z^{-i}$$

$$U(n)=\left[x(n),x(n-1),\ldots x(n-N),y(n-1),y(n-2),\ldots y(n-M)\right]^T$$

$$W(n)=\left[b_0(n),b_1(n),\ldots b_N(n),a_1(n),a_2(n)\ldots a_M(n)\right]^T$$

then

$$y(n)=\sum_{i=0}^{N} b_i(n)x(n-i)+\sum_{i=1}^{M} a_i(n)y(n-i) \tag{13.1a}$$

If $d(n)$ is the desired signal which is the output of the plant/system, the *output error* $e(n)$ is given by

$$e(n)=d(n)-y(n) \tag{13.1b}$$

Presently, it is assumed that the observation noise $v(n)=0$. The transfer function from $X(z)$ to $Y(z)$ is obtained by taking the z-transform of the expression for $y(n)$ (Eqn. (13.1a)).

$$y(n)=\sum_{i=0}^{N} b_i(n)x(n-i)+\sum_{i=1}^{M} a_i(n)y(n-i)$$

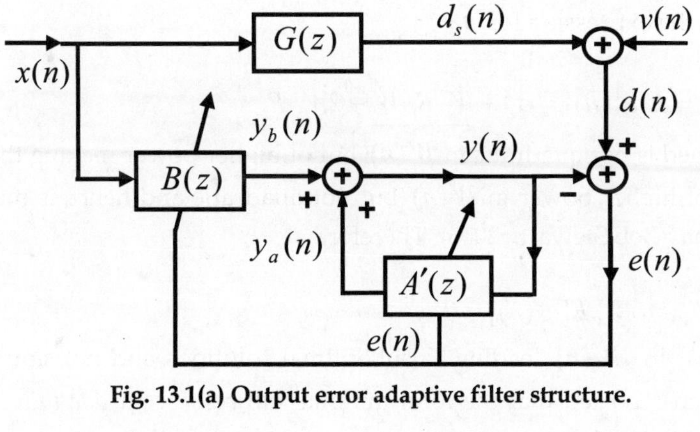

Fig. 13.1(a) Output error adaptive filter structure.

That is,

$$Y(z) = X(z)B(z) + Y(z)A'(z)$$

$$W(z) = \frac{Y(z)}{X(z)} = \frac{B(z)}{1 - A'(z)} \qquad (13.1c)$$

If $Y(z) = D(z)$, then $W(z)$ identifies the plant $P(z)$. Hence, $W(n)$ is found by minimizing the mean *output error* energy $\varepsilon = E[e^2(n)]$.

$$\varepsilon = E[e^2(n)] = E\{[d(n) - y(n)]^2\} = E\{[d(n) - W^T(n)U(n)]^2\}$$

$$= E\{d^2(n)\} + W^T E\{U(n)U^T(n)\}W - 2W^T E\{d(n)U(n)\}$$

$$\varepsilon = E[e^2(n)] = \sigma_d^2 + W^T R_{uu} W - 2W^T P_{du} \qquad (13.1d)$$

$$R_{uu} = \begin{bmatrix} R_{xx} & R_{xy} \\ R_{xy} & R_{yy} \end{bmatrix} = \begin{bmatrix} E\{X(n)X^T(n)\} & E\{X(n)Y^T(n-1)\} \\ E\{X(n)Y^T(n-1)\} & E\{Y(n)Y^T(n)\} \end{bmatrix}$$

$$P_{du} = \begin{bmatrix} P_{dx} \\ P_{dy} \end{bmatrix} = \begin{bmatrix} E\{d(n)X^T(n)\} \\ E\{d(n)Y^T(n-1)\} \end{bmatrix} \qquad (13.1e)$$

For the FIR filter, the $\varepsilon = E[e^2(n)]$ performance surface is quadratic in $W(n)$. But for IIR filter, as $Y(n)$ is a function of $W(n)$ and from Eqn. (13.1e), R_{xy} and R_{yy} are functions of $W(n)$. Hence, in the expression for ε,

$$\varepsilon = E[e^2(n)] = \sigma_d^2 + \underline{W^T R_{uu} W} - 2W^T P_{du} \qquad (13.1f)$$

The term underlined is not quadratic in $W(n)$ but of higher power making the mean square error a function of higher power in $W(n)$ but not quadratic and hence is multimodal with many local minima (Bob Stewart, 2006). Therefore,

$$\nabla_W \varepsilon = \nabla_W E[e^2(n)] = 0 \qquad (13.2a)$$

becomes nonlinear in $W(n)$ leading to suboptimal solutions and not simple to solve for $W(n)$. It has been reported that the local minima will also occur *due to insufficient weight vector order.* Considering instantaneous error energy, for the global/ local minimum Eqn. (13.2a) becomes

$$\nabla_W\left[e^2(n)\right] = -2e(n)\left[\frac{\partial y(n)}{\partial b_0(n)}\frac{\partial y(n)}{\partial b_1(n)}\cdots\frac{\partial y(n)}{\partial b_N(n)}\frac{\partial y(n)}{\partial a_1(n)}\cdots\frac{\partial y(n)}{\partial a_M(n)}\right]^T$$

But from Eqn.(13.1a),

$$y(n) = \sum_{i=0}^{N} b_i(n)x(n-i) + \sum_{l=1}^{M} a_l(n)y(n-l)$$

Therefore,

$$\frac{\partial y(n)}{\partial b_i(n)} = x(n-i) + \sum_{l=1}^{M} a_l(n)\frac{\partial y(n-l)}{\partial b_i(n)}, \quad i = 0,1,\ldots N \tag{13.2b}$$

As $b_i(n)$ does not have any dependence on $a_i(n)$

$$\frac{\partial y(n)}{\partial a_i(n)} = \frac{\partial}{\partial a_i(n)}\left[\sum_{l=1}^{M} a_l(n)y(n-l)\right] \qquad i = 1,\ldots M$$

Further, treating the R.H.S as derivative of the product,

$$\frac{\partial y(n)}{\partial a_i(n)} = y(n-i) + \sum_{l=1}^{M} a_l(n)\frac{\partial y(n-l)}{\partial a_i(n)}, \quad i = 1,\ldots M \tag{13.2c}$$

This is due to the fact that $a_i(n)$ is related to $y(n-1), \ldots, y(n-M)$; the previous outputs. Defining,

$$\beta_i(n) = \frac{\partial y(n)}{\partial b_i(n)}, \quad i = 0,1,\ldots,N \qquad \alpha_i(n) = \frac{\partial y(n)}{\partial a_i(n)}, \quad i = 1,\ldots,M$$

Assuming $\alpha_i(n)$, $\beta_i(n)$ vary slowly with time,

$$\frac{\partial y(n-l)}{\partial b_i(n)} \approx \frac{\partial y(n-l)}{\partial b_i(n-l)} = \beta_i(n-l), \qquad \frac{\partial y(n-l)}{\partial a_i(n)} \approx \frac{\partial y(n-l)}{\partial a_i(n-l)} = \alpha_i(n-l)$$

Eqns. (13.2b) and (13.2c) become

$$\beta_i(n) = x(n-i) + \sum_{l=1}^{M} a_l(n)\beta_i(n-l) \quad or \quad \beta_i(z,n) = \frac{X(z)z^{-i}}{1-A'(z,n)} \tag{13.2d}$$

$$\alpha_i(n) = y(n-i) + \sum_{l=1}^{M} a_l(n)\alpha_i(n-l) \quad or \quad \alpha_j(z,n) = \frac{Y(z)z^{-j}}{1-A'(z,n)} \tag{13.2e}$$

Then

$$W(n+1) = W(n) + 2\mu\, e(n)\eta(n)$$

where,

$$\eta(n) = \left[\beta_0(n), \beta_1(n)\ldots \beta_N(n), \alpha_1(n)\ldots\alpha_M(n)\right]$$

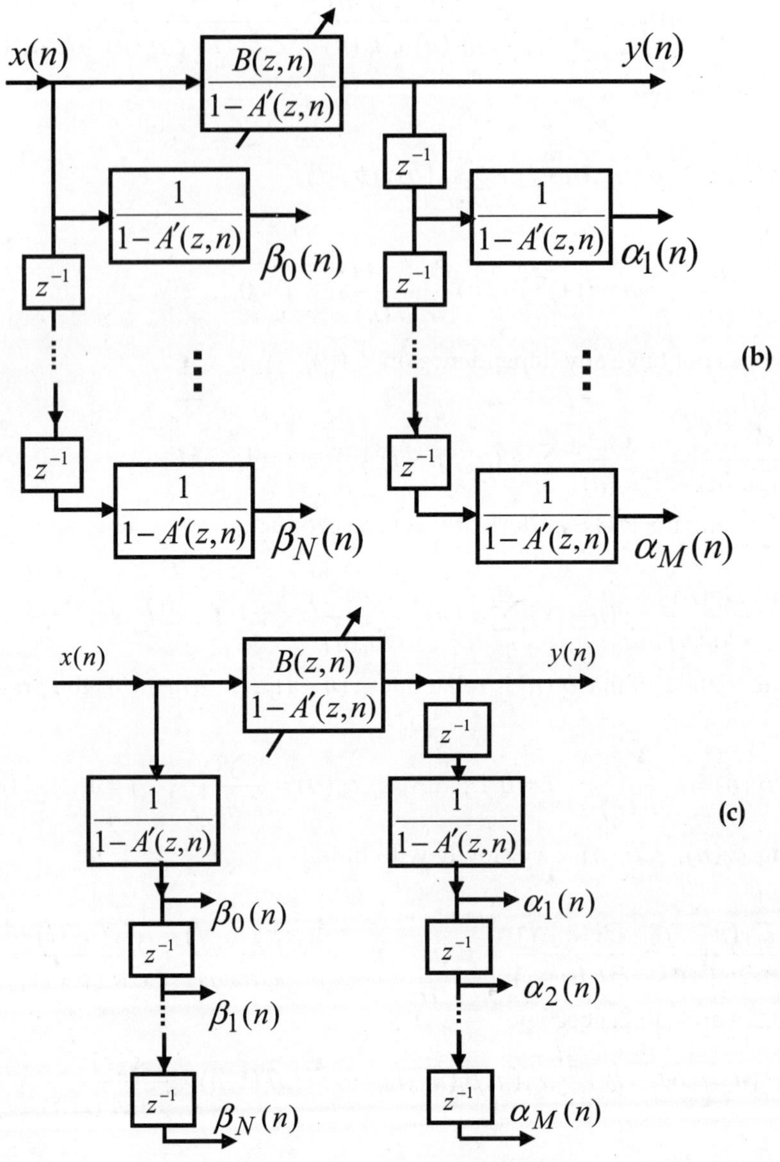

Fig. 13.1(b, c): (b) The OE structure based on Eqns. (13.2d) and (13.2e)
(c) Simplified structure for OE adaptive filter.

The structure based on the Eqns. (13.2d) and (13.2e) are shown in Fig. (13.1b). It is important to note that in generating $\beta_i(n)$ $i = 0, 1, \ldots, N$ by

$$\beta_i(z,n) = \frac{X(z)z^{-i}}{1 - A'(z,n)},$$

the delayed inputs $x(n-i), \quad i = 0, 1, \ldots, N$ are filtered by the same transfer function $1/[1 - A'(z,n)]$. This is true for the generation of $\alpha_i(n), \quad i = 1, \ldots, M$ where the delayed outputs $y(n-i), \quad i = 1, \ldots, M$ are filtered by the same transfer function $1/[1 - A'(z,n)]$.

If $1/[1 - A'(z,n)]$ varies very slowly with time compared to the maximum delay involved, then

$$\frac{1}{1 - A'(z,n)} = \frac{1}{1 - A'(z)}$$

That is, this transfer function is time invariant. *Hence, instead of filtering the delayed signals individually by the transfer function, the input signals $x(n)$ and $y(n)$ can be filtered by $1/[1 - A'(z,n)]$ and then can be delayed.* Thus, the delay chain can be pushed beyond the transfer function $1/[1 - A'(z,n)]$ both at the input and output side resulting in a structure shown in the Fig. 13.1(c). Therefore, from Eqn. (13.2d),

$$\beta_i(n) = x(n-i) + \sum_{l=1}^{M} a_l(n)\beta_i(n-l)$$

$$\beta_0(n) = x(n) + \sum_{l=1}^{M} a_l(n)\beta_0(n-l)$$

$$\beta_0(n-i) = x(n-i) + \sum_{l=1}^{M} a_l(n-i)\beta_0(n-i-l)$$

$$\beta_i(n) = \beta_0(n-i), \qquad i = 1, 2, \ldots N \tag{13.2f}$$

Similarly, Eqn. (13.2e),

$$\alpha_i(n) = y(n-i) + \sum_{l=1}^{M} a_l(n)\alpha_i(n-l)$$

$$\alpha_1(n) = y(n-1) + \sum_{l=1}^{M} a_l(n)\alpha_1(n-l)$$

$$\alpha_2(n) = \alpha_1(n-1)$$

$$\alpha_3(n) = \alpha_1(n-2)$$

$$\vdots$$

$$\alpha_i(n) = \alpha_i(n-i+1), \quad i = 2, 3, \ldots, M \tag{13.2g}$$

This is comparatively simpler as it involves only one filtering operation for each of the signals $x(n)$ and $y(n)$ and respectively saves N and $(M-1)$ filtering operations. Therefore, the LMS adaptation for the coefficients are given by

$$b_i(n+1) = b_i(n) + 2\mu\, e(n)\beta_0(n-i), \quad i = 0,1,2,\ldots,N \tag{13.2h}$$

$$a_i(n+1) = a_i(n) + 2\mu\, e(n)\alpha_1(n-i+1), \quad i = 1,2,\ldots,M \tag{13.2i}$$

Algorithm for output error method:

1. Compute

$$y(n) = W^T(n)\,U(n) \tag{13.3a}$$

$$W(n) = \left[b_0(n), b_1(n), \ldots b_N(n), a_1(n), a_2(n) \ldots a_M(n)\right]^T$$

$$U(n) = \left[x(n), x(n-1), \ldots x(n-N), y(n-1), y(n-2), \ldots y(n-M)\right]^T$$

2. Compute the error

$$e(n) = d(n) - y(n) \tag{13.3b}$$

3. Adaptation of $\eta(n)$

$$\beta_0(n) = x(n-i) + \sum_{l=1}^{M} a_l(n)\beta_0(n-l) \tag{13.3c}$$

$$\alpha_1(n) = y(n-i) + \sum_{l=1}^{M} a_l(n)\alpha_1(n-l) \tag{13.3d}$$

$$\eta(n) = \left[\beta_0(n), \beta_0(n-1) \ldots \beta_0(n-N), \alpha_1(n) \ldots \alpha_1(n-M+1)\right] \tag{13.3e}$$

4. Weight vector update

$$W(n+1) = W(n) + 2\mu\, e(n)\eta(n) \tag{13.3f}$$

13.2.1 Feintuch's IIR Algorithm

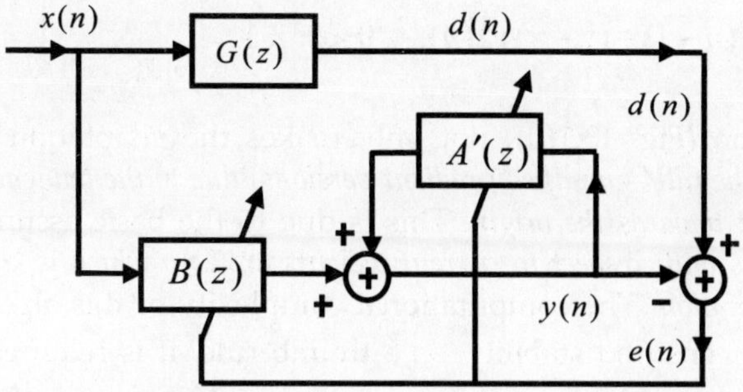

Fig. 13.1(d) Feintuch's IIR LMS algorithm.

From Eqns. (13.2d) and (13.2e)

$$\beta_i(n) = x(n-i) + \sum_{l=1}^{M} a_l(n)\beta_i(n-l)$$

$$\alpha_i(n) = y(n-i) + \sum_{l=1}^{M} a_l(n)\alpha_i(n-l)$$

where

$$\beta_i(n-l) = \frac{\partial y(n-l)}{\partial b_i(n)}, \quad \alpha_i(n-l) = \frac{\partial y(n-l)}{\partial a_i(n)}$$

Here, $\beta_i(n-l)$ and $\alpha_i(n-l)$ represent *the derivatives of the past outputs with respect to current weights and these are assumed to be zero*. Therefore,

$$\beta_i(n) = x(n-i) \text{ and } \alpha_i(n) = y(n-i) \qquad (13.3g, h)$$

Therefore, the adaptation of coefficients,

$$b_i(n+1) = b_i(n) + 2\mu\, e(n)\beta_i(n), \quad \beta_i(n) = \beta_0(n-i), \quad i = 0,1,2,\ldots,N$$

$$a_i(n+1) = a_i(n) + 2\mu\, e(n)\alpha_i(n), \quad \alpha_i(n) = \alpha_1(n-i+1), \quad i = 1,2,\ldots,M$$

Become (13.3i, j)

$$\left.\begin{aligned} b_i(n+1) &= b_i(n) + 2\frac{\mu}{(N+1)P_x(n)} e(n)x(n-i), \quad i = 0,1,2,\ldots,N \\ P_x(n) &= \gamma P_x(n-1) + (1-\gamma)x^2(n), \quad 0 < \gamma < 1 \end{aligned}\right\} \qquad (13.3k)$$

$$\left.\begin{aligned} a_i(n+1) &= a_i(n) + 2\frac{\mu}{MP_y(n)}\, e(n)y(n-i), \quad i = 1,2,\ldots,M \\ P_y(n) &= \gamma P_y(n-1) + (1-\gamma)y^2(n), \quad 0 < \gamma < 1 \end{aligned}\right\} \tag{13.31}$$

This Feintuch's modification (Fig. 13.1(d)) not only makes the adaptation *simple but also inherently more stable (than the full/ simplified gradient versions) due to the tendency to adapt poles away from the unit circle but towards the origin.* This is due to the basic assumption that the *derivatives of the past outputs with respect to current weights are zero which is known as pseudo-linear regression (PLR) assumption.* The computational complexity of this algorithm is of the order $(N + M)$. Further, for ensured stability, as a thumb rule, it is required that the step size for the feedback filter should be kept one tenth that used for the feedforward filter. It has been found that this algorithm is applicable for a system having up to about 20 poles. Full gradient and simplified gradient algorithms are rarely used in circumstances where their performance improvement is marginal over that of Feintuch's algorithm. They are preferred only for systems where the level of feedback is severe and in practice such situations are generally not common (Bob Stewart, 2006).

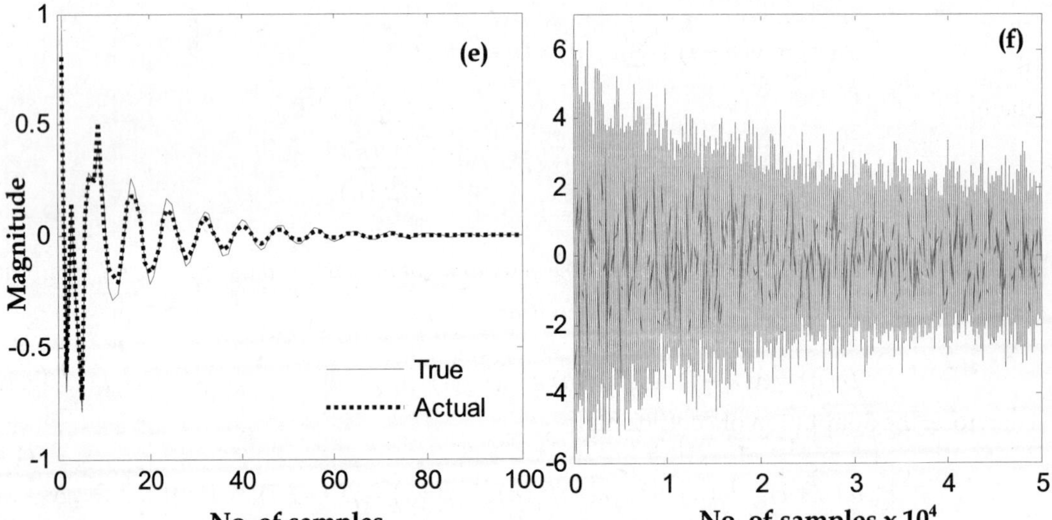

Fig. 13.1 (e, f) Performance of Feintuch's IIR LMS algorithm.
(e) Estimated impulse response and (f) Convergence error.

The performance of Feintuch's algorithm is illustrated for system identification. The systems is given by

$$P(z) = \frac{1-1.289z^{-1}+1.0058z^{-2}-0.7617z^{-3}}{1-0.59z^{-1}+0.455z^{-2}-0.0754z^{-3}+0.504z^{-4}}$$

which is driven by a Gaussian white noise of zero mean and unit variance. The step size for the used for the numerator and denominator adaptation are $\mu = 0.0001$ and $\mu = 0.00001$, respectively. The step size for the denominator adaptation is one tenth that for numerator. The normalized LMS is used for adaptation of numerator and denominator coefficients. The forgetting factor for input power estimations used is $\beta = 0.99$. The lengths of numerator and denominator coefficients used are 9 and 14, respectively. It is seen that the convergence takes place after 25,000 samples (Fig.13.1(f)). The impulse response estimated after 50,000 samples using the numerator and denominator coefficients is shown in comparison with the true one. It is seen that there is a good match of the estimated impulse response with the true one (Fig. 13.1(e)).

13.3 EQUATION ERROR (EQE) METHOD

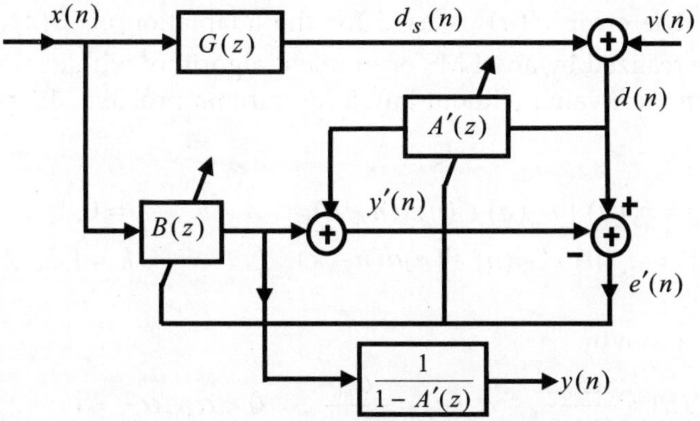

Fig. 13.2(a) Schematic of equation error method.

The output error method suffers from local minima problem and may converge to a local minimum, which is suboptimal compared to global minimum. Further, the OE method requires stability monitoring as during adaptation some of the poles of the adaptive filter may cross the unit circle and hence, may converge slowly due to limit on stepsize. These problems are solved by the equation error method.

In the output error method, the energy of error $e(n) = d(n) - y(n)$, is minimised to match the output $y(n)$ with the desired signal $d(n)$ so that adaptive filter identifies the plant $G(z)$. In this process, the input to the filter $A'(z)$ used is $y(n)$ though it is desired to make $y(n)$ close to $d(n)$ and $d(n)$ is available. This may make one to question, for $A'(z)$ input, why $d(n)$ is not used instead of $y(n)$ though the former is readily available? The answer to this is yes and equation error method of IIR adaptive filter uses $d(n)$ as input to $A'(z)$ instead of $y(n)$.

The error used in this method is

$$e'(n) = d(n) - y'(n) \tag{13.4a}$$

$$y'(n) = \sum_{i=0}^{N} b_i(n)x(n-i) + \sum_{i=1}^{M} a_i(n)\underline{d(n-i)} \tag{13.4b}$$

and $e'(n)$ is different from the output error, $e(n) = d(n) - y(n)$.

It is important to note that in computing the output, $y'(n)$, *the desired signal* $d(n)$ *is used instead of* $y'(n)$. Since the way error is defined determines the method, it is called equation error method. The equation error method is unimodal and this is due to the *nonrecursive* nature of the adaptive filter. Hence, it is a pseudo IIR filter. The output $y'(n)$ is the *output of a linear combiner* with weight vector $W(n)$ and does not involve any feedback weights (poles). The error $e'(n)$ is used for the adaptation of $B(z)$ and $A'(z)$, and the adaptation can be realized by any LMS or gradient algorithm with guaranteed convergence to the unique minimum value without any local minima problem. The adaptation is given by

$$b_k(n+1) = b_k(n) + \alpha_b(n)\, e'(n)x(n-k), \qquad k = 0, 1, \ldots, N-1 \tag{13.4c}$$

$$a_k(n+1) = a_k(n) + \alpha_a(n)\, e'(n)d(n-k), \qquad k = 1, 2, \ldots, M \tag{13.4d}$$

The step sizes are given by

$$\alpha_b(n) = \frac{\alpha_{b0}}{P_x(n)}, \quad \alpha_a(n) = \frac{\alpha_{a0}}{P_d(n)}, \quad 0 < \alpha_{b0}, \alpha_{a0} < 1 \tag{13.4e, f}$$

$$P_x(n) = \gamma P_x(n-1) + (1-\gamma)x^2(n), \quad 0 < \gamma < 1 \tag{13.4g}$$

$$P_d(n) = \gamma P_d(n-1) + (1-\gamma)d^2(n) \tag{13.4h}$$

The use of $d(n)$ instead of $y'(n)$ is justified since goal of the adaptation or error energy minimization is to make $y'(n)$ close to $d(n)$. That is,

$$y'(n) = d(n) \text{ or } Y'(z) = D(z)$$

$$B(z)X(z) + A'(z)D(z) = D(z)$$

$$\frac{D(z)}{X(z)} = \frac{B(z)}{1 - A'(z)}$$

$$D(z) = B(z)X(z)\left[\frac{1}{1 - A'(z)}\right]$$

Therefore, the output $y(n)$ is obtained by passing the output of $B(z)$ filter through the transfer function $1/[1 - A'(z)]$ as shown in Fig. (13.2a) as

$$Y(z) = B(z)X(z)\left[\frac{1}{1 - A'(z)}\right] \tag{13.4i}$$

13.3.1 The Relation between Equation Error $e'(n)$ and Output Error $e(n)$

The error for the OE method is given by (Fig.13. 1a)

$$E(z) = D(z) - Y(z) \tag{13.5a}$$

where,

$$D(z) = X(z)G(z) + V(z), \quad Y(z) = \frac{X(z)B(z)}{1 - A'(z)}$$

Therefore, $E(z)$ is given by

$$E(z) = X(z)G(z) + V(z) - \frac{X(z)B(z)}{1 - A'(z)}$$

$$E(z) = X(z)\left[G(z) - \frac{B(z)}{1 - A'(z)}\right] + V(z) \tag{13.5b}$$

The error for the EQE method is given by (Eqn. (13.3(a))

$$E'(z) = D(z) - Y'(z) \tag{13.5c}$$

where,

$$D(z) = G(z)X(z) + V(z)$$

$$Y'(z) = B(z)X(z) + D(z)A'(z)$$

Therefore, substituting for $Y'(z)$ in Eqn.(13.5c),

$$E'(z) = D(z) - \left[D(z)A'(z) + X(z)B(z)\right]$$

$$= D(z)\left[1 - A'(z)\right] - X(z)B(z)$$

Now substituting for $D(z)$,

$$E'(z) = \left\{G(z)X(z) + V(z)\right\}\left(1 - A'(z)\right) - X(z)B(z)$$

$$E'(z) = G(z)X(z)(1 - A'(z) - X(z)B(z) + V(z)\left(1 - A'(z)\right)$$

$$E'(z) = \left[X(z)\left(G(z) - \frac{B(z)}{1 - A'(z)}\right) + V(z)\right]\left(1 - A'(z)\right) \tag{13.5d}$$

Comparing Eqn.(13.5d) with the output error derived in Eqn.(13.5(b)),

$$E(z) = X(z)\left[G(z) - \frac{B(z)}{1 - A'(z)}\right] + V(z)$$

It is seen that

$$E'(z) = E(z)\left(1 - A'(z)\right) \tag{13.5e}$$

This shows that the equation error $e'(n)$ is related to the output error $e(n)$, through $\left(1 - A'(z)\right)$. The output error $E(z)$ filtered by the transfer function $\left(1 - A'(z)\right)$ results in equation error $E'(z)$. Minimization of the mean square values of the output error $e(n)$ and the equation error $e'(n)$ may lead to different weight vector for the IIR filter. However, in some cases they may converge to nearly same solutions.

The performance of EQE method is illustrated for the example of system identification for the system

$$P(z) = \frac{1 - 1.289z^{-1} + 1.0058z^{-2} - 0.7617z^{-3}}{1 - 0.59z^{-1} + 0.455z^{-2} - 0.0754z^{-3} + 0.504z^{-4}}$$

driven by a Gaussian white noise of zero mean and unit variance. No observation noise is added at the output of the system. The parameters used are step size $\mu = 0.01$ and forgetting factor $\beta = 0.999$ for power estimation for step size normalization. The model order for numerator and denominator have 4 and 4 (excluding the first which is unity), respectively. It is seen that the error converges by about 18000 samples and its magnitude

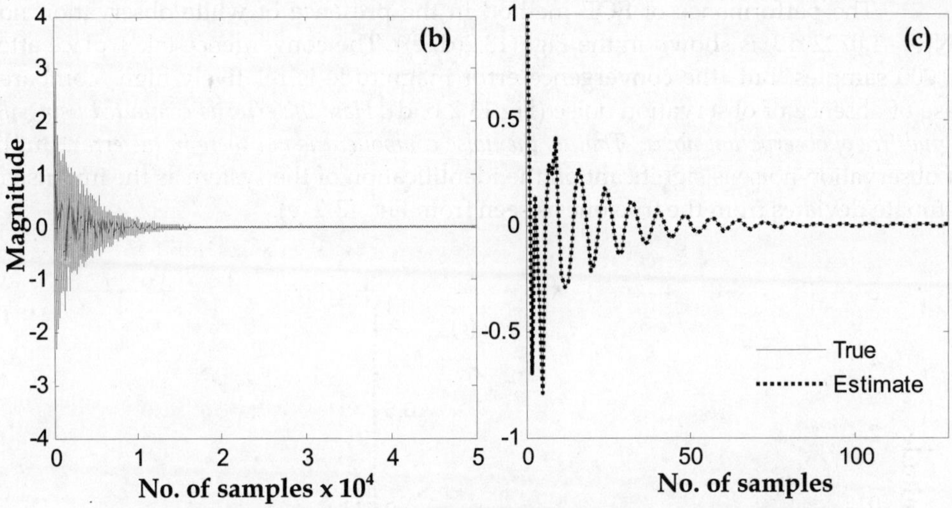

Fig. 13.2 (b, c) Equation error method: (b) convergence error and (c) estimate of impulse response in the absence observation noise.

becomes small after convergence. The impulse response estimated at the 50000 sample instant is almost identical to the true one.

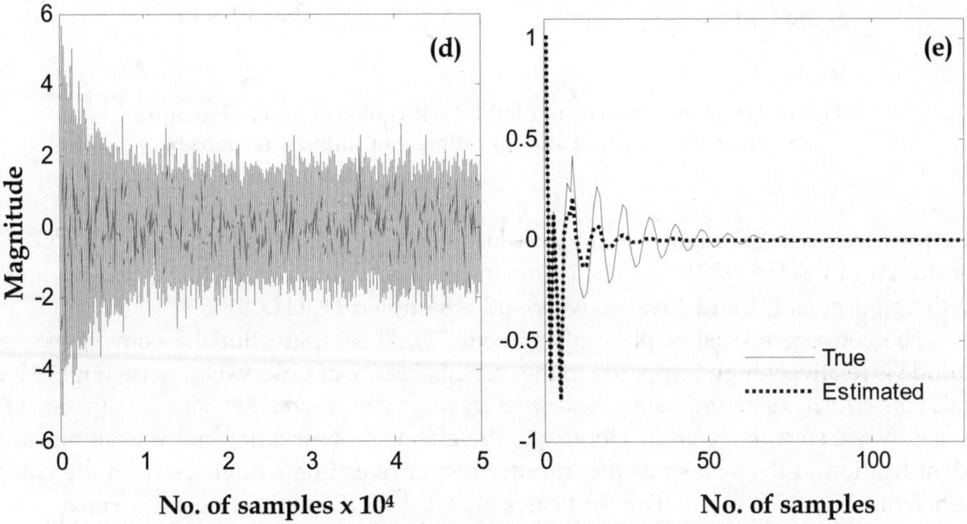

Fig. 13.2 (d, e) Equation error method with white observation noise (d) convergence error and (e) estimate of impulse response.

The performance of EQE method in the presence of white observation noise with SNR= -1.0722 dB, is shown in the Fig. (13.2(d, e)). The convergence takes place after about 20,000 samples. But the convergence error magnitude is relatively high compared to the case of absence of observation noise (Fig.13.2 b, d). *Here the error is computed using the desired signal free of observation noise. That is, the noise component is not there in the error.* But the effect of observation noise is significant on the identification of the system as the impulse response estimate deviates from the true one as seen from Fig. 13.2 (e).

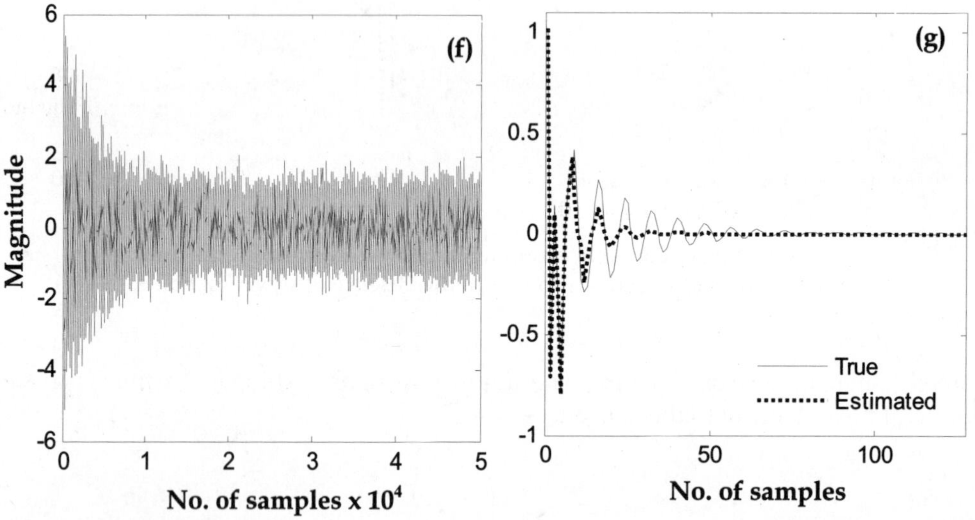

Fig: 13.2 (f, g) Equation error method with colour observation noise (f) error convergence and (g) estimate of impulse response.

The performance of EQE method in the presence of colour observation noise with SNR= 3.19 dB, is shown in the Fig.(132(f, g). The colour noise has Sinusoids of 210Hz, 300 Hz and 400 Hz with amplitudes 1, 1 and 1 respectively and is sampled 1K Hz.

The convergence takes place after about 20,000 samples. But the convergence error magnitude is relatively high compared to the case of absence of observation noise (Fig. 13.2 b, f). *Here also the error is computed using the desired signal free of observation noise. That is, the noise component is not there in the error.* Here also, the effect of observation noise is significant on the identification of the system as the impulse response estimate deviates from the true one as seen from Fig. 13.2 (g). This indicates that the EQE method gets affected by the observation noise irrespective of whether it is white or colored and it is necessary to overcome the effect of observation noise to get proper system identification.

13.4 MONIC NORMALIZATION BIAS REMOVAL FOR EQE METHOD

In the equation error method, by connecting the input to the $A'(z)$ filter to $d(n)$ instead of $y'(n)$, the problem of local minima is solved. However, when the observation noise $v(n)$ is present, the input to the filter $A'(z)$, i.e., the desired signal is corrupted by this noise. Since the desired signal $d(n)$ corrupted by the noise $v(n)$ drives the adaptive filter $A'(z)$, the $A'(z)$ estimate will differ from the true one and is a biased estimate. When the observation noise is white, the coefficients of the filter $A'(z)$ are scaled by a constant factor. Hence, by normalizing all the coefficients of $A(z) = a_0 - A'(z)$ by its first coefficient *in every iteration,* the bias is removed. That is,

$$a(n+1) = a(n) + \alpha_a(n)\, e'(n)D(n)$$

$$a(n+1) = \frac{1}{a_0(n+1)} a(n+1) \tag{13.6a}$$

where,

$$D(n) = [d(n),\, d(n-1),\, d(n-2), \ldots,\, d(n-M)]^T,\; d(n) = d'(n) + v(n),$$

$$a(n) = [a_0(n),\, -a_1(n),\, -a_2(n), \ldots\, a_M(n)]$$

This method to overcome the bias problem is known as *Monic-normalization. Such a normalization is valid for white noise and slow convergence conditions. However this method will not remove bias effect of coloured observation noise on the estimate* $A'(z)$. *After normalization, the denominator part of the coefficient is copied and used for filtering* (H. N. Kim, W.J. Song, 1999).

The adaptation equation for $A(z)$ filter in vector form is given by

$$a(n+1) = a(n) + \alpha_a(n)\, e'(n)D(n),$$

$$e'(n) = d(n) - \sum_{k=0}^{N} b_k(n)x(n-k) - \sum_{l=1}^{M} a_l(n)d(n-l)$$

$$e'(n) = d'(n) + v(n) - \sum_{k=0}^{N} b_k(n)x(n-k) - \sum_{l=1}^{M} a_l(n)d'(n-l) - \sum_{l=1}^{M} a_l(n)v(n-l)$$

$$e'(n) = d'(n) - \sum_{k=0}^{N} b_k(n)x(n-k) - \sum_{l=1}^{M} a_l(n)d'(n-l) - \sum_{l=0}^{M} a_l(n)v(n-l), \qquad (as\ a_0(n) = 1)$$

Noise-free error $e_s(n)$ is given by

$$e_s(n) = d'(n) - \sum_{k=0}^{N} b_k(n)x(n-k) - \sum_{l=1}^{M} a_l(n)d'(n-l)$$

Therefore,

$$e'(n) = e_s(n) - \sum_{l=0}^{M} a_l(n)v(n-l)$$

If

$$V(n) = [v(n), v(n-1), \ldots, v(n-M)]^T,$$

$a^T(n) = [a_0(n), a_1(n), \ldots, a_M(n)]^T$

Then

$$e'(n) = e_s(n) - a^T(n)V(n)$$

Therefore

$$a(n+1) = a(n) + \alpha_a(n)\, e'(n)D(n),$$

becomes

$$a(n+1) = a(n) + \alpha_a(n)\left[\, e_s(n) - a^T(n)V(n)\right]\left[D'(n) + V(n)\right],$$

$$a(n+1) = a(n) + \alpha_a(n)\left[\, e_s(n)D'(n) - a^T(n)V(n)D'(n) + e_s(n)V(n) - a(n)V^T(n)V(n)\right]$$

Taking expectations both sides,

$$E\{a(n+1)\} = E\{a(n)\} + \alpha_a(n)\,[E\{e_s(n)D'\ (n)\} - E\{a^T(n)\}E\{V(n)D'\ (n)\} + E\{e_s(n)V(n)\} - E\{a(n)\}E\{V^T(n)V(n)\}],$$

As $V(n)$ is uncorrelated with $e_s(n)$ and $D(n)$,

$$E\{e_s(n)V(n)\} = 0 \quad \text{and} \quad E\{V(n)D(n)\} = 0$$

Therefore,

$$E\{a(n+1)\} = E\{a(n)\} + \alpha_a(n)\ E\{e_s(n)D'\ (n)\} - \alpha_a(n)E\{a(n)\}E\{V^T(n)V(n)\} \quad (13.6b)$$

The last term on the right hand side of the above Eqn. (13.6b) causes the biasing problem in the conventional equation error method. If the observation noise is white then

$$E\{V^T(n)V(n)\} = \sigma_v^2, \quad (13.6c)$$

Therefore, Eqn. (13.6b),

$$E\{a(n+1)\} = E\{a(n)\} + \alpha_a(n)\ E\{e_s(n)D'\ (n)\} - \alpha_a(n)E\{a(n)\}E\{V^T(n)V(n)\}$$

becomes

$$E\{a(n+1)\} = E\{a(n)\} + \alpha_a(n)\left[\, E\{e_s(n)D'(n)\} - E\{a(n)\}\sigma_v^2 \,\right]$$

$$E\{a(n+1)\} = E\{a(n)\}\left\{1 - \alpha_a(n)\sigma_v^2\right\} + \alpha_a(n)\left[\, E\{e_s(n)D'(n)\}\,\right] \tag{13.6d}$$

That is, *the noise effect is same for all the coefficients as the coefficient vector* $E\{a(n)\}$ *is multiplied by* $\left\{1 - \alpha_a(n)\sigma_v^2\right\}$. In view of this, $E\{a_0(n)\} \neq 1$. On normalization by $E\{a_0(n)\}$, the first coefficient of the vector $E\{a(n)\}$ becomes unity at the nth iteration. Using $E\{a_0(n)\} = 1$ and $(1 - \alpha_a(n)\sigma_v^2) >> \alpha_a E[e_s(n)D'(n)]$, the above equation for $a_0(n)$ becomes

$$E\{a_0(n+1)\} \approx (1 - \alpha_a(n)\sigma_v^2) \tag{13.6e}$$

$$E\{a(n+1)\} = E\{a(n)\}(1 - \alpha_a(n)\sigma_v^2) + \alpha_a(n)E[e_s(n)D'(n)]$$

Dividing each coefficient by

$$E[a_0(n+1)] \approx (1 - \alpha_a(n)\sigma_v^2)$$

known as Monic-normalization and is

$$E\{a(n+1)\} = E\{a(n)\} + \frac{\alpha_a}{(1 - \alpha_a\sigma_v^2)} E\{e_s(n)D'(n)\} \tag{13.6f}$$

At convergence,

$$E\{a(n+1)\} = E\{a(n)\}$$

Hence

$$E\{e_s(n)D'(n)\} = [0]$$

where $[0]$, is a zero vector. This implies that no bias exists at convergence by using Monic Normalization. Due to division of $\alpha_a(n)$ by $(1 - \alpha_a(n)\sigma_v^2)$, there can be slight increase in the value of the stepsize. But as $\alpha_b(n)\sigma_v^2 << 1$, the increase in stepsize is not significant and will not affect the convergence process. In conventional equation error method

$$E\{a(n+1)\} = E\{a(n)\}$$

forces the underlined terms in

$$E\{a(n+1)\} = E\{a(n)\} \underline{- \alpha_a(n)\sigma_v^2 E\{a(n)\} + \alpha_a(n)E\{e_s(n)D'(n)\}}$$

to become zero and this leads to

$$E\{e_s(n)D'(n)\} = \sigma_v^2 E\{a(n)\}$$

This is different from that obtained by Monic normalization.

To implement the method, in each adaptation of weight vector the following procedure is used.

Step 1

Initially, the output is computed by

$$y'(n) = \sum_{k=0}^{N} b_k(n)x(n-k) + \sum_{l=1}^{M} a_l(n)d(n-l)$$

(Invariably the coefficients both $b_k(n)$ and $a_l(n)$ are set to zero value). Further, the error by

$$e'(n) = d(n) - y'(n)$$

This $y'(n)$ and the normalized coefficient vector form valid estimates of output and $b(n)$.

Step 2

The coefficients are updated by

$$a(n+1) = a(n) + \alpha_a(n)\, e'(n)D(n),$$

where

$$D(n) = [d(n), d(n-1), d(n-2), \ldots, d(n-M)]^T,$$

Then the coefficients are normalized as

$$a(n+1) = \frac{a(n)}{a_0(n)}$$

Step 3

With these new coefficients go to step 1 for a next data samples of $x(n)$ and $d(n)$.

Example: Consider the system given by

$$P(z) = \frac{1 - 1.289z^{-1} + 1.0058z^{-2} - 0.7617z^{-3}}{1 - 0.59z^{-1} + 0.455z^{-2} - 0.0754z^{-3} + 0.504z^{-4}}$$

which is driven by a Gaussian white noise of zero mean and unit variance. A Gaussian white noise of zero mean is added to the system output as observation noise to get an SNR =6.4 dB. The numbers of numerator and denominator coefficients used are 4 and 5 respectively. The adaptation used is of fixed stepsize LMS (not normalized LMS) with step size value 0.001.

The norm squared error in dB is given by

$$\varsigma(n) = 10\log_{10}\left(\frac{1}{30}\sum_{i=1}^{30}\left\{\sum_{k=1}^{M}\left[\hat{a}_{ik}(n) - a_{ik}(n)\right]^2 + \sum_{l=0}^{N}\left[\hat{b}_{il}(n) - bi_l(n)\right]^2\right\}\right)$$

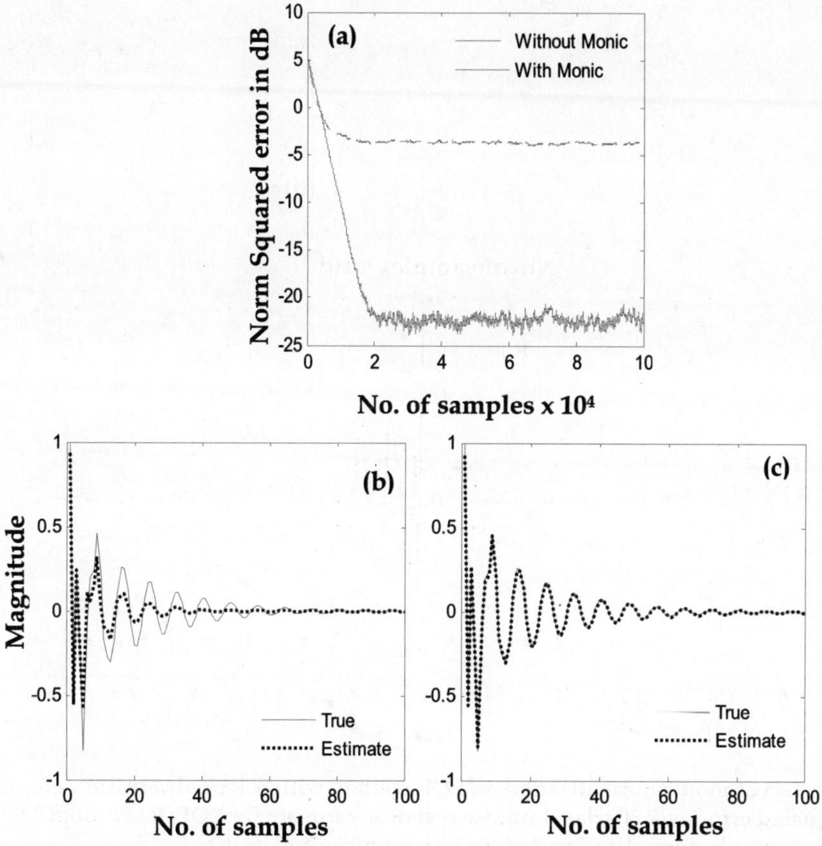

Fig. 13.3 Effect of monic-normalization on EQE method with white observation noise. (a) Norm squared error over 30 trials impulse response estimate for EQE; (b) without monic-normalization and (c) with monic-normalization.

and this indicates that with monic-normalization convergence occurs after about 20,000 samples and the norm square error is deceased significantly by about 28dB (Fig. 13.3a). But for without any monic-normalization, the decrease in error is only by 8dB. Thus, there is an improvement of 20 dB in error which is very insignificant. This is reflected in the impulse response estimate (for the first single trial) obtained after convergence (1,00,000th sample). For monic-normalization, there is a good match with the true impulse response (Fig. 13.3c). But for the case of without monic-normalization, the estimate match with the true impulse response is very poor (Fig. 13.3b)

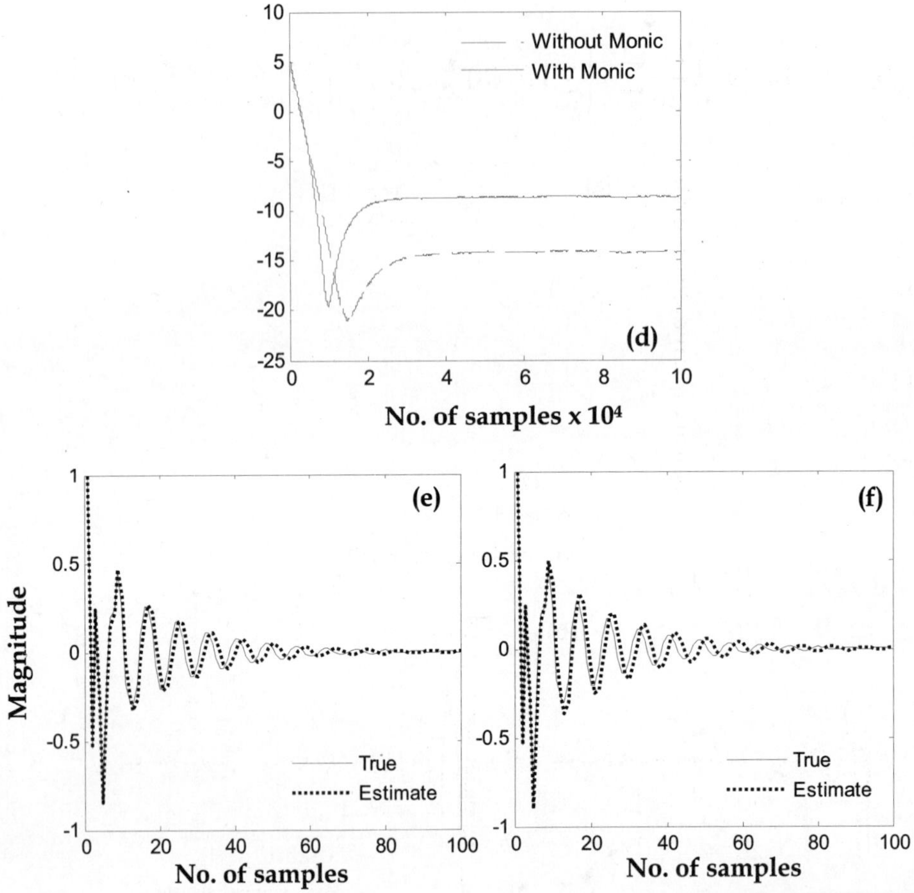

Fig. 13.3(d, e, f) Effect of monic-normalization on EQE method with colour observation noise; (d) Norm squared error over 30 trials Impulse response estimate for EQE; (e) without monic-normalization and (f) with monic-normalization.

For the colored observation noise, sinusoids of 95Hz, 135 Hz and 175 Hz with amplitudes 1, 0.5 and 0.4; respectively are used to get an SNR= 6.4 dB. The sampling rate used is 1 KHz. For the colored observation noise, the norm squared error with monic-normalization is poorer than that for without monic-normalization by about 5dB (Fig. 13.3d). This indicates that the monic normalization is not a solution for the poor performance of EQE in the presence of colored noise though it provides good results for white observation noise. This is reflected in the impulse response estimate also as seen form Fig. 13.3(e, f), where Fig.13.3(e) corresponds to without monic-normalization and Fig.13.3(f) corresponds to with monic-normalization.

13.5 COMPOSITE SQUARE ERROR ALGORITHM

The OE method has the advantage of unbiased parameter estimation in the presence of observation noise and also provides global solution for sufficient order under certain conditions. However, it generally suffers from local minima and specifically when the order is not sufficient and requires stability monitoring. The EQE method is simple and provides unimodal solution but this also requires stability monitoring. However, it suffers from bias in denominator estimation in the presence of observation noise. Hence, it is of interest to combine these two algorithms to achieve the advantages of both (S.L. Netto and P. Agathoklis, 1995). Here, it is worth noting the relation between EQE error $E'(z)$ and the OE error $E(z)$ given by Eqn. (13.5e)

$$E'(z) = E(z)\big(1 - A'(z)\big)$$

The composite error algorithm is achieved by a weighted combination of both the errors as

$$e_C^2(n) = \gamma\left[e'(n)\right]^2 + (1-\gamma)e^2(n) + K \tag{13.7a}$$

The constant $K > 0$ assures the non-negativity of $e_C^2(n)$, the composite square error. The adaptation of the parameter vector $W(n)$ is given the by the LMS gradient algorithm as

$$W(n+1) = W(n) - \mu\,\nabla_W[e_C^2(n)] \tag{13.7b}$$

$$\nabla_W[e_C^2(n)] = 2\gamma e'(n)U'(n) + (1-\gamma)2e(n)U(n) \tag{13.7c}$$

where $U'(n)$ and $U(n)$ are the input vectors for EQE and OE methods, respectively. The EQE algorithm has good initial features like good stability and unique minimum. On the other hand, the OE has unbiased global optimum solution. These can be achieved by making the parameter γ time varying taking initial (prior to convergence) value unity and as convergence progresses, a value close to zero. That is, the composite error algorithm works as EQE prior to convergence and as OE after convergence. This can be realized by adapting γ as

$$\gamma(n+1) = \gamma(n) - \mu_\gamma \nabla_\gamma[e_C^2(n)] \tag{13.7d}$$

where μ_γ is the step size for the adaptation of $\gamma(n)$. Since (Eqn.13.7a))

$$e_C^2(n) = \gamma\left[e'(n)\right]^2 + (1-\gamma)e^2(n) + K,$$

the gradient with respect to γ is given by

$$\nabla_\gamma[e_C^2(n)] = \left[e'(n)\right]^2 - e^2(n) \tag{13.7e}$$

Therefore,

$$\gamma(n+1) = \gamma(n) - \mu_\gamma \left\{ [e'(n)]^2 - e^2(n) \right\}, \quad 0 \le \gamma(n) \le 1 \tag{13.7f}$$

The above procedure is based on the minimization of mean composite error, On convergence/ minimization of composite error energy, γ will tend to unity if the minimum mean energy of EQE error is less than that of OE, that is, $\left[E\left\{ [e'(n)]^2 \right\}\right]_{\min} < \left[E\left\{ e^2(n) \right\}\right]_{\min}$, and otherwise $\gamma(n) = 0$. This is evident from the above adaptation of $\gamma(n)$, as $\left[E\left\{ [e'(n)]^2 \right\}\right]_{\min} < \left[E\left\{ e^2(n) \right\}\right]_{\min}$ makes $\gamma(n)$ much higher than zero. In order to make the parameter $\gamma(n) \to 0$ in all cases, it is required to modifying the adaptation of $\gamma(n)$ as

$$\gamma(n+1) = \begin{cases} \gamma(n) - \mu_\gamma \left| [e'(n)]^2 - e^2(n) \right| & if \;\; \gamma(n+1) > 0 \\ 0, & \end{cases} \tag{13.7g}$$

with the constraint on the range of $\gamma(n)$ as $0 \le \gamma(n) \le 1$. The constraint on the range of $\gamma(n)$ provides the mathematical and physical meaning of the composite square error. This is acceptable in cases where after the first convergence, there is no possibility of divergence from the optimum neighborhood due to any disturbance. However, for a case where convergence gets affected frequently which is true in practice, the composite parameter $\gamma(n)$ should sense the error magnitude and change its value, i.e., prior to convergence $\gamma(n)$ close to unity and after convergence $\gamma(n)$ close to zero. Further, $\gamma(n)$ must be robust to observation noise that will be present invariably.

13.6 STEIGLITZ-MCBRIDE (SM) ALGORITHM

The SM method is also an attempt in the direction of the bias removal problem of EQE IIR filters. But its importance is due the to the fact that *it minimizes the energy of the output error and not that of equation error.* Further, its behavior prior to and after convergence is similar to EQ and OE, respectively. However, *its bias removal ability is valid only for white observation noise and not for the colored noise. Further, it requires the desired signal and in its absence, its generation involves inaccuracies.*

Derivation of SM structure:

Let the original system considered be

$$G(z) = \frac{B(z)}{1 - A'(z)} = \frac{\sum_{k=0}^{N} b_k z^{-k}}{1 - \sum_{j=1}^{M} a_j z^{-j}} \tag{13.8a}$$

and its output $d(n)$ for the input $x(n)$ is given by

$$d(n) = \sum_{k=0}^{N} b_k(n)x(n-k) + \sum_{j=1}^{M} a_j(n)y(n-j) \tag{13.8b}$$

If the estimates of b_k and a_k are $\hat{b}_k(n)$ and, $\hat{a}_k(n)$ respectively, the output of the system model $y(n)$ is given by

$$y(n) = \sum_{k=0}^{N} \hat{b}_k(n)x(n-k) + \sum_{j=1}^{M} \hat{a}_j(n)d(n-j) \tag{13.8c}$$

Therefore, the error is

$$e(n) = d(n) - y(n) \tag{13.8d}$$

Substituting for $d(n)$ and $y(n)$,

$$e(n) = \sum_{k=0}^{N} b_k(n)x(n-k) + \sum_{j=1}^{M} a_j(n)d(n-j) - \sum_{k=0}^{N} \hat{b}_k(n)x(n-k) - \sum_{j=1}^{M} \hat{a}_j(n)y(n-j)$$

Adding and subtracting, $\sum_{j=1}^{M} \hat{a}_j(n)d(n-j)$ on right hand side of the above equation,

$$e(n) = \sum_{k=0}^{N} b_k(n)x(n-k) + \sum_{j=1}^{M} a_j(n)d(n-j) - \sum_{k=0}^{N} \hat{b}_k(n)x(n-k) - \sum_{j=1}^{M} \hat{a}_j(n)y(n-j)$$
$$+ \sum_{j=1}^{M} \hat{a}_j(n)d(n-j) - \sum_{j=1}^{M} \hat{a}_j(n)d(n-j)$$

Rearranging,

$$e(n) = \sum_{j=1}^{M} \hat{a}_j(n)d(n-j) - \sum_{j=1}^{M} \hat{a}_j(n)y(n-j) + \sum_{k=0}^{N} b_k(n)x(n-k) + \sum_{j=1}^{M} a_j(n)d(n-j)$$
$$- \sum_{j=1}^{M} \hat{a}_j(n)d(n-j) - \sum_{k=0}^{N} \hat{b}_k(n)x(n-k)$$

$$= \sum_{j=1}^{M} \hat{a}_j(n)[d(n-j) - y(n-j)] + d(n) - \sum_{j=1}^{M} \hat{a}_j(n)d(n-j) - \sum_{k=0}^{N} \hat{b}_k(n)x(n-k)$$

$$e(n) = \sum_{j=1}^{M} \hat{a}_j(n)e(n-j) + d(n) - \sum_{j=1}^{M} \hat{a}_j(n)d(n-j) - \sum_{k=0}^{N} \hat{b}_k(n)x(n-k)$$

$$e(n) - \sum_{j=1}^{M} \hat{a}_j(n)e(n-j) = d(n) - \sum_{j=1}^{M} \hat{a}_j(n)d(n-j) - \sum_{k=0}^{N} \hat{b}_k(n)x(n-k) \quad (13.8e)$$

Taking Z-transform both sides,

$$E(z) - \hat{A}'(z)E(z) = D(z) - \hat{A}'(z)D(z) - \hat{B}(z)X(z)$$

$$E(z)[1 - \hat{A}'(z)] = D(z) - \hat{A}'(z)D(z) - \hat{B}(z)X(z)$$

$$E(z) = \frac{1}{[1 - \hat{A}'(z)]}\left[D(z) - \hat{A}'(z)D(z) - \hat{B}(z)X(z)\right]$$

$$E(z) = D'(z) - \hat{A}'(z)D'(z) - \hat{B}(z)X'(z) \quad (13.8f)$$

where

$$D'(z) = \frac{D(z)}{[1 - \hat{A}'(z)]}, \quad X'(z) = \frac{X(z)}{[1 - \hat{A}'(z)]}$$

From this expression for $E(z)$, the schematic for SM method can be drawn as in Fig. 13.4 (J.R. Treichler, C.R. Johnson and M. G. Larimore, 2001). *It is important to note that the above equation which decides the SM structure has been derived by using the error used in OE method only.* This also can be seen from the description that follows.

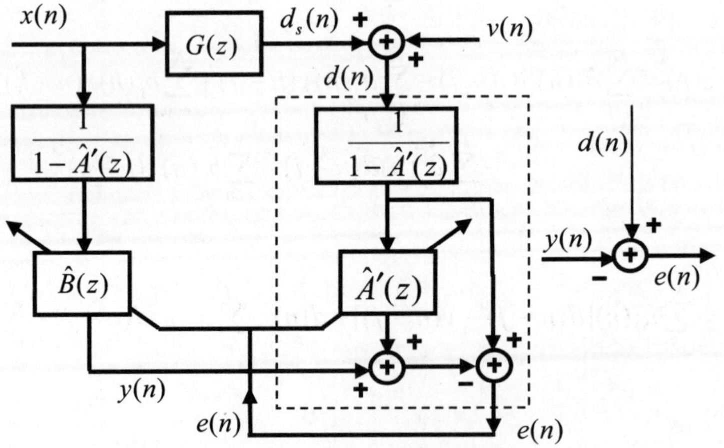

Fig. 13.4 Steiglitz McBride algorithm.

Consider

$$E(z) = D'(z) - \hat{A}'(z)D'(z) - \hat{B}(z)X'(z)$$

$$E(z) = \frac{D(z)}{[1-\hat{A}'(z)]} - \frac{D(z)\hat{A}'(z))}{[1-\hat{A}'(z)]} - \frac{\hat{B}(z)X(z)}{[1-\hat{A}'(z))]}$$

$$E(z) = \frac{D(z)[1-\hat{A}'(z)]}{[1-\hat{A}'(z)]} - \left[\frac{\hat{B}(z)}{[1-\hat{A}'(z)]}X(z)\right],$$

$$E(z) = D(z) - Y(z)$$

Thus in $E(z)$, the contribution of $D(z)$ branch is $D(z)$ only and the operation on $D(z)$ from $D(z)$ to $E(z)$, is unity operation. The other part of $E(z)$ is $Y(z)$ which is due to input branch. Thus, this also indicates that the error considered is *the output error* even though this is not evident directly from the schematic (Fig. 13.4).

Adaptation of the coefficient vectors:

$$e(n) = d'(n) - \sum_{j=1}^{M} \hat{a}_j(n)d'(n-j) - \sum_{k=0}^{N} \hat{b}_k(n)x'(n-k)$$

$$e(n) = d'(n) - \hat{a}'^T D'(n) - \hat{b}^T X'(n) \quad (13.9a)$$

where

$$\hat{b} = [\hat{b}_0, \hat{b}_1, \cdots\cdots \hat{b}_N]^T, \quad \hat{a}' = [\hat{a}_1, \cdots\cdots \hat{a}_M]^T \quad (13.9b)$$

$$X'(n) = [x'(n), x'(n-1), \cdots\cdots x'(n-N)]^T, \quad (13.9c)$$

$$D'(n) = [d'(n-1), d'(n-2)\cdots\cdots d'(n-M)]^T \quad (13.9d)$$

Therefore, the mean square error is given by

$$E\left[e^2(n)\right] = E\left[\left\{d'(n) - \hat{a}'^T D'(n) - \hat{b}^T X'(n)\right\}^2\right]$$

$$\frac{\partial E[e^2(n)]}{\partial \hat{b}} = -2\, E[e(n)\, X'(n)] \quad , \quad \frac{\partial E[e^2(n)]}{\partial \hat{a}'} = -2\, E[e(n)\, D'(n)],$$

The LMS adaptation rule for the estimated coefficient vectors $\hat{b}(n)$ and $\hat{a}'(n)$; are obtained replacing the expected values of the gradients by their respective instantaneous values as

$$\hat{b}(n+1) = \hat{b}(n) - \alpha_b\, e(n)X'(n) \quad (13.9e)$$

$$\hat{a}'(n+1) = \hat{a}'(n) - \alpha_a\, e(n)D'(n) \quad (13.9f)$$

α_a and α_b are the step sizes (and they can be normalized) used for adaptation. Let

$$D'_a(n) = [\, d'(n), d'(n-1), d'(n-2)\cdots\cdots\, d'(n-M)]^T,$$

$$\hat{a} = [1, -\hat{a}']^T \text{ or } A(z) = \left(1 - A'(z)\right)$$

The mean square error is given by

$$E\left[\, e^2(n)\right] = E\left[\left\{\, d'(n) - \hat{a}'^T D'(n) - \hat{b}^T X'(n)\right\}^2\right]$$

$$E[e^2(n)] = E\left[\left\{\hat{a}^T D'_a(n) - \hat{b}^T X'(n)\right\}^2\right]$$

$$E[e^2(n)] = (\hat{a}^T)E\left\{D'_a(n)D'^T_a(n)\right\}\,(\hat{a}\,)$$
$$-2(\hat{a}^T)E\left\{D'_a(n)X'(n)\right\}\hat{b} + \hat{b}^T E\left\{X'(n)X'^T(n)\right\}\hat{b}$$

$$E[e^2(n)] = \hat{a}^T R_{d'd'}\hat{a} - 2\hat{a}^T R_{d'x'}\hat{b} + \hat{b}^T R_{x'x'}\hat{b} \tag{13.9g}$$

where $R_{d'd'} = E\left\{D'_a(n)D'^T_a(n)\right\}$ and $R_{x'x'} = E\left\{X'(n)X'^T(n)\right\}$ are the autocorrelation matrices of $d'(n)$ and $x'(n)$ respectively and $R_{d'x'} = E\left\{D'(n)X'^T(n)\right\}$ is the cross-correlation matrix between $\dot{d}(n)$ and $\hat{x}(n)$. As seen above the MSE is a quadratic function of the adaptive filter weights. Further, as $R_{d'd'}$, $R_{x'x'}$ and $R_{d'x'}$ are not functions of $\hat{b}\,(n)$ and $\hat{a}\,(n)$, the Steiglitz-McBride algorithm is free from local minimum problem. However, in OE method, as $R_{xy}(n)$ and $R_{yy}(n)$ are functions of the weight vector, the MSE is not a quadratic function of $W(n)$ and hence suffers from local minimum problem.

The SM method can be considered composite square error method (G. Simon and G. Peceli, 1995) where its gradient is defined as

$$\nabla_W[e_C^2(n)] = e(n)\left\{\frac{1}{1-\hat{A}'(z)}\begin{bmatrix} x(n) \\ d(n) \end{bmatrix}\right\} \tag{13.9h}$$

Where $e(n) = d(n) - y(n)$. The difference between OE and SM is that in the gradient, $y(n)$ in OE has been replaced by $d(n)$ so that the SM provides the in between properties of EQE and OE methods.

13.6.1 Bias Removal in SM Method due to White Observation Noise

The mean square error is given by (Eqn.(13.9a)

$$E\left[e^2(n)\right] = E\left[\left\{ d'(n) - \hat{a}'^T D'(n) - \hat{b}^T X'(n)\right\}^2\right]$$

and

$$e(n) = d'(n) - \hat{a}'^T D'(n) - \hat{b}^T X'(n)$$

$$E(z) = \frac{D(z)}{[1-\hat{A}'(z)]} - \frac{D(z)\hat{A}'(z)}{[1-\hat{A}'(z)]} - \frac{\hat{B}(z)X(z)}{[1-\hat{A}'(z)]}$$

$$= \frac{D(z) - D(z)\hat{A}'(z)}{1-\hat{A}'(z)} - \frac{\hat{B}(z)X(z)}{[1-\hat{A}'(z)]}$$

$$E(z) = D(z) - \frac{\hat{B}(z)X(z)}{[1-\hat{A}'(z)]}$$

But

$$D(z) = G(z)X(z) + V(z), \qquad \hat{G}(z) = \frac{\hat{B}(z))}{[1-\hat{A}'(z)]}$$

Therefore,

$$E(z) = G(z)X(z) + V(z) - \hat{G}(z)X(z)$$

The adaptation equation for the denominator coefficient $\hat{a}_k(n)$ is given by

$$\hat{a}_k(n+1) = \hat{a}_k(n) - \mu\frac{1}{2}\frac{\partial E[e^2(n)]}{\partial \hat{a}_k}$$

Where

$$e(n) = d'(n) - \hat{a}'^T D'(n) - \hat{b}^T X'(n)$$

Now,

$$\frac{\partial E[e^2(n)]}{\partial \hat{a}_k} = E\left[\left\{ d'(n) - \hat{a}'^T D'(n) - \hat{b}^T X'(n)\right\}^2\right]$$

$$= 2E\left[e(n)\frac{\partial}{\partial \hat{a}_k}\left\{ d'(n) - \hat{a}'^T D'(n) - \hat{b}^T X'(n)\right\}\right]$$

$$= 2E\left[e(n)\frac{\partial}{\partial \hat{a}_k}\left\{ -\hat{a}'^T D'(n)\right\}\right]$$

$$\frac{\partial E[e^2(n)]}{\partial \hat{a}_k} = -2E\left[e(n)d'(n-k)\right]$$

$$\frac{\partial E[e^2(n)]}{\partial \hat{a}_k} = -2E\left[e(n)\left\{\left(\frac{z^{-k}}{1-\hat{A}'(z)}\right)d(n)\right\}\right]$$

Since

$$D(z) = G(z)X(z) + V(z) \qquad E(z) = \left\{G(z) - \hat{G}(z)\right\}X(z) + V(z)$$

$$d(n) = G(z)\left[x(n)\right] + v(n) \qquad e(n) = \left\{G(z) - \hat{G}(z)\right\}\left[x(n)\right] + v(n)$$

Therefore,

$$\frac{\partial E[e^2(n)]}{\partial \hat{a}_k} = -2E\left[e(n)\left\{\left(\frac{z^{-k}}{1-\hat{A}'(z)}\right)d(n)\right\}\right]$$

becomes

$$\frac{\partial E[e^2(n)]}{\partial \hat{a}_k} = -2E\left[\left\{\left(G(z) - \hat{G}(z)\right)\left[x(n)\right] + v(n)\right\}\left\{\left(\frac{z^{-k}}{1-\hat{A}'(z))}\right)\left(G(z)\left[x(n)\right] + v(n)\right)\right\}\right]$$

$$\frac{\partial E[e^2(n)]}{\partial \hat{a}_k} = -2E\left[\left\{e_s(n) + v(n)\right\}\left\{\left(\frac{z^{-k}}{1-\hat{A}'(z)}\right)\left(d_s(n) + v(n)\right)\right\}\right]$$

where,

$$e_s(n) = \left(G(z) - \hat{G}(z)\right)\left[x(n)\right], \quad d_s(n) = G(z)\left[x(n)\right]$$

Therefore,

$$\frac{\partial E[e^2(n)]}{\partial \hat{a}_k} = -2E\left[e_s(n)\left(\frac{z^{-k}}{1-\hat{A}'(z)}\right)[d_s(n)]\right] - 2E\left[v(n)\left(\frac{z^{-k}}{1-\hat{A}'(z)}\right)[d_s(n)\right]$$

$$-2E\left[e_s(n)\left(\frac{z^{-k}}{1-\hat{A}'(z))}\right)[v(n)]\right] - 2E\left[v(n)\left(\frac{z^{-k}}{1-\hat{A}'(z))}\right)[v(n)]\right]$$

$$\frac{\partial E[e^2(n)]}{\partial \hat{a}_k} = -2E\left[e_s(n)d_s'(n-k)\right] - 2E\left[v(n)d_s'(n-k)\right]$$

$$-2E\left[e_s(n)v'(n-k)\right] - 2E\left[v(n)v'(n-k)\right] \tag{13.9i}$$

Note that in the above equation, the second and third terms on right hand side are ensemble averages of uncorrelated products and become zero. Further, $v'(n-k)$ is a linear combinations of $v(n-k), v(n-k-1), \ldots$ which are uncorrelated with $v(n)$ Hence, if $v(n)$ is a white noise sequence, the last term also becomes zero. However if $v(n)$ is colored, the samples of $v(n)$ will be correlated and hence $E\left[v(n)v'(n-k)\right] \neq 0$ and the last term contributes to the gradient and the coefficients $\hat{a}_k$ get biased due to the observation noise. Therefore, the adaptation of $\hat{a}_k$ for white observation noise is given by

$$\hat{a}_k(n+1) = \hat{a}_k(n) + \alpha E\left\{d_s'(n-k)e_s(n)\right\}$$

where, $d_s'(n)$ is the filtered desired signal $d_s(n)$. The above equation is same as the adaptation equation when white noise is absent. Therefore, Steiglitz-McBride algorithm eliminates the biasing problem in the presence of white noise (P.A. Regalia, 1995, Chapter 8).

13.7. BIAS FREE STRUCTURE (BFS) FOR EQE IIR ADAPTIVE FILTERS FOR COLOUR OBSERVATION NOISE

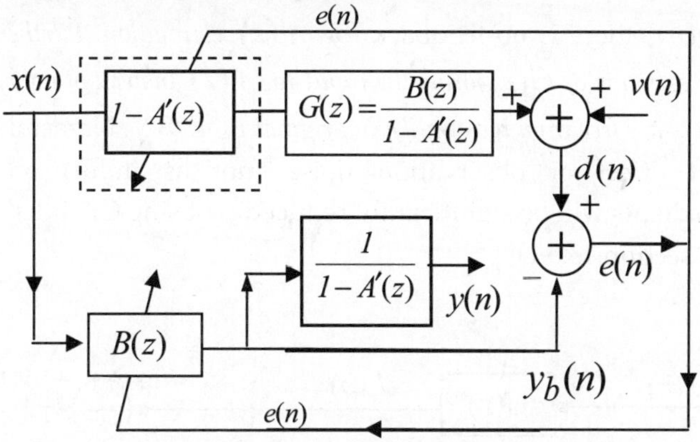

Fig. 13.5(a) Pole cancellation error adaptive filter.

The structures proposed to remove the bias in the denominator polynomial estimate due to observation noise only when it is white. Presently, the structure based on EQE that removes the bias even for the coloured observation noise, will be considered. Since the observation noise appears at the output of the system, it intuitively implies that the input for $A'(z)$ in Fig. 13.2a (also shown in Fig. 13.4b in dotted line) should not be from $d(n)$ but from the input $x(n)$, which is free from the observation noise $v(n)$. Based on this concept of using the system input which is free from noise (unlike $d(n)$) for estimation of $A'(z)$, pole-cancellation error adaptive filtering has been proposed (H.C. Shin and W.J. Song, 2003).

Here both adaptive filters for the numerator and denominator are *driven by the input* $x(n)$ (which is noise free) (Fig.13.5a). Further, the FIR adaptive filter that estimates ***the poles is forced (imposed) to cancel the system*** $G(z)$ ***poles*** and the other FIR adaptive filter identifies the pole cancelled unknown system. This pole cancellation is achieved by using the error $e(n)$ between the pole cancelled system output (plus observation noise) and its estimate $y_b(n)$ (obtained by filtering input $x(n)$ by $B(z)$) to adapt both $1-A'(z)$ and $B(z)$. Theoretically, this method provides unbiased estimates of plant parameters both for white and colored observation noises. However, it requires *an adaptive filter to be introduced between the source and the plant* (indicated by dotted box), hence, its utility will be limited and may not be practically feasible for applications like acoustic echo cancellation and active noise control (where the input and system cannot be separated like sound waves (input) and acoustic path (system). Therefore, for using the input for estimating the pole coefficients, the original conditions of the physical scenario have to be preserved. For this, it requires $x(n)$ to be passed through $G(z)$ and its output can be used as input for $A'(z)$. However, as $G(z)$ is unknown, its output estimate $\hat{d}(n)$ is computed using the adaptively estimated values of $A'(z)$ and $B(z)$ of the previous iteration. *Since the filter structure is realized only by FIR filters, similar to EQE, it is also free from local minima problem.* Unlike in the output error method, as $A'(z)$ input is not $y(n)$, there is no feedback for $A'(z)$ *estimation. Further, as* $\hat{d}(n)$ *derived from* $x(n)$ *through the estimate of* $G(z)$ *forms the input to* $A'(z)$ *instead of noisy* $d(n)$ *forming its input,* $A'(z)$ *parameters are free from bias unlike in original EQE* (S.V. Narasmhan and S. Veena, 2013). (The indirect effect of observation noise (not as input) on the estimation of $A'(z)$ and $B(z)$ gradients will be significantly reduced by using Griffiths' algorithm which will be considered in section 13.5.1).

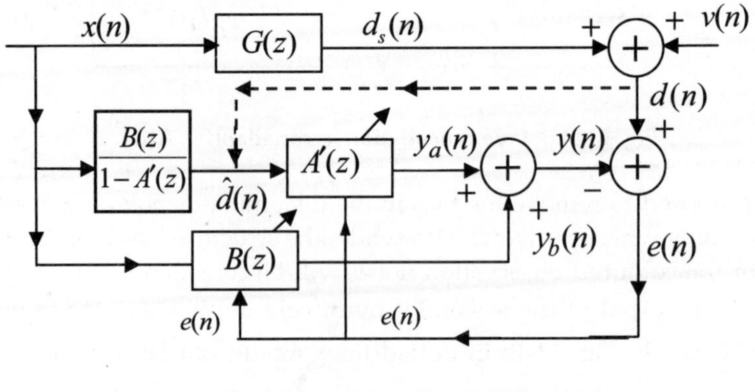

Fig. 13.5(b) Bias-free adaptive EQE IIR filter.

The BFS IIR adaptive filter in the system identification configuration is shown in Fig. 13.5b. Considering an unknown system $G(z)$ as in EQE, the adaptive part from input $x(n)$ to $y(n)$ results in

$$\hat{G}(z) = \frac{\hat{B}(z)}{1-\hat{A}'(z)}\hat{A}'(z) + \hat{B}(z) = \hat{B}(z)\left[\frac{\hat{A}'(z)}{1-\hat{A}'(z)} + 1\right]$$

$$\hat{G}(z) = \frac{\hat{B}(z)}{1-\hat{A}'(z)} \tag{13.10a}$$

and this is same as the plant system estimate used for removing the bias for $\hat{B}(z)$*. This implies that the transfer function between* $X(z)$ *and* $Y(z)$ *provides the estimate of the plant* $G(z)$. Therefore,

$$\hat{d}(n) = \sum_{k=0}^{M} \hat{b}_k(n)x(n-k) + \sum_{l=1}^{N} \hat{a}_l(n)\hat{d}(n-l) \tag{13.10b}$$

The output of the adaptive IIR filter $y(n)$ is given by

$$y(n) = y_a(n) + y_b(n)$$

$$y(n) = \sum_{k=0}^{M} b_k(n)x(n-k) + \sum_{l=0}^{N} a_l(n)\hat{d}(n-l) \tag{13.10c}$$

In this, $A'(z)$ is a FIR filter. The BFS error (Fig. 13.5b) is

$$e(n) = d(n) - y(n) \tag{13.10d}$$

$$e(n) = d(n) - \sum_{k=0}^{N} b_k(n)x(n-k) - \sum_{l=1}^{M} a_l(n)\hat{d}(n-l) \tag{13.10e}$$

which is used for the adaptation of $A'(z)$ and $B(z)$. The NLMS adaptation equations for the coefficients $b_k(n)$ and $a_l(n)$ are,

$$b_k(n+1) = b_k(n) + \frac{\mu}{MP_x(n)}e(n)x(n-k),\ k = 0, 1, \ldots N \tag{13.11a}$$

$$a_l(n+1) = a_l(n) + \frac{\mu}{NP_{\hat{d}}(n)}e(n)\hat{d}(n-l),\quad l = 1, 2, \ldots M \tag{13.11b}$$

$P_x(n)$ and $P_{\hat{d}}(n)$ are the powers of the input signals and they are computed recursively as,

$$P_x(n+1) = \beta P_x(n) + (1-\beta)\, x^2(n), \qquad 0 < \beta < 1 \tag{13.11c}$$

and

$$P_{\hat{d}}(n+1) = \beta P_{\hat{d}}(n) + (1-\beta)\, \hat{d}^2(n) \tag{13.11d}$$

From the adaptation point of view of $B(z)$, there are two paths between input $X(z)$ and output $Y(z)$. One path is where $A(z)$ is adapted and other is its copy used for computation of $\hat{d}(n)$. In view of this, the values of weights $B(z)$ will be half of their true values.

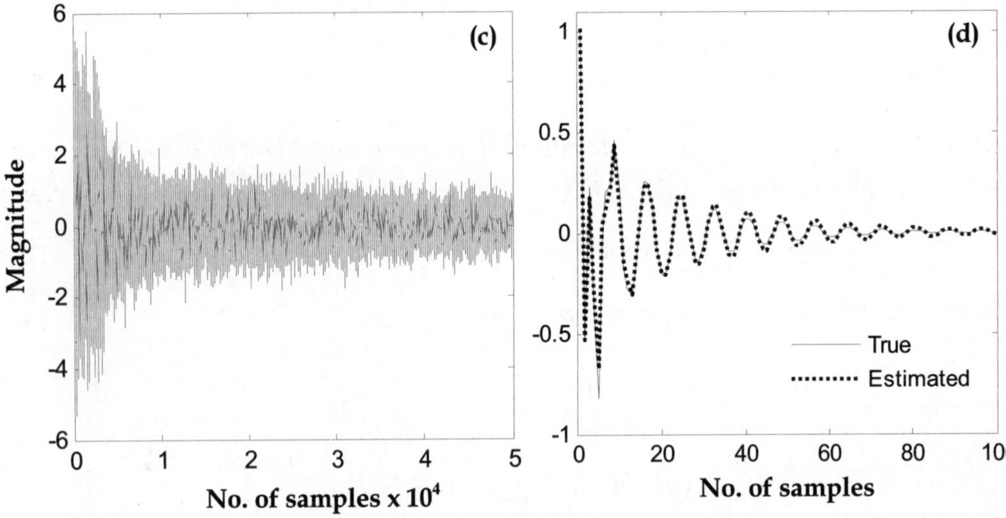

Fig. 13.5 (c, d) Performance of bias free EQE with white observation noise (c)Convergence error and (d) Impulse response.

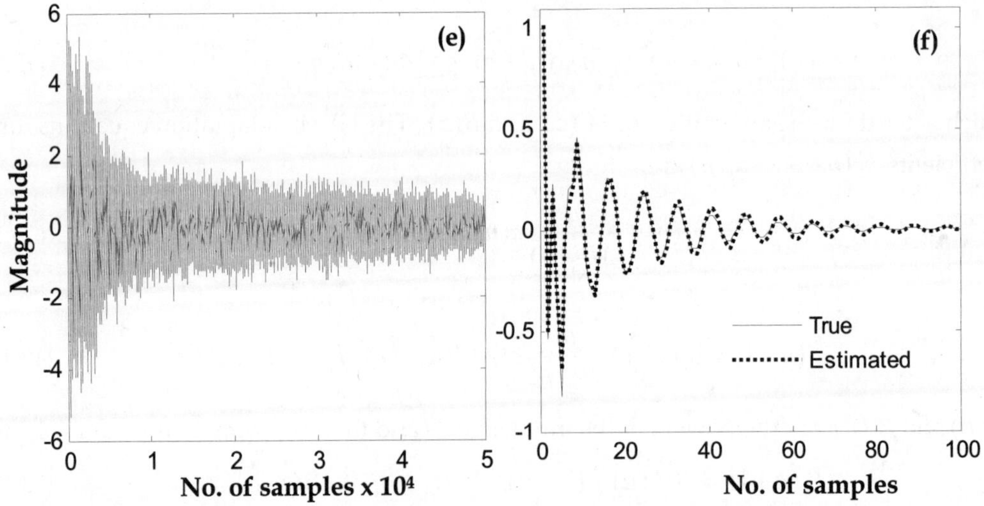

Fig. 13.5 (e, f) Performance of bias free EQE with coloured observation noise (e) Convergence error and (f) Impulse response.

Example: To bring out the performance of bias free EQE method, a system

$$P(z) = \frac{1 - 1.289z^{-1} + 1.0058z^{-2} - 0.7617z^{-3}}{1 - 0.59z^{-1} + 0.455z^{-2} - 0.0754z^{-3} + 0.504z^{-4}}$$

driven by a zero mean unit variance Gaussian white noise, is considered. The observation noise at the system output is a zero mean Gaussian white noise with an SNR= -1.0982 dB. The step size $\mu = 0.001$ and the forgetting factor $\beta = 0.999$ for the input power estimation; are used for the normalized LMS algorithm. The number of coefficients used for the numerator and denominator are 4 for each.

For the colored observation noise, sinusoids of 210Hz, 300 Hz and 400 Hz with amplitudes 1, 1 and 1 respectively are considered to give an SNR of 3.1912 dB. It is seen that the bias free EQE gives almost similar results to both for white and colour observation noise. In both cases, the error converges to a reduced value (Fig.13.5(c, e)) and the estimated impulse responses have good match to the true impulse response (Fig. 13.5(d, f)).

13.7.1 Bias-free Griffiths' Algorithm (BFSG) for EQE

Applying the Griffiths' algorithm developed in chapter-10 to Eqns. (13.11a) and (13.11b) using

$$e(n) = d(n) - y(n),$$

$$b_k(n+1) = b_k(n) + \frac{\mu}{NP_x(n)}(n)(n)\left[G_{bk}(n) - y(n)x(n-k)\right] \tag{13.12a}$$

$$a_l(n+1) = a_l(n) + \frac{\mu}{MP_{\hat{d}}(n)}\left[G_{al}(n) - y(n)\hat{d}(n-l)\right] \tag{13.12b}$$

where

$$G_{bk}(n+1) = \beta G_{bk}(n) + (1-\beta)d(n)x(n-k), \quad 0 < \beta < 1 \tag{13.12c}$$

$$G_{al}(n) = \beta G_{al}(n-1) + (1-\beta)d(n)\hat{d}(n-l) \tag{13.12d}$$

μ is the step-size used to update the coefficients. This algorithm is referred to as *BFS Griffiths'* (BFSG) algorithm.(S.V. Narasimhan and S. Veena, 2013).

13.7.2 Bias-free Griffiths' Variable Step Size Algorithm (BFSGV)

Based on the variable step size Griffiths' LMS (VGLMS) algorithm considered in Chapter 10, the step-sizes μ_b and μ_a are updated as,

$$\mu_b(n+1) = \alpha_b\,\mu_b(n) + \sigma_b \sum_{k=0}^{N}\left[G_{bk}(n) - y(n)x(n-k)\right]^2, \quad 0 < \alpha_b, \sigma_b < 1 \tag{13.13a}$$

$$\mu_a(n+1) = \alpha_a\,\mu_a(n) + \sigma_a \sum_{l=1}^{M}\left[G_{al}(n) - y(n)\hat{d}(n-l)\right]^2, \quad 0 < \alpha_a, \sigma_a < 1 \tag{13.13b}$$

Thus, the adaptation rules developed for both $b_k(n)$ and $a_l(n)$ not only have *variable step sizes, but also gradients, which are immune to observation noise* (S.V. Narasimhan and S. Veena, 2013). In accordance with the NLMS algorithm, for stability the upper limit for step sizes, $\mu_{\max}$ is set close to unity and the lower limit $\mu_{\min}$ is fixed so as to prevent adaptation stalling and to provide minimum misadjustment tolerable.

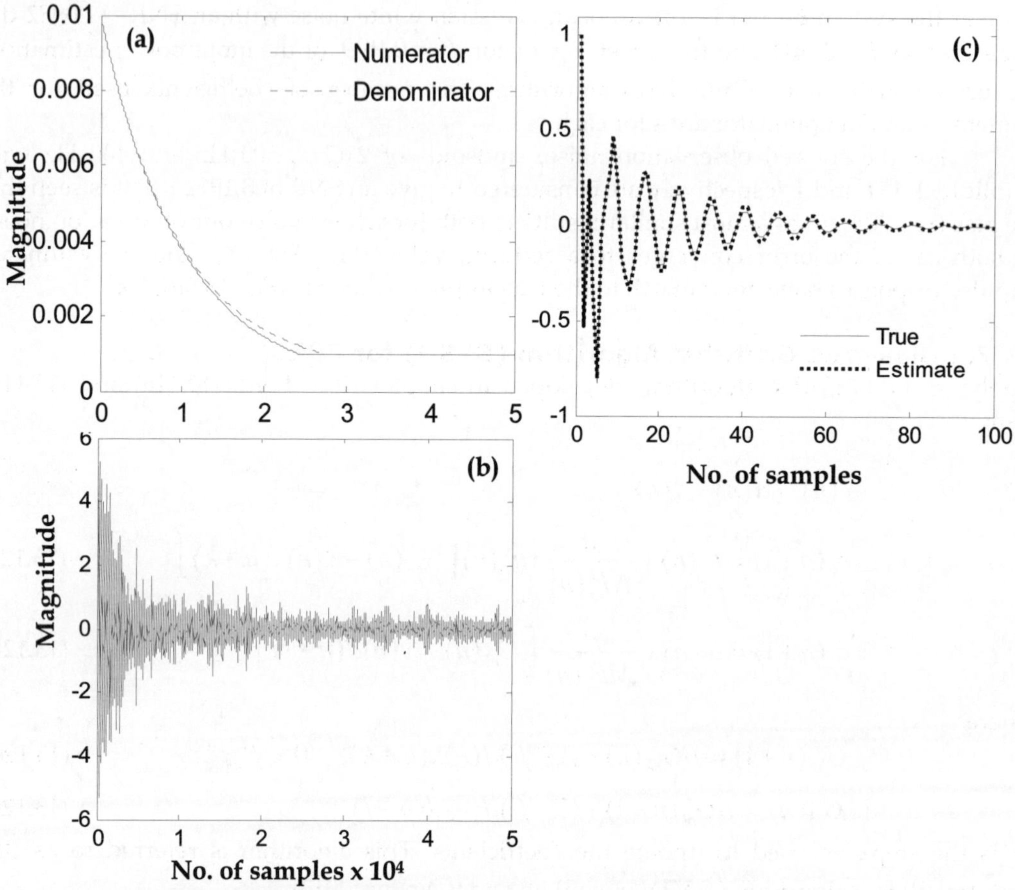

Fig. 13.6 (a, b and c) Bias-free EQE Griffiths' variable step size algorithm for system identification with white observation noise. (a) step size variation; (b) Convergence error and (c) Impulse response estimate.

The performance of BFSGV algorithm is illustrated for the system

$$P(z) = \frac{1-1.289z^{-1}+1.0058z^{-2}-0.7617z^{-3}}{1-0.59z^{-1}+0.455z^{-2}-0.0754z^{-3}+0.504z^{-4}}$$

driven by a white zero mean, unit variance Gaussian white noise. The plant output is corrupted by a zero mean Gaussian white noise to get an SNR= -1.0982 dB . For the bias-free

EQEGV, the step size variation was limited to between a minimum and maximum of 0.001and 0.01, respectively; both for numerator and for denominator. The initial step size is set to the maximum value $\mu = 0.01$. The forgetting factor $\beta = 0.999$ is used for power estimation for step size normalization. The number of numerator and denominator coefficients used are 3 and 4, respectively. The forgetting factor for Griffiths' cross-correlation computation used is 0.999 . The values of α_a, α_b =0.9999 and $\sigma_a, \sigma_b = 1\times10^{-9}$ are used.

The variation of step size for both numerator and denominator from maximum starting value of 0.01 to minimum value of 0.001 takes place over 25000 samples in an almost similar manner (Fig.13.6a). The error convergence also supports this (Fig.13.6b). The impulse response obtained after 5000 samples, shows close resemblance to the true one.

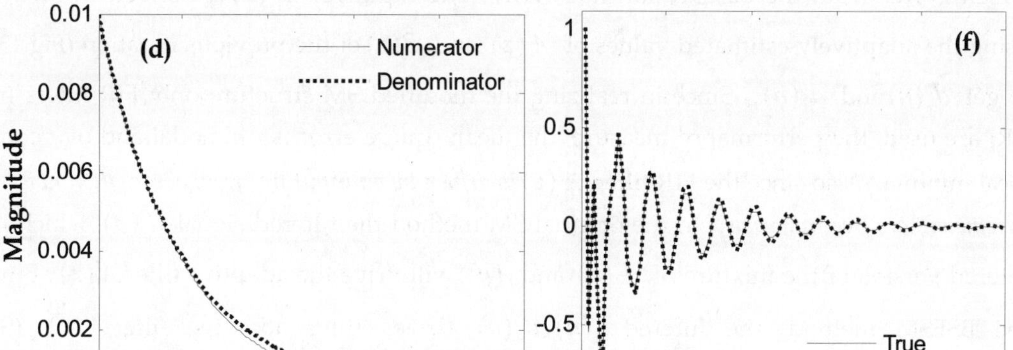

Fig. 13.6 (d, e and f) Bias-free EQE Griffiths' variable step size algorithm for system identification with colour observation noise. (d) Stepsize variation; (e) Convergence error and (f) Impulse response estimate.

The corresponding results for the color observation noise made up of sinusoids 210, 300 and 400 Hz each of amplitude unity for an SNR= 3.1652 dB are shown in Fig, 13.6(d, e, f). The sampling frequency used is 1KHz. The same parameters used for white observation noise are used for colored observation noise case also. The results are almost similar to those obtained for white observation noise case.

13.8 BIAS-FREE STEIGLITZ-MCBRIDE METHOD (BSFSM)

A modification has been incorporated for the SM adaptive IIR filter to avoid the bias problem when the observation noise is colored and not white. Similar to bias free EQE method, here also the input for $\hat{B}(z)$ should not be from $d(n)$ but from the input $x(n)$ which is free from the observation noise $v(n)$. The input for $\hat{A}'(z)$ is derived from $x(n)$ using the adaptively estimated values of $\hat{A}'(z)$ and $\hat{B}(z)$ of the previous iteration (Fig.13.7) to get $\hat{d}_s(n)$ and $\hat{d}_s'(n)$. Since in realizing the modified SM structure only FIR filters (not IIR) are used, the performance measure, the mean square error is unimodal and overcomes local minima. Also since the FIR filter $\hat{A}'(z)$ is *driven by an input free from observation noise, its parameters are free from bias.* In the original SM method the filtered signal $d'(n)$ which is a filtered version of the mixture of $d_s(n)$ and $v(n)$ will drive the adaptive filter $\hat{A}'(z)$. But in the BSFSM method, the filtered $\hat{d}_s(n), \hat{d}_s'(n)$ drives filter adaptive filter $\hat{A}'(z)$ (S.V. Narsimhan and C. Nagaveni, 2011).

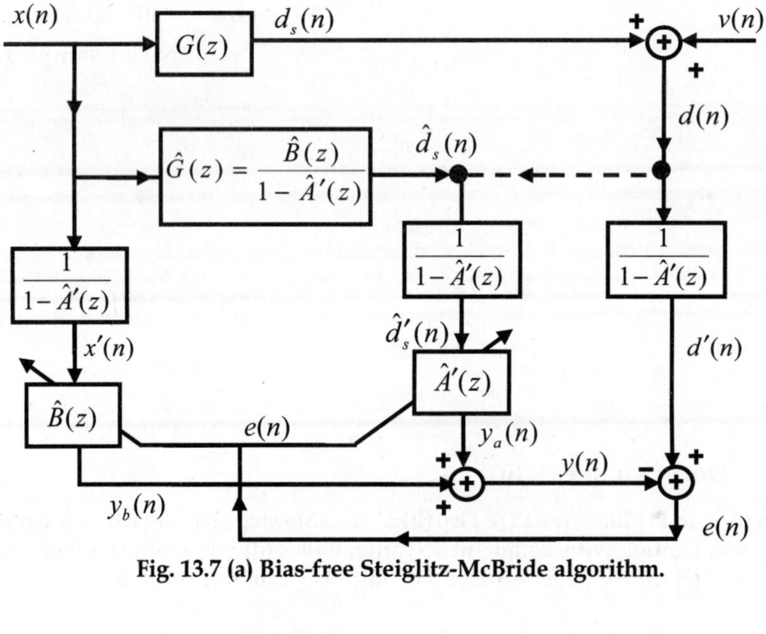

Fig. 13.7 (a) Bias-free Steiglitz-McBride algorithm.

The error is given by

$$e(n) = d'(n) - y_a(n) - y_b(n)$$

$$e(n) = \frac{1}{1-\hat{A}'(z)}[d(n)] - \frac{\hat{A}'(z)}{1-\hat{A}'(z)}[\hat{d}_s(n)] - \frac{\hat{B}(z)}{1-\hat{A}'(z)}[x(n)]$$

where

$$y_a(n) = \frac{\hat{A}'(z)}{1-\hat{A}'(z)}[\hat{d}_s(n)] \quad \text{and} \quad y_b(n) = \frac{\hat{B}(z)}{1-\hat{A}'(z)}[x(n)]$$

Since

$$\hat{d}_s(n) = d_s(n) + \in(n), \quad d(n) = d_s(n) + v(n) \tag{13.14a, b}$$

$\in(n)$ is the error in the estimation of $d_s(n)$ by $\hat{d}_s(n)$. Therefore,

$$e(n) = \frac{1}{1-\hat{A}'(z)}[d_s(n) + v(n)] - \frac{\hat{A}'(z)}{1-\hat{A}'(z)}[d_s(n) + \in(n)] - y_b(n)$$

$$= \frac{1}{1-\hat{A}'(z)}[d_s(n)] - \frac{\hat{A}'(z)}{1-\hat{A}'(z)}[d_s(n)] - y_b(n) + \frac{1}{1-\hat{A}'(z)}[v(n)] - \frac{\hat{A}'(z)}{1-\hat{A}'(z)}[\in(n)]$$

$$= d_s(n) - y_b(n) - \frac{A'(z)}{1-A'(z)}[\in(n)] + \frac{1}{1-A'(z)}[v(n)]$$

$$e(n) = e_s(n) + \in''(n) + v'(n) \tag{13.14c}$$

where,

$$e_s(n) = d_s(n) - y_b(n), \; v'(n) = \frac{1}{1-A'(z)}[v(n)] \;\text{ and }\; \in''(n) = -\frac{A'(z)}{1-A'(z)}[\in(n)]$$

In the error (Eqn.13.14c)), there are three components. The first error component is $e_s(n)$ in between $d_s(n)$ and $y_a(n)$. The second is $\in''(n)$ due to difference between the estimated system parameters $\{\hat{B}(z), \hat{A}'(z), \}$ and the actual parameters $\{B(z), A'(z), \}$ used in estimating $d_s(n)$ by $\hat{d}_s(n)$. As the $\{\hat{B}(z), \hat{A}'(z), \}$ used in deriving $\hat{d}_s(n)$ by filtering $x(n)$ correspond to previous iteration of the adaptation (as explained in the SM method), the $\hat{d}_s(n)$ and the error component $\in''(n)$ is also independent of present $\{\hat{B}(z), \hat{A}'(z), \}$ and hence should be treated as constants for minimization of the energy of the error $e(n)$. The third is due to observation noise $v'(n)$. Again as $v'(n)$ is the filtered $v(n)$ using $\hat{A}'(z)$ of previous iteration, $v'(n)$ is also treated as constant with respect to present $\hat{A}'(z)$. In view of this, $\in''(n)$ can be treated as a part of observation noise. Hence, the error minimized can be considered output error as

$$e(n) = d_s(n) - y_b(n) + u(n), \qquad u(n) = \in''(n) + v'(n) \tag{13.14d, e}$$

In this, $\{d_s(n) - y_b(n)\}$ forms the output error and $u(n)$ is the observation noise. As the error is minimized with respect to the parameters $\{\hat{B}(z), \hat{A}'(z), \}$, $\{\hat{B}(z), \hat{A}'(z), \}$ approach $\{B(z), A'(z), \}$ and hence the error component $\in''(n)$ decreases resulting in $\hat{d}_s(n)$ approaching $d_s(n)$ on convergence.

It is important to note that the component in the error due to observation noise $v(n)$ does not remain as observation noise as in the original SM method but it is a filtered version of observation noise and is $v'(n)$. Now,

$$\hat{d}_s(n) = \sum_{k=0}^{N} \hat{b}_k(n)x(n-k) + \sum_{l=1}^{M} \hat{a}_l(n)\hat{d}_s(n-l)$$

The output of the adaptive IIR filter is $y(n)$ given by

$$y(n) = y_b(n) + y_a(n) \tag{13.15a}$$

$$y_b(n) = \sum_{k=0}^{N} \hat{b}_k(n)x'(n-k) \tag{13.15b}$$

$$y_a(n) = \sum_{l=1}^{M} \hat{a}_l(n)\hat{d}'_s(n-l) \tag{13.15c}$$

where $x'(n)$ and $\hat{d}'_s(n)$ are the filtered forms of $x(n)$ and $\hat{d}_s(n)$.

In the SM method, $y(n) = y_b(n)$. However, in the BSFSM method this is not so. This is due to the fact that $v(n)$ and $\in(n)$ are not common to both $\hat{d}_s(n)$ and $d(n)$ branches. But as $d_s(n)$ is common to both the branches, for $d_s(n)$, the BSFSM method behaves as SM method. The error is

$$e(n) = d'(n) - y(n) \tag{13.16a}$$

where, $d'(n)$ is the filtered version of the desired signal and is

$$d'(n) = d(n) + \sum_{l=1}^{M} a_l(n)d'(n-l) \tag{13.16b}$$

Substituting Eqns. (13.15a), (13.15b) and (13.15c) in Eqn.(13.16a)

$$e(n) = d'(n) - \sum_{k=0}^{N} \hat{b}_k(n)x'(n-k) - \sum_{l=1}^{M} \hat{a}_l(n)\hat{d}'_s(n-l) \tag{13.16c}$$

and this error $e(n)$ is used for the adaptation of $\hat{B}(z)$ and $\hat{A}'(z)$ The NLMS adaptation equations for the coefficients $\hat{b}_k(n)$ and $\hat{a}_l(n)$ are,

$$\hat{b}_k(n+1) = \hat{b}_k(n) + \frac{\mu}{NE_b(n)} e(n)x'(n-k), \quad k=0,1,2,\ldots.. N \tag{13.17a}$$

$$\hat{a}_l(n+1) = \hat{a}_l(n) + \frac{\mu}{ME_a(n)} e(n)\hat{d}'_s(n-l)\ ,\ l=1,2,\dots M \tag{13.17b}$$

$E_b(n)$ and $E_a(n)$ are the powers of the input signals and they are computed recursively as,

$$E_b(n) = \beta_b E_b(n-1) + (1-\beta_b)\{x'(n)\}^2, \qquad 0 < \beta_b < 1 \tag{13.17c}$$

$$E_a(n) = \beta_a E_a(n-1) + (1-\beta_a)\{\hat{d}'_s(n)\}^2, \quad 0 < \beta_a < 1 \tag{13.17d}$$

From the adaptation point of view of $\hat{B}(z)$, there are two paths between input $X(z)$ and output $Y(z)$. One path is where $\hat{B}(z)$ is adapted and other is its copy used for computation of $\hat{d}_s(n)$. In view of this, the values of weights $\hat{B}(z)$ will be half of their true values.

SM method: The performance of SM method and modified SM (MSM) algorithms are illustrated for system identification. The systems is given by

$$P(z) = \frac{1 - 1.289z^{-1} + 1.0058z^{-2} - 0.7617z^{-3}}{1 - 0.59z^{-1} + 0.455z^{-2} - 0.0754z^{-3} + 0.504z^{-4}}$$

which is driven by a Gaussian white noise of zero mean and unit variance.

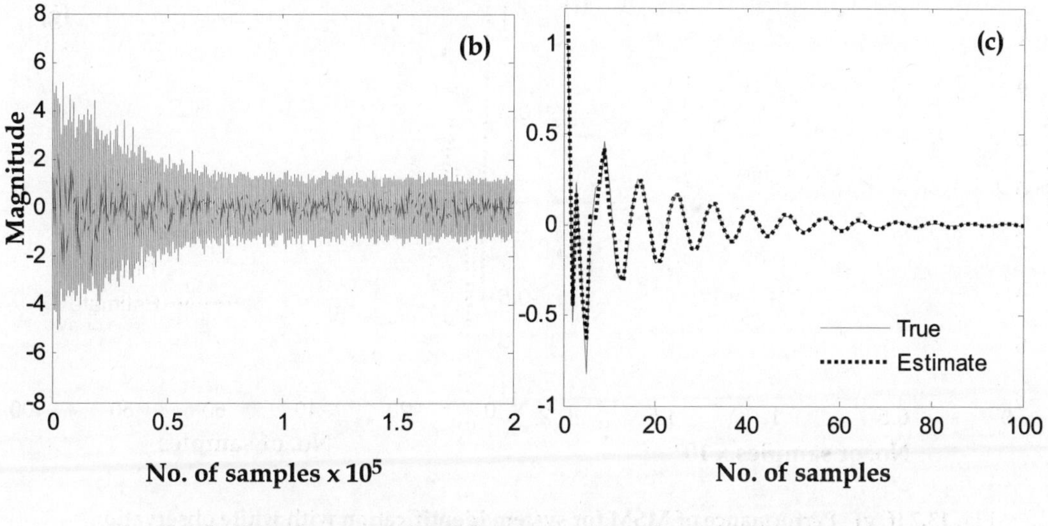

Fig. 13.7 (b, c) Performance of SM for system identification with white observation noise (b) Convergence of error (c) estimate of impulse response.

The step size for the used for the numerator and denominator adaptation are $\mu = 0.002$ and $\mu = 0.0002$, respectively. The stepsize for the denominator adaptation is one tenth that for numerator. The normalized LMS is used for adaptation of numerator and

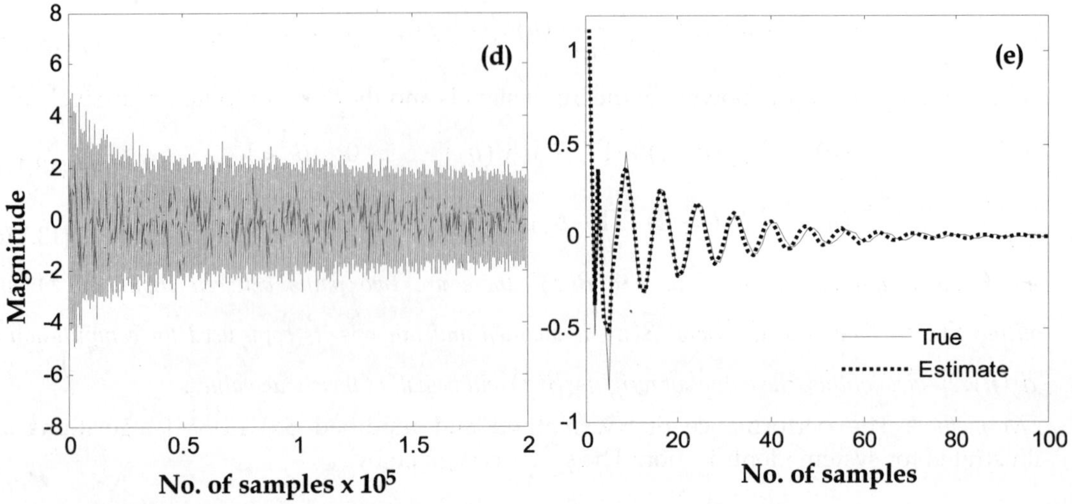

Fig. 13.7 (d, e) Performance of SM for system identification with color observation noise (d) Convergence of error (e) estimate of impulse response.

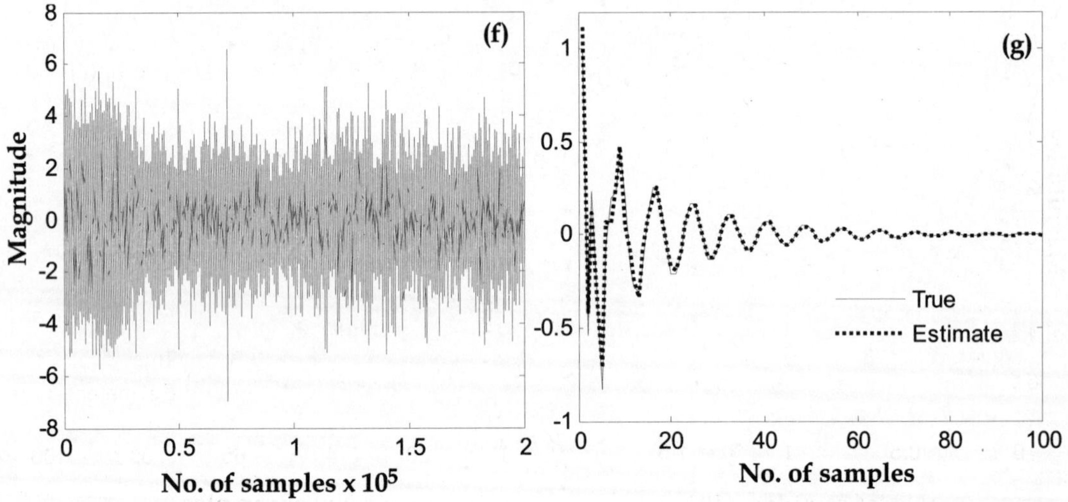

Fig. 13.7 (f, g) Performance of MSM for system identification with white observation noise (f) Convergence of error (g) estimate of impulse response.

denominator coefficients. The forgetting factor for input power estimations used is $\beta = 0.999$.

The lengths of numerator and denominator coefficients used are 3 and 4, respectively. A zero mean white Gaussian noise is added to the system output to get an SNR = -1.07 dB. At every sample instant, from the denominator coefficients, the reflection coefficients are computed and checked for their magnitude to be less than unity and it was found to satisfy the requirement. It is seen that the convergence takes place after 50,000 samples (Fig. 13.7 (b)). The impulse response estimated after 2,00,000 samples using the numerator and denominator coefficients is shown in comparison with the true one. It is seen that there is good match of the estimated impulse response with the true one (Fig.13.7(c)). This is because, in the SM method the filter $1/(1-A'(z))$ *rejects the white observation noise component in the desired signal* (which forms the input to $A'(z)$) and hence its bias effect on the estimation $A'(z)$ is not felt. However, the observation noise effect in the error will affect the estimation $A'(z)$ and $B(z)$.

For the color observation noise with sinusoids of 100Hz, 150Hz and 200Hz with amplitudes 1.5, 1 and 1 respectively. with SNR= -0.3264 dB (with respect to system output). The sampling frequency used is 1KHz. Even though the observation color noise power is less than that for white noise (-1db), the convergence error magnitude is more for the former (Fig. 13.7(b) and (d)).

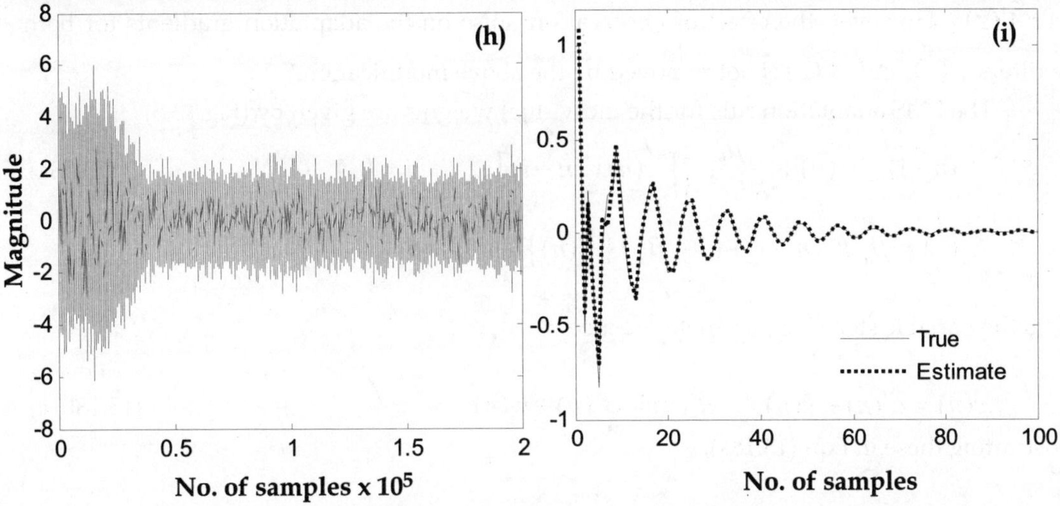

Fig. 13.7(h, i) Performance of MSM for system identification with colour observation noise (h) Convergence of error (i) estimate of impulse response.

This is reflected in the estimate of impulse response. Specifically, the maximum negative dip (in the initial part of impulse response) is not followed by the estimate (Fig.13.7(e)) when compared to that for white noise case (Fig.13.7(c)). Further, for white noise, the overall match of the estimate of impulse response to true one is better than that obtained for colour observation noise case. This is due to the fact that the filtering by $1/(1-A'(z))$ *in SM method as it is only a one step predictor, it cannot reject the noise having correlations at higher lags and can only reject white noise with one step prediction.* This indicates that the bias due to color observation noise is not removed as it happens for white noise case. In this simulation, the other parameters used are kept same as for the white noise case.

For the modified SM method, its performance is evaluated for the same example considered for SM method. The convergence plot for the white observation noise, is somewhat spiky nature compared to that of color observation noise (Fig. 13.7(f, h)) and this may be due poor SNR for white noise (-1.7dB) case compared to that of color noise (-0.3dB). But the impulse response estimates corresponding to the last sample are almost similar to that of actual one (Fig.13.7(g, i)) indicating that the modified SM provides unbiased impulse response estimate even for coloured observation noise (unlike SM). This is because, the desired signal estimate derived from the system estimate is used in generating the input for $A'(z)$ and this desired signal estimate is free from colored observation noise.

13.8.1 Bias-free Steiglitz Mc-Bride Griffiths' Algorithm (BSFSMG)

The modification in BSFSM method overcomes the bias effect on the denominator polynomial of the system function considered for identification. This is due to the fact the modification avoids the presence of observation noise at the input of the adaptive filter $\hat{A}'(z)$. However, the effect of observation noise on the adaptation gradients for both the filters $\hat{B}(z)$ and $\hat{A}'(z)$ is not removed by the above modification.

The LMS adaptation rule for the individual weights are given by (Fig.13.5)

$$\hat{b}_i(n+1) = \hat{b}_i(n) + \frac{\mu_b}{NE_b(n)}\left[e(n)x'(n-i) \right], \qquad i = 0, 1, \cdots, N \tag{13.18a}$$

$$E_a(n) = \beta_a\, E_a(n-1) + (1-\beta_a)\{x'(n)\}^2, \quad 0 < \beta_a < 1$$

μ_b is the variable step size parameter. Since

$$e(n) = d'(n) - y(n), \qquad d'(n) = d_s'(n) + v'(n) \tag{13.18b, c}$$

Substituting these in Eqn.(13.18a),

$$\hat{b}_i(n+1) = \hat{b}_i(n) + \frac{\mu_b}{ME_b(n)}\left[d_s'(n) - y(n) + v'(n) \right] x'(n-i)$$

$$\hat{b}_i(n+1) = \hat{b}_i(n) + \frac{\mu_b}{NE_b(n)}\left[\{d_s'(n) - y(n)\} x'(n-k) \right] + \frac{\mu_b}{NE_b(n)} v'(n)x'(n-k) \tag{13.18d}$$

The last term is due to observation noise and hence is not desirable. As $x'(n)$ and $v'(n)$ are uncorrelated, the expected value of $v'(n)x'(n-k)$, $E[v'(n)x'(n-k)] = 0$. However, the instantaneous product, $[v'(n)x'(n-k)] \neq 0$. Hence, to avoid this undesirable effect of the observation noise, the Griffiths' LMS adaptation rule is used (S.V. Narasimhan and C. Nagaveni, 2011) and is given by

$$\hat{b}_i(n+1) = \hat{b}_i(n) + \frac{\mu_b}{ME_b(n)}\left[\ G_{bi}(n) - y(n)x'(n-i)\right], \tag{13.19a}$$

$$\hat{G}_{bi}(n+1) = \gamma_{cb}\ \hat{G}_{bi}\ (n) + \left(1-\gamma_{cb}\right) d'(n)x(n-k), \quad 0 < \gamma_{cb} < 1, k = 0, 1, \cdots, M-1 \tag{13.19b}$$

$$E_b(n+1) = \beta_b\ E_b(n) + \left(1-\beta_b\right)\left\{x'(n)\right\}^2, \quad 0 < \beta_b < 1 \tag{13.19c}$$

The Griffiths' LMS adaptation for the denominator coefficients are given by

$$\hat{a}_l(n+1) = \hat{a}_l(n) + \frac{\mu_a}{ME_a(n)}[G_{al}(n) - y(n)\hat{d}'_s(n-l)] \tag{13.19d}$$

$$G_{al}(n) = \gamma_{ca} G_{al}(n-1) + (1-\gamma_{ca}) d'(n)\hat{d}'_s(n-l), \quad 0 < \gamma_{ca} < 1 \tag{13.19e}$$

$$E_a(n+1) = \beta_a\ E_a(n) + \left(1-\beta_a\right)\left\{\hat{d}'_s(n)\right\}^2, \quad 0 < \beta_a < 1 \tag{13.19f}$$

μ_a, μ_b are the step sizes used to update the numerator and denominator coefficients.

13.8.2 Bias-free Steiglitz-McBride Griffiths' Variable Step Size Algorithm (BSFSMGV)

To meet the requirement of both faster convergence and small misadjustment, it is essential to have a variable step size, which is a function of true error. The Griffiths' correlation reduces the effect of observation noise on the measure of the error. Therefore, the variable step size version of the BFSSM can be obtained by applying Griffiths' variable step size algorithm for the adaptation step sizes of the numerator and denominator μ_a and μ_b, respectively (S.V. Narasimhan and C. Nagaveni, 2011). These are given by

$$\mu_b(n+1) = \alpha_b \mu_b(n) + \sigma_b \sum_{k=0}^{N} [G_{bk}(n) - y(n)x'(n-k)]^2, \quad 0 < \alpha_b, \sigma_b < 1 \tag{13.20a}$$

$$\mu_a(n+1) = \alpha_a \mu_a(n) + \sigma_a \sum_{l=1}^{N} [G_{al}(n) - y(n)\hat{d}'_s(n-l)]^2, \quad 0 < \alpha_{a,} \sigma_a < 1 \tag{13.20b}$$

The first term on the RHS, $\alpha_a(n)\mu_a(n)$ or $\alpha_b(n)\mu_b(n)$ overcomes the adaptation stalling and the second term as it is positive, avoids the negative step size problem. To track fast changes σ_a or σ_b should be close to unity and α_a or α_b should be small. But for stationary signals, σ_a or σ_b should be small and α_a or α_b can be large. The coefficient adaptation rule is given by

$$\hat{b}_i(n+1) = \hat{b}_i(n) + \frac{\mu_b(n)}{ME_b(n)}\left[\ G_{bi}(n) - y(n)x'(n-i)\right], \quad i = 0, 1, \cdots, N \tag{13.20c}$$

$$\hat{a}_l(n+1) = \hat{a}_l(n) + \frac{\mu_a(n)}{NE_a(n)}[G_{al}(n) - y(n)\hat{d}'_s(n-l)],\ l = 1, \cdots, M \tag{13.20d}$$

Thus the coefficient adaptation for $b_k(n)$ and $a_l(n)$ are not only have robust variable step sizes but also robust gradients, which are immune to observation noise. In accordance with the NLMS algorithm for stability, the upper limit for step sizes, $\mu_{\max}$ can be set close to unity and the lower limit $\mu_{\min}$ is fixed so as to prevent adaptation stalling and to provide minimum misadjustment tolerable.

Example:

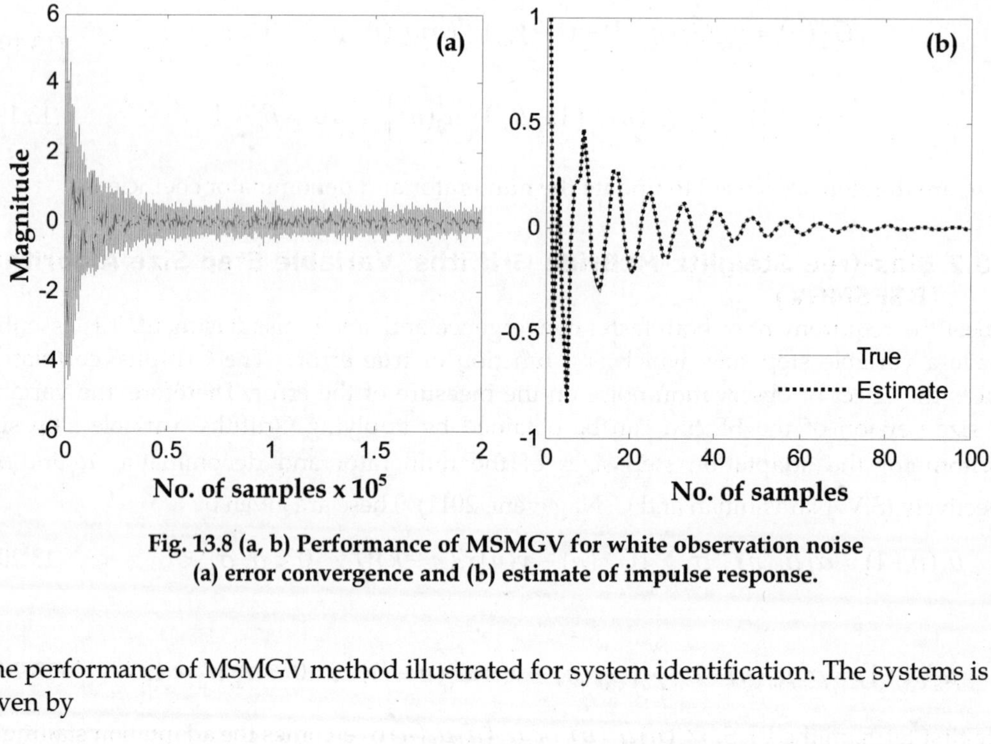

Fig. 13.8 (a, b) Performance of MSMGV for white observation noise (a) error convergence and (b) estimate of impulse response.

The performance of MSMGV method illustrated for system identification. The systems is given by

$$P(z) = \frac{1 - 1.289z^{-1} + 1.0058z^{-2} - 0.7617z^{-3}}{1 - 0.59z^{-1} + 0.455z^{-2} - 0.0754z^{-3} + 0.504z^{-4}}$$

which is driven by a Gaussian white noise of zero mean and unit variance. Further, a Gaussian white observation Noise with $\text{SNR} = -1.0709\text{dB}$ is considered. The system identification is done by using MSMGV algorithm. The parameters selected are: the step-size variation was limited between a minimum and maximum of 10^{-6} and 2×10^{-3} for numerator, and for denominator 10^{-6} and 2×10^{-4} respectively. The forgetting factor of 0.999 is considered for input power normalization and 0.9 for Griffiths' cross-correlation estimation. The values of α_a, α_b =0.999 and $\sigma_a, \sigma_b = 1\times10^{-3}$ are used. The numerator and denominator order used are 10 and 12, respectively.

The convergence of the MSMGV (Fig. (13.8(a)) is significantly far improved compared to that of MSM method (Fig. 13.7f). The estimated impulse response by MSMGV algorithm has an excellent match with the true one (Fig. 13.8b)compared to that obtained by MSM (Fig. 13.7(g)).

For colored observation noise, sinusoids of 100Hz, 150 Hz and 200 Hz with amplitudes 1.5, 1 and 1 respectively, sampled at 1KHz is considered and further the SNR = −0.3264dB.

The convergence performance of MSMGV for color noise is similar to that of white noise with a good error with a smaller magnitude (Fig. 13.8(c)) compared to that of MSM (Fig. 13.7(h). In this case also, the impulse response estimate has a good match with that of true one(Fig. 13.8(d)).

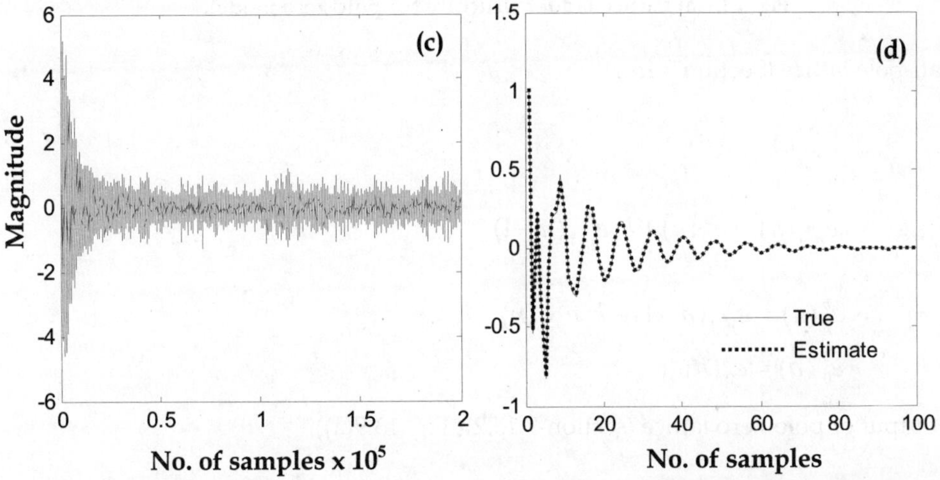

Fig. 13.8 (c, d) Performance of MSMGV for colour observation noise (c) error convergence and (d) estimate of impulse response.

In both types of noise cases, the effect of noise both on the gradient and step size is effectively reduced. Further, as the step size reduces to its set minimum value resulting in a small misadjustment leading to good match between the estimated and true impulse response.

13.9 POLE-ZERO ADAPTIVE FILTER BASED ON ALL-POLE LATTICE STRUCTURE

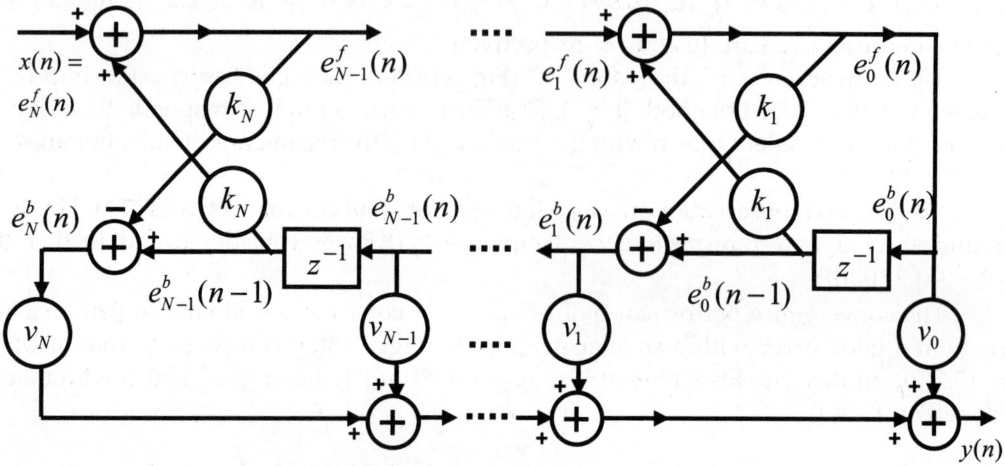

Fig. 13.9(a) Lattice ladder structure for pole-zero model.

For the all-pole lattice (Section 7.15.1),

$$e_N^f(n) = x(n),$$

$$e_{p-1}^f(n) = e_p^f(n) + k_p e_{p-1}^b(n-1)\ ,$$

$$e_p^b(n) = e_{p-1}^b(n-1) - k_p e_{p-1}^f(n),$$
$$e_0^f(n) = e_0^b(n)$$

and the output of pole-zero lattice (section-7.15.2)(Fig. 13.9(a)),

$$y(n) = \sum_{p=0}^{N} v_p(n) e_p^b(n)$$

To find the parameters of the pole-zero lattice adaptive filter, the error to be minimized is

$$e^2(n) = \left[d(n) - y(n)\right]^2$$

where, $d(n)$ and $y(n)$ are the desired signal and the output of the pole-zero adaptive filter. Adaptation of the coefficients $k_p(n)$ and $v_p(n)$ by the LMS gradient method are given by

$$k_p(n+1) = k_p(n) + \mu_k e(n)\Psi_p(n) \tag{13.21a}$$

$$v_p(n+1) = v_p(n) + \mu_v e(n)\phi_p(n) \tag{13.21b}$$

where,

$$\psi_p(n) = \frac{\partial y(n)}{\partial k_p(n)} \text{ and } \phi_p(n) = \frac{\partial y(n)}{\partial v_p(n)} \tag{13.21c, d}$$

For gradient $\phi_p(n)$: Since

$$y(n) = \sum_{p=0}^{N} v_p(n) e_p^b(n)$$

$$\phi_p(n) = \frac{\partial y(n)}{\partial v_p(n)} = e_p^b(n)$$

$$\phi_p(n) = e_p^b(n) \tag{13.22a}$$

For gradient $\psi_p(n)$

Defining

$$\alpha_{p,m}(n) = \frac{\partial e_m^f(n)}{\partial k_p(n)}, \quad \beta_{p,m}(n) = \frac{\partial e_m^b(n)}{\partial k_p(n)}$$

The backward prediction error $e_m^b(n)$ for the mth order predictor is given by

$$e_m^b(n) = e_{m-1}^b(n-1) - k_m e_{m-1}^f(n)$$

$$\beta_{p,m}(n) = \frac{\partial e_m^b(n)}{\partial k_p(n)} = \beta_{p,m-1}(n-1) - k_m \alpha_{p,m-1}(n) - \delta_{p,m} e_{m-1}^f(n)$$

$$\beta_{p,m}(n) = \beta_{p,m-1}(n-1) - k_m \alpha_{p,m-1}(n) - \delta_{p,m} e_{m-1}^f(n) \tag{13.22b}$$

The forward prediction error $e_{m-1}^f(n)$ for the $(m-1)^{th}$ order *all-pole predictor* is given by

$$e_{m-1}^f(n) = e_m^f(n) + k_m e_{m-1}^b(n-1)$$

$$\alpha_{p,m-1}(n) = \frac{\partial e_{m-1}^f(n)}{\partial k_p(n)} = \alpha_{p,m}(n) + k_m \beta_{p,m-1}(n-1) + \delta_{pm} e_{m-1}^b(n-1)$$

$$\alpha_{p,m-1}(n) = \alpha_{p,m}(n) + k_m \beta_{p,m-1}(n-1) + \delta_{pm} e_{m-1}^b(n-1) \tag{13.22c}$$

Further,

$$\alpha_{p,N}(n) = \frac{\partial e_N^f(n)}{\partial k_p(n)} =, \frac{\partial x(n)}{\partial k_p(n)} = 0 \quad \because e_N^f(n) = x(n)$$

$$\alpha_{p,N}(n) = 0 \tag{13.22d}$$

and

$$\alpha_{p,0}(n) = \frac{\partial e_0^f(n)}{\partial k_p(n)} = \frac{\partial e_0^b(n)}{\partial k_p(n)} = \beta_{p,0}(n), \quad \because e_0^f(n) = e_0^b(n)$$

$$\alpha_{p,0}(n) = \beta_{p,0}(n) \tag{13.22e}$$

Since

$$\psi_p(n) = \frac{\partial y(n)}{\partial k_p(n)} \quad \text{and} \quad y(n) = \sum_{m=0}^{N} v_m(n) e_m^b(n)$$

Therefore,

$$\psi_p(n) = \sum_{m=0}^{N} v_m(n) \frac{\partial e_m^b(n)}{\partial k_p(n} \tag{13.22f}$$

Since

$$\beta_{p,m}(n) = \frac{\partial e_m^b(n)}{\partial k_p(n)}$$

$$\psi_p(n) = \sum_{m=0}^{N} v_m(n) \beta_{p,m} \tag{13.22g}$$

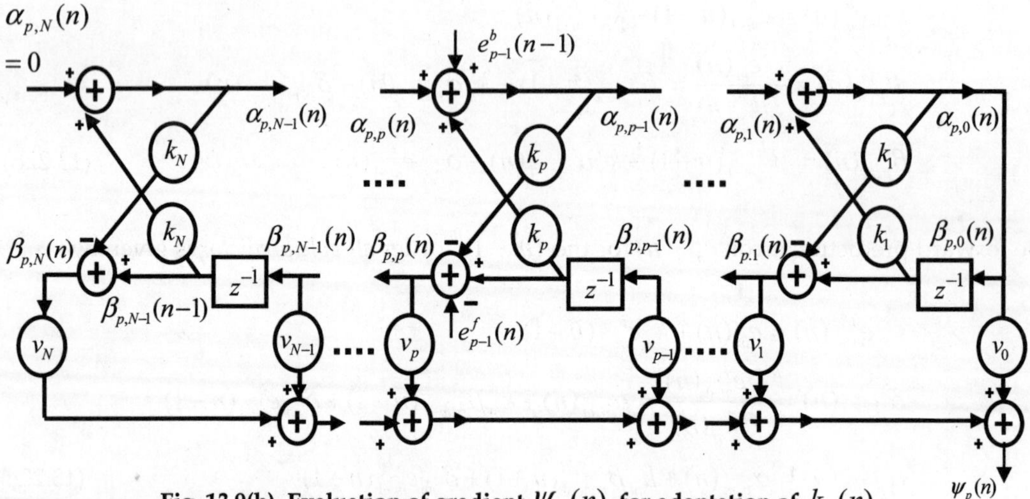

Fig. 13.9(b) Evaluation of gradient $\psi_p(n)$ for adaptation of $k_p(n)$.

The Eqns. (13.22b), (13.22c) and (13.22g)

$$\beta_{p,m}(n) = \beta_{p,m-1}(n-1) - k_m \alpha_{p,m-1}(n) - \delta_{p,m} e_{m-1}^f(n)$$

$$\alpha_{p,m-1}(n) = \alpha_{p,m}(n) + k_m \beta_{p,m-1}(n-1) + \delta_{pm} e_{m-1}^b(n-1)$$

$$\psi_p(n) = \sum_{m=0}^{N} v_m(n)\beta_{p,m}$$

determine the gradient $\psi_p(n)$ required for the adaptation of $k_p(n)$ and are realized as shown in Fig. (13.9(b). It is important to note that for $\psi_p(n)$, to get $\beta_{p,p}(n)$,$e^f_{p-1}(n)$ and to get $\alpha_{p,p-1}(n)$, $e^b_{p-1}(n-1)$ are required and are used as shown in the Fig. 139(b). *For the adaptation of each* $k_p(n),\ p = 1,2,\ldots,N$, *the above lattice scheme indicated has to be repeated and thus its computational complexity becomes* N^2 (John. J. Shynk, 1988) .

.

The IIR OE or EQE filters can be realized either by cascade or parallel structures as an alternative to direct structure. It has been found that for the parallel structure, the gradient computation is simple. On the other hand, for the cascade structure, second order sections can be used whose roots can be easily found for stability monitoring. However, these structures may result in *nonlinearity of the EE error signal with respect to adaptive filter coefficients* and hence their use is not desirable. Hence, the direct structure using lattice is the only desirable solution. The above lattice version and the implementation of OE and EQE methods using lattice structure enable the stability monitoring by checking magnitude of the reflection coefficient to be within its bound, viz., less than unity. Further, during adaptation, in case their magnitudes are not less than unity, the adaptation can be stopped (i.e., that coefficient is not adapted and its previous estimate is used or set to some preset value whose magnitude is less than unity) and restarted. Though this type of stability monitoring may not be convenient for real time implementations, it is a necessity for utilising the advantages of IIR adaptive filters. In spite of the theoretical instability suspicion, the IIR adaptive algorithms have been found to be stable for sufficiently small stepsizes. Thus, the direct implementation of either EE or OE IIR adaptive filters using lattice (two multiplier) structure is of importance.

With such a transversal filter realized by an equivalent lattice structure, as the filters used are effectively transversal filter with known input and the desired signal, it is possible to overcome the bias problem due to observation white/colour noise using Griffiths' algorithm.

13.10 LATTICE BASED EQE IIR ADAPTIVE FILTERS

It has been shown that a pole-zero transfer function can be expressed as

$$H(z) = \frac{\sum_{p=0}^{N} v_p(n) G_p^b(z)}{G_N^f(z)} = \frac{C_N(z)}{A_N(z)}$$

where $A(z) = G_N^f(z)$ is the forward prediction error filter , $B_N(z) = G_N^b(z)$ is the backward prediction error filter with $b_m^{(m)} = 1$. Further,

$$C_N(z) = \sum_{p=0}^{N} v_p G_p^b(z)$$

where $v_p(n),\ p = 0, 1, \ldots, N$ are linear combiner weights and are given by

$$C^{(m-1)}(z) = C^{(m)}(z) - v_m(n) G_m^b(z)$$
$$v_m = C_m^{(m)}$$

The relation between forward and backward prediction error filter is given by

$$G_{p-1}^f(Z) = \frac{G_p^f(Z) + k_p G_p^b(Z)}{\left[1 - k_p^2\right]}$$

If

$$G_N^f(z) = 1 + a_1^{(N)} z^{-1} + a_2^{(N)} z^{-2} + \ldots + a_N^{(N)} z^{-N} = A_N(z)$$
$$G_N^b(z) = 1 + \sum_{i=1}^{N} b_i^{(N)}(i) z^{-i} = B_N(z)$$

then

$$G_N^b(z) = Z^{-N} G_N^f(z^{-1}) = B_N(z) = Z^{-N} A_N(z^{-1})$$

$$b_i^{(N)} = a_{N-i}^{(N)} \text{ and } b_N^{(N)} = 1 \text{ or } b_m^{(m)} = 1$$

$$A_{p-1}(z) = \frac{A_p(z) + k_p B_p(z)}{\left[1 - k_p^2\right]} \quad \text{or}$$

Therefore,

$$A_{p-2}(z) = \frac{A_{p-1}(z) + k_{p-1} B_{p-1}(z)}{\left[1 - k_{p-1}^2\right]}$$

Further as $G_p^b(z) = B_p(z) = Z^{-p} A_p(z^{-1})$,

$$A_{p-2}(z)\left[1 - k_{p-1}^2\right] = A_{p-1}(z) + k_{p-1} z^{-(p-1)} A_{p-1}(z^{-1})$$

$$z^{-(p-1)} A_{p-1}(z^{-1}) = \frac{1}{k_{p-1}} \left\{ A_{p-2}(z)\left[1 - k_{p-1}^2\right] - A_{p-1}(z) \right\} \tag{13.23a}$$

Multiplying by $k_p z^{-1}$ both sides

$$k_p z^{-1} z^{-(p-1)} A_{p-1}(z^{-1}) = \frac{k_p}{k_{p-1}} \left\{ A_{p-2}(z)\left[1 - k_{p-1}^2\right] - A_{p-1}(z) \right\} z^{-1}$$

$$k_p z^{-p} A_{p-1}(z^{-1}) = \frac{k_p}{k_{p-1}}\left\{A_{p-2}(z)\left[1-k_{p-1}^2\right] - A_{p-1}(z)\right\}z^{-1} \tag{13.23b}$$

Since

$$A_{p-1}(z) = \frac{A_p(z)+k_p B_p(z)}{\left[1-k_p^2\right]} = \frac{A_p(z)+k_p A_p(z^{-1})z^{-p}}{\left[1-k_p^2\right]}$$

Replacing z by z^{-1},

$$A_{p-1}(z^{-1}) = \frac{A_p(z^{-1})+k_p A_p(z)z^{p}}{\left[1-k_p^2\right]}$$

Substituting this value in Eqn.(13.23b),

$$k_p z^{-p}\left\{\frac{A_p(z^{-1})+k_p A_p(z)z^{p}}{\left[1-k_p^2\right]}\right\} = \frac{k_p}{k_{p-1}}\left\{A_{p-2}(z)\left[1-k_{p-1}^2\right] - A_{p-1}(z)\right\}z^{-1}$$

$$k_p\left\{\frac{A_p(z^{-1})z^{-p}+k_p A_p(z)}{\left[1-k_p^2\right]}\right\} = \frac{k_p}{k_{p-1}}\left\{A_{p-2}(z)\left[1-k_{p-1}^2\right] - A_{p-1}(z)\right\}z^{-1} \tag{13.23c}$$

Further adding and subtracting $A_p(z)/k_p$ to the numerator of LHS,

$$LHS = \frac{k_p}{\left[1-k_p^2\right]}\left\{A_p(z^{-1})z^{-p}+k_p A_p(z)+\frac{A_p(z)}{k_p}-\frac{A_p(z)}{k_p}\right\}$$

$$= k_p\left\{\frac{A_p(z^{-1})z^{-p}}{\left[1-k_p^2\right]}+\frac{k_p A_p(z)}{\left[1-k_p^2\right]}+\frac{A_p(z)}{k_p\left[1-k_p^2\right]}-\frac{A_p(z)}{k_p\left[1-k_p^2\right]}\right\}$$

$$= k_p\left\{\frac{k_p A_p(z^{-1})z^{-p}+A_p(z)}{k_p\left[1-k_p^2\right]}+\frac{k_p^2 A_p(z)-A_p(z)}{k_p\left[1-k_p^2\right]}\right\}$$

$$LHS = \left\{\frac{A_p(z)+k_p A_p(z^{-1})z^{-p}}{\left[1-k_p^2\right]} - A_p(z)\right\}$$

But

$$A_{p-1}(z) = \frac{A_p(z)+k_p A_p(z^{-1})z^{-p}}{\left[1-k_p^2\right]}$$

Using this,

$$LHS = \{A_{p-1}(z) - A_p(z)\}$$

Therefore, Eqn. (13.23c) becomes,

$$\{A_{p-1}(z) - A_p(z)\} = \frac{k_p}{k_{p-1}}\{A_{p-2}(z)[1-k_{p-1}^2] - A_{p-1}(z)\}z^{-1}$$

$$A_p(z) = A_{p-1}(z) - \frac{k_p}{k_{p-1}}\{A_{p-2}(z)[1-k_{p-1}^2] - A_{p-1}(z)\}z^{-1} \tag{13.23d}$$

Also the linear combining weights v_m are given by

$$C^{(p-1)}(z) = C^{(p)}(z) - v_p G_p^b(z)$$

Substituting for $G_p^b(z)$,

$$C^{(p)}(z) = C^{(p-1)}(z) + v_p A_p(z^{-1})z^{-p} \tag{13.23e}$$

From Eqn. (13.23a),

$$z^{-(p-1)}A_{p-1}(z^{-1}) = \frac{1}{k_{p-1}}\{A_{p-2}(z)[1-k_{p-1}^2] - A_{p-1}(z)\}$$

Hence,

$$z^{-p}A_p(z^{-1}) = \frac{1}{k_p}\{A_{p-1}(z)[1-k_p^2] - A_p(z)\}$$

Substituting this in Eqn.(13.23e),

$$C^{(p)}(z) = C^{(p-1)}(z) + \frac{v_p}{k_p}\{A_{p-1}(z)[1-k_p^2] - A_p(z)\} \tag{13.23f}$$

and Eqn.(13.23d)

$$A_p(z) = A_{p-1}(z) - \frac{k_p}{k_{p-1}}\{A_{p-2}(z)[1-k_{p-1}^2] - A_{p-1}(z)\}z^{-1}$$

These equations (13.23d) and (13.23f) are represented in a slightly different way for the adaptive case and for convenience,

$$A_p(z,n) = A_{p-1}(z,n) - \frac{k_p(n)}{k_{p-1}(n)}\{A_{p-2}(z,n)[1-k_{p-1}^2(n)] - A_{p-1}(z,n)\}z^{-1}$$

$$C_p(z,n) = C_{p-1}(z,n) + \frac{v_p(n)}{k_p(n)}\{A_{p-1}(z,n)[1-k_p^2(n)] - A_p(z,n))\}$$

for $p = 2, 3, \ldots, N$.

with

$$A_0(z,n) = 1,\ A_1(z,n) = 1 - k_1(n)z^{-1},$$

$$C_1(z,n) = \{v_0(n) - v_1(n)k_1(n)\} + v_1(n)z^{-1}$$

Using these equations, the EE error signal can be calculated as

$$e_{EE}(n) = d(n) - y(n)$$

$$e_{EE}(n) = d(n) - \sum_{k=1}^{N} a_k(n)d(n-k) - \sum_{l=0}^{N} c_l(n)x(n-l)$$

$$\theta(n+1) = \theta(n) + \mu\, e_{EE}(n)\, \varphi_E(n)$$

$$\theta(n) = [k_1(n), k_2(n), \ldots k_N(n), v_0(n), v_1(n), \ldots v_N(n)]^T$$

$$\varphi_E(n) = [d(n-1), d(n-2), \ldots d(n-N), \quad x(n), x(n-1), \ldots x(n-N)]^T$$

It is important to note that for the adaptation of $k_i(n), i = 1,2,\ldots,N,$ the inputs used are $d(n-i), i = 1,2,\ldots,N$ (unlike the forward and backward errors used in conventional lattice algorithm) and the error used is $e_{EE}(n)$. Further, for the adaptation of $v_i(n), i = 1.2.\ldots, N,$ the inputs used are $x(n-i), i = 0,1,2.\ldots, N$ and the error used is $e_{EE}(n)$. This is based on the valid assumption that the IIR lattice and the direct form of IIR are equivalent in the sense that both methods have same stationary points corresponding to same input-output descriptions of the adaptive filter and this also implies that the residual errors are equal (S.L. Netto and P. Agathoklis, 1998)

Though joint lattice, with IIR lattice for the denominator can be used in realising the IIR (pole-zero) adaptive filter but it suffers from bias in the presence of observation noise and in this structure, it is not possible to use the Griffiths' algorithm as the actual inputs to the lattice structure are different from those used for adaptation.

The lattice EE adaptive algorithm:

Step 1: Initialize
The convergence parameter μ
The adaptive filter coefficient vector: $\theta(n) = [k_1(n), k_2(n), \ldots k_N(n), v_0(n), v_1(n), \ldots v_N(n)]^T$
The auxiliary vectors

$$Ax^T = [a_0(z,n) \ldots, a_{N-1}(z,n)],\quad Ad^T = [a_0(z,n) \ldots, a_{N-1}(z,n)]$$

$$Atx^T = [a_1(z,n) \ldots a_N(z,n)],\quad Atd^T = [a_1(z,n) \ldots a_N(z,n)]$$

Step 2: Compute

$$Atd_p(z,n) = Atd_{p-1}(z,n) - \frac{k_p(n)}{k_{p-1}(n)}\left[Ad_{p-2}(z,n)\left[1-k_{p-1}^2(n)\right] - Ad_{p-1}(z,n)\right]$$

for. $p = 2, 3, \ldots, N$ with

$$Atd_0(z,n) = d(n),\ Atd_1(z,n) = d(n) - k_1(n)d(n-1),$$

$$Atx_p(z,n) = Atx_{p-1}(z,n) - \frac{k_p(n)}{k_{p-1}(n)}\left[Ax_{p-2}(z,n)\left[1-k_{p-1}^2(n)\right] - Ax_{p-1}(z,n)\right]$$

for. $p = 2, 3, \ldots, N$ with

$$Atx_0(z,n) = x(n), \quad Atx_1(z,n) = x(n) - k_1(n)x(n-1),$$

$$Cx_p(z,n) = Cx_{p-1}(z,n) + \frac{v_p(n)}{k_p(n)}\left\{ Atx_{p-1}(z,n)\left[1-k_p^2(n)\right] - Atx_p(z,n) \right\}$$

for . $p = 2, 3, \ldots, N$ with

$$Cx_1(z,n) = \left\{ v_0(n) - v_1(n)k_1(n) \right\}\left\{x(n)\right\} + v_1(n)\left\{x(n-1)\right\}$$

Step 3 : Obtain

$$e_{EE}(n) = Atd_N(z,n) - Cx_N(z,n)$$

Step 4: Form

$$\phi_E(n) = [d(n-1), d(n-2), \ldots d(n-N), x(n), x(n-1), \ldots x(n-N)]^T$$

Step 5: Update

The adaptive filter coefficient vector:

$$\theta(n+1) = \theta(n) + \mu\, e_{EE}(n)\, \varphi_E(n)$$

If $|\, k_i(n) \,| \geq 1,\ i = 1, 2, 3, \ldots, N$

$$\theta(n+1) = \theta(n)\,; \quad e_{EE}(n) = e_{EE}(n-1)$$

else

the auxiliary vectors defined in step -1

$$Ax_{p-1}(z,n) = Atx_p(z,n)\ p = 1, 2, 3, \ldots, N$$

$$Ad_{p-1}(z,n) = Atd_p(z,n)\ p = 1, 2, 3, \ldots, N$$

$$\theta(n) = \theta(n+1)$$

Assign $k_i(n)$ and $v_i(n)$ values

Update the time counter *n*, return to step 2

In this treatment the transfer function polynomials of the direct form are related to the lattice structure. This method requires only $O(N)$ computations (*N* is the order of the filter) unlike the other method based on gradients which requires $O(N^2)$ computations. Further, the stability can be monitored by checking whether $|k_i(n)| < 1,\ i = 1, 2, \ldots, N$.

However, this suffers from observation noise problem as will be shown by the example that will be considered.

Example : To illustrate this method the system

$$P(z) = \frac{0.0154 + 0.0462z^{-1} + 0.0462z^{-2} + 0.0154z^{-3}}{1 - 1.99z^{-1} + 1.572z^{-2} - 0.4583z^{-3}}$$

driven by zero mean, unit variance white Gaussian noise is considered [Netto,]. For this system N=3 and step size used is $\mu = 0.06$. For this system, the parameter vector from the system transfer function,

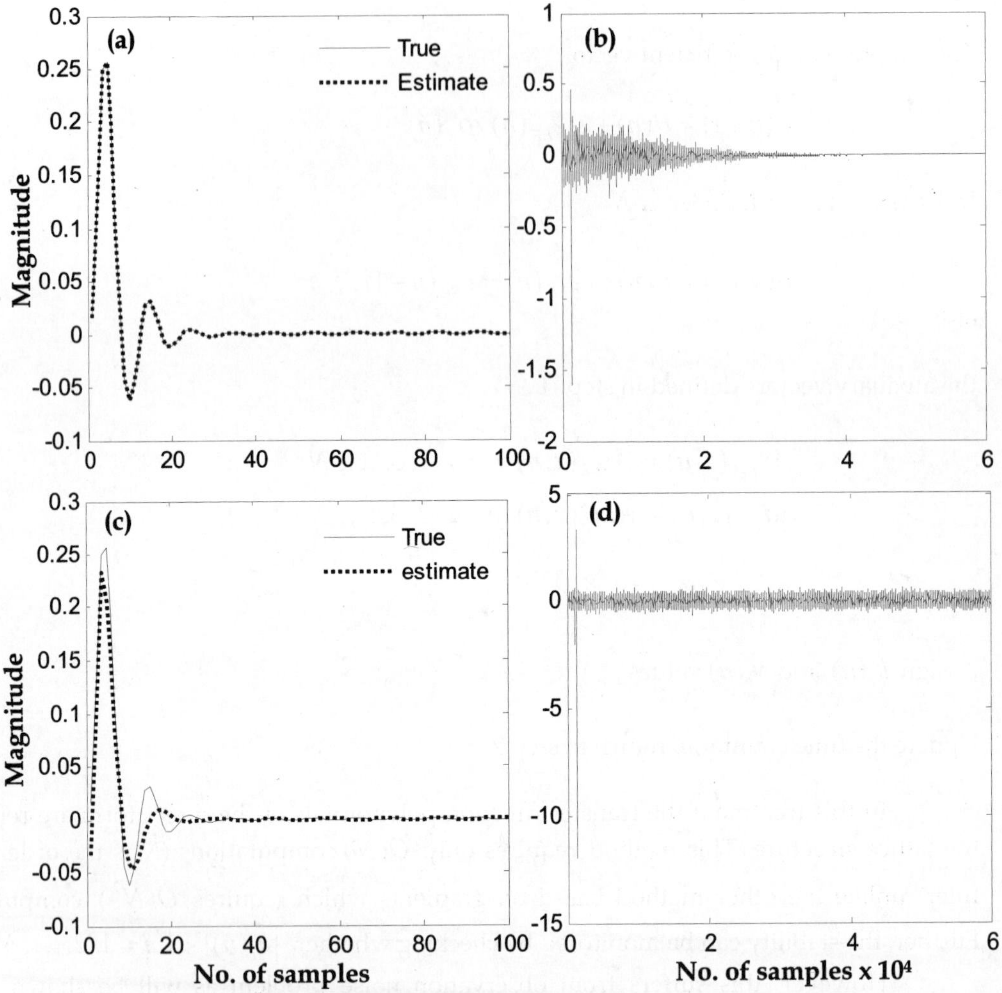

Fig. 13.10 Lattice equation error IIR adaptive system identification, (a, b) without observation noise: (a) Impulse response estimate; (b) Error convergence; (c, d) with observation noise (SNR=15 dB) : c) Impulse response estimate, (d) Error convergence.

$$[a_1(n), a_2(n), a_3(n), c_0(n), c_1(n), c_2(n), c_3(n)]^T$$
$$= [-1.9900, 1.5720, -0.4583, 0.0154, 0.0462, 0.0462, 0.0154]$$

The corresponding vector is (the method of getting $k_i(n), i = 1, 2, \ldots, N$ and $v_i(n), i = 0, 1, 2, \ldots, N$ from numerator and denominator polynomial is explained in chapter 7)

$$[k_1(n), k_2(n), k_3(n), v_0(n), v_1(n), v_2(n), v_3(n)]^T$$
$$= [0.8756, -0.8355, 0.4583, 0.0856, 0.1455, 0.0768, 0.0154]$$

It is seen that the impulse response of the system obtained from $k_i(n), i = 1, 2, \ldots, N$ and $v_i(n), i = 0, 1, 2, \ldots, N$ is shown in the Fig. 13.10. In the absence of observation noise, the algorithm converges and the error decreases significantly (Fig. 13.10(b)) hence, there is a good match between estimated and ideal impulse response (Fig. 13.10(a)). However, in the presence of white Gaussian observation noise (SNR=15dB), though the algorithm converges (Fig. 13.10d), the estimated impulse response deviates from the ideal (Fig.13.10(c)) and this is due to bias resulting from the observation white noise in the desired signal which forms one of the component of the input to the adaptation of the reflection coefficients.

In view of the above discussion, it is preferable to use unbiased EQE or the unbiased SM method for IIR adaptive filter application. Between the two, as *further the SM method as it minimizes the output error, the unbiased SM is better than unbiased EQE method.* Further, the numerator and denominator polynomials can be realized by FIR filters. However, the stability can be ensured by computing $k_i(n), i = 1, 2, \ldots, N$ for the denominator polynomial and checking for $|k_i(n)| < 1, i = 1, 2, \ldots, N$ for the adaptation of the denominator polynomial to be valid and in case this is not satisfied, then this adaptation is skipped and the values of $k_i(n), i = 1, 2, \ldots, N$ and $v_i(n), i = 0, 1, 2, \ldots, N$ for the previous iteration are preserved for the subsequent iteration. In most of the cases, $|k_i(n)| < 1, i = 1, 2, \ldots, N$, will be satisfied. Generally, the implementation of IIR adaptive algorithm can be realized by using tapped delay line structure with stability monitoring and the lattice structure is not necessary unless the round off errors due to limited number bits used for number representation by the processor becomes significant. In such numerical stability cases, the tapped delay line filter can be implemented by lattice for filtering purposes.

Chapter 14

BLOCK/FREQUENCY DOMAIN LMS ALGORITHM

In scenarios like echo cancellation where the impulse response will be very long (greater than 1000 samples) the time domain convolution involved in estimating the output estimate demands high computational complexity. However, the frequency domain convolution based on FFT needs significantly less computational load. But such an efficient convolution requires block processing and hence, the existence of block LMS algorithm (BLMS). In BLMS, the adaptive filter coefficients are adapted on block-to-block rather than sample-to-sample basis. Initially, the time domain BLMS is introduced. As the block of data is available, it can be exploited for averaging over the samples of the block, to reduce the misadjustment. For the same step size, compared to LMS, BLMS has a lower convergence error by a factor of block size and for the same misadjustment, the convergence rate of BLMS and LMS are same.

The BLMS can be implemented in the frequency domain using FFT and is known as frequency domain BLMS (FBLMS). Here, the output estimate, the convolution of the input with the adaptive filter is realized by FFT using the overlap and save method. To compute the gradient, the FT of the input block can be multiplied by using the FT of the error of same size. Further, as the input is in frequency domain, it is orthogonalized and can be used to improve the speed of convergence by scaling the step size by the power of each FT bin of input. As the size of the FFT used in finding the gradient is more than the filter order required, the extra filter length has to be set to zero to reduce the errors and this is achieved by taking the IFFT of the gradient setting excess length to zero and again taking the FFT of the modified time domain gradient sequence and this is known as constrained FBLMS algorithm.

The FBLMS though attractive from computational and convergence performance point of view, there is a delay corresponding to the long impulse response length as there is waiting for the input to form a block of same size. To reduce this delay, the adaptive filter is portioned so that only input block length will be same as that of the filter partition and is known as portioned FBLMS (PFBLMS). Further, attempt has been made to eliminate the delay completely by resorting to the time domain LMS for the first partition of data and is known as delayless PFBLMS. The performance for the first partition and of other partitions can be different as they are realized in different domains.

Further, the same block of data available can be recycled (iterated) to improve the convergence performance as if they are similar to regular input blocks and this is known as Iterated FBLMS algorithm.

The FBLMS algorithm assumes stationarity over the block length hence, valid for stationary or slow nonstationary signals.

14.1 INTRODUCTION

In the earlier chapters, the LMS algorithm and its performance improvement were considered. The convergence rate was improved by using orthogonalization using unitary transforms or Gram- Schmidt orthogonalizers like regular and escalator lattice algorithms. However, this improvement may result in higher misadjustment as higher step sizes are permitted for different orthogonal components. The forward-backward LMS predictor had an improved misadjustment by a factor of 2 but had the same speed of convergence as that of LMS predictor which used only the forward prediction error. Further, the use of variable step size for LMS algorithm and its extension to transform domain, enabled to achieve both convergence rate and the misadjustment. However, these algorithms are computationally expensive compared to the transversal LMS algorithm. Hence, there is a necessity to have adaptive algorithms, which possess good convergence rate, good misadjustment and are also computationally efficient. In situations where, the signal is not varying very fast and can be considered stationary over small segments, the LMS adaptation instead of from sample to sample can be extended to block to block. Here, instead of considering the instantaneous gradient, an averaged gradient over the block can be considered. The use of such averaged gradient reduces the misadjustment by a factor corresponding the block length, which is much higher than the adaptive algorithms considered could achieve. The block LMS can be realized in the frequency domain due to block nature of the data and this also enables faster convergence rate as adaptation is done in frequency domain, which is similar to transform domain operation. Further, the block nature of the data enables filtering by block convolution using computationally efficient Fast Fourier Transform algorithm. In view of this, often the BLMS realized in frequency domain is referred to as fast LMS algorithm. As the BLMS provides all the desirable features, it can be advantageously used in situations where its application is permissible.

14.2 BLOCK LMS ALGORITHM (BLMS)

The LMS algorithm which adapts the filter weights sequentially on sample-to-sample basis is given by the following relations. The error is given by

$$e(n) = d(n) - y(n) = d(n) - \sum_{i=0}^{N-1} w_i \; x(n-i) \tag{14.1a}$$

The error gradient $\nabla_i(n)$ is

$$\nabla_i(n) = \frac{d[e^2(n)]}{dw_i} = -2e(n)x(n-i) \tag{14.1b}$$

The adaptation of the weight $w_i(k)$ by

$$w_i(n+1) = w_i(n) - \mu \, \nabla_i(n)$$

Therefore,

$$w_i(n+1) = w_i(n) + \alpha\, e(n)x(n-i), \;\; \alpha = 2\mu,\; i = 0,1,2,\ldots N-1 \tag{14.1c}$$

N is the adaptive filter length and α is the adaptation step size.

Similar to the LMS instantaneous gradient $\nabla_i(n)$ for the ith element of the weight vector $w_i(n)$, at the nth **sample**; for the BLMS with the block length of L samples, for the ith element of the weight vector for the nth **block**, the required BLMS gradient $\nabla_i\varepsilon$ is given by

$$\nabla_i\varepsilon = \frac{1}{L}\sum_{k=0}^{L-1} x(nL+k-i)\, e(nL+k) \tag{14.1d}$$

This gradient is *not instantaneous* like in LMS algorithm, but it is the *average of L such LMS gradients corresponding to L samples* of the nth block. The gradient estimation can be interpreted as the *cross-correlation between the error signal block with the input signal block with a lag corresponding to filter weight indices.* Hence, for the BLMS algorithm, the weight update *once in every block* of data is

$$w_i(n+1) = w_i(n) + \frac{\alpha}{L}\sum_{k=0}^{L-1} x(nL+k-i)e(nL+k).\;\; i = 0,1,2,\ldots N-1 \tag{14.1e}$$

α is the step size parameter. Comparison of Eqn. (14.1c) and (14.1e) shows that for the weight adaptation, *the gradient noise for the BLMS is reduced by a factor* L, the number of samples constituting the nth block.

14.2.1 Matrix Representation of BLMS Algorithm

The gradient vector for BLMS is,

$$\nabla\varepsilon = \left[\nabla_0\varepsilon, \nabla_1\varepsilon, \ldots \nabla_{N-1}\varepsilon\right]^T \tag{14.2a}$$

where, $\nabla_i\varepsilon$ is the block gradient for the ith weight and is given by Eqn. (12.1d),

$$\nabla_i\varepsilon = \frac{1}{L}\sum_{k=0}^{L-1} e(nL+k)x(nL+k-i)$$

This gradient vector can be expressed as

$$\nabla\varepsilon = X(nL)^T E(nL) \tag{14.2b}$$

$X(nL)$ is a $L\times N$ matrix, $E(nL)$ is a $L\times 1$ column vector and are given by

$$X(nL) = \begin{bmatrix} x(nL) & x(nL+1) & \cdots & \cdots & x(nL+N-1) \\ x(nL-1) & x(nL) & \cdots & \cdots & x(nL+N-2) \\ \vdots & \vdots & \cdots & \cdots & \vdots \\ x(nL-L+1) & x(nL-L+2) & \cdots & \cdots & x(nL+N-L) \end{bmatrix}, \tag{14.2c}$$

and

$$E^T(nL) = \left[e_{nL} \quad e_{nL+1} \qquad \ldots e_{nL+L-1} \right] \tag{14.2d}$$

with

$$e(nL+k) = d(nL+k) - y(nL+k), \quad k = 0,1,2,\ldots L-1, \tag{14.2e}$$

$$y(nL+k) = \sum_{i=0}^{N-1} w_i(n)x(nL+k-i), \quad n = 0,1,2,\ldots L-1 \tag{14.2f}$$

The vector form of update equation is

$$W(n+1) = W(n) + \frac{\alpha}{L} X(nL)^T \, E(nL) \tag{14.2g}$$

14.2.2 Misadjustment and convergence rate of BLMS

It was mentioned that the gradient used for the BLMS algorithm is the average of the L instantaneous gradients used in the LMS algorithm and further, the comparison of Eqns. (12.1c) and (12.1e) indicated that the gradient noise for the BLMS is reduced by a factor L. This implies that the misadjustment must be less, as it is dependent on the gradient noise which is due to the step size. This is true and will be explained.

For the LMS, the misadjustment is (Section-10.4.1, Eqn. (10. 17),

$$M \approx \alpha N \sigma_x^2$$

For the block LMS, the averaging of the gradient over L samples of the block *will reduce its noise or the step size* by the same factor L. Considering the mechanism involved in both LMS and the BLMS are similar except for block-to-block adaptation and the averaging effect in the latter, its misadjustment M_B can be given by an expression similar to that of LMS by

$$M_B \approx \frac{\alpha}{L} N \sigma_x^2 \tag{14.2h}$$

This implies that for the same step size, the BLMS is having a better misadjustment than that of the LMS algorithm. In other words, *for the same misadjustment,* the BLMS can have a step size L times larger than the LMS and its stepsize α_B is given by

$$\alpha_B = L\alpha \tag{14.2i}$$

For the LMS, the time constant, τ_i associated with the ith eigenvalue (eigen mode) is given by

$$\tau_i = \frac{1}{\alpha \, \lambda_i} \tag{14.2j}$$

Therefore, for the BLMS algorithm, the corresponding time constant, τ_{Bi} is given by

$$\tau_{Bi} = \frac{1}{\alpha_B \lambda_i} \tag{14.2k}$$

Here τ_{Bi} *is in terms of block interval.* Substituting for α_B form Eqn. (14.2i),

$$\tau_{Bi} = \frac{1}{L\alpha\,\lambda_i}$$

Since there are L samples in one block, the *time constant in terms of sample intervals* is given by

$$\tau_{Bi} = \left(\frac{1}{L\alpha\,\lambda_i}\right).L$$

$$\tau_{Bi} = \frac{1}{\alpha\,\lambda_i} \qquad (14.2l)$$

Then comparing Eqn. (14.2*l*) with Eqn. (14.2k), the time constant τ_{Bi} has the same value as that of LMS algorithm (in terms of sample intervals). *Hence for the same misadjustment (convergence error), convergence rate of BLMS is same as that of the LMS algorithm.* This is illustrated for a system identification example. The impulse response considered is that of an acoustic path of length 128 samples (Fig. 14 (c, d)). The input is zero mean unit variance white Gaussian noise. The observation noise added at the system output is a zero mean unit variance white Gaussian noise to get an SNR=20dB. The LMS algorithm used is of fixed step size with its value 0.1. For the block LMS, the block size L used is 10 samples and hence to get same misadjustment as that of LMS, the stepsize used is L times 0.1, equal to 1. The smoothed ensemble averaged square error (SEASE) is used as performance index. The ensemble averaging is done over 50 trials and the averaged square error $\xi(n)$ is smoothed by a recursive lowpass filter $\xi_S(n) = \lambda\,\xi_s(n-1) + (1-\lambda)\xi(n)$ with $\lambda = 0.99$.

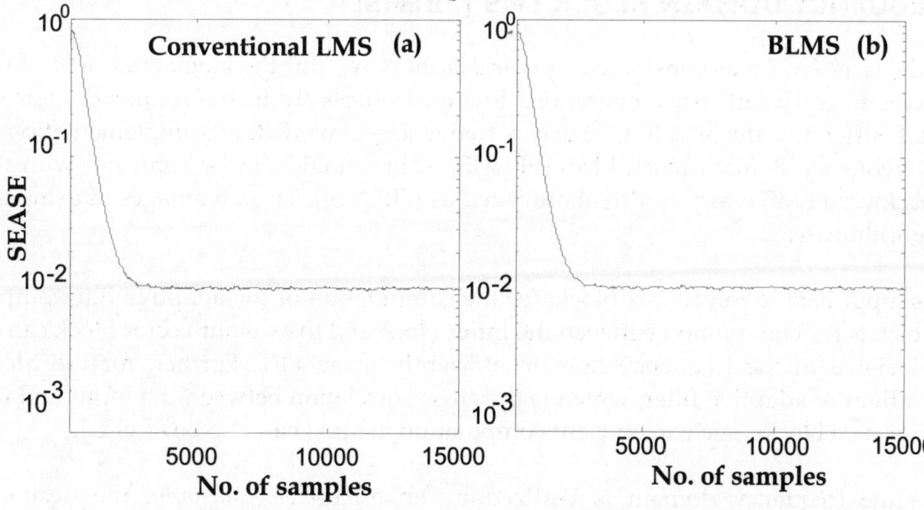

Fig. 14.1 Comparison of the convergence of conventional LMS and block LMS for the same misadjustment. SEASE for (a) Conventional LMS, (b) Block LMS.

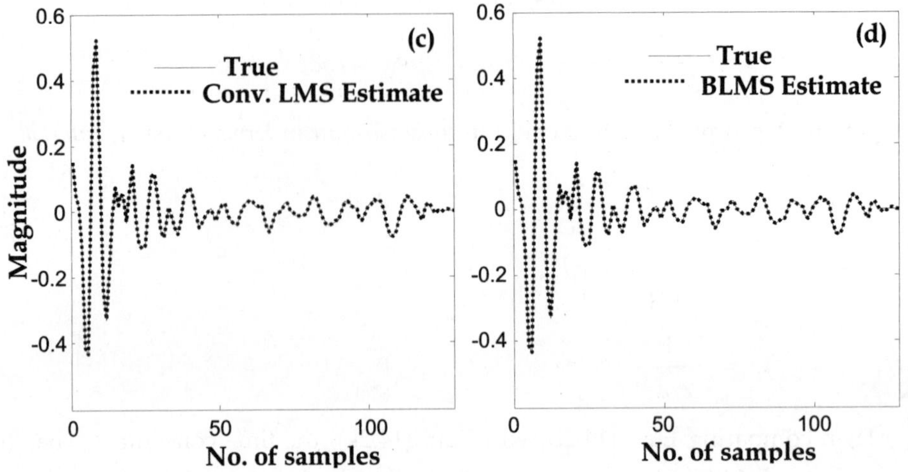

Fig. 14.1(c, d) Comparison of impulse response estimates (c) Conventional LMS and (d) by block LMS.

It is seen that from Fig. 14.1(a, b), for both conventional LMS and BLMS the SEASE plots are almost identical both in terms of convergence rate and convergence error. The convergence takes place at about 4000 samples and as the SEASE plots being constant after convergence, they are shown up to 15,000 samples. Further, the impulse response estimate obtained by the two methods (Fig. 14.1(c, d) computed after 50,000 samples) have similar match to the true impulse response and this indicates that the misadjustments for both cases are same.

14.3 FREQUENCY DOMAIN BLOCK LMS (FBLMS)

Till now, the *block LMS* was considered in *time domain* only. But the block treatment of the input enables to carry out many operations involved efficiently in the frequency domain using fast Fourier transform (FFT). Such a frequency-domain BLMS implementation is known as frequency domain block LMS (FBLMS). (This should not be confused with the forward-backward LMS which is also abbreviated as (FBLMS)). The advantages in using the FBLMS algorithm are:

1. As input data involved is a block, for the computation of the adaptive filter output which is the convolution between the input block and the weight vector block can be calculated in the frequency domain efficiently using FFT. Further, for the **block gradient** of adaptive filter, which is the cross-correlation between the input and the error data blocks, such an efficient computation method can also be applied.

2. As the frequency domain is realized by an *orthogonal transform*, the desirable properties of the transform can be advantageously utilized.

The adaptive filter output computation using the filter impulse response as a block and the input data block by FFT is straightforward. However, the computation of the gradient is not so and hence will be considered. The BLMS weight vector adaptation by gradient method in time domain is given by (Eqn. 14.1e)

$$\begin{bmatrix} w_0(n+1) \\ w_1(n+1) \\ \vdots \\ w_{N-1}(n+1) \end{bmatrix} = \begin{bmatrix} w_0(n) \\ w_1(n) \\ \vdots \\ w_{N-1}(n) \end{bmatrix} + \alpha \begin{bmatrix} (1/L)\sum_{k=0}^{L-1} e(nL+k)x(nL+k) \\ (1/L)\sum_{k=0}^{L-1} e(nL+k)x(nL+k-1) \\ \vdots \\ (1/L)\sum_{k=0}^{L-1} e(nL+k)x(nL+k-N+1) \end{bmatrix} \tag{14.3a}$$

Let the cross-correlation for the nth block be given by

$$r(n,i) = \frac{1}{L}\sum_{k=0}^{L-1} e(nL+k)x(nL+k-i) \tag{14.3b}$$

Then

$$\begin{bmatrix} w_0(n+1) \\ w_1(n+1) \\ \vdots \\ \vdots \\ w_{N-1}(n+1) \end{bmatrix} = \begin{bmatrix} w_0(n) \\ w_1(n) \\ \vdots \\ \vdots \\ w_{N-1}(n) \end{bmatrix} + \alpha \begin{bmatrix} r(n,0) \\ r(n,1) \\ \vdots \\ \vdots \\ r(n,N-1) \end{bmatrix} \tag{14.3c}$$

If

$$W_f(n) = F[W(n)]$$

Then taking the Fourier transforms on both sides of Eqn. (14.3c),

$$W_f(n+1) = W_f(n) + \alpha F\left[\{r(n,0),\, r(n,1),\ldots\ldots r(n,N+L-1)\}\right]$$

$$W_f(n+1) = W_f(n) + \alpha\, F\left\{F^{-1}\left[X_f^*(n)E(n)\right]\right\}$$

or

$$W_f(n+1) = W_f(n) + \alpha\, X_f^*(n)E_f(n) \tag{14.3d}$$

where

$$X_f(n) = F[X(n)],\;\; E_f(n) = F[E(n)]$$

Hence, $\{r(n,0),\, r(n,1),\ldots r(n,L+N-2)\}$ cross-correlation vector between the input and the error can be obtained in frequency domain using FFT by taking the inverse *FT* of the product of the conjugate of *FT* of the input data block and *FT* of the corresponding error data block. Thus, the above derivation enables the implementation of BLMS algorithm in the frequency domain.

The reduction in computation of the output of adaptive filter is possible if *FFT* length is more than 64 that is if $N+L-1 > 64$. Further for the reduction in computation, the block length should be comparable to that of the filter length of adaptive filter. To take care of time domain aliasing, the lengths of $W_f(n)$ and $X_f(n)$ should be $(L+N-1)$, as lengths of $W(n)$ and $X(n)$ are N and L respectively. The frequency domain computation can be done by either *overlap and save method* or *overlap and add method.* But for BLMS, as overlap and save method is efficient, hence it is preferred.

14.3.1 Overlap and Save Method

In this method, input samples are divided into blocks in which the nth block data is having an overlap of $(N-1)$ points with $(n-1)^{th}$ block and the L data points are new. The initial $(N-1)$ points of convolved output are incorrect and only the later L points are valid and hence, later L points of each block are concatenated to get the correct convolved output $y(n)$.

Hence the output of the adaptive filter for the nth block is given by

$$y_n(k) = \sum_{i=0}^{N-1} w_i(n)x(k-i) = Last\ L\ values\ F^{-1}\left[W_f(n)X_f(n)\right] \tag{14.3e}$$

The error is obtained *in time domain* by taking the difference between the desired signal and the output,

$$e(n) = d(n) - y(n)$$

For the block operation, the L output points obtained per block are compared with the corresponding L points of $d(n)$ to get a L point error sequence. But as it is required to do the multiplication in frequency domain with the conjugate of the FT of the input, the error block should be made same as that of the input, i.e., $(L+N-1)$ by padding $(N-1)$ zeros initially to the L error points.

14.3.2 Improving Convergence Rate by the Frequency Domain Orthogonalization

As studied in the previous chapter, to improve the convergence rate, input orthogonalization and adaptation step size scaling are essential. The input orthogonalization, readily available in the frequency domain implementation of BLMS, can be used. Further, by scaling the step size by the input power corresponding to that frequency bin, the convergence rate can be improved.

The adaptation of the ith bin weight at the nth block is given by (14.3d)

$$W_{fi}(n+1) = W_{fi}(n) + \alpha_i(n)\, X^*_{fi}(n)E_{fi}(n) \tag{14.3f}$$

Here, $W_{fi}(n)$, $X_{fi}(n)$ and $E_{fi}(n)$ correspond to the ith bin of the Fourier transforms of weight vector, input and the error, respectively.

The step size $\alpha_i(n)$ to be used for the ith bin is given by

$$\alpha_i(n) = \frac{\alpha}{\sigma^2_{xf\,i}(n)}, \quad 0 < \alpha < 1 \tag{14.3g}$$

The input power for the LMS adaptive filter for the ith bin, $\sigma^2_{xf\,i}(n)$ is given by

$$\sigma^2_{xf\,i}(n) = \beta\, \sigma^2_{xf\,i}(n-1) + (1-\beta)\left|X_{f\,i}(n)\right|^2, \quad 0 < \beta < 1 \tag{14.3h}$$

It is important to note that for the adaptation of the ith bin weight $W_{fi}(n)$, only the ith component of the transform of error $E_{fi}(n)$ is used and not the total error $e(n)$. If the total error is used then the adaptation becomes

$$W_{fi}(n+1) = W_{fi}(n) + \alpha_i(n)\, X^*_{fi}(n) E_{fi}(n) + \sum_{l\neq i} X^*_{fi}(n) E_{fl}(n)$$

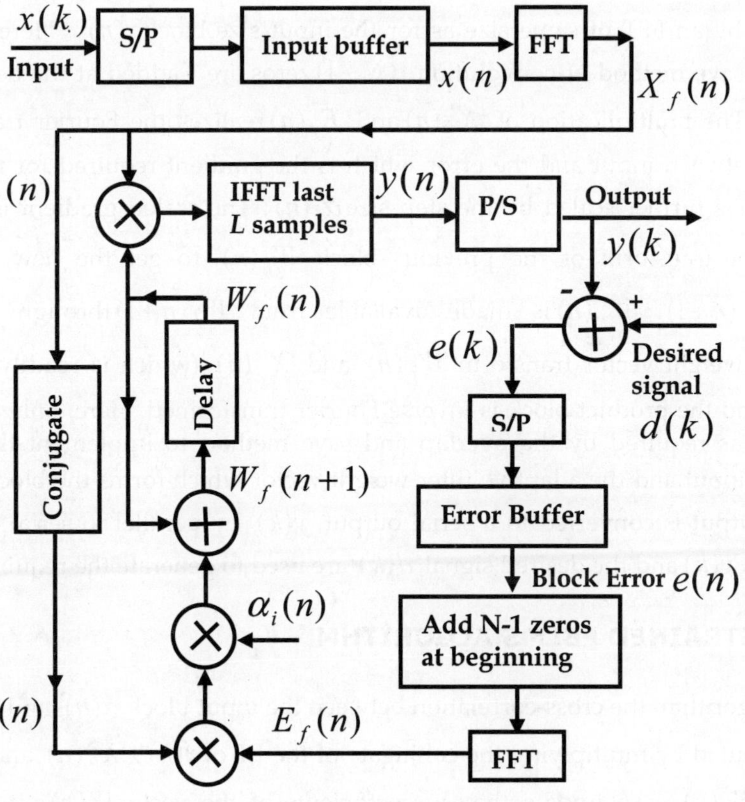

Fig. 14.2 Implementation of BLMS algorithm.

The summation term may become zero in the expectation (average) sense provided $X_{fi}(n)$ and $E_{fl}(n), l \neq i$ are uncorrelated. (This situation is similar to the transform domain LMS algorithm considered without the desired signal decomposition.) However, in the absence of such an uncorrelated property and expectation operation, the summation term is not zero. However, such a nonzero term plays the role of an observation noise hindering the adaptation of that particular bin weight towards convergence and in the worst case may lead to divergence also. Due to the error being decomposed by the transform in the FBLMS, it is possible to use the required bin of the transform of error, i.e., for the ith bin weight adaptation only the ith component of the transform of the error can be used and this enables faster adaptation towards convergence.

The implementation of the BLMS algorithm is shown in Fig. 14.2 The $x(k)$ input L samples are stored in the input buffer to get the block input $x(n)$ and is transformed by an FFT of size $(N+L-1)$ and conjugated to get $X_f^*(n)$. Also L samples of the error $e(k)$ derived as the difference between the adaptive filter output $y(k)$ and the desired signal $d(k)$ is stored in the error buffer to get the block error $e(n)$. This error block $e(n)$ is transformed by an FFT of same size as for the input size block $x(n)$. Here, to satisfy the overlap and save method of convolution, $(N-1)$ zeros are added at the beginning of the error block. The multiplication of $X_f^*(n)$ and $E_f(n)$ realizes the Fourier transform of the correlation between input and the error which is the gradient required for the adaptation. The gradient is further scaled by the step size $\alpha_i(n)$. The scaled gradient is added to the weight vector transform of the previous block $W_f(n)$ to get the new weight vector transform $W_f(n+1)$. $W_f(n)$ is made available from $W_f(n+1)$ through a delay unit. Further, the weight vector transform $W_f(n)$ and $X_f(n)$ (which is readily available) are multiplied and the product block is inverse Fourier transformed. Here, only last L samples are selected as required by the overlap and save method to implement the convolution between the input and the adaptive filter weight vector which forms the block output $y(n)$. The block output is converted to a serial output $y(k)$ by parallel to serial operation. The serial output $y(k)$ and the desired signal $d(k)$ are used to generate the required error $e(k)$.

14.4 CONSTRAINED FBLMS ALGORITHM

In FBLMS algorithm, the cross-correlation between the input block $x(n)$ and the error block $e(n)$ is computed by multiplying the conjugate of the FT of $x(n)$, $X_f^*(n)$ and the FT of the error $e(n)$, $E_f(k)$. To update the N coefficients of the vector $W(n)$, it is required to consider the cross-correlation between $x(n)$ and $e(n)$ *for N lags only.* However, multiplication of $X_f^*(n)$ and $E_f(n)$ correspond to $\{r(n,0), r((n,1),......r(n,L+N-2)\}$. Hence, *the lags beyond* $0,1,\ldots(N-1)$ *also contribute to the cross-correlation resulting in an incorrect weight vector adaptation.* Further, these correlations at *higher lags have high variance and also may suffer from time domain aliasing*. The aliasing is due to insufficient FFT length as the length of *FT's* are only $(L+N-1)$ and not $(2L-1)$ for cases $(N<L)$ as $(L+N-1)<(2L-1)$. *Hence these higher lags should not be included and this is achieved by setting the cross-correlation values to zero for these lags.*

14.4.1 Implementation of Constrained FBLMS Algorithm

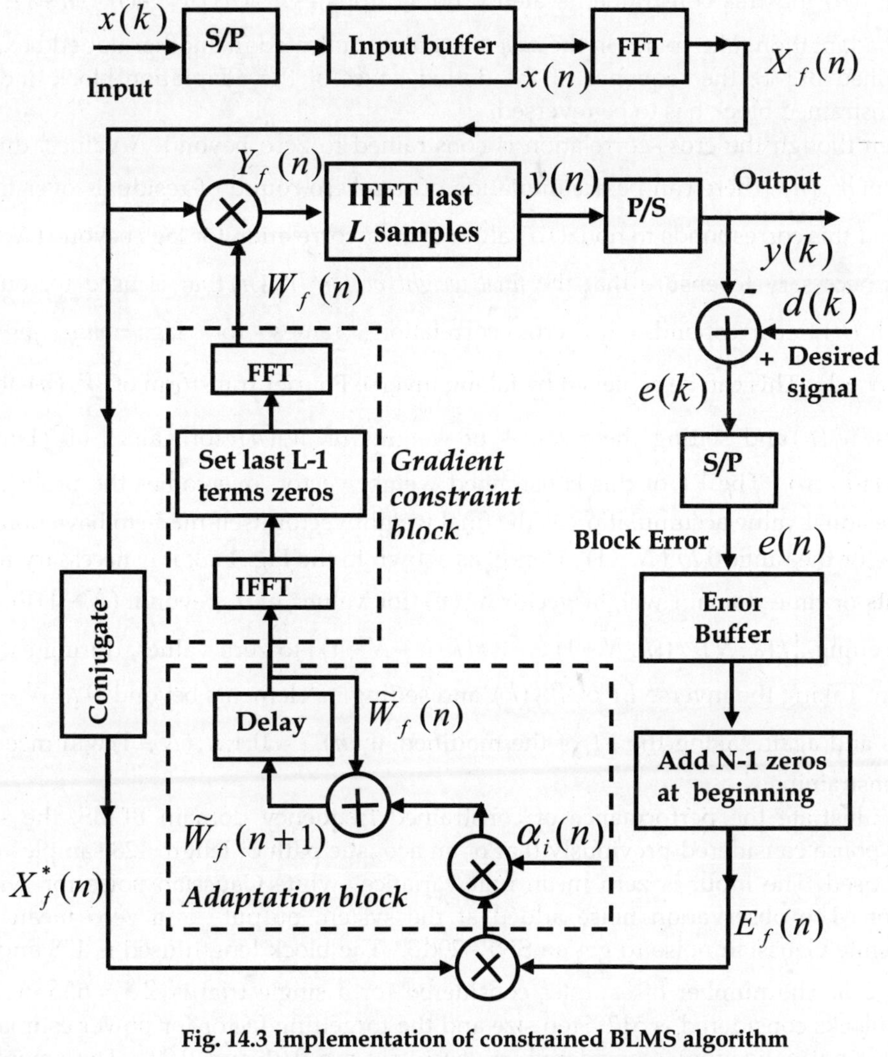

Fig. 14.3 Implementation of constrained BLMS algorithm accounting for round-off noise accumulation.

This can be achieved by taking the inverse FT of $X_f^*(k)E_f(k)$ and setting each element of the vector $\{r(k,N), r((k,N+1),......r(k,L+N-2)\}$ to *zero value* and again taking the *FT* of the modified cross-correlation vector. This modification operation of the cross-correlation is shown in the dotted box and the inclusion of this operation is known as *constrained* FBLMS (CFBLMS) algorithm. The reason is that the cross-correlation at higher lags, *are constrained to zero value.* The CFBLMS provides a good performance than the unconstrained FBLMS algorithm *as it removes the unwanted filter coefficients that contribute to the error.* In Fig.14.3, this corresponds to the product $X_f^*(k)E_f(k)$ being passed through the *gradient constraint block* indicated as a dotted box. The gradient *constraint block* output is scaled by $\alpha_i(n)$ and this constrained scaled product updates $W_f(n)$ to get $W_f(n+1)$. The scaling and adaptation takes place in the *adaptation block* indicated in another dotted box. For the constrained BLMS, the sequence of the dotted boxes of the adaptation block and the gradient constrained block has to be reversed.

Even though the cross-correlation is constrained to zero beyond N values, during adaptation of $W_f(n)$, there can be accumulation of non-zero round of residuals over many iterations and this corresponds to nonzero values of crosscorrelation for lags beyond $(N-1)$. Hence it is necessary to ensure that the *final weight vector* $W_f(n)$ that is used for output generation $Y_f(n)$, corresponds to cross-correlation values for lags *only for the range* 0 *to* $(N-1)$. This can be achieved by taking inverse Fourier transform of $W_f(n)$ to get the weights $w_i(n)$ and setting their values of weights of $w_i(n)$ for values of i beyond 0 *to* $(N-1)$ to zero. The FT of this constrained weight vector, overcomes the problem of round-off residual value accumulation as the final weight vector itself made to have nonzero values only for the range 0 *to* $(N-1)$. Hence, as shown in the Fig. 14.3, it is necessary to set the elements of time domain weight vector $w_i(n)$ for values of i beyond $(N-1)$ to zero instead of setting $\{r(k,N), r((k,N+1),......r(k,L+N-1)\}$ to zero values, during each of the iteration. Taking the inverse *FT* of $W_f(k)$ and setting its elements beyond 0 *to* $N-1$ to zero values and again taking the *FT* of the modified, $w_i(n), i = 0,1,...(N-1)$ will meet the required constraining.

To illustrate the performance of constrained frequency domain BLMS, the same impulse response considered previously that of an acoustic path of length 128 samples (Fig. 14.4 (a)) is used. The input is zero mean unit variance white Gaussian noise for system identification. The observation noise added at the system output is a zero mean unit variance white Gaussian noise to get an SNR=20dB. The block length used is 128 and the FFT size is 256. The number of samples considered for a single trial is $2^{16} = 65536$. The number of blocks considered is 512. Step size and the forgetting factor for power estimation (for step size normalization) considered respectively are 0.01 and 0.99. The smoothed ensemble averaged square error (SEASE) is used as performance index. The ensemble averaging is done over 50 trials and the averaged square error $\xi(n)$ is smoothed by a recursive low-pass filter $\xi_s(n) = \lambda\ \xi_s(n-1) + (1-\lambda)\xi(n)$ with $\lambda = 0.99$.

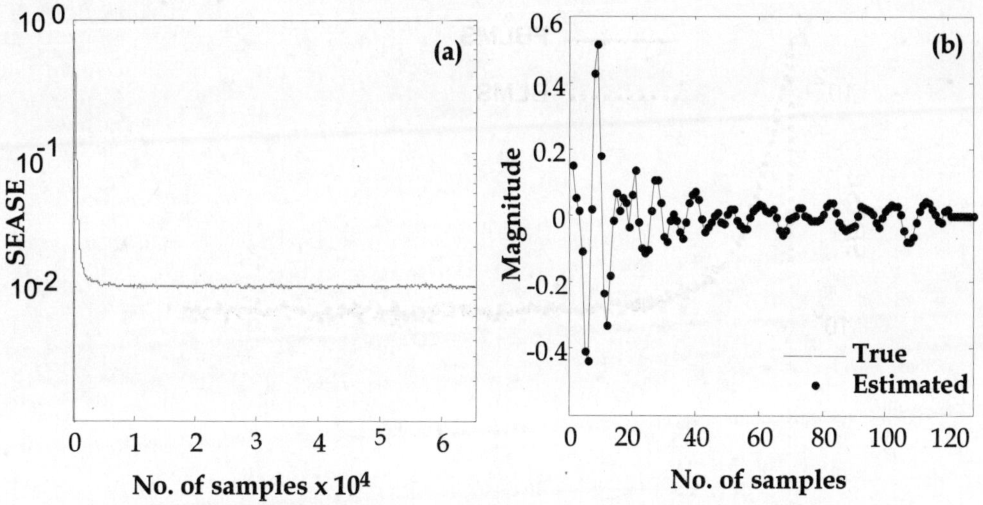

Fig. 14.4 System identification by constrained frequency domain BLMS (a) SEASE (b) Comparison of estimated impulse response with the ideal.

It is seen from Fig. 14.4 (a) that the convergence takes place at about 5000 samples and the convergence error variance magnitude is 0.01and is same that of observation noise variance magnitude 0.01 which is expected. Also the estimate of the impulse response has a close match to the true one and the quality of the match is decided by the variance magnitude of the observation noise.

In the previous example, for the purpose of *complete* system identification, white noise input was considered. However, this will not bring out the orthogonality property of the frequency domain BLMS over that of the time domain BLMS. For this purpose, the input considered is a signal comprising three sinusoids at 50 Hz, 100Hz and 200Hz with amplitudes 2, 1 and 0.5; respectively plus Gaussian white noise to get an SNR of 25 dB. The block length used is 128, the number of samples used for each trial is $2^{16} = 65536$ and 50 trials are used for computing the performance index SEASE. In computing SEASE, the smoothing factor λ used is 0.99. The system impulse response used is same as in the previous case with 128 samples. The step size used for time domain BLMS is 0.001 which is normalized by the filter length and total input power and a forgetting factor of 0.99 is used for recursive power estimation. For the FBLMS, the power input for each bin is recursively computed using a forgetting factor of 0.99. For each bin, the step size 0.001 is normalized by its corresponding bin input power.

It is seen that the SEASE converges faster for FBLMS compared to that for time domain BLMS algorithm (Fig.14.4(c)). This is due to built in orthogonalization and step size normalization by the individual powers of different transform bins which are comparatively smaller than the total power of the input signal (prior to orthogonal transformation).

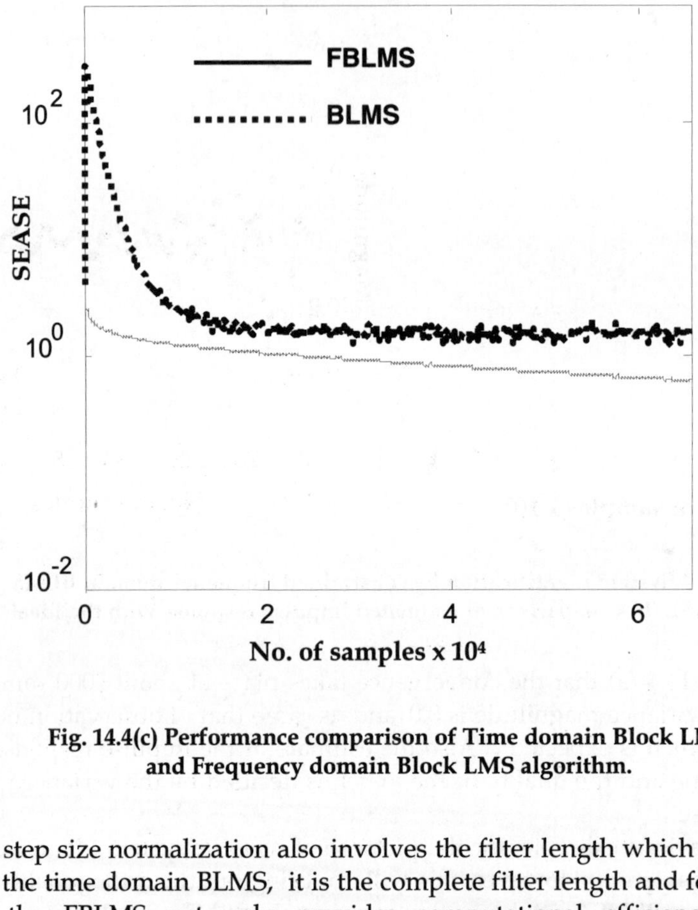

Fig. 14.4(c) Performance comparison of Time domain Block LMS and Frequency domain Block LMS algorithm.

Further, the step size normalization also involves the filter length which is unity for FBLMS whereas for the time domain BLMS, it is the complete filter length and for this example it is 128. Thus, the FBLMS not only provides computational efficiency but also faster convergence.

14.5 PARTITIONED FBLMS ALGORITHM

In applications like echo cancellation, at a sampling rate of 10 kHz, the filter length N required may be very large like 2500 samples. *To get good computational efficiency by frequency domain convolution, the two signal blocks involved should be of similar length.* Therefore, the large impulse response length, forces to use a similar large length input data block to achieve a computationally efficient convolution operation between the two. This results *in a long waiting time to collect input samples to have a required large input data block which delays the output* and such a delay is not desirable. The only way to reduce the delay to the acceptable value is to reduce the input block length. Hence, to achieve the computational efficiency, the length of impulse response of the adaptive filter taking part in convolution, has to be reduced. Such a reduction of the impulse response length can be achieved by *partitioning of the impulse response into blocks of smaller length which is decided by the tolerable limit on the delay instead of using the complete impulse response.* Therefore, *to avoid delay and to have computational efficiency, the FBLMS requires partitioning of the impulse response.*

The output of the adaptive filter which is convolution of its impulse response with the input with the estimate is given by,

$$y(k)=\sum_{i=0}^{N-1} w_i\, x(k-i)$$

In Partitioned FBLMS algorithm (PFBLMS), *filter impulse response of length N is divided into P filters each of length M so that* $N = PM$.

$$\{w_0, w_1, \ldots, w_{M-1}\}, \{w_M, w_{M+1}, \ldots, w_{2M-1}\}, \ldots\ldots\ldots \{w_{M(P-1)}, w_{M(P-1)+1}, \ldots, w_{(MP-1)}\}$$

Here, each partition of the overall filter can be considered an *independent filters for convolution.* The convolution equation can also be split as (for a single value output at the instant n),

$$\begin{aligned} y(k)=\{ & w_0x(k)+w_1x(k-1)+w_2x(k-2)+\ldots\ldots\ldots+w_{M-1}x(k-M+1)\}+. \\ & \{w_Mx(k-M)+w_{M+1}x(k-M-1)+\ldots\ldots+w_{2M-1}x(k-2M+1)\}+\ldots\ldots \\ & +\{w_{(P-1)M}x(k-(P-1)M)+\ldots\ldots+w_{PM-1}x(k-PM+1)\} \end{aligned}$$

This can be written as,

$$y(k)=\sum_{l=0}^{P-1} y_l(k), \quad y_l(k)=\sum_{i=0}^{M-1} w_{i+lM}\, x(k-lM-i)$$

Let

$$\begin{aligned} x_0(n) &= [x(k-M+1)\ x(k-M)\ \ldots\ldots x(k)] \\ x_1(n) &= [x(k-2M+1)\ x(k-2M)\ \ldots\ldots x(k-M)] \\ x_2(n) &= [x(k-3M+1)\ x(k-3M)\ \ldots\ldots x(k-2M)] \\ &\vdots \\ x_l(n) &= [x(k-(l+1)M+1)\ x(k-(l+1)M)\ \ldots\ldots x(k-lM)] \\ &\vdots \\ x_{P-1}(n) &= [x(k-PM+1)\ x(k-PM)\ \ldots\ldots x(k-(P-1)M)] \end{aligned}$$

Here, x_0 , x_1 , ..., *are delayed by a unit block of M samples and are convolved with the corresponding partitioned weight vectors.* Also here, the *block length is considered to be equal to partition length. That is,* $L = M$ (generally any value of L can be chosen and usually L is small). To carry out block convolution, the input data should have $2M$ samples and the weight vector with its M elements is padded with M zeros making its length also $2M$. For the lth partition, the transform of the data and weight vector are given by

$$X_{f,l} = FT\begin{bmatrix} x((n-l)M-M)\ x((n-l)M-(M-1)\ldots\ldots x((n-l)M-1) \\ x((n-l)M\ldots\quad \ldots x((n-l)M+(M-1) \end{bmatrix}^T$$

and,

$$W_{f,l} = FT\begin{bmatrix} w_{lM}(n)\ w_{lM+1}(n)\ \ldots\ldots\ w_{lM+M-1}(n)\ \overset{M\ values}{0,0,\ldots 0} \end{bmatrix}^T$$

Now consider,

$$X_{f,0}(n) = FT\left[x(nM-M)\ x(nM-M+1)\ \ldots\ldots x(nM-1)\ x(nM)\ldots\ldots x(nM+M-1)\right]^T$$

$$X_{f,1}(n) = FT\begin{bmatrix} x(nM-2M)\ x(nM-2M+1)\ \ldots\ldots x(nM-M+1)\ x(nM-M) \\ \ldots\ldots x(nM-1) \end{bmatrix}^T$$

$$\vdots \qquad\qquad \vdots$$

$$X_{f,P-1}(n) = FT\begin{bmatrix} x(nM-PM)\ x(nM-PM+1)\ \ldots.x(nM-PM+M-1) \\ (nM-PM+M)\ldots x(nM-PM+2M-1) \end{bmatrix}^T$$

Consider

$$X_{f,0}(n) = FT\left[x(nM-M)\ x(nM-M+1)\ \ldots\ldots x(nM-1)\ x(nM)\ldots\ldots x(nM+M-1)\right]^T$$

Replacing n by $(n-1)$ in $X_{f,0}(n)$,

$$X_{f,0}(n-1) = FT\begin{bmatrix} x((n-1)M-M)\ x((n-1)M-M+1)\ \ldots\ldots x((n-1)M-1) \\ x((n-1)M)\ldots x((n-1)M+M-1) \end{bmatrix}^T$$

$$X_{f,0}(n-1) = FT\begin{bmatrix} x(nM-2M)\ x(nM-2M+1)\ \ldots\ldots x(nM-M-1) \\ x(nM-M)\ldots x(nM-1) \end{bmatrix}^T$$

But

$$X_{f,1}(n) = FT\begin{bmatrix} x(nM-2M)\ x(nM-2M+1)\ \ldots\ldots x(nM-M+1)\ x(nM-M) \\ \ldots\ldots x(nM-1) \end{bmatrix}^T$$

It is seen that $X_{f,1}(n)$ and $X_{f,0}(n-1)$ are identical. Hence,

$$X_{f,1}(n) = X_{f,0}(n-1)$$

In general,

$$X_{f,l}(n) = X_{f,0}(n-l)$$

This is valid under the assumption that $L = M$.*To realize the convolution between input block with the complete impulse response of the adaptive filter, all the partitions of the impulse response have to be convolved with that input block.* That is, with reference to the impulse response partitions, that data block has to be delayed as the block length and partition length is same. The above relation states this. Thus to realize this convolution, the FT of the lth partition of impulse response has to be multiplied with the FT of data partition (which is a same single block), $X_{f,l}(n)$ which is given by the FT of 0th data partition delayed by l. The filter output is

$$y_l(n) = \text{Last } M \text{ elements of } IFT[W_{f,l}(n) \otimes X_{f,l}(n)]$$

$$y_l(n) = \text{Last } M \text{ elements of } IFT[W_{f,l}(n) \otimes X_{f,0}(n-l)]$$

$\otimes$ denotes the element-by-element multiplication. The nth block output is

$$y(n) = \left[\, y(nM)\ y(nM+1) \ldots\ldots y(nM+M-1)\right]$$

But as

$$y(n) = \sum_{l=0}^{P-1} y_l(n)$$

Therefore,

$$y(n) = \text{Last } M \text{ elements of } IFT\left[\sum_{l=0}^{P-1} W_{f,l}(n) \otimes X_{f,0}(n-l)\right]$$

$$y(n) = \text{Last } M \text{ elements of } IFT \sum_{l=0}^{P-1} Y_l(n)$$

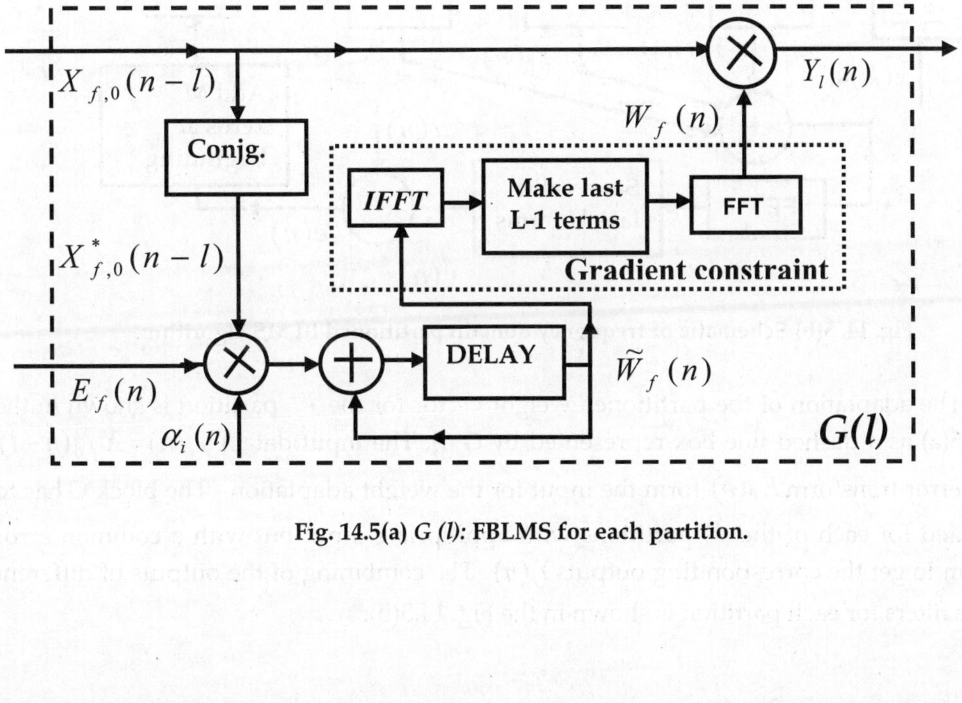

Fig. 14.5(a) ***G (l)*****: FBLMS for each partition.**

The weight update equation is same as that of the FBLMS algorithm and is given as,

$$W_{f,l}(n+1)=W_{f,l}(n)+\mu(n)X_{f,0}(n-l)E_f(n),\ l=0,1,\ldots,P-1$$

Here, $\mu(n)$ is the step size vector used for the nth block and $E_f(n)$ is given by,

$$E_f(n)=FFT[\overset{M}{0,0,\ldots 0}\, e(n)]^T$$

$e(n)$: $M\times 1$ error signal vector and $e(n)=d(n)-y(n)$ and $d(n)$: $M\times 1$ desired signal vector is

$$d(n)=\left[d(nM)\ d(nM+1)\ldots\ldots d(nM+M-1)\right]^T$$

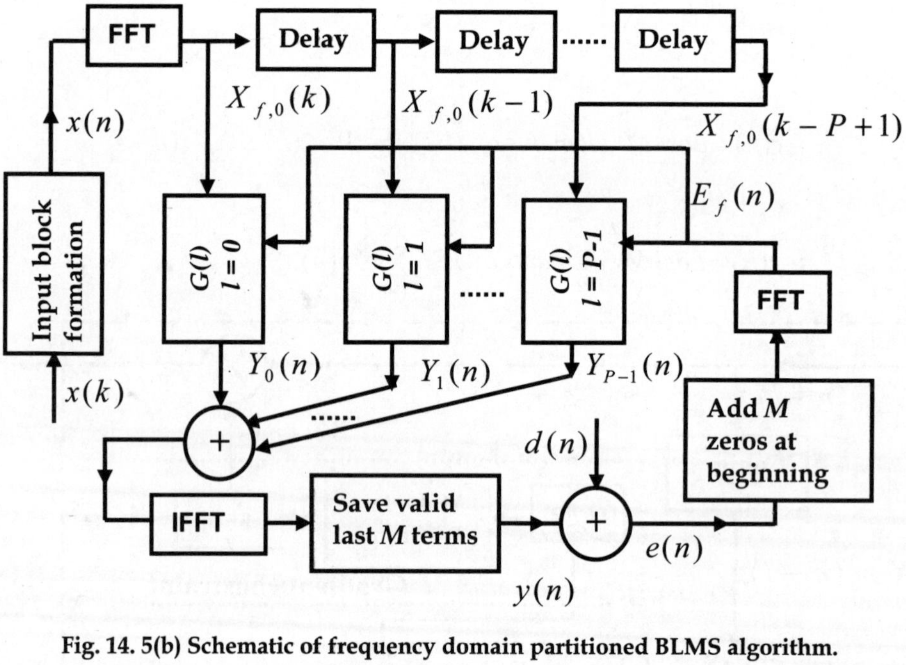

Fig. 14. 5(b) Schematic of frequency domain partitioned BLMS algorithm.

The adaptation of the partitioned weight vector for the lth partition is shown in the Fig. 14.5(a) as a dashed line box represented by G (*l*). The input data $X_{f,l}(n)=X_{f,0}(n-l)$ and the error transform $E_f(n)$ form the input for the weight adaptation. The block G has to be repeated for each of the P partitions with appropriate input but with a common error transform to get the corresponding outputs $Y_l(n)$. The combining of the outputs of different adaptive filters for each partition is shown in the Fig. 14.5(b).

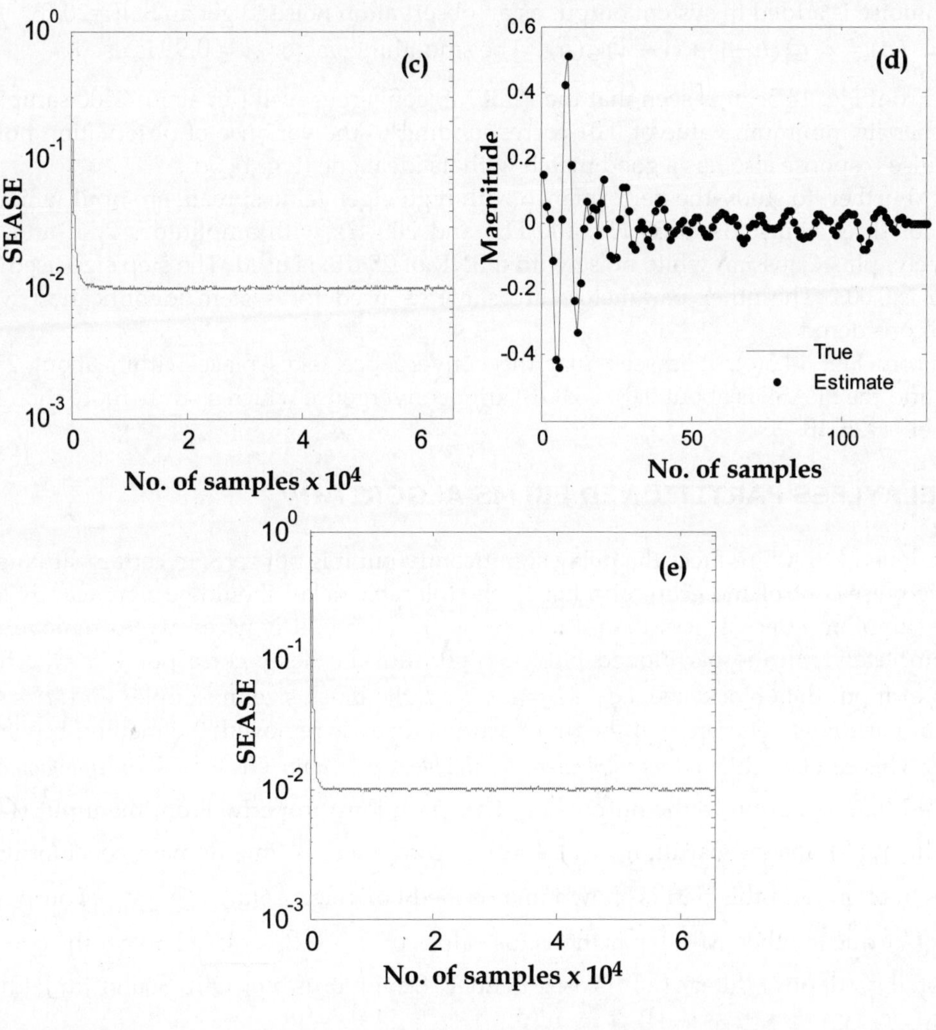

Fig. 14.5 (c, d, e) System identification by PFBLMS: (c) SEASE with white noise input to the system; (d) System impulse response estimate and (e) SEASE with sinusoidal input to the system.

Here also, the previously used impulse response of length 128 samples is considered. The impulse response is partitioned to 4 partitions each of length 32 samples. The data length used for each trial is $2^{16} = 65536$ and the input block data length is also 32 and the number of input data blocks are 2048. The FFT length used is 64. The step size used is 0.01 and for power normalization, the forgetting factor used is 0.99. For system identification, the input signal used is a zero mean, unit variance white Gaussian noise and zero mean white

Gaussian noise is added to system output as an observation noise to get an SNR=20dB. The SEASE $\xi_s(n) = \lambda\ \xi_s(n-1) + (1-\lambda)\xi(n)$, The smoothing factor $\lambda = 0.99$ is used.

From Fig. 14.5c, it is seen that the PFBLMS converges well (in about 2500 samples) and reaches its minimum value of 0.01 corresponding to the variance of observation noise. The impulse response also has a good match with its ideal one (Fig. 14.5d).

Further, to study the performance with high eigenvalue spread, an input which is a sum of three sinusoids at 50 Hz, 100 Hz, and 200 Hz with amplitudes 2, 1, and 0.5 respectively, plus Gaussian white noise with a SNR of 25 dB, is used. The step size used for this case is.0.003. The other parameters are same as used for system identification with PFBLMS considered.

From Fig. 14.5(e), it is seen that the convergence takes place within about 2500 samples and the SEASE is about 0.01 (- 20dB) after convergence which slightly more than the noise level of -25 dB.

14.6 DELAYLESS PARTITIONED FBLMS ALGORITHM

The partitioned FBLMS reduces the delay significantly but it is not zero. In certain situations like active noise control and echo cancellation the tolerable delay should be significantly less than the sampling interval (less than a millisecond). Hence, it is necessary to remove the delay completely. In the partitioned FBLMS algorithm the delay corresponds to the first partition of input data block size, i.e., $M = L/P$, L the block size in samples and P is the number of partitions. Therefore, if the first partition delay is removed, the method becomes delayless. This can be achieved *by implementing the block partition* $G(l), l = 0$ *in time domain.* Further, adding its output to the outputs of other partitions properly. From the input $x(k)$, the output $y_0(k)$ for the partition $G(l), l = 0$ is computed in time domain considering a filter length of M and this part is shown in a dashed box (Fig.14.6(a)). The $y_0(k)$ output is added $y'(k)$ due to other $M-1$ partitions for values of $l = 1. 2, \ldots M-1$ to get the overall output of the adaptive filter $y(k)$ (Yosef Bendel, David Brustein, Ofir Shalvi and Ehud Weinstein, 2001).

$$y'(n) = \text{Last } M \text{ elements of } IFT\left[\sum_{l=1}^{P-1} W_{f,l}(n) \otimes X_{f,0}(n-l)\right]$$

$$y'(n) = \text{Last } M \text{ elements of } IFT \sum_{l=1}^{P-1} Y_l(n)$$

It is important to note that in computing $y'(n)$, for the summation, the lower limit starts as $l = 1$ and not as $l = 0$. For $l = 0$ partition, the adaptation of the filter weights are given by

$$w_{0,i}(k+1) = w_{0,i}(k) + \mu_0(k)x(k-i)e(k), \quad i = 0, 1, \ldots, M-1$$

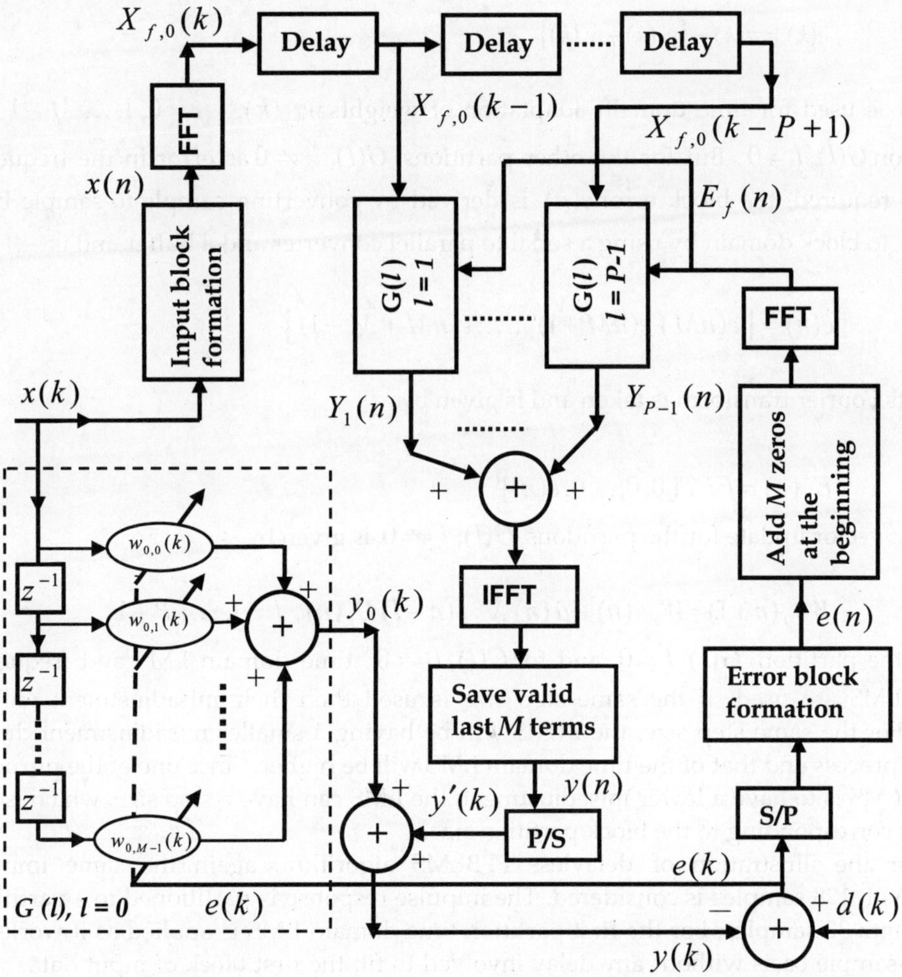

Fig. 14.6(a) Delayless frequency domain partitioned BLMS algorithm.

The output of the adaptive filter is given by

$$y_0(k) = \sum_{i=1}^{M-1} w_{0,i}(k)x(k-i)$$

and the overall output is given by

$$y(k) = y_0(k) + y'(k)$$

To get the error on sample to sample basis $e(k)$, the output $y(k)$ is subtracted from $d(k)$ which is also on sample to sample basis.

$$e(k) = d(k) - \{y_0(k) + y'(k)\}$$

This error is used for time domain adaptation of weights $w_{0,i}(k)$, $i = 0, 1, \ldots M-1$, for the partition $G(l), l = 0$. But for the other partitions $G(l), l \neq 0$ as error in the frequency domain is required, the block error $e(n)$ is derived by converting sample-to-sample basis error $e(k)$ to block domain by using a serial to parallel converter and a buffer and is

$$e(n) = \left[e(nM)\; e(nM+1) \ldots\ldots e(nM+M-1)\right]^T$$

Further, its Fourier transform is taken and is given by

$$E_f(n) = FFT[\overset{M}{0, 0, \ldots 0}, e(n)]^T$$

The weight vector update for the partitions $G(l), l \neq 0$ is given by

$$W_{f,l}(n+1) = W_{f,l}(n) + \mu(n) X_{f,0}(n-l) E_f(n), \; l = 1, \ldots, P-1$$

Since for the partition $G(l), l = 0$ and for $G(l), l \neq 0$, time domain LMS and frequency domain BLMS are used, if the same step size is used then their misadjustment will be different. For the same step size, the BLMS will be having a smaller misadjustment due to averaging process and that of the time domain LMS will be higher. Since one of the purposes of using BLMS is to have a lower misadjustment, the LMS can have a step size, which is less by a factor corresponding to the block partition size.

For the illustration of delayless PFBLMS algorithm, again the same impulse response with 128 samples is considered. The impulse response is partitioned to 4 partitions each of length 32 samples. For the first partition time domain LMS is applied as it works on sample-to-sample basis without any delay involved to fill the first block of input data.. The block data length is same as that of impulse response and is 32. The data length used for each trial is $2^{16} = 65536$ and hence, the number of input data blocks are 2048. The FFT length used is 64. For both time domain LMS for the first partition and for FBLMS for subsequent partitions, the step size used is 0.01and for power normalization, the forgetting factor used is 0.99. For the time domain LMS part for step size normalization the filter length 32 and the total input power are considered. But for PFBLMS, filter length is unity and the input power is the power of particular transform output bin. Also for system identification, the input signal used is a zero mean, unit variance white Gaussian noise and zero mean white Gaussian noise is added to system output as an observation noise to get an SNR=20dB. The SEASE $\xi_s(n) = \lambda\, \xi_s(n-1) + (1-\lambda)\xi(n)$. The smoothing factor $\lambda = 0.99$ and number of trials used is 50.

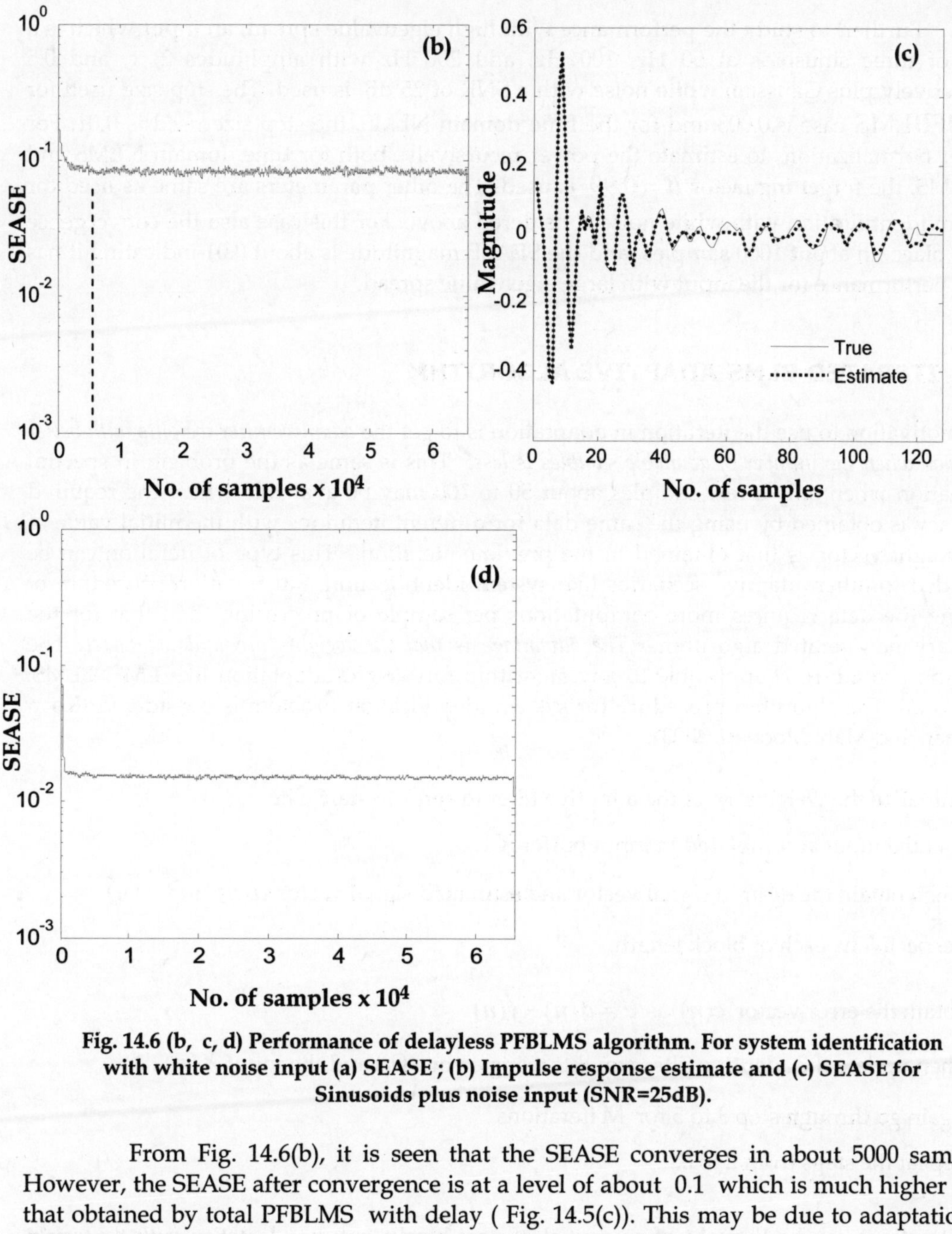

Fig. 14.6 (b, c, d) Performance of delayless PFBLMS algorithm. For system identification with white noise input (a) SEASE ; (b) Impulse response estimate and (c) SEASE for Sinusoids plus noise input (SNR=25dB).

From Fig. 14.6(b), it is seen that the SEASE converges in about 5000 samples. However, the SEASE after convergence is at a level of about 0.1 which is much higher than that obtained by total PFBLMS with delay (Fig. 14.5(c)). This may be due to adaptation in different domains which may result in different performances both with respect to convergence rate and error magnitude. It may be difficult to match the performance in both the domains completely.

Further, to study the performance with high eigenvalue spread, an input which is a sum of three sinusoids at 50 Hz, 100 Hz, and 200 Hz with amplitudes 2, 1, and 0.5 respectively plus Gaussian white noise with a SNR of 25 dB, is used. The step size used for the PFBLMS case is.0.003 and for the time domain NLMS, the step size used is 0.01. For power normalization to estimate the power recursively both for time domain NLMS and PFBLMS, the forgetting factor $\beta = 0.99$ is used. The other parameters are same as used for system identification with white noise considered above. For this case also the convergence takes place in about 1000 samples and the SEASE magnitude is about 0.01 indicating it has good performance for the input with large eigenvalue spread.

14.7 ITERATED BLMS ADAPTIVE ALGORITHM

The motivation to use the iteration in adaptation is to get the *adaptive filter weights with better accuracy when the number of available samples is less.* This is same as the problem in spectral estimation when only a few samples about 50 to 100 may be available where the required accuracy is obtained by using the same data for different iterations with the initial value of the weight vector as that obtained in the previous iteration. This type of iteration can be extended to other adaptive scenarios like system identification. But iteration procedure of reusing the data requires more computations per sample of processing than that for the ordinary non-iterated algorithm. *The advantage is that the weights are updated faster.* The iteration procedure is applicable to any algorithm for weight adaptation like LMS, BLMS, and so on. The algorithm procedure for system identification problem is considered (Koen Eneman and Marc Moonen, 2003).

1. Initialize the weights W of the adaptive filter to zeros to start with.
2. Get the input accumulated in input buffer X.
3. Then obtain the desired signal vector and estimated signal vector $d(n)$ and $y(n)$ respectively, each of block length.
4. Obtain the error vector $e(n)$ as $e = d(n) - y(n)$.
5. Then update the adaptive filter weights using error $e(n)$ and the input $X(n)$.
6. Again go through step 3 to 5 for M iterations.
7. Repeat the steps from 2 to 6.

The iterations can be thought of as arrival of new blocks *but the adaptation with the weight vector of previous iteration.* The weight vector of last iteration forms the weight vector for that block. That is, in between normal block sequence, the iterations take place to improve the quality of that partition of weight vectors.

The FBLMS algorithm is computationally advantageous where the impulse responses are of long length (order of 2000 samples) where the time domain convolution becomes inefficient and frequency domain convolution is appropriate due to use of FFT. The FBLMS provides significant computational efficiency for convolution operation both for the output and the correlation between input and error.

But the FBLMS is valid for slow nonstationary signals and provides better misadjustment due to averaging over the samples in the block. The convergence rate can be improved by scaling the step size by the power of that particular frequency bin. For long impulse responses, partitioned BLMS can be used. The partition delay can be removed by using the sample to sample basis LMS adaptation for the first partition of the impulse response and BLMS for other partitions of the impulse response. If the computational load is not a limit, the iterated PFBLMS where the iterations in between input blocks of data can be used to improve the quality (convergence rate) of adaptation of the weight vector. Computationally fast iterated PFBLMS have also been developed.

Chapter 15

RECURSIVE LEAST SQUARE ADAPTIVE FILTERS

In block processing, block of data of specific size is acquired and the processing is done to get the least square filter on offline basis. Though this has immense offline utility, it is not suitable for real time processing as it involves delay in acquiring data and computing its output. Further, in general, it involves covariance / autocorrelation matrix inversion which is essential and is difficult to perform in real time on sample-to-sample basis. If the inverse of the autocorrelation matrix can be computed recursively on sample to sample basis, the real time least square solution is possible, using exact data available. The autocorrelation matrix of the input and the cross-correlation vector between input and desired signal can be updated recursively using a forgetting factor. Most importantly the inverse of the auto-correlation matrix can computed recursively by using the ABCD lemma and this makes the recursive least square (RLS) solution possible. The RLS does the minimization of the energy of the new error at every sample and hence provides the close to the optimal solution on sample to sample basis. The forgetting factor used prevents in achieving exact optimal solution. For small forgetting factor less than unity, the misadjustment is high but the convergence rate is better and vice versa. Compared to NLMS, the RLS algorithms provide fast initial convergence with a better misadjustment.

Even though the RLS computes its inverse autocorrelation matrix recursively, the computational load involved is N^2 (N is the order of the filter) and is very high. Hence, fast algorithms which have a computational load linearly proportional to N are considered. These fast algorithms achieve this by exploiting both the forward and backward prediction errors. Though the fast transversal RLS (FTRLS) algorithm requires a computational complexity of 8N, its stability is sensitive to finite word length effects. Hence, the lattice structure which is immune to finite word length effects has been exploited for the fast RLS algorithm. Further, the fast lattice RLS (FLRLS) with error feedback is less sensitive to compared to that without error feedback. In terms of practical applicability, FLRLS with error feedback using a priori errors are preferred to that based on a posteriori errors. Though the computational complexity of FLRLS is about 18N, compared to that of FTRLS which is 8N, the former is preferred due to its stability to finite word length effects.

In an attempt to reduce the computational complexity of even the fast RLS algorithms and to get better convergence rate than that of NLMS, the affine projection has come into existence. Instead of a single input vector used by the NLMS algorithm, it forces the a posteriori errors to become zero for L previous input vectors by the same weight vector and this is ensured by minimizing the norm difference between consecutive weight vectors. Since L vectors are simultaneously considered at a single sample, instead of over L samples as in NLMS, the convergence rate will be improved but as it is not possible to make the L errors zero by a single weight vector, the misadjustment will be more *as L increases. It gives a convergence rate in between those of NLMS and RLS but its computational load will increase with L.*

As the Kalman filter is a recursive way of realizing the Wiener filter, it has been considered in a simple form to estimate the signal from a noisy signal.

15.1 INTRODUCTION

The optimum solutions considered use the *statistical error measure, namely, the mean square error or its variations as their basis.* Though they are based on statistical error measure, their implementation utilizes *exact data* available. In block processing, block of data of specific size is acquired and the processing is done to get the least square filter *on offline basis.* Though this has immense offline utility, it is not suitable for *real-time processing* as it involves delay in acquiring data and computing its output. Therefore, if the same least square processing of block method is carried out in a sequential or recursive way which avoids delay and storage, it will be suitable for real-time implementation. Thus, *it is appropriate to consider adaptive filters based on error measure derived from the exact data acquired on sequential basis and perform least square filtering as in block method without any delay and storage problems which is useful for real-time processing.* Such an algorithm is known as *Recursive Least Square algorithm.*

In this, at time *n*, the observations $x(1), x(2)........x(n)$ and the desired responses $d(1), d(2)........d(n)$ are available. It is assumed that for the observations and desired responses, the least square estimator $W_N(n)$ has been estimated. But for the new observation $x(n+1)$ and new desired response $d(n+1)$, *it is required to update the existing least square estimator for time instant "n" instead of re-estimating the least square solution for the (n+1) instant.*

Let the *N* tap estimator at time *n* be given by

$$W_N(n) = [w_0(n), w_1(n), ... w_{N-1}(n)]^T \tag{15.1a}$$

Further, let the input be given by

$$X_N(i) = x(i),\ x(i-1),\ ......,\ x(i-N+1) \tag{15.1b}$$

The error between the desired response and the filter output $e(i)$ is given by

$$e(i) = d(i) - W_N^T(n)X(i) \tag{15.1c}$$

It is important to note that that the index for the tap weight vector, *n* is used and for input and desired response, the index "*i* " is used. This indicates that at each time instant *n*, a multiple regression is made on the input and desired response up to time *n*. That is, $W_N(n)$ depends on $X_N(i)$ and $d(i)$, $i = 1, 2, \dots, n$. This results in the determination $W_N(n)$ by the minimization of the ***cumulative squared error measure*** index given by

$$\varepsilon\,(n) = \sum_{i=1}^{n} \lambda^{\,n-i} e^2(i), \qquad 0 < \lambda \le 1 \tag{15.1d}$$

Here, λ is known as forgetting factor. The factor λ^{n-i} weighs heavily the recent errors compared to the distant past errors. This will be of significance when the input signal is nonstationary for which its statistics will be varying with time. The value of λ to be used depends on the degree of nonstationarity. A value of $\lambda = 1$ weighs equally all the errors/data and this is valid for a stationary input whose statistics does not change with time and this is known as *pre-windowed* case. The value of λ is not very critical as its value in the range $(0.95 < \lambda < 0.9995)$ can track most of the local nonstationarities.

It is important to note that $\varepsilon(n)$ is a function of actual data vectors $X(n), X(n-1), \ldots, X(1)$. Further, the upper limit of the summation is the current sample n, which indicates that *a new error criterion is recomputed and is* ***re-optimized for every time iteration,*** n. That is, for the recursive least square estimator, the optimal filter for the exact data acquired, is computed *for every time iteration n*. The minimization of $\varepsilon(n)$ is achieved by setting its gradient to zero, i.e.,

$$\frac{\partial\, \varepsilon(n)}{\partial W_N(n)} = 0 \tag{15.2a}$$

Substituting for $\varepsilon(n)$ from Eqn. (15.1d),

$$\frac{\partial}{\partial W_N(n)}\left[\sum_{i=1}^{n} \lambda^{n-i} e^2(i)\right] = 0$$

$$\frac{\partial}{\partial W_N(n)}\left[\sum_{i=1}^{n} \lambda^{n-i}\left\{ d(i) - X_N^T(i) W_N(n)\right\}^2\right] = 0$$

$$\sum_{i=1}^{n} \lambda^{n-i} 2\left[d(i) - X_N^T(i) W_N(n)\right]\cdot\left(-X_N(n)\right) = 0$$

$$\left[\sum_{i=1}^{n} \lambda^{n-i} X_N^T(i) X_N(n)\right] W_N(n) = \sum_{i=1}^{n} \lambda^{n-i} d(i) X_N(n)$$

$$R_{NN}(n) W_N(n) = P_N(n) \tag{15.2b}$$

where,

$$R_{NN}(n) = \sum_{i=1}^{n} \lambda^{n-i} X_N(i) X_N^T(i) \tag{15.2c}$$

is the sample autocorrelation matrix and

$$P_N(n) = \sum_{i=1}^{n} \lambda^{n-i} d(i) X_N(i) \tag{15.2d}$$

is the sample cross-correlation vector.

For an ergodic data, $\lambda = 1$ and hence,

$$R_{NN}(n) = \sum_{i=1}^{n} X_N(i)X_N^T(i) \text{ and } P_N(n) = \sum_{i=1}^{n} d(i)X_N(i)$$

Assuming $R_{NN}(n)$ invertible,

$$W_N(n) = R_{NN}^{-1}(n)P_N(n) \tag{15.2e}$$

As the Eqn. (15.2e) is valid for any n,

$$W_N(n-1) = R_{NN}^{-1}(n-1)P_N(n-1) \tag{15.2f}$$

As already stated, in the recursive approach, it is required to compute the next least square set of coefficients $W(n)$ from the knowledge of the quantities in Eqn. (15.2f) and the new data $x(n)$ and $d(n)$ acquired at the time n. This can be started right from the starting time $n = 0$. In such a case, it is required to assume an initial value of $R_{NN}^{-1}(0)$ and it has been found that the choice of $R_{NN}^{-1}(0)$ is not critical for RLS algorithm. The only requirement is that $R_{NN}(0)$ must be non-singular for $R_{NN}^{-1}(0)$ to exist.

15.2 STANDARD RECURSIVE LEAST SQUARES (RLS) ALGORITHM

From Eqn. (15.2e),

$$W_N(n) = R_{NN}^{-1}(n)P_N(n)$$

For recursion of $R_{NN}(n)$ in terms of $R_{NN}(n-1)$, Eqn. (15.2c),

$$R_{NN}(n) = \sum_{i=1}^{n} \lambda^{n-i} X_N(i)X_N^T(i)$$

can be written as

$$R_{NN}(n) = \sum_{i=1}^{n-1} \lambda^{n-1-i} X_N(i)X_N^T(i) + X_N(n)X_N^T(n)$$

$$R_{NN}(n) = \lambda\, R_{NN}(n-1) + X_N(n)X_N^T(n) \tag{15.3a}$$

For recursion of $P_N(n)$ in terms of $P_N(n-1)$, Eqn. (15.2d)

$$\mathrm{P_N}(n) = \sum_{i=1}^{n} \lambda^{n-i} \mathrm{X}_N(i)\, d(i)$$

can be written as

$$\mathrm{P_N}(n) = \lambda \sum_{i=1}^{n-1} \lambda^{(n-i)-1} \mathrm{X_N}(i) d(i) + \mathrm{X}_N(n) d(n)$$

$$\mathrm{P_N}(n) = \lambda\, \mathrm{P_N}(n-1) + \mathrm{X}_N(n) d(n) \tag{15.3b}$$

Matrix inversion lemma

If A and B positive definite, $M \times M$ matrices are related by,

$$A = B^{-1} + C\, D^{-1}\, C^T \tag{15.3c}$$

Then,

$$A^{-1} = B - BC\left(D + C^T BC\right)^{-1} C^T B \tag{15.3d}$$

Comparing Eqn. (15.3c) with Eqn. (15.3a)

$$R_{NN}(n) = \lambda\, R_{NN}(n-1) + X_N(n) X_N^T(n),$$

results in

$$A = \mathrm{R}_{NN}(n), \quad B^{-1} = \lambda\, \mathrm{R}_{NN}(n-1), \quad C = \mathrm{X}_N(n), \quad D = 1 \tag{15.3e}$$

Therefore, from Eqn. (15.3d),

$$\mathrm{R}_{NN}^{-1}(n) = \lambda^{-1} \mathrm{R}_{NN}^{-1}(n-1) - \lambda^{-1} \mathrm{R}_{NN}^{-1}(n-1) \mathrm{X}_N(n) \left[1 + \mathrm{X_N^T}(n) \lambda^{-1} \mathrm{R}_{NN}^{-1}(n-1) \mathrm{X}_N(n)\right]^{-1} \lambda^{-1} \mathrm{X_N^T}(n) \mathrm{R}_{NN}^{-1}(n-1)$$

$$\mathrm{R}_{NN}^{-1}(n) = \lambda^{-1} \mathrm{R}_{NN}^{-1}(n-1) - \frac{\lambda^{-2} \mathrm{R}_{NN}^{-1}(n-1) \mathrm{X}_N(n) X_N^T(n) \mathrm{R}_{NN}^{-1}(n-1)}{1 + \mathrm{X}_N^{\mathrm{H}}(n) \lambda^{-1} \mathrm{R}_{NN}^{-1}(n-1) \mathrm{X}_N(n)} \tag{15.3f}$$

For convenience let,

$$C(n) = \mathrm{R}_{NN}^{-1}(n) \tag{15.4a}$$

Then,

$$C(n) = \lambda^{-1}C(n-1) - \frac{\lambda^{-2}C(n-1)X_N(n)X_N^T(n)C(n-1)}{1+\lambda^{-1}X_N^T(n)C(n-1)X_N(n)} \qquad (15.4b)$$

Further, let

$$K(n) = \frac{\lambda^{-1}C(n-1)X_N(n)}{\left[1+\lambda^{-1}X_N^T(n)C(n-1)X_N(n)\right]} \qquad (15.4c)$$

$K(n)$ is referred to as gain vector. Therefore, Eqn. (15.4b) can be written as

$$C(n) = \lambda^{-1}C(n-1) - \lambda^{-1}K(n)X_N^T(n)C(n-1) \qquad (15.4d)$$

Now consider Eqn. (15.4c)

$$K(n)\left[1+\lambda^{-1}X_N^T(n)C(n-1)X_N(n)\right] = \lambda^{-1}C(n-1)X_N(n)$$

$$K(n)+\lambda^{-1}K(n)X_N^T(n)C(n-1)X_N(n) = \lambda^{-1}C(n-1)X_N(n)$$

$$K(n) = \lambda^{-1}C(n-1)X_N(n) - \lambda^{-1}K(n)X_N^T(n)C(n-1)X_N(n)$$

$$K(n) = X_N(n)\left[\lambda^{-1}C(n-1) - \lambda^{-1}K(n)X_N^T(n)C(n-1)\right]$$

From Eqn. (15.4d)

$$K(n) = C(n)X_N(n) \qquad (15.4e)$$

$$K(n) = R_{NN}^{-1}(n)X_N(n) \qquad (15.4f)$$

Recursive update for tap weight vector: Consider Eqn. (15.2e),

$$W_N(n) = R_{NN}^{-1}(n)P_N(n)$$

Substitute for $R_{NN}^{-1}(n)$ from Eqn. (15.4a),

$$C(n) = R_{NN}^{-1}(n)$$

Then

$$W_N(n) = C(n)P_N(n)$$

Substitute for $P_N(n)$ from Eqn. (15.3b)

$$P_N(n) = \lambda\, P_N(n-1) + X_N(n)d(n)$$

Then,

$$W_N(n) = C(n)\left[\lambda\, P_N(n-1) + X_N(n)d(n)\right]$$
$$= \lambda\, C(n)\, P_N(n-1) + \underline{C(n)X_N(n)}d(n)$$

Since from Eqn. (15.4e) $K(n) = C(n)X_N(n)$,

$$W_N(n) = \lambda\, C(n)\, P_N(n-1) + K(n)d(n)$$

Substitute for $C(n)$ from Eqn. (15.4d)

$$C(n) = \lambda^{-1}C(n-1) - \lambda^{-1}K(n)X_N^T(n)C(n-1)$$

Then,

$$W_N(n) = \lambda\left[\lambda^{-1}C(n-1) - \lambda^{-1}K(n)X_N^T(n)C(n-1)\right] P_N(n-1) + K(n)d(n)$$

$$W_N(n) = C(n-1)P_N(n-1) - K(n)X_N^T(n)C(n-1)P_N(n-1) + K(n)d(n)$$

$$W_N(n) = W_N(n-1) - K(n)X_N^T(n)W_N(n-1) + K(n)d(n)$$
$$= W_N(n-1) + K(n)\left[d(n) - X_N^T(n)W_N(n-1)\right]$$

$$W_N(n) = W_N(n-1) + K(n)\,\alpha(n) \tag{15.4g}$$

where,

$$\alpha(n) = d(n) - X_N^T(n)W_N(n-1)$$
$$\alpha(n) = d(n) - W_N^T(n-1)X_N(n) \tag{15.4h}$$

$\alpha(n)$ is the *a priori* estimation error, since it is updated using the weight vector of the *previous instant* $W_N(n-1)$. Further, $\alpha(n)$ is different from the *a posteriori* estimation error given by

$$e(n) = d(n) - W_N^T(n)X_N(n) \tag{15.4i}$$

Computation of $e(n)$ involves the current *least-squares estimate of filter tap weights,* i.e., $W_N(n)$ in getting the filter output. The error $\alpha\ (n)$ can be viewed as a tentative value of $e(n)$ before updating the tap weight vector. *It is important to note that for the least square problem, the minimization of cumulative error measure is of* $e(n)$ *and not that of* $\alpha\ (n)$.

15.2.1 Initializations for the RLS Algorithm

To start the recursion, the following initializations are required.

- Initialization for $\mathrm{C}(0)$: $\mathrm{R_{NN}}(0)$ should be nonsingular. This is done by evaluating, the value of $\mathrm{C}(0) = \mathrm{R_{NN}^{-1}}(0)$ by

$$C(0) = \left[\sum_{i=-N}^{0} \lambda^{-i} \mathrm{X}_N(i)\mathrm{X}_N^T(i) \right]^{-1},\ P_N(0) = \left[\sum_{i=-N}^{0} \lambda^{-i} d(i)\mathrm{X}_N^T(i) \right]^{-1} \tag{15.5a}$$

With this, $\mathrm{W_N}(0)$ can be initialized as $\mathrm{W_N}(0) = \mathrm{C}(0)P_N(0)$. This preserves the optimality at every step as the RLS is initialized at $n = 0$ by a $\mathrm{W_N}(0)$ which minimizes $\varepsilon\ (0)$. However, this requires direct inversion of $\mathrm{R_{NN}}(0)$ which requires operations of the order of $(N+1)^3)$. Further, it involves delay of $(N+1)$ samples prior to adaptation. A simple initialization procedure is to have

$$\mathrm{R}_{NN}(0) = \delta\ \mathrm{I} \text{ or } \mathrm{C}(0) = \mathrm{R}^{-1}_{NN}(0) = \delta^{-1}\ \mathrm{I} \tag{15.5b}$$

where δ is a small positive constant. This on recursion leads to

$$\mathrm{R_{NN}}(n) = \sum_{i=1}^{n} \lambda^{n-i} \mathrm{X}_N(i)\mathrm{X}_N^T(i) + \delta\ \lambda^{n}\mathrm{I}$$

To start with though this introduces *a bias in the weight vector* $\mathrm{W_N}(n)$, as $\lambda < 1$, the effect of $\mathrm{R}_{NN}(0)$ exponentially decreases with increase of n and this initialization will have little effect in steady state , i.e., as n takes large values. Further, this bias effect can be reduced by choosing a small value for δ. In practice, the value of $\delta \geq 0.01\sigma_x^2$ has been found to be satisfactory. σ_x^2 is the variance of the input.

- Initialization for the tap weight vector

$$\mathrm{W_N}(n) = 0 \tag{15.5c}$$

where 0 is a $M \times 1$ null vector. The initialization procedure using Eqns. (15.5b) and (15.5c) is referred to as *self-constrained* initialization.

15.2.2 RLS ALGORITHM

Initialization:

$$C(0) = \delta^{-1} I, \ \delta = 0.001, \quad W_N(n) = 0$$

Algorithm:

- Update the gain vector,

$$K(n) = \frac{\lambda^{-1} C(n-1) X_N(n)}{\left[1 + \lambda^{-1} X_N^H(n) C(n-1) X_N(n)\right]}$$

- Compute the *a priori* estimation error,

$$\alpha(n) = d(n) - W_N^T(n-1) X_N(n)$$

- Update the tap weight vector,

$$W_N(n) = W_N(n-1) + K(n)\,\alpha(n)$$

- Update the inverse of correlation matrix,

$$C(n) = \lambda^{-1} C(n-1) - K(n) X_N^T(n) C(n-1)$$

15.3 MINIMUM SUM OF WEIGHTED ERROR SQUARES $\varepsilon_{\min}(n)$

From Eqn. (15.1d) the cumulative squared error measure is

$$\varepsilon(n) = \sum_{i=1}^{n} \lambda^{n-i} e^2(i), \qquad 0 < \lambda \le 1$$

From Eqn. (15.4i)

$$\varepsilon(n) = \sum_{i=1}^{n} \lambda^{n-i} \left\{ d(i) - X_N^T(i) W_N(n) \right\}^2$$

$$\varepsilon(n) = \sum_{i=1}^{n} \lambda^{n-i} d(i)^2 - 2 W_N^T(n) \sum_{i=1}^{n} \lambda^{n-i} d(i) X_N(i)$$

$$+ W_N^T(n) \left\{ \sum_{i=1}^{n} \lambda^{n-i} d(i) X_N(i) X_N^T(i) \right\} W_N(n)$$

Using Eqns. (15.2c) and (15.2d),

$$R_{NN}(n) = \sum_{i=1}^{n} \lambda^{n-i} X_N(i) X_N^T(i), \quad P_N(n) = \sum_{i=1}^{n} \lambda^{n-i} d(i) X_N(i)$$

$$\varepsilon_d(n) = \sum_{i=1}^{n} \lambda^{n-i} d(i)^2,$$

$$\varepsilon(n) = \varepsilon_d(n) - 2W_N^T(n)P_N(n) + W_N^T(n)R_{NN}(n)W_N(n) \tag{15.6a}$$

Since least square solution is (Eqn. 15.2b)

$$R_{NN}(n)W_N(n) = P_N(n)$$

Therefore,

$$\varepsilon_{\min}(n) = \varepsilon_d(n) - W_N^T(n)P_N(n)$$

Further, since (Eqn. (15.3b)),

$$\mathrm{P}_N(n) = \lambda\ \mathrm{P}_N(n-1) + X_N(n)d(n),$$

and

$$\varepsilon_d(n) = \lambda\ \varepsilon_d(n-1) + d(n)^2$$

Also from Eqn. (15.4g),

$$\mathrm{W_N}(n) = \mathrm{W_N}(n-1) + \mathrm{K}(n)\ \alpha(n)$$

Therefore,

$$\varepsilon_{\min}(n) = \lambda \varepsilon_d(n-1) + d^2(n) - \left[\mathrm{W_N}(n-1) + \mathrm{K}(n)\alpha(n)\right] \left[\lambda\ \mathrm{P}_N(n-1) + X_N(n)d(n)\right]^T$$

$$= \lambda \varepsilon_d(n-1) + d^2(n) - \left[\mathrm{W_N}(n-1) + \mathrm{K}(n)\alpha(n)\right]\left[\lambda\ \mathrm{P}^{\mathrm{T}}(n-1) + X_N^T(n)d(n)\right]$$

$$= \lambda \varepsilon_d(n-1) + d^2(n) - \left[\lambda\ P_N^T(n-1)\mathrm{W_N}(n-1) + \lambda\ P_N^T(n-1)\mathrm{K}(n)\alpha(n) + X_N^T(n)\mathrm{W_N}(n-1)d(n) + X_N^T(n)\mathrm{K}(n)\alpha(n)d(n)\right]$$

$$\varepsilon_{\min}(n) = \lambda\left[\varepsilon_d(n-1) - P_N^T(n-1)\mathrm{W_N}(n-1)\right]$$
$$+ d(n)\left[d(n) - X_N^T(n)\mathrm{W_N}(n-1)\right]$$
$$\downarrow \alpha(n)$$
$$-\left[\lambda\, P_N^T(n-1) + X_N^T(n)d(n)\right]\mathrm{K}(n)\alpha(n)$$
$$\downarrow P_N^T(n)$$

$$\varepsilon_{\min}(n) = \lambda\,\varepsilon_{\min}(n-1) + d(n)\alpha(n) - P_N^T(n)\mathrm{K}(n)\alpha(n)$$

$$\varepsilon_{\min}(n) = \lambda\,\varepsilon_{\min}(n-1) + \alpha(n)[d(n) - P_N^T(n)\mathrm{K}(n)] \tag{15.6b}$$

Since from (Eqn. 15.2b)

$$\mathrm{R}_{NN}(n)\mathrm{W}_N(n) = P_N(n)$$

and from Eqns. (15.4e, f)

$$\mathrm{K}(n) = \mathrm{C}(n)\mathrm{X}_N(n), \quad \mathrm{K}(n) = \mathrm{R_{NN}^{-1}}(n)\mathrm{X}_N(n)$$

Therefore,

$$\mathrm{P}_N^T(n)\mathrm{K}(n) = \mathrm{P}_N^T(n)\mathrm{R_{NN}^{-1}}(n)\mathrm{X}_N(n)$$
$$= \left[\mathrm{R_{NN}^{-1}}\mathrm{P}_N(n)\right]^{\mathrm{T}}\mathrm{X}_N(n)$$

$$\mathrm{P}_N^T(n)\mathrm{K}(n) = \mathrm{W}_N^{\mathrm{T}}(n)\mathrm{X}_N(n)$$

Hence Eqn. (15.6b) becomes

$$\varepsilon_{\min}(n) = \lambda\,\varepsilon_{\min}(n-1) + \alpha(n)\left[d(n) - W_N^T(n)\mathrm{X}_N(n)\right]$$

Since

$$e(n) = d(n) - W_N^T(n)\mathrm{X}_N(n)$$

$$\varepsilon_{\min}(n) = \lambda\,\varepsilon_{\min}(n-1) + \alpha(n)e(n) \tag{15.6c}$$

This expression enables to estimate the minimum error energy recursively from its previous sample instant value using the *a priori* estimation error $\alpha\,(n)$ and the *a posteriori* estimation error $e\,(n)$.

15.4 CONVERGENCE BEHAVIOR OF RLS ALGORITHM

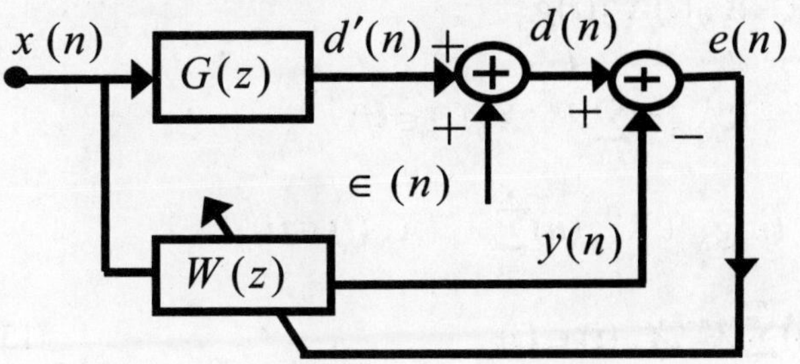

Fig. 15.1(a) System identification for convergence analysis.

The convergence behavior of the RLS algorithm is studied with respect to system identification (Fig. 15.1(a)). Let the system impulse response be $g_N(i), \quad i = 0,1,2,\ldots N-1$. The system output is given by

$$d'(n) = g_N^T(n)\mathrm{X}(n)$$

$$d(n) = g_N^T(n)\mathrm{X}(n) + \in(n) \tag{15.7a}$$

$$\in(n) = \begin{bmatrix} \in(1) & \in(2) & \cdots \in(n) \end{bmatrix}^T \tag{15.7b}$$

where, $\in(n)$ is the *zero mean* white observation noise uncorrelated with the input $x(n)$. The desired signal $d(n)$ is the sum of system output $g_N^T\mathrm{X}(n)$ plus the observation noise $\in(n)$.

15.4.1 Mean-weight Vector Behavior of the RLS Algorithm

It is known that

$$\mathrm{W_N}(n) = \mathrm{R}_{NN}^{-1}(n)P_N(n)$$

$$R_{NN}(n) = \sum_{i=1}^{n} \lambda^{n-i} X_N(i) X_N^T(i)$$

$$P_N(n) = \sum_{i=1}^{n} \lambda^{n-i} d(i) X_N(i)$$

Therefore,

$$W_N(n) = \mathrm{R}_{NN}^{-1}(n)\sum_{i=1}^{n} \lambda^{n-i} d(i) X_N^T(i)$$

Using Eqn. (15.7a)

$$W_N(n) = R^{-1}_{NN}(n)\sum_{i=1}^{n}\lambda^{n-i}X_N(i)X^T_N(i)g_N + R^{-1}_{NN}(n)\sum_{i=1}^{n}\lambda^{n-i}X_N(i)\in(i)$$

$$W_N(n) = R^{-1}_{NN}(n)R_{NN}(n)g_N + R^{-1}_{NN}(n)\sum_{i=1}^{n}\lambda^{n-i}X_N(i)\in(i)$$

$$W_N(n) = g_N + R^{-1}_{NN}(n)\sum_{i=1}^{n}\lambda^{n-i}X_N(i)\in(i) \tag{15.7c}$$

Taking expectations,

$$E\left[W_N(n)\right] = g_N + R^{-1}_{NN}(n)\sum_{i=1}^{n}\lambda^{n-i}E\left[X_N(i)\in(i)\right]$$

Further, using the information that $\in(n)$ is the *zero mean* white observation noise uncorrelated with the input $x(n)$,

$$E\left[W_N(n)\right] = g_N + R^{-1}_{NN}(n)\sum_{i=1}^{n}\lambda^{n-i}E\left[X_N(i)\right]E\left[\in(i)\right]$$

Since

$$E[\in(n)] = 0,$$

$$E\left[W_N(n)\right] = g_N \tag{15.7d}$$

This shows that the RLS results in unbiased estimate of the system impulse response. In practice due to initialization, the $R_{NN}(n)$ used is not a true one and this results in a bias and this becomes unbiased as n increases. That is, the estimate $W_N(n)$ is asymptotically unbiased.

15.4.2 Convergence Behavior in Terms of Convergence Rate and Misadjustment

The weight deviation vector is

$$V_N(n) = W_N(n) - g_N \tag{15.7e}$$

Since from Eqn. (15.7c)

$$W_N(n) = g_N + R^{-1}_{NN}(n)\sum_{i=1}^{n}\lambda^{n-i}X_N(i)\in(i)$$

$$V_N(n) = W_N(n) - g_N = R^{-1}_{NN}(n)\sum_{i=1}^{n}\lambda^{n-i}X_N(i)\in(i)$$

Therefore, the covariance of $V_N(n)$ is

$$\theta_V = \mathrm{E}\left[V_N(n)V_N^T(n)\right] =$$

$$\mathrm{R}_{NN}^{-1}(n)E\left[\sum_{i=1}^{n}\sum_{j=1}^{n}\lambda^{n-i}\lambda^{n-j}X_N(i)X_N^T(j)\in(i)\in(j)\right]\mathrm{R}_{NN}^{-1}(n)$$

Since $\in(n)$ is the *zero mean* white observation noise with power $\sigma_\in^2$,

$$\mathrm{E}\left[V_N(n)V_N^T(n)\right] =$$

$$\sigma_\in^2\mathrm{R}_{NN}^{-1}(n)E\left[\sum_{i=1}^{n}\lambda^{2(n-i)}X_N(i)X_N^T(j)\right]\mathrm{R}_{NN}^{-1}(n) \qquad (15.7f)$$

Here two cases will arise for values of λ, $\lambda = 1$ valid for a stationary signal and $\lambda \neq 1$ valid for a nonstationary signal. Though there is some similarity in the analysis treatment, they require different type of mathematical manipulations hence, will be considered separately. For $\lambda = 1$,

$$\mathrm{E}\left[V_N(n)V_N^T(n)\right] = \sigma_\in^2\mathrm{R}_{NN}^{-1}(n)\mathrm{R}_{NN}(n)\mathrm{R}_{NN}^{-1}(n)$$

$$\mathrm{E}\left[V_N(n)V_N^T(n)\right] = \sigma_\in^2\mathrm{R}_{NN}^{-1}(n) \qquad (15.7g)$$

The a priori error of estimation is

$$\alpha(n) = d(n) - W_N^T(n-1)\mathrm{X}_N(n)$$

Since (Eqn.15.7a)

$$d(n) = g_N^T\mathrm{X}_N(n) + \in(n)$$

$$\alpha(n) = \in(n) - \mathrm{X}_N^T(n)\left[g_N - W_N(n-1)\right]$$

$$\alpha(n) = \in(n) - \mathrm{X}_N^T(n)V_N(n-1) \qquad (15.7h)$$

$$E\left[\alpha^2(n)\right] = \sigma_\in^2 + E\left[\mathrm{X}_N^T(n)E\left\{V_N(n-1)V_N^T(n-1)\right\}\mathrm{X}_N(n)\right]$$

$$\because \quad E[\in(n)\mathrm{X}_N^T(n)V_N(n-1)] = 0, \quad E[\in(n)] = 0$$

Since from Eqn.(15.7g),

$$\mathrm{E}\left[V_N(n)V_N^T(n)\right] = \sigma_\in^2 \mathrm{R}_{NN}^{-1}(n),$$

$$\mathrm{E}\left[V_N(n-1)V_N^T(n-1)\right] = \sigma_\in^2 \mathrm{R}_{NN}^{-1}(n-1)$$

Therefore

$$E\left[\alpha^2(n)\right] = \sigma_\in^2 + \sigma_\in^2 E\left\{\mathrm{X}_N^T(n)R_{NN}^{-1}(n-1)\mathrm{X}_N(n)\right\} \tag{15.7i}$$

Further, the trace of a scalar is equal to scalar. Therefore ,

$$\mathrm{X}_N^T(n)R_{NN}^{-1}(n-1)\mathrm{X}_N(n)$$

$$= trace\left[\mathrm{X}_N^T(n)R_{NN}^{-1}(n-1)\mathrm{X}_N(n)\right]$$

$$trace\left[\mathrm{X}_N^T(n)R_{NN}^{-1}(n-1)\mathrm{X}_N(n)\right]$$

$$= trace\left[R_{NN}^{-1}(n-1)\mathrm{X}_N(n)\mathrm{X}_N^T(n)\right]$$

$$\mathrm{X}_N^T(n)R_{NN}^{-1}(n-1)\mathrm{X}_N(n)$$

$$= trace\left[R_{NN}^{-1}(n-1)\mathrm{X}_N(n)\mathrm{X}_N^T(n)\right]$$

Therefore, Eqn.(15.7i)

$$E\left[\alpha^2(n)\right] = \sigma_\in^2 + \sigma_\in^2 \mathrm{E}\left\{\mathrm{X}_N^T(n)R_{NN}^{-1}(n-1)\mathrm{X}_N(n)\right\}$$

becomes

$$E\left[\alpha^2(n)\right] = \sigma_\in^2 + \sigma_\in^2 \mathrm{trace}\left[R_{NN}^{-1}(n-1)E\left\{\mathrm{X}_N(n)\mathrm{X}_N^T(n)\right\}\right] \tag{15.7j}$$

Consider the definition of $R_{NN}(n)$ given by

$$\mathrm{R}_{NN}(n) = \sum_{i=1}^{n} \lambda^{n-i}\left[X_N(i)X_N^T(i)\right]$$

Since $\lambda = 1$,

$$\mathrm{R}_{NN}(n) = \sum_{i=1}^{n} \left[X_N(i)X_N^T(i)\right]$$

Since

$$\frac{1}{n}\sum_{i=1}^{n}\left[X_N(i)X_N^T(i)\right] \approx R, \quad R_{NN}(n) \approx nR$$

where, R is the ensemble average and due to ergodicity, the ensemble average is equated to time average on the L.H.S. of the above equation. Therefore,

$$R_{NN}^{-1}(n-1) \approx \frac{R^{-1}}{(n-1)}$$

Therefore, by substituting for $R_{NN}^{-1}(n-1)$ in Eqn. (15.7j)

$$E\left[\alpha^2(n)\right] = \sigma_\in^2 + \sigma_\in^2 \text{trace}\left[R_{NN}^{-1}(n-1)E\left\{X_N(n)X_N^T(n)\right\}\right]$$

becomes

$$E\left[\alpha^2(n)\right] = \sigma_\in^2 + \frac{\sigma_\in^2}{(n-1)}\text{trace}\left[R^{-1}E\left\{X_N(n)X_N^T(n)\right\}\right]$$

$$E\left[\alpha^2(n)\right] = \sigma_\in^2 + \frac{\sigma_\in^2}{(n-1)}\text{trace}\left[R^{-1}R\right],$$

$$\because \left\{X_N(n)X_N^T(n)\right\} = R$$

$$E\left[\alpha^2(n)\right] = \sigma_\in^2 + \frac{\sigma_\in^2}{(n-1)}\text{trace}\left[I\right]$$

Since

$$trace[I_N] = N$$

$$E\left[\alpha^2(n)\right] = \sigma_\in^2\left[1 + \frac{N}{(n-1)}\right], \tag{15.7k}$$

As $n \to \infty$,

$$E\left[\alpha^2(n)\right] = \sigma_\in^2 \tag{15.7l}$$

This is the minimum possible mean square error possible that any algorithm can achieve as it is solely due to observation noise $\in(n)$. Hence, RLS algorithm theoretically produces zero misadjustment.

Further, Eqn. (15.7k)

$$E\left[\alpha^2(n)\right] = \sigma_\in^2\left[1 + \frac{N}{(n-1)}\right]$$

indicates that the convergence of mean square error (MSE) does not depend on eigenvalue spread of the input autocorrelation matrix. But in the method of steepest or LMS algorithm, as it is already seen their convergence rate depends on the eigenvalue spread and the largest time constant is given by $1/2\mu\ \eta_{\min}$, where $\eta_{\min}$ is the minimum eigenvalue. The independency of MSE on eigenvalue spread is due to use of $C(n) = R_{NN}^{-1}(n)$ and $P_N(n)$ at every sample instant which involves additional computational load.

For $\lambda \neq 1$,

$$\begin{aligned} R_{NN}(n) &= \sum_{i=1}^{n} \lambda^{n-i} X_N(i) X_N^T(i) \\ &= X_N(n)X_N^T(n) + \lambda X_N(n-1)X_N^T(n-1) \\ &\quad + \lambda^2 X_N(n-2)X_N^T(n-2) + \ldots \lambda^{n-1} X_N(1)X_N^T(1) \end{aligned}$$

For a stationary process since,

$$\begin{aligned} X_N(n)X_N^T(n) &= X_N(n-1)X_N^T(n-1) \\ &= X_N(n-2)X_N^T(n-2) = \ldots = X_N(1)X_N^T(1) = R \end{aligned}$$

Therefore,

$$\begin{aligned} R_{NN}(n) &= R + \lambda R + \lambda^2 R + \ldots\ldots \lambda^{n-1} R \\ &= R(1 + \lambda + \lambda^2 + \ldots\ldots \lambda^{n-1}) \\ &= R + \lambda R + \lambda^2 R + \ldots\ldots \lambda^{n-1} R \end{aligned}$$

$$R_{NN}(n) = R\left[\frac{1-\lambda^n}{1-\lambda}\right]$$

$$R_{NN}^{-1}(n) = R^{-1}\left[\frac{1-\lambda}{1-\lambda^n}\right] \tag{15.8a}$$

Using similar reasoning as above,

$$\mathrm{R}_{NN}^{(2\lambda)}(n) = \left[\sum_{i=1}^{n} \lambda^{2(n-i)} X_N(i) X_N^T(i)\right]$$
$$= X_N(n)X_N^T(n) + \lambda^2 X_N(n-1)X_N^T(n-1)$$
$$+ \lambda^4 X_N(n-2)X_N^T(n-2) + \ldots \lambda^{2(n-1)} X_N(1)X_N^T(1)$$

$$R_{NN}^{(2\lambda)}(n) = R\left[\frac{1-\lambda^{2n}}{1-\lambda^2}\right] \tag{15.8b}$$

Therefore, the Eqn. (15.7f),

$$\mathrm{E}\left[V_N(n)V_N^T(n)\right] = \sigma_\in^2 \mathrm{R}_{NN}^{-1}(n) E\left[\sum_{i=1}^{n} \lambda^{2(n-i)} X_N(i)X_N^T(j)\right] \mathrm{R}_{NN}^{-1}(n)$$

Using Eqns. (15.8a) and (15.8b),

$$\mathrm{E}\left[V_N(n)V_N^T(n)\right] = \sigma_\in^2 R^{-1}\left[\frac{1-\lambda}{1-\lambda^n}\right] R\left[\frac{1-\lambda^{2n}}{1-\lambda^2}\right]\left[\frac{1-\lambda}{1-\lambda^n}\right] R^{-1}$$
$$= \sigma_\in^2 \left[\frac{1-\lambda}{1-\lambda^n}\right]^2 \left[\frac{1-\lambda^{2n}}{1-\lambda^2}\right] R^{-1}$$

$$\mathrm{E}\left[V_N(n)V_N^T(n)\right] = \sigma_\in^2 \left[\frac{1-\lambda}{1+\lambda}\right]\left[\frac{1+\lambda^n}{1-\lambda^n}\right] R^{-1} \tag{15.8c}$$

As $n \to \infty$,

$$\mathrm{E}\left[V_N(\infty)V_N^T(\infty)\right] = \sigma_\in^2 \left[\frac{1-\lambda}{1+\lambda}\right] R^{-1} \tag{15.8d}$$

The *a priori* error is given by Eqn. (15.7h)

$$\alpha(n) = \in(n) - \mathrm{X}_N^T(n)V_N(n-1)$$

$$E\left[\alpha^2(n)\right] = \sigma_\in^2 + E\left\{ V_N^T(n-1)\mathrm{X}_N(n)\mathrm{X}_N^T(n)V_N(n-1)\right\} \tag{15.8e}$$

$$\because E[\in(n)] = 0$$

Since $V_N(n-1)$ is determined by past values of $X_N(n)$ and for large n, $X_N(n)$ and $V_N(n-1)$ can be considered independent. Therefore,

$$E\left\{ V_N^T(n-1)X_N(n)X_N^T(n)V_N(n-1)\right\}$$
$$= E\left[V_N^T(n-1)E\left\{ X_N(n)X_N^T(n)\right\} V_N(n-1)\right]$$
$$= E\left[V_N^T(n-1)RV_N(n-1)\right]$$

This term $E[V_N^T(n-1)RV_N(n-1)]$ is a scalar. Therefore

$$E\left[V_N^T(n-1)RV_N(n-1)\right] = Trace\left\{E\left[V_N^T(n-1)RV_N(n-1)\right]\right\}$$

$$E\left[V_N^T(n-1)RV_N(n-1)\right] = Trace\left\{E\left[V_N(n-1)V_N^T(n-1)R\right]\right\}$$

$$= Trace\left\{E\left[V_N(n-1)V_N^T(n-1)\right] R\right\}$$
$$E\left[V_N^T(n-1)RV_N(n-1)\right] = Trace\left\{\theta_V(n-1)R\right\}$$

Therefore, Eqn.(15.8e)

$$E[\alpha^2(n)] = \sigma_\epsilon^2 + E\left\{V_N^T(n-1)X_N(n)X_N^T(n)V_N(n-1)\right\}$$
$$= \sigma_\epsilon^2 + E[V_N^T(n-1)RV_N(n-1)]$$
$$E[\alpha^2(n)] = \sigma_\epsilon^2 + Trace\left\{\theta_V(n-1)R\right\}$$

Further since

$$\theta_V(\mathrm{n}) = \mathrm{E}\left[V_N(n)V_N^T(n)\right] = \sigma_\epsilon^2\left[\frac{1-\lambda}{1+\lambda}\right]\left[\frac{1+\lambda^n}{1-\lambda^n}\right]R^{-1}$$

$$\theta_V(\mathrm{n\text{-}1}) = \mathrm{E}\left[V_N(n-1)V_N^T(n-1)\right] = \sigma_\epsilon^2\left[\frac{1-\lambda}{1+\lambda}\right]\left[\frac{1+\lambda^{n-1}}{1-\lambda^{n-1}}\right]R^{-1}$$

Therefore

$$E\left[\alpha^2(n)\right] = \sigma_{\in}^2 + Trace\left\{\theta_V(n-1)R\right\}$$

becomes

$$E\left[\alpha^2(n)\right] = \sigma_{\in}^2 + Trace\left\{\sigma_{\in}^2\left[\frac{1-\lambda}{1+\lambda}\right]\left[\frac{1+\lambda^{n-1}}{1-\lambda^{n-1}}\right]R^{-1}R\right\}$$

$$E\left[\alpha^2(n)\right] = \sigma_{\in}^2 + \sigma_{\in}^2\left[\frac{1-\lambda}{1+\lambda}\right]\left[\frac{1+\lambda^{n-1}}{1-\lambda^{n-1}}\right]Trace\left\{R^{-1}R\right\}$$

$$E\left[\alpha^2(n)\right] = \sigma_{\in}^2 + \sigma_{\in}^2\left[\frac{1-\lambda}{1+\lambda}\right]\left[\frac{1+\lambda^{n-1}}{1-\lambda^{n-1}}\right]Trace\left\{I\right\}$$

$$E\left[\alpha^2(n)\right] = \sigma_{\in}^2 + \sigma_{\in}^2\left[\frac{1-\lambda}{1+\lambda}\right]\left[\frac{1+\lambda^{n-1}}{1-\lambda^{n-1}}\right]N$$

As already stated $\sigma_{\in}^2$ is the $\varepsilon_{\min}$, hence

$$E\left[\alpha^2(n)\right] = \varepsilon_{\min} + \varepsilon_{\min}\left[\frac{1-\lambda}{1+\lambda}\right]\left[\frac{1+\lambda^{n-1}}{1-\lambda^{n-1}}\right]N$$

Representing $E[\alpha^2(n)] = \varepsilon_{n-1}(n)$ as it is due to *a priori* error $\alpha(n)$ estimated using $W_N(n-1)$ [and not using $W_N(n)$],

$$\varepsilon_{n-1}(n) = \varepsilon_{\min} + \varepsilon_{\min}\left[\frac{1-\lambda}{1+\lambda}\right]\left[\frac{1+\lambda^{n-1}}{1-\lambda^{n-1}}\right]N \qquad (15.8f)$$

This represents the learning curve for the RLS algorithm. Since the second term on the right-hand side is positive quantity, $\varepsilon_{n-1}(n)$ deviates from $\varepsilon_{\min}$ and further $\varepsilon_{n-1}(n) > \varepsilon_{\min}$. But as n increases $W_N(n)$ converges to g_N and the rate of convergence depends on the factor λ^n. Therefore, the time constant which is the time taken to decay to $(1/e)$ *of its initial value (in this case unity)* that decides the convergence rate is given by

$$\lambda^{\tau} = e^{-1}, \quad \tau\ln(\lambda) = -1, \quad \tau = \frac{-1}{\ln(\lambda)}$$

But

$$\ln(\lambda) = \ln[1-(1-\lambda)]$$

Since $(1-\lambda)$ is small, as λ is close to unity,

$$\ln(\lambda) = -(1-\lambda)$$

Therefore,

$$\tau = \frac{-1}{\ln(\lambda)} = \frac{-1}{-(1-\lambda)}$$

$$\tau = \frac{1}{(1-\lambda)} \tag{15.8g}$$

Thus, the time constant τ depends on the value of λ and smaller the value of λ, the time constant is small making the convergence faster. This is natural as for smaller value of λ, the previous information in the data is forgotten quickly. However, a larger value of λ close to unity can only forget the previous information slowly. Thus, it has been stated that λ close to unity is valid for stationary signals and $\lambda << 1$ is useful for nonstationary signals.

It is important to note that there is only one time constant or one mode of convergence which is different from that of LMS algorithm where many time constants/ eigenvalues/ modes of input autocorrelation matrix decide the convergence rate resulting in a slow convergence rate. As there is only one time constant that decides the convergence rate, it is independent of eigenvalue spread. Further, this point can be made clear by considering the weight vector adaptation given by (Eqn. 15.4g)

$$\mathrm{W}(n) = W(n-1) + \mathrm{K}(n)\,\alpha(n)$$

Since from Eqn. (15.4e)

$$\mathrm{K}(n) = \mathrm{R}_{\mathrm{NN}}^{-1}(n)\mathrm{X}_N(n)$$

Therefore,

$$\mathrm{W}(n) = W(n-1) + R_{NN}^{-1}\ X_N(n)\alpha(n)$$

But

$$R_{NN}^{-1}(n) = R^{-1}\left[\frac{1-\lambda}{1-\lambda^n}\right]$$

Therefore,

$$\mathrm{W}(n) = W(n-1) + R^{-1}\left[\frac{1-\lambda}{1-\lambda^n}\right]\alpha(n)X_N(n)$$

As $n \to \infty$

$$\mathrm{W}(n) = W(n-1) + \left[1-\lambda\right]R^{-1}\alpha(n)X_N(n)$$

If $\lambda = 1 - 2\mu,$

$$W(n) = W(n-1) + 2\mu\, R^{-1}\alpha(n) X_N(n) \tag{15.8h}$$

This is nothing but LMS-Newton algorithm. The Newton algorithm is interpreted as application of method of steepest descent algorithm to the orthogonalized input and the orthogonalization overcomes the eigenvalue spread problem. This does not mean that both RLS and Newton algorithms are same, but the behavior of RLS is similar to that of the latter for large value of n. However, their initial process of convergence may be entirely different.

Again consider Eqn. (15. 8f)

$$\varepsilon_{n-1}(n) = \varepsilon_{\min} + \varepsilon_{\min}\left[\frac{1-\lambda}{1+\lambda}\right]\left[\frac{1+\lambda^{n-1}}{1-\lambda^{n-1}}\right] N$$

As $n \to \infty$,

$$\varepsilon_{n-1}(\infty) = \varepsilon_{\min} + \varepsilon_{\min}\left[\frac{1-\lambda}{1+\lambda}\right] N$$

Therefore, the excess error is given by

$$\varepsilon_{excess} = \varepsilon_{n-1}(\infty) - \varepsilon_{\min} = \varepsilon_{\min}\left[\frac{1-\lambda}{1+\lambda}\right] N$$

Therefore, the misadjustment M is given by

$$M = \frac{\varepsilon_{excess}}{\varepsilon_{\min}} = \frac{\varepsilon_{\min}\left[\frac{1-\lambda}{1+\lambda}\right] N}{\varepsilon_{\min}}$$

$$M = \left[\frac{1-\lambda}{1+\lambda}\right] N \tag{15.8i}$$

Thus, from the above expression for misadjustment, a large value of λ close to unity results in smaller misadjustment. However, from the Eqn. (15.8g), the large value λ close to unity makes the time constant large and convergence becomes slow. But even this convergence rate is better than that obtained by LMS algorithm.

As stated already, if the statistics of the signal are changing with time, a smaller time constant is preferred to track the signal statistical changes. However, this will increase the misadjustment as it is proportional to $(1-\lambda)$ and hence λ value can be decided by the misadjustment that can be tolerated using Eqn. (15.8i) for a particular filter length N.

For RLS, initially, for the error energy to decrease significantly, it may take about (Eqn. 15.8g) 3 filter lengths and this is required to remove the bias effect due to initialization , $C(0) = R^{-1}_{NN}(0) = \delta^{-1}\,I$. Further, the error energy to settle to the steady state value, it may take large number of samples (say *5N*). But the LMS algorithm may take 20N samples. The choice between RLS and LMS depends upon computational load that is permissible. If tracking property is of concern with low computational complexity, then LMS is preferred. But if initial speed of convergence and misadjustment are of concern, RLS is the only solution. It is necessary to note that the accuracy of the weight vector estimate depends on the plant observation noise level and the estimate can match the true plant only if the observation noise is zero.

The complexity of an adaptive algorithm for real-time operation is determined by the number of multiplications (with divisions counted as multiplications) per iteration and the precision required to perform arithmetic operations. For the LMS algorithm, $2N + 1$ multiplications are required and the load , increases linearly with *N*. For RLS algorithm, $3N(3 + N)/2$ multiplications are required and the load increases as the square of *N*. For example, for $N = 21$, LMS algorithm requires only 43 multiplications and RLS algorithm requires 756 multiplications and this is about a factor of 17.5.

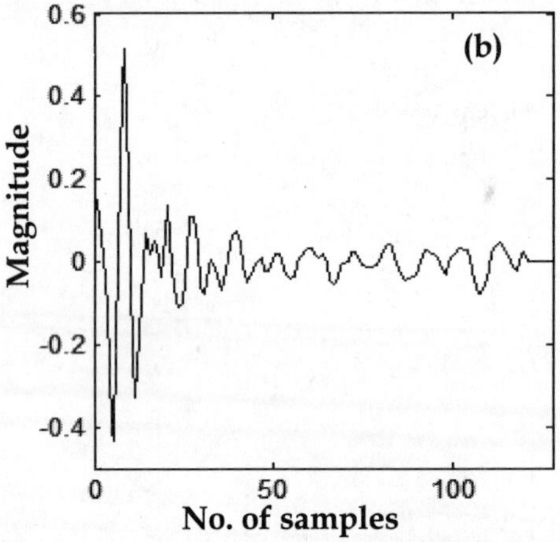

Fig. 15.1(b) Impulse response used for the illustration of RLS performance.

Performance of RLS algorithm for system identification for both types of inputs, viz., the white Gaussian noise (with zero mean and unit variance) and color noise with sinusoids at 100Hz, 150Hz, 200Hz, 250Hz, 300Hz, and 350Hz having amplitudes 0.2, 0.2, 0.1, 0.1, 0.05 and 0.05 respectively. The system impulse response (Fig. 15.1b) is of an acoustic path having 128 samples . The adaptive filter length used is also of length 128 samples. For the LMS (fixed step size), the maximum possible stepsize of value 0.01 is used. For the RLS initialization $C(0) = R^{-1}_{NN}(0) = \delta^{-1}\,I$, $\delta = 1$ and the forgetting factor $\lambda = 0.999$, are used.

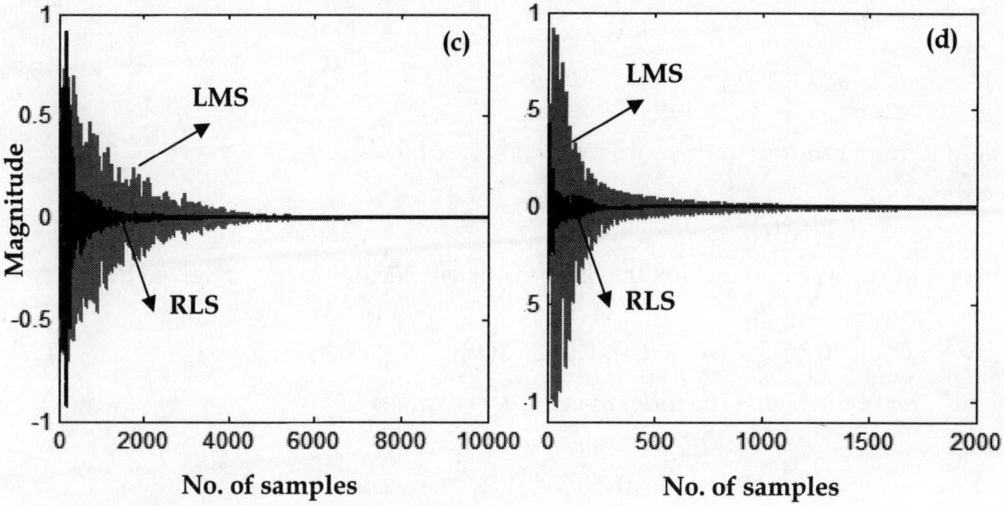

Fig. 15.1(c, d). RLS error convergence for white noise (c) and color noise (d) inputs.

From Figs. 15.1(c, d), it is seen that for both types of inputs, white and color, for the scenario considered, the convergence of RLS is about 5 times faster than LMS algorithm and after convergence, both the algorithms have errors of same order magnitude.

15.5 BACKGROUND THEORY FOR FAST RECURSIVE ALGORITHMS

The standard approach for RLS algorithm exploits transversal structure and its computational complexity is of the order of N^2, where N is the order of the filter. Hence, the complexity has to be reduced. *The algorithm is said to be fast if its computational complexity increases linearly with the order of the filter.* A fast algorithm is therefore, similar to LMS algorithm in its computational requirement.

The mathematical background for fast algorithms is based on adaptive forward and backward predictions, optimized in the least square sense. Basically, the expressions for RLS gain and *a priori* error required for the weight vector adaptation which will be different for different cases, have to be derived. Further, the performance index; the error energy expressions will also be different which have to be derived for different cases.

15.5.1 RLS for Forward Linear Prediction

Let the input vector be $X_N(i-1)$ be given by

$$X_N(i-1) = \left[x(i-1) \quad x(i-2) \quad \ldots \quad x(i-N)\right]^T \tag{15.9a}$$

The desired response $d(i)$ for forward prediction case is $x(i)$, i.e.,

$$d(i) = x(i) \tag{15.9b}$$

Further for the forward prediction, the filter coefficient vector be $W_N^f(n)$ given by

$$W_N^f(n) = [w_1(n), w_2(n), \ldots\ldots, w_N(n)]^T \tag{15.9c}$$

N is the order of the filter. The forward error $e_N^f(i)$ s given by

$$e_N^f(i) = x(i) - W_N^{fT}(n)X_N(i-1) \tag{15.9d}$$

The RLS forward prediction minimizes the prediction error energy $e_N^f(n)$ given by

$$\varepsilon_N^f(n) = \sum_{i=1}^{n} \lambda^{n-i}[e_N^f(i)]^2 \tag{15.9e}$$

Further, the optimal predictor is given by

$$W_N^f(n) = R_{NN}^{-1}(n-1)P_N^f(n) \tag{15.9f}$$

where,

$$R_{NN}(n-1) = \sum_{i=1}^{n} \lambda^{n-i} X_N(i-1)X_N^T(i-1) \tag{15.9g}$$

$$P_N^f(n) = \sum_{i=1}^{n} \lambda^{n-i} x(i)X_N(i-1) \tag{15.9h}$$

The minimum forward prediction error energy is given by

$$\left.\begin{aligned} \varepsilon_N^f(n) &= \sigma_f^2(n) - P_N^{fT}(n)W_N^f(n) \\ \sigma_f^2(n) &= \sum_{i=1}^{n} \lambda^{n-i} x^2(i) \end{aligned}\right\} \tag{15.9i}$$

The forward prediction error filter is given by

$$a_N(n) = \begin{bmatrix} 1 \\ -W_N^f(n) \end{bmatrix} \tag{15.10a}$$

If $X_{N+1}(i)$ indicates the vector of $(N+1)$ most recent samples, then it can be written as

$$X_{N+1}(i) = \begin{bmatrix} x(i) \\ X_N(i-1) \end{bmatrix} \tag{15.10b}$$

and this can be compared with $X_N(i-1)$ given by

$$X_N(i-1) = \begin{bmatrix} x(i-1) & x(i-2) & \dots & x(i-N) \end{bmatrix}^T \tag{15.10c}$$

Therefore, the forward prediction error can be written as

$$e_N^f(i) = a_N^T(n) X_{N+1}(i) \tag{15.10d}$$

The correlation matrix corresponding to the vector $X_{N+1}(i)$ is

$$R_{N+1,N+1}(n) = \sum_{i=1}^{n} \lambda^{n-i} X_{N+1}(i) X_{N+1}^T(i)$$

Consider $X_{N+1}(i) X_{N+1}^T(i)$,

$$X_{N+1}(i) X_{N+1}^T(i) = \begin{bmatrix} x(i) \\ X_N(i-1) \end{bmatrix} \begin{bmatrix} x(i) & X_N^T(i-1) \end{bmatrix}$$

$$= \begin{bmatrix} x^2(i) & x(i) X_N^T(i-1) \\ x(i) X_N(i-1) & X_N(i-1) X_N^T(i-1) \end{bmatrix}$$

Now,

$$R_{N+1,N+1}(n) = \sum_{i=1}^{n} \lambda^{n-i} X_{N+1}(i) X_{N+1}^T(i)$$

$$= \begin{bmatrix} \sum_{i=1}^{n} \lambda^{n-i} x^2(i) & \sum_{i=1}^{n} \lambda^{n-i} x(i) X_N^T(i-1) \\ \sum_{i=1}^{n} \lambda^{n-i} x(i) X_N(i-1) & \sum_{i=1}^{n} \lambda^{n-i} X_N(i-1) X_N^T(i-1) \end{bmatrix}$$

Using Eqns. (15.9g, h, i))

$$R_{N+1,N+1}(n) = \begin{bmatrix} \sigma_f^2(n) & P_N^{fT}(n) \\ P_N^f(n) & R_{NN}(n-1) \end{bmatrix} \tag{15.10d1}$$

Now as (Eqn. (15.10a))

$$a_N(n) = \begin{bmatrix} 1 \\ -W_N^f(n) \end{bmatrix},$$

$$R_{N+1,N+1}(n)a_N(n) = \begin{bmatrix} \sigma_f^2(n) & P_N^{fT}(n) \\ P_N^f(n) & R_{NN}(n-1) \end{bmatrix} \begin{bmatrix} 1 \\ -W_N^f(n) \end{bmatrix}$$

$$R_{N+1,N+1}(n)a_N(n) = \begin{bmatrix} \sigma_f^2(n) - P_N^{fT}(n)W_N^f(n) \\ P_N^f(n) - R_{NN}(n-1)W_N^f(n) \end{bmatrix}$$

From Eqns.(15 .9i) and (15 .9f)

$$\varepsilon_N^f(n) = \sigma_f^2(n) - P_N^{fT}(n)W_N^f(n)$$

$$R_{NN}(n-1)W_N^f(n) = P_N^f(n)$$

Therefore,

$$R_{N+1,N+1}(n)a_N(n) = \begin{bmatrix} \varepsilon_N^f(n) \\ 0 \end{bmatrix} \tag{15.10e}$$

This is known as augmented normal equation for forward LS prediction. The *a priori* forward prediction error $\alpha_N^f(n)$ as given by RLS algorithm is

$$\alpha_N^f(n) = x(n) - W_N^{fT}(n)X_N(n-1) \tag{15.10f}$$

$$\alpha_N^f(n) = \begin{bmatrix} 1 & -W_N^{fT}(n) \begin{bmatrix} x(n) \\ X_N(n-1) \end{bmatrix} \end{bmatrix}$$

Further, as

$$X_{N+1}(n) = \begin{bmatrix} x(n) \\ X_N(n-1) \end{bmatrix}$$

and from (Eqn. (15.10a))

$$a_N(n) = \begin{bmatrix} 1 \\ -W_N^f(n) \end{bmatrix}, \qquad a_N(n-1) = \begin{bmatrix} 1 \\ -W_N^f(n-1) \end{bmatrix}$$

Therefore,

$$\alpha_N^f(n) = a_N^T(n-1)X_{N+1}(n) \tag{15.10g}$$

For the forward predictor $W_N^f(n)$ since the input is $X_N(n-1)$, the corresponding gain has to be used for its adaptation in applying the RLS adaptation rule. Hence, the gain to be used has to be derived from the general RLS case expression given by (Eqn. 15.4e)

$$K_N(n) = R_{NN}^{-1}(n)X_N(n)$$

Therefore, the required gain for the forward predictor adaptation is

$$K_N(n-1) = R_{NN}^{-1}(n-1)X_N(n-1)$$

For the forward prediction, the gain index has $(n-1)$ indicating that it corresponds to the input $X_N(n-1)$. The predictor weight vector adaptation using RLS rule can be applied as the required gain $K_N(n-1)$ and the forward a priori error $\alpha_N^f(n)$ are available and given by

$$W_N^f(n) = W_N^f(n-1) + K_N(n-1)\alpha_N^f(n)$$

Also since prediction error filter $a_N(n)$ is

$$a_N(n) = \begin{bmatrix} 1 \\ -W_N^f(n) \end{bmatrix}$$

The prediction error filter adaptation using RLS rule is

$$a_N(n) = \begin{bmatrix} 1 \\ -W_N^f(n) \end{bmatrix} = \begin{bmatrix} 1 \\ -W_N^f(n-1) \end{bmatrix} - \begin{bmatrix} 0 \\ K_N(n-1) \end{bmatrix} \alpha_N^f(n)$$

$$a_N(n) = a_N(n-1) - \begin{bmatrix} 0 \\ K_N(n-1) \end{bmatrix} \alpha_N^f(n) \tag{15.10h}$$

The a posteriori forward prediction error $e_N^f(n)$ is given by

$$e_N^f(n) = x(n) - W_N^{fT}(n)X_N(n-1)$$

$$e_N^f(n) = a_N^T(n)X_{N+1}(n) \tag{15.10i}$$

Hence, as a posteriori error $e_N^f(n)$ and a priori error $\alpha_N^f(n)$ are available, the forward prediction error energy adaptation using RLS rule is given by

$$\varepsilon_N^f(n) = \lambda\ \varepsilon_N^f(n-1) + \alpha_N^f(n)e_N^f(n) \tag{15.10j}$$

15.5.2 RLS for Backward Linear Prediction

The input vector be $X_N(i)$ consists of N most recent samples of the input and is

$$X_N(i) = \left[x(i) \quad x(i-1) \quad \ldots \quad x(i-N+1)\right]^T \tag{15.11a}$$

The desired response $d(i)$ is at $(i-N)$ for backward prediction is $x(i-N)$, i.e.,

$$d(i) = x(i-N) \tag{15.11b}$$

Further, for the backward prediction, the filter coefficient vector $W_N^b(n)$ is given by

$$W_N^b(n) = [w_0(n), w_1(n), \ldots\ldots, w_{N-1}(n)]^T \tag{15.11c}$$

N is the order of the filter. The backward error $e_N^b(i)$ s given by

$$e_N^b(i) = x(i) - W_N^{bT}(n)X_N(i) \tag{15.11d}$$

The RLS backward prediction minimizes the prediction error energy $\varepsilon_N^b(n)$ given by

$$\varepsilon_N^b(n) = \sum_{i=1}^{n} \lambda^{n-i}[e_N^b(i)]^2 \tag{15.11e}$$

Further, the optimal predictor is given by

$$W_N^b(n) = R_{NN}^{-1}(n)P_N^b(n) \tag{15.11f}$$

where,

$$R_{NN}(n) = \sum_{i=1}^{n} \lambda^{n-i} X_N(i)X_N^T(i) \tag{15.11g}$$

$$P_N^b(n) = \sum_{i=1}^{n} \lambda^{n-i} x(i-N) X_N(i) \tag{15.11h}$$

The minimum forward prediction error energy is given by

$$\left.\begin{aligned} \varepsilon_N^b(n) &= \sigma_b^2(n) - P_N^{bT}(n) W_N^b(n) \\ \sigma_b^2(n) &= \sum_{i=1}^{n} \lambda^{n-i} x^2(i-N) \end{aligned}\right\} \tag{15.11i}$$

The backward prediction error filter is given by

$$b_N(n) = \begin{bmatrix} -W_N^b(n) \\ 1 \end{bmatrix} \tag{15.12a}$$

If $X_{N+1}(i)$ indicates the vector of $(N+1)$ most recent samples, then it can be written as

$$X_{N+1}(i) = \begin{bmatrix} X_N(i) \\ x(i-N) \end{bmatrix} \tag{15.12b}$$

Therefore, the backward prediction error can be written as

$$e_N^b(i) = b_N^T(n) X_{N+1}(i)$$

The correlation matrix corresponding to the vector $X_{N+1}(i)$ is

$$R_{N+1,N+1}(n) = \sum_{i=1}^{n} \lambda^{n-i} X_{N+1}(i) X_{N+1}^T(i)$$

Consider $X_{N+1}(i) X_{N+1}^T(i)$,

$$X_{N+1}(i) X_{N+1}^T(i) = \begin{bmatrix} X_N(i) \\ X_N(i-N) \end{bmatrix} \begin{bmatrix} X_N^T(i) & x(i-N) \end{bmatrix}$$

$$= \begin{bmatrix} X_N(i) X_N^T(i) & X_N(i) x(i-N)) \\ X_N^T(i) x(i-N) & x^2(i-N) \end{bmatrix}$$

Therefore,

$$R_{N+1,N+1}(n) = \sum_{i=1}^{n} \lambda^{n-i} X_{N+1}(i) X_{N+1}^{T}(i)$$

$$= \begin{bmatrix} \sum_{i=1}^{n} \lambda^{n-i} X_N(i) X_N^T(i) & \sum_{i=1}^{n} \lambda^{n-i} X_N(i) x(i-N) \\ \sum_{i=1}^{n} \lambda^{n-i} X_N^T(i) x(i-N) & \sum_{i=1}^{n} \lambda^{n-i} x^2(i-N) \end{bmatrix}$$

Using Eqns. (15.11g, h, i)

$$R_{N+1,N+1}(n) = \begin{bmatrix} R_{NN}(n)) & P_N^b(n) \\ P_N^{bT}(n) & \sigma_b^2(n) \end{bmatrix} \tag{15.12c1}$$

Now as

$$b_N(n) = \begin{bmatrix} -W_N^b(n) \\ 1 \end{bmatrix}$$

$$R_{N+1,N+1}(n) b_N(n) = \begin{bmatrix} R_{NN}(n)) & P_N^b(n) \\ P_N^{bT}(n) & \sigma_b^2(n) \end{bmatrix} \begin{bmatrix} -W_N^b(n) \\ 1 \end{bmatrix}$$

$$R_{N+1,N+1}(n) b_N(n) = \begin{bmatrix} P_N^b(n) - R_{NN}(n) W_N^b(n) \\ \sigma_b^2(n) - P_N^{bT}(n) W_N^b(n)) \end{bmatrix}$$

From Eqns. (15.11i, f),

$$\varepsilon_N^b(n) = \sigma_b^2(n) - P_N^{bT.}(n) W_N^b(n)$$

$$R_{NN}(n) W_N^b(n) = P_N^b(n)$$

Therefore,

$$R_{N+1,N+1}(n) b_N(n) = \begin{bmatrix} 0 \\ \varepsilon_N^b(n) \end{bmatrix} \tag{15.12d}$$

This is known as augmented normal equation for backward LS prediction. The a priori backward prediction error $\alpha_N^b(n)$ as given by RLS algorithm is

$$\alpha_N^b(n) = x(n-N) - W_N^{bT}(n-1) X_N(n) \tag{15.12e}$$

Since

$$b_N(n) = \begin{bmatrix} -W_N^b(n)1 \\ 1 \end{bmatrix}$$

$$b_N(n-1) = \begin{bmatrix} -W_N^b(n-1) \\ 1 \end{bmatrix}$$

Further as

$$X_{N+1}(i) = \begin{bmatrix} X_N(i) \\ x(i-N) \end{bmatrix}$$

Therefore,

$$\alpha_N^b(n) = b_N^T(n-1)X_{N+1}(n) \tag{15.12f}$$

From the general RLS adaptation gain, for the backward prediction case

$$K_N(n) = R_{NN}^{-1}(n)X_N(n)$$

Further adaptation of the weight vector can be written using the general RLS as the required gain $K_N(n)$ and a priori error $\alpha_N^b(n)$ for the backward prediction are available. It is important to note that adaptation gain is indicated by $K_N(n)$ and not by $K_N(n-1)$ and this indicates that the input associated is $X_N(n)$ and not $X_N(n-1)$. The weight vector adaptation rule is

$$W_N^b(n) = W_N^b(n-1) + K_N(n)\alpha_N^b(n) \tag{15.12g}$$

Also since backward prediction error filter $b_N(n)$ is

$$b_N(n) = \begin{bmatrix} -W_N^b(n)1 \\ 1 \end{bmatrix}$$

The prediction error filter adaptation using RLS rule is

$$\begin{bmatrix} -W_N^b(n) \\ 1 \end{bmatrix} = \begin{bmatrix} -W_N^b(n-1) \\ 1 \end{bmatrix} - \begin{bmatrix} K_N(n) \\ 0 \end{bmatrix} \alpha_N^b(n)$$

$$b_N(n) = b_N(n-1) - \begin{bmatrix} K_N(n) \\ 0 \end{bmatrix} \alpha_N^b(n) \tag{15.12h}$$

The a posteriori forward prediction error $e_N^f(n)$ is given by

$$e_N^b(n) = x(n-N) - W_N^{bT}(n)X_N(n)$$

Or

$$e_N^b(n) = b_N^T(n)X_{N+1}(n) \tag{15.12i}$$

Hence the backward prediction error energy adaptation using RLS rule is given by

$$\varepsilon_N^b(n) = \lambda\ \varepsilon_N^b(n-1) + \alpha_N^b(n)e_N^b(n) \tag{15.12j}$$

15.5.3 Gain Relation for Filter Order Update

The weight vector adaptation for RLS algorithm is

$$W_N(n) = W_N(n-1) + K_N(n)\alpha_N(n) \tag{15.13a}$$

It was seen that the Levinson Durbin algorithm plays an important role in obtaining the next higher order filter from its previous lower order and results in an efficient /fast algorithm compared to that of matrix inversion for every next order. For the RLS, the key parameter is the gain and hence, a relationship for gain from the Nth to order to $(N+1)$ order will be important in deriving a fast /efficient RLS algorithm.

The a posteriori gain for the RLS for the Nth order is given by

$$K_N(n) = R_{NN}^{-1}(n)X_N(n)$$

Therefore, the RLS gain for the $(N+1)^{th}$ order is

$$K_{N+1}(n) = R_{N+1N+1}^{-1}(n)X_{N+1}(n)$$

But it is required to find the relation between $K_N(n)$ and $K_{N+1}(n)$. Consider Eqn. (15.10d1),

$$R_{N+1,N+1}(n) = \begin{bmatrix} \sigma_f^2(n) & P_N^{fT}(n) \\ P_N^f(n) & R_{NN}(n-1) \end{bmatrix}$$

If

$$A = \begin{bmatrix} a & b \\ c & d \end{bmatrix}$$

Then

$$A^{-1} = \frac{1}{(ad-bc)}\begin{bmatrix} d & -b \\ -c & a \end{bmatrix}$$

Therefore,

$$R^{-1}_{N+1,N+1}(n) = \frac{\begin{bmatrix} R_{NN}(n-1) & -P_N^{fT}(n) \\ -P_N^f(n) & \sigma_f^2(n) \end{bmatrix}}{\left\{\sigma_f^2(n)R_{NN}(n-1) - P_N^{fT}(n)P_N^f(n)\right\}}$$

Considering the denominator,

$$\begin{aligned}
&\left\{\sigma_f^2(n)R_{NN}(n-1) - P_N^{fT}(n)P_N^f(n)\right\} \\
&= R_{NN}(n-1)\left\{\sigma_f^2(n) - P_N^{fT}(n)R_{NN}^{-1}(n-1)P_N^f(n)\right\} \\
&= R_{NN}(n-1)\left\{\sigma_f^2(n) - P_N^{fT}W_N^f(n)\right\} \\
&= R_{NN}(n-1)\varepsilon_N^f(n)
\end{aligned}$$

Therefore,

$$R^{-1}_{N+1,N+1}(n) = \frac{\begin{bmatrix} R_{NN}(n-1) & -P_N^{fT}(n) \\ -P_N^f(n) & \sigma_f^2(n) \end{bmatrix}}{\left\{R_{NN}(n-1)\varepsilon_N^f(n)\right\}}$$

$$R^{-1}_{N+1,N+1}(n) = \begin{bmatrix} 1/\varepsilon_N^f(n) & -P_N^{fT}(n)/\left\{R_{NN}(n-1)\varepsilon_N^f(n)\right\} \\ -P_N^f(n)/\left\{R_{NN}(n-1)\varepsilon_N^f(n)\right\} & \sigma_f^2(n)/\left\{R_{NN}(n-1)\varepsilon_N^f(n)\right\} \end{bmatrix} \quad (15.13b)$$

From Eqn. (15.9i)

$$\varepsilon_N^f(n) = \sigma_f^2(n) - P_N^{fT}(n)W_N^f(n), \quad \sigma_f^2(n) = \varepsilon_N^f(n) + P_N^{fT}(n)W_N^f(n)$$

Therefore, the term in the above matrix,

$$\frac{\sigma_f^2(n)}{\{R_{NN}(n-1)\varepsilon_N^f(n)\}} = \frac{\varepsilon_N^f(n) + P_N^{fT}(n)W_N^f(n)}{\{R_{NN}(n-1)\varepsilon_N^f(n)\}}$$

$$= \frac{\varepsilon_N^f(n) + W_N^{fT}(n)R_{NN}(n-1)W_N^f(n)}{\{R_{NN}(n-1)\varepsilon_N^f(n)\}}$$

$$= \frac{1}{\varepsilon_N^f(n)}\left[R_{NN}^{-1}(n-1) + W_N^f(n)W_N^{fT}(n)\right]$$

Therefore, Eqn. (15.13b)

$$R_{N+1,N+1}^{-1}(n)$$

$$= 1/\varepsilon_N^f(n)\begin{bmatrix} 1 & -\{R_{NN}(n-1)\}P_N^{fT}(n) \\ R_{NN}^{-1}(n-1)P_N^f(n) & R_{NN}^{-1}(n-1)\varepsilon_N^f(n) + W_N^f(n)W_N^{fT}(n) \end{bmatrix}$$

becomes

$$R_{N+1,N+1}^{-1}(n)$$

$$= \frac{1}{\varepsilon_N^f(n)}\begin{bmatrix} 1 & -W_N^{fT}(n) \\ -W_N^f(n) & R_{NN}^{-1}(n-1)\varepsilon_N^f(n) + W_N^f(n)W_N^{fT}(n) \end{bmatrix}$$

$$\because R_{NN}^{-1}(n-1)P_N^{fT}(n) = W_N^{fT}$$

This can be written as

$$R_{N+1,N+1}^{-1}(n)$$

$$= \begin{bmatrix} 0 & 0 \\ 0 & R_{NN}^{-1}(n-1) \end{bmatrix} + \frac{1}{\varepsilon_N^f(n)}\begin{bmatrix} 1 & -W_N^{fT}(n) \\ -W_N^f(n) & W_N^f(n)W_N^{fT}(n) \end{bmatrix}$$

$$R_{N+1,N+1}^{-1}(n)$$

$$= \begin{bmatrix} 0 & 0 \\ 0 & R_{NN}^{-1}(n-1) \end{bmatrix} + \frac{1}{\varepsilon_N^f(n)} \begin{bmatrix} 1 \\ -W_N^f(n \end{bmatrix} \begin{bmatrix} 1 & -W_N^{fT}(n) \end{bmatrix}$$

Since

$$a_N(n) = \begin{bmatrix} 1 \\ -W_N^f(n) \end{bmatrix}$$

$$R_{N+1,N+1}^{-1}(n) = \begin{bmatrix} 0 & 0 \\ 0 & R_{NN}^{-1}(n-1) \end{bmatrix} + \frac{1}{\varepsilon_N^f(n)} a_N(n) a_N^T(n) \tag{15.13c}$$

For the backward prediction errors, from Eqn.(15.12c1)

$$R_{N+1,N+1}(n) = \begin{bmatrix} R_{NN}(n)) & P_N^b(n) \\ P_N^{bT}(n) & \sigma_b^2(n) \end{bmatrix}$$

Therefore,

$$R_{N+1,N+1}^{-1}(n) = \frac{\begin{bmatrix} \sigma_b^2(n) & -P_N^b(n) \\ -P_N^{bT}(n) & R_{NN}(n) \end{bmatrix}}{\left\{ \sigma_b^2(n) R_{NN}(n) - P_N^{bT}(n) P_N^b(n) \right\}}$$

Considering the denominator,

$$\left\{ \sigma_b^2(n) R_{NN}(n) - P_N^{fT}(n) P_N^f(n) \right\}$$

$$= R_{NN}(n) \left\{ \sigma_b^2(n) - P_N^b(n) R_{NN}^{-1}(n-1) P_N^{bT}(n) \right\}$$

$$= R_{NN}(n) \left\{ \sigma_f^2(n) - P_N^{bT} W_N^f(n) \right\}$$

$$= R_{NN}(n) \varepsilon_N^b(n)$$

Therefore,

$$R_{N+1,N+1}^{-1}(n) = \begin{bmatrix} \sigma_b^2(n) / R_{NN}(n)\varepsilon_N^b(n) & -P_N^b(n) / R_{NN}(n)\varepsilon_N^b(n) \\ -P_N^{bT}(n) / R_{NN}(n)\varepsilon_N^b(n) & R_{NN}(n) / R_{NN}(n)\varepsilon_N^b(n) \end{bmatrix}$$

$$R_{N+1,N+1}^{-1}(n)$$

$$= \begin{bmatrix} \sigma_b^2(n)/\left\{R_{NN}(n)\varepsilon_N^b(n)\right\} & -P_N^b(n)/\left\{R_{NN}(n)\varepsilon_N^b(n)\right\} \\ -P_N^{bT}(n)/\left\{R_{NN}(n)\varepsilon_N^b(n)\right\} & 1/\varepsilon_N^b(n) \end{bmatrix}$$

From Eqn. (15.11i)

$$\varepsilon_N^b(n) = \sigma_b^2(n) - P_N^{bT}(n)W_N^b(n)\,,$$
$$\sigma_b^2(n) = \varepsilon_N^b(n) + P_N^{bT}(n)W_N^b(n)$$

Therefore, the term in the above matrix,

$$\frac{\sigma_b^2(n)}{\left\{R_{NN}(n)\varepsilon_N^b(n)\right\}} = \frac{\varepsilon_N^b(n) + P_N^{bT}(n)W_N^b(n)}{\left\{R_{NN}(n)\varepsilon_N^b(n)\right\}}$$

$$= \frac{\varepsilon_N^b(n) + W_N^{bT}(n)R_{NN}(n)W_N^b(n)}{\left\{R_{NN}(n)\varepsilon_N^b(n)\right\}}$$

$$= \left[R_{NN}^{-1}(n) + \frac{W_N^b(n)W_N^{bT}(n)}{\varepsilon_N^b(n)}\right]$$

$$R_{N+1,N+1}^{-1}(n)$$

$$= \begin{bmatrix} R_{NN}^{-1}(n) + \dfrac{W_N^b(n)W_N^{bT}(n)}{\varepsilon_N^b(n)} & -R_{NN}(n)W_N^b(n)/\left\{R_{NN}(n)\varepsilon_N^b(n)\right\} \\ -R_{NN}(n)W_N^{bT}(n)/\left\{R_{NN}(n)\varepsilon_N^b(n)\right\} & 1/\varepsilon_N^b(n) \end{bmatrix}$$

$$R_{N+1,N+1}^{-1}(n)$$
$$= \begin{bmatrix} R_{NN}^{-1}(n) + \dfrac{W_N^b(n)W_N^{bT}(n)}{\varepsilon_N^b(n)} & -W_N^b(n)/\varepsilon_N^b(n) \\ -W_N^{bT}(n)/\varepsilon_N^b(n) & 1/\varepsilon_N^b(n) \end{bmatrix}$$

This can be written as

$$R_{N+1,N+1}^{-1}(n)$$
$$= \begin{bmatrix} R_{NN}^{-1}(n) & 0 \\ 0 & 0 \end{bmatrix} + \frac{1}{\varepsilon_N^b(n)} \begin{bmatrix} W_N^b(n)W_N^{bT}(n) & -W_N^b(n) \\ -W_N^{bT}(n) & 1 \end{bmatrix}$$

$$R_{N+1,N+1}^{-1}(n)$$
$$= \begin{bmatrix} R_{NN}^{-1}(n) & 0 \\ 0 & 0 \end{bmatrix} + \frac{1}{\varepsilon_N^b(n)} \begin{bmatrix} -W_N^b(n) \\ 1 \end{bmatrix} \begin{bmatrix} -W_N^b(n) & 1 \end{bmatrix}$$

Since

$$b_N(n) = \begin{bmatrix} -W_N^b(n) \\ 1 \end{bmatrix}$$

$$R_{N+1,N+1}^{-1}(n) = \begin{bmatrix} R_{NN}^{-1}(n) & 0 \\ 0 & 0 \end{bmatrix} + \frac{1}{\varepsilon_N^b(n)} b_N(n) b_N^T(n) \tag{15.13d}$$

By definition,

$$K_{N+1}(n) = R_{N+1N+1}^{-1}(n) X_{N+1}(n)$$

and

$$X_{N+1}(n) = \begin{bmatrix} x(n) \\ X_N(n-1) \end{bmatrix}$$

Multiplying Eqn. (15.13c) by $X_{N+1}(n)$,

$$R_{N+1,N+1}^{-1}(n) X_{N+1}(n)$$
$$= \begin{bmatrix} 0 & 0 \\ 0 & R_{NN}^{-1}(n-1) \end{bmatrix} X_{N+1}(n) + \frac{1}{\varepsilon_N^f(n)} a_N(n) a_N^T(n) X_{N+1}(n)$$

$$K_{N+1}(n) = \begin{bmatrix} 0 & 0 \\ 0 & R_{NN}^{-1}(n-1) \end{bmatrix}\begin{bmatrix} x(n) \\ X_N(n-1) \end{bmatrix} + \frac{1}{\varepsilon_N^f(n)} a_N(n) a_N^T(n) X_{N+1}(n)$$

$$\because K_{N+1}(n) = R_{N+1N+1}^{-1}(n) X_{N+1}(n)$$

$$K_{N+1}(n) = \begin{bmatrix} 0 \\ R_{NN}^{-1}(n-1) X_N(n-1) \end{bmatrix} + \frac{1}{\varepsilon_N^f(n)} a_N(n) a_N^T(n) X_{N+1}(n)$$

Since

$$K_N(n-1) = R_{NN}^{-1}(n-1) X_N(n-1)$$

$$a_N^T(n) X_{N+1}(n) = e_N^f(n)$$

$$K_{N+1}(n) = \begin{bmatrix} 0 \\ K_N(n-1) \end{bmatrix} + \frac{e_N^f(n)}{\varepsilon_N^f(n)} a_N(n) \tag{15.13e}$$

This provides the relation between the gain at order $(N+1)$, $K_{N+1}(n)$ at time n to the gain $K_N(n-1)$ of previous order N at time $(n-1)$ and this does not involve any matrix inversion. Similarly, for the backward prediction errors, the gain order update relation will be derived. For this consider Eqn. (15.13d)

$$R_{N+1,N+1}^{-1}(n) = \begin{bmatrix} R_{NN}^{-1}(n) & 0 \\ 0 & 0 \end{bmatrix} + \frac{1}{\varepsilon_N^b(n)} b_N(n) b_N^T(n)$$

By definition,

$$K_{N+1}(n) = R_{N+1N+1}^{-1}(n) X_{N+1}(n), \quad X_{N+1}(n) = \begin{bmatrix} X_N(n) \\ x(n-N) \end{bmatrix}$$

Multiplying Eqn. (15.13c) by $X_{N+1}(n)$,

$$R_{N+1,N+1}^{-1}(n) X_{N+1}(n) = \begin{bmatrix} R_{NN}^{-1}(n) & 0 \\ 0 & 0 \end{bmatrix} X_{N+1}(n) + \frac{1}{\varepsilon_N^b(n)} b a_N(n) b_N^T(n) X_{N+1}(n)$$

$$K_{N+1}(n) = \begin{bmatrix} R_{NN}^{-1}(n) & 0 \\ 0 & 0 \end{bmatrix} \begin{bmatrix} X_N(n) \\ x(n-N) \end{bmatrix} + \frac{1}{\varepsilon_N^b(n)} b_N(n) b_N^T(n) X_{N+1}$$

$$K_{N+1}(n) = \begin{bmatrix} R_{NN}^{-1}(n) X_N(n) \\ 0 \end{bmatrix} + \frac{1}{\varepsilon_N^b(n)} b_N(n) b_N^T(n) X_{N+1}(n)$$

Since

$$K_N(n) = R_{NN}^{-1}(n) X_N(n)$$

$$b_N^T(n) X_{N+1}(n) = e_N^b(n)$$

$$K_{N+1}(n) = \begin{bmatrix} K_N(n) \\ 0 \end{bmatrix} + \frac{e_N^b(n)}{\varepsilon_N^b(n)} b_N(n) \tag{15.13f}$$

It is important to note that for the forward error the gain is updated both for the order and time. However, for the backward error the update of gain is only with respect to order and not for time.

15.5.4 A Priori Gain

Let the a priori gain be denoted by $U_N(n)$ and be defined as

$$R_{NN}(n-1) U_N(n) = X_N(n)$$

Therefore,

$$U_N(n) = R_{NN}^{-1}(n-1) X_N(n) \tag{15.13g}$$

From Eqns. (15.13c)

$$R_{N+1,N+1}^{-1}(n) = \begin{bmatrix} 0 & 0 \\ 0 & R_{NN}^{-1}(n-1) \end{bmatrix} + \frac{1}{\varepsilon_N^f(n)} a_N(n) a_N^T(n)$$

Therefore,

$$R_{N+1,N+1}^{-1}(n-1) = \begin{bmatrix} 0 & 0 \\ 0 & R_{NN}^{-1}(n-2) \end{bmatrix} + \frac{1}{\varepsilon_N^f(n-1)} a_N(n-1) a_N^T(n-1)$$

By definition

$$U_{N+1}(n) = R_{N+1N+1}^{-1}(n-1) X_{N+1}(n), \quad X_{N+1}(n) = \begin{bmatrix} x(n) \\ X_N(n-1) \end{bmatrix}$$

Therefore,

$$U_{N+1}(n) = \begin{bmatrix} 0 & 0 \\ 0 & R_{NN}^{-1}(n-2) \end{bmatrix} \begin{bmatrix} x(n) \\ X_N(n-1) \end{bmatrix} + \frac{1}{\varepsilon_N^f(n-1)} a_N(n-1) a_N^T(n-1) X_{N+1}(n)$$

$$U_{N+1}(n) = \begin{bmatrix} 0 \\ R_{NN}^{-1}(n-2) X_N(n-1) \end{bmatrix} + \frac{1}{\varepsilon_N^f(n-1)} a_N(n-1) a_N^T(n-1) X_{N+1}(n)$$

From Eqn. (15.13g)

$$U_N(n) = R_{NN}^{-1}(n-1) X_N(n)$$

Therefore,

$$U_N(n-1) = R_{NN}^{-1}(n-2) X_N(n-1)$$

Also,

$$a_N^T(n-1) X_{N+1}(n) = \alpha_N^T(n)$$

Hence,

$$U_{N+1}(n) = \begin{bmatrix} 0 \\ U_N(n-1) \end{bmatrix} + \frac{\alpha_N^f(n)}{\varepsilon_N^f(n-1)} a_N(n-1) \tag{15.13h}$$

For the backward prediction, a similar relation will be derived. For this consider Eqn. (15.13d)

$$R_{N+1,N+1}^{-1}(n) = \begin{bmatrix} R_{NN}^{-1}(n) & 0 \\ 0 & 0 \end{bmatrix} + \frac{1}{\varepsilon_N^b(n)} b_N(n) b_N^T(n)$$

$$R_{N+1,N+1}^{-1}(n-1) = \begin{bmatrix} R_{NN}^{-1}(n-1) & 0 \\ 0 & 0 \end{bmatrix} + \frac{1}{\varepsilon_N^b(n-1)} b_N(n-1) b_N^T(n-1)$$

From Eqn. (15.13g)

$$U_N(n) = R_{NN}^{-1}(n-1) X_N(n)$$

$$U_{N+1}(n) = R_{N+1N+1}^{-1}(n-1) X_{N+1}(n)$$

Also

$$X_{N+1}(n) = \begin{bmatrix} X_N(n) \\ x(n-N) \end{bmatrix}$$

$$U_{N+1}(n) = R^{-1}_{N+1N+1}(n-1)X_{N+1}(n)$$

$$R^{-1}_{N+1,N+1}(n-1)X_{N+1}(n)$$

$$= \left\{ \begin{bmatrix} R^{-1}_{NN}(n-1) & 0 \\ 0 & 0 \end{bmatrix} + \frac{1}{\varepsilon^b_N(n-1)} b_N(n-1)b^T_N(n-1) \right\} X_{N+1}(n)$$

$$= \begin{bmatrix} R^{-1}_{NN}(n-1) & 0 \\ 0 & 0 \end{bmatrix} \begin{bmatrix} X_N(n) \\ x(n-N) \end{bmatrix} + \frac{1}{\varepsilon^b_N(n-1)} b_N(n-1)b^T_N(n-1)X_{N+1}(n)$$

$$= \begin{bmatrix} R^{-1}_{NN}(n-1)X_N(n) \\ 0 \end{bmatrix} + \frac{1}{\varepsilon^b_N(n-1)} b_N(n-1)b^T_N(n-1)X_{N+1}(n)$$

Since

$$U_N(n) = R^{-1}_{NN}(n-1)X_N(n), \quad \alpha^b_N(N) = b^T_N(n-1)X_{N+1}(n)$$

$$U_{N+1}(n) = \begin{bmatrix} U_N(n) \\ 0 \end{bmatrix} + \frac{\alpha^b_N(n)}{\varepsilon^b_N(n-1)} b_N(n-1) \tag{15.13i}$$

15.5.5 Relation between $U_N(n)$ and $K_N(n)$

The RLS gain $K_N(n)$ is given by

$$K_N(n) = R^{-1}_{NN}(n)X_N(n), \quad R_{NN}(n)K_N(n) = X_N(n)$$

But

$$U_N(n) = R^{-1}_{NN}(n-1)X_N(n)$$

Therefore, substituting for $X_N(n)$,

$$U_N(n) = R^{-1}_{NN}(n-1)R_{NN}(n)K_N(n)$$

But $R_{NN}(n)$ adaptation with respect to time is given by

$$R_{NN}(n) = \lambda R_{NN}(n-1) + X_N(n)X_N^T(n)$$

Hence, substituting for $R_{NN}(n)$

$$U_N(n) = R_{NN}^{-1}(n-1)\left[\lambda R_{NN}(n-1) + X_N(n)X_N^T(n)\right] K_N(n)$$

$$= \lambda K_N(n) + \underline{R_{NN}^{-1}(n-1)X_N(n)}X_N^T(n)K_N(n)$$

$$U_N(n) = \lambda K_N(n) + U_N(n)K_N(n)X_N^T(n)$$

$$U_N(n)\left[1 - K_N(n)X_N^T(n)\right] = \lambda K_N(n)$$

$$U_N(n) = \frac{\lambda}{\left[1 - K_N(n)X_N^T(n)\right]} K_N(n) \tag{15.14}$$

15.5.6 Error Conversion Factor

The conversion of a priori error to a posteriori error is necessary for the derivation of fast RLS algorithm. This conversion factor $\varphi(n)$ is defined as

$$\phi(n) = \frac{e(n)}{\alpha(n)} \tag{15.15a}$$

$$\alpha(n) = d(n) - W_N^T(n-1)X_N(n)$$

and

$$e(n) = d(n) - W_N^T(n)X_N(n)$$

Further, the adaptation of $W_N(n)$ is

$$W_N(n) = W_N(n-1) - K_N(n)\alpha(n)$$

Therefore,

$$e(n) = d(n) - [W_N(n-1) - K_N(n)\alpha(n)]^T X_N(n)$$

$$e(n) = d(n) - W_N^T(n-1)X_N(n)) - K_N^T(n)\alpha(n)X_N(n)$$
$$e(n) = \alpha(n) - K_N^T(n)X_N(n)\alpha(n)$$
$$e(n) = \alpha(n)\left[1 - K_N^T(n)X_N(n)\right]$$

Therefore,

$$\phi(n) = \frac{e(n)}{\alpha(n} = \left[1 - K_N^T(n)X_N(n)\right]$$

But

$$K_N(n) = R_{NN}^{-1}(n)X_N(n)$$

Hence,

$$\phi(n) = \left[1 - X_N^T(n)R_{NN}^{-1}(n)X_N(n)\right]$$

Since $\phi(n)$ is the ratio of two estimation errors and as if a priori error $\alpha(n)$ is known, the a posteriori error $e(n)$ can be obtained, it is known as conversion factor. Therefore, knowing $\alpha(n)$, $e(n)$ is given by $e(n) = \phi(n)\alpha(n)$. For the forward prediction error,

$$\frac{e_N^f(n)}{\alpha_N^f(n} = \left[1 - K_N^T(n-1)X_N(n-1)\right]$$

This is represented by $\phi_N(n-1)$ as

$$\phi_N(n-1) = \frac{e_N^f(n)}{\alpha_N^f(n)} = \left[1 - K_N^T(n-1)X_N(n-1)\right] \tag{15.15b}$$

The notation $\phi_N(n-1)$ indicates that the conversion factor involves $K_N^T(n-1)$ and $X_N(n-1)$. For the backward prediction error,

$$\left.\begin{aligned} \phi_N(n) &= \frac{e_N^b(n)}{\alpha_N^b(n)} \\ \phi_N(n) &= \left[1 - K_N^T(n)X_N(n)\right] \end{aligned}\right\} \tag{15.15c}$$

The notation used indicates that the conversion factor $\phi_N(n)$ involves $K_N^T(n)$ and $X_N(n)$.

As the *a priori* error $\alpha(n)$ and the a posteriori error $e(n)$ are related, their corresponding gains $U_N(n)$ and $K_N(n)$, can be related respectively. The a priori adaptation gain is given by Eqn. (15.14)

$$U_N(n) = \frac{\lambda}{\left[1 - K_N^T(n)X_N(n)\right]} K_N(n)$$

Since from Eqn. (15.15c),

$$\phi_N(n) = \left[1 - K_N^T(n)X_N(n)\right], \quad U_N(n) = \frac{\lambda}{\varphi_N(n)} K_N(n)$$

$$K_N(n) = \frac{\phi_N(n)}{\lambda} U_N(n) \tag{15.15d}$$

15.5.7 Order Updating for Error Conversion Factor

The adaptation of the conversion factor as filter order increases from N to $(N+1)$ is necessary for fast RLS algorithm. The conversion factor for the $(N+1)$ order is given by

$$\phi_{N+1}(n) = \frac{e_{N+1}^f(n+1)}{\alpha_{N+1}^f(n+1)}$$

$$\phi_{N+1}(n) = \left[1 - K_{N+1}^T(n)X_{N+1}(n)\right]$$

But $K_{N+1}(n)$ is given by (Eqn.15.13e)

$$K_{N+1}(n) = \begin{bmatrix} 0 \\ K_N(n-1) \end{bmatrix} + \frac{e_N^f(n)}{\varepsilon_N^f(n)} a_N(n), \quad X_{N+1}(n) = \begin{bmatrix} x(n) \\ X_N(n-1) \end{bmatrix}$$

Therefore,

$$K_{N+1}^T(n)X_{N+1}(n) = \left\{ \begin{bmatrix} 0 \\ K_N(n-1) \end{bmatrix} + \frac{e_N^f(n)}{\varepsilon_N^f(n)} a_N(n) \right\}^T \begin{bmatrix} x(n) \\ X_N(n-1) \end{bmatrix}$$

$$K_{N+1}^T(n)X_{N+1}(n) = \begin{bmatrix} 0 & K_N^T(n-1) \end{bmatrix} \begin{bmatrix} x(n) \\ X_N(n-1) \end{bmatrix} + \frac{e_N^f(n)}{\varepsilon_N^f(n)} a_N^T(n)X_{N+1}(n)$$

Since

$$a_N^T(n)X_{N+1}(n) = e_N^f(n),$$

$$K_{N+1}^T(n)X_{N+1}(n) = K_N^T(n-1)X_N(n-1) + \frac{[e_N^f(n)]^2}{\varepsilon_N^f(n)}$$

Substituting this in

$$\phi_{N+1}(n) = \left[1 - K_{N+1}^T(n)X_{N+1}(n)\right]$$

$$\phi_{N+1}(n) = \left\{1 - K_N^T(n-1)X_N(n-1)\right\} - \frac{[e_N^f(n)]^2}{\varepsilon_N^f(n)}$$

Since

$$\phi_N(n-1) = \left\{1 - K_N^T(n-1)X_N(n-1)\right\},$$

$$\phi_{N+1}(n) = \phi_N(n-1) - \frac{[e_N^f(n)]^2}{\varepsilon_N^f(n)} \tag{15.15e}$$

Similarly, for the backward prediction,

$$\phi_{N+1}(n) = \phi_N(n) - \frac{[e_N^b(n)]^2}{\varepsilon_N^b(n)} \tag{15.15f}$$

To prove this, consider (Eqn. 15.15c),

$$\phi_N(n) = \frac{e_N^b(n)}{\alpha_N^b(n)}$$

From Eqn. (15.15c)

$$\phi_N(n) = \left[1 - K_N^T(n)X_N(n)\right],$$

Also

$$K_{N+1}(n) = \begin{bmatrix} K_N(n) \\ 0 \end{bmatrix} + \frac{e_N^b(n)}{\varepsilon_N^b(n)} b_N(n) \quad \text{and} \quad X_{N+1}(n) = \begin{bmatrix} X(n) \\ x(n-N) \end{bmatrix}$$

Consider

$$K_{N+1}^T(n)X_{N+1}(n) = \left\{ \begin{bmatrix} K_N(n) \\ 0 \end{bmatrix} + \frac{e_N^b(n)}{\varepsilon_N^b(n)} b_N(n) \right\}^T \begin{bmatrix} X_N(n) \\ x(n-N) \end{bmatrix}$$

$$K_{N+1}^T(n)X_{N+1}(n) = \begin{bmatrix} K_N^T(n) & 0 \end{bmatrix} \begin{bmatrix} X_N(n) \\ x(n-N) \end{bmatrix} + \frac{e_N^b(n)}{\varepsilon_N^b(n)} b_N^T(n)X_{N+1}(n)$$

$$K_{N+1}^T(n)X_{N+1}(n) = K_N^T(n)X_N(n) + \frac{e_N^b(n)}{\varepsilon_N^b(n-1)} b_N^T(n)X_{N+1}(n)$$

Therefore,

$$1 - K_{N+1}^T(n)X_{N+1}(n) = 1 - K_N^T(n)X_N(n) - \frac{e_N^b(n)}{\varepsilon_N^b(n-1)} e_N^b(n)$$

$$\because e_N^b(n) = b_N^T(n)X_{N+1}(n)$$

But

$$\phi_N(n) = 1 - K_N^T(n)X_N(n)$$

$$\phi_{N+1}(n) = 1 - K_{N+1}^T(n)X_{N+1}(n)$$

Therefore,

$$\phi_{N+1}(n) = \phi_N(n) - \frac{[e_N^b(n)]^2}{\varepsilon_N^b(n-1)}$$

15.6 FAST TRANSVERSAL FILTER RLS ALGORITHM (FTFRLS)

In this algorithm, the posteriori error $e(n)$ and a priori gain $U_N(n)$ are used. The a priori forward and backward errors are given by

$$\alpha_N^f(n) = a_N^T(n-1)X_{N+1}(n) \tag{15.16a}$$

$$\alpha_N^b(n) = b_N^T(n-1)X_{N+1}(n) \tag{15.16b}$$

The posteriori and a priori gains $K_N(n)$ and $U_N(n)$, respectively are related by

$$K_N(n) = \lambda^{-1}\phi_N(n)U_N(n)$$

Hence,

$$K_N(n-1) = \lambda^{-1}\phi_N(n-1)U_N(n-1)$$

The forward prediction error filter is given by Eqn. (15.10h)

$$a_N(n) = a_N(n-1) - \begin{bmatrix} 0 \\ K_N(n-1) \end{bmatrix} \alpha_N^f(n)$$

Therefore,

$$a_N(n) = a_N(n-1) - \begin{bmatrix} 0 \\ \lambda^{-1}\phi_N(n-1)U_N(n-1) \end{bmatrix} \alpha_N^f(n)$$

$$a_N(n) = a_N(n-1) - \begin{bmatrix} 0 \\ U_N(n-1) \end{bmatrix} \lambda^{-1}\varphi_N(n-1)\alpha_N^f(n) \qquad (15.16c)$$

The a priori gain $U_N(n)$ order adaptation is given by Eqn. (15.13h)

$$U_{N+1}(n) = \begin{bmatrix} 0 \\ U_N(n-1) \end{bmatrix} + \frac{\alpha_N^f(n)}{\varepsilon_N^f(n-1)} a_N(n-1)$$

This can be portioned as

$$U_{N+1}(n) = \begin{bmatrix} q(n) \\ Q(n) \end{bmatrix} = \begin{bmatrix} 0 \\ U_N(n-1) \end{bmatrix} + \frac{\alpha_N^f(n)}{\varepsilon_N^f(n-1)} a_N(n-1) \qquad (15.16d)$$

$Q(n)$ is a $(N \times 1)$ vector and $q(n)$ is a scalar.

$$\begin{bmatrix} q(n) \\ Q(n) \end{bmatrix} = \begin{bmatrix} 0 \\ U_N(n-1) \end{bmatrix} + \frac{\alpha_N^f(n)}{\varepsilon_N^f(n-1)} \begin{bmatrix} 1 \\ -W_N^f(n-1) \end{bmatrix}$$

$$= \begin{bmatrix} \alpha_N^f(n) / \varepsilon_N^f(n-1) \\ U_N(n-1) - \dfrac{\alpha_N^f(n) W_N^f(n-1)}{\varepsilon_N^f(n-1)} \end{bmatrix} = \begin{bmatrix} q(n) \\ U_N(n-1) - q(n) W_N^f(n-1) \end{bmatrix}$$

Hence,

$$q(n) = \frac{\alpha_N^f(n)}{\varepsilon_N^f(n-1)}, \quad Q(n) = U_N(n-1) - q(n) W_N^f(n-1) \tag{15.16e}$$

For the backward prediction, the *a priori* gain order updating is given by (Eqn. (15.13i)

$$U_{N+1}(n) = \begin{bmatrix} U_N(n) \\ 0 \end{bmatrix} + \frac{\alpha_N^b(n)}{\varepsilon_N^b(n-1)} b_N(n-1)$$

$$U_{N+1}(n) = \begin{bmatrix} Q(n) \\ q(n) \end{bmatrix} = \begin{bmatrix} U_N(n) \\ 0 \end{bmatrix} + \frac{\alpha_N^b(n)}{\varepsilon_N^b(n-1)} \begin{bmatrix} -W_N^b(n-1) \\ 1 \end{bmatrix}$$

$Q(n)$ is a $(N \times 1)$ vector and $q(n)$ is a scalar.

$$\begin{bmatrix} Q(n) \\ q(n) \end{bmatrix} = \begin{bmatrix} U_N(n) - \dfrac{\alpha_N^b(n)}{\varepsilon_N^b(n-1)} W_N^b(n-1) \\ \alpha_N^b(n) / \varepsilon_N^b(n-1) \end{bmatrix}$$

$$\begin{bmatrix} Q(n) \\ q(n) \end{bmatrix} = \begin{bmatrix} U_N(n) - q(n) W_N^b(n-1) \\ q(n) \end{bmatrix}$$

Therefore,

$$q(n) = \frac{\alpha_N^b(n)}{\varepsilon_N^b(n-1)}, \quad Q(n) = U_N(n) - q(n) W_N^b(n-1)$$

$$U_N(n) = Q(n) + q(n) W_N^b(n-1)$$

Since $Q(n)$ and $q(n)$ are known from Eqn. (15.16e), the a priori gain $U_N(n)$ can be computed.

15.6.1 Error Conversion Factor Order Update

The conversion factor order update is given by (Eqn. (15.15e))

$$\phi_{N+1}(n) = \phi_N(n-1) - \frac{[e_N^f(n)]^2}{\varepsilon_N^f(n)}$$

From Eqn. (15.15b)

$$\phi_N(n-1) = \frac{e_N^f(n)}{\alpha_N^f(n)}$$

Therefore,

$$\phi_{N+1}(n) = \frac{e_N^f(n)}{\alpha_N^f(n)} - \frac{[e_N^f(n)]^2}{\varepsilon_N^f(n)}$$

$$\phi_{N+1}(n) = \frac{e_N^f(n)\varepsilon_N^f(n) - \alpha_N^f(n)[e_N^f(n)]^2}{\alpha_N^f(n)\varepsilon_N^f(n)}$$

$$\phi_{N+1}(n) = \left(\frac{e_N^f(n)}{\alpha_N^f(n)}\right)\frac{\left\{\varepsilon_N^f(n) - \alpha_N^f(n)e_N^f(n)\right\}}{\varepsilon_N^f(n)}$$

Since

$$\phi_N(n-1) = \frac{e_N^f(n)}{\alpha_N^f(n)},$$

$$\phi_{N+1}(n) = \phi_N(n-1)\frac{\left\{\varepsilon_N^f(n) - \alpha_N^f(n)e_N^f(n)\right\}}{\varepsilon_N^f(n)}$$

The forward energy time update is given by

$$\varepsilon_N^f(n) = \lambda\ \varepsilon_N^f(n-1) + \alpha_N^f(n)e_N^f(n)$$

$$\lambda\ \varepsilon_N^f(n-1) = \varepsilon_N^f(n) - \alpha_N^f(n)e_N^f(n)$$

Hence,

$$\phi_{N+1}(n) = \phi_N(n-1)\frac{\lambda\ \varepsilon_N^f(n-1)}{\varepsilon_N^f(n)} \tag{15.16f}$$

Similarly, for backward prediction,

$$\phi_{N+1}(n) = \phi_N(n)\frac{\lambda\ \varepsilon_N^b(n-1)}{\varepsilon_N^b(n)}$$

To derive this, consider from Eqn.(15.15d)

$$\phi_{N+1}(n) = \frac{e_{N+1}^b(n)}{\alpha_{N+1}^b(n)},\quad \phi_N(n) = \frac{e_N^b(n)}{\alpha_N^b(n)}$$

Further, from Eqn. (15. 15f)

$$\phi_{N+1}(n) = \phi_N(n) - \frac{[e_N^b(n)]^2}{\varepsilon_N^b(n)}$$

Substituting for $\phi_N(n)$,

$$\phi_{N+1}(n) = \frac{e_N^b(n)}{\alpha_N^b(n)} - \frac{[e_N^b(n)]^2}{\varepsilon_N^b(n)}$$

$$\phi_{N+1}(n) = \frac{\varepsilon_N^b(n)e_N^b(n) - \alpha_N^b(n)[e_N^b(n)]^2}{\alpha_N^b(n)\varepsilon_N^b(n)}$$

$$= \frac{e_N^b(n)}{\alpha_N^b(n)}\,\frac{\varepsilon_N^b(n) - \alpha_N^b(n)e_N^b(n)}{\varepsilon_N^b(n)}$$

$$= \varphi_N(n)\frac{\varepsilon_N^b(n) - \alpha_N^b(n)e_N^b(n)}{\varepsilon_N^b(n)}$$

But the backward energy time update is given by Eqn. (15.12j)

$$\varepsilon_N^b(n) = \lambda\ \varepsilon_N^b(n-1) + \alpha_N^b(n)e_N^b(n)$$
$$\varepsilon_N^b(n) - \alpha_N^b(n)e_N^b(n) = \lambda\ \varepsilon_N^b(n-1)$$

Hence,

$$\phi_{N+1}(n) = \phi_N(n)\frac{\lambda\ \varepsilon_N^b(n-1)}{\varepsilon_N^b(n)} \qquad (15.16g)$$

or $$\phi_N(n) = \phi_{N+1}(n)\frac{\varepsilon_N^b(n)}{\lambda\ \varepsilon_N^b(n-1)}$$

Again substituting for backward error energy,

$$\phi_N(n) = \phi_N(n)\frac{\lambda\ \varepsilon_N^b(n-1) + \alpha_N^b(n)e_N^b(n))}{\lambda\ \varepsilon_N^b(n-1)}$$

$$\phi_N(n) = \phi_{N+1}(n)\left[1 + \frac{\alpha_N^b(n)e_N^b(n)}{\lambda\ \varepsilon_N^b(n-1)}\right] \tag{15.16h}$$

Using the representation

$$q(n) = \frac{\alpha_N^b(n)}{\varepsilon_N^b(n-1)},$$

$$\phi_N(n) = \phi_{N+1}(n)\left[1 + \lambda^{-1}q(n)e_N^b(n)\right]$$

For backward prediction, as the a posteriori error can be obtained from the *a priori* error by

$$e_N^b(n) = \phi_N(n)\alpha_N^b(n)$$

Then

$$\phi_N(n) = \phi_{N+1}(n)\left[1 + \lambda^{-1}q(n)\phi_N(n)\alpha_N^b(n)\right]$$

$$\phi_N(n)\left[1 - \phi_{N+1}(n)\lambda^{-1}q(n)\alpha_N^b(n)\right] = \phi_{N+1}(n)$$

$$\phi_N(n) = \frac{\phi_{N+1}(n)}{\left[1 - \phi_{N+1}(n)\lambda^{-1}q(n)\alpha_N^b(n)\right]} \tag{15.16i}$$

For backward prediction error filter from Eqn. (15.12h)

$$b_N(n) = b_N(n-1) - \begin{bmatrix} K_N(n) \\ 0 \end{bmatrix}\alpha_N^b(n)$$

But from Eqn. (15.15d)

$$K_N(n) = \frac{\phi_N(n)}{\lambda} U_N(n)$$

Substituting this,

$$b_N(n) = b_N(n-1) - \begin{bmatrix} \frac{\phi_N(n)}{\lambda} U_N(n) \\ 0 \end{bmatrix} \alpha_N^b(n)$$

$$b_N(n) = b_N(n-1) - \begin{bmatrix} U_N(n) \\ 0 \end{bmatrix} \lambda^{-1} \phi_N(n) \alpha_N^b(n) \tag{15.16j}$$

Here ends the prediction part of the algorithm. The LS filtering and updating can be done as the a priori gain is available. The a priori estimation error is given by

$$\alpha_N(n) = d(n) - W_N^T(n-1) X_N(n)$$

The LS weight vector adaptation is given by

$$W_N(n) = W_N(n-1) + K_N(n) \alpha_N(n)$$

Since the gain $K_N(n)$ is related to a priori adaptation gain $U_N(n)$ by

$$K_N(n) = \lambda^{-1} \phi_N(n) U_N(n)$$

The LS weight vector adaptation using a priori adaptation gain $U_N(n)$ is given by

$$W_N(n) = W_N(n-1) + \lambda^{-1} \phi_N(n) U_N(n) \alpha_N(n) \tag{15.16k}$$

This forms the end of the ***fast transversal filter RLS*** algorithm.

15.6.2 Fast Transversal Filter RLS (FTFRLS) Algorithm

It starts with the initialization of $\varepsilon_N^f(0), a_N(0), b_N(0), W_N(0)$.

Further, the variables $a_N(n-1), b_N(n-1), U_N(0), \varepsilon_N^f(n-1)$ are available for every n.

Prediction adaptation gain updation: By Eqns. (15.16a, b)

$$\alpha_N^f(n) = a_N^T(n-1)X_{N+1}(n)$$
$$\alpha_N^b(n) = b_N^T(n-1)X_{N+1}(n)$$

By Eqn. (15.16c)

$$a_N(n) = a_N(n-1) - \begin{bmatrix} 0 \\ U_N(n-1) \end{bmatrix} \lambda^{-1}\varphi_N(n-1)\alpha_N^f(n)$$

By Eqn. (15.16d)

$$U_{N+1}(n) = \begin{bmatrix} q(n) \\ Q(n) \end{bmatrix} = \begin{bmatrix} 0 \\ U_N(n-1) \end{bmatrix} + \frac{\alpha_N^f(n)}{\varepsilon_N^f(n-1)} a_N(n-1)$$

By Eqn. (15.16e)

$$q(n) = \frac{\alpha_N^b(n)}{\varepsilon_N^b(n-1)}, \quad Q(n) = U_N(n) - q(n)W_N^b(n-1)$$

Knowing $q(n)$ and $Q(n)$, $U_N(n)$ can be found using

$$U_N(n) = Q(n) + q(n)W_N^b(n-1)$$

Forward error energy is found by

$$\varepsilon_N^f(n) = \lambda\ \varepsilon_N^f(n-1) + \phi_N(n-1)\left[\alpha_N^f(n)\right]^2$$

(Since $\varepsilon_N^f(n) = \lambda\ \varepsilon_N^f(n-1) + \alpha_N^f(n)e_N^f(n)$, $\quad \phi_N(n-1) = \dfrac{e_N^f(n)}{\alpha_N^f(n)}$)

Further by Eqn. (15.16f)

$$\phi_{N+1}(n) = \phi_N(n-1)\frac{\lambda\ \varepsilon_N^f(n-1)}{\varepsilon_N^f(n)}$$

By Eqn. (15.16i)

$$\phi_N(n) = \frac{\phi_{N+1}(n)}{\left[1-\phi_{N+1}(n)\lambda^{-1}q(n)\alpha_N^b(n)\right]}$$

By Eqn. (15.16j)

$$b_N(n) = b_N(n-1) - \begin{bmatrix} U_N(n) \\ 0 \end{bmatrix}\lambda^{-1}\phi_N(n)\alpha_N^b(n)$$

Filtering and weight vector adaptation:

$$\alpha_N(n) = d(n) - W_N^T(n-1)X_N(n)$$

$$W_N(n) = W_N(n-1) + K_N(n)\alpha_N(n), \text{ where } K_N(n) = \lambda^{-1}\phi_N(n)U_N(n)$$

The computational load of the FTF RLS algorithm involves $8N$ out of which $6N$ multiplications for gain updating and $2N$ multiplications for filtering are used. The old fast RLS algorithm, known as Fast Kalman algorithm requires $10N$ multiplications and this achievement in computational reduction is significant compared to the direct RLS requirement of N^2 multiplications. For the fast Kalman algorithm, adaptation of gain computations are based on a priori prediction errors. By using a posteriori errors, the computational load can be reduced and such an algorithm is known as fast a posteriori errors sequential algorithm (FAEST). The FAEST and FTF-RLS require same computational load but they differ in gain conversion updating. The FTF-RLS requires $8N$ instead of $10N$ multiplications which is 20% better than that of Fast Kalman and hence, the present discussion was confined only to the FTF algorithm.

15.7 RECURSIVE LEAST SQUARE LATTICE ALGORITHM

In FTF-RLS, the adaptation is done indirectly through auxiliary variables which lead to round off error accumulation. This makes the FTF-RLS algorithm sensitive to finite word length effects. Hence, the FTF–RLS demands high numerical precision and this is met by the floating point arithmetic. However, in some situations, even the floating point arithmetic may not be sufficient and this has led to development of new algorithms like the one based on lattice structure which is known for *its better insensitivity to round off errors*. Further, the

FTF-RLS is only time recursive. A change in filter length results in a filter with totally different set of coefficients. But as the lattice filter is order recursive, all intermediate order filters are available and this will be useful when correct filter order required is not known. These make lattice structure attractive for RLS and therefore presently the lattice structure based RLS (RLSL) will be considered.

The augmented normal equations for forward (Eqn. 15.10e) and backward (Eqn. 15.12d) LS predictions are

$$R_{N+1N+1}(n)a_N(n) = \begin{bmatrix} \varepsilon_N^f(n) \\ 0 \end{bmatrix}, \quad R_{N+1,N+1}(n)b_N(n) = \begin{bmatrix} 0 \\ \varepsilon_N^b(n) \end{bmatrix}$$

Therefore,

$$R_{N+1N+1}(n)a_{N-1}(n) = \begin{bmatrix} \varepsilon_{N-1}^f(n) \\ 0 \end{bmatrix}, \; R_{N+1N+1}(n)b_{N-1}(n) = \begin{bmatrix} 0 \\ \varepsilon_{N-1}^b(n) \end{bmatrix}$$

Further, from Eqns. (15.10d1) and (15.12c1),

$$R_{N+1,N+1}(n) = \begin{bmatrix} \sigma_f^2(n) & P_N^{fT}(n) \\ P_N^f(n) & R_{NN}(n-1) \end{bmatrix}, \quad R_{N+1,N+1}(n) = \begin{bmatrix} R_{NN}(n) & P_N^b(n) \\ P_N^{bT}(n) & \sigma_b^2(n) \end{bmatrix}$$

Consider the product,

$$R_{N+1,N+1}(n)\begin{bmatrix} a_{N-1}(n) \\ 0 \end{bmatrix} = \begin{bmatrix} R_{NN}(n) & P_N^b(n) \\ P_N^{bT}(n) & \sigma_b^2(n) \end{bmatrix}\begin{bmatrix} a_{N-1}(n) \\ 0 \end{bmatrix}$$

$$R_{N+1,N+1}(n)\begin{bmatrix} a_{N-1}(n) \\ 0 \end{bmatrix} = \begin{bmatrix} R_{NN}(n)a_{N-1}(n) \\ P_N^{bT}(n)a_{N-1}(n) \end{bmatrix}$$

Also

$$R_{N+1,N+1}(n)\begin{bmatrix} 0 \\ b_{N-1}(n-1) \end{bmatrix} = \begin{bmatrix} \sigma_f^2(n) & P_N^{fT}(n) \\ P_N^f(n) & R_{NN}(n-1) \end{bmatrix}\begin{bmatrix} 0 \\ b_{N-1}(n-1) \end{bmatrix}$$

$$R_{N+1,N+1}(n)\begin{bmatrix} 0 \\ b_{N-1}(n-1) \end{bmatrix} = \begin{bmatrix} P_N^{fT}(n)b_{N-1}(n-1) \\ R_{NN}(n-1)b_{N-1}(n-1) \end{bmatrix}$$

Consider

$$R_{N+1,N+1}(n)\left\{\begin{bmatrix} a_{N-1}(n) \\ 0 \end{bmatrix} + \mathbf{\Gamma}_N^f \begin{bmatrix} 0 \\ b_{N-1}(n-1) \end{bmatrix}\right\}$$

$$= \begin{bmatrix} R_{NN}(n)a_{N-1}(n) \\ P_N^{bT}(n)a_{N-1}(n) \end{bmatrix} + \mathbf{\Gamma}_N^f \begin{bmatrix} P_N^{fT}(n)b_{N-1}(n-1) \\ R_{NN}(n-1)b_{N-1}(n-1) \end{bmatrix}$$

$$= \begin{bmatrix} R_{NN}(n)a_{N-1}(n) + \mathbf{\Gamma}_N^f(n)P_N^{fT}(n)b_{N-1}(n-1) \\ \\ P_N^{bT}(n)a_{N-1}(n) + \mathbf{\Gamma}_N^{f(n)} R_{NN}(n-1)b_{N-1}(n-1) \end{bmatrix}$$

Comparing the RHS of this with that of

$$R_{N+1N+1}(n)a_N(n) = \begin{bmatrix} \varepsilon_N^f(n) \\ 0 \end{bmatrix}$$

$$\varepsilon_N^f(n) = R_{NN}(n)a_{N-1}(n) + \mathbf{\Gamma}_N^f(n)P_N^{fT}(n)b_{N-1}(n-1) \tag{15.17a}$$

$$P_N^{bT}(n)a_{N-1}(n) + \mathbf{\Gamma}_N^f(n)R_{NN}(n-1)b_{N-1}(n-1) = 0 \tag{15.17b}$$

If this is satisfied then,

$$a_N(n) = \begin{bmatrix} a_{N-1}(n) \\ 0 \end{bmatrix} + \mathbf{\Gamma}_N^f(n)\begin{bmatrix} 0 \\ b_{N-1}(n-1) \end{bmatrix} \tag{15.17c}$$

and

$$\varepsilon_N^f(n) = \varepsilon_{N-1}^f(n) + \mathbf{\Gamma}_N^f(n)P_N^{fT}(n)b_{N-1}(n-1)$$

$$\because R_{NN}(n-1)a_{N-1}(n) = \varepsilon_{N-1}^f(n)$$

$$\mathbf{\Gamma}_N^f(n) = -\frac{P_N^{bT}(n)a_{N-1}(n)}{R_{NN}(n-1)b_{N-1}(n-1)}$$

as

$$\varepsilon_{N-1}^f(n-1) = R_{NN}(n-1)b_{N-1}(n-1)$$

$$\mathbf{\Gamma}_N^f(n) = -\frac{P_N^{bT}(n)a_{N-1}(n)}{\varepsilon_{N-1}^f(n-1)} \tag{15.17d}$$

Thus, the value of $\mathbf{\Gamma}_N^f(n)$ which satisfies the above conditions, results in a solution for $a_N(n)$. Also the forward prediction error energy is order updated by

$$\varepsilon_N^f(n) = \varepsilon_{N-1}^f(n) + \mathbf{\Gamma}_N^f(n)P_N^{fT}(n)b_{N-1}(n-1) \tag{15.17e}$$

For the backward prediction, consider

$$R_{N+1,N+1}(n)\left\{\begin{bmatrix} 0 \\ b_{N-1}(n-1) \end{bmatrix} + \mathbf{\Gamma}_N^f(n)\begin{bmatrix} a_{N-1}(n) \\ 0 \end{bmatrix}\right\}$$

$$= \begin{bmatrix} P_N^{fT}(n)b_{N-1}(n-1) \\ R_{NN}(n-1)b_{N-1}(n-1) \end{bmatrix} + \mathbf{\Gamma}_N^f(n)\begin{bmatrix} R_{NN}(n)a_{N-1}(n) \\ P_{NN}^{bT}(n)a_{N-1}(n) \end{bmatrix}$$

$$= \begin{bmatrix} P_N^{fT}(n)b_{N-1}(n-1) + \mathbf{\Gamma}_N^f R_{NN}(n)a_{N-1}(n) \\ R_{NN}(n-1)b_{N-1}(n-1) + \mathbf{\Gamma}_N^f P_N^{bT}(n)a_{N-1}(n) \end{bmatrix}$$

Comparing the RHS of this with that of

$$R_{N+1N+1}(n)b_{N-1}(n-1) = \begin{bmatrix} 0 \\ \varepsilon^b(n) \end{bmatrix},$$

$$P_N^{fT}(n)b_{N-1}(n-1) + \mathbf{\Gamma}_N^b(n)R_{NN}(n)a_{N-1}(n) = 0$$

$$R_{NN}(n-1)b_{N-1}(n-1) + \mathbf{\Gamma}_N^b(n)P_N^{bT}(n)a_{N-1}(n) = \varepsilon_N^b(n)$$

If this is satisfied then

$$b_N(n) = \begin{bmatrix} 0 \\ b_{N-1}(n-1) \end{bmatrix} + \mathbf{\Gamma}_N^b(n)\begin{bmatrix} a_{N-1}(n) \\ 0 \end{bmatrix} \tag{15.18a}$$

and

$$\varepsilon_N^b(n) = R_{NN}(n-1)b_{N-1}(n-1) + \mathbf{\Gamma}_N^b(n)P_N^{bT}(n)a_{N-1}(n)$$

$$\varepsilon_N^b(n) = \varepsilon_{N-1}^b(n-1) + \Gamma_N^b(n) P_N^{bT}(n) a_{N-1}(n)$$

$$\because R_{NN}(n-1) b_{N-1}(n-1) = \varepsilon_{N-1}^b(n-1)$$

$$\Gamma_N^b(n) = -\frac{P_N^{fT}(n) b_{N-1}(n-1)}{R_{NN}(n) a_{N-1}(n)}$$

As

$$\varepsilon_{N-1}^f(n-1) = R_{NN}(n) a_{N-1}(n),$$

$$\Gamma_N^b(n) = -\frac{P_N^{fT}(n) b_{N-1}(n-1)}{\varepsilon_{N-1}^f(n)} \tag{15.18b}$$

Thus, the value of $\Gamma_N^b(n)$ which satisfies the above conditions, results in a solution for $b_N(n)$. Also the backward prediction error energy is order updated by

$$\varepsilon_N^b(n) = \varepsilon_{N-1}^b(n-1) + \Gamma_N^b(n) P_N^{bT}(n) a_{N-1}(n) \tag{15.18c}$$

Again consider L.H.S of the forward error augmented normal equation

$$R_{N+1,N+1}(n) a_N(n) = R_{N+1,N+1}(n) \begin{bmatrix} a_{N-1}(n) \\ 0 \end{bmatrix} = \begin{bmatrix} R_{NN}(n) a_{N-1}(n) \\ P_N^{bT}(n) a_{N-1}(n) \end{bmatrix}$$

Since

$$R_{NN}(n) a_{N-1}(n) = \begin{bmatrix} \varepsilon_{N-1}^f(n) \\ 0 \end{bmatrix}$$

$$R_{N+1,N+1}(n) \begin{bmatrix} a_{N-1}(n) \\ 0 \end{bmatrix} = \begin{bmatrix} \begin{bmatrix} \varepsilon_{N-1}^f(n) \\ 0 \end{bmatrix} \\ P_N^{bT}(n) a_{N-1}(n) \end{bmatrix}$$

Pre-multiplying both sides by $[0 \quad b_{N-1}^{T}(n-1)]$

$$\begin{bmatrix}0 & b_{N-1}^{T}(n-1)\end{bmatrix} R_{N+1,N+1}(n)\begin{bmatrix}a_{N-1}(n)\\0\end{bmatrix} = \begin{bmatrix}0 & b_{N-1}^{T}(n-1)\end{bmatrix}\begin{bmatrix}\begin{bmatrix}\varepsilon_{N-1}^{f}(n)\\0\end{bmatrix}\\P_{N}^{bT}(n)a_{N-1}(n)\end{bmatrix}$$

$$\begin{bmatrix}0 & b_{N-1}^{T}(n-1)\end{bmatrix} R_{N+1,N+1}(n)\begin{bmatrix}a_{N-1}(n)\\0\end{bmatrix} = \begin{bmatrix}0 & -W_{N-1}^{bT}(n-1) & 1\end{bmatrix}\begin{bmatrix}\varepsilon_{N-1}^{f}(n)\\0\\P_{N}^{bT}(n)a_{N-1}(n)\end{bmatrix}$$

$$\begin{bmatrix}0 & b_{N-1}^{T}(n-1)\end{bmatrix} R_{N+1,N+1}(n)\begin{bmatrix}a_{N-1}(n)\\0\end{bmatrix} = P_{N}^{bT}(n)a_{N-1}(n) \qquad (15.18d)$$

By transposition property for

$$R_{N+1N+1}(n)\begin{bmatrix}0\\b_{N-1}(n-1)\end{bmatrix} = \begin{bmatrix}0 & b_{N-1}^{T}(n-1)\end{bmatrix}R_{N+1N+1}(n)$$

$$\begin{bmatrix}P_{N}^{fT}(n)b_{N-1}(n-1)\\0\\\varepsilon_{N-1}^{b}(n)\end{bmatrix} = \begin{bmatrix}P_{N}^{f}(n)b_{N-1}^{T}(n-1) & 0 & \varepsilon_{N-1}^{b}(n)\end{bmatrix}$$

Post-multiplying both sides by $\begin{bmatrix}a_{N-1}(n)\\0\end{bmatrix}$

$$\begin{bmatrix}0 & b_{N-1}(n-1)\end{bmatrix}R_{N+1N+1}(n)\begin{bmatrix}a_{N-1}(n)\\0\end{bmatrix} = \begin{bmatrix}P_{N}^{f}(n)b_{N-1}^{T}(n-1) & 0 & \varepsilon_{N-1}^{b}(n)\end{bmatrix}\begin{bmatrix}a_{N-1}(n)\\0\end{bmatrix}$$

$$\begin{bmatrix} 0 & b_{N-1}(n-1) \end{bmatrix} R_{N+1N+1}(n) \begin{bmatrix} 1 \\ -W^f_{N-1}(n) \\ 0 \end{bmatrix}$$

$$= \begin{bmatrix} P^f_N(n) b^T_{N-1}(n-1) & 0 & \varepsilon^b_{N-1}(n) \end{bmatrix} \begin{bmatrix} 1 \\ -W^f_{N-1}(n) \\ 0 \end{bmatrix}$$

$$\begin{bmatrix} 0 & b_{N-1}(n-1) \end{bmatrix} R_{N+1N+1}(n) \begin{bmatrix} a_{N-1}(n) \\ 0 \end{bmatrix} = P^f_N(n) b^T_{N-1}(n-1) \tag{15.18e}$$

From Eqns. (15. 18d) and (15.18e) as their L.H.S are equal,

$$P^f_N(n) b^T_{N-1}(n-1) = P^{bT}_N(n) a_{N-1}(n) = \Delta_N(n) \tag{15.18f}$$

From Eqns. (15. 17d) and (15.18b),

$$\Gamma^f_N(n) = -\frac{P^{bT}_N(n) a_{N-1}(n)}{\varepsilon^b_{N-1}(n-1)}, \quad \Gamma^b_N(n) = -\frac{P^{fT}_N(n) b_{N-1}(n-1)}{\varepsilon^f_{N-1}(n)} \tag{15.18g}$$

$$P^{bT}_N(n) a_{N-1}(n) = \Delta_N(n), \qquad P^{fT}_N(n) b_{N-1}(n-1) = \Delta_N(n)$$

$$\Gamma^f_N(n) = -\frac{\Delta_N(n)}{\varepsilon^b_{N-1}(n-1)}, \quad \Gamma^b_N(n) = -\frac{\Delta_N(n)}{\varepsilon^f_{N-1}(n)} \tag{15.18h}$$

From Eqn. (15.17e)

$$\varepsilon^f_N(n) = \varepsilon^f_{N-1}(n) + \Gamma^f_N(n) P^{fT}_N(n) b_{N-1}(n-1)$$

Therefore,

$$\varepsilon^f_N(n) = \varepsilon^f_{N-1}(n) - \frac{\Delta_N(n)}{\varepsilon^b_{N-1}(n-1)} P^{fT}_N(n) b_{N-1}(n-1)$$

Since from Eqn. (15.18f)

$$P^{fT}_N(n) b_{N-1}(n-1) = \Delta_N(n)$$

Therefore,

$$\varepsilon_N^f(n) = \varepsilon_{N-1}^f(n) - \frac{\Delta_N^2(n)}{\varepsilon_{N-1}^b(n-1)} \tag{15.18i}$$

Again for the forward error energy (Eqn. 15.18c)

$$\varepsilon_N^b(n) = \varepsilon_{N-1}^b(n-1) + \mathbf{\Gamma}_N^b(n)P_N^{bT}(n)a_{N-1}(n)$$

Since from Eqn. (15.18h)

$$\mathbf{\Gamma}_N^b(n) = -\frac{\Delta_N(n)}{\varepsilon_{N-1}^f(n)},$$

$$\varepsilon_N^b(n) = \varepsilon_{N-1}^b(n-1) - \frac{\Delta_N(n)}{\varepsilon_{N-1}^f(n)}P_N^{bT}(n)a_{N-1}(n)$$

Since from Eqn. (15.18f)

$$P_N^{bT}(n)a_{N-1}(n) = \Delta_N(n)$$

Therefore

$$\varepsilon_N^b(n) = \varepsilon_{N-1}^b(n-1) - \frac{\Delta_N^2(n)}{\varepsilon_{N-1}^f(n)} \tag{15.18j}$$

The forward prediction error is given by

$$e_N^f(n) = a_N^T(n)X_{N+1}(n)$$

$$a_N(n) = \begin{bmatrix} a_{N-1}(n) \\ 0 \end{bmatrix} + \mathbf{\Gamma}_N^f(n)\begin{bmatrix} 0 \\ b_{N-1}(n-1) \end{bmatrix}, (\textit{from Eqn. } (15.17c)$$

$$X_{N+1}(n) = \begin{bmatrix} x(n) \\ X_N(n-1) \end{bmatrix}, \qquad X_{N+1}(n) = \begin{bmatrix} X_N(n) \\ x(n-N) \end{bmatrix}$$

$$e_N^f(n) = a_N^T(n)X_{N+1}(n)$$

$$e_N^f(n) = \begin{bmatrix} a_{N-1}^T(n) & 0 \end{bmatrix} \begin{bmatrix} X_N(n) \\ x(n-N) \end{bmatrix}$$

$$+\boldsymbol{\Gamma}_N^f(n)\begin{bmatrix} 0 & b_{N-1}^T(n-1) \end{bmatrix} \begin{bmatrix} x(n) \\ X_N(n-1) \end{bmatrix}$$

$$e_N^f(n) = a_{N-1}^T(n)X_N(n) + \boldsymbol{\Gamma}_N^f(n)b_{N-1}^T(n-1)X_N(n-1)$$

$$e_N^f(n) = e_{N-1}^f(n) + \boldsymbol{\Gamma}_N^f(n)e_{N-1}^b(n-1) \tag{15.19a}$$

The backward posteriori prediction error is given by

$$e_N^b(n) = b_N^T(n)X_{N+1}(n)$$

$$b_N(n) = \begin{bmatrix} 0 \\ b_{N-1}(n-1) \end{bmatrix} + \boldsymbol{\Gamma}_N^b(n)\begin{bmatrix} a_{N-1}(n) \\ 0 \end{bmatrix}, \quad (\textit{from Eqn.}\ (15.18a)$$

Therefore,

$$e_N^b(n) = \begin{bmatrix} 0 & b_{N-1}^T(n-1) \end{bmatrix} \begin{bmatrix} x(n) \\ X_N(n-1) \end{bmatrix}$$

$$+\boldsymbol{\Gamma}_N^b(n)\begin{bmatrix} a_{N-1}^T(n) & 0 \end{bmatrix} \begin{bmatrix} X_N(n) \\ x(n-N) \end{bmatrix}$$

$$e_N^b(n) = b_{N-1}^T(n-1)X_N(n-1) + \boldsymbol{\Gamma}_N^b(n)a_{N-1}^T(n)X_N(n)$$

$$e_N^b(n) = e_{N-1}^b(n-1) + \boldsymbol{\Gamma}_N^b(n)e_{N-1}^f(n) \tag{15.19b}$$

(Here instead of $e_N^b(n-1) = b_N^T(n-1)X_{N+1}(n-1)$, $e_{N-1}^b(n-1) = b_{N-1}^T(n)X_N(n)$ is present.)

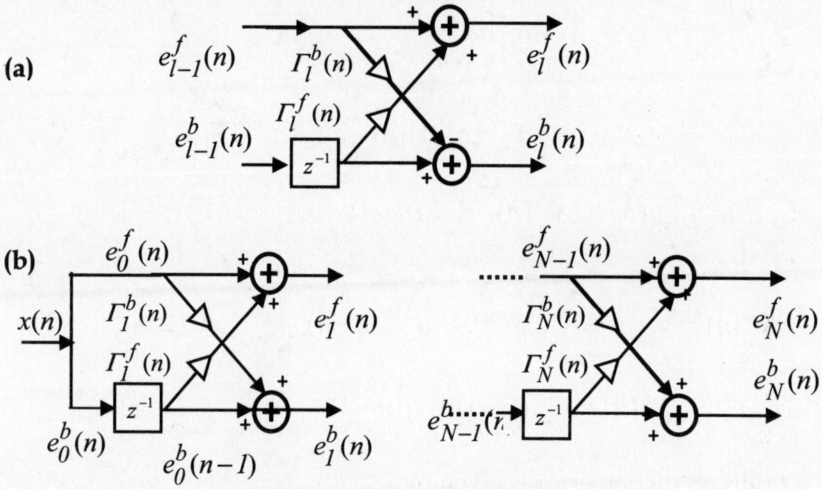

Fig. 15.2 (a) Single stage and (b) Multiple stage lattice structures.

Using the Eqns. (15.19a, b) for the forward and backward errors, respectively,

$$e^f_N(n) = e^f_{N-1}(n) + \Gamma^f_N(n) e^b_{N-1}(n-1)$$
$$e^b_N(n) = e^b_{N-1}(n-1) + \Gamma^b_N(n) e^f_{N-1}(n)$$

the lattice structure can be drawn (Fig. 15.2). (Note the change in sign associated with $\Gamma^f_N(n)$ and $\Gamma^b_N(n)$ compared to the previous lattice structure and this is only due to change in notation)

A priori forward and backward errors:

Similar to the above a posteriori forward and backward errors, the corresponding a priori errors are required for the lattice RLS algorithm.

$$\alpha^f_N(n) = a^T_N(n-1) X_{N+1}(n), \qquad \text{(from Eqn.(15.16a)}$$
$$\alpha^b_N(n) = b^T_N(n-1) X_{N+1}(n), \qquad \text{(from Eqn.(15.16b)}$$

$$a_N(n-1) = \begin{bmatrix} a_{N-1}(n-1) \\ 0 \end{bmatrix} + \Gamma^f_N(n-1) \begin{bmatrix} 0 \\ b_{N-1}(n-2) \end{bmatrix}, \qquad \text{(from Eqn.(15.17c)}$$

$$X_{N+1}(n) \begin{bmatrix} x(n) \\ X_N(n-1) \end{bmatrix}, \qquad X_{N+1}(n) \begin{bmatrix} X_N(n) \\ x(n-N) \end{bmatrix}$$

$$\alpha_N^f(n) = \begin{bmatrix} a_{N-1}^T(n-1) & 0 \end{bmatrix} \begin{bmatrix} X_N(n) \\ x(n-N) \end{bmatrix}$$

$$+\mathbf{\Gamma}_N^f(n-1)\begin{bmatrix} 0 & b_{N-1}^T(n-2) \end{bmatrix} \begin{bmatrix} x(n) \\ X_N(n-1) \end{bmatrix}$$

$$\alpha_N^f(n) = a_{N-1}^T(n-1)X_N(n) + \mathbf{\Gamma}_N^f(n-1)b_{N-1}^T(n-2)X_N(n-1)$$

Since

$$\alpha_{N-1}^f(n) = a_{N-1}^T(n-1)X_N(n), \quad \alpha_{N-1}^b(n-1) = b_{N-1}^T(n-2)X_N(n-1),$$

$$\alpha_N^f(n) = \alpha_{N-1}^f(n) + \mathbf{\Gamma}_N^f(n-1)\alpha_{N-1}^b(n-1) \tag{15.19c}$$

The backward a priori prediction error is given by

$$\alpha_N^b(n) = b_N^T(n-1)X_{N+1}(n)$$ (from Eqn.(15.16b)

$$b_N(n) = \begin{bmatrix} 0 \\ b_{N-1}(n-1) \end{bmatrix} + \mathbf{\Gamma}_N^f(n) \begin{bmatrix} a_{N-1}(n) \\ 0 \end{bmatrix},$$ (from Eqn.(15.18a)

$$b_N(n-1) = \begin{bmatrix} 0 \\ b_{N-1}(n-2) \end{bmatrix} + \mathbf{\Gamma}_N^f(n-1) \begin{bmatrix} a_{N-1}(n-1) \\ 0 \end{bmatrix}$$

Therefore,

$$\alpha_N^b(n) = \begin{bmatrix} 0 & b_{N-1}^T(n-2) \end{bmatrix} \begin{bmatrix} x(n) \\ X_N(n-1) \end{bmatrix}$$

$$+\mathbf{\Gamma}_N^b(n-1)\begin{bmatrix} a_{N-1}^T(n-1) & 0 \end{bmatrix} \begin{bmatrix} X_N(n) \\ x(n-N) \end{bmatrix}$$

$$\alpha_N^b(n) = b_{N-1}^T(n-2)X_N(n-1) + \mathbf{\Gamma}_N^b(n-1)a_{N-1}^T(n-1)X_N(n)$$

Since

$$\alpha_{N-1}^b(n-1) = b_{N-1}^T(n-2)X_N(n-1), \quad \alpha_{N-1}^f(n) = a_{N-1}^T(n-1)X_N(n),$$

$$\alpha_N^b(n) = \alpha_{N-1}^b(n-1) + \mathbf{\Gamma}_N^b(n-1)\alpha_{N-1}^f(n) \tag{15.19d}$$

$$\alpha_N^f(n) = \alpha_{N-1}^f(n) + \mathbf{\Gamma}_N^f(n-1)\alpha_{N-1}^b(n-1)$$ (Eqn. (15.19c)

Eqns. (15.19c) and (15.19d) decide the structure of the lattice based on a priori errors (Fig. 15.3a). From Eqns (15.18d), (15.18e) and (15.18f),

$$\begin{bmatrix} 0 & b_{N-1}(n-1) \end{bmatrix} R_{N+1N+1}(n) \begin{bmatrix} a_{N-1}(n) \\ 0 \end{bmatrix} = \Delta_N(n)$$

Since the leading element of $a_{N-1}(n-1)$ is equal to unity,

$$\Delta_N(n) = \begin{bmatrix} 0 & b_{N-1}^T(n-1) \end{bmatrix} R_{N+1N+1}(n) \begin{bmatrix} a_{N-1}(n-1) \\ 0 \end{bmatrix}$$

$$R_{N+1\,N+1}(n) = \lambda R_{N+1\,N+1}(n-1) + X_{N+1}(n) X_{N+1}^T(n)$$

Therefore,

$$\Delta_N(n) = \begin{bmatrix} 0 & b_{N-1}^T(n-1) \end{bmatrix} \begin{bmatrix} \lambda R_{N+1\,N+1}(n-1) + X_{N+1}(n) X_{N+1}^T(n) \end{bmatrix} \begin{bmatrix} a_{N-1}(n-1) \\ 0 \end{bmatrix}$$

$$= \begin{bmatrix} 0 & b_{N-1}^T(n-1) \end{bmatrix} \lambda R_{N+1\,N+1}(n-1) \begin{bmatrix} a_{N-1}(n-1) \\ 0 \end{bmatrix}$$

$$+ \begin{bmatrix} 0 & b_{N-1}^T(n-1) \end{bmatrix} \begin{bmatrix} x(n) \\ X_N(n-1) \end{bmatrix} \begin{bmatrix} X_N^T(n) & x(n-N) \end{bmatrix} \begin{bmatrix} a_{N-1}(n-1) \\ 0 \end{bmatrix}$$

$$\Delta_N(n) = \lambda\, \Delta_N(n-1) + \left\{ b_{N-1}^T(n-1) X_N(n-1) \right\} \left\{ X_N^T(n) a_{N-1}(n-1) \right\}$$

Since

$$b_{N-1}^T(n-1) X_N(n-1) = e_{N-1}^b(n-1)$$

$$X_N^T(n) a_{N-1}(n-1) = \alpha_{N-1}^f(n)$$

$$\Delta_N(n) = \lambda\, \Delta_N(n-1) + e_{N-1}^b(n-1) \alpha_{N-1}^f(n)$$

Therefore

$$\Delta_N(n) = \lambda\, \Delta_N(n-1) + e_{N-1}^b(n-1) \alpha_{N-1}^f(n) \qquad (15.19e)$$

Similarly,

$$\Delta_N(n) = \lambda\,\Delta_N(n-1) + \begin{bmatrix} x\,(n) & X_N^T(n-1) \end{bmatrix} \begin{bmatrix} 0 \\ b_{N-1}(n-1) \end{bmatrix} \begin{bmatrix} a_{N-1}^T(n) & (0) \end{bmatrix} \begin{bmatrix} X_N(n) \\ x(n-N) \end{bmatrix}$$

$$= \lambda\,\Delta_N(n-1) + \left\{X_N^T(n-1)b_{N-1}(n-1)\right\}\left\{a_{N-1}^T(n-1)X_N(n)\right\}$$

$$= \lambda\,\Delta_N(n-1) + \left\{b_{N-1}^T(n-1)X_N(n-1)\right\}\left\{a_{N-1}^T(n)X_N(n)\right\}$$

Since

$$b_{N-1}^T(n-1)X_N(n-1) = \alpha_{N-1}^b(n-1), \quad a_{N-1}^T(n)X(n) = e_{N-1}^f(n)$$

$$\Delta_N(n) = \lambda\,\Delta_N(n-1) + \alpha_{N-1}^b(n-1)e_{N-1}^f(n) \tag{15.19f}$$

The lattice predictor developed will be used for joint process estimation. From the general RLS algorithm theory, for the desired signal $d(n)$, Eqn. (15.4h) gives

$$\alpha_N(n) = d(n) - W_N^T(n-1)X_N(n)$$

Further

$$R_{NN}(n)W_N(n) = P_N(n), \qquad P_N(n) = \sum_{i=1}^{n} \lambda^{n-i} d(i)X_N(i)$$

For the $(N+1)$ order filter,

$$\alpha_{N+1}(n) = d(n) - W_{N+1}^T(n-1)X_{N+1}(n)$$

For this,

$$R_{N+1N+1}(n)W_{N+1}(n) = P_{N+1}(n), \; P_{N+1}(n) = \sum_{i=1}^{n} \lambda^{n-i} d(i)X_{N+1}(i)$$

Since $X_{N+1}(i)$ can be portioned as

$$X_{N+1}(i) = \begin{bmatrix} X_N(i) \\ x(i-N) \end{bmatrix}$$

$$P_{N+1}(n) = \begin{bmatrix} \sum_{i=1}^{n} \lambda^{n-i} d(i) X_N(i) \\ \sum_{i=1}^{n} \lambda^{n-i} d(i) x(i-N) \end{bmatrix} = \begin{bmatrix} P_N(n) \\ \sum_{i=1}^{n} \lambda^{n-i} d(i) x(i-N) \end{bmatrix}$$

and (Eqn. (15.12c1)

$$R_{N+1}(n) = \begin{bmatrix} R_N(n) & P_N^b(n) \\ P_N^{bT}(n) & \sigma_b^2(n) \end{bmatrix}$$

Consider

$$R_{N+1}(n) \begin{bmatrix} W_N(n) \\ 0 \end{bmatrix} = \begin{bmatrix} R_N(n) & P_N^b(n) \\ P_N^{bT}(n) & \sigma_b^2(n) \end{bmatrix} \begin{bmatrix} W_N(n) \\ 0 \end{bmatrix}$$

$$R_{N+1}(n) \begin{bmatrix} W_N(n) \\ 0 \end{bmatrix} = \begin{bmatrix} R_N(n) W_N(n) \\ P_N^{bT}(n) W_N(n) \end{bmatrix} = \begin{bmatrix} P_N(n) \\ P_N^{bT}(n) R_{NN}^{-1}(n) P_N(n) \end{bmatrix}$$

$$R_{N+1}(n) \begin{bmatrix} W_N(n) \\ 0 \end{bmatrix} = \begin{bmatrix} P_N(n) \\ W_N^{bT}(n)) P_N(n) \end{bmatrix}$$

Also

$$R_{N+1N+1}(n) W_{N+1}(n) = P_{N+1}(n),$$

Consider

$$R_{N+1}(n) \left\{ W_{N+1}(n) - \begin{bmatrix} W_N(n) \\ 0 \end{bmatrix} \right\} = P_{N+1}(n) - \begin{bmatrix} P_N(n) \\ W_N^{bT}(n)) P_N(n) \end{bmatrix}$$

$$= \begin{bmatrix} P_N(n) \\ \sum_{i=1}^{n} \lambda^{n-i} d(i) x(i-N) \end{bmatrix} - \begin{bmatrix} P_N(n) \\ W_N^{bT}(n)) P_N(n) \end{bmatrix}$$

$$\because P_{N+1}(n) = \begin{bmatrix} P_N(n) \\ \sum_{i=1}^{n} \lambda^{n-i} d(i) x(i-N) \end{bmatrix}$$

$$R_{N+1}(n)\left\{W_{N+1}(n)-\begin{bmatrix}W_N(n)\\0\end{bmatrix}\right\}=$$

$$=\begin{bmatrix}0\\ \sum_{i=1}^{n}\lambda^{n-i}d(i)x(i-N)-W_N^{bT}(n)\sum_{i=1}^{n}\lambda^{n-i}d(i)X_N(i)\end{bmatrix}$$

$$\because P_N(n)=\sum_{i=1}^{n}\lambda^{n-i}d(i)X_N(i)$$

$$R_{N+1}(n)\left\{W_{N+1}(n)-\begin{bmatrix}W_N(n)\\0\end{bmatrix}\right\}=\begin{bmatrix}0\\ \sum_{i=1}^{n}\lambda^{n-i}d(i)\left\{x(i-N)-W_N^{bT}(n)X_N(i)\right\}\end{bmatrix}$$

Let

$$\xi_N^f(n)=\sum_{i=1}^{n}\lambda^{n-i}d(i)\left\{x(i-N)-W_N^{bT}(n)X_N(i)\right\} \quad (15.19g)$$

$\xi_N^f(n)$ is the time averaged crosscorrelation between the backward prediction error and the desired signal. Therefore

$$R_{N+1}(n)\left\{W_{N+1}(n)-\begin{bmatrix}W_N(n)\\0\end{bmatrix}\right\}=\begin{bmatrix}0\\ \xi_N^f(n)\end{bmatrix} \quad (15.19h)$$

Multiplying both sides by $[\varepsilon_N^b(n)/\xi_N^f(n)]$

$$R_{N+1N+1}(n)\left\{W_{N+1}(n)-\begin{bmatrix}W_N(n)\\0\end{bmatrix}\right\}\frac{\varepsilon_N^b(n)}{\xi_N^f(n)}=\begin{bmatrix}0\\ \varepsilon_N^b(n)\end{bmatrix}$$

Comparing this with the augmented normal equation for backward prediction (Eqn. (15.12d)),

$$R_{N+1N+1}(n)b_N(n)=\begin{bmatrix}0\\ \varepsilon_N^b(n)\end{bmatrix}$$

$$b_N(n)=\left\{W_{N+1}(n)-\begin{bmatrix}W_N(n)\\0\end{bmatrix}\right\}\frac{\varepsilon_N^b(n)}{\xi_N^f(n)}$$

$$W_{N+1}(n) = \begin{bmatrix} W_N(n) \\ 0 \end{bmatrix} + \frac{\xi_N^f(n)}{\varepsilon_N^b(n)} b_N(n) \tag{15.19i}$$

The *a priori* estimation error for the $(N+1)^{th}$ order filter is given by (Eqn. (15.4h))

$$\alpha_{N+1}(n) = d(n) - W_{N+1}^T(n-1) X_{N+1}(n) \tag{15.19j}$$

Here consider the term $W_{N+1}^T(n-1)X_{N+1}(n)$ and from the above Eqn.(15.19i),

$$W_{N+1}^T(n-1)X_{N+1}(n) = \left\{ \begin{bmatrix} W_N(n-1) \\ 0 \end{bmatrix} + \frac{\xi_N^f(n-1)}{\varepsilon_N^b(n-1)} b_N(n-1) \right\}^T X_{N+1}(n)$$

$$W_{N+1}^T(n-1)X_{N+1}(n) = \left\{ \begin{bmatrix} W_N^T(n-1) & 0 \end{bmatrix} \begin{bmatrix} X_N(n) \\ x(n-N) \end{bmatrix} + \frac{\xi_N^f(n-1)}{\varepsilon_N^b(n-1)} b_N^T(n-1) \right\} X_{N+1}(n)$$

$$W_{N+1}^T(n-1)X_{N+1}(n) = \left\{ \begin{bmatrix} W_N^T(n-1) & 0 \end{bmatrix} \begin{bmatrix} X_N(n) \\ x(n-N) \end{bmatrix} + \frac{\xi_N^f(n-1)}{\varepsilon_N^b(n-1)} b_N^T(n-1) \right\} X_{N+1}(n)$$

$$W_{N+1}^T(n-1)X_{N+1}(n) = W_N^T(n-1)X_N(n) + \frac{\xi_N^f(n-1)}{\varepsilon_N^b(n-1)} b_N^T(n-1) X_{N+1}(n)$$

$$W_{N+1}^T(n-1)X_{N+1}(n) = W_N^T(n-1)X_N(n) + \frac{\xi_N^f(n-1)}{\varepsilon_N^b(n-1)} b_N^T(n-1) X_{N+1}(n)$$

Substituting this in Eqn.(15.19j)

$$\alpha_{N+1}(n) = d(n) - W_{N+1}^T(n-1)X_{N+1}(n)$$

$$\alpha_{N+1}(n) = d(n) - W_N^T(n-1)X_N(n) - \frac{\xi_N^f(n-1)}{\varepsilon_N^b(n-1)} b_N^T(n-1)X_{N+1}(n)$$

$$\alpha_{N+1}(n) = \alpha_N(n) - \frac{\xi_N^f(n-1)}{\varepsilon_N^b(n-1)} \alpha_N^b(n) \qquad (15.19k)$$

Order recursion for error $e_N(n)$:

Consider Eqn.(15.19i)

$$W_{N+1}(n) = \begin{bmatrix} W_N(n) \\ 0 \end{bmatrix} + \frac{\xi_N^f(n)}{\varepsilon_N^b(n)} b_N(n)$$

The error is given by

$$e_{N+1}(n) = d(n) - W_{N+1}^T(n-1)X_{N+1}(n)$$

$$e_{N+1}(n) = d(n) - \left\{ \begin{bmatrix} W_N(n) \\ 0 \end{bmatrix} + \frac{\xi_N^f(n)}{\varepsilon_N^b(n)} b_N(n) \right\}^T X_{N+1}(n)$$

$$e_{N+1}(n) = d(n) - \left\{ \begin{bmatrix} W_N(n) \\ 0 \end{bmatrix}^T \begin{bmatrix} X_N(n) \\ x(n-N) \end{bmatrix} + \frac{\xi_N^f(n)}{\varepsilon_N^b(n)} b_N^T(n) \right\} X_{N+1}(n)$$

$$e_{N+1}(n) = d(n) - \left\{ \begin{bmatrix} W_N^T(n) & 0 \end{bmatrix} \begin{bmatrix} X_N(n) \\ x(n-N) \end{bmatrix} + \frac{\xi_N^f(n)}{\varepsilon_N^b(n)} b_N^T(n) \right\} X_{N+1}(n)$$

$$e_{N+1}(n) = d(n) - W_N^T(n)X_N(n) - \frac{\xi_N^f(n)}{\varepsilon_N^b(n)} b_N^T(n)X_{N+1}(n)$$

Since

$$e_N^b(n) = b_N^T(n)X_{N+1}(n), \quad e_N(n) = d(n) - W_N^T(n)X_N(n)$$

$$e_{N+1}(n) = e_N(n) - \frac{\xi_N^f(n)}{\varepsilon_N^b(n)} e_N^b(n) \tag{15.19l}$$

Recursion in time for $\xi_N^f(n)$: From Eqn. (15.19g), $\xi_N^f(n)$ is given by

$$\xi_N^f(n) = \sum_{i=1}^{n} \lambda^{n-i} d(i)\left\{x(i-N) - W_N^{bT}(n) X_N(i)\right\}$$

$$\xi_N^f(n) = \sum_{i=1}^{n} \lambda^{n-i} d(i)x(i-N) - W_N^{bT}(n)\sum_{i=1}^{n} \lambda^{n-i} d(i) X_N(i)$$

Since

$$P_N(n) = \sum_{i=1}^{n} \lambda^{n-i} d(i) X_N(i)$$

Therefore,

$$\xi_N^f(n) = \sum_{i=1}^{n} \lambda^{n-i} d(i)x(i-N) - W_N^{bT}(n) P_N(n)$$

It is known that (Eqn. 15.12g)

$$W_N^b(n) = W_N^b(n-1) + K_N(n)\alpha_N^b(n)$$

Therefore,

$$\xi_N^f(n) = \sum_{i=1}^{n} \lambda^{n-i} d(i)x(i-N) - \left[W_N^b(n-1) + K_N(n)\alpha_N^b(n)\right]^T P_N(n)$$

$$\xi_N^f(n) = \sum_{i=1}^{n} \lambda^{n-i} d(i)x(i-N) - W_N^{bT}(n-1)P_N(n) - K_N^T(n)\alpha_N^b(n)P_N(n)$$

But

$$P_N(n) = \lambda P_N(n-1) + d(n)X_N(n)$$

$$\xi_N^f(n) = \sum_{i=1}^{n} \lambda^{n-i} d(i)x(i-N) - W_N^{bT}(n-1)\{\lambda P_N(n-1) + d(n)X_N(n)\} - K_N^T(n)\alpha_N^b(n)P_N(n)$$

$$\xi_N^f(n) = \lambda \sum_{i=1}^{n-1} \lambda^{n-i-1} d(i)x(i-N) - W_N^{bT}(n-1)\{P_N(n-1)$$
$$+ d(n)W_N^{bT}(n-1)X_N(n)\} - K_N^T(n)\alpha_N^b(n)P_N(n)$$

$$\xi_N^f(n) = \lambda\ \xi_N^f(n-1) + \alpha_N^b(n)\left\{\ d(n) - K_N^T(n)P_N(n)\right\}$$

$$\xi_N^f(n) = \lambda\ \xi_N^f(n-1) + \alpha_N^b(n)\left[d(n) - X_N^T(n)R_{NN}^{-1}(n)P_N(n)\right],$$
$$\because K_N(n) = R_{NN}^{-1}X_N(n)$$

$$\xi_N^f(n) = \lambda\ \xi_N^f(n-1) + \alpha_N^b(n)\left\{\ d(n) - X_N^T(n)W_N(n)\right\},$$
$$\because e_N(n) = d(n) - X_N^T(n)W_N(n)$$

$$\xi_N^f(n) = \lambda\ \xi_N^f(n-1) + \alpha_N^b(n)e_N(n) \tag{15.19m}$$

15.7.1 Recursive Least Square Joint Lattice Algorithm Based on a Priori Errors with Error Feedback

Recursion in time for reflection coefficients:

For forward reflection coefficient: From Eqn. (15.18h),

$$\Gamma_N^f(n) = -\ \frac{\Delta_N(n)}{\varepsilon_{N-1}^b(n-1)},$$

$$\varepsilon_{N-1}^b(n-1)\Gamma_N^f(n) = -\ \Delta_N(n), \tag{15.19n}$$

But $\Delta_N(n)$ is given by Eqn. (15.19e)

$$\Delta_N(n) = \lambda\ \Delta_N(n-1) + e_{N-1}^b(n-1)\alpha_{N-1}^f(n)$$

$$\Delta_N(n+1) = \lambda\ \Delta_N(n) + e_{N-1}^b(n)\alpha_{N-1}^f(n+1)$$

$$\Delta_N(n) = \lambda^{-1}\left[\ \Delta_N(n+1) - e_{N-1}^b(n)\alpha_{N-1}^f(n+1)\right] \tag{15.19o}$$

The backward prediction error energy is given by Eqn. (15.12j),

$$\varepsilon^b_{N-1}(n) = \lambda\ \varepsilon^b_{N-1}(n-1) + e^b_{N-1}(n)\alpha^b_{N-1}(n)$$

$$\varepsilon^b_{N-1}(n-1) = \lambda^{-1}\left[\varepsilon^b_{N-1}(n) - e^b_{N-1}(n)\alpha^b_{N-1}(n)\right] \tag{15.19p}$$

Substituting for $\Delta_N(n)$ (Eqn.(15.19o) and for $\varepsilon^b_{N-1}(n-1)$ (Eqn. (15.19p) in Eqn. (15.19n)

$$\varepsilon^b_{N-1}(n-1)\Gamma^f_N(n) = -\Delta_N(n)$$

$$\lambda^{-1}\left[\varepsilon^b_{N-1}(n) - e^b_{N-1}(n)\alpha^b_{N-1}(n)\right]\Gamma^f_N(n)$$
$$= -\lambda^{-1}\left[\Delta_N(n+1) - e^b_{N-1}(n)\alpha^f_{N-1}(n+1)\right]$$

$$\left[\varepsilon^b_{N-1}(n) - e^b_{N-1}(n)\alpha^b_{N-1}(n)\right]\Gamma^f_N(n)$$
$$= -\left[\Delta_N(n+1) - e^b_{N-1}(n)\alpha^f_{N-1}(n+1)\right]$$

Dividing both sides by $\varepsilon^b_{N-1}(n)$

$$\Gamma^f_N(n) - \frac{e^b_{N-1}(n)\alpha^b_{N-1}(n)}{\varepsilon^b_{N-1}(n)}\Gamma^f_N(n) = -\frac{\Delta_N(n+1)}{\varepsilon^b_{N-1}(n)} + \frac{e^b_{N-1}(n)\alpha^f_{N-1}(n+1)}{\varepsilon^b_{N-1}(n)}$$

$$-\frac{\Delta_N(n+1)}{\varepsilon^b_{N-1}(n)} = \Gamma^f_N(n)$$
$$-\Gamma^f_N(n)\frac{e^b_{N-1}(n)\alpha^b_{N-1}(n)}{\varepsilon^b_{N-1}(n)} - \frac{e^b_{N-1}(n)\alpha^f_{N-1}(n+1)}{\varepsilon^b_{N-1}(n)}$$

$$-\frac{\Delta_N(n+1)}{\varepsilon^b_{N-1}(n)} = \Gamma^f_N(n) - \frac{e^b_{N-1}(n)}{\varepsilon^b_{N-1}(n)}\left[\alpha^f_{N-1}(n+1) + \Gamma^f_N(n)\alpha^b_{N-1}(n)\right]$$

Since from Eqn. (15.18h),

$$\Gamma^f_N(n+1) = -\frac{\Delta_N(n+1)}{\varepsilon^b_{N-1}(n)},$$

Therefore,

$$\Gamma_N^f(n+1) = \Gamma_N^f(n) - \frac{e_{N-1}^b(n)}{\varepsilon_{N-1}^b(n)}\left[\alpha_{N-1}^f(n+1) + \Gamma_N^f(n)\alpha_{N-1}^b(n)\right] \tag{15.20a}$$

But from Eqn. (15.19c)

$$\alpha_N^f(n) = \alpha_{N-1}^f(n) + \Gamma_N^f(n-1)\alpha_{N-1}^b(n-1)$$

$$\alpha_N^f(n+1) = \alpha_{N-1}^f(n+1) + \Gamma_N^f(n)\alpha_{N-1}^b(n)$$

Therefore,

$$\Gamma_N^f(n+1) = \Gamma_N^f(n) - \frac{e_{N-1}^b(n)\alpha_N^f(n+1)}{\varepsilon_{N-1}^b(n)}$$

$$\Gamma_N^f(n) = \Gamma_N^f(n-1) - \frac{e_{N-1}^b(n-1)\alpha_N^f(n)}{\varepsilon_{N-1}^b(n-1)} \tag{15.20b}$$

Recursion for the backward reflection coefficient:

$$\Gamma_N^b(n) = -\ \frac{\Delta_N(n)}{\varepsilon_{N-1}^f(n)},$$

$$\varepsilon_{N-1}^f(n)\Gamma_N^b(n) = -\ \Delta_N(n),$$

But $\Delta_N(n)$ is given by

$$\Delta_N(n) = \lambda\,\Delta_N(n-1) + e_{N-1}^f(n)\alpha_{N-1}^b(n-1)$$

$$\Delta_N(n+1) = \lambda\,\Delta_N(n) + e_{N-1}^f(n+1)\alpha_{N-1}^b(n)$$

$$\Delta_N(n) = \lambda^{-1}\left[\Delta_N(n+1) - e_{N-1}^f(n+1)\alpha_{N-1}^b(n)\right] \tag{15.20c}$$

The forward prediction error energy is given by Eqn.(15.10j)

$$\varepsilon_{N-1}^f(n) = \lambda\ \ \varepsilon_{N-1}^f(n-1) + e_{N-1}^f(n)\alpha_{N-1}^f(n)$$

$$\varepsilon_{N-1}^f(n-1) = \lambda^{-1}\left[\varepsilon_{N-1}^f(n) - e_{N-1}^f(n)\alpha_{N-1}^f(n)\right]$$

$$\varepsilon^f_{N-1}(n) = \lambda^{-1}\left[\, \varepsilon^f_{N-1}(n+1) - e^f_{N-1}(n+1)\alpha^f_{N-1}(n+1) \right] \tag{15.20d}$$

Substituting for $\Delta_N(n)$ (Eqn. (15.20c)) and $\varepsilon^f_{N-1}(n)$ (Eqn. (15.20d)) in

$$\varepsilon^f_{N-1}(n)\boldsymbol{\Gamma}^b_N(n) = -\boldsymbol{\Delta}_N(n)$$

$$\lambda^{-1}\left[\, \varepsilon^f_{N-1}(n+1) - e^f_{N-1}(n+1)\alpha^f_{N-1}(n+1) \right] \boldsymbol{\Gamma}^b_N(n)$$
$$= -\lambda^{-1}\left[\, \boldsymbol{\Delta}_N(n+1) - e^f_{N-1}(n+1)\alpha^b_{N-1}(n) \right]$$

$$\left[\, \varepsilon^f_{N-1}(n+1) - e^f_{N-1}(n+1)\alpha^f_{N-1}(n+1) \right] \boldsymbol{\Gamma}^b_N(n)$$
$$= e^f_{N-1}(n+1)\alpha^b_{N-1}(n) - \boldsymbol{\Delta}_N(n+1)$$

Dividing both sides by $\varepsilon^f_{N-1}(n+1)$

$$\boldsymbol{\Gamma}^b_N(n) - \frac{e^f_{N-1}(n+1)\alpha^b_{N-1}(n+1)}{\varepsilon^f_{N-1}(n+1)}\boldsymbol{\Gamma}^b_N(n)$$
$$= -\frac{\boldsymbol{\Delta}_N(n+1)}{\varepsilon^f_{N-1}(n+1)} + \frac{e^f_{N-1}(n+1)\alpha^b_{N-1}(n)}{\varepsilon^f_{N-1}(n+1)}$$

$$-\frac{\boldsymbol{\Delta}_N(n+1)}{\varepsilon^b_{N-1}(n+1)} = \boldsymbol{\Gamma}^b_N(n)$$
$$-\boldsymbol{\Gamma}^b_N(n)\frac{e^f_{N-1}(n+1)\alpha^b_{N-1}(n+1)}{\varepsilon^f_{N-1}(n+1)} - \frac{e^f_{N-1}(n+1)\alpha^b_{N-1}(n)}{\varepsilon^f_{N-1}(n+1)}$$

$$-\frac{\boldsymbol{\Delta}_N(n+1)}{\varepsilon^f_{N-1}(n+1)} = \boldsymbol{\Gamma}^b_N(n) - \frac{e^f_{N-1}(n+1)}{\varepsilon^f_{N-1}(n+1)}\left[\, \alpha^b_{N-1}(n) + \boldsymbol{\Gamma}^b_N(n)\alpha^b_{N-1}(n+1) \right]$$

Since

$$\mathbf{\Gamma}_N^b(n+1) = -\frac{\Delta_N(n+1)}{\varepsilon_{N-1}^f(n+1)},$$

Therefore,

$$\mathbf{\Gamma}_N^b(n+1) = \mathbf{\Gamma}_N^b(n) - \frac{e_{N-1}^f(n+1)}{\varepsilon_{N-1}^f(n+1)}\left[\alpha_{N-1}^b(n) + \mathbf{\Gamma}_N^b(n)\alpha_{N-1}^b(n+1)\right]$$

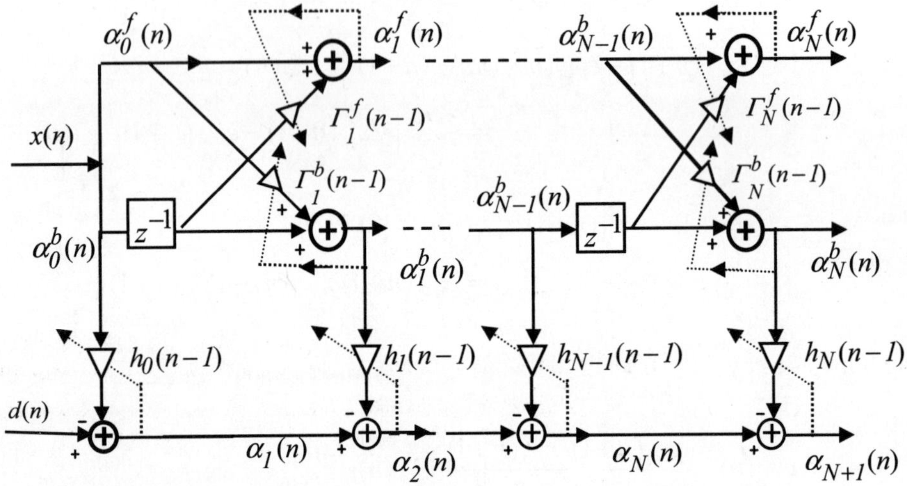

Fig. 15.3(a) Joint lattice with a priori errors with error feedback.

But from Eqn. (15.19d)

$$\alpha_N^b(n) = \alpha_{N-1}^b(n-1) + \mathbf{\Gamma}_N^b(n-1)\alpha_{N-1}^b(n)$$

$$\alpha_N^b(n+1) = \alpha_{N-1}^b(n) + \mathbf{\Gamma}_N^b(n)\alpha_{N-1}^b(n+1)$$

Therefore,

$$\mathbf{\Gamma}_N^b(n+1) = \mathbf{\Gamma}_N^b(n) - \frac{e_{N-1}^f(n+1)\alpha_N^b(n+1)}{\varepsilon_{N-1}^f(n+1)}$$

$$\mathbf{\Gamma}_N^b(n) = \mathbf{\Gamma}_N^b(n-1) - \frac{e_{N-1}^f(n)\alpha_N^b(n)}{\varepsilon_{N-1}^f(n)} \tag{15.20e}$$

For a joint lattice algorithm, it is required to derive the time recursion for the output section, that is the ladder section. For this, let the ladder weight vector be defined by

$$H_N(n) = \frac{\xi_N^f(n)}{\varepsilon_N^b(n)} \tag{15.21a}$$

The vector $H_N(n) = [h_0(n), h_1(n), \ldots\ldots, h_N(n)]$ corresponds to weights to be associated with the ladder network for which the orthogonal a priori backward prediction errors $\alpha_N^b(n) = [\alpha_0^b(n), \alpha_1^b(n), \ldots\ldots, \alpha_N^b(n)]$ form their inputs which are generated by the lattice from its input $x(n)$ (Fig.15.3a).

The backward prediction errors

$$B\ (i) = [e_0^b(i), e_1^b(i)), \ldots\ldots, e_N^b(i)i]^T$$

can be obtained from the input vector $X_{N+1}(i)$ by the relation

$$B(i) = \Omega(n) X_{N+1}(i)$$

where, $\Omega(n)$ is matrix consisting of all backward prediction filters of all intermediate orders and is given by

$$\Omega(n) = \begin{bmatrix} 1 & 0 & \cdots & 0 \\ -w_{1,1}^b(n) & 1 & \cdots & 0 \\ \vdots & \vdots & \vdots & \vdots \\ -w_{N,,1}^b(n) & -w_{N,,2}^b(n) & \cdots & 1 \end{bmatrix}$$

The time averaged autocorrelation matrix of $B(n)$ is given by $\Psi(n)$ and is

$$\Psi(n) = \sum_{i=1}^{n} \lambda^{n-i} B(i) B^T(i)$$

From the definition of Ω

$$\Psi(n) = \Omega(n) R_{N+1,N+1}(n) \Omega^T(n)$$

It can be shown that

$$\Psi(n) = \Omega(n) R_{N+1,N+1}(n) \Omega^T(n)$$

$$= \begin{bmatrix} \varepsilon_0^b(n) & 0 & 0 & \cdots & 0 \\ 0 & \varepsilon_1^b(n) & & \cdots & 0 \\ 0 & 0 & \varepsilon_2^b(n) & \cdots & 0 \\ \vdots & \vdots & \vdots & \ddots & \vdots \\ 0 & 0 & 0 & & \varepsilon_N^b(n) \end{bmatrix}$$

Hence, $e_i^b(n)$, $i = 0,1,2,\ldots N$ are uncorrelated to each other and also to the input as $e_0^b(n), = x(n)$.

The weight vector $H_N(n) = [h_0(n), h_1(n), \ldots\ldots, h_N(n)]$ for the backward a posteriori errors is obtained such that the cost function

$$J(n) = \sum_{i=1}^{n} \lambda^{n-i} \left\{ d(i) - H_N^T(n) B(i) \right\}^2$$

is minimized for each n. This is achieved by solving the normal equation

$$\left[\sum_{i=1}^{n} \lambda^{n-i} B(i) B^T(i) \right] H_N(n) = \sum_{i=1}^{n} \lambda^{n-i} d(i) B(i)$$

As already stated the backward prediction errors $B(i) = [e_0^b(i), e_1^b(i)), \ldots\ldots, e_N^b(i)]^T$ are obtained from the input vector $X_{N+1}(i)$ by the relation

$$B(i) = \Omega(n) X_{N+1}(i)$$

$$\Psi(n) H_N(n) = \Omega(n) P_{N+1}(n)$$

where,

$$\Psi(n) = \sum_{i=1}^{n} \lambda^{n-i} B(i) B^T(i),$$

$$\Omega(n) \sum_{i=1}^{n} \lambda^{n-i} d(i) X_{N+1}(n) = \Omega(n) P_{N+1}(n)$$

Therefore,

$$H_N(n) = \Psi^{-1}(n)\Omega(n)P_{N+1}(n)$$

Further as

$$\Psi(n) = \Omega(n)R_{N+1\,N+1}(n)\Omega^T(n)$$

The solution for $H_N(n)$ is

$$H_N = \left[\Omega(n)R_{N+1\,N+1}(n)\Omega^T(n)\right]^{-1}\Omega\ (n)P_{N+1}$$

$$H_N = \left\{\Omega^T(n)\right\}^{-1} R^{-1}_{N+1\,N+1}(n)\Omega^{-1}(n)\Omega\ (n)P_{N+1}$$

Since determinant of $\Omega(n)$ is unity, it is definitely invertible, therefore

$$H_N = \left\{\Omega^T(n)\right\}^{-1} R^{-1}_{N+1\,N+1}(n)P_{N+1}$$

This ensures that the coefficient vector $H_N(n) = [h_0(n), h_1(n), \ldots\ldots, h_N(n)]$ can be chosen for the uncorrelated backward prediction errors derived from the input $x(n)$ using the lattice filter. From Eqn. (15.21a),

$$H_N(n-1) = \frac{\xi_N^f(n-1)}{\varepsilon_N^b(n-1)} \tag{15.21b}$$

$$\xi_N^f(n-1) = H_N(n-1)\varepsilon_N^b(n-1) \tag{15.21c}$$

But from Eqns. (15.19m) and (15.12j)

$$\xi_N^f(n) = \lambda\ \xi_N^f(n-1) + e_N(n)\alpha_N^b(n)$$

$$\varepsilon_N^b(n) = \lambda\ \varepsilon_N^b(n-1) + e_N^b(n)\alpha_N^b(n)$$

$$\xi_N^f(n-1) = \lambda^{-1}\left[\xi_N^f(n) - e_N(n)\alpha_N^b(n)\right]$$

$$\varepsilon_N^b(n-1) = \lambda^{-1}\left[\varepsilon_N^b(n) - e_N^b(n)\alpha_N^b(n)\right]$$

Now substituting for $\xi_N^f(n-1)$ and $\varepsilon_N^b(n-1)$ in Eqn. (15.21c)

$$\xi_N^f(n-1) = H_N(n-1)\varepsilon_N^b(n-1)$$

$$\lambda^{-1}\left[\xi_N^f(n) - e_N(n)\alpha_N^b(n)\right] = H_N(n-1)\lambda^{-1}\left[\varepsilon_N^b(n) - e_N^b(n)\alpha_N^b(n)\right]$$

$$\left[\xi_N^f(n) - e_N(n)\alpha_N^b(n)\right] = H_N(n-1)\left[\varepsilon_N^b(n) - e_N^b(n)\alpha_N^b(n)\right]$$

$$\xi_N^f(n) = H_N(n-1)\left[\varepsilon_N^b(n) - e_N^b(n)\alpha_N^b(n)\right] + e_N(n)\alpha_N^b(n)$$

Dividing both sides by $\varepsilon_N^b(n)$,

$$\frac{\xi_N^f(n)}{\varepsilon_N^b(n)} = \frac{H_N(n-1)}{\varepsilon_N^b(n)}\left[\varepsilon_N^b(n) - e_N^b(n)\alpha_N^b(n)\right] + \frac{e_N(n)\alpha_N^b(n)}{\varepsilon_N^b(n)}$$

From Eqn. (15.21a)

$$H(n) = H_N(n-1) - \frac{e_N^b(n)\alpha_N^b(n)}{\varepsilon_N^b(n)}H_N(n-1) + \frac{e_N(n)\alpha_N^b(n)}{\varepsilon_N^b(n)}$$

$$H_N(n) = H_N(n-1) + \frac{\alpha_N^b(n)}{\varepsilon_N^b(n)}\left[e_N(n) - e_N^b(n)H_N(n-1)\right]$$

Since from Eqn. (15.15a, c),

$$e_N(n) = \varphi_N(n)\alpha_N(n), \quad e_N^b(n) = \varphi_N(n)\alpha_N^b(n)$$

$$H_N(n) = H_N(n-1) + \frac{\alpha_N^b(n)}{\varepsilon_N^b(n)}\left[\varphi_N(n)\alpha_N(n) - \varphi_N(n)\alpha_N^b(n)H_N(n-1)\right]$$

$$H_N(n) = H_N(n-1) + \frac{\alpha_N^b(n)\varphi_N(n)}{\varepsilon_N^b(n)}\left[\alpha_N(n) - \alpha_N^b(n)H_N(n-1)\right] \qquad (15.21d)$$

From Eqn. (15.19k)

$$\alpha_{N+1}(n) = \alpha_N(n) - \frac{\xi_N^f(n-1)}{\varepsilon_N^b(n-1)}\alpha_N^b(n)$$

Using Eqn. (15.21b)

$$\alpha_{N+1}(n) = \alpha_N(n) - H_N(n-1)\alpha_N^b(n)$$

Therefore Eqn. (15.21d) becomes

$$H_N(n) = H_N(n-1) + \frac{\alpha_N^b(n)\phi_N(n)\alpha_{N+1}(n)}{\varepsilon_N^b(n)} \qquad (15.21e)$$

In this lattice RLS algorithm, *the a priori errors* (prediction and filtering) *are fed back in time updating the reflection and ladder coefficients* and hence it is called *RLS lattice with error feedback algorithm.*

Recursive least square lattice algorithm based on a priori with error feedback:
Initializations:

$$\Gamma_i^f(0) = \Gamma_i^b(0) = h_i(0) = 0$$
$$\varepsilon_i^f(0) = \varepsilon_i^b(0) = \delta \text{ (its value is not critical)}, \quad \varphi_0(0) = 1,$$
$$\alpha_0^f(n) = \alpha_0^b(n) = x(k), \alpha_0(n) = d(n)$$
$$\varepsilon_0^f(n) = \varepsilon_0^b(n) = \lambda \ \varepsilon_0^f(n-1) + x^2(n), \varphi_0(n) = 1$$

Compute for $i = 1, 2, \ldots\ldots, N$:

$$\alpha_i^f(n) = \alpha_{i-1}^f(n) + \Gamma_i^f(n-1)\alpha_{i-1}^b(n-1)$$
$$\alpha_i^b(n) = \alpha_{i-1}^b(n-1) + \Gamma_i^b(n-1)\alpha_{i-1}^f(n)$$

$$\Gamma_i^f(n) = \Gamma_i^f(n-1) - \frac{\phi_{i-1}(n-1)\alpha_{i-1}^b(n-1)\alpha_i^f(n)}{\varepsilon_{i-1}^b(n-1)}$$

$$\varepsilon_{i-1}^f(n) = \lambda \ \varepsilon_{i-1}^f(n-1) + \phi_{i-1}(n-1)[\alpha_{i-1}^f(n)]^2$$
$$\varepsilon_{i-1}^b(n) = \lambda \ \varepsilon_{i-1}^b(n-1) + \phi_{i-1}(n-1)[\alpha_i^b(n)]^2$$

$$\mathbf{\Gamma}_i^b(n) = \mathbf{\Gamma}_i^b(n-1) - \frac{\phi_{i-1}(n)\alpha_{i-1}^f(n)\alpha_i^b(n)}{\varepsilon_{i-1}^f(n)}$$

$$\varphi_i(n) = \varphi_{i-1}(n) - \frac{[\phi_{i-1}(n)\alpha_{i-1}^b(n)]^2}{\varepsilon_{i-1}^f(n)}$$

For filtering for $i = 1, 2, \ldots\ldots, N$

$$\alpha_{i+1}(n) = \alpha_i(n) - h_i(n-1)\alpha_i^b(n)$$

$$h_i(n) = h_i(n-1) + \frac{\phi(n)\alpha_i^b(n)\alpha_{i+1}(n)}{\varepsilon_i^b(n)}$$

The RLS Lattice algorithm based on a priori errors with error feedback considered, *is less sensitive to numerical errors* compared to those RLS lattice based on a posteriori errors and that based on a priori errors; which do not use error feedback for the adaptation of reflection coefficients and the ladder coefficients. The RLS lattice based on a posteriori errors and that based on a priori errors *without feedback* will be considered to bring out the difference with the one that uses error feedback, presented.

15.7.2 Recursive Least Square Lattice Algorithm Based on a Posteriori Errors

The required equations are already derived and hence they will be collected here to form the algorithm (Fig. 15. 3b).

Prediction section:
From Eqn. (15.19f)

$$\Delta_i(n) = \lambda\,\Delta_i(n-1) + \alpha_{i-1}^b(n-1)e_{i-1}^f(n), \quad i = 1,2,3,\ldots\ldots,N$$

$$\Delta_i(n) = \lambda\,\Delta_i(n-1) + \frac{e_{i-1}^b(n-1)e_{i-1}^f(n)}{\varphi_{i-1}(n-1)},$$

$$\because e_{i-1}^b(n-1) = \varphi_{i-1}(n-1)\alpha_{i-1}^b(n-1)$$

From Eqn. (15.18h)

$$\mathbf{\Gamma}_i^f(n) = -\frac{\Delta_i(n)}{\varepsilon_{i-1}^b(n-1)}, \quad \mathbf{\Gamma}_i^b(n) = -\frac{\Delta_i(n)}{\varepsilon_{i-1}^f(n)}$$

(Compare these with Eqns. (15.20b) and (15.20e) which have error feedback)

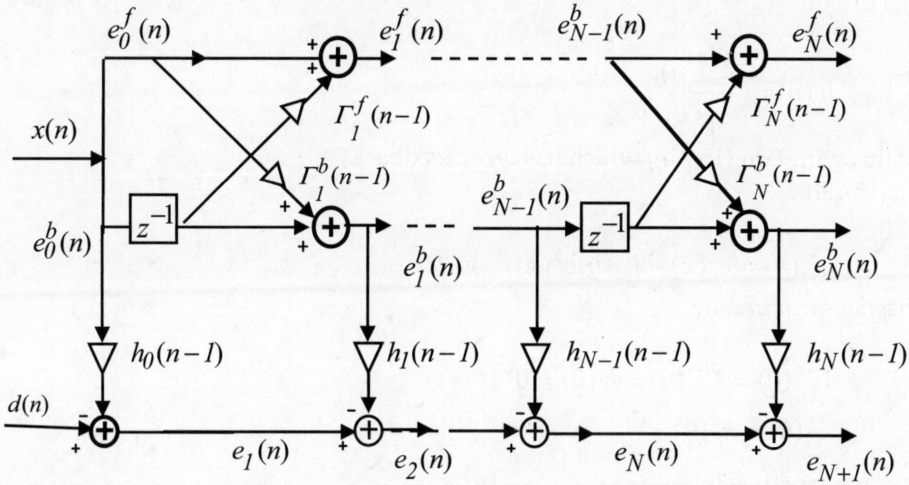

Fig. 15.3(b) Joint lattice with a posteriori errors.

From Eqn. (15.19a, b)

$$e_i^f(n) = e_{i-1}^f(n) + \boldsymbol{\Gamma}_i^f(n)e_{i-1}^b(n-1)$$

$$e_i^b(n) = e_{i-1}^b(n-1) + \boldsymbol{\Gamma}_i^f(n)e_{i-1}^f(n)$$

From Eqn. (15.18i, k)

$$\varepsilon_i^f(n) = \varepsilon_{i-1}^f(n) - \frac{\boldsymbol{\Delta}_i^2(n)}{\varepsilon_{i-1}^b(n-1)}, \qquad \varepsilon_i^b(n) = \varepsilon_{i-1}^b(n) - \frac{\boldsymbol{\Delta}_i^2(n)}{\varepsilon_{i-1}^f(n)}$$

From Eqn. (15.15f),

$$\phi_i(n) = \phi_{i-1}(n) - \frac{[e_{i-1}^b(n)]^2}{\varepsilon_{i-1}^b(n)}$$

From Eqn. (15.19m),

$$\xi_i^f(n) = \lambda\ \xi_i^f(n-1) + \alpha_i^b(n)e_i(n)$$

$$\xi_i^f(n) = \lambda\ \xi_i^f(n-1) + \frac{e_i^b(n)e_i(n)}{\phi_i(n)}, \quad \because e_i^b(n) = \phi_i(n)\alpha_i^b(n)$$

From Eqn. (15.21a)

$$h_i(n) = \frac{\xi_i^f(n)}{\varepsilon_i^b(n)},$$

(Compare this with Eqn.(15,21e) which has error feedback)
From Eqn. (15.19*l*)

$$e_{i+1}(n) = e_i(n) - h_i(n)e_i^b(n)$$

The initializations required are

$$\Gamma_i^f(0) = \Gamma_i^b(0) = h_i(0) = 0$$

$$\varepsilon_i^f(0) = \varepsilon_i^b(0) = \delta$$

$$\varphi_0(0) = 1,$$

$$\alpha_0^f(n) = \alpha_0^b(n) = x(n)$$

$$\alpha_0(n) = d(n),$$

$$\varepsilon_0^f(n) = \varepsilon_0^b(n) = \varepsilon_0^f(n-1) + x^2(n)$$

$$\phi_0(k) = 1, \quad e_0(n) = d(n)$$

15.7.3 Lattice RLS Algorithm Based on a Priori Errors

The required equations for joint lattice based on a priori errors are already derived (Fig.15.3c) and hence they will be collected here to form the algorithm

Prediction section:

From Eqn. (15.19c, d), for $i = 1, 2, 3, \ldots\ldots, N$

$$\alpha_i^f(n) = \alpha_{i-1}^f(n) + \Gamma_i^f(n-1)\alpha_{i-1}^b(n-1), \because$$

$$\alpha_i^b(n) = \alpha_{i-1}^b(n-1) + \Gamma_i^f(n-1)\alpha_{i-1}^f(n)$$

From Eqn. (15.19f)

$$\Delta_i(n) = \lambda\,\Delta_i(n-1) + \alpha_{i-1}^f(n)e_{i-1}^b(n-1),$$

$$\Delta_i(n) = \lambda\,\Delta_i(n-1) + \alpha_{i-1}^f(n)\alpha_{i-1}^b(n-1)\phi_{i-1}(n-1),$$

$$\because e_{i-1}^b(n-1) = \phi_{i-1}(n-1)\alpha_{i-1}^b(n-1)$$

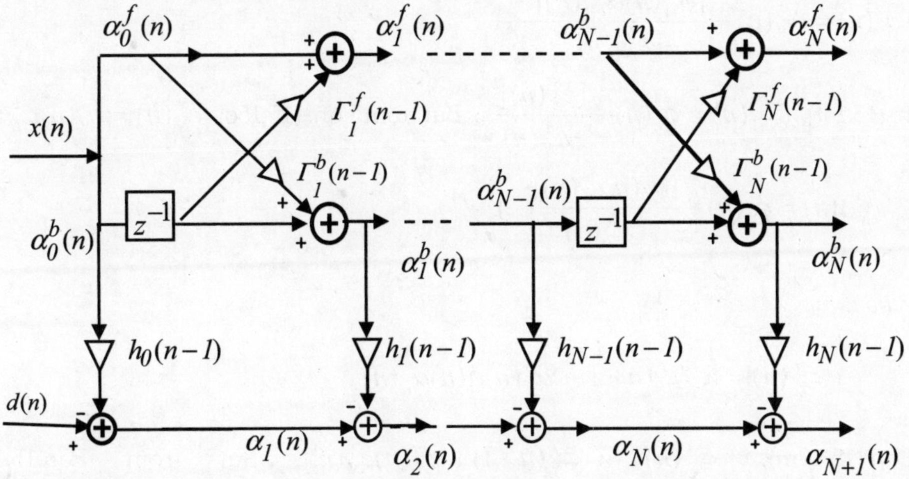

Fig. 15.3(c) Joint lattice with a priori errors.

From Eqn. (15.18h)

$$\Gamma_i^f(n) = -\frac{\Delta_i(n)}{\varepsilon_{i-1}^b(n-1)}, \quad \Gamma_i^b(n) = -\frac{\Delta_i(n)}{\varepsilon_{i-1}^f(n)}$$

(Compare these with Eqns. (15.20b) and (15.20e) which have error feedback)

$$\varepsilon_{i-1}^f(n) = \lambda\ \varepsilon_{i-1}^f(n-1) + \phi_i\ (n-1)[\alpha_{i-1}^f(n)]^2$$

(From Eqn. (15.10j), $\varepsilon_N^f(n) = \lambda\ \varepsilon_N^f(n-1) + \alpha_N^f(n)e_N^f(n)$. But from Eqn. (15.15b) $e_N^f(n) = \phi_N(n-1)\alpha_N^f(n)$. Therefore, $\varepsilon_N^f(n) = \lambda\ \varepsilon_N^f(n-1) + \phi_N(n-1)[\alpha_N^f(n)]^2$.)

$$\varepsilon_{i-1}^b(n) = \lambda\ \varepsilon_{i-1}^b(n-1) + \phi_{i-1}(n)[\alpha_i^b(n)]^2$$

(From Eqn. (15.12j), $\varepsilon_N^b(n) = \lambda\ \varepsilon_N^b(n-1) + \alpha_N^f(n)e_N^b(n)$. But from Eqn. (15.15c), $e_N^b(n) = \phi_N(n)\alpha_N^b(n)$. Therefore, $\varepsilon_N^b(n) = \lambda\ \varepsilon_N^b(n-1) + \phi_N(n)[\alpha_N^b(n)]^2$.)

$$\phi_i(n) = \phi_{i-1}(n) - \frac{[\phi_{i-1}(n)\alpha_{i-1}^b(n)]^2}{\varepsilon_{i-1}^b(n)}$$

(From Eqn. (15.15f), $\phi_{N+1}(n) = \phi_N(n) - \dfrac{[e_N^b(n)]^2}{\varepsilon_N^b(n)}$. But from Eqn. (15.15c), $e_N^b(n) = \phi_N(n)\alpha_N^b(n)$.

Therefore, $\phi_{N+1}(n) = \phi_N(n) - \dfrac{[\phi_N(n\alpha_N^b(n)]^2}{\varepsilon_N^b(n)}$.)

Filtering section:

$$\xi_i^f(n) = \lambda\ \xi_i^f(n-1) + \alpha_i^b(n)\phi_i(n)\alpha_i(n)$$

(From Eqn.(15.19m), $\xi_N^f(n) = \lambda\ \xi_N^f(n-1) + \alpha_N^b(n)e_N(n)$. But from Eqn.15.15c) $e_N^b(n) = \phi_N(n)\alpha_N^b(n)$. Therefore, $\xi_N^f(n) = \lambda\ \xi_N^f(n-1) + \alpha_N^b(n)\phi_N(n)\alpha_N(n)$.)

From Eqn. (15.19*l*)

$$\alpha_{i+1}(n) = \alpha_i(n) - h_i(n-1)\alpha_i^b(n)$$

From Eqn. (15.18j)

$$\varepsilon_i^b(n) = \varepsilon_{i-1}^b(n) - \frac{\Delta_i^2(n)}{\varepsilon_{i-1}^f(n)}$$

From Eqn. (15.21a)

$$h_i(n) = \frac{\xi_i^f(n)}{\varepsilon_i^b(n)},$$

(Compare this with Eqn. (15,21e) which has error feedback)

The initializations required are

$$\Gamma_i^f(0) = \Gamma_i^b(0) = h_i(0) = 0,$$

$$\varepsilon_i^f(0) = \varepsilon_i^b(0) = \delta$$

$$\varphi_0(0) = 1,$$

$$\alpha_0^f(n) = \alpha_0^b(n) = x(n)\ , \quad \alpha_0(n) = d(n)$$

$$\varepsilon_0^f(n) = \varepsilon_0^b(n) = \lambda\ \varepsilon_0^f(n-1) + x^2(n),\ \phi_0(n) = 1$$

In the a posteriori and *a priori* algorithms, the reflection and ladder coefficients are updated using $\Delta_N(n)$ (the cross-correlations between forward and delayed backward prediction errors of the lattice) and $\xi_N^f(n)$ (the cross-correlations between the backward errors of the lattice and the ladder errors). This makes them *sensitive to numerical errors.* But in lattice RLS based on *a priori errors with error feedback,* these coefficients are directly recursively updated without using the auto- (with the exception of zeroth lag) and cross-correlations and this makes the algorithm *less sensitive to numerical errors* compared to those of a posteriori and a priori error based algorithms without error feedback. The RLS lattice algorithm based on the a posteriori errors with error feedback also exists.. But in applications like channel equalization and echo cancellation *only a priori version can be used* and the a posteriori form cannot be used (J.G. Proakis and D.K. Manolakis. 2006, Chapter 13). This is due to the fact the present ladder weight vector corresponding to time instant *n* cannot be computed prior to computation of $d(n)$. In view of this, the a priori version with error feedback is most often used RLS joint lattice algorithm hence, the a posteriori version with error feedback is not discussed.

It is important to note that the RLS lattice ladder/ joint lattice has the same speed of convergence as that of the transversal structure, the FTRLS. This is due to the fact that both are optimum in the least square sense. Further, compared to the gradient lattice, the RLS lattice is superior as it is optimum in the least square sense. The gradient lattice though has some optimal characteristics, it is not optimum in least square sense and hence its convergence rate is poor compared to that of RLS lattice.

The computational complexity of joint lattice algorithms is in the range of 15*N* to 20*N*. Though this computational load is significantly less compared to that of direct implementation of RLS of N^2, it is still relatively high compared to that of fast transversal RLS implementation (8*N*). However, the fast transversal RLS algorithms are highly sensitive to finite word length effects for a fixed point processor implementation and it can be sensitive even for a floating point processor implementation. Hence, the fast lattice RLS algorithm which is robust to numerical errors plays an important role in utilizing the desirable features of RLS algorithm even though its computational load is somewhat high compared to that of transversal RLS algorithm.

15.8 AFFINE PROJECTION ALGORITHM

Among the many attempts for improving the convergence of an adaptive algorithm, reusing the old data available is of importance and such a data-reusing algorithm is known as Affine Projection algorithm (APA). The APA adaptation plays a significant role in increasing the convergence rate specifically for correlated inputs. However, the APA results in increased misadjustment. Hence, there is a trade-off between convergence speed and misadjustment which can be controlled by the step size parameter of the adaptation process (P.S. R. Diniz, 2002). The APA can be considered the generalization of normalized LMS (NLMS) algorithm. With this interpretation, each weight vector update of NLMS can be considered a one-dimensional APA. In APA, the projections are made on multiple dimensions (4 or 5) and as the projection dimension increases, the convergence speed of the weight vector also increases (S.L. Gay and Sanjeev Tavathia,1995) For this, the reason can be thought of as instead of considering a single projection on each dimension sequentially, considering the $L+1$ projections simultaneously at every sample instant provides a better approximation of the estimate to the desired one resulting in an increased convergence speed. In other words, for many sample instants, the APA weight vector has to match its estimates to their corresponding desired signal samples unlike the matching at only one sample of NLMS. *This simultaneous matching at more than one sample instant makes the algorithm to converge faster. Also as the single weight vector has to satisfy matching at many sample instants, it is required that the change due to adaptation in the weight vector should be minimum.* As the matching between NLMS estimate to its desired signal is only at one sample instant and can be regarded what APA does, is done sequentially and hence, the NLMS takes more samples to reach convergence. Thus, APA has a better convergence rate than that of NLMS algorithm. However, *matching at different sample instants by the same weight vector may not be perfect and results in a higher error, in matching.* Also, the use of more than one projection at every sample, increases the computational complexity compared to that of NLMS. The convergence rate of APA is inferior to that of RLS algorithm as it considers effectively more number of projections decided by the forgetting factor used. Using techniques similar to those used for RLS, fast APA has been achieved (S.L Gay and Sanjeev Thavathia, 1995). Also the computational complexity of APA lies between that of NLMS and RLS algorithms. A fast version of the APA (FAPA) requires *2N+20(L+1)* multiplications (*N* is the filter order). For applications like echo cancellation, the filter length will be generally larger than the number of projections (*L+1*) required and this makes computational complexity of FAPA comparable to that of normalized LMS algorithm.

For APA, the last $L+1$ input signal vectors are considered and put in a matrix as

$$X_{af}(n) = \begin{bmatrix} x(n) & x(n-1) & \dots & x(n-L) \\ x(n-1) & x(n-2) & \dots & x(n-l+1) \\ \vdots & \vdots & \vdots & \vdots \\ x(n-N) & x(n-N-1) & \dots & x(n-L-N) \end{bmatrix} \tag{15.22a}$$

Representing the column vectors of the above matrix by

$$X(n) = \begin{bmatrix} x(n) \\ x(n-1) \\ \vdots \\ x(n-N) \end{bmatrix},$$

$$X_{af}(n) = [X(n) \quad X(n-1) \quad \ldots \quad X(n-L)] \tag{15.22a1}$$

The corresponding outputs are given by $y_{af}(n)$ as

$$y_{af}(n) = X_{af}^T(n)W(n) = \left[y_0(n) \quad y_1(n) \quad \ldots \quad y_L(n)\right]^T \tag{15.22b}$$

The desired signal vector $d_{af}(n)$

$$d_{af}(n) = \left[d(n) \quad d(n-1) \quad \ldots \quad d(n-L)\right]^T \tag{15.22c}$$

The error vector $e_{af}(n)$

$$e_{af}(n) = \begin{bmatrix} e_0(n) \\ e_1(n) \\ \vdots \\ e_L(n) \end{bmatrix} = \begin{bmatrix} d(k) - y_0(n) \\ d(k-1) - y_1(n) \\ \vdots \\ d(k-L) - y_L(n) \end{bmatrix} = d_{af}(n) - y_{af}(n) \tag{15.22d}$$

The above considerations mentioned lead to an objective of the affine projection algorithm which has to minimize the cost function

$$J\left[W(n+1)\right] = \left\{\frac{1}{2}\left\|W(n+1) - W(n)\right\|^2\right\}$$

with constraint :

$$d_{af}(n) - X_{af}^T(n)W(n+1) = 0 \tag{15.22e}$$

That is, the APA makes next coefficient vector $W(n+1)$ as close as possible to the present current vector $W(n)$. Ideally, a single weight vector at $L+1$ consecutive samples should ensure that the a posteriori error is zero and this can be achieved by minimizing the norm square of the difference of successive weight vectors. Such a constrained minimization is achieved by using the Lagrange multiplier approach. The cost function is given by

$$J\left[w(n+1)\right] = \frac{1}{2}\left\|W(n+1) - W(n)\right\|^2 + \lambda_{af}^T(n)[d(n) - X_{af}^T(n)W(n+1)] \tag{15.23a}$$

where $\lambda_{af}(n)$ is an $(L+1)\times 1$ vector of Lagrange multipliers. By setting the gradient of $J[W(n+1)]$ with respect to $W(n+1)$ equal to zero.,

$$\nabla\{J\,[W(n+1)]\} = \frac{\partial}{\partial W(n+1)}\left\{\frac{1}{2}\|W(n+1)-W(n)\|^2 + \lambda_{af}^T(n)[d(n)-X_{af}^T(n)W(n+1)]\right\} = 0$$

$$W(n+1)-W(n)-X_{af}(n)\lambda_{af}(n)=0$$

$$W(n+1)=W(n)+X_{af}(n)\lambda_{af}(n) \tag{15.23b}$$

By substituting for $W(n+1)$ in the constraint, Eqn. (15.22e)

$$d_{af}(n)-X_{af}^T(n)W(n+1)=0$$

$$d_{af}(n)-X_{af}^T(n)\{W(n)+X_{af}(n)\lambda_{af}(n)\}=0$$

$$X_{af}^T(n)X_{af}(n)\lambda_{af}(n)=d_{af}(n)-X_{af}^T(n)W(n)$$

$$\lambda_{af}(n)X_{af}^T(n)X_{af}(n)=e_{af}(n), \quad \because d_{af}(n)-X_{af}^T(n)W(n)=e_{af}(n)$$

Therefore,

$$\lambda_{af}(n)=\left[X_{af}^T(n)X_{af}(n)\right]^{-1}e_{af}(n) \tag{15.23c}$$

Substituting this value of $\lambda_{af}(n)$ in the weight vector update Eqn. (15.23b)

$$W(n+1)=W(n)+X_{af}(n)\lambda_{af}(n)$$

$$W(n+1)=W(n)+X_{af}(k)\left[X_{af}^T(k)X_{af}(k)\right]^{-1}e_{af}(n) \tag{15.23d}$$

This is an affine projection algorithm with unity step size and has maximum permissible convergence rate with maximum misadjustment. But by introducing the step size other than unity, a trade-off between the misadjustment and convergence speed can be achieved and is

$$W(n+1)=W(n)+\mu X_{af}(k)\left[X_{af}^T(k)X_{af}(k)\right]^{-1}e_{af}(n) \tag{15.23e}$$

With the introduction of a step size μ, it is not possible to make the a posteriori errors zero (Eqn.(15.22e)) as it disturbs the weight vector solution (Eqn. (15.23d)). For a step size μ, the misadjustment is given by [S.G. Sankaran and A.A. Beex, 2000]

$$M = \frac{\mu}{(2-\mu)} E\left\{\frac{1}{\|X(n)\|^2}\right\} Trace[R_{NN}]$$

Here, the misadjustment is independent of L. Also, in the presence of observation noise uncorrelated with the input, forcing the posteriori error to zero, results in a compensation for the noise resulting in a misadjustment. Further, this misadjustment becomes maximum when the step size is unity (i.e., $\mu = 1$). For a single input vector, from Eqn. (15.22a1), Eqn. (15.23e) becomes

$$W(n+1) = W(n) + \mu X(n)\left[X^T(n)X(n)\right]^{-1} e_0(n)$$

where, $d(n) - X^T(n)W(n) = e_0(n)$

$$W(n+1) = W(n) + \frac{\mu}{X^T(n)X(n)} X(n)e_0(n)$$

$$W(n+1) = W(n) + \frac{\mu}{\|X(n)\|^2} e_0(n)X(n)$$

That is, APA with a single input vector is nothing but NLMS algorithm.

To overcome the numerical matrix inversion problems, an identity matrix I multiplied by a small constant γ is added to the matrix $X_{af}^T(n)\, X_{af}(n)$ as

$$W(n+1) = W(n) + \mu X_{af}(n)(X_{af}^T(n)X_{af}(n) + \gamma\, I)^{-1} e_{af}(n) \qquad (15.23f)$$

The computational complexity of the affine projection algorithm is decided by the number of data vectors $(L+1)$ used, as it determines the order of the matrix to be inverted.

Here, the performance of affine projection algorithm is illustrated for the system identification and the considered system has the impulse response

$$h(n) = [0.1,\ 0.3,\ 0.0,\ 0.2,\ -0.4,\ -0.7,\ -0.4,\ -0.2]$$

The adaptive filter used has an order 10, i.e., has 10 coefficients. The input used is a sum of three sinusoidal signals of frequencies 100, 200 and 300 Hz plus the white Gaussian noise with SNR=26dB, sampled at 1000 Hz. The convergence factor $\mu = 0.01$ and the constant $\gamma = 10^{-6}$.

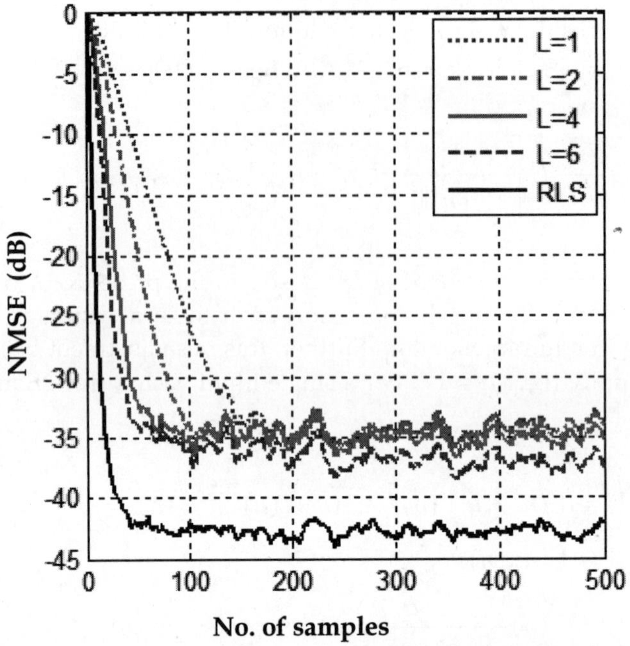

Fig. 15.4 Performance of affine projection algorithm for different number of input vectors *L*.

Fig. 15.4 depicts the NMSE learning curves obtained by averaging the results of 500 simulation runs for different values of *L*. When *L*=0, affine projection algorithm becomes the normalized LMS algorithm. Increasing *L* makes the algorithm to converge faster. However, beyond certain number of input vectors, the improvement in convergence rate is not significant. Here, for *L*=0 and for *L*=3 there is a significant difference in convergence rate and the latter is better than the former. But for *L*=3 and 5, the difference in convergence rate is small indicating that a high value of L may not be necessary and it can be in the range of 4 to 9. The NMSE for the RLS algorithm for the same input with forgetting factor 0.999 and the initialization $C(0) = R^{-1}_{NN}(0) = \delta^{-1}\,I$, $\delta = 1$ is also shown in the Fig. 15.4. This indicates that the error convergence of APA though slower than that of RLS but it closer to that of the RLS and is significantly superior to that of NLMS which corresponds to *L*= 0.

15.9 SIGNAL ESTIMATION BY KALMAN FILTERING

Signal estimation from observations in which signal is corrupted by white noise was dealt with by Wiener filtering, a block/off line approach. A recursive approach which dynamically estimates signal from observations on a sample to sample online basis in its general vector form is known as Kalman filtering. To bring out the concept, a simplified version of Kalman filtering, its scalar form will be considered (C. Therrien, 1992). This involves estimation of a random process at any specific point in its evolution course using observations from its starting point $n = 0$ till the present instant n. This requires a filter which changes its parameters recursively with the arrival of each new observation which is used to filter the observations to get signal estimate.

Here an estimation of a random signal $s(n)$ with autocorrelation $R_s(l)$ in additive white noise $v(n)$ having variance σ_v^2 and uncorrelated with $s(n)$ is considered. It is required to find the optimal linear mean square estimate of $s(n)$ from the observations

I $$x(k) = s(k) + v(k), \quad k = 0,1,2,\ldots n \tag{15.24a}$$

It is required to have a recursive linear mean square estimate of $s(n)$ as

$$\hat{s}(n) = C_n \hat{s}(n-1) + K_n x(n) \tag{15.24b}$$

where,

$\hat{s}(n-1)$: Estimate of $s(n-1)$

$x(n)$: Present observation

C_n, K_n : Dynamically varying parameters to be found.

In general such a representation may not be always possible but due to its utility, it is necessary to find the conditions under which such a recursive representation is possible. For this, let the possible error be given by

$$e(n) = s(n) - \hat{s}(n) \tag{15.24c}$$

and $\sigma_{e(n)}^2$ be its variance. Since the noise $v(n)$ is white and uncorrelated with the signal $s(n)$,

$$E\left[s(n)x(n-l)\right] = R_s(l) \tag{15.24d}$$

$$E\left[x(n)x(n-l)\right] = R_s(l) + \sigma_v^2 \delta(l) \tag{15.24e}$$

$$E\left[x(n)v(n)\right] = E\left[\{s(n)+v(n)\}v(n)\right] = \sigma_v^2 \tag{15.24f}$$

For the recursive estimation to be optimal, its parameters C_n and K_n are found by applying the orthogonality principle, the input and the error are orthogonal to each other.

That is,

$$E\left[e(n)x(n-l)\right]=0, \quad l=0,1,2,\ldots n \tag{15.25a}$$

For $l=0$,

$$E\left[e(n)x(n)\right]=E[e(n)\{s(n)+v(n)\}=0$$

$$E\left[e(n)s(n)\right]+E\left[e(n)v(n)\right]=0 \tag{15.25b}$$

But

$$E\left[e(n)s(n)\right]=E\left[e(n)e(n)\right]=\sigma_{e(n)}^2$$

Therefore, Eqn. (15.25b) becomes,

$$\sigma_{e(n)}^2+E\left[e(n)v(n)\right]=0 \tag{15.25c}$$

But from Eqns. (15.24b, c),

$$e(n)=s(n)-\hat{s}(n) \text{ and } \hat{s}(n)=C_n\hat{s}(n-1)+K_n x(n)$$

Hence,

$$\sigma_{e(n)}^2+E\left[\left\{s(n)-C_n\hat{s}(n-1)+K_n x(n)\right\}v(n)\right]=0$$

$$\sigma_{e(n)}^2+E\left[s(n)v(n)\right]-C_nE\left[\hat{s}(n-1)v(n)\right]-K_nE\left[x(n)v(n)\right]=0$$

Since the signal $s(n)$ is orthogonal to noise $v(n)$, $E\left[s(n)v(n)\right]=0$. Further $v(n)$ is white and $\hat{s}(n-1)$ is a function of only signal values and noise values up to $(n-1)$, $E\left[\hat{s}(n-1)v(n)\right]=0$. Hence,

$$\sigma_{e(n)}^2-K_nE\left[x(n)v(n)\right]=0$$

$$\sigma_{e(n)}^2-K_nE\left[\left\{s(n)+v(n)\right\}v(n)\right]=0$$

$$\sigma_{e(n)}^2-K_nE\left[v(n)v(n)\right]=0$$

Therefore,

$$K_n=\frac{\sigma_{e(n)}^2}{\sigma_v^2}, \quad \sigma_v^2=E\left[v(n)v(n)\right] \tag{15.25d}$$

Here, as $\sigma_{e(n)}^2$ and σ_v^2 are real and positive, K_n is real and positive. Again considering orthogonality principle for $l > 0$,

$$E\left[e(n)x(n-l)\right] = 0$$

From Eqn. (15.24c)

$$e(n) = s(n) - \hat{s}(n)$$

Therefore,

$$E\left[\left\{s(n) - \hat{s}(n)\right\}x(n-l)\right] = 0,$$

From Eqn. (15.24b)

$$\hat{s}(n) = C_n\hat{s}(n-1) + K_n x(n)$$

Hence,

$$E\left[\left\{s(n) - C_n\hat{s}(n-1) - K_n x(n)\right\}x(n-l)\right] = 0,$$

$$E\left[s(n)x(n-l)\right] - C_n E\left[\hat{s}(n-1)x(n-l)\right] - K_n E\left[x(n)x(n-l)\right] = 0 \qquad (15.25e)$$

Since

$$E\left[s(n)x(n-l)\right] = R_s(l),\; E\left[x(n)x(n-l)\right] = R_s(l) + \sigma_v^2\delta(l)$$

Further, for $l > 0$,

$$E\left[x(n)x(n-l)\right] = R_s(l)$$

Therefore, Eqn. (15.25e) becomes,

$$R_s(l) - C_n E\left[\hat{s}(n-1)x(n-l)\right] - K_n R_s(l) = 0$$

$$R_s(l)\left(1 - K_n\right) - C_n E\left[\hat{s}(n-1)x(n-l)\right] = 0 \qquad (15.25f)$$

Consider

$$e(n) = s(n) - \hat{s}(n)$$
$$e(n-1) = s(n-1) - \hat{s}(n-1)$$
$$\hat{s}(n-1) = s(n-1) - e(n-1)$$

Therefore, Eqn. (15.25f) becomes,

$$R_s(l)(1-K_n)-C_nE\left[\{s(n-1)-e(n-1)\}x(n-l)\right]=0$$

$$R_s(l)(1-K_n)-C_nE\left[s(n-1)x(n-l)\right]+C_nE\left[e(n-1)x(n-l)\right]=0$$

Since error and input are orthogonal to each other, $E\left[e(n-1)x(n-l)\right]=0$. Therefore,

$$R_s(l)(1-K_n)-C_nE\left[s(n-1)x(n-l)\right]=0$$

$$R_s(l)(1-K_n)-C_nR_s(l-1)=0,\because R_s(l-1)=E\left[s(n-1)x(n-l)\right]$$

$$R_s(l)=\frac{C_n}{(1-K_n)}R_s(l-1),\ \text{ for } l>0$$

$$R_s(l)=\alpha R_s(l-1),\ \text{for } l>0 \quad \text{where} \quad \alpha=\frac{C_n}{(1-K_n)}$$

For $l=1$, $R_s(1)=\alpha R_s(0)$,

For $l=2$, $R_s(2)=\alpha R_s(1)=\alpha\left[\alpha R_s(0)\right]=\alpha^2R_s(0)$

For $l=3$, $R_s(3)=\alpha R_s(2)=\alpha\left[\alpha^2R_s(0)\right]=\alpha^3R_s(0)$

Therefore,

$$R_s(l)=R_s(0)\alpha^l,\quad l>0 \text{ with } \quad \alpha=\frac{C_n}{(1-K_n)} \tag{15.25g}$$

For the recursive signal estimation given by Eqn.(15.24b) to be optimal, the condition to be satisfied is

$$\hat{s}(n)=C_n\hat{s}(n-1)+K_nx(n)$$

Consider the signal,

$$s(n)=\alpha s(n-1)+w(n) \tag{15.25h}$$

generated by a white noise process $w(n)$ with variance σ_w^2. The driving noise $w(n)$ is different from the observation noise $v(n)$. Now,

$$R_{sw}(l)-\alpha R_{sw}(l-1)=\sigma_w^2\delta(l)$$

The solution of this difference equation is found recursively as below:

$$R_{sw}(0) = \sigma_w^2$$

$$R_{sw}(1) - \alpha R_{sw}(0) = 0, R_{sw}(1) = \alpha R_{sw}(0) = \alpha\ \sigma_w^2,$$

$$R_{sw}(2) - \alpha R_{sw}(1) = 0, R_{sw}(2) = \alpha R_{sw}(1) = \alpha\left(\alpha\ \sigma_w^2\right),$$

$$\vdots$$

$$R_{sw}(l) = \alpha^l \sigma_w^2 (\alpha) u(l), \quad l \geq 0$$

The solution is,

$$R_{sw}(l) = \sigma_w^2 \alpha^l u(l), \quad u(l) = 1, \quad l \geq 0, \tag{15.26a}$$

Also

$$E[s(n)s(n-l)] = \alpha E[s(n-1)s(n-l)] + E[w(n)s(n-l)]$$

$$R_S(l) - \alpha R_S(l-1) = R_{ws}(l)$$

$$R_{ws}(l) = R_{sw}(-l) = \sigma_w^2 \alpha^{-l} u(-l) \tag{15.26b}$$

The input to the difference equation is an exponential for $l \leq 0$, zero for $l > 0$. The solution for the difference equation for $l \leq 0$ can be obtained by assuming the solution in the form

$$R_S(l) = p_1 \alpha^{-l}$$

and the constant p_1 is found by substituting the solution in the difference equation. That is,

$$p_1 \alpha^{-l} - \alpha p_1 \alpha^{(l-1)} = \sigma_w^2 \alpha^{-l}, \ p_1 - p_1 \alpha^2 = \sigma_w^2, \ p_1 = \frac{\sigma_w^2}{1-\alpha^2}$$

Therefore, the solution for $l \leq 0$ is

$$R_S(l) = \frac{\sigma_w^2}{1-\alpha^2} \alpha^{-l} \tag{15.26c}$$

For $l > 0$, the input is zero and the difference equation has homogeneous solution of the form,

$$R_S(l) = p_2 \alpha^l, \ l > 0$$

For $l = 1$,

$$R_S(1) - \alpha R_S(0) = 0$$

This gives the boundary condition and then substituting the assumed solution in the this equation,

$$p_2\alpha - \alpha R_S(0) = 0 \ , \ p_2 = R_S(0)$$

Since

$$R_S(0) = \frac{\sigma_w^2}{1-\alpha^2},$$

Therefore, the solution for $l > 0$ is

$$R_S(l) = \frac{\sigma_w^2}{1-\alpha^2}\alpha^l$$

Hence, the correlation of this signal is

$$R_s(l) = \sigma_s^2 \alpha^l, \qquad l > 0$$

where,

$$\sigma_s^2 = R_s(0) == \frac{\sigma_w^2}{\left(1-|\alpha|^2\right)}$$

The above derivation shows that for the correlation for the signal generated by Eqn. (15.25h)

$$s(n) = \alpha s(n-1) + w(n),$$

satisfies the *orthogonality condition required for the estimate of the form (Eqn. (15.24b))*

$$\hat{s}(n) = C_n \hat{s}(n-1) + K_n x(n)$$

and therefore is an optimal estimate. Since from Eqn. (15.25g)

$$C_n = \alpha(1-K_n)$$

Eqn. (15.24b) becomes

$$\hat{s}(n) = \alpha(1-K_n)\hat{s}(n-1) + K_n x(n)$$

$$\hat{s}(n) = \alpha\hat{s}(n-1) + K_n x(n) - \alpha K_n \hat{s}(n-1)$$

$$\hat{s}(n) = \alpha\hat{s}(n-1) + K_n\left[x(n) - \alpha\hat{s}(n-1)\right] \qquad (15.26d)$$

Further, from Eqn. (15.25d)

$$K_n = \sigma_{e(n)}^2 / \sigma_v^2$$

These Eqns. (15.26d) and (15.25d), together form the required result. Thus, the estimator is given by

$$\hat{s}(n) = \alpha\hat{s}(n-1) + K_n\left[x(n) - \alpha\hat{s}(n-1)\right]$$

$$K_n = \frac{\sigma^2_{e(n)}}{\sigma^2_v}, \quad \sigma^2_v = E\left[v(n)v(n)\right] \quad \alpha = \frac{C_n}{(1-K_n)},$$

For the estimator to be complete, it is required to find dynamically the error variance $\sigma^2_{e(n)}$ which determines the value of K_n. Hence, an expression for $\sigma^2_{e(n)}$ will be derived. The error is given by (Eqn. (15.24c))

$$e(n) = s(n) - \hat{s}(n),$$

$$\sigma^2_{e(n)} = E[e(n)e(n)] = E[\{s(n) - \hat{s}(n)\}e(n)]$$

$$= E[s(n)e(n)]$$

$$= E[s(n)\{s(n) - \alpha(1-K_n)\hat{s}(n-1) - K_n x(n)\}]$$

$$= \sigma^2_S - \alpha(1-K_n)E[s(n)\hat{s}(n-1)] - K_n E[s(n)x(n)]$$

$$= \sigma^2_S - \alpha(1-K_n)E[s(n)\hat{s}(n-1)] - K_n E[s(n)\{s(n)+v(n)\}]$$

$$= \sigma^2_S(1-K_n) - \alpha(1-K_n)E[s(n)\hat{s}(n-1)]$$

$$\sigma^2_{e(n)} = (1-K_n)\left\{\sigma^2_S - \alpha E[s(n)\hat{s}(n-1)]\right\}$$

$$\sigma^2_{e(n)} = (1-K_n)\left\{\sigma^2_S - \alpha E[\{\alpha s(n-1) + w(n)\}\hat{s}(n-1)]\right\}$$
$$= (1-K_n)\left\{\sigma^2_S - \alpha E[\{\alpha s(n-1) + w(n)\}\hat{s}(n-1)]\right\}$$

$$\sigma^2_{e(n)} = (1-K_n)\left\{\sigma^2_S - \alpha^2 E[s(n-1)\hat{s}(n-1)]\right\} \quad (15.27a)$$

In the first approach, for $n = 0$, $\hat{s}(n-1) = \hat{s}(-1)$ and $\hat{s}(-1)$ is taken as $\hat{s}(-1) = x(-1)$. Therefore,

$$E[\hat{s}(-1)x(-1)] = R_{Sx}(0) = \sigma^2_S$$

Hence,

$$\sigma_{e(0)}^2 = (1-K_0)\left\{\sigma_S^2 - \alpha^2\sigma_S^2\right\}$$

But

$$K_n = \frac{\sigma_{e(n)}^2}{\sigma_v^2} \text{ and } K_0 = \frac{\sigma_{e(0)}^2}{\sigma_v^2}$$

Therefore,

$$\sigma_{e(0)}^2 = \left(1 - \frac{\sigma_{e(0)}^2}{\sigma_v^2}\right)\sigma_S^2(1-\alpha^2)$$

$$\sigma_{e(0)}^2\left(1 + \frac{\sigma_S^2(1-\alpha^2)}{\sigma_v^2}\right) = \sigma_S^2(1-\alpha^2)$$

$$\sigma_{e(0)}^2 = \frac{\sigma_S^2(1-\alpha^2)}{\left(1 + \dfrac{\sigma_S^2(1-\alpha^2)}{\sigma_v^2}\right)} \tag{15.27b}$$

In the second approach, at $n = 0$, it is assumed $\hat{s}(-1) = 0$. Therefore,

$$\sigma_{e(n)}^2 = (1-K_n)\left\{\sigma_S^2 - \alpha^2 E[s(n-1)\hat{s}(n-1)]\right\}$$

becomes

$$\sigma_{e(0)}^2 = (1-K_0)\sigma_S^2$$

$$\sigma_{e(0)}^2 = (1 - \frac{\sigma_{e(0)}^2}{\sigma_v^2})\sigma_S^2$$

$$\sigma_{e(0)}^2 = \frac{\sigma_S^2}{\left(1 + \dfrac{\sigma_S^2}{\sigma_v^2}\right)} \tag{15.27c}$$

Now consider $E[s(n)s(n)]$ and since $s(n) = \alpha s(n-1) + w(n)$,

$$E[s(n)s(n)] = \alpha^2 E[s(n-1)s(n-1)] + 2\alpha E[s(n-1)w(n)] + E[w(n)w(n)]$$

$$\sigma_S^2 = \alpha^2\sigma_S^2 + \sigma_w^2$$

$$\sigma_w^2 = \sigma_S^2(1-\alpha^2)$$

$$\sigma^2_{e(0)} = \frac{\sigma_S^2}{\left(1+\dfrac{\sigma_S^2}{\sigma_v^2}\right)}$$

Multiplying and dividing by $(1-\alpha^2)$,

$$\sigma^2_{e(0)} = \frac{\sigma_S^2(1-\alpha^2)}{(1-\alpha^2)\left(1+\dfrac{\sigma_S^2}{\sigma_v^2}\right)}$$

Therefore substituting for $\sigma_S^2(1-\alpha^2) = \sigma_w^2$ (Eqn. (15.26c))

$$\sigma^2_{e(0)} = \frac{\sigma_w^2}{\left((1-\alpha^2)+\dfrac{\sigma_w^2}{\sigma_v^2}\right)}$$

$$\sigma^2_{e(0)} = \frac{\sigma_w^2\,\sigma_v^2}{\left((1-\alpha^2)\sigma_v^2+\sigma_w^2\right)} \tag{15.27d}$$

This relation can be used if σ_w^2 is given instead of σ_S^2.

Estimation of $\sigma^2_{e(n)}$ for $n > 0$:

$$\sigma^2_{e(n)} = E[e(n)e(n)] = E[\{s(n)-\hat{s}(n)\}e(n)]$$
$$\sigma^2_{e(n)} = E[s(n)e(n)]$$
$$\sigma^2_{e(n-1)} = E[s(n-1)e(n-1)]$$
$$\sigma^2_{e(n-1)} = E\left[s(n-1)\{s(n-1)-\hat{s}(n-1)\}\right]$$
$$\sigma^2_{e(n-1)} = \sigma_S^2 - E\left[s(n-1)\hat{s}(n-1)\right]$$

$$E\left[s(n-1)\hat{s}(n-1)\right] = \sigma_S^2 - \sigma^2_{e(n-1)}$$

From Eqn. (15.27a),

$$\sigma^2_{e(n)} = (1-K_n)\left\{\sigma_S^2 - \alpha^2 E[s(n-1)\hat{s}(n-1)]\right\}$$

Substituting for $E\left[s(n-1)\hat{s}(n-1)\right]$,

$$\sigma_{e(n)}^2 = (1-K_n)\left\{\sigma_S^2 - \alpha^2\left(\sigma_S^2 - \sigma_{e(n-1)}^2\right)\right\}$$

$$\sigma_{e(n)}^2 = (1-K_n)\left\{\sigma_S^2(1-\alpha^2) + \alpha^2\sigma_{e(n-1)}^2\right\}$$

$$\sigma_{e(n)}^2 = (1-K_n)\left\{\sigma_w^2 + \alpha^2\sigma_{e(n-1)}^2\right\}$$

But from Eqn. (15.25d)

$$K_n = \frac{\sigma_{e(n)}^2}{\sigma_v^2}$$

Therefore,

$$\sigma_{e(n)}^2 = \left(1 - \frac{\sigma_{e(n)}^2}{\sigma_v^2}\right)\left\{\sigma_w^2 + \alpha^2\sigma_{e(n-1)}^2\right\}$$

$$\sigma_{e(n)}^2\left[1 + \frac{\sigma_w^2 + \alpha^2\sigma_{e(n-1)}^2}{\sigma_v^2}\right] = \left\{\sigma_w^2 + \alpha^2\sigma_{e(n-1)}^2\right\}$$

$$\sigma_{e(n)}^2 = \frac{\left\{\sigma_w^2 + \alpha^2\sigma_{e(n-1)}^2\right\}\sigma_v^2}{\left[\sigma_v^2 + \sigma_w^2 + \alpha^2\sigma_{e(n-1)}^2\right]}, \qquad \text{for } n > 0 \tag{15.27e}$$

This the recursive expression for the error variance required for the estimation of K_n. Further,

$$\sigma_{e(n)}^2 = \left[\frac{\sigma_v^2 + \sigma_w^2 + \alpha^2\sigma_{e(n-1)}^2}{\sigma_w^2 + \alpha^2\sigma_{e(n-1)}^2}\right]^{-1}\sigma_v^2$$

$$\sigma_{e(n)}^2 = \left[\frac{\sigma_v^2}{\sigma_w^2 + \alpha^2\sigma_{e(n-1)}^2} + 1\right]^{-1}\sigma_v^2$$

$$\sigma_{e(n)}^2 = \left[\sigma_v^2\left\{\frac{1}{\sigma_w^2 + \alpha^2\sigma_{e(n-1)}^2} + \frac{1}{\sigma_v^2}\right\}\right]^{-1}\sigma_v^2$$

$$\sigma_{e(n)}^2 = \left[\frac{1}{\sigma_w^2 + \alpha^2\sigma_{e(n-1)}^2} + \frac{1}{\sigma_v^2}\right]^{-1}, \quad n > 0 \tag{15.27f}$$

If $\sigma_w^2 >> \sigma_v^2$, then the SNR is large and

$$\sigma_{e(n)}^2 = \left[\frac{1}{\sigma_v^2}\right]^{-1} = \sigma_v^2 \quad n > 0, \quad \sigma_{e(n)}^2 = \sigma_v^2 \quad n > 0$$

Also as

$$K_n = \frac{\sigma_{e(n)}^2}{\sigma_v^2} = \frac{\sigma_v^2}{\sigma_v^2}, \quad K_n = 1$$

The signal $s(n)$ is same as the observations as $x(n)$ as $v(n)$ is very small. The error variance and filter gain given respectively by

$$\sigma_{e(n)}^2 = \left[\frac{1}{\sigma_w^2 + \alpha^2\sigma_{e(n-1)}^2} + \frac{1}{\sigma_v^2}\right]^{-1}, \quad K_n = \frac{\sigma_{e(n)}^2}{\sigma_v^2}$$

are not functions of $x(n)$. Hence these can be computed off-line prior to filtering and can be stored for later use. The estimator (Eqn. (15.26d))

$$\hat{s}(n) = \alpha\hat{s}(n-1) + K_n\left[x(n) - \alpha\hat{s}(n-1)\right]$$

is generally represented by two equations as,

$$\hat{s}(n/n-1) = \alpha\hat{s}(n-1) \tag{15.28a}$$

$$\hat{s}(n) = \alpha\hat{s}(n-1) + K_n\left[x(n) - \hat{s}(n/n-1)\right] \tag{15.28b}$$

for the signal model and observations given by

$$s(n) = \alpha s(n-1) + w(n) \text{ and } x(k) = s(k) + v(k), \quad k = 0,1,2,\ldots,n$$

According to signal model, the Eqn. (15.28a),

$$\hat{s}(n/n-1) = \alpha\hat{s}(n-1)$$

approximates $s(n)$ and can be treated *an* ***approximate prediction*** (not actual prediction) since here for the prediction, the previous actual signal value $s(n-1)$ is not used but only its estimate $\hat{s}(n-1)$ is used. The second Eqn.(15.5b) gives the signal estimate (not prediction) at instant n using all the observations including $x(n)$ at the nth instant. This estimate can be expressed as prediction plus correction.

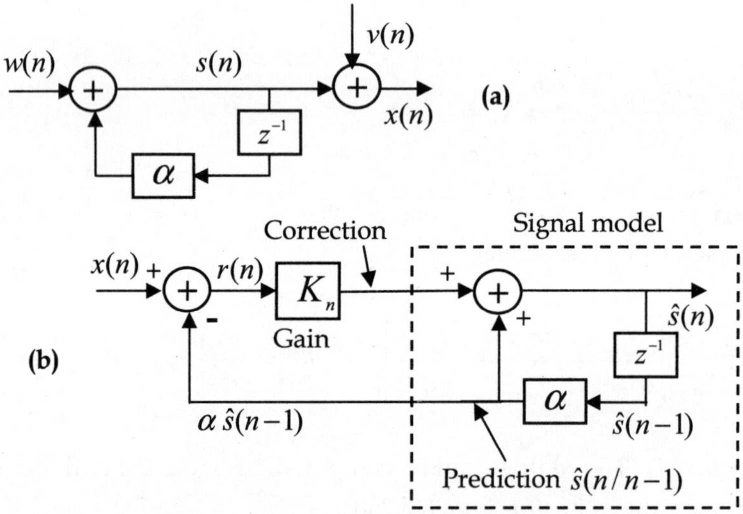

Fig. 15.5. (a) Signal model, (b) Kalman Signal estimator.

$$\textit{estimate} = \textit{prediction} + \textit{correction}$$

$$\hat{s}(n) = \alpha\hat{s}(n-1) + K_n\left[x(n) - \alpha\hat{s}(n-1)\right]$$

Using Eqn. (15.28a),

$$\hat{s}(n) = s(n/n-1) + K_n\left[x(n) - s(n/n-1)\right]$$

If

$$r(n) = \left[x(n) - s(n/n-1)\right],$$

$$\hat{s}(n) = s(n/n-1) + K_n r(n) \tag{15.28c}$$

The quantity $r(n)$ is the error or residual and is a non-stationary process (as its variance depends n) which represents new information in the observation. As $n \to \infty$, the variance of $r(n)$ becomes a constant and $r(n)$ becomes an innovation process. This means that $r(n)$ becomes a white noise process which when passed through a suitable filter produces the observation process $x(n)$. The signal model and the Kalman signal estimator are shown in the Fig. 15.5(a) and 15.5(b), respectively. It is important to notice that the signal model appears in estimator. Further, the driving noise $w(n)$ and the observation noise $v(n)$ are different. If $w(n) = 0$, there is no signal. The observation noise $v(n)$ is not desirable but its presence is inevitable. If $v(n) = 0$, then no estimator is required and the signal is available directly.

In addition to estimating the present signal value $s(n)$ from the observations including the present observation $x(n)$, it is possible to predict the next value of the signal, i.e., $s(n+1)$ (from observation up to time index n). This is given by

$$\hat{s}(n+1/n) = \alpha\hat{s}(n) \tag{15.29a}$$

The corresponding error is

$$e(n+1/n) = s(n+1) - \hat{s}(n+1/n) \tag{15.29b}$$

To show that the prediction given by Eqn. (15.6a) is optimal, it is necessary to show that the input and the error are orthogonal to each other. That is,

$$E\left[x(n-l)e(n+1/n)\right] = 0$$

For this, consider

$$E\left[x(n-l)e(n+1/n)\right] = E[x(n-l)\{s(n+1) - \hat{s}(n+1/n)\}]$$

Substituting for $s(n+1)$ and $\hat{s}(n+1/n)$

$$E[x(n-l)e(n+1/n) = E[x(n-l)\{\alpha s(n) + w(n+1) - \alpha\hat{s}(n)\}]$$

As $w(n)$ is white noise, for $l = 0,1,2,\ldots,n$

$$E[x(n-l)w(n+1)] = 0$$

Therefore,

$$E[x(n-l)e(n+1/n) = \alpha E[x(n-l)\{s(n) - \hat{s}(n)\}$$

$$E[x(n-l)e(n+1/n) = \alpha E[x(n-l)e(n),] \qquad \because e(n) = \{s(n) - \hat{s}(n)\}$$

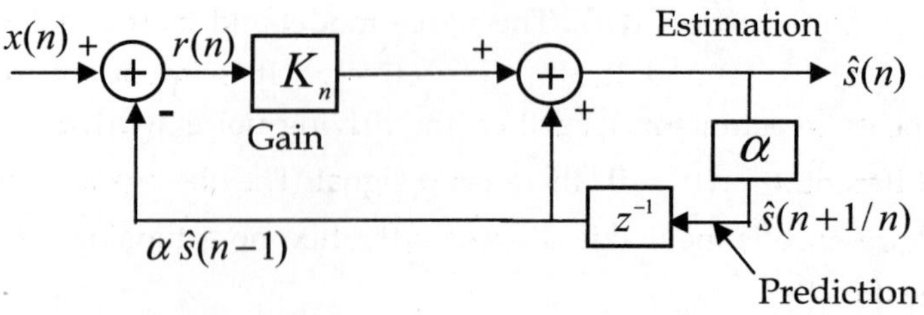

Fig. 15.5 (c) System achieving both signal estimation and prediction.

Since $x(n)$ and $e(n)$ are orthogonal to each other,

$$E[x(n-l)e(n)] = 0$$

Therefore,

$$E[x(n-l)e(n+1/n) = 0$$

This indicates that the estimate $\hat{s}(n+1/n) = \alpha\hat{s}(n)$, is optimum. The *optimum prediction* in addition to the optimum estimate of the signal can be obtained by just by interchanging the delay and the factor α of the system which provides only the signal estimate (Fig. 15.5(b)) and is shown in Fig. 15.5 (c). The error variance for this prediction estimate is

$$\sigma^2_{e(n+1/n)} = E[e(n+1/n)e(n+1/n)]$$

But from Eqn. (15.29b),

$$e(n+1/n) = s(n+1) - \hat{s}(n+1/n)$$

Therefore,

$$\sigma^2_{e(n+1/n)} = E[e(n+1/n)\{s(n+1) - \hat{s}(n+1/n)\}]$$

$$\sigma^2_{e(n+1/n)} = E[e(n+1/n)s(n+1)], \qquad \because E[e(n+1/n)\hat{s}(n+1/n)] = 0$$

$$\sigma^2_{e(n+1/n)} = E[\{s(n+1) - s(n+1/n)\}s(n+1)]$$

From the signal model and from Eqn. (15.29a), respectively,

$$s(n+1) = \alpha s(n) + w(n+1) \text{ and } \hat{s}(n+1/n) = \alpha\hat{s}(n)$$

Hence,

$$\sigma^2_{e(n+1/n)} = E[\{\alpha s(n) + w(n+1) - \alpha\hat{s}(n)\}s(n+1)]$$

$$= E[\{\alpha e(n) + w(n+1)\}s(n+1)], \quad \because e(n) = s(n) - \hat{s}(n)$$

$$= E[\{\alpha e(n) + w(n+1)\}\{\alpha s(n) + w(n+1)\}]$$

$$= \alpha^2 E[s(n)e(n)] + E[w(n+1)w(n+1)]$$

$$\sigma^2_{e(n+1/n)} = \alpha^2\sigma^2_{e(n)} + \sigma^2_w \tag{15.29c}$$

From Eqn. (15.26d) the signal estimate is given by

$$\hat{s}(n) = \alpha\hat{s}(n-1) + K_n\left[x(n) - \alpha\hat{s}(n-1)\right]$$

Also the prediction is given by Eqn. (15.29a)

$$\hat{s}(n+1/n) = \alpha\hat{s}(n)$$

$$\hat{s}(n+1/n) = \alpha\hat{s}(n) = \alpha\left\{\alpha\hat{s}(n-1) + K_n\left[x(n) - \alpha\hat{s}(n-1)\right]\right\}$$

Using the definition of prediction (Eqn. 15.29a)),

$$\hat{s}(n/n-1) = \alpha\hat{s}(n-1)$$

Therefore,

$$\hat{s}(n+1/n) = \alpha\hat{s}(n/n-1) + \alpha K_n\left[x(n) - \hat{s}(n/n-1)\right] \tag{15.29d}$$

At high SNR, $\sigma^2_v \rightarrow 0$ and the error variance given by (Eqn. (15.4e))

$$\sigma^2_{e(n)} = \frac{\left\{\sigma^2_w + \alpha^2\sigma^2_{e(n-1)}\right\}\sigma^2_v}{\left[\sigma^2_v + \sigma^2_w + \alpha^2\sigma^2_{e(n-1)}\right]}$$

becomes $\sigma^2_{e(n)} = \sigma^2_v$. Therefore since $K_n = \sigma^2_{e(n)} / \sigma^2_v$, $K_n \rightarrow 1$. In view of this, Eqn. (15.6d) becomes

$$\hat{s}(n+1/n) = \alpha x(n)$$

At high SNR, as $x(n) \approx s(n)$, $\hat{s}(n+1/n) = \alpha s(n)$ which is same as linear prediction.

For large value of n, the filter gain K_n and the error variance $\sigma^2_{e(n)}$ converge to constant steady state values. Also, the recursive filter is expected to approach the causal Wiener filter and their steady state mean square errors are also expected to become equal. In the above treatment, only first order AR process was considered, but an extension of this to P^{th} order AR process can be done using state variables.

Example: Consider he signal be generated by

$$s(n) = \alpha s(n-1) + w(n)$$

With $\alpha = 0.6$ and $w(n)$ is zero mean white Gaussian noise with variance 0.98. Therefore,

$$x(n) = s(n) + v(n)$$

$s(n)$ has a variance $\sigma_s^2 = 1.5$ and $v(n)$ is zero mean white Gaussian noise and has a variance $\sigma_v^2 = 1.4741$ to give an SNR=0.07 dB. From Eqn. (15.26c),

$$R_S(l) = \frac{\sigma_w^2}{1-\alpha^2}\alpha^{-l} \text{ and } \sigma_s^2 = R_S(0) = \frac{\sigma_w^2}{1-\alpha^2}$$

$$\sigma_w^2 = \sigma_s^2(1-\alpha^2) = 1.5\{1-(0.6)^2\} = 0.96$$

Therefore,

$$\sigma_s^2 = 1.5, \quad \sigma_v^2 = 1.474, \quad \sigma_w^2 = 0.96,$$

From Eqn. (15.27c),

$$\sigma^2_{e(0)} = \frac{\sigma_S^2}{\left(1+\dfrac{\sigma_S^2}{\sigma_v^2}\right)}, \qquad \sigma^2_{e(0)} = \frac{1.5}{\left(1+\dfrac{1.5}{1.474}\right)} = 0.74344$$

From Eqns. (15.27f) and (15.25d),

$$\sigma^2_{e(n)} = \left[\frac{1}{\sigma_w^2 + \alpha^2\sigma^2_{e(n-1)}} + \frac{1}{\sigma_v^2}\right]^{-1}, \quad n > 0\ , \quad K_n = \frac{\sigma^2_{e(n)}}{\sigma_v^2}$$

The values of $\sigma^2_{e(n)}$ and K_n for successive values of $n = 1, 2, 3\dots.$ are listed in Table 15.1. It is seen that the error variance $\sigma^2_{e(n)}$ and the Kalman gain converge after four iterations. The signal estimate is given by

$$K_n = \frac{\sigma^2_{e(n)}}{\sigma_v^2}, \quad \hat{s}(n) = \alpha\hat{s}(n-1) + K_n\left[x(n) - \alpha\hat{s}(n-1)\right]$$

Table- 15.1 $\sigma^2_{e(n)}$ and K_n for different iterations

N	$\sigma^2_{e(n)}$	K_n
0	0.74344	0.5044
1	0.6700	0.4544
2	0.6620	0.4490
3	0.6611	0.4484
4	0.6610	0.4483
5	0.6610	0.4483
6	0.6610	0.4483

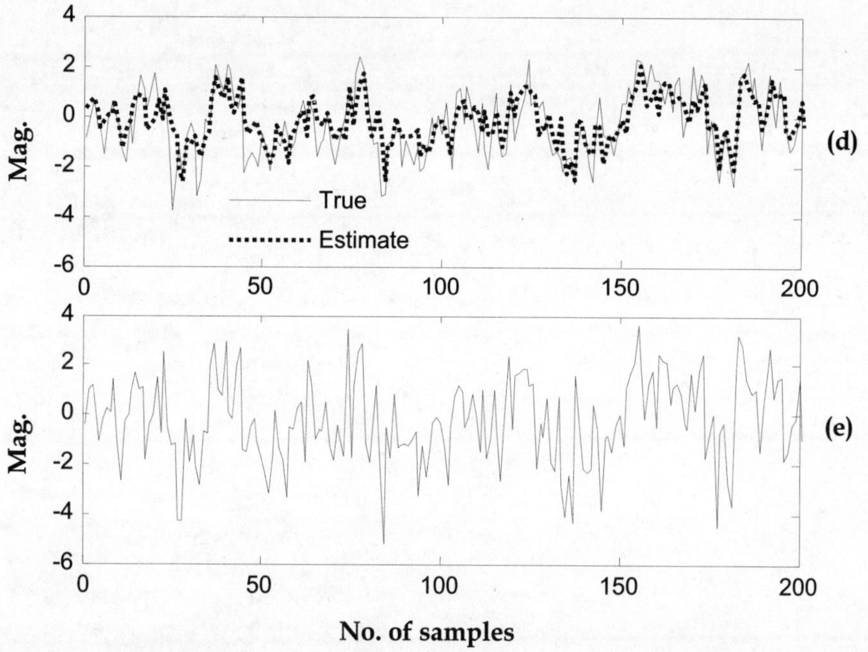

Fig. 15.5 (d, e) Kalman filtering. (d) Signal and its Kalman estimate and (e) Noisy signal.

The signal estimate and the noisy signal (last 200 samples) are shown in Fig. 15.5(d, e). It is seen that the signal estimate follows the true signal reasonably and its magnitude matches with that of true signal. But the noisy signal has a larger magnitude compared to the true signal. The error signal power is 0.6610 and the noise power is 1.474. That is,

$$10\log_{10}[\sigma^2_{e(n)} / \sigma^2_{v(n)}] = 10\log_{10}\left[0.6610/1.474\right] = -3.483dB$$

Therefore, for the filtered signal, the reduction in noise is $3.483dB$.

Chapter 16

NONLINEAR ADAPTIVE FILTERS

In all the adaptive algorithms considered so for, the signals or systems involved are assumed to be linear. However, in practice, most of the signals/systems have a nonlinear component in addition to linear component. Therefore, use of linear adaptive filters, only represents the actual signals/systems only partially and the nonlinear part will be left out resulting in a huge error in their estimation. To overcome this, the adaptive algorithms also should have nonlinear elements. Such a feature is inbuilt for neural network and in Volterra filters. Generally, the neural network with many nodes operates in three stages viz., the input layer, hidden layer and output layer and there can be many hidden layers. In generating the output, activation function which are nonlinear like sigmoid/hyperbolic tangent. This is equivalent to many transfer functions in cascade like an efficient cascade lattice structure (but with single node), instead of a single transfer function like a inefficient transversal structure used in simple signal processing algorithms. This multiple node cascade structure enables efficient feature extraction from the signal/system. But as it is required to adapt to the changes in the signal /system, the intermediate stages have to be adapted using the local gradients which involve local errors and inputs. However, as they are not directly available they are generated by feeding back the final error by using derivative chain rule and is known as back propagation algorithm. Thus, the neural network which not only takes into account the nonlinearity but also extracts information efficiently due to its structure is considered. Generally, the computational complexity will be high and it depends on the input and output nodes and number of hidden layers, considered.

The Volterra model represents the output as a linear combination of nonlinear functions of the input and the output is a linear function of model (filter) parameters. Further, as the input-output relation is nonrecursive and depends only on the input, the linear adaptive filter approaches can be extended even to nonlinear cases. For a nonlinear system, the Volterra LMS algorithm converges significantly better than linear LMS as the latter cannot account for nonlinearity. The extension of Volterra to lattice is considered and due to orthogonality of the input, it provides significantly better convergence rate compared to that of transversal Volterra algorithm. Further, for Volterra algorithm, the orthogonalization of input by an orthogonal transform like, DCT is also considered.

16.1 INTRODUCTION

The adaptive filters considered so far are valid only for linear signals/systems but not for nonlinear signals/systems. The linear models are appropriate for Gaussian noise environment. However, in certain practical scenarios, not only the nonlinearity exists but also plays a major role as it characterizes the processes/systems. In many applications, linear Gaussian model is not proper as nonlinearity becomes prominent in applications like signal companding, amplifier saturation, multiplicative interaction between Gaussian signals, digital storage medium, high data rate transmissions as copper line, satellite links and wireless communication. Hence, in spite of the fact that the nonlinearity creates uneasiness, there can be situations where its handling may become absolutely necessary. In view of this, the linear model performance like bit error rate in communication becomes not satisfactory. This involves nonlinear features extraction and their use (towards the goal) for the complete treatment. In such cases, the linear estimation becomes only partial treatment or even may not suffice. Artificial neural network structure and Voltera filter are used as adaptive filters mainly to cater to the presence of nonlinearities in a system.

The ability of the neural network to perform nonlinear filtering is due to the nonlinear operation called *activation function* that is associated with each processing node/ processing element (PE). However, the neural network, being a distributed one unlike a transversal filter, cannot provide the impulse response of a system. The neural network (with many layers) is similar to a lattice (joint lattice) structure with many stages where for the adaptation of intermediate stages, the required errors are generated from the final error. The field of neural network is very wide, but here its consideration is only from the point of view of temporal adaptive signal processing. That is, the manner in which neural network concepts aid the adaptation process taking into account the existing nonlinearities. The back propagation algorithm realizes this and hence plays a crucial role for neural network based adaptive filters.

The neural network approach has drawbacks like initialization of variance and weights, selecting parameters like input dimension and number of network units. Among polynomial filters, the truncated Volterra filter is popular as it fits a large class of nonlinear systems and is appealing from mathematical point of view also. But it requires knowledge about how many kernels and orders can be handled with the limited available storage and computational resource.

This chapter deals with the basic structure of an artificial neuron, a multi-layer feed-forward network with its training algorithm, viz., the back propagation algorithm and the neural network adaptive filter for identification of a nonlinear system. It also deals with the Volterra adaptive filter based on transversal, transform domain and lattice approaches.

16.2 BASIC MODEL OF AN ARTIFICIAL NEURON

A neuron is an information processing unit and it derives its name from its equivalence to its biological unit. An artificial neural network is made up of many neurons, each referred to as a processing element (PE), perceptron or node.

An artificial neuron model receives input signals x_i, from n different sources (Fig. 16.1). These signals are weighted by weights w_{ji}, to generate net_j an *internal activation* (output), which is a linear weighted aggregation of the input signals, modified by an internal threshold, θ_j.

$$net_j = \sum_{i=1}^{n} w_{ji} x_i + \theta_j \tag{16.1a}$$

To generate the output of the neuron out_j, the internal activation is transformed by $f(.)$, referred to as *signal /activation function* and is

$$out_j = f(net_j) \tag{16.1b}$$

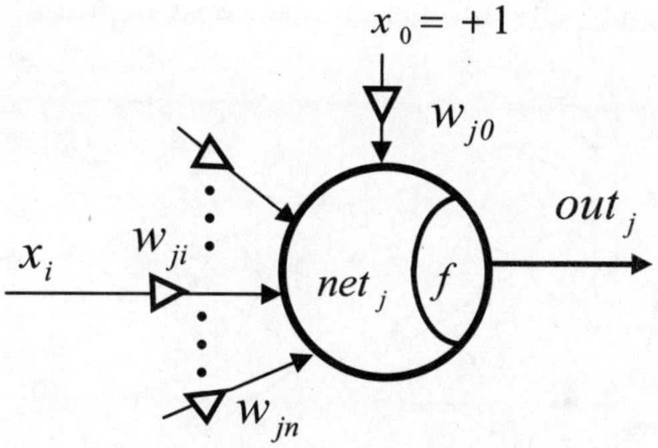

Fig. 16.1 Artificial neuron model / processing element.

A signal function can be a binary threshold, linear threshold, Sigmoid, Gaussian, or probabilistic. For the neuron, *the threshold θ_j is considered an additional weight $w_{j0} = \theta_j$ with its input $x_0 = +1$, a constant.* This treatment will enable the notational convenience. With this, the activation becomes simply an inner product between the input vector $X = (x_{0,} ....x_n)^T$, and the weight vector $W = (w_{j0},....., w_{jn})^T$

$$net_j = X^T W_j = \sum_{i=0}^{n} w_{ji} x_i \tag{16.1c}$$

where $x_0 = +1$ and $w_{j0} = \theta_j$. The activation (inner product) is a measure of similarity or match between the input signal vector and the weight vector. Since the inner product is also discrete convolution of a linear filter, the neuron can be treated as a linear *filter*. Thus the neuron activation measures the similarity between input and the weight vector. The neuron becomes an *adaptive filter* when it acquires the ability of learning.

16.3 MULTI-LAYERED NETWORK ARCHITECTURE

Neural network with one or more layers of neurons hidden in between the input and the output neuron layers represents a large class of feed-forward neural networks, called the multi-layer perceptron networks. Generally, all neurons in a layer are connected to all neurons in the adjacent layers through unidirectional links with their associated weights. The structure of the MLP characterized by its connections/ links within or in between layers is often referred to as its *topology*. The architecture of a multi-layer perceptron (MLP) network with one hidden layer is shown in Fig.16.2 (a). The MLP feed-forward network consists of simple processing elements arranged in layers. Each element takes as its input, the weighted sum of all the outputs of the previous layer and passes it through a nonlinear activation function. *The feed forward nature is due to absence of connections within layers and from outer layers towards the input.* A three-layer network is sufficient to design any kind of non-linear classifier, but there can be any number of layers also.

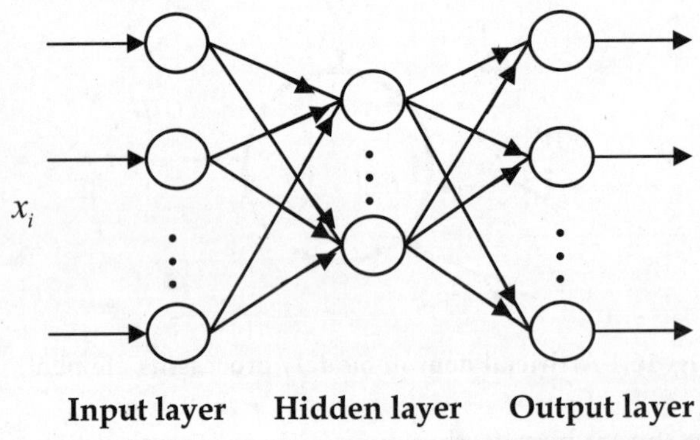

Fig. 16.2 (a) An MLP network with one hidden layer.

The input layer is the first layer and it feeds its input signals to the second layer (hidden layer) without any modification. The input layer neuron uses single input and a *linear activation function*. The layer of processing elements which is not connected directly to the external world, is the hidden layer. The last layer, from which the response of the network comes, is the output layer. *The hidden layer and the output layer use nonlinear activation function.* Usually, the activation function can be either unipolar or bipolar sigmoid function. The number of processing elements in the input and the output layers is governed by the size of the input and output vectors, respectively. There can be any number of processing elements in the hidden layer. Adjusting the weights using some learning rule, like back propagation so that the network produces the desired nonlinear mapping from the input to output, is known as *network training*.

16.4 BACK PROPAGATION ALGORITHM

Back propagation algorithm is a systematic method for training multi-layer feed forward neural networks. Figure 16.2 (b) shows a three-layer feedforward neural network considered for training with back propagation.

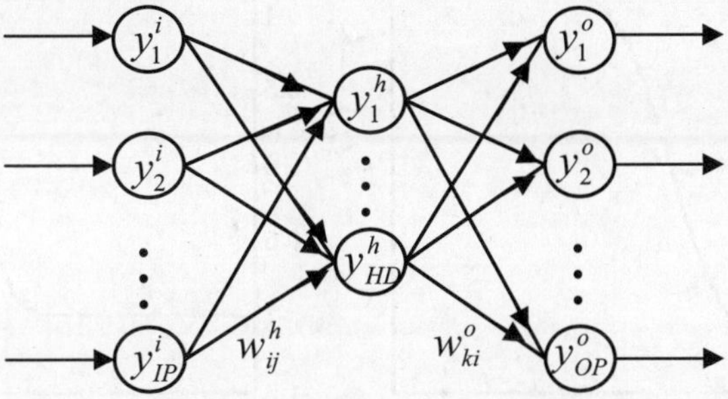

Fig. 16.2 (b) Multi-layer feedforward neural network.

A neuron in the input layer, hidden layer and output layer are respectively denoted by $y_j^i, j = 1,\ldots, IP$, y_i^h, $i = 1,\ldots, HD$ and y_k^o, $k = 1,\ldots, OP$. The number of neurons in input, hidden and output layers are IP, HD and OP; respectively.

The input layer neuron can be considered a single input single output neuron with a linear activation function. The weights for inputs into the hidden layer are denoted by $w_{ij}^h, i = 1, \ldots, HD, j = 1, \ldots, IP$. The weights for inputs from the hidden layer into the output layer are denoted by w_{ki}^o, $k = 1, \ldots, OP, i = 1, \ldots, HD$.

16.4.1 Activation Function

The most commonly used nonlinear activation functions are the *Sigmoid* and *Hyperbolic tangent* functions.

Sigmoid function:

$$f(x) = \frac{1}{1+e^{-x}}, \tag{16.2a}$$

Hyperbolic tangent:

$$f(x) = \frac{1-e^{-2x}}{1+e^{-2x}} \tag{16.2b}$$

These functions are also known as squashing functions as they compress the range of *net* so that the output lies within the range (0, 1) and (-1, 1) for the sigmoid and hyperbolic tangent functions, respectively (Fig. 16.3a). This compression is due to their nonlinearity which is the desired property to represent the nonlinear system/signal. These functions satisfy the requirement of back propagation algorithm, viz., the function is differentiable

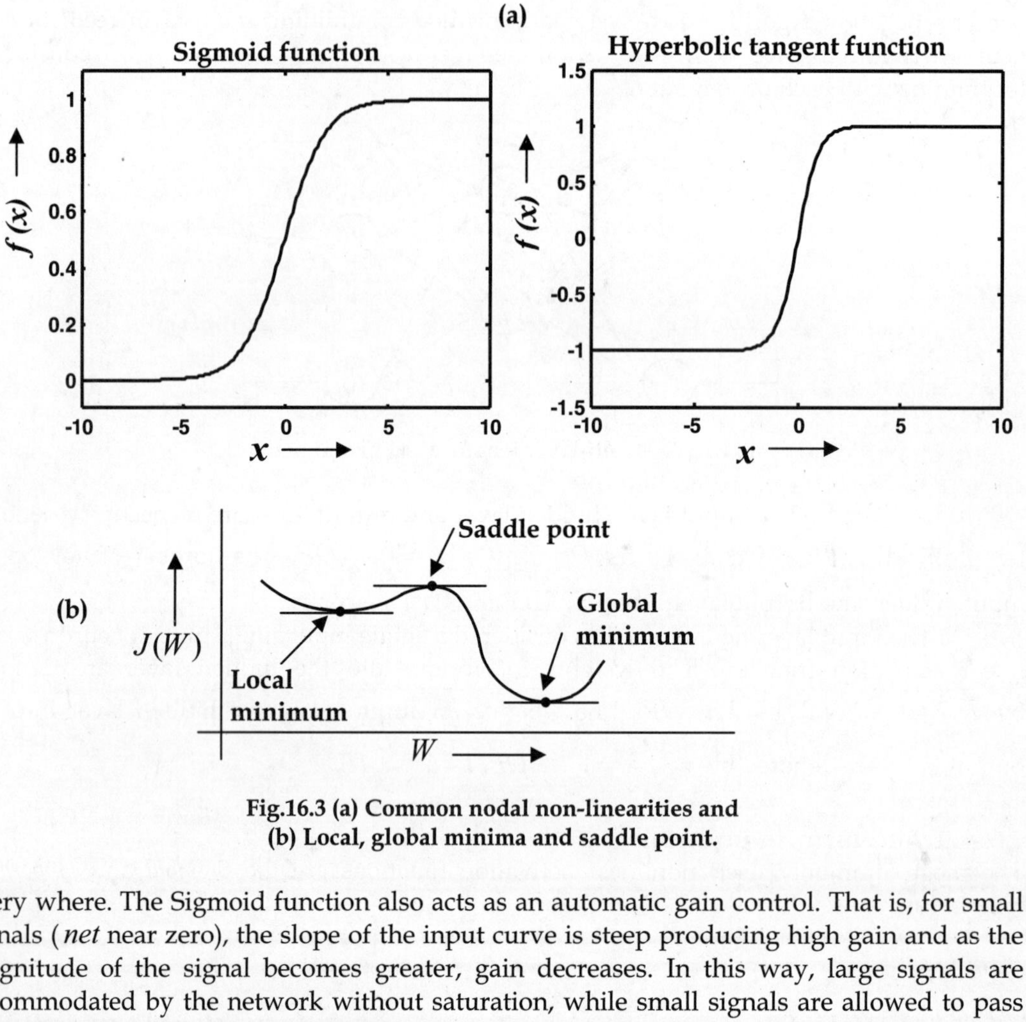

Fig.16.3 (a) Common nodal non-linearities and (b) Local, global minima and saddle point.

every where. The Sigmoid function also acts as an automatic gain control. That is, for small signals (*net* near zero), the slope of the input curve is steep producing high gain and as the magnitude of the signal becomes greater, gain decreases. In this way, large signals are accommodated by the network without saturation, while small signals are allowed to pass through without excessive attenuation.

The nonlinear activation makes the MSE performance and weights relation non-linear hence, there can be many local minima instead of single global minimum. The gradient can become zero where MSE surface is flat, called the *saddle point* and at this point, the adaptation may stop/stall (Fig. 16.3b). Thus, the solution may get caught at local minima or at a saddle point resulting in a suboptimal solution. But if the local minima/ saddle point problem is solved and the global minimum is reached, then this minimum will be lower than that provided by the linear solution as it gives better fit to the data than that by linear approach.

16.4.2 Necessity for Error Back Propagation

In multi-layer feed-forward networks, in addition to input and output layers, there are hidden layers which are not accessible. In view of this, the corresponding desired signals at their outputs hence, the errors (which are the differences between the desired and the actual output) are not known. But to update the weights for hidden neurons, it is necessary to know the error at each output of neuron. This necessitates propagating the error backwards to the hidden layer and hence the algorithm name, back propagation.

The weights are updated using the gradient LMS algorithm. The adaptation is done on a distribution basis depending up on the sensitivity of the output to each weight. To minimize the output error, it is required to make larger changes in the weights which affect the output value most and this is measured by the sensitivity. This sensitivity is obtained by applying the *chain rule* to the network structure (topology). For this, the output error is propagated back to intermediate points in the network and will be scaled according to elemental transfer functions which will be found. This is a powerful approach as it does not require explicitly the error at the intermediate nodes but it derives the contribution to the error at these nodes.

The error criterion for the adaptive filtering problem is the minimization of the cost function J, which is the mean square value of the error signal

$$\text{Minimize} \qquad J = E\left[\sum_{k=1}^{OP} e_k^{\,2}(n)\right] \tag{16.3a}$$

The error criterion is based on OP multiple outputs of the neural network. The gradient descent algorithm adjusts the weights of the neural network to minimize this performance measure by adding to each weight a portion of the negative gradient of error with respect to weight of interest,

$$w(n+1) = w(n) - \mu\,\nabla_w J = w(n) - \mu\frac{\partial J}{\partial w} \tag{16.3b}$$

where $\nabla_w J$ is the gradient of the cost function with respect to the weight w and μ is the scaling factor/adaptation stepsize. To implement this algorithm, an expression for the gradient with respect to each weight in the neural network, is required.

For practicality, the stochastic approximation is made (i.e., expectation is removed) and is

$$J \approx \frac{1}{2}\sum_{k=1}^{OP} e_k^2(n) = \frac{1}{2}\sum_{k=1}^{OP} (d_k^o(n) - y_k^o(n))^2 \tag{16.3c}$$

$$y_k^o(n) = \sum_i w_i(n) x_i(n) \tag{16.3d}$$

where $y_k^o(n)$ is the kth PE output and $d_k(n)$ is the kth PE desired signal. For a single output neuron, as $d_k^o(n)$ is not a function of the weights of neural network,

$$\Delta w = \frac{\partial J_k}{\partial w_i} \approx \frac{1}{2}\frac{\partial e_k^{\,2}(n)}{\partial w_i(n)} = e_k(n)\frac{\partial e_k(n)}{\partial w_i(n)} = -e_k(n)x_i(n) \tag{16.3e}$$

The goal is to find a solution to Eqn. (16.3e) for each weight in the network. This solution is used in the gradient descent algorithm (Eqn. (16.3b)) to adapt each weight in the network.

$$\frac{\partial J_k}{\partial y_k^o(n)} = \frac{\partial}{\partial y_k^o(n)}\left[\frac{\left\{d_k^o(n) - y_k^o(n)\right\}^2}{2}\right] = -\left\{d_k^o(n) - y_k^o(n)\right\} = -e_k(n) \tag{16.4a}$$

Using Eqn. (16.3d),

$$\frac{\partial y_k^o(n)}{\partial w_i(n)} = \frac{\partial}{\partial w_i(n)}\left[\sum_i w_i(n)x_i(n)\right] = x_i(n) \tag{16.4b}$$

Therefore,

$$\Delta w = \frac{\partial J_k}{\partial w_i} = -e_k(n)x_i(n) = \frac{\partial J_k}{\partial y_k^o(n)}\frac{\partial y_k^o(n)}{\partial w_i(n)} \tag{16.4c}$$

That is,

$$\Delta w = \frac{\partial J_k}{\partial w_i} = \frac{\partial J_k}{\partial y_k^o(n)}\frac{\partial y_k^o(n)}{\partial w_i(n)} \tag{16.4d}$$

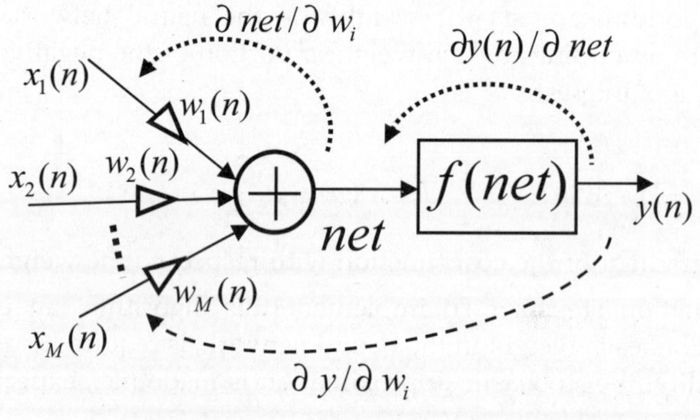

Fig. 16.4 Sensitivity of PE with nonlinear threshold.

This is nothing but the chain rule and the gradient has been expressed in terms of sensitivities. The sensitivity of the cost function with respect to individual weights is equal to the product of the sensitivity of the cost with respect to the output and the sensitivity of the output to the individual weight. This plays a central role in the neural network learning. This concept can be further extended to a PE (processing element/perceptron/node) followed by threshold nonlinearity shown in Fig. 16.4. Here x_i are inputs to PE (from M

different sources) and w_i are their associated weights. The output of PE is *net* and the output of PE after nonlinear threshold is y. From the chain rule, the sensitivity of the output y with respect to the weight w_i is

$$\frac{\partial y}{\partial w_i} = \frac{\partial y}{\partial\, net} \frac{\partial\, net}{\partial w_i(n)} \tag{16.5a}$$

But

$$y = f(net) \qquad \text{and} \qquad net = \sum_{i=1}^{M} w_i(n)x_i(n) + \theta$$

Therefore,

$$y = f(net) = f\left(\sum_{i=1}^{M} w_i x_i(n) + \theta \right)$$

and

$$\frac{\partial y}{\partial\, net} = f'(net), \qquad \frac{\partial\, net}{\partial w_i(n)} = x_i(n)$$

$$\frac{\partial y}{\partial w_i(n)} = \frac{\partial y}{\partial\, net} \frac{\partial\, net}{\partial w_i(n)} = f'(net)\, x_i(n) \tag{16.5b}$$

From Eqn.(16.4a),

$$\frac{\partial J_k}{\partial y_k^o(n)} = -e_k(n)$$

Therefore the sensitivity of the cost function with respect to the individual weight is given by

$$\frac{\partial J_k}{\partial w_i(n)} = \frac{\partial J_k}{\partial y_k^o(n)} \frac{\partial y_k^o(n)}{\partial net_k(n)} \frac{\partial\, net_k(n)}{\partial w_i(n)} = -e_k(n) f'(net)\, x_i(n) \tag{16.5c}$$

The gradient of the cost function with respect to $w_i(n)$ is the product of the derivative of the cost with respect to PE state (output), $\partial J_k / \partial y_k^o(n)$ scaled by the derivative of the nonlinearity of the PE and the input activation. Comparing the above Eqn. (16.5c) with the standard LMS adaptation gradient which is the product of the error and the input, $e_k(n) f'(net)$ corresponds to the error and $x_i(n)$ to the input. Since again from Eqn. (16.4a)

$$\frac{\partial J_k}{\partial y_k^o(n)} = -e_k(n)$$

Eqn.(16.5c) can be written as

$$\frac{\partial J_k}{\partial w_i(n)} = \left[\frac{\partial J_k}{\partial y_k^o(n)} f'(net) \right] x_i(n) \tag{16.6a}$$

In this, the term in the square bracket corresponds to the error and is called the *local error* $\delta_i(n)$ defined by

$$\delta_i(n) = \frac{\partial J_k}{\partial y_k^o(n)} f'(net) \tag{16.6b}$$

Therefore,

$$\frac{\partial J_k}{\partial w_i(n)} = \delta_i(n) x_i(n) \tag{16.6c}$$

Further,

$$\frac{\partial J_k}{\partial w_i(n)} = \frac{\partial J_k}{\partial y_k^o(n)} \frac{\partial y_k^o(n)}{\partial\, w_i(n)} = -\, e_k(n) \frac{\partial y_k^o(n)}{\partial\, w_i(n)} \qquad \text{(Using Eqn.16.4a)}$$

where,

$$\frac{\partial y_k^o(n)}{\partial\, w_i(n)} = f'(net)\, x_i(n)$$

Therefore, adaptation is given by

$$w_i(n+1) = w_i(n) - \mu \frac{\partial J_k}{\partial w_i(n)}$$

$$w_i(n+1) = w_i(n) + \mu\, e_k(n) f'(net_k(n)) x_i(n) \tag{16.6d}$$

The weight adaptation in terms of local error is

$$w_i(n+1) = w_i(n) + \mu\, \delta_k(n) x_i(n) \tag{16.6e}$$

Here, it is to be noted that the quantities involved for the weight $w_i(n)$ adaptation are the input $x_i(n)$ and the error $\delta_k(n)$ are local to the weight $w_i(n)$. That is, irrespective of the number of PEs and the way they are interconnected, its learning is dependent only on its local input and local error.

16.4.3 Derivation of Back Propagation Algorithm

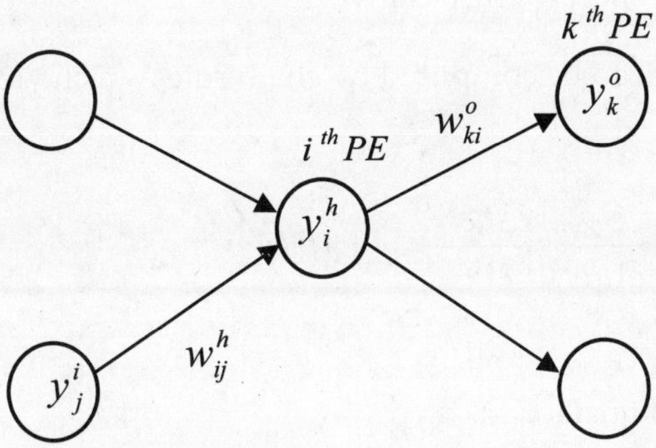

Fig. 16.5 Details of a hidden-layer network.

To start with, a simple model which traces the data through one possible route through the neural network with one hidden layer and an output layer (Fig 16.5) is considered. Here, it is assumed that the kth processing element is the only output PE in the network. Applying Eqns. (16.5c) and (16.6c) ,
Eqn. (16.5c) :

$$\frac{\partial J_k}{\partial w_i(n)} = \frac{\partial J_k}{\partial y_k^o(n)} \frac{\partial y_k^o(n)}{\partial net_k(n)} \frac{\partial\, net_k(n)}{\partial\, w_i(n)} = -e_k(n) f'(net)\, x_i(n)$$

Eqn. (16.6c) :

$$\frac{\partial J_k}{\partial w_i(n)} = -\delta_i(n) x_i(n)$$

to the present model, for the weight w_{ij}^h with input y_j^i and the output y_i^h

$$\frac{\partial J}{\partial w_{ij}^h} = \frac{\partial J}{\partial y_i^h} \frac{\partial y_i^h}{\partial net_i^h} \frac{\partial net_i^h}{\partial w_{ij}^h} = \frac{\partial J}{\partial y_i^h} f'(net_i^h) y_j^i = -\delta_i^h y_j^i$$

$$\frac{\partial J}{\partial w_{ij}^h} = \frac{\partial J}{\partial y_i^h} f'(net_i^h) y_j^i = -\delta_i^h y_j^i \qquad (16.7a)$$

where

$$\frac{\partial J}{\partial y_i^h} f'(net_i^h) = \delta_i^h$$

where the local error δ_i^h is the partial derivative of the cost function with respect to the PE state (output) scaled by the nonlinearity of the PE. The sensitivity of the cost with respect to weights is decomposed into the sensitivity of the cost with respect to the state y_i^h times the sensitivity of the state with respect to the local weights.

The sensitivity $\frac{\partial J}{\partial y_i^h}$ can be computed by chain rule, which yields from the Fig 16.5,

$$\frac{\partial J}{\partial y_i^h} = \frac{\partial J}{\partial y_k^o}\frac{\partial y_k^o}{\partial net_k^o}\frac{\partial net_k^o}{\partial y_i^h} = -e_k f'(net_k^o)w_{ki}^o \qquad (16.7b)$$

Substituting for $\frac{\partial J}{\partial y_i^h}$ in the Eqn. (16.7a),

$$\frac{\partial J}{\partial w_{ij}^h} = \frac{\partial J}{\partial y_i^h} f'(net_i^h)y_j^i$$

results in

$$\frac{\partial J}{\partial w_{ij}^h} = -e_k f'(net_k^o)w_{ki}^o f'(net_i^h)y_j^i \qquad (16.7c)$$

This gives the gradient of the cost function with respect to the weight w_{ij}^h and for this derivation, only single PE was assumed. But there can be many output PEs connected to the ith PE each contributing additively to the gradient of the cost with respect to PE state. That is, Eqn.(16.7b) is summed over the number of output PEs as

$$\frac{\partial J}{\partial y_i^h} = \sum_k \frac{\partial J}{\partial y_k^o}\frac{\partial y_k^o}{\partial net_k^o}\frac{\partial net_k^o}{\partial y_i^h}) = -\sum_k e_k f'(net_k^o)w_{ki}^o \qquad (16.7d)$$

Substituting for $\frac{\partial J}{\partial y_i^h}$ from Eqn. (16.7d) in the Eqn. (16.7a),

$$\frac{\partial J}{\partial w_{ij}^h} = \frac{\partial J}{\partial y_i^h}\frac{\partial y_i^h}{\partial net_i^h}\frac{\partial net_i^h}{\partial w_{ij}^h} = \frac{\partial J}{\partial y_i^h} f'(net_i^h)y_j^i = -\delta_i^h y_j^i$$

results into

$$\frac{\partial J}{\partial w_{ij}^h} = \left\{-\sum_k e_k f'(net_k^h)w_{ki}^o\right\} f'(net_i^h)(y_j^i) \qquad (16.7e)$$

This provides the required gradient of the cost function with respect to the weight w_{ij}^h, hence the w_{ij}^h adaptation by the gradient descent learning is achieved by changing the weight proportionally to the negative of the gradient given by Eqn. (16.7e) and is

$$w_{ij}^h(n+1) = w_{ij}^h(n) + \eta\, f'(net_i^h(n))\left[\sum_k e_k(n) f'(net_k^h(n)) w_{ki}^o(k)\right] y_j^i(n) \qquad (16.7f)$$

This completes the derivation of back propagation algorithm (J.C. Principe, N.R. Euliano, W. Curt Lefebvre, 2000). The summation in the Eqn. (16.7f) is the sum of local errors δ_k^o at each network output PE, scaled by the weights connecting the output PEs to the ith PE. Thus, the term in the square brackets, effectively computes the total error reaching the ith PE from the output layer. When this passes through the ith PE nonlinearity, its local error is obtained and is

$$\delta_i^h(n) = f'(net_i^h(n)) \sum_k \delta_k w_{ki}^o(n) \qquad (16.7g)$$

Thus, all the weights in the gradient descent learning are updated by multiplying the *local error* $\delta_i^h(n)$ by the *local activation* $y_j^i(n)$ according to the instantaneous gradient LMS rule,

$$\Delta w_{ij}^h(n) = \eta\, \delta_i^h(n) y_j^i(n) \qquad (16.7h)$$

The calculation of the local error has to take into account whether the PE is linear or nonlinear and whether the weight is attached to an output PE or a hidden layer PE.

Case 1: If the PE is linear and it is at the output layer,

$$\delta_k^o(n) = -e_k(n) \qquad (16.8a)$$

since $f'(.)$ is equal to a constant value of 1.

Case 2: If the PE is nonlinear and it is at the output,

$$\delta_k^o(n) = -e_k(n) f'(net_k^o(n)) \qquad (16.8b)$$

where $e_k(n)$ is the error associated with the kth output.

Case 3: If the PE is nonlinear and is hidden, the local error is computed by summing all the contributions of the local errors in the output layer, scaled by the corresponding weights, which yields

$$\delta_i^h(n) = f'(net_i^h(n)) \sum_k \delta_k^o w_{ki}^o(n) \qquad (16.8c)$$

To get a better understanding of the back propagation algorithm, it is considered for a network with two neurons in hidden and input layers. Fig. 16.6a shows a three-layer perceptron with the input layer, a single hidden layer and an output layer having two neurons each in the input and the hidden layer and a single neuron at the output layer. Further, inner details of the neural network are shown in Fig. (16.6b).

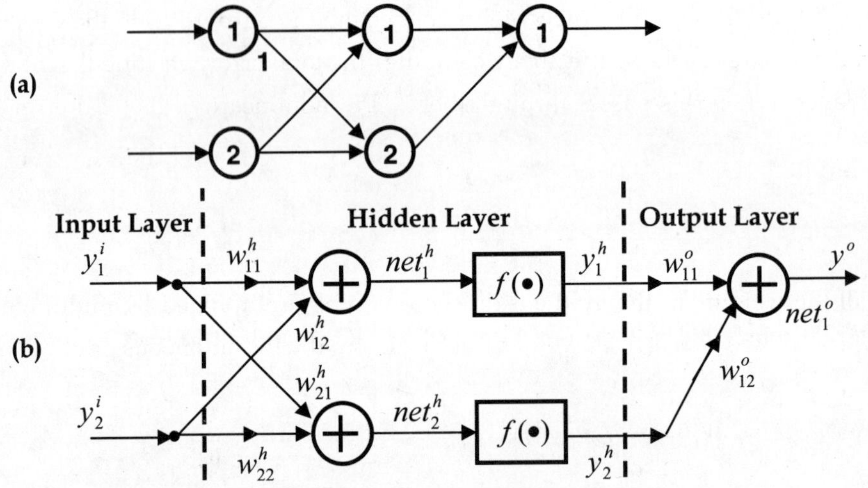

Fig. 16.6 (a) A simple three-layer perceptron network and (b) Detailed diagram of a three-layer network.

I) ***Computing the signals for the forward propagation***

For the input layer: The input vector y^i can be written as,

$$y^i = \begin{bmatrix} y_1^i \\ y_2^i \end{bmatrix} \qquad (16.9a)$$

For the hidden layer: The hidden layer activation vector is the inner product of input to hidden weight w_{ij}^h and the input vector y^i and is calculated as,

$$net^h = \begin{bmatrix} net_1^h \\ net_2^h \end{bmatrix} = \begin{bmatrix} w_{11}^h y_1^i + w_{12}^h y_2^i \\ w_{21}^h y_1^i + w_{22}^h y_2^i \end{bmatrix} = \begin{bmatrix} w_{11}^h & w_{12}^h \\ w_{21}^h & w_{22}^h \end{bmatrix} \begin{bmatrix} y_1^i \\ y_2^i \end{bmatrix}$$

$$net^h = w^h y^i \tag{16.9b}$$

The output of hidden layer

$$y^h = f(net^h) \tag{16.9c}$$

For the output layer: Output layer activation is given by,

$$net^o = \begin{bmatrix} w_{11}^o & w_{12}^o \end{bmatrix} \begin{bmatrix} y_1^h \\ y_2^h \end{bmatrix} = w^{oT} y^h \tag{16.9d}$$

Since a linear activation function is used, $f(net^0) = 1$. Therefore, the output of an output layer neuron is,

$$y_1^o = net_1^o \tag{16.9e}$$

II) Computing the Error

$$e(n) = d(n) - y_1^o(n) \tag{16.9f}$$

where $d(n)$ is the desired signal at time instant n.

III) Updating the Weights

The cost function J to be minimized is,

$$J = \frac{1}{2}\left[d(n) - y_1^o(n) \right]^2$$

$$\frac{\partial J}{\partial w^h} = \frac{1}{2} \times (2) \times \left[d(n) - y_1^o(n) \right] \times \left[\frac{-\partial y_1^o}{\partial w^h} \right]$$

$$\frac{\partial J}{\partial w^h} = -e(n) \frac{\partial y_1^o}{\partial w^h} \tag{16.9g}$$

The derivation below illustrates how to update the weight, for example, the weight connecting the first input neuron to the first hidden neuron i.e., w_{11}^h. Weight update equation for a weight w_{11}^h is given by,

$$w_{11}^h(n+1) = w_{11}^h(n) - \eta \frac{\partial J}{\partial w_{11}^h} \quad (16.9h)$$

Substitute Eqn. (16.9g)in Eqn. (16.9h)

$$w_{11}^h(n+1) = w_{11}^h(n) - \eta \left[-e(n) \frac{\partial y_1^o}{\partial w_{11}^h} \right]$$

$$w_{11}^h(n+1) = w_{11}^h(n) + \eta \left[e(n) \frac{\partial y_1^o}{\partial w_{11}^h} \right] \quad (16.9i)$$

The sensitivity $\frac{\partial y_1^o}{\partial w_{11}^h}$ can be computed by chain rule, which yields,

$$\frac{\partial y_1^o}{\partial w_{11}^h} = \frac{\partial y_1^o}{\partial net_1^o} \frac{\partial net_1^o}{\partial y_1^h} \frac{\partial y_1^h}{\partial net_1^h} \frac{\partial net_1^h}{\partial w_{11}^h} \quad (16.9j)$$

The sensitivities are calculated as follows:

Computing $\frac{\partial y_1^o}{\partial net_1^o}$:

$$y_1^o = f(net_1^o) \quad , \quad \frac{\partial y_1^o}{\partial net_1^o} = f'(net_1^o)$$

Computing $\frac{\partial net_1^o}{\partial y_1^h}$:

$$net_1^o = \sum_{i=1}^{HD} w_{1i}^h y_i^h \quad , \quad \frac{\partial net_1^o}{\partial y_1^h} = w_{11}^o$$

Computing $\dfrac{\partial y_1^h}{\partial net_1^h}$:

$$y_1^h = f(net_1^h), \qquad \frac{\partial y_1^h}{\partial net_1^h} = f'(net_1^h)$$

Computing $\dfrac{\partial net_1^h}{\partial w_{11}^h}$:

$$net_1^h = \sum_{j=1}^{IP} w_{1j}^h y_j^i, \qquad \frac{\partial net_1^h}{\partial w_{11}^h} = y_1^i$$

Substituting the sensitivities in Eqn. (16.9j),

$$\frac{\partial y^o}{\partial w_{11}^h} = f'(net_1^o) w_{11}^o f'(net_1^h) y_1^i \tag{16.9k}$$

Substituting Eqn. (16.9k) in Eqn. (16.9i),

$$w_{11}^h(n+1) = w_{11}^h(n) + \eta e(n) f'(net_1^o) w_{11}^o f'(net_1^h) y_1^i \tag{16.9l}$$

Eqn. (16.9*l*) shows how to update the weight connecting first neuron of the input layer to the first neuron of hidden layer. Similarly, other weights can be updated.

16.4.4 Temporal Processing using Neural Networks

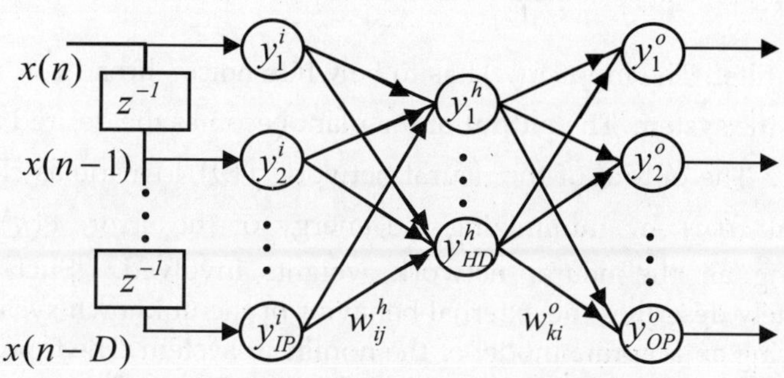

Fig. 16.6(c) Time delay neural network.

Time plays an important role in the learning process and time index will enable the assessment of speed of learning. A neural network which is static (e.g., multi-layer perceptron) can be made dynamic by involving time as an index. Time can be brought into the structure of a neural network by using time delays and is implemented at the synaptic level inside the network or at the input layer of the network.

The multi-layer neural network with a delay line incorporated at the input is shown in Fig. 16.6(c) and is known as time delay neural network (TDNN). The input signal flows into the delay line and the PE receives not only the present input but also (D-1) past samples. Even though the variables involved were indicated as functions of time (n), they were treated as constants and some times as iterations which need not necessarily refer to time.

16.5 SYSTEM IDENTIFICATION USING NEURAL NETWORK

The general problem of non-linear system identification is shown in Fig. 16.7(a) and is similar to linear system identification except for the nonlinearity to be considered. For this, a common signal is injected as input both to the unknown plant / system and to the neural network adaptive filter.

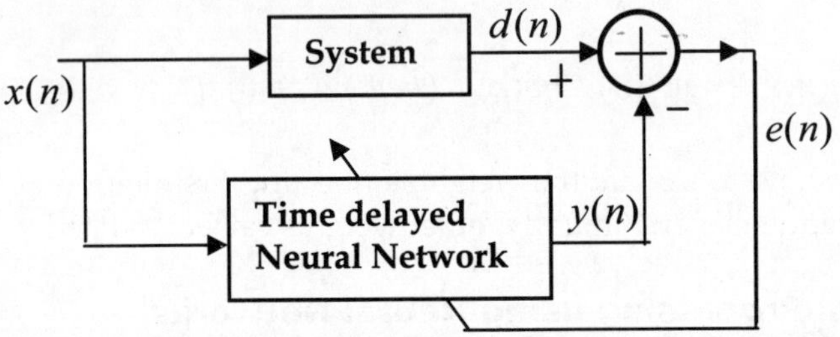

Fig. 16.7(a)System identification by time delayed neural network.

The input $x(n)$ has to be white noise/ broadband noise to excite all the frequencies of the system. The output of the plant becomes the desired response $d(n)$ for the adaptive filter. The output of the neural network $y(n)$ is made as close as possible to the desired signal $d(n)$ by minimizing the energy of the error, $e(n) = d(n) - y(n)$ by optimally choosing all the neural network weights involved. Such an input-output modeling completely describes the external behavior of the unknown system and should be capable of producing an accurate model of the nonlinear system. The topology of the neural network is decided by the model used for identification. The models used can be in general autoregressive (AR) or autoregressive moving average (ARMA) models also but here for simplicity, only moving average model (MA) will be considered. Further, for the adaptive filter, the time delayed neural network is well suited as the learning is in temporal domain and not for applications like classification. Neural network model used is the multi-layer perceptron network trained with the back propagation algorithm (BPA), which is explained in section 16.4. A nonlinear system having the input output relation given below is considered for illustration.

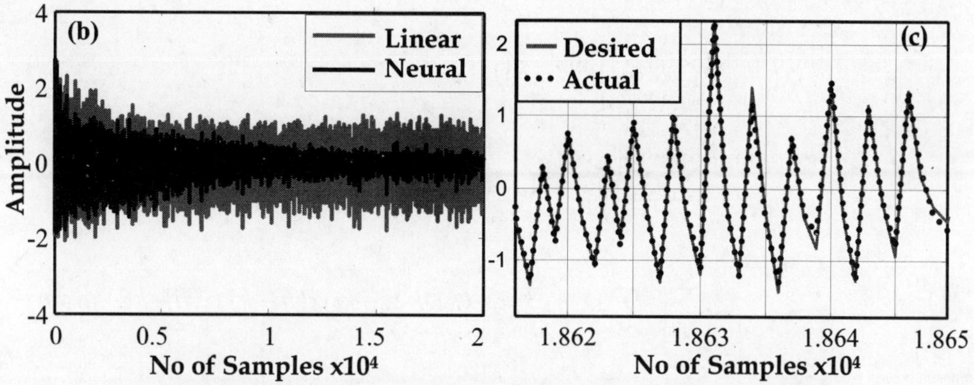

Fig. 16.7 Performance of neural network for system identification (b) Convergence (c) Match between actual desired signal and its estimate.

$$d(n) = -0.78x(n) - 1.48x(n-1) + 1.39x(n-2) + 0.04x(n-3) + 0.54x^2(n)$$
$$-1.62x^2(n-1) + 1.41x^2(n-2) - 0.13x^2(n-3) + 3.72x(n)x(n-1)$$
$$+1.86x(n)x(n-2) - 0.76x(n)x(n-3) + 0.76x(n-1)x(n-2)$$
$$-0.12x(n-1)x(n-3) - 1.52x(n-2)x(n-3).$$

For the simulation of a MLP network, a single hidden layer is considered. Further the numbers of inputs, hidden and output neurons chosen are 10, 30 and 1, respectively. The step sizes for the output layer and hidden layers of the neural network used are 0.005 and 0.05 , respectively. For FIR LMS filter, the order is chosen as 10 and the stepsize used is 0.05. The system input is a white Gaussian noise with unit variance. The convergence characteristics of neural network trained by BPA with the linear FIR filter using LMS algorithm is compared in Fig. 16.7 (b). It is seen that the neural network can converge faster and has a smaller error compared to that of the FIR filter. The matching between the desired and the actual response of neural network is also shown in Fig. 16.7 (c).

16.6 ADAPTIVE VOLTERRA FILTERS

The Volterra model is attractive and most widely used for modeling nonlinear systems as it represents the output *as a linear combination of nonlinear functions of the input.* The output is a linear function of model (filter) parameters. Also, as this input output relation is *nonrecursive, the output only depends on the input.* Hence the classical formulation of linear adaptive filters can be easily extended for nonlinear case also. The Volterra series expansion of a nonlinear system consists of a non-recursive series in which the output signal is related to the input signal as follows:

$$y(n) = \sum_{k1=0}^{N-1} w_{01}(k1)x(n-k1)$$

$$+\sum_{k1=0}^{N-1}\sum_{k2=0}^{N-1} w_{02}(k1,k2)x(n-k1)x(n-k2)$$

$$+\sum_{k1=0}^{N-1}\sum_{k2=0}^{N-1}.......\sum_{ki=0}^{N-1} w_{0i}(k1,k2.....ki)x(n-k1)x(n-k2)....x(n-ki)$$

$$+\ldots\ldots \tag{16.10}$$

Here, $w_{01}(k_1)$, $w_{02}(k_1,k_2)$, ... $w_{0P}(k_1,k_2,\ldots k_P)$ are the weights of the 1st, 2nd, ... P^{th} *order kernels*

$$x(n-k_1),$$
$$x(n-k_1)x(n-k_2),$$
$$\vdots$$
$$x(n-k_1)x(n-k_2)\ldots x(n-k_P),$$

respectively, of truncated Volterra series for N input samples. From Eqn. (16.10), it is seen that the Volterra series consists of second and higher orders terms in addition to the first order term. It is because of these higher order terms, the filter is able to take care of the nonlinearities in the system. However, the Volterra series representation is not suitable to model systems containing discontinuities on their models. Further, Volterra series representation becomes computationally complex if the complete series, i.e., $P=\infty$, is employed. The permutation of indices of the Volterra series results in symmetry of the kernels as all permutations of any number of coefficients multiply the same combinations of input samples and this symmetry results in reduction in the number of Volterra series coefficients. By truncating the series, the computational complexity can be reduced but this is at the cost of accuracy of the series expansion. But for most of the practical nonlinear systems, the second order nonlinearity is sufficient. Therefore, presently, adaptive Volterra filter of second-order will be considered.

16.6.1. Nonlinear System Identification by Adaptive Second-order Volterra Filter

Here, the adaptive filter acts on the input sample $x(n)$ to generate an estimate $y(n)$ of the desired signal $d(n)$. The difference between the desired and the estimated gives the error, which is to be minimized by adapting the Volterra filter coefficients according to adaptive filtering algorithms. The Volterra LMS algorithm is used for nonlinear system identification and other applications due to its simplicity and low computational complexity.

The unit impulse response is not sufficient to represent and identify all kernel elements of a general nonlinear system (polynomial system) modeled by the Volterra series. This is because, the unit impulse response can represent only the first order kernel and not the other higher order kernels.

Volterra LMS adaptive algorithm

The LMS algorithm minimizes the instantaneous square error between the output of the adaptive filter $y(n)$ and the desired signal $d(n)$,

$$J = e^2(n) = [d(n) - y(n)]^2$$

$$J = d^2(n) - 2d(n)y(n) + y^2(n) \tag{16.11a}$$

The output is

$$y(n) = w^T(n)X(n)$$

Further, the input and coefficient vectors are given by (Fig.16.8)

$$X^T(n) = \big[\ x(n), x(n-1), \ldots x(n-N+1), x^2(n), x(n)x(n-1), \ldots, x(n)x(n-N=1), x^2(n-1), \ldots, x^2(n-N+1)\ \big] \tag{16.11b}$$

$$w^T(n) = \big[w_0(n), w_1(n), \ldots w_{N-1}(n), w_{0,0}(n), w_{0,1}(n), \ldots, w_{0,N-1}(n), w_{1,1}(n) \ldots, w_{N-1,N-1}(n) \big] \tag{16.11c}$$

Therefore,

$$y(n) = \sum_{k1=0}^{N-1} w_{k1}(n)x(n-k1) + \sum_{k1=0}^{N-1}\sum_{k2=k1}^{N-1} w_{k1,k2}(n)x(n-k1)x(n-k2) \tag{16.11d}$$

Therefore, the instantaneous error is

$$e(n) = [d(n) - y(n)]$$

$$e(n) = d(n) - \sum_{k1=0}^{N-1} w_{k1}(n)x(n-k1) + \sum_{k1=0}^{N-1}\sum_{k2=k1}^{N-1} w_{k1,k2}(n)x(n-k1)x(n-k2) \tag{16.11e}$$

The dimension of the input and the coefficient vector for a second-order Volterra filter is $\frac{N(N+3)}{2}$, where N is the length of the Volterra filter. The instantaneous square error is

$$e^2(n) = d^2(n) - 2d(n)w^T(n)X(n) + w^T(n)X(n)X^T(n)w(n) \tag{16.12}$$

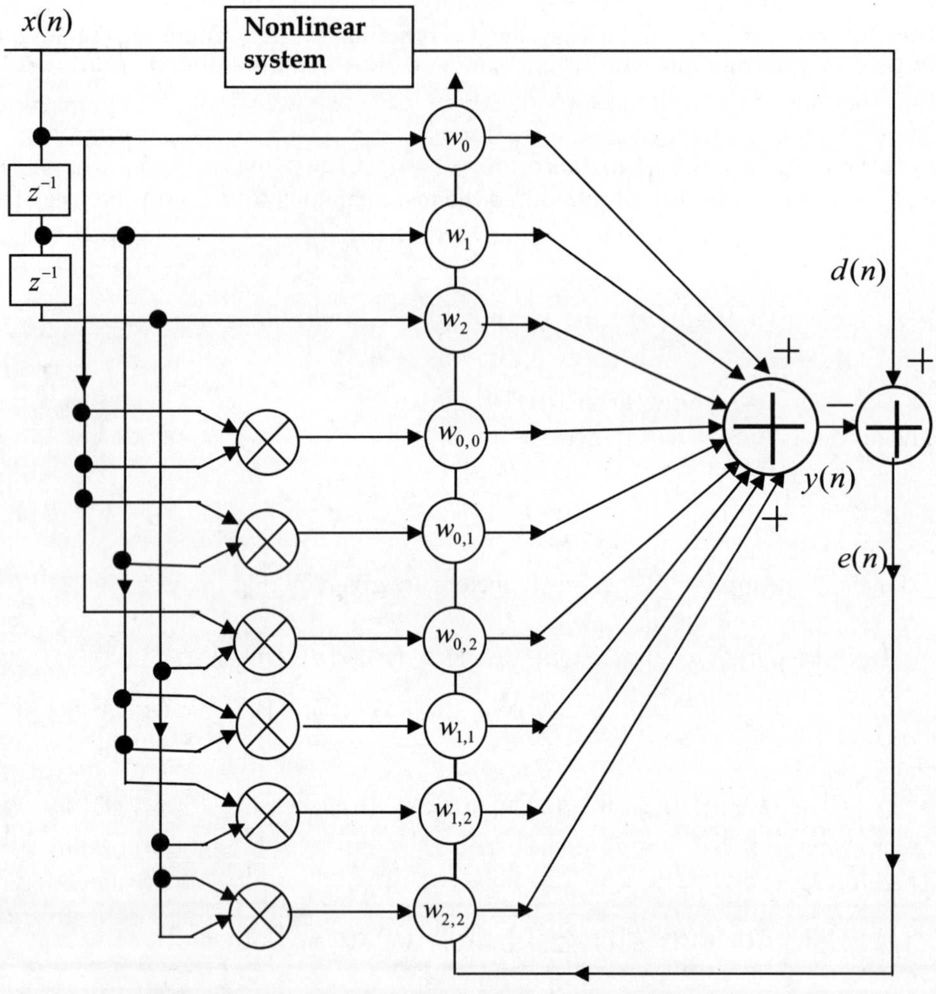

Fig. 16.8 Adaptive 2nd order Volterra LMS filter for N=3.

According to the LMS algorithm, the adaptation of the coefficients are given by

$$w(n+1) = w(n) - 2\mu\ e(n)\frac{\partial e(n)}{\partial w(n)}$$

Therefore,

$$w_{k1}(n+1) = w_{k1}(n) - 2\mu_1 e(n)\frac{\partial e(n)}{\partial w_{k1}(n)}$$

From Eqn. (16.11e),

$$e(n) = d(n) - \sum_{k1=0}^{N-1} w_{k1}(n)x(n-k1) + \sum_{k1=0}^{N-1}\sum_{k2=k1}^{N-1} w_{k1,k2}(n)x(n-k1)x(n-k2)$$

$$\frac{\partial e(n)}{\partial w_{k1}(n)} = -x(n-k1)$$

Therefore,

$$w_{k1}(n+1) = w_{k1}(n) + \alpha_1 e(n)x(n-k1),\ \alpha_1 = 2\mu_1,\ \ k1 = 0,1,\ldots,N-1 \tag{16.13a}$$

Also

$$w_{k1,k2}(n+1) = w_{k1,k2}(n) - 2\mu_2 e(n)\frac{\partial e(n)}{\partial w_{k1,k2}(n)}$$

Again from Eqn. (16.11e),

$$\frac{\partial e(n)}{\partial w_{k1,k2}(n)} = -x(n-k1)x(n-k2)$$

Therefore,

$$w_{k1,k2}(n+1) = w_{k1,k2}(n) + \alpha_2 e(n)x(n-k1)x(n-k2), \quad \alpha_2 = 2\mu_2$$
$$k1 = 0,1,\ldots,N-1 \text{ and } k2 = k1,\ k1+1,\ ,\ldots..N-1. \tag{16.13b}$$

It has been assumed that the input signal DC component (mean value) is zero or removed. If it is not so, a weight $w0(n)$ is associated with the mean value and is adapted as

$$w0(n+1) = w0(n) + \alpha_0 e(n), \quad \alpha_0 = 2\mu_0, \quad e(n) = [d(n) - y(n)]$$

and

$$y(n) = w0(n) + \sum_{k1=0}^{N-1} w_{k1}(n)x(n-k1) + \sum_{k1=0}^{N-1}\sum_{k2=k1}^{N-1} w_{k1,k2}(n)x(n-k1)x(n-k2)$$

This new error which takes into account the mean value should be used in Eqns.(16.13a) and (16.13b) for adapting $w_{k1}(n)$ and $w_{k1,k2}(n)$, respectively (T. Ogunfunmi, 2007 and V.J. Mathews and G.L. Sicuranza, 2000).

To ensure convergence of Volterra LMS algorithm coefficients in the mean, the step sizes of the must be chosen as

$$0 < \alpha_1 < \frac{1}{tr(R)} < \frac{1}{\lambda_{\max}}, \tag{16.13c}$$

$$0 < \alpha_2 < \frac{1}{tr(R)} < \frac{1}{\lambda_{\max}} \tag{16.13d}$$

$$0 < \alpha_0 < \frac{1}{tr(R)} < \frac{1}{\lambda_{\max}} \tag{16.13e}$$

where $\lambda_{\max}$ is the largest eigenvalue of the input signal vector autocorrelation matrix $R = E[x(n)x^T(n)]$. For example, for a second order , 3 memory Volterra series, the elements of input vector

$$\left[x(n), x^2(n), x(n-1), x^2(n-1), x(n)x(n-1), x(n-2), x^2(n-2), x(n-1)x(n-2), x(n)x(n-2) \right]$$

are not mutually orthogonal even when $x(n)$ is white. In view of this, eigenvalue spread of the matrix R of the input becomes large. For example, the eigenvalue spread of 2 ($\lambda_{\max} / \lambda_{\min} = 1.4 / 0.7$) of a pseudo-white noise becomes 67 ($\lambda_{\max} / \lambda_{\min} = 52 / 0.8$) due to Volterra nonlinearity. Also the eigenvalue spread of 800 ($\lambda_{\max} / \lambda_{\min} = 64 / 0.08$) of a colored noise becomes 660000 ($\lambda_{\max} / \lambda_{\min} = 6600 / 0.01$) due to Volterra nonlinearity. Thus, the Volterra nonlinearity enhances the eigenvalue spread. As a consequence in general, the LMS algorithm used for Volterra type nonlinear system has a significantly slow convergence rate than that of the LMS algorithm used for linear systems. In view of this, it is better to use two step sizes for the linear and nonlinear terms. This enables to use a higher value of μ_1 for linear terms which are orthogonal to each other compared to the nonlinear which may not be orthogonal.

Now, the improvement in the performance achieved by an adaptive second-order Volterra filter in identifying a nonlinear system over the linear FIR LMS algorithm has been presented by considering different examples. For these, the input is a white Gaussian noise with unit variance. The step sizes chosen for the linear part and nonlinear part of Volterra filter are 2×10^{-3} and 5×10^{-4} respectively and 2×10^{-3} for a linear FIR filter. The order is chosen as N=6 for both Volterra filter and FIR filter and the series order for a Volterra filter chosen is 2

***Experiment 1*:** In this, the unknown system to be identified has an input-output relation

$$\begin{aligned} d(n) = &-0.78x(n) - 1.48x(n-1) + 1.39x(n-2) + 0.04x(n-3) + 0.54x^2(n) \\ &-1.62x^2(n-1) + 1.41x^2(n-2) - 0.13x^2(n-3) + 3.72x(n)x(n-1) \\ &+1.86x(n)x(n-2) - 0.76x(n)x(n-3) + 0.76x(n-1)x(n-2) \\ &-0.12x(n-1)x(n-3) - 1.52x(n-2)x(n-3)\,. \end{aligned}$$

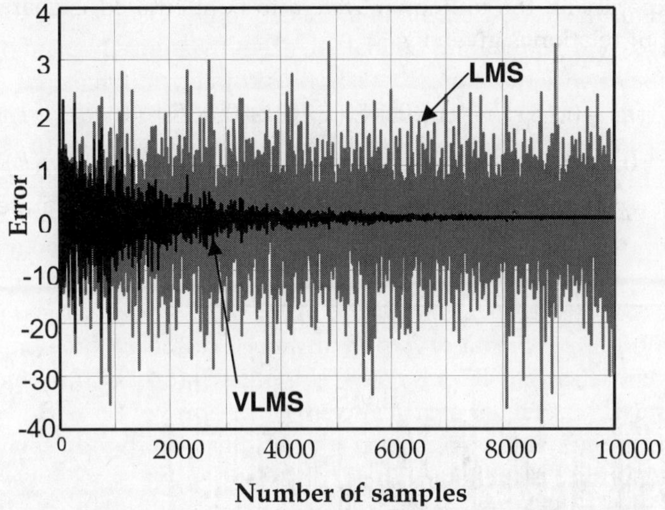

Fig. 16.9 (a) Co-convergence co-comparison for VolVLMS and LMS Co-algorithms.

Experiment 2: In this experiment, the unknown system to be identified has an input-output relation is

$$d(k) = -0.76x(k) - x(k-1) + x(k-2) + 0.5x^2(k) + 2x(k)x(k-2)$$
$$-1.6x^2(k-1) + 1.2x^2(k-2) + 0.8x(k-1)x(k-2) + n(k).$$

The desired signal $d(n)$ is obtained by adding the measurement noise to the output signal of unknown system to be identified. The measurement noise $\eta(n)$ is a Gaussian white noise with variance 0.01.

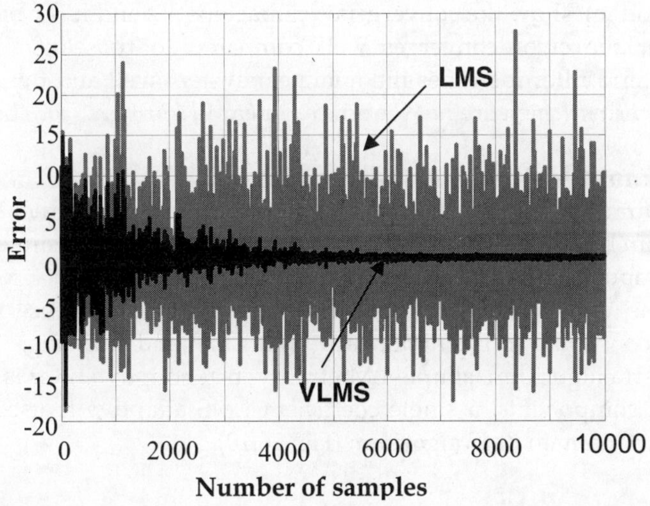

Fig. 16.9 (b) Co-convergence co-comparison for VolVolterra LMS and LMS Co-algorithms.

***Experiment* 3:** In this experiment, the unknown system to be identified is characterized by the nonlinear input-output relation is given by,

$$d(k) = -0.08x(k) - 0.15x(k-1) + 0.14x(k-2) + 0.055x^2(k)$$
$$+0.3x(k)x(k-2) - 0.16x^2(k-1) + 0.14x^2(k-2) + n(k)$$

The measurement noise $\eta(n)$ in this case is also a Gaussian white noise, independent of input with variance 10^{-3}.

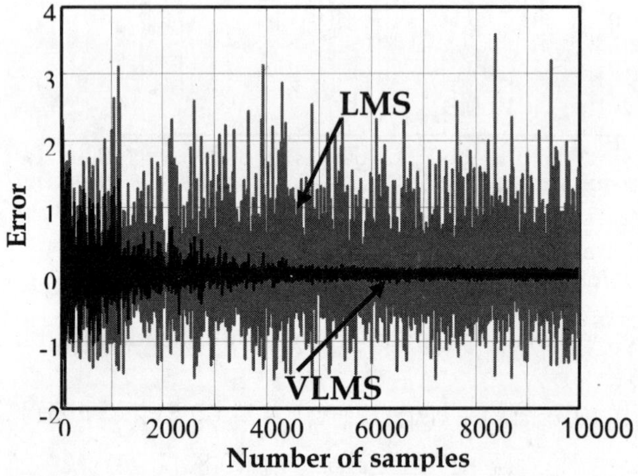

Fig. 16.9 (c) Co-convergence co-comparison for VolVLMS and LMS algorithms Co-algorithms.

Figs. 16.9 (a), (b) and (c), show the convergence characteristics and it is observed that the for the Volterra filter the error converges well compared to those of linear LMS algorithm. Hence, the Volterra filter identifies the nonlinear system well and this is due to the presence of s 2nd order terms which are not found in a linear LMS filter.

16.6.2 Transform Domain Volterra Adaptive Filter

To simplify the Volterra adaptive algorithm, by reducing the number of kernel terms, the transform domain algorithm has been explored. Here, basically, it has been assumed that the orthogonalization of the input by the transform removes the contribution of cross second order terms and only the auto-second order terms (square terms) of the orthogonal components significantly contribute to the nonlinearity (H. Zhao and J. Zhang, 2007). For the linear part of the Volterra algorithm, simple LMS transversal adaptive filter is used. But for the auto-second order components, a single coefficient LMS adaptive is used for each component as in a transform domain LMS algorithm (Fig. 16.10).

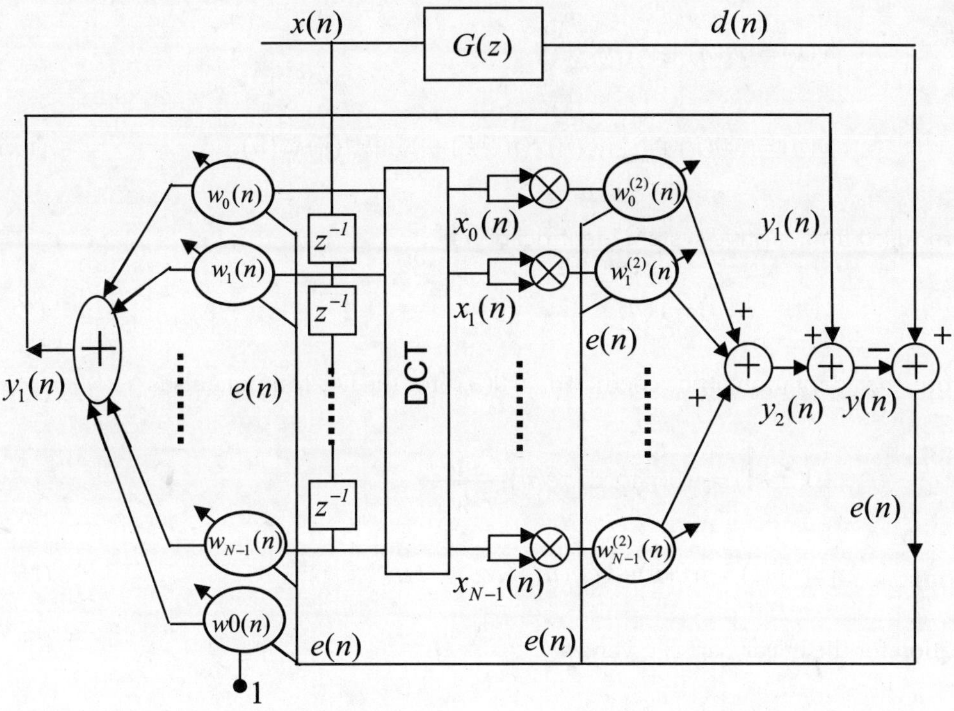

Fig. 16.10 Transform domain Volterra adaptive system identification.

Let

$$w^T(n) = [w_0(n), w_1(n), \ldots, w_{N-1}(n)] \tag{16.14a}$$

Therefore, the output due to linear part plus due to DC is given by

$$y_1(n) = w0(n) + \sum_{k=0}^{N-1} w_k(n)x(n-k) \tag{16.14b}$$

where, $w_k(n)$ are the adaptation weights of the transversal filter for the linear part and $w0(n)$ is weight to cater to the DC component associated with the input.

Let the orthogonal components of $x(n)$ obtained by DCT be

$$x_O{}^T(n) = [x_0(n), x_1(n), \ldots, x_{N-1}(n)] \tag{16.14c}$$

Therefore, the output due to squared auto-components is given by

$$y_2(n) = \sum_{k=0}^{N-1} w_k^{(2)}(n)x_k^2(n) \tag{16.14d}$$

where, $w_k^{(2)}(n)$ is the adaptation weight associated with the orthogonal component $x_k(n)$.

Therefore, the output due to all components is

$$y(n) = y_1(n) + y_2(n)$$

$$y(n) = w0(n) + \sum_{k=0}^{N-1} w_k(n)x(n-k) + \sum_{k=0}^{N-1} w_k^{(2)}(n)x_k^2(n) \tag{16.14e}$$

The error used for adaptive system identification is

$$e(n) = d(n) - y(n) \tag{16.14f}$$

Therefore, the adaptation of the coefficient $w0(n)$, for the DC components is

$$w0(n+1) = w0(n) - 2\mu_0 e(n)\frac{\partial y(n)}{\partial w0}$$

$$w0(n+1) = w0(n) + \alpha_0 e(n), \quad \alpha_0 = 2\mu_0 \tag{16.15a}$$

Adaptation for the linear part is given by

$$w_k(n+1) = w_k(n) - 2\mu_1 e(n)\frac{\partial y(n)}{\partial w_k}$$

$$w_k(n+1) = w_k(n) + \frac{\alpha_1}{NE\ (n)} e(n)x(n-k), \quad \alpha_1 = 2\mu_1, \tag{16.15b}$$

$$k = 0.1 \ldots, N-1$$

$$E(n) = \beta\, E(n-1) + (1-\beta)x^2(n), \qquad 0 < \alpha_1 < 1, \quad 0 < \beta < 1$$

Adaptation for the nonlinear part (due to squared auto terms) is given by

$$w_k^{(2)}(n+1) = w_k^{(2)}(n) - 2\mu_2 e(n)\frac{\partial y(n)}{\partial w_k^{(2)}(n)}$$

$$w_k^{(2)}(n+1) = w_k^{(2)}(n) + \frac{\alpha_2}{E_k(n)} e(n)x_k^2(n), \quad \alpha_2 = 2\mu_2 \tag{16.15c}$$

$$E_k(n) = \gamma E_k(n-1) + (1-\gamma)x_k^2(n), \quad 0 < \alpha_2 < 1, \quad 0 < \gamma < 1,$$

$$k = 0, 1, \ldots, N-1$$

16.6.3 Second Order Volterra Adaptive Filter Based on Lattice Predictor

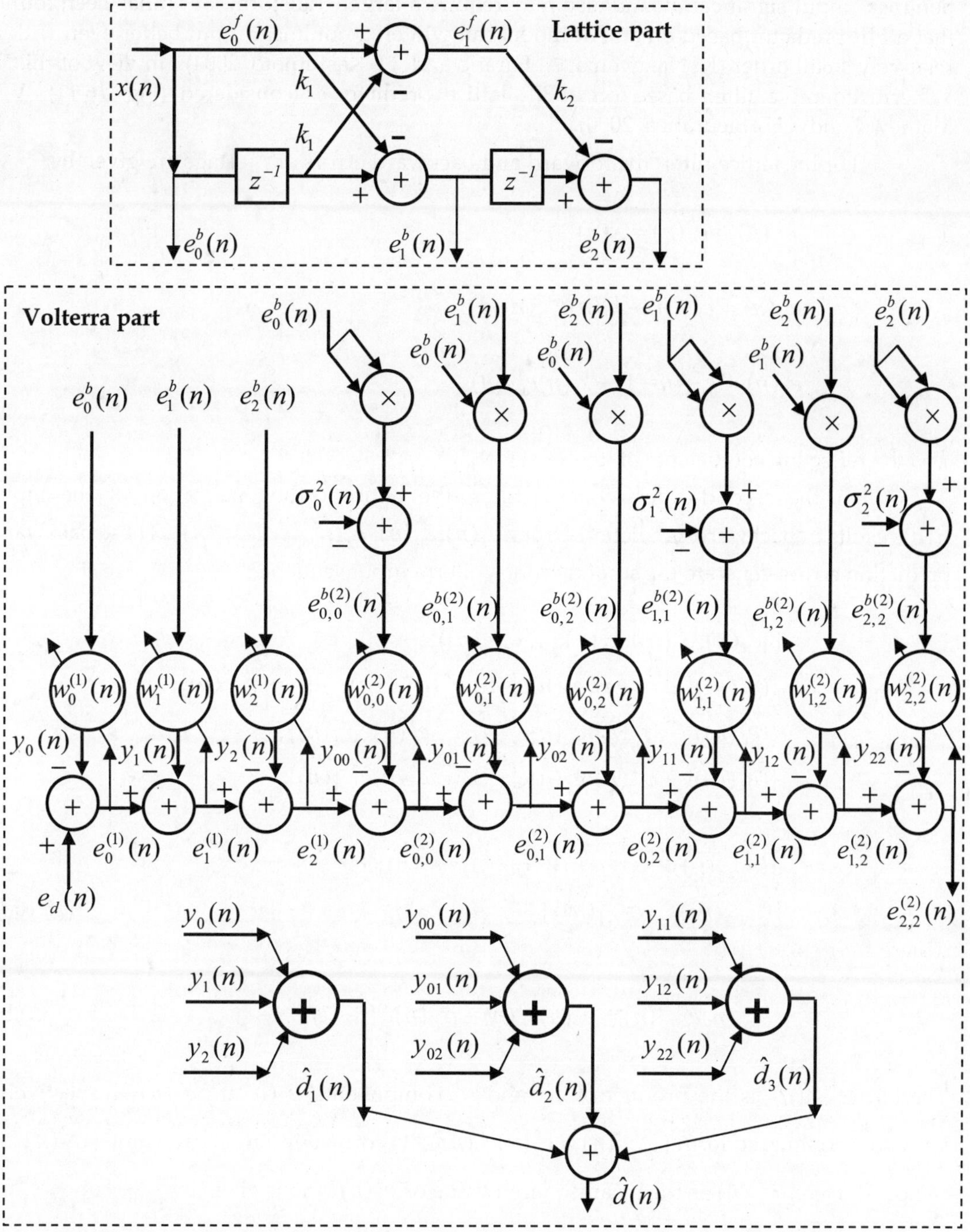

Fig. 16.11 Adaptive second order Volterra filter based on lattice predictor (N=3).

The orthogonal transform considered for Volterra adaptive filter cannot whiten the nonlinear input significantly and also may require a larger transform. But it has been found that a FIR predictor based on an AR model can whiten a nonlinear input better even with a relatively small order (K. Nakayama, A. Hirano and H. Kashimoto, 2004). In view of this, a Volterra adaptive filter based on a FIR lattice predictor is considered (Fig. 16.11) (V.J. Mathews and G.L. Sicuranza, 2000).

For the lattice filter, the forward and backward errors at l^{th} stage are given by

$$e_0^f(n) = e_0^b(n) = x(n)$$

$$e_l^f(n) = e_{l-1}^f(n) - k_l(n)e_{l-1}^b(n-1)$$

$$e_l^b(n) = e_{l-1}^b(n-1) - k_l(n)e_{l-1}^f(n)$$

k_l is the reflection coefficient of the lth stage.

For the second-order Volterra filter, the lattice orthogonalization of the input $x(n)$ results in backward prediction errors $e_0^b(n), e_1^b(n), e_2^b(n), \ldots, e_{N-1}^b(n)$,. These backward prediction errors generate the second order Volterra input sequence

$$\begin{aligned}
&[\{e_0^b(n), e_1^b(n)), e_2^b(n), \ldots\, e_{N-1}^b(n))\},\\
&\{(e_0^b(n))^2 - \sigma_0^2(n), e_0^b(n)e_1^b(n), e_0^b(n)e_2^b(n)\ldots e_0^b(n)e_{N-1}^b(n)\},\\
&\{(e_1^b(n))^2 - \sigma_1^2(n), e_1^b(n)e_2^b(n), \ldots\, e_1^b(n)e_{N-1}^b(n)\},\\
&\{(e_2^b(n))^2 - \sigma_2^2(n), e_2^b(n)e_3^b(n), \ldots\, e_2^b(n)e_{N-1}^b(n)\},\\
&\vdots\\
&\{(e_{N-2}^b(n))^2 - \sigma_{N-2}^2(n), e_{N-2}^b(n)e_{N-1}^b(n)\},\\
&\{(e_{N-1}^b(n))^2 - \sigma_{N-1}^2(n)\}]
\end{aligned} \tag{16.16a}$$

where

$$b_{i,j}^{(2)}(n)e_{i,j}^{b(2)}(n) = e_i^b(n)e_j^b(n) - \sigma_i^2(n)\delta(i-j) \tag{16.16b}$$

The term $e_{i,j}^{b(2)}(n)$ is the product of ith and j^{th} components $e_i^b(n)$ and $e_j^b(n)$, respectively. Only for the squared term ($e_{i,j}^{b(2)}(n) = e_i^b(n)e_j^b(n)$), its corresponding mean value ($\sigma_i^2(n)$) is removed. Here, $\sigma_i^2(n)$ is the mean-squared value of $e_i^b(n)$ and is given by

$$\sigma_i^2(n) = \lambda\ \sigma_i^2(n-1) + (1-\lambda)[e_i^b(n)]^2, \quad 0 < \lambda < 1 \tag{16.16c}$$

But $e_{i,j}^{b(2)}(n) = e_i^b(n)e_j^b(n), i \neq j$, are used as inputs directly without removing any mean value.

The linear components, the backward prediction errors $e_i^b(n), i = 0, 1, 2, \ldots, N-1$ are orthogonal to each other and are used in the joint lattice structure. In addition to the linear components $e_i^b(n)$, the product components $e_{i,j}^{b(2)}(n)$ are also considered orthogonal to each other hence, the weights associated with each of the components are of single coefficient and are adapted using LMS algorithm for the joint lattice structure. However, with the product terms, the orthogonality may not be complete. But for the purpose of estimation, such an approximation can be considered as it enables simplicity in considering nonlinearity. Let

$$e_d(n) = d(n) - \overline{d(n)}, \tag{16.17a}$$

$\overline{d(n)}$ is the mean value of the desired response signal $d(n)$ and is given by

$$\overline{d(n)} = \lambda \overline{d(n-1)} + (1-\lambda)d(n), \quad 0 < \lambda < 1 \tag{16.17b}$$

Let $e_i^{(1)}(n)$ represent the errors in estimating the mean-removed desired response signal $e_d(n)$ using the first i backward prediction error signals and are given by

$$e_i^{(1)} = e_d(n) - \sum_{k=0}^{i} w_k^{(1)}(n)e_k^b(n); \qquad i = 0, 1, 2, \ldots, N-1 \tag{16.17c}$$

This can also be expressed as

$$e_i^{(1)}(n) = e_{i-1}^{(1)}(n) - w_i^{(1)}(n)e_i^b(n),\ for\ \ e_{-1}^{(1)}(n) = e_d(n) \tag{16.17d}$$

$w_i^{(1)}(n)$ is the coefficient to be associated with the component $e_i^b(n)$ to estimate its contribution in $e_d(n)$.

The errors required for nonlinear components are given by

$$e_{i,j}^{(2)}(n) = e_{i,j-1}^{(2)}(n) - w_{i,j}^{(2)}(n)e_{i,j}^{b(2)}(n) \tag{16.18a}$$

with

$$\begin{aligned} e_{i,-1}^{(2)}(n) &= e_{i-1,N-1}^{(2)}(n), \quad i = 1, 2, \ldots N-1 \\ &= e_{N-1}^{(1)}(n), \quad i = 0 \end{aligned} \tag{16.18b}$$

$w_{i,j}^{(2)}(n)$ is the coefficient to be associated with the input component m $e_{i,j}^{b(2)}(n)$. The generation of errors for the joint lattice is shown in Fig.16.11. For the lattice predictor of order N=3. Having defined the inputs and the errors, their corresponding coefficient adaptation can be achieved.

As already stated, in addition to the linear part, the nonlinear components are also considered orthogonal to each other hence, their corresponding adaptive filters will also have single coefficient.

For the linear part:
The coefficients are now updated as

$$w_i^{(1)}(n+1) = w_i^{(1)}(n) + \frac{\mu}{\sigma_i^2(n)} e_i^{(1)}(n) e_i^b(n), \quad i = 0, 1, 2, \quad 0 < \mu < 1 \tag{16.19a}$$

$$\sigma_i^2(n) = \beta\ \sigma_i^2(n-1) + (1-\beta)[e_i^b(n)]^2\ , \quad 0 < \beta < 1 \tag{16.19b}$$

The error $e_i^{(1)}(n)$ is given by Eqn. (16.17d).

For the non-linear part:
The coefficients are now updated as

$$w_{i,j}^{(2)}(n+1) = w_{i,j}^{(2)}(n) + \frac{\mu}{\sigma_{2,i,j}^2(n)} e_{i,j}^{(2)}(n) e_{i,j}^{b(2)}(n), \quad 0 < \mu < 1 \tag{16.19c}$$

$$\sigma_{2,i,j}^2(n) = \gamma\ \sigma_{2,i,j}^2(n-1) + (1-\gamma)[e_{i,j}^{b(2)}(n)]^2, \quad 0 < \gamma < 1 \tag{16.19d}$$

The error $e_{i,j}^{(2)}(n)$ is given by Eqns. (16.18a) and (16.18b).

The performance of lattice Volterra based adaptive filter is illustrated for the nonlinear system identification. The desired signal is generated based on the following third-order polynomial model given by

$$d(n) = t(n-2) + 0.08 * t^2(n-2) - 0.04 * t^3(n-1)$$

where $t(n)$ is obtained from the following linear convolution $t(n) = x(n) * f(n)$ and $f(n)$ is the impulse response of the transfer function

$$F(z) = z^{-3} - 0.3z^{-4} + 0.5z^{-5}.$$

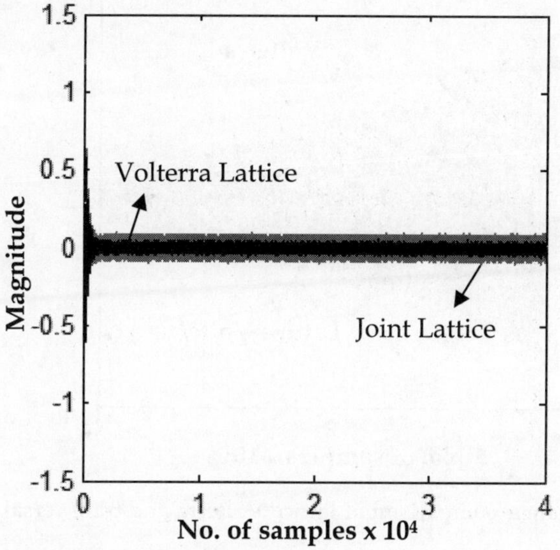

Fig. 16.12(a) Error convergence comparison between lattice Volterra and joint lattice.

The reference signal is the sum of a sinusoidal wave of 500 Hz sampled at the rate of 8000 samples/sec, that is,

$$x(n) = \sqrt{2} * \sin(2\pi * 62.5 / 1000 * n) + v(n)$$

and $v(n)$ is a white noise process with the Gaussian distribution giving a signal to noise ratio (SNR) of 40dB.

The performance of adaptive lattice Volterra filter is compared with that of joint lattice. The parameters used are : number of lattice stages=10, forgetting factor $\lambda = 0.99$, stepsize for regression coefficient adaptation is 0.0009 and step size used for lattice filter reflection adaptation is 0.00001.

The lattice Volterra reduces the error better than that by joint lattice (Fig. 16.12a). This is due to the fact that the joint lattice is a linear filter whereas the lattice Volterra is nonlinear. However, the convergence speed in both cases is same.

The performance of lattice Volterra is also compared with that of transversal Volterra. The parameters used are: number of lattice stages 10, the forgetting factor $\lambda = 0.99$, regression coefficient adaptation step size for is 0.001 and step size for lattice reflection coefficient is 0.00001.

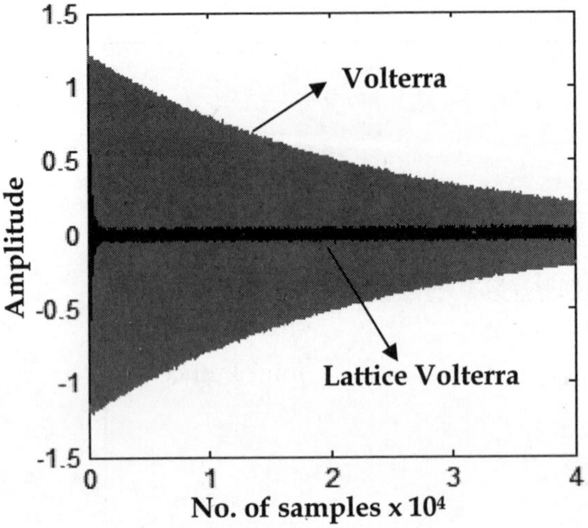

Fig. 16.12(b) Performance comparison of lattice Volterra and transversal Volterra filter.

Fig. 16.12 (b) indicates that the lattice Volterra has a very significantly faster convergence compared to that of transversal Volterra filter. Thus, for a nonlinear path, lattice Volterra filter can achieve a better error reduction than the joint lattice filter and its convergence is significantly faster than that of transversal Volterra filter.

16.6.4 Volterra Adaptive Filter Based on Two Joint Lattice Preprocessors

In an attempt to reduce the eigenvalue spread for the nonlinear squared input terms, a separate joint lattice filter has been proposed in addition to the joint lattice used for the linear input terms (X.M. Zhang and Z.J. Zhao, 2009). Thus, there are two backward prediction error sets and there will be two associated combiners (Fig.16.13). The cross-kernels are treated in the usual way as in a transversal Volterra adaptive filter.

Here also it is based on the fact that the square terms have higher correlation and their eigenvalue spread will get enhanced significantly compared to the cross-terms. The two lattice predictor structures which generate the backward prediction errors for the linear terms and for the squared input terms are similar. If $e_k^b(n)$ and $e_k^{b(2)}(n)$ represent the backward prediction errors and $w_k(n)$ and $w_{k,k}^{(2)}(n)$ are linear combiner coefficients of the joint lattice for the linear and squared inputs, respectively, then the output of the adaptive filter is given by

$$y(n) = w0(n) + \sum_{k=0}^{N-1} w_k(n)e_k^b(n) + \sum_{k=0}^{N-1} w_{k,k}^{(2)}(n)e_k^{b(2)}(n) + \sum_{k=0}^{N-1} \sum_{\substack{l=1 \\ l\neq k}}^{N-1} w_{k,l}^{(2)}(n)x(n-k)x(n-l) \tag{16.20a}$$

where, $w_{k,l}^{(2)}(n)$ are the weights of the transversal adaptive filter associated with the cross-kernels. If

$$y_1(n) = w0(n) + \sum_{k=0}^{N-1} w_k(n) e_k^b(n) \tag{16.20b}$$

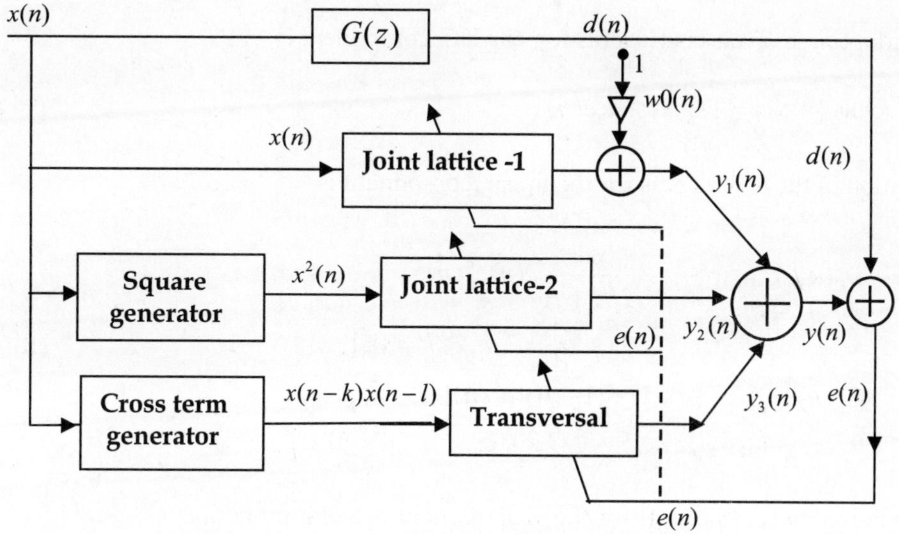

Fig. 16.13 Second order Volterra adaptive filter based on two joint lattice processors.

This forms the output of the joint lattice-1 for linear part which does not include the first term due to DC/mean component. The backward prediction errors are $e_k^b(n),\ k = 0, 1, \ldots, N-1$. Further, if

$$y_2(n) = \sum_{k=0}^{N-1} w_{k,k}^{(2)}(n) e_k^{b(2)}(n) \tag{16.20c}$$

This forms the output of the joint lattice-2 for nonlinear squared input terms. The backward prediction errors are $e_k^{b(2)}(n),\ k = 0, 1, \ldots, N-1$.

If

$$y_3(n) = \sum_{k=0}^{N-1} \sum_{\substack{l=1 \\ l \neq k}}^{N-1} w_{k,l}^{(2)}(n) x(n-k) x(n-l) \tag{16.20d}$$

This forms the third part of the output due to second order cross-kernel terms. Therefore,

$$y(n) = y_1(n) + y_2(n) + y_3(n) \tag{16.20e}$$

The error for a desired signal $d(n)$ is

$$e(n) = d(n) - y(n) \tag{16.20f}$$

Then, the adaptation of the coefficients for the DC component is

$$w0(n+1) = w0(n) + \alpha_0 e(n) \tag{16.21a}$$

The adaptation of the coefficients for the linear component is

$$\begin{aligned} w_k(n+1) &= w_k(n) + \frac{\alpha_1}{NE_k(n)} e(n) e_k^b(n), \\ &\qquad \alpha_1 = 2\mu_1, \quad k = 0.1\dots, N-1 \\ E_k(n) &= \beta\, E_k(n-1) + (1-\beta)[e_k^b(n)]^2, \\ &\qquad 0 < \alpha_1 < 1, \quad 0 < \beta < 1 \end{aligned} \tag{16.21b}$$

The adaptation of the coefficients for the squared nonlinear component is

$$\begin{aligned} w_{k,k}^{(2)}(n+1) &= w_{k,k}^{(2)}(n) + \frac{\alpha_2}{E_{kk}(n)} e(n) e_k^{b(2)}(n), \quad k = 0.1\dots, N-1 \\ E_{kk}(n) &= \beta\, E_{kk}(n-1) + (1-\beta)[e_k^{b(2)}(n)]^2, \qquad 0 < \alpha_2 < 1, \quad 0 < \beta < 1 \end{aligned} \tag{16.21c}$$

The adaptation of the coefficients for the nonlinear crossterm component is

$$\begin{aligned} w_{k,l}^{(2)}(n+1) &= w_{k,l}^{(2)}(n) + \frac{\alpha_3}{E_{kk}(n)} e(n) x(n-k) x(n-l), \quad 0 < \alpha_3 < 1 \\ &\qquad k \neq l,\ k,l = 0.1\dots, N-1, \end{aligned} \tag{16.21d}$$

It is important to note that the overall error $e(n)$ is used in the two joint lattice structures only for their linear combiners of backward prediction errors but the adaptation of the lattice predictor reflection coefficients are based on the forward and backward errors in the usual manner. The lattice predictor can be a gradient lattice or least square lattice. Certainly with the use of least square lattice, the orthogonalization will be better but its computational requirement will be relatively higher than that for the gradient lattice.

Chapter 17

APPLICATIONS OF ADAPTIVE FILTERS

Any real-time processing should be sequential as the processing has to take place as data arrives successively one sample after the other sample. Further not only the processing but also any action if any to be taken should be over within the same sample interval. To provide desirable solution, the processing should be adaptive, i.e., it should learn and adjust to the changing environment. To meet this requirement to the best extent, generally, the adaptive filters/processing systems should have fast convergence rate and small misadjustment/ error. However, there can be some relaxations depending on the application scenario. These relaxations may come due to computational load that can be handled by the processor specifications on which it is realized and the cost that is affordable for the considered application. A wide variety of applications belonging to different fields like, Communication (ADPCM based on lattice predictor and pole zero predictor), sigma-delta modulation with an adaptive line enhancer, decision feedback equalization based on joint lattice filter, adaptive time delay estimation, adaptive coherence estimation, telephone line and data network echo cancellation, optimal and adaptive beamformers for communication), speech audio and acoustics (noise cancellation for speech, crosstalk suppression, acoustic echo and Hearing aid feedback cancellation, active low frequency noise control), adaptive inverse modeling and inverse control, biomedical (enhancement of fetus ECG in the presence of mothers ECG, muscle noise suppression from EEG, suppression of EOG from EEG, magnitude square coherence function, adaptive filtering of evoked potentials). The adaptive notch filter(ANF) is considered in detail covering FIR, IIR ANF, adaptive IIR lattice notch filter, notch filter realization by Steiglitz-McBride system identification method. The lattice notch filter has application in spread spectrum suppressing multiple jammers with good convergence rate. It can be used as a line enhancer with good frequency tracking ability. The line enhancer based on notch filter has the ability to detect the notch frequency of the sinusoid (fundamental frequency the pitch) in a broadband noise like speech and hence, useful for pitch estimation. The ANF effectively removes the power line interference from ECG.

17.1 INTRODUCTION

In chapter 1, to motivate the study of adaptive filters some of the applications were considered briefly. But now having covered the subject in depth, it will be of interest to consider those and other applications in detail bringing out some of the important aspects relevant to the specific application. These applications are adaptive differential pulse code modulation (ADPCM), sigma-delta modulation with adaptive line enhancer, adaptive inverse modeling and inverse control, adaptive channel equalization, adaptive time delay estimation, adaptive magnitude square coherence function estimation, adaptive notch filter and its application, feedback cancellation (FBC) in hearing aids, adaptive noise cancellation for speech, adaptive echo cancellation, Active noise control (ANC), Applications of adaptive filtering for biosignals and adaptive beam formers.

17.2 ADPCM-ADAPTIVE DIFFERENTIAL PULSE CODE MODULATION

The pulse code modulation (PCM) is nothing but converting each sample value of the signal to a binary number. Presently, this operation is done by analog to digital converter (ADC). To represent both a small and a large value, the number of bits required is large. The variance of a signal represents this range hence, in PCM transmission, *the number of bits required depends on the variance.* For a given channel bandwidth, for efficient transmission, the signal variance should be as small as possible for the bit rate to be within the channel bandwidth. The ADPCM achieves such a variance reduction by removing the redundancy/ correlation in the signal. Here, the difference / error between the present sample and its estimate (based on a small number of previous samples) is transmitted instead of complete signal. Hence, a better estimate leads to a smaller variance of the difference signal resulting in an efficient transmission.

An adaptive predictor is used to get the estimate of the input signal $\hat{x}(n)$ from the quantized error $e_q(n)$ and the predictor input $\tilde{x}(n)$ is obtained by adding $\hat{x}(n)$ to $e_q(n)$ (Fig. 17.1a).

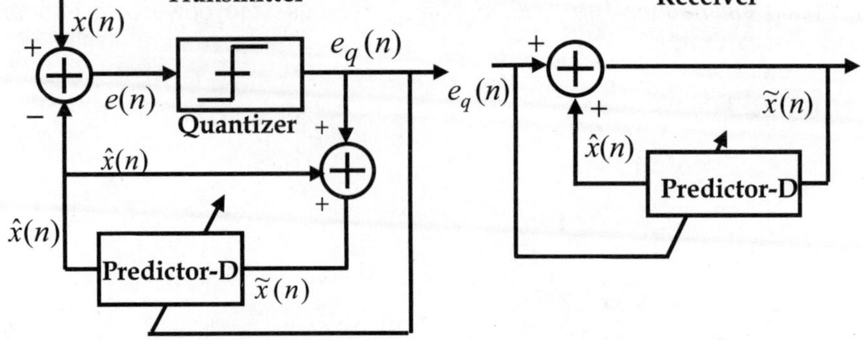

Fig. 17.1(a) Schematic of ADPCM.

$$\hat{x}(n) = \sum_{k=1}^{M} a_k(n)\tilde{x}(n-k), \qquad \tilde{x}(n) = \hat{x}(n) + e_q(n)$$

The coefficients $a_k(n)$ are adapted by LMS algorithm using error $e_q(n)$,

$$a_k(n+1) = a_k(n) + \mu_a(n)\, e_q(n)\tilde{x}(n-k),\ 1 \le k \le M$$

$$\mu_a(n) = \alpha_a / ME_{xa}(n), \quad 0 < \alpha_a < 1$$

$$E_{xa}(n) = \beta_a\, E_{xa}(n-1) + (1-\beta_a)\tilde{x}^2(n), \quad 0 < \beta_a < 1$$

Even on receiver side, such a predictor estimates $\hat{x}(n)$ using $\tilde{x}(n) = \hat{x}(n) + e_q(n)$ as its input and $a_k(n)$ are adapted by $e_q(n)$. Thus, the transmission of the quantized error itself is sufficient to generate the signal at the receiver side making the transmission significantly efficient. In the CCITT ADPCM standard for 32000 bps system, the sign data-sign error LMS algorithm has been used.

17.2.1 ADPCM Based on Lattice Predictor

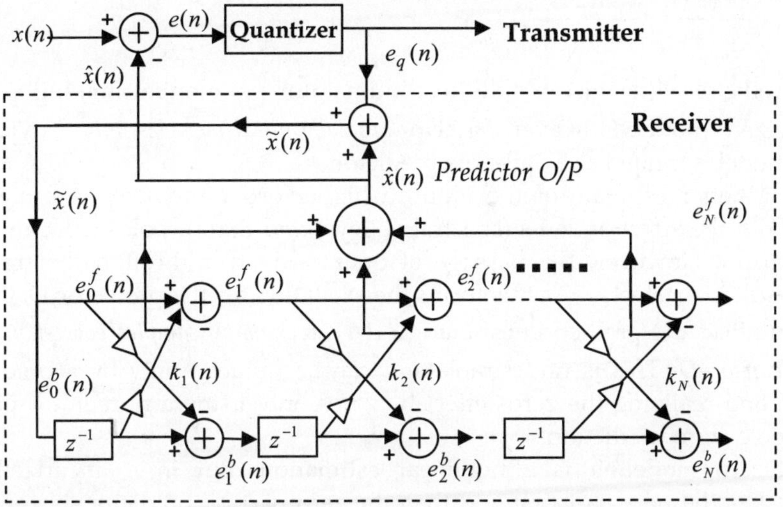

Fig. 17.1 (b) Lattice predictor based ADPCM.

In the lattice filter (Fig. 17 (b)), the lth order filter errors are given by

$$e_l^f(n) = e_{l-1}^f(n) - k_l(n)e_{l-1}^b(n-1)$$

$$e_l^b(n) = e_{l-1}^b(n-1) - k_l(n)e_{l-1}^f(n), \quad l = 1, \cdots 2, \ldots, N-1$$

The reflection coefficients are adapted by Burg recursive approach,

$$k_l(n+1) = \frac{C_l(n+1)}{D_l(n+1)}, \qquad l = 1, \cdots 2, \ldots, N$$

$$C_l(n+1) = \lambda C_l(n) + \sigma\, e_l^f(n) e_l^b(n-1),$$

$$D(n+1) = \lambda D(n) + e_l^{f2}(n), \qquad 0 < \lambda < 1, \quad 0 < \sigma < 1$$

λ: fading memory parameter, σ : damping factor to improve noisy channel performance. [R.C. Reininger and J.D.Gibson, 1986] The lattice predictor provides a better performance than that by transversal predictor both for noisy and noise-free channels.

The transfer function between $\hat{X}(z)$ and $E_q(z)$,

$$\hat{x}(n) = \sum_{k=1}^{M} a_k(n)\tilde{x}(n-k)$$

$$\hat{X}(z) = A(z)\tilde{X}(z) = A(z)[\hat{X}(z) + E_q(z)]$$

$$\frac{\hat{X}(z)}{E_q(z)} = \frac{A(z)}{[1 - A(z)]},$$

Thus, the predictor $A(z)$ also results in a numerator polynomial making the transfer function a pole-zero model. However, *the zeros are determined by poles only*. In view of this, this pole-zero model is treated as an all-pole predictor.

To get a better estimate of the input, a higher order predictor is preferred as it approximates both the effect of zeros (spectral valleys) and that of poles (spectral peaks) in the input spectrum. However, the passage of quantized error through the transmission channel may result in artefacts. As this forms the excitation for the predictor-D, the artefact in the error can affect the prediction estimate $\hat{x}_D(n)$ *over many samples corresponding to the length of the predictor-D*. This is undesirable and can be reduced only by a predictor-D of reduced length and realizing the zeros effect by zeros only using a numerator polynomial that leads to a pole -zero prediction.

A pole-zero modeling is a nonlinear estimation. But in a linearized version, predictor-D models the pole part first and then the zero part by predictor -N. A pole-zero predictor not only reduces the channel error problem but also provides better input estimate than that is provided by a higher order all-pole predictor and reduces the error variance [B.Koo and J.D. Gibson, 1986]. The coefficients $b_k(n)$ of the predictor-N are adapted by the LMS algorithm using error $e_q(n)$ (Fig. 17.1(c)) and is

$$b_k(n+1) = b_k(n) + \mu_b(n)\, e_q(n) e_q(n-k), \; o \le k \le N$$

$$\mu_b(n) = \frac{\alpha_b}{NE_{xb}(n)},, \quad E_{xb}(n) = \beta_b\, E_{xb}(n-1) + (1-\beta_b)e_q^{\,2}(n), \quad 0 < \alpha_b, \beta_b < 1$$

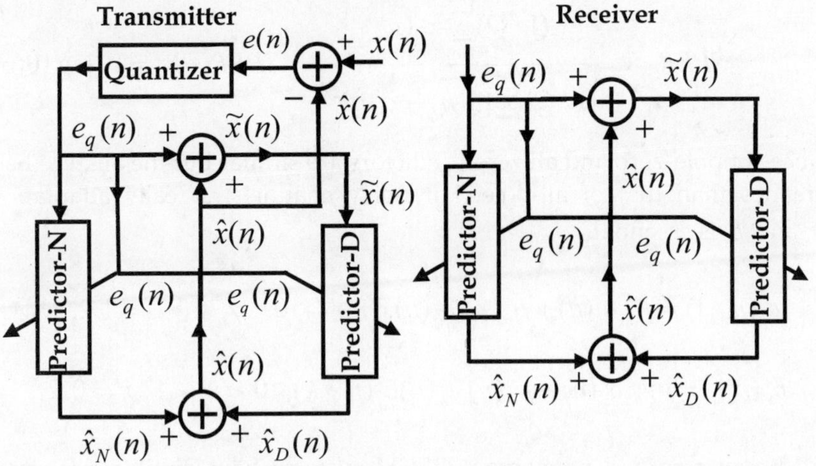

Fig. 17.1(c) ADPCM based on pole-zero predictor.

For the pole zero predictor,

$$\hat{x}(n) = \sum_{k=1}^{M} a_k(n)\tilde{x}(n-k) + \sum_{k=0}^{N-1} b_k(n)e_q(n-k)$$

$$\hat{X}(z) = A(z)\tilde{X}(z) + B(z)E_q(z),$$

$$\hat{X}(z) = A(z)\hat{X}(z) + A(z)E_q(z) + B(z)E_q(z)$$

$$\frac{\hat{X}(z)}{E_q(z)} = \frac{A(z)+B(z)}{[1-A(z)]}$$

Here, $B(z)$ controls effect of zeros and this is independent of poles.

For ideal channels, the pole-zero predictor performs better than all-zero and all-pole predictors in terms of average SNR and segmented SNR.

$$SNR = \frac{(1/L)\sum_{n=0}^{L-1} x^2(n)}{\sum_{n=0}^{L-1}[x(n)-\tilde{x}(n)]^2}, \quad SNR(dB) = 10\log 10(SNR)$$

$$SNRSEG = \frac{1}{K}\sum_{j=1}^{K} SNR\mathbf{B}_j(dB)$$

The $SNRSEG$ represents the subjective performance and for this the data is segmented into non-overlapping blocks and $SNR\mathbf{B}_j(dB)$, the $SNR(dB)$ for the jth block. The predictor performance is indicated by $SPER(dB)$,

$$SPER = \frac{(1/Q)\sum_{n=0}^{Q-1} x^2(n)}{(1/Q)\sum_{n=0}^{Q-1} [x(n) - \hat{x}(n)]^2}, \quad SPER(dB) = 10\log 10(SPER)$$

The performances of pole-zero and all zero predictors are similar but the all-zero has a much better performance than that of all-pole. For noisy channels, a leaky adaptation of the coefficients a_k and b_k is essential.

$$a_l(n+1) = \gamma_a\ a_l(n) + \mu_a(n)\ e_q(n)\tilde{x}(n-l),\ 0 < \gamma_a < 1,\ l = 1,2,\ldots M$$

$$b_k(n+1) = \gamma_b\ b_k(n) + \mu_b(n)\ e_q(n)e_q(n-k),\ \ 0 < \gamma_a < 1,\ k = 0,2,\ldots N-1$$

With this, the all-zero and pole-zero ADPCM systems have similar performance and they outperform ADPCM based on the all-pole predictor.

In adaptive quantization (AQ), the quantization step size is updated in accordance with the changes in the input signal variance. This is made possible by using the short time estimate of variance of the input. An AQ is realized by scaling the input and output levels of a fixed quantizer that is matched to unit variance input by the standard deviation of the input. This indicates the importance of variance for AQ. In adaptive quantization, instantaneous, syllabic and hybrid approaches exist. In instantaneous, the step size is updated at every sample instant according to changes in the input signal. In syllabic step size is updated at a slow rate (not at every sample). These two methods assume that the step size range required is same for all segments of speech. However, speech being quasi-stationary, the optimum step size required can vary for different segments of speech. This has been solved in the hybrid approach in which the adaptation range of the instantaneous method is varied according to the envelope of the input signal that is available in the syllabic method. The use of exponential power estimation (EPE) introduced for normalizing the adaptation step size of LMS adaptive filter has been proposed quantization. But the quantizer realized by EPE, will suffer from limited number of quantization step sizes. But the use of hybrid approach enables to make best use of finite number of step sizes. Here, the EPE is used to find the standard deviations of the input for both instantaneous and syllabic methods. Also EPE quantizer is modified to have intermediate step sizes preserving the simplicity of EPE.

The use of EPE for an adaptive filter (predictor) has already been explained in chapter 10. Thus, the EPE can be used to estimate the powers of input speech and prediction error signals, for use in the adaptive predictor and adaptive quantizer, respectively of a ADPCM. Such a ADPCM based on EPE (E-ADPCM) has been studied for different types of inputs and channel conditions at a bit rate of 16Kbits per second comparing its performance with those of ADPCM which uses block power estimation (B-ADPCM) and ADPCM based on sign-sign LMS predictor (S-ADPCM) (G. Mathew, S.V. Narasimhan and A.P. Shivaprasad, 1992). Both B-ADPCM and S-ADPCM use robust adaptive quantizer. The E-ADPCM has shown a better performance than those of B-ADPCM and S-ADPCM.

The E-ADPCM has a computational advantage of *(3N + B + 1)* multiplications and one division (*N*: order of the predictor and B : number of bits used for quantizing the prediction error) over that of B-ADPCM. Further, the E-ADPCM has a computational advantage of *(B+1)* multiplications over the S-ADPCM. The novelty of the E-ADPCM is that same EPE approach is used both for the quantizer and the predictor and it provides superior performance with computational advantage.

17.3 SIGMA-DELTA MODULATION WITH ADAPTIVE LINE ENHANCER

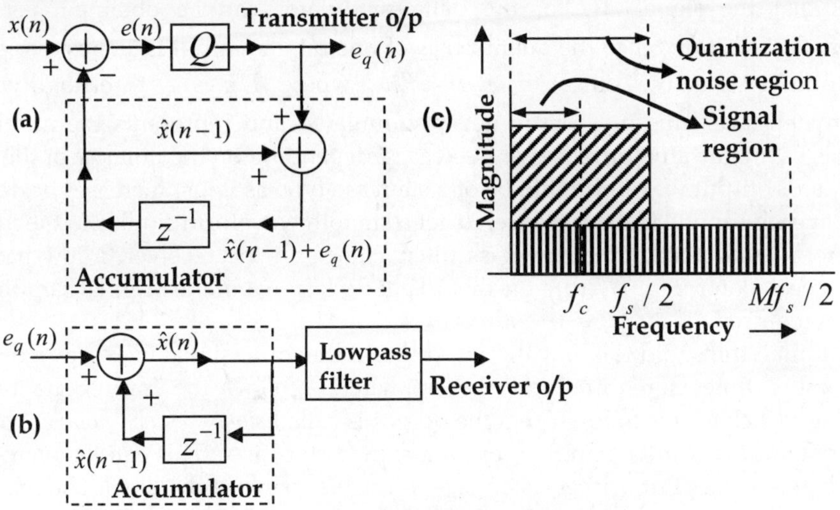

Fig. 17.2 (a) DM transmitter; (b) DM receiver and (c) Effect of oversampling on quantization noise.

This section considers the application of a adaptive line enhancer (ALE) to sigma-delta modulator to extend its performance of high resolution to wide band scenario even with low oversampling factor. For the purpose of understanding and bringing out the importance of ALE in this scenario, the delta and delta-sigma modulators are briefly described with relevant details.

In ADPCM, as the predictor provides a better estimate of the input signal, the variance of the error between the original signal and its estimate gets reduced. Hence it is possible to have low bit quantization (4bits) ADPCM compared to that of in PCM (8 bits). The complexity of the ADPCM can be reduced by replacing the predictor by a delay unit but this increases the number of bits required as the error signal variance increases. The variance of the error can be reduced by increasing the correlation between adjacent samples and this is achieved by *increasing the input sampling rate* by a factor of 4 *and using the previous sample as the estimate of the present sample.* As the error variance is decreased to a large extent, even the minimum one bit quantization which has only two levels, known as delta modulator is possible. For the delta modulator though the number of bits per sample is reduced, due to oversampling the effective bit rate remains same as that of ADPCM. Hence, delta modulation can be thought of as a simplified version of ADPCM.

$$\hat{x}(n) = \hat{x}(n-1) + e_q(n)$$
$$\hat{x}(n-1) = \hat{x}(n-2) + e_q(n-1)$$
$$\hat{x}(n) = \hat{x}(n-2) + e_q(n-1) + e_q(n)$$

Therefore, iterating like this and assuming $\hat{x}(0) = 0$,

$$\hat{x}(n) = \sum_{n=0}^{n} e_q(n)$$

This shows that the signal $\hat{x}(n)$ of the delta modulator can be obtained just by an accumulator (Fig. 17.2(a)). Since the quantizer is of one bit, its output is 1 (positive pulse) if $e(n) > TH$, and 0 (negative pulse) if $e(n) < TH$, where TH is a threshold which is quantization step size. This pulse train gets accumulated and represents the input signal estimate. This estimate value increases or decreases, depending on the polarity of the pulses. The accumulator output waveform will be of a staircase type as it just increases or decreases by one step at each sample instant. Hence, to get a smooth waveform similar to the input, its steps have to be smoothed, by a low-pass filter. Hence, in the receiver, a low-pass filter follows the accumulator (Fig. 17.2(b)). Further to get back signal in the original sampling rate decimation is required.

For a small threshold/quantization step size, if the input signal increases / decreases faster then more time is required for the estimate to follow the input. The resulting distortion due to lack of this follow-up in the output is called *slope overload noise* as the input slope is more than that can be handled for an assigned step size. This can be overcome by increasing the step size. But a larger step size is unable to sense the small changes in the slope and results in a *granular noise*. Generally, the slope overload noise results in a severe distortion compared to that by the granular noise.

The threshold (the step size) and the amplitude of the pulse that the quantizer generates are related. If the pulse amplitude is A, then the maximum value of the amplitude of the slope that can be handled is $A / T_s = Af_s$, where T_s and f_s are the sampling interval and sampling frequency, respectively. Therefore,

$$\left|\frac{dx(t)}{dt}\right| = \frac{e(n)}{T_s} < \frac{A}{T_s} = Af_s$$

For a sinusoidal signal,

$$x(t) = K\sin(\omega t), \quad \left|\frac{dx(t)}{dt}\right| = \omega K\cos\omega t$$

Therefore, the maximum value of the slope is

$$\left|\frac{dx(t)}{dt}\right|_{\max} = \omega K < Af_s$$

The maximum value of the signal permissible without slope overload is $K < [(A\, f_s)/\omega]$.

For a given quantization stepsize, the maximum amplitude of the input signal is inversely proportional to the input signal frequency and for its higher frequencies, overload occurs even for smaller amplitudes. But the limit on the input amplitude can be raised by increasing the sampling frequency. To take care of both the types of noise, it is required to change the threshold/ step size depending on the error signal variance. The step size has to be large when $e(n)$ is high and vice versa and such a system is an *adaptive delta modulator.*

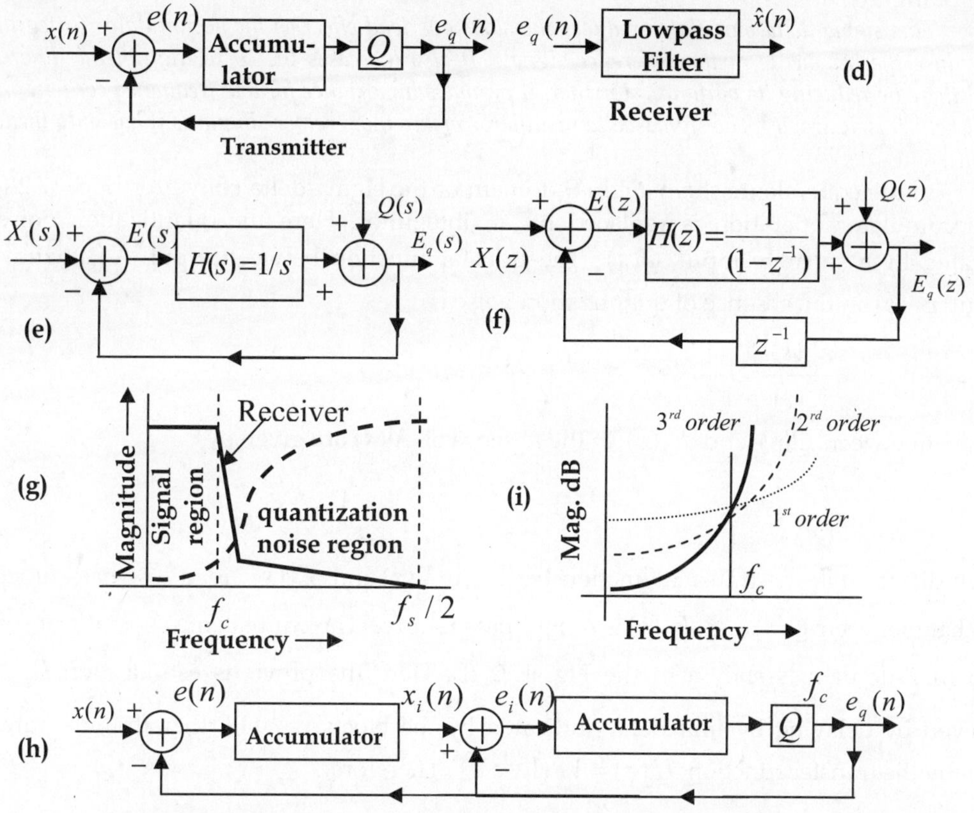

Fig. 17. 2(d, e, f, g, h) (d) Sigma-Delta Modulator, Its equivalent schematic; (e) in s-domain; (f) in z-domain; (g) Its frequency response for the signal and quantization noise; (h) Second order delta-sigma modulator transmitter and (i) Characteristics of 1st, 2nd and 3rd order quantizers.

Oversampling will distribute the noise spectrum confined to $f_s / 2$ to over a wider frequency range $Mf_s / 2$ decided by the oversampling factor M (Fig. 17.2(c)). *Thus, the noise energy in the signal frequency band gets reduced by the same factor. This can be attributed to increase in correlation between adjacent samples due to higher sampling rate.*

In the delta modulator receiver, the accumulator used not only accumulates the signal but also the noise picked up during transmission while reaching the receiver. To overcome this, the accumulator at the receiver is shifted to the input side of the transmitter prior to summing point and quantizer Q. Further, this shifted accumulator and the existing accumulator of the transmitter (in the feedback path) together are replaced by a single accumulator as shown in the Fig. 17.2 (d) and is known as *sigma-delta modulation* (B. P. Lathi and Z. Ding, 2009).

This sigma-delta modulator reduces the slope overload effect as the accumulator (integrator) will reduce the transients in the difference signal. It also increases the correlation of the difference signal there by reducing its variance. Further, it provides importance for low frequency content of the signal, which is desirable. The sigma-delta modulator makes the receiver simple as it has only the low-pass filter.

In the equivalent schematic in S-domain of the sigma-delta converter (Fig. 17.2 (e)), the accumulator operation is replaced by an integrator. Here the quantization noise is indicated by a separate input $Q(s)$. The transfer function between input $X(s)$ and the output $E_q(s)$ in the absence of quantization noise is

$$\frac{E_q(s)}{X(s)} = \frac{1}{1+s},$$

and that between $Q(s)$ and $E_q(s)$ in the absence of $X(s)$ are given by

$$\frac{E_q(s)}{Q(s)} = \frac{s}{1+s}$$

This indicates that the transfer function between $X(s)$ and $E_q(s)$, is a low-pass function and that between $Q(s)$ and $E_q(s)$ is of highpass nature. The equivalent schematic of sigma delta in Z-domain is shown in the Fig. 17.2 (f). Here the previous estimate of $E_q(z)$ is achieved by delaying by unit delay indicated by the block z^{-1}. Further the integrator is given by the transfer function $H(z) = 1/(1 - z^{-1})$. Therefore,

$$E_q(z) = Q(z) + E(z)H(z)$$

But

$$E(z) = X(z) - E_q(z)z^{-1}, \quad E_q(z) = Q(z) + \left[X(z) - E_q(z)z^{-1}\right] H(z)$$

$$E_q(z)\left[1 + H(z)z^{-1}\right] = Q(z) + X(z)H(z)$$

$$E_q(z) = \frac{Q(z) + X(z)H(z)}{\left[1 + H(z)z^{-1}\right]}$$

Substituting for

$$H(z)=\frac{1}{1-z^{-1}},\quad E_q(z)=\frac{Q(z)+X(z)\dfrac{1}{1-z^{-1}}}{\left[1+\dfrac{1}{1-z^{-1}}z^{-1}\right]}$$

$$E_q(z)=Q(z)(1-z^{-1})+X(z)$$

This indicates that the transfer function between $x(n)$ and $e_q(n)$, the input $x(n)$ does not get affected but the quantization noise $q(n)$ is filtered by high-pass filter. That is, the quantization noise $q(n)$ is of high-pass nature.

That is, the *sigma delta modulator has pushed out the quantization noise into high frequency region and thus it does not enter into the signal frequency band.* Thus, the sigma-delta modulator enables separation of the signal and quantization noise (Fig. 17.2(g)). Hence, the low-pass filter in the receiver will remove the quantization noise. A second order sigma delta modulator is shown in Fig. 17.2(h). The in-band noise power n_0 in the signal frequency range $(0,\ f_0)$ for a Pth order sigma -delta modulator is given by

$$n_0=q_{rms}\left(\frac{\pi^P}{\sqrt{2P+1}}\right)\left(\frac{2f_0}{f_s}\right)^{\left(P+\frac{1}{2}\right)}$$

Here, $q_{rms}^2=\delta^2/12$ is the RMS quantization noise power for a quantization step-size δ which is uniformly distributed. This indicates that the quantization noise magnitude decreases with increase of oversampling factor $\left(2f_0/f_S\right)$ and with the increase of the order of the modulator P. For a first order modulator, doubling f_S will result in decrease in inband noise by 9dB and for a second order its decrease is 15dB [INTERSIL, 1995]. Thus, for order P, the decrease in in-band noise is $3(2P+1)$ dB. This performance characteristic of the quantizer for different orders is schematically shown in Fig. 17.2 (i)).

For wideband applications, the oversampling factor required may be very high (as large as 256) resulting in a very high sampling rate (which may be limited by the existing electronics technology). Further, with the oversampling factor being fixed, a very high order modulator may be required to meet the quantization noise limit and this may suffer from stability problems. Thus, though the sigma delta modulators provide *very good resolution* required for applications like asynchronous subscriber digital line (ASDL), sensors, wireless and wireline communication, its frequency range is only suitable to low and medium speed applications like audio transmission. But its low cost, good dynamic range and its flexibility in converting low bandwidth signals have enabled its use in consumer and professional audio and communication systems, industrials weight scales and in precision measuring instruments.

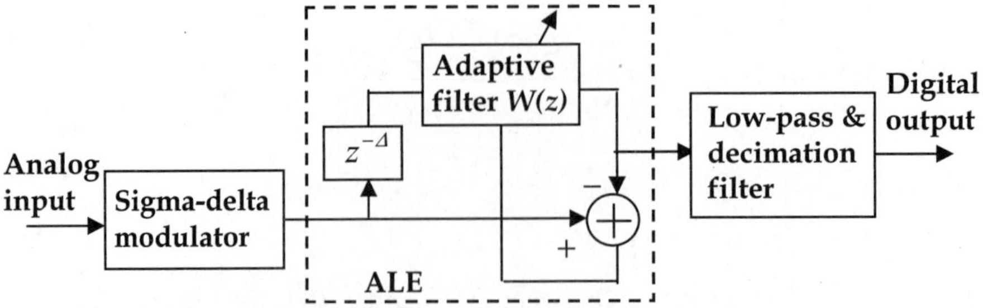

Fig. 17.2 (j) Sigma-Delta modulator with ALE.

This drawback of the sigma-delta modulator can be overcome by using an adaptive line enhancer (ALE) based on prediction filtering (Fig. 17.2(j)) and its utility can be extended to wideband applications [B.J. Farahani and M. Ismail, 2007]. The ALE will not add any complexity to the modulator and it will be a part of digital post-processing. The ALE basically removes the broadband quantization noise and allows only the narrowband signal. This is possible as there is only correlation between the input and the reference (delayed version of the input by Δ samples) only for the signal component but not for the quantization noise component. The adaptive filter uses this correlation to estimate its output.

The ALE almost whitens the high-pass quantization noise and this relaxes the severity on the specifications of the low-pass and decimation filter which are essential for a conventional sigma delta modulator. Specifically the low-pass filter realized by a FIR filter should be of long length to provide the required out of band attenuation. But as the ALE adapts to its input signal, its performance is far superior to using fixed coefficient low-pass and decimation filters. Depending on the application, the ALE may even do away with such low-pass filters.

The quantization noise performance for a given sigma-delta modulation order and oversampling frequency factor, *depends on the length of the adaptive filter of ALE* and improves with increase in its length. The performance of ALE sigma delta is independent of the order of the sigma delta modulator. The performance of ALE is more effective when the oversampling ratio is low as there will be significant amount of inband quantization noise and the ALE effectively removes this noise. However, an excess length may attenuate the signal also and also adaptation of large number of coefficients may result in a poor misadjustment. But this can be overcome by using IIR type of filters or Laguerre type of structures for the adaptive filters. For a Wide Local area network (WLAN), a 7 level quantizer has an SNR of 33.7 dB. But the use of ALE realized by a IIR filter of 20th order, improves its performance by 11 dB as SNR becomes 44dB. Thus, the use of ALE very effectively improves the SNR performance of a sigma-delta modulator with low over sampling frequency factor irrespective of its order and is highly useful for wideband applications which otherwise would have been limited to medium speed applications like audio transmission.

17.4 ADAPTIVE INVERSE MODELING AND INVERSE CONTROL

In inverse modeling, the plant inverse model estimate $Q(z)$ will be in cascade with the unknown actual plant $P(z)$ and the plant input $s(n)$ will form the desired signal in the estimation of the inverse model (Fig. 17.3(a)). On convergence of the adaptive filter $Q(z)$ which is the inverse model of the plant $P(z)$, its output $y(n)$ will be approximately matching the plant input $s(n)$. Then,

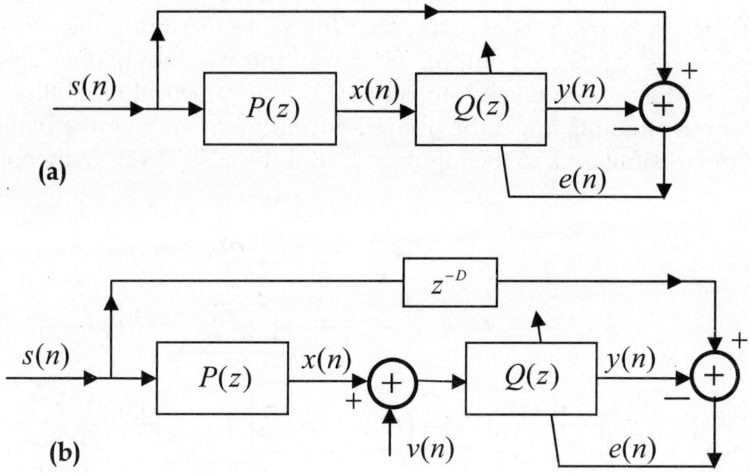

Fig. 17.3 (a) Inverse modeling concept and (b) Inverse modeling with delay and noise.

$$Y(z) = S(z)P(z)Q(z) \approx S(z)$$

$$P(z)Q(z) \approx 1 \text{ or } \quad Q(z) \approx \frac{1}{P(z)}$$

Inverse modeling realized by a FIR filter $Q(z)$ assumes that the plant $P(z)$ is an all-pole system and the model will be of the same length as that of the plant. If this is not so, then the model will only approximate the inverse of the input-output relation of the plant, i.e., $Q(z) \approx 1/P(z)$ and this is possible only by an inverse model of higher order than that of the actual pant. To get the complete inverse model of the plant, it is required to excite the plant over the complete desired range of frequencies equally. This can be realized by adding a small amount of white noise to the plant input.

The plant generally will introduce a delay D and also the plant output which forms the input to the inverse model adaptive filter can get corrupted by noise $v(n)$, the plant noise (Fig. 17.3 (b)). There can also be a drift (due to integration operation in a feedback loop) in the plant output. The delay will also be required if the plant is a non-minimum phase system, i.e., it has zeros outside the unit circle. Such zeros may make the inverse model unstable (as zero become poles which lie outside the unit circle). To account for the delay D, the plant input $s(n)$ should be correspondingly delayed as $s(n-D)$ and used as the desired signal for the adaptive filter. Also, the plant output noise $v(n)$ affects the inverse model estimate and this noise effect may become severe if its magnitude is high and may require additional precautions to make the inverse model estimate robust to noise.

In engineering problems, it is expected that for a specific input, the system chosen for a certain purpose gives the desired output. However, this may not happen since various aspects which may not have been taken into account due to practical feasibility problems. The control engineering attempts to get the desired output by modifying the plant input by sensing its output and feeding back to the input side. But this also always may not be easily

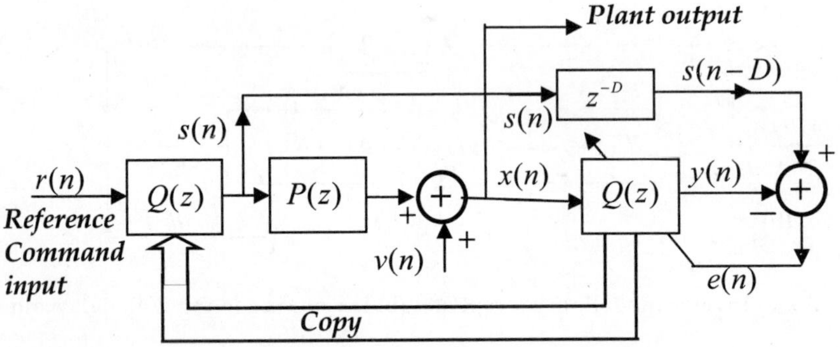

Fig. 17.3 (c) Adaptive inverse model control.

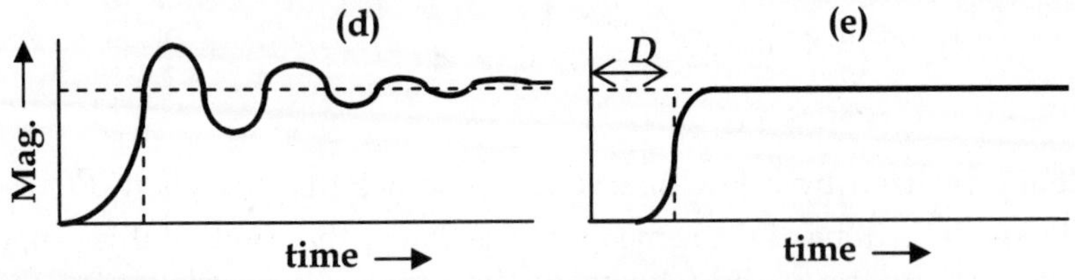

Fig. 17.3 (d, e) Typical step response of (d) closed loop feedback control and (e) adaptive inverse control.

feasible and other approaches like inverse control based on inverse modeling can be a simpler solution. For inverse control, in between the command input and the plant; a controller is introduced. This controller is nothing but the adaptive filter that estimates inverse model of the plant. This is realized by copying the weights of the adaptive filter as shown in Fig. 17.3(c). With this controller, the plant output obeys its command input properly. This is so because, the command input to the plant is *predistorted* by the controller and this is what the plant expects for its output to follow the command input (J.C. Principe, N.R. Euliano, W. Curt Lefebvre, 2000, chapter 9). That is, the plant's undesired behaviour is compensated by putting its inverse (of the undesired part of the plant which is identified by using a reference plant model. This explanation is covered towards the end) at plant's input. For example, if the plant introduces undesired gain over a frequency range, then the inverse model will attenuate this gain and thus, the undesired effect of the plant in obeying its input is removed. The Fig. 17.3(d, e) shows the typical performance of an inverse control system for a plant and that with a conventional closed loop feedback control.

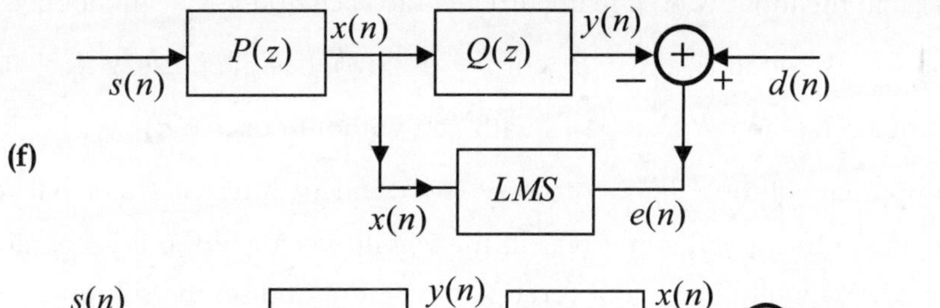

Fig. 17.3 (f, g, h) Development of FXLMS algorithm for inverse control.

For the inverse control there is a delay as expected due to delay introduced in deriving the inverse model corresponding to plant delay but it settles down to its steady state neatly without any over- and undershoots hence, its settling time is negligible after the delay. But for the conventional closed loop feedback control though the response starts without any delay, it has over- and undershoots and takes significant time to settle to its steady state. Thus, the performance of a conventional feedback system is poorer compared to that of an inverse control system except for the delay.

For the proper performance of a practical inverse control system, the noise at the plant output including its drift and the plant's non-minimum phase zeros, have to be taken into account. In the absence of plant output noise, the inverse modeling adaptive filter will converge to its optimal solution, $W_N^o = R_{xx}^{-1} R_{dx}$, Here, the cross-correlation between $d(n)$ and $x(n)$ is $R_{dx} = E[d(n)X(n)]$ and the autocorrelation matrix $R_{xx} = E[X(n)X^T(n)]$ and $X^T(n) = [x(n), x(n-1), \ldots x(n-N+1)]$. In the presence of plant noise, as the desired signal $d(n)$ and the input $x(n)$ are uncorrelated to each other, R_{dx} will not get affected. But the plant noise will certainly affect R_{xx} or R_{xx}^{-1} as $x(n)$ has the noise $v(n)$ as its component hence, the autocorrelation R_{xx} of $x(n)$ with and without noise $v(n)$ will be different. This makes the optimal solution W_N^o for noisy $x(n)$ much different from that for noise free $x(n)$. This noise effect results in a bias in the weight vector of the inverse model adaptive filter, when $v(n)$ is white noise. But for a plant coloured noise, the solution will be not just biased but can be far from the solution with that of noise-free case. This plant noise ill effect can be removed by shifting the inverse model adaptive filter prior to the plant and this is achieved by the filtered input LMS (FXLMS) algorithm and its development (B. Widrow and S.D. Stearns,1985, chapter 11) is shown in the Fig. 17.3 (f, g, h).

In Fig. 17.3(f), the plant and the adaptive filter in the absence of plant noise is shown. As $P(z)$ is common to both the inputs of $Q(z)$ and LMS block, it is moved to each branch and its effect is same as it was prior to shifting $P(z)$ to each branch (Fig. 17.3 (g)). Now, the order of $P(z)$ and $Q(z)$ is interchanged and $Q(z)$ *is brought prior to* $P(z)$ (Fig. 17.3 (h)) and the plant noise $v(n)$ is included at the output of $P(z)$. This order interchanging is under the assumption that $P(z)$ and $Q(z)$ are either time invariant or very slowly varying so that the delay introduced by them is negligible compared to their rate of variation. *Thus, the effect of noise* $v(n)$ *on the inverse model* $Q(z)$ *is removed.* But this requires a filtered version of $Q(z)$ input to the LMS block (of the adaptive filter $Q(z)$), the filter being the plant $P(z)$. Hence, the plant $P(z)$ should be known or it has to be estimated to overcome plant noise effect on the inverse modeling adaptive filter. Use of FXLMS for removing the effect of plant noise on the inverse modeling adaptive is shown in the Fig. 17.3(i). For estimating the plant $P(z)$, a small amount of dither (white noise) $\xi(n)$ can be added to the reference input $r(n)$ to ensure persistent excitation.

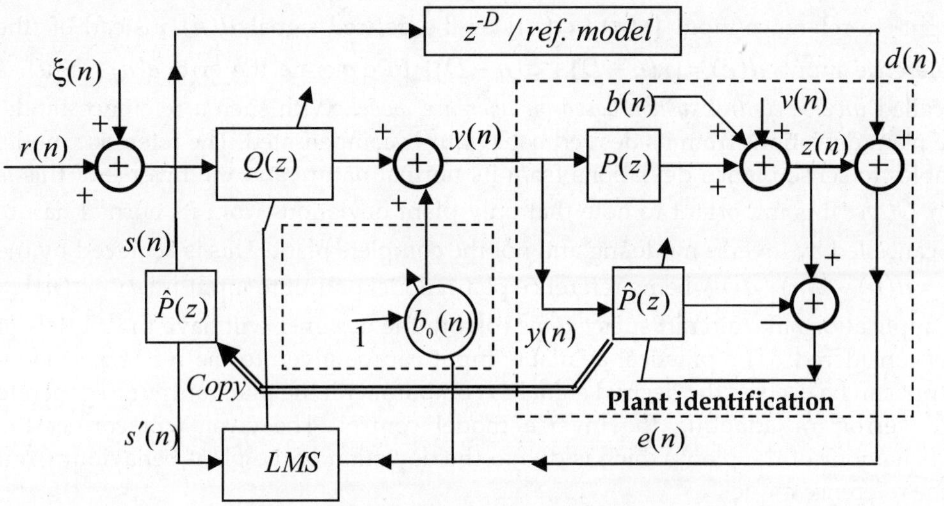

Fig. 17.3 (i) Adaptive inverse control with reference model using FXLMS algorithm.

Here, the plant noise and its drift both are considered in realising the inverse control system. The drift is indicated by $b(n)$ and it is cancelled by a single coefficient LMS adaptive filter $b_0(n)$ (using the error $e(n)$). The plant $P(z)$ is estimated by the adaptive filter $\hat{P}(z)$ using $y(n)$ as their inputs. The required error for the adaptation of $\hat{P}(z)$ is generated by comparing its output with that of the plant to which the plant noise and its drift are added. To adapt the inverse adaptive filter $Q(z)$ using the FXLMS algorithm, $s'(n)$, a filtered version of its input $s(n)$ is required. For this filtering operation, the coefficients of the estimate $\hat{P}(z)$ are copied and used for filtering the input $s(n)$. Further, the error $e(n)$ for FXLMS algorithm is generated by comparing the delayed reference input $[r(n-D)+\xi(n-D)]$ with the plant output (including the drift and plant noise). The delay is necessary to take into account the transport delay and or the non-minimum phase nature of the plant. (Transport delay is time duration required for the plant output to attain nonzero value from the instant the input is applied to it.) Thus, for the inverse control of the plant, the problem of plant noise and its drift on the inverse model are solved by using FXLMS algorithm. It is to be noted that the use of FXLMS algorithm for the inverse modeling only makes its mean weight vector to reach the noise-free optimum solution. But the plant noise effect on the misadjustment of the inverse model adaptive filter continues to exist. The accuracy of the plant identification is not very critical and it is sufficient that the estimate represents the transport delay of the plant. In Fig. 17.3 (i), the plant output $z(n)$ follows the delayed input $d(n)=s(n-D)$. However, *it is the deviations of the plant from its desired behaviour which are required to be removed.* For this, it is required to use the plant reference model output to reference input $[r(n)+\xi(n)]$ as the desired signal $d(n)$ instead of the delayed reference input $d(n)=[r(n-D)+\xi(n-D)]$ to generate the error $e(n)$. Such a system is called *inverse control system based on reference model.* With such a reference model output, the plant deviations from its desired behaviour is compensated. The reference model output enables to sense plant's deviations from its normal nature and the inverse of this is modeled in $Q(z)$. It is important to note that only plant deviations from its normal nature have to be cancelled by inverse modeling and not the complete plant. This is achieved by the error $e(n)=d(n)-z(n)$, $d(n)$ being reference plant model output to input $[r(n)+\xi(n)]$.

In applications like aircraft subsystems (plant), the designer will have the models of subsystems considered. The practical available input can be given to the reference model and its output can be used as the desired signal to compare with the plant output to generate the required error for adapting the inverse model control. The error will correct the undesired behaviour of the plant which is due to the deviation of the plant behaviour from the ideal, the reference model.

17.5 ADAPTIVE CHANNEL EQUALIZATION

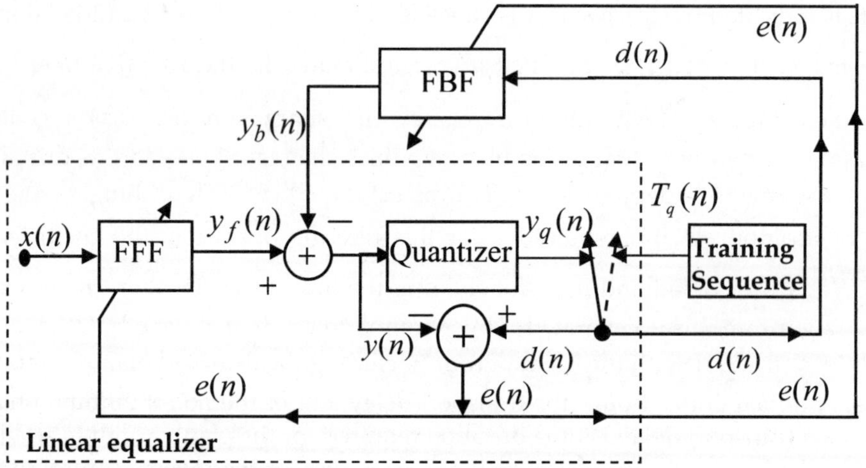

Fig. 17.4 (a) Decision feedback equalization.

In digital communication, if the bandwidth of the channel is larger than the data rate (signal bandwidth), the transmission is acceptable. In this case, the impulse response of the channel $h(n)$ is confined to less than one period of the bit duration and the received bit stream represents the input bit stream even after it passes through the channel. However, if the channel bandwidth is smaller than the data rate, the impulse response of the channel will be of longer duration than the bit period. In such a case, the received bit is a weighted sum of many bits in its neighbourhood and may not represent the transmitted bit stream. This phenomenon is called *inter-symbol interference* (ISI) as symbols in the neighbourhood of any symbol will have influence on it and this is due to limited channel bandwidth. One solution for this is channel equalization where the channel output $x(n) = r(n) * h(n)$ is passed through a transfer function that is reciprocal of the channel, $F(z) = [1/H(z)]$.

$$Y(z) = X(z)F(z) = R(z)H(z)\left[1/H(z)\right] = R(z)$$

The output $Y(z) = R(z)$, is free from channel effect due to $H(z)$, the channel. Thus, the equalization compensates for the deviations of the channel from having a constant gain and linear phase, with respect to frequency. The equalization effectively makes the combined impulse response of the channel and the equalizer, to be an impulse at the desired instant by forcing its values at other instants to zero value. Hence, such an equalizer is called *zero forcing equalizer.* For an FIR channel $H(z)$, since the equalizer transfer function $F(z)$ becomes an IIR, equalization will be proper only when the impulse response of the equalizer is of infinite length (ideally).

The channel transfer function can be estimated using least means square gradient adaptive method and is known as a *linear equalizer / feedforward filter* (FFF). To identify the channel, a known (at the receiver end) training signal $T_q(n)$ is transmitted through the channel. With this training signal as the desired signal, the inverse of the channel is identified. Since the input transmitted is a sequence involving 0 and 1, the output of the adaptive filter FFF is quantized depending on whether it is above or below a set threshold (Fig. 17.4(a)). However, for time varying channels, frequent transmission of a training signal is necessary and this makes the transmission inefficient. To avoid this, the output of the quantizer $y_q(n)$ (the decision made about 1 or 0) is assumed to represent the channel input and is used as a desired signal. Since, here the decision made by the quantizer decides the performance, it is known as *decision directed mode.* Generally, the decision directed mode takes more time to converge compared to that of the training mode (20% more time), as the initial decisions generally will not be correct.

For the adaptive channel linear equalizer the use of exponential power estimator (EPE) for its gradient step size normalization of FFF has been found to be advantageous both for stationary and time-varying channels (M. Girish Chandra and S.V. Narasimhan, 1996). This is due to the fact that the equalizer with FFF based on EPE, better or equal convergence rate, and convergence error resulting in a better or equal bit-error performance (over the RPE, BPE and constant step size based equalizers). Further, the new equalizer has a computational simplicity of 2(N + 1) multiplications plus one division and (N + 5) multiplications plus one division per sample; over that of the block and recursive methods, respectively. For channels with large eigenvalue spread, the performance advantage is by a factor of about 2 and this is valid even at SNR of about 15 dB. The performance advantage is due to the exponential nature of the EPE and the computational simplicity is due to the input power or magnitude being expressed as a power of 2.

The linear equalizer is not suitable when the ISI is severe. Also for a channel with spectral nulls, the equalizer introduces large gain to compensate them and this amplifies the noise. To avoid this, decision feedback equalizer (DFE) is used. The DFE avoids the zero forcing operation allowing the equalized channel impulse response to have nonzero values at other instants also in addition to that at the desired instant. This is achieved by removing the effect of previous decisions on the current output by using another adaptive filter called the feedback filter (FBF) (Fig. 17.4(a)). Here, the FBF output (which is the weighted sum of previous outputs) is subtracted from the output of FFF. This removes the requirement of exact *inverse modeling* (zero forcing) hence, does not require high gain at spectral nulls of the channel thereby avoids the noise enhancement. Also as the quantized output is input for the FBF, its performance is free from noise effects. The FBF and FFF weights are adapted such that their effective output matches the quantizer output, in a minimum mean square error sense.

DFE based on joint process lattice:

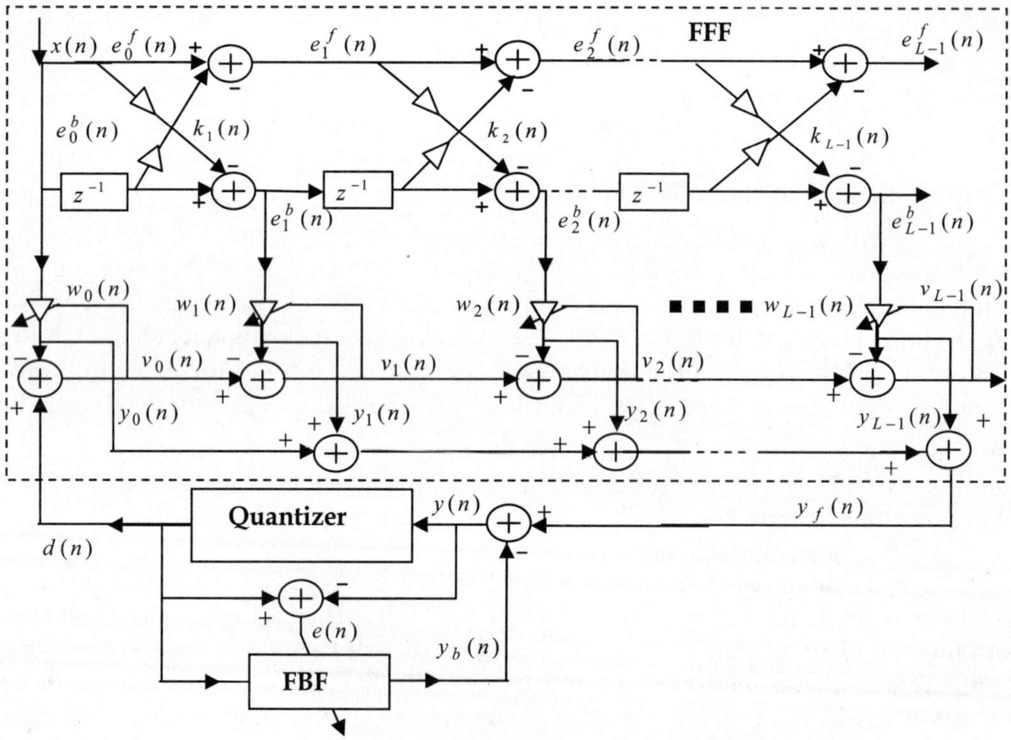

Fig. 17.4 (b) Decision feedback equalization based on joint lattice filter.

For channels severely distorted or with deep spectral nulls, *convergence is slow due to large eigenvalue spread* . The large eigenvalue spread is reduced by orthogonalising the input signal using a lattice filter into orthogonal backward prediction errors $e_l^b(n)$, $0 \le l \le L-1$

(Fig. 17.4(b)) (E. H. Satorius and S. T. Alexander, 1979). For each $e_l^b(n)$, an LMS algorithm is applied and it updates the corresponding weight $w_l(n)$.

$$w_0(n+1) = w_0(n) + \alpha_0(n)v_0(n)e_0^b(n),$$
$$v_0(n) = d(n) - y_0(n) = d(n) - w_0(n)e_0^b(n),$$

$d(n)$: quantizer output, the desired signal and $y_b(n)$: feedback filter output.

$$w_l(n+1) = w_l(n) + \alpha_l(n)v_l(n)e_l^b(n)$$

$$v_l(n) = v_{l-1}(n) - y_l(n) = v_{l-1}(n) - w_l(n)e_l^b(n), \quad 1 \le l \le L-1$$

$$\alpha_l(n) = \mu / E_l(n), \quad E_l(n) = \beta E_l(n-1) + (1-\beta)\left[e_l^b(n)\right]^2, \quad 0 < \beta < 1, \quad 0 < \mu < 1$$

The summation of all the outputs $y_l(n)$, $0 \le l \le L-1$ gives the output $y_f(n)$ of the adaptive filter FFF. The FBF filter is a transversal adaptive filter with input as quantizer output $d(n)$ and the error

$$e(n) = d(n) - y(n) = d(n) - \left\{ y_f(n) - y_b(n) \right\}$$

$y_f(n)$ and $y_b(n)$ are the outputs of the filter FFF and FBF respectively.

Equalizer based on transform domain variable step size Griffiths LMS algorithm (TVGLMS)

The TVGLMS which is robust to observation noise for a system identification setup, was considered in chapter 11. This algorithm used not only robust step size but also robust gradient for the weight adaptation (compared to using only robust step size and not also robust gradient by other algorithms). The performance of TVGLMS was superior to that of transform domain variable step size algorithm (TDVSS) as the latter does not take into account the effect of noise. The application of TVGLMS to linear equalizer and DFE is possible except for the difference that the noise is associated with the input and not with the output in system identification (S. Roopa and S.V. Narasimhan, December 2013 (on online)). Thus, for the equalizer, the Griffiths correlation involves cross-correlation between the desired signal and with each of the components of input which have noise components. These cross-correlations will be free of noise as the components of noise input and the desired signal are uncorrelated. Thus, one term of Griffiths gradient (out of two terms) will be significantly free from input noise effect. Hence, the variable step size derived using the Griffiths gradient will also be robust to noise. For linear equalizer, for a channel with eigenvalue spread of 11, the bit error performance of TVGLMS is 50% that of TDVSS algorithm. For DFE, for a channel with eigenvalue spread of 46, the bit error performance of TVGLMS is 70% that of TDVSS algorithm. The input SNR range considered is from -10dB to 15 dB. This performance is due to noise robustness of the gradient and that of the variable step size of TVGLMS algorithm.

In the above approaches, the large eigenvalue spread due to deep spectral channel nulls was solved by orthogonalising the channel output as backward prediction errors of lattice or orthogonal components due to orthogonal transform. The eigenvalue spread can be reduced by decomposing the adaptive filter input into subbands/ wavelet scales/wave packets. A DFE system where the different orthogonal scales of the input (obtained by using wavelet transform) form the input to FFF filter however, for FBF, the quantizer output is directly given (without any decomposition), has about 3 times faster convergence compared to that of using FFF without any input decomposition. The type of wavelet transform is not of concern as even Harr wavelet provides similar results as with DB4 wavelet. In view of this, for the wavelet based DFE, the increase in computation is very small. Thus, for the DFE, use of simple wavelet transform can improve the convergence rate with a small increase in computational load (Liu Feng and Wang Xinmei, 1997).

In the above discussion, it has been treated that the equalizer works only in decision directed mode by considering the decision is correct. But the training and decision directed modes both can be used together. First for learning (acquisition) mode, the training mode and then for tracking, the decision directed mode. The channel equalization can be realized in a computationally efficient manner by using IIR adaptive filters considered in chapter 13. In the training mode, the decision feedback equalizer is essentially an equation error adaptive IIR filter and in the decision feedback mode, it is an output error IIR adaptive filter. If both the modes are used, it is like using both the equation error and output error IIR adaptive filters.

17.6 ADAPTIVE TIME DELAY ESTIMATION

The time delay between two processes plays an important role in the fields like, radar, sonar, geophysics and biomedical applications. The model used for this is (Fig. 17.5)

$$r_1(n) = s(n-D) + u(n), \quad r_2(n) = s(n) + v(n), \qquad n = 0, 1, \dots, N-1$$

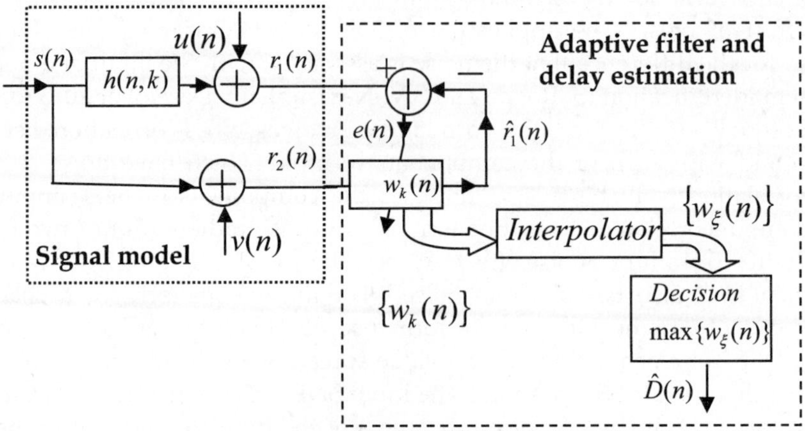

Fig. 17.5 Adaptive delay estimation.

Here, $s(n)$ is the stationary signal and is delayed by the system $h(n,k)$. This delayed signal $s(n-D)$ and $s(n)$ are associated with white Gaussian noises $v(n)$ and $u(n)$ which are uncorrelated to each other to give signals $r_1(n)$ and $r_2(n)$ respectively. Further, $v(n)$ and $u(n)$ are also independent of $s(n)$. The cross-correlation between two signals provides the delay information but it is valid for stationary signals but not for non-stationary signals.

For a delayed impulse in discrete domain is given by

$$h(n) = \frac{1}{2\pi}\int_{-\pi}^{\pi} e^{-j\omega D} e^{j\omega n} d\omega$$

$$= \frac{1}{2\pi}\int_{-\pi}^{\pi} e^{j\omega(n-D)} d\omega = \frac{1}{2\pi}\frac{e^{j\omega(n-D)}}{j(n-D)}\Bigg|_{-\pi}^{\pi} = \frac{1}{2\pi}\left[\frac{e^{j\pi(n-D)} - e^{-j\pi(n-D)}}{j(n-D)}\right]$$

$$h(n) = \frac{\sin[\pi(n-D)]}{[\pi(n-D)]}, \qquad \because e^{j\theta} - e^{-j\theta} = 2j\sin\theta$$

Since

$$\sin c(x) = \frac{\sin(\pi x)}{\pi x},$$

Therefore,

$$h(n) = \sin c(n-D)$$

Now, the Fourier transform of $s(n-D)$ is

$$F[s(n-D)] = S(\omega)e^{-j\omega D}$$

If

$$H(\omega) = e^{-j\omega D},$$

Then

$$F[s(n-D)] = S(\omega)H(\omega)$$

Therefore, $s(n-D)$ can be expressed as convolution of $s(n)$ with $h(n)$, that is,

$$s(n-D) = h(n) * s(n)$$

that is, the delayed signal is the output of a filter whose impulse response is $h(n)$ and its input is the original signal (undelayed).

Since

$$h(n) = \sin c(n-D),$$

$$s(n-D) = \sum_{k=-\infty}^{\infty} h(k)s(n-k)$$

$$s(n-D) = \sum_{k=-\infty}^{\infty} \sin c(k-D)s(n-k)$$

This indicates that a *delayed signal can be realized as the output of a non-causal FIR filter of infinite length.* But as $\sin c(n-D)$ approaches zero as $|n|$ increases, the $\sin c(n-D)$ can be approximated to be of finite length which is sufficient for many practical situations. Therefore,

$$s(n-D) = \sum_{k=-M}^{M} \sin c(k-D)s(n-k),$$

for sufficiently large value of M. If

$$w_k(n) = \sin c(k-D,n)\ , \qquad k = -M, -M+1, \ldots, M$$

$$W(z,n) = \sum_{k=-M}^{M} w_k(n)z^{-k}$$

$$s(n-D) = \sum_{k=-M}^{M} w_k(n)s(n-k)$$

The value of k for which $\{w_k(n)\}$ is maximum gives the non-integer delay estimate $\hat{D}(n)$.

$$\hat{D} = \max_k \left[w_k(n) \right]$$

For finite values of M, the interpolator is biased but this bias decreases as the filter length increases. Further, the precision with which the delay is estimated by locating the maximum of the estimated filter impulse response $\{w_k(n)\}$ will be poor. Hence, to get a better precision, its continuous version $\{w_\xi(n)\}$, where ξ is continuous; has to be derived from $\{w_k(n)\}$. This can be achieved by interpolating the discrete time impulse response $\{w_k(n)\}$. For this, any interpolation method can be used. But it is known that for signals whose spectrum monotonically deceases with frequency, the sinc interpolation is most accurate. This sinc interpolation operation on $\{w_k(n)\}$ is given by

$$w_\xi(n) = \sum_{k=-M}^{M} w_k(n) \sin c(\xi - k)$$

This sinc function used for interpolation should not be confused with the one used for representation of the delayed signal. The sinc interpolation can be achieved by *zero padding in the frequency domain.*

The implementation of the above concept of delay estimation by the adaptive filter will be considered (H. Youn, Nasir Ahmed and G.C. Carter,1982). Here,

$$e(n) = r_1(n) - \sum_{k=-M}^{M} w_k(n)r_2(n-k)$$

Therefore, the LMS algorithm for the time delay estimation is,

$$w_k(n+1) = w_k(n) + \mu(n)\frac{\partial e^2(n)}{\partial w_k(n)}$$

$$w_k(n+1) = w_k(n) + 2\mu(n)e(n)r_2(n-k), \quad k = -M, -M+1, \ldots M$$

$$\mu(n) = \alpha / E(n), \quad E(n) = \beta E(n-1) + (1-\beta)r_2^2(n), \quad 0 < \beta < 1$$

The $\{w_k(n)\}$ achieved by the adaptive filtering can be interpolated by zero padding as explained already and the time delay estimate at time n can be computed by finding the ξ value for which $\{w_\xi(n)\}$ is maximum.

17.7 ADAPTIVE MAGNITUDE SQUARE COHERENCE FUNCTION (MSC) ESTIMATION

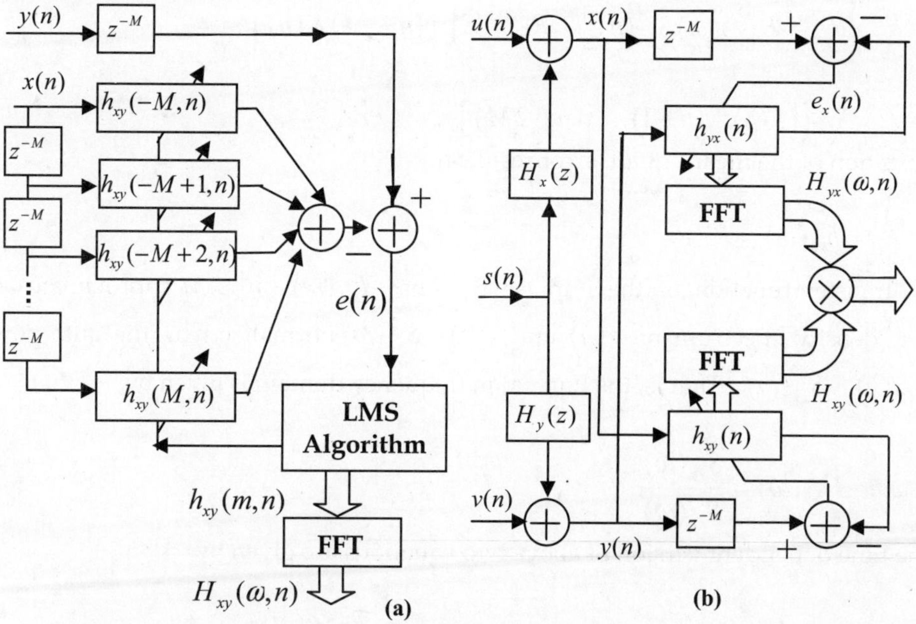

Fig. 17.6 (a) Schematic for adaptive $H_{xy}(\omega, n)$ estimation by LMS algorithm and (b) Schematic for adaptive MSC estimation by LMS algorithm.

The magnitude square coherence function (MSC) estimation was introduced in section 5.4.12. The MSC is a measure of linearity between two processes. The MSC $|\gamma_{xy}(\omega)|^2$

estimation can be made by computing the auto- and cross-power density spectrum using FFT. For this, the data $x(n)$ and $y(n)$ can be partitioned into overlapping segments and the auto-periodograms / cross-periodograms can be averaged to reduce their bias and variance.

$$|\gamma_{xy}(\omega)|^2 = \frac{|S_{xy}(\omega)|^2}{S_{xx}(\omega)S_{yy}(\omega)}, \quad 0 \le |\gamma_{xy}(\omega)|^2 \le 1$$

The MSC can be estimated based on the Wiener filter concept. Here, for the stationary signals $x(n)$ and $y(n)$, the output of the Wiener filter $H_{xy}(n)$ for the input $x(n)$, $\hat{y}(n)$ is compared with $y(n)$ and the mean square error $E[e^2(n)] = E[\{y(n) - \hat{y}(n)\}^2]$ is minimised.

If

$$h_{xy}(m) = \left[h_{xy}(-M), \ldots h_{xy}(0), \ldots h_{xy}(M)\right]^T,$$

$$R_{xx} = E\left[X(n)X^T(n)\right], \; R_{xy} = E\left[y(n-M)X(n)\right]$$

$$X = \left[x(n), x(n-1) \ldots x(n-2M)\right]^T,$$

The minimization of the mean square error results in

$$h_{xy}(m) = R_{xx}^{-1}R_{xy}$$

Here, the transfer function of the impulse response $h_{xy}(m), |m| \le M$ approximates the cross-power density spectrum of $x(n)$ and $y(n)$, $S_{xy}(\omega)$ normalized by the auto-power density of $x(n)$, $S_{xx}(\omega)$. That is, the Eqn. (a) in frequency domain is given by

$$H_{xy}(\omega) = \frac{S_{xy}(\omega)}{S_{xx}(\omega)}$$

Further, the time dependent version of above two equations are given by

$$h_{xy}(m,n) = R_{xx}^{-1}(n)R_{xy}(n), \quad H_{xy}(\omega,n) = \frac{S_{xy}(\omega,n)}{S_{xx}(\omega,n)}$$

Here, $h_{xy}(m,n)$ can be estimated using LMS algorithm (Fig. 17. 6 (a)) (H. Youn, Nasir Ahmed and G.C. Carter, 1983).

$$h_{xy}(m,n+1) = h_{xy}(m,n) + \mu(n)e(n)x(n-m), \qquad -M \le m < M$$

$$\mu(n) = \frac{\alpha}{E_x(n)\,(2M+1)}, \qquad 0 < \alpha < 1$$

$$E_x(n) = \beta\, E_x(n-1) + \left(1-\beta\right) x^2(n), \quad 0 < \beta < 1$$

The value of $\alpha = (1-\beta)/2$, *provides a better dynamic adaptation in a nonstationary environment.*

If $x(n)$ and $y(n)$ are the input signals whose MSC is required (Fig.17. 6 (b)). The MSC is

$$|\gamma_{xy}(\omega,n)|^2 = \frac{|S_{xy}(\omega,n)|^2}{S_{xx}(\omega,n)S_{yy}(\omega,n)} = \frac{S_{xy}(\omega,n)}{S_{xx}(\omega,n)}\frac{S_{yx}(\omega,n)}{S_{yy}(\omega,n)}$$

Since

$$H_{xy}(\omega,n) = \frac{S_{xy}(\omega,n)}{S_{xx}(\omega,n)} \quad \text{and} \quad H_{yx}(\omega,n) = \frac{S_{yx}(\omega,n)}{S_{yy}(\omega,n)},$$

$$|\gamma_{xy}(\omega)|^2 = H_{xy}(\omega,n)\,H_{yx}(\omega,n)$$

Here, $H_{xy}(\omega,n)$ and $H_{yx}(\omega,n)$ are the Fourier transforms of $h_{xy}(m,n)$ and $h_{yx}(m,n)$ which are obtained for the inputs $x(n)$ and $y(n)$ by adaptive filtering using the LMS algorithm as explained with reference to Fig. (17. 6(a)). The method of estimating the MSC using the adaptive filters $h_{xy}(m,n)$ and $h_{yx}(m,n)$ estimated by the LMS algorithm is given in Fig. (17. 6(b)).

17.8 ADAPTIVE NOTCH FILTER AND ITS APPLICATION

In many scenarios, the presence of interference which plays the role of noise obscures the signal and it may be difficult to get the desired information from the noisy signal. This is specifically true with biomedical signals, the interference being the power line noise. To remove the interference, band stop filter which can be realized using low-pass and high-pass filters, can be used. However, this will remove the signal components within the stopband and introduces signal distortion. To overcome this signal distortion, fixed notch filter which has maximum attenuation only at the interference frequency within a narrow band, was proposed. However, this fixed notch filter is valid only if the interference frequency remains stable. But the interference frequency can vary hence, the fixed notch filter will not be suitable as its notch frequency does not match with that of interference.

For a notch filter as it should provide zero gain, it is characterised by *the zeros and should be located on the unit circle* at the desired frequency of rejection like the power line frequency ω_0. For a signal sampled at frequency ω_S, the zeros are given by

$$z_{1,2} = r\left\{\cos(\beta_0) \pm j\sin(\beta_0)\right\}$$

Here, as zeros locations should be on the unit circle to provide zero gain, $r = 1$. Further,

$$\beta_0 = \left(\frac{\omega_0}{\omega_S}\right)2\pi == \omega_0\left(\frac{2\pi}{\omega_S}\right) = \omega_0 T_S$$

The FIR transfer function is

$$\begin{aligned} H(z) &= (z - z_1)(z - z_2) \\ &= \{z - [\cos(\beta_0) + j\sin(\beta_0)]\}\{z - [\cos(\beta_0) - j\sin(\beta_0)]\} \\ &= z^2 - z[\cos(\beta_0) + j\sin(\beta_0)] \\ &\quad - z[\cos(\beta_0) - j\sin(\beta_0)] + \left[\cos^2(\beta_0) + \sin^2(\beta_0)\right] \\ &= \left\{ z^2 - z[e^{j\beta_0} + e^{-j\beta_0}] + 1\right\} \end{aligned}$$

$$H(z) = \left\{ z^2 - 2z\cos(\beta_0) + 1\right\}$$

Dividing by z^2 to make filter causal,

$$H(z) = 1 - 2z^{-1}\cos(\beta_0) + z^{-2}$$

If $H(z)$ is represented by

$$H(z) = a_0 + a_1 z^{-1} + a_2 z^{-2}$$

then

$$a_0 = 1,\ a_1 = -2\cos\beta_0,\ a_2 = 1$$

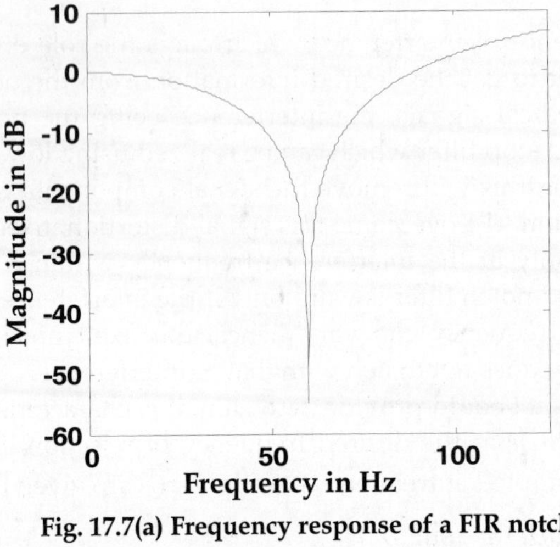

Fig. 17.7(a) Frequency response of a FIR notch filter.

The sharpness of the notch is determined by 3dB bandwidth of the magnitude of the transfer function $H(z)$ and is given by

$$BW = \frac{1}{\sqrt{2}\left|\sin\beta_0\right|}$$

It is important to note that *the bandwidth of the notch/ notch width is solely decided by the notch frequency only and there is no other control to make it narrow which is desirable.*

In time domain, the FIR notch filter output $y(n)$ for the input $x(n)$ containing the power line frequency ω_0, is given by

$$H(z) = Y(z)/X(z) = [a_0 + a_1 z^{-1} + a_2 z^{-2}]$$

$$y(n) = a_0 x(n) + a_1 x(n-1) + a_2 x(n-2)$$

This output $y(n)$ is supposed to have an attenuated power line component of frequency ω_0. But due to its bandwidth decided by ω_0 itself, some of the frequency components in the neighbourhood of the frequency ω_0, also may get attenuated. This may cause some signal distortion in addition to suppression of the power line frequency and forms the main drawback of the FIR notch filter. The frequency response of FIR notch filter with notch at 60Hz for a sampling frequency of 360Hz using an FFT of 2048 points is shown in Fig. 17.7(a). To overcome/reduce this distortion, the notch width can be controlled by introducing poles at the notch frequency but inside the unit circle. The poles location inside the unit circle is necessary from the point of view of stability of the notch filter (Fig. 17.7(b)). But it should be recalled that at the notch frequency, the *zeros are on the unit circle.* Thus, the required poles are given by

$$p_{1,2} = r\left[\cos(\beta_0) \pm j\sin(\beta_0)\right], \quad r = \alpha < 1$$

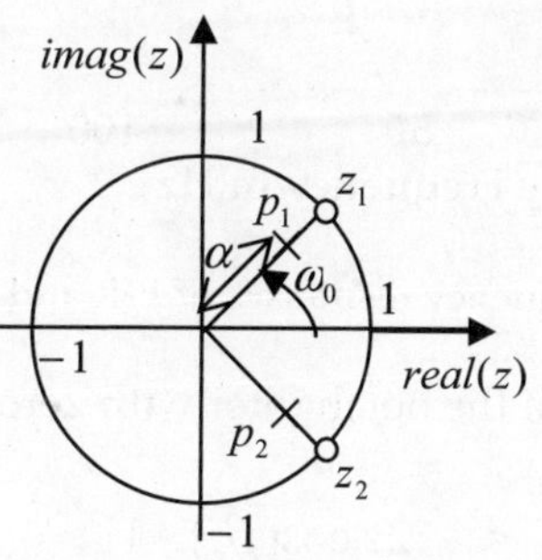

Fig. 17.7(b) Pole-zero locations for the notch filter.

The denominator of the transfer function is

$$D(z) = (z - p_1)(z - p_2)$$

$$= \{z - \alpha[\cos(\beta_0) + j\sin(\beta_0)]\}\{z - \alpha[\cos(\beta_0) - j\sin(\beta_0)]\}$$

$$= z^2 - z\alpha[\cos(\beta_0) + j\sin(\beta_0)]$$

$$- z\alpha[\cos(\beta_0) - j\sin(\beta_0)] + \alpha^2\left[\cos^2(\beta_0) + \sin^2(\beta_0)\right]$$

$$= z^2 - z\alpha[e^{j\beta_0} + e^{-j\beta_0}] + \alpha^2$$

$$D(z) = [z^2 - 2z\alpha\cos(\beta_0) + \alpha^2]$$

and the zeros are same as used previously,

$$z_{1,2} = \left[\cos(\beta_0) \pm j\sin(\beta_0)\right]$$

hence,

$$N(z) = [z^2 - 2z\cos(\beta_0) + 1]$$

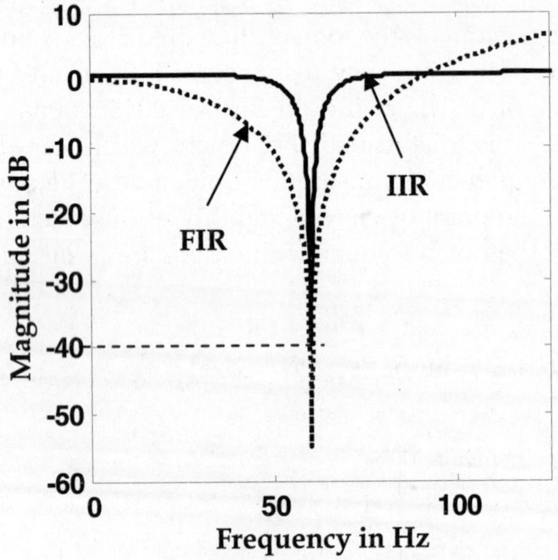

Fig. 17.7(c) Frequency responses of FIR and IIR notch filters.

Therefore, the transfer function of the notch filter with zeros and poles is

$$H(z) = \frac{N(z)}{D(z)} = \frac{z^2 - 2z\cos(\beta_0) + 1}{z^2 - 2\alpha z\cos(\beta_0) + \alpha^2}$$

$$H(z) = \frac{1 - 2\cos(\beta_0)z^{-1} + z^{-2}}{1 - 2\alpha\cos(\beta_0)z^{-1} + \alpha^2 z^{-2}}$$

Here, the parameter $\alpha < 1$ controls the notch width, the closer the value of α to unity, the smaller is the notch width. The notch width is given by

$$BW = \frac{2\sqrt{2}\,(1-\alpha^2)\pi}{\left[16 - 2\alpha(1+\alpha)^2\right]^{1/2}}$$

As $\alpha \approx 1$, $\alpha^2 = \alpha$,,

$$BW = \frac{2\sqrt{2}\,(1-\alpha)\pi}{\left[16 - 2\alpha(1+\alpha)^2\right]^{1/2}} \approx \frac{2\sqrt{2}\,(1-\alpha)\pi}{\left[16 - 2(2)^2\right]^{1/2}}, \qquad \because \alpha \approx 1$$

$$\approx \frac{2\sqrt{2}\,(1-\alpha)\pi}{2\sqrt{2}}$$

$$BW \approx (1-\alpha)\pi$$

The frequency response of FIR notch filter and that of IIR with $\alpha = 0.9$, are shown in Fig. 17.7 (c) for the notch frequency of 60Hz. It is seen that the IIR notch filter has a narrower notch compared to that of FIR (Fig. 17.7(c)). The rising pole frequency response in the neighbourhood of the notch frequency compensates for the fall in gain due to zeros of FIR hence, makes the overall response flatter over a wider region except at the notch frequency. Further, frequency response value at the notch is not zero as with FIR hence, the notch depth of IIR (-40dB) is less than that of FIR (-55dB). For the IIR notch filter, the closer the poles to the unit circle, the narrower is the notch width and but lesser is the notch depth.

17.8.1 Adaptive FIR Notch Filter

It has been pointed out that in many scenarios, the interference frequency varies and this calls for an adaptive notch filter. For this,

$$H(z) = 1 - 2\cos(\beta_0)z^{-1} + z^{-2} = a_0 + a_1 z^{-1} + a_2 z^{-2}$$

The coefficients $a_0 = 1$, $a_2 = 1$ are fixed and the only coefficient $a_1 = -2\cos(\beta_0)$ varies with the power line frequency has to be adapted.

$$H(z) = 1 + a_1 z^{-1} + z^{-2}$$

The output which is the error $e(n)$, is given by

$$e(n) = x(n) + a_1 x(n-1) + x(n-2)$$

The minimization of error energy estimates optimum value for a_1 which is found by

$$\frac{\partial e^2(n)}{\partial a_1} = 0$$

To achieve this by gradient method, the gradient of the error energy is given by [M. Ferdjallah and R. E. Barr, 1994]

$$\frac{\partial e^2(n)}{\partial a_1} = 2e(n)x(n-1)$$

Therefore, the coefficient adaptation is

$$a_1(n+1) = a_1(n) - \mu \frac{\partial e^2(n)}{\partial a_1}$$

$$a_1(n+1) = a_1(n) - \gamma(n)e(n)x(n-1), \quad \gamma = 2\mu$$

$$\gamma(n) = \gamma_0 / 3E_x(n), \; E_x(n) = \beta E_x(n-1) + (1-\beta)x^2(n)$$

$$0 < \gamma_0, \beta < 1$$

17.8.2 IIR Notch Filter

$$H(z) = \frac{1 - 2\cos(\beta_0)z^{-1} + z^{-2}}{1 - 2\alpha\cos(\beta_0)z^{-1} + \alpha^2 z^{-2}} = \frac{1 + a_1 z^{-1} + z^{-2}}{1 + \alpha a_1 z^{-1} + \alpha^2 z^{-2}}$$

Here, the parameter α is assumed to be known and is a constant . For this, the error/ output is given [M. Ferdjallah and R.E. Barr, 1994] by

$$e(n) = x(n) + a_1 x(n-1) + x(n-2) - \alpha a_1 e(n-1) - \alpha^2 e(n-2)$$

Therefore,

$$\frac{\partial e^2(n)}{\partial a_1} = 2e(n)\{x(n-1) - \alpha e(n-1)\}$$

Adaptation of the coefficient a_1 is given by

$$a_1(n+1) = a_1(n) - \mu \frac{\partial e^2(n)}{\partial a_1}$$

$$a_1(n+1) = a_1(n) - \gamma(n)e(n)\{x(n-1) - \alpha e(n-1)\}, \quad \gamma = 2\mu$$

$$\gamma(n) = \gamma_0 / 6E_x(n), \; E_x(n) = \beta E_x(n-1) + (1-\beta)x^2(n), \; 0 < \gamma_0, \beta < 1$$

Depending on the value of α, the notch width will be narrow. Since value of α is chosen in advance and the value of a_1 is estimated, the denominator does not have any unknown variables to be estimated by adaptation. Further, as $\alpha < 1$, it ensures stability as it fixes the roots of the denominator polynomial are within the unit circle.

IIR notch filter with the adaptation of α :
The error is

$$e(n) = x(n) + a_1 x(n-1) + x(n-2) - \alpha a_1 e(n-1) - \alpha^2 e(n-2)$$
$$e(n) = x(n) + a_1 x(n-1) + x(n-2) - b_1 e(n-1) - b_2 e(n-2),$$
$$b_1 = \alpha a_1, \quad b_2 = \alpha^2$$

Hence, a_1 and b_2 are required to be estimated as $b_1 = \alpha a_1 = a_1\sqrt{b_2}$

$$\frac{\partial e^2(n)}{\partial a_1} = 2e(n)\{x(n-1) - \alpha e(n-1)\}, \quad \frac{\partial e^2(n)}{\partial b_2} = -2e(n)\{e(n-2)\}$$

Therefore, the adaptation of a_1 and b_2 are given by [M. Ferdjallah and R.E. Barr, 1994]

$$a_1(n+1) = a_1(n) - \gamma(n)e(n)\{x(n-1) - \alpha e(n-1)\}, \quad \gamma_0 = 2\mu$$
$$b_2(n+1) = b_2(n) + \gamma(n)e(n)\{e(n-2)\}$$
$$\gamma(n) = \gamma_0 / 6E_x(n), \; E_x(n) = \beta E_x(n-1) + (1-\beta)x^2(n), 0 < \gamma_0, \beta < 1$$

In the above, it has been taken as total filter length is 6 taking into account both numerator and denominator. Further for stability, the roots of the denominator polynomial

$$D(z) = 1 + b_1 z^{-1} + b_2 z^{-2}, \qquad p_1, p_2 = \frac{1}{2}\left[-b_1 \pm \sqrt{b_1^2 - 4b_2}\right]$$

The magnitude of the roots p_1, p_2 must be less than unity. In terms of filter coefficients, $b_2 = \alpha^2 < 1, \; b_1^2 < 4$. If any pole is outside the unit circle, it is reflected inside the unit circle (reciprocal conjugate) or the adaptation is considered invalid and values of the coefficients of previous instant are used.

17.8.3 Adaptive IIR Lattice Notch Filter

Consider the pole zero lattice structure shown in the Fig. 17.7(d). For the upper lattice branch,

$$\xi_1^b(n) = x(n-1) - \rho_1 x(n),$$
$$\xi_1^f(n) = u(n) + \rho_2 \xi_1^b(n-1)$$
$$x(n) = \xi_1^f(n) + \rho_1 x(n-1)$$

Therefore,

$$x(n) = u(n) + \rho_2 \xi_1^b(n-1) + \rho_1 x(n-1)$$

Substituting for $\xi_1^b(n-1)$

$$\begin{aligned} x(n) &= u(n) + \rho_2 \{x(n-2) - \rho_1 x(n-1)\} + \rho_1 x(n-1) \\ &= u(n) - \rho_1\rho_2 x(n-1) + \rho_2 x(n-2) + \rho_1 x(n-1) \\ &= u(n) + \rho_1(1-\rho_2)x(n-1) + \rho_2 x(n-2) \end{aligned}$$

$$U(z) = X(z)\{1 - \rho_1(1-\rho_2)z^{-1} - \rho_2 z^{-2}\}$$

$$\frac{X(z)}{U(z)} = \frac{1}{D(z)} = \frac{1}{\{1 - \rho_1(1-\rho_2)z^{-1} - \rho_2 z^{-2}\}}$$

$$\begin{aligned} e_1^f(n) &= x(n) - k_1 x(n-1) \\ e_1^b(n) &= x(n-1) - k_1 x(n) \\ e_2^f(n) &= y(n) = e_1^f(n) - k_2 e_1^b(n-1) \\ y(n) &= x(n) - k_1 x(n-1) - k_2\{x(n-2) - k_1 x(n-1)\} \\ &= x(n) - k_1 x(n-1) + k_1 k_2 x(n-1) - k_2 x(n-2) \\ &= x(n) - k_1(1-k_2)x(n-1) - k_1 x(n-2) \end{aligned}$$

$$Y(z) = X(z) - k_1(1-k_2)X(z)z^{-1} - k_1 X(z)z^{-2}$$

$$\frac{Y(z)}{X(z)} = N(z) = 1 - k_1(1-k_2)z^{-1} - k_2 z^{-2}, \quad \frac{X(z)}{U(z)} = \frac{1}{D(z)} = \frac{1}{\{1 - \rho_1(1-\rho_2)z^{-1} - \rho_2 z^{-2}\}}$$

Therefore,

$$H(z) = \frac{Y(z)}{U(z)} = \frac{X(z)}{U(z)}\frac{Y(z)}{X(z)} = \left(\frac{1}{1 - \rho_1(1-\rho_2)z^{-1} - \rho_2 z^{-2}}\right)\left(1 - k_1(1-k_2)z^{-1} - k_2 z^{-2}\right)$$

$$H(z) = \frac{N(z)}{D(z)} = \frac{1 - k_1(1-k_2)z^{-1} - k_2 z^{-2}}{1 - \rho_1(1-\rho_2)z^{-1} - \rho_2 z^{-2}}$$

If

$$\rho_1(1-\rho_2) = \alpha k_1(1-k_2), \quad \rho_2 = \alpha^2 k_2,$$

then

$$H(z) = \frac{1 - k_1(1-k_2)z^{-1} - k_2 z^{-2}}{1 - \alpha k_1(1-k_2)z^{-1} - \alpha^2 k_2 z^{-2}}$$

$$H(z) = \frac{1 + w_1 z^{-1} + w_2 z^{-2}}{1 + \alpha w_1 z^{-1} + \alpha^2 w_2 z^{-2}}$$

where,

$$w_1 = -k_1(1 - k_2) \text{ and } w_2 = -k_2$$

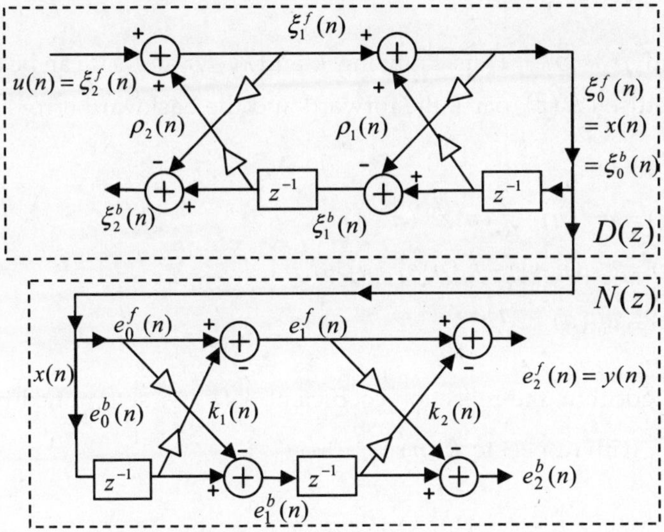

Fig. 17. 7 (d) IIR notch filter using lattice structure.

Further, if $w_1 = -2\cos(\beta_0)$ and $w_2 = 1$,

$$H(z) = \frac{1 - 2\cos\beta_0 z^{-1} + z^{-2}}{1 - 2\alpha\cos\beta_0 z^{-1} + \alpha^2 z^{-2}}$$

This is nothing but the transfer function of an IIR notch filter. From the already considered equations,

$$\rho_1 = \frac{\alpha k_1(1 - k_2)}{(1 - \rho_2)}, \quad \rho_2 = \alpha^2 k_2$$

This indicates that ρ_1 and ρ_2 can be computed from k_1 and k_2 knowing the value for α. Further, the coefficients k_1 and k_2 can be adapted using FIR lattice algorithm (N.K. Cho, C.H.

Choi and S.U. Lee, 1989). Generally, the value of α chosen will be close to unity and with this $\alpha^2 \approx \alpha$ and $\rho_2 = \alpha k_2$. Further,

$$\rho_1(1-\rho_2) = \alpha k_1(1-k_2), \quad \rho_1(1-\alpha k_2) = \alpha k_1(1-k_2) = k_1(\alpha - \alpha k_2)$$

$$\rho_1(1-\alpha k_2) = k_1(1-\alpha k_2), \quad \because \alpha \approx 1$$

Therefore, $\rho_1 = k_1$ and $\rho_2 = \alpha k_2$. Thus, knowing k_1 and k_2, ρ_1 and ρ_2 can be easily found.

For the FIR lattice ($N(z)$ part), the forward and the backward errors at the stage l are given by

$$e_l^f(n) = e_{l-1}^f(n) - k_l(n)e_{l-1}^b(n-1)$$
$$e_l^b(n) = e_{l-1}^b(n-1) - k_l(n)e_{l-1}^f(n), \quad l = 1, 2, \ldots, N$$
$$e_0^f(n) = e_0^f(n) = x(n)$$

For the Burg lattice algorithm, the reflection coefficient $k_l(n)$ is determined by minimising the cost function $\varepsilon_l(n)$ with respect to $k_l(n)$.

$$\varepsilon_l(n) = E\left[\left\{e_l^f(n)\right\}^2\right] + E\left[\left\{e_l^b(n)\right\}^2\right]$$

This results in

$$k_l(n) = \frac{\frac{1}{L}\sum_{n=0}^{L-1} e_{l-1}^f(n)e_{l-1}^b(n-1)}{\frac{1}{L}\sum_{n=0}^{L-1}\left\{ e_{l-1}^f(n)\right\}^2 + \left\{ e_{l-1}^b(n-1)\right\}^2}$$

where, L is the data length in block processsing. Further this ensures the stability requirement $|k_l| \leq 1, l = 1, 2, \ldots, N$.

For sequenctial processing this is realized recursively as

$$k_l(n) = \frac{C_l(n)}{D_l(n)}, \quad C_l(n) = \beta C_l(n-1) + e_{l-1}^f(n)e_{l-1}^b(n-1)$$

$$D_l(n) = \beta D_l(n-1) + \left\{e_{l-1}^f(n)\right\}^2 + \left\{e_{l-1}^b(n-1)\right\}^2, \quad 0 < \beta < 1$$

Here, β is a forgetting factor that enables adaptation. Since for notch filter, the nuemarator, uses only k_1 and k_2, the above equations can be expressed in terms of input $x(n)$ as

For k_1:

$$C_1(n) = \beta C_1(n-1) + x(n)x(n-1)$$

$$D_1(n) = \beta D_1(n-1) + \left\{x^2(n) + x^2(n-1)\right\}, \quad k_1(n) = \frac{C_1(n)}{D_1(n)}$$

For k_2:

$$C_2(n) = \beta C_2(n-1)\left\{x(n) - k_1(n-1)x(n)x(n-1)\right\}\left\{x(n-2) - k_1x(n-1)\right\}$$

$$D_2(n) = \beta D_2(n-1) + [\left\{x(n) - k_1(n-1)x(n)x(n-1)\right\}^2$$

$$+\left\{x(n-2) - k_1(n-1)x(n-1\right\}^2], \quad k_2(n) = \frac{C_2(n)}{D_2(n)}$$

For the notch filter, $k_2 = 1$ and only k_1 has to be adapted.

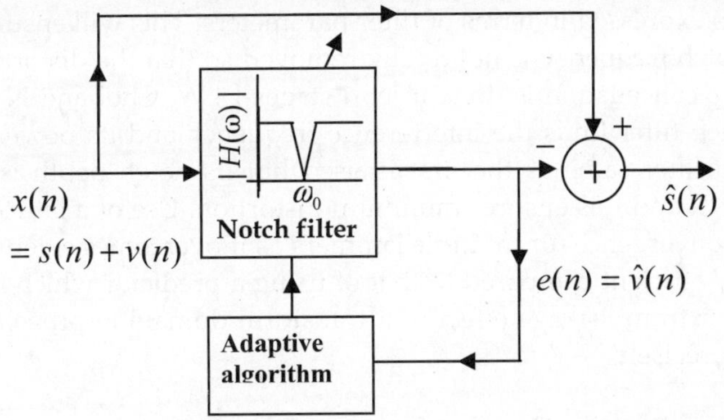

Fig. 17.7(e) Adaptive line enhancer based on notch filter.

The notch filter can be used as a line enhancer in a scenario where a sinusoid is buried in broadband noise. It may be required to extract the sinusoid from its associated noise even when the frequency of the sinusoid is changing with time. That is, the sinusoid extraction also should have the ability to track the frequency changes of the sinusoid. The notch filter will block the sinusoid $s(n)$ with frequency ω_0 and pass all others. Also some of the components in the neighbourhood of frequency ω_0 will get attenuated to some extent depending on the notch width. In the adaptive notch filter, the notch frequency is varied by the adaptive algorithm and the notch filter output forms the required error signal. The error minimization occurs when the notch frequency matches that of sinusoid and the output of notch filter $\hat{v}(n)$ is an estimate of broadband noise (Fig. 17.7(e)). The subtraction of $\hat{v}(n)$ from the input $x(n)$ results in $\hat{s}(n)$ which is devoid of noise and therefore enhanced/ brought out clearly. Further, as the frequency of $s(n)$ changes, the adaptive filter is able to track these changes in frequency of the sinusoid. Thus, the line enhancement has the ability of frequency tracking also. For the line enhancement as the sinusoid has to be totally removed, $|k_2|$ is equal to unity and the zeros lie on the unit circle. This line enhancement is different from the one considered previously where again a narrowband interference associated with broadband noise is enhanced by using a linear predictor with a decorrelation delay. The delay enables rejection of the broadband noise and allows only correlated spectral line to its output. In some cases, the available reference can be used for the estimation of the interference present in the input. By subtracting the interference estimate, the broadband signal can be recovered. The interference canceller as a whole can be considered a notch filter as it removes a narrowband signal from the input.

The spread spectrum technique used in wireless communication has the inbuilt immunity to interference/jammers due to its wideband nature compared to its original signal. But for very strong interferences/ jammers, additional cancellation is required. In such cases, the sinusoidal interference has to be attenuated depending on its power and when the interference is weak, then a higher attenuation than really required may result in a signal distortion. To avoid this, the optimum notch depth is found by maximising the output SNR which is expressed in terms of filter parameters. This will ensure that the signal component at the notch frequency is not totally removed so that the distortion is minimum. Here the interference cancellation is done in two stages [J.W. Choi and N.I. Cho, 2002]. In the first stage, a notch filter finds the interference frequency and its power. In the second stage, another notch filter same as the first is used but its notch depth is adjusted to get maximum output SNR and this ensures minimum distortion. Use of a lattice notch filter not only enables faster convergence for multiple jammers as it overcomes eigenvalue spread but also computationally efficient compared to that of using a predictor which is of long length. It also does not suffer from leakage effect of a transform domain approach which may not localize the jammer precisely.

17.8.4 Notch Filter Realization by System Identification Approach

The notch filter can also be formulated as a system identification problem. This enables extension of system identification algorithms for notch filter realization. The notch filter transfer function for input $x(n)$ and its output $e(n)$ (Fig. 17.7(f)) is given by

$$H(z) = \frac{E(z)}{X(z)} = \frac{A(z)}{A(\alpha z)} = 1 - \frac{A(\alpha z) - A(z)}{A(\alpha z)}$$

If $$B(z) = z\left[A(\alpha z) - A(z)\right],$$

$$\frac{E(z)}{X(z)} = 1 - \frac{z^{-1}B(z)}{A(\alpha z)}, \quad E(z) = X(z) - \left[z^{-1}X(z)\right]\left[\frac{B(z)}{A(\alpha z)}\right]$$

This realization of notch filter is shown in the Fig. 17.7(g).

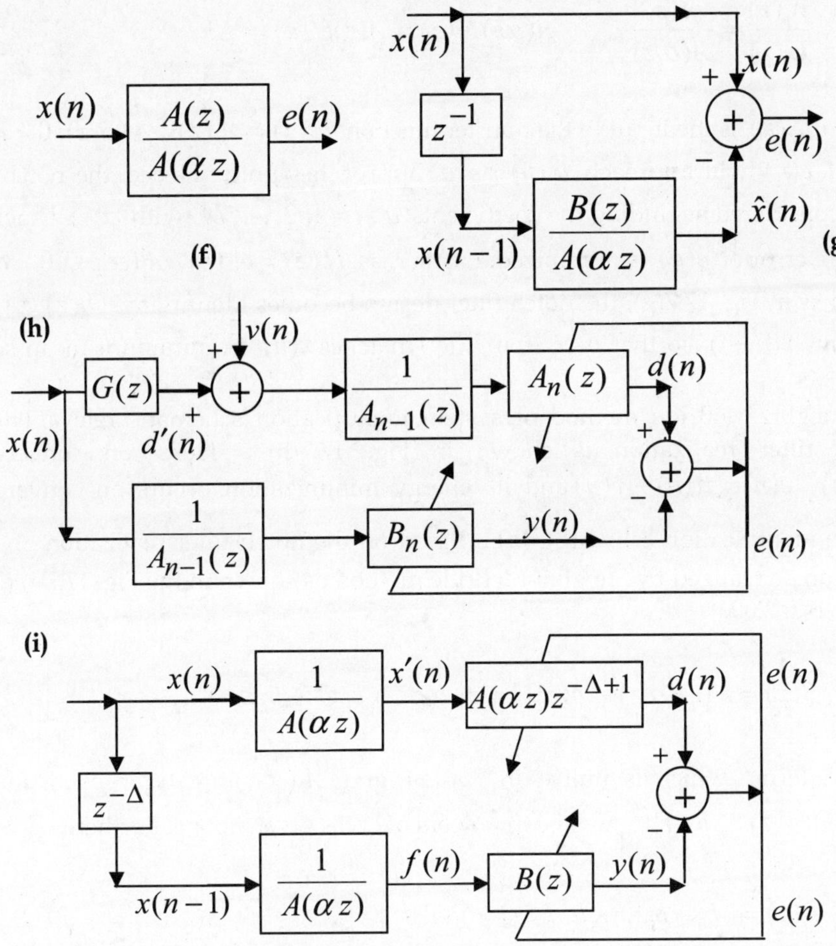

Fig. 17.7 (f-i): (f) Notch filter transfer function for an input and its output; (g) Adaptive notch filter realization by system identification; (h) Steiglitz-McBride system identification and (i) Adaptive notch filter realization by Steiglitz-McBride method.

For a signal with M interfering sinusoids with a broad band signal $u(n)$,

$$x(n) = u(n) + \sum_{i=1}^{M} c_i(n)\sin(\omega_i n + \varphi_i)$$

The signal $x(n)$ can be modeled as the output of an ARMA system with its input as noise $u(n)$ as

$$\frac{X(z)}{U(z)} = \frac{A(z)}{A(\alpha z)}, \qquad A(\alpha z)X(z) = A(z)U(z)$$

In such a case, if $x(n)$ is the input to the transfer function $H(z) = A(z)/A(\alpha z),\ 0 < \alpha < 1$, then its output $e(n)$ will approach $u(n)$ as α approaches unity. Hence, the notch filter design can be considered as finding the coefficients $a_i,\ i = 1, 2, \ldots, M$ with $\alpha \to 1$ such that the mean square error of $e(n)$ is minimized. Further, as $H(z) = A(z)/A(\alpha z),\ 0 < \alpha < 1$; can be realized as in Fig. 17.7(g), the notch filter design becomes identifying $Q(z)/A(\alpha z)$ using the input $x(n-1)$ so that its output $\hat{x}(n)$ matches $x(n)$ in minimum mean square sense.

The Steiglitz-McBride method of system identification schematic relevant to the present notch filter realization is shown in Fig. 17.7(h). Here, on convergence $A_n(z) = A_{n-1}(z)$, $e(n) = d(n) - y(n)$ and its energy minimization occurs on convergence resulting in the identification estimate $\hat{G}(z)$. Based on the notch filter realization given in Fig. 17.7(g), it can be realized by Steiglitz-McBride method as shown in the Fig. 17.7(i) (M. H. Cheng and J.L. Tsai, 2006).

$$B(z) = z\left[A(\alpha z) - A(z)\right] = z\left[b_0 z^{-1} + b_1 z^{-2} + b_2 z^{-3} + \ldots b_M z^{-(M+1)}\right]$$

Here, the first term which is unity in $A(z)$ and $A(\alpha z)$ get cancelled and the polynomial $\{A(\alpha z) - A(z)\}$ will be in powers of z as given above. Further,

$$B(z) = \left[b_0 + b_1 z^{-1} + b_2 z^{-2} + \ldots b_M z^{-M}\right]$$

Thus, the adaptation of $B(z)$ results in a polynomial $b_0 + b_1 z^{-1} + b_2 z^{-2} + \ldots b_M z^{-M}$ and

$$y(n) = \sum_{k=0}^{M} b_k f(n-k), \quad f(n) = \frac{1}{A(\alpha z)}\left[x(n-1)\right]$$

As the eigenvalue spread for $A(\alpha z) = C(z)$ is large, the LMS algorithm will be slow and is not suitable . A fast adaptation for $C(z)$ is necessary and this can be a variable step size algorithm (S. Roopa, S.V. Narasimhan and B. Babloo, 2014).

$$c_k(n+1) = c_k(n) + \frac{\mu(n)}{NP_{x'}(n)} e(n)x'(n-k), \quad k = 1, \ldots, N$$

$$P_{x'}(n) = \beta P_{x'}(n-1) + (1-\beta)\{x'(n)\}^2, 0 < \beta < 1$$

$$Q_k(n) = \gamma Q_k(n-1) + (1-\gamma)\left[e(n)x'(n-k)\right], \quad 0 < \gamma < 1$$

$$\mu(n) = \rho\, \mu(n-1) + \sigma \sum_{k=1}^{N} Q_k^2(n), \qquad 0 < \rho, \sigma < 1, \quad \mu_{\min} \le \mu(n) \le \mu_{\max}$$

The adaptation of $B(z)$ is given by

$$b_k(n+1) = b_k(n) + \eta(n)e(n)f(n-k), \quad k = 0, 1, \ldots, M$$

$$\eta(n) = \frac{\eta_0}{NP_f(n)}, \qquad 0 < \eta_0 < 1$$

$$P_f(n) = \beta f(n-1) + (1-\beta)\{f(n)\}^2$$

To reduce the bias problem due to white noise associated with the sinusoids, $\Delta = 1$ is used But for colour noise, an appropriate value for the delay Δ can be used (M. H. Cheng and J.L. Tsai, 2006).

The performance of the notch filter realized by the SM method based on VSS-LMS algorithm is illustrated. The input signal is the sum of three sinusoids of frequencies 80, 160, and 200 Hz with equal amplitudes (which acts as interference) and a zero-mean unit-variance white Gaussian noise (acts as signal) to get an interference-to-signal ratio, ISR =10 dB. The sampling frequency used is 500 Hz. The VSS-LMS algorithm is used only for adaptation of the denominator coefficients. For the numerator coefficient adaptation, the NLMS algorithm is used. The number of denominator and numerator polynomials considered is 15 (as each spectral peak requires about 5 coefficients for estimation). For the denominator polynomial, the forgetting factor $\beta = 0.99$, $\gamma = 0.9$, $\sigma = 1e-11$, $\mu_{\max} = 0.01$, $\mu_{\min} = 0.0001$, and delay $\Delta = 1$. The step size for numerator adaptation is 0.01, and $\beta = 0.99$.

The step size is varied from a maximum value of 0.01 to a minimum value of 0.0001. The convergence of the step size is reached in about 50 samples, indicating significantly fast convergence. The frequency response of the notch filter shows an attenuation of about 25 dB and the spectral flatness is within 5 dB (peak-to-peak) (Fig. 17.7j).

The step size is varied from a maximum value of 0.01 to a minimum value of 0.0001. The convergence of the step size is reached in about 50 samples, indicating significantly fast convergence. The frequency response of the notch filter shows an attenuation of about 25 dB and the spectral flatness is within 5 dB (peak-to-peak) (Fig. 17.7j).

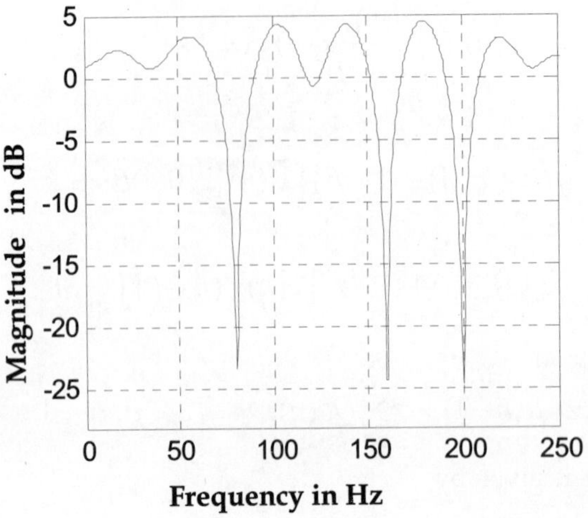

Fig. 17.7(j) Frequency response of ANF.

To evaluate the performance of the ANF, an index that indicates the amount of residual interference in the output is considered and is given by

$$R_{\mathrm{i}} = 10 * \log_{10}\left[\frac{P_{(v-e)}}{P_{\mathrm{s}}}\right].$$

$P_{(v-e)}$ is the power of the difference between input Gaussian white noise (v) and output error signal (e), P_{s} is the power of sinusoids, and R_{i} is the residual interference. The index R_{i} is found to be −13 dB which indicates that the sinusoids (colored noise) are reduced in the output to a better extent.

17.9 FEEDBACK CANCELLATION (FBC) IN HEARING AIDS

As already introduced in chapter 1, in a hearing aid, a microphone picks up the audio signal and amplifies it and delivers through a loudspeaker (earphone) for better hearing. But the microphone in addition to picking up the actual input, it also picks up the sound signal given out by the speaker earphone and is known as the acoustic feedback. This acoustic feedback in a hearing aid is due to the acoustical coupling between its earphone (loudspeaker) and the microphone. A severe distortion in the desired signal, an annoying howling sound and system instability; either any of these combinations or all of them may be the result of this coupling on increasing the gain. With the growing use of open fittings to improve listening comfort and the decreasing distance between the loudspeaker and the microphone for the purpose of miniaturization, the demand for the effective feedback cancellation techniques is growing. The acoustic feedback limits the maximum gain that can be achieved in most of the hearing aids. The computer simulation of feedback cancellation provides gain improvement in the range of 6 to 20dB. However, feedback cancellation in built into *behind the ear hearing aid* for severely impaired subjects, improves the gain by 8-10dB. In the hearing aid due to the limited processing power available as it operates on a battery, the simple LMS algorithm is generally preferred compared to the more sophisticated algorithms. To generate the feedback signal picked up by the microphone for cancellation from microphone pickup signal, it is necessary to identify the feedback path. For this identification, the output of the hearing aid is used as input instead of using a separate excitation and this has become a standard practice as it does not increase the output noise level. For a broadband input the feedback path identification will be fast and will be of good quality but for a narrowband input, it will not be so due to the slow convergence of LMS algorithm for a large eigenvalue spread. The schematic of feedback cancellation is shown in the Fig. (17.8(a)).

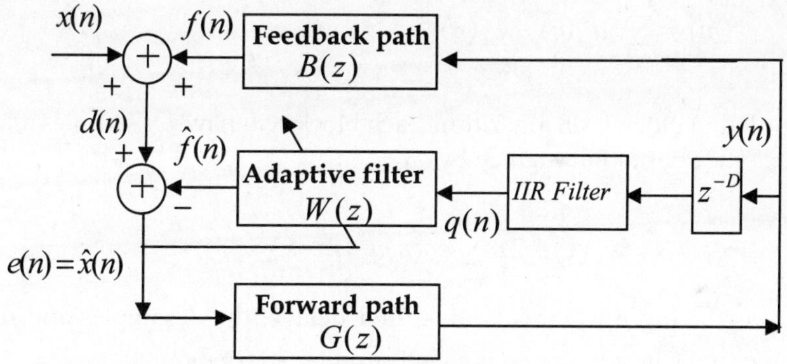

Fig. 17.8(a) Schematic of feedback cancellation by running adaptation.

To account for the delay introduced by the feedback path, it is essential to introduce the delay D in the cancellation path. The feedback canceller estimates the feedback path based on the hearing aid input as the *desired signal* $d(n)$ and its output $y(n)$ as the *reference input signal.* Since the desired signal $d(n)$ is sum of feedback signal $f(n)$ and actual input $x(n)$, the correlation between these two components will result in a *biased estimation* of the feedback path, especially for narrowband inputs. The delay used can be used in solving this bias problem. An appropriate delay decorrelates the input $x(n)$ and feedback signal $f(n)$. It is required to estimate $f(n)$ and not $f(n)+x(n)$ and the delay enables the estimation of $f(n)$ effectively though $f(n)$ and $x(n)$ are together. The input $x(n)$ acts as observation noise for the estimation of $f(n)$ and the delay reduces this ill effect. A higher value of delay may result in better decorrelation, but a higher delay slows the convergence of the adaptation. Hence, it is required to use a minimum value of the delay that gives best decorrelation (Y.C. Park, I.Y. Kim, S.M. Lee, 2003).

In the feedback path, there are two components one which changes slowly and can be considered stationary and the other which changes faster with respect to time. The slow part is due to changes in microphone, amplifier and earphone and can be due to their ageing. The fast changes in the feedback path can be due to the position of hearing aid in the ear canal and the surrounding environment (like wearing and removing the hat or putting the palm on cheek) taking into account the echo/reflections of the room, etc. The slow part can be determined and can be approximated by an IIR filter having five poles which can also be initialized. The second fast changing part has to be adapted and can be a FIR filter of sufficient length to accommodate the effect of the changes in the environment. The degree of feedback cancellation is finally decided by the environment changes including acoustic reflections inside a room and to what extent the feedback cancellation filter can accommodate these changes.

For the adaptation of the FBC filter, the error energy is minimized on block-to-block basis instead of sample to sample basis and is given by

$$\varepsilon(m) = \sum_{n=0}^{N-1} \{d_m(n) - \hat{f}_m(n)\}^2$$

As studied in block LMS algorithm, each block will have N samples and m refers to the block index. The adaptation is given by

$$w_{m+1}(k) = w_m(k) + 2\mu \sum_{n=0}^{N-1} e_m(n) q_m(n-k)$$

Here, $q_m(n-k)$ is the input to the adaptive filter delayed by k samples and filtered by IIR filter for the block m. This adaptation is called running adaptation.

For the speech sampled at 16KHz, it has been found that a 3 millisecond delay gives minimum bias. Use of time varying delays using second order all-pass filters whose frequency is varied by a low frequency modulation in addition to the small fixed delay, has also been explored to reduce the bias.

The FBC though reduces the acoustic feedback, but for a narrowband / tonal input signal, it may result in signal cancellation and coloration artefacts. This process can result in large deviation of the feedback path estimate from its actual one. Generally, this results in large magnitude values for the feedback cancellation filter which in association with the hearing aid gain can lead to system instability or coloration of the hearing aid output due to large changes in the frequency response of the system.

The bias or deviation of the feedback path estimate from its actual one that can be tolerated is decided by stability of the hearing aid system as a whole. For the complete system to be stable, the gain of hearing aid $G(\omega)$ has to satisfy (J.M. Kates, 1999, J. Benessty and Y. Huang, 2003).

$$|G(\omega)||B(\omega) - W(\omega)| < 1, \qquad |G(\omega)| < \frac{1}{|B(\omega) - W(\omega)|}$$

Therefore, the maximum stable gain (MSG) possible is

$$MSG = \min_{\omega}\left\{\frac{1}{|B(\omega) - W(\omega)|}\right\}$$

The maximum gain that the hearing gain can provide is determined by the maximum deviation of the feedback path estimate from its actual value which can occur at any frequency with in the audio frequency range considered. The MSG is used as performance measure as the feedback estimate is closer to the actual path, this gain increases and it is desirable to have this as high as possible. In the absence of feedback path cancellation, i.e., $W(\omega) = 0$, then the MSG is determined by the maximum peak value of $B(\omega)$. The maximum deviation of the feedback path estimate will occur at the peaks of feedback path frequency response.

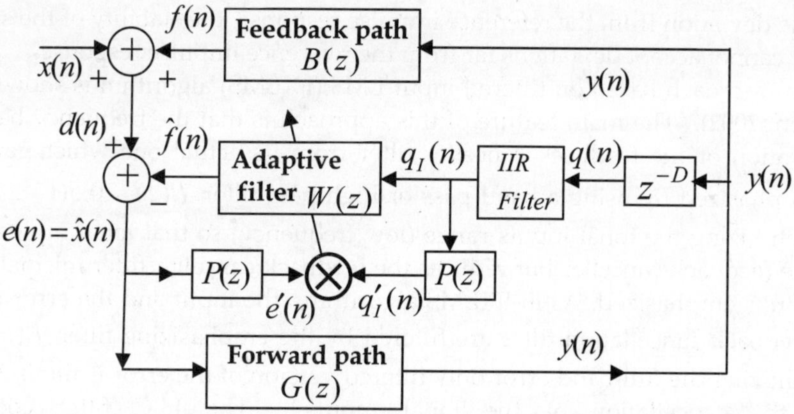

Fig. 17.8(b) Schematic of feedback cancellation using FXLMS adaptation.

It is desirable to have maximum possible gain for the hearing aid but at the same time the system should be stable. Hence, there is a necessity to limit the deviations of the feedback path estimate from its actual or its approximate. Ideally, it is desirable to have the actual feedback path impulse response at any instant of time but as it is unknown, it is attempted to keep it close to a safe reference FIR impulse response. This is achieved by minimizing a cost function based on block-to-block adaptation. The cost function is given by

$$\varepsilon(m) = \sum_{n=0}^{N-1}\{d_m(n) - \hat{f}_m(n)\}^2 + \beta\sum_{k=0}^{K-1}\{w_m(k) - w_0(k)\}^2$$

and this is in place of the generally used cost function

$$\varepsilon(m) = \sum_{n=0}^{N-1} \{d_m(n) - \hat{f}_m(n)\}^2$$

Here each block has N samples, m refers to the block index and β is the weighting factor. The new cost function permits the FBC freely to adapt in the neighbourhood of the reference impulse response $w_0(k)$ but hinders going far from the reference. The weight adaptation of the feedback path canceller is given by

$$w_{m+1}(k) = w_m(k) - 2\mu\ \beta[w_m(k) - w_0(k)] + 2\mu \sum_{n=0}^{N-1} e_m(n) q_m(n-k)$$

This algorithm enables exponential decay of the feedback path estimate towards its reference. Also the constraint on the weights may hinder the adaptation reaching the actual optimum solution. But this adaptation significantly reduces the processing artifacts which become severe due to unconstrained growth of the feedback path estimate. Due to the constraint on the deviation from the reference impulse response, the stability of the system is improved but it cannot accept deviations far from the reference impulse response.

Another approach based on filtered input LMS (FXLMS) algorithm is shown in Fig. 17.8(b) (Hellgren, 2002). The main feature of this approach is that the frequency band over which the deviation of the feedback canceller filter from its actual one which causes the instability is emphasized by using a band-pass or high-pass filter $P(z) = 0.5(1 - z^{-1})$. This enables de-emphasizing the tonal inputs range (low frequency) so that tonal inputs are not cancelled by the feedback canceller but reduces the feedback canceller filter mismatch as its frequency region is emphasized. As in FXLMS algorithm, the input and the error required for adapting feedback cancellation filter are filtered by the emphasizing filter $P(z)$. Thus, instead of minimizing the full-band error only filtered version of the error is minimized and this reduces the cancellation of the tonal inputs. The FBC filter coefficients $w_k(n), k = 0, 1, 2, \ldots, M-1$; are adapted on sample to sample basis as

$$w_k(n+1) = w_k(n) + \frac{\alpha}{ME(n)} e'(n) q'_I(n-k), \quad 0 < \alpha < 1$$

$$E(n) = \beta E(n-1) + (1-\beta)[q'_I(n)]^2, \quad 0 < \beta < 1$$

It has been reported that the FXLMS approach with a sampling rate of 16KHz, using a 5 pole IIR filter , 8-tap FIR filter and $P(z) = 0.5(1 - z^{-1})$; provides similar maximum stable gain (MSG) as that given by running adaptation by block approach. The filter $P(z) = 0.5(1 - z^{-1})$ is very effective in reducing the amplitude of 1kHz sinusoid so that it is not cancelled by the FBC. But for a 4kHz sinusoid the filter $P(z)$ does not have good attenuation hence, it only reduces its cancellation.

The FXLMS also reduces the trade-off between slow adaptation for tonal signals and fast adaptation required for large changes which occur in the feedback path. The above constrained cost minimization and the FXLMS methods do not compromise on the hearing aid gain to reduce the processing artefacts.

17.10 ADAPTIVE NOISE CANCELLATION FOR SPEECH AND ITS SPECTRAL ESTIMATION

17.10.1 Noise Cancellation for Speech by Suppressing the Crosstalk using Two Adaptive Noise Cancellers

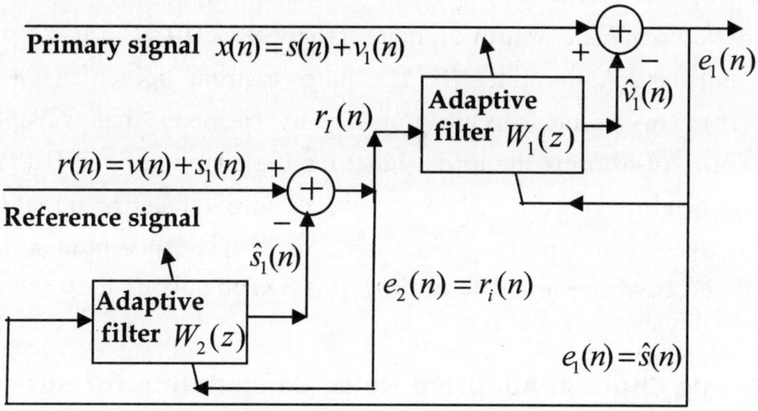

Fig. 17.9(a) Noise cancellation for speech by suppressing the crosstalk using two adaptive noise cancellers.

The general noise cancellation scheme requires noise reference that is free from signal and the primary noisy signal. In an aircraft cockpit scenario, it is possible to get the primary and reference signals as the pilot oxygen mask attenuates the speech considerably and hence the reference will not get corrupted as the microphone picking this can be kept sufficiently away from the pilot. But in many cases in practice, it may be difficult to get such a reference as it is not possible to keep the reference microphone close to noise source and away from the primary microphone. Even a few meters of separation of these microphones, requires either the estimation of the delay involved between the two signals for processing purpose or a long adaptive filter impulse response to get good noise cancellation. The length of the impulse response may be of the order of 1500 samples and this may lead to a high adaptive filter misadjustment resulting in a poor noise cancellation. Further, the long impulse response demands a high computational load and to solve this, frequency domain adaptive filtering method (considered in chapter 14) is used. The long filter length may also introduce echo in the processed speech. If the two microphones, primary and the reference are close to each other, there will be *crosstalk*. That is, there will be strong mutual leakage between primary and reference signals. *The ill-effect of such a crosstalk is that the speech also may get suppressed to some extent along with the noise.*

This is solved by estimating that speech part $s_1(n)$ that corrupts the reference noise signal $r(n)$ as $\hat{s}_1(n)$ using a second adaptive filter $W_2(z)$ and subtracting it from $r(n)$ (Fig. 17.9(a)) (G. Mirchandani, R.C. Gaus and L.K. Bechtel, 1986; J.R. Deller, J.H.L. Hansen and J.G. Proakis, 2000, Chapter 8).

Here, it is assumed that in the reference $r(n)$, the strength of speech $s_1(n)$ is relatively low compared to noise $v(n)$ and requires a good reference speech. This is in addition to the normal first adaptive filter $W_1(z)$ which estimates and removes the noise part $v_1(n)$ that corrupts the signal in the primary signal $x(n)$. Basically, the second adaptive filter enables to create an improved noise reference $r_I(n)$ as it has a reduced speech component. But the adaptive filter $W_2(z)$ requires a strong speech signal as its input for estimating $s_1(n)$. This strong speech can only be the cleaned signal available $e_1(n) = \hat{s}(n)$. This signal $e_1(n)$ also forms the error signal for the adaptive filter $W_1(z)$. On the other hand, the improved reference signal $e_2(n) = r_I(n)$ forms the error signal for the adaptive filter $W_2(z)$. Such a system has been used for spoken speech in a helicopter noise environment and has been found to improve the SNR by about 10dB which resulted in a better intelligibility.

17.10.2 Single Channel Adaptive Noise Cancellation for Speech

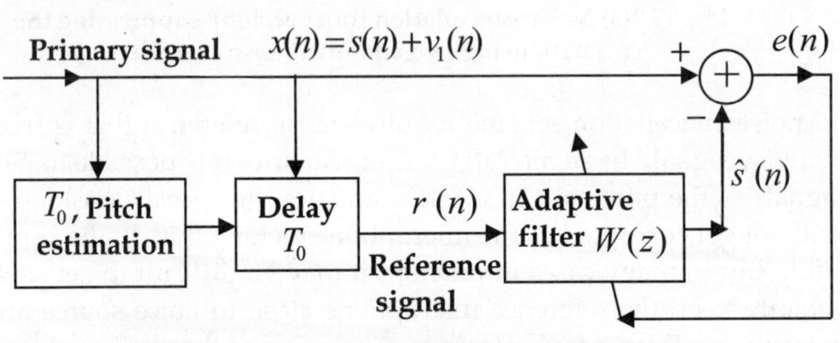

Fig. 17.9(b) Noise cancellation for speech by generating the reference signal using pitch information.

The success of dual channel adaptive noise cancellation depends on the availability of a good correlated noise reference. However, in a single channel approach, such a noise reference is not at all available. Therefore, there is a necessity to get the noise estimate and one such is to get it during silence durations and use it assuming that it will be stationary during the subsequent speech duration. Such an approach is used in block processing methods like spectral subtraction where a scaled noise magnitude spectrum is subtracted from the magnitude spectrum of the noisy signal and for the resulting magnitude spectrum, the noisy phase is associated to get the enhanced speech. However such an approach requires detection of silent zone and also may not be applicable to cases where noise and speech always exist simultaneously as in the case of quantization process. But as the speech is quasi-periodic with pitch period T_0 , it is possible to create the reference $r(n)$ by delaying the primary signal $x(n)$ by one or two pitch periods (M.R.Sambur,1978). Such a reference $r(n)$ is given as input to the adaptive filter $W(z)$ and its output $\hat{s}(n)$ is compared with the primary noisy signal to generate the error $e(n) = x(n) - \hat{s}(n)$ required for the adaptive filter (Fig. 17.9(b)).

Here $\hat{s}(n)$ is given by

$$\hat{s}(n) = \sum_{k=1}^{M} w_k(n)x(n-k-T)$$

The adaptation of $W(z)$ can be done by using the standard normalized LMS algorithm. This approach is based on the predictive filtering which exploits distinction between correlation property of the signal and broadband noise. As the input and the desired signal have to be correlated, the adaptive filter output estimates only the signal component and not the noise component. This is so as only the signal component in the primary signal and that in the reference (which is the delayed version of the primary signal) are correlated but not their noise components. By using such a reference, there will be a correlation between the primary noisy signal and the reference signal only for the speech component and for the noise component there will not be any correlation when the noise is white or broadband compared to that of speech. Thus, the output of the adaptive filter $\hat{s}(n)$ is the estimate of speech signal and the error $e(n)$ corresponds to that of noise. This method provides good performance for low SNR (< 5dB) signals and the performance becomes better as input SNR decreases.

This method has been found to perform well even for the case of quantization noise of a variable step size delta modulator (already considered) and reduces the granular noise component which is signal independent and broadband. However, it cannot reduce the slope overload noise (due to small quantization step size) which is signal dependent and hence, correlated.

This approach of generation of reference requires quite accurate estimation of pitch. Attempts to eliminate the pitch estimation for the noise cancellation are made by using an approximate estimate of speech component by an adaptive predictor and subtracting it from the primary signal to get reference signal. This reference is used for adaptive filtering of the noise from the primary signal.

Pitch period estimation is not only for the purpose of enhancement but also forms an important pre-processing step in speech compression. Many techniques like autocorrelation, cepstrum, average magnitude difference function; are used in practice for pitch estimation but these are based on estimation using multiples of the actual pitch period. A speech codec based on such pitch information has a poor performance. Further, accurate estimation of pitch in a noisy environment like mobile communication will be very difficult. The voiced speech of short duration can be well approximated by a stationary sinusoidal model. That is the speech is considered weighted sum of fundamental and its harmonics given by

$$s(n) = \sum_{l=0}^{M-1} A_l \sin\left(l\omega_0 n + \phi_l \right)$$

where, ω_0 is the fundamental frequency (pitch frequency), A_l and ϕ_l are the amplitudes and phases of lth harmonic. Further, the signal is approximated by $(M-1)$ harmonics.

The properties of an adaptive notch filter (considered already) will enable the adaptive line enhancer to detect at its notch frequency of a single sinusoid embedded in a broadband input like speech. The adaptive lattice notch filter due to its ability to detect a sinusoid in noise, forms an appropriate tool (J.H. Chang, N.S. Kim and S.K'Mitra, 2005). If this notch frequency is at $\hat{\omega}_0$, then the pitch period is given by $\hat{p}_0 = 2\pi / \hat{\omega}_0$ and is assumed to be constant over a short interval. *It is important to note that this pitch estimation is done on a sequential sample to sample basis and is unlike by the other methods which are on block basis. This enables a better estimation of a time varying pitch without any averaging over many pitch periods.* The pitch period can be used to realize a comb filter and can be used to remove broadband noise from a noisy speech.

17.10.3 Spectral Estimation of Noisy Speech

The linear prediction spectrum computed using the adaptive predictor coefficients $w_k(n), 1 \le k \le M$ will have a better match to that of the clean signal compared to that of the noisy speech. Due to noise as the correlation gets disturbed, the spectral peaks of noisy speech may be broad and may not represent the true narrow spectral peaks and there can be large shifts in peak locations from their true positions of the clean speech. But the use of delayed input to the adaptive filter enables noise rejection as explained already (under predictive filtering) and the coefficients correspond closer to that of the clean signal hence, provides better spectral estimation than that is obtained by feeding the noisy signal without delaying as input to the predictor.

$$\hat{x}(n) = \sum_{k=1}^{M} w_k(n) x(n-k-T),$$

$$e(n) = x(n) - \hat{x}(n);$$

$$X(z) = \frac{\sigma_e}{1 + \sum_{k=1}^{M} a_k z^{-k}}, a_k = -w_k(n), \sigma_e^2 = \frac{1}{L}\sum_{n=0}^{L-1} e^2(n)$$

The frequency response of such an IIR filter provides the spectrum of the signal $s(n)$.

17.11 ADAPTIVE ECHO CANCELLATION

One of the important applications of system identification is in acoustic/line echo cancellation. The acoustic echo occurs when an incoming signal gets reverberated resulting in a signal which has original signal plus its attenuated delayed copies. This signal picked up by the microphone reaches the sender as echo of his own speech. In the acoustic echo cancellation, the acoustic path called the echo path to be considered is between the loudspeaker and the microphone of a receiver set. This path depends on the geometry, volume and the material present in the enclosure where the handset is placed. The reverberation time of the enclosure depends on its size and application. For example, the reverberation time for a car of size 2.5 cubic meters is 60 meters. For video conferencing room of size 50 cubic meters and for teleconferencing room of size 90 cubic meters, it is 500 meters and 400 meters, respectively. The reverberation time represents the order of delays involved. If the delay involved is small, then the echo almost overlaps the original speech of the sender and his ear cannot make out the presence of the echo. However, if the delay involved for the echo is large say more than 20 meters, then the sender ear can feel the presence of the echo and larger the delay , higher is ill effect of the echo. The acoustic path to be considered also covers multiple reflection paths due to walls, floor and roof (r_1, r_2) which cause reverberations and

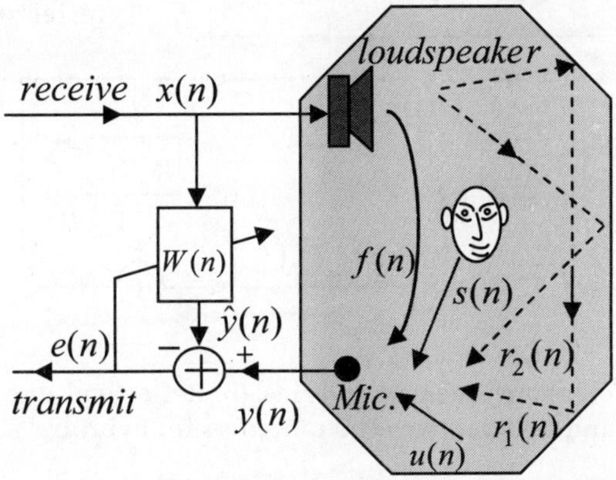

Fig. 17.10 Acoustic echo cancellation.

The feedback path (resulting in $f(n)$) (Fig. 17.10). As the overall acoustic path may be changing due to variations in temperature, humidity and geometry (physical placement of the objects, etc.), it has to be identified adaptively using the incoming speech. The microphone is supposed to pick up only the speech $s(n)$ to be transmitted but it will also acquire the reflections/echoes and the ambient noise $u(n)$. To avoid the echo being transmitted to the other side, the estimate of the echo $\hat{y}(n)$ is derived by filtering the incoming speech $x(n)$ by the acoustic path estimate $W(n)$ and this $\hat{y}(n)$ is subtracted from the microphone signal $y(n)$ (Fig. 17.10a). The echo path estimate is obtained by minimising the energy of the error signal $e(n) = y(n) - \hat{y}(n)$ by an adaptive algorithm like gradient LMS. The performance of the echo canceller is measured by echo return loss enhancement (*ERLE*) given by

$$ERLE = 10\log 10[E_y / E_e],$$

where

$$E_y = \sum_{n=0}^{L-1} y^2(n), \quad E_e = \sum_{n=0}^{L-1} e^2(n)$$

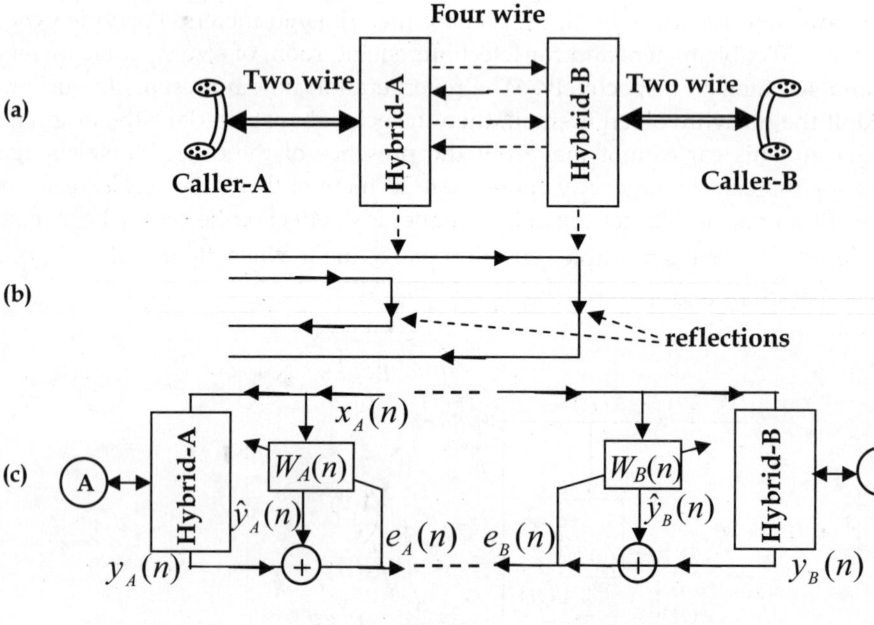

Fig. 17.11(a) Telephone line with hybrids; (b) Reflections due to hybrids and (c) Adaptive echo cancellers for hybrids.

The impulse response length of the room echo path can be of the order of 3500 samples for a sampling frequency of 8 kHz and requires an adaptive filter of 1200 coefficients. Such an echo canceller can provide ERLE = 15dB. Since the adaptive filter length is very large and the speech is having a large eigenvalue spread, it is preferable to use subband/ wavelet based adaptive filter which will improve the convergence speed. This is because as signal gets decomposed into orthogonal subbands/scales reduce the eigenvalue spread and for each subband/scale, the filter length required is also reduced. Further, the adaptation step size is decided by the input power to the subadaptive filter and its length. The wavelet based adaptive filter in addition to convergence speed also provides perception performance. A wavelet based delayless subband adaptive filter can improve the performance of the acoustic echo canceller with computational efficiency.

Similar to acoustic echo, there is a line echo also in communication. This is due to channel not being terminated with matching impedance and this mismatch can cause reflection as the signal travels from source to destination. The mismatch can also be due to the variations in the line characteristics. If the delay between the original speech and the echo is small, then the echo almost overlaps the signal, its effect is negligible. But if this delay is large, the speaker can distinguish between the echo and his own speech and it will be very annoying for the speaker and also reduces the intelligibility of the speech. This type of echo often occurs with long distance international telephone communication. Here also, to overcome this problem echo cancellers are used.

The reason for line echo is the transformer called the hybrid which receives a two-wire input and converts into a four-wire output to facilitate multiplexing for the purpose of transmission (Fig. 17.11 (a)). Due to this, when the caller-A calls, he gets two reflections, i.e., one from the near end hybrid-A and the other from the far end hybrid-B (Fig. 17.11(b)). The near end echo is not serious as its delay is small. However, the far end echo is serious as the delay involved is large. For the caller-B, the reflection from hybrid-A is of serious concern as it is having large delay. The delay can be of the order of 50 meters for few thousand miles and it can be as high as 500 meters for satellite communication (satellite being positioned at a distance of 23000 miles). But the delay due to dispersion and leakage are tolerable as it is about 20 meters.

The Fig. 17.11(c) shows the cancellation of line echo due to the hybrids-A and B. Here also, the echo path is modeled and the echo estimate is subtracted from the echo plus the incoming signal. Similar to acoustic echo problem, due to the large eigenvalue spread of speech, the identification of the echo path will suffer from slow convergence. Here also, the subband/wavelet transform/wave packet based adaptive filter can be exploited to reduce the eigenvalue spread thereby improving the convergence speed. This also enables implementation of the algorithm using low speed (low clock rate) multi DSP processors as the subband/wave packet enables decimation/increased sampling interval.

The data network also suffers from echo problem. That is, the modems used for data transmission suffer from echo due to reflections in the local subscriber loop. This problem becomes serious especially when the data rate is high and the data can get corrupted. Hence, all the modems operating above 4800 baud (9600 bits/sec) will have built-in adaptive echo cancellers. Generally, the network echo path will be sparse in nature. That is, only few samples will have significant values and others will be very small. This characteristic has been exploited to get faster convergence and reduce computational load by using proportionate normalized least mean square algorithm (PNLMS). Here, the step size is changed depending upon the magnitude of the coefficient estimate (previous) so that significant coefficients converge faster. The PNLMS is given by

$$\hat{y}(n) = \sum_{k=0}^{N-1} w_k x(n-k)\,, \quad e(n) = y(n) - \hat{y}(n)$$

$$w_{\max}(n) = \max\left[\,|w_0(n)|, |w_1(n)|, \ldots |w_{N-1}(n)|\,\right], \quad \overline{w}_{\max}(n) = \max\left[\delta, w_{\max}(n)\right]$$

$$g_k(n) = \max\left[\rho \bar{w}_{\max}(n), |w_k(n)|\right], \quad g_{AV}(n) = \frac{1}{N}\sum_{k=0}^{N-1} g_k(n)$$

$$w_k(n+1) = \left[\frac{g_k(n)}{g_{AV}(n)}\right]\mu(n)e(n)x(n-k), \quad \mu(n) = \frac{\alpha}{N\sigma_x^2(n)}, \quad 0 < \alpha < 1$$

The term in square bracket decides whether the kth coefficient is large or small as $g_k(n)$ is compared with average $g_{AV}(n)$. Further, based on their ratio $[g_k(n)/g_{AV}(n)]$, the step size $\mu(n)$ is scaled for the used for the kth coefficient. The parameter δ prevents stalling when the coefficients are initialized to zero value. The parameter ρ prevents stalling of coefficients which are much smaller than the largest coefficient. Typical values for these parameters is of the order of 0.01. Thus, as the PNLMS assigns larger step size to the active coefficients, it converges faster than the LMS algorithm. But after the initial fast convergence, it slows down and this is due to slow convergence of small coefficients (P. Loganathan, A.W.H. Khong, P. A. Naylor, 2008). If $\rho = 1$, then the PNLMS becomes NLMS algorithm. A small value for ρ makes the adaptation faster as $g_{AV}(n)$ becomes small.

For a dispersive channel (channel with deep spectral nulls/large eigenvalue spread), the value of $g_k(n)$ becomes very large for most of the N coefficients and hence $g_{AV}(n)$ will become large reducing the step size for all the large coefficients. This results in slower convergence. A transform domain adaptive filter can be used to reduce the eigenvalue spread as it orthogonalizes the input. The implementation of the transform domain adaptive filter demands the echo path estimate to be in the transform domain. Since the echo path *which is sparse in time domain becomes nonsparse in transform domain*, the benefits of PNLMS algorithm cannot be utilized (H. Deng, 2007). But the wavelet transform (even after transformation) preserves the sparsity of the impulse response in time due to its time frequency localization. Thus, the input orthogonalization by wavelet transform enables use of PNLMS algorithm hence, gets the benefit of faster convergence and reduced computational load. The use of wavelet packet will bring down the effective number of scales and the coefficients of the adaptive filter for each wave packet scale again can be activated or deactivated depending on their magnitude (compared to some threshold) (Odair A. Noskoski, 2007). This will further reduce the computations as the wave packets are reduced in addition to reducing the number of coefficients adapted.

Thus, in different echo cancellation scenarios, the use of wavelet transform or its variations provide advantages like faster convergence, reduction in computational load and algorithm realization even with slow DSP processors.

17.12 ACTIVE NOISE CONTROL (ANC)

The high noise levels not only impede the conversation and communication but also results in a serious health hazard to a human being on long exposures hence, some type of noise reduction is absolutely necessary. For the passive acoustic absorbers to be useful their size should be the order of wavelength for the frequency considered. At 2000 Hz, the wavelength is 7 inches but at 100Hz, it is 12 feet. Hence, the passive absorbers are suitable for only high frequencies and are not suitable for low frequencies (less than 500Hz) due to their size and weight for mobile vehicles. ANC is a suitable technology for the suppression of low frequency noise (below 500 Hz) in mobile vehicles like automobiles, aircraft cabins and cockpits and industrial sites where the noise levels are high (above 80 dB).

Basically, in active noise control (ANC), an antinoise is generated in acoustical domain and superimposed on the existing noise field so that the latter's effect is reduced. The generation of the antinoise is achieved by adaptive signal processing. With reference to a duct (exhaust or ventilator) shown in Fig. 17.12(a), as the signal propagates from reference source point/input microphone to the point of interest/ error microphone, it undergoes both magnitude and phase changes depending on the acoustic path. For this, the acoustic path $P(z)$ (between the input microphone and loudspeaker) called the main path (MP) is identified and the antinoise generated $\hat{y}(n)$ is given out through the loudspeaker. The two noise fields interact over a physical space and the resulting residual noise field is expected to

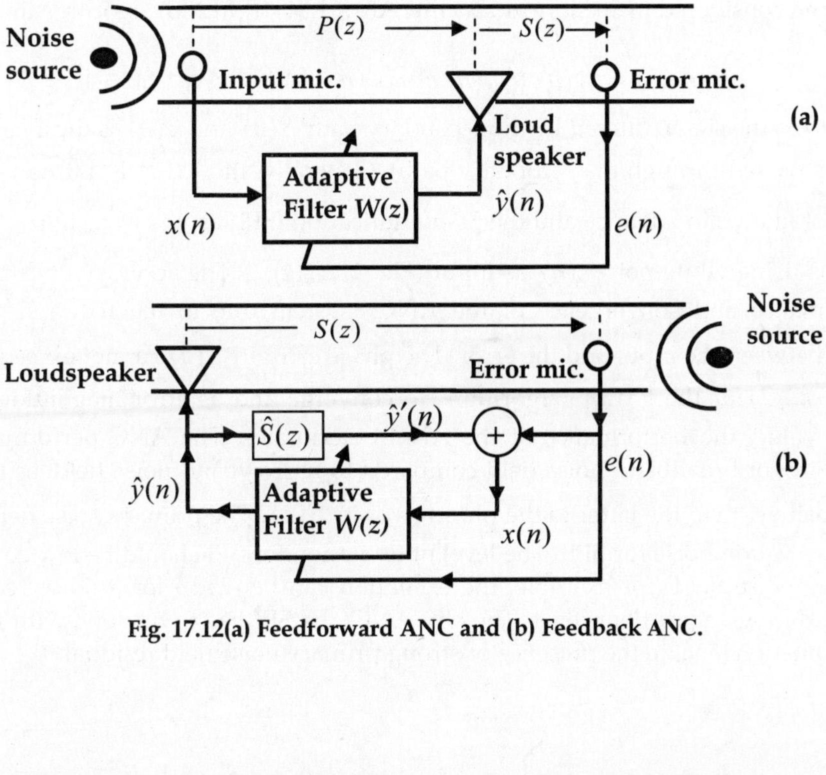

Fig. 17.12(a) Feedforward ANC and (b) Feedback ANC.

have a smaller magnitude compared to that of primary. The residual field, the error is effectively available only after it has passed through an acoustic path, the secondary/error path, $S(z)$ (Fig. 17.12(a)). In this case, the correlated reference $x(n)$ (correlated to the noise field at the error microphone) is available and this ANC system is known as feedforward ANC (FFANC). In scenarios like boundary layer noise and noise generated from different sources having many propagation paths, as it is not practically possible to sense a correlated reference signal FFANC cannot be used and the solution is feedback active noise control (FBANC). For generating the reference input $x(n)$, the MP adaptive filter output is filtered by the SP estimate $\hat{S}(z)$ and the resulting signal is added to the error signal $e(n)$ (residual noise) (Fig. 17.12(b)). The major part of the input $x(n)$ to the MP adaptive filter forms delayed (and modified magnitude) version of its output. *Hence, the MP adaptive filter has to predict the noise field in advance corresponding to the delay (and changes) introduced by SP. Thus, the MP adaptive filter will be playing the role of a predictor.* In view of this, the FBANC is suitable for predictable /correlated (narrowband) noise fields but not for broadband noise fields.

In the LMS adaptation for the weight $w_k(n)$,

$$w_k(n+1) = w_k(n) + \mu(n)\,\underline{e(n)x(n-k)}, \quad 0 \le k \le M-1$$

there is a time alignment between the error $e(n)$ and the input $x(n)$ and this is crucial for the algorithm to converge. Therefore, error/ secondary path $S(z)$ after which only the error is available cannot be neglected and has to be accounted for applying the LMS algorithm. This has been considered in Section-10.12 Filtered -X LMS (FXLMS) algorithm and is

$$w_k(n+1) = w_k(n) + \mu(n)\,\underline{e(n)x'(n-k)}, \quad 0 \le k \le M-1$$

where $x'(n)$ is the signal filtered by the secondary path $S(z)$ and $e(n)$ is the available error (after it has passed through the secondary path). Generally, the $S(z)$ is a time varying one which is not known in advance and the SP identification (SPI) as $\hat{S}(z)$ has to be done on an online basis using white noise $r(n)$ as input (Fig. 17.12(g)). The accuracy of $\hat{S}(z)$ decides the performance and convergence of the ANC system due to the required strict time alignment between the input and the error. The phase error in $\hat{S}(z)$ cannot exceed the range of $(\pi/2, -\pi/2)$ for the FXLMS algorithm to converge and as error magnitude increases from zero value, the performance of the ANC deteriorates. The ANC performance gets affected more for broadband noise field compared to narrowband noise field, as the former is more sensitive than the latter to the phase error in $\hat{S}(z)$. The primary noise field residual acts as observation noise for SPI. The level of excitation for SPI should be low compared to the primary noise field (for example, the excitation can be -25dB for a noise reduction of 20dB) as it decides noise floor level. Thus, in ANC, the SPI has to be done with the lowest possible excitation level in the presence of strong primary noise field residual.

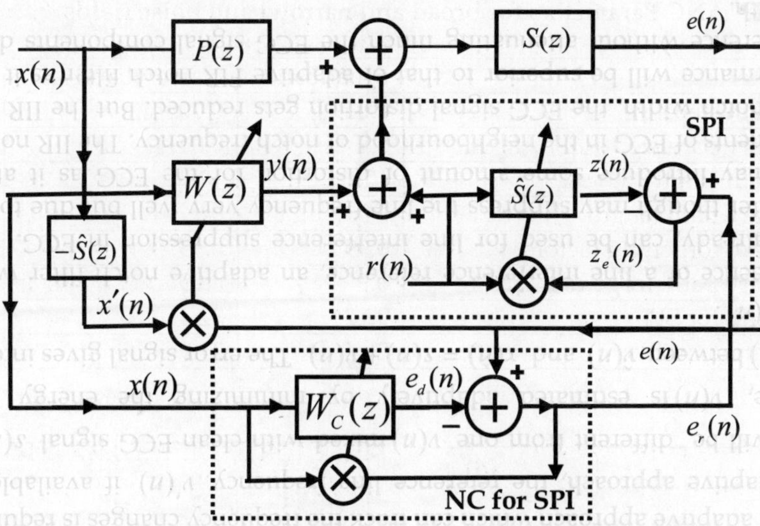

Fig. 17.12 (c) Feedforward ANC with on line SPI and NC for SPI.

The ANC error $e(n)$ is made up of two components and is given by $e(n) = e_r(n) + e_d(n)$. The component $e_r(n)$ is due to input $r(n)$ for SP and forms the required desired signal for SPI. However, the component $e_d(n)$ due to primary noise acts as an observation noise for SPI and is not required. A noise canceller (NC) has been used to extract $e_r(n)$ from $e(n)$ by estimating $e_d(n)$ using the input noise field signal $x(n)$ (Fig. 17.12(c)). The component $e_r(n)$ not only forms the desired signal for SPI but also the error signal for the adaptation of the weights of $W_C(z)$ of the NC. In the case of feedback ANC as good reference is not available, the estimate of $e_d(n)$ component from $e(n)$ is obtained using adaptive predictive filtering which exploits correlation properties of $e_d(n)$ and $e_r(n)$ as the former will be more correlated compared to the latter. The use of NC for SPI improves the accuracy of $\hat{S}(z)$ which is used for filtering the input $x(n)$ for FXLMS results in improved MP estimate leading to better noise reduction. With the FBANC use of NC for SPI has twin benefits as it not only improves FXLMS performance but also the generation of the input $x(n)$ using the error $e(n)$ and the estimate $\hat{y}'(n)$ derived by filtering $\hat{y}(n)$ by $\hat{S}(z)$.

To illustrate the performance of a duct ANC system both for narrowband and wideband noise fields., the impulse responses of main path, secondary path corresponds to duct of length 12 inches and secondary source is kept at a distance of 10 inches from the reference microphone (source end). Further, the error microphone is separated from the secondary source (loudspeaker) by a distance of 2 inches. The other system parameters used are listed in Table 17.1.

Table 17.1 : ANC Parameters for broad and narrowband noise fields

Parameter	Broadband	Narrowband
Main path step size	0.05	0.005
Secondary path step size	0.1	0.01
Main path length	128	128
Secondary path length	64	64
Main path forgetting factor	0.999	0.99
Secondary path forgetting factor	0.9	0.99
SNR in dB for SPI	27.2	32.4
Attenuation in dB	26.4	31.9

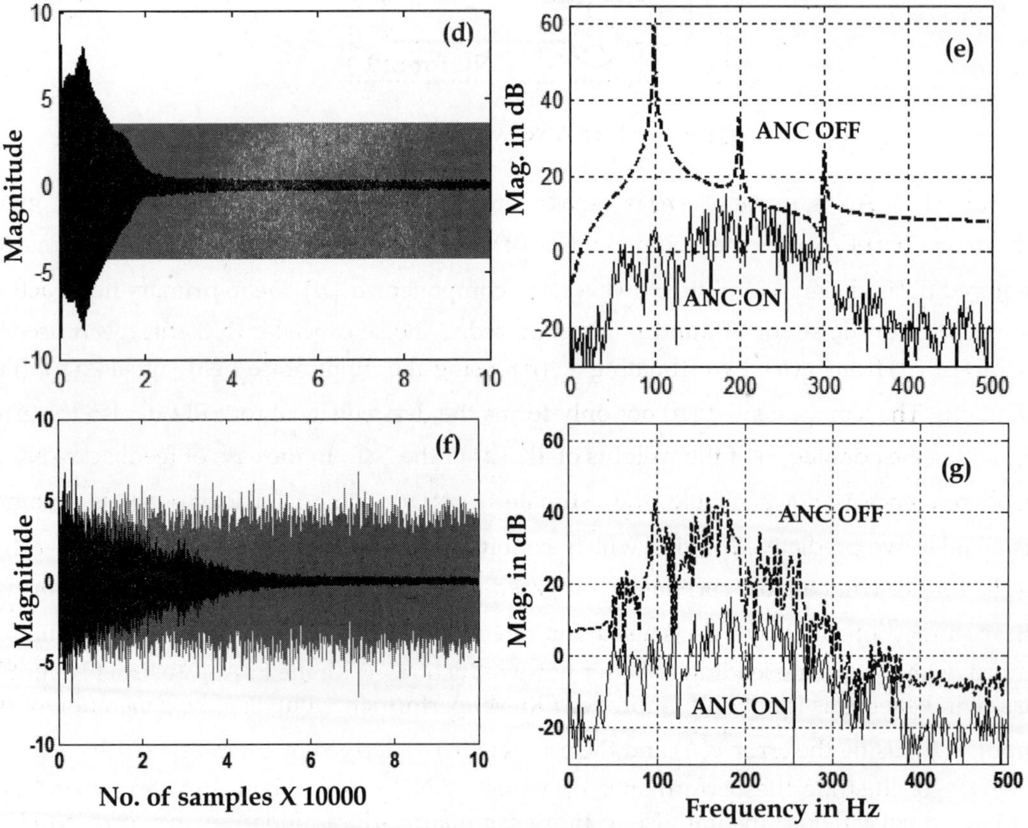

Fig. 17. 12(d-g) For narrowband ANC: (d) convergence error; (e) spectrum of error For wideband ANC; (f) convergence error and (g) spectrum of error.

For the narrowband noise field, 3 sinusoids at 100 Hz, 200Hz and 300Hz plus white Gaussian noise with SNR=30dB and the white noise injected for SPI is 32.4 dB below the primary noise field. The attenuation achieved is 31.9dB. It is seen that the residual noise decreases and converges to a level slightly above the injected noise for SPI (Fig. 17.12d). This is also seen from the PSD of residual noise after convergence as the magnitude of spectral peaks are reduced(Fig. (17.12e)).

For the broadband noise field, white Gaussian noise is considered and the white noise injected for SPI is 27.2 dB below the primary noise field.. The attenuation achieved is 26.4dB. It is seen that the residual noise decreases and converges to a level slightly above the injected noise for SPI (Fig. 17.12f). This is also seen from the PSD of residual noise after convergence as the magnitude of spectral envelope is reduced (Fig.(17.12g)). The performance of ANC for broadband noise will be generally poorer than for narrowband noise, as it is more sensitive to phase error of SP estimate.

The different aspects which improve the performance of ANC system both for narrowband and wideband noise fields are investigated. For a narrowband noise field, due to the large eigenvalue spread of the input autocorrelation matrix, the convergence is slow. To overcome, this lattice orthogonalization for the input based on Laguerre lattice has been proposed. The input orthogonalization has bee used not only for the MP adaptation but also for NC for SPI (S. Veena and S.V. Narasimhan, 2004) . The Laguerre structure not only reduces the adaptive filter length by about 50% but also provides better modeling efficiency as Laguerre is a combination of both FIR and IIR structures where only the FIR part is adapted and IIR (all-pass filter) is fixed. Further, to handle nonstationary noise field, the Laguerre escalator lattice orthogonalization which is based only on the backward error (unlike regular lattice which uses both forward and backward errors) energy minimization has been proposed. Here again, the orthogonalization has been used both for MP identification and for the NC for SPI (S.Veena and S.V.Narasimhan, 2007). In this direction, the notch filter based on Steiglitz-Mc bride method has been used as a noise canceller for SPI (S.Roopa, S.V. Narasimhan and B. Babloo, 2014). To improve the SPI without using the NC, variable step size algorithm based on Griffiths gradient which is robust to observation noise (to both white and coloured) has been proposed. Here not only the step size but also the gradient is robust to observation noise (S.V. Narasimhan, S. Veena and H. Lokesha, 2009). Further unbiased adaptive IIR equation error filter based on Griffiths gradient has been used for the SPI (S. Veena and S.V. Narasimhan, 2011). Unbiased Steiglitz-Mcbride method has been used to improve the SPI for narrowband ANC (S.V. Narasimhan and Nagaveni, 2012). To improve the performance of FBANC, transversal predictor based on both forward and backward error energy minimization (FBLMS) which has 50% reduction in misadjustment has been explored both for the MP predictor and the NC predictor (S. Pavithra and S.V. Narasimhan, 2011). Further, FBLMS predictor realization in transform domain has been used both for MP and NC predictors to improve their convergence rate in addition to the misadjustment reduction (S. Pavithra and S.V. Narasimhan, 2012). Further, delayless subband adaptive filter based on DCT harmonic wavelet transform has been used for MP identification has been explored for narrowband nonstationary noise fields (S.V. Narasimhan and S. Veena, 2010).

17.13 APPLICATIONS OF ADAPTIVE FILTERING FOR BIOSIGNALS

There is a wide variety of applications of adaptive filtering to process biosignals. Only a few of them will be considered here to bring out their importance.

17.13.1 Enhancement of Foetus ECG in the Presence of Mother's ECG

In ECG, the P-wave, QRS-complex and T-wave; their amplitude, relative positions and their durations are of significance for proper diagnosis. Generally, the magnitude of foetus ECG will be smaller compared to that of mother ECG. Further, as their QRS complexes occur closely side by side, the foetus ECG QRS complex will get distorted by Mother's ECG (Fig 17.13 (a, c))and the heart rate by the combination the two ECGs may give incorrect information about the foetus ECG. Hence, there is a necessity to enhance the fetus ECG from the background of mother's ECG. For this, the ECG picked up by the mother's chest lead signals $v_0(n), v_1(n), \ldots v_{M-1}(n)$ form the reference signals (Fig. 17.13 (a)).

The mother's ECGs at the chest leads get modified both in magnitude and phase as they propagate from the chest to abdomen. Hence, the abdominal leads (Fig 17.13 (a)) contain not only foetus ECG $s(n)$ but also $v(n)$ a modified version of mother's ECG available at the chest leads. The mother's ECG component $v(n)$ in the abdomen lead signal $x(n)$ can be considered a weighted sum of chest lead ECGs $v_0(n), v_1(n), \ldots v_{M-1}(n)$ (Fig 17.13 (b)). Hence, it is possible to get $\hat{v}(n)$, the estimate of mother's ECG component in the abdomen lead signal $x(n) = s(n) + v(n)$ by minimising energy of the difference signal $e(n)$ between the mother's ECG estimate $\hat{v}(n)$ and abdomen lead ECG signal $x(n)$. Thus, by using the LMS algorithm, the weights $w_0(n), w_1(n), \ldots w_{M-1}(n)$ can be adapted and the estimate of mother ECG component in the abdominal lead can be achieved (E.C. Ifeachor and B.W. Jervis, 1993).

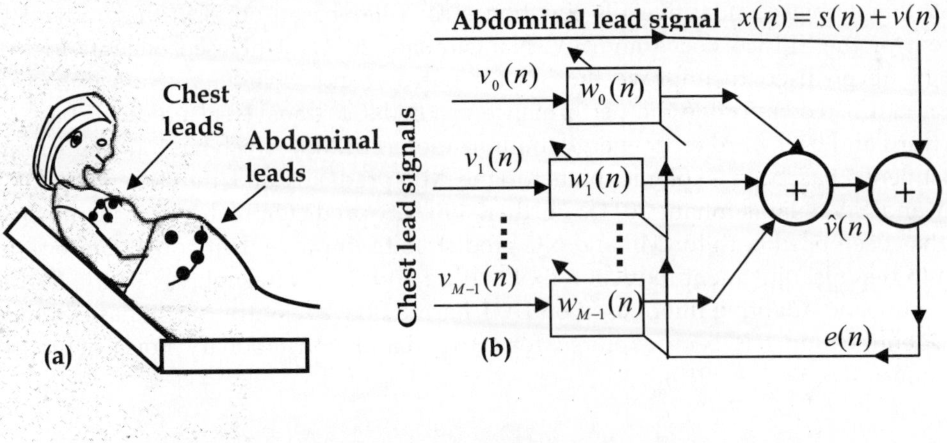

Fig. 17.13 Foetus ECG enhancement by adaptive filtering: (a) Electrode positioning and (b) Adaptive filtering.

The weight adaptation is

$$w_k(n+1) = w_k(n) + \alpha_k(n)e(n)v_k(n)$$

$$e(n) = x(n) - \hat{v}(n), \quad \hat{v}(n) = \sum_{k=0}^{M-1} w_k(n)v_k(n), \quad \alpha_k(n) = \frac{\mu}{P_k(n)}, \; 0 < \mu < 1$$

$$P_k(n) = \beta P_k(n-1) + (1-\beta)v_k^2(n), \quad 0 < \beta < 1$$

Once convergence is reached, in the error signal the mother's ECG component gets suppressed in an optimal sense hence, the fetus ECG will be enhanced so that a proper in formation about it is available.

17.13.2 Adaptive Cancellation of Power Line Interference from ECG

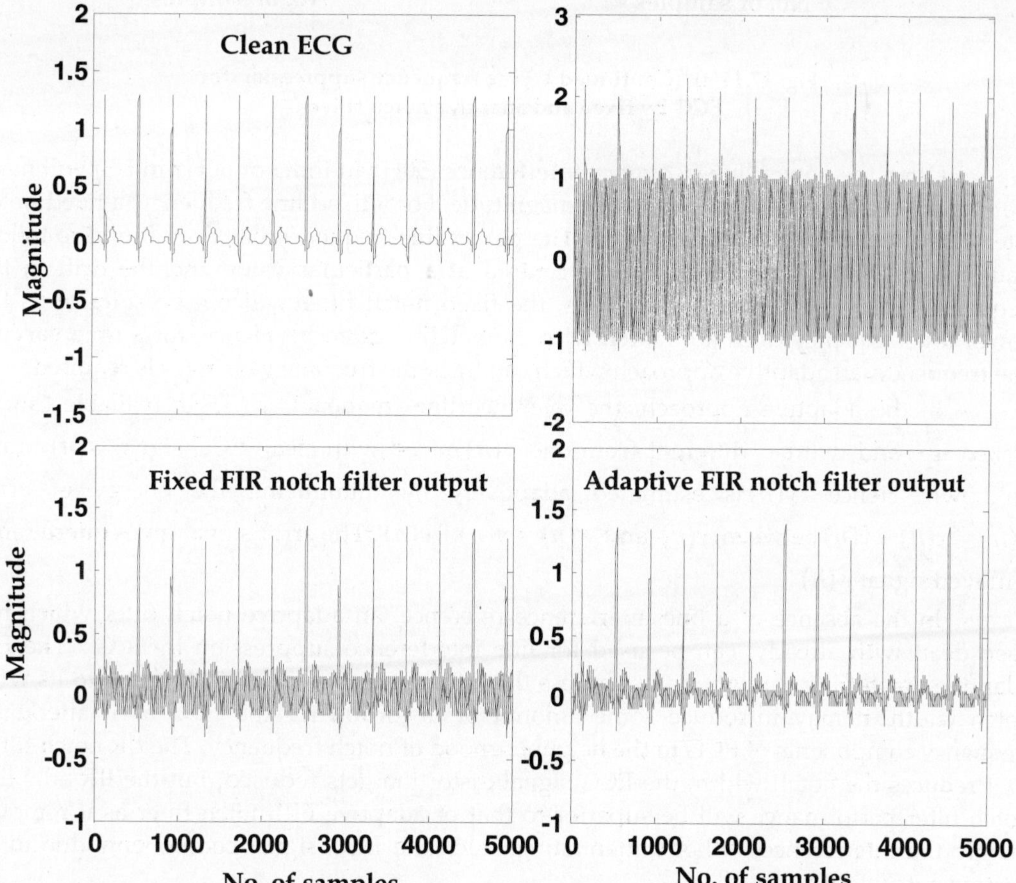

Fig. 17.14(a) Line frequency suppression for ECG by fixed and adaptive notch filters.

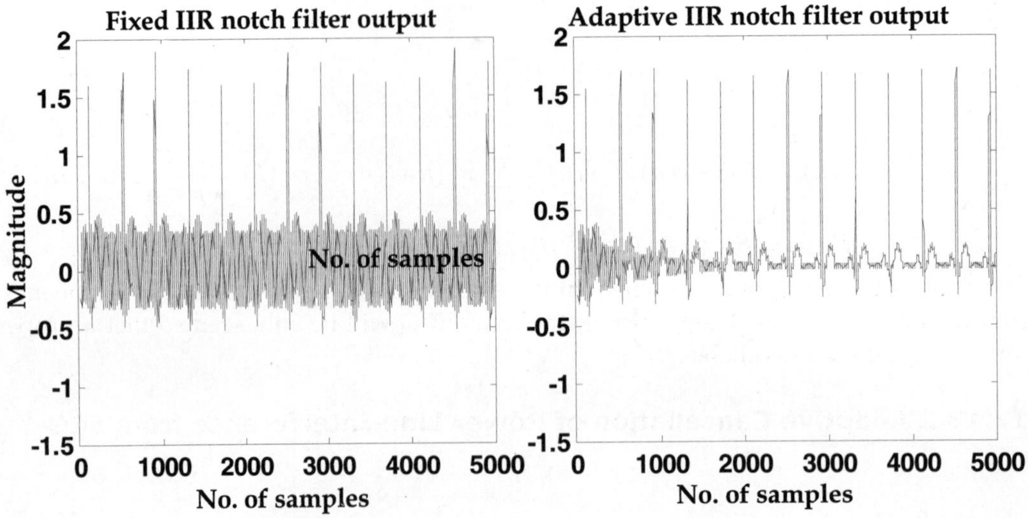

Fig. 17.14(a) (Continued) Line frequency suppression for ECG by fixed and adaptive notch filters.

Often the power line frequency interference (50Hz in India or 60Hz in US) will mask the P and T waves which are of smaller magnitude. For a fixed line frequency, a fixed notch filter at the line frequency can be used. The power line frequency has been found to follow Gaussian distribution and may not be centred at a particular value and the drift in the frequency may not be one sided. Thus, the fixed notch filter will not be suitable as the power line frequency does not match with the notch frequency. Hence, for a time varying line frequency, an adaptive approach which can track the frequency changes is required.

In the adaptive approach, the reference line frequency $v'(n)$ if available can be picked up and will be different from one $v(n)$ mixed with clean ECG signal $s(n)$ and is unknown. Hence, $v(n)$ is estimated adaptively by minimizing the energy of error $e(n) = x(n) - \hat{v}(n)$ between $\hat{v}(n)$ and $x(n) = s(n) + v(n)$. The error signal gives interference removed signal $\hat{s}(n)$.

In the absence of a line interference reference, an adaptive notch filter which has been dealt with already, can be used for line interference suppression in ECG. The FIR adaptive notch filter though may suppress the line frequency very well but due to its wide notch width, it may introduce some amount of distortion for the ECG as it attenuates frequency components of ECG in the neighbourhood of notch frequency. The IIR notch filter, as it reduces the notch width, the ECG signal distortion gets reduced. But the IIR adaptive notch filter performance will be superior to that of adaptive FIR notch filter as it removes mainly the interference without attenuating much the ECG signal components due to its narrow notch width.

The clean ECG sampled at $f_S = 360Hz$ and the ECG with a simulated line interference $v(n) = \cos[2\pi * \{60 + f(n)\} n / f_S]$ having time varying frequency $f(n)$ about 60Hz as $f(n) = 5\cos[2\pi * 0.0005n / f_S]$ are shown in the Fig.17.14(a). Further, the interference magnitude is such that with respect to clean ECG, it has an SNR = -12.5 dB.

The parameters used for adaptive filters are step size $\mu = 0.01$, forgetting factor $\beta = 0.999$, for power computation. For the notch filter

$$H(z) = \frac{1 + a_1(n)z^{-1} + a_2(n)z^{-2}}{1 + b_1(n)z^{-1} + b_2(n)z^{-2}},$$
$$a_1(n) = -2\cos[2\pi\{\omega_I(n)/\omega_S\}, a_2(n) = 1,$$
$$b_1(n) = \alpha a_1(n), b_2(n) = \alpha^2(n)$$

The coefficients $a_1(n)$ and $b_2(n)$ are adapted and hence, $b_1(n)$ is computed. For the adaptation, for $a_1(n)$, $\omega_I(n) = 2\pi * 60$ and for $b_2(n)$, $\alpha = 0.7$ are taken as initial values.

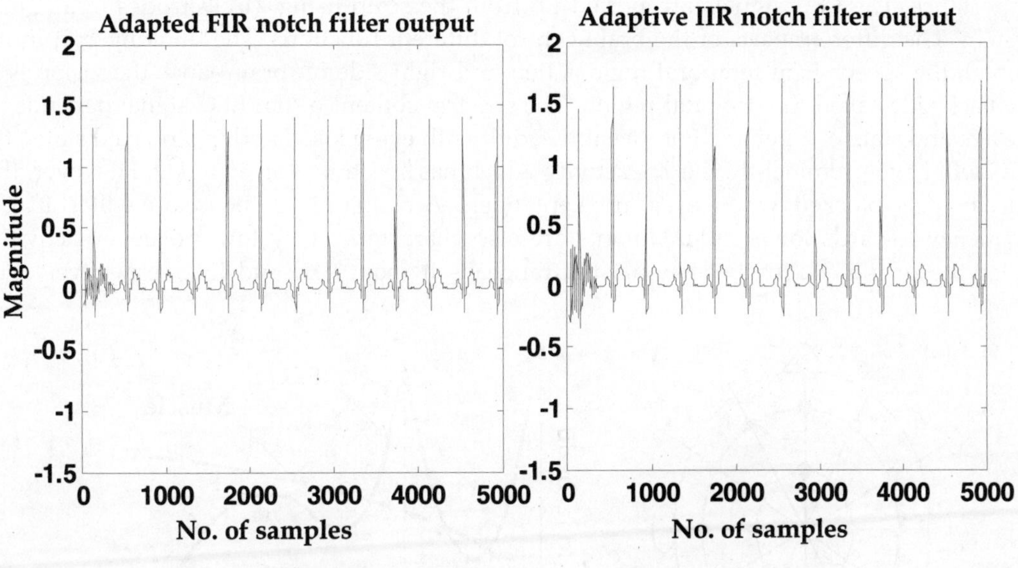

Fig. 17.14(b) Adaptive interference removal by FIR and IIR notch filters using sign error LMS algorithm.

The fixed FIR and IIR (with $\alpha = 0.7$) notch filters with notch at 60 Hz though reduce the interference magnitude but they are unable to bring out the ECG signal (specifically S and T waves) from the interference. As expected, the line suppression by fixed FIR notch filter is better than that by fixed IIR notch filter (Fig. 17.14(a)). Among the adaptive notch filters compared to FIR, the IIR very significantly reduces the interference by clearly bringing out the S and T waves also.

For sinusoidal frequency estimation in the presence of impulsive noise, the L_1 norm minimization is preferred and this leads to sign error LMS algorithm. The sign error LMS adaptation provides better results (S.C. Pei and C.C.Tseng,1993). With the above initialization with fixed step size equal to 0.001, the interference cancellation obtained is shown in the Fig. 17.14(b).

$$a_1(n+1) = a_1(n) - \mu \; sign[e(n)]\{x(n-1) - \alpha e(n-1)\},$$
$$b_2(n+1) = b_2(n) + \mu \; sign[e(n)]\{e(n-2)\}$$

It is clear that the sign error adaptive algorithm both FIR and IIR remarkably remove the interference compared to those of fixed notch filters. The sign algorithm results are even better than those of LMS algorithms both in terms of convergence rate and convergence error.

17.13.3 Suppression of Muscle Artefact (noise) from EEG

In EEG unlike the time domain features of ECG, the frequency content is of clinical importance. The EEG signals are picked up from the scalp using 21 electrodes in a (10-20) system. The different parts of the brain control different functions. The thinking part in the frontal, the speech is in temporal regions (left and right side of brain) and the vision is in the back side called the occipital region. The spectral content of the EEG signal depends on the age and state of a person. For a healthy adult with eyes closed and in a relaxed state , the EEG will have prominently the α - activity which has spectral peak at 10 Hz. However, this activity gets blocked when eyes are kept open. Any injury to the brain will shift the frequency towards lower side. Infants are also characterised by low frequency activities called the δ and θ activities having spectral peaks at about 1 Hz and 4Hz, respectively. For

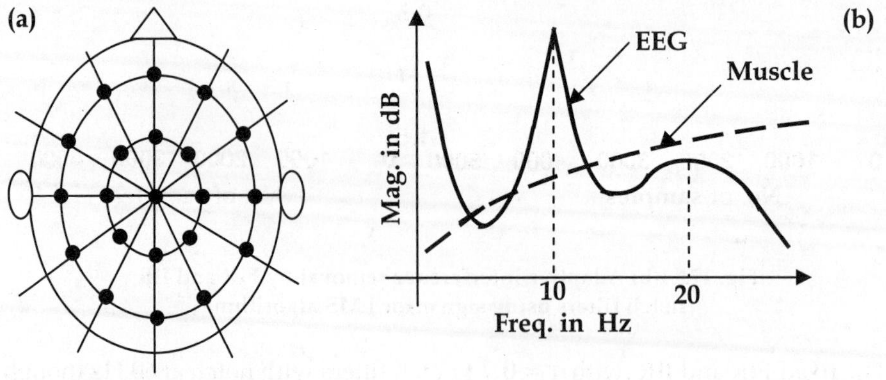

Fig. 17.15 (a) 10-20 EEG electrode system positions and (b) Typical EEG spectrum and muscle noise spectrum.

old persons or with mental activity, the β-activity having spectral peak at about 20 Hz will be prominent. Thus, the spectral content of the EEG signal can indicate the brain activity. Further, it is possible to compare the signals recovered from two hemispheres of the brain which will indicate the malfunction of the specific part of the brain. Thus, an EEG signal is characterised by δ, θ, α and β activities with corresponding spectral peaks of different magnitudes at different frequencies. The spectrum of an EEG signal can indicate functional disability of the brain. The EEG is stationary for about 10 seconds and is sampled at 100 Hz. The spectrum can be represented by an AR spectrum with about 15 coefficients. Hence, assuming 1.5 seconds of data is modeled by a 15th order AR model, a compression factor of 10 is achieved.

Similar to ECG, EEG also suffers from unwanted signals/noise called the artefacts. The major artefact for EEG is the muscle noise which is due to contraction of neck and scalp muscles which is common with tense subjects. Its magnitude may be many times larger than that of EEG and may affect EEG picked up by all the electrodes (21) and is very serious as it affects the processing of EEG. The muscle noise can be modeled as a bandpass activity with 70Hz centre frequency and 70Hz bandwidth. The muscle noise is an impulsive type of noise following Poisson distribution. The spectra of muscle and the EEG have maximum overlap around 30Hz and the muscle spectrum extends up to folding frequency, 50 Hz. Hence, muscle noise has two components, in-band (with in 30 Hz) and out-of-band component (30-50Hz). The muscle noise affects EEG above 14 Hz, i.e., the β-activity (20Hz). The out-of-band component is removed by a 4th order IIR Butterworth digital low-pass filter with a cut-off frequency equal to 35 Hz as EEG spectrum is confined to 30Hz. Compared to the in-band muscle noise, the EEG is of narrowband nature hence, by exploiting the difference in their correlation properties, the muscle noise can be suppressed from the muscle noise contaminated EEG using predictive filtering (S.V. Narasimhan, D. Narayana Dutt, 1996). The correlation length of the muscle noise, $v(n)$ is less compared to that of EEG, $s(n)$. For this, the low-pass filtered noisy EEG, $x(n) = s(n) + v(n)$ is used as the desired signal and its delayed version $x(n-D)$ is used as input to the adaptive predictor $W(n)$. The weight vector $W(n)$ is achieved by minimising the energy of the error $e(n)$ between $x(n)$ and the EEG estimate $\hat{s}(n)$.

$$e(n) = x(n) - \hat{s}(n), \quad \hat{s}(n) = \sum_{k=1}^{M} w_k(n) x(n-k-D),$$

Only for the EEG component of $x(n)$, there will be correlation with the delayed signal $x(n-D)$. However, for the noise component there will not be any correlation between $v(n)$ and $v(n-D)$. Hence, the adaptive filter output estimates only the EEG component $\hat{s}(n)$ and not the noise component $v(n)$ as $W(n) = R_{NN}^{-1} P_N$, where R_{NN} is the autocorrelation of the input $x(n-D)$ and P_N is the cross-correlation between $x(n)$ and $x(n-D)$. This is due to the fact that P_N will have nonzero value only for EEG component $s(n)$ and it is zero for noise component $v(n)$.

$$\hat{s}(n) = \sum_{k=1}^{M} w_k(n)s(n-k-D) + \sum_{k=1}^{M} w_k(n)v(n-k-D)$$

Therefore, the second term on the R.H.S of this equation is zero and

$$\hat{s}(n) = \sum_{k=1}^{M} w_k(n)s(n-k-D)$$

The adaptation of the weights of the predictive filter is

$$w_k(n+1) = w_k(n) + \alpha(n)e(n)x(n-k-D), \qquad \alpha(n) = \frac{\mu}{E(n)}, \quad 0 < \mu < 1$$

$$E(n) = \beta E(n-1) + (1-\beta)x^2(n-D), \quad 0 < \beta < 1$$

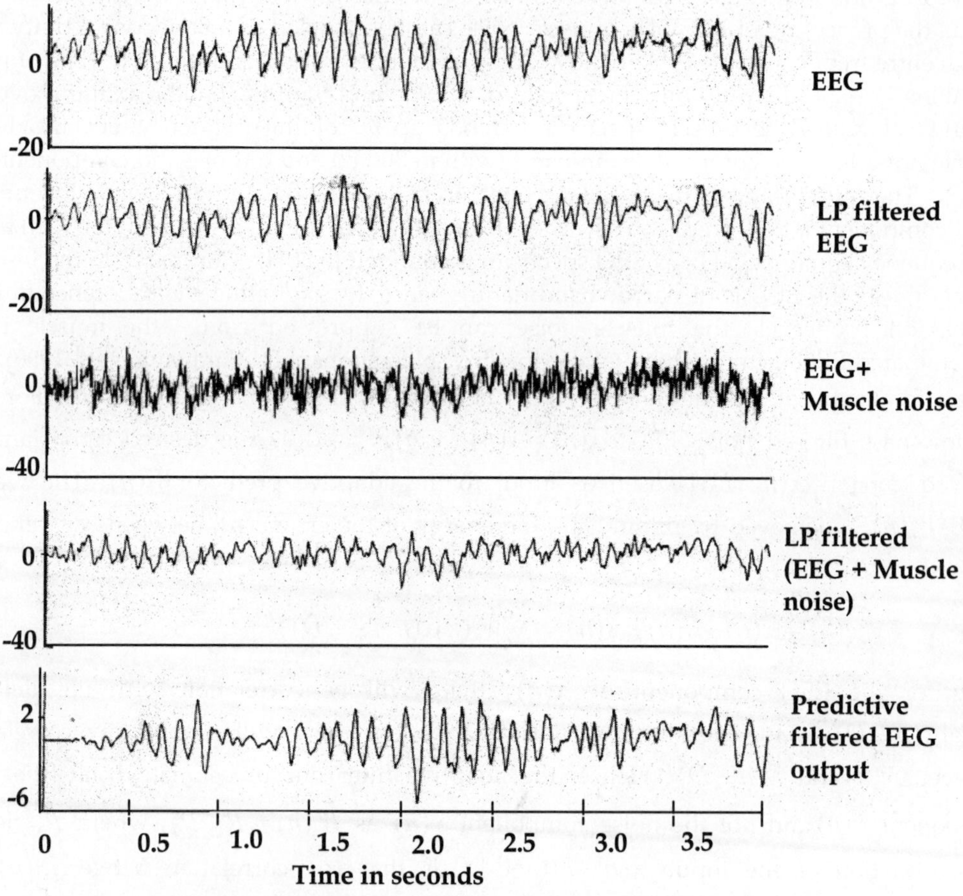

Fig. 17.16 (a) Muscle noise suppression from muscle noise contaminated EEG by predictive filtering.

The noise suppression for a simulated EEG with a sampling rate of 200 Hz is shown in the Fig.17.16 (a). The low-pass filter does not distort the signal as its output is same as the input. The muscle noise simulated with a sampling rate of 170 Hz is added to the EEG to give a SNR of -2.5 dB (Fig. (17.16. (a)). The low-pass (LP) filtered EEG plus muscle noise signal is shown indicates the effect of out-of-band noise removal beyond 35 Hz. But still this signal is of different nature from that of the clean EEG due to the presence of inband muscle noise. A 12th order adaptive predictor is used with a delay $D = 20$ to remove the in-band noise. The muscle noise cancelled signal after convergence is similar in nature to that of original EEG.

For EEG as the spectral information is of importance, to find the improvement achieved by the predictive filtering on it, the original EEG, the muscle artefact contaminated EEG, the low-pass filtered (EEG + muscle noise) and the adaptive filtered output (EEG estimate); are subjected to Burg's block spectral estimation. For this purpose, a 15th order prediction filter is applied to the 512 samples of the predictive filtered output and to ensure that the convergence of the adaptive filter has taken place, the 512 samples are considered after leaving the initial 240 samples.

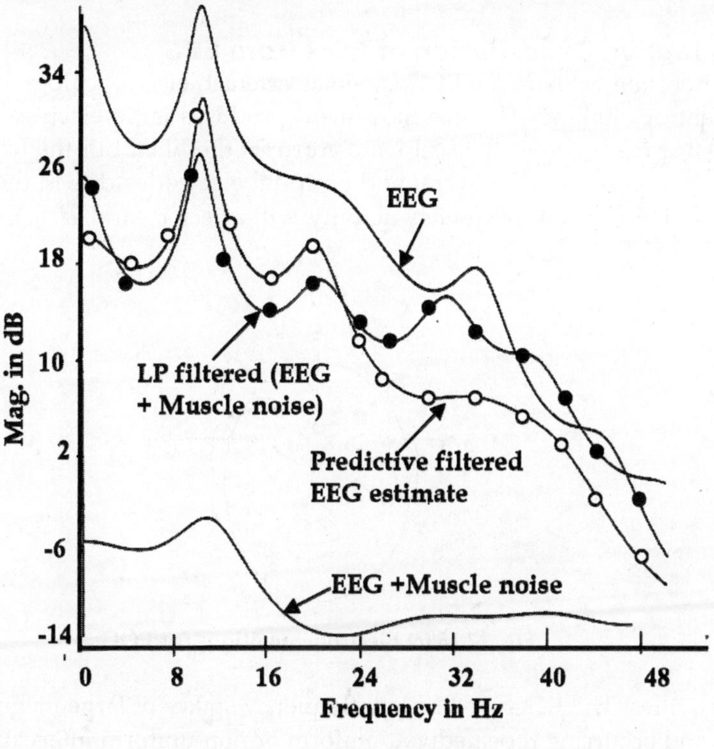

Fig. 17.16(b) Burg spectrum of predictive filtered Muscle noise contaminated EEG in comparison with other signals.

For the signals at other stages of filtering, the corresponding data blocks of 512 points are used. Fig. 17.16(b) shows the respective Burg's spectra. It is seen that for the muscle artefact contaminated EEG signal, the Burg's spectrum does not indicate the peak corresponding to β-activity at all and does not indicate the spectral peaks but only shows a broad peak roughly around the α-activity frequency range. Thus, even the peak position for the α-activity is not accurate.

For the signals at other stages of filtering, the corresponding data blocks of 512 points are used. Fig. 17.16(b) shows the respective Burg's spectra. It is seen that for the muscle artefact contaminated EEG signal, the Burg's spectrum does not indicate the peak corresponding to β-activity at all and does not indicate the spectral peaks but only shows a broad peak roughly around the α-activity frequency range. Thus, even the peak position for the α-activity is not accurate. For the low-pass filtered signal, though there is a peak in the spectrum corresponding to β-activity, this peak is almost at the same level as that of noise peak. But for the adaptive filtering, all the peaks are clearly brought out at reasonably proper positions by the Burg's method and specifically β-activity peak which is affected most by the muscle noise brought out well which is also much above the noise level. This indicates that the predictive filtering enables in getting proper spectral information of the EEG when it is contaminated with muscle noise.

17.13.4 Adaptive Cancellation of EOG from EEG

The other important artefact for EEG is ocular artefact/electrooculogram or (EOG) due to eye movement or blinking. The eye movement produces large electrical potentials around the eye which spreads across the scalp and corrupts the EEG. But the frontal electrodes get significantly compared to the central and occipital electrodes due to their proximity. The ocular artefact being a low frequency activity will affect δ and θ activities compared to

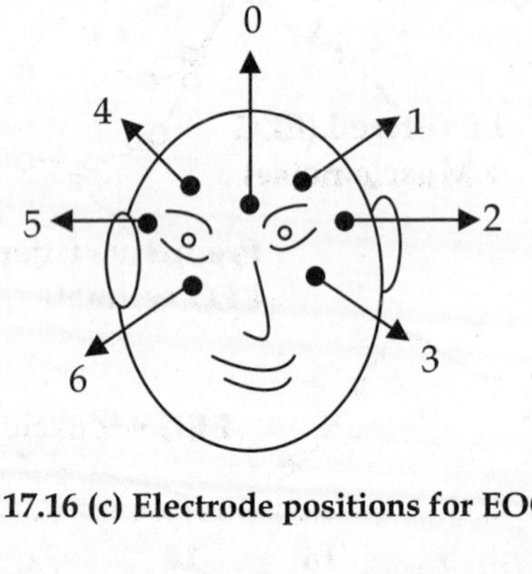

Fig. 17.16 (c) Electrode positions for EOG.

α and β-activities. The EOG are like low frequency spikes of large magnitude compared to that of EEG and occurring repeatedly at uniform or non-uniform intervals depending on the eye movement/blinking nature. Thus, normal EEG due to EOG appears to be discontinuous as EEG in between separated by EOG spike. This low frequency spiky signal is not desirable. Further, the patients with frontal tumour which are of low frequency nature are similar to EOG and this makes both clinical and computer analysis tedious. For the removal of this artefact, the reference is picked up by placing electrodes in and around each eye

(Fig. 17.16(c)). Using these reference signals $v_k(n), k = 0, 1, \ldots, M-1$ as inputs and the noisy EEG $x(n) = s(n) + v(n)$ as the desired signal, the component of EOG $v(n)$ in the noisy EEG $x(n)$ is estimated as $\hat{v}(n)$ by the adaptive filter $W(n)$ by minimising the energy of error $e(n)$ (E.C. Ifeachor and B.W. Jervis, 1993).

$$e(n) = x(n) - \hat{v}(n), \qquad \hat{v}(n) = \sum_{k=0}^{M-1} w_k(n) v_k(n)$$

In the Fig. 17,16 (c), $M = 7$. The weights of the adaptive filter are updated by

$$w_k(n+1) = w_k(n) + \alpha_k(n) e(n) v_k(n),$$

$$\alpha_k(n) = \frac{\mu}{P_k(n)}, \quad 0 < \mu < 1$$

$$P_k(n) = \beta P_k(n-1) + (1-\beta) v_k^2(n), \quad 0 < \beta < 1$$

Compared to the performance of linear adaptive filter, the second order Volterra nonlinear adaptive filter performs better in cancelling the EOG from EEG.

17.13.5 Adaptive Filtering of Evoked Potentials

The evoked potentials (EP) can be visual, auditory and somatosensory responses obtained by giving light (optical), auditory and muscle excitation as stimulus to eyes, ears and muscle of the part of the body, respectively and the response is picked up by a specific electrode on the scalp. The background on going EEG acts as noise for the evoked response. Generally, as the SNR will be poor, an improvement is essential for the evoked potential signal analysis. The evoked response signals are assumed to occur synchronously for all the repeated excitations and to improve the SNR, ensemble averaging is used. The averaging only suppresses the noise as the signal remains same in all ensembles/trials. As the SNR is poor and also for N trials the improvement in SNR is by a factor $\sqrt{N}$, the averaging may involve few hundreds to thousand ensembles to arrive at a reasonable SNR. In the process of getting the required number of ensembles, the properties of the part of the human body under investigation may deteriorate due to strain resulting in changes in the inherent signal in the ensembles and the averaging of such ensembles does not provide its true characteristics. Hence, it is important to achieve the signal enhancement of required SNR with a minimum number of ensembles.

To utilize adaptive filtering, two successive ensemble members $x_k(n)$ and $x_{k+1}(n)$ are taken as primary and reference signals, respectively (Fig.17.16(d)). The reference forms the input to the adaptive filter and the primary forms the desired signal. For the adaptive filter, a tapped delay line filter with M weights are used. Further, to take care of the DC component of the EP, a bias weight $w_{dc}(n)$ with a constant input unity is added to the adaptive filter output. The desired signal $x_k(n)$ and the output $y_k(n)$ result in the error $e_k(n) = x_k(n) - y_k(n)$ which is used for the adaptation of the weight vector $W(n)$ and the

DC bias weight $w_{dc}(n)$. For several such pairs, the adaptive filter outputs which provide the estimates of EP, are averaged as

$$y(n) = \frac{\sum_{i}^{k} y_i(n)}{[(k+1)/2]}, \quad i = 1, 3, \ldots, k$$

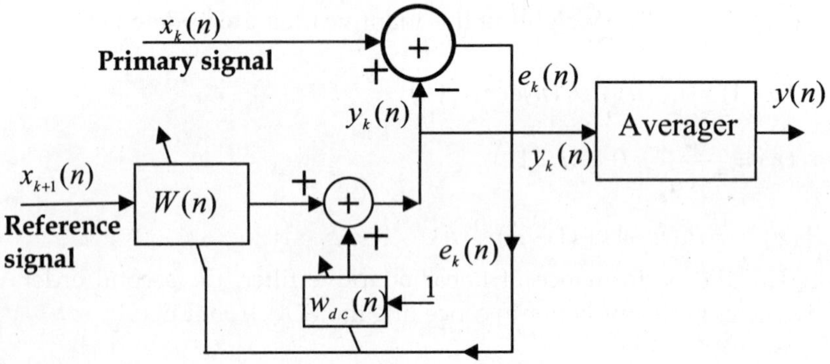

Fig. 17.16 (d) Adaptive filter based averaging for EP.

For example, if $k = 5$, then the pairs (1, 2), (3, 4) and (5, 6) form the pairs for the desired signal and input for the adaptive filter and the outputs $y_1(n)$, $y_3(n)$ and $y_5(n)$ have to be averaged and hence the averaging factor is $(k+1)/2 = (5+1)/2 = 3$. Therefore, the averaged output is

$$y(n) = [y_1(n) + y_3(n) + y_5(n)]/3$$

For a simulated EP signal, a one and a half cycle of sinusoid in which the half cycle is an attenuated version of full cycle is used. Further, to this EP signal, a filtered noise which is similar to actual EP noise is added to get the required SNR. For such a EP signal, it has been shown that the mean square error obtained by averaging the adaptive filter outputs (with sufficiently large filter length say 24) with 10 ensembles is about half that of mere averaging with 60 ensembles. This indicates that the number of ensemble members required has been reduced by a factor of 6 and the strain on the human body part is reduced by the same factor. This reduction in strain in the part of the body makes its response consistent as it will not change due to absence of extra strain and this results in a proper EP output. This method works well for auditory and somatosensory data bring out the amplitude and latency of the primary wave (largest peak with shortest latency (delay)).

An alternate approach is to use the accumulated average as the reference /input to the adaptive filter given by

$$r_k(n) = \frac{1}{k}\sum_{i=1}^{k} x_i, \qquad i = 1, 2, \ldots, k$$

and the desired signal is $x_k(n)$. The output of the adaptive filter $y(n) = y_k(n)$. The use of accumulated average as reference input to the adaptive filter improves the SNR of the reference signal. The adaptive filter operates on a better reference on consideration of successive ensemble members. This method for brainstem auditory EP, even with 256 trials provides an output similar to that obtained by normal averaging of 1700 trials. Thus, the combination of adaptive filtering and averaging is significantly beneficial for EP processing as it reduces the number of trials required for the same SNR improvement achieved by normal averaging (N.V. Thakor, 1987).

17.14 OPTIMAL AND ADAPTIVE BEAMFORMERS

A single symmetrically spherical element antenna/sensor generally provides omni directionality and receives radiation from all directions. To receive radiation from a particular direction, an array of antenna/sensors is used. If the antenna/sensors are located at a uniform spacing d along a straight line, it is called a linear array. If the wavefront is coming from a direction inclined at an angle θ (less than 90°) to the array with a speed c, then the sensor on one end (top) of the array (Fig. 17.17 (a)) will receive the signal first compared to those which are on the other end (bottom) and the time delay τ involved for successive elements to receive the signal is $\tau = (d \sin\theta)/c$. Hence, by delaying the signal from sensors which receive signal earlier compared to its successor, all the received signals can be synchronized in time and their summation amplifies the signal strength by a factor equal to the number of sensor elements. Such a summation rejects the noise picked up by the element antennas (from other directions) as they will not be in synchrony. Such a sensor array system is called *delay and sum beamforming*.

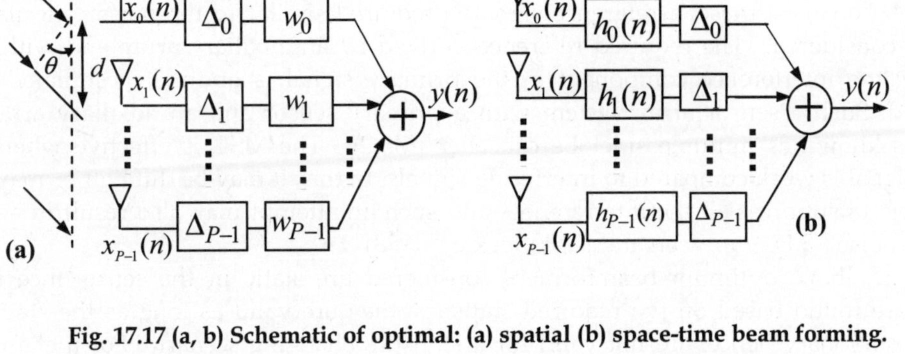

Fig. 17.17 (a, b) Schematic of optimal: (a) spatial (b) space-time beam forming.

The disadvantage of such a beam former is that a large number of sensors are required to improve the SNR and each doubling of sensors result in only additional 3dB increase in SNR provided the incoming signals are completely uncorrelated between the sensors and with the desired signal. The linear array response is characterised by a mainlobe and sidelobes whose level is sufficiently close to mainlobe (just -13dB below mainlobe level). Hence, the interfering / jamming signals can be received through the sidelobes of the array which can distort the desired signal from the look direction. But the delay and sum beam former cannot place any nulls in the jammimg /sidelobe directions. In addition to delaying sensor element signals, if they are weighed properly by the weight vector and then summed, it will not only enhance the beam shape in the desired direction but also reduces the sidelobe levels improving the rejection in other directions. This type of weighing is similar to the design of a FIR filter. Thus, there can be necessity to reduce/cancel the sidelobes (sidelobe canceller) depending up on the application. Such a beamformer which only does spatial filtering is known as narrowband beamformer. This beamformer is independent of the data and data statistics. In order to select signals in specific frequency bands by different sensor elements, they have to be passed through $h_0(n), h_1(n), \ldots , h_{P-1}(n)$ different proper linear FIR filters prior to delaying. Such a beamforming which not only does spatial filtering but also temporal filtering (frequency selection) is known as *space-time or broadband beamformer*. The spatial filtering provides additional discrimination to cancel the signals occupying the frequency band which is not possible by fixed linear filters like low-pass, high-pass, band-pass and band-reject filters. (But for the signals with overlapping spectra the mean square error minimization enables their optimal separation.) The passband characteristics of linear filters should correspond to that of the desired spectrum. The performance of such a beamformer is optimized by using different criteria (G.J. Miao, 2007). In the minimum mean square error criterion, the weights are chosen so that the energy of the difference between the desired output and actual array output is minimised. In the linearly constrained minimum variance algorithm (LMCV), the beamformer output power is minimized so that it contains minimum contribution due to noise and signals from other directions other than the desired direction. In the maximum signal to noise/interference power ratio criterion, the weights are chosen to maximize the signal to noise/interference power ratio and this corresponds to finding maximum eigenvalue.

The interferences to the primary sensor/main channel can be from different directions. To cancel these, a reference signal is required similar to the noise cancellation scenario, considered. The required reference derived by an auxiliary array and with this reference, the interference component in the primary signal is optimally estimated and subtracted. Such a sensor array system with a primary sensor and an auxiliary array of sensors is known as multiple sidelobe canceller (MSC). The MSC is effective when the desired signal is weak compared to interfering signals. In this, it may be difficult or may not be possible to suppress all the interferences and such an attempt may also result in a gain for white noise (B.D. Van Veen and K.M. Buckley, 1988).

The above optimum beamformers considered are static in the sense once their weights are found based on the required statistics, they are valid as long as the statistics used are stationary hence, are not valid for a statistics changing scenario. For a changing scenario, the weights of the beamformer have to be *adapted* to maintain the optimal performance. For this purpose, the adaptive algorithms like LMS and its variations or recursive least square algorithms are used and from computational simplicity point of view, the LMS algorithm is preferred. The Griffiths and Frost beamformers based on the LMS algorithm are very popular beamformers. Further, most of the beamformers can be treated as different versions of the LCMV beamformer.

17.14.1 Adaptive Griffiths Beamformer

The Griffiths beamformer achieves its optimal performance by minimising the energy of the error between the output of the beamformer and the desired signal. In this beamformer (Fig. 17.17(c)), after the bulk delays Δ_i the time aligned signals are available and hence they steer the receiving beam in the look direction. The delays $\Delta_i, i = 0, 1, \ldots, P-1$ are decided by the array geometry. The time aligned signals $x_i(n-\Delta_i)$ are fed to tapped delay line filters $w_i^G(n,k), k = 0, 1, \ldots,, M-1$. Further, based on the delayed inputs, $x_i(n-\Delta_i)$, the desired signal $d(n)$ and output $y(n)$, the Griffiths gradients are given by

$$G_{i,k}(n) = E[d(n)x_i(n-\Delta_i-k)] - x_i(n-\Delta_i)y(n)$$
$$G_{i,k}(n) = P_{i,k} - x_i(n-\Delta_i)y(n), \quad P_{i,k} = E[d(n)x_i(n-\Delta_i-k)]$$

Here, $P_{i,k}$ the first term on the R.H.S is the expected value between $d(n)$ and $x_i(n-\Delta_i-k)$ and is a crosscorrelation constant for the weight $w_i^G(n,k)$ that can be computed and used for adaptation of the weight by LMS algorithm (B. Widrow and S.D. Stearns, 1985, chapter 14)

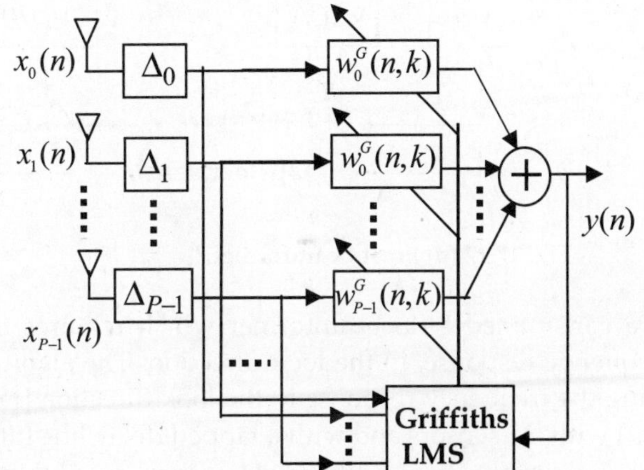

Fig. 17.17 (c) Griffiths LMS beamformer.

$$w_i^G(n+1,k) = w_i^G(n,k) + \alpha(n)[P_{i,k}(n) - x_i(n-\Delta_i - k)],$$

$$\alpha(n) = \mu / E_{xi}(n),\ 0 < \mu < 1,\ \ E_i\,(n+1) = \beta E_i(n) + (1-\beta)\left|x_i\,(n-\Delta_i)\right|^2,$$

$$i = 0, 1, \ldots, P-1, \quad k = 0, 1, \ldots, M-1$$

Thus, for Griffiths beamformer, the actual desired signal is not required and only its cross-correlation with the different inputs corresponding to each weight $w_i^G(n,k)$ are sufficient. The Griffiths algorithm provides unbiased estimates of the weight vectors due to the crosscorrelation between the noisy input and the desired signal which suppress the noise.

17.14.2 Frost Adaptive Beamformer

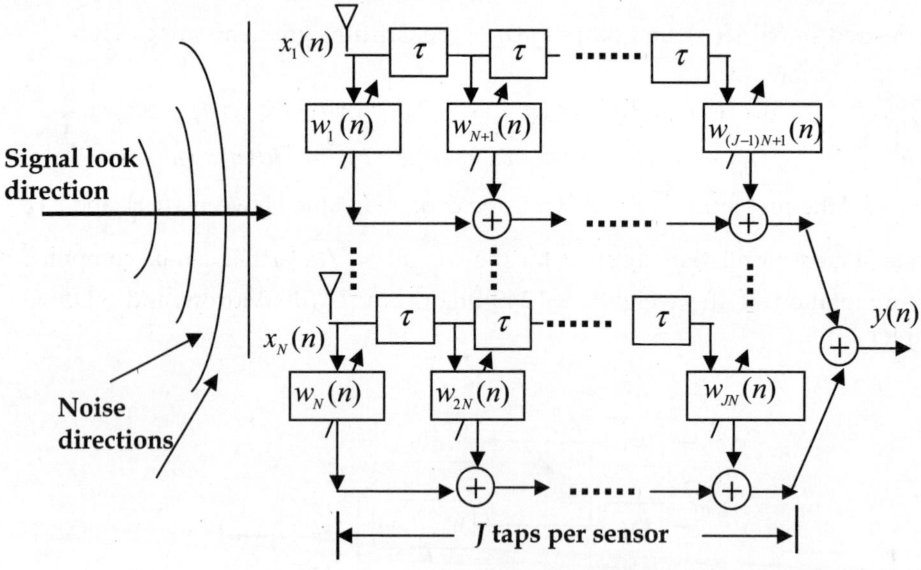

Fig. 17.17 (d) Frost beamformer.

The Frost beamformer is a constrained LMS beamformer which requires the direction of arrival and the desired frequency response, in the look direction. The algorithm minimizes the output power preserving the frequency response in the look direction [Frost, 1972]. For a beamformer (Fig. 17.17(d)) with N sensors and with a tapped delay line filter with J taps per sensor, requires in total NJ weights to be adapted. However, out of these NJ weights, only J of them determine the look direction frequency response and the remaining $J(N-1)$ weights minimize the array total output power. Such minimization provides best possible reduction of the power of the non-look direction signals which play the role of noise and are assumed to be uncorrelated with the signal in the look-direction. As in delay sum beamformer, proper delays after each sensor steer the beam in the look direction and these delays are not shown in the figure. The tap outputs at n^{th} sample are given by

$$X^T(n) = [x_1(nT), x_2(nT), \ldots x_{NJ}(nT)]$$

The time interval T may be a multiple of delay τ between the taps of the delay line.

$$X(n) = S(n) + V(n)$$

The component $S(n)$ is due to look direction and $V(n)$ is due to other than look direction and noise. Further,

$$S^T(n) = [\underline{s(nT)\ldots s(nT)},\ \underline{s(nT-\tau),\ldots s(nT-\tau)}\ldots\ldots \underline{s(nT-(J-1)\tau)\ldots s(nT-(J-1)\tau)}]$$

Each underlined set has N elements and

$$V^T(n) = [v_1(nT), v_2(nT), \ldots v_{NJ}(nT)]$$

The weight vector at each tap is given by

$$W^T(n) = [w_1(n), w_2(n), \ldots w_{NJ}(n)]$$

$$E\left[X(n)X^T(n)\right] = R_{XX},\quad E\left[V(n)V^T(n)\right] = R_{VV},\quad E\left[S(n)S^T(n)\right] = R_{SS}$$

As the signal in look direction is uncorrelated with those of non-look direction noises, $E\left[V(n)S^T(n)\right] = 0$. The array output at n^{th} sample is given by

$$y(n) = W^T X(n) = X^T W$$

$$E\left[y^2(n)\right] = E\left[W^T X(n) X^T W\right] = W^T R_{XX} W$$

For the array matched to the look direction, the plane wavefront will be parallel to the line of sensors (Fig. 17.17(d)) and produce equal amplitude signals at any taps in a vertical column. But noise arriving in other directions, will not generate equal amplitude signals at any taps in a vertical column. Thus, for the signal in the look direction, the beam former is equivalent to a single tapped delay line filter with its each weight equal to the sum of the weights in the vertical column of the beamformer (Fig. 17.17d). This implies that the summation of weights in the equivalent tapped delay line has to be chosen to get the desired frequency response in the look direction. Therefore the constraint that the sum of the weights of the taps of the jth column add to a chosen constant f_j is

$$y(n) = c_j^T W = f_j,\quad j = 1, 2, \ldots, J$$

The NJ dimensional c_j is

$$c_j^T = [\underbrace{0\ldots0}_{N},\ldots\ldots\ldots,\underbrace{0\ldots0}_{N},\ \underbrace{1\ldots1}_{j^{th}\ N\ elements},0\ldots0\ldots\ldots0\ldots0]$$

The constraint matrix C is defined as

$$C = \left[\, c_1 \ldots\ldots c_j \ldots\ldots c_J \,\right]$$

This is of dimension J columns and NJ rows. Defining the equivalent tapped delay line in look direction as F with J weights as

$$F^T = \left[\, f_1 \ldots\ldots f_j \ldots\ldots f_J \,\right]$$

The constraint becomes

$$C^T W = F$$

Therefore, the optimization can be expressed as minimization of $W^T R_{XX} W$ subject to the constraint $C^T W = F$. This is solved by Lagrange multipliers approach considering the cost function as

$$\varepsilon(W) = \frac{1}{2} W^T R_{XX} W + \lambda^T (C^T W - F)$$

The gradient of $\varepsilon(W)$ with respect to W,

$$\nabla_W \left[\, \varepsilon(W) \,\right] = R_{XX} W + \lambda\, C$$

Therefore,

$$W_{opt} = -R_{xx}^{-1} \lambda\, C$$

Substituting this in the constraint equation $C^T W = F$,

$$C^T W_{opt} = F = -C^T R_{xx}^{-1} \lambda\, C$$

Therefore,

$$\lambda = -\left[\, C^T R_{xx}^{-1} C \,\right]^{-1} F$$

Therefore,

$$W_{opt} = -R_{xx}^{-1} \lambda\, C = R_{xx}^{-1} C \left[\, C^T R_{xx}^{-1}\, C \,\right]^{-1} F$$

$$W_{opt} = R_{xx}^{-1} C \left[\, C^T R_{xx}^{-1}\, C \,\right]^{-1} F$$

This is the constrained optimum weight vector for the beam former which is filter both in space and frequency. The output of the beamformer is given by

$$y_{opt}(n) = W_{opt}^T X(n)$$

By making different suitable choice for F, a variety of optimal beamformers can be derived. Such an optimal beamformer valid for a stationary environment and requires the knowledge of R_{xx}. Further, the inverse of R_{xx} demands high computational load of the order of cube of number of weights used. But for the LMS algorithm, not only R_{xx} is not required but also the computations are reduced to the order of number of weights. For this, the weight vector is initialized as $W(0) = C\left[C^T C\right]^{-1} F$. The adaptation of weight vector by of steepest descent is

$$W(n+1) = W(n) - \mu \nabla_W \left[\varepsilon(W(n))\right]$$

But

$$\nabla_W \left[\varepsilon(W(n))\right] = R_{XX}W(n) + \lambda\ (n)C$$

$$W(n+1) = W(n) - \mu[R_{XX}W(n) + C\lambda\ (n)]$$

The weight vector $W(n+1)$ is chosen to satisfy the constraint,

$$F = C^T W(n+1) = C^T W(n) - \mu[C^T R_{XX} W(n) + C^T C\lambda\ (n)]$$

$$F = C^T W(n) - \mu[C^T R_{XX} W(n) + C^T C\lambda\ (n)]$$

$$C^T W(n) - \mu C^T R_{XX} W(n) - F] = \mu C^T C\lambda\ (n)$$

$$\mu\ \lambda\ (n) = [C^T C]^{-1}[C^T W(n) - \mu C^T R_{XX} W(n) - F]$$

Substituting for $\mu\ \lambda$ in

$$W(n+1) = W(n) - [\mu R_{XX} W(n) + C\mu\ \lambda\ (n)]$$

$$= W(n) - \mu R_{XX} W(n) - C[C^T C]^{-1}[C^T W(n) - \mu C^T R_{XX} W(n) - F]$$

$$W(n+1) = W(n) - \mu[I - C\left\{C^T C\ \right\}^{-1} C^T]R_{XX} W(n) + C\left\{C^T C\right\}^{-1}[F - C^T W(n)]$$

If $\tilde{F} = C\left\{C^T C\right\}^{-1} F$, the NJ dimensional vector and $P = I - C\left\{C^T C\ \right\}^{-1} C^T$ the $NJ \times NJ$ matrix, then

$$W(n+1) = W(n) - \mu P R_{XX} W(n) + \tilde{F} - C\left\{C^T C\right\}^{-1} C^T W(n)$$

$$= W(n) - C\left\{C^T C\right\}^{-1} C^T W(n) - \mu P R_{XX} W(n) + \tilde{F}$$

$$W(n+1) = [I - C\{C^T C\}^{-1} C^T]W(n) - \mu P R_{XX} W(n) + \tilde{F}$$

Substituting for $P = I - C\{C^T C\}^{-1} C^T$,

$$W(n+1) = PW(n) - \mu P R_{XX} W(n) + \tilde{F}$$
$$W(n+1) = P[W(n) - \mu R_{XX} W(n)] + \tilde{F}$$

For the LMS algorithm, $R_{XX} \approx X(n)X^T(n)$ and using $y(n) = X^T(n)W(n)$,

$$W(n+1) = P[W(n) - \mu X(n)X^T(n)W(n)] + \tilde{F}$$

$$W(n+1) = P[W(n) - \mu\, y(n)X(n)] + \tilde{F}$$

with $W(0) = \tilde{F}$. The vector F is chosen to have all one's in the entries for the desired signal not to have any distortion. The step size μ should be less than the reciprocal of the eigenvalue of the sample correlation matrix of the received data vector $X(n)$.

The time domain equations for the weights are given by

$$w_1(n+1) = w_1(n) - \mu y(n)x_1(n) - \frac{1}{N}\sum_{k=1}^{N} w_k(n) - \mu y(n)x_k(n) + \frac{f_1}{N}$$

$$\vdots$$

$$w_N(n+1) = w_N(n) - \mu y(n)x_N(n) - \frac{1}{N}\sum_{k=1}^{N} w_k(n) - \mu y(n)x_k(n) + \frac{f_1}{N}$$

$$w_{N+1}(n+1) = w_{N+1}(n) - \mu y(n)x_{N+1}(n) - \frac{1}{N}\sum_{k=N+1}^{2N} w_k(n) - \mu y(n)x_k(n) + \frac{f_2}{N}$$

$$\vdots$$

$$w_{2N}(n+1) = w_{2N}(n) - \mu y(n)x_{2N}(n) - \frac{1}{N}\sum_{k=N+1}^{2N} w_j(n) - \mu y(n)x_k(n) + \frac{f_2}{N}$$

$$\vdots$$

$$w_{JN}(n+1) = w_{JN}(n) - \mu y(n)x_{JN}(n) - \frac{1}{N}\sum_{k=(J-1)N+1}^{JN} w_k(n) - \mu y(n)x_k(n) + \frac{f_J}{N}$$

The computational complexity of the Frost algorithm described is quite low and can be implemented on a single chip DSP. But its performance is sensitive to array deformations.

The generalised sidelobe canceller (GSC) is an alternate formulation of LCMV and simplifies the LMCV beamformer implementation. The reason for this is that the GSC basically converts the constrained minimization to an unconstrained one. The maximum SNR beamformer cannot be realized in the standard adaptive filter configuration having the primary and reference signals. But the maximum SNR and LMCV beamformers become equivalent, for a narrowband desired signal.

The beamforming has applications in wireless/ satellite communication, radar (like phased array radar, synthetic aperture radar and air traffic control) seismology (for earth crust mapping and oil exploration), imaging (ultrasonic, optical and tomography) speech enhancement in an ambient noise environment.

17.15 CURRENT TOPICS

Few current topics of research on adaptive filters are briefly indicated.

17.15.1 State Space LMS Algorithm for Single Channel and for Multi-channels

The LMS adaptive algorithm can be generalized by incorporating linear time-varying state-space model of the underlying environment and this is known as state-space LMS (SSLMS) (M.B. Malik, M. Salman, 2008). This facilitates significant improvement in its tracking performance over the standard LMS. In control applications, the use of SSLMS for state estimation is common. The performance of SSLMS depends on the model and time-varying nature for which it is applied. The SSLMS with adaptive memory and variable step-size can handle such scenarios. Further, using adaptive memory, variable optimum step size is achieved by gradient method which effectively minimizes the mean square value of the prediction error. This SSLMS algorithm provides significant performance improvement and it can be modified to many application scenarios.

The use of State space model (SSM) provides a degree of flexibility in model parameterization and this improves the performance significantly. This is due to the fact that SSM facilitates the use of numerically robust methods like eigenvalue decomposition and well-conditioned matrices to obtain desired results. However, the transfer function models achieve the final results only by fitting the polynomials. The state space approach provides estimates of the model order from the data even though it is not known a priori. However, the transfer function lacks this advantage. The SSM uses minimum parameters and this enables to avoid ill-effects of finite precision errors for implementation on digital signal processors. Thus, the reduction in minimum number of parameters by use of SSM not only reduces the computational load but also make the computations numerically exact. These benefits are due to replacing the single input single output FIR models by multiple input multiple output form in SSM . Instead of having $(p \times m)$ FIR models for a system with p inputs and m outputs, SSM requires only one model *having a common state* (G. Nijsse, M.

Verhaegen, B. De Schutter, D. Westwick, and N. Doelman, 1999). Further, though SSM is different from IIR model, both are similar in having infinite impulse response. The main feature of SSM is that it uses minimum number of coefficients but still provides same I/O representation of FIR model which uses large number of coefficients. The reduction in the number of coefficients and computational reduction can be of the order of 90% and 80%, respectively. There is a direct relation between SSM and FIR model but; finding an SSM from FIR model is not an identification type of problem, but it represents the transformation of one model representation into another model.

17.15.2 Bayesian Adaptive Filtering

The block Bayesian processing/estimation utilises the available prior information about the parameter to be estimated and this prior information utilization by Bayesian approach, gives better results. The adaptive filtering aims at tracking non-stationary systems. But most of the adaptive filtering algorithms are evaluated for convergence to an unknown fixed system. But it is important to note that improving the convergence time results in better tracking. Hence, a nonstationary system can be handled by increasing the step size/decreasing the forgetting factor (for RLS) which results in better tracking. However, such algorithms have only one parameter which makes them to handle better, a nonstationary system. Hence, it is possible that an orthogonalization of the input and use of individual variable step sizes which improves tracking ability of a nonstationary system. The Kalman filter, is an exception to common adaptive filters as in it the time-varying optimal filter is modeled as a vector AR(1) process. But Kalman filter is not used like an adaptive filter due to its complexity and other difficulties. In the optimal Bayesian approach, adaptive filtering is considered a stationary optimal filter evolution. This takes into account not only parameter variation bandwidth but also a priori parameter size and parameter dynamics (Tayeb Sadiki, Dirk T.M. Slock, 2004). Further, some constrained versions of Bayesian adaptive filters which have modest complexity have come into existence.

17.15.3 Adaptive Set-membership Filtering (ASMF)

Under the concept of data reusing, adaptive filtering based on data selection using set membership is known as Adaptive Set-Membership Filtering (ASMF). They may involve a data-dependent step size that provides fast convergence. Some of the algorithms of this class may result in a faster convergence for correlated inputs but at the cost of large convergence error. To overcome this, the ASMF with a new cost function which puts a bound on the magnitude of the estimation error over a model space of interest, has been proposed (Li Gu o and Y.F. Huang. 2005 and Sridhar Gollamudi, Shirish Nagaraj, Samir Kapoor and Y.F Huang, 1998). The model space is made up of input vector-desired output pairs and the error bound criterion is applied for this space. Any parameter estimate that satisfies the error bound for all data pairs space considered, forms an acceptable solution. Therefore, the solution for SMF is a set in parameter space and not a point estimate. A proper error bound results in non-empty solution set and any point in it is a valid solution estimate. The variations of this algorithm with some benefits have come into existence. Adaptive SMF has applications in speech coding, adaptive equalization and in solving multiple access interference in wireless communication. To overcome the problems of fixed error bound which may be small or large than actual, results in a poor performance and hence error bound has been derived from the weights of the filter vector.

17.15.4 Partial-Update and Sparse Adaptive Filtering

Partial/sparse update adaptive filtering is a technique suitable for applications where the order of the adaptive filter is so high that it may impair even the implementation of low computational complexity algorithms, such as the NLMS algorithm. Partial-update adaptive filters reduce the algorithm complexity by properly decreasing the number of filter coefficients that is updated in each iteration so that the filter order may be kept fixed (K. Dogancay and P.A. Nylor 2005, K. Dogancay, 2008). Initially developed partial update algorithms suffered from slow convergence compared to that of their full update algorithms. This was mainly due to improper updating approach used. The proportionate partial update has resulted in better performance for network echo cancellation. Many variations in this direction have come into existence.

17.15.5 Kernel and Kernelized Affine Projection Adaptive Filters

The mathematical problems of nonlinear adaptive filters have been avoided by using Volterra filters. But the complexity of Volterra series *exponentially increases* with increase in its modeling ability. The multilayer neural network known for handling nonlinearities also due to their non-convex optimization nature, are not suitable for online applications. The alternative to these is Kernel adaptive filters (W. Liu, J. C. Principe and Simon Haykin, 2010. The basic goal is to perform adaptive filtering in linear space that is related to the nonlinearity of the original input space. Here, the input data is transformed into a high dimensional feature space through a reproducing kernel such that the inner product operation can be computed efficiently in the feature space through kernel evaluations. After that, appropriate linear methods are applied on the transformed data. As long as an algorithm can be formulated in terms of equivalent kernel evaluation (or inner products), there is no necessity to perform computations in the high dimensional feature space. This is the main advantage when compared to the traditional methods. This methodology includes kernel principal component analysis, recursive least squares, Hebbian algorithm, Adaline, and support vector machines (SVM), etc. The SVM based approach has improved the accuracy of speech recognition activity. As an improvement, ***kernelized affine projection algorithms*** (W. Liu and J. C. Principe, 2008) are achieved by reformulating the conventional affine projection algorithm in general, reproducing kernel Hilbert spaces. These new algorithms are online, simple, and significantly reduce the gradient noise compared with the kernel LMS hence, provide improved performance.

17.15.6 Wavenets and Wavelet Network

This is a new field of signal processing which exploits the multiresolution property of wavelets and the efficiency of neural networks in extracting the signal features. This finds application in the representation of nonstationary signals in terms of wavelet functions and in classification. The wavelets can be used only for multiresolution decomposition and the neural network only for efficient learning. That is, the wavelet parameters and the weights to be associated with each scale are treated separately. Only the weights associated with each scale is adapted in neural learning process and this category is called wavenet. It is possible to combine both wavelets and neural networks into a single method without defining their roles independently and this may be advantageous in some situations.

Here, the scale and the translation parameters of the wavelet transform is also optimized in addition to the weights associated with that scale and this category is often known as wavelet network. Thus, the wavenet is a subset of wavelet network.

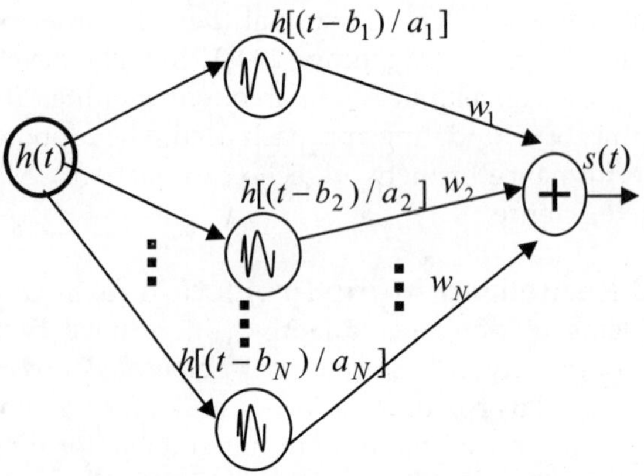

Fig. 17. 18 Wavelet network structure.

The basic structure of a wavenet is shown in the Fig. 17.18. Here, the signal $s(t)$ is approximated by a weighted sum of N modified versions of the wavelet function $h(t)$ or wavelet nodes and is given by

$$\hat{s}(t) = \sum_{k=1}^{N} w_k h\left(\frac{t-b_k}{a_k}\right)$$

The shift parameter b_k decides its position for the node k. The parameter a_k decides the centre frequency and the bandwidth for the node k. The weight w_k represents the contribution of the node k to the signal. The parameters b_k, a_k and w_k are chosen to minimize the cost function ε given by

$$\varepsilon = \sum_{i=1}^{L} \left(s(t_i) - \hat{s}(t_i) \right)^2$$

In the process of getting good and efficient approximation, the well known neural network techniques can be utilized (H. Heinrich and H. Dickhaus, 1998).

17.15.7 Complex Valued (nonlinear) Adaptive Filters

Complex valued signals play a major role in communications, array signal processing, power, environmental, biomedical signal processing and related fields. The nature of signals can be complex due to their design as in communication like QPSK. It can also be due to signal representation which is preferred due to ease of handling as analytic signal, in applications like magnetic source imaging, interferometric radar, direction of arrival estimation and smart antennas and mobile communication etc. Linear adaptive filters can be tailored accurately for processing a wider class of signals called ***complex noncircular signals with performance advantages.*** Complex circularity of signals refers to their invariant statistical nature with respect to rotation. That is, a complex random variable Z is circular, if Z and $Ze^{j\phi}$ (ϕ is angle of rotation) have the same probability distribution. It has been found that in wind forecasting which involves wind speed and its direction, use of complex adaptive filtering is advantageous from the point of view of performance. Thus, adaptive processing of noncircular and nonlinear signals form a current and future important research topic (D.P. Mandic and V.S.L. Goh, 2009).

Bibliography

1. Communication Systems, B.P. Lathi, B.S Publishers (India), 2001.

2. The Fast Fourier Transform and its Applications, E .O. Brigham, Prentice-Hall, 1988.

3. Random signals and communication theory, B.P. Lathi, International book company, 1968.

4. Probability and Random process, Donald Childers, Irwin Book Team, McGra Hill Book Companies Inc, 1997.

5. Introduction to Random signal analysis and Kalman Filtering, R.G. Brown. John Wiley & Sons Inc., New York, 1983.

6. Digital Signal Processing, A.V. Oppenheim and R.W. Schafer, Prentice Hall, India, 1975.

7. Signal Processing, Principles and Implementation, S.V. Narasimhan and S. Veena, Revised edition, 2008, Narosa Publishing House, India, Alpha Science Publishers, Harrow, U.K.

8. Two dimensional signal image processing, J.S. Lim, Prentice Hall, 1990.

9. Nikias, C.L. and Raghuveer, M.R.: Bispectrum estimation: a digital signal processing framework. Proc. IEEE 75(7), 869–89 (1987).

10. B. Yegnanarayana, Hema A. Murthy, "Significance of Group delay functions in Spectral Estimation", *IEEE transactions on Signal Processing, Vol. 40, No. 9, September 1992.* pp. 2281-2289.

11. S.V. Narasimhan, E.I. Plotkin and M.N.S. Swamy, "Power Spectrum Estimation of Complex Signals: Group delay approach", *Electronics letters, Vol. 35, No. 25,* December 1999. pp. 2182-2184.

12. S.V. Narasimhan, M. Harish, "Spectral estimation based on Discrete cosine transform and Modified group delay", Signal processing, Vol. 86, No. 7, 2006, pp. 1586-1596.

13. P. Sandeep, B.K. Shreyamsha kumar and S.V. Narasimhan, 2D-Spectral estimation based on DCT and Modified group delay", Signal Image and Video Processing, Vol. 7, No. 5, 2013, pp. 959-973.

14. D.J. Thomson, Spectrum estimation and harmonic analysis, Proeedings of IEEE, Vol. 70, No. 9, 1982, pp. 1055-1096.

15. B.Babadi and E.N. Brown, A review of mutitaper spectral analysis, IEEE Trnsactions on Biomedical Engineering, Vol. 61, No. 5, May 2014.

16. Maria Hansson-Sandsten, Mean square error optimal weighting fot multitaper cepstrum estimation, EURASIP Journal on advances in Signal Processing, 2013, 2013:158 (11 pages)

17. P. Moulin, Wavelet thresholding techniques for power spectrum estimation, IEEE Transactions on Signal processing, Vol. 42, Nov. 1994, pp. 3126-3136.

18. Narasimhan, S.V., Reddy, G.R., Plotkin, E.I., Swamy, M.N.S, Phase estimation by bispectrum : A group delay approach. Signal Process. 27 (1), 79–86 (1992).

19. Holambe, R.S., Ray, A.K., Basu, T.K.: Signal phase recovery using the bispectrum. Signal Process. 55, 321–337 (1996).

20. S.V. Narasimhan, Nandini, B., Chaitanya, R., Improved phase estimation based on complete bispectrum and modified group delay. Signal Image Video Process. 2, (2008) pp.261–274.

21. Nandini Basumallick, S.V. Narasimhan, Improved bispectrum estimation based on modified group delay, Signal Image and Video Processing , 2011

22. Modern filters, S. S. Haykin, Mac Millan, NewYork, 1989.

23. Statistical Digital Signal Processing and Modeling, M.H. Hayes, John Wiley & Sons Inc., New York, 1996 .

24. Digital spectral analysis: with applications, S.L. Marple Jr., Prentice Hall Inc., Upper Saddle River , NJ, U.S.A, 1986.

25. Algorithms for Statistical Signal processing, J.G. Proakis, C.M. Rader, F.Ling and C.L. Nikias, Pearson Education Inc, 2002.

26. Nasir Ahmed and D.H. Youn, On a realization and related algorithm for adaptive prediction, IEEE Trans. On Acoustics, Speech and Signal Processing, Vol. ASSP-28, No.5, 1980, pp. 493-497.

27. A. Harma, Evaluation of warped linear predictive coding scheme, IEEE Int. Conf. Acoustics, speech and Signal Processing, Vol. 2, 2000, pp. 11897-11900

28. Kari Roth, I.Kauppinen, P.A.A. Esquef and V. Valimaki, Frequency warped Burg's method of AR modeling, IEEE Workshop on Applications of signal processing to audio and acoustics, October 19-22, 2003, New Paltz, New York.

29. S. Veena and S. V. Narasimhan, "Laguerre escalator lattice and Feedforward/ feedback active noise control", Signal Processing, Volume 87 , pp. 725-738 , Issue 4, April 2007.

30. Keneth Steiglitz, On the simultaneous estimation of poles and zeros in speech analysis, IEEE transctions on Acoustics, speech and signal processing, 1977, Acoustics, speech and signal processing, Vol. ASSP_25, No. 3, 1977, pp. 229-234.

31. W.J. Done and C.K. Rushforth, Evaluation of Steiglitz algorithm for estimating parameters of an ARMA process, IEEE International Conf. on Acoustics, speech and signal processing, 1979, pp. 415-418.

32. T.H. Joo, J.H. Mc Clellan, R.A. Pole, G.S. Mayers and R.S. Lees, Pole-zero modeling and classification of phonocardiograms, IEEE transctions on Biomedical Engineering, Vol. BME_30, No. 2, 1983, pp. 110-118.

33. B. Yegnanarayana, Speech analysis by pole-zero decomoposition of short time spectra, Signal Processing , Vol. 3 1981, pp. 5-17.

34. S.V. Narasimhan, Pole-zero spectral modeling of EEG, Signal Processing , Vol. 8 1981, pp.17-32.

35. S.V. Narasimhan, G.R. Reddy, E.I. Plotkin and M.N.S. Swamy, Group delay based magnitude square coherence estimation by an ARMA model, Signal Processing, Vol. 46 1995, pp. 285-296.

36. Y.T. Chen and R.K. Miskowicz, "Estimation of coherence and time delay with ARMA models", IEEE Trans. Acoustics Speech Signal Processing, Vol. ASSP-32, April 1984, pp. 295-303.

37. J.A. Cadzow and O.M. Solomon, Jr., "Linear modeling and coherence function", IEEE Trans. Acoustics Speech Signal Processing, Vol. ASSP-35, January 1987, pp. 19-28.

38. S.V. Narasimhan, G.R. Reddy, E.L. Plotkin and M.N.S. Swamy, Bispectrum based mixed phase system identification by AR/ARMA model : A group delay approach, IEEE Transactions on Circuits and Systems-II, vol. 39, No.1, Sept. 1992.

39. Mysore R. Raghuveer and C.L. Nikias, Bispectrum estimation: A parametric approach, IEEE Transactions on Acoustics, Speech an Signal Processing, Vol. 33, No.4, 1985, pp. 1213-1230.

40. Mysore R. Raghuveer and C.L. Nikias, Bispectrum estimation via AR modelling, Signal Processing, Vol. 10, 1986, pp. 35-48.

41. S.V. Narasimhan, M. Hazarathaiah and P.V.S. Giridhar, Channel Blind Identification based on Cyclostationarity and Group Delay, Signal Processing, Vol. 85, July 2005, pp. 1275-1286.

42. Adaptive Signal Processing S.T. Alexander, Springer -Verlag, 1986 .

43. Linear Algebra and its applications, 3rd edition by D. C Lay, 2003, Pearson Education Inc.

44. Adaptive Filter: Theory and Applications, B. Farhang-Boroujeny, John Wiley & Sons,1998.

45. S.Vasehgi, Multimedia signal Processing, Theory and applications, in speech, music and communications, 2007, John Wiley & Sons.

46. T.J. Shan and T.Kailath,1988, Adaptive algorithms with an automatic gain control feature, IEEE Trans. On Circuits and Systems,, Vol. 35, No. 1, 1988, pp. 122-127.

47. T. Aboulnasr and K. Mayyas, "A robust variable step-size LMS-type algorithm: analysis and simulation," IEEE Trans. Signal Processing,, Vol. 45¸ 1997, pp. 631–639.

48. Theory and Design of Adaptive Filters, John R. Treichler, C. Richard Johnson Jr., Michael G. Larimore, Prentice Hall, New Delhi, 2000.

49. L. J. Griffiths, "A simple adaptive algorithm for real time processing of in antenna arrays", Proc. IEEE, Vol. 57, pp. 1696-1704, October 1967.

50. James Okello et. al., "A new modified variable step-size for the LMS algorithm", IEEE International Symposium on circuits and systems, 5, pp. 170-173, Vol. 5, 1998.

51. S.V. Narasimhan, S. Veena and Lokesha. H, Variable step-size Griffiths algorithm for improved performance of feedforward/ feedback active noise control, *Signal, Image and Video Processing,* Vol. 4., No. 3, 2010, pp. 309-317.

52. Snyder S.D. and Hansen. C.H., Influence of transducer transfer functions and acoustic time delays on the implementation of the LMS algorithm in active noise control systems, Journal of Sound and Vibration, Vol. 141, No. 31990, pp. 409-424.

53. M. Kamenetsky and B. Widrow, A variable leaky LMS adaptive algorithm, IEEE Asilomar Conf. On Signals, Systems and Computers, ACSSC 2004, pp. 125-128.

54. L.J. Eriksson, development of filtered-X and filtered-U LMS algorithms, Journal of Acoustical Society of America, Vol. 89, No. 1, 1991, pp. 257-265.

55. Widrow, et. al., On adaptive inverse control,, Proceedings of 15th Asilomar Conf., 1981, pp. 185-189.

56. J.C. Burgess, Active adaptive control in a duct, A computer simulation, Journal of Acoustical Society of America, , 70, Sept. 1981, pp. 715-726.

57. Y.C. Lim and C.C. Ko. "Forward-backward LMS Adaptive Line Enhancer", IEEE trans. On circuits and systems Vol., 37, July 1990, PP. 936-940.

58. M.A. Masnadi-Shirazi*, M. Ghasemi, Adaptive Laguerre network realization, Signal Processing, Vol. 80 , 2000, pp. 2169-2186.

59. Rajesh Sharma W.A. Sethares and J.A. Bucklew, Analysis of momentum adaptive filtering algorithms, IEEE Transactions on Signal Processing, Vol. 46, No. 5, 1998, pp. 1430-1434.

60. M.A. Tugay and Y. Tanik, Properties of momentum LMS algorithm, Electro technical conference, 11-13 April, 1989, Lisbon, pp. 197-200.

61. P.M. Clarkson and T.I. Haweel, Median LMS algorithm, Electronics Letters, Vol. 25, No. 8, 13th April, 1989, pp. 520-522.

62. Optimal and Adaptive signal processing, P.M. Clarkson, CRC press, 1993.

63. Arghya Deb, Lecture-18,19, Conjugate Gradient Method, National Programme on enhanced learning, 2012, Dept. of Civil Engineering, Indian Institute of Technology, Kharagpur, India

64. Dragica Vasileska, (2010) Methods of Conjugate gradient tutorial, https// connanohub.org/resources/9127 .

65. Introduction to the Conjugate Gradient method, Martin Zachariasen, DIKU, WWW. diku.dk/OLD/undervisning 2001e/e01.448/intro_qp.ps

66. Sumit Roy and J.J. Shynk, Analysis of momentum LMS algorithm, IEEE Transactions on Acoustics, Speech and Signal Processing, Vol. 38, No. 12, 1990, pp. 2088-2098.

67. Y.H. Chung, C.K. Tzou and N.J. Bershad, "Postsmoothing for the LMS algorithm and a fixed point round off error analysis," IEEE Trans. on Sig. Process.,Vol. 39, No. 4,1991, pp. 959-962.

68. Saleh A. Alshebeili, A. Enis Cetin andA. N. Venetsanopoulos, An Adaptive System Identification Method Based on Bispectrum, IEEE transactions on Circuits and Systems, VOL. 38, NO. *8*, 1991, pp. 967-969.

69. P.S. Chang and A.N. Wilson, Analysis of conjugate gradient algorithms for adaptive filtering, IEEE Trans. on Signal Processing, Vol. 48, No. 2, Feb. 2000, pp. 409-418.

70. S. Shankara Narayan, A.M. Peterson and M.J. Narasimha, Transform domain LMS algorithm, IEEE Transactions on Acoustics, Speech and Signal Processing, Vol. 31, No. 3, 1983, pp. 609-615.

71. D.F. Marshall, W.K. Jenkins and J.J. Murphy, The use of orthogonal transforms for improving performance of adaptive filters, IEEE Transactions on Circuits and Systems, Vol. 36, No. 4, 1989, pp. 474-484.

72. Beaufays F. and Widrow B. Transform-domain adaptive filters: an analytical approach, IEEE Transactions on Signal Processing,vol. 43, 1995 pp. 422–431.

73. A.O. Ogunfunmi and C. Pham, "The Transform-Domain Forward-backward LMS Adaptive with Applications. *IEEE-CAS*, May 9-10, 1991.

74. S. Roopa and S.V. Narasimhan, Transform domain variable step-size Griffiths Least Mean Square adaptive algorithm and its applications, Computers and Electrical Engineering, Vol. 40, Issue-4, 2014, pp. 1025-1041.

75. Radu Ciprian Bilcu, Pauli Kuosmanen and Karen Egiazarian, "A Transform domain LMS adaptive filter with variable Step-size", IEEE Signal processing Letters, Vol. 9, No. 2, pp. 51-53, February 2002.

76. V.N. Parihk, A.Z. Baraniecki, Use of the modified escalator algorithm to improve the performance of transform domain LMS adaptive filters, IEEE Trans. Signal Processing, Vol. 46 , No.3, 1998, pp. 625–626.

77. M.L. Honig and D.G. Messerschmitt, Adaptive filters, structures, algorithms and applications1, KLuwer, 984.

78. Zoran Fejzo, Hanoch Lev-Ari, Adaptive Laguerre-lattice filters, IEEE Trans. Signal Processing. 45 (12) (1997) pp. 3006–3016.

79. S. Veena and S.V. Narasimhan, Improved active noise control performance based on Laguerre lattice, Signal Processing, Vol. 84, (2004), 695–707.

80. S. Veena and S.V. Narasimhan, Laguerre escalator lattice and feedforward /feedback active noise control, Signal processing, Vol. 87, 2007, pp. 725-738.

81. P. Padroni and M.Vetterli, A FIR cascade structure for adaptive linear prediction, IEEE Trans. Signal Processing. 46, No.9, 1998, pp. 2566-2571.

82. H.G. Yeh and Martin Vaklev, An adaptive predictor with cascaded forward-backward structure, IEEE Trans. Circuit and Systems-II, 53, No. 8, 2006, pp. 773-776.

83. S. , Forward-backward adaptive Predictors and their application in Active Noise control, M. Tech Thesis, 20011, National Institute of Technology, Karnataka (NITK), Surathkal, India.

84. Zou Yuexian, S.C. Chan and T.S. Ng, Transform domain Conjugate gradient algorithm for adaptive filtering, Journal of Electronics, Vol. 17, No. 1,2000, pp. 69-76.

85. Multi resolution signal decomposition, Transforms, Subbands an Wavelets, A.N. Akansu and R.A. Haddad , Academic Press Inc., 1992.

86. Digital Signal Processing and statistical Classification, George. J. Miao and M.A. Clements, Artech House, 2002.

87. Multirate Digital Signal Processing,- Multirate systems. Filter banks and wavelets, N.J. Fliege, John Wiley & Sons, 1994.

88. Multirate systems and Filter banks, P.P. Vaidyanathan, Prentice –Hall, 1993.

89. Introduction to Wavelet transform: A signal processing approach, S.V.Narasimhan, Nandini BasuMallik and S. Veena, 2011, Narosa Publishing House, India, Alpha Science Publishers, Harrow, U.K .

90. Advanced topics in Signal Processing, Editors : J.S. Lim and Oppenheim, Prentice –Hall, 1988.

91. Subband adaptive filtering, Theory and implementation, K.A. Lee, W.S. GAn and S.M. Kuo, John Wiley and Sons, 2009.

92. Introduction to Wavelet transform: Theory and Applications, Raghuveer M. Rao and Ajit. S. Bopardikar, Addison Wesley, 1998.

93. An Introduction to Random Vibrations, Spectral and wavelet analysis, Newland .D.E , (Chapter- 3), New York Longman Publication, 1993.

94. D. R. Morgan and J. C. Thi, A Delayless Subband Adaptive Filter Architecture:' IEEE Transactions on Signal Processing Vol. 43, No. 8 1995, pp. 1819-1830.

95. L.-O. Larson, J.M. de Haan and I. Claesson, A New Subband Weight Transform for Delayless Subband Adaptive Filtering Structures, *IEEE Digital Signal Processing Workshop 2002*, October 2002, pp. 201–206.

96. Newland, D.E. Harmonic wavelet analysis, Proceedings of Royal Society London-A, 443, 1993, pp. 203-225.

97. S.V. Narasimhan, M. Harish and A.R. Haripriya and Nandini Basumallick, Discrete Cosine Harmonic Wavelet Transform: Application to signal compression and subband spectral estimation using modified group delay, *Signal, Image and Video Processing,* Vol. 3., No.1, Feb. 2009, pp. 85-99.

98. Erdol N, Basbug F. "Wavelet transform based adaptive filters: analysis and new results," IEEE Trans, Signal Processing, 1996, Vol. 44, pp. 1156–1167.

99. Hosur S, Tewfik H. "Wavelet domain adaptive FIR filtering," IEEE Trans. Signal Processing, 1997, Vol. 45, pp. 617–630.

100. Liu changyun, Chen changxing, Jia gui, "The application of multi-scale wavelet transform in adaptive filtering," Journal of Engineering, Univ. of Air Force. 2002, Vol. 3 pp. 50--52.

101. Cunqian Feng, Linrang Zhang and Xiaoping Hui, A New Adaptive Filtering Algorithm Based on Discrete Wavelet Transforms,2010, 3rd International Congress on Image and Signal Processing (CISP2010), pp. 3284-3286.

102. S.V. Narasimhan and S. Veena, A new Delayless Subband Adaptive Filter based on Discrete Cosine Harmonic Wavelet Transform (CHWT), Inter National Conf. on Signal Processing and Communication SPCOM-2010, Bangalore, India.

103. Z. Lin, J.Z. Chen and R.W. McCallum, Transform domain adaptive filtering using recursive running DCT and its application in surface recording of small intestine, South East conf. 4-7 April, 1993, IEEE proceedings 1993.

104. Subband adaptive filtering: The mutual wavelet packets Approach, C.B. Lambrecht and M. Karrakchou, Int. Conf. On Acoustics, Speech and Signal processing, Vol. 3, 1993, pp. 316-318.

105. S. Rao and Pearlman, Analysis of linear prediction, coding and spectral estimation from subbands, IEEE Transactions on Information Theory, 42(4), 1996, pp. 1160-1178.

106. S.V. Narasimhan and M. Harish, "Spectral estimation based on subband decomposition by harmonic wavelet transform and modified group delay", *Proceedings of IEEE International Conference* SPCOM-2004, December 11-14, Bangalore, India.

107. Adaptive IIR filtering, Bob Stewart, 5th IEE Adaptive Signal Processing Professional Development Course, March 2006, Birmingham, UK.

108. H.N. Kim, W.J. Song, Unbiased Equation Error Adaptive IIR Filtering Based on Monic Normalization, IEEE signal Processing letters, Vol. 6, No. 2, 1999, pp. 35-37.

109. Hyun-Chool Shin and Woo-Jin Song ,"Pole-Cancellation Error Adaptive IIR Filtering", IEEE Signal Processing Letters, No. 9, Vol. 10, Sept. 2003, pp. 280-282.

110. S.L. Netto and P. Agathoklis, A new composite IIR adaptive filter algorithm, , IEEE 28th Asilmoar Conference on Signals, Systems and Computers, Vol. 2, 1995, pp. 1506-1510.

111. Theory and Design of Adaptive Filters, J.R. Treichler, C.R. Johnson and M.G. Larimore, Prentice Hall, 2001.

112. G. Simon and G. Peceli, A New Composite Gradient Algorithm to achieve Global Convergence, IEEE Trans. On Circuits and Systems=II, Analog and Digital Signal Processing, Vol. 42, No.10, 1995, pp. 681-684.

113. Adaptive IIR filtering in signal processing and Control, P.A. Regalia, 1995, Marcel dekker Inc.

114. S.V. Narasimhan and S. Veena, New unbiased IIR adaptive filter: Its robust and Variable step-size versions and application to active noise control, *Signal, Image and Video Processing*, (2013) 7, pp. 197-207, January 2013.

115. S.V. Narasimhan and C. Nagaveni, New unbiased Steiglitz-McBride IIR adaptive filter: Its Robust and Variable stepsize versions and applications to Active noise control, Proceedings EE Centanary Conference, IISc Bangalore, 14-17, December 2011, pp. 327-332.

116. John. J. Shynk, On the lattice algorithms for IIR filtering, IEEE International Conference on Acoustics, speech and Signal processing, 1988, pp. 1554-1557.

117. S.L.Netto, P.Agathoklis, Efficient lattice realization of adaptive IIR algorithm., IEEE Trans. on Signal Processing, Vol. 46, No. 1, 1998, pp. 223-227.

118. Yosef Bendel, David Brustein, Ofir Shalvi and Ehud Weinstein, Delayless frequency domain acoustic echo cancellation, IEEE Transactions on Speech and Audio Processing, Vol. 9, No. 5, 2001, pp. 589-597.

119. Koen Eneman and Marc Moonen, Iterated portioned block frequency domain adaptive filtering for echo cancellation, IEEE Transactions on Speech and Audio Processing, Vol. 11, No. 2, 2003, pp. 143158.

120. Advanced Digital signal Processing: Theory and applications, G. Zelniker and F.J. Taylor, Marcel Dekker Inc., New York, 1994 .

121. Digital Signal Processing: Principles, Algorithms and applications, (4th edition, 2006, Chapter-13, J.G. Proakis and D.K. Manolakis. 2006,

122. Adaptive Filtering: Algorithms and practical implementation (Second Edition), 2002, P.S.R. Diniz, Publishers, Kluwer Academic Publishers.

123. S.L Gay and Sanjeev Thavathia, The fast affine projection algorithm, IEEE Inter. Conf. on Acoustics, Speech and Signal Processing, 1995, ICASSP-95, (Vol.5) pp. 3023-3026.

124. Discrete Random Signals, and Statistical Signal Processing, Charles Therrien, Prentice-Hall, 1992.

125. Neural and Adaptive Systems: Fundamentals through simulations, J.C. Principe, N.R. Euliano, W. Curt Lefebvre, 2000, John Wiely and Sons.

126. Polynomial Signal Processing, V.J. Mathews and G.L. Sicuranza, 2000, John Wiely & Sons.

127. Adaptive nonlinear system identification: The Volterra and Wiener model approaches, T. Ogunfunmi , 2007, Springer Science.

128. A novel Adaptive equalizer with DCT domain Second order Volterra series for non linear channel, H. Zhao and J. Zhang, 5-7 Sept. 2007, International Conf. on Innovative computing , Information and Control, 2007 (ICICIC), Chengdu, China.

129. A lattice predictor based adaptive Volterra filter and its convergence analysis, K.Nakayama, A.Hirano and H.Kashimoto, IFAC workshop on Adaptation and Learning in Control and Signal Processing, Yokohoma, Japan 30th August-1 Sept. 2004, pp. 415-420.

130. Adaptive Volterra filtering algorithm based on lattice preprocessing, X.M. Zhang and Z.J. Zhao, IEEE International Congress on Image and Signal Processing (CSIP-2009), 17-19, Oct. 2009, Hangzhou, China.

131. B.Koo and J.D. Gibson, Experimental comparison of allpole, all-zero and pole-zero predictors for ADPCM speech coding, IEEE transactions on Communication, Vol. COM-34, No. 3, 1986, pp. 285-290.

132. R.C. Reininger and J.D. Gibson, Backward adaptive lattice and transversal predictors in ADPCM, IEEE trans. on Communication, Vol. COM-33, No1, 1985, pp. 74-82.

133. M. Bellanger and C.C. Evci, An efficient stepsize adaptation technique for LMS adaptive Filters, Proceedings of IEEE Int. Conf. on Acoustics, Speech and Signal Processing, 1985, pp. 1153-1156.

134. G. Mathew, S.V. Narasimhan and A.P. Shivaprasad, Computationally simple ADPCM based on exponential power estimation (EPE), I.E.E. Proceedings-I, Vol.139, No.1, 1992.

135. B.J. Farahani and M. Ismail, Adaptive noise cancellation techniques in Sigma-Delta Analog to Digital Converters, IEEE trans. On Circuits and systems-1, Regular paper, Vol. 54, No. 9, 2007, pp. 1891-1899.

136. Application note on "A brief introduction to sigma-delta conversion", 1-888-INTERSIL, May 1995, AN 9504.

137. M. Girish Chandra and S.V. Narasimhan, Exponential power estimation (EPE) based daptive equalization for stationary and time varying channels, Signal Processing 54 (1996) pp. 49-58.

138. E. H. Satorius and S. T. Alexander, Channel Equalization Using Adaptive Lattice Algorithms, IEEE Trans. on Communication, Vol.27, No.6, 1979, pp. 895-905.

139. Liu Feng and Wang Xinmei, Wavelet based decision feedback equalizer, Electronics Letters, 27th March 1997 Vol. 33, no. 7, pp. 565-567.

140. P. Loganathan, A.W.H. Khong, P.A. Naylor, A sparseness controlled proportionate algorithm for acoustic echo cancellation, 16th European signal Processing Conf. (EUROSIPCO 2008), Lausanne, Switzerland, Aug. 25-29, 2008.

141. H. Deng and M. Doroslovacki, Wavelet based MPNLMS adaptive algorithm for network echo cancellation, EURASIP Journal on audio, speech and music processing, vol. 2007, Article ID 96101.

142. Odair A. Noskoski and Jose C.M. Bermudez, Wavelet packet based adaptive algorithm for sparse impulse response identification, IEEE Int. Conf. on Acoustics, Speech and Signal Porcessing, 2007, Vol. III, pp. 1321-1324.

143. M. Ferdjallah and R.E. Barr, Adaptive notch filter design on unit circle for the removal of power line noise from biomedical signals, IEEE Trans. on Biomedical Engineering, Vol. 41, No. 6, 1994, pp. 529-536.

144. N.K. Cho, C.H. Choi and S.U. Lee, Adaptive line enhancement by using an IIR lattice notch filter, IEEE trans. On Acoustics, speech and Signal Processing, Vol. 37, No. 4, 1989, pp. 585-589.

145. J.W. Choi and N.I. Cho, Suppression of narrow band interference in DS-spread spectrum system using IIR notch filter, Signal processing, Vol. 82, 2002, pp. 2003-2013.

146. M.H. Cheng and J.L. Tsai, A New IIR Adaptive notch filter, Signal processing, Vol. 86, 2006, pp. 1648-1655.

147. Mirchandani, R.C. Gaus and L.K. Bechtel, Performance characteristics of a hardware implementation of the crosstalk resistant adaptive noise canceller. IEEE Inter. Conf. On Acoustics, Speech and Signal processing, Tokyo, Japan, pp. 93-96, 1986.

148. H. Chang, N.S. Kim and S.K. Mitra, Pitch estimation of speech based on adaptive notch filter, Signal processing (85), 2005, pp. 637-641.

149. M.R. Sambur, Adaptive noise cancelling for speech signals, IEEE Trans. on Acoustics, Speech and Signal processing, Vol. ASSP_26, No.5, 1978, pp. 419-423.

150. J.R. Deller, J.H.L. Hansen and J.G. Proakis, Discrete time processing of speech signals, IEEE Press, 2000, Chapter-8.

151. S.C. pei and C.C. Tseng, Adaptive IIR notch Filter based on least mean P^{th} power error criterion, IEEE trans. On Circuits and Systems-II, Analog and digital signal processing, Vol. 40, No. 8, 1993, pp. 525-529.

152. Y.C. Park, I.Y. Kim, S.M. Lee, An efficient adaptive feedback cancellation for hearing aids, Proceedings of the International Conf. Of the IEEE EMBS, Sept. 17-21, 2003, pp. 1647-1650.

153. S. Roopa, S.V. Narasimhan and B. Babloo, Steiglitz-McBride adaptive notch filter based on variable step-size algorithm and its application to active noise control, International Journal of Adaptive control and Signal Processing, John Wiley & Sons, (under third review), 2014.

154. J.M. Kates, Constrained adaptation of feedback cancellation in hearing aids, Journal of Acoustical Society of America, Vol. 106, 1999, pp. 1010-1019.

155. J Hellgren, Analysis of feedback cancellation in hearing aids with filtered-X LMS and t.he direct method of closed loop identification, IEEE Trans. on Speech and Audio processing, Vol. 10, 2002, pp. 119-131.

156. J. Benessty and Y. Huang, Adaptive Signal Processing: Applications to real world problems (Chapter-2), Springer Publication, 2003.

157. D.H. Youn, Nasir Ahmed and G.C. Carter, On using the LMS algorithm for time delay estimation, IEEE Trans. Acoustics, Speech and Signal processing, Vol. 30, No. 5, 1982, pp. 798-801.

158. D.H. Youn, Nasir Ahmed and G.C. Carter, Magnitude square coherence function estimation: An adaptive approach, IEEE Trans. Acoustics, Speech and Signal processing, Vol. 31, No.1, 1983, pp. 137-142.

159. B. Widrow and S.D. Stearns, Adaptive Signal processing , Pearson Education Inc., 1985.

160. E.C. Ifeachor and B.W. Jervis, Digital signal processing: A practical approach, (Chapter 9) Addison-Wesley, 1993.

161. S.V. Narasimhan and D. Narayana dutt, Application of LMS adaptive predictive filtering for muscle artifact (noise) cancellation from EEG signals, Computers Electrical Enginering Vol. 22, No. 1, pp. 13-30, 1996.

162. N.V. Thakor, Adaptive filtering of evoked potentials, IEEE Transactions on Biomedical Engineering, Vol. BME-34, No. 1, 1987.

163. S. Veena and S. V. Narasimhan "Improved active noise control performance based on Laguerre lattice", Signal processing, 84(2004), pp. 695-707.

164. S. Veena and S.V. Narasimhan, "Laguerre escalator lattice and Feedforward/feedback active noise control", Signal Processing, Volume 87, pp. 725-738, Issue 4, April 2007.

165. S.V. Narasimhan, S. Veena and Lokesha, Variable step-size Griffiths' algorithm for improved performance of Feedforward/Feedback Active Noise Control.

166. S.V. Narasimhan and S. Veena, New unbiased adaptive IIR filter: its robust and variable step-size versions and application to active noise control, Signal, Image and video Processing, June 2009.

167. S. Pavithra and S.V. Narasimhan, Feedback active noise control based on forward-backward LMS predictor, Signal, Image and video Processing, Vol. 7, 2013, pp.1083-1091.

168. S. Pavithra and S.V. Narasimhan, Feedback active noise control based on transform domain forward-backward LMS predictor, Signal, Image and video Processing, Vol. 8, No. 3, 2014, pp. 479-487.

169. G.J. Miao , Signal processing in digital communication, Chapter-5, Artech House, 2007.

170. O.L. Frost, An algorithm for linearly constrained adaptive array processing, Proceedings of IEEE, Vol. 60 Aug. 1972, pp. 926-935.

171. B.D. Vanveen and K.M. Buckley, Beam forming: A versatile approach to spatial filtering, IEEE ASSP Magazine, April 1988, pp. 4-24.

172. M.B. Malik, M. Salman, State space least mean square, Digital signal Processing, Vol. 18, No. 3, 2008, pp. 334-345.

173. G. Nijsse, M. Verhaegen, B. De Schutter, D. Westwick, and N. Doelman , State space modeling in multichannel active control systems, International symposium on active control of sound and vobration (ACTIVE 99), Fort Lauderdale, Florida, December 1999, pp. 909-920.

174. Tayeb Sadiki, Dirk T.M. Slock, Bayesian adaptive filtering: Principles and practical approaches, EURASIP 2004.

175. Li Gu o and Y.F. Huang, Adaptive set-membership filtering with parameter error bound tunining, IEEE ICASSP 2005, pp. IV-369-372.

176. Sridhar Gollamudi, , Shirish Nagaraj, Samir Kapoor, and Y.F Huang, Set-Membership Filtering and a Set-Membership Normalized LMS Algorithm with an Adaptive Step Size, IEEE signal processing letters, Vol. 5, No. 5, May 1998, pp. 111-114.

177. K. Dogancay and P.A. Nylor, Recent advances in partial update and sparse adaptive filters. EURASIP-2005.

178. K. Dogancay Partial Update Adaptive Signal Processing: Design, analysis and implementation, Academic Press, 2008.

179. W.Liu, J.C, Princpe and Simon Haykin. Kernal adaptive filtering: A comprehensive introduction, John Wiley, 2010.

180. W. Liu and J.C. Principe, Kernel affine projection algorithms, EURASIP Journal on Advances in Signal Processing, Volume 2008, Article ID 784292,12 pages, doi: 10.1155 /2008/784292.

181. H. Heinrich and H. Dickhaus, Analysis of evoked potentials using wavelet networks, Chapter-27, Time frequency and wavelets in biomedical signal processing, Edited by Metin Akay, IEEE Press, 1998.

182. D.P. Mandic and V.S.L. Goh, Complex valued nonlinear adaptive filters: Noncircularity, widely linear and neural models, John Wiley & Sons Ltd., 2009.

INDEX

A

B

S

T

U

V

W